SOFTWARE RADIO TECHNOLOGIES

IEEE Press
445 Hoes Lane, P.O. Box 1331
Piscataway, NJ 08855-1331

Cover design: Sharon Klein, *Sharon Klein Graphic Design*

Technical Reviewers

Javier R. Fonollosa, *Universitat Politecnica de Catalunya, Barcelona, Spain*
Thierry Turletti, *Project RODEO – INRIA, Sophia Antipolis, France*
Neiyer Correal, *University of Florida, Gainesville, FL*
John B. Anderson, *University of Lund, Sweden*

Books of Related Interest from IEEE Press

REVIEW OF RADIO SCIENCE: 1996–1999
Edited by W. Ross Stone
Published for the International Union of Radio Science by Oxford University Press
2000 Hardcover 1008 pp IEEE Order No. PC5838 ISBN 0-7803-6003-6

ELEMENTS OF SOFTWARE PROCESS ASSESSMENT AND IMPROVEMENT
Khaled El Emam and Nazim Madhavji
1999 Hardcover 384 pp IEEE Order No. BP8523 ISBN 0-8186-8523-9

SOFTWARE ENGINEERING PROJECT MANAGEMENT, Second Edition
Edited by Richard H. Thayer
1997 Softcover 400 pp IEEE Order No. BP8000 ISBN 0-8186-8000-8

MOBILE RADIO COMMUNICATIONS: Second and Third Generation Cellular and WATM Systems,
Second Edition
Edited by Raymond Steele and Lajos Hanzo
Published in cooperation with John Wiley & Sons, Ltd.
1999 Hardcover 1064 pp IEEE Order No. PC5820 ISBN 0-471-97806-X

RADIO FREQUENCY PRINCIPLES AND APPLICATIONS: The Generation, Propagation, and Reception of Signals and Noise
Albert A. Smith, Jr.
1998 Hardcover 240 pp IEEE Order No. PC5704 ISBN 0-7803-3431-0

SOFTWARE RADIO TECHNOLOGIES

Selected Readings

Edited by

Joseph Mitola III
Consulting Scientist,
Fairfax, VA

Zoran Zvonar
Systems Development Manager,
Communications Division
Analog Devices
Wilmington, MA

A Selected Reprint Volume

IEEE PRESS

The Institute of Electrical and Electronics Engineers, Inc., New York

This book and other books may be purchased at a discount
from the publisher when ordered in bulk quantities. Contact:

IEEE Press Marketing
Attn: Special Sales
445 Hoes Lane, P.O. Box 1331
Piscataway, NJ 08855-1331
Fax: +1 732 981 9334

For more information on IEEE Press products, visit the
IEEE Online Catalog & Store: http://www.ieee.org/store.

Printed in the United States of America.

10 9 8 7 6 5 4 3 2 1

ISBN 0-7803-6022-2

IEEE Order No. PC5871

Library of Congress Cataloging-in-Publication Data

Software radio technologies : selected readings / edited by Joseph Mitola III, Zoran Zvonar.
p. cm.
Includes bibliographical references and index.
ISBN 0-7803-6022-2
1. Broadband communication systems. 2. Cellular telephone systems. 3. Radio frequency modulation—Data processing. 4. Digital-to-analog converters. 5. Adaptive antennas. I. Mitola, Joseph. II. Zvonar, Zoran.

TK5103.4 .S64 2000
384.5—dc21

00-050540

Contents

Preface ix

Chapter 1 Foundations 1

Software Radios: Survey, Critical Evaluation and Future Directions **3**
J. Mitola (*IEEE National Telesystems Conference*, May 1992).
The Software Radio Architecture **12**
J. Mitola (*IEEE Communications Magazine*, May 1995).
Technical Challenges in the Globalization of Software Radio **25**
J. Mitola (*IEEE Communications Magazine*, February 1999).
Software-Defined Radios: Facets of a Developing Technology **31**
W. H. W. Tuttlebee (*IEEE Personal Communications Magazine*, April 1999).
Software Radio Architecture: A Mathematical Perspective **38**
J. Mitola (*IEEE Journal on Selected Areas of Communications*, April 1999).
Beyond Software Radio, Towards Re-configurability Across the Whole System and Across Networks **63**
J. M. Pereira (*IEEE 50th Vehicular Technology Conference*, Fall 1999).

Chapter 2 Enabling Technologies 67

CMOS Wireless Transceivers: The New Wave **69**
A. A. Abidi (*IEEE Communications Magazine*, August 1999).
Analog-to-Digital Converters and Their Applications in Radio Receivers **75**
J. A. Wepman (*IEEE Communications Magazine*, May 1995).
Analog-to-Digital Converter Survey and Analysis **82**
R. H. Walden (*IEEE Journal on Selected Areas of Communications*, April 1999).
Power Consumption of A/D Converters for Software Radio Applications **94**
P. B. Kennington and L. Astier (*IEEE Transactions on Vehicular Technology*, March 2000).
An Overview of Sigma-Delta Converters **102**
P. M. Aziz, H. V. Sorensen, and J. Van DerSpiegel (*IEEE Signal Processing Magazine*, January 1996).
The Theory of Bandpass Sampling **126**
R. G. Vaughan, N. L. Scott, and D. R. White (*IEEE Transactions on Signal Processing*, September 1991).
RSFQ Front-end for a Software Radio Receiver **138**
E. B. Wikborg, V. K. Semenov, and K. K. Likharev (*IEEE Transactions on Applied Superconductivity*, June 1999).

On Sampling Rate, Analog Prefiltering, and Sufficient Statistics for Digital Receivers **142**
H. Meyr, M. Oerder, and A. Polydoros (*IEEE Transactions on Communications*, December 1994).
Digital IF Filter Technology for 3G Systems: An Introduction **148**
D. B. Chester (*IEEE Communications Magazine*, February 1999).
The DSP Bottleneck **154**
R. Baines (*IEEE Communications Magazine*, May 1995).
VLSI Design and Implementation Fuels the Signal-Processing Revolution **163**
T. Chen (*IEEE Signal Processing Magazine*, January 1998).
Digital Signal Processors in Cellular Radio Communications **179**
Z. Kostic and S. Seetharaman (*IEEE Communications Magazine*, December 1997).
Recent Developments in Enabling Technologies for Software Defined Radio **193**
D. Efstathiou, J. Fridman, and Z. Zvonar (*IEEE Communications Magazine*, August 1999).
FPGA in the Software Radio **199**
M. Cummings and S. Haruyama (*IEEE Communications Magazine*, February 1999).
The Flexibility of Configurable Computing **204**
J. Villasenor and B. Hutchings (*IEEE Signal Processing Magazine*, September 1998).

Chapter 3 Systems and Architectures 223

Programmable Channelized Digital Radio **225**
R. S. Prill and M. Antonesco (*IEEE National Telesystems Conference*, 1992).
Speakeasy: The Military Software Radio **229**
R. J. Lackey and D. W. Upmal (*IEEE Communications Magazine*, May 1995).
Advanced Digital Receiver Principles and Technologies for PCS **235**
H. Meyr and R. Subramanian (*IEEE Communications Magazine*, January 1995).
Broadband RF Stage Architecture for Software-Defined Radio in Handheld Terminal Applications **246**
H. Tsurumi and Y. Suzuki (*IEEE Communications Magazine*, February 1999).
Trends in Silicon Radio Large Scale Integration: Zero IF Receiver! Zero I & Q Transmitter! Zero Discrete Passives! **252**
J. Sevenhans, B. Verstraeten, and S. Taraborrelli (*IEEE Communications Magazine*, January 2000).
Advanced Base Station Technology **258**
R. Arnott, S. Ponnekanti, C. Taylor, and H. Chaloupka (*IEEE Communications Magazine*, February 1998).
Advanced Software Radio Architecture for 3rd Generation Mobile Systems **265**
H. Erben and K. Sabatakakis (*IEEE Vehicular Technology Conference*, September 1998).
A Soft Radio Architecture for Reconfigurable Platforms **270**
S. Srikanteswara, J. H. Reed, P. Athanas, and R. Boyle (*IEEE Communications Magazine*, February 2000).
Software Radio Issues in Cellular Base Stations **278**
K. C. Zangi and R. D. Koipillai (*IEEE Journal on Selected Areas of Communications*, April 1999).
A Low-Power DSP Core-Based Software Radio Architecture **291**
J. E. Gunn, K. S. Barron, and W. Ruczczyk (*IEEE Journal on Selected Areas of Communications*, April 1999).
DSP-Based Architectures for Mobile Communications: Past, Present and Future **307**
A. Gatherer, T. Stetzler, M. McMahan, and E. Auslander (*IEEE Communications Magazine*, January 2000).
Virtual Radios **314**
V. Bose, M. Ismert, M. Welborn, and J. Guttag (*IEEE Journal on Selected Areas of Communications*, April 1999).
Architectural Overview of SPEAKeasy System **326**
P. G. Cook and W. Bonser (*IEEE Journal on Selected Areas of Communications*, April 1999).
An Architecture for Radio-Independent Wireless Access Networks **337**
R. E. Schuh, C. Schuler, and M. Mateescu (*IEEE Vehicular Technology Conference*, 1999).
Software-Defined Radio Architectures for Interference Cancellation in DS-CDMA Systems **342**
I. Seskar and N. B. Mandayam (*IEEE Personal Communications Magazine*, August 1999).

Chapter 4 Software-Defined Radio Emerging Technologies 351

Direction Finding and "Smart Antennas" Using Software Radio Architectures **353**
J. Kennedy and M. C. Sullivan (*IEEE Communications Magazine*, May 1995).

Software Radio Architecture with Smart Antennas: A Tutorial on Algorithms and Complexity **360**
J. Razavilar, F. Rashid-Farrokhi, and K. J. Ray Liu (*IEEE Journal on Selected Areas of Communications*, April 1999).
The Rapidly Deployable Radio Network **375**
J. B. Evans, G. J. Minden, K. S. Shanmugan, G. Prescott, V. S. Frost, B. Ewy, R. Sanchez, C. Sparks, K. Malinimohan, J. Roberts, R. Plumb, and D. Petr (*IEEE Journal on Selected Areas of Communications*, April 1999).
Sample Rate Conversion for Software Radio **389**
T. Hentschel and G. Fettweiss (*IEEE Communications Magazine*, August 2000).
Mobile Middleware for the Reconfigurable Software Radio **398**
A. Munro (*IEEE Communications Magazine*, August 2000).
Engineering the Embedded Software Radio **408**
R. Shepherd (*IEEE Communications Magazine*, November 1999).
Cognitive Radio: Making Software Radios More Personal **413**
J. Mitola and G. Q. Maguire Jr. (*IEEE Personal Communications Magazine*, August 1999).

Chapter 5 Software Defined Radio Applications and Economics 419

Mode Switching and Software Download for Software Defined Radio: The SDR Forum Approach **421**
M. Cummings and S. Heath (*IEEE Communications Magazine*, August 1999).
Toward the Software Realization of a GSM Base Station **424**
T. Turletti, H. J. Bentzen, and D. Tennenhouse (*IEEE Journal on Selected Areas of Communications*, April 1999).
Real-Time Implementation of a Reconfigurable IMT-2000 Base Station Channel Modem **434**
D. Murotake, J. Oates, and A. Fuchs (*IEEE Communications Magazine*, February 2000).
Code-Division Multiplexing of a Sensor Channel: A Software Implementation **439**
J. R. Smith, C. Salthouse, and N. Gershenfeld (*IEEE Journal on Selected Areas of Communications*, April 1999).
Software Radios for Airborne Platforms **445**
J. P. Cummings (*IEEE Journal on Selected Areas of Communications*, April 1999).
Receiver Dimensioning in a Hybrid Multicarrier GSM Base Station **461**
H. Posti, R. Järvelä, and P. A. Leppänen (*IEEE Personal Communications Magazine*, August 1999).
Software Radio Economics **470**
A. S. Margulies and J. Mitola (*IEEE Milcom*, 1998)

Author Index 475

Subject Index 477

About the Editors 483

Preface

AT the beginning of each volume, there should be an interesting story. So, we start this book with an anecdote about the origins of software radio.

In the early 1990s, U.S. defense contractors were concerned that the end of the Cold War also meant the end of the Department of Defense (DoD) radio business. For example, Joe Mitola's general manager asked for ideas to replace business lost to the "peace dividend." Joe suggested that they merge their electronic warfare receiver product line with their programmable communications jamming product line to produce multiband multimode cell sites for the emerging wireless industry. First-generation cellular products were such a success that the proliferation of global system for mobile communications (GSM) and other digital standards might create a global marketplace where such products would be needed "within a few years."

"What are we going to call this new product?" asked the marketing vice president.

"I want to call it software radio," Joe replied.

The director of receiver products declared: "There will never be such a thing as a software radio. You will need wideband, highly linear radio frequency (RF) that is so far beyond the state of the art that you will never get there."

However, after discussing whether we could sell "software radio" technology to our current DoD customers to bootstrap the commercial thrust, the general manager said, "Okay, get going."

This interoffice exchange led to the dissemination of software radio theory and concepts to an ever-widening audience.

By 1992, the company had created several software radio precursors but could not get permission from our DoD sponsors to describe them, so the first paper to appear on software radio technology, "Software Radios: Survey, Critical Evaluation and Future Direction," presented at the May 1992 IEEE National Telesystems Conference, was largely theoretical. Fortunately, Bob Prill of GEC-Marconi had received permission to describe the DoD's experimental Pave Pillar programmable digital radio (PDR) in that same conference. Bob's opening words were: "Joe described the theory of the software radio, and we have developed one." Bob then described a multichannel digital radio, another of the software radio precursors. His paper is included as the first in the "Systems and Architecture" section of this volume. His radio was also among the first of a rapid succession of PDRs to emerge during the next few years. By 1995, the U.S. Defense Advanced Research Projects Agency (DARPA), the U.S. Army, and the Air Force Rome Laboratories had collaborated on SPEAKeasy, "the military software radio," also described in Upmal and Lackey's May 1995 paper.

Looking back at the last decade, many of the issues that seemed to be roadblocks to software radio development have been solved to some degree. A major wireless infrastructure supplier, for example, recently introduced a second-generation base station that uses wideband analog-to-digital converters (ADCs), digital interconnect, and a pool of digital signal processors. The design, which is not yet fully in the public domain, implements the software radio architecture in an appropriate compromise among the ideal, technical limitations of current devices and business considerations. So there are successes, but most areas of software-defined radio (SDR) need further work. Thanks to Moore's law, the size, weight, and power needed for giga-floating point operations per second (GFLOPS) processing capacities in affordable formats have yielded high-performance digital signal processors (DSPs) and application-specific integrated circuits (ASICs) embedded in cell phones and base stations. Advanced work in filtering algorithms, RF circuits, ADCs, and field-programmable gate arrays (FPGAs) has transformed cell site technology from mostly analog to mostly digital, with SDR cell sites not far off, depending on the details of the third-generation rollout. But linear RF, wideband, high-dynamic-range ADCs, affordable smart antennas, integrated CDMA/TDMA/FDMA architectures, and other pragmatics required to transfer technology to the marketplace remain daunting. In addition, software radio technology propels radio engineering from an enterprise concerned with mostly RF and small-scale DSP software (e.g., <100,000 lines of code [kLOC]) to a large-scale software systems enterprise (>500 kLOC) that employs analog antennas, RF, intermediate frequency (IF), and heterogeneous signal processing as its "platform." This volume therefore provides both a historical perspective that identifies the chronic problems and a cutting-edge view of software radio derived technology challenges that indicate the path ahead.

The special issue of the *IEEE Communications Magazine* of May 1995 was, in retrospect, a landmark event. The issue also included papers on architecture (Mitola), ADCs (Wepman), DSP chips (Baines), and smart antennas (Sullivan and Kennedy). It provided a "cutting-edge tutorial" of the software radio. By this time, the multiband RF access, wideband ADCs and digital-to-analog converters (DACs), and multimode software programmability had become the identifying features of the software radio. Joao daSilva of the European Community has said that that issue sparked additional focus on software radio technology, including the First International Workshop on Software Radio in 1996, organized by Joao. The European ACTS program included a project for Flexible Integrated Radio Systems Technology (FIRST) and other software radio–related projects. Jorge Pereira's paper "Beyond Software Radio, Towards Re-Configurability Across the Whole System and Across Networks" presented at IEEE VTC'99 summarizes European Union projects in this area. We regret that the rich literature from that venue can only be highlighted in that text. Other contributions of the work on SDR within EC projects in this volume are provided in papers by Tuttlebee, Kennington, Arnott, Erben, Hentschel, Shepherd, and Posti.

In March 1996 in conjunction with the award of the SPEAKeasy II program, the U.S. DoD served as catalyst for the formation of what has become the SDR Forum. Stephen Blust of BellSouth published the Request for Information on the SDR at the same meeting. The SDR represented the compromises of many of the research ideals (e.g., digital RF) necessary to bring the benefits of the software radio architecture into the marketplace sooner rather than later. BellSouth's RFI contemplated migration from the "Velcro" radio (one RF/baseband chip set per air interface standard) to the DSP-enabled radio and ultimately to the multipersonality SDR. This notion of a migration path for reduced cost of ownership has helped unite commercial and military program managers and technologists behind the vision of the SDR Forum. Among the SDR-driven contributions to this text are the radio device applications programmer's interface (API) from SPEAKeasy II (Cook). Mark Cummings, then chair of the technical committee of the SDR Forum, and Shinichiro Haruyama, the organizer of the first Asian workshop on software radio, jointly contributed the paper on the use of reconfigurable computing in SDR, "FPGA in the Software Radio" (Cummings and Haruyama). Walter Tuttlebee's paper "Software-Defined Radios: Facets of a Developing Technology" addressed issues then being hotly debated in the SDR Forum, among other venues. The debates in the SDR Forum also focused work on over-the-air downloads, highlighted in "Mode Switching and Software Download for Software Defined Radio: The SDR Forum Approach" (Cummings and Heath). Other SDR Forum topics beyond the scope of this text seem to be of growing importance in the globalization of the software radio. These include smart antenna interfaces, object-oriented modeling of radio systems, and CORBA-based middleware for open architecture plug-and-play.

In parallel, Asian interest in software radio technology has been gaining momentum. In addition to the first Asian workshop at Keio University in April 1998, Professor Masami Araki, Science University of Tokyo, for example, chaired a session of the Union of Radio Scientists International on software radio in 1999. In addition, the IEICE, the IEEE's sister organization in Japan, has formed a special study group on software radio technology. For example, we include in this volume the paper by Tsurumi and Suzuki, "Broadband RF Stage Architecture for Software-Defined Radio in Handheld Terminal Applications." They address the issue that continues to haunt practical SDR implementations: The RF and IF processing of the SDR has to be more linear over more bandwidth and more dynamic range with less noise than with conventional radios. CMOS transceivers (Abidi) and Zero IF ("homodyne") receivers (Sevenhans, Verstraeten, and Taraborrelli) offer near-term SDR trade-offs. But with a wideband linear RF/IF section, one gets to employ a wideband ADC and digital IF processing, which can substantially improve system performance.

Some say the critical technology of the software radio is the RF hardware. Although that is true in many ways, software radio also places extreme demands on ADCs. Thus, the discussion of enabling technologies in this text addresses Nyquist sampling (Wepman, Walden, Kennington and Astier), oversampling (Aziz, Sorensen and DerSpiegel), bandpass sampling (Vaughan, Scott, and White), cryogenic ADCs (Wikborg, Semenov and Likharev), and related issues (Meyr, Oerder and Polydoros). Given the gigabyte per second digitized streams of wideband ADC(s), one needs very efficient, wideband high-performance digital filter chips to extract the subscriber signals and manage the data rate (Meyr, Oerder and Polydoros, Chester). Alternatively, a W-CDMA despreader may extract a baseband data stream more or less directly from the wideband signal. If the despreader is an ASIC, a little added silicon may turn the despreader into an oversampling ADC for narrowband standards, facilitating backward compatibility with second- and first-generation cellular standards via DSP software. Such hybrid front-end components are in the definition stages as this text goes to press.

Today's digital processing engines for both research and product development of software radios consist of a mix of ASICs, FPGAs, DSPs, and/or general-purpose processors. This text surveys the key technologies in a set of complementary papers. Rupert Baines paper "The DSP Bottleneck" exposes the necessity of wideband I/O and memory architecture to complement raw DSP horsepower in software radio applications on DSPs. Related approaches are described in "VLSI Design and Implementation Fuels the Signal-Processing Revolution." Kostic and Seetharaman's, "Digital Signal Processors in Cellular Radio Communications" and Efstathiou, Fridman, and Zvonar's "Recent Developments in Enabling Technologies for Software Defined Radio" focus on the enabling DSP technologies. Two papers address FPGAs, the "higher speed alternative" to DSPs that have a reputation for being harder to program and of limited spatial efficiency. There is no set answer to whether a DSP is better than an FPGA because their strengths are application dependent. The papers that examine the FPGA alternative include Cummings and Haruyama's, "FPGA in the Software Radio" and Villasenor and Hutchings's, "The Flexibility of Configurable Computing."

Chapter 3: Systems and Architectures provides an overview of what can be done with the enabling technologies.

Chapter 4: Software-Defined Emerging Technologies explores those that result from the evolution of software radio. These include smart antennas (Sullivan and Kennedy; Razavillar et al.), dynamic networks (Evans et al.), sample rate conversion (Hentschel and Fetweiss), middleware (Munro, Shepherd), and cognitive radio (Mitola).

Finally, the volume concludes with a survey of software radio applications and economics.

Joseph Mitola III
Consulting Scientist

Zoran Zvonar
Systems Development Manager, Communications Division Analog Devices

Chapter 1

Foundations

WITHOUT mathematical foundations, systems architecture discussions quickly devolve into whether "my" approach is better than "your" approach. With mathematical foundations, both "you" and "I" have an outside reference point, some quantitative criteria by which to assess the quality of both of our designs. Structured programming abolished the GOTO statement and set criteria for module size, for example. The banishment of the GOTO was not just good engineering practice. Mathematicians proved that one could express any Turing-computable expression without the GOTO, so there was no need to clutter up software with random branches when the "structured" primitives were easier to understand and were provably capable of computing whatever is needed. A software radio contains a lot of software, so the software design principles apply. In addition, however, the software radio supports isochronous processes, tasks that have to be accomplished within a short time window that can be specified tightly in advance. Speech signal processing tasks in a software radio, for example, cannot process more than a few hundred milliseconds of data in a block without introducing time delays that the speakers find objectionable. Software radio code is therefore isochronous code. This means that the code must run to complete within a short time window that may be tightly specified in advance. The second paper shows that this constraint actually rules out the use of Turing-complete constructs in isochronous radio software. Although this sounds like a limitation, it really is not. This paper shows that it is an advantage because in this case, one can write an algorithm to answer the general question "Is a given program an algorithm—that is, will it terminate?" This classic problem is provably unsolvable in the Turing-computable case, per the famous Goedel incompleteness theorem. But this question is provably computable for isochronous software. Not only that, but the algorithm yields a tight upper bound on how long the program will require to run to complete. This algorithm can then be used to determine automatically whether software radio plug-and-play modules will support isochronous streams even if they have never been used together before. The mathematical foundations of the software radio, then, offer a relatively untapped reserve of tools to constrain and manipulate the performance of the software radio.

In general, architecture is about functions, components, and design rules. A good architecture prescribes design rules for applying a family of hardware and/or software components to achieve a specified set of functions within specified constraints that may also be considered high-level design rules. The design rules include interface specifications but also include performance constraints and management-oriented rules. The mandatory reuse of existing hardware or software components is an example of a management-oriented design rule that is typically part of a software radio architecture. The mathematical foundations of architecture employ the methods of point-set topology. The first paper introduces topological techniques, and the second extends and applies these techniques to plug-and-play goals of software radio. The third paper emphasizes the need for performance management of the signal processing resources, particularly of multithreaded multiprocessing pooled processors. These architecture considerations also have an impact on the business case, which is also introduced in the third paper. The two papers also offer precise definitions of (ideal) software radio, software-defined radio (SDR), and programmable digital radio (PDR).

The hallmarks of software radio—multiband multimode capability, wideband RF, high-speed ADCs and DACs, and software realization of signal processing functions—imply additional foundations besides mathematics. Engineering practical software-defined radios therefore require cutting-edge analog, ADC/DAC, frequency standards, timing, and related engineering subsystems. The last three papers emphasize the engineering foundations.

Software Radios
Survey, Critical Evaluation and Future Directions

J. Mitola III, Chief Scientist, Electronic Systems
E-Systems, Melpar Division
11225 Waples Mill Road, Fairfax, VA 22030

Abstract

A software radio is a set of Digital Signal Processing (DSP) primitives, a meta-level system for combining the primitives into communications systems functions (transmitter, channel model, receiver...) and a set of target processors on which the software radio is hosted for real-time communications. Typical applications include speech/music, modems, packet radio, telemetry and High Definition Television. Low cost high performance DSP chips promote delivery of enhanced communications services as software radios. Time to incorporate a new service into a product is reduced dramatically using this approach. Low costs and new services will continue to increase demand for software radio tool sets and CAD environments.

This paper relates performance of enabling hardware technologies to software radio requirements, portending a decade of shift from hardware radios toward software intensive approaches. Such approaches require efficient use of computational resources through topological consistency of radio functions and host architectures. This leads to a layered topology oriented design approach encapsulated in a canonical Open Architecture Software Radio model. This model underscores challenges in simulation and computer aided design tools for radio engineering. It also provides a unified mathematical framework for quantitative analysis of algorithm structures, host architectures and system performance for radio engineering CAD environments of the 90's.

I. Introduction

An ideal software radio transceiver is illustrated in Figure 1. D/A and A/D converters at the transmit/ receive antenna and at handset allow all radio transmit, receive, signal generation, modulation/ demodulation, timing, control, coding and decoding functions to be performed in software.

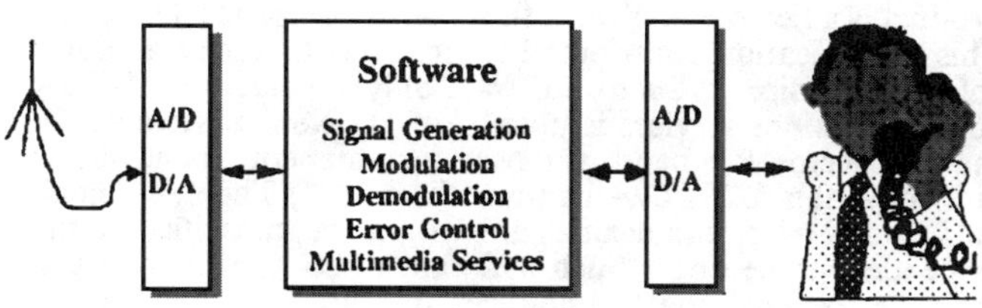

Figure 1, An Idealized Software Radio

This "software" radio of course includes many non DSP hardware components like RF conversion, RF distribution, anti-aliasing filters, power handling, etc. But the increased performance and continually dropping costs of the enabling technologies of A/D and D/A converters, high speed digital signal distribution, DSP chips and embedded computing are facilitating a shift toward software intensive approaches especially in large scale telesystems applications.

A. Software Radios Expand Multimedia Services

The ideal software radio interoperates with any communications service in its RF preselector band and A/D bandwidth. By running a different algorithm, the software radio instantly reconfigures itself to the appropriate signal format. This opens interesting possibilities for expanded radio services. A future software radio might autonomously select the best transmission mode (Personal Communication Network, Mobile Cellular Network, etc), send probing signals to establish a link, explore communications protocols with the remote end and adapt to the remote signal format. It could select the mode for lowest cost, service availability or best signal quality. The software radio reconfigures itself on the fly to support the required services.

This kind of flexibility opens opportunities for reduced costs and improved services for military as well as civilian applications. Prior generation military radios used single signal formats. Comms centers require different radios for different modes: SINGCARS for voice, TACFIRE for data, etc. Radios under development embed computing resources for a wider range of data formats. The Commanders' Tactical Terminal Hybrid (CTT-H), for example, will interoperate with several signal families in UHF, providing voice, data and imagery data relay [1]. CTT radios are multi-media capable today using a separate image compression unit. In the future image compression services may be embedded in a software radio service of CTT. Such software radio technology will also offer expanded services. Instead of calling target coordinates to a missile battery by voice, a future forward observer might uplink a video frame of the target from his night vision goggles to an airborne relay and thence to the terminal guidance of a missile. Later, he might send a battle damage assessment video frame to headquarters using the same airborne relay but a different protocol. Such advanced services require flexible signal generation, wideband Intermediate Frequency (IF) Analog to Digital (A/D) conversion, adaptive signal processing in the radio relay and data-driven routing.

B. Software Radio Telesystems Architecture

Such advanced services are on an evolutionary path which began in the early 1980's. The data links, mobile radios and LAN's of Figure 2 illustrate the system architecture of software radio oriented wide area telesystems.

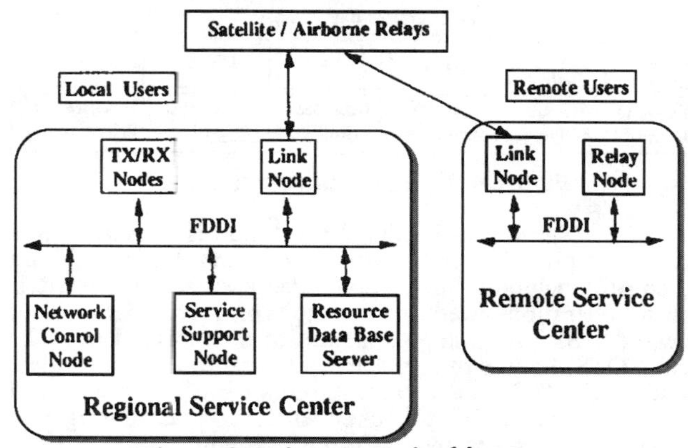

Figure 2 Telesystems Architecture

A regional service center provides central control while local service centers provide statistical multiplexing, bandwidth management and ancillary data. These telesystems have used message passing for distributed remote control since 1982. Layering and message passing have provided a robust, extensible applications architecture through several generations of hardware and operating systems. Figure 3 shows how telesystems connectivity services are layered according to the International Standards Organization/ Open Systems Interconnect (ISO/OSI) model (a) within a service center, (b) on line of sight radio data links and (c) on wide area remote satellite links. With this approach 44 thousand Lines of Code (KLOC) allocated to connectivity layers insulates over half a

Reprinted from *IEEE National Telesystems Conference*, pp. 13-15–13-23, 1992.

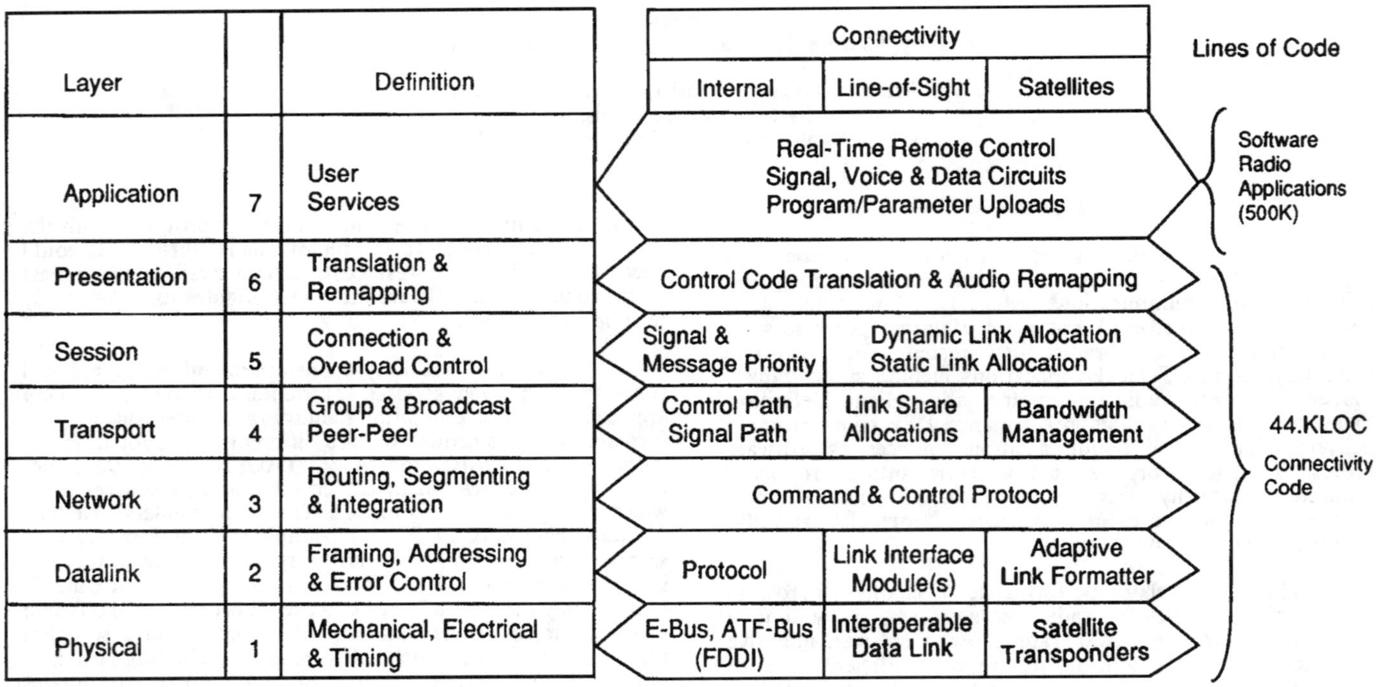

Figure 3. Layered Connectivity Architecture

million lines of software radio applications code from changes in the implementations. This open architecture now extends to server nodes based on the VME backplane (Figure 4).

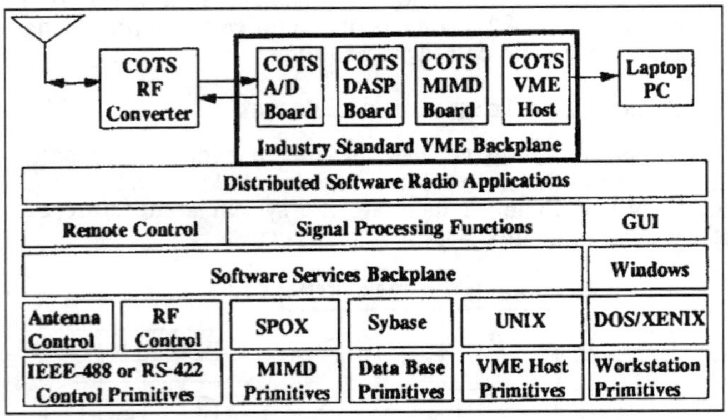

Figure 4 Open Architecture Node

The Digital Multimedia Workstation is a typical VME node [2]. It provides remote voice, video and data acquisition and control including full motion video and facsimile services. Earlier generations were limited to voice and data. But increased DSP processing capacities of Commercial Off The Shelf (COTS) boards have made it practical to integrate video, extending the ISO model to multimedia. Earlier systems also required dedicated PCM backbones for signal routing and separate audio intercoms. Current systems integrate these functions on a COTS FDDI network.

Each VME node has a signal flow architecture. But closer examination reveals the <u>VME Open Architecture Myth</u>. Although the multiple DSP boards conform to VME at the physical level, that is where open-ness ends. With no organizing paradigm at the applications layer, the integration of useful primitives like digital filters into radio systems is highly labor intensive. The system developer manually structures the data sets and signal and control flows. SPOX provides non-application specific support, but lacks a radio engineering paradigm for applications layer interoperability. The Software Services Backplane developed by E-Systems provides interoperability for signal processing data bases between VAX/VMS-ORACLE and Sun/UNIX-Sybase. But the Fast Fourier Transform (FFT) output from a DASP/ Fast Fourier Transform (FFT) board is not structured for the TMS320 MIMD board. If VME board and software vendors followed a common signal flow model, the open-ness of the VME architecture would begin to extend to software radio applications. The open architecture software radio offers insights into necessary applications layer interoperability.

The Celltap node [3] illustrates the reductions in product development time realized through software radios. This law enforcement node was conceived in 1990. An initial AMPS based product was delivered in a laptop computer with a COTS DSP board in 1991. Product extensions from AMPS to TACS to Nordic Mobile required only a few months of software development with a small Independent R&D staff. The product is a flexible receiver rather than a complete transceiver. This simplification contributed to product success as a "pure" software radio. Such software-only evolution of new communications services is possible if (1) New services are in the preselector RF band, (2) New instantaneous bandwidths are within the A/D - D/A bandwidths and (3) The new signal, data and service complexities are within the capacities of the embedded processing. Such software radio applications will grow as the enabling technologies increase in instantaneous bandwidth and capacity of embedded processing.

II Enabling Technologies

Technologies which support A/D, DSP/embedded computing and high speed digital interconnect define feasible applications and evolution paths for software radios.

A. A/D and D/A Conversion

Placement of the A/D converter in the software radio architecture drives sampling rate and dynamic range. Baseband A/D samples a single subscriber bandwidth exclusive of hopping or spreading. Agility band IF A/D samples the predetected, hopped or spread spectrum bandwidth. Wideband IF A/D samples the instantaneous bandwidth of a network of users. RF A/D samples filtered antenna signals without RF conversion. RF/IF-A/D needs greatest dynamic range because of the "near-far" situation in which strong nearby signals and weak distant signals must be processed by one A/D. Table 1 shows the relationship between potential software radio applications and representative A/D converter requirements. HF-IF requires large dynamic range due to propagation, fluctuating noise and small coherence bandwidth [4]. Dynamic range requirements diminish with increasing frequency [5] and increase for multiple users due to the near-far problem.

Table 1 A/D Conversion Requirements

Software Radio	Sampling Rate	Dynamic Range (dB)
HF Baseband	.5-8 kHz	24-64
Modems	8-32 kHz	48-64
Music	20-100 kHz	60-96
VHF-UHF BB	50-150 kHz	20-60
UHF-SHF FDM	.1-25 MHz	48-96
HF-IF	.2-10 MHz	72-120
Cellular Radio	.2-75 MHz	48-90
UHF Air Nav	2-25 MHz	48-90
SHF QAM	12-100 MHz	30-72
LVHF-IF (FH)	12-200 MHz	66-108
VHF/UHF-IF	25-500 MHz	60-96
HF RF	75 MHz	130
SHF CDMA	125-500 MHz	60-90
VHF RF	650 MHz	96
SHF Agility	.5-3 GHz	48-72
SHF-IF	1-8 GHz	48-90
UHF RF	5.4-10 GHz	48-90

Figure 5 shows dynamic range as a function of sampling rate for representative A/D converters. Phillips described a 650 MHz 7.8 bit A/D with input signal bandwidths up to 150 MHz [6]. A 1 GHz optical A/D with 1.8 bits of noise free dynamic range has been demonstrated [7]. The 8 GHz 8 bit converter goal was announced by DARPA [8]. Resolution in dB is not equal to linear dynamic range. Generally, A/D resolution is 3 to 6 dB greater than spurious free noise free linear dynamic range. For example, the Burr Brown 10 MHz 12 bit model ZPB1603 has a spurious free dynamic range of 72 dB, but a Signal to Noise Ratio of 67 dB [9].

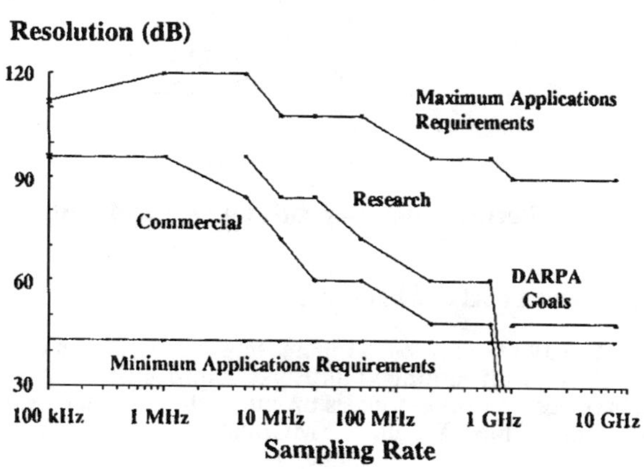

Figure 5. A/D Converter Products/Plans

Comparing A/D requirements to capabilities shows the feasibility of software radio applications in speech, IF and agility band VHF and UHF; and wideband applications of limited instantaneous dynamic range (e.g. microwave QAM with constellations < 64 and symbol rates < 30 MHz).

Critical Assessment: The noise free spurious free linear dynamic range of A/D converters is limited by aperture uncertainty and device noise of the sample and hold circuit. The aperture uncertainty tolerated by a B bit converter is approximated for a signal V(t)= A cos(wt), dV/dt= Aw sin(wt) and [dV/dt]*Dt = A/2**(B+1). Thus, aperture uncertainty Dt is approximately 1/(w*2**(B+1)).

Table 2 Aperture Uncertainty Requirements

f (MHz)	B (bits)	Dt
250	8	.64 ps
500	10	.16 ps
1000	12	20 fs

Aperture uncertainties increase linearly with frequency and exponentially with the number of bits (Table 2). Thus, an 8 GHz 8 bit converter requires aperture uncertainty < 100 fs. Hundred femtosecond stabilities have been reported in conjunction with optoelectronic RF sampling circuits with bandwidths of 275 GHz [10]. Thus an 8 GHz x 8 bit A/D is within reach in terms of aperture uncertainty. Conversion of the sample and hold signal depends on switching speeds. An electron-gas GaAs Schottky diode with 3 THz cutoff frequency has been reported [11]. Such switching device technology is sufficient to evolve toward 300 GHz A/D rates but aperture uncertainty requirements will limit dynamic range. There are many untapped software radio applications within this growing envelope of evolving A/D technology.

Software radio algorithms extend the linear dynamic range of A/D devices. Oversampling with interpolation extends effective SNR and linear dynamic range. Thao and Vetterli [12] reported an optimal 4th order projection filter which improves SNR by 80 dB vs 12 db from oversampling. This extends the theory of reconstruction of signals from level crossings [13]. Such techniques are useful if the bandwidth allocation to a set of users is large but each user baseband is small. For example 200 cellular radio channels of 25 kHz occupy a 5 MHz IF requiring a 12.5 MHz sample rate which limits dynamic range to 60 dB. Using [12], each user's SNR may be improved by 38 dB to 98 dB. Such advanced algorithms improve overall software radio performance beyond the capabilities of the A/D devices at the expense of additional DSP capability.

B. Digital Signal Processing Hosts

DSP hosts embed software radio functions with the rough order of magnitude computational demands of Table 3.

Table 3 Illustrative Processing Demands

Radio Function	Bandwidth	Nominal MIPS/MFLOPS
T1 Modem	1.544 Mb/s	2/0
GSM Radio	60 MHz agility	5/5
Modem	9.6 kb/s	6/8
Adaptive Relay	25 MHz	5/750
HDTV	50 MHz	2000/500
SDH Equalizer	620 Mb/s	20,000

Processing demands shown in the table vary by up to three orders of magnitude. For example, the T1 multiplexer has been benchmarked for variations of language, programming style and host hardware (Table 4). Ada packages promote software reuse but are inefficient with subroutine calls in the inner multiplexing loop. In-line Ada improves throughput, but in-line C and hand coded assembler provide the best throughput margins. Nominal processing requirements of Table 3 may be compared to processing capacities of chip-sets, multi processors and dedicated hardware of Table 5. SPECMark, Linpack, DRHYSTONE and other instruction mix benchmarks are more representative than peak performance but are not always available.

Critical Assessment: With careful interpretation, peak capacity is adequate for rough order of magnitude analyses. Processing efficiency is the ratio of peak capacity to achieved performance. Efficiencies of 80% are rare. Efficiencies less than 100% occur because of (1) inadequate interconnect bandwidth, (2) inability of an algorithm to access available capacity and (3) statistical

Table 5 Nominal Processing Capacities

Technology	Device	Parallelism General/DSP	MIPS/MFLOPS (Peak)
[14]IBM PC/AT	i80486	1/0	5/2
[15]DSP Chip	TMS320	0/1	25/50
[16]DSP Board	i960/i860	1/1	40/80
[17]DSP Chip	PA-RISC 7000	1/1	120/120*
[18]SIMD	AMT/DAP	1024/	20000x1/39
[19]MIMD	PC/MC860	1/4	5/320
[20]SuperCPU	Convex-210	1	200/200
[21]DSP Chip	Fujitsu VPU	0/1	70/289*
[22]SIMD	MASPAR	1/4096	6400/325
[23]SuperMIMD	CSP	1/4	36/600
[24]MIMD	GT860	1/8	10/640
[25]Hardware	G-FEC	N/A	1250
[26]MIMD	DSP-3	1/64	10/1280
[27]Super MIMD	Convex-3800	8	2000/2000
[28]DSP MIMD	VASP	1/16	5000/5000*
[29]Super MIMD	Cray YMP90	16	/16000*
[30]MIMD	Kendall	/1088	21760/43520*
[31]SPARC MIMD	TMI CM-5	16384	/128000*
[32]i860 MIMD	iParagon	4096	/327000*

* Announced but not yet delivered

patterns of demand. Interconnect includes all memory and system busses and interprocessor communications paths. Interconnect utilization is the ratio of interconnect demand to bandwidth. If this is < 0.1, the system is processor limited, but if it is > 0.6 on any path, the system is interconnect-limited. Software radios are reliable with interconnect utilizations < 0.5.

Table 4 Software T1 Multiplexors

Software Style	Host	MIPS per T1
Ada/Package	VAX	2000
Ada/Inline	VAX	200
Ada/Inline	Gould	40
C/Inline	Gould	8
Assembler	Gould	2

Efficiency also depends on the statistical pattern of service demand. Figure 6 shows how achieved software radio performance depends on aggregate demand on a typical MIMD host. A function which requires 280 msec to run to complete in a non-loaded environment requires over 700 msec during peak loads. The general shape of this curve is well approximated by a multiple server queue latency model [33]. As demand exceeds 60% of resources, service time grows exponentially. A robust software radio host therefore provides capacity in excess of the average loads by an amount predicted by queueing theory, typically 3 times the average demand.

To compare the nominal demands of software radio applications to the nominal peak capacities of hosts, one must account for interconnect utilization, algorithm partitioning to the host architecture, software implementation efficiency on each processor and statistical patterns of service demand. Generally an order of magnitude excess capacity is required to reliably support software radio functions. Thus, comparing Tables 3 and 5, software radio products for DS0 voice channel modems, T1 multiplexing and cellular radio appear within the capabilities of current hosts. High Definition TV research is conducted on Convex supercomputers and SIMD image processors with real time signal acquisition and non real time signal processing [34]. But wider bandwidth real time systems seem to be on the horizon due to the MIPS/MFLOPS of planned massively parallel systems. The pace of transition to massive parallelism and related software intensive approaches will be accelerated if the new supercomputing architectures provide the interconnect bandwidth required for software radio topological flexibility.

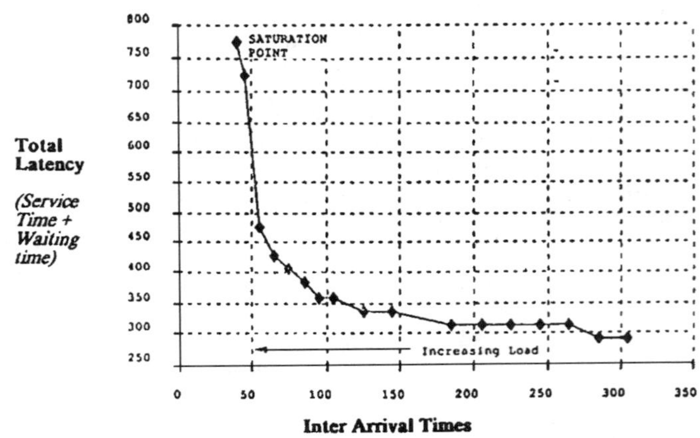

Figure 6 Performance Depends on Demand Patterns

C. Interconnect and Packaging Technology

The 1980's focused on increasing the capacities of DSP chips. Along with continued thrusts to improve chip capacities, there has been recent emphasis on integrating unpackaged die into Multi Chip Modules (MCM's). MCM's increase functionality per unit volume by replacing board level interconnect with substrate interconnect. The traditional board, backplane and system level interconnects persist, resulting in the interconnect hierarchy of Table 6.

Table 6 Simplified Interconnect Hierarchy

Level	Width	Layers	Clock Limits
Die	micron	3	GHz
MCM	50-10 um	4-10	500 MHz
Board	50 mil	5-15	100 MHz
Backplane	100 mil	N/A	20 MHz
FDDI	N/A	N/A	100 Mbit
HiPPi	N/A	N/A	800 Mbit
Fiber	N/A	N/A	2 Gigabit

Differences in speeds across the interconnect levels may be overcome through parallelism. The functional transparency desired for software radios requires ultra wideband N x N non-blocking switching among all levels of system interconnect. Massively parallel processors strive for flexible interconnect with desirable theoretical properties [35], functional transparency [36] and/or general malleability [37]. But physical limits to the interconnect technology are at hand. MCM line widths with 50 ohm resistance require thirty layers to interconnect a typical 4" MCM. Eden [38] has shown that when copper interconnects reach 1 um x .5 um the line resistance on a 4" MCM is 7000 ohms. Thus reduction of line widths to reduce the number of interconnect layers for large MCM's is impractical.

High Temperature Super Conductors reduce the 1 um line resistance to near zero, reducing the number of layers to 2. In addition, CMOS integrated circuits have a 2X speed improvement at HTSC temperatures. This technology is at the research stages, reaching demonstrations in 1994 [39]. Chip-to-chip integrated optical interconnect circuits have been studied with GHz bandwidths [40]. Free space or fiber optic paths could provide interconnect at the MCM and board level.

Critical Assessment: Although the focus on improved interconnect devices is important, little is known about interconnect topologies for software radios. A pyramidal architecture based on vision theory and algorithm structure has emerged for computer vision. But the topological structure of radios is not guided by a strong theory. Instead, it depends on bandwidths, signal complexity and service demands in an application dependent way. As the A/D converter moves inexorably from baseband towards RF, increasingly wideband front end interconnect is needed. If the interconnect bandwidth is three orders of magnitude faster than the signal flow, then the interconnect network is functionally transparent to the application [36]. Such transparency in future hardware architectures would help hosts provide the flexibility required for software radio service growth.

D. Technology and Applications Trends

Figure 7 illustrates communications technology trends relevant to software radios: analog instantaneous bandwidth in research environments, bandwidth of RF sampling circuits (e.g. in sampling oscilloscopes), equivalent bit rate of digital communications products, 8 bit commercial A/D rates, DSP chip capacities and massively parallel processing capacities. Fiber optics research achieves an order of magnitude more bandwidth than contemporary sampled RF circuits, while digital communications products lag by two orders of magnitude. Synchronous Digital Hierarchy (SDH) radio modems will reach 2.4 Gbit/sec this year. Signal complexities have increased to pack the higher bit rate signals more efficiently into limited bandwidth allocations. PSK has given way to 16 and 64 QAM for production radios, with successful research in 256 and 1024 QAM. Such signal complexities create demand for increased DSP capacities in ASICs, gate arrays and general purpose DSP chips. Gigabit FEC Codec chips and programmable 289 MFLOP DSP chips for HDTV receivers have been announced [41]. Such low cost high performance modules are ideal hosts for software radio functions. Increased demand thus fuels the device markets, creating continued investments in these enabling technologies. Software radios clearly ride this technology wave.

As the figure shows, during 1992 the enabling technologies will cross the frontier from baseband/ audio DSP applications to the IF/video range of software radio applications. Along this trend line even wideband SDH and HDTV products and systems will become software intensive soon. The communications industry is in transition from digital (hardware) radios to software intensive radio systems. The open architecture software radio provides a framework for capitalizing on the advantages of this trend.

III Computational Models and Architectures

The importance of topology in radio engineering is evident by examining signal flows, structure of algorithms and related systems architectures.

A. Computational Models Shape System Architectures

Software radios are beginning to embed extensive computational models to achieve high performance with a high degree of autonomy. Today's voice mail systems use dial tones for interactive control, but tomorrows will respond to spoken

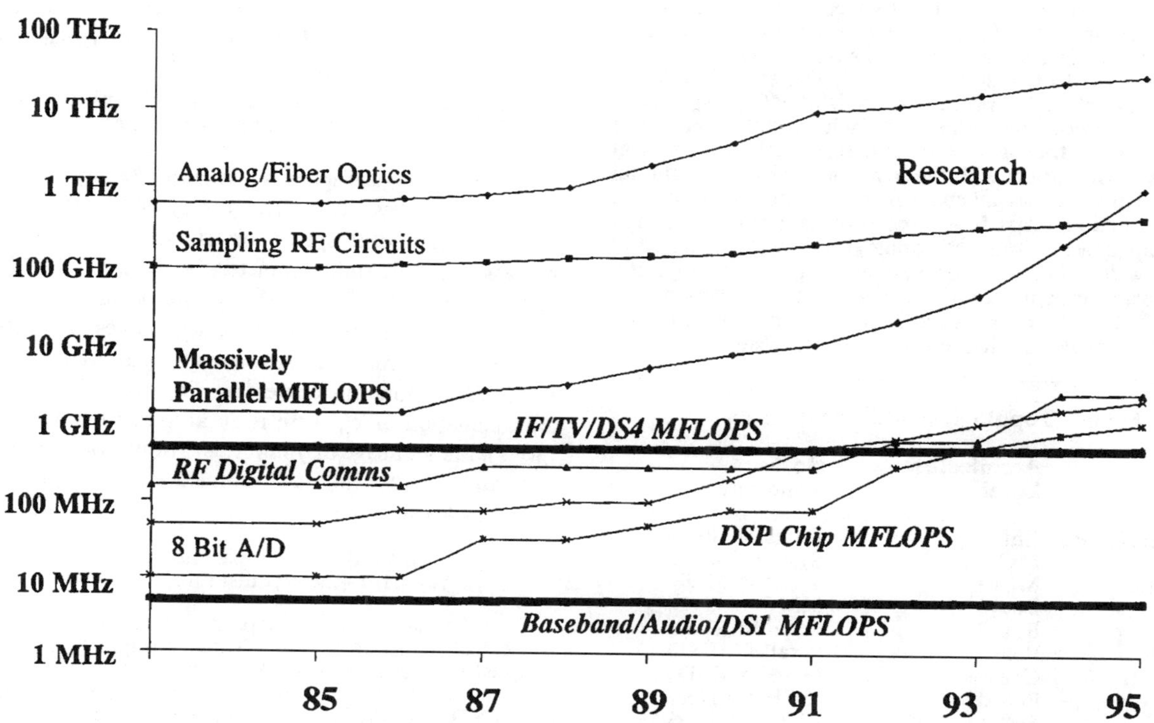

Figure 7 Enabling Technology Trends

commands. If DARPA's WHISPER program is successful, the necessary speech understanding technology will be available soon [42]. Hearsay and Harpy were exemplary speech understanding systems of the 1970's. Both embedded a priori characteristics of speech and language in computational models. Hearsay encapsulated information in Knowledge Sources (KS's). KS expert system rules, attached procedures and data transformed signal information on a global hierarchical Blackboard. Hearsay's high computational demands led to a coarse grain multiprocessors, C.mmp, with a Blackboard hosted on a global memory and KS's hosted on parallel processors, providing an isomorphism between algorithm and processor structures [43]. In software radios, SIMD architectures provide bulk signal processing (equalization, demultiplexing, FEC coding/decoding...). MIMD architectures provide multiple asynchronous channel processing. Neither SIMD nor MIMD architectures alone are adequate for large scale applications because the mix of signal flows mandates a mix of architectures as illustrated in Figure 8.

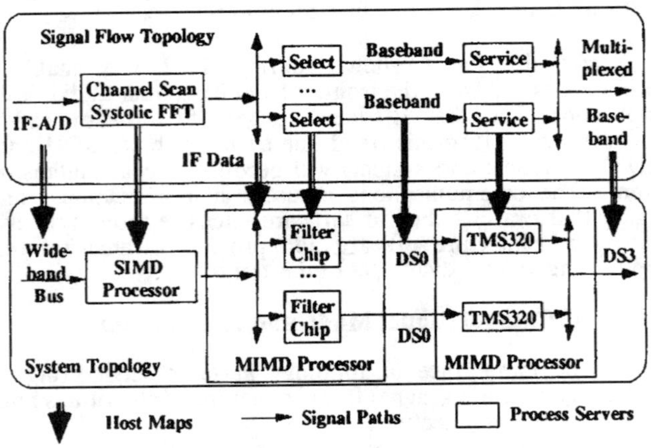

Figure 8 Effective Capacity Is High If Host System Topology Matches Radio Function Topology

The figure shows a data-driven radio relay signal flow mapped to a notional multiprocessor. The software radio algorithm topology is evident in signal flows among signal spaces (IF-A/D, Baseband, etc). Each radio function is a transform or map from one signal space to another. Signal streams partition radio functions naturally into functionally cohesive transforms coupled by signal streams. The multi user IF signal is hosted on a 200 MHz bus. The scan function is hosted in a systolic FFT processor. A DSP MCM hosts each subscriber's service algorithms. Each signal space and transformation in the signal flow is mapped 1:1 to a physical host which is compatible with the bandwidth or computational demands. Thus each signal space maps to a host interconnect and each function maps to a host processor providing high effective capacities. Study of topological structure therefore leads to a well balanced radio system design. Topologically efficient hosts for representative software radio functions are shown in Table 7. Large scale radio systems rely on such topological efficiency at all levels from chip to system.

Table 7 Topologically Efficient Hosts

Radio Function	Architecture Level	Host Topology
HDTV FEC Codec	Chip	Function Specific
FFT	MCM	Systolic [44]
Digital Filter	MCM	Serial MIMD
Switching	Board	SIMD
Equalizer	Board	Serial MIMD
FAX Modems	Board	Parallel MIMD
Bulk Modem	Chassis	Mixed MIMD
TX Control	Board	IEEE-488 Network
Voice Mail	System	FDDI Network

IV. The Software Radio Model

The need to efficiently map radio functions to hardware architectures led the author to apply the topological theory of algorithms [45] to radio engineering [46]. The resulting study of the set-theoretic topological properties of analog hardware radios, DSP algorithms and computational architectures provides organizing principles for telesystems design and insights relevant to future radio CAD environments. Signal processing theory, DSP algorithms, problem oriented languages and the hosting of algorithms onto DSP chips all share a common mathematical structure of mappings and operators on topological spaces. This section provides a brief overview of this approach to radio system engineering.

A. The Topological Model

A topological model of a software radio is a formal system {X, Y, P, H, M, F} (Figure 9). X is a space of radio system abstractions: infinite dimensional signal spaces, control spaces and transformations. X is organized by a radio engineering taxonomy consisting of information sources, transmission facilities, radio channels, reception facilities and information sinks. Y is a space of finite approximations to the signal and control spaces of X. P is a set of primitive maps from Y onto Y. P are the transforms and control algorithms of digital signal processing. H is a set of hosts for Y and P. H consists of physical channels, users, DSP chips, memories, etc from which one configures a radio system. M is a meta-level system for extending P and for mapping primitives in P onto hosts. F is a set of constraints among X, Y, P, M and H.

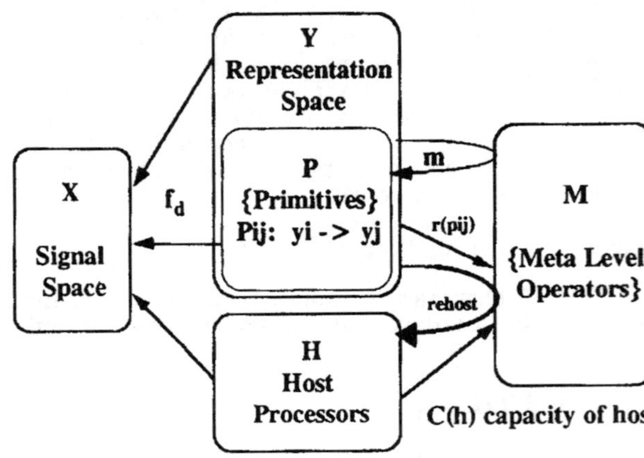

Figure 9 The Topological Model Unifies Concept, Design and Fabrication

A topological space is a set and a family of subsets with certain properties [47]. The theory of iterative algorithms induces topological spaces on subsets of X (e.g. random processes and decision parameters [48]). In {X,Y,F} for each y in Y there is an x in X such that y approximates x with acuity d and there is a map f_d in F which makes this relationship explicit. For example, an A/D stream y_k in Y could represent a band limited random process x_t in X. If y_k is a "good" approximation of x_t, there is an accuracy bound d such that $f_d(y_k)$, the set image of y_k in X, is "always within d of x". If x_t has bandwidth W and y_k is sampled at the Nyquist rate, then there exists d such that:

$$\|[k+1]-[k]\| < d \text{ - the sample interval and} \quad (1)$$
$$(2*W) < 1/d \text{ - the Nyquist rate.} \quad (2)$$

The map f_d: y->x maps subsets of y_k to corresponding subsets of xt according to Nyquist sampling and makes the Nyquist accuracy explicit. The distance d induces a topology of "open balls" for x_t. The map f_d is continuous, One To One and ONTO and the inverse image of open sets are open sets; so f_d is a homeomorphism, a topology preserving map [49].

B. Topological Constraints Encode Semantics

The creation of formal maps f_d and the proofs that "y approximates x with accuracy d" is the science of digital signal processing. Nyquist sampling is just one example of a theory which implicitly establishes such maps. The maps f_d are not DSP functions. They are the formal definitions by which the approximations of DSP are linked to the abstractions of communications theory. In the topological model, the constraint map f_d:Y->X expresses these relationships explicitly. Suppose, for example, p_{ij} is a digital filter. Domain $D(p_{ij}) = y_i$ is the filter input space. It includes an A/D converter stream with resolution and sampling rate and inherited information such as the original RF carrier. Control inputs like 3 db bandwidth W of the filter are also in y_i:

y_i = {Continuous Sampling;
RF = 240 MHz;
Rate =100 Ksa/sec;
Resolution = 8 bits;
W = 350 Hz;
Stream @ FIFO.1;}

The abstraction in X which yi represents is:

x_i = {Bandlimited Signal;
UHF Signal;
Nyquist Bandwidth = 50 kHz;
Nominal Dynamic Range = 48 db;
Information Bandwidth = 350 Hz;
Stream = 'From Antenna Element #1';}

The map f_i: y_i -> x_i establishes the semantic constraints of y_i, e.g. mapping the approximation "Continuous Sampling" to the abstraction "Bandlimited Signal", and so forth for all the corresponding entries in y_i and x_i above. All input and output spaces in a well formulated software radio system have such definitive maps whether expressed in documents, data bases or computational models. Expert system rules and attached procedures are computational models of such constraints as research in qualitative physics, for example, has shown [50]. A spreadsheet model of radar design used a rule base to encode such constraints [51]. In traditional development environments, the maps are expressed in documentation but seldom in computational models. In future CAD environments based on the software radio paradigm, such maps would be made explicit through topological constraint maps.

C. Topological Information Hiding

P is the set $\{p_{ij}\}$ of predefined primitive signal processing functions. In typical DSP design tools, 50 to 200 primitives are supplied. Any DSP function may be analyzed as a potential topological map. One first explicates the topology of the function's domain and range. From these topological spaces, one tests p_{ij} for homeomorphism by determining whether p_{ij} is continuous, 1:1, ONTO and has a set inverse. The analysis leads to refinement of $\{y_i, x_i, y_j, x_j\}$ and p_{ij} towards simple topological structures easily mapped to data structures, algorithms and host hardware.

The designer constructs radio functions from the primitive set through function composition. Composition provides a signal oriented model of information hiding. Suppose P = $\{p_{ij}, p_{jk}\}$. From p_{ij}: y_i->y_j and p_{jk}: y_j->y_k, the meta-level composition operator m in M yields:

$m(p_{ij}, p_{jk}) = p_{ik}$, where

p_{ik}: y_i-> y_k,

Domain, $D(p_{ik}) = D(p_{ij}) = y_i$

Range, $R(p_{ik}) = R(p_{jk}) = y_k$

but $R(p_{ij}) = D(p_{jk}) = y_j$ is *hidden*, embedded in p_{ik}.

The new primitive p_{ik} hides the embedded space y_j. The composition operator propagates constraints from y_j and p_{jk} onto the output space y_k. The information hiding of composition of primitives induces structure in software radios. As a primitive, the composite map p_{ik} is a virtual machine embedding lower level machines p_{ij} and p_{jk}. This virtual machine hierarchy extends from an initial set of primitives at the lowest level to complete software radios. These machines define a virtual machine hierarchy the key levels of which generalize the layers of the ISO/OSI model (Figure 10). The physical layer contains the host hardware. A host imbued with a software radio function is a hosted primitive. This corresponds to the ISO link layer because a the virtual machine map from y_i to y_j is analogous to a point to point link. The highest level is the Radio System, the Application Layer. Within each layer, composition of primitives and information spaces in Y define signal oriented information hiding. The layered virtual machine architecture isolates the software investments in the communications functions of layers 3 - 7 from the rapidly evolving implementation technology of the lower layers.

Layer	OSI	Radio	*Software Radio Virtual Machines*
7	Application	Radio System	User, Transmit, Channel, Receive, User
6	Presentation	Control	Communications Events and Controls
5	Session	Resource	Radio Functions Mapped to Hosts
4	Transport	Signal Flow	Signal Streams & Connectivity
3	Network	Function	Signal Scripts / Composition of Primitives
2	Link	Primitive	Lower Level DSP Primitives
1	Physical	ISA	Instruction Set Architectures & Hardware

Figure 10 Topological Virtual Machine Hierarchy

D. Host Models Facilitate System Design

The set H provides the physical hosts for the communications functions of P and for the interconnect spaces of Y. In the topological software radio, device properties are encapsulated in topological maps as in Table 8. Such maps are complete specifications of input, output, function, resources (supplied/demanded) and design constraints of the device. The widespread use of such formal computational models of radio system devices would make it easier to assess relationships between resource requirements of radio services and resource capabilities of potential system components. Electronic publication of such models would make it easier to find the right component for a given job.

E. Computability And The Meta-Level Model M

The set M contains meta-level operators m:P -> P by which to combine primitives to create new radio functions which themselves are primitives. The operators of M are sufficient for primitive recursion. Partial recursion is the most powerful computational model, equivalent to Turing Computability [52]. Partial recursion employs unbounded minimalization and is not guaranteed to terminate for all inputs. Total recursion is defined everywhere but may use unbounded resources. Primitive recursion uses the bounded resources of bounded minimalization. It allows iteration only over a bounded search space and returns the result "it can't be determined within these resources" if such is the case. Primitive recursion is defined for all inputs, even if the result is "out of resources". This computational model is compatible with a central design issue in software radios: fixed real-time computational resources.

M includes structured control primitives IF-THEN, DO-WHILE, etc. with system upper bounds on resources. M

Table 8. A/D Converter Topological Resource Map

```
Resource h: yi -> yj:
        {hij=pij: A/D Converter;
        Sample Rate = 750 kHz;
        Resolution = 8 bits;
        Dynamic Range = ?;
        Size = 3 x 5 x 2 in;
        Weight = 11 oz;
        Power max = 100 mW;
        Temp-Low = 0 C;
        Temp-Hi = 27 C;
        Shock = 'Office;
        Vibration = 'Office;
        Board-cost = $500 US;
        fi: yi -> xi:
                yi = {RF = ? MHz;
                        IF = ? MHz;
                        W < 350 kHz;
                        Analog Stream @ ?;};
                xi: {? RF Signal;
                        Predetected Signal;
                        Bandlimited signal;
                        Nyquist Bandwidth = 350 kHz
                        Analog BNC Connector Signal;};
        fj: yj -> xj:
                yj = {RF = ? MHz;
                        Continuous Sampling;
                        Rate =700 Ksa/sec;
                        Resolution = 8 bits;
                        W = 350 Hz;
                        Stream@ FIFO.?;};
                xj = {? RF Band;
                        Sampled Signal;
                         AT Bus Signal;
                        Band Limited Signal;}; }
        resource(h,*) = {Serves Analog-W = 350 kHz;
                        Demands Digital-W = 750 kB/s}
```

provides explicit composition through procedural languages. It also provides signal scripts in which a representation in X is interpreted by a script interpreter to create a sequence of primitives which will yield the desired signal. Script macros support interactive synthesis by analysis in which a signal processing procedure is the trace of user operations on a signal.

The formal system {X, Y, P, H, M, F} can be used in a system design paradigm to help a design team define a modular, extensible, computationally efficient radio system architecture. It can also be used this way to analyze an existing design. It may also provide some mathematical structure to guide the evolution of future radio CAD environments.

V. Future Radio Simulation and CAD Environments

Consider the process of designing and developing large scale software radio systems. The transitions from (1) service concept to (2) system definition to (3) simulation and validation to (4) delivery invariably require a mix of radio engineering disciplines. In one vision of the future, an ideal radio CAD environment would facilitate such transitions as suggested in Figure 11. In the software radio paradigm the environment would integrate systems analysis, software definition and system design disciplines. One technical basis for the integration of such disparate disciplines would be the consistent representation of algorithms, software components, existing chips, MCM's, boards, chassis and subsystems as maps among topological spaces. Thus a comprehensive computational model would be associated with each entity in the CAD environment. The CAD system would automatically hand off end item specifications to software, chip, MCM, and board level CAD tools from different vendors for detailed definition of components. As-built computational models (e.g. test results) of hardware and software would be integrated back into the larger CAD environment as items are designed and manufactured. Thus, difficulties in one area could be compensated by other areas early in the project (e.g. an A/D dynamic range limitation could be overcome by a more advanced filter algorithm using spare MFLOPS). These as-built computational models would persist, available for design analyses in future systems. Such environments could accelerate pace at which new services transition from the conceptual design and analysis stage into products.

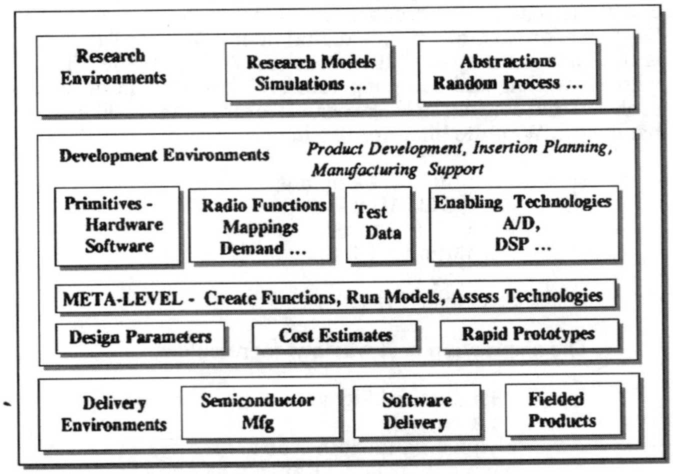

Figure 11 Future Radio CAD Environments

The applications scope of such an environment is shown in the radio system taxonomy of Figure 12. There is a tradeoff among (a) open architectures with large commercial bases, (b) high design coherence and (c) architecture efficiency with respect to the specific radio service requirements. Early software radios were designed and developed by an integrated team, were delivered on closed architecture hardware from a few vendors, had limited bandwidth and were relatively expensive. Yet the high design coherence of these systems contributed to the rapid and continued evolution of flexible services. These systems are in the field and their design coherence has supported more than a decade of product evolution including fairly radical hardware changes.

But recent projects which involved the use of commercial tools to the greatest extent possible underscored the lack of design coherence across tools from multiple vendors. While this is only natural, it is nevertheless an impediment to getting the most from the commercial base. Figure 12 also lists widely available tools related to radio system simulation and design. The marketplace is replete with general purpose hardware, software and simulation tools for signal processing. These tools are inadequate for large scale applications, lacking bandwidth and support for mixed MIMD/SIMD hosts. There are few widely available simulation tools which include communications specific functions. Computer aided design of the 80's provided hardware CAD tools for device designers and software CASE tools for software engineers. Extending these tools to the systems level and integrating them into networks of interoperable tool sets is the challenge of the 90's. Greater CAD focus on specific applications like radio engineering seems necessary to gaining the next plateau of enhanced productivity.

Considering radio systems and the rapidly evolving technologies from the perspective of the software radio is revealing. Reduced costs and increasing capacities of the enabling technologies are moving us rapidly towards software radios. Their reduced product cycles and software-only services create many opportunities and challenges. The topological model and resulting layered virtual machine model consitute an open architecture software radio model. When such models guide implementations, the investments in applications are insulated from the rapidly evolving host technology. The study of radio function and systems topology reveals unifying principles for better integrating the related hardware, software and systems engineering disciplines.

Radio System Taxonomy	Simulation & CAD Tools
System Level	
Structures	
Systems	-
Discrete Event Models	Many eg BONES[53]
Properties	
Information Capacities,...	-
Subsystems Level	
Users	
Sources...	-
Source Access (A/D...)	Many < 100 KHz
Transmission Facilities	
Source Coding - Audio	LPC Waves+[54]
- Video	Convex [55]
Bitstreams	
Error Control - BCH, RS	SPW [56]
Multiplexing	-
Channel Modulation	
Families - FM, QAM, PSK	SPW [56]
Structures	
Large Bandwidth	-
Large TW Product	MODAC [57]
Short Duration	-
Multiple Access	-
Timing Control	-
RF	-
Antennas	Ohio State [58]
Channels	
Models	Few e.g. SPW [56]
Noise Models	Few e.g. SPW [56]
Reception Facilities	
Channel Scanning	< 100 KHz
Demodulation	
Carrier Recovery	-
Timing Recovery	-
Analysis Facilities	
General	Many
Specialized	Few e.g.Elvira [59]
High Level Programming Facilities	
Signal Flow Languages	HyperSignal [60]
Script Interpreters	-
Targeting/ Rehosting Facilities (CAD)	
Hardware Description Languages	VHDL [56]
Microcode/PAL Downloading	SPROC, Xylinx[61]

Figure 12 Simulation and CAD Tools

References

[1] *Commanders' Tactical Terminal,* Farifax, VA: E-Systems Melpar, 1992.

[2] *Digital Multimedia Workstation,* Fairfax, VA: E-Systems Melpar, 1992.

[3] *Celltap,* Farifax, FA: E-Systems Melar, 1992.

[4] Cai, Phan and O'Connor, "Energy Detector Performance in A Noise Fluctuating Environment", *CH2681,* New York: IEEE Press, 1989.

[5] *Reference Data for Engineers,* Indianapolis: H.W.Sams 1986, 34-9.

[6] "At ISSCC: signals, SRAMs and systems", NY: *EE Times,* 24 Feb 92.

[7] Boeing Corp, "Advanced Optical A/D Converter," *DARPA Optics Review,* Arlington, VA: DARPA, 1992.

[8] "DARPA Critical Technologies", EE Times, 24 Feb 92

[9] *ZPB1603 Product Specification,* Tuscon, AZ: Burr Brown Inc, 1992

[10] Kamegawa et al, "Picosecond GaAs Monolithic Optoelectronic Sampling Circuit," NY: *IEEE Photonic Technology Letters,* Vol 3 No 6, Jun 91.

[11] Brown, "Schottky Heterojunction Sets Records," NY: *EE Times,* 24 Feb 92.

[12] Thao and Vetterli, "Optimal MSE Signal Recovery in Oversampled A/D Conversion Using Convexity," *Proceedings of the ICASSP-92.*

[13] Zakhor and Oppenheim, "Reconstruction of Two-Dimensional Signals from Level Crossings,' *Proc IEEE* Vol 78, No 1, Jan 90.

[14] *The Intel 80486,* Santa Monica: Intel Corp, 1991.

[15] *Guide to the TMS320C4x Parallel Processing Development System,* Dallas,TX:Texas Instruments,91.

[16] *Introducing the i860,* Santa Monica: Intel Corp, 1991.

[17] *PA RISC 7000,* Palo Alto, CA: Hewlett-Packard, 1992.

[18] *DAP,* LaJoya, CA:Active Memory Technologies, 1992.

[19] *MC860VS,* Lowell, MA: Mercury Corp, 1992.

[20] *Convex C210,* Greenbelt, MD: Convex Computer Corp, 1991.

[21] "Fujitsu has single chip VPU" *EE Times,* 24 Feb 92.

[22] *Maspar,* Gaithersburg, MD: Maspar Inc.

[23] *Common Signal Processor,*Manassas,VA:IBM Corp,91.

[24] "640 MFLOPS hits desktop," *EE Times,* 20 May 91.

[25] "Gigabit Codec" *EE Times,* 24 Jan 92.

[26] *DSP-3 Users' Guide,* Greensboro, NC: AT&T, 1991.

[27] *Convex C3800,* Greenbelt, MD: Convex Corp, 1991.

[28] "VASP integrates arrays of 32-bit DSP's", *EE Times,* 16 Nov 90.

[29] *Cray YMP90,* Greenbelt, MD: Cray Research, 92.

[30] "Kendall Readies Parallel System," *EE Times,* 9 Dec 91.

[31] "Thinking Machines CM-5 Boasts Scaleable Software," *EE Times,* 4 Nov 91.

[32] *Paragon Massive Parallel Systems,* Intel Corp, 1992.

[33] Kleinrock, *Queueing Systems, Vol II,* NY: Wiley 76.

[34] *HDTV Research,* Greenbelt, MD: Convex Corp, 1992.

[35] Weems, "Architectural Requirements of Image Understanding with Respect to Parallel Processing," *PROC IEEE,* Apr 91.

[36] Reiger, et al, *ZMOB: A New Computing Engine for AI,* College Park, MD: U. Maryland, 1981.

[37] *DSP-3 Users' Guide,* Greensboro, NC: AT&T, 1991.

[38] Eden, R.

[39] *High Temperature Super Conductuctivity,* Falls Church, VA: E-Systems Melpar Division, 1991.

[40] "Physics, Dynamics and Optoelectronic Integration of Surface Emitting Laser Diodes," *DARPA Optics Review,* 1992.

[41] "Fujitsu has single chip VPU" *EE Times,* 24 Feb 92.

[42] "DARPA Whisper Program", *EE Times,* 17 Feb 92.

[43] Lesser et al,"Organization of the Hearsay II Speech Understanding System," *IEEE Trans ASSP* 23 Jan 75.

[44] Whelchel et al "The Systolic Phase Rotation FFT - A New Algorithm and Parallel Processor Architecture," *Proc IEEE ICASSP 91.*

[45] Meyer, "Asymptotic Properties of Sequences Iteratively Generated by Point-to-Set Maps," *Mathematical Programming Study 10,* North Holland, 1979.

[46] Mitola, *Modular Signals Analysis Interactive Control System,* NY: ITT Corp, 1976.

[47] Ono, *Introduction to Topology,* Balt, MD: Johns Hopkins University Course Notes, 1974.

[48] Van de Linde, V.D., *Stochastic Optimal Control,* Balt, MD: Johns Hopkinks University Course Notes, 1974.

[49] Mitola, *Software Radios,* Fairfax, VA: E-Systems Melpar Division, 1992.

[50] "Research In Qualitative Physics," *Handbook of Artificial Intelligence,* Vol IV, 1990.

[51] *Assistant for Scientific and Technical Analysis,* Palo Alto, CA: Advanced Decision Systems, 1987.

[52] Machtey and Young, *An Introduction to the General Theory of Algorithms* NY: Elsevier, 1978.

[53] *BONES,* Foster City, CA: Comdisco Inc., 1992.

[54] *Waves+,* Washington, DC: Entropic Research Laboratory, Inc, 1992.

[55] *HDTV Research,* Greenbelt, MD: Convex Corp, 1992.

[56] *Signal Processing Workbench,* Foster City, CA: Comdisco Systems, Inc., 1992.

[57] *MODAC,* San Diego, CA: Pacific Sierra Research 1990.

[58] *Antenna Model,* Dayton, OH:Ohio State Univ, 1983.

[59] *Elvira,* Sunnyvale,CA: Applied Signal Technology 1992.

[60] *HyperSignal Macros,* Dallas, TX: Hyperception, 1992.

[61] *SprocLab,* Warren, NJ: Star Semiconductor, 1992.

SPOX, Santa Barbara, CA: Spectron Microsystems, Inc.

UNIX is a trademark of AT&T Corp.

The Software Radio Architecture

As communications technology continues its rapid transition from analog to digital, more functions of contemporary radio systems are implemented in software, leading toward the software radio. What distinguishes software radio architectures? What new capabilities are more economically accessible in software radios than digital radios? What are the pitfalls? And the prognosis?

Joe Mitola

JOE MITOLA is a consulting scientist with The MITRE Corporation.

SPW is a registered trademark of the Alta Group.

TMS320 is a registered trademark of Texas Instruments, Inc.

Matlab is a registered trademark of The MathWorks Inc.

SPOX is a registered trademark of Spectrum Microsystem.

COSSAP is a registered trademark of SYNOPSIS, Inc.

Matcad is a registered trademark of MathSoft, Inc.

We are poised on the threshold of another revolution in radio systems engineering. Throughout the '70s and '80s radio systems migrated from analog to digital in almost every respect from system control to source and channel coding to hardware technology. And now the software radio revolution extends these horizons by liberating radio-based services from chronic dependency on hard-wired characteristics, including frequency band, channel bandwidth, and channel coding. This liberation is accomplished through a combination of techniques that includes multi-band antennas and RF conversion; wideband Analog to Digital (A/D) and Digital to Analog (D/A) conversion (A/D/A conversion); and the implementation of IF, baseband, and bitstream processing functions in general-purpose programmable processors. The resulting software-defined radio (or "software radio") in part extends the evolution of programmable hardware, increasing flexibility via increased programmability. And, in part, it represents an ideal that may never be fully implemented but that nevertheless simplifies and illuminates tradeoffs in radio architectures that seek to balance standards compatibility, technology insertion, and the compelling economics of today's highly competitive marketplaces.

Radio Architecture Evolution

Of the many possible definitions of architecture, the one that best relates systems, technology and economics is best suited to this discussion. We shall therefore define "architecture" as the comprehensive, consistent set of functions, components and design rules according to which systems of interest may be organized, designed, and constructed. A specific architecture entails a partitioning of functions and components such that functions are assigned to components and interfaces among components correspond to interfaces among functions.

When such functions and interfaces are defined in formal design rules via a public forum, the resulting architectures are called "open." The full economic benefits of open architectures require the existence of a large commercial base which sometimes fails to emerge in spite of openness. As system complexity increases, architecture becomes more critical because of its power to either simplify and facilitate system development (a "powerful" architecture) or to complicate development and impede progress (a "weak" architecture).

Radio architectures may be plotted in the phase space of network organization versus channel data rate, as shown in Fig. 1. These architectures have evolved from early point-to-point and relatively chaotic peer networks (e.g., citizens band and push-to-talk mobile military radio networks) toward more hierarchical structures with improved service quality. In addition, channel data rates continue to increase through multiplexing and spectrum spreading. In a multiple-hierarchy application, a single radio unit, typically a mobile terminal, participates in more than one network hierarchy. A software radio terminal, for example, could operate in a GSM network, an AMPS network, and a future satellite mobile network. Multiband multimode military radios and future Personal Communications Systems (PCS) that seamlessly integrate multimedia services across such diverse access modes represent the high end of that evolution and the focus of this discussion. The complexity of functions, components and design rules of these architectures continues to increase with each generation, as shown in Table 1. In particular, future seamless multimode networks will require radio terminals and base stations with agile RF bands, channel access modes, data rates, Bit Error Rates (BERs), power, and functionality. Software radios have emerged to increase quality of service through such agility. At the same time, software radio architectures simplify hardware component tradeoffs and provide new ways of managing the complexity of rapidly emerging standards.

The Canonical Software Radio Architecture

The components of the canonical software radio consist of a power supply, an antenna, a multiband RF converter, and a single chip containing

Reprinted from *IEEE Communications Magazine*, Vol. 33, No. 5, pp. 26–37, May 1995.

Generation	Illustrative Functions	Typical Components	Key Design Rules
Analog	Transmit and receive, channel select, squelch	Power, antenna, packaging (discrete analog baseband)	Channel allocations, power limits, standard modulations (AM, FM)
Early digital microwave	Transmit or receive protected modes, BER control	Analog + quadrature modems + Forward Error Control (FEC)	Analog + operations/management and bitstream multiplex interfaces, adjacent channel power envelopes
Analog Mobile Cellular Radio (MCR)	Analog + signaling and control	Analog + digital modems + embedded control processors	Analog + early digital + cell site and frequency plan + handoff protocol
Spread spectrum (CDMA/FH)	Code synchronization, code management, BER control	Analog + early digital + de-/spreading devices + embedded control processors	Analog + code design + peer network protocols + digital voice channel
TDMA MCR	Analog + early digital +analog MCR + spread spectrum + diversity + directivity	Analog + analog MCR + spread spectrum + multibeam antennas	Analog + early digital + analog MCR + digital voice channel + privacy and authentication
Future seamless multimode multimedia networks	Digital MCR + agile directional power management + data rate management + BER agility + mode handover + location reporting	Digital MCR + high programmability + agile modulators + multiband antennas + multiband RF	Digital MCR + mode handover criteria and protocols + end-to-end encryption + software defined services

Digital MCR = CDMA, TDMA, Frequency Hop (FH), and related hybrids.

■ **Table 1.** *Key elements of radio architecture evolution. Complexity of functions, components, and design rules increases with successive generations.*

A/D/A converters with an on-chip general purpose processor and memory that perform the radio functions and required interfaces illustrated in Fig. 2. The canonical mobile software radio terminal interfaces directly to the user (e.g., via voice, data, fax, and/or multimedia). The canonical base station interfaces to the public switched telephone network (PSTN). Fully instrumented base stations support operations and maintenance, developers and researchers via services development workstation(s). The placement of the A/D/A converters as close to the antenna as possible and the definition of radio functions in software are the hallmarks of the software radio. Thus, although software radios use digital techniques, software-controlled digital radios are generally not software radios. The key difference is the total programmability of software radios, including programmable RF bands, channel access modes, and channel modulation.

Contemporary radio designs mix analog hard-

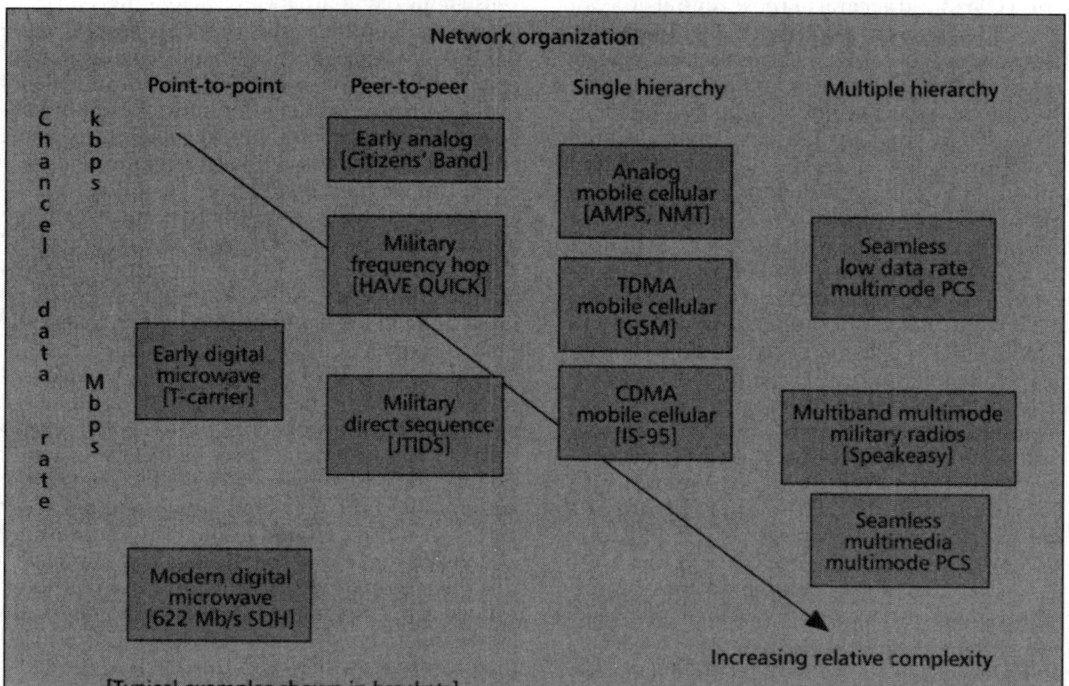

■ **Figure 1.** *Radio architectures are evolving towards high levels of complexity.*

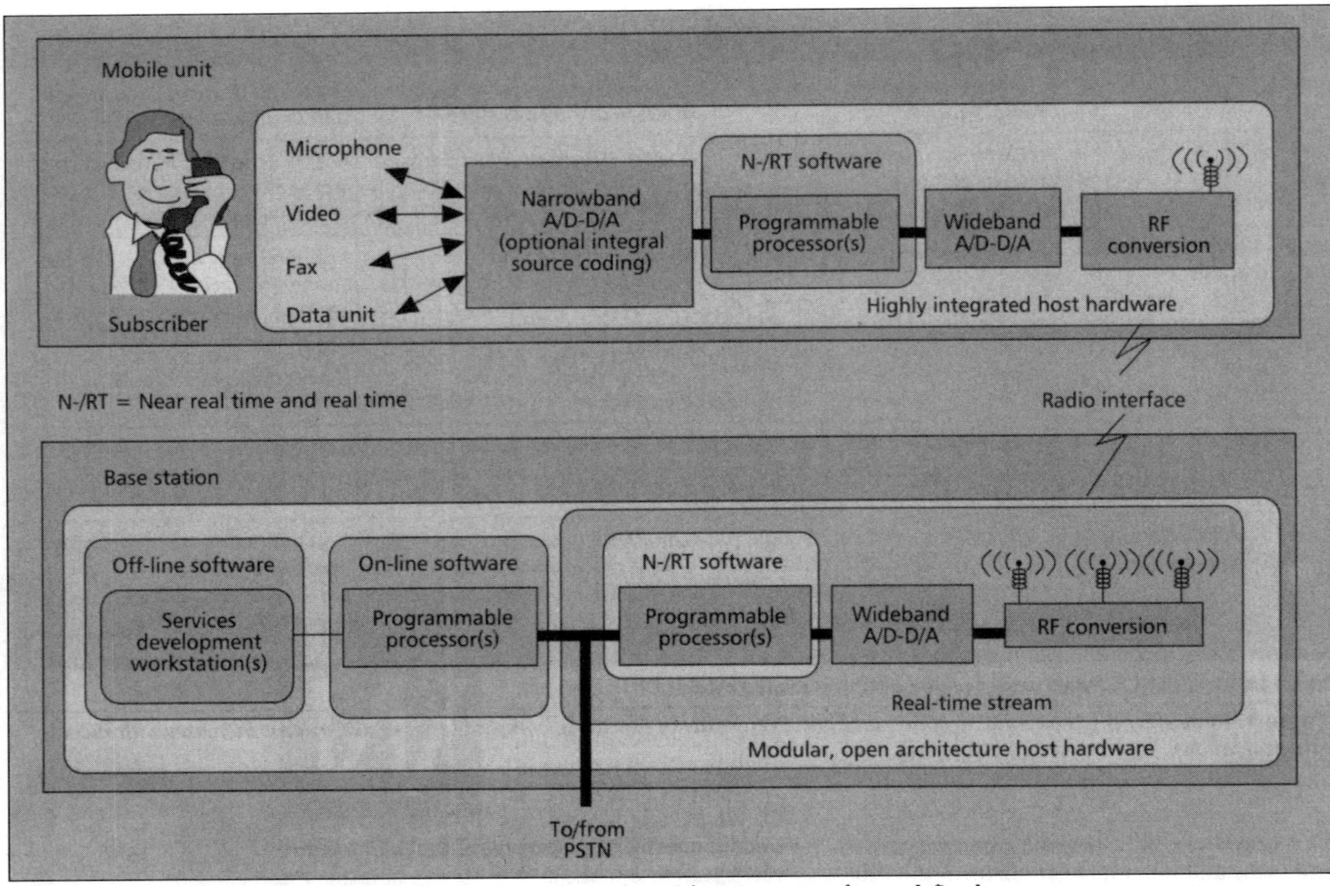

Figure 2. *In the canonical software radio, hardware is simple and functions are software-defined.*

ware, digital hardware, and software technologies. It is instructive to consider software radios *per se* to better understand benefits, pitfalls and relationships to other technologies. Software radios have become practical as costs per millions of instructions per second (MIPS) of digital signal processors (DSPs) and general purpose central processor units (CPUs) have dropped below U.S. $10 per MIPS. The economics of software radios become increasingly compelling as demands for flexibility increase while these costs continue to drop by a factor of two every few years. At the same time, absolute capacities continue to climb into the hundreds of millions of floating-point operations per second (MFLOPS) per chip, at which point software radios are compatible with commercial TDMA and CDMA applications. In addition, A/D/A converters available in affordable single-board open architecture configurations offer bandwidths of tens of MHz with the dynamic range required for software radio applications [1]. Multimedia requirements for the desktop and palmtop continue to exert downward pressure on parts count and on power consumption of such chip sets, pushing the software radio technology from the base station to the mobile terminal. Although the tradeoffs among analog devices, low-power ASICs, DSP cores and embedded microprocessors in handsets remain fluid, cutting-edge base stations employ software radio architectures. Finally, the multiband, multimode flexibility of software radios appears central to the goal of seamless integration of PCS, land mobile and satellite mobile services (including truly nomadic computing) toward which many of us aspire.

But software radio engineering is fraught with pitfalls. It is difficult to engineer wideband, low-loss antennas and RF converters. It is also difficult to accurately estimate processing demand of applications and processing capacity of reprogrammable DSP/CPU configurations. In addition, sustaining required data rates across interprocessor interfaces is problematic. There are several vendor-unique high capacity wideband signal buses, but open architecture standards for this critical element of the software radio architecture have not yet emerged. DSP function libraries continue to expand, and block diagram-based "integrated environments" exist. But we do not yet have the ability to mix and match real-time software tools and modules from different software suppliers as we can mix and match VME boards today. Most of the pitfalls can be avoided, however. And continued progress in these difficult areas will further reduce costs and time to market.

This article therefore provides a tutorial review of software radio architectures and technology, highlighting benefits, pitfalls, and lessons learned. This includes a closer look at the canonical functional partitioning of channel coding into antenna, RF, IF, baseband, and bitstream segments. A more detailed look at the estimation of demand for critical resources is key. This leads to a discussion of affordable hardware configurations, the mapping of functions to component hardware, and related software tools. This article then concludes with a brief treatment of the economics and likely future directions of software radio technology. Companion articles in this issue address the critical technologies of A/D convert-

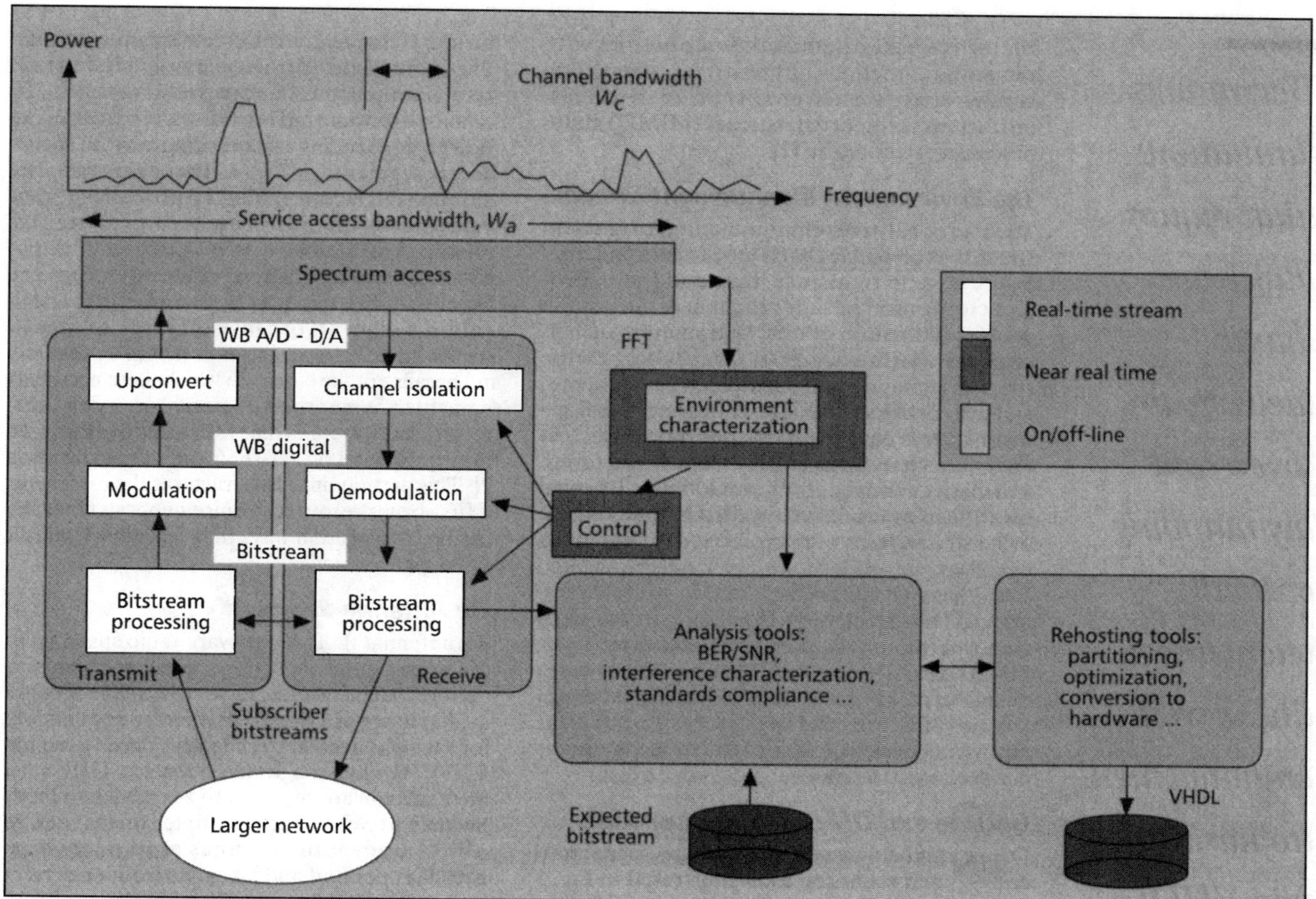

■ **Figure 3.** *Software radio functional architecture in a mobile cellular base station application.*

ers and DSP cores and cutting-edge applications in greater detail.

Software Radio Overview

The software radio architecture is widely applicable to trunk radios, peer networks, air and sea traffic management, mobile military communications, and satellite mobile systems. For simplicity, this overview describes the software radio architecture in a mobile cellular/PCS setting.

In an advanced application, a software radio does not just transmit: it characterizes the available transmission channels, probes the propagation path, constructs an appropriate channel modulation, electronically steers its transmit beam in the right direction, selects the appropriate power level, and then transmits. Again, in an advanced application, a software radio does not just receive: it characterizes the energy distribution in the channel and in adjacent channels, recognizes the mode of the incoming transmission, adaptively nulls interferers, estimates the dynamic properties of desired-signal multipath, coherently combines desired-signal multipath, adaptively equalizes this ensemble, trellis decodes the channel modulation, and then corrects residual errors via forward error control (FEC) decoding to receive the signal with lowest possible BER.

Finally, the software radio supports incremental service enhancements through a wide range of software tools. These tools assist in analyzing the radio environment, defining the required enhancements, prototyping incremental enhancements via software, testing the enhancements in the radio environment, and finally delivering the service enhancements via software and/or hardware.

The Real-Time Channel Processing Stream

The canonical software radio architecture includes the channel processing stream, the environment management stream and associated software tools illustrated in Fig. 3. The real-time channel processing stream incorporates channel coding and radio access protocols. Channel processing is characterized by discrete time point-operations such as the translation of a baseband signal to an intermediate frequency (IF) by multiplying a discrete time-domain baseband waveform by a discrete reference carrier to yield a sampled IF signal. The time between samples is on the order of tens of microseconds to hundreds of nanoseconds. Such point-operations require hundreds of MIPS and/or MFLOPS to Giga-FLOPS with strictly isochronous performance. That is, sampled data values must be computationally produced and consumed within timing windows on the order of the time between samples in order to maintain the integrity of the signals represented therein. Input/output (I/O) data rates of this stream approach a gigabit per second per A/D converter. Although these data rates are decimated through processing, it is challenging to sustain isochronism through I/O interfaces and hard real-time embedded software in this stream. Multiprocessing is

Technology limitations that require hardware-based delivery are overcome by mapping critical elements of the service enhancement to hardware via VHDL.

therefore best organized as a pipeline with sequential functions of the stream assigned to serially interconnected processors, i.e., a multiple instruction multiple data-stream (MIMD) multi-processing architecture [2].

The Environment Management Stream

The near-real-time environment management stream continuously characterizes radio environment usage in frequency, time and space. This characterization includes channel identification and the estimation of other parameters such as channel interference levels (depending on the specific signaling and multiple access scheme) and subscriber locations. The environment management stream employs block operations such as fast Fourier transforms (FFTs), wavelet transforms, and matrix multiplies for beam forming. Channel identification results are needed in times on the order of hundreds of microseconds to hundreds of milliseconds, while power levels may be updated in milliseconds and subscriber locations may be updated less frequently. The block structure of such operations is readily accommodated by a MIMD parallel processor. The interface between this highly parallel environment management stream and the pipelined channel processing streams must synchronize the environment management parameters to the channel processing streams.

On-Line and Off-Line Software Tools

On-line and off-line systems analysis, signal processing, and rehosting tools illustrated in Fig. 3 allow one to define incremental service enhancements. For example, an enhanced beamformer, equalizer and trellis decoder may be needed to increase subscriber density. These enhancements may be prototyped and linked into the channel processing stream, allowing one to debug the algorithm(s), to experiment with parameter settings, and to determine the service value (e.g., in improved subscriber density) and resources impact (e.g., on processing resources, I/O bandwidth, and time delays).

Software-based enhancements may be organized around managed objects, collections of data and associated executable procedures that work with object resource brokers and conform to related open architecture software interface standards such as the Common Object Resource Broker (CORBA) [3]. Enhancements may then be delivered over the air to other software radio nodes, as contemplated in the future software-defined telecommunications architectures being considered by ITU-T and embraced by NTT [4] and others [5, 6]. A well integrated set of analysis and rehosting tools leads to the creation of incremental software enhancements relatively quickly, with service upgrades provided over-the-air as software-defined networks proliferate. Technology limitations that require hardware-based delivery are overcome by mapping critical elements of the service enhancement to hardware via VHDL.

Partitioning of The Channel Processing Stream

The classical canonical model of communications concatenates source encoder, channel encoder, channel (which adds noise, interference, and distortion), channel decoder, and source decoder. The channel encoder/decoder and related radio access functions constitute the real-time stream. The canonical software radio architecture partitions classical channel coding and decoding into the channel access segments of Fig. 4. These segments are: antennas, RF conversion, IF processing, baseband processing and bitstream processing. This canonical partitioning is useful because of the significant differences in functionality between segments, because of the strong cohesion among functions within a segment; because of large changes in bandwidth due to decimation within a segment; and because of the ease with which these particular segments are mapped to affordable open-architecture hardware. Further rationale for this partitioning is based on set-theoretic considerations [7]. This partitioning also structures the estimation of first-order resource requirements so that they may be combined in ways that accurately predict system performance.

The Antenna Segment

The antenna(s) of the software radio span multiple bands, up to multiple octaves per band with uniform shape and low losses to provide access to available service bands. In military applications, for example, a mobile terminal may need to employ VHF/UHF line of sight frequencies, UHF satellite communications, and HF as a backup mode. Switched access to such multiple bands requires octave bandwidth antennas and/or multiple antennas per band and an agile frequency reference in the RF Segment. In addition, multiple antenna elements may be part of a beam forming network for interference reduction or space division multiple access (SDMA).

The relationship between interference cancellation capacity and the number of antenna elements varies. A single auxiliary element, for example, can reduce interference of a large number of interferers (e.g., those in the "back lobe"). Algorithms that reduce interference through non-spatial techniques (e.g., cyclostationary algorithms) can also reduce a large number of interferers with one or with no auxiliary antenna elements. Beam forming of N antenna elements can place $N-1$ adaptive nulls on interferers sufficiently separated in azimuth, but coherent multipath may require a distinct null for each distinct multipath direction, reducing the number of interferers accordingly. Polarization scrambling from nearby reflecting surfaces may require $2N + 1$ elements for N paths. The structure of the antenna array(s) determines the number of distinct physical and logical signal processing paths in the RF conversion and IF processing segments. As a result, the competing demands for directional selectivity, multipath compensation and interference suppression versus wideband low-loss antennas versus affordability define the tradeoffs of the antenna segment.

The RF Conversion Segment

RF conversion includes output power generation, preamplification, and conversion of RF signals to and from standard intermediate frequencies (IFs) suitable for wideband A/D/A conversion. In most radio bands, RF conversion will be analog. Certain critical RF problems are exacerbated in the software radio. These include the need for amplifier

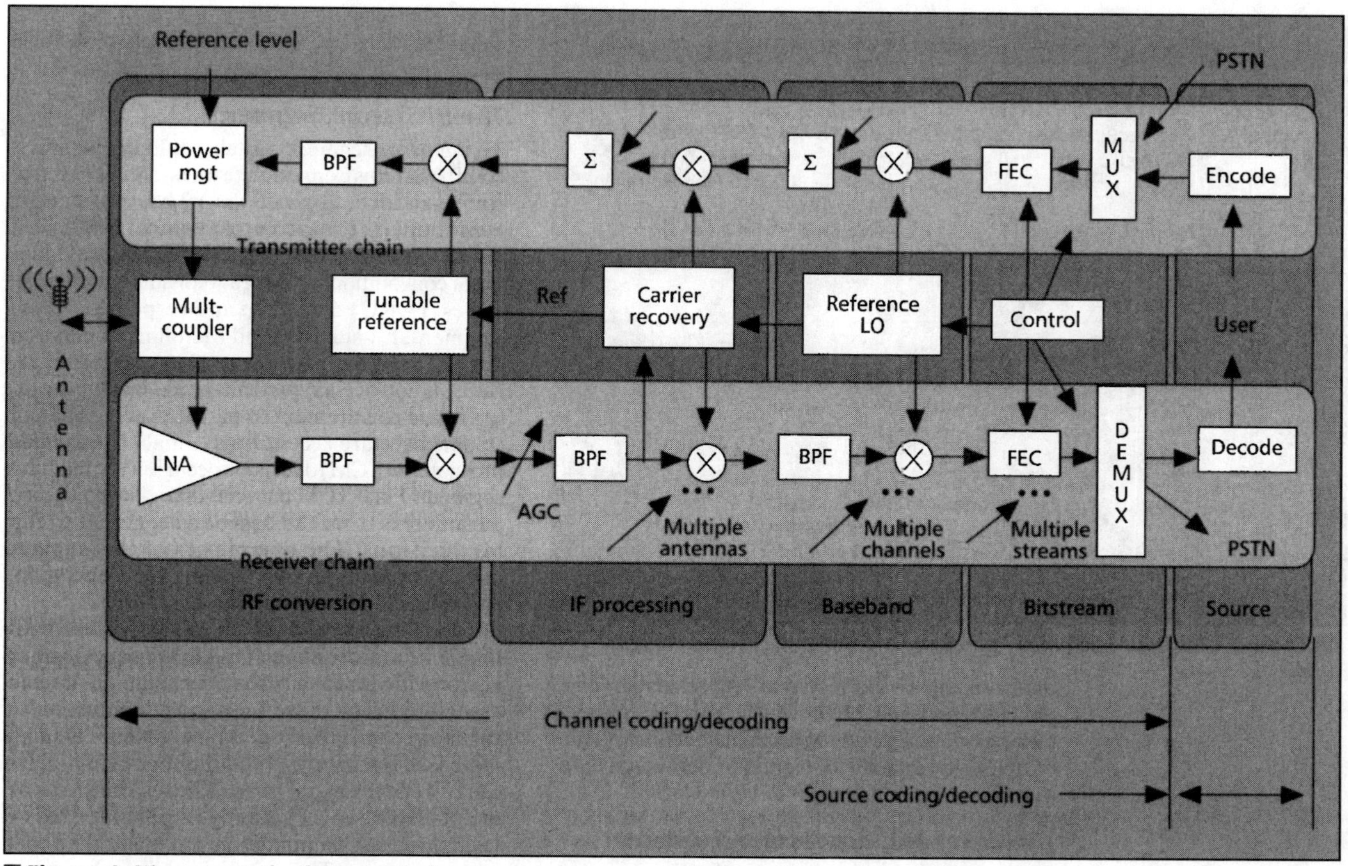

■ **Figure 4.** *The canonical software radio functional architecture maximizes cohesion and minimizes coupling.*

linearity and efficiency across the access band. RF shielding of processors may also be necessary to avoid the introduction of processor clock harmonics into the analog RF/IF circuits. Cositing of multiple transmitters also creates electromagnetic interference (EMI) problems, but these are about the same for software radios as for cosite collections of multiple discrete hardware radios.

Placement of the A/D/A Converters Is Key

W_a, the bandwidth of the IF to be digitized, determines what kinds of A/D techniques are feasible. According to the Nyquist criterion for band limited signals f_s, the sampling rate of the A/D converter, must be at least twice W_a. Practical systems typically require modest oversampling:

$$f_s > 2.5\, W_a$$

Wideband A/D/A converters access broad instantaneous segments of spectrum, typically 10 to 50 MHz. Such wide access may also be achieved in parallel subbands of more modest 1 to 10 MHz bandwidths each. The dynamic range of each parallel subband depends on the dynamic range of the A/D/A converters. Since the product of dynamic range times sampling rate is approximately constant for a given A/D/A technology, narrower subbands generally increase the useful dynamic range, albeit at the cost of increased system complexity. The many issues related to filtering and A/D conversion are deferred to the companion article on A/D converters in this issue [1].

The placement of wideband A/D/A conversion before the final IF and channel isolation filters achieves three key architectural objectives:

- It enables digital signal processing before detection and demodulation.
- It reduces the cost of mixed channel access modes by consolidating IF and baseband processing into programmable hardware.
- It focuses the component tradeoffs to a single central issue: providing the computational resources (I/O bandwidth, memory, and processing capacity) critical to each architecture segment, subject to the size, weight, power, and cost constraints of the application.

The IF Processing Segment

The IF processing segment maps the transmit and receive signals between modulated baseband and IF. The IF receiver processing segment includes wideband digital filtering to select a service band from among those available. Furthermore, IF filtering recovers medium band channels (e.g., a 200 kHz TDMA channel in GSM) and/or wideband subscriber channels (e.g., a 2 MHz CDMA channel) and converts the signal to baseband. The complexity of frequency conversion and filtering is the first order determinant of the processing demand of the IF segment. In a typical application, a 12.5 MHz mobile cellular band is sampled at 30.72 MHz (M samples per second). Frequency translation, filtering and decimation requiring 100 operations per sample equates to more than 3000 MIPS of processing demand. Although such microprocessors are on the horizon, contemporary implementations offload this computationally intensive demand to dedicated chips such as the Harris Decimating Down Converter (DDC) or Gray digital receiver chip. Spreading and de-spreading of CDMA, also an IF processing function, creates demand that is proportional to the bandwidth of

Application	Radio function	Segment	First order demand drivers
Analog Voice Channel (Receiver)	Companding	Source	Speech bandwidth (W_v) and sampling rate
	Gap suppression	Bitstream	Gap identification algorithm
	FM modulation	Baseband	Interpolation required (W_{fm}/W_v)
	Up conversion	IF	IF carrier, f_i, and FM bandwidth: $W_i = W_{fm}$
	Band selection	IF	Access bandwidth (W_a)
	Channel selection	IF	Channel bandwidth (W_c)
	FM demodulation	Baseband	f_i, W_{fm}
	DS0 reconstruction	Bitstream	Speech bandwidth
TDMA (TDM)	Voice codec	Source	Choice of voice codec
	FEC coding	Bitstream	Code rate, algorithm complexity
	Framing	Bitstream	Frame rate (R_f)
	MSK modulation	Baseband	Baud rate (R_b)
	Up conversion	IF	f_i, $R_b/2$
(Receiver)	Band selection	IF	Access bandwidth (W_a)
	Channel selection	IF	Channel bandwidth (W_c)
	Demodulation	Baseband	Baud rate (R_b) or channel bandwidth (W_c)
	Demultiplexing	Bitstream	Frame rate (R_f)
	FEC decoding	Bitstream	Code rate, algorithm complexity
	DS0 reconstruction	Source	Voice codec

■ **Table 2.** *Illustrative functions, segments, and resource demand drivers.*

the spreading waveform (typically the chip rate) times the baseband signal bandwidth. This is so computationally intensive that with current technology limitations, it is typically assigned to dedicated chips as well.

The Baseband Processing Segment

The baseband segment imparts the first level of channel modulation onto the signal (and conversely demodulates the signal in the receiver). Predistortion for nonlinear channels would be included in baseband processing. Trellis coding and soft decision parameter estimation also occur in the baseband processing segment. The complexity of this segment therefore depends on the bandwidth at baseband W_b, the complexity of the channel waveform, and the complexity of related processing (e.g., soft decision support). For typical digitally encoded baseband waveforms such as binary phase shift keying (BPSK), quadrature phase shift keying (QPSK), Gaussian minimal shift keying (GMSK), and 8-PSK with channel symbol (baud) rates of R_b:

$$R_b/3 < W_b < 2^*R_b$$

In the transmission side of the baseband segment, such waveforms are generated one sample at a time (a "point operation"). If three samples are generated for the highest frequency component, demand falls between R_b and 6^*R_b. Greater oversampling decreases the transmitted power of spectral artifacts, but also increases transmit power and processing demand. In the receiver, digital baseband modulations require timing recovery which typically includes the integration of baud intervals over time. If baud interval is measured in transitions of a high-speed clock, some timing-sensitive signal structures (e.g., TDMA) and some synchronization algorithms require up to 96 b precision integer arithmetic in the clock recovery loop(s), and such extended precision arithmetic may not be readily available, particularly on newer chips. Analog baseband modulation, such as FM voice, may also be encoded and demodulated in software in the baseband segment, with a processing demand of less than 1 MIPS per subscriber. Such software simulation of analog modulations helps achieve backwards compatibility with analog standards.

The Bitstream Segment

The bitstream segment digitally multiplexes source-coded bitstreams from multiple users (and, conversely, frames and demultiplexes them). The bitstream segment imparts forward error control (FEC) onto the bitstream, including bit interleaving and block and/or convolutional coding and/or automatic repeat request (ARQ) detection and response. Frame alignment, bit-stuffing, and radio link encryption occur in the bitstream segment. Encryption requires the isolation of encrypted bits from clear bits, resulting in the requirement to partition and isolate bitstream hardware accordingly. Final trellis-coded modulation (TCM) decisions occur in the bitstream segment. Final TCM converts soft/delayed decision parameters from the baseband segment to final bit decisions. The complexity of this segment depends on multiplexing, framing, FEC, encryption, and related bit manipulation operations.

Signaling, control and operations, administration and maintenance (OA&M) functions are also provided in the bitstream segment. The demand associated with these functions depends on the signaling, control and operations systems. Demand increases linearly with the number of simultaneously active subscribers. These functions are event-driven and typically impart an order of magnitude less computational demand than baseband processing. These functions, may, however, require access to distributed data bases, not all of which will be local to the base station. Thus, although the processing demand is relatively small, the timing requirements may be severe.

The Source Segment

The source segment differs between the mobile terminal and the base station. In the mobile terminal, the source segment consists of the user and the source encoders and decoders. Here, the relatively narrowband voice and fax A/D/A converters are typically located in the handset, palmtop, or workstation. In the base station, on the other hand, the source segment consists of the interface to the PSTN for access to remote source coding. Conversion of protocols required for interoperability with the PSTN creates processing demand in the base station's source segment. Conversion of DS0 64 kb/s PCM to RPE-LTP (GSM [11]), for example, would create 1 to 2 MIPS of demand per subscriber.

End-to-End Timing Budgets

Time delays are introduced in the IF, baseband, bitstream and source segments due to finite I/O and processing resources that empty and fill buffers in finite but sometimes random amounts of time. The end-to-end accumulation of these delays must be kept within bounds of isochronism at each segment-to-segment interface. These bounds depend on signal type and the larger network architecture. Thus, for example, end-to-end voice delay should be less than 150 ms, but the external network may consume 100 ms of this timing budget, leaving only 50 ms for the software radio. Maintaining such budgets in software radios is complicated (compared to digital radios) by queuing delays internal to and between processors.

Typical application	RF fc	Wa	IF Wi	Channel code	Baseband Wi	Bitstream States	FEC	Multi-plexing	Privacy	Source
FM mobile	VHF	30 kHz	30 kHz	FM	4 kHz	Inf	—	(PTT)	None	Compand
MCR voice	UHF	25 MHz	25 kHz	FM	4 kHz	Inf	—	(CA)	None	Compand
control		"	"	MSK	10 kb/s	4	Conv	TDM	None	Data
GSM	UHF	25 MHz	200 kHz	GMSK	270.83 kb/s 13 kb/s	4	CPF	TDM	Encrypt	RPE-LTP
Mil FH	VHF	60 MHz	30 kHz Agile	FH-QPSK	16 kb/s	4	Conv ARQ	(PTT)	Encrypt	LPC/Delta
CDMA	UHF	175 MHZ	50 MHz	M-PSK	8 kb/s	M	Conv	(CD)	Privacy	LPC
JTIDS [12]	L-Band	250 MHz	3 MHz	MSK	150 kb/s	4	Multiple	TDM	Encrypt	LPC
Microwave	SHF	20 MHz	20 MHz	64 QAM	90 Mb/s	64	Multiple	TDM	Bit Stuff	PCM

(Implicit Multiplexing): (PTT) = Push to Talk; (CA) = Circuit Assigned; (CD) = Code Division
CPF: GSM specifies various modes of Convolutional, Parity and Fire codes [8].

■ **Table 3.** *Critical applications parameters bound offered demand.*

Estimating Resource Requirements

Resources critical to the software radio architecture include I/O bandwidth, memory and processing capacity. Good estimates of the demand for these resources results in a well informed mapping of the above segments to appropriate hardware. Depending on the details of the hardware, the critical resource may be memory, bus, or I/O bandwidth, or a particular embedded processor. Identification of the critical resource can be accomplished quickly using the techniques highlighted in this section. When such critical resources are identified early in the design process, order of magnitude shortfalls in performance can be avoided. By identifying first-order demand drivers; aggregating demand for each major system resource; comparing demand systematically to hardware/software capacity; and then managing the critical resources appropriately, such shortfalls can be avoided.

Standardized Measures of Demand and Capacity

Since many contemporary processors include pipelined floating point arithmetic sections or single instruction butterfly operations, MIPS and MFLOPS are *not* interchangeable. Both types of operations, however, require processor clock cycles, allowing one to express demand in a common measure of millions of operations per second (MOPS) where an operation is the amount of work that can be accomplished by a given resource (CPU, floating point unit, etc.) in a single clock cycle of a standard width (e.g., 32 b). Using this measure, required MIPS and MFLOPS may both be expressed in MOPS. In addition, I/O, direct memory access, auxiliary I/O processor throughput, memory and bus bandwidths may all be expressed in MOPS where the operand is a standard data word and the operation is store or fetch. MOPS may then be accumulated independently for each potentially critical resource (CPU, DSP unit, floating point processor, I/O bus, memory bank, etc.). Finally, software demand may be translated rigorously to equivalent MOPS. Benchmarking is the key to this last step since hand coded assembly language algorithms often out-perform high order language (HOL) code by an order of magnitude and HOL often outperforms code-generating software tools by an order of magnitude. Rigorous analysis of performance is accomplished by comparing demand and capacity in standards MOPS for each resource using queueing theory. This identifies bottlenecks and yields useful estimates of performance.

Estimate Demand In The Context of The Canonical Data Flow

Table 2 shows how functional architecture parameters drive the resource demand of the associated segment in illustrative applications. Although the associated demand may exceed the capacity available with a given generation of devices, the capacity estimates serve to identify the hardware that best supports a given segment. By determining the number of operations required per point-operation (e.g., a filtering stage) and multiplying by the critical parameter (e.g., the data rate of the stream being filtered), one can quickly arrive at demand estimates that frame the related implementation decisions.

Table 3 shows critical parameters of processing demand for illustrative applications. The demand depends to a first-order approximation on the signal bandwidths and on the complexity of key operations within IF, baseband, bitstream, and source segments as follows:

$$D = D_{if} + N^{*}(D_{bb} + D_{bs} + D_{s}) + D_{o}$$

Where D_{if}, the IF processing demand, is proportional to the acquisition bandwidth, and to the complexity of the service band isolation filter and of the subscriber channel isolation filter(s). N is the maximum number of simultaneous subscribers to be serviced by the node. The baseband demand D_{bb} is proportional to the bandwidth of a single subscriber channel and the complexity of demodulation. The bitstream demand D_{bs} is proportional to the data rate and the complexity of the forward error control algorithm, bit stuffing/interleaving, signaling and control and miscellaneous operations functions. The source segment demand D_s depends on the complexity of the interface to the PSTN. And finally, the management overhead processing demand D_o depends on the signaling system, the radio access network overhead and on the degree of instrumentation that is provided to OA&M, developers and/or researchers.

Table 4 shows how key parameters and processing demand are related for an illustrative application. This demand must be met by processors with sufficient capacity to support real-time performance, including isochronism and end-to-end delay budgets. At present, most IF processing is off-loaded to special-purpose digital receiver chips because general purpose processors with sufficient MOPS are not yet cost effective compared to specialized hardware. This tradeoff changes approximately every 18 months, with the trend in favor of the general-purpose processor. In the table, the total baseband plus bitstream processing demand of 4.7 MOPS per user is within the capabilities of nearly all DSP chips. As a result, several subscribers may share a single DSP chip. Aggregate demand (all users, including overhead) of 142.6 MOPS is nominally within the capacity of a Quad TMS320 C50 board, but when multiplexing more than one user's stream into a single processor, memory buffer sizes, bus bandwidth and fan-in/fan-out may be more critical than processing capacity.

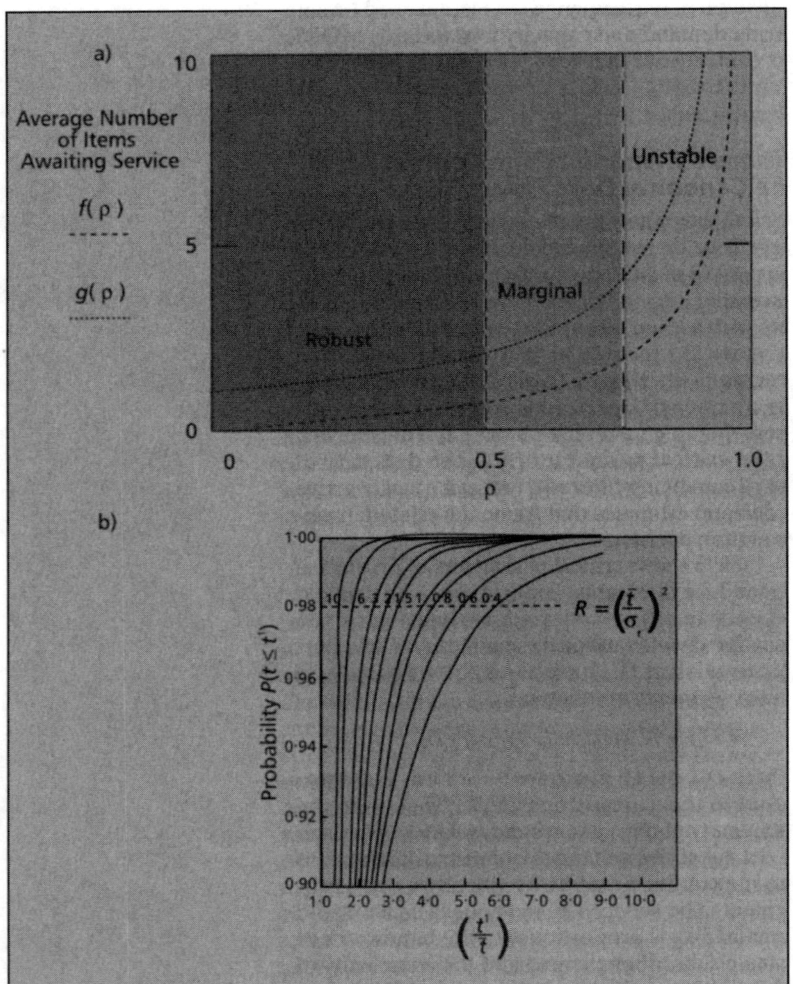

■ **Figure 5.** *a) Facility utilization determines performance, underscoring the criticality of demand estimates; b) ratio of variance to mean determines reliability, $P(t \leq t')$ for specific offered demand [Gamma distribution courtesy of Tebbs & Collins ©1977, McGraw-Hill (UK), reproduced with permissionof McGraw-Hill].*

Facility Utilization Accurately Predicts Performance

The most significant design parameter of the mapping of processing demand to processor capacity is resource utilization (of the CPU, DSP chip, memory, bus, etc.). Resource utilization, ρ, is the ratio of offered demand to available capacity. Figure 5a shows how the average number of items awaiting service at the resource varies as a function of utilization. (See [9] for an accessible treatment of material covered at a deeper level in [8] and [10].) The curve $f(\rho)$ represents exponentially distributed service times, while $g(\rho)$ represents uniform service times; system performance generally lies on or between these two curves. Rarely does usable processing capacity exceed 50 percent of stand-alone capacity as indicated by operating regions. Robust performance occurs when ρ is less than 0.5. The undesired events that result in service degradation will occur with noticeable regularity for $0.5 < \rho < 0.75$. For $\rho > 0.75$, the system is generally unstable, with queue overflows regularly destroying essential information.

An analysis of variance is required to establish the risk that the required time delays will be exceeded. The incomplete Gamma distribution of Fig. 5b characterizes the risk of exceeding a specified delay through the ratio of the specified delay to the average delay (under various relationships of the mean to the variance). Software radios work well if there is a 98 percent probability of staying within specified performance using the curve for $R = 1$ in the figure.

The simplified analysis presented above illustrates the key ideas with the processor as the critical resource. But in many cases, I/O, the backplane bus or memory access will be the critical resource. The approach is readily adapted to any critical resource. Simulation can be used to refine the estimates obtained from this simple model. Management of the system bottlenecks requires the further insights gained through benchmarking and rapid prototyping. But there is no free lunch. Software radios require up to ten times the raw processing power of ASICs and special purpose chips of the same technology. Software radios therefore lag special purpose hardware implementations by about one or two hardware generations. Thus, software radio architectures have appeared first in base stations using heterogeneous multiprocessing hardware, which is discussed briefly in the following section.

Heterogeneous Multiprocessing Hardware

Segment boundaries among antennas, RF, IF, baseband, bitstream, and source segments evolved because they make it easy to build software radios on parallel, pipelined, heterogeneous multiprocessing hardware. Such partitioning maps the segments to affordable open hardware architectures, as illustrated in Fig. 6. In this example, the VME chassis hosts the RF, IF, baseband, and bitstream segments, while the workstation houses the user interface, research tools, development

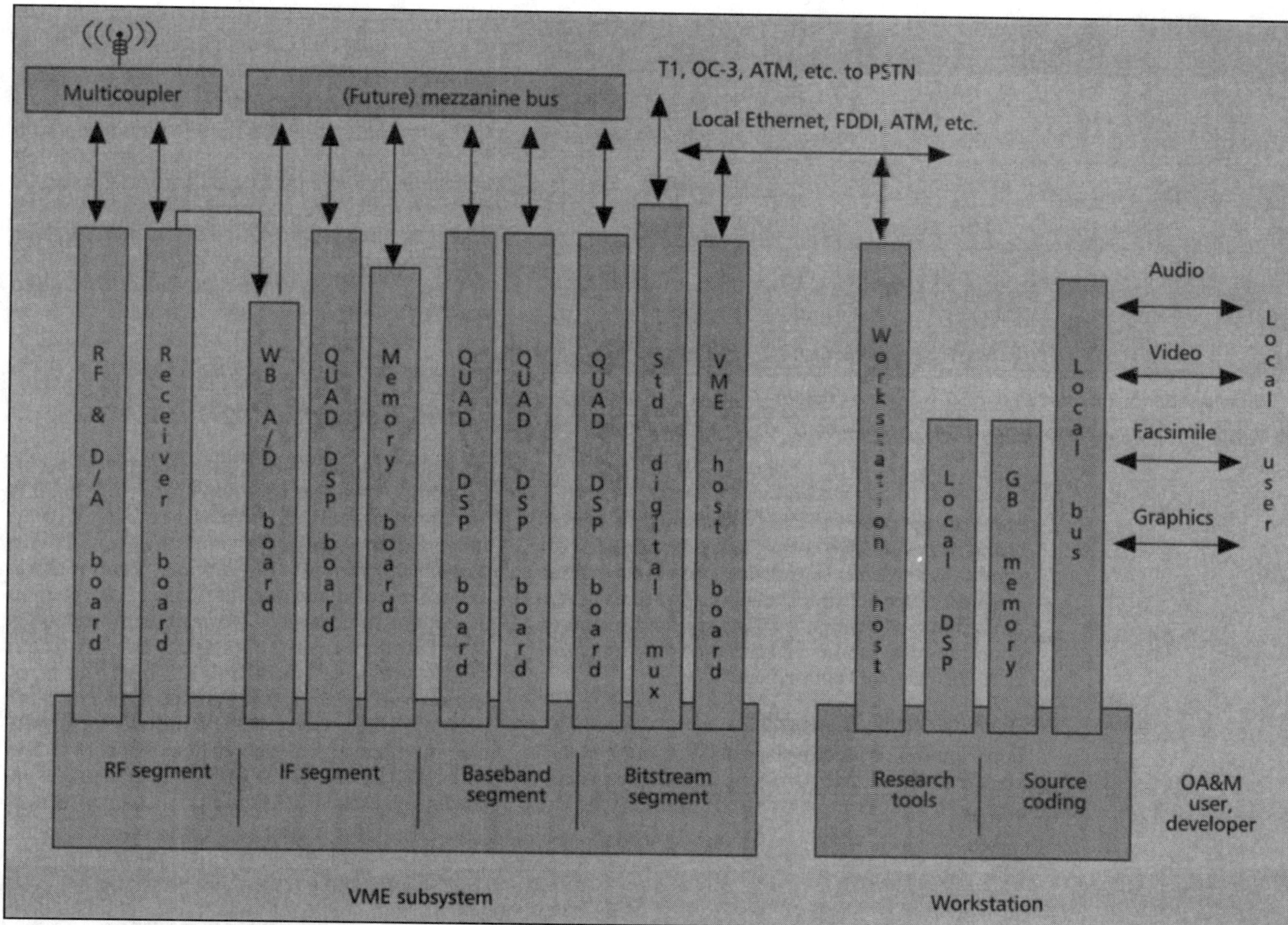

■ **Figure 6.** *Open component architecture supplies processing capacity with affordable technology insertion.*

tools and local source coding/decoding (Table 5).

In the most cost effective leveraging of open architectures, VME systems are configured by experienced developers as heterogeneous multiprocessors. The VME host serves as systems control processor . The DSP processors support the real-time channel processing stream, sometimes configured as one DSP per subscriber channel. The path from A/D to the first filtering/decimation stage typically uses a dedicated point-to-point mezzanine interconnect such as DT Connect™ (Data Translation). Timing the data transfers across this bus with the point operations of the first filtering and decimation stage introduces inefficiencies that reduce throughput. Fan-out from IF processing to multiple baseband processing DSPs is also typically accomplished via a dedicated point-to-point mezzanine path. Current implementations use customized fiber channel, TMS320/C50 or transputer links. As suggested in the figure, however, an open architecture optical mezzanine bus or local ATM switch should supplant today's vendor-unique point-to-point interconnects. An open architecture mezzanine bus with a capacity of 125 MHz x 16 bits per channel with an aggregate capacity of eight simultaneous channels would fill a critical void in contemporary open architecture standards.

Instead of configuring such heterogeneous multiprocessors at the board level, one might chose a preconfigured system. Mercury™, for example, offers a mix of SHARC 21060 (Analog Devices), PowerPC RISC, and Intel i860 chips with a unique Raceway interconnect with nominally three paths at 160 MB/sec interconnect capacity. Arrays of WE32s were used in AT&T's DSP-3 system. Arrays of i860s are available from Sky, CSPI, and others. Of particular note is the militarized TOUCHSTONE system which is also based on the i860. Such turnkey systems integrate block programming languages, automatic translation to C

Segment	Parameter	Illustrative value	Demand estimate
IF	W_a	10 MHz (2.5 oversampling)	
	IF Filter	100 OPS/ Hz	D_{if} = 2500 MOPS*
Users	N	30/ cell site	
Baseband	W_c	30 kHz	
	Demodulator	50 OPS/Hz	D_{bb}= 1.5 MOPS
Bitstream	R_b	32 kb/s	
	FEC, Signaling	100 OPS/b/s	D_{bs} = 3.2 MOPS
Source	CELP Codec	1.6 MOPS/user	D_s = 1.6 MOPS/user
Signaling	SS7	2 MOPS/site	D_o = 2 MOPS
Aggregate	DSP MOPS		D = 142.6 MOPS per cell site

* D_{if} is off-loaded to dedicated digital hardware and is therefore not included in D.

■ **Table 4.** *Illustrative processing demand: analog mobile cellular base station.*

Segment	Module	Characteristics	Illustrative manufacturers
RF	RF/IF	HF, VHF, UHF	Watkins Johnson, Steinbrecher
(A/D/A)	A/D	1 to 70 Msa/sec	Analog Devices, Pentek, DT
IF, BB	DSP	4 x 30 MFLOPS	AMD, TI, Intel, Mercury, Sky
IF	Digital Rx	30.72 MHz filters	Harris, Gray, E-Systems
IF	Memory	64 MB at 30.72 MHz	TI, Harris, TRW
BS, SC	Hosts	M680 x 0	Motorola, Force, Intel
SC	Workstation	50 to 100 Specmarks	Sun, Hewlett-Packard, DEC
(BB = Baseband; BS = Bitstream; SC = Source Coding)			

■ **Table 5.** *Mapping of segments to open architecture VME modules.*

and machine code and extensive debugging support but at relatively high cost of entry. Vendor-unique turnkey systems may be excellent choices when there is a good match between the application and the vendor's standard configured system. Vendor-unique details tend to limit hardware and software choices to the original vendor.

Architecture Tradeoffs

Software radios ideally place all IF, baseband, bitstream and source processing in a single processor. The assessment of the feasibility of the software radio centers on comparing estimated demand to the capacities of the available processors. Implementations back off from the ideal single-CPU implementation where driven to do so by processor, memory, or interconnect technology limitations, or to achieve cost advantages (e.g., of off-loading filtering to a Harris chip so that the DSP is less expensive). Software tasks are then structured into managed objects designed to run on any DSP or CPU with access to the data and sufficient processing capacity.

Open Architecture Software Tools Remain Problematic

Figure 7 shows how the time criticality of performance varies considerably with functionality. No signal processing environments in existence fully span the range from hard real-time IF processing through off-line support. Each tool and environment on the market excels in one or more aspects of the required support, although a few integrated environments approach this ideal. UNIX works well as both VME host and workstation host in laboratory systems. Portable VME systems often use a DOS/Windows laptop for user interface and display in lieu of a workstation, with a DOS/EISA or UNIX/VME signal processing configuration. SPOX and Harris Corporation's

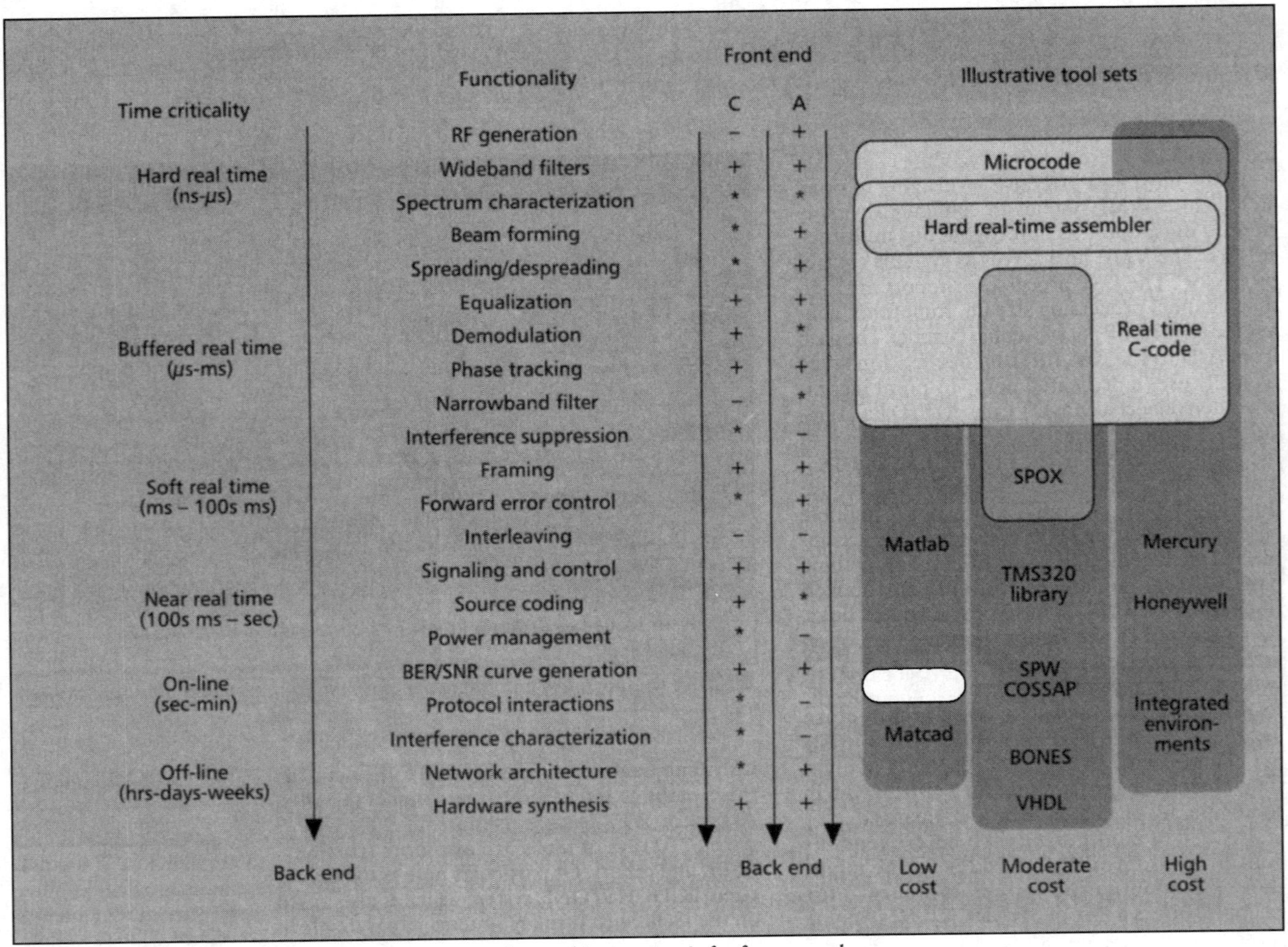

■ **Figure 7.** *Software tools span the function space, but tool integration is far from seamless.*

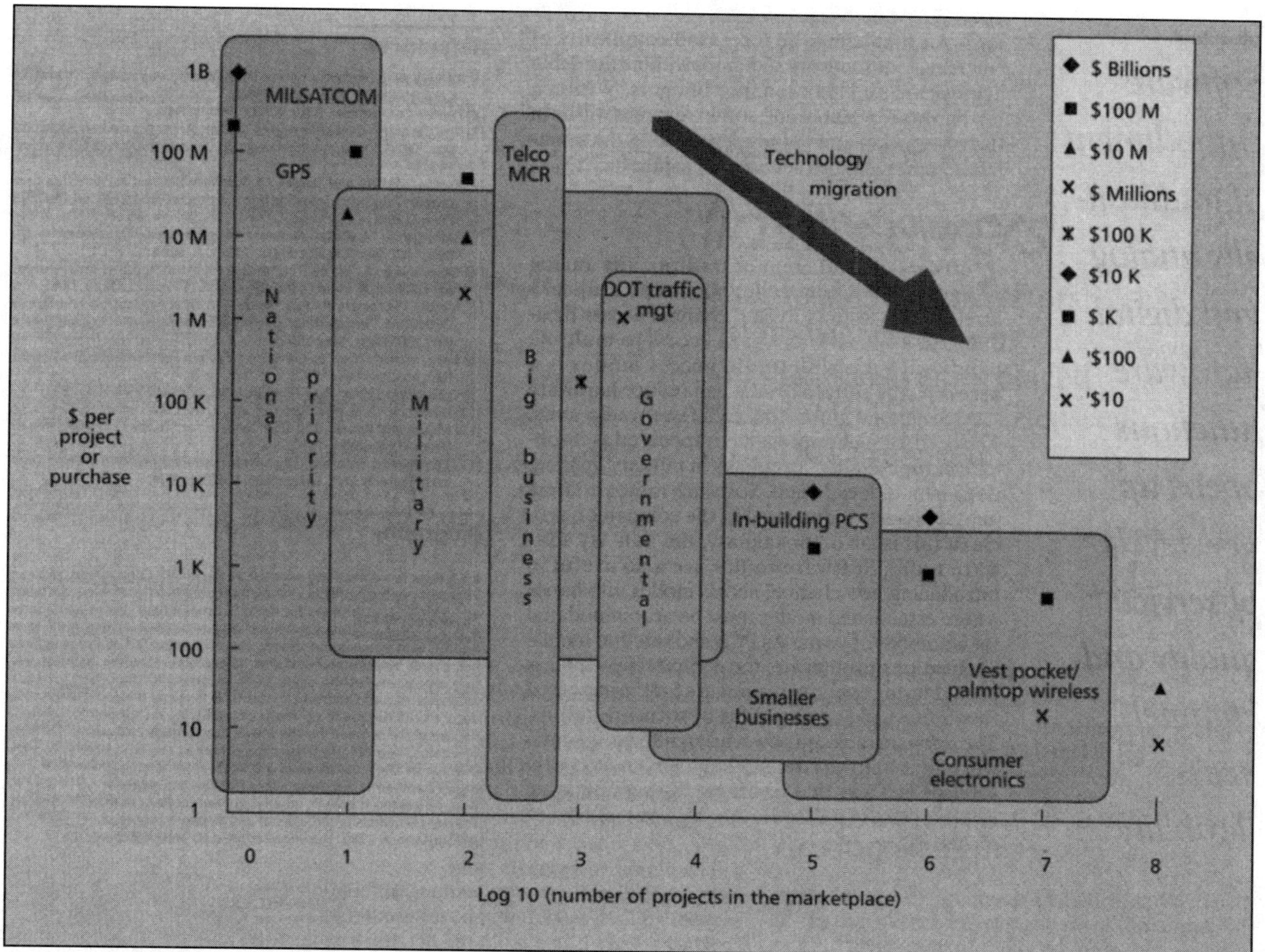

Figure 8. *Economics of project size versus number of projects drive software radio technology migration.*

real-time UNIX excel in near-real-time support. Software tools for heterogeneous distributed processing are developing rapidly, but lack extended precision arithmetic and standardized application-to-application interfaces.

The Signal Processing Workbench (SPW) and COSSAP environments each span a wide range from analysis and design through VHDL code generation. These environments include block diagram languages that allow the designer to express processing in a signal flow representation. Block diagram languages are useful for concept development. The compilers can also map these to an intermediate language (typically C) which can be compiled and downloaded to target DSP chips (with up to an order of magnitude of performance degradation compared to hand coded C). Real-time access to the DSP status and performance is relatively hard to come by, however. Mercury and UNISYS' Militarized Touchstone environments provide counter examples, offering real-time operating systems with integrated performance monitoring and debugging capabilities. Some operating systems require one to build specialized status reporting tasks to send data of interest to the workstation. This is not a big impediment on a small project, but on larger, more complex projects, the integrated environments have significant advantages in reduced development times.

Rehosting tools tend to be aligned by vendor with Vendor-X providing a tool that will download C code to Vendor-X's own DSP hardware. The Signal Processing Workbench (SPW), Alta Corp., is a notable counter-example, with a very complete set of rehosting tools including targeting for several popular DSP cores and translation to VHDL for rapid transition to product hardware. Code libraries of such tool sets will evolve to standards-based suites of managed objets as object resource broker technlogy becomes more widely implemented in DSP tools.

Economics

Figure 8 shows the rough order of magnitude relationship between size of a project (or purchase) and the number of such projects (or potential purchases) in the marketplace. Software radios are now in the segment of the market dominated by the military, big business (e.g., the telephone and wireless service providers) and governments. The companion applications articles describe projects with costs on the order of a few million dollars to more than ten million, for example. Over time, the software radio will continue to move down and to the right as the size, power and cost of general purpose DSP chips, A/D and D/A converters and related interconnect and memory allow. In addition, object management software technology applied to software radios offers powerful

Software embodiment of traditionally analog and digital hardware functions opens up new levels of service quality and channel access flexibility.

tools for managing the increased complexity of emerging radio network standards within affordable acquisition and maintenance budgets. Within a few hardware generations, software radios will be in the vest-pocket and palmtop, providing the seamless communications services we aspire to.

Conclusions

Software embodiment of traditionally analog and digital hardware functions opens up new levels of service quality and channel access flexibility. In applications where access to multiple bands with multiple radio access modes is a necessity, the software radio can reduce hardware size, weight and power through fewer radio units. The need for such improvement appeared as chronic "interoperability" problems in military applications over a decade ago. Software radios address military issues as discussed in the companion article in this issue on Speakeasy, the military software radio. Software radios are also useful in introducing new channel access modes into bands where established modes must be accommodated for a number of years. As PCS and satellite mobile applications proliferate, the commercial sector is encountering continuing pressure for the increased flexibility and seamlessness of software radios. The software radio is a powerful architecture framework that helps us deliver such advanced radio services in a way that leverages the economics of contemporary microelectronics and software technologies.

References

[1] Bensley *et al.*, Introduction to Parallel Supercomputing, The MITRE Corporation, Bedford, MA, 1988.

[2] Pentek Corporation, (Laurel, MD: SPSI, 1994).

[3] The Common Object Resource Broker: Architecture and Specification, OMG 91.12.1, (Maynard, MA: Digital Equipment Corp. *et al.*, 1992).

[4] Kondo, Matsuo and Suzuki, "A Software-Defined Architecture Concept for Telecommunication Information Systems," ICC '94, (NY: IEEE Press, 1994).

[5] Voruganti, "A Global Network Management Framework for the 90's," ICC 94, (NY: IEEE Press, 1994), p. 1094.

[6] Celestino and Claude, "Building a Platform Support Environment for TMN Applications", ICC 94, (NY: IEEE Press, 1994) p.1847.

[7] Mitola "Software Radios: Survey, Critical Evaluation and Future Directions," Proceedings of the National Telesystems Conference, (NY: IEEE Press, May 1992).

[8] Kant, *Introduction to Computer System Performance Evaluation*, (NY: McGraw Hill, 1992).

[9] Collins and Tebbs, *Real Time Systems*, (UK: McGraw Hill, 1977).

[10] Kleinrock, *Queueing Systems*, vol. II, (NY: Macmillan, 1982).

[11] Mouly and Pautet, *The GSM System for Mobile Communications*, (France: Palaiseau, 1992).

[12] Ziemer and Petersen, *Digital Communications and Spread Spectrum Systems*, (NY: Macmillan, 1985), p. 694.

Technical Challenges in the Globalization of Software Radio

Joseph Mitola III, The MITRE Corporation

ABSTRACT During the past five years, the software radio has emerged from demonstrations in military research to become a cornerstone of the third-generation strategy for affordable, ubiquitous, global communications. This article reviews the concepts, architecture, technology challenges, and economics of the continuing productization and globalization of software radio. It provides a point of departure for subsequent articles in this feature topic.

"We made our first 'software radio,' as people are now calling them, in the late 1970s. In those days the analog-to-digital converter (ADC) fed an 8085 microprocessor, and the radio operated at VLF," Dr. Walter Tuttlebee of Roke Manor Research (Romsey, United Kingdom) explained to the First International Workshop on Software Radios [1]. Indeed, the lower the carrier frequency, the more practical the use of contemporary ADC technology at intermediate frequency (IF), which is an aspect of software radios. First described as a military-related technology in 1992 [2], the military applications of software radio were more fully described in terms of the U.S. Defense Advanced Research Projects Agency's (DARPA's) SPEAKeasy I technology pathfinder in 1995 [3]. The SPEAKeasy II program was the catalyst for the Modular Multifunction Information Transfer Systems (MMITS) Forum, founded in March 1996. MMITS' global participation has included Alcatel (France), Ericsson (Sweden), Keio University (Japan), Motorola (United Kingdom), Nokia (Finland), Orange Personal Communications (United Kingdom), Rhode and Schwarz (Germany), Samsung Electronics (South Korea), and Siemens (Germany), among others. In addition, as illustrated in Fig. 1, the European Community has sponsored precompetitive software radio programs in its R&D in Advanced Communications in Europe (RACE) and Advanced Communications Technology and Services (ACTS) programs [4]. Asian thrusts were not widely published until the First Asian Workshop in April 1998 at Keio University, Japan.

The recent renaming of the MMITS Forum to the Software-Defined Radio (SDR) Forum signals a shift from military to commercial emphasis in open architecture standards for software radio. The commercial sector defines an SDR as one that implements a specified range of capabilities through elements that are software-reconfigurable. The SDR, then, includes radios with a limited set of predefined hardware functions such as air interface application-specific integrated circuits (ASICs), among which one may select by software. To the degree that the functions may be redefined in software (e.g., by downloads or subscriber identity cards), the radio is a software radio (SWR). If the vocoder function, for example, is implemented in software or electronically reprogrammable firmware, it may be downloaded over the air. This brings high potential for service differentiation. If the personality of the radio may be totally redefined in software, the implementation would be an ideal SWR. Usually, however, limited-function ASICs cost less than fully flexible digital signal processor (DSP) or field programmable gate array (FPGA) implementations. Early SWR research focused on the military's goal of total flexibility across dozens of radio frequency (RF) bands and air interface modes. Current research, particularly in Europe, focuses more on economic viability in introducing third-generation (3G) handsets and infrastructure. Military research is now attracting relatively few investment resources.

The next section reviews SWR architecture, emphasizing the pragmatic issues that arise in engineering this technology into affordable handsets, mobile nodes, and infrastructure. Subsequent sections address the technology challenges and related economics.

EVOLVING CONCEPTS AND ARCHITECTURE

An ideal SWR is a multiband multimode radio with dynamic capability defined through software in all layers of the protocol stack, including the physical layer. The ideal SWR supports the functions of Fig. 2. Each band and mode is realized in a software-defined personality. In base stations, a wideband ADC may convert an entire service band at IF, with programmable digital filters for channel isolation and DSPs for demodulation. Coupled with a corresponding wideband digital-to-analog converter (DAC) and software-repro-

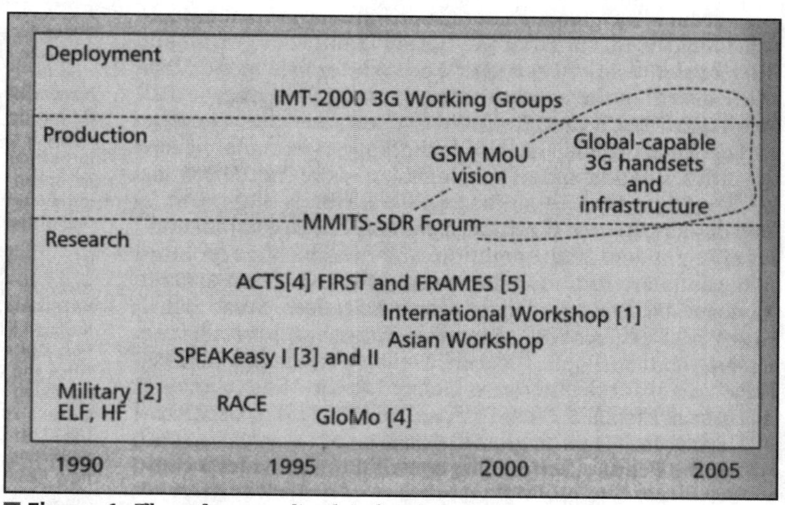

Figure 1. *The software radio globalization timeline.*

Views expressed in this article are the author's and do not necessarily reflect the policy of the MITRE Corporation or the U.S. Department of Defense.

Reprinted from *IEEE Communications Magazine*, Vol. 37, No. 2, pp. 84–89, February 1999.

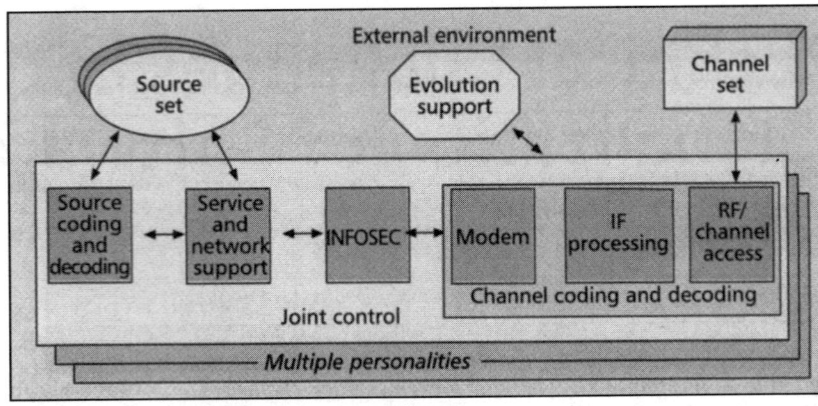

■ **Figure 2.** *Software-defined radio architecture.*

grammable functions from IF to baseband, such a radio supports arbitrary air interfaces within engineering limits. These limits include the bandwidth and dynamic range of the ADC and DAC, timing accuracy, and the processing capacity of the digital processing hardware (including reprogrammable ASICs, FPGAs, DSP chips, and general-purpose processors).

Object-oriented constructs best describe the implementation strategy. Each functional box in the figure corresponds to a class library. Any of the functions of Fig. 2 may be "null" in a specific implementation. In the open architecture vision, system integrators, network operators, and service providers choose from among standards-based interoperable class libraries to define services implemented using reusable classes. A 3G mobile Internet service, for example, might be crafted from an existing Internet Protocol (IP) stack from a service and network support library mixed with a 3G IF modem from a channel coding and decoding library. The resulting 3G mobile class could be implemented in Java-like middleware executing on a Java virtual machine. An instance of this class may be downloaded to an SWR handset, just one of its virtually unlimited personalities. The platform for this personality could include a 4.096 Mb/s despreader ASIC with an embedded DSP core. Such a handset may not be an ideal SWR due to the limited programmability of the despreader. However, if the ASIC supports, say, wideband code-division multiple access (W-CDMA) and Global System for Mobile Communications (GSM), the benefits of open architecture SWR standards apply. This enlarged view represents an evolution of the SWR concept from the ideal academic abstraction to a framework for pragmatic implementations. Commercial products are taking the SWR in this direction. Productization thus calls for a mix of compatible hardware, firmware, and software modules. Such a module-driven strategy of SWR productization could yield a wide variety of hybrid implementations.

The layered virtual machine reference model of Fig. 3 shows how such modules may be organized to promote productization. The aggregation of functions and components into layered virtual machines defines the key interfaces among virtual machines required for cost-effective implementations. The lowest virtual machine layer includes conventional processors plus the instruction set architectures (ISAs) implicit in ASICs and FPGAs. This lowest layer includes device drivers, interrupt service routines, task control, resource allocation, and related operating systems services. The heterogeneous nature of the hardware requires that this layer normalize access to data paths and processing resources. On this layer is built the radio infrastructure that sets up and controls infor-

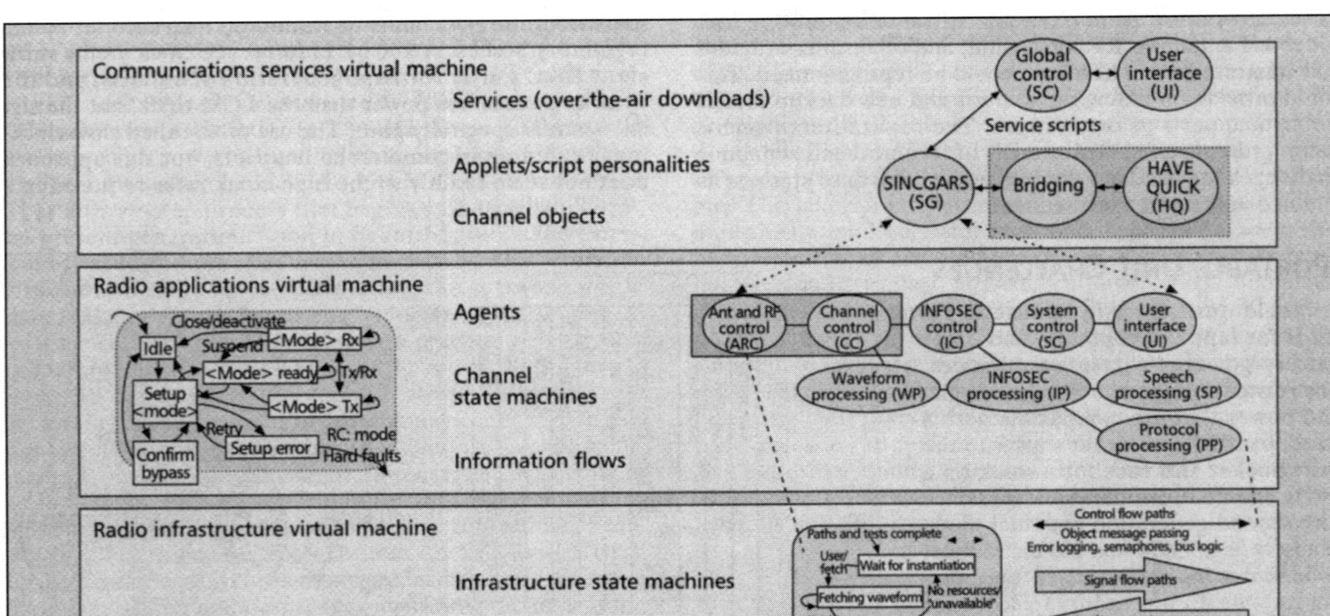

■ **Figure 3.** *Layered virtual machine reference model. (© Mitola's STATISfaction, used by permission)*

Challenge area	Challenges	Approaches
Power management	Sleep vs. paging delay	Power-managed DSP devices
Clocking	Multiple standards	Normalization, low power (LCM[1])
Receiver architecture	Efficiency vs multiple standards	Direct conversion or wideband SWR
Parts count	Off-chip passive devices	MEMS
Software	Computational efficiency	FPGAs, Java engines
[1] Least common multiple		

■ **Table 1.** *Illustrative SWR handset implementation challenges.*

mation transfer threads. This includes voice channels, data pathways, radio control message threads, timing distribution, frequency calibration, and location services. Infrastructure state machines request and allocate the internal resources necessary for a subscriber channel. The radio application layer includes agents, the peer-level control algorithms that maintain a radio mode such as a cellular interface standard or a military standard like Single Channel Ground Air Radio System (SINCGARS). Channel state machines control squelch, synchronize crypto functions, and mediate voice and data transfers. Communications services such as linking a cellular user with a SINCGARS user are defined by scripts that interconnect and invoke agents.

In principle, such a layered virtual machine strategy "future-proofs" the architecture. That is, evolving air interface standards and related services could be delivered on SWR infrastructure and handsets with minimum hardware redeployment. Bugs could be fixed without recalling handsets. Vocoders could be customized to the language of the speaker over the air. Data quality could improve with advances in coding theory quickly and at low cost. In the limit, as new RF bands are shifted from fixed terrestrial to 3G mobile, the wideband antennas, RF conversion, and other infrastructure now immortalized in hardware could be reprogrammed. This could provide both new capabilities and new backup modes for critical parts of the network. The realization of such a vision presents an imposing array of technical and economic challenges for handsets, mobile nodes, and base stations as outlined in the next three sections.

Portable Unit Challenges

Portable units include handsets, PC cards for laptop computers, and other battery-powered wireless appliances. The related designs demand size, weight, and power that are compatible with a sleek, handheld unit that slips into the shirt pocket and runs for a week on a single charge. Considerable experience with second-generation and dual-mode handsets (e.g., GSM and Nordic Mobile) underscores the challenges of transitioning today's digital radios to software radios as suggested in Table 1.

Power Management

Power-managed DSP chips now form the basis of power-managed ASICs with the DSP ISA as the ASIC core [5]. Sleep modes make power available for computationally intensive software-radio tasks by exploiting the low duty cycle of voice and data communications. Sleep modes save power but introduce a probability that the radio will be asleep during a paging message. Structured timing of paging messages reduces the miss probability, further conserving battery life.

Clock Generation and Distribution

Clock generation and distribution consumes a large fraction of handset power. Multimode SWR handsets in particular must generate several different clock rates for each supported standard. One reference oscillator per standard increases parts count, complexity, and therefore cost. A single master clock may use the least common multiple (LCM) of the required clocks. The LCM of GSM and IS-136, for example, is 6.31800 GHz. An integer down counter N_0 derives the clock for the invoked standard from such a master clock. N_0 equals 486 for GSM and 325 for IS-136. Integer decimators derive the channel spacing (N_1) and channel symbol clock (N_2) from these clocks as shown in Table 2. A W-CDMA clock of 2.56 GHz yields the integer channel spacing shown, but a W-CDMA and GSM LCM integer clock runs at 33.28 GHz. The total dissipated power is roughly proportional to the square of the clock rate. The impracticality of such high-speed master clocks illustrates the difficulty of such a conventional approach to the handset. Alternatively, the clocks may be generated using a $\Delta\Sigma$ approach where clocks are exact on the average, but in error instantaneously. As the timing error variance exceeds a few percent of the channel symbol interval, the bit error rate (BER) begins to degrade.

Nearest-fraction clock generation requires high-Q filters that enable the reduction of the master clock rate at the expense of increased circuit complexity. To synthesize the required clocks in software, the spectral purity of the sinusoids from the DAC must be within the tolerance set by the regulatory bodies. A 100 MHz reference clock yields sufficient DAC purity for first-generation standards, and the clock consumes less power than the LCM clock, but the signal is not as spectrally pure. The use of so-called slow clocking is common in commercial handsets, but this approach does not scale readily to the high clock rates required in a

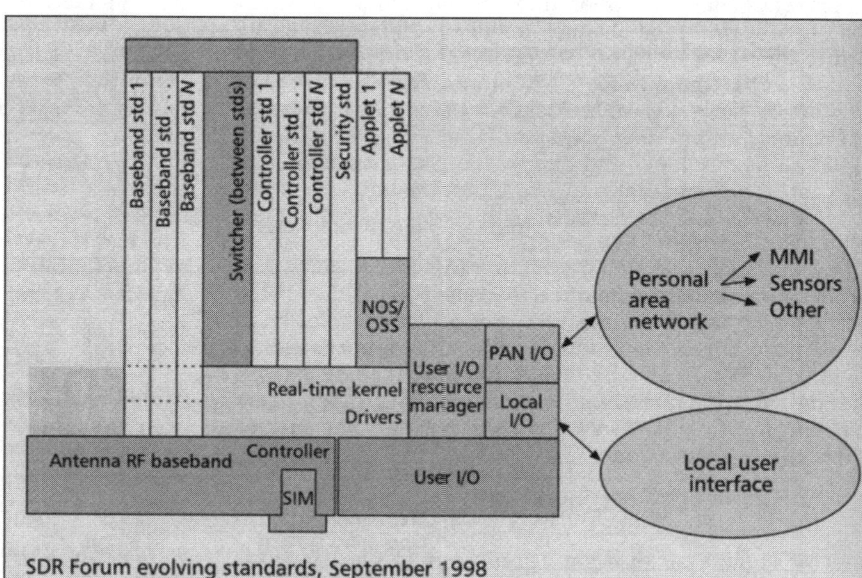

■ **Figure 4.** *Handset architecture emphasizes computational efficiency.*

handset that supports W-CDMA, IS-136, and GSM. A new technology SWR handset clock with low power dissipation seems to be needed so that the spectrally pure signals may be generated in software in multistandard handsets. Thus, in the near term, multimode handsets will be software-defined to the degree that standards are normalized to avoid clock rates with large LCMs. The LCM of the European Telecommunications Standards Institute's (ETSI's) normalized W-CDMA chip rate of 4.096 Mb/s and IS-95's 1.2288 Mb/s is 1.257472 GHz, an order of magnitude too high for power efficiency. Clock choices could make or break 3G, either reducing the cost of multimode handsets to the levels mass markets will accept, or relegating 3G to a small high-end market niche.

Standard	Reference clock (MHz)	Channel spacing (N_1)	Symbol clock (N_2)
GSM	13	200 kHz (65)	270.83333 kHz (48)
D-AMPS/ IS-136	19.440	30 kHz (648)	24.3 kHz (800)
W-CDMA	2560	5000 kHz (512)	4096 kHz (625)

■ **Table 2.** *High clock rates needed for LCMs drain battery power.*

Receiver Architecture

Receiver complexity is typically four or more times transmitter complexity. Thus, the receiver architecture has a first-order impact on handset cost. Direct conversion receivers mix a reference waveform with the RF signal to extract baseband signals without multiple superheterodyne conversion stages [6]. These ASICs efficiently transform RF to baseband, but each different air interface standard requires additional ASIC area. An ideal SWR receiver uses a wideband ADC to access the IF, with channel selection and conversion to baseband implemented in programmable filter ASICs or FPGA code. The peak computational demand of such a wideband reconfigurable SWR for a 4.096 MHz W-CDMA mode is about 400 MIPS per finger or 1600 MIPS. An FPGA multichip module (MCM) or DSP-core-based ASIC could deliver the required computational capacity. Such an SWR could be reprogrammed for GSM, IS-136, and analog standards, but today's DSP and ADC would consume the battery quickly. Alternatively, a software-reconfigurable direct conversion hybrid could digitally demodulate W-CDMA at essentially RF clock rates of 400 MHz to 2.3 GHz. These MCMs would also have high power dissipation compared to hybrid analog receivers. Panasonic has described its handset evolution plan [7] as a three-step process that begins with baseband DSP, the present generation. Soon to be introduced is the extension of software to the channel modem at an IF compatible with today's clock rates. Ultimately, the software would include the IF signal processing, an unspecified number of years hence. Since 400 MHz clocks are practical on-chip, several companies are developing ASIC and/or FPGA-based digital IF despreaders and demodulators. With such high clock rates, a demodulator algorithm may count zero-crossings to decode channel symbols. High power dissipation, low immunity to adjacent channel interference, and sensitivity to timing errors must be overcome for this highly software-defined RF modem technology to become competitive. Panasonic's timing seems to reflect the industry as a whole. Thus, the 3G handset receiver seems constrained to a few modes of direct conversion and despreader hybrids in the near term. The limited programmability of such a near-term approach does not preclude SDR baseband signal processing such as software-defined vocoders.

Handset Production

The cost of handsets in volume production is a nearly linear function of parts count. Emerging micro electro-mechanical systems (MEMS) technology uses three-dimensional bulk micromachining to integrate components into semiconductor substrates [8]. MEMS devices include 400 micron RF switches with 1/10,000 the size and less than 1/100th the average power of conventional PIN-diode RF switches. Cantilever beam filters have quality factors (Q) as high as 80,000. In addition, nickel MEMS devices may be fabricated onto existing complementary metal oxide semiconductor (CMOS) chips without melting them. The mechanical structures that set the tuned frequency of the MEMS devices themselves may be variable, set to specific values by micro-actuators. In addition to trimming parameters of production components, such micro-actuators may have sufficient dynamic range and mechanical life for programmable features of future SDR.

Handset Software Computational Efficiency

Finally, handset software has to be extraordinarily efficient in use of computational resources. The W-CDMA despreader is the critical modem component for 3G. Since on the order of 100 operations may be required on the average to process a CDMA chip, a simple despreader would require on the order of 400 MIPS of processing capacity. Although the new DSP chips can deliver from 400 to 1000 MFLOPS, the trade-off of flexibility versus power dissipation generally favors an ASIC despreader. Software-defined despreaders, however, could encapsulate ASIC personalities as downloadable objects for FPGAs.

In addition, the interfaces among software components have to be streamlined as suggested in the handset architecture [9] of Fig. 4. This architecture is a mapping of the functional interfaces of Fig. 1 (as further characterized by the SDR Forum) to the handset, showing the API interfaces. In principle, these functional components may be implemented in power-efficient ASICs, FPGAs, DSPs, or general-purpose software, provided the data elements and timing are compatible. The layered virtual machine allows one to associate implementation levels with each module of the handset. The handset software architecture of Fig. 4 makes lower-level interfaces more explicit. Drivers, real-time kernels, and mode controllers have simpler application programming interfaces (APIs) than the layered virtual machine might suggest. In addition, fewer layers mean greater computational efficiency in invoking specific handset resources. Some developers have put many man years into the reduction of the few lines of code necessary to reduce parts count by one memory chip. The handset economics are such that this cost is recovered within months on lines that produced about 70 million GSM handsets in 1996.

Mobile Node Challenges

Mobile nodes share many of the challenges of handsets, but with less of a premium on power efficiency. The vehicular radio exemplifies the mobile node. Law enforcement, aviation, and the military all envision mobile nodes with a capability of from four to six simultaneous channels operating generally in high frequency (HF) through ultra high frequency (UHF) (nominally 2 MHz to 3 GHz). In some applications, such as flight safety, there is fixed infrastructure with which the mobile nodes communicate.

Challenge area	Challenge	Illustrative approaches
Protocols	Network mobility	Dynamic networks (GloMo)
Multimode personalities	20–40 different modes	Open architecture API standards
Wideband modes	> 200 MHz agility	Hybrid SDRs with fast LOs
Cosite interference	Noise, high-power artifacts	MEMS roofing filters
Self-generated interference	Time-domain duplexing	Active cancellation

■ **Table 3.** *Mobile node challenges.*

However, the military applications are the most demanding in some ways due to the mobility of the network itself as the military unit moves. As suggested in Table 3, this is an area of research in DARPA's Global Mobile (GloMo) program [10]. Coalition partners who engage in joint humanitarian aid, for example, are generally confronted with an amazingly wide array of bands and modes in just the terrestrial bands. The U.S. Joint Tactical Radio System (JTRS) Program's requirements document identifies over 30 different air interfaces with which JTRS should be compatible [11]. The approach taken in SPEAKeasy, the JTRS technology pathfinder, is the use of open architecture API standards. The hope is that modes implemented in a standards-based API will cost less and become available more quickly than modes without such standards. Referring to the layered virtual machine reference model of Fig. 2, one would like to procure standard hardware engines and infrastructure software as commodity parts. Value-added contributions in the upper layers of the reference model, in principle, should cost less because of the amortization of hardware, distributed processing infrastructure, and radio mode software costs.

The military also has to contend with more complex modes than the civilian sector. With its emphasis on information security, some military waveforms hop the signal over hundreds of hegahertz and spread it using CDMA techniques to avoid detection. JTIDS [12], for example, hops thousands of times per second over about 250 MHz. The ideal software radio would have to convert the signal at > 500 MSPS to preserve the JTIDS frequency hop bandwidth. Although such ADCs exist, they lack dynamic range and consume too much power. Instead, the typical JTIDS radio dehops the signal to its 3 MHz instantaneous direct sequence spread spectrum bandwidth. Dehopping uses a fast-settling tunable local oscillator (LO) to track the received signal in accordance with a prearranged plan. For the foreseeable future, JTIDS will employ the fast tuning LO instead of the wideband ADC, limiting the migration of these radios to the software-defined ideal.

Military users do not have the luxury of siting antennas at ideal locations. Instead, less than ideal whip antennas on vehicles must operate wherever the driver decides to stop. Military units employ software tools and other mechanisms to avoid interference, but often it is unavoidable. A high-power troposcatter system, for example, may be sited near a mobile command post. The troposcatter signal may nonlinearly intermodulate in a metal structure, creating harmonics with high power very high frequency (VHF) or UHF components. Future MEMS narrowband filters may be programmed to suppress such interference. Sometimes, a radio interferes with itself, particularly in the time-domain duplex (TDD) bands (called the "push-to-talk" bands by the military). A four-channel JTRS radio, for example, might need to both listen and transmit in the same ADC band at the same time. Only a few frequency hop modes like SINCGARS, invoked at the same time, can render a node essentially inoperable. In these cases, one may inject a scaled replica of the transmitted signal into the receiver, 180° out of RF phase with the transmitted signal. This technique both reduces the peak interference power and allows the receiver to accept incoming RF in the same ADC band at the same time. Narrowband filters reject the interference and the intended incoming signal, so are inappropriate for suppressing this kind of cosite interference. In the future, such military-driven applications may lag behind handset and base station technology since the compelling economics of 3G do not apply to the military.

Parameter	AMPS	GSM	IS-95	W-CDMA
W_c	30	200	1250	5000
N_c	12500	25000	25000	25000
W_c	416.66667	125	20	5
Nyq	2.5	2.5	2.5	2.5
Rs	31250	62500	62500	62500
R (dB)	87	90	90	90
Smin (dB)	9	12	-10	-10
Rmax (dB)	96	102	80	80
Rmax bits	16	17	13.33333	13.33333
Pad bits	2	2	2	2
B	18	19	15.33333	15.33333
$R_{oversamp}$	416.66667	125	20	5
$G_{oversamp}$	26.197888	20.9691	13.0103	6.9897
RES_{req} (db)	69.802112	81.0309	66.9897	73.0103
RES_{req} bits	14	16	14	15
WADC Best	35000	4000	35000	8000
WACD COTS	17500	2000	17500	4000
N-ADCs	2	32	4	16

■ **Table 4.** *Conservative channelization requirements.*

Base Station Challenges

The Flexible Integrated Radio System and Technology (FIRST) and FRAMES projects of the ACTS program investigated next generation air interfaces using prototype SWRs. Lessons learned in FIRST, FRAMES, and SPEAKeasy reinforce each other. The lack of ADC performance is the economic showstopper for base station SWRs. The number of parallel digital channels required for a base station application depends on the effective dynamic range (resolution accuracy) of the ADC at a given bandwidth. The numbers of parallel RF-ADC channels required of an SWR base station for important cellular standards are derived in Table 4. The total allocated spectrum (W) is processed in N_c parallel RF-ADC wideband channels to yield baseband channels of bandwidth W_c. Some IF oversampling (Nyq) yields a required real sampling rate (R_s). The near-far ratio (R) plus the minimum detectable signal (S_{min}) yields the net dynamic range (R_{max}). In a conservative design, two pad bits are allocated to the automatic gain control ceiling bit and to the least significant bit, yielding B bits required; but the wideband oversampling of $R_{oversamp}$ provides $G_{oversamp}$ gain, which is recoverable in the channel isolation filters. This yields RES_{req} bits actually required of the ADC. The maximum sampling rates of the available (WADC Best) and affordable (WADC COTS) ADCs are divided into the required sampling rates (R_s) to establish the number of channels required.

The cost of the base station is strongly linear with the number of RF-ADC channels. Thus, the GSM base station with 32 channels seems to require too much parallelism to be affordable. The IS-95 and AMPS base stations require only two RF-ADC channels. These may be competitive for a single standard, but the architecture does not address ETSI's basic W-CDMA requirement. Hence, the ADC technology as yet falls short of SWR products with truly global capabilities. Continued investments and innovation, however, may yield viable base station products within a year or two.

Conclusions

Software radio is in the midst of a painful birthing process. Its embryology included the military's low RF bands, leveraging programmable filter ASICs and baseband DSP. Its near-term economic future, however, is tied to 3G. Some technology challenges arise due to an inability to normalize air interfaces across regions of the world. Others arise due to the laws of physics that limit technical progress in power dissipation, clock speeds, dynamic range, and linearity. Over time, both should yield to advancing ADC, DSP, ASIC, FPGA, and software technologies. One key to success is to leverage the progress being made on these fronts through open architecture standards. The balance of this feature topic therefore summarizes the state of globalization of critical SWR technologies, including a variety of perspectives on the drivers, pitfalls, and prognosis.

References

[1] First Int'l. Wksp. Software Radios, Rhodos, Greece, May 1998.

[2] J. Mitola, "The Software Radio," IEEE Nat'l. Telesys. Conf., 1992.

[3] Lackey and Upmall, "SPEAKeasy: The Military Software Radio," *IEEE Commun. Mag.*, May 1995.

[4] *Advanced Communications Technology and Services (ACTS) Overview*, Brussels: European Commission, 1998.

[5] J. Gunn, "Low Power DSP Cores," *IEEE JSAC*, Apr. 1999.

[6] Abidi, "Low power Radio Frequency ICs for Portable Communications," *Proc. IEEE*, New York: IEEE Press, Apr. 1995.

[7] Shinagawa, "Software Radio Technologies, 1st Int'l. Software Radio Wksp., Rhodes, Greece, June 1998.

[8] Panasonic, ACTS, May 1998.

[9] DARPA MEMS Program; http://www.darpa.mil

[10] SDR Handset Architecture; http://www.SDRforum.org

[11] DARPA GloMo Program; http://www.darpa.mil

[12] JTRS Requirements (JTRS SPO); http://www.acq.osd.mil

[13] Petersen and Zeimer, *Digital Communications and Spread Spectrum Systems*, Reading, MA: Addison-Wesley, 1996.

Abstract

Software radio has emerged from obscurity to be heralded by some in the personal communications industry as offering a potential solution to man's historical inability to achieve common global standards. In such a scenario reconfigurable terminals, able to adapt to the differing regional radio interfaces, if feasible in the appropriate timescales, appear a very attractive option. In reality, the future of software-defined radio is perhaps, at the same time, both much less and much more than this. At one level, the advent of the true software-reconfigurable "universal handset" may lie beyond the launch date of third-generation systems. Equally, however, software radio will profoundly impact not only technology but also industry structures. Software-defined radio reflects the convergence of two dynamically developing technological forces of recent years: digital radio and real-time downloadable software. The former has facilitated the wireless revolution, and the latter has both facilitated and ridden the Internet wave. The massive growth and convergence of these two markets is both creating the applications demand for third-generation personal communications and simultaneously threatening some of the existing ground rules of the wireless industry. Software radio is in its infancy — in this article we explore its first steps, describe facets of the technological possibilities and implications, and try to postulate into what will it grow.

Software-Defined Radio: Facets of a Developing Technology

WALTER H. W. TUTTLEBEE
ROKE MANOR RESEARCH LTD. (A SIEMENS COMPANY)

Software-defined radio (SDR) has made the transision from obscurity to commercial credibility in a very short time, due in part to rapid technology development and in part to its perception as a panacea to the difficulties of securing common global standards for mobile communications.

Technology is a poor substitute for political will; however, whatever the motivation, advances in this field have the potential to dramatically influence the complex world of multimedia, multimode, multiband personal communications of the 21st century, offering benefits across the entire industry value chain.

What are the benefits to be derived from the introduction of software radio technologies, and who will derive these benefits? At the top level, examples are:

- For subscribers — easier international roaming, improved and more flexible services, increased personalization and choice
- For mobile network operators — the potential to rapidly develop and introduce new, personalized, and customized services; a tool for increased customer retention and added-value services
- For handset and base station manufacturers — the promise of new scale economies, increased production flexibility, and improved product evolution

This article reviews the status and development of software radio as it relates to the personal communications industry of today and tomorrow.

We first define terminology, explore the enabling technology for software radio, and consider the role of standards in a world of soft, reconfigurable products. Are they needed? What should be standardized? Who will do the job? Handset radio interface reconfigurability has been perceived as a key area of potential impact. Thus, after a cursory look at current European R&D, we specifically address the issue of how manufacturers are pragmatically approaching this issue at present. Alternative options for download mechanisms — static, pseudo-static, and dynamic — are reviewed and their implications discussed. Standardization and network implications are also considered. The article concludes by briefly exploring a number of longer-term issues, namely regulatory aspects, service provision and convergence, the relationship between soft antennas and software base stations, spectrum management in a world of soft radios, and the wider value chain impacts.

This article complements one in the recent *IEEE Communications Magazine* Feature Topic on Globalization of Software Radio [1], which provides a more comprehensive review of current R&D activities in software radio handsets and base stations. For a general introduction to software radio the reader is referred to the classic 1995 Special Issue of *IEEE Communications Magazine* [2] on the topic and to [3, 4].

Proceedings of the ACTS Mobile Communications Summits referenced in this article are available from the European Commission, Directorate General XIII-B, BU-9 4/91, Rue de la Loi 200, B-1049, Brussels, Belgium.

What Is Software Radio?

Definitions and Potential Benefits

In its original "purist" or "academic" form, the term *software radio* refers to reconfigurability of the radio interface by software, possibly using over-the-air (OTA) download, with an often implicit assumption of analog-to-digital (A/D) conversion at the antenna (Fig. 1). While such an implementation is already feasible at very low frequencies (VLF), such an architecture is still a long way in the future for personal communications at 2 GHz and above.

Thus, in recent years a more pragmatic definition has

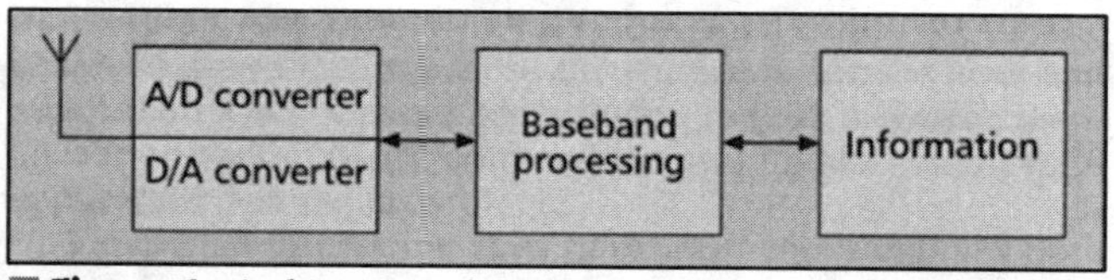

Figure 1. *Architecture of the "purist" software radio implementation [5].*

Reprinted from *IEEE Personal Communications Magazine*, Vol. 6, No. 2, pp. 38–44, April 1999.

emerged in personal communications, which encompasses reconfigurability at any level of the radio protocol stack by software either by OTA download or by other means, together with an acknowledgment that some signal processing will continue to be done in RF circuitry. This wider definition, illustrated in Fig. 2, represents an evolutionary recognition of how the industry is beginning to approach and implement the concepts of software radio reconfigurability.

In Europe, dynamic application download to Javaphones has been trialed, and product developments are in hand. While not in the purist sense software radio, such early activities offer operators a means of service differentiation and a potential source of revenue, and are thus setting a direction for commercial development. Such download techniques can be used to provide rapid introduction of new services, flexible and interactive personalized subscriber services, and operator customization of the handset interface.

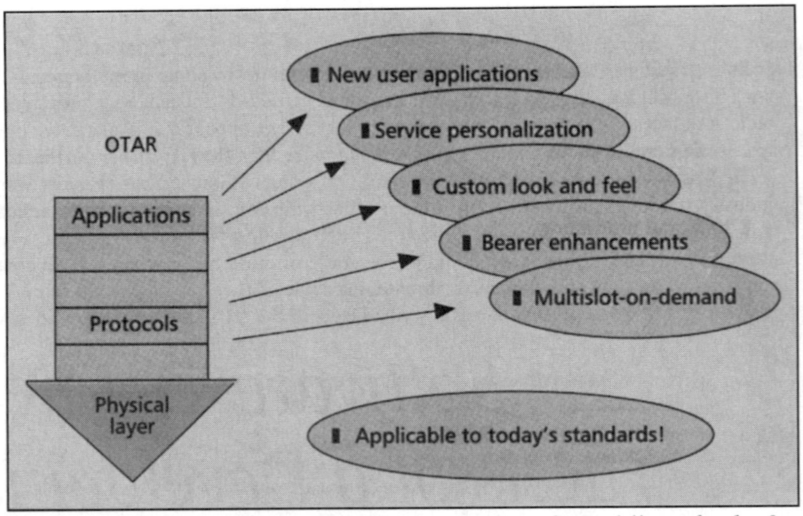

■ **Figure 2.** *Reconfigurability can offer a range of benefits at differing levels of the protocol stack, some already applicable to today's digital standards.*

Moving further down the protocol stack, dynamic bearer enhancements (e.g., bandwidth on demand) could be envisaged. These concepts to some degree are applicable already to today's second-generation digital personal communications. All of these features are highly desirable benefits to the mobile operator who wishes to maximize handset loyalty[1] and sell added-value services to its subscriber base. They provide ways to maximize the return on operator investment.

Within the timeframe of third-generation (3G) systems — Universal Mobile Telecommunications System/International Mobile Telecommunications (UMTS/IMT-2000) — technology will permit software-downloadable reconfiguration of the terminal radio interface. Alternative download options exist, as described below, which may build out from simple first steps. Such radio interface reconfiguration could eventually find important new applications.

Clear market differences exist between North America and Europe and are creating different driving forces for software radio. These are described more fully in [1] and summarized in Fig. 3. At present in Europe 3G is seen as the main driver for product development, where software radio offers the potential for network evolution from Global System for Mobile Communications (GSM) to UMTS, particularly for base station products.

In North America the paucity of seamless nationwide roaming and the existence of multiple standards is creating an apparent demand for multimode handsets. A cautionary note, however, is that, in general, multimode software base station products in North America to date have met with very limited market success.

Enabling Technology for Software Radio

Software radio is being facilitated by the evolution and convergence of several different technologies, illustrated in Fig. 4, which we will very briefly summarize.

Signal Digitization

Extensive development of A/D capability in recent years by the semiconductor industry has been stimulated by the prospects of digital wireless applications. This has resulted in improvements in accuracy, linearity, sampling rates, and resolution; however, the trade-off between A/D performance and sampling rate continues to be a limitation. While the use of multiple A/D channels may be a short-term solution, power consumption limits this approach to base station products.

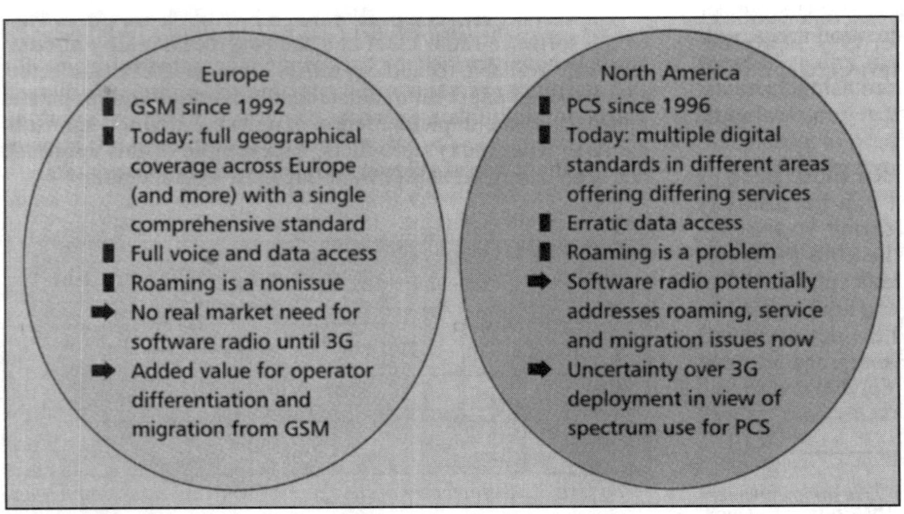

■ **Figure 3.** *Clear market differences between the digital mobile markets in Europe and North America are creating different drivers and imperatives for software radio R&D.*

Silicon: Functionality, Size, and Power Consumption

Silicon geometries continue to shrink, with complexities still doubling every 18–24 mo; 0.18 μm technology is almost commercially here, with significant further reductions already anticipated. Memory sizes continue to increase, as do the investments needed to construct their fabrication facilities! Today 3 V operation is standard in portable products; experimental devices have demonstrated programmable digital signal processors (DSPs) at 1 V operation, consuming a fraction of the power of the equiva-

[1] *Mobile operators often subsidize handset pricing and therefore wish to minimize subscriber "churn" onto a competitor's network.*

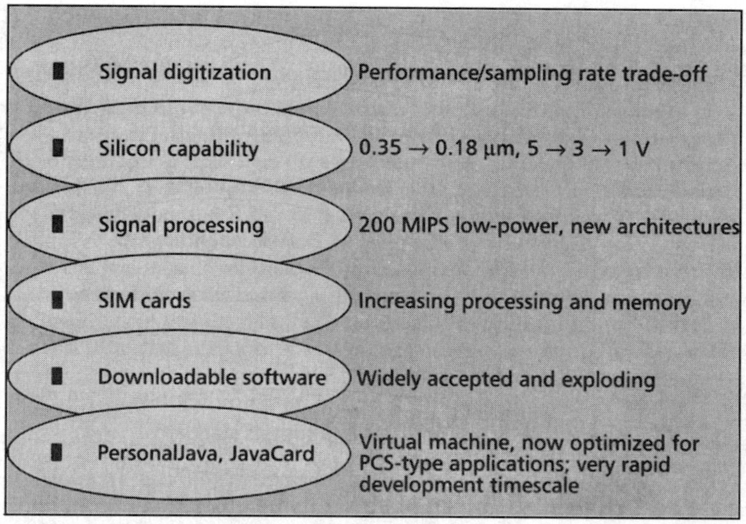

■ **Figure 4.** *Key technology trends enabling software radio handset terminal implementations.*

lent 3 V device. Such low-voltage operation may emerge as standard early in the next century. Such power consumption reductions promise the viability of very-high-performance DSPs in battery-powered products.

DSP Power

Advances in DSP processors have included new architectural concepts and increasingly customized hardware engines (e.g., for equalization, Viterbi multichannel demodulation and correlation); new non-Von-Neumann architectures offer increasingly power efficient implementation. Low-power-consumption single-processor DSP engines already offer 200 MIPS, with power consumption at around 0.5 W. 1 GIPS devices at reasonable handset power consumption and cost should be available within three years; arguably the activity in 3G mobile standardization is already accelerating such DSP developments, as did the advent of GSM.

PersonalJava and the JavaCard

The concept of the Virtual Machine, enabling Java programs to run on any processing platform, has created the "write once, run anywhere" software paradigm. Once a Java program is running, new software components can be imported and dynamically incorporated, allowing functionality enhancement on the fly. Java is playing a significant role in the evolution of software, not least for mobile phones. Since the finalization of the PersonalJava[2] specification in September 1997, new product offerings have emerged; likewise, the JavaCard specification.

Smartcard Technology

Smartcard functionality has traditionally been constrained by the size of integrated circuit (IC) that can be robustly deployed. However, as silicon geometries shrink, so the processing and memory capability within the chip footprint increases. Ratification of the JavaCard specification promises further flexibility enhancement of functionality for a given amount of on-card processing. Together these developments promise a rapid evolution in capability. Smartcard applications are growing fast in many markets — notably e-cash and pay-per-view conditional access to broadcast services, as well as cell phone SIM cards. The wide application domains of the technology, and resultant synergies, will have major implications for personal services of all kinds.

Software Download

Software sales, pricing, and distribution have been revolutionized by the advent of the Internet, with its possibilities for instant impulse buys, easy access try-before-buy, and online upgrades. For the average public switched telephone network (PSTN) modem-connected PC user, downloading is still slow and frustrating, and represents a bottleneck in wider acceptance at the present time. Despite such limitations, downloadable software over wireline is now widely and increasingly accepted — indeed, upgradeable modems, using software download, are already here. Over the wireless medium, the concept is already being deployed in Europe within the context of digital television.

Soft Terminals

Hardware Architectures

European research into software radio architectures and technology, as it relates to handset terminals and base stations, was recently reviewed in [1]. Such work is focused mainly on 3G standards. Table 1 summarizes the relevant collaborative projects, mainly undertaken within the European ACTS and ESPRIT research programs, and gives sources for additional information.

The main issues explored within these public domain research projects include:

- RF hardware architecture innovation and RF semiconductor process technology evolution
- Baseband DSP architecture evolution, including specific accelerators and software-programmable hardware reconfigurability
- Architectural and algorithmic implications of software radio, including combinations with other advanced processing concepts (e.g., beamforming and multiple signal extraction)
- Network implications of software radio — developing a nonconstraining, open evolutionary route from where we are today

In addition, work is beginning on the development of tools, libraries, and environments to support product development and software portability.

In practice, reflecting the wide gulf between the purist view of the software radio ideal and the pragmatic reality of commercially available technology today, some handset manufacturers choose to adopt an understandably cautious and pragmatic approach to the implementation of software-defined handsets, evolving from today's products. An example of this approach has been described in [6] by one Japanese manufacturer.

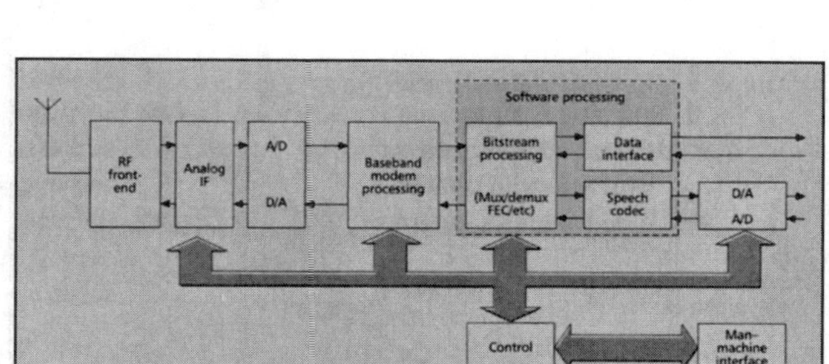

■ **Figure 5.** *Possible handset evolution: phase 1 — one manufacturer's view [6].*

[2] *PersonalJava is a specifically tuned development of Java for low-memory consumer appliances, such as cell phones. For more details visit http://www.javasoft.com/products/personaljava/index.html*

Project name	Objectives and technical emphasis as related to software radio	Website
	ACTS Mobility and Personal Communications Domain	http://www.infowin.org/ACTS/ANALYSIS/CONCERTATION/MOBILITY
FRAMES (ACTS)	UMTS radio interface definition and demonstrator implementation (TD-CDMA) in software on a DSP platform	http://www.de.infowin.org/ACTS/RUS/PROJECTS/FRAMES/
FIRST (ACTS)	Feasibility demonstration of software radio for second- and third-generation mobile systems (DCS1800 / TD-CDMA)	http://www.era.co.uk/first/first.htm
PROMURA (ESPRIT)	Validation of critical building blocks and system concepts for wideband RF aspects	New project, not yet available
MEDIAN (ACTS)	Implementation of very-high-rate wireless ATM on an advanced high-capability programmable processor	http://www.imst.de/mobile/median/median.html
SINUS (ACTS)	Architectures and implementation for hardware-reconfigurable terminal for S-UMTS	http://www.eeng.brad.ac.uk/research/SINUS
SUNBEAM (ACTS)	Exploration of the architectural synergies between smart antenna beamforming and software basestations for second- and third-generation mobile systems	http://www.era.co.uk/sunbeam/sunbeam.htm
SORT (ACTS)	Examination of hardware reconfigurability and software portability, including PC-based terminal demonstration for UTRA and S-UMTS	http://www.ifn.et.tu-dresden.de/~sort/welcome.html
SLATS (ESPRIT)	Development of software libraries, for second- and third-generation systems, suitable for multiple hardware platforms	New project, not yet available
RAINBOW(ACTS)	Network aspects and implications of broadband over wireless, including the implications of having multiple radio interfaces	http://www.cselt.it/sonah/RAINBOW
IBMS	Integrated broadband mobile system (4h generation), exploring dynamic flexible configuration of multimedia radio channels	http://www.comnets.rwth-aachen.de/project/ATMmobil/IBMS.htm

■ **Table 1.** *European software-radio-related R&D projects.*

Their phase 1 architecture, shown in Fig. 5, implements channel coding, source coding. and control functionality in software on DSP/μC/programmable logic. This architecture is already in use for today's digital phones, such as their PDC product, and allows a measure of new service introduction onto the phone in the field; in effect it allows reconfiguration at the very top of the protocol stack, the applications layer.

The extension of this to implement the baseband modem functionality in software is shown in Fig. 6. This step allows the realization of new and adaptive modulation schemes under either self-adaptive or download control.

Further extension of this, involving a major change to the overall architecture to implement the intermediate frequency (IF) signal processing digitally in software, will allow a single terminal to adapt to multiple radio interface standards by software reconfigurability (i.e., reconfigurability at the lowest level of the protocol stack); see Fig. 7.

It is clear that the processing power required to implement a phase 3 handset, perhaps 4 GIPS [6], exceeds that available from generic low-power DSPs in the near future by a large margin. Moreover, new technical requirements for 3G systems — RAKE receivers, soft handover, turbo codes, and so on — would appear to be increasing faster than improvements in traditional DSP technology may be able to deliver. This has led to the concept of augmenting generic DSP platforms with hardware-specific accelerators, as described in [6] and [1], to create in effect the "SDR engine" IC. A further enhancement of this concept, providing increased flexibility, is the implementation of such hardware accelerator modules using reconfigurable programmable logic. While attractive in concept, the divergence between accelerator architectures is such that commercial reality may still be some years away. The uncertain speed at which such new concepts will progress in commercial development represents something of a wild card when attempting to predict the pace of evolution of software handsets. Such approaches illustrate the pragmatic trend toward

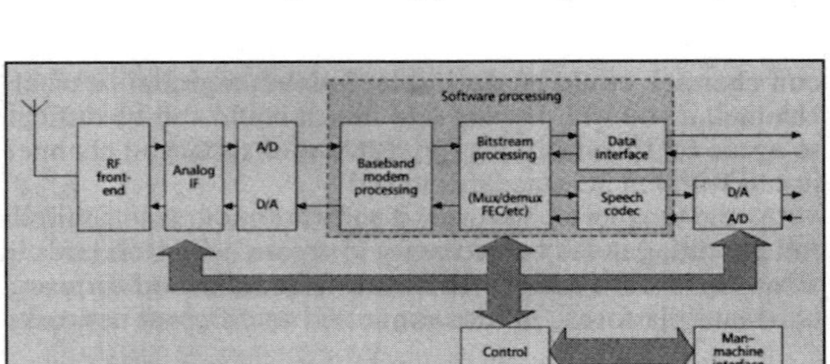

■ **Figure 6.** *Possible handset evolution: phase 2 — one manufacturer's view [6].*

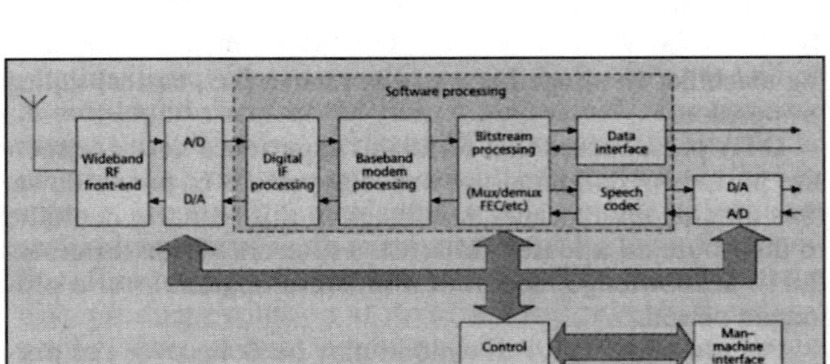

■ **Figure 7.** *Possible handset evolution: phase 3 — one manufacturer's view [6].*

combinations of flexible hardware/software combinations — referred to by Mitola [2] as "the use of CHFS, compatible hardware, firmware and software to achieve productization."

Download Mechanisms

Software radio technology, at the reconfigurable radio interface level, can be used in a handset in different ways. Three generic options exist: static, pseudo-static, and dynamic software download.

Static download refers to the situation where an SDR-enabled handset can support a variety of standards and is preprogrammed in a static manner to address one of these possibilities, by means of a Smartcard, for example. This could be a first step in radio interface reconfigurability.

Pseudo-static download refers to the use of OTA download to preconfigure a terminal to accommodate a defined set of applications, protocols, or radio interface. Clearly this offers increased flexibility over the static download option, in that it puts more control in the hands of the network operator and can appear transparent to the user.

The third option, dynamic reconfiguration, offers further flexibility, allowing OTA reconfiguration to occur during a call, providing, for example, bandwidth on demand for a time-varying multimedia service.

These three approaches correspond, at the same time, to increasing flexibility and increasing complexity. Possibly they also represent evolutionary steps over the coming years.

Smartcard technology is advancing fast in terms of on-card processing and memory. The step to static download by Smartcard is a small one from where we are today, allowing application, protocol, and radio interface upgrade via software downloaded to a "personality card." The potential combination of today's cell phone SIMcards with automatic teller machine (ATM)/credit cards for multiple applications, is a likely scenario for the management of a wide range of personalizable services, including communications and entertainment, which would support integrated payment options.

New terminals could be supplied with the operator's current local standard, and downloaded by wire at the point of sale (PoS) to a Smartcard to equip the phone with the latest release of radio interface, protocol, and applications software. Some time later, as required, the user could download new facilities or services (software) onto the card from a vending machine, with the cost automatically debited to the cardholder's account. When roaming abroad, downloading a new radio interface could appear to the user as simple as buying airtime from a vending machine in an airport arrival lounge, similar to today's rechargeable prepaid card options.

In the longer term, pseudo-dynamic OTA reconfiguration has significant logistical attractions for an operator. OTA upgrade can be transparent to the user, can allow the operator a greater degree of control, and, importantly, will more readily allow new services to be sold to subscribers as impulse buys. In effect it puts the vending machine into the phone, rather than the user having to put his credit card into a vending machine, an altogether more attractive (i.e., easier) selling proposition.

OTA presupposes that terminals experience neither excessive periods of continual use nor long periods of power-down, thus missing an upgrade. A fallback in this scenario could be to distribute an updated Smartcard to users whose terminals fail to acknowledge receipt of a software upgrade over a prolonged period.

Pseudo-static OTA download may be done over the pre-existing radio interface or over a common signaling channel as described below.

The concept of dynamic download, configuring a software radio terminal on the fly to support a time-varying multimedia bearer channel, has been explored within the Integrated Broadband Mobile System (IBMS) project [7, 8], part of the German ATMmobil program. The IBMS project, among other things, has conceived a minimum air interface backbone — a single separate signaling channel designed solely for establishing access to the wireless network and for mobility management across a hierarchy of pico-, micro-, and macro-cell environments capable of supporting different data rates. Such a signaling channel could also be used to set up and tear down a variety of custom radio interface bearers dynamically on demand, to support varying multimedia service requirements at any specific point in time.

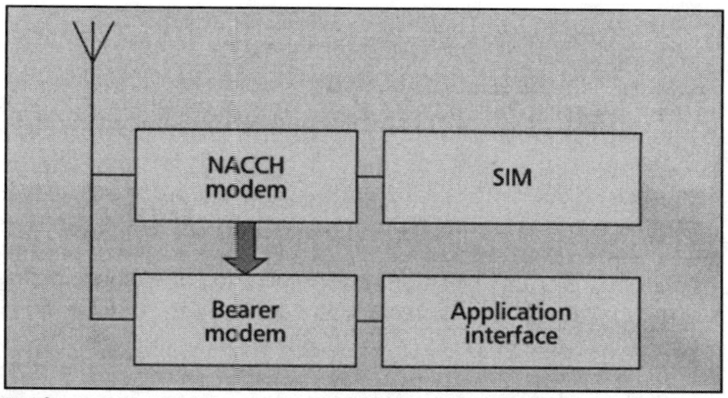

Figure 8. *A minimum-specification radio interface: the NACCH concept from the IBMS project [7].*

This concept of the minimum radio interface, a so-called network access and connectivity channel (NACCH) or global access channel, and its relationship to the handset are illustrated in Fig. 8.

Within this concept, the NACCH holds responsibility for the common (service-independent) functions of registration, authentication, and location management, configuring the terminal's bearer modem as required according to the network service class (degree of mobility) and specific radio interface required [8]. This concept can perhaps be viewed as a dynamic extension to the "switcher" concept of the SDR Forum,[3] in which the terminal is requested to switch between modem functionality to support the appropriate radio interface required by a call.

Standards

In this new world of soft terminals, what should be standardized, and how? The answers to these questions are still emerging, and they are nontrivial.

To the cellular handset manufacturer involved in standards activities, the radio interface is perhaps the first thing that springs to mind. For a soft terminal, as discussed above, it need not be necessary, beyond perhaps a basic spectral mask and other factors affecting mutual interference, to specify radio interface details. However, although not strictly necessary, a global radio interface standard to act as a bearer for local OTA download or to provide switching instructions to handsets containing a set of preloaded options, a global beacon channel, could be desirable; indeed, a global NACCH channel, if you will. Having said this, it could still be difficult to agree on specifications for such a global beacon channel, just as with full 3G standards today!

At the next level, however, if software radio is to fulfill its full potential, it will be necessary to create open standards to allow different vendors' software to function on different hardware platforms, in turn connected to different networks,

[3] The Software Defined Radio Forum, previously the MMITS Forum; for more information see http://www.sdr.forum.org

using different vendors' infrastructure — a nontrivial scenario which will not be achieved overnight! Work to assess the implications of such scenarios is progressing within the context of the SDR Forum. The SDR Forum does not see its role as replacing that of the standardization bodies, but rather as an industry resource to create frameworks and specifications by consensus which can then be input to the relevant existing bodies, in much the same way as has happened with other such fora in recent years. The SDR Forum is thus participating in the current 3G standardization activities, providing input to the International Telecommunication Union (ITU) process and other standards bodies, to ensure the role of SDR within the future 3G scenario.

Extensive work has been undertaken by the SDR Forum in the areas of software download and the creation of an application programming interface (API) framework for compatible software and hardware modules, which could form the basis for intravendor interoperability. Clearly individual companies could seek to create their own proprietary approaches. The full commercial potential of SDR is, however, more likely to be realized through open standards — indeed, the growth of the wireless industry owes much to its endorsement of open standards in recent years.

Network Implications

Little attention has been focused to date on the network implications of software radio; the initial focus has been on handsets and base stations. Recognition of the importance of network issues is now emerging.

Just as soft handsets are envisaged as supporting multiple radio interfaces, so it is anticipated that future mobile multimedia network infrastructures will support multiple radio interface standards; indeed, this is a fundamental aspect of the philosophy of the ITU approach to IMT-2000 standards. This requirement implies the need for separation of radio-dependent and radio-independent functionality within the network. The separation of control and transport functionality, conceived in earlier RACE UMTS work, is now being adopted and extended by the ACTS RAINBOW project [9, 10]. What is not yet fully clear is how an evolution from today's second-generation network infrastructures toward such concepts will be pragmatically implemented; clearly different vendors' different proprietary implementations will be subject to different constraints in this respect.

In migrating operational networks to 3G, interworking with GSM and other preexisting radio interfaces will be required to provide for evolution and allow backward compatibility. In addition, issues of home network service support when roaming impact not only on the radio interface, which can be accommodated by OTA software download, but also on the signaling and infrastructure support — provision of the virtual home environment (VHE). The operational system implications of the broad range of network–terminal interactions possible in a software radio environment still remain to be fully understood and characterized if the full potential of software radio is to be secured.

Long-Term Implications

We conclude by briefly considering a few longer-term issues arising from the prospect of software radio. Assuming that pseudo-dynamic and dynamic download become possible, what are the longer-term implications?

Regulatory Issues

Wireless product standards exist to facilitate, among other things, conformance testing and/or type approval, important issues when one considers the potential for mutual interference from an errant soft terminal product. Both national and international regulatory bodies will need to be educated and convinced that the evolving mechanisms for configuration management (software/hardware/network combinations) and terminal control (e.g., automatic shutdown of errantly configured equipment) are both secure and reliable. In addition, issues of what degree of such external control should exist and where it should reside have yet to be determined and will not be simple to agree on, since many vested interests exist — those of operators, users, national governments, civil liberties groups, handset manufacturers, and software providers, among others. Although these issues are already being considered by the industry, firm proposals have yet to be finalized. The international dimension of securing the necessary agreements and procedures should not be underestimated. For truly *dynamically reconfigurable* products at the radio interface level to become internationally acceptable will require extensive collaboration across the global industry.

Service Provision and Convergence

Applications download and mobile Internet access have already begun. We anticipate that it may not be long before third-party providers begin to offer new services (client/server applications) downloadable over the Internet independent of the mobile network operators. Such providers could be the handset manufacturers, independent applications developers encouraged by them, or today's large commercial Web businesses with a proposition targeted at the mobile consumer. In the long term, just as today anyone can be a Web site publisher providing limited interactive services, so we may anticipate a proliferation of new mobile service provision possibilities. Today's mobile operators may view such a scenario as a threat or an opportunity; their future fortunes could depend on their response. A proactive strategy of embracing new commercial content and service partners could be key to customer retention, growth in added-value offerings, and subscriber growth.

In the United Kingdom, already at least one mobile operator has announced not simply to their investors, but in marketing literature issued with their subscribers' monthly bills, their intention to introduce a videophone service. Once such phones are available, supported by a suitable infrastructure, it will be a small step to dial-up access to niche, narrowcast television services. Relatively low bit rates, ~ 64 kb/s, should be sufficient with MPEG-4 for adequate resolution on a small handset screen 5 cm^2. Once such a service is implemented, further combined service offerings become possible.

Soft Antennas and Soft Base Stations

Implementation of soft (or smart) antennas for 3G systems will not be possible using the traditional 2G approaches, with analog beam steering at radio frequency (RF), but will require digital receiver architectures. Soft antennas will not be deployed early in the life of 3G systems, for the same reasons that we have seen with 2G: the capacity benefits are proven [11], but the cost-benefit equation does not yet square. However, the balance of cost-benefit will change, and at some time early in the next century the added value of soft antennas will become positive. To accommodate an easy product evolution at that time will require the appropriate choice of base station architectures now (i.e., the use of an architecture that can easily evolve to accommodate digital receivers).

A basic software base station architecture is not dissimilar from the phase 3 handset architecture shown in Fig. 7; the key difference, of course, is that the base station suffers from far fewer implementation constraints, in terms of size and power consumption, than a handset product. Thus, such designs can be implemented more readily. In practice, such software-defined multistandard base station architectures have clear synergies with the requirements of soft antennas (and, for that matter, with a number of other advanced functions).

Adaptive Spectrum Management

In most countries large portions of the electromagnetic spectrum, a scarce resource, are at present permanently and inefficiently allocated to systems with low real usage. With the increasing recognition of the economic value of the radio spectrum, such a situation will not in the long term be allowed to continue. Software radio technology offers one tool for harnessing this wasted resource through adaptive spectrum management (ASM).

ASM, at a simple level, is already implemented in some existing wireless systems. In the Digital Enhanced Cordless Telecommunications (DECT) standard, for example, the terminal and base station both monitor their local frequency spectrum/time slot space to identify, reserve, and use spectrum that is currently unused, using dynamic channel assignment [12]. In the United States, the spectrum etiquette used in the unlicensed personal communications services (PCS) band is another example [12]. One can foresee the potential of extending such basic capabilities to negotiate dynamically not only frequency and time space usage, but also optimization of radio interface to match transmission power and modulation scheme, for example, to the required service requirements, operational range, and local RF environment, as well as extending the operation of such negotiation across much wider chunks of radio spectrum. Service priority information could be included in such negotiation schemes in order to ensure that subscribers' terminals defer to those of emergency or defense services needing to use the spectrum when present. While the regulatory implications of such adaptivity will pose headaches for regulators, the economic value arguments are such that they may be forced to understand, encourage, and eventually mandate such techniques.

Value Chain Implications

With any major technology shift comes an associated shift in power base within an industry; software radio is no exception. Many potential benefits of software radio exist at different levels of the value chain. Eventual impact will depend on how commercialization of the technology develops.

Development environments are needed to facilitate the creation of software-defined products. These may be expected to evolve from today's electronic design tools industry, with a focus initially on digital processing. Nonetheless, RF and software aspects must also be addressed.

As more functionality is implemented in software a shift in end product value from hardware to software will occur; indeed, this is already happening and may be expected to accelerate.

As open standards emerge, assuming that the regulatory issues can be adequately addressed in a timely manner, we may expect to see a new base of software providers developing and offering services either directly to subscribers or, perhaps more likely, bundled through network operators.

Conclusions

Software radio today is in its infancy. Already, however, the potential benefits, impacts, and implications are beginning to be understood and recognized. Software radio cannot be ignored: players around the world are busy trying to understand its particular implications for their position in the industry value chain. In the short term its pragmatic manifestations at the application layer will provide network operators with some of the answers they have been seeking for customer loyalty and differentiation. In the longer term, as "pure" software radio becomes commercially feasible, it will have profound impacts in many different ways on personal communications, and indeed on other industries.

Acknowledgments

The author would like to acknowledge the contributions of colleagues from various disciplines at Roke Manor Research to the development of the concepts and views expressed in this article. The significant contribution of industry colleagues within the ACTS community, the European Commission, and the SDR Forum is also gratefully acknowledged.

References

[1] W. H. W Tuttlebee, "Software Radio Technology: A European Perspective," *IEEE Commun. Mag.*, Feb. 1999.

[2] *IEEE Commun. Mag.*, Special Issue on Software Radio, May 1995.

[3] W. H. W. Tuttlebee, "Software Radio — Impacts and Implications," *Proc. IEEE 5th Int'l. Symp. Spread Spectrum Techniques and Apps.*, South Africa, Sept. 1998.

[4] *Proc. 1st Int'l. Software Radio Wksp.*, Rhodes, Greece, June 1998, jointly organized by the EC DG XIII and the SDR Forum; available at http://www.infowin.org/ACTS/ANALYSYS/CONCERTATION/MOBILITY/1intlswr.htm

[5] H. Huomo, "Software Radio, A Manufacturer's Perspective," *Proc. 1st Int'l. Software Radio Wksp.*, Rhodes, Greece, June 1998; http://www.infowin.org/ACTS/ANALYSYS/CONCERTATION/MOBILITY/1intlswr.htm

[6] Y. Shinagawa, "Software Radio Technologies: From the Viewpoint of the Terminal Manufacturer," *Proc. 1st Int'l. Software Radio Wksp*, Rhodes, Greece, June 1998; http://www.infowin.org/ACTS/ANALYSYS/CONCERTATION/MOBILITY/1intlswr.htm

[7] G. Fettweis *et al.*, "Integrated Broadband Mobile System (IBMS) featuring Wireless ATM," 2nd ACTS Mobile Commun. Summit, Aalborg, Denmark, Oct. 1997.

[8] V. Brankovic *et al.*, "Integrated Broadband Mobile System (IBMS) Demonstrator," *Proc. 3rd ACTS Mobile Commun. Summit*, Rhodes, Greece, June 1998.

[9] G. Fleming *et al.*, "A Flexible Network Architecture for UMTS," *IEEE Pers. Commun.*, Apr. 1998, pp. 8–15.

[10] G. Fleming *et al.*, "A Radio Independent Network — The Enabler for Software Radio," *Proc. 3rd ACTS Mobile Commun. Summit*, Rhodes, Greece, June 1998.

[11] M Wells, "Increasing the Capacity of GSM Using Adaptive Antennas," *IEE Proc.*, vol. 143, no. 5,Oct. 1996, pp. 304–10.

[12] W. H. W. Tuttlebee, *Cordless Telecommunications Worldwide*, Springer Verlag, 1997; http://ourworld.compuserve.com/homepages/cordless

Software Radio Architecture: A Mathematical Perspective

Joseph Mitola, III, *Member, IEEE*

Abstract— **As the software radio makes its transition from research to practice, it becomes increasingly important to establish provable properties of the software radio architecture on which product developers and service providers can base technology insertion decisions. Establishing provable properties requires a mathematical perspective on the software radio architecture. This paper contributes to that perspective by critically reviewing the fundamental concept of the software radio, using mathematical models to characterize this rapidly emerging technology in the context of similar technologies like programmable digital radios. The software radio delivers dynamically defined services through programmable processing capacity that has the mathematical structure of the Turing machine. The bounded recursive functions, a subset of the total recursive functions, are shown to be the largest class of Turing-computable functions for which software radios exhibit provable stability in plug-and-play scenarios. Understanding the topological properties of the software radio architecture promotes plug-and-play applications and cost-effective reuse. Analysis of these topological properties yields a layered distributed virtual machine reference model and a set of architecture design principles for the software radio. These criteria may be useful in defining interfaces among hardware, middleware, and higher level software components that are needed for cost-effective software reuse.**

Index Terms— **Computability, digital radio, software radio, topology.**

I. Introduction

It is well established that a communications system is the set of devices by which one employs a communications channel to convey information to a recipient. From a radio engineering perspective, the communications device must first encode the information from the source into a suitable electronic representation. The device then must transform this internal form into a waveform compatible with the radio frequency (RF) communications channel. The channel distorts the RF signal, adds noise, and creates time-delayed distorted replicas of the signal. The process is (imperfectly) reversed in the receiver as illustrated in Fig. 1.

In this venerable historical framework, control of the radio needs little attention, generally limited to power on/off, audio volume control, a noise-riding receiver threshold ("squelch"), and a switch to manually select from among predefined RF channels. Several such transmitter(s) and receiver(s) located and working together comprise a radio "node." The multiband, multimode, multithreaded, multipersonality capabilities of software radios require an expansion of this simple model.

A. Essential Functions of the Software Radio

Technology advances have ushered in a new level of physical-layer flexibility of the radio. *Multiband* technology [1], first of all, accesses more than one RF band of communications channels at once. The RF channel of Fig. 1, then, is generalized to the *channel set* of Fig. 2. This set includes RF channels, but radio nodes like personal communications system (PCS) base stations [2] and portable military radios also interconnect to fiber and cable; therefore, these are also included in the channel set. The channel encoder functions are expanded to programmable *RF/channel access, intermediate frequency (IF) processing, and modem.* Antennas and RF conversion that span multiple RF bands (and fiber-optic interfaces) comprise the RF/channel access function. *IF processing* may include filtering, further frequency translation, space/time diversity processing, beamforming, and related functions. *Multimode radios* [3] generate multiple air interface waveforms ("modes") defined principally in the *modem*, the RF channel modulator–demodulator. These waveforms may be in different bands, and may span multiple bands. The source and channel coders of Fig. 1 therefore become the *multiple personalities* of Fig. 2. A personality combines RF band, channel set (e.g., control and traffic channels), air interface waveform, protocol, and related functions.

Although many applications do not require *information security* (INFOSEC), there are incentives for its use. Authentication reduces fraud. Stream encipherment ensures privacy. Transmission security (TRANSEC) hides the fact of a communications event (e.g., by spread-spectrum techniques [4]). INFOSEC is therefore included in Fig. 2, although the function may be null for many applications.

In addition, the source coder/decoder pair of Fig. 1 must now be expanded to include the data, facsimile, video, and multimedia sources implicit in Fig. 2. Some sources will be physically remote from the radio node, e.g., connected via the synchronous digital hierarchy (SDH) [5], a local area network (LAN) [6], or other network through *service and network support* of Fig. 2.

These functions are implemented in multithreaded, multiprocessor software orchestrated by a *joint control* function. Joint control assures system stability, error recovery, timely data flow, and isochronous streaming of voice and video. As radios become more advanced, joint control becomes more complex, evolving toward autonomous selection of band,

Manuscript received November 11, 1997; revised March 17, 1998 and August 23, 1998.

The author is with The MITRE Corporation, McLean, VA 22102 USA.

Publisher Item Identifier S 0733-8716(99)02974-1.

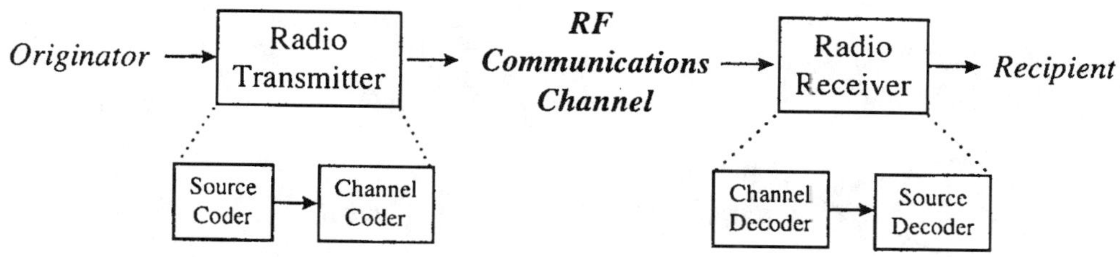

Fig. 1. Traditional model of radio communications.

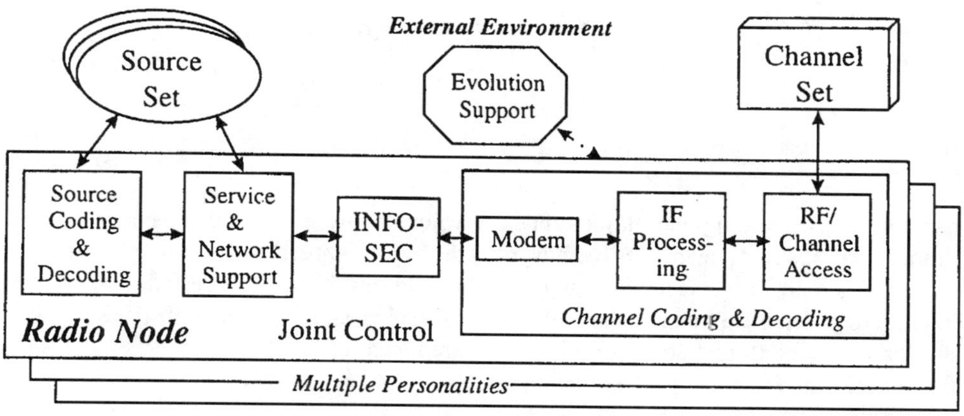

Fig. 2. Functional model of a software radio communications system. © Mitola's STATIS*faction*, used by permission.

mode, and data format. Any of the functions may be singleton (e.g., single band versus multiple bands) or null, further complicating joint control. Agile beamforming supports additional users and enhances quality of service (QoS) [7]. Beamforming today requires dedicated processors, but in the future, joint space–time equalization algorithms may time-share a digital signal processor (DSP) pool along with the Rake receiver [8] and other modem functions. Joint source and channel coding [9] also yields computationally intensive waveforms. Dynamic selection of band, mode, and diversity as a function of QoS [10] introduces large variations into computational-demand, potentially causing conflicts for processing resources. Channel strapping, adaptive waveform selection, and other forms of data rate agility [11] further complicate the statistical structure of the computational demand. In addition, processing resources are occasionally lost through equipment failures [12]. Joint control integrates fault modes, personalities, and support functions on a limited resource of applications-specific integrated circuits (ASIC's), field-programmable gate arrays (FPGA's), DSP's, and general-purpose computers to yield a reliable telecommunications object [13].

A software radio can upload new air interface personalities. These may modify any aspect of the air interface, including whether the waveform is hopped, spread, or otherwise constructed. The required resources (e.g., bandwidth, memory, and processing capacity), of course, must not exceed those available. Some mechanism for *evolution support* is therefore necessary to define the waveform personalities, to download them (e.g., over the air), and to assure that each new personality is safe before being activated. To type certify such a radio, one must guarantee that the properties specified by the regulatory bodies are preserved in spite of this high degree of flexibility. The need for such guarantees motivates the study of the mathematical properties of this architecture.

Traditionally, one may model the statistical demand for computational resources versus processing capacity using queueing theory [14]–[16]. Real-time performance can be assured in a fixed architecture using this approach [17]. Plug-and-play, however, creates a variable architecture as modules are introduced into the environment and removed. This raises the complexity of the statistics, particularly in complex nodes such as a base station cell site in which hundreds of users can invoke dozens of variable-bandwidth services in a pool of shared DSP's. A deeper understanding of the statistical properties of such environments is no doubt needed. Underlying such statistical analyses, there must be a predictable computational-demand relationship between the plug-and-play module and the environment. This calls for a theory of plug-and-play resource bounds for the software radio within which such predictable relationships will exist.

An obvious challenge is to define interface points for plug-and-play hardware and software modules. Industry organizations including the Software-Defined Radio (SDR) Forum (formerly the Modular Multifunction Information Transfer Systems—MMITS—Forum) are in the process of identifying such interface points using generalized applications programmers interfaces (API's) [18]. A less obvious challenge is to define architecture principles that assure that plug-and-play architectures will have the mathematical properties of controllability and predictability necessary for true plug-and-play services. This paper focuses on that challenge. The relevant properties may be characterized in the following mathematical framework.

B. Toward a Mathematical Model of Plug-and-Play Architecture

The thesis of this paper is that the next-generation radio functions of Fig. 2 have a mathematical structure that con-

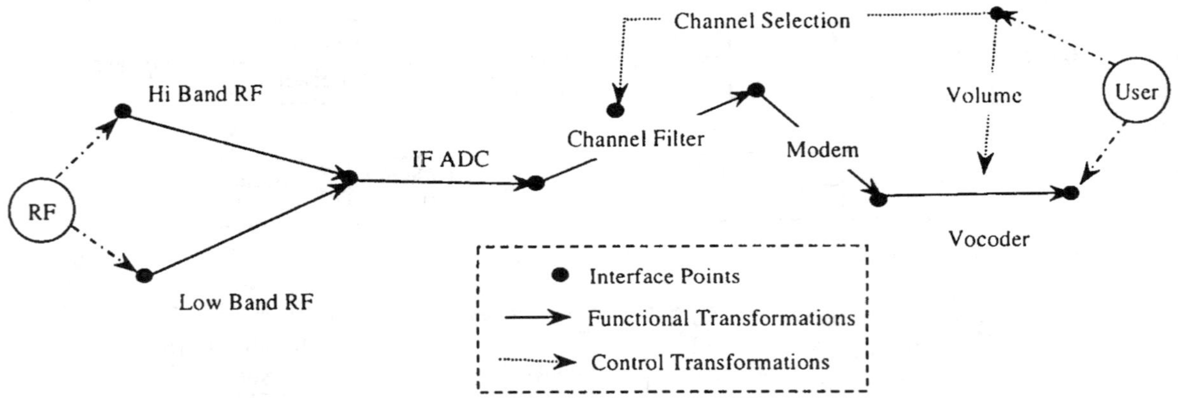

Fig. 3. Topological structure of an illustrative radio.

strains the behavior and interfaces of plug-and-play modules. The mathematical framework abstracts away the details of the radio functions, interfaces, and implementations. It identifies the computational and geometric structure that diverse software, services, and hardware platforms share.

In the proposed framework, modules are modeled as mappings of signals among architecture interface points defined as vertices in a topological space. Topological spaces are very general geometric spaces with a few set-theoretic axioms [19]. Mappings among interface points are represented as edges or arcs in this space. An edge is that which joins two vertices. An RF ASIC, for example, transforms the RF signal at the input vertex into the baseband signal at the output vertex. An arc, on the other hand, is a map that has properties in addition to its vertices, such as the power dissipated by the associated RF ASIC. This model may be used

1) to identify top level plug-and-play interfaces;
2) to predict and control system performance;
3) to define a reference model for standards setting;
4) to derive architecture principles for product evolution.

Mathematical models of multiprocessor task specifications based on topological spaces have been used to prove important properties of shared memory multiprocessors [20]. The model used in this paper is akin to such models.[1] The model in Fig. 3 shows radio hardware and software functional arcs joining the interface points in a topological space. This example shows a dual-band receiver with an analog-to-digital converter (ADC) IF, a channel filter ASIC, and software modem and vocoder (voice coding) functions. Some arcs represent the movement of signals and data from one interface point to another via isochronous, real-time streams (e.g., the high-band RF hardware and modem software of the figure). Others represent the exertion of control over states (e.g., user choice of a channel) and parameters that affect a stream (e.g., the channel filter).

Any important properties of the arcs become dimensions of the topological space. For example, the interface node between the IF ADC and the channel filter would have the value "hardware" in an "implementation" dimension. The interface between the modem and vocoder would have the value "software." Other dimensions may be continuous in $\boldsymbol{R}$, the real numbers, such as the range of a control parameter. In addition, there may be infinite-dimensional subspaces such as analog signals. Each subspace represents an aspect of the architecture.

[1] Rather than present the mathematical ideas formally and at once, they are introduced throughout the paper as needed.

C. Architecture Goals

A successful plug-and-play architecture entails at least the following.

Compatibility: The structure of plug-and-play modules must be compatible with that of the software radio environment—arcs plug into nodes for function composition.

Predictability: Module composition must preserve radio service-defining properties of the system, and when control is exerted, it must not have unintended consequences. Modules must not consume excessive computational resources.

An architecture is a framework in which a set of components is used to achieve specified functions (services) within specified constraints or design rules [21]. Compatibility and predictability are the properties of plug-and-play architecture studied in this paper. A mathematical framework for a plug-and-play architecture should inductively establish the desired properties of a given set of components within specified design rules. Due to the open-ended nature of radio services and technology, the framework must be extensible both to new services and to new implementation platforms. This paper begins the development of this framework with a top-down review of the properties of the software radio. A bottom-up analysis of the computational properties of software radio components then sets the stage for the derivation of the layered virtual machine reference model. Architecture principles that support plug-and-play modularity are developed in the process.

II. Defining the Software Radio

This section defines the software radio from several perspectives: engineering design, topological structure, and computational structure. The programmable digital radio (PDR) is defined and differentiated from the software radio. Topological properties of components are introduced in this section and developed in Sections III and V. In Section IV, the bounded

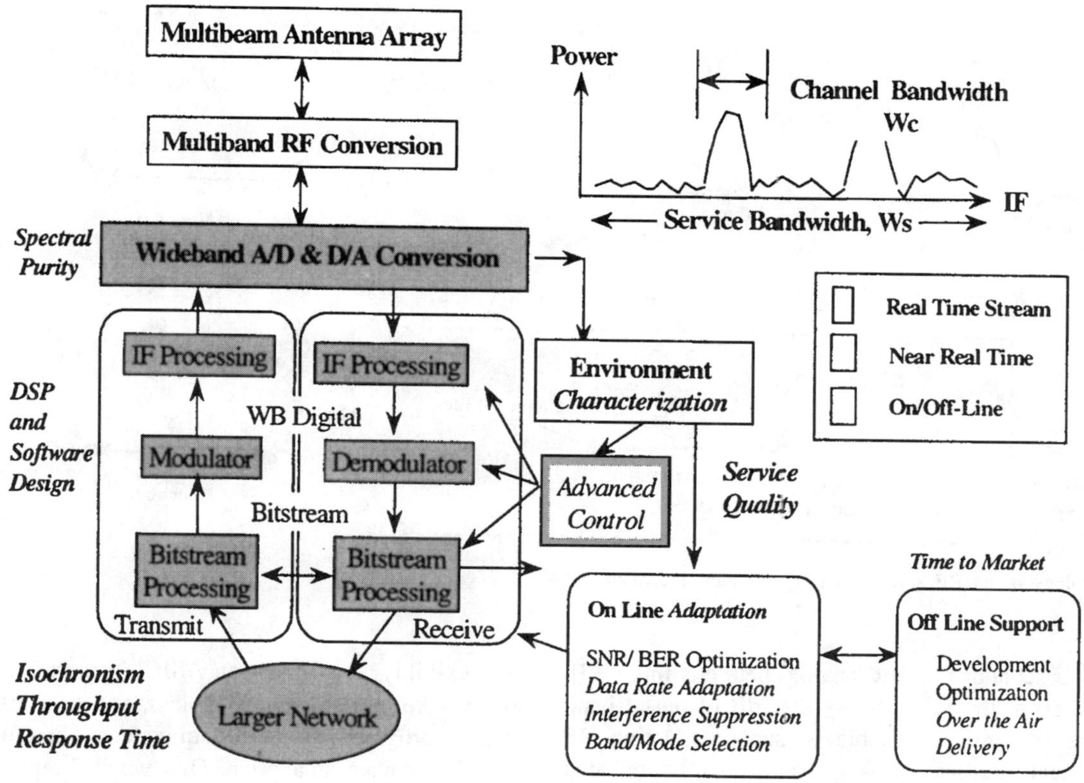

Fig. 4. Key software radio functions and components.

recursive functions are developed. Section VI derives architecture partitions and design principles.

A. Programmable Digital Radios

There are many ways to impart more than one personality to a radio. The term "PDR" applies to those radios that use a hardware-intensive mix of hardware and software techniques to access more than one RF band with a choice of air interface modes [22]. A PDR's programmability may be achieved using baseband DSP's. However, hardware modules or "slices" must be interchanged for the radio to change RF band and air interface mode ("waveform").

The software-defined radio (SDR) was defined by BellSouth to describe an evolution toward greater programmability of a wireless infrastructure [23]. This evolution includes programmable multiband, multimode radios implemented using the PDR approach. An IS-95/AMPS handset, for example, may employ a code-division multiple-access (CDMA) chip set for IS-95; an analog mobile phone system (AMPS) chip set; a dual-mode RF integrated circuit (RFIC) chip set [24]; a DSP chip [25] for filtering, voice coding, and computationally intensive tasks; and a microcontroller for user interface and system control. Many of the functions are programmable, but the CDMA despreader, for example, is defined in an ASIC, and may not be changed in the field. Over time, functions initially implemented in hardware (e.g., the RF modem) will migrate to software and conversely. The mathematical framework must therefore capture SDR migration between hardware and software.

B. The Software Radio

The software radio imparts multiple personalities to a radio in a specific way [26]. A software radio defines all aspects of the air interface including RF channel access and waveform synthesis (not just selection) in software. In the software radio, then, wide-band analog-to-digital and digital-to-analog converters (ADC's [27] and DAC's) transform each RF service band among digital and analog forms at IF as in Fig. 4. The wide-band digitized receiver stream of bandwidth W_s accommodates all subscriber channels, each of which has bandwidth W_c ($W_c \ll W_s$). The mathematical framework differentiates alternative ADC and DAC architectures, with the attendant differences in programmability.

In the software radio of Fig. 4, IF ADC and DAC channels are processed isochronously using programmable digital hardware, and software.[2] IF processing may include filtering to isolate subscriber channels [28]; digital formation of nulls and beams [29]; joint space–time equalization [30]; integration of space diversity [31], polarization, or frequency diversity channels [32]; and other means of acquiring a high-quality waveform. There may be multiple IF frequencies (in the heterodyne receiver) or the IF frequency may be zero (in the homodyne receiver [33]). IF processing may be null, for example, in a direct conversion receiver [34]. Digital downconversion is the process of using the frequency-domain periodicity of sampled bandpass waveforms to translate the

[2]Software here means algorithms stored in random access memory of processors which may, in principle, be downloaded over the air. Firmware is software to the degree that it can be downloaded; otherwise, it is part of the hardware personality.

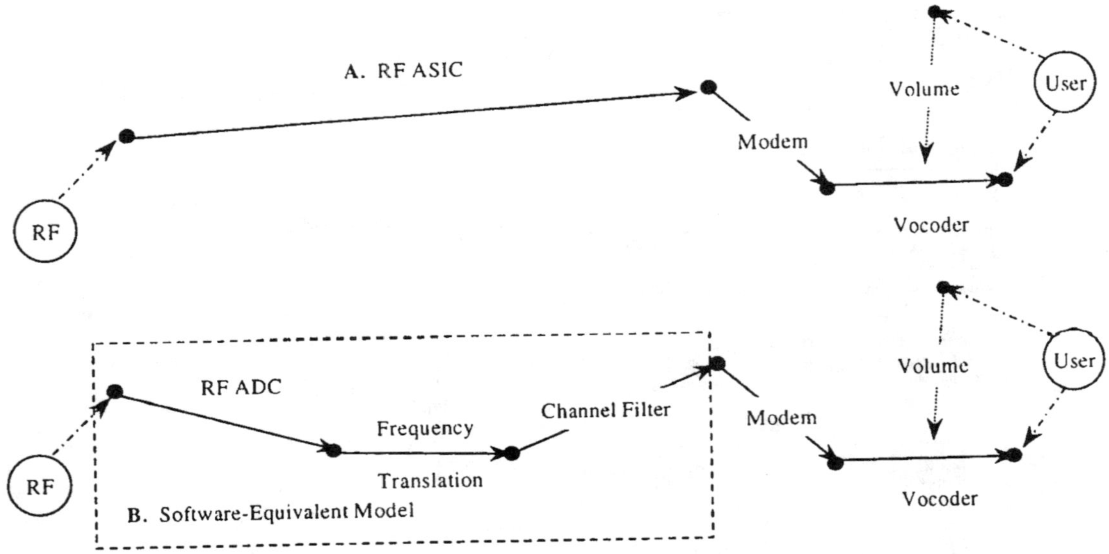

Fig. 5. (A) Topology of an RF ASIC and (B) its digital equivalent.

waveform to baseband without analog heterodyning [35]. Preselection filters with the necessary 80+ dB of out-of-band attenuation are becoming available in superconducting RF filters [36]. Local oscillator leakage may also be reduced proportionally so that digital downconversion is becoming more feasible. The mathematical framework represents plug-and-play aspects of such variations via sets of arcs and nodes with subset relationships.

In the software radio transmitter, baseband signals are transformed into sampled channel waveforms via channel modem functions implemented in software to drive high-performance DAC's. These signals may be preemphasized or nonlinearly precoded [37] by the IF processing software. In some implementations, modem functions, IF processing, and RF channel access may be amalgamated into a single component such as a direct conversion receiver RFIC [34]. In addition, dynamic compilation of software or real-time switching among personalities can allow these discrete functions to be integrated into a single component such as an FPGA [38]. The mathematical framework represents both discrete and integrated implementations. In discrete implementations, each edge in the topological space is also an arc corresponding to a component. In an integrated implementation, a component arc subsumes many topological edges.

C. A Software-Equivalent Model of Hardware

The ADC and DAC define the point at which functions are potentially software defined. We may call this the *digital access point*. Relatively inflexible analog and digital hardware such as ASIC's may be necessary in a handset, e.g., for direct conversion of the channel waveform to a bitstream. Although an ASIC may have a few control parameters, the personality of an ASIC cannot be changed in the field (e.g., from one air interface standard to another). To change the air interface personality, one must replace the ASIC, switch among different ASIC's, or switch among modes of one ASIC. In the tradeoffs among size, weight, power, cost, and flexibility, fixed-function ASIC's generally consume less power, take less space, weigh less, and cost less per device than the programmable DSP equivalent. Therefore, increased flexibility comes at a price. One way of representing these differences mathematically is to attribute equivalent computational capacity to the nonprogrammable devices.

In the topological model of the hypothetical radio illustrated in Fig. 5 (top), an RF ASIC (A) transforms the analog signal from the antenna directly to input for the baseband modem. The software-equivalent model (B, bottom) postulates an RF ADC with exactly the bandpass characteristics of the RF ASIC. The frequency translation behavior of RF ASIC (A) is modeled as the frequency translation software process of model (B). The filtering behavior of the RF ASIC is modeled by the channel filter software of (B). The sequence of RF ADC, frequency translation, and channel filter arcs constitute the software-equivalent model of this ASIC. The equivalent computational capability of RFIC (A) is the total computational demand of the postulated software processes (B). Since analog hardware has variable performance over time and over different devices, the exact filter model of a given RF ASIC will change over time. Thus, each device is represented in a *set* of models. The variations are contained within "open balls" [39] in the parameter subspace of the topological model (Fig. 6). One may also aggregate multiple device models into such a parametric model by taking set unions. The set of all such arcs and open balls comprises the topological model of the ASIC. Each model may also have parameter sets in dimensions that characterize or predict such diverse properties as time delays and power dissipation.

Although the sequel focuses on computational properties, this set-theoretic framework is general, capable of representing local properties of a device including interface details, size, weight, and power. Since topological spaces are not necessarily linear, aggregate properties of the radio system in the higher level spaces may have nonlinear relationships to the properties of the components in the lower level spaces.

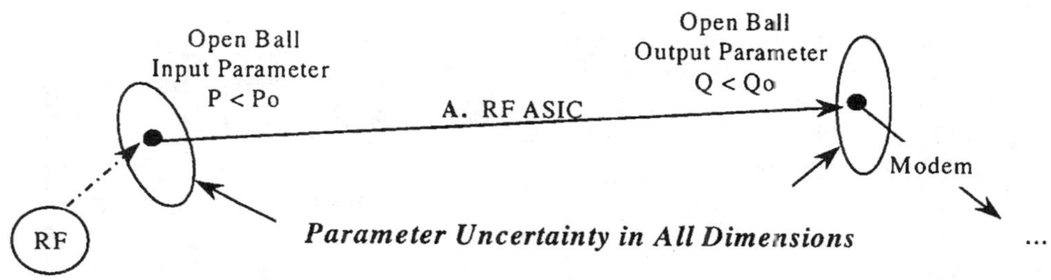

Fig. 6. Parameter uncertainty constrained to "open balls."

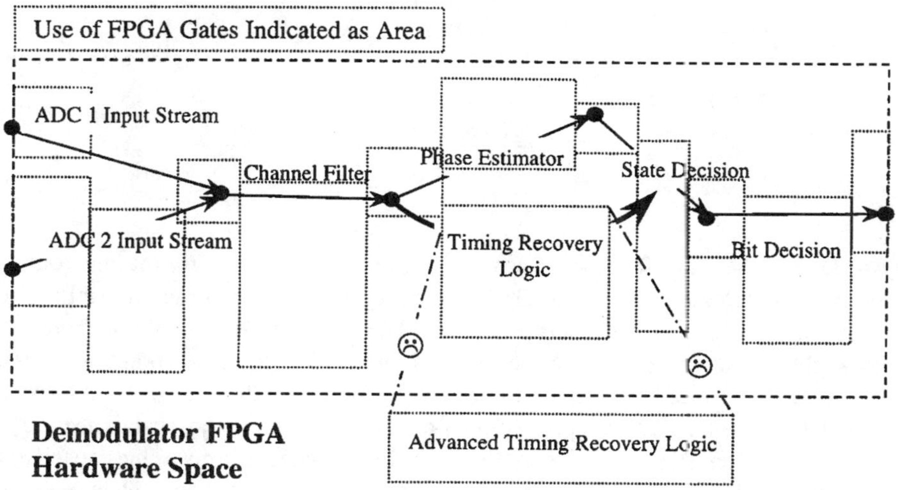

Fig. 7. Hardware topology shows incompatibility.

D. Quantifying Degrees of Programmability

The *degree of programmability* of an implementation is fundamental to software radio architecture. Since contemporary radios use a mix of processor types, one must characterize this mix precisely. Consider the highest level topological model of a radio (a single arc). This arc may be hierarchically divided into its primitive components. Hardware primitives are the discrete devices, while software primitives are single machine instructions. Primitives that may be redefined via software in the field are labeled. The number of labeled primitive arcs divided by the total number of primitive arcs is a measure of programmability. Since an ASIC's programmability is limited to the modification of a few parameters, most of its gate-level arcs will not be labeled.

FPGA's are, in principle, completely programmable. In practice, they are more programmable than ASIC's, but subject to gate and interconnect constraints. Programmable radios have been based almost entirely on reconfigurable FPGA's [40]. The field programmability of an FPGA is more constrained than that of a DSP chip, in part because of the likelihood of running out of usable gates on the FPGA as radio functions expand. The topological model of each type of device allows one to characterize programmability more precisely. In the FPGA topology of Fig. 7, two dimensions (corresponding to states and interconnect) represent the allocation of functions to hardware.

Suppose that an advanced timing recovery algorithm, comprising, say, 10% of the FPGA area is to be downloaded to the radio. As shown in the figure, its needs are incompatible with the gate use of the existing timing recovery logic. Assume that one redefines the personality of the FPGA to accommodate the new logic. The download size increases from the 10% needed for the increment to 100% for the entire personality, a 900% increase in the size of the download. It is also possible that a moderately populated (70%) FPGA will be unable to accommodate the 10% download because of hardware constraints such as the required placement of I/O buffers, lack of registers, etc. A similar topological model of a DSP chip is shown in Fig. 8. The additional 10% of DSP code associated with the advanced timing recovery logic (now implemented in software) is accommodated via space available in the nonvolatile random access memory (RAM). The memory map allocates this logic to RAM hardware. DSP may appear easier to program than FPGA's because RAM allocation is accomplished by loader software while gate allocation in an FPGA generally requires an experienced designer. New waveforms also seem to outgrow the gates of an FPGA more easily than they outgrow the program memory of a DSP chip. On the other hand, the timing constraints of DSP software are more severe than the timing of an equivalent FPGA. Logic in an FPGA runs at nearly the chip's clock rate, while the DSP code runs from one to three orders of magnitude slower than the system clock. The tradeoff of ASIC hardware allocation and DSP task timing is reflected in these subsets of the topological model.

Complex or reduced instruction set computers (CISC/RISC) provide less hardware acceleration than DSP chips. To quantify

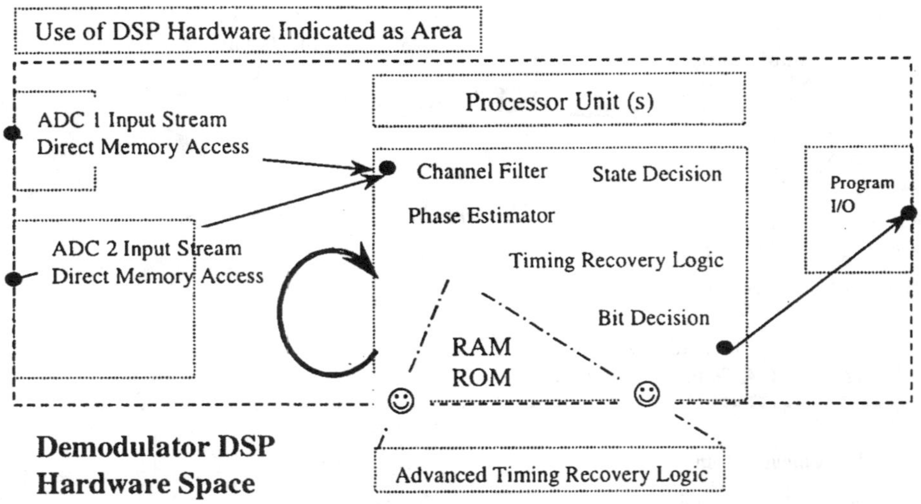

Fig. 8. DSP hardware commitment subspace topology.

the degree of flexibility of DSP, CISC, and RISC processors, one again defines an appropriate topological space. Let {ISA} be the space of single-instruction register-state transformations in a processor with a given instruction set architecture (ISA). From an arbitrary initial state, a DSP has many more edges connecting reachable data states than a CISC processor, which in turn has more arcs than an RISC processor. So, from the perspective of {ISA} topology, the RISC processor is the simplest, and thus in some sense, the most general programmable platform. CISC is more complex, and DSP the most complex of the fixed ISA machines. Performance of high-quality code underscores these differences. DSP code that employs zero-overhead loops with full register stacks and processing elements yields higher throughput than CISC code of a processor with the same system clock, memory, and I/O delays. FPGA's, on the other hand, effectively have a variable ISA that makes them even more computationally efficient than DSP machines, with an attendant loss of flexibility. That is, once we employ an FPGA's gates for a specific task (e.g., a digital filtering algorithm), those gates in that configuration cannot be used for a different task (e.g., a timing recovery algorithm). A reconfigurable FPGA's gates may support both functions, provided reconfiguration delays are tolerable. DSP chips use the same gates for both functions time shared through software.

These relationships define a phase space for the software radio. Physicists use a phase space to represent states of a substance (e.g., solid, liquid, and gas) as a function of parameters such as temperature and pressure. The software radio phase space of Fig. 9 represents the states of radio implementations as a function of the digital access point and the degree of programmability.

Fig. 9 places five types of radios in the phase space. The vertical axis represents the bandwidth at the digital access point. The horizontal axis represents the degree of flexibility, the fraction of functionality that may be changed in the field using plug-and-play software. To place a system in the phase space, one examines the ADC's and DAC's, placing the digital access point where the functionality is fully programmable. One places the radio on the horizontal axis according to the aggregate degree of programmability of the device.

The HF STR-2000, a commercial product of Standard Marine AB, shown at point $[A]$, employs baseband DSP using the TMS320C30. Most commercial off-the-shelf (COTS) cellular telephone handsets fall near $[B]$. ASIC's provide much of the equivalent processing capacity, shifting these designs toward the less programmable end of the axis. Current digital cell site designs $[C]$, similarly, rely heavily on digital filter ASIC's for frequency translation and filtering, but they access the spectrum at IF. SPEAK*easy* II, $[D]$, provides a GFLOP of programmable DSP, shifting this implementation to the right. A product of research at the Massachusetts Institute of Technology, the virtual radio [41] $[V]$ delivers a single channel radio using a general-purpose processor, DEC's alpha. This is the most general-purpose computing platform reported in the literature. Point $[X]$ represents the ideal software radio with the digital access point at RF and all functions programmed on RISC processors. Although maximally flexible, and thus of research interest, such designs tend to be economically impractical. This phase space quantitatively differentiates software radios in the upper right quadrant from the PDR's elsewhere in the figure.

A PDR is not a software radio if any crucial aspect of the channel waveform is implemented using programmable hardware (such as a voltage-controlled oscillator) rather than using software (e.g., sin/cosine lookup table). A joint tactical information dissemination system (JTIDS) radio [42] implementation, with a 3 MHz IF from which hop frequencies are generated using a 250 MHz programmable local oscillator, is not a software radio for that waveform. On the other hand, a 250 MHz digitized IF that could set every hop frequency in software would be a software radio. The radio may meet the software radio criterion for a large class of narrow-band waveforms, while failing for wide-band waveforms. The phase space of the topological model thus unambiguously defines the degree to which an implementation is reprogrammable. Given an unambiguous definition, one may consider the components in more detail.

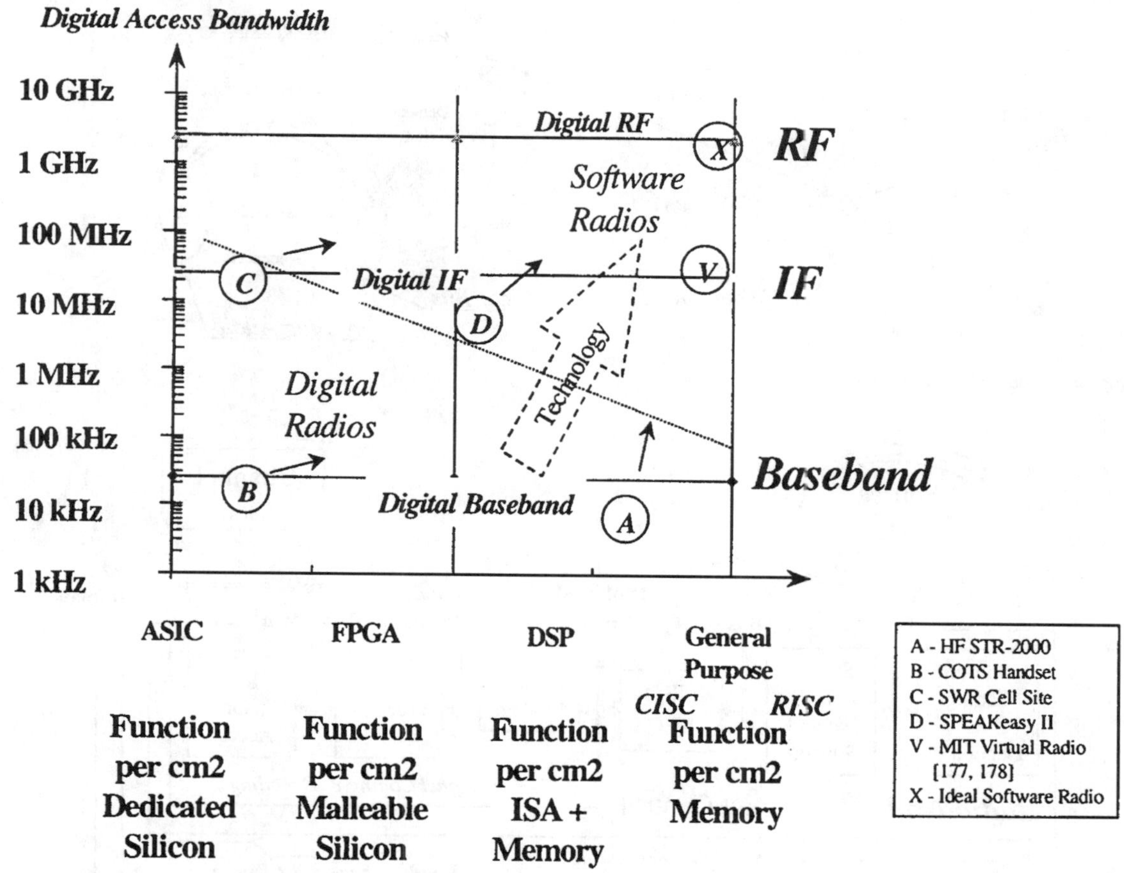

Fig. 9. Software radio phase space. © Mitola's STATIS*faction*, used by permission.

TABLE I
ATTRIBUTES OF TOP-LEVEL SOFTWARE RADIO FUNCTIONAL COMPONENTS

Functional Component	Attributes	Remarks
Source Coding & Decoding	Audio, video, fax and data interfaces	Ubiquitous standard algorithms (e.g. ITU[1], ETSI[2])
Service & Network Support	Multiplexing; setup and control; data services; internetworking	Wireline and Internet standards including mobility [3]
*Information Security**	Transmission security, authentication, non-repudiation, privacy, data integrity	May be null, but is increasingly essential in wireless applications [4]
*Channel Coding & Decoding: Modem**	Baseband modem, timing recovery, equalization, channel waveforms, predistortion, black data processing, etc.	INFOSEC, modem, and IF interfaces are not standardized
*IF Processing**	Beamforming, diversity combining, characterization of all IF channels	Innovative channel decoding for signal and QoS enhancement
RF Access	Antenna, diversity, RF conversion	IF interfaces are not standardized
Channel Set(s)	Simultaneity, multiband propagation, wireline interoperability	Automatically employ multiple channels or modes for managed QoS
*Multiple Personalities**	Multiband, multimode, agile services, interoperable with legacy modes	Multiple *simultaneous* personalities may cause considerable RFI
*Evolution Support**	Define & manage personalities	Local or network support
*Joint Control** (over *Channel Set*)	Joint source/channel coding, dynamic QoS vs. load control, processing resource management	Integrates user and network interfaces; multi-user; multiband; and multimode capabilities

* *Interfaces to these functions have historically been internal to the radio, not plug-and-play*

III. TOP-LEVEL COMPONENT TOPOLOGY

Radio components are represented as arcs in the topological model. An arc may be a union or composition of other arcs, defining a natural encapsulation hierarchy for the radio system. At the top level of the hierarchy (the "context" level in object-oriented design), the radio node is a black box mapping air interface, user, and network events to appropriate responses. The functions of Fig. 2 define the second level of partitioning with the component attributes highlighted in Table I. The attributes listed in this table indicate the allocation of functions

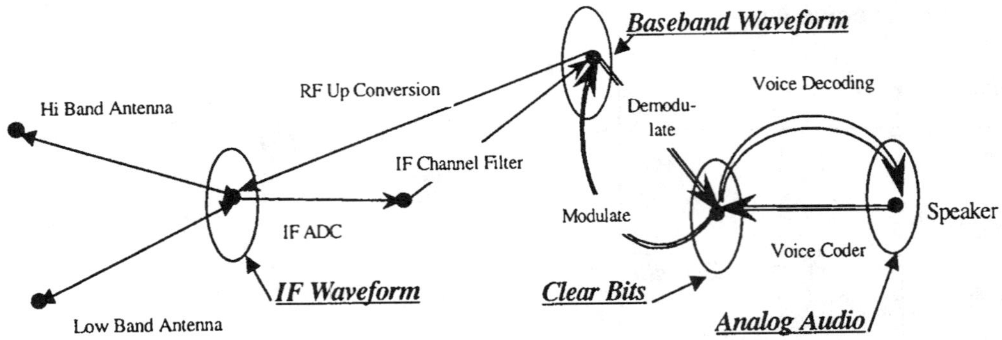

Fig. 10. Topological model of dual band handset streams.

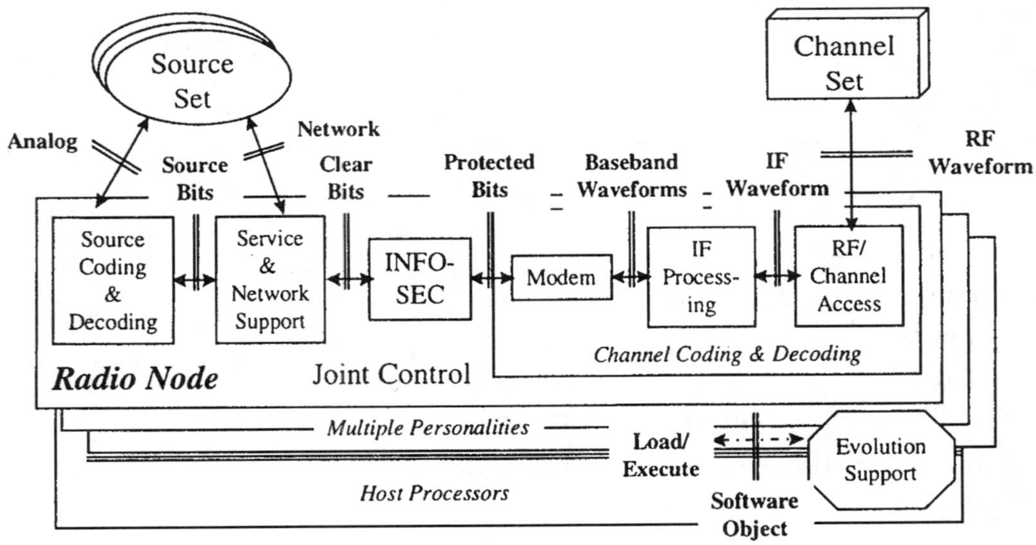

Fig. 11. Interfaces constrain topological properties.

to top-level components. In a topological space, each allocated function may be a singleton, a subset, or the null set. This corresponds to the occurrence of a function implemented in a single component, implemented multiple times (e.g., in parallel channels) in multiple components, or not implemented at all. The remarks column highlights functional diversity leveraged by the software radio architecture.

The stream-oriented functional components are source coding and decoding, service and network support, INFOSEC, RF/channel modem, IF processing, and RF/channel access. Each is represented topologically by a pair of arcs between data interfaces, the topological domain, and range of the maps. A topological model of a specific radio consists of those arcs that correspond to the components. For example, the signal-stream topology of a dual-mode handset is shown in Fig. 10.

The critical interfaces among dual-band antenna, IF ADC, baseband modem, and vocoder of this notional device are readily apparent in the topological model. Additional interfaces associated with the functions of Table I are shown in Fig. 11. Analog (audio and video) waveforms comprise the interface between the source set and the source coding and decoding functions. Source bits are encoded, but services and network support adds forward error control structure. If INFOSEC is null, protected bits become clear bits. Interfaces may be null at one level and visible at another. An RFIC, for example, may subsume the IF waveform interface, exhibiting only baseband and RF waveforms in the higher level of the topological space.

Implementation topologies are constrained by these interfaces and related standards as summarized in Table II. Interfaces are represented in a space which includes the interface signal itself $x(t)$ and related meta-level characteristics: implementation class (hardware or software), impedance, connector type, bandwidth, etc. The analog stream has infinite dimensionality in both time and amplitude, defined over $\boldsymbol{R}$. A digitized analog stream consists of an infinite stream of one-dimensional vectors, the samples. But, given a pair of sample buffers of length N, this stream is reduced to dimension $2*N$.

In addition, as shown in Fig. 12, the topological model structures the meta-level aspects of these interfaces. Each dimension of the space must be identified, and the associated elements that apply must be specified. The syntactic and semantic definition of the topological spaces of such a generic radio would constitute a knowledge representation language (KRL) [47] for radio (an RKRL). The definition of a complete, extensible RKRL is a significant undertaking that is beyond the scope of this introduction. Important properties of plug-and-play interfaces may be defined without a complete RKRL, however. In particular, general properties of the interfaces and

TABLE II
TOP-LEVEL INTERFACE TOPOLOGIES

Interface	Key Characteristics	Topological Properties
Analog Stream	Audio, video, facsimile streams	Infinite dimensional; Filtering constraints comprise open ball(s)
Source Bitstream	Coded bitstreams and packets. ADCs define a finite window into a quantized, discrete-time sampled waveform.	Finite dimensional; frame and data structure defines subspaces. Finite precision defines a Dynamic Range Subspace[2] for the ADC
Clear Bitstream	Framed, Multiplexed, Forward Error Controlled (FEC) bitstreams and packets	Finite dimensional; FEC subspaces have rich algebraic propert es
Protected Bitstream	Random challenge, authentication responses; public key; enciphered bitstreams and packets	Finite dimensional; randomized streams; complex message passing for downloads; If null, interface reverts to clear bits
Baseband Waveform	Discrete time synchronous quantized sample streams (one per carrier)	Digital waveform properties determine fidelity of analytic representation
IF Waveform	Composite, digitally pre-emphasized waveform ready for up-conversion	Analog IF has infinite dimensional topology; Digital IF may have baseband product topology
RF Waveform	Power level, shape, adjacent channel interference, etc. are controlled	Analog RF has infinite dimensional topology; Includes spatial and temporal dimensions
Network Interface	Packaged bitstreams may require ATM, SS7, or ISO protocol stack processing	Synchronous Digital Hierarchy (SDH), Signaling System 7 (SS7) subspaces
Joint Control	Control interfaces to all hardware and software; initialization; fault-recovery	(Not illustrated in the figure) Parameter spaces; non-linear logic subspaces
Software Objects	Download from evolution support systems	Represents binaries, applets; includes self-descriptive language subspaces
Load/Execute	Software object encapsulation	Download topologies are highly nonlinear

[2] A Nyquist-dynamic range subspace has been sampled so as to meet the Nyquist criteria for the bandwidth of the sampled signal; it has been quantized with sufficient accuracy for the two-tone spurious-free dynamic range of the application.

Domain:
Class: Analog-Stream
Implementation: Hardware
Signal-Interface: Coax-DC-Coupled
Impedance: 50 Ohms ...
Carrier Frequency: Baseband

3dB Bandwidth: 350 kHz
Signals: {Predetected-Signal}
Interface-Signal: x(t)
Control Parameters:
Gain: 0dB to 20 dB
Gain Control: AGC ...

Fig. 12. Illustrative meta-level topological space for analog stream interface, a topological domain.

computational components can be studied independently of an RKRL. In addition, air interface specifications and radio API's constitute informal RKRL's.

Such interface domain descriptions could promote the assured delivery of services in the software radio's plug-and-play environment. A plug-and-play interface should effectively separate the module from the rest of the system. Effective separability implies at least the following topological properties.

1) Composition of module functions on the interface topology yields a well-defined system-level function consuming specified resources to deliver the intended service.
2) Performance (e.g., spectral purity, data formats, throughput, response time, etc.) under function composition is within specified bounds.

The separability of modules at plug-and-play interface points is illustrated in Fig. 13. Plug-and-play modules may comprise top-level functional components such as the modem. Or, they may be defined across arbitrary points deep in the hierarchy such as at the vocoder.

The goal is that the system as a whole behave as desired if the plug-and-play modules and the host system each has the prescribed testable local properties. What properties can be guaranteed? The composite system should use bounded computational resources with guarantees against wait-induced fault conditions. This property is necessary for isochronism of voice and video streams and acceptable timing of packet data flows. This property is also essential for type certification by induction.

IV. COMPUTATIONAL PROPERTIES OF FUNCTIONAL COMPONENTS

A fundamental aspect of software radio architecture is the set of conditions under which plugging in a module results in acceptable use of computational resources. The composability of functions defined over general topological spaces is well understood. Homeomorphisms are topology-preserving maps that may be composed to yield other homeomorphisms provided the domain of one map and the range of the other are topologically compatible [39]. From a topological perspective,

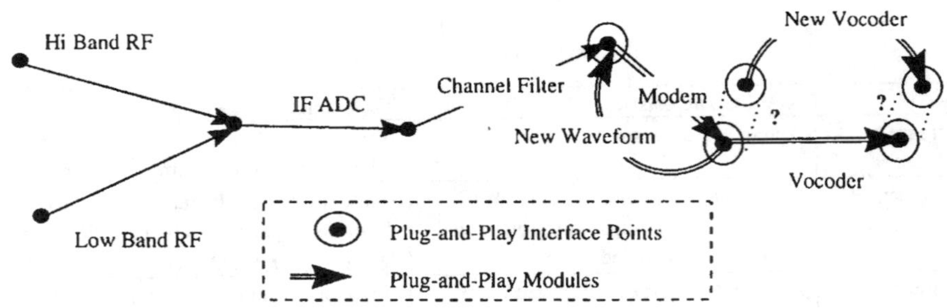

Fig. 13. Topology of plug-and-play interfaces.

the domain of software consists of those inputs over which its results are defined. The range consists of the corresponding results (including side effects and returned values). Under what conditions can well-behaved software be composed with other well-behaved software to yield a well-behaved system? This very general question may be undecideable. But software radios are engineering systems with timing constraints that allow us to prescribe constraints on the topological structure of the software, and to establish conditions under which the composition of software modules consumes finite resources that can be predicted accurately.

The concept in addressing bounded resource use is to construct the largest set of functions for which predictable finite resource consumption may be guaranteed. In addition, the composition of such functions must also be bounded for a proof that defines conditions under which plug-and-play modules will not use excessive resources. Furthermore, any constraints on the radio functions that can be computed with such functions must be specified. The analysis of this section shows that the bounded recursive functions both guarantee predictably bounded resource use and present no constraints on radio functionality. A system constructed using these functions has the ability to maintain predicted throughput across dynamic changes of plug-and-play software modules.

A. Models of Computation

Earlier in this paper, the notion of software-equivalent hardware capability was introduced. To each ASIC and FPGA may be attributed the computational capability of a canonical processor, which may in turn be modeled as a (collection of) von Neuman machine(s). Each procedure encoded in software on a von Neuman processor may be modeled in terms of computability and concrete complexity as a random access machine (RAM) [48]. A RAM consists of a state machine, input, and output arrays; internal registers; the capability to load and store data via direct and indirect addresses; and the capability to increment memory values. The RAM model provides an intuitive but mathematically precise description of a single von Neuman processor. The RAM model has also been proven equivalent to the recursive function and Turing machine models of computing. Important theoretical properties include the ability to simulate in polynomial time capabilities that require exponential time on Turing machines [49]. Let {RAM} represent the RAM model of the instruction set architecture {ISA} of an arbitrary processor. Theorems outlined below use the recursive function and the RAM models to establish tight upper bounds on processing resources, a necessary condition for predictable throughput.

B. The Primitive Recursive Functions

The recursive function model of computing consists of a set of functions from $\boldsymbol{N}$, the natural numbers, onto $\boldsymbol{N}$ with closure properties that define classes of functions. The *primitive recursive functions* consist of the following.

1) The *zero* function z: $\boldsymbol{N} \rightarrow \boldsymbol{N}$: $z(x) = 0$, which yields the constant zero for any natural number x in $\boldsymbol{N}$.
2) The *successor* function s: $\boldsymbol{N} \rightarrow \boldsymbol{N}$: $s(x) = x+1$, which associates a successor to each x in $\boldsymbol{N}$.
3) The *projection* function U_m^n: $\boldsymbol{N}_n \rightarrow \boldsymbol{N}$: $U_m^n(x_1, x_2, \cdots, x_n) = x_m$ by which arguments may be selected, with closure under:
4) *Composition:* for $g_1, g_2 \cdots g_m$: $\boldsymbol{N}^n \rightarrow \boldsymbol{N}$ and h: $\boldsymbol{N}^m \rightarrow \boldsymbol{N}$, $f(\boldsymbol{x}_n) = h(g_1(\boldsymbol{x}_n), g_2(\boldsymbol{x}_n) \cdots g_m(\boldsymbol{x}_n))$, where $(\boldsymbol{x}_n)$ is brief notation for $(x_1, x_2 \cdots x_n)$; equivalently, $f = h^*(\boldsymbol{g}_m)$.

That is, the set is closed under composition of functions. One can compose functions in a natural way that corresponds to function calls and/or selection of data inputs in a random access machine.

5) *Primitive Recursion:* for g: $\boldsymbol{N}^n \rightarrow \boldsymbol{N}$, $\boldsymbol{h}$: $N^{n+2} \rightarrow N$, and finite $y > 0$, $f(\boldsymbol{x}_n, y) = g(\boldsymbol{x}_n)$ if $y = 0$, and

$$\boldsymbol{h}(\boldsymbol{x}_n, (y-1), f(\boldsymbol{x}_n, (y-1))) \qquad \text{for } y > 0.$$

That is, the set of functions is closed under primitive recursion. Primitive recursion for $y = 1$ and $\boldsymbol{h}(\boldsymbol{x}_n, 0, f(\boldsymbol{x}_n, 0)) = h(\boldsymbol{x}_n)$, f is simply the conditional execution of g or h based on the value of y [i.e., since $f(\boldsymbol{x}_n, 0) = g(\boldsymbol{x}_n)$, but $f(\boldsymbol{x}_n, 1) = \boldsymbol{h}(\boldsymbol{x}_n, 0, g(\boldsymbol{x}_n)) = h(\boldsymbol{x}_n)$]. This recursive pattern is equivalent to the if–then–else programming construct in terms of effective computability. Since f appears in its own definition, it requires a pushdown stack of successive arguments. But since initially $y > 0$ and y decreases at each invocation, the procedure will terminate, and a stack which is as large as $|y|$ will not overflow. It is assumed that the stack allocator in a specific {ISA} keeps the finite stack from overflowing, provided an upper bound for $|y|$ is known in advance. The knowledge of such bounds is a feature of the software radio application of this theory.

The *primitive-recursive functions* are the smallest set of such functions closed under composition and primitive recursion. Sequential logic, transversal and recursive filters, bit manipulation, and data packing are common primitive-recursive functions of software radios. They include addition, subtraction,[4] multiplication, and others that do not require iterative search.

C. The Total Recursive Functions

Iterative loops are not primitive recursive, but they are essential to radio software. The simplest model of iteration is bounded minimalization.

6) *Bounded Minimalization*: Let $g\colon \boldsymbol{N}^{n+1} \to \boldsymbol{N}$ be primitive recursive; then

$$f\colon \boldsymbol{N}^{n+1} \to \boldsymbol{N}\colon f(\boldsymbol{x}, y) = \mu z\{z < y\}[g(\boldsymbol{x}, z) = 0]$$

is also primitive recursive.

Read μz as "$f(\boldsymbol{x}, y)$ *is the least* z *less than* y for which $g(\boldsymbol{x}, z) = 0$". The class of functions closed under composition, primitive recursion, and bounded minimalization is *total*. That is, they are defined for all $\boldsymbol{N}$. The Fortran **do** loop exemplifies *bounded minimalization*. Ackerman's function [50] is defined everywhere, but uses resources faster than any known function due to the structure of its recursive calling sequences. It is not considered primitive recursive, although it is a total function. Yet, for small finite parameters of Ackerman's function which are known in advance, one can determine whether the function will exceed an upper resource bound by using a suitable step-counting function. Any primitive recursive function induces a primitive-recursive step-counting function. Let the *bounded primitive recursive* functions[5] be the primitive-recursive functions with the associated step-counting functions and finite resource bounds. In software engineering terms, this model of computing insists that when a module is tested, its resource use in isolation [e.g., millions of instructions per second (MIPS)] be tested and resource bounds be established as a parameter of the module. Such characterization is regularly used software-development practice for real-time systems. Integer division, exponentiation, generation of primes, finite logic predicates, and finite iteration are examples of bounded primitive functions. These functions have properties attractive for software radio implementations as expressed in the following theorems.

Theorem 1 (Primitive Bounded Resources): Any bounded primitive recursive function is equivalent to a RAM program which terminates in a finite number of steps which can be bounded tightly from above, given bounds y_i, of the minimalizations from which the composite functions are created.

Outline of the Proof[6]*:* Associate a sequence of RAM instructions with each bounded primitive function and closure construct. Each corresponding RAM instruction sequence is primitive recursive and bounded. These RAM instruction sequences are like in-line subroutines. Each such sequence has a small finite number of steps (the bound). Compute the maximum concrete complexity of the recursive function structure from the ISA and a suitably chosen step-counting function [51]. Let the number of instructions from RAM be the step-counting function. The resources used by any bounded recursive function with a given set of calling parameters can be computed from the calling parameters and the step-counting function.

Lemma 1 (Primitive Bounded Execution Time): The dedicated-processor RAM-equivalent of any bounded primitive recursive function executes in an amount of time that may be tightly bounded. One determines the maximum execution time of each class of RAM instructions on a specific ISA. The execution time bound is the product of the number of steps from Theorem 1 times the maximum execution time for that class of instruction. The time used by the function tested in isolation on a specified dedicated machine is thus an alternative step-counting function. From a software engineering perspective, this lemma states that the bounds allocated may be specified in terms of execution time on a dedicated processor.

Relevance: Software radio incorporates isochronous streams for which there is a fixed timing window during which specific functions must be accomplished in order for services to be delivered properly and/or for the system to remain stable. Modules whose structure is not constrained as above may use resources that are bounded on test cases, but that "blow up" in other conditions. All primitive recursive functions can be guaranteed to either: 1) complete within a specified window or 2) be *easily computed in advance* to be incapable of meeting that timing window (Lemma 1). If the system control algorithm predicts such a fault, it can signal the network or the user and preclude the (plug-and-play) module from being invoked. Such predictable timing properties provide the foundation necessary to allocate computational resources to plug-and-play modules safely. Queueing theory [14] translates service time measured in isolation into statistical bounds on isochronous performance in the service environment where other tasks are sharing the same processor(s). The control algorithm may employ such techniques to establish a statistical bound on aggregate resource use which, if exceeded, could degrade services. But since the resource bounds may be computed, the system should not become unstable.

D. Bounding the Partial Recursive Functions

Bounded primitive recursive functions do not express all of the programming constructs needed in software radios. Notably absent are the **while** and **until** loops. These loops search for a condition under which to terminate. This condition may never occur, so the RAM programs may loop forever. Hardware and software processes that wait or search for a condition that may not occur are computationally equivalent to while or until loops. Current research in wait-free computation considers the related problem of defining instruction sequences ("protocols") which assure that any process that itself is not in a fault condition will terminate in a finite number of steps [58]. The source code of software radios like SPEAK*easy* I reveals the relatively widespread use of wait-

[4]Unconstrained subtraction and division are not primitive recursive, but appropriately constrained versions are primitive recursive.

[5]The class of total functions includes closure under restricted minimalization (unbounded minimalization defined everywhere). This class of functions is *too big* for software radios because its resource use cannot be guaranteed to be less than a bound specified in advance.

[6]Since these theorems are applications of well-known results from the theory of computing, only the outline of the proof is given.

prone constructs, e.g., waiting for sufficient signal strength to initiate receiver processing. The process-dispatch loop of the kernel operating system may "consume" infinite resources to deliver resources to applications software. Other infinite loops are computationally equivalent to the *partial* recursive functions. The computational structure is called (unbounded) minimalization.

7) *Minimalization:* for g: $N^{n+1} \rightarrow N$,

$$f(\boldsymbol{x}_n, y) = \mu y[g(\boldsymbol{x}_n, y) = 0]$$

that is, y is "the least y for which $g(\boldsymbol{x}_n, y)$ is zero."

The search operator μ will continue to increment y and test $\boldsymbol{g}$ without bound. The partial recursive functions are the smallest set of functions closed under minimalization. Unlike bounded minimalization which consumes finite resources, the extent of which is computable in advance, unbounded minimalization consumes resources until g is satisfied. For many situations, g may never be satisfied. For example, a carrier detection loop that is waiting for a signal that will never be transmitted due to an impossible RF value will never terminate. It will keep using resources until someone steps in to force the release of the resources. The partial recursive functions have been proven equivalent to the Turing computable functions, the Post productions, and to the RAM model [52]. These functions compute essentially anything that we know how to compute.

Theorem 2 (Unbounded Loops): Software radio functions implemented with **while** and/or **until** loops or their equivalents (e.g., implemented using go-to programming styles) cannot be guaranteed to use bounded computational resources.

Outline of Proof: Construct a **while** loop that simulates minimalization:

$$\text{While}(\text{NOT}(g(\boldsymbol{x}_n, y) = 0), \text{ increment } y); \text{ Return } y;$$

One may show the equivalence of **until**, **while**, and **for** loops to unbounded minimalization. They thus may be undefined, and are not total. The equivalent RAM procedure may consume unbounded computational resources, "crashing" the system.

The instruction set {RAM} of a programmable processor always includes instructions by which such unbounded loops may be constructed. There are, however, practical ways that radio engineers and real-time programmers have devised for assuring system stability in spite of the regular use of such constructs. One may assign time limits to each search condition. A time-out condition causes an error state that deals with the excessive time used. Such programming techniques informally convert unbounded minimalization to bounded minimalization. The time limits must include the statistical effects of competition for resources with functions that have higher priority.

Radio applications are distinct from general-purpose computing applications in that there is a rich set of timing constraints imposed on the creation, emission, reception, processing, conversion, network interoperation, and user interface of radio signals. These timing constraints are often represented in message sequence charts and state machines. Timing constraints apply to call setup and control; delivery of isochronous voice, video, and multimedia streams; packet networking; and related control. This is a heavily time-constrained applications domain. The constraints imply timing, and hence resource bounds that can be allocated to individual hardware and software components. These bounds constrain the logic (software and hardware) to tightly boundable computational constructs without the loss of radio functionality. The formal assignment of such bounds is the basis for a theory of plug-and-play software radio stability as follows.

Theorem 3 (Local Feasibility of Bounded Partial Recursion): A single-threaded single processor software radio function may be synthesized in RAM instruction sequences equivalent to a resource-constrained subset of the partial recursive functions.

Outline of Construction (Locally Bounded Partial Recursion): This proof builds on Theorems 1 and 2. Unconstrained **While** and **Until** loops and their "go-to" equivalents which cannot be guaranteed to terminate are precluded. Each such construct from a conventional implementation is allocated a maximum number of iterations (or, equivalently, a maximum processor time). These limits are coded into **bounded-while** or **bounded-until** loops as indivisible operations (in the same way one that a semaphore is an indivisible operation). Hardware that waits until a condition is met is similarly bounded, redesigned to generate a fault interrupt if the condition is not met after a specified time interval. Watchdog timers exemplify this principle.

Hardware and software "read" operations that wait for a response from a port (which may wait forever) are also precluded. Poll and bounded-wait interrupt handshakes are substituted. In these, the system will continue (with the read task suspended if the port has no data available) to be interrupted with a read failure condition if the read timer expires. "Read" is thus allowed in which the reading software task waits a specified time to be interrupted, and then returns an error condition. Write operations generate interrupts in the receiver and acknowledgment to the transmitter. Since there are RAM instruction sequences that implement these time-constrained search/wait loops, they are guaranteed to complete (either successfully or in a time-out state) in an amount of time that can be tightly bounded. This construction constrains the structure of the software away from unbounded minimalization, and hence from partial recursion to the set of functions called bounded partial recursion. Since this construction applies to software within a given von Neuman processor, the proof establishes the *local* feasibility of bounded partial recursion (i.e., within a single processor).

Theorem 4 (Totality): The bounded partial recursive functions are total.

Outline of Proof: Total functions are those that are defined for all $\boldsymbol{N}$. Each of the primitive recursive functions is defined for all $\boldsymbol{N}$. The closure under composition, primitive recursion, and bounded minimalization is defined for all $\boldsymbol{N}$. Bounded minimalization is defined for all $\boldsymbol{N}$ as follows. Associate with each range of arguments of a minimalization a step-counting function with a limit $y_{\max}$. If the search terminates in fewer than $|y_{\max}|$ steps, the loop yields the result of the minimalization. If not, then the result is FAIL for all

$N > |y_{\max}|$. Bounded minimalization is therefore total. The set of functions closed under such operations is therefore total.

Lemma 2 (Bounded Stability): The RAM instruction sequences that are equivalent to the bounded recursive functions are total. (Proof: If not, then the RAM model is not equivalent to the recursive function model, which would be a contradiction.)

Relevance: A software radio has to deliver services predictably. When combinations of inputs and states occur for which responses have not been defined, the system's behavior is unpredictable; hence, it will produce undesired results, and may ultimately go into an unrecoverable "crash" state. Software radios—in particular those with downloaded applications modules—have many modes and control parameters which result in a combinatorial explosion of control states, making it impossible to test every combination of states in advance. However, if the software radio is constructed of modules that are each guaranteed to be total, then any state combinations are also total by induction. Thus, some specific system behavior (noncrash) will be defined under all conditions. The *results* of such stable computations may or may not be as intended (i.e., "correct"), but if some specific state is defined, behavior will be *repeatable*. Unintended results, then, may be more readily diagnosed and corrected. Crash states due to undefined unstable behavior, on the other hand, often require heroic labor-intensive debugging efforts. Radio systems fielded with software that is not total appear unreliable to the user.

Theorem 5 (Global Bounded Recursion): Multithreaded multiprocessor software radio functions implemented using RAM sequences that are equivalent to the bounded recursive functions may be guaranteed to run to complete or to cause a resource fault. The number of instructions before a fault may be tightly bounded using polynomial resources to compute the bound.

Outline of Proof: (Existence of Globally Bounded Recursion): Nielson and Nielson [53] present a process algebra for a polymorphic subset of concurrent ML, a parallel multiprocessing language [54], [55]. With this, they show that the number of processes and interprocessor communications channels associated with a subset of concurrent ML is finite. They point out that their analysis is not complete: some programs with finite resource use will be rejected as potentially infinite due to an inability to infer the finiteness of sequences of (unbounded) recursions which terminate. In addition, their semantics admit sequences in which an infinite number of tasks terminate before a finite number begins, again resulting in an erroneous rejection of a finite resource use case.

Substitute bounded recursion into their schema to preclude the partial recursive constructs, rendering the analysis complete. The formal language (the subset of concurrent ML) and semantics constrain the entire system (all processors, applications, and the concurrent ML) to bounded recursion. In addition, the above guarantees that a step-counting function exists, but not that it will be effective in halting a process that is consuming inordinate resources. Therefore, global bounded recursion requires a two-level priority system in which the step-counting function can interrupt the function being monitored. An infinite regression can occur when step-counting functions themselves could consume excessive resources. This may be curtailed in building software radios by constraining the step-counting function to use the real-time clock. The existence proof is therefore a first-order result (it assumes two levels of priority and only isolated clock faults), not a completely general result. But it isolates the risk of exceeding resources to the reliability of the real-time clock (versus the millions of lines of code of the radio applications). It is much easier to make highly reliable clocks than to make error-free totally defined software.

Generalizations: The degree to which Theorem 5 may be generalized to the asynchronous multithreaded heterogeneous multiprocessors of software radios touches on much current research in computer science. The degree of further generalization is, no doubt, limited. Communications paths, data-sharing mechanisms, and process control protocols (e.g., Compare and Swap, Load Linked/Validate/Store Conditional [56]) would have to be bounded in a more general setting than [53] for a fully general proof. Topological models of computation have produced relevant results including necessary and sufficient conditions for wait-free protocols [57], the asynchronous complexity theorem [58], nonuniform and uniform iterated intermediate snapshot models of wait-free computation [59], and related results. In addition, practical nonblocking primitives can be implemented on contemporary machines like the DEC Alpha, MIPS R4000, and PowerPC [56]. The question of task termination has been proven to be undecideable if even one process can be faulty (e.g., fail-stop [60]). Theorem 5 asserts that a mechanism for ensuring bounded recursive interprocessor communications together with locally bounded-recursive software modules guarantees globally bounded resources provided the step-counting function uses a reliable real-time clock. This analysis yields an architecture principle for the software radio.

Architecture Principle 1—Bounded Modules: Software radio architectures which limit (software and hardware) control structures to those that constrain partial recursion to the ISA-equivalent of bounded recursion will consume bounded resources, possibly entering a resource fault condition. Components that conform to this criterion may be called total bounded modules.

Implications: There are many implications to such an architecture principle. Setting resource bounds so tight that the statistical structure of the demand causes them to be exceeded frequently can drive resource use to zero due to thrashing over resource-use fault conditions. Loss of service from misallocating resource bounds is no less painful on the service provider than other types of resource loss, but it is easy to avoid.

In addition, the theorems above do not preclude combinatorial explosion. That is, using these constrained hardware and software structures, one may create data sets and instruction sequences with concrete complexity of $O(2^N)$. One is guaranteed, however, that N is small enough for the exponential resources to be within those allocated in advance. One *must* know both the structure of the dependency and the bounds on N in advance in order to establish the time-out criteria for each potentially exponential loop in the system. Such

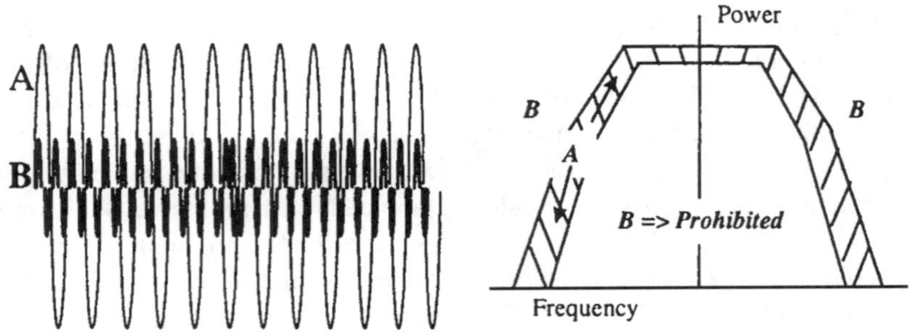

(a) (b)

Fig. 14. Interface waveform topology. A permitted, and B prohibited. (a) Time domain and (b) frequency domain.

controlled combinatorial explosion balances the needs of an algorithm to search a large space (e.g., in an equalizer) against the finite resources of the system. It will also guarantee that such algorithms will not consume more than the allocated resources. In addition, although the type of constructs that yield Ackerman's function cannot, in general, be ruled out, bounded recursion assures that such a construct would yield a resource fault after a specified amount of resource use. Such limits may be proscribed in the type certification process. Thus, any modules that meet the constraints when tested in isolation will not exceed them when inserted into the plug-and-play environment. In addition, the system will not become unstable because of undefined states since all functions are total. Such guarantees help assure reliable, predictable, plug-and-play software radios.

This section has presented an initial examination of the computational properties of software radio functions. The heterogeneous mix of ASIC's, FPGA's, DSP's, and general-purpose computers can be constructed so as to consume predictable, tightly bounded resources. One must constrain the logic to the bounded recursive constructs with prioritized step-counting fault handling. The theorems provide an initial outline of a theory of resource integrity needed for plug-and-play modules. The constructions from the proofs suggest engineering techniques for applying the theory to development projects. This stability theory must then be combined with queuing theory [16], [26] and other resource management techniques that address the statistical nature of the demand for robust management of processing resources.

V. Interface Topologies Among Plug-and-Play Modules

Constraints on the computational structure of modules can ensure bounds on resource use for plug-and-play. Are there constraints on the interface structure among these modules that further facilitate plug-and-play? The answer requires additional aspects of the topological model.

A. Topological Spaces

Definition (Topological Space): A topological space, denoted (X, O_x), is a set X and a family of subsets O_x, the "open sets," which include X and the empty set ϕ, and which are closed under countable union and finite intersection [39]. The topology is the family of subsets O_x which has the geometric and algebraic structure.

In a topological space, one can represent the geometric properties of interfaces among software radio modules. An interface to an analog source, for example, may be modeled as a subset of functions on $\boldsymbol{R}$ that obeys certain constraints. The constraints include bandwidth, adjacent channel interference, and minimum and maximum transmitted power. An uncountable number of such waveforms is possible in an analog interface, but regulatory bodies and the hardware limit the waveform to a structured subset of the possible waveforms. Since the space of randomly perturbed signals is also a metric space [61], one can precisely define distance among waveforms, and thus the topological structure of analog interfaces. Fig. 14, for example, shows two analog waveforms in the time domain (a) and in the frequency domain (b).

A range of waveforms is (A) allowed and prohibited (B). Time-domain waveform A is within the allowed geometry, while waveform B is not, e.g., due to unacceptable power or RF. Such widely used interface constraints are *set theoretic* in that they limit the elements in the interfaces to a subset with specified properties. These constraints are also *geometric,* defining an interior A, the conforming region, and an exterior B, the nonconforming regions. The representations (a) and (b) are homeomorphic because the Fourier transform provides a topology-preserving map among the two.

B. Finite Interface Topologies

If a set X has a finite number of elements $|X|$, all subsets are open sets (and are also closed sets). If $|X| = M$, then the number of topologies that induce a topological space on X is $2^{(2^M-2)}$, a double exponential. Not all of the candidate topologies satisfy closure under union and finite intersection as required for a topological space [62]. The huge number of possible topologies compels one to define finite interface topologies more compactly.

Definition (Basis): A set $\boldsymbol{B} \subseteq X$ is a basis for O_x if the members of O_x are the union of members of $\boldsymbol{B}$. A basis is a smaller set than O_x from which O_x may be induced by taking unions.

From a hardware perspective, a pin in a connector is an interface point $x_i \in X$. The set $\{X = \bigcup_N x_i, \phi\}$ is a basis for O_x. The set $Y = \{\bigcup_{N-1} x_i, \{x_1\}, \phi\}$ which contains three subsets $\{x_1\}$, ϕ, and the union of all the other interface pins works just as well. The basis with the most subsets is $\{\phi, \{x_i\}\}$, the $N+1$ sets that include the empty set and each element of X taken as a singleton set. If all of the pins are needed for a viable connection, then $\{X\}$ is the only subset of X that consummates a connection, and therefore is the only element in the interface topology. In this case, $O_x = \{\{X\}\}$ and the topology is just the fixed set of required inputs X. This may be called the rigid topology. It is not a topological space because it lacks the empty set. The empty set is not a valid member of the interface set if the interface will not "work" if no pins are present. If the system will "work" with the connector unplugged (e.g., resort to a default or fail-soft mode), then the empty set *is* a member of the interface topology.

These principles apply to software. An API may specify a call to the synthesize() function, for example, with arguments RF (frequency) and W (bandwidth). If both are required, then $Y = \{\text{RF}, W\}$ and $O_Y = \{\{Y\}\}$. This is just as inflexible as the equivalent hardware interface. On the other hand, a tagged API with the expressions RF = 859 or W = 30 or both or neither would be defined over the space containing $\{\text{RF}\}$, $\{W\}$, $X = \{\text{RF}, W\}$, and ϕ:

$$O_X = \{\{\text{RF}, W\}, \{W\}, \{\text{RF}\}, \phi\}.$$

The empty set is included in the topology because the interface "works" even if no arguments are provided. Thus, $O_X = \{X, \{\text{RF}\}, \{W\}, \phi\}$, the set of all subsets of X or power set, which is a topological space. This power-set topology is the *discrete* topology; it may also be called the *flexible API* topology. This flexible interface geometry may be implemented if synthesize() uses default values for the missing arguments. Since the power set is the largest set of subsets, the discrete topology maximizes the number of combinations of API parameters over which a function call is valid. The basis for the power set consists of the singleton arguments plus the empty set: $\{\{W\}, \{\text{RF}\}, \phi\}$.

C. Function-Call Parameter Topologies

The geometric structure of interface spaces may be better understood using additional notions.

Simplex: A simplex is an ordered set of points in a topological space that are adjacent in some sense, such as sharing a relation R.[7] Higher dimensionality simplexes induce lower dimensionality simplexes. Simplexes may be embedded in Euclidean space, but need not be [63].

Complex: A simplicial complex is a union of simplexes that includes the union of all of the lower dimensionality simplexes of a given simplex.

Q-Connected: Simplicial complexes that share a $q+1$ face are q-connected [64].

The three vertices of a plane triangle $[A, B, C]$ in Fig. 15 comprise a two-dimensional simplex, adjacent in the sense that they are connected by the points in the plane. Each line segment joining these vertices comprises a one-dimensional

[7] An n-simplex is the convex hull of $n+1$ independent points in m-dimensional Euclidean space subject to completeness and closure axioms. The full geometric treatment is not needed at this point.

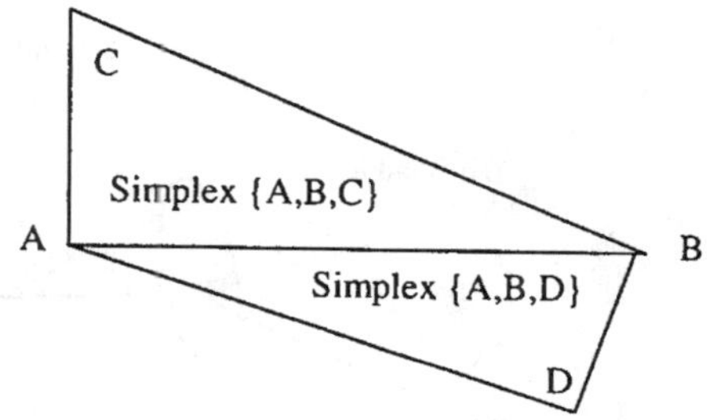

Fig. 15. Simplicial complex consists of two simplexes.

simplex (e.g., $[A, C]$), denoted by the pairs of vertices. A second triangle ABD that shares one line segment with the first is also a simplex. Each triangle together with its edges and vertices comprise a simplicial complex. The two triangles also comprise a simplicial complex in which the simplexes are 1-connected by the line $[A, B]$.

Interface parameter sets may be modeled as simplexes with a large number of dimensions. In the Synthesize() function above, suppose RF is discrete and has as its explicit basis the range [400.001, 2400.000] MHz and resolution 1 kHz. The RF's that are valid parameters of Synthesize() constitute a simplex consisting of the 2 million RF's in this range. Let BW be defined for $\{25, 30, 200\}$ kHz. The parameter space $\{\text{RF}, BW\}$ contains 6 million points, a simplicial complex consisting of the union of three simplexes of 2 million RF points each, indexed by BW. Then the set of RF's reachable from Synthesize(925.000, 30) is exactly one, $\{[925, 30]\}$. Synthesize(BW = 30), on the other hand, may reach any of the 2 million point RF simplex through a global default. Synthesize() reaches all 6 million points through a global default, connecting them in a single simplex.

For the example function Synthesize() to have an explicit basis, its definition must include basis tuples expressed as parameter (Type, Range). Parameter indicates the parameter being defined. Type chooses among predefined subspace types (e.g., Discrete, Continuous). Range establishes the valid values. In the example above, an explicit basis could be BW (Discrete, $\{25, 30, 200\}$) and RF (Discrete, [400.001, 400.002; 2400.000]) using Mathcad® range notation. This explicit basis can be enforced by the environment to yield an error result "out of range" for arguments that violate the basis description. Such semantics define Synthesize() as an effectively computable function over all inputs. A product topology on these bases yields the 6 million points of ($BW\times$ RF) as the input space to this function. However, it is often necessary to limit the parameters to subsets of this space. One might limit the 200 kHz bandwidth to the RF's allocated to GSM, say, 925 to 960 MHz. The simplex

NOT (200, {[925.000, 925.200, 926.000]})
→ "bandwidth-out-of-range"

defines the set-complement subspace for which the 200 kHz bandwidth choice is mapped to a fault condition. To the degree that such subspaces can be changed while the radio node is in the field, the service provider has a "future proof" interface. In order to be changed in the field, the "capability" implicit in the interface constraints must be defined in such a way that

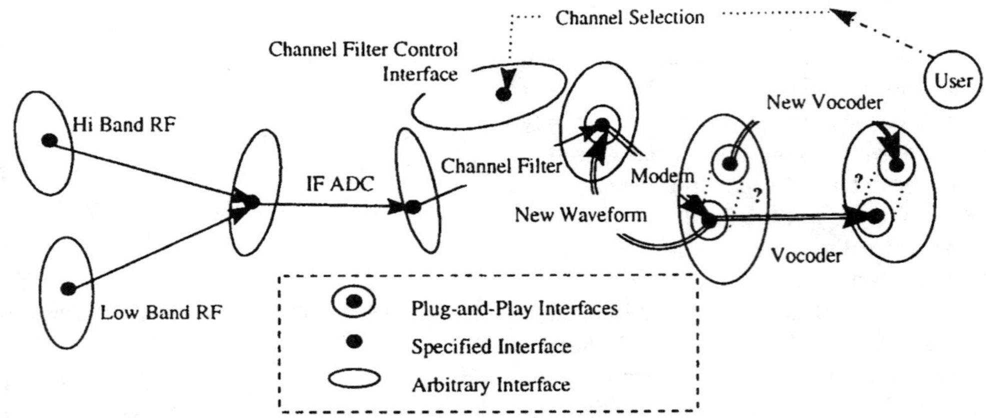

Fig. 16. Interface geometry reflects design decisions.

the control algorithms can test and manipulate the constraints, e.g., to decide whether to accept a download. The geometric structure of the interfaces can be expressed naturally in strongly typed languages to define plug-and-play constraints. More appropriately, for plug-and-play environments, they can be expressed in execution time interface parameter sets.

D. Plug-and-Play Interface Geometry

The implementation decisions made during design, development, integration, and test may constrain an interface to a point in the simplicial complex of interfaces as illustrated in Fig. 16.

Each domain and range of a conventional interface consists of a designated point in its simplicial complex. A plug-and-play interface, however, defines an interoperable subset of the interface space. The physical interface subspaces must change as a function of the hardware in which the service is delivered. The logical interface subspaces also may have to change as a function of the software modules configured to deliver the services. To be fully extensible, plug-and-play modules have to be combined dynamically. To do this in a controlled way, the control algorithm(s) must have a way of comparing the range of one function to the domain of the next to determine whether the functions are compatible.

E. Extensible Capabilities

Capabilities may be represented as levels ("tags") such as the type of video teleconferencing interface in the ITU H.320 Recommendation [65]. The level of call control support from Microsoft's telephony applications programmer interface (TAPI) [66] provides another example of tagged capability levels. The advantage of tags is simplicity. Capability tags are defined during the standards-setting process and documented in the text of the standards. One disadvantage of this approach is the possible misinterpretation of the text of the standard. To counter this chronic problem in telecommunications standardization, ETSI promulgated language that makes the formal specification of the ITU Z.100 specification and description language (SDL) [67] the *normative* expression of the standard [68], with text providing amplification and explanation. Historically, the text has been normative, while the computer-based representations have been auxiliary.

```
Resource:
        Type: ADC
        Set Extension:
        Range (Output-Space):
         {<Buffering>:= {None, Double},
          <Buffer-Size> = N,
          <Buffer>={Register(X[1,N])}
          <Buffer-Flag>:= {A,B}}
        Buffering: None (Null)
        Buffering: Double (
                Buffer-Size: 128
                Buffer (A); Buffer(B)
                Buffer-Flag: Singleton('2A1 = {0,1})
                       )
```

Fig. 17. Resource topology extension: ADC buffering.

Another disadvantage of capability tags is that capability cannot be defined dynamically by the radio itself. Computational manipulation of tags by the software radio joint control function is also limited. For control of dynamically defined software radio-based services, the constraints must be computer processable, and the system must be able to act on them during operations. Automated reasoning over the constraint spaces and capabilities of the host system could enable the radio itself to construct the level of capability necessary for a desired service. SDL falls somewhat short of this vision since the application semantics are left to the designer. Dynamic capability requires designer-independent semantics.

It is easy to define designer-independent semantics in a way that is readable by people. To do this in a way that is processable by a control algorithm requires a radio knowledge-representation language (RKRL). Parts of RKRL that are defined *a priori* may be extended dynamically. For example, to add a control that selects from either the A or B buffer in a double-buffered interface of a notional ADC requires extension of the interface topology, adding sets and subsets to (X, O_X) as suggested in the example of Fig. 17.

The meta-level expressions ⟨Buffering⟩, ⟨Buffer-Size⟩, ⟨Buffer⟩, and ⟨Buffer-flag⟩ introduce the meta-level primitives **Buffering, Buffer-Size, Buffer,** and **Buffer-flag** to an extensible RKRL. The RKRL is assumed to include *a priori*

TABLE III
SPEAKEASY I SOFTWARE MODULES, DMA: DIRECT MEMORY ACCESS

Module	Source	*Ada Module Descriptions/* Functions
At	(127 kB)	C040 interprocessor communications
BIT	(318 kB)	Built-In-Test packages, including CRC, EEPROM, PID, I/O registers, interrupts & DMA
Cm	(1.29 MB)	Configuration Management
ALE	(125 kB)	ALE Receive (Rx) and Transmit (Tx) Functions
ALE_Rx1	(378 kB)	Automatic Link Establishment (ALE) Receive Modules
Hvq	(645 kB)	Have Quick Communications Ensemble
Hvq_Ct	(109 kB)	Control Modules (Initialization, Mode Control, Errors)
Hvq_Glob	(25 kB)	Globals
Hvq_ Rx	(379 kB)	Receive Mode (Synchronize, TOD, Rx, Active...)
Hvq_ Tx	(131 kB)	Transmit Mode
Work	(299 kB)	ALE packages & specs
Hfm	(518 kB)	HF Modem Communications Ensemble
Hfm_ctrl	(58 kB)	Controls Waveform start/stop messages; RTS Events; PM Query; TX/RX Done (local);
Hfm_dc	(22 kB)	Data Control Packages, source messages error checking
Hfm_ rx	(289 kB)	Receiver Bit & message operations, text I/O, Rx utilities, data correlation tables, filters, queues
Hfm_ tx	(149 kB)	Squelch, TX/RX mode, TX templates, RF Control, Timing
Nbg	(334 kB)	Narrowband Frequency Hopping Group
Nbg_ct	(49 kB)	State Machine, Sync Loss, TX/ RX, Waveform, PTT State...
Nbg_glob	(105 kB)	Global parameters for NBG package
Nbg_hp	(57 kB)	Hop Packages – timing, data request/processing, PTT acknowledge, cryptographic processing...
Nbg_rx	(73 kB)	Receiver Packages MFSK, Preamble, Galois (FEC), Dead Bits, Flags, Bitsync, RX flush, Detect/Track
Nbg_t	(49 kB)	TX: Amplitude, Preamble fill, IQ Samples, AM on Voice, Filter, Inter-Process Communications (IPC) Messages, SSB, DSB, QAM, OQPSK, Event & Constraint Checking

semantics for Resource, Type, Set-Extension, Output-Space, Range, None, Double, and Singleton topologies. The set extension augments the interface topology ($X\ O_X$) with new subsets. Range expressions define the new subspaces. Discrete values constrain the subspace to an interface point.

The resource extension indicates that if **Buffering** = none, then the effect is null. This effectively glues this new subspace to the existing subspaces in which there was no buffering. In addition, the new subspace **Buffering(Double)** has the rigid interface topology. Topological spaces represent radio capability in a way that is readily processed by control algorithms. The extensibility of interface topologies may be expressed in the following architecture principle.

Architecture Principle 2—Explicit Extensible Interface Topology: API's and hardware interfaces which exhibit an explicit topological basis for each interface parameter space and which are extensible in the field exhibit the level of flexibility necessary for extensible plug-and-play services.

Implications: The requirement for an explicit basis assures that the interface topology is defined completely, including error states. The full extensibility of the topological bases in the field requires an RKRL-like representation of radio resources. This is an area of current research in wireless computer-communications systems [69].

VI. ARCHITECTURE PARTITIONS

One goal of a plug-and-play software radio architecture is to provide computing resources from as yet undefined hardware modules to support as yet undefined software modules for as yet undefined services. The following architecture analysis applies the computational and topological properties introduced above to begin to identify the mathematically based partitions of a software radio system.

The source code of a widely published military software radio, SPEAKeasy I [70], is considered. A sequence of virtual machines is then derived that both expand the RAM equivalent instruction set model and constrain the computational geometry, yielding a layered virtual machine reference model for the software radio.

A. SPEAKeasy I

SPEAKeasy I resulted in the software radio source code structure summarized in Table III. The as-built code has some strong features—such as real-time performance and accurate handling of timing differences between radio networks. Since an Ada implementation was mandated, the real-time executive is the Ada run-time kernel.

The Ada modules can be viewed as software objects. They include databases, channels, and agents. Databases store personalities, filter parameters, lengthy chunks of compiled code, and data sets to be loaded into a personality at run time.

Channels are abstractions around which modes (e.g., HAVE QUICK) are organized. A channel is supported by several Ada packages that perform the systems-level functions of RF control, modem processing, INFOSEC, and related internetworking. A channel gets the system resources (paths

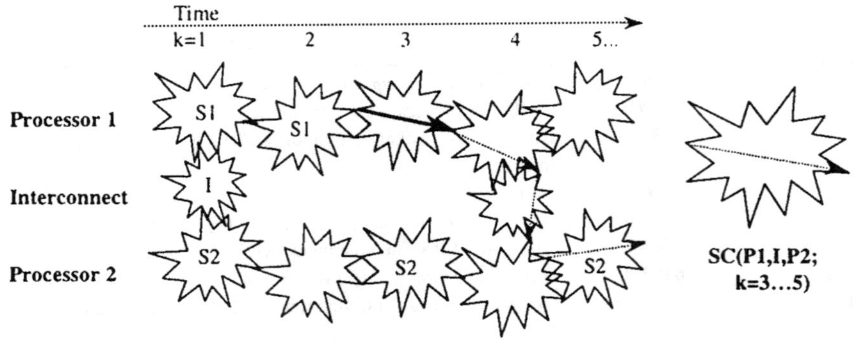

Fig. 18. Processor simplexes ($S1$, $S2$), interconnect simplexes (I), and an equivalent simplicial complex SC($P1$, I, $P2$; $k = 3 \cdots 5$).

or threads through the system). It installs its personality on these resources to implement a mode. It then keeps track of the overall state of the processing thread that delivers the associated services. In SPEAK*easy* I, lower level modules implement the personalities of the channels. They also serve as hosts for buses, manage IO processes, access timing and positioning data, and control the radio. Channels also keep track of the status of the mode, number of resources employed, volume, data rate, throughput, network parameters such as network number, and assigned time slot(s).

Timing packages manipulate the system clock, time of day (ToD), the day/time format, and the timing resolution. The RF control packages determine RF direction (i.e., transmit or receive); the RF mode (e.g., linear or nonlinear amplification); preemphasis for predistortion; and frequency of transmission. The modem packages include modulation (AM, FM, QAM, USB, and MSK), demodulation, automatic gain control (AGC), packing, and unpacking of protected bits.

The system may also perform a loop-back function for network testing or local diagnosis. The back-end functions include message processing, internetworking, managing protocol stacks, and the user interface. Radio control handles system boot up, initial ToD, current hop, calibration, status requests, and security level. Comparing the Nbg, Hvq, and Hfm mode software shows a common pattern of control module(s), global parameters, transmit and receive (modem) modules, along with specialized modules such as hop generation for frequency-hopping modes.

The as-built code includes four distinct types of software. One type delivers voice, data, and channel bridging services. Another type structures the radio applications using state machines. The infrastructure-type software moves data packets and isochronous streams. The fourth type consists of hardware-specific support software including the operating system. About 30–40% of the application-specific code is infrastructure, 30% structures the radio applications, and 30% implements services. These different functions are almost inextricably intertwined in the SPEAKeasy code. Much of the code is hardware dependent. The partitioning analysis that follows is oriented toward reuse. The most primitive level of code is the hardware-specific operating system.

B. The Hardware-Specific Partition

The hardware–software interfaces define the lowest level partition of a radio system. From a topological perspective, processor hardware *connects* the states of data in its registers. The set of data states reachable in one clock cycle from any initial state is the processor's characteristic simplex. A processor with 20 32-bit address registers, cache pointers, general-purpose registers, with multiple direct memory access (DMA) registers and other registers can, in principle, assume any of $2^{(32*20)}$ states $= 10^{192.6} = 0.358846241M2$—see the Appendix for magnitude notation. This is the size of the processor's instantaneous characteristic simplex. Fewer registers yield a smaller simplex. A more complex instruction set architecture yields a larger simplex. An FPGA's simplex is determined by its use of state memory, which will generally be much larger than a DSP chip of the same area and device feature size. The size of the simplex indicates how many different things the processor could accomplish in a single clock cycle.

1) *Paths in Simplexes Induce Partitions:* Sequential simplexes are q-connected to each other by the clock sequence. Over time, a series of RAM instructions traces a path through a sequence of simplexes, the union of which is a simplicial complex (SC). In 1 s, a processor with a 100 MHz instruction clock may assume any of (10^{192})^(10^{8}) or $0.87822M5$ states. In addition, a software radio with distributed multiprocessor hardware and interconnect devices has additional connectedness among processor simplexes as illustrated in Fig. 18.

At each clock cycle, an edge in the simplex is traversed (e.g., the solid arrow in processor 1's simplex at time $k = 3$ in the figure). The path traversed through a sequence of simplexes may be subsumed into a simplicial complex, e.g., SC($P1$, I, $P2$; $k - 3 \cdots 5$) in the figure. The dotted arrow in the SC indicates the path traversed by data sent from $P1$ to $P2$ during clock sequence [3, 5]. The way in which software constrains such paths implicitly defines architecture partitions. A larger number of possible paths indicates greater built-in flexibility, while topological loops in such paths suggest partitions.

2) *Interrupt Service Routine (ISR) Topological Loops:* Consider hardware-dependent software such as operating system services. This software consumes processing resources in a way that is tightly coupled to hardware-related processor states. A typical ISR, for example, might turn off hardware interrupts, push the processor state onto a stack, test the interrupt condition, set a dispatch pointer, restore the registers, and exit. In a

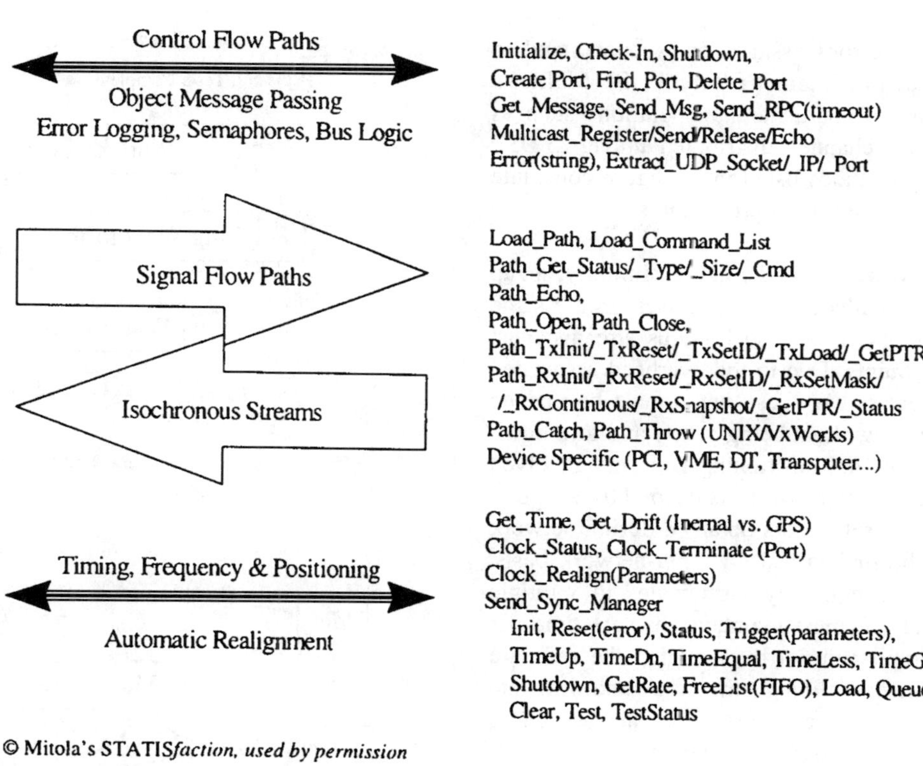

Fig. 19. Infrastructure software function calls.

real-time system, this process could take from 10 to 100 instructions. Windows NT requirements for template data might expand this to 1000 instructions or more. During this sequence, the path through machine's simplicial complex, normally arbitrary, is constrained to a path that involves states of a the interrupt register. In addition, the instruction sequence that is so constrained is defined temporally by the interrupt event and the termination of the ISR. This produces a path in the temporal simplicial complex that begins with a hardware interrupt "set" and a particular configuration of registers and ends with the interrupt "clear" and the identical configuration of registers. Adopt a convenient distance measure on data states such as Hamming distance, the number of bits that are not identical. A topological loop over ([RAM], *d*) at x, a point in the simplicial complex [RAM], is the shortest path in [RAM] that returns to within d of x. An ISR is a topological loop in [RAM] for $d = 1$ in the above example. Input/output drivers, task schedulers, disk access control, real-time clock support software, and the like have short characteristic topological loops.

3) *The Topology of the Kernel Substrate:* The set {kernel} comprises the software most closely aligned to hardware states and having the shortest topological loops. The kernel is the lowest level software substrate. Kernel sequences can be modeled as extensions to the processor's native ISA. The hardware interrupt ISR, for example, could be modeled as the function Service() defined over the set {Interrupt-Registers}. Service() then becomes an instruction in {kernel}. From a topological perspective, Service() defines a subgraph within the processor's SC. A processor with a library of such hardware-dependent modules can be viewed as a machine with an extended instruction set, $\text{ISA}^+ = \{\text{RAM}\} \cup \{\text{kernel}\}$. The set {kernel} may have a distinguished sequence of {RAM} instructions for the idle state. Idle $\Leftrightarrow$ {RAM} in that, during the Idle state, an arbitrary RAM sequence is possible. Let $\{\text{kernel}^-\}$ represent the kernel-less idle. Minimizing the number of {RAM} sequences needed for the {kernel} maximizes the {RAM} capacity left for higher level radio services. The simplex of {RAM} $\cup$ $\{\text{kernel}^-\}$ contains a distinguished subspace corresponding to the paths reachable through {kernel}.

C. Topology of Infrastructure Software

Given a set of kernel software, the next step in a bottom-up definition of virtual machines identifies those functions that allocate, set up, and control physical resources to create logical resources. Fig. 19 lists the function calls required for distributed processing. These functions include the structures recommended in ITU's X.900 open distributed processing reference model [71]. They also include structures unique to radio applications such as frequency distribution. This infrastructure code manages control flow paths; signal flow paths; and timing, frequency, and positioning information. It also includes features needed for isochronous delivery of voice, video, and other real-time streams. The control flow paths mediate message passing among most objects in the system. Error logging, semaphores that manage shared resources, and bus access protocols are examples of control message flows.

The infrastructure functions initialize the system, create and manipulate logical and physical ports, move messages, and perform remote procedure calls (RPC). They also provide multicast services, handle error messages, and support standard protocol interactions such as Internet protocol (IP). Multicast simplifies programming of multichannel operations such as initializing 100 subscriber channels distributed among 25 DSP chips. The control-flow methods listed in the figure constitute a minimum set necessary for multiprocessor software radio infrastructure.

The signal-flow methods listed in the figure, similarly, set up and manage signal flow paths among processes on the same or on different processors. These isochronous streams must meet tight timing constraints. Due to the overhead associated with path setup and tear down, these paths must be opened and closed multiple times without being set up and torn down again. Timing, frequency, and positioning is a very involved process, a complete discussion of which is beyond the scope of this brief treatment. Time references obtained during network synchronization must be maintained on a per-network basis. Since the software radio generally participates in multiple networks simultaneously, it must maintain absolute time per network. This is accomplished not by changing the software radio's clock, but rather by defining time offset and drift rate for each network. Additional ancillary functions related to queueing data messages (e.g., load, clear, test status, and reset) are also part of infrastructure.

Applying the virtual machine paradigm to this set of software yields a new virtual instruction set {Infrastructure} which consists of {Message-Passing}, {Isochronous Paths}, {Timing}, {Frequency}, and {Positioning} instruction subsets. These extensions build on the facilities of {kernel}. The expanded {RAM} simplex contains {Message-Passing}, etc. Each subspace depends on $\{\text{kernel}^{-}\}$ services, the lowest layer of the emerging virtual machine hierarchy.

D. Radio State Machines

State machines control access to many software radio resources. State machines typically control the transmit and receive channels and fault recovery actions. A resource-allocation state machine is illustrated in Fig. 20. Three states are shown in boxes: waiting for instantiation, fetching a waveform, and waiting for a response. The arcs are labeled with conditions that cause one to transition from one state to the next and with actions performed upon such a transition. This control structure may be implemented as a software object with a set of slots (the states) and attached methods. Methods test for state transitions and perform required actions. Recommendations for defining and simulating state machines are provided in the SDL Recommendation Z.100.

Such state machines may operate on several levels. At the top level, each channel (e.g., HAVE QUICK) has a channel-setup state machine that keeps track of resources. The next-level state machine manages the mode, as illustrated in Fig. 21. The channel is initially idle. When it has been set up, it enters the Ready state. If the user (or network) initiates a "talk" sequence, the transmit (Tx) state is entered, upon which the radio will alternate between Tx and Rx states until conditions are met which either suspend or deactivate the channel.

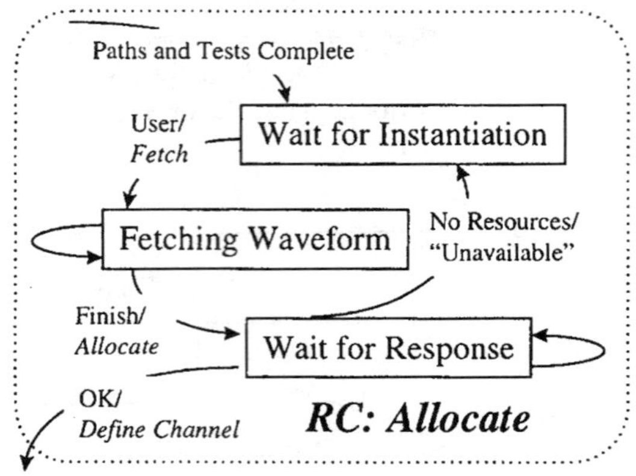

Fig. 20. Infrastructure state machine. © Mitola's STATIS*faction*, used by permission.

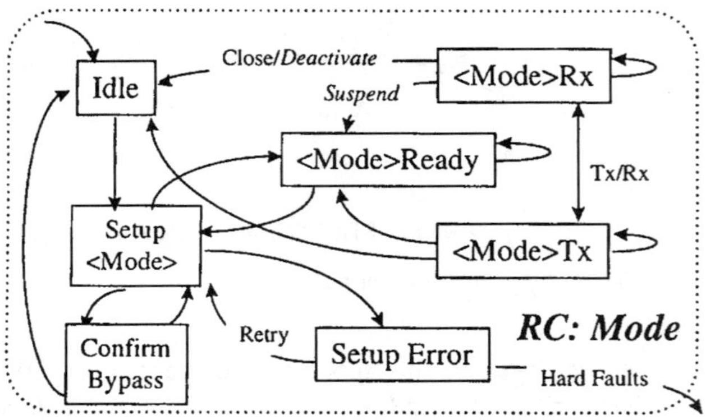

Fig. 21. Channel control state machine. © Mitola's STATIS*faction*, used by permission.

Lower level modem state machines track the states for active plain text receive and transmit, crypto-sync, receiving analog and digital, and bridging fades. The SPEAKeasy state machines also have built-in timing and error-recovery procedures. Automatic gain control (AGC) and squelch (Sq) are adjusted on all transitions in the lower level state machines.

Thus, the state machines schedule both routine processes such as AGC and processes driven by channel conditions such as bridging across fades. States also reflect failure modes. These include sync failure, loss of carrier, and loss of system resources. Such state machines are a central mechanism for controlling system and radio resources in the software radio.

From a geometric perspective, the topological loops of these state machines begin and end on conditions of identical state values. That is, the states "setup," "transmit," "receive," etc., are the constants or fixed points in the evolution of the [RAM] simplicial complex. Periodically, the processor's register set will return to topologically close states for some suitably defined distance. Since the functions called from these state transitions (e.g., "transmit next block") employ facilities from {Message-passing}, {Isochronous stream}, etc., these state machines define the next level of virtual machine built on top of the infrastructure machine. The boundary of this layer of virtual machine is based on the function calls attached to the state-machine objects. This new layer of virtual machine is the {State machine} instruction set, consisting of the {RAM} sequences that set up, execute, and control the state-machine

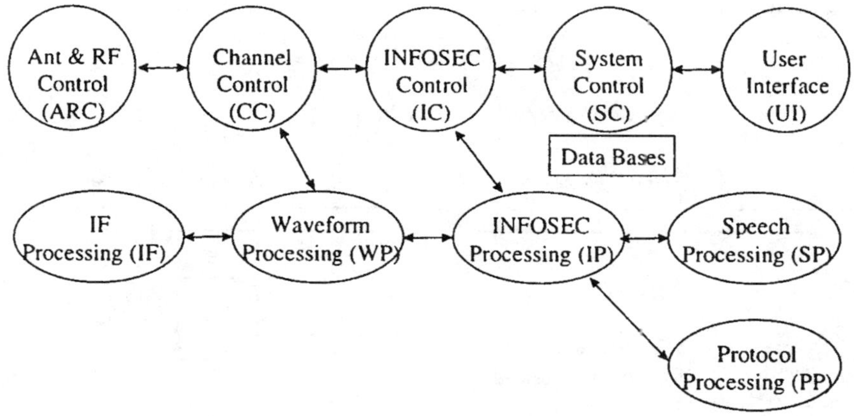

Fig. 22. Channel agent software objects. © Mitola's STATIS*faction*, used by permission.

TABLE IV
FUNCTIONS OF RADIO SOFTWARE AGENTS

Radio Agent	Object Methods and Slots
Antenna & RF Control (ARC)	TX/RX, Power, Polarization
Channel Control (CC)	Allocate Resources, Configure, State Machines
IF Processor (IF)	Diversity combining, Beamforming, Pre-emphasis
Waveform Processor (WP)	Generation, Timing, Fault Detection, Modulator and Demodulator (Modem), link-related data processing
INFOSEC Control (IC)	Key, Control Bridge to Black Side, Authenticate
INFOSEC Processing (IP)	Encrypt, Decrypt, Transec
System Control (SC)	Initialize/ Shutdown, Test, System Status
User Interface (UI)	Commands & Displays
Speech Processing (SP)	Codecs; Echo Cancellation, Voice Channel Modems
Protocol Processing (PP)	Packetization, Routing

objects. The state-machine objects themselves comprise a virtual processor.

E. Channel Agents

The characteristics of SPEAKeasy I and other software radios have been abstracted to the top-level software object diagram of Fig. 22. These high-level objects may be called channel agents. Each software object has been allocated corresponding functions from the software radio functional block diagram of Fig. 2.

Those agents in the top row of the figure control the system, while those in the lower rows implement the traffic channels and related services. The objects of Fig. 22 implement the functions summarized in Table IV.

Each of these high-level software objects has subordinate objects, to the level of primitive functions like filters, modulators, bit decisions, etc. The protocol and speech processing "back end" of the software radio employs a protocol stack such as ATM, TCP/IP, Mobile IP, etc. Consequently, internetworking to the wireline infrastructure consists of a few monolithic predefined software objects.

These radio functions are relatively generic. Thus, each waveform or mode of operation may be expected to have either its own set of agents or a set of parameters and a script that tell the agent how to treat that specific mode. Radio services, like bridging across waveforms, could be constructed as scripts or Java-like applets that interconnect channel agents and other high-level functions. The semantics of such top-level modules are readily represented in the unified modeling language (UML) [72].

The timing and overall behavior of the channel agents are constrained by the lower level state machines that manage the channels. From a geometric perspective, agent subspaces in the [RAM] simplex subsume the {State machine} subspace. The extended instruction set for channel agents {Channel agents} is more of a category than a virtual instruction set. The top-level radio functions {Antenna and RF Control}, {Channel Control}, {IF Processor}, {Waveform Processor}, etc., each constitute a set of virtual instructions. Since these channel agents must be mutually consistent in order to work together, the class {Channel agents} refers to a mutually consistent set of such instruction set extensions comprising a radio mode.

F. Distributed Layered Virtual Machine Reference Model

The preceding sections have described essential software mechanisms employed in the construction of software radios. When viewed in terms of the topological spaces defined by the paths traced in the simplex of an equivalent RAM processor, layers of successively less machine-dependent software emerge. The distributed layered virtual machine is a way of representing the resulting hierarchy. Each layer is a virtual machine built on subordinate layers. The virtual machine interfaces may be held constant so that the implementation details and intellectual property present within a component at a given layer are hidden from all other layers. The software radio architecture that results from this process is illustrated in Fig. 23.

As the capacity of FPGA's and DSP's continues to grow, the processing to support such virtual machines becomes more affordable. The services offered from the top-layer virtual machines thus are independent of the hardware implementations in the bottom-layer virtual machine. The FPGA and DSP hardware vendors or software tool suppliers, for example,

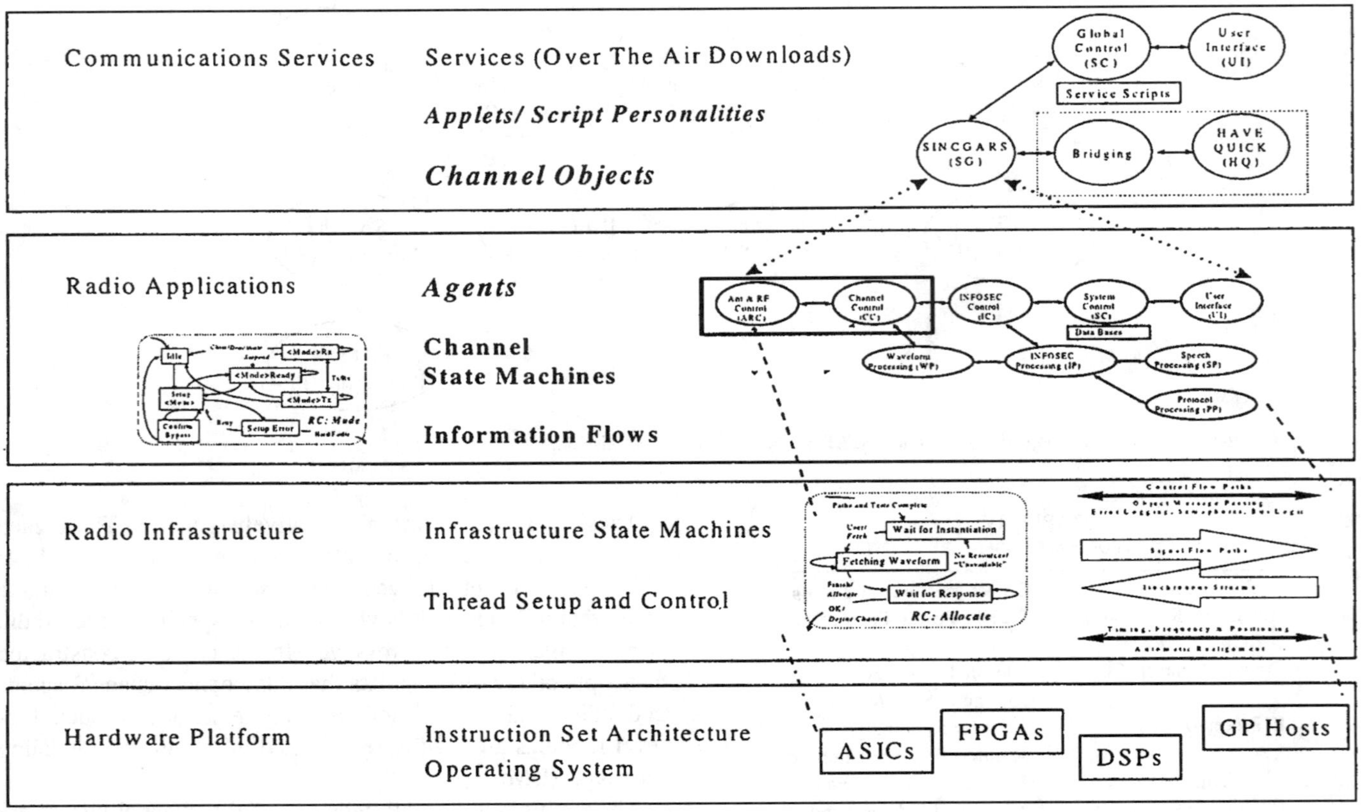

Fig. 23. Software radio distributed layered virtual machine. © Mitola's STATIS*faction*, used by permission.

could offer infrastructure layer software. This layer could be implemented in part using an augmented message-passing interface (MPI) [73]. The state machines in the infrastructure and radio applications layers would naturally fall into the domain of SDL. Third-party waveform vendors could offer middle layers. Interfaces among components may be represented in UML or the interface definition language (IDL) [74], possibly including abstract syntax notation (ASN.1) [75]. Systems integrators and service providers could then define their unique added value in the top layers, building on the broad base of industry support at the lower layers. In addition, computer-aided software/systems engineering (CASE) vendors could embrace the wide range of reusable components to assist developers to encompass the entire software radio distributed virtual machine hierarchy for reduced time to market.

The joint control function is distributed over these layers. Joint control at the top layer sets up channel objects and orchestrates services. In the radio applications layer, it consists of state machines that manage the radio channels. In the infrastructure layer, it establishes paths for signal and data flow. And in the hardware layer, it interacts with the operating system. Each layer accepts commands from higher layers, and returns status including fault conditions. With layering of the joint control function, the layered virtual machine insulates the higher layer applications from the details of the hardware.

There are generally significant computational resource penalties in making the translations among the layers. Two factors mitigate these penalties. Compilers, first of all, have become fairly efficient in the use of machine register sets. As radio API's become more widely used, precompilers should emerge which suppress unnecessary levels of function calls and make compile-time translations, enhancing run-time efficiency. In addition, late bindings that are computationally efficient should also emerge. Until they do, the penalties for the layered architecture implementations will remain high. Second, the computational capacity of the underlying chips continues to double every 18 months or so, making the performance of the layered machines acceptable in spite of the inefficiency. In particular, the internetworking functions that a few years ago were implemented in special-purpose hardware are now supported by general-purpose computers. These observations may be summarized in a third architecture principle.

Architecture Principle 3—Distributed Layered Virtual Machine Reference Model: Modules partitioned according to the distributed virtual machine reference model will insulate lower layer hardware-dependent modules from upper layer service-defining modules.

The economic incentives for such a model center on integrating markets for economy of scale. The global wireless marketplace now consists of dozens of niches defined by unique hardware platforms and waveform-unique infrastructure. Industry is attempting to organize itself to integrate these diverse markets so that the next generation of wireless will offer low-cost plug-and-play services. The process of defining the required industry standards has already begun. The SDR Forum, for example, has defined an architecture framework based on functional threads in which a generic applications programmer interface (API) is being defined across the virtual machine. The forum is integrating the GloMo radio device API

[76] into its own. The object management group (OMG), on the other hand, has requested technology for CORBA real-time objects [77] including multimedia at video data rates. A real-time CORBA might provide software radio infrastructure like an SDR API. The partitioning strategy and layered reference model presented here may provide insights helpful in evolving toward broadly accepted standards for future software-defined radios.

VII. Conclusion

The computational and geometric properties of the software radio will shape the degree to which plug-and-play is realizable for the global marketplace. Greater understanding of the mathematical properties of the software radio architecture should accelerate progress toward cost-effective plug-and-play services to which developers aspire. The architecture principles are particularly informative.

1) *Bounded Recursive Modules:* Construct the software radio system—host environment and plug-and-play modules—of bounded recursive modules. These modules will be defined for all inputs, and they will consume predictable system resources. Although the software may contain errors, those errors will be much less likely to cause faults in the delivery of audio, data, and multimedia services. They also will be less likely to cause system crashes than unconstrained modules.
2) *Explicit Extensible Interface Topologies:* Define software radio interfaces using an explicit basis for the underlying topological spaces. Use extensible languages such as UML, SDL, IDL, and ASN.1 until a more domain-specific language such as a radio knowledge representation language (RKRL) emerges.
3) *Distributed Layered Virtual Machine:* Position plug-and-play modules in the distributed layered virtual machine hierarchy in a way that maximizes the use of existing (lower and higher level) components.

Appendix
Magnitude Notation

A recursive logarithmic number system makes the large size of simplexes and simplicial complexes easier to understand and manipulate. The notation yMn ("y magnitude n") indicates raising y to the power of 10 n times where n is an integer and $0 < y < 1$. The magnitude n of this measure of concrete complexity of a simplicial complex is a strongly nonlinear measure of complexity. For example, $32 = 10^{1.5051}$ which equals $10\wedge 10^{0.1775}$, which can be expressed in magnitude notation as $0.1775M2$, a magnitude 2 number. The size of the processor's simplex above, $10^{192.6}$, equals $0.35884M3$, a magnitude 3 number. The 1 s simplicial complex of the same machine has size $0.87822M5$. This magnitude 5 number is exponentially bigger than a magnitude 3 number by 10^{10}. Put another way, the size of this complex is 10^{10} raised to a power containing 35 million zeros. For a perspective on the size of this number, the number of neutrino-sized places in the known universe (10^{40} m on each side) over the last 10 billion years is only 10^{266} or $0.384836241M3$. Using the size of the simplex ($10^{192.6}$) as a base and raising it to the power $k = 100$ million is not, perhaps, a familiar concept, but it establishes an upper bound on the size of the simplicial complex associated with just 1 s of use of such a processor. This notation is useful in comparing the huge numbers associated with simplexes and large simplicial complexes.

References

[1] McGarth *et al.*, "RFIC technology for wireless consumer products—Trends in GaAs," in *M/A-COM LOUD & Clear.* Lowell, MA: M/A-COM, 1995.
[2] Mouly and Pautet, *The GSM System for Mobile Communications.* Plaiseau, France: Mouley & Pautet, 1992.
[3] D. Upmal and R. Lackey, "SPEAKeasy, The military software radio," *IEEE Commun. Mag.*, 1995.
[4] D. Nicholson, *Spread Spectrum Signal Design.* Rockville, MD: Computer Science Press, 1988.
[5] *Reference Data for Engineers.* Carmel, IN: Sams Publishing, 1993.
[6] Stallings, *Handbook of Computer-Communications Standards—Volume 1: The Open Systems Interconnection (OSI) Model and OSI-Related Standards.* New York: Macmillan, 1987.
[7] R. L. Pickholtz and Hill, *Adaptive Beamforming for Interference Reduction*, GW University PW3312A, 1990.
[8] Zoltowski *et al.*, "Blind 2-D rake receivers based on space-time adaptive MVDR processing for IS-95 CDMA system," presented at MILCOM'96, IEEE, NY, Oct. 1996.
[9] Belzer *et al.*, "Joint source channel coding of images with trellis coded quantization and convolutional codes," UCLA, 1998.
[10] Ferguson and Huston, *Quality of Service.* New York: Wiley, 1998.
[11] Paradells *et al.*, *DECT Multibearer Channels.* New York: IEEE Press, 1994.
[12] Strom and Shaula, "Optimistic recovery in distributed systems," *ACM Trans. Comput Syst.*, 1985.
[13] Pesonen, *Object-Based Design of Embedded Software Using Real-Time Operating Systems.* New York: IEEE Press, 1994.
[14] Tebbs and Garfield, *Real Time Systems.* Berkshire, U.K.: McGraw-Hill, 1977.
[15] K. Ellison, *Developing Real-Time Embedded Software in a Market-Driven Company.* New York: Wiley, 1994.
[16] J. Mitola, "The software radio architecture," *IEEE Commun. Mag.*, May 1995.
[17] ——, *Software Radios: Wireless Architectures for the 21st Century.* Fairfax, VA: Statisfaction, 1999.
[18] MMITS Forum Tech. Rep. (www.mmitsforum.com), 1997.
[19] Milewski, *The Topology Problem Solver.* Piscataway, NJ: Research and Education Association, 1994.
[20] Herlihy and Shavit, "A simple constructive computability theorem for wait-free computation," in *Proc. STOC'94.* Montreal, Canada: ACM, 1994.
[21] "DII strategic enterprise architecture, DISA," *IEEE Commun. Mag.*, Oct. 1995.
[22] *Programmable Digital Radio.* GEC Marconi, 1996.
[23] *The Software Defined Radio Request for Information.* Atlanta, GA: BellSouth, Dec. 1995.
[24] R. Pengelly, "Low cost GaAs MMIC chip set for dual mode AMPS/CDMA phones," in *Proc. WESCON 96*, p. 96.
[25] S. Oh, "DSP technology for a hands-free car cellular phone," in *Proc. WESCON 96* p. 821.
[26] J. Mitola, "Software radios: Technology and prognosis," in *Proc. Nat. Telesyst. Conf.*, IEEE, NY, May 1992.
[27] Walden, "ADC integrated circuits," in *Proc. IEEE GaAs IC Symp.*, Dec. 1995.
[28] K. C. Zangi and R. D. Koilpillai, "Software radio issues in cellular base stations," this issue, pp. 561–573.
[29] R. L. Pickholtz and Hill, *Adaptive Beamforming for Interface Reduction*, GW University PW3312A, 1990.
[30] M. Martone, "Cumulant-based adaptive multichannel filtering for wireless communications systems with multipath RF propagation using antenna arrays," *IEEE Trans. Veh. Technol.*, May 1998.
[31] Loundu *et al.*, *Estimating the Capacity of a Frequency-Selective Fading Mobile Radio Channel with Antenna Diversity.* New York: IEEE Press, 1994, 0-7803-1927-3.
[32] Morgensen and Petersen, *Practical Considerations of Using Antenna Diversity in DECT.* New York: IEEE Press, 1994.
[33] J. Tsui, *Digital Techniques for Wideband Receivers.* Boston, MA: Artech, 1995.

[34] A. Abidi, "Low power RF IC's for portable communications," *Proc. IEEE*, Apr. 1995.

[35] Brown and Sward, *Digital Downconversion Test Results with a Broadband L-Band GPS Receiver.* New York: IEEE Press, 1994.

[36] M. Rolfes, "Field trial results of superconducting products in wireless," in *Proc. WESCON 96*, p. 109.

[37] K. Feher, *Wireless Digital Communications.* Upper Saddle River, NJ: Prentice-Hall, 1995.

[38] B. Fawcett, "Advancements in SRAM-based FPGA technology," in *Proc. WESCON 96*, p. 994.

[39] Ono, *Introduction to Point Set Topology.* Baltimore, MD: Johns Hopkins University Press, 1974.

[40] "Software drives multi-channel radio," *Mil. Aerosp.*, June 1997.

[41] V. Bose, M. Ismert, M. Welborn, and J. Guttag, "Virtual radios," this issue, pp. 591–602.

[42] R. Ziemer and Peterson, *Digital Communications and Spread Spectrum Systems.* New York: Macmillan, 1985.

[43] Recommendation H.320, "Narrow-band visual telephone systems and terminal equipment," /itu-t/ituth13.htm: International Telecommunications Union, 1998.

[44] "Coding of analogue signals by pulse code modulation (G.711–G.712) and by methods other than PCM (G.720–G.729)," International Telecommunications Union, Geneva, Switzerland, 1998.

[45] "Introduction to IP multicast routing," Internet: ds.internic.net: Internet Engineering Task Force, Mar. 1997.

[46] Mouly and Pautet, "Evolution of the GSM system," *IEEE PCS Mag.*, Oct. 1995.

[47] *The Artificial Intelligence Handbook.* Tioga: Menlo Park, CA: 1985.

[48] Hennie, *Introduction to Computability.* Reading, MA: Addison-Wesley, 1997.

[49] Simon and Szegedy, "On the complexity of RAM with various operation sets," presented at the 24th Annu. ACM STOC, ACM, Victoria, B. C., Canada, 1992.

[50] Cutland, *Computability: An Introduction to Recursive Function Theory.* London: Cambridge University Press, 1980.

[51] Machtey and Young, *An Introduction to the General Theory of Algorithms.* New York: North-Holland, 1978.

[52] Hennie, *Introduction to Computability.* Reading, MA: Addison Wesley, 1977.

[53] H. R. Nielson and F. Nielson, "Higher-order concurrent programs with finite communication topology" (ext. Abst.) POPL94-1K14, ACM, Portland, OR, 1994.

[54] Milner, Tofte, and Harper, *The Definition of Standard ML.* Boston, MA: MIT Press, 1990.

[55] Reppy, "CML: A higher-order concurrent language," in *Proc. ACM SIGPLAN'91, (Conf. Programming Language Design and Implementation*, 1991.

[56] Moir, "Practical implementations of nonblocking synchronization primitives," in *Proc. 1997 PODC 97*, ACM, Santa Barbara, CA, 1997, p. 219.

[57] Herlihy and Shavit, "A simple constructive computability theorem for wait-free computation," in *Proc. STOC 94*, ACM, Montreal, Canada, 1994, p. 243.

[58] Hoest and Shavit, "Toward a topological characterization of asynchronous complexity" (preliminary version), in *Proc. 1997 PODC*, ACM, Santa Barbara, CA, 1997, p. 199.

[59] Borowsky and Gafni, "A simple algorithmically reasoned characterization of wait-free computations" (ext. Abstr.), in *Proc. 1h997 PODC*, ACM, Santa Barbara, CA, 1997, p. 188.

[60] M. Fischer, N. Lynch, and M. Paterson, "Impossibility of distributed consensus with one faulty process," *J. ACM*, vol. 32, no. 2, pp. 374–382, 1985.

[61] D. Luenberger, *Optimal Control.* Englewood Cliffs, NJ: Prentice Hall, 1970.

[62] Milewski, *The Topology Problem Solver.* Piscataway, NJ: Research and Education Association, 1994.

[63] Laubenbacher, *Complex Adaptive Systems.* Las Cruces, NM: New Mexico State Univ., 1996.

[64] Johnson, "Design and control of self-organizing complexes," in *Control Mechanisms for Complex Systems.* Las Cruces, NM: New Mexico State University, 1996.

[65] Recommendation H.320, "Narrow-band visual telephone systems and terminal equipment," www.itu.int./publications/itu-t/ituth13.htm: International Telecommunications Union, 1998.

[66] http://www.microsoft.com/ntserver/communications/tapi_wp.htm.

[67] ITU, "Specification and description language, Recommendations Z.100," Geneva, ITU, 1991.

[68] "Methods for testing and specifications (MTS); Strategy for the use of formal SDL for descriptive purposes in ETSI products," TR 101 081 V1.1.1, ETSI, Sophia Antipolis Cedex, France, 1997.

[69] J. Mitola, *Cognitive Radio—Model Based Control of Soft Radios.* Kista, Sweden: KTH, 1998.

[70] D. Upmal and R. Lackey, "SPEAKeasy, The military software radio," *IEEE Commun. Mag.*, 1995.

[71] ITU, "Open distributed processing, X.900–X.999 Recommendations." Geneva, ITU, Nov. 1995.

[72] Eriksson and Penker, *UML Toolkit.* New York: Wiley, 1998.

[73] *MPI-2: Message Passing Interface Extensions.* Knoxville, TN: University of Tennessee, 1998.

[74] Wolfe *et al., Expressing and Enforcing Timing Constraints in Dynamic Real-Time CORBA System.* Kingston, RI: University of Rhode Island, 1998.

[75] ITU-T Recommendation Z.105, "SDL combined with ASN.1, 1994."

[76] Beyer *et al., Radio Device API.* Menlo Park, CA: SRI International, Sept. 1, 1997.

[77] Sunaga and Yamada, "NOSES distributed communication switching software structure based on CORBA," presented at IEEE ICC'97.

Beyond Software Radio, towards Re-configurability across the whole System and across Networks

Jorge M. Pereira*

European Commission, DG XIII F4
Ave. de Beaulieu 9, Office BU-9 04/61
B-1160 Brussels, Belgium
Tel: +32 2 296 1547 Fax: +32 2 296 2178
Email: Jorge.Pereira@bxl.dg13.cec.be

Abstract

From its military origins, the Software Radio concept has evolved slowly, remaining mainly focused on the terminal, and still with considerable military influence. Such narrow perspective does not fit with our long held conviction that Software Defined Radio (SDR) is a far-reaching concept, with much broader application.

We propose a much broader, all encompassing approach. While the SDR approach concentrates on the Terminal side, ours covers the whole System, extending through the Network into Service creation and Applications development with the involvement of new players, thus delving into the resulting revamped Business Model, and paying attention to institutional issues like Standardisation and Regulation.

This paper introduces this novel approach, and establishes the link with ongoing research in the area of Re-configurable Radio Systems and Networks in Europe.

1 Introduction

From its military origins, namely with SPEAKeasy, the first widely publicised military software radio [1], the Software Radio concept evolved slowly, even with the insight of people like Mitola, whom coined the term Software Radio in 1991 and effectively defined the area [2] and Steinbrecher, whom called the attention for the appropriateness of Software Radio for implementing multi-standard, flexible Base Stations [3]. BellSouth's Request for Information [4], focusing on the commercial application of Software Radio, seen as key for the "future-proof" infrastructure needed by wireless service providers, was certainly a step in the right direction, but in the end the focus remained on the terminal, and still with considerable military influence. It was not, as Mitola pointed out in [5], that the technology was not within reach, but the fact that commercial applications are quite cost sensitive.

The MMITS Forum was born in March 96, in the wake of the PCS (2nd Generation) auctions in the US, initially with 7 competing standards fighting in the market, and patching the territory. These special circumstances reinforced the trend to focus on the terminal by "requiring" multi-mode terminals (AMPS plus some version of PCS) just to insure nation-wide coverage.

Even after changing the name to SDR (Software Defined Radio) Forum in Dec 98, the focus hasn't changed significantly, in fact it has become even more short term [6]: "A software defined radio is defined as a radio in which the receive digitisation is performed at some stage downstream from the antenna, typically after wideband filtering, low noise amplification, and down conversion to a lower frequency in subsequent stages, with a reverse process occurring for the transmit digitisation. As technology progresses, a software defined radio can move to an almost total software radio where the digitisation is at the antenna and all of the processing is performed by software residing in high-speed digital signal processors. *The software defined radio will occur in the near term migrating to the software radio in the longer term subject to the progression of core technologies.*"

In the meantime, in Europe, already in 1995, the importance of the area was recognised, and the Second Call for Proposals of the ACTS (Advanced Communication Technologies and Services) Programme resulted in 1996 in a project, FIRST (Flexible Integrated Radio Systems Technology), investigating Intelligent Multimode Terminals. Soon after, the First European Workshop on Software Radio was organised in Brussels [7], launching a debate that facilitated the response to the ACTS Third Call for Proposals, which included novel technological work in Software Radio technologies. Two projects resulted in this area: SUNBEAM (Smart Universal Beamforming) and

* The opinions expressed herein are those of the author, and do not engage the European Commission. The author is on leave of absence from Instituto Superior Técnico, Lisbon Technical University, Portugal.

Reprinted from *IEEE 50th Vehicular Technology Conference*, pp. 2815–2818, Fall 1999.

SORT (Software Radio Technology), one dealing with smart array antenna integration in a "software radio Base Station", the other looking into Base Band issue, and namely channelisation.

A joint workshop was organised with the, at the time, MMITS Forum [8], with the objective of fostering the exchange of experience in the field, and mainly of broadening the debate, extending beyond the terminal. It soon became evident that the circumstances of mobile communications (1st and 2nd Generation) in the US were effectively preventing the evolution of the Software Radio concept beyond the need for multi-mode, multi-band terminals. Another area where the approach is rather restrictive concerns the data rate, with only very low data rates being considered.

Such narrow perspective did not fit with the long held conviction that Software Radio is a far-reaching concept, with much broader application – nor with Europe's clear emphasis on mobile multimedia.

2 From Software Radio ...

In a recent contribution to the ITU [9], the SDR Forum re-stated its mission to evolve towards a Universal Terminal, seen as a Quadruple-mode Terminal implemented in software, initially with four separate RF parts, later with an intelligent wide-band front end (Figure 1). No consideration is given to anything beyond the Terminal.

3 ... towards Re-configurability across the whole System and across Networks

In contrast with this canonical approach, we propose a much broader, all encompassing concept. While the SDR Forum's approach concentrates mainly on the Terminal side (recently a working group has been formed to look into Base Station issues), ours covers the whole System, extending through the Network into Service creation and Applications development (with the involvement of new players), thus delving into the resulting revamped Business Model, and paying attention to institutional issues like Standardisation and Regulation.

Already now, the benefits of re-configurable radio techniques in handset design (commonly known as "software radio") are becoming clear. And these same techniques apply *mutatis mutandis* to Base Station design. In particular, the ability to incorporate enhancements, to deploy new features and to instantiate new air interfaces "on demand", are some of the factors making re-configurable radio techniques so attractive.

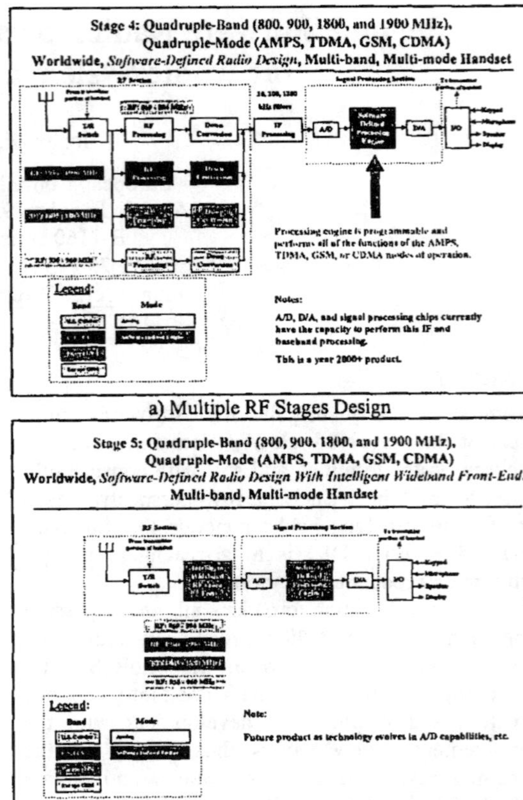

Figure 1: SDR Forum's Universal Terminal [9]

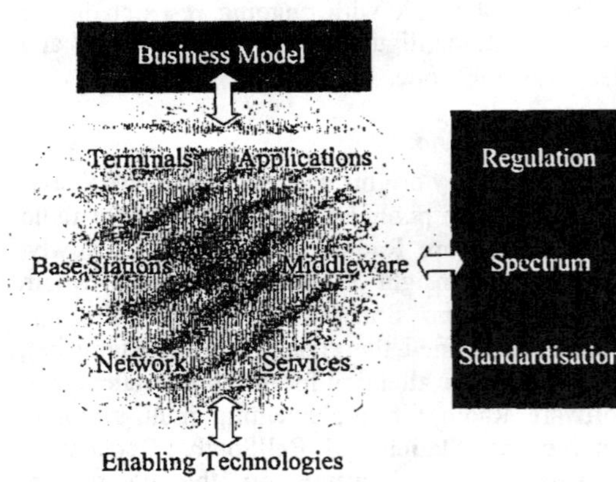

Figure 2: Scope of the Re-configurable Radio Systems and Networks approach

Naturally, the ability of the network to support re-configuration is critical, as is the required amount of signalling, very much a function of the download mechanisms implemented. Re-configurability, on the other hand, will take network management to a new level, by allowing for much more flexible (and dynamic) resource management.

Clearly, were it to stay at the "system" level only, most of the potential of re-configurability would be lost. On the "application" side, re-configurability opens new forays to application developers and service providers (taken the broadest sense), by allowing for the instantiation of processes across intervening networks/network segments (across heterogeneous networks and even irrespective of the serving/visited end network), as well as at the terminals, such that transparent service provision results.

Application updates, service enhancements, and above all personalisation, are effected effortlessly in such environment, with middleware providing the necessary conversion/tunneling mechanisms, taking into account terminal capabilities and user preferences.

In order for all the above to come to fruition, a number of "enabling technologies" needs to be developed, spanning software, hardware and algorithmic aspects.

Many "institutional" implications arise from new capabilities and increased flexibility and from the change of roles of the market players, with particular incidence on standardisation. The more flexible management of radio resources will certainly impact spectrum usage and planning, and will have regulatory implications in terms of spectrum management and monitoring, and through (easier) spectrum sharing also in terms of licensing.

Driving it all, are (quite obviously a posteriori) economic/business considerations, the same that have kept Software Radio so focused on the short term. On the other hand, Re-configurability will create new opportunities, and effectively shape new "business models" and create new value-added chains.

This approach was discussed in the First European Colloquium on Re-configurable Radio Systems and Networks [10], where the most important issues/problems associate with each of the areas in the Functional Diagram above were discussed, together with the interdependencies amongst areas.

3.1 Defining Re-configurability

Re-configurability comprises in this context the pervasive use of "software re-definition" (it is not enough to define once), empowering (possibly live) upgrades/patching of any element in the network, and of all services and applications running on it.

It cuts across all types of radio systems (paging to cellular, W-LAN to MBS, terrestrial to satellite, personal communications to broadcasting) enabling the integration of many systems in the same platform, but more importantly, it also cuts across generations (second to third to fourth).

It is not just a way to implement third generation, ignoring second generation and compromising the potential of fourth generation.

It plays a critical role in maintaining Europe's leadership in the area of Mobile Communications by increasing flexibility; facilitating enhancements and personalisation; reducing deployment, operation and maintenance costs; creating new business opportunities and new jobs; etc.

3.2 Functional Characterisation

Figure 2 does not pretend to suggest the existence of simple one-to-one relationships amongst "functional elements". On the contrary, the complex interdependencies warrant a careful characterisation of the elements. Some of the questions to be answered are:

- What does Re-configurability add?
- Interdependencies with other Elements
- What are the Short/Medium/Long Term issues?
- Essential Technologies involved/required
- Institutional/Market Impact

Due to space limitations, and mainly for illustration purposes, we will focus only on two elements: Business Model and Institutional issues.

3.2.1 Business Model

There are obvious benefits arising from re-configurability in terms of Deployment, Operations and Maintenance, but where re-configurability makes a difference is in opening the field for application developers and service providers (taken the broadest sense), by allowing them to become players in the mobile/wireless market.

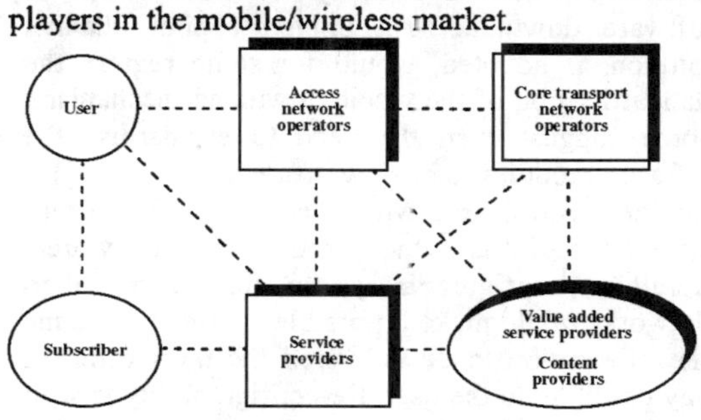

Figure 3: UMTS Forum Technology Aspects Group Business Model

By requiring an Open Architecture (open business model), a need that is being insistently voiced in Europe, it is already changing the mobile landscape.

In terms of interdependencies, if re-configurability makes economic sense by renovating and enlarging the market, empowering new players and creating new opportunities, the resulting new business model raises regulatory questions namely in terms of right-of-access to both information and resources (competitive, open, non-discriminatory access), and of accountability/liability. Standardisation is also impacted by re-configurability and at the same time impacts the business model.

3.2.2 Institutional Issues

In this section we will focus on standardisation and spectrum issues, and a topic of great importance: software download.

Some radical thinkers saw in Software Radio the end of standardisation, as "everything" could be defined (and re-defined) at will, without any need for consensus. The whole standardisation process was seen as unnecessarily forcing the crystallisation of solutions and thus thwarting evolution. The Standardisation bodies would then be the direct victims of re-configurability.

Against this interpretation, standardisation is seen as a protection against proprietary solutions, and the only way to achieve interoperability, without which there is no customer choice.

We have already seen that the existence (or lack) of standards affects both "system" and "application" development, as well as the business model, the "rules of the game". The call for an Open Architecture should be seen in this context.

But here arises the first (absurd, it would seem) question: Is there a need to standardise this Open Architecture? If so, who would do it? Would we "legalise" standardisation de facto?

A concrete case where standardisation is being clearly suggested by many is exactly that of software download. An universal pilot channel solution, if adopted, would for some require the standardisation of the whole download mechanism. Some suggest even the need to standardise the software modules to be downloaded.

Another critical area where re-configurability will make an impact is in the Spectrum Efficiency area. A fully re-configurable system, from Terminal to Network, would make it possible for (many streams of) information to be delivered the most efficient way possible to the user. Re-configurability is seen as the enabler of full *spectrum sharing*, the culmination of this trend towards efficient use of a increasingly scarce and valuable resource.

This, however, raises many issues concerning "ownership" of spectrum, particularly sensitive when spectrum auctions are an option, or when current operators had to pay considerable amount of money for their licenses (associated with a precise portion of spectrum, much like real estate).

4 Conclusions

This paper proposes a broader, all-encompassing approach to the concept of Software Radio. A functional characterisation of Re-configurable Radio Systems & Networks is briefly presented and some of the issues associated with two of the functional elements discussed.

There was no intention to propose solutions to the problem areas identified, just submit for discussion this new approach, in order to jump start the much needed dialogue.

5 Acknowledgements

The author would like to thank his colleagues at the European Commission, namely Dr. João da Silva and Demosthenes Oikonomou, for the long discussions that helped crystallise this vision, and to the members of the Steering Committee of the First European Colloquium on Re-configurable Radio Systems and Networks, Dr. Mark Beach, Prof. Gerhard Fettweis, Prof. Lajos Hanzo, Dr. Enrico Buracchini, and the University of Bristol supporting team, and namely Dr. Alistair Munro – they made possible the Colloquium, and throughout the preparatory discussions helped flesh out the details of the approach.

6 References

[1] Lackey, Upmal, "SPEAKeasy: The Military Software Radio", *Special Issue on Software Radio*, IEEE Communications Magazine, May 95

[2] Joseph Mitola III, Editor, *Special Issue on Software Radio*, IEEE Comm Magazine, May 95

[3] George Gilder, "Auctioning the Airwaves", Forbes ASAP, April 11, 1994

[4] BellSouth, *The Software Defined Radio*, Request for Information, Dec 95

[5] Joseph Mitola III, "Software Radio Technology: Challenges and Opportunities", Proceedings of the First European Workshop on Software Radio, May 97

[6] *Software Defined Radio: A Window of Opportunity in Wireless Communications*, Market Demand Forecast Series, SDR Forum, Nov 98

[7] *Proceedings* of the First Workshop on Software Radio, European Commission, Brussels, May 97

[8] *Proceedings* of the First International Workshop on Software Radio Technologies, EC/MMITS Forum, Rhodes, June 98

[9] "The impact of Step 7 Key Characteristics on Wireless Devices for IMT-2000", SDR Forum contribution to ITU-R TG 8/1, Document 8-1/INFO/25-E, Nov 98

[10] *Re-configurable Radio Systems and Networks*, European Commission, London, Mar 99

Chapter 2

Enabling Technologies

THE first chapter introduces the need for the enabling technologies developed in greater detail in this chapter. The themes of this chapter are RF components; ADC technology; digital filtering; and processor architectures, including the problem-specific architectures made possible by configurable computing in FPGAs. These are the fundamental enabling technologies of effective SDR implementations.

Analog components, first of all, include GaAs and CMOS device technologies. RF Micro Electro-Mechanical Subsystems (MEMS) are quietly revolutionizing the filtering and switching components of software-defined radio platforms. Abidi's paper examines CMOS technology in detail. This technology benefits from the advances in foundries associated with general-purpose computing platforms. These billions of dollars per year of revenue-generation capability result in multibillion dollar investments that competing technologies have yet to muster. Consequently, CMOS RF can leverage the mainstream semiconductor industry substantially. Abidi's paper provides an up-to-date grounding in this critical technology.

Wepman's paper introduced the fundamentals of the ADC in software radio in 1995. This provides the background necessary to tackle the more advanced paper by Walden. Walden's critical look at ADC progress notes that we have only made 1.5 bits of progress in the past 8 years, not Moore's law performance. He also establishes the fundamental physical limits of ADC progress. Kennington and Astier examine the power consumption of ADCs in detail. Low-power ADCs are necessary to approach the ideal software radio in the handset, an elusive goal. Enhanced linearity may be achieved using the Sigma-Delta techniques described by Aziz et al. Vaughan, Scott, and White provide the technical underpinnings of sampling theory required to apply the ADC device technology effectively.

Wikborg et al. describe a technology that could completely revolutionize software radio. Their rapid single flux quantum superconducting ADCs can access wide RF bands with extremely high dynamic range. 200 MHz with 80 dB of true spurious-free dynamic range has been demonstrated in the laboratory. The remarkable aspect of this ADC is that it is implemented in a 3-μm process. The bad news is that today's cryogenic coolers are large and consume substantial power compared with room-temperature silicon. The paper points the way toward digital RF in the not too distant future.

Chester emphasizes the computational efficiency gained when an ASIC is structured to provide programmable digital filters. In the next chapter, Zangi et al. analyze the filtering requirements further, showing how to reduce parts count substantially when the ASIC is designed for a specific cellular standard. Infrastructure implemented exclusively with such optimized ASICs might not be as readily reprogrammable to evolving standards as the more general programmable ASICs. Conversely, however, the optimized ASICs have the potential for a simpler and thus possibly lower cost and more reliable implementation. The April 1999 JSAC companion to this text includes several papers that extend the principles introduced in these papers.

The final set of papers in this chapter address the key issues in digital processors. These include DSP architectures and FPGAs. FPGA advocates point out the advantages of reconfigurability, whereas DSP advocates point out the relative ease of extending DSP code compared with extending an FPGA that is, say, 70% used but that cannot be used further because of race conditions. It seems to be much more labor intensive to reprogram an FPGA that is near full utilization than an analogous DSP. One cannot say unequivocally which approach is best because the trade-offs are architecture, algorithm, and air-interface dependent. We therefore chose a subset of the current literature that addresses the key issues so that the reader can make informed trade-offs.

CMOS WIRELESS TRANSCEIVERS: THE NEW WAVE

ASAD A. ABIDI, UNIVERSITY OF CALIFORNIA, LOS ANGELES

ABSTRACT

A new generation of wireless transceivers is being intergrated into CMOS IC technology, which so far has been used mainly to realize digital and mixed analog–digital baseband circuits. This article reviews some of the recently developed RF CMOS circuit design techniques, and shows how an understanding of the strengths and weaknesses of these circuits influences choice of radio architecture. The CMOS approach to radio design calls for the elimination of discrete components in favor of high levels of on-chip integration which freely use translators and mix analog and digital functionality; in these respects, it departs from traditional RF circuit practices. Successful wireless devices of the future will require that radio system design evolve around these new trends in RF integration.

With the spectacular popularity of mobile telephones, wireless architectures and circuit techniques have experienced a renaissance in the past few years. The art of radio engineering was almost lost in the many decades following the invention of the transistor, when the focus of most electronic engineers turned to analog circuits for over-the-wire telephony or, in the case of logic circuits, to electronic computers. These applications have preoccupied engineers in the integrated circuit (IC) era, culminating in the electronics that enable fiber optic communications and today's microprocessors. Baseband analog circuits and digital signal processing (DSP) have made enormous strides over the years, motivated by consumers and telephony. But radio engineering was somehow always different. Radio circuits were thought to be incompatible with IC technology, because they rely heavily on tuned circuits and discrete filters, and as a result their teaching and practice became less fashionable, indeed even irrelevant to the immediate needs of industry; that is, until the advent of worldwide mobile telephony.

As mobile telephony became popular in the 1990s, it was clear that the number of users and the base station infrastructure require great sophistication in the mobile radio transceivers, the associated DSP and power management, and the software to drive it all. Moreover, the hardware had to be packaged into a very small physical volume and consume as low a battery current as possible. These constraints rekindled research in radio architectures and circuits, although now in a community of engineers steeped in the design of baseband or broadband silicon ICs. This has led to new syntheses and concepts, notably that like all other communications devices, radios too should be highly integrated, while requiring the fewest or preferably no external components; and that by penetrating the core of the radio, digital functionality adds strengths such as multimode operation and adaptation. Complementary metal oxide semiconductor (CMOS) is the technology of choice to permit the free mixture of analog and digital functions.

Starting from a topic initially confined to academic research, the field of radio frequency (RF)-CMOS ICs has now progressed to the point of being on the threshold of important commercial deployment. Progress is on several fronts. New MOS-appropriate circuit techniques have been discovered for many of the RF, intermediate frequency (IF), and baseband blocks required in state-of-the-art wireless transceivers. Transceiver architectures have evolved to take

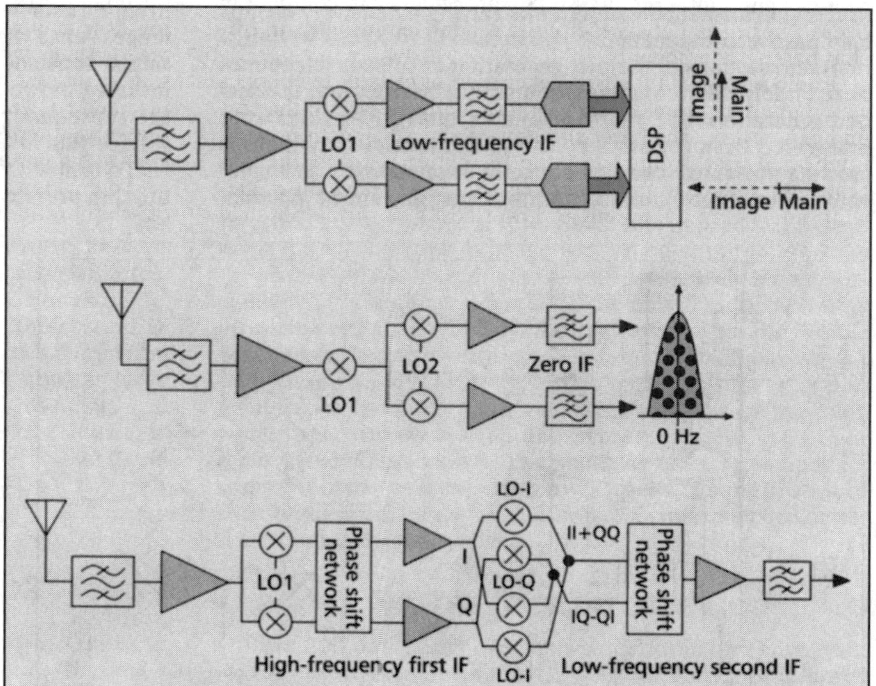

Figure 1. *Three possible receiver architectures suited to CMOS: a) single-conversion with low IF; b) indirect conversion to zero IF (in the case of direct conversion the first mixer is not present); c) dual conversion with low second IF and on-chip image rejection.*

Reprinted from *IEEE Communications Magazine*, Vol. 37, No. 8, pp. 119–124, August 1999.

advantage of the strengths of CMOS, and to circumvent its weaknesses. Recent CMOS implementations of RF and IF blocks combine analog circuits with switched and digital functions in unprecedented ways. Although it is possible to realize individual wireless building blocks such as amplifiers or oscillators with bipolar transistors, often with slightly superior performance to CMOS, the author believes that future wireless devices will increasingly gravitate toward the CMOS paradigm. The object of this article, then, is to summarize the important concepts underlying good practice in the design of RF-CMOS circuits, and how it is availed in integrated wireless transceivers.

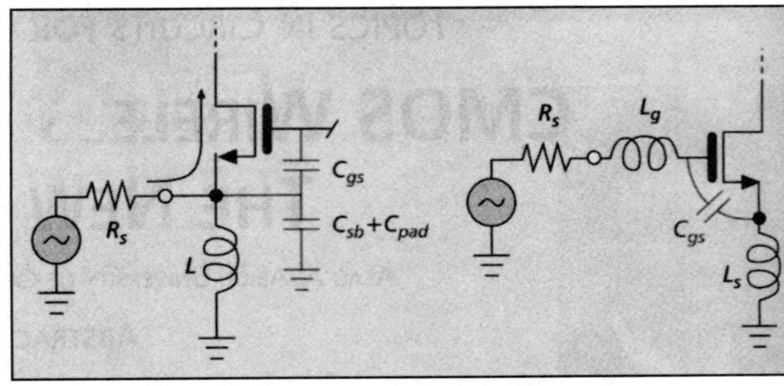

■ **Figure 2.** *LNA topologies: common-gate and common-source.*

CMOS Transceivers Strive for Greater Integration

There is a greater effort in RF-CMOS research to integrate all transceiver components on a single chip [1–4] than is seen in contemporary ICs for cellular and cordless telephones developed in industry, although a 900 MHz CMOS Global System for Mobile Communications (GSM) receiver, which implements a superheterodyne architecture [5] with off-chip inductors and filters, is an exception. Experience with baseband and other mixed analog-digital circuits shows that high levels of integration usually bring benefits such as simple assembly and testing, and lower interference from stray signal pickup. However, integration also compromises the performance of some RF-CMOS building blocks. For example, the close-in phase noise (PN) of oscillators tuned with lossy on-chip inductors and capacitors is inferior to oscillators tuned with less lossy off-chip elements. The dynamic range of active filters is severely limited by large noise and limited signal swing, as compared to the essentially unlimited dynamic range of passive filters. Nevertheless, well-designed integrated CMOS transceivers can compete in overall performance with circuits relying extensively on off-chip passive components.

Radio sections in the first generation of cellular telephones were largely discrete; integration gradually crept into the second generation. The baseband analog and DSP sections were integrated first, followed by the IF sections. Today there is growing pressure to implement cellular telephones on single-chip radios (Fig. 1), mainly to lower cost and shrink physical volume. The experience of integrating CMOS transceivers will now prove valuable, particularly since there is also interest in implementing future cellular telephones in CMOS.

Any systematic exploration of small-signal MOS circuits must start from the basic fact that the transconductance (g_m) of a field effect transistor (FET) relates to the gate overdrive bias voltage ($V_{GS} - V_t$) and the bias current approximately as

$$g_m = 2I/(V_{GS} - V_t)$$

Now in the front end of a receiver, the mean-square input noise voltage depends on 1/gm, whereas the maximum input voltage swing at the onset of clipping, which determines the intercept point, is set by ($V_{GS} - V_t$). Therefore, once the noise figure and intercept point are specified in a simple FET circuit, the current consumption of that circuit is fixed, *regardless* of technology scaling.

The Low-Noise Amplifier

In an integrated CMOS receiver the low-noise amplifier (LNA) drives the capacitive load of the subsequent mixer. To provide gain at 900 MHz or 2 GHz, the amplifier may be designed either as a narrowband amplifier tuned by an *LC* load, or as a wideband amplifier. The latter solution only makes sense when the mixer input is shunted with 50 Ω or a similar resistance, for instance, imposed by a discrete RF filter inserted between the LNA and mixer. In other cases, when the subsequent load is capacitive, the *LC* load gives higher voltage gain at a given bias current (Fig. 2).

To enable the LNA to properly terminate the antenna and off-chip prefilter, its input port must be well matched to 50 Ω in the frequency band of interest — an input return loss of at least 10 dB usually suffices. An impedance match is obtained with either a common-gate input [2] or common-source input with inductor degeneration [6]. Whereas the minimum noise figure of the former is limited in practice to about 3 dB, with the latter approach sub-2 dB noise figures at 900 MHz have been reported in a 0.25 μm CMOS LNA [7]. Second, the on-chip load inductor must be designed to provide high impedance at the frequency of interest. Inductor design will be discussed in a later section, in the context of oscillators.

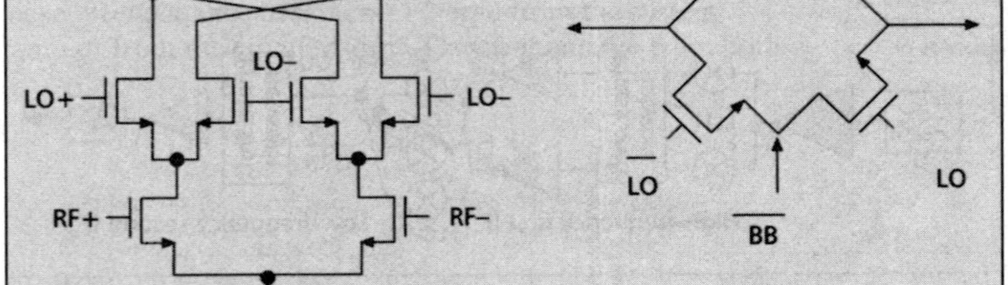

■ **Figure 3.** *Mixer topologies: transconductance mixer and four-FET switch mixer.*

Mixers

Two styles of mixer may be implemented in CMOS: a transconductance mixer similar to the Gilbert cell, or a four-FET switch to commutate voltage or current (Fig. 3). The capacitive input impedance of the transconductance mixer is desirable when it follows the LNA, but in image-reject arrangements where mixer gains must match well, the four-FET commutating switch is preferred. Threshold voltage mismatch is masked when the switches are driven with square-wave local oscillator (LO) signals (usually at the second IF), whereas channel conductance mismatch may be lowered by using FETs of appropriately large geometry with well-matched fixed resistors in series.

In an integrated receiver, the LNA output will usually drive *two* mixers configured in an image-reject arrangement (Fig. 1). The alternative is to obtain the required image rejection from the prefilter and the narrowband characteristic of the LNA. In a superheterodyne receiver the mixer output lies at some IF, possibly as high as hundreds of megahertz, whereas in a direct conversion receiver the IF is zero. In both cases a high-frequency LO drives the MOSFET mixers. Associated with this are some important limitations. In the superheterodyne case, the high-frequency LO soft-switches the mixers, which exposes statistical mismatches between the two mixers. The mixer gain mismatch limits the achievable image rejection. In a direct conversion receiver the channel of interest is downconverted to DC after only a small amplification at RF. It must now compete with the large flicker noise (1/*f* noise) in the CMOS baseband circuits. (Due to the very low levels of flicker noise in bipolar transistors, this concern did not arise prior to the advent of RF-CMOS.) While it is obvious that the active loads contribute flicker noise, what is not evident is that the mixer's switch FETs also contribute a very large, possibly even the dominant, flicker noise. This may significantly raise the cascade noise figure in a direct-conversion receiver. However, if the mixer switches are driven at a lower frequency with sharp edges, the flicker noise contribution of the switches falls. Also, the short transition times mask threshold voltage differences between the mixer FETs, which will otherwise introduce phase skews. At lower LO frequencies, then, the matching between the mixers improves, and so does the image rejection.

Starting from these considerations of transistors and circuits, it becomes apparent that certain transceiver architectures are more "CMOS-friendly" than others. For example, in a dual-conversion superheterodyne with the second IF at zero, the channel of interest is converted away from zero at the first mixer, whose output flicker noise is large as explained above, and then converted to zero at the second mixer driven by a low-frequency LO. This "indirect" conversion to zero IF is therefore preferred to classic direct conversion.

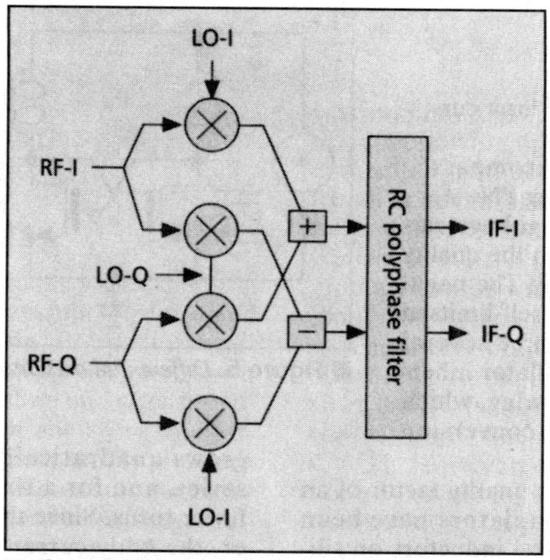

■ **Figure 4.** *A double quadrature downconversion mixer, followed by a polyphase filter to reject the image.*

The Choice of IF

A zero IF receiver has the advantage that the desired channel may be selected by a low-pass active filter (Fig. 1b). However, the mildly amplified downconverted channel competes in CMOS with baseband flicker noise and DC offsets. Baseband DSP methods summarized in [8] have been developed to estimate and remove DC offset, but they are complicated. The offsets in an integrated CMOS zero-IF receiver may be two orders of magnitude higher than the signal, and without perfect cancellation they will be troublesome. Modulations such as wideband frequency shift keying (FSK) are tolerant of DC offsets, which may be removed by capacitor coupling, as has been shown in an integrated CMOS receiver using 4-FSK modulation superimposed on a frequency-hopping carrier [2]. With larger transistor area the circuit designer can push down the frequency of onset of flicker noise to, say, below 1 kHz, although because of the larger capacitance this tends to narrow the bandwidth of the baseband circuits. However to detect signals with more complex modulations, IF well above the flicker noise corner frequency is very likely required.

The receiver must suppress the image channel, which otherwise superimposes on the desired channel being received. This can be done in several ways, depending on the specifics of the situation. The final choice is dictated by how much image rejection is practically achievable in the transceiver. For example, GSM base stations regulate the strength of adjacent channels, and the channel strength mask suggests some good choices of IF. A low IF of 100 kHz has been proposed where the image is the adjacent channel, which is at most 20 dB larger than the minimum channel. However, a single-conversion mixer with a 100 kHz IF output may suffer from large flicker noise. Therefore, a dual-conversion receiver makes better sense. Now the image associated with the higher-frequency first IF may be up to 100 dB larger than the desired channel. The IF must therefore be chosen high enough that most of the image is attenuated in the stopband of the antenna prefilter, and the remaining in a modestly accurate on-chip image reject mixer.

The critical tasks of IF signal processing, namely, amplification of the desired channel to the full scale of a baseband analog–digital (A/D) converter, automatic gain control (AGC), and filtering to remove the adjacent channels, should preferably be implemented at a low IF. These circuits may then realize the required dynamic range at modest power consumption. However, in unregulated wireless systems where adjacent channel power is not known, the problem of image rejection is much more severe. Since it is impractical to suppress the much closer-in image with an on-chip filter *prior* to the second downconversion, this becomes entirely the responsibility of the image reject mixer. The only feasible option to attenuate the image by 50–60 dB appears to be the double quadrature downconversion mixer followed by a passive RC polyphase filter [9] (Fig. 4). In this mixer, mismatch of element values and mixer gains mainly limits the image rejection, while inaccuracies in the quadrature phases of the LO or input signal do not matter. The matching of FETs and passive components improves as the inverse of the square root of their physical area. Passive FET switching mixers match better than transconductance mixers.

The Oscillator

Differential oscillators with integrated inductors have been widely investigated in RF-CMOS (Fig. 5). The current biased oscillator usually gives the same PN at lower power consumption than the voltage biased oscillator, because in the latter topology the FET operates at an unfavorably large $V_{GS} - V_t$. There are two main design issues in RF oscillators:

- How to achieve the lowest PN with the least current consumption
- How to widen the tuning range to encompass process variations without compromising PN

Phase noise depends inversely on the voltage carrier-to-noise ratio in the oscillator, and on the quality factor of the resonant tuning circuit [10]. The negative resistance core of an MOS oscillator self-limits at voltages of about $(V_{GS} - V_t)$, which can be several volts large. This means that a MOS oscillator inherently produces a large output voltage swing, which helps to lower its PN and improve the conversion gain of any mixer it drives.

The on-chip spiral inductor limits the quality factor of an integrated resonator. Customized simulators have been developed to assist in the design of spiral inductors on silicon substrates. There are three important losses associated with a spiral inductor: metal loss due to the nonzero resistance of the coil; loss due to eddy currents induced in planes parallel to the inductor; and loss due to displacement currents flowing perpendicular to the plane of the inductor through the substrate. On the heavily doped substrates used in modern CMOS processes, the first two losses are the most important. It has been found through simulation that a certain inductor with a given self-resonance frequency achieves the highest quality factor when realized with stacked spirals in series on as many layers of metal as are available (assuming that the specific sheet resistance of all metal layers is comparable). The inductance of the stack grows quadratically with the number of metal layers in series, and for a fixed total inductance each spiral needs fewer turns. Since the inductor footprint is now much smaller, the eddy currents it induces in the substrate occupy a smaller volume, and this lowers the total substrate loss. The inductor structure balancing metal loss and eddy current loss achieves the highest Q [11]. On certain medium doped substrates used in bipolar ICs, the displacement current loss may be largest, and then a sliced grounded shield of polysilicon under the inductor is effective in shunting away the displacement currents from the substrate. Best of all would be to fabricate inductors on a lightly doped substrate through which neither displacement currents nor eddy currents flow, and to implement the inductor coils on thick low-resistance metal layers. The problem is that the substrate is intimately linked with the transistors fabricated on its surface, and cannot be selected independently.

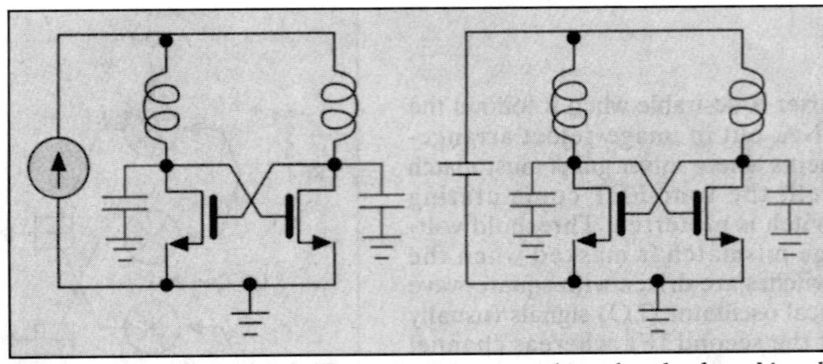

■ **Figure 5.** *Differential oscillator circuits: current biased and voltage biased.*

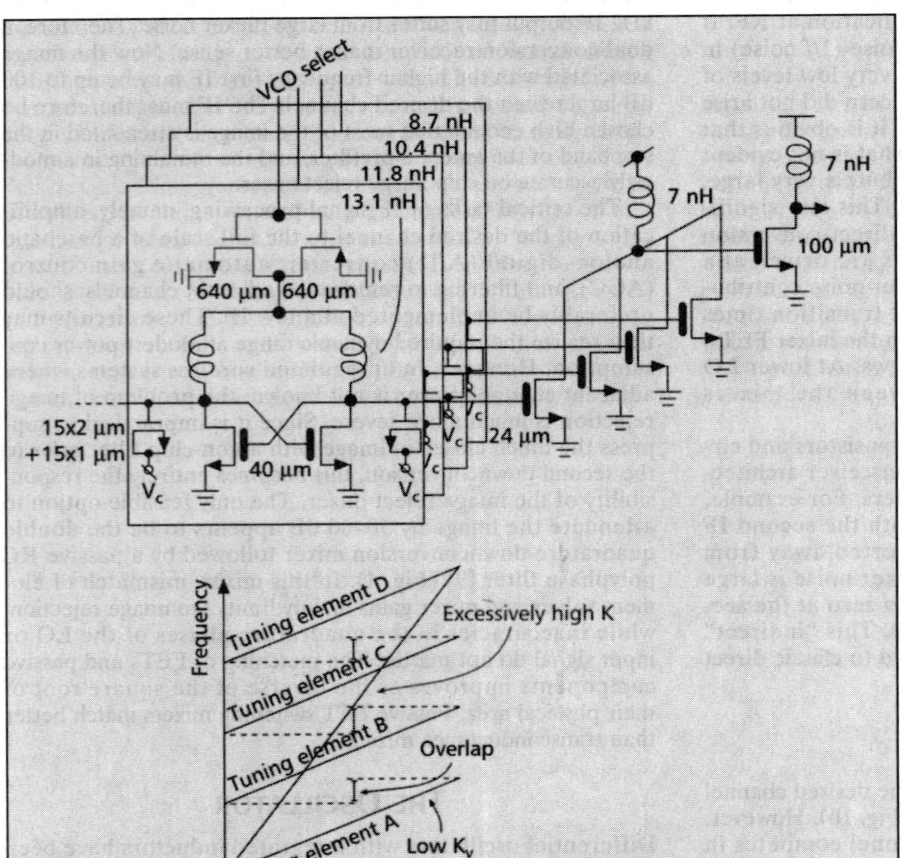

■ **Figure 6.** *An oscillator discretely tuned with inductors and continuously with MOS varactors, and the resulting piecewise characteristics.*

Any oscillator must be continuously tunable in frequency. There is renewed interest in the classic voltage-dependent MOS capacitor as a varactor because, unlike the junction diode, a large oscillation can never force this capacitor into conduction. The MOS capacitor is fully compatible with standard CMOS, and achieves higher Q than a diode junction. In practical oscillators the varactor usually takes up only a fraction of the total tuning capacitance. This fixed capacitance is subject to spreads as large as ±20 percent across wafer lots. To absorb these process variations, the oscillator requires a tuning range much larger than the minimum necessary to cover, say, a cellular band.

CMOS *LC* oscillators may be discretely tuned with switched elements [12] (Fig. 6). One of an array of oscillators, each loaded with different inductors, may be powered on by a digital control loop, and then a varactor continuously tunes the oscillation frequency. In this way the switched array coarsely sets the nominal oscillation frequency over an arbitrary range, and the varactor smoothly interpolates to reach any specific frequency. Fixed capacitors, too, in a binary weighted array may be switched in parallel with the varactor for intermediate tuning. With discrete tuning elements the PN does not degrade compared to a simple varactor-tuned oscillator. In fact, the lower slope of the continuous portions of the f-V characteristics means that noise on the control voltage induces less PN than it would in a single oscillator spanning the same range.

IF Strip Circuits

The variable gain amplifier and active channel-select filter in the IF strip must meet stringent requirements for noise and linearity at modest power consumption. For best overall dynamic range these blocks must be jointly designed, with appropriate allocation of variable gain before and after the filter. A smart IF strip ascertains whether or not strong interferers accompany a desired signal, and appropriately allocates prefilter gain to balance noise and intermodulation. Accuracy of the amplified signal level is also a concern when using complex modulations. A programmable gain amplifier (PGA) with discrete gain settings usually offers a better trade-off between noise, linearity, and current drain than an amplifier with continuously variable gain (VGA). This is because a PGA may be constructed with a bank of switch-selected amplifiers, each optimized for a fixed gain and noise level. Among various candidate MOS variable gain topologies the most promising for this application seems to be a method based on polysilicon resistor arrays in the load and degeneration on differential pairs hard-switched (for PGA) or soft-switched (for VGA) by FETs [13].

The channel select filter must be highly linear, because it accepts large interferers or blocking signals accompanying a weak desired signal, and attenuates the interferers in its stopband without substantial intermodulation which may create in-band spurious tones. Op-amp-based active filters are attractive for high linearity because of their capacity to swing rail to rail. Switched-capacitor filters have been shown to work well in a direct-conversion receiver [2, 3]. An active *RC* filter using op amps, fixed resistors, and programmable capacitor arrays performs as well at lower power, and without the need for an anti-aliasing prefilter. Continuous-time *gm-C* filters are usually limited in voltage swing, but their linearity can be improved with degeneration techniques similar to the circuits for the VGA core described above. They are well suited to compact, low-power implementations of filters with *adaptive* characteristics, for instance, to select a single channel which is 1 or 10 MHz wide always centered at a fixed 10 MHz IF.

Frequency Synthesizer

Under control of a digital word, the frequency synthesizer sets the LO frequency to precisely the value required to tune a sought channel. There are two fundamentally different approaches to frequency synthesis: the closed-loop architecture, using a phase locked loop (PLL) with an integer or fractional frequency divider, which servos the oscillator frequency to a fixed, stable reference frequency derived from a crystal oscillator; and the open-loop architecture, which digitally synthesizes a sinusoidal waveform with a programmable rate accumulator and a lookup table.

PLL-based synthesizers are ubiquitous. The majority use variable integer-modulus dividers programmed by a control word, and a reference frequency equal to the desired channel spacing. In a fractional-N or $\Delta\Sigma$ synthesizer, the PLL output frequency can increment by fractions of the reference frequency, which means that the reference frequency may be much higher than the required channel spacing. This has the advantage that the PLL bandwidth, usually set at roughly 10 percent of the reference frequency to avoid any significant feedthrough of the reference tone, may span several channels. The greater than unity gain of the PLL then proportionally lowers the intrinsic PN of the voltage-controlled oscillator (VCO) over this number of channels, which is advantageous in systems with large nearby channels, such as GSM. In response to a change in frequency control word, the PLL output frequency settles to the programmed value with a time-constant inverse related to the loop bandwidth.

The direct-digital frequency synthesizer (DDFS) is an alternative to implement a fast frequency-hopping synthesizer whose frequency changes instantly in response to a control word. The DDFS logic can only be implemented in CMOS. However, CMOS is also well suited to implement the 10–12-bit digital–analog (D/A) converters (DACs) required after the DDFS. High-frequency sine waves with spurious levels of about –60 dBc have been synthesized with a glitch-free charge redistribution DAC [1] as well as current steering DACs clocked at hundreds of megahertz. The DDFS/DAC is an important example of a mixed analog-digital CMOS alternative to a fractional-N PLL frequency synthesizer.

The PA and Preamplifier

CMOS RF power amplifiers (PAs) capable of producing 1 W output with 40–50 percent efficiency at 1–2 GHz have been demonstrated [14], but it is not yet clear whether it is a good idea to integrate them on the transceiver chip. If the PA is a separate chip, there is no reason it cannot be fabricated in silicon bipolar or GaAs. At high power outputs the PA efficiency is sensitive to small parasitic losses, and in this respect GaAs technology offers compelling advantages. In any case, a preamplifier with an output of 0 to +10 dBm to drive the PA must be integrated on the transceiver; this is a mini-PA in its own right. For constant envelope modulations this can be a simple Class-B-type circuit with peak efficiency of 40 to 50 percent [1].

Conclusions

This article surveys some of the emerging circuit techniques for the building blocks of wireless transceivers in CMOS. Many of these techniques would not be feasible in other than CMOS technologies. It is also shown how transceiver architectures are jointly optimized with knowledge of the signal modulation, transistor characteristics, and known circuit techniques. The pressure for lower cost or the needs for adaptation in some of the emerging wireless applications will force exploitation of these new capabilities. Using a mixture of analog and digital techniques, future wireless transceivers will sense the prevailing channel conditions to adapt their characteristics and power consumption in sophisticated ways. Higher levels of integration will be essential to realize disposable radios, such as the ones expected to be built into common household appliances. The passive filters and resonators which over the years have helped to identify radio circuits will slowly but surely disappear onto the silicon, also obviating the need for the tin-can metal shields that surround them. The silicon radio of the future, while more sophisticated in its functions than ever before, will be indistinguishable in all outward appearances from any other mixed analog-digital IC.

References

[1] A. Rofougaran *et al.*, "A Single-Chip 900 MHz Spread-Spectrum Wireless Transceiver in 1-μm CMOS (Part I: Architecture and Transmitter Design)," *IEEE J. Solid-State Circuits*, vol. 33, no. 4, 1998, pp. 515–34.

[2] A. Rofougaran *et al.*, "A Single-Chip 900 MHz Spread-Spectrum Wireless Transceiver in 1-μm CMOS (Part II: Receiver Design)," *IEEE J. Solid-State Circuits*, vol. 33, no. 4, 1998, pp. 535–47.

[3] J. C. Rudell *et al.*, "A 1.9-GHz Wide-band IF Double Conversion CMOS Receiver for Cordless Telephone Applications," *IEEE J. Solid-State Circuits*, vol. 32, no. 12, 1997, pp. 2071–88.

[4] M. Steyaert *et al.*, "A single-chip CMOS transceiver for DCS-1800 wireless communications," *Int'l. Solid-State Circuits Conf.*, San Francisco, CA, 1998, pp. 48–49, 411.

[5] F. Piazza and Q. Huang, "The Development of a 0.25μm CMOS Receiver for GSM," *URSI Int'l. Symp. Signals, Sys., and Elect.*, Pisa, Italy, 1998, pp. 64–67.

[6] D. K. Shaeffer and T. H. Lee, "A 1.5-V, 1.5-GHz CMOS Low Noise Amplifier," *IEEE J. Solid-State Circuits*, vol. 32, no. 5, 1997, pp. 745–59.

[7] Q. Huang, P. Orsatti, and F. Piazza, "Broadband, 0.25μm CMOS LNAs with Sub-2dB NF for GSM Applications," *Proc. Custom IC Conf.*, Santa Clara, CA, 1998, pp. 67–70.

[8] A. A. Abidi, "Direct-Conversion Radio Transceivers for Digital Communication," *IEEE J. Solid-State Circuits*, vol. 30, no. 12, 1995, pp. 1399–1410.

[9] J. Crols and M. Steyaert, "A 1.5GHz Highly Linear CMOS Downconversion Mixer," *IEEE J. Solid State Circuits*, vol. 30, no. 7, 1995, pp. 736–42.

[10] A. A. Abidi, "How Phase Noise Appears in Oscillators," *Analog Circuit Design-RF Analog-to-Digital Converters; Sensor and Actuator Interfaces; Low-Noise Oscillators, PLLs and Synthesizers*, R. van de Plassche, J. Huijsing, and W. Sansen, Eds. Boston: Kluwer, 1997.

[11] J. Lee *et al.*, "Design of Spiral Inductors on Silicon Substrates with a Fast Simulator," *Proc. Euro. Solid-State Circuits Conf.*, The Hague, The Netherlands, 1998, pp. 328–31.

[12] A. Kral, F. Behbahani, and A. A. Abidi, "RF-CMOS Oscillators with Switched Tuning," *Custom IC Conf.*, Santa Clara, CA, 1998, pp. 555–58.

[13] C. H. J. Mensink, B. Nauta, and H. Wallinga, "'Soft-Switched' Transconductor and its Application in Gain Control and Filters," *IEEE J. Solid-State Circuits*, vol. 32, no. 7, 1997, pp. 989–98.

[14] K.-C. Tsai and P. R. Gray, "A 1.9GHz 1W CMOS Class E Power Amplifier for Wireless Communications," *Euro. Solid-State Circuits Conf.*, The Hague, The Netherlands, 1998, pp. 76–79.

Analog-to-Digital Converters and Their Applications in Radio Receivers

Rapid advances in hardware development of analog-to-digital converters (ADCs) have paved the way for development of radio receivers using digitization at the IF, and in some cases, at the RF. The constraints placed on these receivers due to hardware limitations of these devices are discussed and some examples of high-speed, state-of-the-art ADCs are given.

Jeffery A. Wepman

As advances in technology provide increasingly faster and less expensive digital hardware, more of the traditional analog functions of a radio receiver will be replaced with software or digital hardware. The ultimate goal in radio receiver design is to directly digitize the RF signal at the output of the receive antenna and hence implement all receiver functions in either digital hardware or software. Trends in receiver design have evolved toward this goal by incorporating digitization closer and closer to the receive antenna for systems at increasingly higher frequencies and wider bandwidths. Applications for these receivers are expected to increase rapidly in areas such as mobile cellular, satellite, and personal communications services (PCS) systems.

The analog-to-digital converter (ADC) is a key component in these radio receivers. ADCs most often used for wideband digitization at the RF or IF are organized as shown in Fig. 1. Key parameters of ADCs are effected by specific ADC circuit elements. For example, accuracy and linearity are primarily determined by the sample-and-hold circuitry, while jitter in the sampling clock can introduce noise in the desired output of the ADC. The quantizer establishes the resolution of the ADC. For a given burst rate, the buffers limit the sustainable throughput.

This article discusses some of the key parameters for ADCs used in radio receiver applications. The requirements, practical limitations, and potential problems for ADCs are also discussed. Conversion methods of practical ADCs such as flash, successive approximation, sigma-delta, bandpass sigma-delta, and subranging are not presented here. A discussion of these techniques can be found in [1].

Certain commercial equipment, components, instruments, or materials are identified in this article to provide some examples of current technology. In no case does such identification imply recommendation or endorsement by the National Telecommunications and Information Administration, nor does it imply that the material or equipment identified is necessarily the best available for the purpose. Furthermore, examples of technology identified in this article are not intended to be all inclusive; they represent only a sampling of what is available.

JEFFERY A. WEPMAN is an electronics engineer with the Institute for Telecommunication Sciences (ITS), National Telecommunications and Information Administration (NTIA), U.S. Department of Commerce.

Sampling Methods and Analog Filtering

The sampling process is of critical importance in radio receivers using digitization at the RF or IF. The content of the resulting sampled signal waveform is highly dependent on the relationship between the sampling rate employed and the minimum and maximum frequency components of the analog input signal. Some common sampling techniques that utilize a uniform spacing between the samples include Nyquist sampling, oversampling, quadrature sampling, and bandpass sampling (also called downsampling or direct downconversion). Sampling techniques with nonuniform spacing between the samples do exist, but they are not widely used and therefore are not considered in this article.

Any time a continuous-time analog signal is uniformly sampled, the spectrum of the original signal $F(f)$ is repeated at integer multiples of the sampling frequency (i.e., $F(f)$ becomes periodic). This is an inherent effect of sampling and cannot be avoided. This phenomenon is shown graphically in Fig. 2. Figure 2a shows the spectrum of the original analog signal $F(f)$. Figure 2b shows the spectrum of the sampled signal $F_s(f)$ using a sampling rate of $f_s = 2f_{max}$.

Nyquist Sampling

The general sampling theorem for sampling a bandlimited analog signal (a signal having no frequency components above a certain frequency f_{max}) requires that the sampling rate be at least two times the highest frequency component of the analog signal $2f_{max}$. This ensures that the original signal can be reconstructed exactly from the samples. A sampling rate of two times the highest frequency component of the analog signal is

Reprinted from *IEEE Communications Magazine,* Vol. 33, No. 5, pp. 39–45, May 1995.

called the Nyquist sampling rate. Figure 2b shows an example of sampling a bandlimited signal with a maximum frequency of f_{max} at the Nyquist rate ($f_s = 2f_{max}$). Note that the copies of the spectrum of the analog signal $F(f)$ that are present in the spectrum of the sampled signal $F_s(f)$ do not overlap. As the sampling rate is increased beyond the Nyquist rate, the copies of the spectrum of the analog signal $F(f)$ that are present in the spectrum of the sampled signal $F_s(f)$ are spread even farther apart. This is shown in Fig. 2c. Sampling a bandlimited signal at rates equal to or greater than the Nyquist rate guarantees that spectrum overlap (often called aliasing) does not occur and that the original analog signal can be reconstructed exactly [2, 3].

Out-of-Band Energy

Two practical problems arise when sampling at the Nyquist rate: defining what a bandlimited signal truly is in a practical sense and analog filtering before the ADC stage. A theoretically defined bandlimited signal is a signal with no frequency components above a certain frequency. When considering real signals such as an RF signal at the input of a radio receiver, however, signals of all frequencies are always present. It is a matter of the amplitude of these frequencies that is important. In particular, the relative amplitude of the undesired signals to the desired signal is important. When digitizing an RF or IF signal at the Nyquist rate in a radio receiver, undesired signals (above one-half the sampling rate) of a sufficient amplitude can create spectrum overlap and distort the desired signal. This phenomenon is illustrated in Fig. 3. Figure 3a shows the spectrum of the analog input signal with its desired and undesired components. If this signal is sampled at two times the highest frequency in the desired signal f_d, the resulting spectrum of the sampled signal $F_s(f)$ is shown in Fig. 3b. Note that spectrum overlap has occurred here (i.e., the spectrum of the undesired signal occurs within the spectrum of the desired signal). This causes distortion in the reconstructed desired signal.

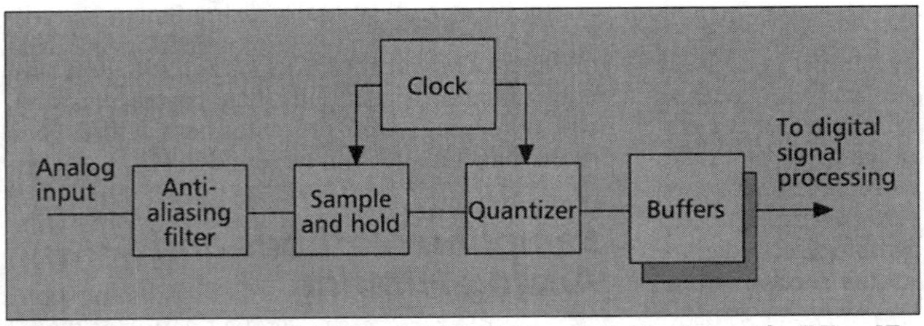

■ **Figure 1.** *Elements of ADCs used for wideband digitization at the RF or IF.*

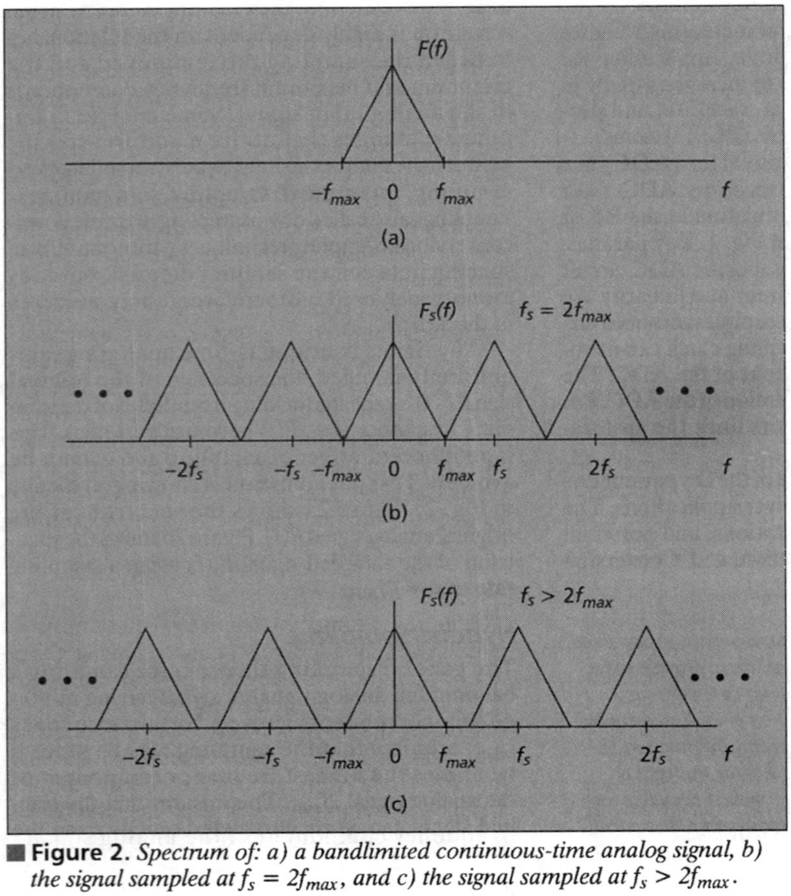

■ **Figure 2.** *Spectrum of: a) a bandlimited continuous-time analog signal, b) the signal sampled at $f_s = 2f_{max}$, and c) the signal sampled at $f_s > 2f_{max}$.*

This effect raises an important question: What is the relative amplitude of the signals occurring above one-half of the sampling rate at which distortion of the desired signal due to spectrum overlap begins to predominate the distortion due to ADC nonlinearities? Nonlinearities in the ADC cause spurious responses in the ADC output spectrum. Distortion due to spectrum overlap can be said to predominate distortion due to ADC nonlinearities when the undesired signals appearing in the Nyquist band (DC to one-half the sampling rate) due to spectrum overlap exceed the largest spurious response of the ADC due to nonlinearities. Therefore, undesired signals appearing in the Nyquist band due to spectrum overlap must be lower in power than the largest spurious response of the ADC. In other words, distortion of the desired signal is predominated by ADC nonlinearities (and not spectrum overlap) if signals higher in frequency than $f_s/2$ are lower in power than the largest spurious response of the ADC. This can be quite a stringent requirement. Depending upon the specific type of radio system, it is possible that this requirement can be relaxed.

To determine ways to "relax" this requirement, the following questions should be asked: How much distortion of the desired signal is tolerable? Does the bandwidth and frequency content of both the desired signal in the Nyquist band and the undesired signals above the Nyquist band effect the distortion of the desired signal? These questions are best answered by considering the details of the specific radio communication system, such as the type of source information (voice, data, video, etc.), desired signal bandwidth, modulation and coding techniques, undesired signal characteristics (bandwidth, power, and type of signal), and finally, the performance criterion used to evaluate the quality of the reception of the desired signal.

Realizable Anti-Aliasing Filters

Analog filtering before the ADC stage is intimately related to the definition of bandlimiting. Where the definition of bandlimiting deals with the content of the signals that may be present, analog filtering before the ADC represents a signal processing stage where certain frequencies can be attenuated. It is important to know both the signals that can be present before filtering and the amount of attenuation that the filter offers for different frequencies. With knowledge of both of these, the true spectrum of the signal to be digitized can be determined. Sampling at the Nyquist rate presents a large and often impractical demand on the filter used before digitization (anti-aliasing filter). Ideally, an anti-aliasing filter placed before an ADC would pass all of the desired frequencies up to some cutoff frequency and provide infinite attenuation for frequencies above the cutoff frequency. Then sampling at the Nyquist rate would be two times the cutoff frequency and no spectrum overlap would occur. Unfortunately, practically realizable filters cannot provide this type of "brickwall" response. The attenuation of real filters increases more gradually from the cutoff frequency to the stop band. Therefore, for a given cutoff frequency on a real filter, sampling at two times this cutoff frequency will produce some spectrum overlap. The steeper the transition from the passband to the stop band and the more attenuation in the stop band, the less the sampled signal will be distorted by spectrum overlap. In general, more complicated filters are required to achieve steeper transitions and higher attenuation in the stop band. Therefore, more complicated filters are required to reduce the distortion in the sampled signal due to spectrum overlap for a given sampling rate. Limitations on the practical implementation of analog filters make high-order, steep rolloff filters difficult to realize. Also, as the steepness of the rolloff is increased, the phase response tends to become more nonlinear. This can create distortion of the desired receive signal since different frequencies within a signal can be delayed in time by different amounts (i.e., phase shifted by amounts not proportional to frequency).

Oversampling Eases Requirements on the Anti-Aliasing Filter

Sampling at rates greater than the Nyquist sampling rate is called oversampling. One of the benefits of oversampling is that the copies of the spectrum of the analog signal $F(f)$ that are present in the spectrum of the sampled signal $F_s(f)$ become increasingly separated as the sampling rate is increased beyond the Nyquist rate. For an analog signal with a given frequency content and a given anti-aliasing filter with a cutoff frequency of f_c, sampling at the Nyquist rate (two times the cutoff frequency) produces a certain amount of distortion due to spectrum overlap. When sampling at a higher rate, a simpler anti-aliasing filter with a more gradual transition from passband to stop band and less stop band attenuation can be used without any increase in the distortion due to spectrum overlap. Therefore, oversampling can minimize the requirements of the anti-aliasing filter. The tradeoff, of course, is that increasingly faster ADCs are required to digitize relatively low frequency signals.

Quadrature Sampling Reduces Required Sampling Rate

In quadrature sampling the signal to be digitized is split into two signals. One of these signals is multiplied by a sinusoid to downconvert it to a zero center frequency and form the in-phase component of the original signal. The other signal is multiplied by a 90-degree phase-shifted sinusoid to downconvert it to a zero center frequency and form the quadrature-phase component of the original signal. Each of these components occupies only one-half of the bandwidth of the original signal and can be sampled at one-half the sampling rate required for the original signal. Therefore, quadrature sampling reduces the required sampling rate by a factor of two at the expense of using two phase-locked ADCs instead of one.

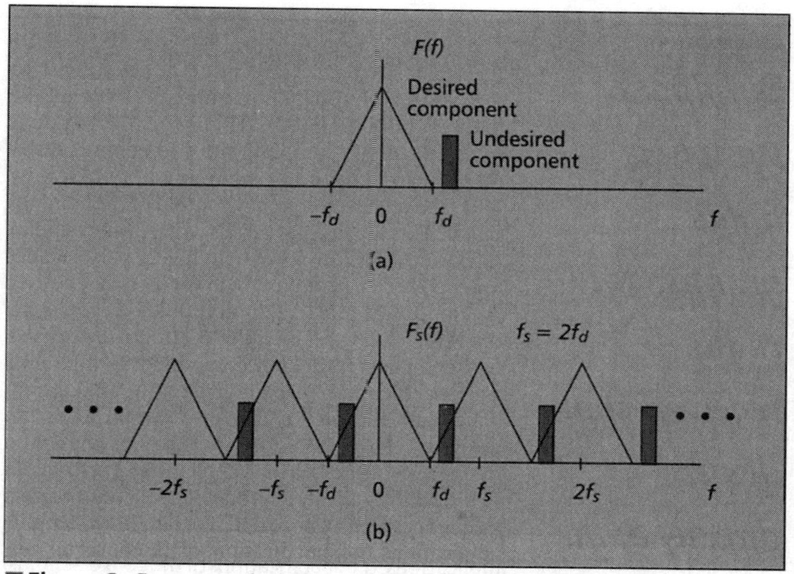

Figure 3. *Spectrum of: a) a continuous-time analog signal with a desired and undesired component and b) the signal sampled at* $f_s = 2f_d$.

Bandpass Sampling for Direct Downconversion

Sampling at rates lower than $2f_{max}$ can still allow for an exact reconstruction of the information content of the analog signal if the signal is a bandpass signal. An ideal bandpass signal has no frequency components below a certain frequency f_l and above a certain frequency f_h. For a bandpass signal, the minimum requirements on the sampling rate to allow for exact reconstruction are that the sampling rate be at least two times the *bandwidth* $f_h - f_l$ of the signal. To ensure that spectrum overlap does not occur, when sampling rates are between two times the bandwidth of the bandpass signal and two times the highest frequency in the bandpass signal, the sampling frequency f_s must satisfy

$$\frac{2f_h}{k} \le f_s \le \frac{2f_l}{(k-1)}$$

where k is restricted to integer values that satisfy

$$2 \le k \le \frac{f_h}{(f_h - f_l)}$$

and $(f_h - f_l) \le f_l$ [3].

Bandpass sampling can be used to downconvert a signal from a bandpass signal at an RF or IF to a bandpass signal at a lower IF. Since the bandpass signal is repeated at integer multiples of the sampling frequency, selecting the appropriate spectral replica of the original bandpass signal provides the downconversion function.

Bandpass sampling holds promise for radio receivers that digitize directly at the RF or IF, since the desired input signals to radio receivers are normally bandpass

Bandpass sampling holds promise for radio receivers that digitize directly at the RF or IF, since the desired input signals to radio receivers are normally bandpass signals. Theoretically, bandpass sampling allows sampling rates to be much lower than those required by sampling at two or more times the highest frequency content of the bandpass signal. This means that ADCs with slower sampling rates (and therefore potentially higher performance, lower power consumption, or lower cost) may be used. An important practical limitation, however, is that the ADC must still be able to effectively operate on the highest frequency component in the signal. This specification is usually given as the analog input bandwidth for the ADC. Conventional ADCs are designed to operate on signals having maximum frequencies of one-half the sampling rate. Performance of the ADC typically degrades with increasing input frequency. When using ADCs for bandpass sampling applications, the specifications of the converter must be examined to determine the behavior at higher frequency inputs. In addition, when bandpass sampling, stringent requirements on analog bandpass filters (steep rolloffs) are needed to prevent distortion of the desired signal from strong adjacent channel signals.

Effects of Quantization Noise, Distortion, and Receiver Noise

This section addresses the relationships among quantization noise, harmonic distortion, and receiver noise. The ADCs best suited to RF and IF processing that have widespread availability use uniform quantization. In uniform quantization, the voltage difference between each quantization level is the same. Other methods of quantization include logarithmic (A-law and μ-law), adaptive, and differential quantization. These methods are currently used in source coding. A discussion of these quantization techniques may be found in [1].

In uniform quantization, the analog signal cannot be represented exactly with only a finite number of discrete amplitude levels. Therefore, some error is introduced into the quantized signal. The error signal is the difference between the analog signal and the quantized signal. Statistically, the error signal is assumed to be uniformly distributed within a quantization level. Using this assumption, the mean squared quantization noise power P_{qn} is

$$P_{qn} = \frac{q^2}{12R}$$

where q is the quantization step size and R is the input resistance of the ADC [4]. In an ideal ADC, this representation of the quantization noise power is accurate to within a dB for input signals that are not correlated with the sampling clock.

If, on the other hand, the analog input into an ADC is periodic, the error signal is also periodic. This periodic error signal includes harmonics of the analog input signal and results in harmonic distortion. Furthermore, harmonics that fall above the Nyquist frequency appear in the Nyquist band due to aliasing. Dithering is commonly used to reduce this harmonic distortion. In one implementation of this technique, thermal noise is added at the ADC input to provide a relatively flat noise power spectrum over the Nyquist bandwidth; the thermal noise plus the quantization noise combine so that the quantization error is uniform. A simple way to implement this technique is to provide amplifier gain to boost the receiver noise to, or several dB above, the level of the quantization noise. The implications of this are examined next.

Commercially available ADCs typically have a full scale range (FSR) of 1 to 20 V. The FSR of the ADC is the difference between the maximum and the minimum analog input voltages to the ADC. Dividing the FSR by the number of quantization levels 2^B, where B is the number of bits of the ADC, provides the quantization step size q. For an 8 b ADC with a FSR of 2.5 V, the quantization step size is 9.77 mV. To compute the quantization noise power, the effective input resistance of the ADC must be known.

Components in radio receivers typically have a 50-ohm input and output impedance. The input impedance of ADCs is usually higher than this and is not well specified. Therefore, when interfacing an RF component with an ADC, as is necessary for digitization at the RF or IF, this impedance mismatch must be considered. A simple method of impedance matching is to place a 50-ohm resistive load at the input of the ADC. This forces the effective input resistance of the ADC to be close to 50 ohms. The quantization noise power can then be computed. Assuming a 50-ohm effective input resistance R to the ADC in this example, the quantization noise power equals –38 dBm. For a noise-limited receiver, the receiver noise power P_{rn} can be computed as the thermal noise power in the given receiver bandwidth (BW) plus the receiver noise figure (NF). This is given as

$$P_{rn} = -174dBm + 10\log_{10}\text{BW}(Hz) + \text{NF}\,(dB).$$

For a receiver with a 10-MHz BW and a 6-dB NF, the receiver noise power is –98 dBm. Therefore, to boost the receiver noise to the quantization noise power level requires a gain of 60 dB. For an ADC of higher resolution, less gain would be needed, since the quantization noise power would be smaller. Also, wider receiver bandwidths and higher receiver noise figures would require less gain since the receiver noise power would be larger. Nevertheless, for most practical receiver and ADC combinations, automatic gain control is necessary to assure that the least significant bit (LSB) represents a uniform noise input while the peak power does not exceed the ADC's FSR.

Important Specifications

In this section, theoretical signal-to-noise ratio (SNR) due to quantization noise and aperture jitter is discussed. Practical specifications for real ADCs are then presented.

Theoretical Signal-to-Noise Ratio Specifications

For radio receiver applications where the amplitude of the desired signal falls within the ADC's FSR, and the bandwidth of the desired signal is equal to the Nyquist bandwidth, the SNR of an ADC is a useful specification. The theoretical SNR of ADCs is generally thought of as $6B$ (dB), where B

is the number of bits of resolution of the ADC. A more precise expression providing the maximum possible theoretical SNR can be derived based on some assumptions about the noise and the input signal. First, it is assumed that the noise present is due to quantization error only. The amplitude of this quantization noise is assumed to be a random variable uniformly distributed over one quantization step. Assuming a sinusoidal input with an amplitude equal to the FSR of the ADC, the maximum possible theoretical SNR is given as

$$\mathrm{SNR} = 6.02B + 1.76 + 10\log_{10}\left(\frac{f_s}{2f_{\max}}\right)\ (dB)$$

where f_s is the sampling frequency and $f_{\max}$ is the maximum frequency of the input analog signal [5, 6]. The commonly stated theoretical SNR of $6B$ (dB) is an approximation to this equation when $f_s = 2f_{\max}$ and the 1.76 dB is neglected. From this equation, note that as the sampling frequency is increased beyond the Nyquist rate of $2f_{\max}$, the SNR increases. This occurs because the quantization noise power, which is fixed and independent of bandwidth, is spread out over an increasingly wider band as the sampling frequency is increased. This lessens the amount of the quantization noise that falls within the Nyquist band (DC to $f_{\max}$). Consequently, this means that oversampling increases the maximum possible SNR. Such oversampling is sometimes used to realize a greater maximum SNR than at first appears possible. An 8 b ADC, with a sampling rate of 20 Msamples/s, for example, can provide 68 dB rather than 48 dB of maximum SNR for 100 kHz signals in the passband if appropriate digital filtering is used to recover the 100 kHz signal.

Besides being limited by the quantization step size (resolution), the SNR of the ADC is also limited by aperture jitter. Aperture jitter is the variation in time of the exact sampling instant. Aperture jitter can be caused externally by jitter in the sampling clock, or internally since the sampling switch does not open at precise times. Aperture jitter causes a phase modulation of the sampled signal and thus results in an additional noise component in the sampled signal [7]. The maximum analog input frequency of the ADC is limited by this aperture jitter since the SNR due to aperture jitter (SNR_{aj}) degrades as the input frequency increases. The SNR_{aj} is given as

$$\mathrm{SNR}_{aj} = 20\log_{10}\left(\frac{1}{2\pi f_{\max}\, t_a}\right)$$

where t_a is the aperture jitter of the ADC [5]. For sampling at the Nyquist rate, where $f_s = 2f_{\max}$, both the SNR due to quantization noise and the SNR due to aperture jitter can be combined to give the overall SNR [8]. Distortion of the ADC output signal can also be introduced because each sample is taken over a finite time period in real sample and hold circuits. This distortion occurs as the input signal varies over this time period.

Practical Specifications for Real ADCs

The SNR in a real ADC can be determined by measuring the residual error. Residual error is the combination of quantization noise, random noise, and nonlinear distortion (i.e., all of the

Specification	Application	Definition
Signal-to-noise ratio (SNR)	Desired signal BW equal to Nyquist BW	MS signal power / MS power of residual error
Spurious free dynamic range (SFDR)	Desired signal BW less than Nyquist BW	MS signal Power / MS power of the largest spurious product
Noise power ratio (NPR)	Desired signal spectrum contains many narrowband channels	Power spectral density of noise outside frequency band of notch filter[1] / Power spectral density of noise inside frequency band of notch filter
Full-power analog input BW	Bandpass sampling	Range from frequency where output amplitude falls to 3 dB less than maximum[2]

[1] With an input signal having a bandlimited, flat noise spectrum and a narrow band of frequencies removed by a notch filter.

[2] For a full-scale input signal.

Table 1. *Summary of ADC specifications for radio receiver applications.*

undesired components of the output signal from the ADC). The residual error for an ADC is found by using a sinusoidal input into the ADC. An estimate of the input signal is subtracted from the output of the ADC and the remaining signal is the residual error. The mean squared power of the residual error is then computed. The SNR is then found by dividing the mean squared (MS) power of the input signal by the mean squared power of the residual error.[1]

A specification sometimes used for real ADCs instead of the SNR is the effective number of bits. This specification is defined as *the number of bits required in an ideal ADC so that the mean squared noise power in the ideal ADC equals the mean squared power of the residual error in the real ADC.*

One definition of the spurious free dynamic range (SFDR) assumes a sinusoidal input to the ADC. In this case, the SFDR is the ratio of the sinusoidal signal power to the peak power of the largest spurious signal in the ADC output spectrum. SFDR allows one to assess how well an ADC can simultaneously detect a very small signal in the presence of a very large signal. Hence, it is an important specification for ADCs used in radio receiver applications. A common misconception is that the SFDR of the ADC is equivalent to the SNR of the ADC. In fact, there is typically a large difference between the SFDR and the SNR of an ADC. The SNR is the ratio between the signal power and the power of the residual error. The SFDR, however, is the ratio between the signal power and the peak power of only the largest spurious product. Since the power of the residual error includes quantization noise, random noise, and nonlinear distortion within the entire Nyquist band, the power of the residual error can be much higher than the peak power of the largest spurious product. Hence, the SFDR can be much larger than the SNR [4].

The SFDR specification is useful for applications when the desired signal bandwidth is smaller than the Nyquist bandwidth. In this case, a wide band of frequencies is digitized and results in a

[1] *The SNR is often (and more accurately) called the Signal-to-Noise Plus Distortion Ratio when distortion is included with the noise, as in this case.*

Resolution (number of bits)	Sampling rate (Msamples/s)	Manufacturer
6	4000	Rockwell International
8	750	Signal Processing Technology
8	2000	Hewlett-Packard
8	3000	*
10	70	Pentek
12	50	Hughes Aircraft
12	100	*
14	24	Hughes Aircraft
18	10	Hewlett-Packard

* Device in development. Work is being sponsored by the Advanced Research Projects Agency (ARPA) of the U.S. Department of Defense.

■ **Table 2.** *Examples of current high-speed ADC technology.*

given SNR. The desired signal is then obtained by using a narrowband digital bandpass filter on this entire band of frequencies. The SNR is improved by this digital filtering process since the power of the residual error is decreased by filtering. The SFDR specification for the ADC becomes very important since a spurious component may still fall within the bandwidth of the digital filter, and hence the SFDR, unlike the SNR, does not necessarily improve by the digital filtering process. However, several techniques are available to improve the SFDR. Dithering (discussed previously) improves the SFDR of ADCs. Additionally, post-digitization processing techniques such as phase-plane compensation [9], state variable compensation [10], and projection filtering [11] have been used to improve SFDR.

For an ideal ADC, the maximum SFDR occurs at a full-scale input level. In practical ADCs, the maximum SFDR occurs at input levels at least several dB below the full-scale input level. This occurs because as the input levels approach full-scale, the response of the ADC becomes more nonlinear and more distortion is exhibited. Additionally, due to random fluctuations in the amplitude of real input signals, as the input signal level approaches the FSR of the ADC, the probability of the signal amplitude exceeding the FSR increases. This causes additional distortion from clipping. Therefore, it is extremely important to avoid input signal levels that closely approach the full-scale level in ADCs. Prediction of the SFDR for practical ADCs is difficult, therefore measurements are usually required to characterize the SFDR of the ADC.

In the preceding discussion on SFDR, a sinusoidal ADC input signal was assumed. However, intermodulation distortion (IMD) due to multi-tone inputs is important in ADCs used for wideband radio receiver applications. To characterize this IMD due to multi-tone inputs, another definition of the SFDR could be used. In this case, the SFDR is the ratio of the combined signal power of all of the multi-tone inputs to the peak power of the largest spurious signal in the ADC output spectrum. A current example of test equipment to generate multi-tone inputs produces up to 48 tones.

The noise power ratio (NPR) specification is useful in applications such as mobile cellular radio, where the spectrum of a signal to be digitized consists of many narrowband channels and where adjacent channel interference can degrade system performance. Particularly, the NPR provides information on how well an ADC can limit crosstalk between channels [9].

The NPR is measured by using a noise input signal into the ADC. This noise signal has a flat spectrum that is bandlimited to a frequency that is less than one-half the sampling frequency. Additionally, a narrow band of frequencies is removed from the noise signal using a notch filter. This noise spectrum is used as the input signal to the ADC. The frequency spectrum of the output of the ADC is then determined. The NPR is then computed by dividing the power spectral density of the noise outside the frequency band of the notch filter by the power spectral density of the noise inside the frequency band of the notch filter [4].

When using an ADC in a bandpass sampling application where the maximum input frequency into the ADC is actually higher than one-half the sampling frequency, the full-power analog input bandwidth is an important specification. A common definition of full-power analog input bandwidth (although not universal) is the range from DC to the frequency where the amplitude of the output of the ADC falls to 3 dB below the maximum output level. This assumes a full-scale input signal to the ADC. Typically, the ADC is operated at input frequencies below this bandwidth. Aside from full-power analog input bandwidth, it is important to examine the behavior of the other specifications such as SNR, SFDR, and NPR at the desired operating frequencies since these specifications may vary with frequency. Table 1 provides a summary of the important ADC specifications for radio receiver applications.

ADC Applications Tradeoffs

The performance of ADCs continues to improve at a rapid rate. For radio receiver applications using digitization at the RF or IF, ADCs with both high sampling rates and high performance are desired. Unfortunately, there is a tradeoff between these two requirements. As a general trend, although not always true, the higher the performance of the ADC, the lower its maximum sampling rate will be. Table 2 shows some examples of current high-speed ADC technology for various ADC resolutions.

When selecting an ADC for a specific radio receiver application, in addition to the sampling rate, one must consider critical specifications that characterize the ADC performance such as the SNR, SFDR, and NPR. The ADC specifications most important for various applications are listed in Table 3. In certain applications such as channelized PCS, the Universal Mobile Telecommunication System (UMTS), the Future Public Land Mobile Telecommunication System (FPLMTS), and mobile cellular systems, instead of digitizing the entire band with a single high-speed ADC, parallel ADCs used to digitize narrower bandwidths are often practical ADC architectures. In this case, ADCs with better performance can be used since the demands of a high sampling rate are relieved.

Summary

This article provides an introduction to some of the important factors that must be considered when using ADCs in radio receiver applications. These factors include the choice of sampling method, the amount and effects of out-of-band energy, the analog filtering required, the effects of quantization noise, receiver noise, and distortion, and the critical ADC specifications for radio receiver applications. The differences between Nyquist sampling, oversampling, and bandpass sampling were discussed. It was shown that sampling at the Nyquist rate presents a large and often impractical demand on the anti-aliasing filter. Oversampling eases the requirements on the anti-aliasing filter. A very steep rolloff bandpass filter is required for bandpass sampling when there are strong signals present in adjacent channels. Quadrature sampling was shown to reduce the required sampling rate by a factor of two at the expense of using two phase-locked ADCs instead of one. It was also shown that, in general, some sort of gain control is required for proper digitization of signals at the RF or IF in a receiver. ADC specifications of particular importance to radio receiver applications such as SNR, SFDR, full-power bandwidth, and NPR were examined and some examples of high-speed, state-of-the-art ADCs were given. As ADC performance continues to improve, digitization at the RF and IF in radio receivers, at increasingly higher frequencies, will be used in an increasingly broad range of applications.

Acknowledgments

The author would like to thank Daniel Moulin of The MITRE Corporation for providing the table on critical ADC specifications and performance issues for typical applications.

<table>
<tr><th>Typical applications</th><th>Critical ADC specifications</th><th>Performance issues</th></tr>
<tr><td>Spread spectrum</td><td>SNR
SFDR
NPR</td><td>SNR for quantization of small signals in an environment with strong interference
SFDR for spatial filtering
NPR for interchannel crosstalk</td></tr>
<tr><td>Wideband digital receivers</td><td>SFDR</td><td>SFDR for accurate detection of low-level signals in an environment with strong interference</td></tr>
<tr><td>Radar</td><td>SNR
SFDR
Overvoltage recovery</td><td>SNR for clutter cancellation
SFDR for Doppler processing</td></tr>
<tr><td>Cellular mobile and PCS</td><td>SNR
SFDR
NPR</td><td>SNR and SFDR for wide bandwidth channelized receivers
NPR for interchannel crosstalk</td></tr>
<tr><td>Spectrum analysis</td><td>SNR
SFDR</td><td>SNR and SFDR for high fidelity measurements</td></tr>
<tr><td>Digital sampling oscilloscopes</td><td>SNR
DNL*</td><td>SNR for better amplitude resolution
DNL for accurate representation of waveform</td></tr>
</table>

* Differential nonlinearity (DNL) is the maximum amount of deviation of any quantization step in the ADC from the theoretical quantization step size of $FSR/2^B$.

Table 3. *Critical ADC specifications and performance issues for typical applications.*

References

[1] J. A. Wepman, J. R. Hoffman, and J. E. Schroeder , "An initial study of RF and IF digitization in radio receivers," NTIA Report 95-xxx (in preparation), 1995.

[2] F. G. Stremler, *Introduction to Communication Systems*, (Reading, MA: Addison-Wesley, Inc., 1977), pp. 112-120.

[3] E. O. Brigham, *The Fast Fourier Transform and its Applications*, (Englewood Cliffs, NJ: Prentice Hall, Inc., 1988), pp. 83-86 and 320-337.

[4] D. Asta, "Recent dynamic range characterization of analog-to-digital converters for spectral analysis applications," Massachusetts Institute of Technology, Lincoln Laboratory, Lexington, MA, Project Report AST-14, July, 1991.

[5] R. M. Lober, "A DSP-based approach to HF receiver design: Higher performance at a lower cost," *RF Design*, vol. 16, no. 8, 1993, pp. 92-100.

[6] M. Amarandos and S. Andrezyk, "Considerations in the development of a low cost, high performance receiver based on DSP techniques," *DSP Applications*, vol. 2, no. 12, Dec., 1993, pp. 1- 14.

[7] R. Groshong and S. Ruscak, "Undersampling techniques simplify digital radio," *Electronic Design*, vol. 39, no. 10, May, 1991, pp. 67-78.

[8] Harris Semiconductor Corporation, *Digital Signal Processing Databook*. (Melbourne, FL: Harris Semiconductor Corporation, 1994), pp. 6-3, 8-7.

[9] N. W. Spencer, "Comparison of state-of-the-art analog-to-digital converters," Massachusetts Institute of Technology, Lincoln Laboratory, Lexington, MA, Project Report AST-4, March, 1988.

[10] F. H. Irons and T. A. Rebold, "Characterization of high-frequency analog-to-digital converters for spectral analysis applications," Massachusetts Institute of Technology, Lincoln Laboratory, Lexington, MA, Project Report AST-2, Nov., 1986.

[11] N. T. Thao and M. Vetterli, "Optimal MSE signal reconstruction in oversampled A/D conversion using convexity," *Proc. ICASSP '92*, 1992, vol. 4, pp. 165-168.

Analog-to-Digital Converter Survey and Analysis

Robert H. Walden, *Member, IEEE*

***Abstract*—Analog-to-digital converters (ADC's) are ubiquitous, critical components of software radio and other signal processing systems. This paper surveys the state-of-the-art of ADC's, including experimental converters and commercially available parts. The distribution of resolution versus sampling rate provides insight into ADC performance limitations. At sampling rates below 2 million samples per second (Ms/s), resolution appears to be limited by thermal noise. At sampling rates ranging from ~2 Ms/s to ~4 giga samples per second (Gs/s), resolution falls off by ~1 bit for every doubling of the sampling rate. This behavior may be attributed to uncertainty in the sampling instant due to aperture jitter. For ADC's operating at multi-Gs/s rates, the speed of the device technology is also a limiting factor due to comparator ambiguity. Many ADC architectures and integrated circuit technologies have been proposed and implemented to push back these limits. The recent trend toward single-chip ADC's brings lower power dissipation. However, technological progress as measured by the product of the ADC resolution (bits) times the sampling rate is slow. Average improvement is only ~1.5 bits for any given sampling frequency over the last six–eight years.**

***Index Terms*—Analog-to-digital converters, aperture jitter, comparator ambiguity, input-referred noise, signal-to-noise ratio, spurious-free dynamic range.**

I. Introduction

During the past two decades, the rapid evolution of digital integrated circuit technologies has led to ever more sophisticated signal processing systems. These systems operate on a wide variety of continuous-time signals including speech, medical imaging, sonar, radar, electronic warfare, instrumentation, consumer electronics, and telecommunications (terrestrial and satellite). One of the keys to the success of these systems has been the advance in analog-to-digital converters (ADC's) which convert the continuous-time signals to discrete-time, binary-coded form. As an example, in the telecommunications arena, advances in software radio development [1]–[3] have provided impetus for ADC performance improvements, especially for sampling rates of approximately 100 million samples per second (Ms/s). More generally, the large number of signal types to be digitized has led to a diverse selection of data converters in terms of architectures, resolution, and sampling rates.

Despite the variety in ADC's, their performances can be summarized by a relatively small number of parameters: stated resolution (number of bits per sample), signal-to-noise ratio (SNR), spurious-free dynamic range (SFDR), and power dissipation P_{diss} [4]. Two-tone intermodulation distortion is also important for ADC's to be used in receivers. Fig. 1 shows ADC resolution, as stated by the manufacturer versus sampling rate, f_{samp}. Over 150 converters (listed in Appendix I), including experimental systems and commercially available parts, are represented in the graph. Strictly speaking, this data does not represent measured performance. It does, however, show two important features. First, approximately one bit of resolution is lost for every doubling of the sampling rate. This is indicated by the state-of-the-art line on the graph. Second, the highest Nyquist sampling rate attained is 8 giga samples per second (Gs/s) [5]. An analysis of SNR shows that the 1-bit per octave slope is related to the sample-to-sample variation of the instant in time at which sampling occurs. This variation is called aperture jitter or aperture uncertainty.[1] In addition, the speed of sampling is limited by the ability of the comparator(s) to make an unambiguous decision regarding the relative amplitude of the input voltage due to comparator ambiguity. This is related to the speed of the device technology used to fabricate the ADC. Device speed is measured as the frequency f_T, at which there is unity current gain.

Section II of this paper discusses how ADC's are evaluated, then Section III deals with the performancae limitations in more detail. ADC architectures that are presently under investigation are presented in Sections IV and V. Performance trends are discussed in Section VI.

II. ADC Characterization

There are a number of ways to measure and compare ADC performance. This paper focuses on determining the resolution in bits for a given sampling rate. In an increasing number of applications, the power consumption is also important. Resolution can be determined both quasistatically and dynamically. Quasi-static measures include differential nonlinearity (DNL) and integral nonlinearity (INL). Dynamic measures include SNR, SFDR, and noise power ratio (NPR). These quantities are determined from spectral analysis, usually in the form of a fast Fourier transform (FFT) of a sequence of ADC output samples. This study focuses on SNR and SFDR because dynamic performance is most important for high-speed applications and SNR and SFDR provide a more accurate measure of ADC performance than the stated-number-of-bits. In addition, SNR and SFDR are universally accepted performance measures.

SNR is the ratio of the root-mean-square (rms) signal amplitude to the square-root of the integral of the noise power

Manuscript received November 17, 1997; revised June 3, 1998 and August 9, 1998.

The author is with HRL Laboratories LLC, Malibu, CA 90265 USA (e-mail: walden@hrl.com).

Publisher Item Identifier S 0733-8716(99)02983-2.

[1] In this paper, the terms aperture jitter and aperture uncertainty are synonymous. Another term, aperture time, relates to the fact that sampling is, in fact, not instantaneous, but is actually the result of a weighted averaging of the sample over a period of time. This effect does not limit SNR to the degree that aperture jitter does, however variations in aperture time can be thought of as being included in the jitter effect.

Reprinted from *IEEE Journal on Selected Areas of Communications,* Vol. 17, No. 4, pp. 539–550, April 1999.

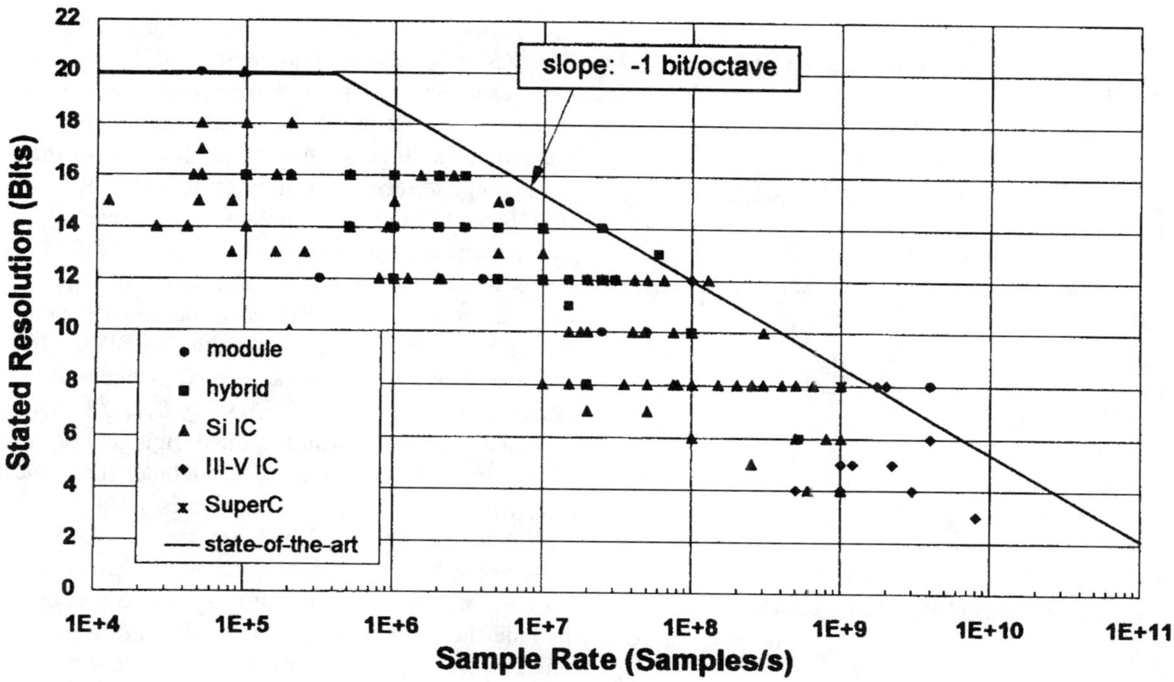

Fig. 1. Survey of ADC's.

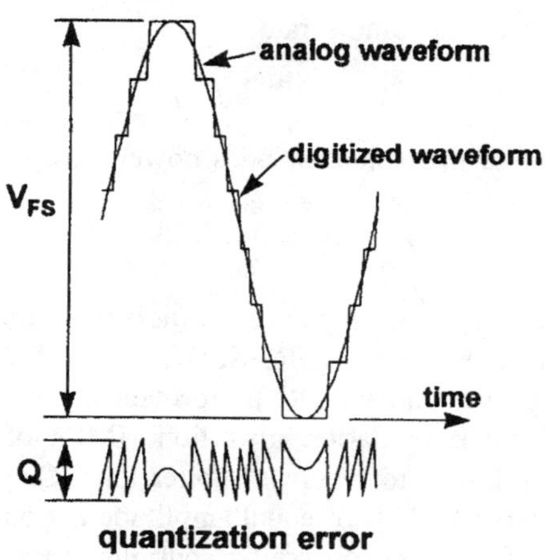

Fig. 2. Example of quantization error. V_{FS} is the full-scale voltage range, and Q is the size of the LSB.

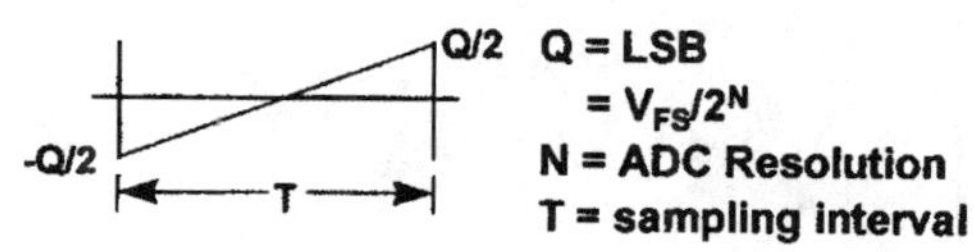

$$e(t) = Q \cdot \left(\frac{t}{T} - \frac{1}{2}\right)$$

$$NP_0(rms) = \sqrt{\frac{1}{T}\int_0^T [Q \cdot (\frac{t}{T} - \frac{1}{2})]^2 dt} = \frac{Q}{\sqrt{12}}$$

$$SNR(dB) = 20 \cdot \log_{10}\left(\frac{V_{FS}(rms)}{NP_0(rms)}\right)$$

$$= 20 \cdot \log_{10}\left(\frac{V_{FS}}{2\sqrt{2}} \frac{1}{\frac{V_{FS}}{2^N\sqrt{12}}}\right) = 6.02 \cdot N + 1.76$$

Fig. 3. Random approximation for quantization error. All errors within the range $\pm Q/2$ are equally likely. The resulting SNR is linear in the number of bits of resolution N.

spectrum over the frequency band of interest. For a Nyquist converter the frequency band of interest ranges from 0 to $f_{\text{samp}}/2$ Hz. The noise spectrum contains contributions from all the error mechanisms present. These include quantization noise, circuit noise, aperture uncertainty, and comparator ambiguity.

The only error mechanism present in an ideal ADC is quantization. This error arises because the analog input signal may assume any value within the input range of the ADC while the output data is a sequence of finite precision samples [6]. The example of Fig. 2 compares a sinusoidal waveform and its (reconstructed) digitized representation. The difference is the quantization error. Q is the size of the elementary quantization step, which is the least significant bit (LSB) of a binary representation of that value. In this case, the quantization error waveform and the analog waveform are strongly correlated. In a more typical case, the analog input contains frequency content due to the simultaneity of complicated signals and noise. In this situation, the quantization error is approximately random. The common white noise approximation is to assume that the probabilities of quantization errors are equal. This random error process is described in Fig. 3 with the equations of SNR due solely to quantization noise. T, the sampling interval, equals $1/f_{\text{samp}}$. N is the resolution of the converter in bits. The SNR (in dB) of an ideal ADC is shown in the lower portion of the figure ($\text{SNR} = 6.02N + 1.76$). It can be improved only by increasing N.

In physical ADC devices, additional error mechanisms are present. Some of these other errors may also be characterized as white noise with the same expression for SNR as in Fig. 3, except that N represents N_{eff}, an effective number of bits. The notation SNR-bits refers to N_{eff}. SNR-bits is given by

$$\text{SNR bits} = (\text{SNR(dB)} - 1.76)/6.02. \tag{1}$$

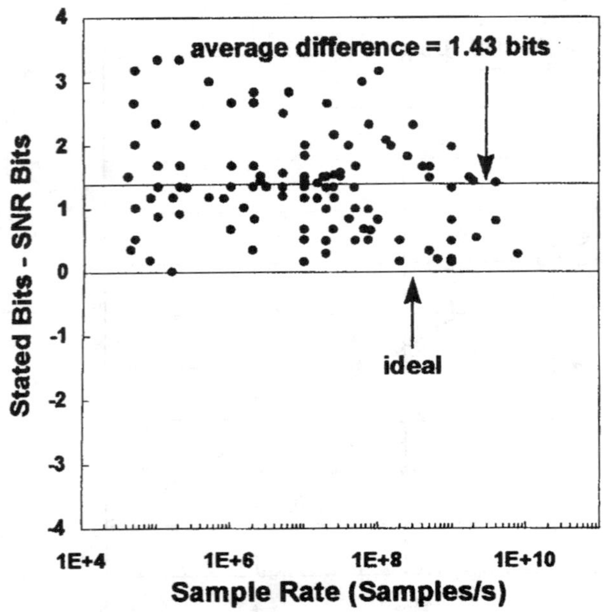

(a)

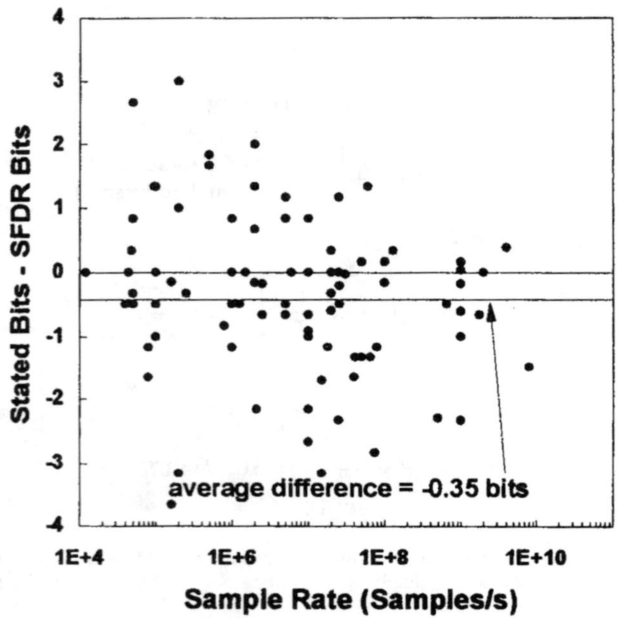

(b)

Fig. 4. Comparisons of stated bits (number of output leads) with (a) SNR-bits and with (b) SFDR-bits.

The difference between stated resolution and SNR bits for a given ADC indicates the degradation in SNR due to all other error sources. Fig. 4(a) exhibits this difference with a degradation of approximately 1.5 bits for a given sampling rate, with scatter in the data.

The effective number of bits associated with SFDR is

$$\text{SFDR bits} = \text{SFDR(dBc)}/6.02. \tag{2}$$

SFDR is the ratio of the single-tone signal amplitude to the largest nonsignal component within the spectrum of interest. Fig. 4(b) shows the difference between stated resolution and SFDR-bits. Although the average difference is less than .5 LSB, there is more scatter in this plot than for the SNR data of Fig. 4(a). There are many reasons for such a wide variation. The design emphasis may render SNR more important in some cases and SFDR more important in others. Other factors include how well the design overcomes noise, aperture jitter, comparator ambiguity, and the nonlinearity of the transistors.

A complete characterization of an ADC includes the values of SNR and SFDR as a function of frequency f_{sig}, with f_{samp} as a parameter. For low values of f_{sig}, the SNR is constant. It decreases as f_{sig} increases. The value of f_{sig} at which the SNR decreases to 3 dB below the low-frequency value is the effective resolution bandwidth (ERBW). This important characteristic implies the range of frequencies over which the converter may be used. If $\text{ERBW} \geq f_{\text{samp}}/2$, then the ADC is a Nyquist converter, which is the design goal of many ADC's. The characterization of an ADC includes the highest value of f_{samp} for which Nyquist operation is sustained.

Not all widely published reports on ADC's include the conditions for Nyquist conversion. However, some still achieve noteworthy sampling speed, SNR or SFDR. To include these, the criterion for inclusion in this study is that $\text{ERBW} > \sim f_{\text{samp}}/4$. Furthermore, the low-frequency values of SNR and SFDR are used.

P, a universal measure of ADC performance, is the product of the effective number of quantization levels, $ql_{\text{eff}} = 2^{\text{SNRbits}}$, times the sample rate

$$P = 2^{\text{SNRbits}} f_{\text{samp}}. \tag{3}$$

F, a figure of merit that includes power dissipation [4], is

$$F = \frac{2^{\text{SNRbits}} f_{\text{samp}}}{P_{\text{diss}}}. \tag{4}$$

This figure of merit emphasizes efficiency with respect to dissipated power, P_{diss}. SNR, SFDR, P, and F are used subsequently to quantify ADC performance.

Two-tone intermodulation distortion (IMD) of ADC's is particularly relevant to receiver applications. One excites an ADC with two sinusoids of equal amplitude but with different frequencies, f_1 and f_2, observing spurious tones in the FFT spectrum of the ADC output. The strongest such tone is usually either second- $(\pm f_1 \pm f_2)$ or third-order $(\pm f_1 \pm 2f_2$ or $\pm 2f_1 \pm f_2)$. Unfortunately, IMD data reported in the literature is minimal. In addition, there is no standard set of conditions for IMD evaluation, making comparisons between ADC's more difficult. Hence, IMD's must be evaluated by the prospective user for the intended application.

III. Performance Analysis

For a better understanding of ADC performance limits, it is helpful to plot the effective resolution as determined from SFDR and SNR. Fig. 5 shows the reported SFDR (where available). Comparing Figs. 1 and 5 indicates that the effective resolution expressed as SFDR-bits is roughly the same as the stated resolution for the population is taken as a whole.

This is somewhat misleading because the difference Stated-bits minus SFDR-bits [see Fig. 4(b)], for a given converter is $\sim \pm 3$ bits, a wide variation.[2]

[2]The terms stated bits and stated resolution are synonymous.

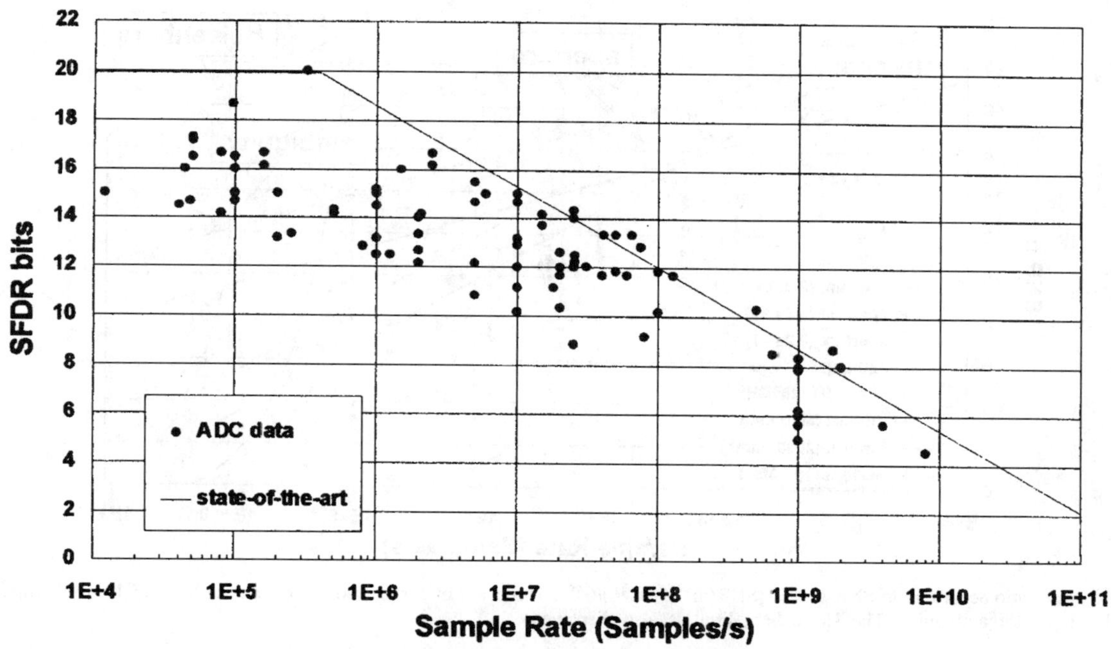

Fig. 5. Spur-free dynamic range expressed as effective number of bits according to SFDR-bits = SFDR(dBc)/6.02.

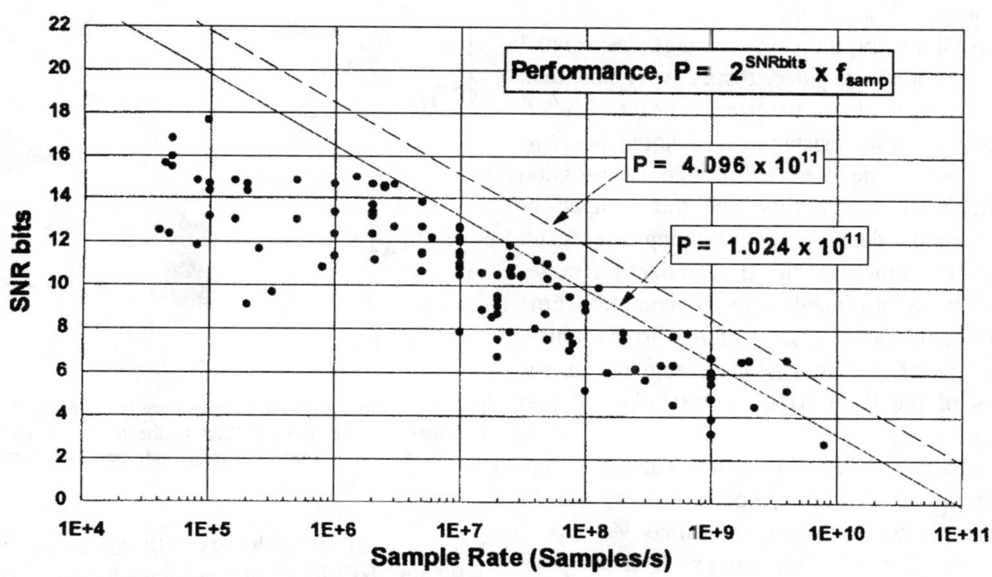

Fig. 6. Signal-to-noise ratio expressed as effective number of bits according to SNR-bits = (SNR(dB) − 1.76)/6.02. The two values of P bracket the current state of the art.

A similar graph of SNR-bits versus sample rate is shown in Fig. 6. Comparing with Fig. 1 shows that the distribution of ADC's in the SNR-bits plane is approximately 1.5 to 2 bits lower than the distribution of stated resolutions. This conclusion is consistent with Fig. 4(a). From the figure, the state-of-the-art corresponds to the range $1 \times 10^{11} < P < 4 \times 10^{11}$.

The data of Fig. 6 reveal global performance factors for the ADC population. Many factors and loss mechanisms affect ADC performance. Aside from quantization noise, three mechanisms limit achieved SNR: input-referred circuit noise (equivalent thermal noise), aperture uncertainty, and comparator ambiguity. The equations that calculate the associated maximum achievable resolutions, in SNR-bits, are

thermal noise (referred to the input:)

$$B_{\text{thermal}} = \log_2\left(\frac{V_{FS}^2}{6kTR_{\text{eff}}f_{\text{samp}}}\right)^{1/2} - 1 \tag{5}$$

aperture uncertainty:

$$B_{\text{aperture}} = \log_2\left(\frac{2}{\sqrt{3}\pi f_{\text{samp}}\tau_a}\right) - 1 \tag{6}$$

comparator ambiguity:

$$B_{\text{ambiguity}} = \frac{\pi f_T}{6.93 f_{\text{samp}}} - 1.1. \tag{7}$$

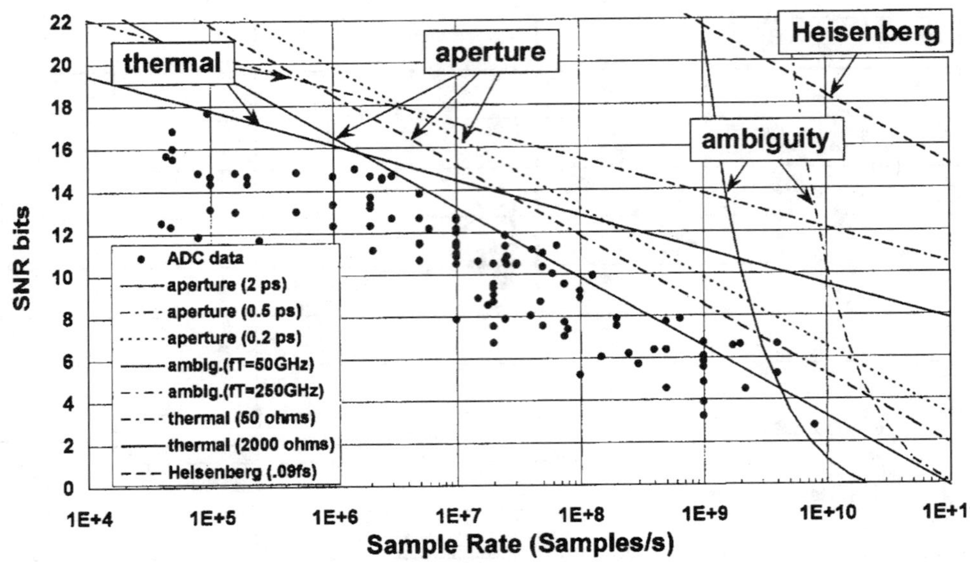

Fig. 7. Signal-to-noise ratio according to SNR-bits = (SNR(dB) − 1.76)/6.02. Three sets of curves show performance limiters due to thermal noise, aperture uncertainty, and comparator ambiguity. The Heisenberg limit is also displayed.

The derivations of these three equations are given in Appendix II. To summarize, B_{thermal} and B_{aperture} were derived by developing expressions for the noise voltages associated with thermal noise and aperture jitter, respectively and then equating each with the equivalent quantization noise $Q/\sqrt{12}$. The equivalent thermal noise resistance is denoted as R_{eff}. The rms aperture jitter is denoted as τ_a. The expression for $B_{\text{ambiguity}}$ reflects the probability that the comparator will make an ambiguous decision [7], treating the result as additive noise to the otherwise ideal quantization noise. The ambiguity probability is related to the regeneration time constant t_{reg} of the comparator. This is related to the unity-current-gain frequency of the transistors employed in the circuit. An analysis of the flash ADC in [5] indicated that $t_{\text{reg}} \sim 2.5/\pi f_T$ [8].

From these expressions SNR curves are calculated for values of input-referred thermal noise, aperture jitter, and f_T which measures comparator ambiguity. Assume $V_{FS} = 1$ V, $T = 300$ K for the thermal noise curves. Aperture jitter is calculated for Nyquist sampling, i.e., $f_{\text{sig}} = f_{\text{samp}}/2$. Comparator ambiguity is determined from the regeneration time constant of the IC technology. These are included in Fig. 7, which contains the same SNR data as Fig. 6. The current state-of-the-art is limited by the equivalent of thermal noise associated with a ~2 kΩ resistor for sampling rates under 2 Ms/s. Aperture jitter, in the range 0.5 ps to 2 ps, limits SNR for the sampling frequency range of ~2 Ms/s to 4 Gs/s. Comparator ambiguity is limited via the regeneration time constant corresponding to a value of $f_T \sim 50$ GHz for ADC's at the highest sampling rates. The 3-bit, 8 Gs/s Nyquist ADC [5] was fabricated with an $f_T \sim 80$ GHz InP process. The aperture uncertainty results in the −1 bit/octave slope; this is the dominant factor because the range of affected f_{samp} values is so large.

To continue to advance the state-of-the-art requires low-noise designs that achieve less than 0.5 ps of aperture uncertainty and/or technologies with $f_T \gg 50$ GHz. Experimental HBT and HEMT IC technologies have been reported that have devices with f_T and $f_{\max}$ ranging from ~150 GHz to ~260 GHz [9]–[11]. Hence, one can envision an eventual increase in sampling rates of a factor of two to about four beyond today's 8 Gs/s.

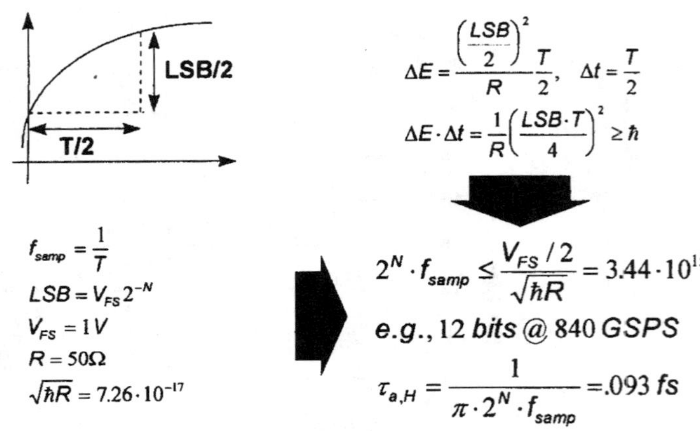

Fig. 8. Applying the Heisenberg uncertainty principle to ADC performance indicates that the ultimate limit in the resolution–sampling rate product is approximately four orders of magnitude beyond the current state of the art.

The ultimate limit to the ADC resolution-sampling rate product P, may be estimated using the Heisenberg uncertainty principle. Let $\Delta E \Delta t > h/2\pi$, where ΔE is the energy of the smallest resolvable signal, equivalent to .5 LSB, Δt is .5 sampling period $(T/2)$, and $h = 6.626\,17 \times 10^{-34}$ J-s is Planck's constant. The analysis is summarized in Fig. 8. For 50 Ω impedance and a 1 V peak-to-peak input signal, that limit is approximately four orders of magnitude beyond the state-of-the-art, which is aperture jitter limited (see the curve labeled Heisenberg in Fig. 7). There are probably other limiting factors between aperture jitter and the uncertainty principle. Although these are worthy of study, it is more urgent to develop a thorough understanding of aperture jitter.

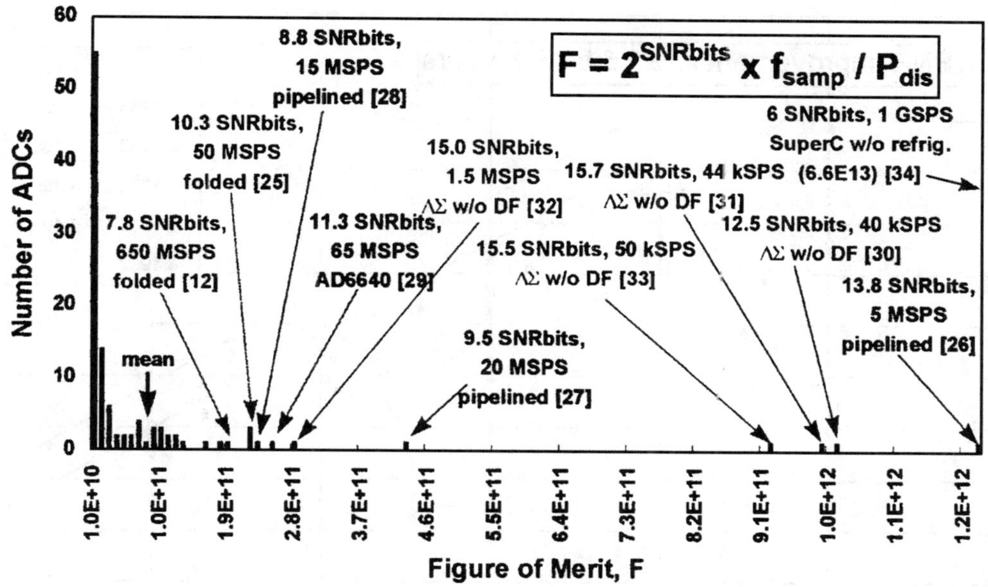

Fig. 9. Histogram of the figure of merit F. The most power-efficient ADC's have been reported within the past six years.

IV. High-Performance ADC Architectures

The ADC's of the preceding figures include architectures ranging from flash, a parallel technique, which is the fastest, through integrating which is probably the most accurate but which also is the slowest. Most of the converters have been fabricated in silicon, while a few have been realized in gallium arsenide (GaAs) and indium phosphide (InP).

The flash architecture uses $2^N - 1$ comparators, where N is the stated resolution. Flash converters often include one or two additional comparators to measure overflow conditions. All comparators sample the analog input voltage simultaneously. This ADC is thus inherently fast. The fastest ADC reported is the 3-bit, 8 Gs/s Nyquist flash converter [5] cited above. This ADC had a maximum sampling rate of 14 Gs/s.

The parallelism of the flash architecture has drawbacks for high-resolution applications. The number of comparators grows exponentially with N. In addition, the separation of adjacent reference voltages grows smaller exponentially. Consequently, this architecture requires very large IC's. It has high power dissipation. It is difficult to match components in the parallel comparator channels. Finally, increasingly large input capacitance reduces analog input bandwidth. Most flash converters available today have $\leq$ 8-bit resolution. In order to overcome these problems, variations on the flash architecture have been developed which use relatively few comparators yet retain good speed. Examples capable of Gs/s rates are the folded-flash [12]–[14]; and pipelined [15], [16] architectures.

Another approach to high-speed conversion is to time-interleave two or more converters [17]. The reported ADC achieves <1 ps aperture jitter, but requires two hybrids, each with five LSI chips. The total P_{diss} is 40 W, roughly an order of magnitude larger than single-chip converters.

An architecture that trades speed for resolution combines delta–sigma ($\Delta\Sigma$) modulation with digital decimation filtering [18]. Delta–sigma converters sample the analog input signal at a rate which is many times the Nyquist output rate. Integration and feedback suppress the quantization noise in the lower portions of the spectrum relative to the delta–sigma clock frequency. This technique requires few analog components.

The challenge is that a high speed IC technology is needed for RF applications. Recently, near ideal performance was reported with an InP HBT second-order $\Delta\Sigma$ modulator with a sampling rate of 3.2 Gs/s and an over-sampling ratio of 32 for a Nyquist rate of 100 Ms/s [19]. This converter technology has $f_T = 70$ GHz and $f_{\max} = 90$ GHz.

Delta–sigma modulators may be designed with a bandpass characteristic [20]–[22]. This is useful when a relatively narrow band of intermediate frequencies contains the signal to be digitized. Furthermore, the center frequency of the converter is tunable. Finally, in receiver applications, down conversion stages are eliminated. Recently two bandpass delta–sigma modulators were reported with a 60 MHz center frequency [21] and an 800 MHz center frequency [22]. Both of these sample at 4 GHz. These are the fastest bandpass $\Delta\Sigma$ modulators yet built. Further discussion of bandpass sampling for RF applications can be found in [2].

V. Low-Power ADC Architectures

Another facet of ADC performance is power dissipation P_{diss}. Generally the highest performing converters also dissipate the most power. A convenient way to include P_{diss} in the performance comparison is to use the figure of merit, F, defined above. Fig. 9 shows a histogram of F for the ADC population under study. Most of the ADC's have values of $F \leq 7.9 \times 10^{10}$ (the mean).

Two fictitious examples of converter specifications that would correspond approximately to this value of F are: 1) 13 SNR-bits, 10 Ms/s, 1.1 W, and 2) 7 SNR-bits, 1 Gs/s, 1.75 W. These two examples would represent present-day state-of-the-art performance and correspond to an aperture jitter of ~2 ps (see Fig. 7).

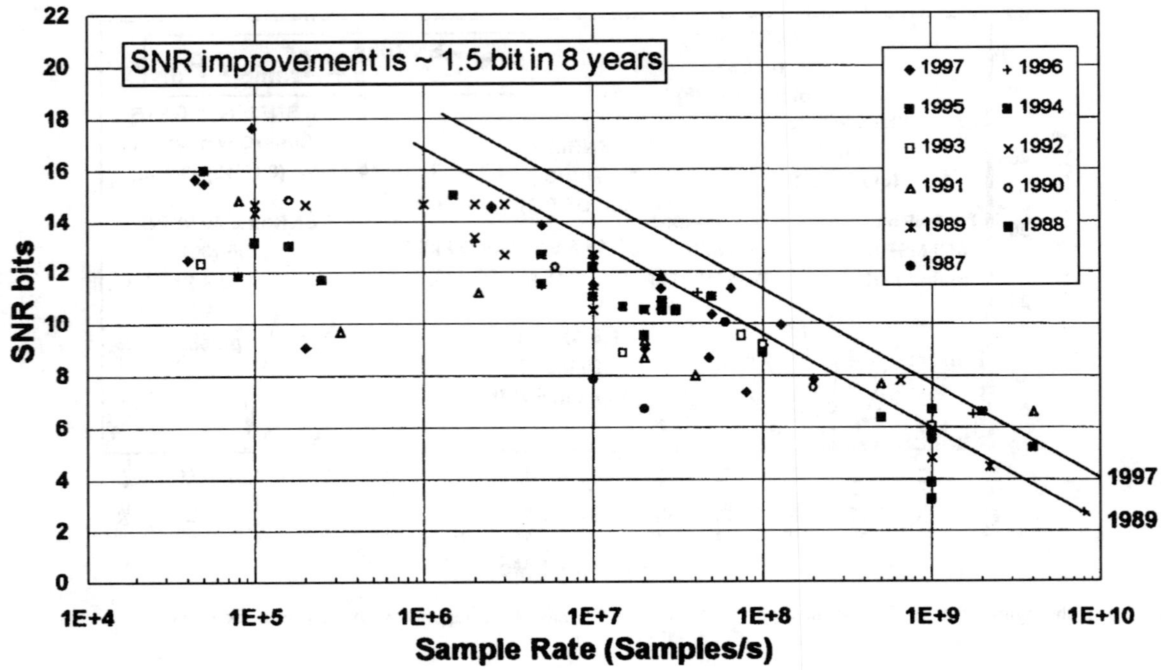

Fig. 10. Trend in SNR bits over time.

There are a few ADC's with F values significantly above 7.9×10^{10} and some of these are pointed out in the figure. These power-efficient converters utilize four architectures: flash (low resolution only) [23], [24], folded-flash [12], [25], pipelined [26]–[29], and $\Delta\Sigma$ modulation [30]–[33]. These particular $\Delta\Sigma$ modulators, in contrast to the GHz circuits mentioned above, were designed in CMOS using switched-capacitors and are oriented toward lower frequency applications. In addition, the accompanying digital decimation filters for these modulators have not been included in the power dissipation, so the actual complete ADC's will have somewhat lower values for F. The highest value of F is 6.6×10^{13}, and corresponds to a superconducting (denoted by S.C. in Fig. 1) ADC [34]. The refrigeration overhead was not included in the determination of F for this circuit. Most of these very efficient converters have been reported within the last six years.

VI. ADC Performance over Time

It is revealing to examine the trends in ADC performances during recent years. As an example, the data in Fig. 9 show that excellent progress has been made recently in developing power-efficient designs. However, the same is not true for the advancement of the resolution-speed product P. To show this, the SNR data in Fig. 6 were sorted according to the year in which the ADC's were reported. The results are given in Fig. 10, and it is evident that relatively little improvement has been made over the last six–eight years or so. From the scatter in the data it is also evident that the improvement is quite sporadic. A similar lack of advancement for SFDR-bits also holds.

If it is assumed that aperture uncertainty is the performance limiter for the best converters and if it is further assumed that all of the ADC's represented in Fig. 10 are Nyquist converters (optimistic) then, for each converter a value of τ_a can be derived from each value of P using the relation

$$\tau_a = \frac{1}{\pi\sqrt{3}P}.$$

Using this relation a graph of τ_a as a function of time (year) can be generated and is shown in Fig. 11. In this figure only the best results ($\tau_a \leq 6$ ps) are shown and the very best aperture jitter values achieved in each year are connected by a line. The scatter in the data emphasizes the sporadic nature of improvement in P, however, a least squares fit through the logs of the very best yearly data values indicates a gradual improvement over time. It can be conjectured that there may be an aperture uncertainty barrier of ~0.5 ps. Some other reasons for the stagnation in ADC performance improvement may be: 1) that much of the recent research has been aimed at monolithic, and therefore, power-efficient ADC's (c.f. Fig. 9); 2) a recent and general de-emphasis on research and development; and 3) few application drivers that push the state-of-the-art. Although software radios (~100 Ms/s) and satellite communications (>1 Gs/s) may provide the incentives for a breakthrough.

VII. Summary

The state-of-the-art for ADC's has been reviewed and analyzed. Data for SNR and SFDR as functions of f_{samp} has been discussed. The SNR data show that converter performance is limited by input-referred noise, aperture uncertainty and comparator ambiguity. The best results have been achieved for flash, folded-flash, pipelined, and time-interleaved architectures.

It is clear from the data presented above, that in order to improve upon the present state-of-the-art in ADC performance, significant technical challenges must be met. Specifically 1) a reduction in aperture uncertainty to well below 1 ps, 2) an

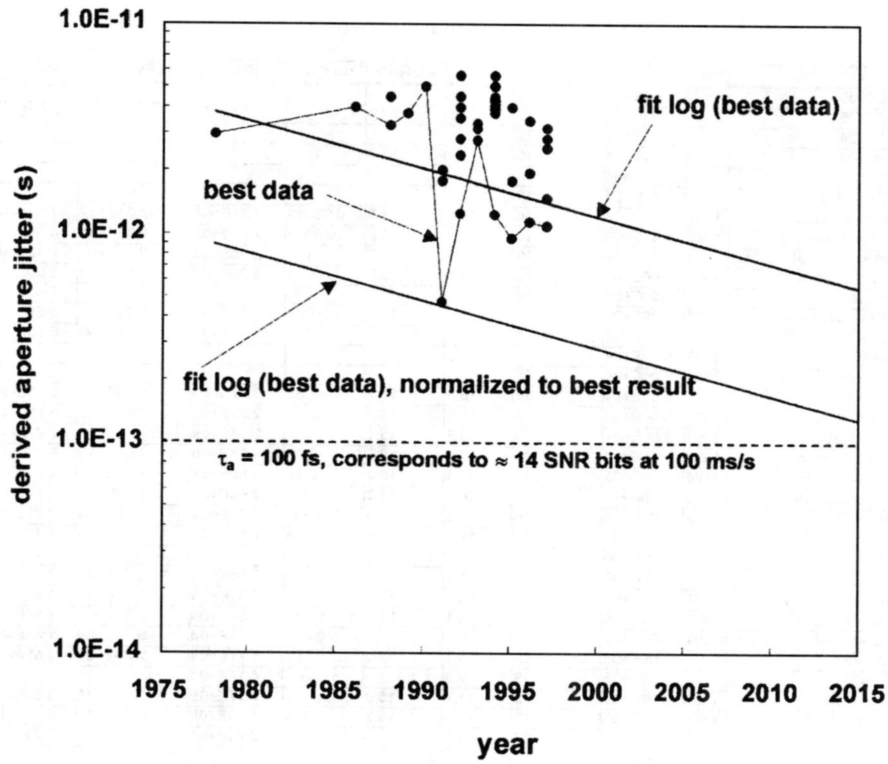

Fig. 11. Derived aperture jitter for the best ADC performances as a function of the year of introduction. Converter performances are gradually improving, although actual progress is sporadic.

increase in the maximum sampling frequency to beyond 8 Gs/s, and 3) accomplishing both 1) and 2) while maintaining low power consumption, e.g., < 5 W.

With respect to aperture uncertainty, only about 1.5 bits of overall improvement has been achieved over the last six–eight years in SNR (and only one bit in SFDR). The best effort was the time-interleaved ADC which achieved $\tau_a < 1$ ps [17]. In addition, while significant progress has been made in achieving power-efficient ADC designs (high F), none of these efforts has gone below ~0.5–2 ps of aperture jitter.

Appendix I

See Table I.

Appendix II

This appendix contains the derivation of the three equations which calculate the maximum ADC resolutions in SNR-bits for input-referred thermal noise, aperture uncertainty, and, comparator ambiguity. The first two equations are obtained by developing expressions for the noise voltage due to each mechanism and then equating to an equivalent quantization noise. The ambiguity equation is developed by generating an expression for the probability that an ambiguous decision will be made by a particular comparator, then summing the probabilities for all comparators, then adding the resulting noise to an otherwise ideal quantization noise voltage. All three equations are developed as if each error (thermal, aperture, ambiguity) is acting alone.

Thermal Noise Derivation: The spectral noise density seen at the ADC input consists of various contributions such as thermal noise, shot noise, $1/f$ noise, and input-referred noise. The thermal and shot components are white while the $1/f$ and input-referred components are frequency dependent. The resulting input-referred noise voltage is obtained by integrating these spectra over the full-Nyquist band $f_{\text{samp}}/2$ and can be expressed as

$$\langle v_n^2 \rangle = 4kTR_{\text{eff}} f_{\text{samp}}/2$$

where k is Boltzmann's constant $= 1.380\,658 \times 10^{-23}$ J/K, $T =$ temperature in K (assumed $= 300$ K in Fig. 7), and R_{eff} is an effective thermal resistance which lumps together the effects of all noises. The equivalent quantization noise voltage is given by

$$\langle v_Q^2 \rangle = \frac{Q^2}{12} = \frac{V_{FS}^2 2^{-2B_{\text{thermal}}}}{12}$$

where B_{thermal} is the maximum resolution in SNR-bits (for a given value of R_{eff}). Equating these two expressions leads directly to the desired result

$$B_{\text{thermal}} = \log_2 \left(\frac{V_{FS}^2}{6kTR_{\text{eff}} f_{\text{samp}}} \right)^{1/2} - 1.$$

Aperture Uncertainty Derivation: This effect comes about because an ADC does not sample the input signal at precisely equal time-intervals, $T = 1/f_{\text{samp}}$. Instead the sampling process can be characterized by a mean and a standard deviation with regard to the location in time of when sampling

TABLE I

Institution	Author/Part No.	Year	fsamp(Hz)	Stated Bits	SNRbits	SFDRbits	Pdiss(W)	P(LSBs-Hz)	F(LSBs-Hz/W)	Derived τa	Vinput(V)	BWinput(Hz)
Crystal	CS5506	1990	2.00E+01	20								
Analog Devices	AD1175		2.00E+01	22	21.8	22.2		7.31E+07		2.52E-09		
Crystal	CS5324		1.00E+03	20								
UC Berkeley	Lee, et al.	1984	1.20E+04	15		15.0	0.02					
Harris	ICL 7115		2.50E+04	14								
Analogic	ADC 20048		3.00E+04	18								
TI & Waterloo	Chen & Leung	1997	4.00E+04	14	12.5	14.5	0.00023	2.31E+08	1.01E+12	7.94E-10	2.0	
van der Zwan	Philips Research	1997	4.41E+04	16	15.7	16.0	0.0023	2.27E+09	9.89E+11	8.08E-11		2.0E+04
TI & Vitesse	Fattaruso, et al.	1993	4.80E+04	15	12.3	14.7	0.1	2.48E+08	2.48E+09	7.42E-10		2.4E+04
Stanford	Rabbii & Wooley	1997	5.00E+04	16	15.5	16.5	0.0025	2.30E+09	9.19E+11	8.00E-11	3.2	
Fujitsu	MB87020		5.00E+04	16						3.68E-06		
Stanford	Williams & Wooley	1994	5.00E+04	17	16.0	17.3	0.047	3.25E+09	6.91E+10	5.66E-11		2.5E+04
Analog Devices	AD1879		5.00E+04	18	16.0	17.2	0.9	3.25E+09	3.61E+09	5.66E-11	6.0	2.2E+04
Analog Devices	Adams	1989	5.00E+04	18								
Hitachi	Matsumoto, et al.	1988	5.00E+04	18								
Analogic	ADC5120		5.00E+04	20	16.8	17.3	2.38	5.77E+09	2.43E+09	3.18E-11	10.0	1.4E+05
Bell Northern R	Longo, et al.	1988	8.00E+04	13	11.8	14.2	0.02	2.92E+08	1.46E+10	6.29E-10		
GE + Yokogawa	Ribner, et al.	1991	8.00E+04	15	14.8	16.7		2.32E+09		7.92E-11		
Leung, et al,	Crystal Semic.	1997	9.60E+04	20	17.6	18.7	0.76	1.97E+10	2.59E+10	9.32E-12	4.0	2.2E+04
Analog Devices	Fernandes, et al.	1988	1.00E+05	14	13.1	15.0	0.48	8.96E+08	1.87E+09	2.05E-10	10.0	6.0E+05
Motorola	DSP56ADC16	1992	1.00E+05	16	14.7	14.7	0.3	2.58E+09	8.62E+09	7.11E-11	3.5	5.0E+04
Micro Networks	MN6500	1992	1.00E+05	16	14.3	16.0	0.685	2.05E+09	3.00E+09	8.95E-11	10.0	5.0E+04
Analog Devices	AD1876		1.00E+05	16	14.7	16.5	0.235	2.58E+09	1.10E+10	7.11E-11	10.0	1.0E+06
Crystal	CS5126		1.00E+05	16								
Atmel	AT76		1.00E+05	18	14.7		1	2.58E+09	2.58E+09	7.11E-11	1.0	
AT&T Bell L	Norsworthy, et al.	1988	1.60E+05	13	13.0	16.7	0.075	1.31E+09	1.74E+10	1.41E-10		
Motorola	Rebeschini, et al.	1990	1.60E+05	16	14.8	16.2	0.076	4.64E+09	6.11E+10	3.96E-11		8.0E+04
Swiss Federal Institute	Hammerschmied, Huang	1997	2.00E+05	10	9.1	13.2	0.012	1.09E+08	9.10E+09	1.68E-09		
Analogic	ADC4357		2.00E+05	16	14.7	15.0	2.2	5.17E+09	2.35E+09	3.55E-11	10.0	
Burr-Brown	PCM78		2.00E+05	16	14.3			4.11E+09		4.48E-11	6.0	
Burr-Brown	PCM1750	1992	2.00E+05	18	14.7	15.0	0.21	5.17E+09	2.46E+10	3.55E-11	5.5	5.0E+05
UC Berkeley	Sutarja, et al.	1988	2.50E+05	13	11.7	13.3	0.015	8.13E+08	5.42E+10	2.26E-10		
Micro Networks	MN5420	1991	3.20E+05	12	9.7	20.0	6.5	2.61E+08	4.02E+07	7.03E-10		
Sipex	SP9478		5.00E+05	14								
Alcatel	Mietec (ISSCC)		5.00E+05	14								
Analog Devices	AD1382		5.00E+05	16	14.8	14.2	2.8	1.45E+10	5.18E+09	1.27E-11	10.0	2.0E+05
Datel	ADS 930		5.00E+05	16	13.0	14.3	3.4	4.09E+09	1.20E+09	4.50E-11	10.0	2.0E+06
Analog Devices	AD7886		8.00E+05	12	10.8	12.8	0.25	1.46E+09	5.85E+09	1.26E-10	5.0	1.0E+06
Harris	ICL 7135		9.00E+05	14								
Datel	ADS 112		1.00E+06	12	11.3	12.5	1.3	2.58E+09	1.99E+09	7.11E-11	10.0	1.0E+07
Univ. of Illinois	Song, et al.	1988	1.00E+06	12		13.2	0.4					
Analog Devices	AD7586		1.00E+06	12			0.2				4.0	
Datel	ADS941		1.00E+06	14	12.3	14.5	2.8	5.16E+09	1.84E+09	3.56E-11	10.0	6.0E+06
MIT & Harris	Karanicolas, et al.	1993	1.00E+06	15		15.0	1.8					
Analogic	ADC 4344		1.00E+06	16	13.3	15.2	3.4	1.03E+10	3.03E+09	1.79E-11	5.0	4.0E+06
Edge Technology	ET1661	1992	1.00E+06	16	14.7		5.5	2.58E+10	4.70E+09	7.11E-12	10.0	
Sipex	SP9490		1.00E+06	16								
Analog Devices	Mercer	1991	1.25E+06	12		12.5	0.6				5.0	8.5E+06
Katholieke Univ. Leuven	Yin & Sansen	1994	1.50E+06	16	15.0	16.0	0.18	4.88E+10	2.71E+11	3.76E-12	0.7	7.5E+05
Signetics	Kolluri	1989	2.00E+06	12		12.2	0.65				10.0	
Analogic	ADC3110		2.00E+06	14	13.7	12.7	4.1	2.59E+10	6.32E+09	7.09E-12	10.0	2.0E+07
Datel	ADS942		2.00E+06	14	12.3	13.3	2.9	1.03E+10	3.56E+09	1.78E-11	10.0	6.0E+06
Datel	ADS932	1996	2.00E+06	16	13.2	14.0	1.85	1.83E+10	9.91E+09	1.00E-11	5.5	4.0E+06
Edge Technology	ET1662	1992	2.00E+06	16	14.7		5.5	5.17E+10	9.40E+09	3.55E-12	10.0	
Analog Devices	AD1388	1992	2.00E+06	16	13.3		6.5	2.06E+10	3.17E+09	8.93E-12	10.0	9.0E+06
Stanford	Brandt, et al.	1991	2.10E+06	12	11.2	14.2	0.041	4.84E+09	1.18E+11	3.80E-11		1.0E+06
Analog Devices	Brooks, et al.	1997	2.50E+06	16	14.5	16.2	0.55	5.76E+10	1.05E+11	3.19E-12		
Analog Devices	AD9260	1997	2.50E+06	16	14.6	16.7	0.6	6.10E+10	1.02E+11	3.01E-12		1.3E+06
Edge Technology	ET1463	1992	3.00E+06	14	12.7		2.17	1.95E+10	8.98E+09	9.44E-12	2.0	
Edge Technology	ET1663	1992	3.00E+06	16	14.7		5.5	7.75E+10	1.41E+10	2.37E-12	10.0	
Raytheon	was Hughes	1978	4.00E+06	12								
Datel	ADS 118		5.00E+06	12	10.7	10.8	1.9	8.15E+09	4.29E+09	2.25E-11	2.0	6.5E+07
UC Berkeley	Cline & Gray	1996	5.00E+06	13	11.5	12.2	0.166	1.40E+10	8.44E+10	1.31E-11	6.6	5.0E+07
Burr-Brown	ADC614	1994	5.00E+06	14	12.7	14.7	6.1	3.25E+10	5.32E+09	5.66E-12	2.5	3.0E+07
Datel	ADS944	1994	5.00E+06	14	11.5		3.37	1.45E+10	4.30E+09	1.27E-11	2.5	
Edge Technology	ET1465	1992	5.00E+06	14	12.7		2.17	3.25E+10	1.50E+10	5.66E-12	2.0	
Univ of Illinois, Harris	Kwak et al.	1997	5.00E+06	15	13.8	15.5	0.06	7.18E+10	1.20E+12	2.56E-12		
Raytheon	Boyko, GSG	1990	6.00E+06	15	12.2	15.0	22	2.76E+10	1.25E+09	6.66E-12		
Raytheon	was Hughes	1984	1.00E+07	8	7.8	10.2	0.4	2.30E+09	5.76E+09	7.98E-11		
TRW	THC1202		1.00E+07	12	10.8	11.2	4.5	1.83E+10	4.07E+09	1.00E-11	2.0	7.0E+07

(a)

occurs. The mean is the average position of the sampling time and the standard deviation is a measure of the variation of the sampling point and is defined as the rms aperture jitter, τ_a. Assuming that τ_a is known, an expression for the voltage error due to τ_a can be derived. The worst case situation corresponds to sampling a sinusoidal waveform with the highest frequency in the Nyquist band, which is $f_{\rm samp}/2$, i.e., $v(t) = (V_{FS}/2)\ sin(\pi f_{\rm samp} t)$. The maximum rms voltage error will occur when attempting to sample the sinusoid at its zero-crossing, and is given by the product of the maximum slope of the wave and the aperture uncertainty

$$v_{\rm rms} = \pi f_{\rm samp} V_{FS} \tau_a / 2.$$

Equating this to the square root of the expression for $\langle v_Q^2 \rangle$ given above (here $B_{\rm thermal}$ is replaced by $B_{\rm aperture}$), leads to

TABLE I *(Continued.)*

Institution	Author/Part No.	Year	fsamp(Hz)	Stated Bits	SNRbits	SFDRbits	Pdiss(W)	P(LSBs-Hz)	F(LSBs-Hz/W)	Derived τa	Vinput(V)	BWinput(Hz)
Datel	ADS130		1.00E+07	12	10.5	11.2	3.85	1.45E+10	3.77E+09	1.26E-11	2.5	6.5E+07
Burr Brown	ADC603	1992	1.00E+07	12	10.5	12.0	6.1	1.45E+10	2.38E+09	1.26E-11	2.5	4.0E+07
Analog Devices	AD9220	1996	1.00E+07	12	11.3	12.9	0.28	2.58E+10	9.23E+10	7.11E-12		3.5E+07
Analog Devices	AD9220	1997	1.00E+07	12	11.5	14.7	0.25	2.90E+10	1.16E+11	6.34E-12		6.0E+07
TI + Army	Hamlett, et al.	1990	1.00E+07	12			1.6					
Univ. of Illinois	Shu et al.	1995	1.00E+07	13	11.0		0.36	2.05E+10	5.70E+10	8.95E-12		
Datel	ADS945	1994	1.00E+07	14	12.0	13.2	4.2	4.10E+10	9.75E+09	4.49E-12	2.5	5.0E+07
Analog Devices	AD9014	1992	1.00E+07	14	12.2	14.7	12.8	4.60E+10	3.59E+09	4.00E-12	2.0	6.0E+07
Analog Devices	AD9240	1997	1.00E+07	14	12.6	15.0	0.285	6.13E+10	2.15E+11	3.00E-12		7.0E+07
Edge Technology	ET1471	1992	1.00E+07	14	12.7		5.7	6.49E+10	1.14E+10	2.83E-12	2.0	
Harris	CA 3318C		1.50E+07	8								
Matsushita	Kusumoto, et al.	1993	1.50E+07	10	8.8		0.03	6.89E+09	2.30E+11	2.67E-11		3.0E+06
Raytheon	was Hughes	1986	1.50E+07	11		14.2	90					
Comlinear	CLC935B	1994	1.50E+07	12	10.6	13.7	4.75	2.34E+10	4.92E+09	7.87E-12	2.0	8.0E+07
Analog Devices	AD773		1.80E+07	10	8.5	11.2	1.2	6.57E+09	5.47E+09	2.80E-11	1.0	1.0E+08
Raytheon	was Hughes	1986	2.00E+07	7	6.7		0.64	2.10E+09	3.29E+09	8.73E-11		
Harris	HI 5700		2.00E+07	8	7.5			3.66E+09		5.02E-11		9.0E+06
Datel	ADC208		2.00E+07	8								
Analog Devices	AD9200	1997	2.00E+07	10	9.0	10.3	0.08	1.03E+10	1.29E+11	1.78E-11	2.0	9.0E+07
Analog Devices	Real, et al.	1991	2.00E+07	10	8.7	10.3	1	8.19E+09	8.19E+09	2.24E-11		1.5E+08
Univ. of Cal. Berk.	Cho & Gray	1995	2.00E+07	10	9.5		0.035	1.47E+10	4.21E+11	1.25E-11	2.0	
Sig Proc Tech	SPT7912	1991	2.00E+07	12	9.3	11.7	1.8	1.30E+10	7.21E+09	1.42E-11	2.0	1.2E+08
HP	Jewett, et al.	1992	2.00E+07	12	10.5	12.0	3.5	2.91E+10	8.30E+09	6.32E-12		9.5E+07
Comlinear	CLC936C	1994	2.00E+07	12	10.5	12.6	5.28	2.91E+10	5.50E+09	6.32E-12	2.0	9.0E+07
Raytheon	was Hughes	1982	2.00E+07	12								
Analog Devices	CAV1220		2.00E+07	12			20					
TRW	TAC1025		2.50E+07	10	7.8	8.8	7.8	5.76E+09	7.38E+08	3.19E-11	1.0	6.0E+07
Analog Devices	AD9032	1992	2.50E+07	12	10.7	12.0	5	4.08E+10	8.15E+09	4.51E-12	2.0	1.0E+07
Analog Devices	AD9225	1997	2.50E+07	12	11.3	14.3	0.3	6.46E+10	2.15E+11	2.84E-12		2.0E+08
Raytheon	was Hughes	1991	2.50E+07	14	11.8	14.0	30	9.13E+10	3.04E+09	2.01E-12		
Comlinear	CLC937B`	1994	2.56E+07	12	10.5	12.2	7.35	3.64E+10	4.95E+09	5.06E-12	2.0	1.0E+08
Comlinear	CLC950	1994	2.56E+07	12	10.8	12.5	2	4.68E+10	2.34E+10	3.92E-12	2.0	1.8E+08
Comlinear	CLC938C`	1994	3.07E+07	12	10.4	12.0	6.57	4.26E+10	6.49E+09	4.31E-12	2.0	1.0E+08
Analog Devices	AD9026	1994	3.10E+07	12	10.5		1.46	4.50E+10	3.09E+10	4.08E-12	2.0	
Sony	CXD1179Q		3.50E+07	8			0.1				1.8	
Sig Proc Tech	SPT7824	1991	4.00E+07	10	8.0	11.7	1.8	1.03E+10	5.74E+09	1.78E-11	1.8	1.2E+08
Analog Devices	AD9042	1996	4.10E+07	12	11.2	13.3	0.595	9.44E+10	1.59E+11	1.95E-12		1.4E+09
Broadcom Corp., UCLA	Bult, et al.	1997	4.80E+07	10	8.7		0.24	1.97E+10	8.19E+10	9.35E-12	2.0	3.2E+07
Raytheon	was Hughes	1989	5.00E+07	7			0.5					
Philips Research Labs	Van de Plaasche	1979	5.00E+07	7								
TRW	TDC1025		5.00E+07	8	7.5			9.15E+09		2.01E-11	2.0	1.3E+07
Datel	ADC9060		5.00E+07	10								
Philips Components, FR	Vorenkamp, Roovers	1997	5.00E+07	12	10.3	11.8	0.3	6.48E+10	2.16E+11	2.84E-12		2.6E+05
Analog Devices	Murden & Gosser	1995	5.00E+07	12	11.0	13.3	0.575	1.03E+11	1.78E+11	1.79E-12	1.0	2.0E+08
Raytheon	was Hughes	1978	6.00E+07	13	10.0	11.7	25	6.17E+10	2.47E+09	2.98E-12		3.0E+07
Analog Devices	AD6640	1997	6.50E+07	12	11.3	13.3	0.695	1.68E+11	2.42E+11	1.09E-12		2.5E+07
Sony	CXA1386P		7.50E+07	8	7.0		0.58	9.71E+09	1.67E+10	1.89E-11		1.5E+08
UCLA	Colleran & Abidi	1993	7.50E+07	10	9.5	12.8	0.8	5.46E+10	6.83E+10	3.36E-12		5.0E+07
Analog Devices	AD9060		7.50E+07	10	7.7		2.8	1.54E+10	5.50E+09	1.19E-11	4.0	1.8E+08
Analog Devices	AD9057-80	1997	8.00E+07	8	7.3	9.2	0.175	1.30E+10	7.45E+10	1.41E-11	2.5	1.2E+08
TRW	TDC1029		1.00E+08	6	5.2			3.65E+09		5.04E-11	1.0	5.0E+07
Philips Research Labs	Van de Plaasche	1988	1.00E+08	8			0.8				1.0	4.0E+07
NEC	Sone, et al.	1993	1.00E+08	10	9.2	10.2	0.95	5.78E+10	6.09E+10	3.18E-12		1.5E+07
Analog Devices	AD9070	1997	1.00E+08	10	9.2		0.7	5.78E+10	8.26E+10	3.18E-12		2.3E+08
Raytheon	was Hughes	1994	1.00E+08	10			1.8					
HRL Labs	Jensen, et al.	1995	1.00E+08	12	8.8	11.8	1	4.59E+10	4.59E+10	4.00E-12		
Hewlett Packard	Jewett, et al.	1997	1.28E+08	12	9.9	11.7	5.7	1.24E+11	2.18E+10	1.48E-12	0.5	2.5E+08
Sig Proc Tech	HADC77100		1.50E+08	8	6.0		2.6	9.73E+09	3.74E+09	1.89E-11	2.0	1.8E+08
Analog Devices	AD9054-200	1997	2.00E+08	8	7.8		0.5	4.61E+10	9.21E+10	3.99E-12		3.8E+08
Analog Devices	Mangelsdorf	1990	2.00E+08	8	7.5		2	3.66E+10	1.83E+10	5.02E-12		4.0E+08
Raytheon	was Hughes	1982	2.50E+08	5			2.2					
Sony	CXA1176K		2.50E+08	8	6.2		1.4	1.82E+10	1.30E+10	1.01E-11		2.5E+08
NTT LSI Lab	Akazawa, et al.	1987	3.00E+08	8								
Analog Devices	AD9028		3.00E+08	8	5.7		2.2	1.55E+10	7.03E+09	1.19E-11	2.0	2.5E+08
Matsushita		1992	3.00E+08	10			4					
Sony	CXA1276K		4.00E+08	8	6.4		3.1	3.27E+10	1.05E+10	5.62E-12		2.5E+08

(b)

the desired relation for B_{aperture}

$$B_{\text{aperture}} = \log_2\left(\frac{2}{\sqrt{3}\pi f_{\text{samp}}\tau_a}\right) - 1.$$

Comparator Ambiguity Derivation: This effect is due to the finite speed with which the transistors in the comparators are able to respond to a (small) voltage difference (~.5 LSB). The probability that the ith comparator will produce an ambiguous decision as to whether the input signal is above or below the reference voltage associated with said comparator is given by [7]

$$P_i = \frac{V_{FS}e^{-t/\tau_{\text{reg}}}}{A_0 Q}$$

where $t = .5$ clock period $= (2f_{\text{samp}})^{-1}$, $\tau_{\text{reg}} =$ regeneration time constant $= 2.5/\pi f_T$ [8], $A_0 =$ comparator gain, and Q is the effective LSB voltage $= V_{FS}2^{-B_{\text{ambiguity}}}$. Making the implied substitutions yields

$$P_i = \frac{e^{-\pi f_T/5f_{\text{samp}}}}{A_0 2^{-B_{\text{ambiguity}}}}.$$

TABLE I *(Continued.)*

Institution	Author/Part No.	Year	fsamp(Hz)	Stated Bits	SNRbits	SFDRbits	Pdiss(W)	P(LSBs-Hz)	F(LSBs-Hz/W)	Derived τa	Vinput(V)	BWinput(Hz)
TRW	L. Tran et al.	1993	5.00E+08	4			1					2.5E+08
Analog Devices	AD9006		5.00E+08	6	4.5		2	1.15E+10	5.76E+09	1.60E-11	2.0	5.5E+08
Micro Networks	MN6900	1991	5.00E+08	8	7.7	10.3	7.5	1.03E+11	1.37E+10	1.79E-12	0.5	1.2E+09
Raytheon	was Hughes	1988	5.00E+08	8	6.4		11	4.08E+10	3.71E+09	4.50E-12		2.5E+08
Raytheon	was Hughes	1984	5.20E+08	6			4.1					
Raytheon	was Hughes	1986	6.00E+08	4			1.8					
Philips, Netherl	van Valburg,et al.	1992	6.50E+08	8	7.8	8.5	0.81	1.46E+11	1.81E+11	1.26E-12	2.0	1.5E+08
Raytheon	was Hughes	1986	8.00E+08	6			7.9					3.5E+08
HRL Labs	W. & Scnmitz, HRL	1988	1.00E+09	4	3.2	5.0	0.1	9.16E+09	9.16E+10	2.01E-11	1.4	1.0E+07
Inst. fur Electr., Ruhr Uni	Daniel et al.	1988	1.00E+09	4	3.9	6.0	2.4	1.45E+10	6.05E+09	1.27E-11	1.3	8.0E+08
TRW & Honeywell	Kieks et al.	1987	1.00E+09	4								1.0E+09
TRW	Oki et al.	1987	1.00E+09	4			2.25					5.0E+08
Rockwell	Wang et al.	1987	1.00E+09	4			0.6					
Fraunhofer & TriQuint	Hagelauer et al.	1992	1.00E+09	5	4.8	5.6	3.4	2.80E+10	8.24E+09	6.56E-12	1.8	5.0E+08
Raytheon	was Hughes	1986	1.00E+09	6	5.5	6.2	2.6	4.59E+10	1.77E+10	4.00E-12		5.0E+08
NTT LSI Lab	Wakimoto, et al.	1988	1.00E+09	6	5.8	8.3	2	5.59E+10	2.79E+10	3.29E-12	2.0	1.7E+09
Signal Processing Tech	SPT7760A	1995	1.00E+09	8	6.7	7.8	5.5	1.03E+11	1.87E+10	1.79E-12	2.0	9.0E+08
HYPRES	Stebbins & Bradley	1993	1.00E+09	8	6.0	8.0	0.001	6.56E+10	6.56E+13	2.80E-12	0.3	
Raytheon	was Hughes	1989	1.20E+09	5			3					
TRW	B. Wong, et al	1996	1.75E+09	8	6.5	8.7		1.60E+11		1.15E-12		3.0E+09
Rockwell	Nary et al.	1995	2.00E+09	8	6.6	8.0	5.3	1.92E+11	3.62E+10	9.57E-13	0.6	3.0E+09
Lab d'Electr & de Phys A	DuCourant, et al.	1989	2.20E+09	5	4.5		1.05	4.89E+10	4.66E+10	3.76E-12		4.0E+08
Lab d'Electr & de Phys A	DuCourant, et al.	1986	3.00E+09	4			0.15					
HP & Rockwell	Poulton(HP),Wang(R).	1994	4.00E+09	6	5.2	5.6	5.7	1.48E+11	2.59E+10	1.24E-12		1.8E+09
Hewlett Packard	Schiller & Byrne	1991	4.00E+09	8	6.6		39	3.88E+11	9.96E+09	4.73E-13		
HRL Labs	Baringer, et al.	1996	8.00E+09	3	2.7	4.5	3.5	5.31E+10	1.52E+10	3.46E-12	0.6	1.2E+10

(c)

An equation for the quantization noise plus the contributions from all comparators due to the ambiguity probabilities is [7]

$$\langle v_{Q+}^2 \rangle = \frac{Q^2}{12}\left(1 + P_i(2^N)\right)$$

where 2^N is the number of comparators in the ADC which holds for a flash converter with one overflow comparator (this is the fastest architecture). Substituting for P_i and defining the quantity $\Delta N = N - B_{\text{ambiguity}}$ gives

$$\langle v_{Q+}^2 \rangle = \frac{Q^2}{12}\left(1 + \frac{2^{\Delta N}}{A_0}\, e^{-\pi f_T/5 f_{\text{samp}}}\left(2^{2B_{\text{ambiguity}}}\right)\right).$$

The second term in the parentheses can be considered as an excess noise and can be equated with a fraction, α, of the quantization noise, i.e.,

$$\alpha = \frac{2^{\Delta N}}{A_0}\, e^{-\pi f_T/5 f_{\text{samp}}}\left(2^{2B_{\text{ambiguity}}}\right).$$

Solving for $B_{\text{ambiguity}}$ yields the following:

$$B_{\text{ambiguity}} = \frac{\pi f_T}{10\,\ln 2 f_{\text{samp}}} + \frac{1}{2}\left(\log_2(\alpha A_0) - \Delta N\right).$$

If the following assumptions are made: 1) the additive term should be no more than .5 LSB, then $\alpha = .25$, 2) a value for $A_0 = 2.5$ [5], and 3) $\Delta N = 1.5$ [see Fig. 4(a)], then the final expression for $B_{\text{ambiguity}}$ is obtained

$$B_{\text{ambiguity}} = \frac{\pi f_T}{6.93 f_{\text{samp}}} - 1.1.$$

REFERENCES

[1] J. Mitola, "The software radio architecture," *IEEE Commun. Mag.*, vol. 33, pp. 26–38, May 1995.

[2] J. A. Wepman, "Analog-to-digital converters and their applications in radio receivers," *IEEE Commun. Mag.*, vol. 33, pp. 39–45, May 1995.

[3] R. Baines, "The DSP bottleneck," *IEEE Commun. Mag.*, vol. 33, pp. 46–54, May 1995.

[4] R. H. Walden, "Analog-to-digital converter technology comparison," in *IEEE GaAs IC Symp. Tech. Dig.*, Oct. 1994, pp. 217–219.

[5] C. Baringer, J. F. Jensen, L. Burns, and R. H. Walden, "A 3-bit, 8 GSPS flash ADC," in *Proc. Indium Phos. Rel. Mater. Conf.*, Apr. 1996, pp. 64–67.

[6] A. V. Oppenheim and R. W. Schafer, *Discrete-Time Signal Processing*. Englewood Cliffs, NJ: Prentice-Hall, ch. 3, p. 114.

[7] L. E. Larson, "High-speed analog-to-digital conversion with GaAs technology: Prospects, trends, and obstacles," presented at the 14th European Solid-State Circuits Conf., Manchester, England, Sept. 1988.

[8] J. F. Jensen, private communication.

[9] Y. Matsuoka and E. Sano, "High-speed AlGaAs/GaAs HBT's and their applications to 40-Gbit/s-class IC's," in *IEEE GaAs IC Symp. Tech. Dig.*, Oct. 1994, vol. 16, pp. 185–188.

[10] T. Enoki, T. Kobayashi, and Y. Ishii, "Device technologies for InP-based HEMT's and their application to IC's," in *IEEE GaAs IC Symp. Tech. Dig.*, Oct. 1994, vol. 16, pp. 337–340.

[11] S. Yamahata *et al.*, "Ultra-high $f_{\max}$ and f_T InP/InGaAs double-heterojunction bipolar transistors with step-graded InGaAsP collectors," in *IEEE GaAs IC Symp. Tech. Dig.*, Oct. 1994, vol. 16, pp. 345–348.

[12] J. van Valberg and R. J. van de Plassche, "An 8-bit 650 MHz folding ADC," *IEEE J. Solid-State Circuits*, vol. 27, pp. 1662–1666, Dec. 1992.

[13] K. Poulton, K. L. Knudsen, J. Corcoran, K.-C. Wang, R. B. Nubling, R. L. Pierson, M.-C. F. Chang, and P. M. Asbeck, "A 6-bit, 4 Gsa/s ADC fabricated in a GaAs HBT process," in *GaAs IC Symp. Tech. Dig.*, Oct. 1994, vol. 16, pp. 240–243.

[14] K. Nary, R. Nubling, S. Beccue, W. Colleran, J. Penney, and K. Wang, "An 8-bit, 2 gigasample per second analog to digital converter," in *GaAs IC Symp. Tech. Dig.*, Oct. 1995, vol. 17, pp. 303–246.

[15] K. Sone, N. Naotoshi, Y. Nishida, M. Ishida, Y. Sekine, and M. Yotsuyanagi, "A 10b 100 Ms/s pipelined subranging BiCMOS ADC," in *IEEE ISSCC Dig. Tech. Papers*, Feb. 1993, vol. 36, pp. 66–67.

[16] W. T. Colleran, T. H. Phan, and A. A. Abidi, "A 10b 100 Ms/s pipelined A/D converter," in *IEEE ISSCC Dig. Tech. Papers*, Feb. 1993, vol. 36, pp. 68–69.

[17] C. Schiller and P. Byrne, "A 4-GHz 8-b ADC system," *IEEE J. Solid-State Circuits*, vol. 26, pp. 1781–1789, Dec. 1991.

[18] J. C. Candy and G. C. Temes, Eds., *"Oversampling Delta-Sigma Data Converters.* New York: IEEE Press, 1992 (see Introduction).

[19] J. F. Jensen, G. Raghavan, A. E. Cosand, and R. H. Walden, "A 3.2 GHz 2nd order delta-sigma modulator implemented in InP HBT technology,"

IEEE J. Solid-State Circuits, vol. 30, pp. 1119–1127, Oct. 1995.

[20] F. W. Singor and W. M. Snelgrove, "Switched capacitor bandpass delta-sigma A/D modulation at 10.7 MHz," *IEEE J. Solid-State Circuits*, vol. 30, pp. 184–192, Mar. 1995.

[21] G. Raghavan, J. F. Jensen, R. H. Walden, and W. P. Posey, "A bandpass $\Delta\Sigma$ modulator with 92 dB SNR and center frequency continuously programmable from 0 to 70 MHz," in *ISSCC Tech. Dig.*, Feb. 1997, vol. 40, pp. 214–215.

[22] A. Jayaraman *et al.*, "Bandpass delta-sigma modulator with 800 MHz center frequency," in *IEEE GaAs IC Symp. Tech. Dig.*, Oct. 1997, vol. 19, pp. 95–98.

[23] R. H. Walden, A. E. Schmitz, A. R. Kramer, L. E. Larson, and J. Pasiecznik, "A deep-submicrometer analog-to-digital converter using focused-ion-beam implants," *IEEE J. Solid-State Circuits*, vol. 25, pp. 562–571, Apr. 1990.

[24] T. Ducourant, M. Binet, J.-C. Baelde, Ch. Rocher, and J.-M. Gibereau, "3 GHz 150 mW, 4-bit GaAs analog to digital converter," *IEEE GaAs IC Symp. Tech. Dig.*, Oct. 1986, p. 209.

[25] P. Vorenkamp and R. Roovers, "A 12b 50Msample/s cascaded folding and interpolating ADC," in *IEEE ISSCC Dig. Tech. Papers*, Feb. 1997, vol. 40, pp. 134–135.

[26] S.-U. Kwak, B.-S. Song, and K. Bacrania, "A 15b 5M sample/s low-spurious CMOS ADC," in *IEEE ISSCC Dig. Tech. Papers*, Feb. 1997, vol. 40, pp. 146–147.

[27] T. B. Cho and P. R. Gray, "A 10b, 20 Msample/s, 35 mW pipeline A/D converter," *IEEE J. Solid-State Circuits*, vol. 30, pp. 166–172, Mar. 1995.

[28] K. Kusumoto, A. Matsuzawa, and K. Murata, "A 10-b, 20-MHz, 30 mW pipelined interpolating CMOS ADC," *IEEE J. Solid-State Circuits*, vol. 28, pp. 1200–1206, Dec. 1993.

[29] Analog Devices, Inc., part number AD6640, 12-bit, 65 Msps ADC, 1997.

[30] F. Chen and B. Leung, "A 0.25-mW low-pass passive sigma-delta modulator with built-in mixer for a 10-MHz IF input," *IEEE J. Solid-State Circuits*, vol. 32, pp. 774–782, June 1997.

[31] E. van der Zwan, "A 2.3 mW CMOS SD modulator for audio applications," *IEEE ISSCC Dig. Tech. Papers*, Feb. 1997, vol. 40, pp. 220–221.

[32] G. Yin and W. Sansen, "A high-frequency and high-resolution fourth-order SD A/D converter in BiCMOS technology," *IEEE J. Solid-State Circuits*, vol. 29, pp. 857–865, Aug. 1994.

[33] S. Rabii and B. A. Wooley, "A 1.8 V digital-audio sigma-delta modulator in 0.8 mm CMOS," *IEEE J. Solid-State Circuits*, vol. 32, pp. 783–796, June 1997.

[34] E. Stebbins and P. Bradley, "Hypres flash ADC program report," Hypres, Inc., 1993.

Power Consumption of A/D Converters for Software Radio Applications

Peter B. Kenington, *Member, IEEE,* and Luc Astier

***Abstract*—An "ideal" software radio receiver is often considered as an analog/digital (A/D) converter connected directly to an antenna. This paper analyzes this ideal approach in terms of the power consumption of the A/D converter, based on a theoretical analysis of its minimum possible value.**

The results indicate that for the number of bits required to cope with the slow and fast fading environments and consequent large dynamic range encountered in a mobile radio system, the power consumption of the A/D converter is a fundamental limitation.

***Index Terms*—A/D converters, mobile radio, power consumption, radio transmitters and receivers, software radio.**

I. Introduction

A. Background

ALTHOUGH there is an increasing trend towards regional (and to some degree worldwide) air interface standardization, there are still many current and proposed standards and consequently different handset designs. The European trend towards harmonization, in the shape of the GSM,[1] DECT,[2] DCS1800, and TETRA[3] standards, has shown subscribers the benefits of a single handset with Europe-wide (and to some degree, worldwide) mobility. Even if these standards become dominant throughout the world (which is unlikely, given the strong penetration of U.S.-derived systems, such as AMPS,[4] DAMPS,[5] CDMA,[6] and PDC[7]—which is similar to DAMPS in many respects), there will still be many different standards for different applications. Even in Europe, there are different standards for cellular (GSM and DCS1800), cordless phone (DECT), and private mobile radio (TETRA). A single handset capable of interworking with all of these systems would therefore require either three or four different circuit cards (with a common case, display and keyboard), or a *flexible architecture radio* (often referred to as a "software" radio, although there are reconfigurable hardware options available for its implementation).

Manuscript received July 7, 1997; revised September 8, 1998. This work was supported by the European Commission.

P. B. Kenington is with Wireless Systems International Ltd., Clifton, Bristol BS8 1EJ, U.K.

L. Astier is with Gemplus, France.

Publisher Item Identifier S 0018-9545(00)02576-7.

[1]Groupe special mobile (or global system for mobile communications, as it has become known).

[2]Digital equipment cordless telephone.

[3]Trans-European trunked radio.

[4]American mobile phone system.

[5]Digital American mobile phone system.

[6]Code division multiple access.

[7]Personal digital cellular (formerly known as Japanese digital cellular, JDC).

The increasing convergence of mobile satellite and terrestrial cellular technologies inevitably leads to the requirement for a combined handportable terminal. At present, this is envisaged as a dual-mode or even a dual radio handset due to the current difficulties in achieving a satisfactory performance from shared radio frequency (RF) hardware and flexible digital signal processing (DSP). Considerable research is underway in order to solve these problems and such items should be available in the medium, if not short term.

B. Software Radio

A software radio (also called a "flexible architecture radio") is a transceiver in which, ideally, all aspects of its operation are determined using reconfigurable elements [1], [2]. This is usually thought of in terms of baseband DSP's, but FPGA's and other techniques are also possible [3]. It is also usually assumed to be broadband in nature, as one of its principal applications is perceived to be in replacing the numerous handsets currently required to guarantee cellular (and in the future, satellite) operation worldwide. This is strictly speaking an extension of the basic "software radio" concept into that of a broad-band flexible architecture radio, since the reprogrammability and adaptability aspects of operation do not depend upon multiband coverage. It would be possible, for example, to construct a useful software radio which operated in the 800/900-MHz area of spectrum and which could adapt between TACS, AMPS, GSM, DAMPS, CT2, and PDC. The trend is, however, for multifrequency operation (there are emerging handsets which cover both GSM and DECT, generally by employing separate hardware for each, in a common case).

There are many issues which must be addressed in determining if a software radio is realistic and also to what extent it is flexible. For example, it is possible to create a single-band software radio with a narrow-band channel restriction relatively easily (e.g., [4]). Coping with wider channel bandwidths and operating in multiple bands in differing parts of the spectrum is much more difficult, but nevertheless essential for a combined terrestrial/satellite telephone.

This paper addresses one of these issues, namely, the analog/digital (A/D) converter and its likely effect on the power consumption of future software radios.

C. "Ideal" Software Radio Architecture

An idealized software radio is shown in Fig. 1. It has the following features.

- The modulation scheme, protocols, equalization, etc., for transmit and receive are all determined in software within the DSP.

Reprinted from *IEEE Transactions on Vehicular Technology*, Vol. 49, No. 2, pp. 643–650, March 2000.

- The "ideal" circulator, in conjunction with ideal (perfect) matching between it and the antenna and power amplifier impedances, results in the elimination of the requirement for a diplexer in the radio. Since the diplexer is very much a fixed-frequency component within a radio, its elimination is a key element in any multiband or even multistandard radio. Note that the circulator would also have to be very broadband, which most current designs are not.
- The linear power amplifier ensures an ideal transfer of the RF modulation from the DAC to high power suitable for transmission, with low, ideally no, adjacent channel emissions.

D. A/D Performance

The specifications for an A/D converter appropriate for use in the "ideal" software radio could be as follows.

Resolution: 20 b (= 121.76-dB dynamic range). This value results from assuming that >100-dB signal range is required (e.g., from −20 to −120 dBm) and that a 12-dB signal-to-noise ratio is required at minimum sensitivity.

Sample Rate: 48 MSPS (based on 8 × oversampling of the proposed FRAMES mode 2 CDMA scheme [5], with alias downconversion). Although this sample rate could be used to effectively realize an extra bit of resolution, it is assumed that oversampling is required as a method of relaxing the analog antialias and–or IF filtering requirements. These would otherwise be onerous, particularly when the gain and phase flatness requirements of most digital schemes are considered.

RF Input Bandwidth: DC/100 MHz–2.2 GHz (to cover PMR, cellular, PCN/PCS, mobile satellite, and the UMTS satellite band).

The above are clearly very exacting specifications and are way in excess of those achievable with present technology. They do, however, serve to illustrate what would be required in a software radio receiver of the type illustrated in Fig. 1 for a reasonable level of RF performance (i.e., reasonable resistance to blocking coupled with good sensitivity).

Converter technology is continually improving, and, hence, it is not unrealistic to predict that further improvements in resolution and sampling rate will continue to be made. In addition, the trend from bipolar or BiCMOS implementations [6]–[10] to CMOS [11], [12] brings further potential benefits in terms of lower power consumption. In particular, the converter described in [12] incorporates a sample-and-hold device which does not reset between samples, and this may be one method of reducing the power consumed by the mechanism described in the following section. It does, however, also indicate that the sample-and-hold device consumes a fraction of the power of the complete chip (10–15%), indicating that the other technology areas have a long way to go before the sample-and-hold circuit dominates power consumption.

E. Generic Form of an A/D Converter

A generic form of A/D converter is shown in Fig. 2 [13] and consists of an antialias filter, a sample-and-hold, a method of converting the (now constant level) analog voltage into a digital word (the "quantizer"), and finally a digital buffer. The quantizer element may be implemented in a wide variety of ways,

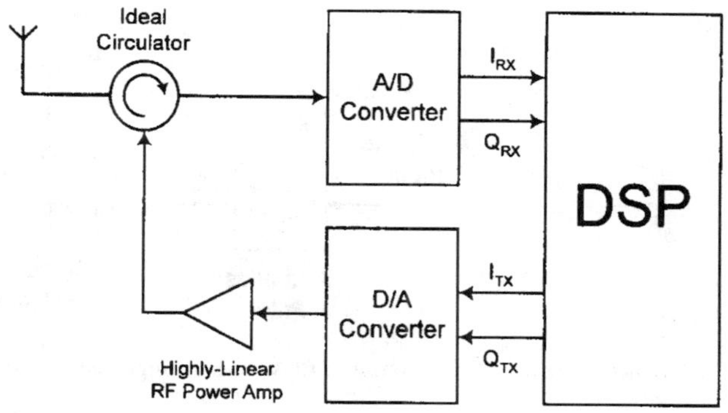

Fig. 1. "Ideal" software radio architecture.

including flash, successive approximation, sigma-delta, bandpass sigma-delta, and subranging. The analysis presented below is independent of the implementation technology and is based on the requirements of the sample-and-hold element.

Note that although the antialias filter of Fig. 2 is shown as a bandpass element, it may be a low-pass device in many applications.

II. Derivation of Minimum Power Consumption

This analysis is based on the assumption that the A/D converter itself consumes no power and that the only power supplied to it is in the form of the input signal. This is obviously a very simplistic argument, as the digital output circuitry and the actual conversion circuitry will clearly need to consume power in a practical design. It does, however, serve to illustrate the theoretical minimum power which a converter could possibly consume and hence provides a power limit below which is it potentially not possible to go.

Current A/D converters are very far from meeting this limit (by over four orders of magnitude), hence indicating that technology has a long way to go before the figures presented here may be approached. They do, however, serve to indicate that the "ideal" software radio architecture discussed above may never be a practical reality for a handportable (i.e., battery-powered) design.

The analysis assumes that a sample-and-hold device is employed in the converter to be used and hence may not be valid for all converter types (e.g., flash converters of the type reported in [14]). Given the complexity of this type of converter however, it is unlikely that 18+ bit devices will be available in a reasonably foreseeable timescale.

It is also assumed that it is the input signal to the converter which is supplying the "power" in order to charge the capacitance in the sample-and-hold. It is common practice for the input signal to be buffered within the converter and hence it is this buffer which would, in practice, supply this power (deriving its power in turn from the power supply to the converter circuit). There are, however, benefits in not using a buffer in very high-speed applications, for example, in order to eliminate offsets, such a converter is described in [15].

For an A/D converter to take full advantage of the resolution available to it, the quantization noise power must be lower than (or possibly equal to) the thermal noise power present at the converter input (within the required converter bandwidth).

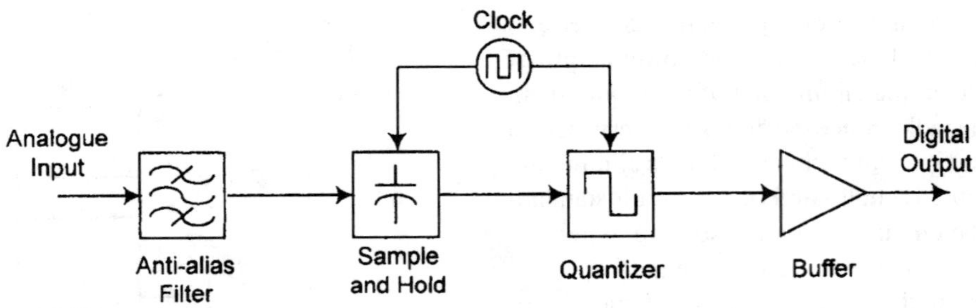

Fig. 2. Generic form of an A/D converter for wide-band digitization at IF or RF.

This noise level then sets the minimum possible peak input signal level (i.e., the minimum possible full-scale voltage) for the converter (see Fig. 3). Based on these levels, it is possible to calculate the converter power consumption from the minimum charging current of the converter input capacitance, which is in turn based upon the thermal noise floor (kT/C_i).

The signal to quantization noise ratio (dynamic range) of an A/D converter is given by

$$D_C = 6n + 1.76 \text{ dB} \tag{1}$$

where n is the resolution (number of bits) of the converter.

The converter will have a full-scale input voltage level V_{fs}, and, hence, the converter noise floor must be at a maximum level of D_C decibels below this value in order for the converter resolution to be fully realized. Hence

$$D_C = 20 \log_{10}\left(\frac{V_{fs}}{e_{nq}}\right) \tag{2}$$

where e_{nq} is the noise voltage level of the quantization noise floor.

Combining (1) and (2) gives

$$e_{nq} = \frac{V_{fs}}{10^{(6n+1.76)/20}}. \tag{3}$$

Hence, the mean-square quantization noise voltage is

$$e_{nq}^2 = \frac{V_{fs}^2}{10^{(6n+1.76)/10}}. \tag{4}$$

For the converter to take full advantage of this dynamic range, the quantization noise level must be greater than or equal to the thermal noise floor of the converter. This thermal noise floor is given by [16]

$$e_{nt}^2 = \frac{kT}{C_i} \tag{5}$$

where k is Boltzmann's constant ($k = 1.38 \times 10^{-23}$ J/K), T is the device temperature (in Kelvin), and C_i is the converter input capacitance (in farads).

Setting the two noise floors to be equal yields [from (4) and (5)]

$$e_{nq}^2 = e_{nt}^2 = \frac{V_{fs}^2}{10^{(6n+1.76)/10}} = \frac{kT}{C_i}. \tag{6}$$

Hence

$$C_i = kT\,\frac{10^{(6n+1.76)/10}}{V_{fs}^2}. \tag{7}$$

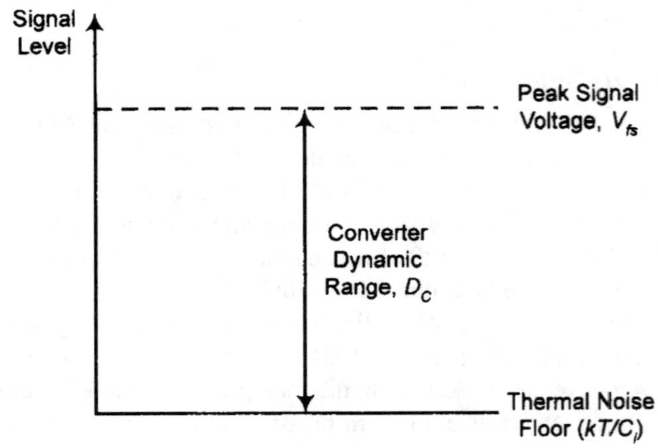

Fig. 3. Dynamic range of an A/D converter.

For the converter to function correctly, in terms of it accurately converting the input voltage it is presented with, this capacitance must be capable of being charged to the full-scale voltage of the converter, within its sampling interval (preferably well within). A charge of Q_i coulombs must therefore be stored in the input capacitance within the sampling interval t_s, giving

$$Q_i = I_i t_s \tag{8}$$

and

$$Q_i = C_i V_{fs}. \tag{9}$$

Hence

$$C_i = \frac{I_i t_s}{V_{fs}}. \tag{10}$$

Combining (7) and (10) gives

$$I_i = \frac{kT}{t_s}\,\frac{10^{(6n+1.76)/10}}{V_{fs}}. \tag{11}$$

Finally, the power consumed in this process may be found from

$$P_i = I_i V_{fs}. \tag{12}$$

Giving

$$P_i = \frac{kT}{t_s}\,10^{(6n+1.76)/10} \text{ W} \tag{13}$$

from (11). Note that this is independent of V_{fs}, the full-scale voltage for the converter.

Note that in many systems, the quantization noise may be artificially reduced to a level below the thermal noise floor (e.g.,

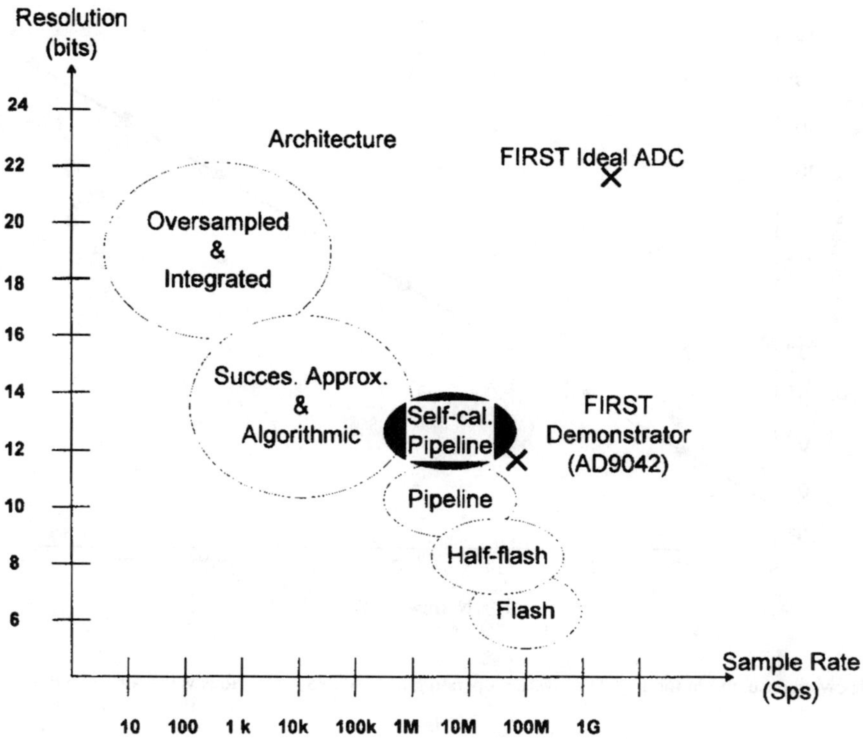

Fig. 4. ADC technology map.

by decimation), and hence it is the thermal noise floor which is the ultimate limit. Systems employing a sampling rate too low to allow decimation can, however, also be assessed using this analysis. Results for both are included below.

III. ADC Technology

The different A/D converter technologies are outlined in Fig. 4, with the "ideal" requirement identified within the ACTS FIRST[8] project highlighted, along with the actual converter chosen for the demonstrator. Pipeline and self-calibrated pipeline converters are highlighted as these are potentially the most appropriate for wireless transceiver applications.

A. *Factor of Merit*

1) Converter Efficiency: Current A/D converters are very far from meeting the theoretical limit discussed in the above sections (by over four orders of magnitude), hence indicating that technology has a long way to go before the figures presented above may be approached. They do, however, serve to indicate that the "ideal" software radio architecture discussed above may never be a practical reality for a handportable (i.e., battery-powered) design, without a revolution in A/D converter technology.

The power consumption performance of an ADC can be quantified by the energy per conversion and per unit of resolution calculated as follows. Combining (5) and (10) gives

$$(\mathrm{S/N})^2 = \frac{V_{fs}^2}{e_{nt}^2} = \frac{V_{fs} I_i t_s}{KT} = \frac{P_i}{KT f_s}. \tag{14}$$

[8]ACTS FIRST (flexible integrated radio system technology) is a European Commission-funded project examining the issues surrounding the implementation of a software radio for UMTS.

Hence, for a nonideal converter in which the conversion energy E_{CR} exceeds the ideal value of kT

$$E_{CR} = \frac{P_i}{f_s(\mathrm{S/N})^2}\ \mathrm{J} \tag{15}$$

where f_s is the sampling frequency and (S/N) is the signal to noise ratio of the converter.

For an ADC, the ratio E_{CR}/kT represents an excess power consumption factor. This can therefore be used to define an A/D power consumption factor of merit

$$M = \frac{E_{CR}}{kT}. \tag{16}$$

An ideal converter would have a factor of merit of unity.

IV. Results

For the results presented in Fig. 5, the sampling rate of 48 MSPS was chosen as potentially representative for alias down-conversion sampling of a future UMTS standard (FRAMES mode 2 [5], with 8 × oversampling). In addition, it was assumed that a 3-dB margin would be necessary between the quantization noise floor and the thermal noise floor for the converter to operate over its full usable dynamic range (as defined by its resolution). Alternatively, this could be considered as a 3-dB noise figure for the converter, assuming that decimation is employed to reduce the quantization noise floor. The results were derived from (13) using a sampling frequency of 48 MSPS.

It can be seen from this graph that for the 20-b resolution chosen above, the theoretical minimum power consumption would be around 600 mW when operating at 48 MSPS. This is clearly a high value for consideration to be included in a handset [in addition to the fast DSP processor(s), memory,

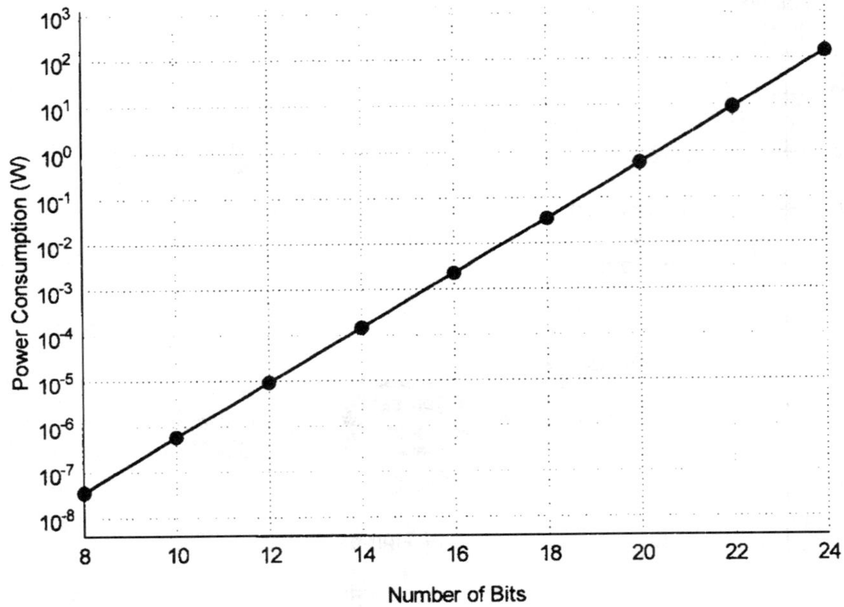

Fig. 5. Minimum theoretical power consumption for an A/D converter, operating at 48 MSPS, for various values of resolution.

linear transmitter, etc.] and would probably be considered excessive in most designs. Any resolution above this value would obviously be far too inefficient for handset applications and would probably be considered excessive for a base-station use (particularly the 24-b device).

Fig. 6 shows how the situation alters with sample rate and hence what could be expected to happen if direct sampling of the RF waveform (as opposed to alias downconversion) were to be contemplated. This would have the advantage of making the antialias filter design more realistic and hence would be significantly closer to the ideal scenario.

Again, considering the example of a 20-b converter, but now examining the case where the sample rate is chosen to (just) Nyquist sample the complete bandwidth outlined in Section I-D (DC to 2.2 GHz), the resulting power consumption is around 50 W. This is clearly not realistic for a handset application and hence indicates that this mechanism will not be realistic as a solution within a software radio without a radical (and currently unknown) advance in technology. Reducing the sampling rate by an order of magnitude, to accommodate, for example, a VHF software radio or wide-band IF sampling for a higher frequency radio, still results in a power consumption of around 5 W. This is again clearly out of the question for a handportable unit.

It is also worth examining how far current A/D converter solutions fall short of the limits identified in this analysis. As an example, the AD773JD converter from analog devices is a 10-b 10-MSPS device, with an input bandwidth of 100 MHz. To give this performance, it consumes a maximum power of 1.5 W. Comparing this with the theoretical minimum power consumption for this specification yields a figure of 1.2×10^{-7} W, which is around seven orders of magnitude smaller. While this does not represent the state-of-the-art in terms of power consumption versus speed and resolution, it does give an indication of how far above the ideal power consumption level current devices are operating.

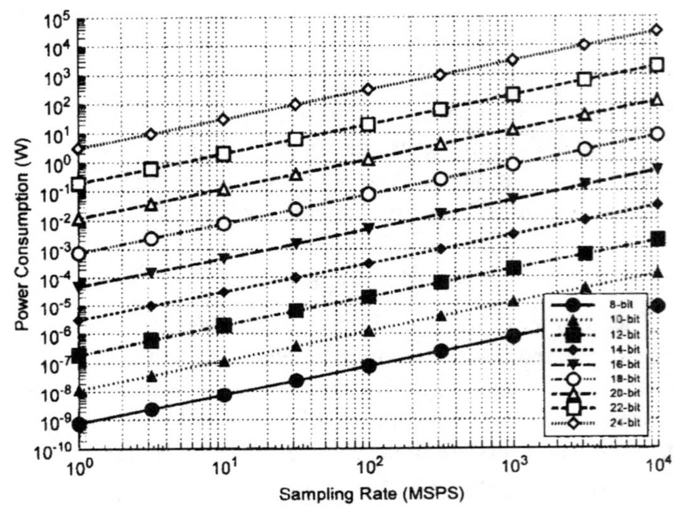

Fig. 6. Minimum theoretical power consumption for an A/D converter over a range of sampling rates.

Finally, the effect of different assumptions regarding the margin required between the thermal and quantization noise floors (or practical noise figure values for the converter) and the amount of the sample time allowed for input capacitance charging is shown in Figs. 7 and 8. In Fig. 7, the margin between the thermal and quantization noise floors is increased to 10 dB [giving the converter a theoretical noise figure of $10\,log(1.1) = 0.4$ dB compared to a noise figure of 1.8 dB for the assumption used in deriving Fig. 6]. This will allow a greater margin for added noise from, for example, practical input buffer stages.

If Figs. 6 and 7 are compared, taking for example the 20-b converter discussed previously, it can be seen that the power consumption has increased from 600 mW to almost 3 W at 48 MSPS. This is a 7-dB increase in power consumption, which is

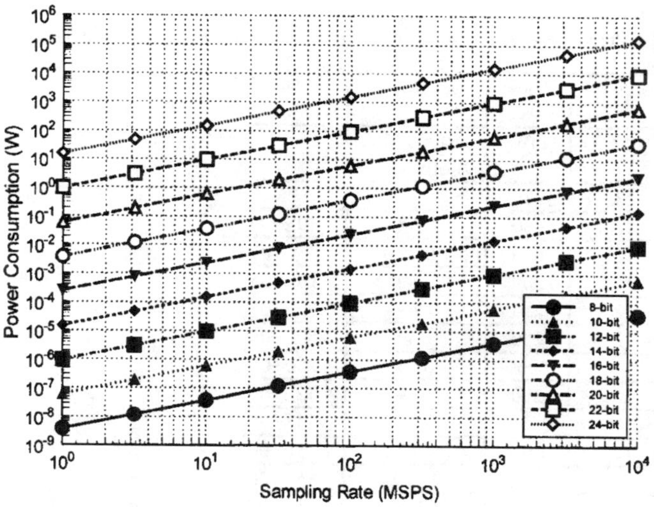

Fig. 7. Minimum theoretical power consumption for an A/D converter over a range of sampling rates, assuming that a 10-dB margin is required between the thermal and quantization noise floors.

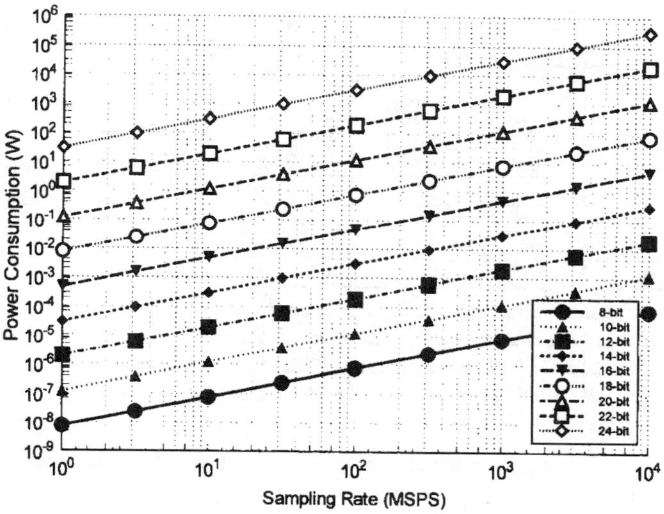

Fig. 8. Minimum theoretical power consumption for an A/D converter over a range of sampling rates, assuming that only 10% of the sample time is available for charging of the input capacitance.

directly equivalent to the 7-dB change in the noise floor differential. Clearly, this relaxation in noise performance for the converter is paid for by a substantial increase in power consumption and hence it is worthwhile utilizing the minimum possible differential between the two noise floors.

Examining Figs. 6 and 8, indicates a similar picture if the time allowed for capacitor charging is reduced to one tenth of the sample time (with the remainder then being available for the conversion operation itself). Again, the increase in power consumed is in direct proportion to the change (in this case, the change in time), as the power required for the 20-b 48-MSPS converter has increased from around 600 mW to around 6 W. Again, this is significant and indicates that other aspects of conversion should occupy the minimum amount of time possible in order to optimize power consumption.

It is possible, of course, to trade sampling rate for resolution, i.e.. to achieve a higher effective resolution using oversampling techniques. The benefit which may be gained by this technique, in the form of an effective increase in the number of bits of A/D resolution N_{eb}, is given by

$$N_{eb} = \frac{10 \log(M/2)}{6} \tag{17}$$

where M is the oversampling factor, given by

$$M = \frac{f_s}{B} \tag{18}$$

and B is the RF channel bandwidth. Thus, using a 200-MSPS converter on a GSM channel (200-kHz bandwidth) effectively provides an extra 4.5 b of A/D resolution. Thus, a 16-b converter could be used in place of the 20-b unit proposed in Section I-D. The increase in sampling rate also leads to a proportionate increase in power consumption [from (13)], and this negates the power consumption benefits of the extra resolution. This method may well, however, permit high-resolution converters to be realized at usable effective sampling rates within a few years, whereas conventional methods of increasing resolution may take longer to achieve.

A. Factor of Merit

Using (16), it is possible to compare the factors of merit for a range of current ADC devices. Table I shows a comparison between the factors of merit for a range of high-performance high-speed converters currently available. It can be seen from this that current devices are many orders of magnitude above the theoretical performance level. Table II shows the same for a range of current audio ADC's; some of these are closer to the theoretical level, with the best parts being just under four orders of magnitude above.

V. Trends

A. Power Consumption

Since all existing converters are still many orders of magnitude above the minimum theoretical limit, the power consumption reduction observed during the last 10–15 yr is mainly due to process evolution and design optimization. It is therefore most likely to continue during the next 5–10 yr when the minimum limit will be approached (excess power consumption in the order of 10–100 times).

Fig. 9 illustrates the fast decrease of the energy spent by conversion per unit of resolution (E_{CR}) by commercial high-speed ADC's during the last five years. It shows the position of all high-speed ADC's with output sample rates of at least 20 MSPS listed in Table I. By this method, it can be shown that the factor of merit of ADC's has decreased almost consistently by a factor of ten every five years during the last 12 years, based upon research carried out within the ACTS FIRST project.

B. Analog Bandwidth

The analog bandwidth of a converter is proportional to the reciprocal of the square of the gate length of the active devices used in the sample-and-hold, i.e.,

$$B = \frac{k_a}{L^2}\ \text{Hz} \tag{19}$$

TABLE I

FACTOR OF MERIT FOR A SELECTION OF CURRENT HIGH-SPEED A/D CONVERTER IC'S. THESE ARE MAINLY PIPELINE AND SELF-CALIBRATED PIPELINE DEVICES

Year	*Company*	*Reference*	*Sample rate*	*Resolution*	*Power cons.*	E_{CR}	*M*
			MSPS	bits	mW	$\times 10^{-15}$ J /unit of SNR	-
1997	Comlinear	CLC952	41	12	?	?	?
1997	Burr Brown	ADS800	40	12	390	0.58	1.40E+05
1997	ADI	AD9042	50	12	600	0.72	1.73E+05
1996	Comlinear	CLC949	20	12	300	0.89	2.16E+05
1996	ADI	AD9220	10	12	250	1.49	3.60E+05
1996	BB	ADS802	10	12	250	1.49	3.60E+05
1995	Analogic	ADC3120	20	14	5000	0.93	2.25E+05
1995	Harris	HI5808	10	12	300	1.79	4.32E+05
1995	Harris	HI5810	20	10	150	7.15	1.73E+06
1994	ADI	AD9023	20	12	1500	4.47	1.08E+06
1994	Comlinear	CLC938	30	12	6570	13.05	3.15E+06
1992	ADI	AD9020	60	10	3400	54.04	1.31E+07
1992	ADI	AD9014	10	14	12800	4.77	1.15E+06

TABLE II

FACTOR OF MERIT FOR A SELECTION OF CURRENT AUDIO A/D CONVERTER IC'S. THESE ARE MAINLY OVERSAMPLED DEVICES

Year	*Company*	*Reference*	*Sample rate*	*Resolution*	*Power cons.*	E_{CR}	*M*
			KSPS	bits	mW	$\times 10^{-15}$ J /unit of SNR	-
1997	Crystal	CS7870	100	12	75	44.70	1.08E+07
1997	Crystal	CS7875	100	16	260	0.61	1.46E+05
1996	ISSCC	J. Rabii	50	16	6.3	0.03	7.09E+03
1994	MAXIM	MAX120	50	12	210	250.34	6.05E+07
1994	MAXIM	MAX176	25	12	148	352.86	8.52E+07
1994	MAXIM	MAX191	100	12	30	17.88	4.32E+06
1992	ADI	AD676	100	16	350	0.81	1.97E+05
1992	ADI	AD679	128	14	750	21.83	5.27E+06
1992	ADI	AD678	200	12	750	223.52	5.40E+07
1992	Crystal	CS5389	48	18	550	0.17	4.03E+04

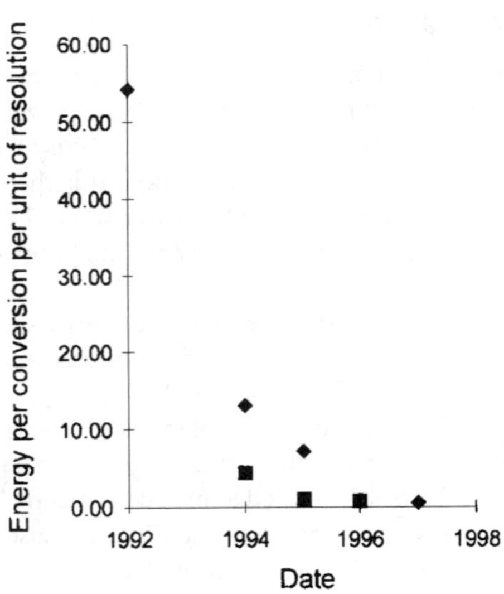

Fig. 9. Trends in energy per conversion per unit of resolution (E_{CR}); the square markers indicate 20-MSPS devices, while the diamond markers indicate devices with the highest sample rate available in that year (or earliest introduced at that sample rate for 1995).

where L is the gate length of the active device and k_a is a constant.

Examining the trend over recent years indicates that the minimum gate length reduction factor is two every four years. Therefore, this implies an analog bandwidth increase by a factor of four during the same time period. Based on current devices, it can therefore be predicted that within the UMTS timeframe (2005–2008) an analog input bandwidth in the mid 100's of the megahertz region will be possible with a resolution of around 16-b and an output sample rate of 50 MSPS. This device could be predicted to consume around 100 mW.

VI. CONCLUSIONS

This paper has derived a theoretical minimum power consumption for an analog-to-digital converter with a view to examining the practical realization of an "ideal" software radio of the type shown in Fig. 1. The results of the analysis appear to indicate that the conversion rates and resolutions required to implement a complete software radio by this "ideal" means are unlikely ever to be realistic with a power consumption commensurate with handportable operation. It is therefore concluded that in the absence of a revolution in ADC technology, alternative software radio architectures and, in particular, RF receiver architectures, are required in order to make the software radio concept a reality, with good radio performance, in a handportable design.

The performance levels required for IF and baseband sampling of high-bandwidth signals (e.g., UMTS) are, however, a considerably more realistic proposition, and it is therefore this type of architecture which shows the most promise for implementation in handportable software radio designs.

ACKNOWLEDGMENT

The authors would like to thank their colleagues on the ACTS FIRST project for helpful discussions on the subject of this paper.

REFERENCES

[1] A. Bateman and D. M. Haines, "Direct conversion transceiver design for compact low-cost portable mobile radio terminals," in *39th IEEE Veh. Technol. Conf.*, vol. 1, San Fransisco, CA, May 1–3, 1989, pp. 57–62.

[2] A. Bateman, D. M. Haines, and R. J. Wilkinson, "Direct conversion linear transceiver design," in *IEE 5th Int. Conf. Mobile Radio and Personal Communications*, Warwick, U.K., Dec. 1989, pp. 53–56.

[3] J. M. Guez and M. Mazzella, "SINUS: Satellite integration into network for UMTS services," in *IEE Colloquium on Integration of Satellite and Terrestrial PCS*, London, U.K., Apr. 7, 1997, pp. 2/1–2/5.

[4] P. B. Kenington, "Dynamic channel multicarrier architecture (DC/MA)—Digital PMR in an analog world," in *Microwaves and RF Conf.*, London, U.K., Sept. 1998.

[5] "Basic description of multiple access scheme," Future Radio Wideband Multiple Access System—FRAMES Rep., CEC Deliverable Number: AC090/NOK/A10/DS/R/008/b1, Nov. 15, 1996.

[6] R. Petschacher, B. Zojer, B. Astegher, H. Jessner, and A. Lechner, "A 10 b 75 MSPS subranging A/D converter with integrated sample and hold," *IEEE J. Solid-State Circuits*, vol. 25, no. 6, pp. 1339–1346, 1990.

[7] P. Vorenkamp and J. P. M. Verdaasdonk, "A 10 b 50 MS/s pipelined ADC," in *Proc. Int. Solid-State Circuits Conf.*, San Fransisco, CA, 1992, pp. 32–33.

[8] W. T. Colleran and A. A. Abidi, "A 10-b 75-MHz two-stage pipelined bipolar A/D converter," *IEEE J. Solid-State Circuits*, vol. 28, no. 12, pp. 1187–1199, 1993.

[9] K. Sone, Y. Nishida, and N. Nakadai, "A 10-b 100-Msample/s pipelined subranging BiCMOS ADC," *IEEE J. Solid-State Circuits*, vol. 28, no. 12, pp. 1180–1186, 1993.

[10] F. Murden and R. Gosser, "12 b 40 M Sample/s two-stage A/D converter," in *Proc. Int. Solid-State Circuits Conf.*, San Fransisco, CA, 1995, pp. 216–217.

[11] K. Nagaraj, H. S. Fetterman, J. Anidjar, S. H. Lewis, and R. G. Renninger, "A 250-mW, 8-b, 52-Msamples/s parallel-pipelined A/D converter with reduced number of amplifiers," *IEEE J. Solid-State Circuits*, vol. 32, no. 3, pp. 312–320, 1997.

[12] K. Y. Kim, N. Kusayanagi, and A. A. Abidi, "A 10-b, 100-MS/s CMOS A/D converter," *IEEE J. Solid-State Circuits*, vol. 32, no. 3, pp. 302–311, 1997.

[13] J. A. Wepman, "Analog-to-digital converters and their applications in radio deceivers," *IEEE Commun. Mag.*, vol. 33, pp. 39–45, May 1995.

[14] H. Reyhani and P. Quinlan, "A 5V 6-b 80 Ms/s BiCMOS flash ADC," *IEEE J. Solid-State Circuits*, vol. 29, pp. 873–878, Aug. 1994.

[15] J. Yuan and C. Svensson, "A 10-bit 5-MS/s successive approximation ADC cell used in a 70-MS/s ADC array in 1.2-μm CMOS," *IEEE J. Solid-State Circuits*, vol. 29, pp. 866–872, Aug. 1994.

[16] J. Smith, *Modern Communication Circuits*. New York: McGraw-Hill, 1986, ch. 3.

An Overview of Sigma-Delta Converters

© Gary S. Chapman/The Image Bank

How a 1-bit ADC achieves more than 16-bit resolution

PERVEZ M. AZIZ,
HENRIK V. SORENSEN, and
JAN VAN DER SPIEGEL

Although real world signals are analog, it is often desirable to convert them into the digital domain using an analog to digital converter (ADC). Motivating designers to apply this conversion is the efficient transmission and storage of digital signals. Digital representation of an audio signal, for example, allows CD players to achieve virtually error free storage using optical disks [1]. Intricate processing of the signal may also necessitate analog to digital (A/D) conversion, since such processing is only feasible in the digital domain using either conventional digital computers or special purpose digital signal processors (DSPs). Signal processing in the digital domain is also extremely useful in such areas as biomedical applications, providing the needed accuracy for tasks such as ultrasound imaging.

One technique, sigma-delta modulation, has become quite popular for achieving high resolution. One significant advantage of the method is that analog signals are converted using only a 1-bit ADC and analog signal processing circuits having a precision that is usually much less than the resolution of the overall converter.

Although sigma-delta concepts have existed since the middle of the century, it is only in the last two decades that this method has become more attractive [2]. One reason is that recent advances in VLSI technology, focused towards realizing high speed densely packed digital circuits, have made feasible the adequate digital processing of the bit stream. Using sigma-delta A/D methods, high resolution can be obtained for only low to medium signal bandwidths.

This article briefly describes conventional A/D conversion, as well as its performance modeling. We then look at the technique of *oversampling*, which can be used to improve the resolution of classical A/D methods. We discuss how sigma-delta converters use the technique of *noise shaping* in addition to oversampling to allow high resolution conversion of relatively low bandwidth signals. Next, we examine the use of sigma-delta converters to convert narrowband bandpass signals with high resolution. Several parallel sigma-delta converters, which offer the potential of extending high resolution conversion to signals with higher bandwidths, are also described.

Reprinted from *IEEE Signal Processing Magazine*, Vol. 13, No. 1, pp. 61–84, January 1996.

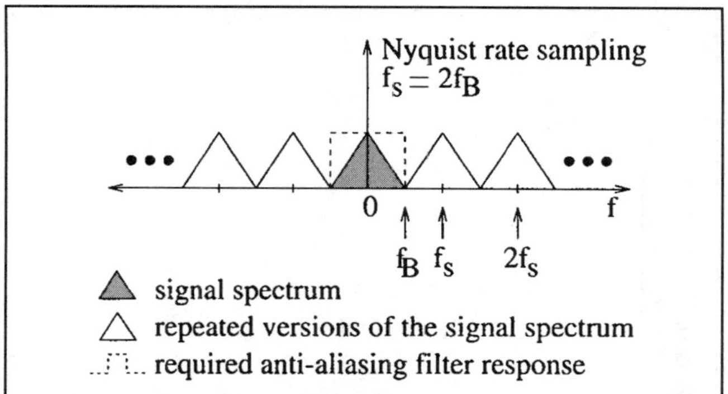

1. Nyquist rate sampling showing the original band limited signal spectrum, periodically repeated versions of the signal spectrum due to sampling, and the anti-aliasing filter response needed to band limit the signal.

PCM A/D Conversion

Nyquist Rate Conversion

Sampling

Analog to digital conversion of a signal is traditionally described in terms of two separate operations: uniform sampling in time, and quantization in amplitude. In the sampling process, a continuous time signal is sampled at uniformly spaced time intervals, T_s. The samples, $x[n]$, of the continuous time signal, $x(t)$ can be represented as $x[n] = x(nT_s)$. The effect, in the frequency domain, of the sampling process is to create periodically repeated versions of the signal spectrum at multiples of the sampling frequency $f_s = 1/T_s$ [3, pp. 80-87]. This relationship is written in Eq. 1, where $X_s(f)$ represents the spectrum of the sampled signal, and $X(f)$ is the spectrum of the original continuous time signal.

$$X_s(f) = \frac{1}{T_s} \sum_{k=-\infty}^{\infty} X(f - kf_s) \tag{1}$$

The sampling process is shown graphically in Fig. 1 for the case where $f_s = 2f_B$, and f_B is the bandwidth of the signal. In general, the signal can be reconstructed back to continuous time if the repeated versions of the signal spectrum do not overlap. Thus, the signal must be band limited to half the sampling rate, i.e., a signal with bandwidth f_B must be sampled at a rate greater than twice the bandwidth, $f_s \geqslant 2\ f_B$. Interference between the repeated versions of the signal spectrum is known as aliasing and it prevents reconstruction of the signal.

Even if a signal is nominally band limited to $f_s/2$, an anti-aliasing filter is often used to ensure that the signal is indeed band limited. For example, speech has a nominal bandwidth of 4 kHz and so in principle can be sampled at 8 kHz. However, there is some residual signal energy above 4 kHz, which results in aliasing if a 8 kHz sampling rate is used. The anti-aliasing filter is a continuous time analog filter preceding the sampler.

The case where $f_s = 2f_B$ is known as Nyquist rate sampling, and clearly the anti-aliasing filter here must have a very sharp cutoff at frequency $f_B = f_s/2$ as shown in Fig 1. Later, we will discuss how the sharp cutoff requirement on this filter can be relaxed.

The discretization or quantization in time as a result of the sampling is an invertible operation, since no signal information is lost and the original continuous time signal can be perfectly reconstructed [3, pp. 80-91]. Note that although Fig. 1 shows the sampling process for the case where the signal is a baseband signal, i.e., the spectrum has a bandwidth centered at DC frequency, Eq. 1 still describes the sampled spectrum even if the signal spectrum is centered at some higher frequency f_c. In this case, for a signal bandwidth f_B, the signal spectrum occupies the region $[f_c - f_B/2, fc + f_B/2]$, and it will still be possible to avoid aliasing and reconstruct the signal provided that $f_s \geqslant 2\ f_B$.

Quantization

Once sampled, the signal samples must also be quantized in amplitude to a finite set of output values. Typical transfer characteristics of quantizers or A/D converters with an input signal sample, $x[n]$, and an output, $y[n]$, are shown in Fig. 2.

Quantization is a non-invertible process, since an infinite number of input amplitude values are mapped to a finite number of output amplitude values. The quantized output amplitudes are usually represented by a digital code word composed of a finite number of bits. For example, for the 1 bit A/D converter of Fig. 2c, the output levels V and $-V$ can be mapped to digital codes "1" and "0." The digital code words are often said to be in pulse code modulation (PCM) format.

Another way of looking at this would be to plot the digital

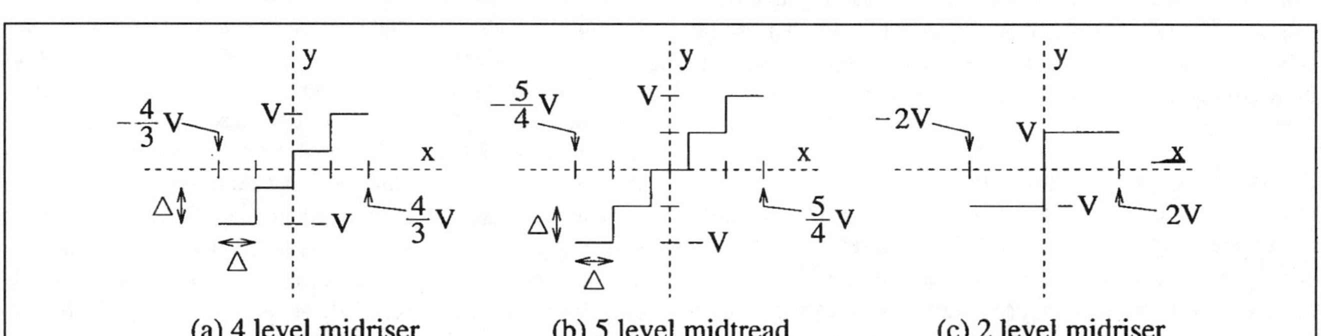

2. Transfer characteristics of typical A/D converters (ADCs) or quantizers.

code words instead of the quantized amplitude values for y in Fig 2. The quantized output amplitude values can also be considered the output of an ideal digital to analog converter (DAC) whose inputs are the corresponding digital code words.

An ADC or quantizer with Q output levels is said to have N bits of resolution where $N = \log_2(Q)$. As should be clear from Fig 2, for an ADC with Q quantization levels, only input values separated by at least $\Delta = 2V/(Q-1)$ can be distinguished or resolved to different output levels. N digital bits are needed to encode the Q codewords corresponding with each output level. The difference between the binary digital codes for two adjacent output levels is one least significant bit (LSB) of the overall N bit codeword. Consequently, a difference in input amplitudes corresponds to a one LSB difference in the digital output code words.

We now point out some general properties of the quantization transfer curves shown in Figs. 2a-2b for a four level (2 bit) "midriser" and a five level (roughly 2 bit) "midtread" ADC. Unlike the midtread ADC, the midriser ADC does not contain a zero output level for a zero input value, effectively creating a DC offset that may be undesirable in some applications. Note that all the transfer characteristics shown in Fig. 2 are symmetric. The midriser needs to have an even number of output levels to produce a completely symmetric transfer curve, whereas the midtread needs an odd number of output levels.

The midriser ADC's symmetric characteristic, with an even number of levels, is an advantage because the number of output levels, Q, can be made a power of two and encoded with exactly $N = \log_2(Q)$ bits. However, the number of output levels, Q, for a symmetric midtread ADC must be odd, and so cannot be made a power of two and encoded as efficiently. The number of bits needed will be $N = \log_2(Q-1) + 1$, where Q-1 is chosen a power of two. If the number of levels for the midtread ADC is forced to be a power of two by using only Q-1 levels, it will no longer have a symmetric transfer characteristic and will distort large amplitude symmetric input signals (e.g., a sinusoid). This distortion, of course, may be negligible when the number of output levels is very large.

These issues may play a role in choosing whether a midriser or midtread quantizer transfer characteristic should be used. For the special case of a 2 level quantizer, a midtread characteristic will not be able to represent both positive and negative output levels, and so will severely distort a signal containing samples of both polarities. For this 2 level case, a midriser characteristic, shown in Fig. 2c, will almost always be used.

Trading Resolution for Bandwidth

Most conventional A/D converters, such as the successive approximation, subranging, and flash converter types quantize signals sampled at, or slightly above, the Nyquist rate. Consequently, these converters are often referred to as Nyquist rate PCM converters. These, and other converters, provide tradeoffs among signal bandwidth, output resolution, and the complexity of the analog and digital hardware.

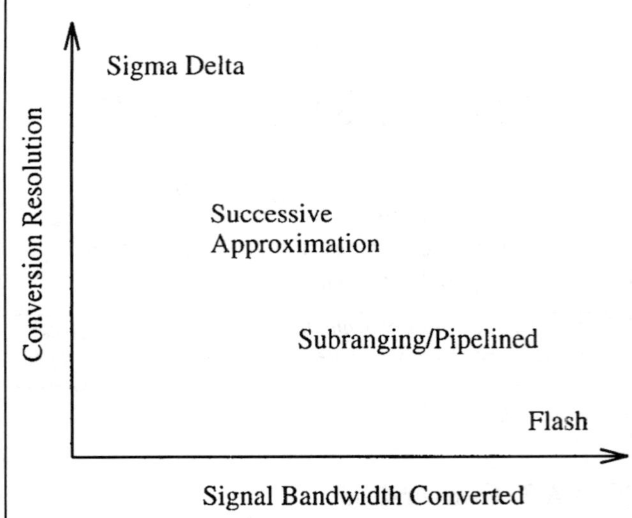

3. Bandwidth resolution tradeoffs.

Qualitative bandwidth and resolution tradeoffs of some of these A/D techniques, as well as sigma-delta conversion, is shown in Fig 3. As is evident from Fig 3, sigma-delta A/D converters attain the highest resolution for relatively low signal bandwidths. Consequently, sigma-delta techniques are often used in speech applications where the signal bandwidth is only 4 kHz and where up to 14 bits of resolution may be needed. Similarly, sigma-delta ADCs are popular for digital audio applications, where the signal bandwidth is 20-24 kHz and where high fidelity audio requires 16-18 bits of resolution. Flash converters, on the other hand, may be used for broadcast video applications where the signal band is about 5 MHz, but the resolution required is only about 8 bits.

Performance Modeling

Having looked at the sampling and quantization processes, we now examine the A/D converter and characterize its performance. The diagrams in Fig. 2 show the transfer characteristic for typical quantizers with input x and output y.

Let the maximum and minimum quantized output values always be V and $-V$. The least significant bit (LSB) of an ADC with Q quantization levels is equivalent to $2V/(Q-1)$. For both the midriser and midtread type of ADCs of Fig 2, the magnitude of the quantization error ($e = y - x$) between the output and input does not exceed half a LSB, i.e., $|e| \leq \Delta/2$, provided that $|x| \leq V + \Delta/2$. Under these circumstances, the quantizer or ADC is said to be not overloaded. For $|x| > V + \Delta/2$ (hence, $|e| > \Delta/2$), the ADC is said to be overloaded.

The quantizer embedded in any ADC is a non-linear system, which makes its analysis difficult. To make the analysis tractable, the quantizer is often linearized and modeled by a noise source, $e[n]$, added to the signal $x[n]$, to produce the quantized output signal $y[n]$:

$$y[n] = x[n] + e[n] \tag{2}$$

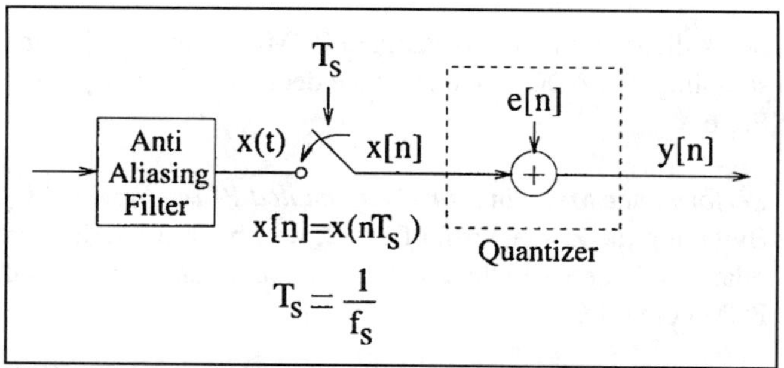

4. Block diagram and model of a conventional A/D converter (ADC) system.

A block diagram of an A/D system showing the sampling process and the quantizer model is shown in Fig 4. To further simplify the analysis of the noise from the quantizer, the following assumptions about the noise process and its statistics are traditionally made [3, p. 120]:

- The error sequence, $e[n]$, is a sample sequence of a stationary random process.
- $e[n]$ is uncorrelated with the sequence $x[n]$.
- The probability density function of the error process is uniform over the range of quantization error, i.e., over $\pm \Delta/2$.
- The random variables of the error process are uncorrelated, i.e., the error is a white noise process.

Under certain conditions, such as when the quantizer is not overloaded, N is large, and the successive signal values are not excessively correlated, these assumptions are reasonable [4]. Consider an N bit ADC with $Q = 2^N$ quantization levels, i.e., with $\Delta = 2V/ Q\text{-}1 = 2V/(2^N\text{-}1)$. For a zero mean $e[n]$, its variance σ_e^2 or power is

$$\sigma_e^2 = \frac{\Delta^2}{12} = \left(\frac{2V}{2^N - 1}\right)^2 / 12 \cong \left(\frac{2V}{2^N}\right)^2 / 12 \tag{3}$$

If the signal is treated as a zero mean random process and its power is σ_x^2, then the signal to quantization noise ratio is:

$$SNR = 10\log\left(\frac{\sigma_x^2}{\sigma_e^2}\right) = 10\log\left(\frac{\sigma_x^2}{V^2}\right) + 4.77 + 6.02\ (dB) \tag{4}$$

Note that for each extra bit of resolution in the ADC, i.e., for every increment in N, there is about a 6 dB improvement in the SNR. Thus, there is a direct relationship between the resolution of an ADC and its SNR, and it is common to equate differences in SNR in dB to bits by dividing the dB value by 6. For example, if an ADC has a SNR that is 3 dB better than that of another ADC, the better ADC will be said to have 1/2 bit higher resolution. Also, note that for a given N, the SNR in dB is linearly related to the signal power, σ_x^2, in dB.

Let us now examine the dynamic range of the ADC, which is a measure of the range of input amplitudes for which the ADC produces a positive SNR. For sinusoidal inputs, the dynamic range of the A/D converter is defined as the ratio of the signal power of a full scale sinusoid to the signal power of a small sinusoidal input that results in a SNR of 1 (or 0 dB) [5]. The signal power of a full scale sinusoid is $V^2/2$. A sinusoid with signal power $\sigma_x^2 = \sigma_e^2 = \Delta^2/12$ will result in an SNR of 1, or 0 dB.

The dynamic range, by definition, is

$$\left(\frac{V^2}{2} / \frac{\Delta^2}{12}\right) \cong \left(\frac{V^2}{2} / \frac{(2V/2^N)^2}{12}\right)$$

This expression reduces to a dynamic range value given by:

$$R = 6.02N + 1.76\ (dB) \tag{5}$$

Note that the ratio of $V^2/2$ to $\Delta^2/2$ is just the peak SNR of the ADC for a sinusoidal input. Consequently, the dynamic range of the Nyquist rate ADC is the same as its peak SNR. Sigma-delta converters do not necessarily have their peak SNR equal to their dynamic range. However, by using the dynamic range of a sigma-delta converter in Eq. 5 and calculating the corresponding N, we will be able to determine the resolution of a Nyquist rate PCM converter that would be required to produce the same dynamic range.

Limitations of Nyquist Rate ADCs

For Nyquist rate converters, each signal sample is quantized at the full precision or resolution of the converter. The resolution of such converters implemented on VLSI chips is limited by the technology in which these chips are fabricated. For example, some successive approximation A/D techniques rely on matching of two capacitors to perform a repeated division of a reference voltage by 2.

If such a converter is to convert signal values to N bits of resolution, the required matching on the capacitor components needs to be at least one part in 2^N. Matching of components to greater than 10 bits (one part in 2^{10}) or equivalently to more than 0.1% is difficult in VLSI.

High resolution Nyquist rate converters are extremely difficult to attain in current integrated circuit technology without the use of techniques such as laser trimming of components or calibration. Furthermore, if the signal is sampled too close to the Nyquist rate, the anti-aliasing filter must have a very sharp cutoff, which is a non-trivial design requirement for analog filters.

Oversampled PCM Conversion

System Description

Oversampled PCM conversion is a technique that improves the resolution obtained from straightforward Nyquist rate PCM conversion. This improvment is achieved by oversampling the signal, i.e., samples are acquired from the analog waveform at a rate significantly faster than the Nyquist rate. Each of these samples is quantized by an N bit ADC. Since quantization is described by Eq. 2, the total amount of noise power injected into the sampled signal, $x[n]$, is σ_e^2 and is given by Eq. 3.

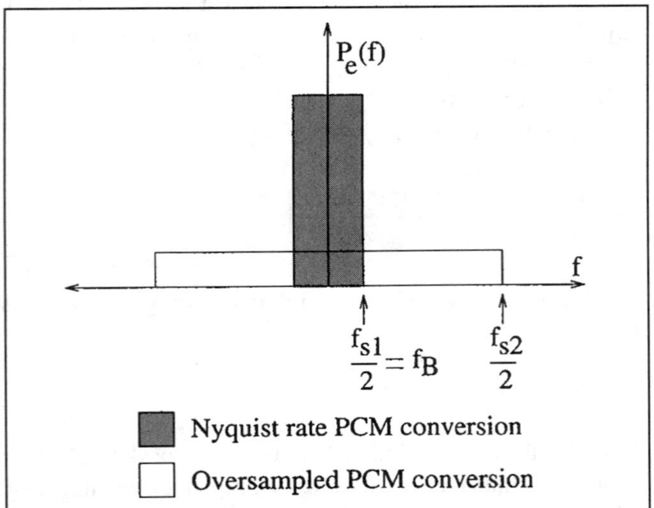

5. Quantization noise power spectral density for Nyquist rate PCM and oversampled PCM conversion.

Obviously, this is exactly the same noise power produced by a Nyquist rate converter, but its frequency distribution is different because of the higher sampling rate. The performance modeling criteria designated the noise process as white, which means that the noise power is uniformly distributed between $-f_s/2$ to $f_s/2$, where f_s is the sampling frequency.

Fig 5 shows the power spectral density, $P_e(f)$, of the quantization noise for Nyquist rate sampling with rate f_{s1} and oversampling rate f_{s2}. For Nyquist rate sampling where the signal band, $f_B = f_{s1}/2$, all the quantization noise power, represented by the area of the tall shaded rectangle, occurs across the signal bandwidth.

In the oversampled case, the same noise power, represented by the area of the unshaded rectangle has been spread over a bandwidth equal to the sampling frequency, f_{s2}, which is much greater than the signal bandwidth, f_B. Only a relatively small fraction of the total noise power falls in the band $[-f_B, f_B]$, and the noise power outside the signal band can be greatly attenuated with a digital low-pass filter following the ADC.

After the low-pass filtering is performed, the signal can be downsampled to the Nyquist rate without affecting the signal to noise ratio. The collective operation of low-pass filtering and downsampling is known as decimation. The low-pass filter and downsampler are collectively called a decimator. A block diagram of an oversampled PCM system showing the sampling, the ADC model, and the decimator is presented in Fig 6.

Performance Modeling for Oversampled PCM Converters

By taking the Z transform of Eq. 2, we obtain the Z domain relationship between the input and output of an oversampled PCM converter:

$$Y(z) = X(z) + E(z) \tag{6}$$

where Y, X and E are the Z transforms of the output, input signal and the quantization error process, respectively. Based on our two-input linear system model for the quantizer, Eq. 6 states that in the Z domain, the output is the input plus the quantization error or noise. We can observe that X and E both experience a unity transfer function.

A more general way of writing Eq. 6, where X and E do not necessarily experience a unity transfer function, is

$$Y(z) = X(z)H_x(z) + E(z)H_e(z) \tag{7}$$

The output is now the input signal modulated by a signal transfer function, denoted by $H_x(z)$, plus the quantization noise modulated by a noise transfer function, denoted by $H_e(z)$. To evaluate the performance of such a converter, we need to find the total signal and noise power at the output of the converter. To do this, we need to evaluate the power spectral densities, $P_{xy}(f)$ and $P_{ey}(f)$, of the signal and noise at the output of the converter, based on the power spectral densities, $P_x(f)$ and $P_e(f)$ of the signal and noise at the input of the converter. We can make use of the fact that if a stationary random process with power spectral density $P(f)$ is the input to a linear filter with transfer function $H(f)$, the power spectral density of the output random process is $P(f)|H(f)|^2$. Consequently,

$$P_{xy}(f) = P_x(f)|H_x(f)|^2$$
$$P_{ey}(f) = P_e(f)|H_e(f)|^2$$

For the oversampled PCM converter, $|H_x(f)| = |H_e(f)| = 1$, and our white noise assumption for $e[n]$ states that $P_e(f) = \sigma_e^2/f_s$, which implies $P_{ey}(f)$ is also σ_e^2/f_s.

Assuming an *ideal* low-pass filter with cutoff frequency

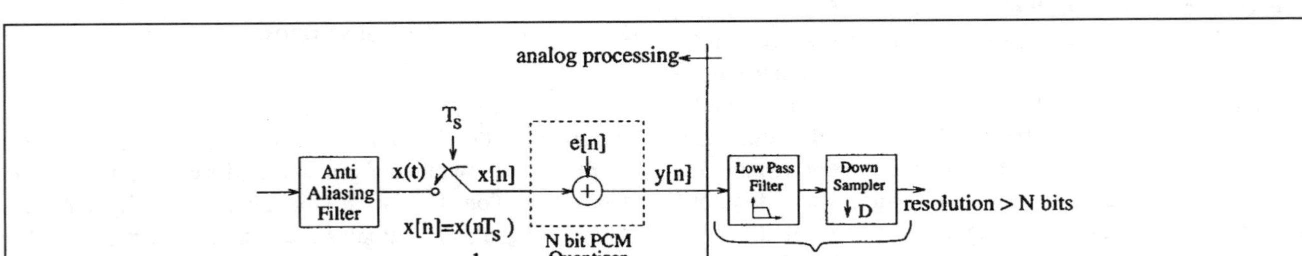

6. Oversampled PCM conversion system.

f_B following the oversampled quantizer, the *in-band* noise power, σ_{ey}^2 at the output of the A/D is

$$\sigma_{ey}^2 = \int_{-f_B}^{f_B} P_{ey}(f)df = 2\int_0^{f_B} P_{ey}(f)df$$

$$= \int_0^{f_B} \frac{2\sigma_e^2}{f_s} df = \sigma_e^2\left(\frac{2f_B}{f_s}\right)$$

Note that some of the noise power is now located outside of the signal band as a result of the oversampling, and so the in-band power σ_{ey}^2 is less than what it would have been without any oversampling (σ_e^2). Since the signal power is assumed to occur over the signal band only, it is not modified in any way and the signal power at the output (σ_{xy}^2) is the same as the input signal power σ_x^2. The maximum achievable SNR in dB is then:

$$SNR = 10\log\left(\frac{\sigma_x^2}{\sigma_{ey}^2}\right)$$

$$= 10\log(\sigma_x^2) - 10\log(\sigma_e^2) + 10\log\left(\frac{f_s}{2f_B}\right) \quad (dB) \tag{8}$$

For the case of Nyquist rate sampling where $f_s = 2 f_B$, this formula reduces to Eq. 3, which is the SNR for the Nyquist rate PCM quantizer. Letting the oversampling ratio $f_s/2f_B = 2^r$, we obtain,

$$SNR = 10\log(\sigma_x^2) - 10\log(\sigma_e^2) + 3.01r \quad (dB) \tag{9}$$

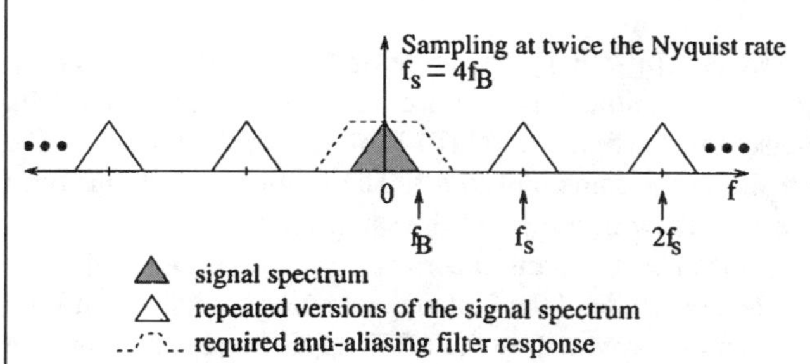

7. Sampling at twice the Nyquist rate, showing the original band limited signal spectrum, repeated versions of the signal spectrum due to sampling, and an anti-aliasing filter that is sufficient to band limit the signal.

For every doubling of the oversampling ratio, i.e., for every increment in r, the SNR improves by about 3 dB, or the resolution improves by one-half bit.

Note that in this scheme, we are trading speed for resolution. The higher resolutions are obtained at the expense of requiring the internal PCM quantizer to quantize samples at the oversampled rate. Analog circuit complexity has also been traded for digital circuit complexity. The analog circuit complexity is simplified, since we have said the resolution of the internal N bit quantizer, an analog circuit, is lower than that of the overall conversion resolution.

Another benefit, which is a direct consequence of the oversampling, is that the analog anti-aliasing filter does not need as sharp a cutoff. This can be seen from Fig. 7, where a signal is sampled at four times the nominal signal bandwidth (twice the Nyquist rate). In this case, the anti-aliasing filter can have a transition band between f_B and $f_s/2$ as long as it provides very good attenuation beyond $f_s/2$. However, a price is paid in the digital domain, since the digital filter must attenuate the remaining quantization noise power (beyond f_B) as much as possible. In the process of filtering out-of-band quantization noise, any other noise that existed in the transition band of the anti-aliasing filter prior to sampling will be attenuated further. The closer the low-pass filter approximates an ideal low-pass filter, the more resources it will need.

Finally, note that for high resolution conversion, one needs, $f_s >> f_B$, and the signal bandwidth must be small so that f_s does not exceed the maximum circuit speed attainable in the given technology. Let us use Eq. 8 for a simple calculation. Suppose we apply a full scale sinusoid with amplitude $V = 1$ corresponding with signal power $V^2/2 = 0.5$, as the input to an oversampled PCM ADC with a 20 kHz audio range signal band. Let the final desired resolution be 16 bits (for CD quality audio), which corresponds to a 98 dB SNR according to Eq. 4. Now, if we use an 8 bit A/D in an oversampled PCM scheme, i.e., if we use $N = 8$ in the expression for σ_e^2 in Eq. 8, we can calculate the f_s required for SNR = 98 dB and f_B = 20 kHz. The needed f_s is 2.64 GHz! Eight bit ADCs implemented in current CMOS technology certainly can not operate at such a high speed.

Suppose a 12 bit internal ADC is used instead. In this case, the required f_s is about 10 MHz, an operating speed that is not extremely high, per se, but which is still not trivial for a 12 bit ADC to attain. We will later see how sigma-delta modulation A/D conversion allows the use of internal ADCs with as low as 1 bit (i.e., $N = 1$) of resolution to achieve an overall resolution of 16 bits for a 20 kHz audio bandwidth.

Sigma-Delta Modulation A/D Conversion

A general way of writing the Z domain output of an A/D converter was given in Eq. 7, as $Y(z) = X(z)H_x(z) + E(z)H_e(z)$, where H_x is the signal transfer function (STF) and H_e is the noise transfer function (NTF). For oversampled PCM conversion, we saw that $H_x(z) = H_e(z) = 1$. This need not be the case and, in fact, oversampled A/D converters can be designed to incorporate noise shaping, where H_e is designed to be different from H_x such that H_x usually leaves the signal undisturbed but H_e shapes the noise to allow a high resolution output [2, 6].

Although the term delta-sigma ($\Delta\Sigma$) was used by some of the earliest researchers in the field [7], the term sigma-delta ($\Sigma\Delta$) has also become almost synonymous with noise shaping ADCs. We will use the term sigma-delta to describe noise shaping ADCs. As noted, oversampling reduces the quanti-

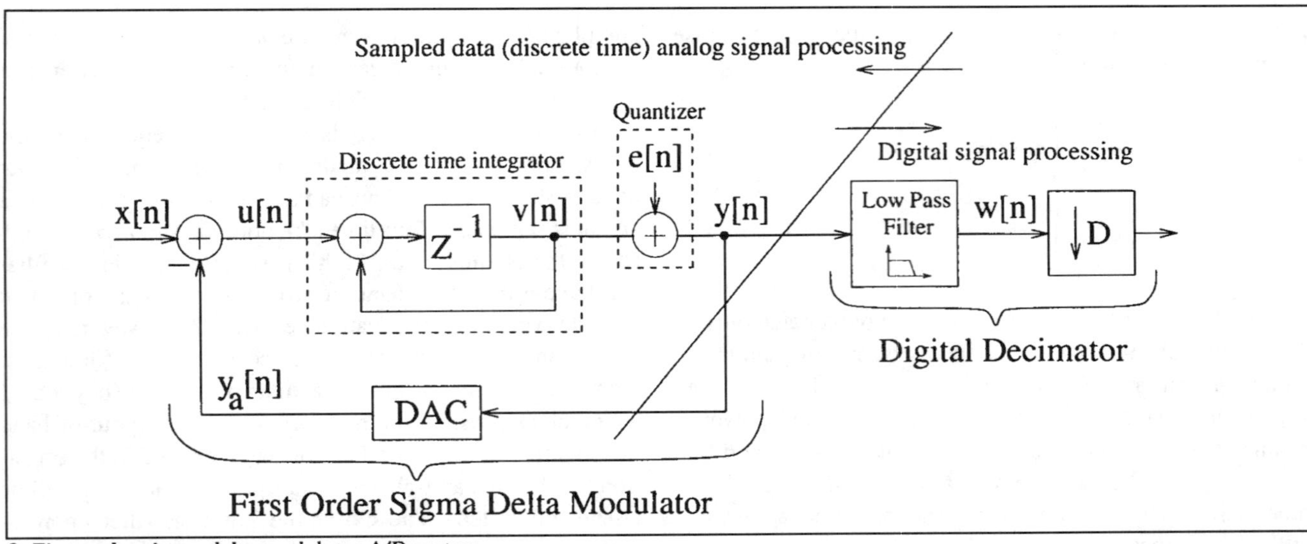

8. First order sigma-delta modulator A/D system.

zation noise power in the signal band by spreading a fixed quantization noise power over a bandwidth much larger than the signal band. Noise shaping or modulation further attenuates this noise in the signal band and amplifies it outside the signal band. Consequently, this process of noise shaping by the sigma-delta modulator can be viewed as pushing quantization noise power from the signal band to other frequencies. The modulator output can then be low-pass filtered to attenuate the out-of-band quantization noise and finally can be downsampled to the Nyquist rate.

The price of attaining high resolution is again a penalty in speed, as the hardware has to operate at the oversampled rate, and an increased complexity of the digital hardware. For high resolution conversion, the sampling frequency f_s must still be much greater than the signal bandwidth f_B, but, of course, not as great as needed by oversampled PCM conversion.

First Order Sigma-Delta Modulation

Operation and Performance Modeling

A block diagram of a first order sigma-delta modulator A/D system is shown in Fig 8. The system consists of an analog sigma-delta modulator, followed by a digital decimator. The modulator consists of an integrator, an internal A/D converter or quantizer, and a D/A converter (DAC) used in the feedback path.

The signal that is quantized is not the input $x[n]$ but a filtered version of the difference between the input and an analog representation, $y_a[n]$, of the quantized output, $y[n]$. The filter, often called the feedforward loop filter, is a discrete time integrator whose transfer function is $z^{-1}/(1-z^{-1})$.

The integrator and the rest of the sigma-delta analog circuit are typically implemented in sampled data switched capacitor technology. Consequently, the sampling operation is not shown explicitly in Fig. 8 or any other modulator architectures to be described in the rest of this article. Continuous-time versions of the modulator have also been considered [8], but this aspect of the modulators will not be discussed here. The linearized model replaces the quantizer with a noise source, $e[n]$, as shown in Fig 8.

If the DAC is ideal, it is replaced by a unity gain transfer function. The modulator output $Y(z)$ is then given by:

$$Y(z) = X(z)z^{-1} + E(z)(1 - z^{-1}) \tag{10}$$

so that $H_x(z) = z^{-1}$ and $H_e(z) = (1\text{-}z^{-1})$. The output is just a delayed version of the signal plus quantization noise that has been shaped by a first order Z domain differentiator or high-pass filter. The corresponding time domain version of the modulator output is

$$y[n] = x[n-1] + e[n] - e[n-1] \tag{11}$$

where the $e[n]$-$e[n$-$1]$ term is the first order difference of $e[n]$.

The magnitude spectrum of a first order sigma-delta noise transfer function (NTF) is plotted in Fig 9a, while Fig 9b shows the same plot in dB. The frequency axis has been normalized with respect to the sampling frequency, f_s.

Since $H_e(z)$ contains a zero at $z = 1$, i.e., at DC frequency on the unit circle of the Z plane, note the zero gain or infinite attenuation provided by the NTF at DC frequency. Note the large attenuation at lower frequencies and relative amplification at higher frequencies. For comparison, the oversampled PCM NTF, which has unity gain, is shown in Fig 9(a). The vertical bar demarcates the extent of the signal band, f_B, where f_B = 0.05 f_s. Quantization noise to the left of the bar that contributes to the finite resolution of the modulator is greatly attenuated while noise to the right of the bar is ~~not~~ attenuated as much or is actually amplified. However, noise to the right of the bar can mostly be removed with a digital low-pass filter.

Let us now examine the DAC before going on to examine the performance of the converter. The DAC is required to be nearly as linear as the overall conversion resolution. Any D/A non-linearity can be modeled as an error source that adds directly to the input. This error source benefits from the

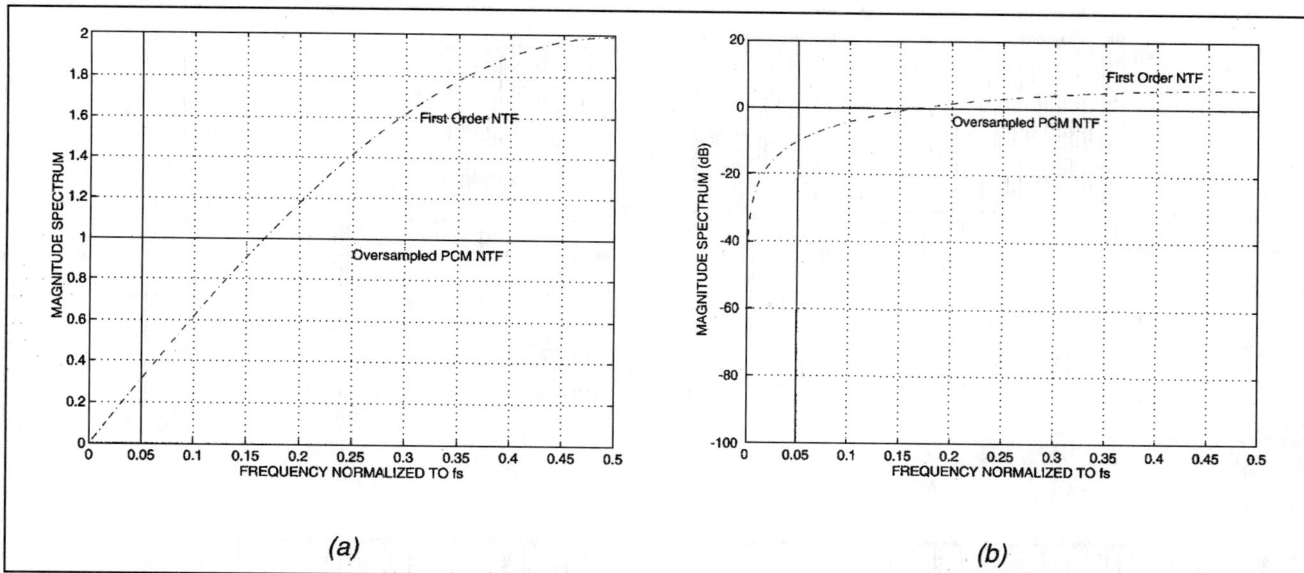

9. NTF for a first order sigma-delta modulator (a) magnitude spectrum on linear scale. For comparison, the oversampled PCM NTF, which has unity gain, is shown; (b) magnitude spectrum in dB.

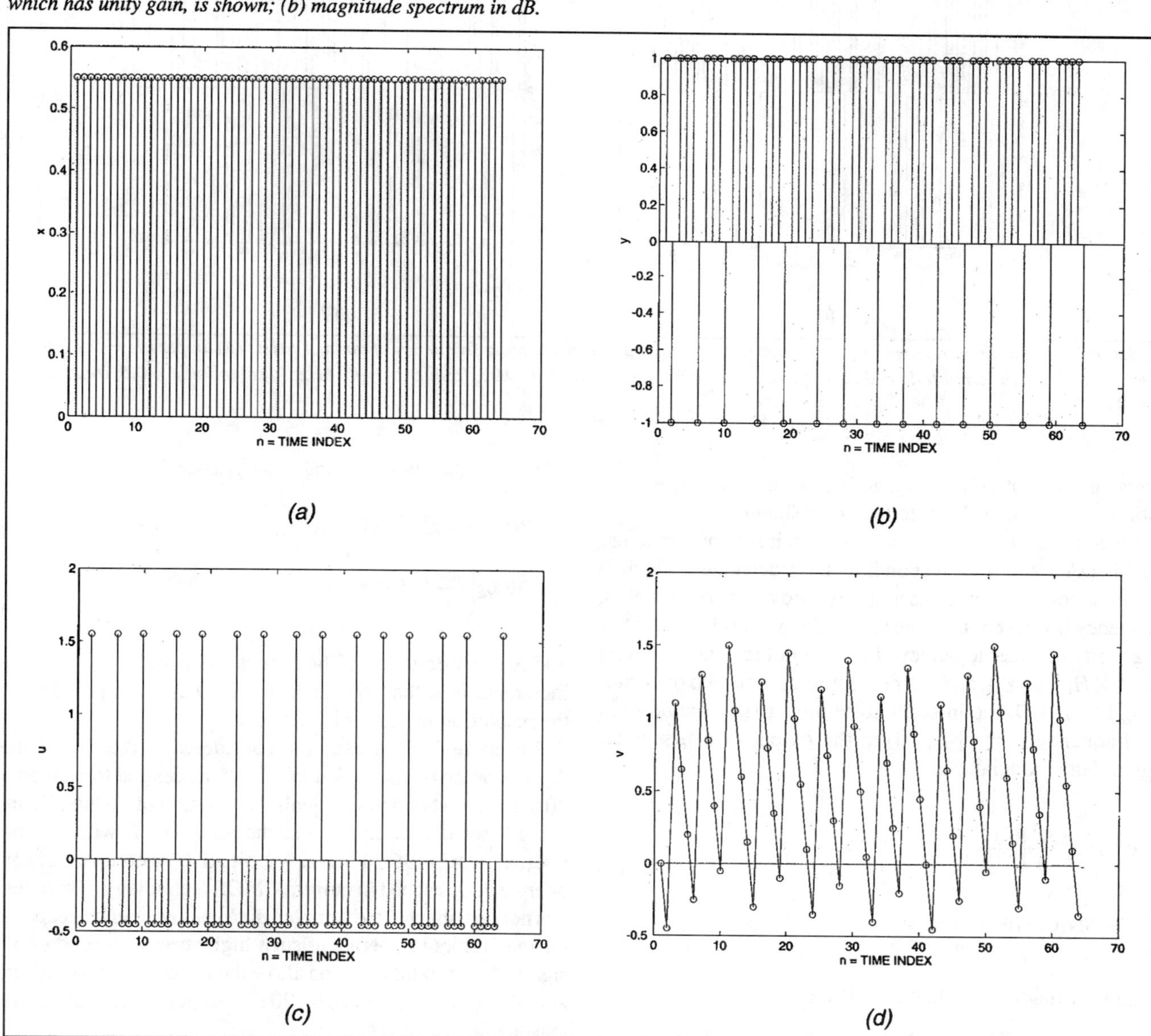

10. 1st order sigma-delta responding to a DC input: (a) DC input x[n] = 0.55 = 11/20. (b) modulator output y[n]; (c) "error" signal u[n]; (d) integrator output v[n]

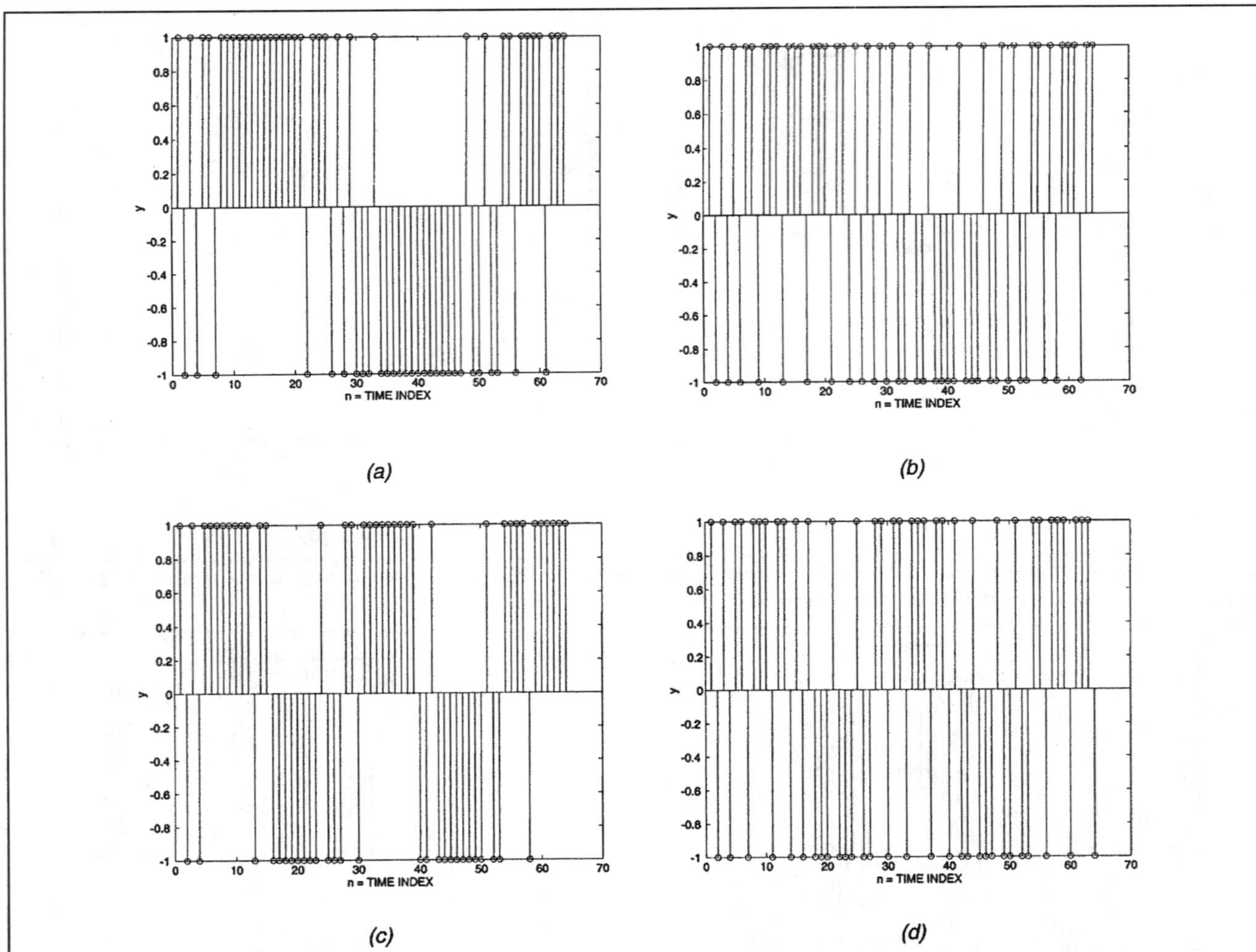

11. 1st order sigma-delta responding to various sinusoidal inputs with sampling frequency of about 1 MHz: (a) amplitude = 0.95, frequency= 20 kHz; (b) amplitude = 0.5, frequency = 20 kHz; (c) amplitude = 0.95, frequency = 40 kHz; (d) amplitude = 0.5, frequency = 40 kHz.

oversampling but unlike $e[n]$, which models the A/D quantization error, is not subject to the noise shaping.

Since a 1 bit DAC is perfectly linear, it is common to use a 1 bit DAC and a corresponding 1 bit quantizer, which is simply a comparator. Consequently, provided the sampling frequency is high enough, the sigma-delta A/D allows the use of a 1 bit quantizer to achieve high overall resolution. Using $H_x = z^{-1}$, $H_e = (1\text{-}z^{-1})$, and the procedure used for the oversampling PCM A/D, the in-band noise power (i.e., the noise in the frequency range $[-f_B, f_B]$) at the output of a first order sigma-delta modulator is

$$\sigma_{ey}^2 = \sigma_e^2 \frac{\pi^2}{3}\left(\frac{2f_B}{f_s}\right)^3 \qquad (12)$$

The SNR in dB is then:

$$SNR = 10\log(\sigma_x^2) - 10\log(\sigma_e^2) - 10\log\left(\frac{\pi^2}{3}\right) + 30\log\left(\frac{f_s}{2f_B}\right)\ (dB) \qquad (13)$$

Letting the oversampling ratio, $f_s/2f_B = 2^r$, we obtain:

$$SNR = 10\log\left(\sigma_x^2\right) - 10\log(\sigma_e^2) - 10\log\left(\frac{\pi^2}{3}\right) + 9.03r\ (dB) \qquad (14)$$

For every doubling of the oversampling ratio i.e., for every increment in r, the SNR improves by 9 dB, or equivalently, the resolution improves by 1.5 bits.

Let us revisit the example considered at the end of the discussion on oversampled PCM. It was desired to convert a 20 kHz audio band to CD quality resolution of 16 bits. Using Eq. 13, we can compute that the required f_s with a 1 bit internal ADC is 96.78 MHz. A 1 bit ADC or comparator can operate at this speed in current CMOS technology. However, it is not possible for the sampled data analog switched capacitor integrators to operate at such high speeds. Second order sigma-delta modulation can allow the use of a 1 bit quantizer, allowing us to meet the 16 bit 20 kHz target with a much more reasonable f_s.

Qualitative Time Domain Behavior

The sigma-delta modulator can be thought of as a PCM converter with feedback, which attempts to force the output signal $y[n]$ to be equal to the input signal, $x[n]$. Consider the case where a 1 bit A/D converter or comparator is used. The transfer characteristics of this 1 bit ADC with output levels V and $-V$ is shown in Fig 2c.

Assume that $V = 1$, such that the comparator's digital output is 1 or -1, so that $y[n]$ and $y_a[n]$ can be used interchangeably. In the time domain, referring to Fig. 8, we have,

$$v[n] = u[n-1] + v[n-1] \tag{15}$$

$$y[n] = \begin{cases} 1 & v[n] \geq 0 \\ -1 & v[n] < 0 \end{cases} \tag{16}$$

$$u[n] = x[n] - y[n] \tag{17}$$

The "error" between the modulator output and input is $u[n]$. Note that this is not the quantization error, which is given by $e[n] = y[n] - v[n]$.

Since $y[n]$ can take on values of 1 or -1 only, it can never equal the input unless the input happens to be one of these two values exactly. Consequently, except for the mentioned cases, there will always be an error, $u[n] \neq 0$. Consider a DC input for $x[n]$. When $y[n] = 1$, $y[n]$ is greater than the input $x[n]$ and the error $u[n]$ is negative, and so negative values are accumulated by the integrator to produce $v[n]$. After a number of clock cycles, enough negative values will have accumulated to cause the quantizer to produce $y[n] = -1$, thereby changing the sign of the error $u[n]$ to be positive. The error between the output and input has been reduced, in some sense, because the positive errors will now cancel the prior negative errors when averaged over a period of time.

Now with $y[n]=-1$, the errors will be positive, and positive values of the error will be accumulated again until the quantizer output changes, this time back to $y[n] = 1$. Over a period of time, the proportion (or density) of 1's and -1's will be related to the DC input value—the larger the input, the more 1's will be present in the output, and vice versa, for smaller inputs. For this reason, the output of a sigma-delta modulator using a 1 bit quantizer is often said to be in pulse density modulated (PDM) format.

Let us now illustrate the time domain behavior using a few examples. Fig 10a shows a DC input $x[n] = 0.55 = 11/20$, while Fig 10b shows the corresponding modulator output, $y[n]$. Roughly, three fourths of the output values are 1's, and the others are -1's. Fig 10c shows the error signal, $u[n]$, and Fig 10d shows the accumulated error signal or integrator output $v[n]$, whose sign change forces the quantizer output to change. For a DC input of $x[n] = 1$, all the modulator output values will be 1's. For a zero DC input, half the modulator output values will be 1's, half -1's. For a DC input of -1, all the values will be -1's.

By averaging the modulator output over a period of time, we can approximate the input. This averaging operation represents the low-pass filter block in Fig. 8, since averaging is a crude low-pass filtering operation. Using a better low-pass filter will result in the modulator output being a better approximation to the input for a given oversampling ratio.

Finally, let us look at some time domain examples of the modulator output for sinusoidal inputs. Figures 11a-d show the modulator outputs for various sinusoidal inputs. As for the DC input case, the sinusoid amplitude information is encoded in the relative number of 1's vs -1's. The modulator output pulse pattern has periodic components, and the fundamental period encodes the sinusoid frequency. This is particularly clear in Figs. 11a-c.

Implementation Imperfections

The results presented thus far have not considered imperfections in the analog hardware. Let us now discuss the consequences of imperfections in some of the main circuit parameters.

The integrator in the modulator may have a gain of g instead of unity, and may be leaky. For an input $u[n]$, an integrator with gain g and leakage factor α has an output $v[n] = g\, u[n-1] + \alpha v[n-1]$ instead of $v[n] = u[n-1] + v[n-1]$, and the integrator transfer function is $g\, z^{-1}/1-\alpha z^{-1}$ instead of $z^{-1}/1-z^{-1}$.

The D/A gain may also not be perfectly unity, and assuming a gain of d, we find the STF and NTF are $H_x = g\, z^{-1}/(1+(g\, d-\alpha)z^{-1})$ and $H_e = 1-\alpha z^{-1}/(1+(g\, d-\alpha)z^{-1})$.

The original NTF $(1-z^{-1})$, which had a Z domain zero at $z = 1$ (on the unit circle and at DC), now has a zero which is still at DC but is moved inside the unit circle. This degrades the NTF noise attenuation in the signal band and can thus affect the noise performance significantly. The term "leaky" comes from the fact that there is charge leakage in the switched capacitor implementation of the integrator circuit and only a portion of the charge from the input capacitor is transferred to the integrating capacitor.

The leakage factor α is related to the open loop gain, A, of the operational amplifier (opamp) used to implement the switched capacitor integrator such that $1 - \alpha \simeq 1/A$. Ignoring the denominator (which will add a slight ripple to the numerator) of the degraded H_e, i.e., by considering the degraded NTF to be $(1-\alpha z^{-1})$, we find the in-band quantization noise power with integrator leakage is:

$$\sigma_{ey}^2 = \sigma_e^2 \left(\frac{2f_B}{f_s}\right)\frac{1}{A^2} + \alpha\sigma_e^2 \frac{\pi^2}{3}\left(\frac{2f_B}{f_s}\right)^3 \tag{18}$$

instead of Eq. 12. Note that the noise power now contains a term that is inversely proportional to the oversampling ratio, $f_s/2f_B$, as well as a term inversely proportional to the oversampling ratio cubed. However, the first term is divided by A^2, and it has been found that if the opamp open loop gain, A, exceeds the oversampling ratio, $f_s/2f_B$, leakage causes no significant degradation of the SNR [2]. Consequently, the circuit constraint required to implement good integrators is not that difficult to meet unless the oversampling ratio is extremely high.

From the linearized analysis, the STF and NTF pole is stable only for $0< gd <2$. Consequently, there is a relatively wide margin over which the two gains may vary from this point of view. It has been reported [2] that variations of g by up to 10% from unity does not degrade the SNR significantly. The gain is implemented in practice as a ratio of two capacitors and so the corresponding precision on the capacitor matching is minimal—one part in 10, or three and a quarter bits.

Now consider an imperfect DAC gain, d, that is slightly different from unity. This can be modeled as a gain of $1/d$ at the input of the modulator. To see this, consider that a gain of $1/d$ is inserted at the input of the integrator. This gain can be moved past the summing node at the modulator input in Fig 8. The result is that the DAC gain d is cancelled by the gain $1/d$ but the input now experiences a gain of $1/d$ before the modulator. Consequently, the STF experiences a slight gain change but there is no great impact on the modulator SNR.

Finally, consider imperfections in the quantizer. Any nonlinearity in the quantizer can be modeled as another noise source which adds to $e[n]$, the quantization error. However, the noise from this extra source is subject to noise shaping by the modulator and so its affect on SNR degradation is not significant.

If the 1 bit quantizer or comparator has a non-zero threshold, v_{th}, its output is given by:

$$y[n]=\begin{cases}1 & v[n]-v_{th}\geq 0\\ -1 & v[n]-v_{th}<0\end{cases}$$

This is equivalent to an offset at the input of the comparator, i.e., at the output of the integrator. However, an offset v_{th} at the output $v[n]$ of the integrator corresponds to an impulse at its input $u[n] = x[n]- y[n]$ which amounts to one incorrect output $y[n]$ for a given $x[n]$. One such incorrect value will have a negligible impact on the overall performance of the modulator. The offset can also be modeled as an error source, with mean v_{th} and zero variance, that adds to $e[n]$.

If for any reason, the offset is input dependent or changes with time, this new error source will have non-zero variance. However, it will be subject to the noise shaping property of the modulator just as $e[n]$ and its presence will not degrade the performance of the modulator significantly.

Non-linear Behavior

A sigma-delta modulator is a non-linear system incorporating feedback. Not surprisingly, the modulator may display limit cycle oscillations that result in the presence of periodic (tone) components in the output. This phenomenon is analogous to limit cycles that occur in digital IIR filters operating with finite precision arithmetic, because, like a sigma-delta modulator, such a filter is a non-linear system that employs feedback. The quantizer error spectrum is not white, which is not surprising as the conditions for the white noise assumption are not perfectly satisfied—the quantizer has only two output levels, and due to the oversampling, successive quantizer input samples may be correlated.

Now consider the existence of limit cycles in the modulator, as has been done in [9] for the simple case of a DC input, $x[n] = x$. For a limit cycle of period T, $v[n]$ should be periodic with period T, i.e., $v[n] = v[n+T]$. This clearly implies that $y[n] = y[n+T]$.

For the DC input, the input to the integrator, $u[n] = x - y[n]$, will likewise be periodic with period T. Thus, the modulator behavior can be represented by T equations.

Combining Eqs. 15 and 17, we obtain: $v[n] - v[n-1] = x[n-1] - y[n-1]$, which for a DC input becomes $v[n] - v[n-1] = x - y[n-1]$. Writing this equation for T different time instances starting (arbitrarily) at $n = 1$, and adding up all these equations we obtain,

$$v[T]-v[0]=\sum_{l=0}^{T-1}x-\sum_{l=0}^{T-1}y[l] \tag{19}$$

However, $v[T]=v[0]$ by assumption, and consequently:

$$x=\frac{1}{T}\sum_{l=0}^{T-1}y[l]=\frac{(P-M)}{T}V \tag{20}$$

where P is the number of positive quantizer outputs over T output samples, and M is the number of negative quantizer outputs over T output samples. Since $(P - M)$ is an integer as is T by assumption, we have $x = bV/a$, with a and b integers. Thus, the output y consists of a limit cycle with period T, provided that x is a rational multiple of V.

The limit cycle with period T will manifest itself in the output spectrum as tones present at frequency f_s/T and its harmonics. The period is $T = 2a$ if a or b is even. $T = a$ if both a and b are odd [11]. For the prior example of $x = 11/20$ in Fig 10, $a = 20$ and $b = 11$, and as can be seen in the figure, $u[n]$, $v[n]$, and $y[n]$ do have the expected period of 40.

For the special case of $x = 0$, the output oscillates between V and $-V$ and the output spectrum consists of a pure tone at $f_s/2$. More complete results, which are independent of the integrator initial condition, $v[0]$, are provided in [9, 10, 12]. In fact, it is shown in [10] and [12] that even if the DC input is an irrational multiple of V, the quantization noise will not be white and the spectrum at the output of the modulator will be discrete, consisting of tones.

Even for sinusoidal inputs, the quantization error is not white and strong tone components are observed in the output and the strength of the tone distortion components depends on the input amplitudes in a complicated way [14]. The tone structure present at the output of the modulator for very low DC or sinusoidal input amplitudes is often called idle-channel noise.

One other point is worth noting. It can be shown that for comparator output levels of $\pm V$, the output of the integrator can have magnitude of at most $2V$ if the input to the modulator is bounded by $\pm V$ [14]. This is easily seen from Eqs. 15 and

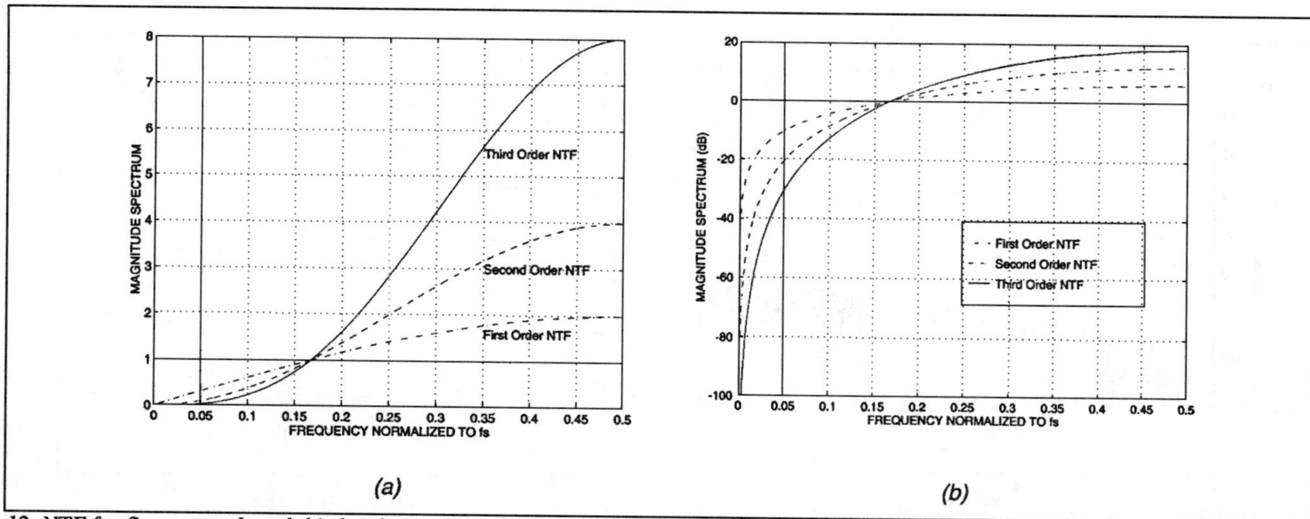

12. NTF for first, second, and third order sigma-delta modulators. (a) Magnitude spectra on linear scale. For comparison, the oversampled PCM NTF, which has unity gain, is shown; (b) magnitude spectra in dB.

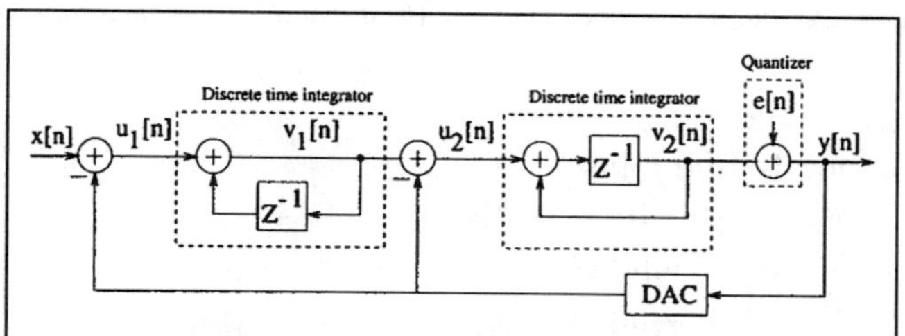

13. Second order sigma-delta modulator.

17, from which we have $v[n] = x[n\text{-}1] - (y[n\text{-}1] - v[n\text{-}1])$, which is $v[n] = x[n\text{-}1] - e[n\text{-}1]$. If we assume $|v[n\text{-}1]| \leq 2V$, then from the transfer characteristic of the 1 bit quantizer (Fig. 2c), $|e[n\text{-}1]| \leq V$, i.e., the quantizer is not overloaded. We then have, $|v[n]| = |x[n\text{-}1] - e[n\text{-}1]| \leq |x[n\text{-}1]| + |e[n\text{-}1]| \leq V + V = 2V$. Thus if $|x| \leq V$ and $|v[n\text{-}1]| \leq 2V$, then $|v[n]| \leq 2V$. This can be guaranteed by ensuring that $|v[0]| \leq 2V$, so that $|v[1]| \leq 2V$, and so on.

In practice, because of the significant tone structure present at the output of a first order sigma-delta ADC, it is rarely used in applications such as speech or audio, where the presence of such tones is objectionable even if the oversampling ratio, $f_s/2f_B$, is high enough to provide a good overall SNR based on the linearized white noise model.

Higher Order Sigma-Delta Modulation

The fundamental ideas presented can be extended to create sigma-delta architectures in a variety of ways that provide different tradeoffs among resolution, bandwidth, circuit complexity, and modulator stability. Our discussion will include higher order, multi-bit, and multi-stage (cascaded) architectures. In general, to obtain a performance improvement, most of these converters require analog circuits that need to be more complex and precise than those used in the 1st order sigma-delta modulator. Of course, the precision required must still be significantly less than the overall high conversion resolution.

Second Order Modulation

Operation and Performance Modeling

The standard 2nd order sigma-delta modulator A/D is widely used. This modulator realizes $H_x(z) = z^{-1}$ and $H_e(z) = (1\text{-}z^{-1})^2$, so that

$$Y(z) = X(z)z^{-1} + E(z)(1 - z^{-1})^2 \tag{21}$$

The second order modulator noise transfer function (NTF) is shown in Fig 12 along with the NTFs for the first order modulator and the third order modulator (a logical extension of the second order modulator which will be discussed later).

Note that, compared with the first order sigma-delta NTF, the second order NTF provides more quantization noise suppression over the low frequency signal band, and more amplification of the noise outside the signal band. Compared with a first order sigma-delta, more noise power is pushed outside the signal band.

A block diagram of the modulator is shown in Fig. 13. The structure now contains two integrators. The transfer function of the first one is $1/1\text{-}z^{-1}$ and that of the second one is $z^{-1}/1\text{-}z^{-1}$. Assuming the modulator output is filtered by an ideal lowpass filter, the linearized white noise model yields the following for the in-band SNR:

$$SNR = 10\log(\sigma_x^2) - 10\log(\sigma_e^2)$$
$$-10\log\left(\frac{\pi^4}{5}\right) + 50\log\left(\frac{f_s}{2f_B}\right) \ (dB) \tag{22}$$

Again, letting $f_s/2f_B = 2^r$, we obtain:

$$SNR = 10\log(\sigma_x^2) - 10\log(\sigma_e^2)$$
$$-10\log\left(\frac{\pi^4}{5}\right) + 15.05r \ (dB) \tag{23}$$

Thus, for every increment in r or for every doubling of the oversampling ratio, $f_s/2f_B$, the SNR improves by 15 dB, or the

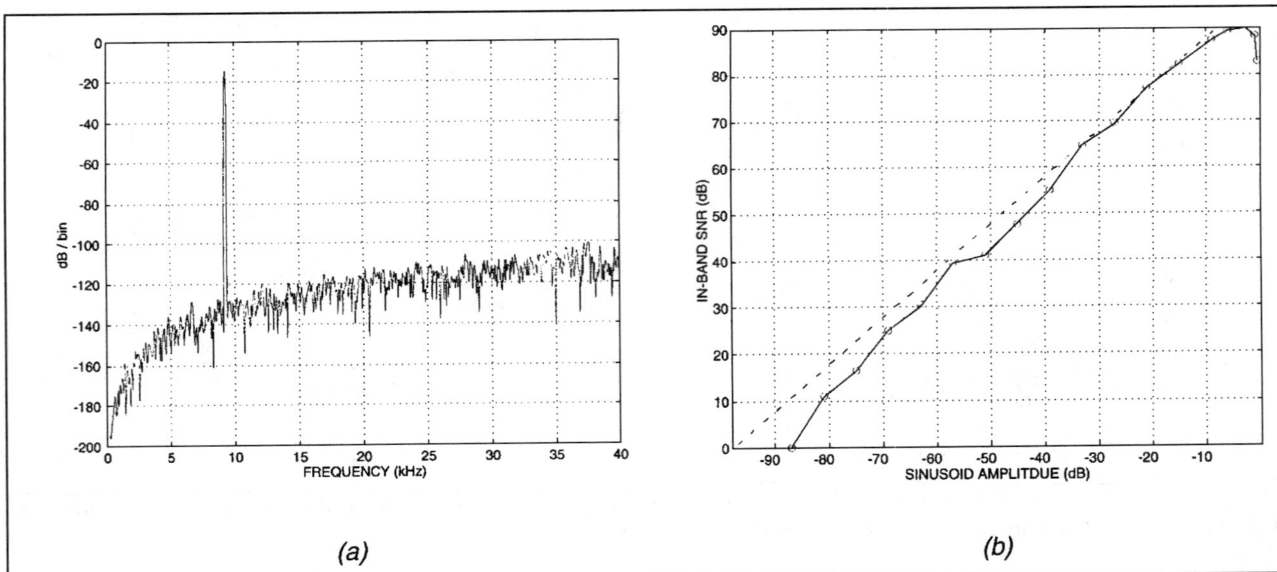

14. Simulation of a 2nd order modulator (a) power spectrum of the modulator output before decimation; (b) in-band SNR vs amplitude of input sinusoid.

equivalent resolution by 2.5 bits, which is 1 bit better than the improvement achieved by a first order sigma-delta.

Now consider the example used previously where a 20 kHz audio band needs to be converted to a resolution of 16 bits. The f_s needed by a 2nd order sigma-delta modulator using only a 1 bit quantizer is, from Eq. 22, only 6.12 MHz in contrast with the 96 MHz needed by the 1st order sigma-delta. This circuit speed is very reasonable in current CMOS technology.

Fig 14a shows the low frequency portion (0 to 40 kHz) of an FFT based power spectral density estimate of the output of a 2nd order modulator with a sinusoidal input frequency of 9.3 kHz and f_s of 6.20 MHz. The tall peak is, of course, the sinusoidal signal. Notice the noise shaping whereby the noise at lower frequencies is greatly attenuated. Such power spectra are often used to numerically calculate the in-band SNR. Here, the in-band SNRs are computed over the 20 kHz audio band. The SNR computation can be repeated for various sinusoid amplitudes to obtain a plot of in-band SNR vs amplitudes. This plot is shown in Fig 14b.

The dashed curve corresponds to the values which are obtained from the linearized white noise model SNR formula (Eq. 22). The discrepancy between the two plots will be discussed later. The dynamic range, R, can be determined from the in-band SNR vs amplitude plot by looking at the amplitude value for which the SNR is 0 dB. The peak SNR is about 90 dB and R is about 88 dB which, from Eq. 5, yields an equivalent resolution, N, of 14.32 bits.

Qualitative Time Domain Behavior

In the time domain, referring to Fig. 13, we have,

$$u_2[n] = v_1[n] - y[n]$$
$$v_2[n] = u_2[n-1] + v_2[n-1]$$
$$y[n] = \begin{cases} 1 & v_2[n] \geq 0 \\ -1 & v_2[n] < 0 \end{cases}$$
$$u_1[n] = x[n] - y[n]$$
$$v_1[n] = u_1[n] + v_1[n-1]$$

Figures 15a and b show plots of a DC input, $x[n] = 0.55$, and the resulting output, $y[n]$, of a second order sigma-delta modulator. The input and output of the first integrator are $u_1[n]$, and $v_1[n]$, respectively, while the input and output of the second integrator are $u_2[n]$, and $v_2[n]$, respectively. The "error" between the modulator input and output is $u_1[n]$, which again is not the quantization error (given by $e[n] = y[n] - v_2[n]$). Looking at $u_2[n]$, we see that it is the difference between an integrated (or low-pass filtered) version, $v_1[n]$, of the modulator "error," $u_1[n]$, and the output, $y[n]$. Thus, $u_2[n]$ can be considered to be a more fine or accurate version of the modulator error. The signal which is quantized, $v_2[n]$, is an integrated version of the "fine error," $u_2[n]$. Consequently, $u_1[n]$, and $v1[n]$ are analogous to the $u[n]$ and $v[n]$ of the 1st order modulator. Outputs $u_2[n]$ and $v_2[n]$ are more accurate representations of $u_1[n]$ and $v_1[n]$, and thus produce an output $y[n]$, which is more accurate than the output of a 1st order modulator. This should be clear from comparing Fig 15c with Fig 15e and Fig 15d with Fig 15f.

In comparing $y[n]$ of the second order sigma-delta of Fig. 15b to the $y[n]$ of the first order sigma-delta of Fig. 10b, the key point is that the distribution of 1's and -1's in Fig. 15b is such their average provides a more accurate representation of the input than the corresponding average of the first order modulator output. In other words, for a given block of output samples, the second order modulator uses its allocation of samples more efficiently to represent the input.

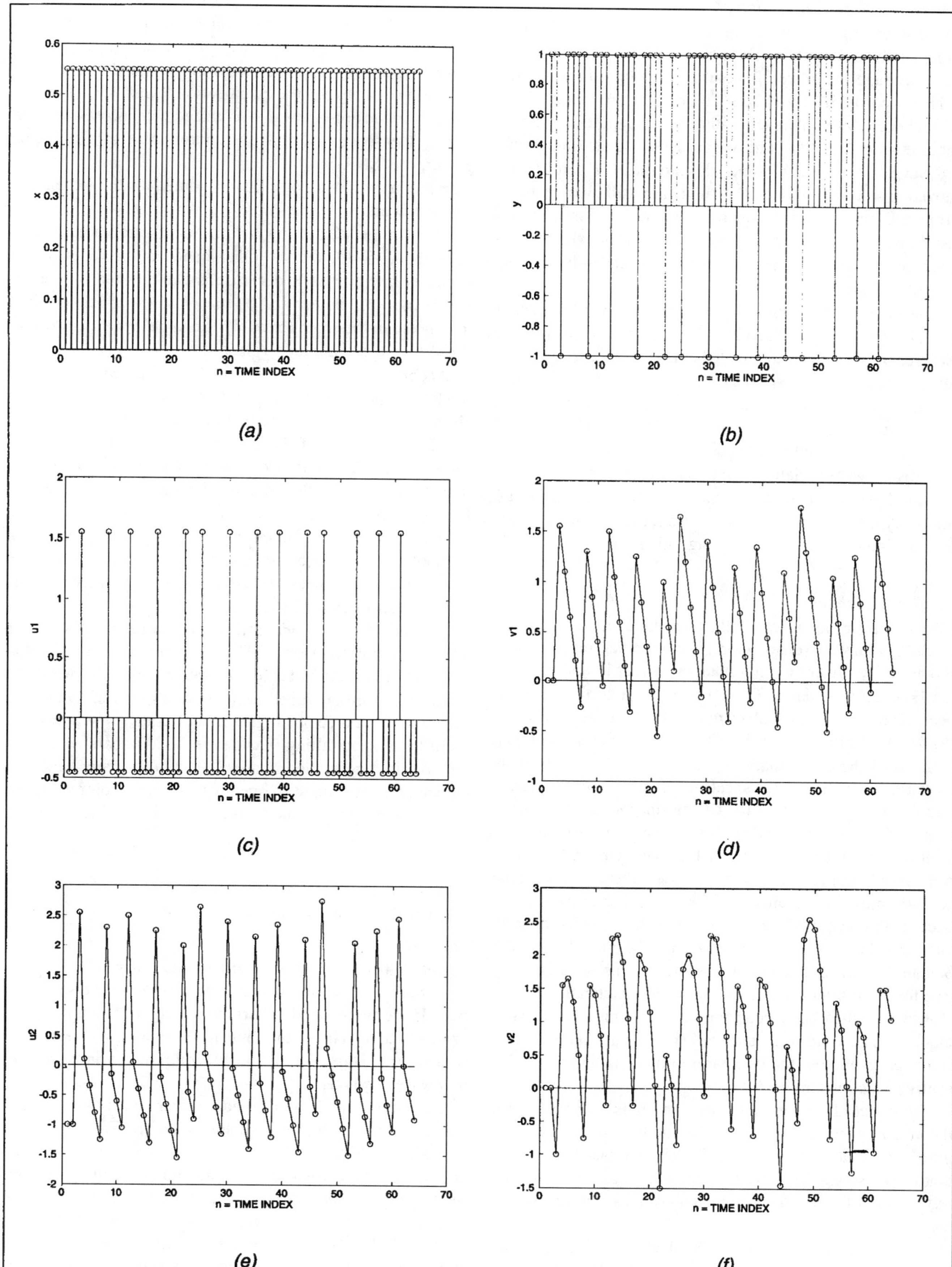

15. Second order sigma-delta responding to a DC input: (a) DC input $x[n] = 0.55$; (b) output $y[n]$; (c) "error" signal $u_1[n]$; (d) first integrator output $v_1[n]$; (e) more accurate "error" signal $u2[n]$; (f) second integrator output $v_2[n]$.

Implementation Imperfections

Compared with a first order sigma-delta modulator, a second order modulator contains an extra integrator. Assuming the same leakage factor, α , for both integrators, the degraded NTF is approximately $(1- \alpha z^{-1})^2$. The leakage factor of the second order integrators can satisfy somewhat less stringent requirements than that of the first order modulator integrator.

One might expect this since despite the NTF zeros being moved inside the unit circle, there are still two of them providing attenuation of quantization noise. The reduced requirement on the second order modulator integrator leakage factors can also be seen if one calculates the in-band noise power, σ_{ey}^2, from the $(1-\alpha z^{-1})^2$ NTF.

The noise term inversely proportional to the oversampling ratio, fs/2fB, is now divided by A4 rather than A2, as was the case for the first order modulator in Eq. 18. Consequently, a lower op-amp gain can suffice for the second order modulator.

The other parameter to consider is the gain of the second integrator. From the linearized analysis, the STF and NTF poles are stable for integrator gains up to 4/3, and so still allow a relatively wide variation from this point of view.

Simulations indeed show that the integrator gains are relatively insensitive to deviations from their nominal values over a wide range of oversampling ratios [8].

Non-linear Behavior

Like the 1st order sigma-delta modulator, a second order modulator may also display limit cycle oscillations [2, 9], and this is easily illustrated in a manner similar to that used for the 1st order modulator. The nature of these limit cycles has been investigated and unlike the first order modulator depend on the initial conditions of the integrator outputs [9, 11].

Most of the exact analyses, e.g., [13-15], which provide an exact description of the spectrum of the quantization error and the modulator output for DC or sinusoidal inputs, have been performed for second order modulators using quantizers with 2 or more bits. In fact, a 1 bit quantizer used in a 2nd order modulator can become overloaded, thereby making the analysis much more difficult. The quantizer is overloaded because the output of the second integrator can significantly exceed values of $\pm 2V$ [2] even with the modulator input bounded by the quantization levels $\pm V$. This is particularly true for large modulator inputs near the quantization levels. However, it has been determined from simulations that a modified 2nd order architecture [5] using a 1 bit quantizer can operate without the integrator outputs having to significantly exceed values of $\pm 2V$.

The spectral properties of overload noise have not been theoretically characterized but simulations have been reported in the literature [16, 17]. The simulations demonstrate that the noise significantly manifests itself as harmonic distortion tones for sinusoidal inputs as well as significant tone components near $f_s/2$.

As with the first order sigma-delta, idle channel tones may be observed for small DC or low amplitude tone inputs. According to the linearized model (Eq. 22), σ_e^2 is fixed and the SNR should increase linearly with signal power σ_x^2. However, due to the presence of overload or idle channel tones, the SNR of a second order modulator, using a 1 bit quantizer, increases linearly with signal power only over a certain range of signal power even though the modulator input may be between the quantization levels $\pm V$. More and more overload noise power is produced with increasing input values. Consequently, above a certain value of high signal power, the SNR will actually start to decrease when the increase in overload noise power dominates the increase in signal power.

This can be seen in Fig 14b for input amplitudes greater than about -5 dB. On the other hand, as the input power becomes low, the SNR decrease is caused both by the decrease in signal power and by the presence of idle channel tone noise in the signal band. This can be clearly be seen in Fig. 14 for low amplitude values.

According to Eq. 22 based on the linearized white noise model, the peak SNR and dynamic range should have been about 98 dB. Thus, idle channel tone problems, resulted in a 10 dB degradation of the dynamic range predicted by the linearized model while overload noise prevented the modulator from reaching the peak SNR predicted by the model. Dithering techniques [17] will often break up tone structures including overload and idle channel tones, thereby producing a smoother power spectrum output, a more linear dependence of SNR on signal power, and a dynamic range, which is closer to the value predicted by Eq. 22.

Another factor which affects the in-band SNR is the sinusoid input frequency. This is particularly true if the sinusoid amplitude is such that strong harmonic distortion components are produced in the power spectrum output. In this case, the choice of a higher input frequency will result in these harmonics falling outside the 20 kHz bandwidth and not contributing to the in-band SNR. On the other hand, the choice of a lower input frequency will yield poorer values of in-band SNR because the harmonics will now fall inside the 20 kHz signal band.

Other Types of Higher Order Sigma-Delta Modulation

Sigma-delta converters realizing higher order NTFs achieve even higher resolution by pushing even more noise power outside the signal band. Alternatively, a lower sampling rate can be used to achieve the same resolution for a given signal bandwidth. In this case, the speed requirements on the analog hardware is relaxed.

An order L modulator based on a straightforward extension of the first order sigma-delta realizes a STF given by $H_x = z^{-1}$ and a NTF given by $H_e = (1-z^{-1})^L$, which contains L zeros at $z = 1$ or at DC frequency on the unit circle. A third order modulator structure can be created from the second order structure of Fig. 13 by inserting an integrator with transfer function $1/(1-z^{-1})$ between the the summing node of the modulator input and the first integrator of the second order modulator. The input to this new integrator is now $x[n]$-$y[n]$ and the output of this integrator minus $y[n]$ is fed as the input to

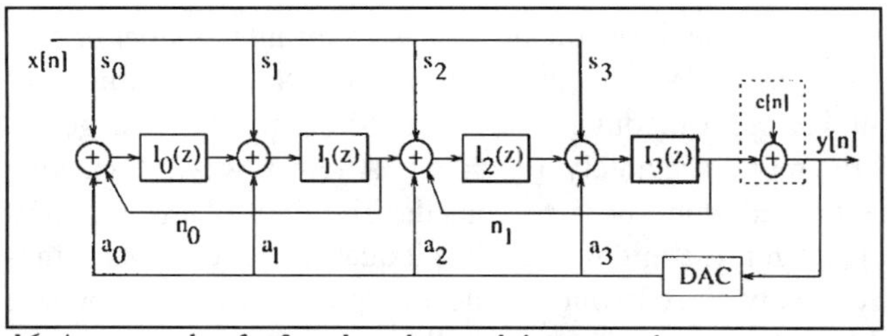

16. An example of a fourth order modulator topology.

what used to be the first integrator of the second order modulator.

The magnitude spectra for a third ($L = 3$) order NTF plotted on a linear scale and in dB are shown in Figs. 12a and b. Note that over the signal band, which is $0.05f_s$ in the figure, the 3rd order NTF provides more attenuation of the quantization noise than the second or first order NTFs and so is capable of pushing more noise power outside the signal band than the second or first order modulators. The ideal in-band SNR achieved by an Lth order modulator is given by

$$SNR = 10\log(\sigma_x^2) - 10\log(\sigma_e^2) - 10\log\left(\frac{\pi^{2L}}{2L+1}\right)$$
$$+(20L+10)\log\left(\frac{f_s}{2f_B}\right) \quad (dB) \tag{24}$$

Thus, for every doubling of the oversampling ratio, this modulator provides an extra ($6L + 3$) dB of SNR, or an extra (L+ 1/2) bits of resolution. For the CD example used throughout this article, a 3rd order modulator will need a f_s of 1.92 MHz to convert a 20 kHz band to 16 bits of resolution.

Modulator topologies are not constrained to always realize FIR NTFs as we have seen until now or to realize all the NTF zeros which provide attenuation at DC frequency. In fact, distributing the zeros over the signal band rather than placing them all at DC frequency can be more efficient in pushing quantization noise outside the signal band.

Examples of higher order topologies are described in [18-21]. One such fourth order topology described in [19] is shown in Fig 16, where $I_k(z)$ denotes the kth integrator. This structure realizes Eq. 7 in the form

$$Y(z) = X(z)\frac{B_x(z)}{A(z)} + E(z)\frac{B_e(z)}{A(z)} \tag{25}$$

Note that the STF and NTF are IIR transfer functions in this case. The feedforward coefficients s_k realize B_x, which contains the Z domain zeros of the STF. The feedback coefficients n_k between every second integrator realize B_e, which contains the zeros of the NTF. Finally, the feedback coefficients a_k realize A, which contains the poles of both the STF and the NTF.

In order to implement Eq. 25, based on a sigma-delta modulator that incorporates feedback, the z^0 coefficient in B_e and A must be equal to ensure a causal feedback loop (in other words to avoid non-computable delays, [3, pp. 308-309]). If the NTF is FIR, i.e., if $A(z) = 1$, then the causality constraint will require the z^0 coefficient of $B(z)$ to be 1.

Note that the performance will be limited by the degree to which the analog coefficients a_k, s_k, and n_k, match their desired values. For the modulator to be useful, the degree of required matching should be significantly less than the overall resolution of the converter.

Higher order architectures also alleviate some of the tone problems mentioned earlier [2, 18]. The main difficulty with such higher order structures is that such modulators are only conditionally stable when an one bit quantizer is used. Stability may for example depend on the input signal being kept below a certain value or on precise circuit matching needed to satisfy a stability criterion. Stability is often described in the sense of the quantizer not being overloaded. This is useful because any higher order modulator structure can be transformed into an equivalent modulator in the form of the 1st order modulator of Fig 8 with the integrator being replaced by a general "loop filter" $H(z)$. If the quantizer is not overloaded, its input is bounded by $\pm 2V$, and this implies that the loop filter is also operating in a stable manner and all internal signals will be bounded. Stability is, in general, difficult to determine for a modulator using an 1 bit quantizer. One reason is the difficulty in characterizing the gain of the 1 bit quantizer. The gain of the 1 bit quantizer of Fig. 2c is variable—it depends on the quantizer input. The smaller the quantizer input, the larger is its gain (e.g., if the input is zero, the output is V and the gain is infinite). The larger the quantizer input, the smaller is its gain (e.g., if the input is infinite, the output is V and the gain is zero). Another way of looking at this is to observe that if one attempts to linearize the quantizer transfer curve of Fig. 2c by trying to fit a straight line to the curve, the correct slope of the straight line is arbitrary.

As we have seen earlier, the input to the quantizer clearly changes with time, even if the input to the modulator is a DC signal. The manner in which the quantizer gain changes over time will also depend on the type of input applied to the modulator. Consequently, performing a linear system analysis of the modulator signal and noise transfer functions in terms of modulator parameters is inadequate because the poles of the transfer function, which determine stability, depend on a time varying and input dependent quantizer gain. An attempt to characterize the quantizer gain more accurately for DC and sinusoidal inputs has been made for several modulators [22].

The phenomenon of limit cycle oscillations is also connected to stability. This is because the structure of limit cycles may be such that the amplitude of internal modulator variables is large, causing the quantizer to overload. The frequency of such a deleterious limit cycle oscillation can correspond to the point on the unit circle where the modulator transfer function pole crosses the unit circle into the unstable region. Properties of limit cycles in the context of stability have been investigated in [11]. A limit cycle that corresponds to the modulator transfer function pole moving from the inside to the outside of the unit circle may not necessarily result in unstable behavior in the long term, provided integrator outputs do not saturate before stability is restored [23].

Suppose an unstable limit cycle, corresponding to poles moving outside the unit circle, results when the quantizer gain is too high, i.e., when the input to the quantizer is small. If this is the case, growing signal values in the modulator which result from this instability will eventually increase the input to the quantizer thereby reducing its gain and so moving the poles back inside the unit circle [23].

One way to guarantee stability would be to reset the integrators if it was detected, by additional circuitry, that their values were becoming too large. However, this approach may cause a significant decrease in the SNR [24]. Similarly, allowing integrator outputs to clip or saturate may also cause degradations in the SNR performance. In particular, low frequency limit cycles which may introduce distortion components in the signal band may occur (albeit with reduced values of integrator output) and may persist for a long time thereby reducing the SNR [23]. An alternative approach uses local feedback loops in an attempt to gracefully return integrator outputs to their normal operating region [24]. Since the linearized transfer function of the system is modified due to the local feedback loops, the effect of these loops is cancelled in the digital domain.

Several quantitative criteria have also been proposed to characterize stability. The l_l norm criterion [14, 25, 26] relates the sum of the magnitudes of the modulator NTF impulse response coefficients, the number of bits in the quantizer, and the modulator input level to the no overload stability requirement. This is a sufficient but not necessary condition for stability, and it has unfortunately been found to be too conservative for practical use. An ad-hoc stability criterion which has been proposed [18] and found to be useful [21] is to design the NTF to possess less than 2 to 6 dB of out-of-band gain.

Multi-bit Sigma-Delta Modulation

Until now, we have assumed that the quantizer and DAC inside our sigma-delta modulator were 1 bit devices. However, converters using a multi-bit internal quantizer offer more potential resolution from the internal quantizer. A 2nd order multi-bit sigma-delta converter would look exactly the same as the modulator shown in Fig. 13, except that $e[n]$ in the figure would be the model for an N bit quantizer instead of a 1 bit quantizer, and the DAC would be an N bit DAC instead of a 1 bit DAC. The use of a multi-bit quantizer affects the σ_e^2 term in the expressions for the SNR, where each additional bit used in the quantizer will yield a 6 dB improvement in the SNR. Using Eq. 3 without the approximation, it is easy to see that if a 5 bit internal quantizer is used instead of a 1 bit quantizer, a 30 dB improvement in SNR is possible. Alternatively, the sampling frequency can be reduced by a factor of 4, while keeping the resolution the same. For our CD example, a 2nd order modulator using a 5 bit internal quantizer can use a f_s of 1.53 MHz, rather than the f_s of 6.12 MHz needed by a modulator using a 1 bit quantizer.

The behavior of multi-bit sigma-delta systems more closely follow that predicted by the linearized model (in the extreme case, if the quantizer has an infinite number of bits, there is no non-linearity). Consequently, the stability of higher order modulators using multi-bit quantizers is generally more accurately predicted. Another way of viewing enhanced stability is to consider the gain of the multi-bit quantizer. If a midtread multi-bit quantizer is used, its gain is relatively close to one for most output values (even for zero input, a midtread quantizer will have a zero output and hence unity gain unlike the midriser 1 bit quantizer, which has infinite gain for a zero input). This is because a straight line drawn through the multi-bit quantizer transfer characteristic (e.g., Fig. 2b) can no longer be arbitrarily drawn. Of course, if for any reason the quantizer does start to overload, its gain will start to deviate more and more from unity. Even though the l_l stability criterion mentioned earlier may be too conservative, it may allow one to obtain an initial idea about the number of bits needed for stable operation for a given NTF and input signal level.

Modulators using multi-bit quantizers also display less of the tone problems associated with the 1st and 2nd order sigma-delta converters using a 1 bit internal quantizer. The main disadvantage is that the multi-bit DAC cannot be easily fabricated in VLSI with sufficient linearity needed for high resolution conversion. Various techniques, examples of which can be found in [27-33], have been proposed to reduce the linearity required for the DAC. The multi-bit output also complicates the digital low-pass filter hardware following the modulator, because for multi-bit processing, the filter requires multi-bit hardware multipliers.

Multi-stage (Cascaded) Sigma-Delta Modulation

Higher order NTFs can also be created by cascading independent modulator stages. This cascading does not adversely affect the stability of the overall modulator, provided the individual stages are stable. They may also suffer fewer of the tone problems than a first or second order sigma-delta alone [2, 14].

An example of a second order modulator obtained by cascading two 1st order modulators is shown in Fig 17. The signal $x[n]$ is the input to the first modulator in the cascade but the quantization error, $e_l[n]$, of the first modulator is used as the input to the second modulator. Finally, the outputs of the first stage and second stage modulators are added in the digital domain after passing through a digital delay (z^{-1}) and a digital differentiator ($1-z^{-1}$), respectively.

From the Z domain analysis of the linear system model with the DACs replaced by unity gains, we have,

$$Y_1(z) = X(z)z^{-1} + E_1(z)(1-z^{-1}) \tag{26}$$

$$Y_2(z) = E_1(z)z^{-1} + E_2(z)(1-z^{-1}) \tag{27}$$

The output is computed as $Y_1 z^{-1} - Y_2(1-z^{-1})$. This sum results in a cancellation of the first order noise term $E_1(z)$ to produce the overall output,

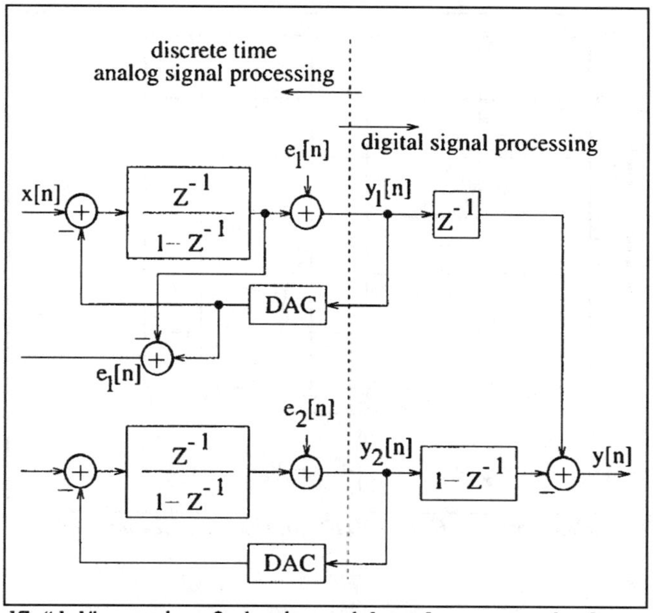

17. "1-1" cascade: a 2nd order modulator from a cascade of two 1st order modulators.

$$Y(z) = X(z)z^{-2} - E_2(z)(1-z^{-1})^2 \tag{28}$$

Except for the sign on the noise that is irrelevant and an extra delay experienced by the input, the modulator realizes the same output as the standard second order sigma-delta modulator. One advantage of using this structure over the second order modulator is the fact that the quantizer in either of the first order modulator sections will never overload for $x[n]$ bound by $\pm V$. However, the cascaded structure requires matching between the analog and digital transfer functions as well matching among the D/A output levels among various stages [2, 34]. In fact, mismatch effects and integrator leakage can lead to the propagation of unshaped or poorly shaped noise from an earlier section to the final output [36, 40].

Assume there are circuit imperfections in the "1-1" cascade of Fig. 17, such that the transfer function of the integrator in the first section is $g\,z^{-1}/(1-\alpha z^{-1})$ instead of $z^{-1}/(1-z^{-1})$. Even if we assume the integrator in the second stage is ideal, the output is then

$$Y(z) = \frac{X(z)gz^{-2}}{1-(g-\alpha)z^{-1}} + \frac{(1-\alpha)z^{-2}E_1(z)}{1-(g-\alpha)z^{-1}} + \frac{(g-\alpha)z^{-2}E_1(z)(1-z^{-1})}{1-(g-\alpha)z^{-1}} - E_2(z)(1-z^{-1})^2 \tag{29}$$

The first term of the equation contains the signal, which is no longer a pure delay but will have a ripple to it determined by the factor $g/(1-(g-\alpha)z^{-1})$. However, since the signal is oversampled, this additional signal transfer function will mostly be flat at lower baseband frequencies. Ignoring the ripple due to the factor $1/(1-(g-\alpha)z^{-1})$, the second term in Eq. 29 is the unshaped noise from the first stage, the third term in Eq. 29 is the first order shaped noise from the first stage, and the fourth term in Eq. 29 is the desired second order shaped noise term. Clearly, for large imperfections, the unshaped noise term might dominate the noise term subject to second order noise shaping. If more than two stages are cascaded, the cumulative effects of such quantization error leakage effects will yield diminishing returns in performance improvement.

Architectures using only first order modulators have been realized [35], as have architectures using second order modulators [37-39]. A comparison of some architectures can also be found in [40]. Finally, note that due to the addition of various single bit intermediate outputs, the architecture has a multi-bit final output, which complicates the decimation filter hardware.

Band-pass Sigma-Delta Systems

Thus far, we have assumed that the sampling frequency f_s is much greater than the Nyquist rate, which is twice the highest frequency component in the input signal. For low-pass signals, the highest frequency component is also the signal bandwidth f_B. If a signal with bandwidth f_B is band-pass but is located at a center frequency, f_c, its highest frequency is $f_c + f_B/2$. If f_c is large, choosing f_s to be much greater than the highest frequency will lead to an unreasonably large f_s, and does not take advantage of the band-pass nature of the signal.

Band-pass sigma-delta modulation [41] allows high resolution conversion of band-pass signals if f_s is much greater than the signal bandwidth f_B, rather than the highest signal frequency. Band-pass sigma-delta modulators can be used in AM digital radios [41] or receivers for digital cellular mobile radios [45].

Real Band-pass Modulation

Unlike low-pass sigma-delta modulators, which realize NTF zeros at DC or low frequencies on the unit circle of the Z plane, band-pass modulators have NTFs that realize zeros or notches in the signal band of interest, $[f_c - f_B/2,\ f_c + f_B/2]$. Consequently, quantization noise that occurs over the signal band is attenuated, and noise power is pushed outside this band. Regardless of where the signal band is centered, the smaller the signal band, f_B, relative to the sampling frequency, f_s, there is less in-band noise power for a given NTF. Noise outside the signal band can be attenuated with a digital decimation filter and so high resolution conversion is possible for large oversampling ratios $f_s/2f_B$. The modulator STF and the decimation filter will typically have a band-pass characteristic, providing unity gain over the signal band.

As an example, the NTF and STF magnitude spectra for the design in [42] are shown in Fig 18a. In this example, the signal band has a center frequency of $f_c = 455$ kHz, $f_s = 3$ MHz, $f_B = 20$ kHz, so the oversampling ratio is 75. Figure 18a shows the magnitude spectra in dB of H_x, and H_e. Figure 18b shows a closeup view of the magnitude spectrum in dB of H_e over the 435 kHz to 475 kHz region. The vertical bars delineate the 20 kHz signal band centered at 455 kHz. The NTF is sixth order, and as should be clear from Fig. 18b, contains three notches or zeros, which provide attenuation of the quantization noise over the 20 kHz signal bandwidth (the other three

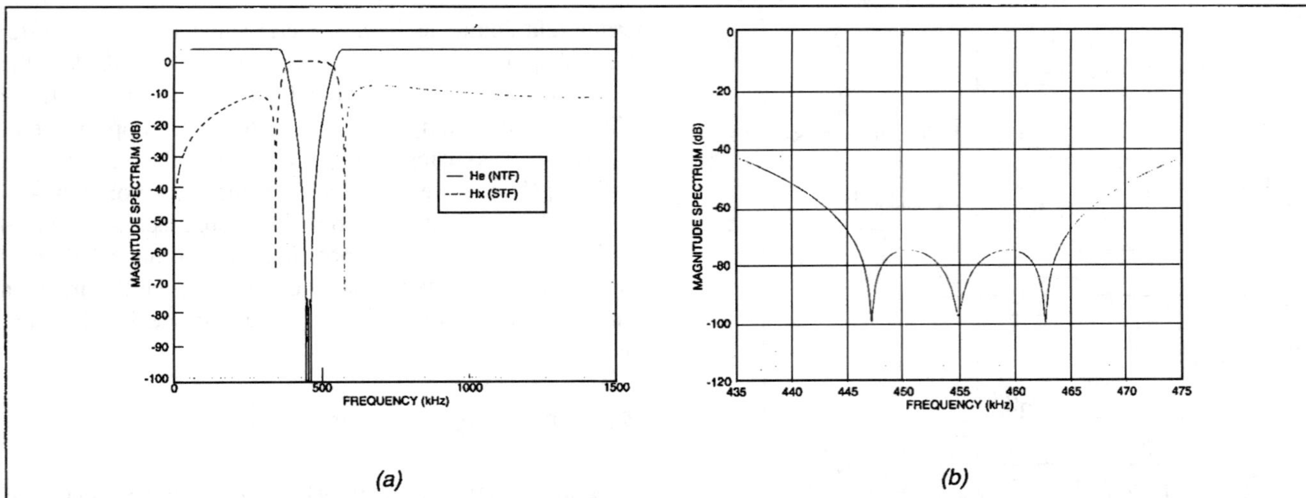

18. Band-pass sigma-delta: (a) magnitude spectra in dB of H_x (the STF, dashed plot), and H_e (the NTF); (b) closeup of the magnitude spectrum in dB of H_e.

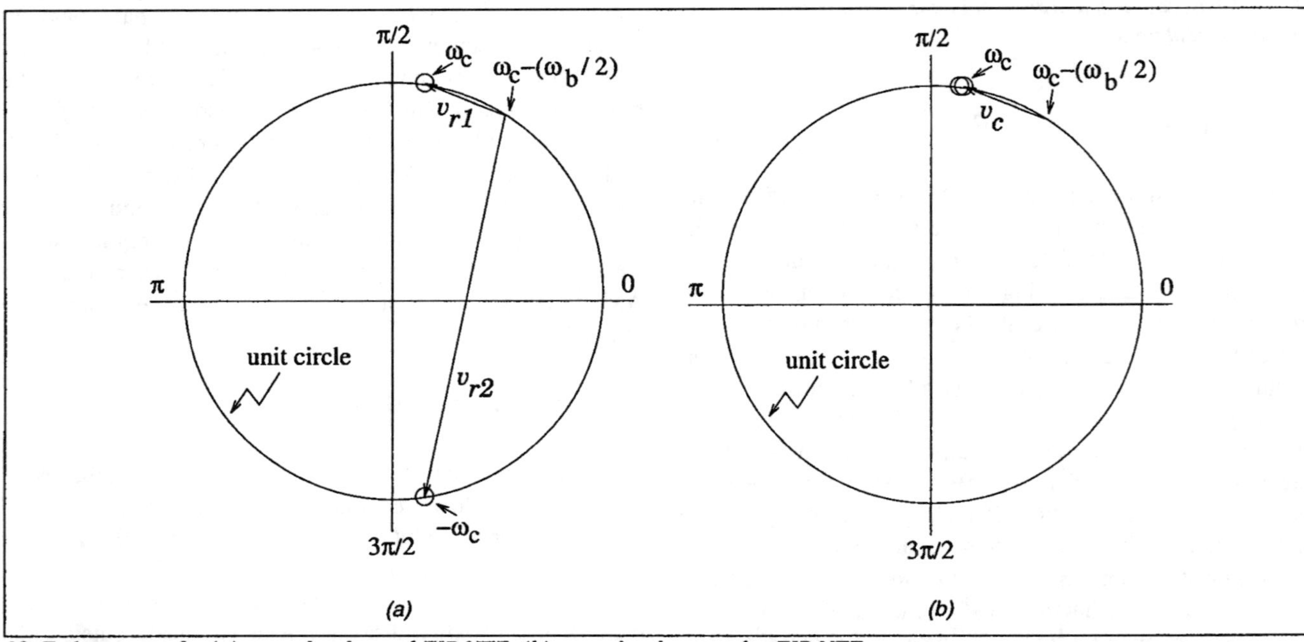

19. Z plane zeros for (a) second order real FIR NTF; (b) second order complex FIR NTF.

zeros are complex conjugates of these). The STF is band-pass and has minimal ripple and is approximately linear phase in the signal band. Band-pass converters employing Lth order modulators display a SNR performance that improves at the rate of $(3L + 3)$ dB per octave increment with the oversampling ratio $f_s/2f_B$ [42].

Complex Band-pass Modulation

Most band-pass sigma-delta converters [43-45] use real NTFs. Thus, all the coefficients in the Z domain transfer function are real. Let us now discuss the idea of complex band-pass NTFs, which have been proposed independently in [46] and [47]. The use of complex NTFs can improve the resolution that can be obtained for real band-pass signals [47]. The reason for this improvement is best illustrated through an example using second order FIR NTFs.

Figure 19a shows the unit circle on the Z plane containing zeros for a real second order FIR NTF. We find it more convenient to use the discrete time frequency $\omega = 2\pi f/f_s$ instead of f in our discussion. A real NTF realizing a zero at center frequency $\omega = \omega_c$ is constrained to also realize one at $-\omega_c$.

For a signal bandwidth ω_b consider the attenuation of the NTF at half the bandwidth away from the zero, that is, at $\omega_c - \omega_b/2$. This attenuation is given by the product of the magnitude of the vectors v_{r1} and v_{r2}, i.e., by $|v_{r1}| \times |v_{r2}|$. The smaller this product, the better the attenuation provided by the NTF, and the better the quantization noise suppression over the signal band, hence a better SNR will be obtained.

Vector v_{r1} represents the contribution from the zero at ω_c, and v_{r2} represents the contribution from the zero at $-\omega_c$. In the figure, the distance between ω_c and $\omega_c - \omega_b/2$ has been exaggerated—in practice, high resolution can be obtained only if

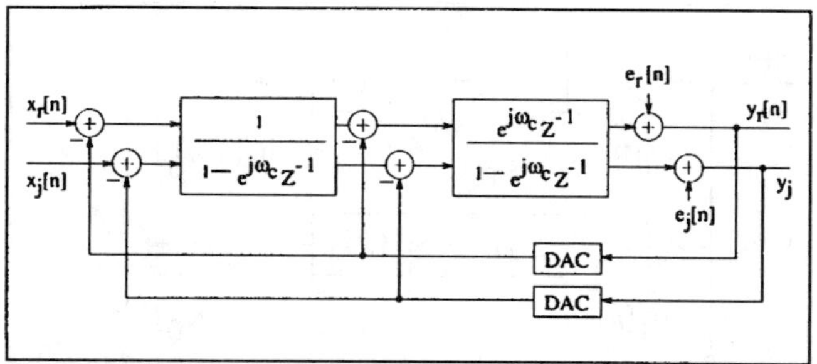

20. Complex second order modulator.

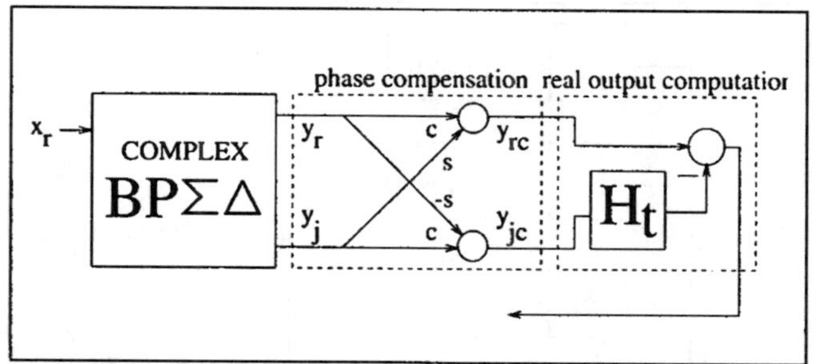

21. Complex BP sigma-delta modulation of a real input signal followed by phase compensation and computation of a real output.

the signal is narrowband and this distance is small. As the center frequency ω_c increases, v_{r2} increases as well, and the attenuation provided by the NTF becomes worse and so there will be less quantization noise suppression.

For the complex case, the zeros of a second order FIR NTF with both zeros located at a center frequency ω_c, are shown in Fig. 19b. The attenuation at ω_c-$\omega_b/2$ is given by the quantity $|v_c| \times |v_c|$, which depends only on ω_b and not on ω_c, as there is no influence from a zero at -ω_c. Consequently, the attenuation provided by the complex NTFs does not suffer any degradation with increasing center frequency. Therefore, at higher center frequencies, complex NTFs can provide more attenuation in the signal band, i.e., better quantization noise suppression and so a better SNR than a real transfer function. Complex band-pass converters employing Lth order modulators display a SNR performance which increases at the rate of $(6L + 3)$ dB per octave increment in the the oversampling ratio $f_s/2f_B$. This is in contrast to the $(3L + 3)$ dB rate of improvement for Lth order real band-pass modulators.

Having discussed the complex NTF, we now discuss the implementation of a complex 2nd order modulator with a band centered at ω_c. Such a modulator can be generated by modulating the Z domain NTFs and STFs of a standard 2nd order modulator with the transformation $z^{-1} \rightarrow z^{-1}\exp(j\omega_c) = z^{-1}\exp(j\,2\pi f_c/f_s)$. A block diagram of such a system obtained from Fig. 13 with the above transformation is shown in Fig. 20. Notice, for example, that the integrator $z^{-1}/(1 - z^{-1})$ of Fig. 13 is replaced with the integrator $(z^{-1}e^{j\omega_c})/(1 - e^{j\omega_c}z^{-1})$, which will have complex inputs and outputs consisting of real and imaginary parts. Such a complex integrator can be physically realized in switched capacitor technology using two cross coupled integrators [48]. Note there also need to be two physical quantizers, E_r and E_j, one for the "real" channel and the other for the "imaginary" channel. The Z domain output of the complex modulator is given by

$$Y(z) = [X_r(z) + jX_j(z)]z^{-1}\exp(j\omega_c) + [E_r(z) + jE_j(z)][1 - z^{-1}\exp(j\omega_c)]^2 \tag{30}$$

The STF is no longer a pure delay but contains the phase factor $e^{j\omega_c}$ However, this phase factor can be compensated digitally by multiplying the output of the complex modulator with the complex constant $e^{-j\omega_c}$ to obtain the phase compensated output. The phase compensated output is $Y_c(z) = Y(z)e^{-j\omega_c}$. For a real input signal, we have $x_j[n] = 0$ and $X_j(z) = 0$ so that $x[n] = x_r[n]$ and $X(z) = X_r(z)$. In this case,

$$Y_c(z) = X^r(z)z^{-1} + [E_r(z) + jE_j(z)] \cdot\exp(-j\omega_c)[1 - z^{-1}\exp(j\omega_c)]^2 \tag{31}$$

where $Y_c(z) = Y_{rc}(z) + j\,Y_{jc}(z)$.

$Y_c(z)$ consists of a signal term $X_r(z)\,z^{-1}$, which we have assumed results from a real signal, plus the shaped quantization noise term. From Eq. 31, we can also see that in the time domain, the signal component in $y_c[n]$ is $x_r[n\text{-}1]$. However, $y_c[n]$ will still be complex because of the complex noise and, accordingly, the spectrum of $y_c[n] = y_{rc}[n] + j\,y_{jc}[n]$ will not be symmetric. A real output can be obtained without disturbing the signal or altering the SNR by considering the final output to be $y_{rc}[n]$ $-y_{jc}[n]*h_t[n]$ where h_t is the impulse response of an ideal Hilbert transformer, $H_t(z)$. In the Z domain, the final output is then $[Y_{rc}(z)\text{-}Y_{jc}(z)H_t(z)]$. This operation amounts to keeping only positive frequencies (discarding negative frequencies) with the Hilbert transformer and then taking the real part to make the spectrum symmetric by folding the positive frequencies on to the negative frequency axis.

The phase compensation and computation of the real output are shown in Fig. 21 where $c = \cos(\omega_c) = \cos(2\pi f_c/f_s)$ and $s = \sin(\omega_c) = \sin(2\pi f_c/f_s)$. Decimation of the sigma-delta output and demodulation of the band-pass signal to baseband are not shown here.

Note that in the case of a real input signal, the imaginary channel of the modulator is not directly connected to the input signal. One could feed $x_r[n]$ into the imaginary channel input as well. This has the benefit of resulting in a $\sqrt{2}$ gain of the STF which in principle will result in a 3 dB improvement in signal power. However, simulations show that with the input being fed to the imaginary channel, the quantizers overload much more often and the increase in quantization noise does not merit the gain in the signal power.

Parallel Sigma-Delta Systems

The use of parallelism for PCM A/D conversion has been considered in [49] and [50]. This section very briefly discusses several schemes that use architectural parallelism to improve the performance of sigma-delta modulators A/D

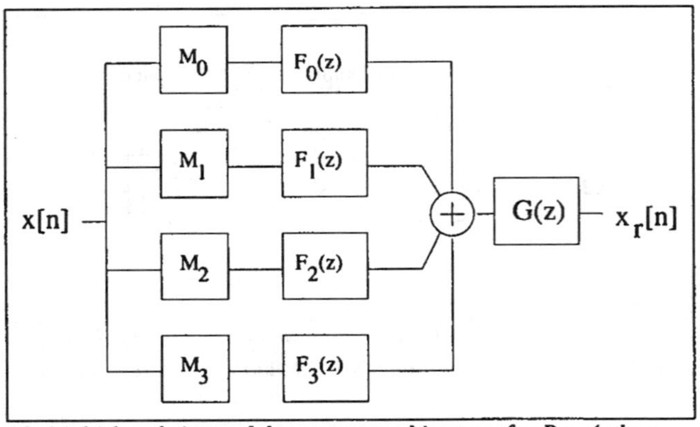

22. Multi-band sigma-delta system architecture for P = 4 channels.

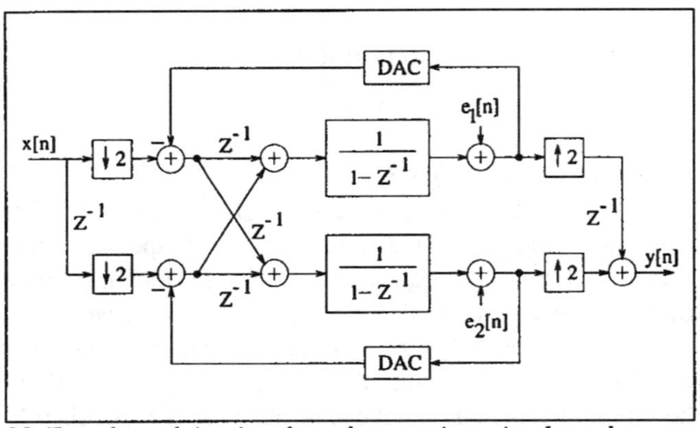

23. Two channel time interleaved conversion using 1st order sigma-delta modulators.

converters. For a given signal bandwidth, modulator order, and sampling frequency, these architectures can attain higher resolution. The cost of realizing the improved performance lies clearly in the extra hardware needed for each parallel channel.

Multi-band Sigma-Delta Modulation

One architecture [51] uses modulators that realize different band reject NTFs for different portions of the signal band. Each band (channel) is converted in parallel. A bank of FIR filters attenuates the out-of-band noise for each band and can achieve perfect reconstruction of the signal component assuming that the modulator STF is a simple delay [52].

A block diagram of the system architecture for four channels is shown in Fig. 22, where M_k denote the modulators, and $F_k(z)$ and $G(z)$ comprise the digital filter bank. Using Lth order complex band-pass modulators, assuming equal sized bands, and assuming the quantization errors from the different channels to be mutually uncorrelated, the total in-band noise power, σ_{er}^2, at the output of the reconstructed signal is

$$\sigma_{er}^2 = \frac{4\sigma_e^2}{2^{2L+1}} \frac{\pi^{2L}}{2L+1}\left(\frac{2f_C}{f_s}\right)^{2L+1}\left(\frac{1}{P}\right)^{2L} \quad (32)$$

where f_B is the total signal bandwidth converted, $f_C = f_B/P$ is the bandwidth per channel, P is the number of channels, and

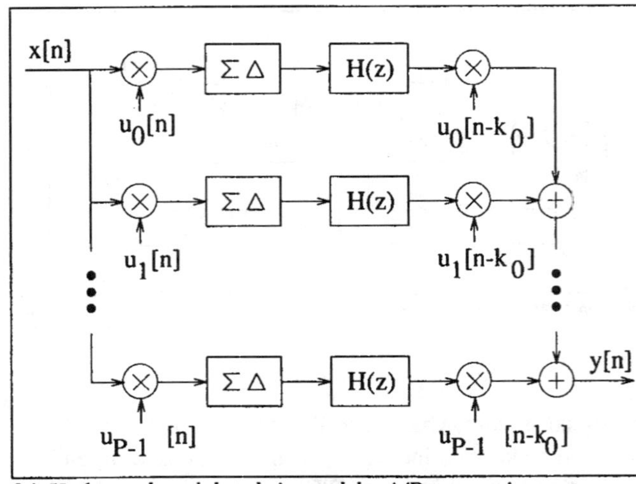

24. Hadamard modulated sigma-delta A/D conversion system.

σ_e^2 is the quantizer noise power in any one modulator. The SNR improves at a rate of $(6L + 3)$ dB, or the resolution by $(L + 1/2)$ bits per octave increment in the oversampling ratio per channel, $f_s/2f_C$. The SNR and resolution also improve at rates of $(6L)$ dB, and L bits, respectively, per octave increment in the number of channels, P.

Time Interleaved Sigma-Delta Modulation

Another method for incorporating parallelism into sigma-delta converters is through time interleaving [53]. This architecture employs ideas of block filter theory to use P identical, mutually cross coupled, modulators running at a sampling rate f_s to generate the same modulator transfer function, which runs at an equivalent sampling rate of Pf_s. The block diagram for a two channel system using 1st order modulators running in each individual channel is shown in Fig 23. Note that downsampling the signal can result in aliasing, but the choice of the cross coupling terms ensure that the aliasing is cancelled in the final output for the ideal system.

If Lth order low-pass sigma-delta modulators are used in each channel, the ideal in-band SNR improves by $(6L + 3)$ dB, or the resolution by $L + 1/2$ bits per octave increment in the number of channels, P, since each octave increment in P amounts to an octave increment in the oversampling ratio of the modulators.

Hadamard System

Another parallel sigma-delta system has been described in the literature [54] recently. Here, each channel contains a Hadamard modulator, which multiplies the input signal by a ± 1 sequence, $u_k[n]$. This operation is called Hadamard modulation.

The Hadamard sequences are obtained from repeating the rows of the Hadamard transform matrix. The Hadamard modulated sequence of each channel is then quantized using a standard sigma-delta modulator. The output of each sigma-delta modulator is filtered to attenuate out-of-band noise, and again multiplied by a Hadamard sequence before all the channel outputs are added to provide the final output.

The block diagram of the system is shown in Fig 24, where k_0 is assumed to be the delay in the STF of the sigma-delta modulators. The overall effect of the system on the signal is to filter it with a subset of the filter coefficients of $H(z)$. The relevant $H(z)$ coefficients affecting the signal can be chosen such that the STF experiences a delay, while the other coefficients can be chosen to maximize the quantization noise attenuation.

The quantization error does not see the first Hadamard sequence and the effect of the second Hadamard sequence modulation is to frequency shift the filtered quantization error power spectral density. Using Lth order low-pass sigma-delta modulators, the SNR using this approach improves by ($6L$) dB or the resolution improves by L bits per octave increment in the number of channels P.

Applications Using Actual Sigma-Delta Converters

We now present some applications using actual sigma-delta converters that have been fabricated in VLSI. The purpose here is to provide a sense of the final performance achieved by the converters as to resolution and bandwidth, rather than to compare them with respect to theoretical performance or to the multitude of other performance criterion (power, area, topology, technology, etc.). To aid the resolution/bandwidth evaluation, we use the abbreviation *osr* for the oversampling ratio, $f_s/2f_B$, where f_B is the signal bandwidth and f_s is the sampling frequency. To provide an overview of the converters discussed here, we have included all the converters in a common format in the Table.

Data conversion for instrumentation applications may require resolutions up to 19-20 bits, albeit at low bandwidths. One such converter used in instrumentation transducers [55] uses a 5th order modulator topology with a f_s = 128 kHz (*osr* ≃ 128) to a achieve a 118 dB dynamic range, or more than 19 bits of resolution over a 492 Hz bandwidth. Similarly, a converter that is used for seismic activity measurements uses a 128 kHz sampling rate with an *osr* of 128, achieving more than 120 dB peak SNR, or almost 20 bits of resolution over about a 500 Hz bandwidth [56]. A 4th order topology is used. Note that for extremely high resolution such as reported here, the quantization noise floor approaches the level of circuit noise for state of the art technologies. Thus, very careful circuit design and optimization is required to fully take advantage of the potential performance that can be realized by the sigma-delta ADC.

Sigma-delta converters are good candidates for voiceband (speech) applications where the signal bandwidth is 4 kHz and 13-14 bits of resolution is desirable. One such converter [57] actually used a single bit 1st order modulator with f_s = 4 MHz, or an *osr* of about 500 to achieve a 79 dB dynamic range, i.e., about 13 bits of resolution. Dithering was required to alleviate the tone problem associated with the first order modulator. Another converter [58] also achieved 13 bits resolution using a f_s = 1.024 MHz with an *osr* of 128. The modulator used is a standard second order modulator employing a 1 bit quantizer.

Digital audio applications such Hi-Fi CD and DAT systems often use sigma-delta A/D converters. Consumer quality Hi-Fi audio needs to be digitized at 16 bits of resolution, and audiophiles prefer up to 18-20 bits of resolution. Many converters have been reported in the literature for audio bandwidths of 20-24 kHz. A high resolution audio range converter was reported as early as 1986 and achieved a dynamic range of 106 dB, or almost 18 bits of resolution over a 24 kHz bandwidth using a fourth order modulator and 4 bit internal A/D and D/A converters [59]. The sampling frequency used was 6.144 MHz with an *osr* of 128. Another converter that also uses a fourth order modulator to convert a 24 kHz bandwidth achieves near 16 bit dynamic range using single bit quantizers with an *osr* of 64 or f_s = 3.072 MHz [60]. Audio band conversion has also been performed by a standard 2nd order modulator with a 4 bit internal ADC and DAC. The converter achieved nearly 16 bit peak SNR for a 20.5 kHz

Table: Sigma Delta Converters in Use Today

Signal Band (f_B)	Sampling Frequency (f_s)	OSR ($f_s/2f_B$)	Overall Resolution (# bits)	Modulator Structure	Internal Quantizer	Application	Reference
492 Hz	128 kHz	128	20	4th order	1 bit	instrumentation	[55]
500 Hz	128 kHz	128	20	4th order	1 bit	seismic	[56]
4 kHz	4 MHz	500	13	1st order	1 bit	speech	[57]
4 kHz	1.024 MHz	128	13	2nd order	1 bit	speech	[58]
20.5 kHz	5.25 MHz	128	16	2nd order	4 bits	audio	[30]
24 kHz	6.144 MHz	128	18	4th order	4 bits	audio	[59]
24 kHz	3.072 MHz	64	16	2nd order	1 bit	audio	[60]
25 kHz	6.4 MHz	128	17	"2-1" cascade	1 bit	audio	[39]
40 kHz	10.24 MHz	128	14	2nd order	1 bit	ISDN	[16]
40 kHz	2.56 MHz	32	13	"2-1" cascade	1 bit	ISDN	[37]
100 kHz	3.25 MHz	16	15	"2-2-2" cascade	3 level	digital cellular radio	[61]
160 kHz	20.48 MHz	24	16	"2-1" cascade	1 bit	-	[38]
250 kHz	32 MHz	32	14	4th order	1 bit	-	[20]
1 MHz	50 MHz	20	12	"2-1" cascade	1 and 3 bit	ultrasound	[62]

bandwidth using an *osr* of 128, or $f_s = 5.25$ MHz [30]. Finally, an architecture consisting of a cascade of second order and a first order modulator and employing 1 bit quantizers achieves nearly 17 bit performance for a 25 kHz bandwidth, also using an *osr* of 128 with a $f_s = 6.4$ MHz [39].

A sigma-delta converter has also been used as a receiver input of an ISDN U-interface 2B1Q access rate receiver [16]. The converter attains a dynamic range of 89 dB, or a resolution of 14 bits, for a 40 kHz bandwidth. A standard 2nd order modulator was used with a 1 bit quantizer running at a 10.24 MHz sampling frequency, and an *osr* of 128. Another converter [37] used for a similar ISDN U-interface consisted of a cascade of a 2nd order modulator, followed by a 1st order modulator. The resolution was 13 bits, using a f_s of 2.56 MHz, i.e., an *osr* of 32.

A sigma-delta ADC has been used as the baseband converter for a digital cellular radio that required a moderately higher bandwidth of 100 kHz [61]. The converter produced 15 bit peak resolution by using a three stage cascade employing three 2nd order modulators running at $f_s = 3.25$ MHz. The *osr* was about 16.

At somewhat higher bandwidths, high resolution conversion has been attained for 160 kHz and 250 kHz bandwidths. The 160 kHz bandwidth converter achieves a dynamic range of 96 dB, or nearly 16 bit performance using a cascade of second and first order modulators that employ 1 bit quantizers [38]. The sampling rate is $f_s = 20.48$ MHz and the resulting *osr* is 24. The 250 kHz bandwidth converter achieves 14 bit resolution using a 4th order modulator using a 1 bit quantizer [20]. The *osr* is 32 and thus the f_s is 32 MHz.

At high conversion bandwidths of about 1 MHz, a 12 bit converter has been realized with an *osr* of about 24, and a f_s of about 50 MHz [62]. The converter uses a cascade of second order and first order modulators. The second order modulator used in the first stage employs a 1 bit quantizer, while the 1st order modulator used in the second stage utilizes a 3 bit internal ADC and DAC. Such a converter can find use in data acquisition for ultrasound imaging systems.

Finally, sigma-delta ADCs have been integrated with digital signal processor (DSPs) to provide a single chip data conversion/ computation engine solution. Sigma-delta ADCs require relatively imprecise analog circuits and digital decimation filtering, thus making them good candidates for fabrication using digital technology such as CMOS. The task of decimation can be handled entirely by a DSP or shared by a DSP and some extra digital hardware dedicated to performing a portion of the decimation.

Conclusion

We have reviewed the basic principles of A/D conversion with sigma-delta modulators. The techniques of oversampling and noise shaping allows the use of relatively imprecise analog circuits to perform high resolution conversion using only a 1 bit A/D converter. Oversampling reduces the amount of quantization noise power present in the signal band, and noise shaping further attenuates quantization noise in the signal band, thereby pushing noise power to out-of-band frequencies. The use of analog filtering combined with feedback around the 1 bit A/D can be used to implement the noise shaping sigma-delta modulator. The noise power that is pushed outside the signal band can be attenuated by a digital filter such that it has no further effect on the signal.

Various sigma-delta architectures exist and many of these have been used in applications such as instrumentation, speech and Hi-Fi audio digitization, ISDN and digital cellular radio. Sigma-delta techniques are also applicable to the high resolution A/D conversion of narrowband band-pass signals using band-pass sigma-delta modulators. Parallel sigma-delta systems offer the potential for extending high resolution operation to larger signal bandwidths than currently possible with single channel systems.

Pervez M. Aziz is a Member of Technical Staff at AT&T Bell Laboratories, Murray Hill, NJ. Henrik V. Sorensen is Senior Researcher at Ariel Corporation, Highland Park, NJ. Jan Van der Spiegel is Professor, Department of Electrical Engineering, University of Pennsylvania, Philadelphia. This work was completed while the first two authors were at the University of Pennsylvania.

References

1. S. Renukunta and D. Wells, "Optical memory and blue lasers," *IEEE Potentials*, pp. 14-18, Oct/Nov 1994.

2. J. Candy and G. Temes, "Oversampling methods for A/D and D/A conversion" in *Oversampling Delta-Sigma Data Converters*, pp. 1-25, IEEE Press, 1992.

3. A. Oppennheim and R. Schafer:, *Discrete Time Signal Processing*, (Prentice-Hall, 1989).

4. W. Bennett, "Spectra of quantized signals," *Bell System Technical Journal*, pp. 446-472, July 1948.

5. Boser and B. Wooley,"The design of sigma-delta modulation analog-to-digital converters," *IEEE Journal of Solid State Circuits*, pp. 1298-1308, December, 1988.

6. B. Leung, "Theory of sigma-delta analog to digital converter," *IEEE International Symposium on Circuits and Systems Tutorials*, pp. 196-223, 1994.

7. H. Inose and Y. Yasuda, "A unity bit coding method by negative feedback," *Proceedings of the IEEE*, pp. 1524-1535, November, 1963.

8. J. Candy, "A use of double integration in sigma delta modulation," *IEEE Transactions on Communications*, pp. 249-258, March, 1985.

9. V. Friedman, "The structure of limit cycles in sigma delta modulation," *IEEE Transactions on Communications*, pp. 972-979, August, 1988.

10. J. Candy, O. Benjamin, "The structure of quantization noise from sigma-delta modulation," *IEEE Transactions on Communications*, pp. 1316-1323, September, 1981.

11. S. Hein and A. Zakhor, "On the stability of sigma delta modulators," *IEEE Transactions on Signal Processing*, pp. 2322-2348, July, 1993.

12. R. Gray, "Spectral analysis of quantization noise in a single-loop sigma-delta modulator with dc input," *IEEE Transactions on Communications*, pp. 588-599, June, 1989.

13. N. He, F. Kuhlmann, A. Buzo, "Double-loop sigma-delta modulation with dc input," *IEEE Transactions on Communications*, pp. 487-495, April, 1990.

14. R. Gray, "Quantization noise spectra," *IEEE Transactions on Information Theory*, pp. 1220-1244, November, 1990.

15. S. Rangan and B. Leung, "Quantization noise spectrum of double-loop sigma-delta converter with sinusoidal input," *IEEE Transactions on Circuits and Systems II*, pp. 168-173, February, 1994.

16. S. Norsworthy, I. Post, H. Fetterman,"A 14-bit 80kHz sigma-delta A/D converter: modeling, design, and performance evaluation," *IEEE Journal of Solid State Circuits*, pp. 256-266, April, 1989.

17. S. Norsworthy and D. Rich, "Idle channel tones and dithering in delta sigma modulators," *95th Convention of the Audioengineering Society*, preprint 3711, October, 1993.

18. K. Chao, S. Nadeem, W. Lee, and C. Sodini, "A higher order topology for interpolative modulators for oversampling A/D converters," *IEEE Transactions on Circuits and Systems*, pp. 309-318, March, 1990.

19. P. Ferguson, A. Ganesan, R. Adams, "One bit higher order sigma-delta A/D converters," *Proceedings, IEEE International Symposium on Circuits and Systems*, pp. 890-893, 1990.

20. F. Op't Eynde, G. Yin, W. Sansen, "A CMOS fourth-order 14b 500k-sample/s sigma-delta ADC converter," Digest of Technical Papers, *International Solid State Circuits Conference*, pp. 62-63, 1991.

21. R. Adams, "Design aspects of high-order delta-sigma A/D converters," *IEEE International Symposium on Circuits and Systems Tutorials*, pp. 235-259, 1994.

22. S. Ardalan and J. Paulos, "An analysis of nonlinear behavior in delta-sigma modulators," *IEEE Transactions on Circuits and Systems*, pp. 593-603, June, 1987.

23. R. Baird and T. Fiez, "Stability analysis of high-order delta-sigma modulation for ADCs," *IEEE Transactions on Circuits and Systems II*, pp. 59-62, January, 1994.

24. S. Moussavi and B. Leung, "High-order single-stage single-bit oversampling A/D converter stabilized with local feedback loops," *IEEE Transactions on Circuits and Systems II*, pp. 19-25, January, 1994.

25. D. Anastassiou, "Error diffusion coding for A/D conversion," *IEEE Transactions on Circuits and Systems*, pp. 1175-1186, September, 1989.

26. R. Schreier and Y. Yang, "Stability tests for single-bit sigma-delta modulators with second-order FIR noise transfer functions," *Proceedings, IEEE International Symposium on Circuits and Systems*, pp. 1316-1319, 1992.

27. L. Larson, T. Cataltepe, and G. Temes, "Multibit oversampled sigma-delta A/D convertor with digital error correction," *Electronics Letters*, pp. 1051-1052, August 4, 1988.

28. T. Leslie and B. Singh, "An improved sigma-delta modulator architecture," *Proceedings, IEEE International Symposium on Circuits and Systems*, pp. 372-375, 1990.

29. A. Hairapetian, G. Temes, and Z. Zhang, "Multibit sigma-delta modulator with reduced sensitivity to DAC nonlinearity," *Electronics Letters*, pp. 990-991, May 23, 1991.

30. M. Sarhang-Nejad and G. Temes, "A high-resolution multibit sigma-delta ADC with digital correction and relaxed amplifier requirements," *IEEE Journal of Solid State Circuits*, pp. 648-660, June, 1993.

31. L. Carley, "A noise-shaping coder topology for 15+ bit converters," *IEEE Journal of Solid State Circuits*, pp. 267-273, April, 1989.

32. B. Leung and S. Sutarja, "Multibit sigma-delta A/D converter incorporating a novel class of dynamic element matching technique," *IEEE Transactions on Circuits and Systems II*, pp. 35-51, January, 1992.

33. F. Chen and B. Leung, "A high resolution multi-bit sigma-delta modulator with individual level averaging," Digest of Technical Papers, *IEEE Symposium on VLSI Circuits*, pp. 101-102, June, 1994.

34. Y. Matsuya, K. Uchimura et al, "A 16-bit oversampling A-to-D conversion technology using triple-integration noise shaping," *IEEE Journal of Solid State Circuits*, pp. 921-929, December, 1987.

35. K. Uchimura, T. Hayashi, T. Kimura and A. Iwata, "Oversampling A-to-D and D-to-A converters with multistage noise shaping modulators," *IEEE Transactions on Acoustics, Speech, and Signal Processing*, pp. 1899-1905, December, 1988.

36. M. Rebeschini, N. van Bavel, P. Rakers, et al, "A 16-b 160-kHz CMOS A/D converter using sigma-delta modulation," *IEEE Journal of Solid State Circuits*, pp. 431-440, April, 1990.

37. L. Longo and M. Copeland, "A 13 bit ISDN-band oversampled ADC using two-stage third order noise shaping," *Proceedings, IEEE Custom Integrated Circuits Conference*, pp. 21.2.1-21.2.4, 1988.

38. G. Yin, F. Stubbe, W. Sansen, "A 16-b 320-kHz CMOS A/D converter using two-stage third-order sigma-delta noise shaping," *IEEE Journal of Solid State Circuits*, pp. 640-647, June, 1993.

39. L. Williams and B. Wooley, "Third-order sigma-delta modulator with extended dynamic range," *IEEE Journal of Solid State Circuits*, pp. 193-202, March, 1994.

40. D. Ribner, "A comparison of modulator networks for high-order oversampled sigma-delta analog-to-digital converters," *IEEE Transactions on Circuits and Systems*, pp. 145-159, February, 1991.

41. R. Schreier and M. Snelgrove, "Bandpass sigma-delta modulation," *Electronics Letters*, pp. 1560-1561, November 9, 1989.

42. S. Jantzi, R. Schreier and M. Snelgrove, "Bandpass sigma-delta analog-to-digital conversion," *IEEE Transactions On Circuits and Systems*, pp. 1406-1409, November, 1991.

43. L. Longo and B. Horng, "A 15b 30kHz bandpass sigma delta modulator," Digest of Technical Papers, *International Solid State Circuits Conference*, pp. 226-227, 1993.

44. S. Jantzi, M. Snelgrove, P. Ferguson, "A 4th-order bandpass sigma-delta modulator," *IEEE Journal of Solid State Circuits*, pp. 282-291, March, 1993.

45. G. Tröster, H Dreβler, et al, "An interpolative bandpass converter on a 1.2-um BiCMOS analog/digital array," *IEEE Journal of Solid State Circuits*, pp. 471-477, April, 1993.

46. S. Jantzi, K. Martin, M. Snelgrove, A. Sedra, "Complex bandpass sigma-delta converter for digital radio," *Proceedings, IEEE International Symposium on Circuits and Systems*, pp. 453-456, 1994.

47. P. Aziz, H. Sorensen, J. Van der Spiegel, "Performance of complex noise transfer functions in bandpass and multi band sigma delta systems," *Proceedings, IEEE International Symposium on Circuits and Systems*, pp. 641-644, 1995.

48. Q. Liu, M. Snelgrove, A. Sedra, "Switched-capacitor implementation of complex filters," *Proceedings, IEEE International Symposium on Circuits and Systems*}, pp. 1121-1124, 1986.

49. W. Black and D. Hodges, "Time interleaved converter arrays," *IEEE Journal of Solid State Circuits*, pp. 1022-1029, December, 1980.

50. A. Petraglia and S. Mitra, "High speed A/D conversion using QMF banks," *Proceedings, IEEE International Symposium on Circuits and Systems*, pp. 2797-2800, 1990.

51. P. Aziz, H. Sorensen, J. Van der Spiegel, "Multiband sigma-delta modulation," *Electronics Letters*, pp. 760-762, April 29, 1993.

52. P. Aziz, H. Sorensen, J. Van der Spiegel, "Multiband sigma delta analog to digital conversion," *Proceedings, IEEE International Conference on Acoustics, Speech, and Signal Processing*, pp. 249-252, 1994.

53. R. Khoini-Poorfard, L. Lim, D. Johns, "Time-interleaved oversampling convertors," *Electronics Letters*, pp. 1673-1674, September 16, 1993.

54. I. Galton and H. Jensen, "Delta-sigma modulator based A/D conversion without oversampling," to appear, *IEEE Transactions on Circuits and Systems*.

55. C. Thompson, S. Bernadas, "A digitally corrected 20b delta-sigma modulator," Digest of Technical Papers, *International Solid State Circuits Conference*, pp. 194-195, 1994.

56. D. Kerth, D. Kasha, et al, "A 120 dB linear switched-capacitor delta-sigma modulator," Digest of Technical Papers, *International Solid State Circuits Conference*, pp. 196-197, 1994.

57. B. Leung, R. Neff, P. Gray, R. Broderson, "Area-Efficient Multichannel Oversampled PCM Voice-Band Coder," *IEEE Journal of Solid State Circuits*, pp. 1351-1357, December, 1988.

58. V. Friedman, D. Brinthaupt, et al, "A dual-channel voice-band PCM codec using sigma-delta modulation technique," *IEEE Journal of Solid State Circuits*, pp. 274-280, April, 1989.

59. R. Adams, "Design and implementation of an audio 18-bit analog-to-digital converter using oversampling techniques," *Journal of the Audioengineering Society*, pp. 153-166, March, 1986.

60. D. Welland, B. Del Signore, et al, "A stereo 16-bit delta-sigma A/D converter for digital audio," *Journal of the Audioengineering Society*, pp. 476-486, June, 1989.

61. I. Dedic, "A sixth-order triple-loop sigma-delta CMOS ADC with 90 dB SNR and 100 kHz bandwidth," Digest of Technical Papers, *International Solid State Circuits Conference*, pp. 188-189, 1994.

62. B. Brandt and B. Wooley, "A CMOS oversampling A/D converter with 12b resolution at conversion rates above 1 MHz," Digest of Technical Papers, *International Solid State Circuits Conference*, pp. 64-65, 1991.

The Theory of Bandpass Sampling

Rodney G. Vaughan, *Member, IEEE*, Neil L. Scott, and D. Rod White

Abstract—**The sampling of bandpass signals is discussed with respect to band position, noise considerations, and parameter sensitivity. For first-order sampling, the acceptable and unacceptable sample rates are presented, with specific discussion of the practical rates which are nonminimum. The minimum sampling rate is pathological in that any imperfection in the implementation will cause aliasing. The susceptibility of the acceptable sampling rates to aliasing decreases with increasing guard-band size from the minimum sampling rate.**

In applying bandpass sampling to relocate signals to a baseband position, the signal-to-noise ratio is not preserved owing to the out-of-band noise being aliased. The degradation in signal-to-noise ratio is quantified in terms of the position of the bandpass signal.

For the construction of a bandpass signal from second-order samples, the cost of implementing the interpolant (dynamic range and length) depends on Kohlenberg's sampling factor k, the relative delay between the uniform sampling streams. An elaboration on the disallowed discrete values of k shows that some allowed values are better than others for implementation. Optimum values of k correspond to the quadrature sampling values with the band being either integer or half-integer positioned.

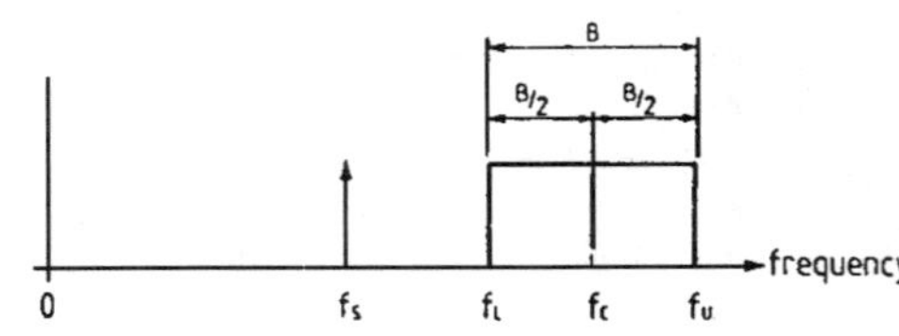

Fig. 1. The bandpass situation as an analog signal spectrum. The sampling rate is expressed as f_s Hz and the band is located at (f_L, f_u). Only the positive frequencies are shown.

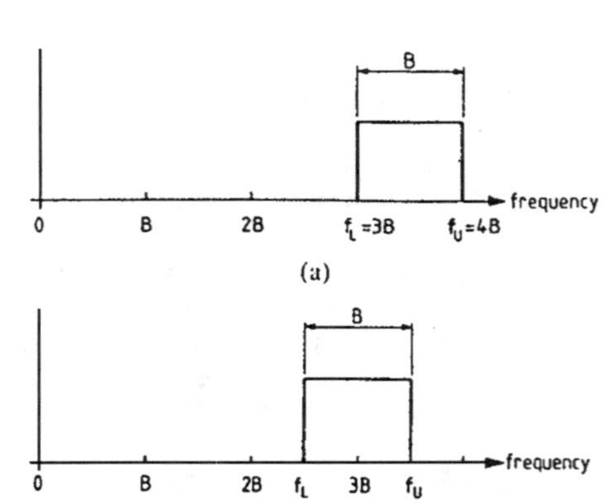

Fig. 2. Examples of (a) integer band positioning for $c = 3$, and (b) half-integer band positioning for $c = 3$.

I. Introduction

The subject of bandpass sampling is an aspect of digital signal processing which has relevance to a variety of disciplines including, for example, optics [1], radar [2], sonar [3], communications [4], biomedical signals [5], power measurement [6], and general instrumentation, such as sampling oscilloscopes.

Historically, Cauchy seems to have been the first to hypothesize the bandpass sampling requirement [7]. Nyquist [8] and Gabor [9] also alluded to the bandpass case. Kohlenberg [10] introduced second-order sampling and provided the basic interpolation formula, which was then reported in the texts by Goldman [11] and Middleton [12].

The bandpass situation is depicted in Fig. 1, where the sampling rate is expressed as f_s Hz and the bandpass signal is located between f_L Hz and f_u Hz. The signal bandwidth B is less than $f_u - f_L$, and the positive frequency band can be expressed as the interval (f_L, f_u).

The *band position* refers to the fractional number of bandwidths from the origin at which the lower band edge resides. A special case is *integer band positioning*, which holds when the band is located at an integral number of bandwidths from the origin, i.e., $f_L = c(f_u - f_L)$, $c = 0, +1, +2, \cdots$; and $c = 0$ is the low-pass case. Fig. 2(a) indicates integer band positioning, for $c = 3$.

Another special case is *half-integer positioning*, when $f_L = ((2c + 1)/2)(f_u - f_L)$. Fig. 2(b) shows an example of half-integer positioning for $c = 2$. These cases of band positioning are useful for illustrating aspects of aliasing, and are particularly important in the application of sampling to the effective relocation of signals between bandpass and low-pass positions.

The classical bandpass theorem for uniform sampling states that the signal can be reconstructed if the sampling rate is at least $f_s^{(\min)} = 2f_u/n$, where n is the largest integer within f_u/B, denoted by $n = I_g[f_u/B]$.

This minimum rate for uniform sampling is of interest from a theoretical viewpoint, but for practical applications this rate is often presented in a misleading way. Fig. 3 from Feldman and Bennett [16], also reproduced in several texts (e.g., [13]–[15]), illustrates the minimum uniform sampling rate for bandpass signals. The theoretical minimum rate $f_s = 2B$ is seen to apply only for integer band positioning. The skew lines represent the minimum uniform rate required when the passband is not located in an integer position. The vertical lines, which represent a

Manuscript received July 5, 1989; revised June 24, 1990.

R. G. Vaughan and N. L. Scott are with Industrial Development, DSIR, Lower Hutt, New Zealand.

D. R. White is with Physical Sciences, DSIR, Lower Hutt, New Zealand.

IEEE Log Number 9101475.

Reprinted from *IEEE Transactions on Signal Processing*, Vol. 39, No. 9, pp. 1973–1984, September 1991.

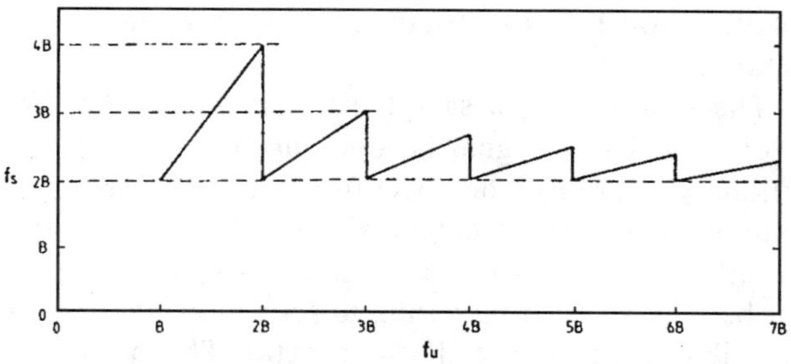

Fig. 3. Minimum sampling frequency for band of width B [16].

discontinuity, traverse *disallowed* values of the sampling frequency, a point elucidated in Section II. Gregg's [17] version of Fig. 3 displays more clearly the discontinuity in the function by using dotted lines for the vertical sections.

Note that $f_s = 2B$ is a valid uniform sampling rate only if there is no signal component at the frequencies f_L or f_u. This is analogous to the anomaly of the low-pass case (cf. Shannon's theorem) where the equality in $f_s \geq 2f_u$ can hold only if there is no signal component at f_u, i.e., the signal is band limited to *less than* B and the signal occupies the interval $(0, f_u)$.[1] Kida and Kuroda [18] discuss the disallowed signal components for higher order sampling.

The theoretical minimum sampling rate is pathological in the sense that any engineering imperfections in an implementation will cause aliasing. This illustrates the need to sample at above the theoretical minimum rate or, alternatively stated, guard-bands need to be included. Gregg [17] states without elaboration that rates between $f_s^{(\min)}$ and $2f_u$ (the low-pass case rate) are not, in general, valid. These rates correspond to the region above the function in Fig. 3. Nevertheless, some authors mistakenly interpret figures such as Fig. 3 to mean that no aliasing will occur as long as $f_s > 4B$. Gaskell [1] gives an expression (see (16)) for the valid rates, but otherwise there is little information regarding bandpass sampling at rates above the theoretical minimum but still well below the frequencies of the signal being sampled. In Section II, this subject is discussed and results are depicted graphically. The existing information is clarified and the cases for bandpass sampling at above the minimum rate are reviewed. The discussion includes sensitivity of sampling rates and offers a practical guide to bandpass uniform sampling.

The theoretical minimum uniform sampling rate $f_s = 2B$ can be applied only when the band is integer positioned. Kohlenberg's [10] second-order sampling theorem allows the minimum rate, in the form of an *average* rate, to be applied independent of the band position. The technique is second order in that two interleaved uniform sample sequences are applied. Each sequence is of rate B (i.e., sampling every $1/B$ s), and the sequence separation is denoted by k s.

Using the samples to construct a signal $f(t)$ at (f_L, f_u), the interpolation is given by

$$f(t) = \sum_p \left[f\left(\frac{p}{B}\right) S\left(t - \frac{p}{B}\right) + f\left(\frac{p}{B} + k\right) S\left(-t + \frac{p}{B} + k\right) \right] \quad (1)$$

in which p is the sample number index and the interpolant is

$$S(t) = S_0(t) + S_1(t) \quad (2)$$

where

$$S_0(t) = \frac{\cos\left[2\pi(rb - f_L)t - r\pi Bk\right] - \cos\left[2\pi f_L t - r\pi Bk\right]}{2\pi Bt \sin r\pi Bk} \quad (3)$$

$$S_1(t) = \frac{\cos\left[2\pi(f_L + B)t - (r+1)\pi Bk\right] - \cos\left[2\pi(rB - f_L)t - (r+1)\pi Bk\right]}{2\pi Bt \sin (r+1)\pi Bk} \quad (4)$$

and r is an integer given by the band position such that

$$\frac{2f_L}{B} \leq r < \frac{2f_L}{B} + 1. \quad (5)$$

The right-hand side of (5) can be expressed in terms of the center frequency f_c,

$$\frac{2f_L}{B} + 1 = \frac{f_L + f_U}{B} = \frac{2f_c}{B}. \quad (6)$$

Geometrically, there are at least $r - 1$ bandwidths between the two-sided spectrum, i.e., between $-f_L$ and $+f_L$. The interpolant has the properties that $S(0) = 1$; $S(p/B) = 0$ for $p \neq 0$, and $S(p/B + k) = 0$, and in general it has extrema which are larger than unity.

In the general case when the equality of (5) does not hold, the value of k may not take on the discrete values

$$kBr,\ kB(r+1) = 0, 1, 2, \cdots \quad (7)$$

for which the interpolant blows up. k is otherwise unrestricted, but implementation difficulty using finite word-length machines is affected by the choice of k. This sensitivity is discussed in Section III.

When the equality in (5) holds, the first term of the interpolant, $S_0(t)$, is zero, and the restrictions $kBr \neq 0, 1, 2, \cdots$ on k are removed. The situation now corresponds to either integer band positioning (r even) or half-integer band positioning (r odd) and for these conditions there are exactly r bandwidths between $-f_L$ and $+f_L$.[2] Three special cases are noteworthy:

[1]The low-pass case is exceptional in that the component at $f_l = 0$ can be recovered because it contains trivial phase information.

[2]An alternative definition for r in (5) is to place the equality with the right-hand term instead of the left-hand term. This causes a swap in the above meaning of odd and even r, and a swap of the term that drops out of the interpolant when the equality holds. For this alternative definition, r cannot be zero and the low-pass case corresponds to $r = 1$.

i) The low-pass case is $f_L = 0$ and $r = 0$ for which the interpolant reduces to

$$S_{LP}(t) = \frac{\cos(2\pi Bt - \pi Bk) - \cos \pi Bk}{2\pi Bt \sin \pi Bk} \tag{8}$$

ii) The special case of $k = 1/2B$ corresponds to uniform rate sampling, with its well-known sinc $(2\pi Bt)$ interpolant, which, as mentioned above, requires integer band positioning.

iii) The quadrature sampling case has

$$k = \frac{1}{4f_c} \pm \frac{m}{2f_c}, \qquad m = 0, 1, 2 \cdots \tag{9}$$

and the simplified interpolant

$$S_Q(t) = \frac{\sin \pi Bt}{\pi Bt} \cos 2\pi f_c t. \tag{10}$$

Quadrature sampling is important because, as the name indicates, the in-phase and quadrature components are sampled explicitly from the bandpass signal. This is seen by denoting the bandpass signal in the usual way:

$$f(t) = I(t) \cos \omega_c t - Q(t) \sin \omega_c t \tag{11}$$

where $I(t)$ and $Q(t)$ are quadrature components of the center angular frequency ω_c rad/s.

To extract $I(t)$ directly from $f(t)$, i.e.,

$$f(t_p) = I(t_p) \tag{12}$$

requires sampling at times

$$\omega_c t_p = p\pi, \qquad \text{i.e.,} \quad t_p = \frac{p}{2f_c}, \qquad p = 0, \pm, \pm 2, \cdots. \tag{13}$$

Similarly, to obtain $Q(t)$

$$\omega_c t_p' = (2p + 1)\frac{\pi}{2}, \qquad \text{i.e.,} \quad t_p' = \frac{2p+1}{4f_c} \tag{14}$$

whence

$$f(t_p') = -Q(t_p'). \tag{15}$$

The signals $I(t)$ and $Q(t)$ are band limited to the interval $(0, B/2)$ so that each stream requires sampling at a rate B.

Grace and Pitt [3] drew attention to this special case of Kohlenberg's sampling scheme. Rice and Wu [19] applied the special case of uniform sampling combined with the band being integer positioned, which is a form of quadrature sampling in that $k = 1/2B = 1/4f_c + m/2f_c$ where $m = 2/r (= 2$ for integer band positioning). They used the fact that $I(t)$ and $Q(t)$ are a Hilbert transform pair to derive the required interpolation weights used for time aligning the samples. Their innovation was to "decimate by two" the samples before applying the Hilbert transform interpolator. Waters and Jarrett [4] implemented quadrature sampling using the $m = 0$ case and also used decimation before the interpolation filter. Jackson and Matthewson [2] also discussed implementation of quadrature sampling.

The spectrum of a sampled signal is periodic. Whenever a band-pass signal is reproduced at a baseband position by sampling, the noise from all the aliased bands is combined into the baseband. Even with an ideal antialiasing filter, the signal-to-noise ratio (SNR) is not preserved for bandpass sampling owing to (postfilter) thermal noise contributions from the aliased spectra. This has clear implications for communications and radar receivers where bandpass sampling is applied to relocate the signal effectively to a low(er) pass position. This is in contrast to analog complex mixing (i.e., using an image rejecting mixer), in which the signal-to-noise ratio is ideally preserved. The degradation in signal-to-noise ratio does not seem to have been discussed to date, despite its ubiquitous presence in less critical applications, such as sampling oscilloscopes. Section IV describes and quantifies the signal-to-noise ratio degradation, with a simple experiment to demonstrate the effect. Also, the aliasing noise resulting from using typical bandpass filters is presented in terms of a signal-to-(aliasing) distortion ratio. This estimates the effect of applying "acceptable" sampling rates with practical, instead of ideal, rectangular bandpass filters.

II. Uniform Sampling

The conditions for acceptable uniform sampling rates can be written [1]

$$\frac{2f_u}{n} \le f_s \le \frac{2f_L}{n-1} \tag{16}$$

where n is the integer given by

$$1 \le n \le I_g \left[\frac{f_u}{B}\right]. \tag{17}$$

The equality conditions of (16) hold only if the band is confined to the interval (f_L, f_u): a frequency component at $f = f_L$ or $f = f_u$ will be aliased, as discussed in Section I and by Kida and Kuroda [18].

Equations (16) and (17) are depicted graphically in Fig. 4. The sampling frequency, normalized by B, is on the ordinate, and the abscissa represents the band position f_u/B. Fig. 4 is a corrected and extended version of Fig. 3. The areas inside the wedges are the allowed zones for sampling without aliasing. The shaded area represents uniform sampling rates that result in aliasing. The low-pass case $f_s \ge 2f_u$, given by $n = 1$, corresponds to the large wedge to the left of the figure. Each wedge across the figure corresponds to a successive value of n.

The theoretical minimum sampling rate $f_s = 2B$, corresponding to integer band positioning, occurs at the tips of the wedges, and these rates are clearly dangerous points for implementation, since any engineering imperfection will move the sampling rate into the disallowed area. In a sampling system, the allowed sampling rates are represented by a vertical line within the allowed areas and above the band position given on the abscissa.

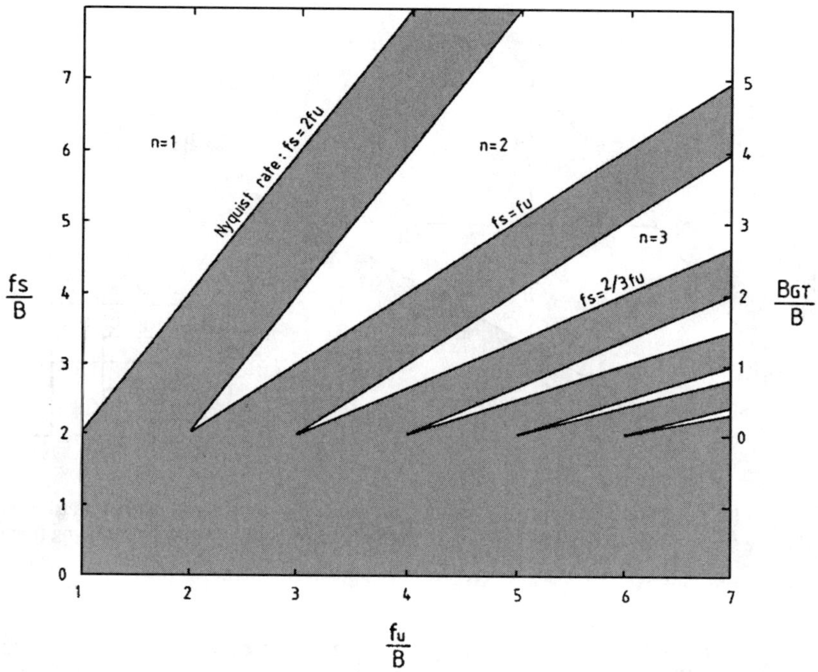

Fig. 4. The allowed and disallowed (shaded areas) uniform sampling rates versus the band position. The figure is an extension of Fig. 3, to show the practical areas of nonminimum sampling rates. The right-hand ordinate is the total guard-band (sum of upper and lower guard-bands). f_s is the sampling rate, B is the bandwidth, and the band is located at (f_L, f_u).

Brown [20] has pointed out that for symmetric double sideband signals, the spectra can "fold over" each other without loss of information, allowing a theoretical minimum sampling rate of $1/B$ for a band-limited signal where the total bandwidth (both sidebands) is B. This case would have an allowed operating "region" in Fig. 4 of straight lines bisecting the disallowed areas between the parallel borders. The case also represents a pathological condition, because any sampling rate variation will not only cause aliasing, but will also cause the signal to become folded incorrectly with a likely loss of otherwise retrievable information.

Sampling at nonminimum rates is equivalent to augmenting the signal band with a guard-band. In fact, any allowed operating point away from the tips of the wedges of Fig. 4 can be interpreted as having a guard-band between positions where the spectrum will be aliased. The total guard-band is given by $B_{GT} = f_s - 2B$ Hz, which is labelled on the right-hand ordinate of Fig. 4.

To find the relation between the guard-bands and the sampling frequency, the order of the wedge n is required. This is found from the overall bandwidth (signal bandwidth plus total guard-band)

$$W = B + B_{GT} \tag{18}$$

and its location (f_L', f_u'), where

$$f_L' = f_L - B_{GL} \tag{19}$$

$$f_u' = f_u + B_{Gu} \tag{20}$$

and B_{GL} and B_{Gu} are the lower and upper guard-bands, respectively, with $B_{GT} = B_{GL} + B_{Gu}$. The lower order wedge is then given by

$$n' = I_g\left[\frac{f_u'}{W}\right]. \tag{21}$$

Fig. 5 is an expansion of Fig. 4, showing the nth wedge with the guard-bands and the sampling frequency tolerance.

The allowable range of sampling frequencies is

$$\Delta f_s = \frac{2f_L'}{n'-1} - \frac{2f_u'}{n'} \tag{22}$$

and this range is divided into values above and below a nominal operating point (see Fig. 5)

$$\Delta f_s = \Delta f_{su} + \Delta f_{sL} \tag{23}$$

giving the lower and upper guard-bands as

$$B_{GL} = \Delta f_{su} \frac{n'-1}{2} \tag{24}$$

$$B_{Gu} = \Delta f_{sL} \frac{n'}{2}. \tag{25}$$

Note that symmetric guard-bands imply asymmetric sampling frequency tolerances. However, the asymmetry decreases with increasing n.

If the operating point is the vertical midpoint of the wedge, then the sampling rate is

$$f_s = \frac{1}{2}\left(\frac{2f_u'}{n'} + \frac{2f_L'}{n'-1}\right) \tag{26}$$

and the guard-bands become

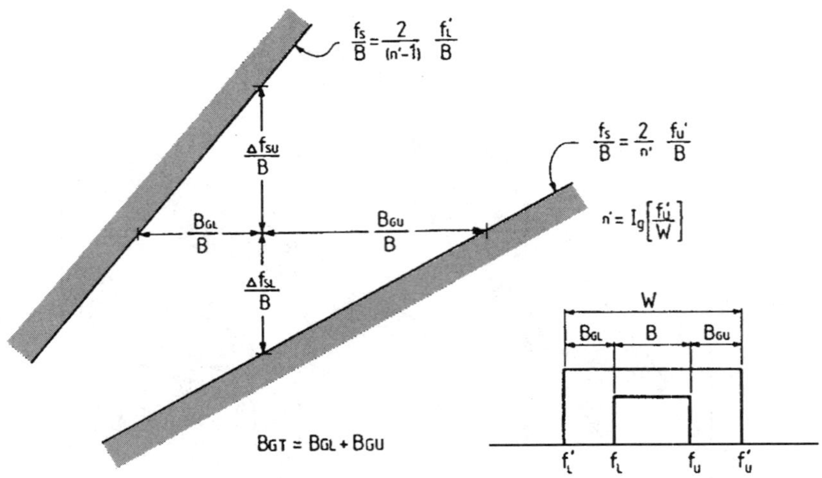

Fig. 5. Parameters in the nth wedge from Fig. 4. B_{GL} and B_{Gu} represent guard-bands against aliasing of the signal occupying bandwidth B, and Δfs_u and Δfs_L represent the allowable upward and downward range of the sampling frequency from its nominal value.

$$B_{GL} = \Delta f_s \frac{n' - 1}{4} \tag{27}$$

$$B_{Gu} = \Delta f_s \frac{n'}{4}. \tag{28}$$

An example illustrates the use of the equations. If a 25-kHz signal is allocated symmetric 2.5-kHz guard-bands so that the overall band is (10.7, 10.703 MHz), then $n' = I_g[10703/30] = 356$ and the sampling rate has limits of $2(10703)/356 = 60.130$ kHz to $2(10700)/355 = 60.282$ kHz, i.e., an allowable range of $\Delta f_s = 152$ Hz. The theoretical minimum sampling rate (for the signal without guard-bands), corresponding to a zero allowable sampling rate range, is on the lower edge of the nth wedge, where $n = I_g[10702.5/25] = 428$, and is $f_s^{(\min)} = 2(10702.5)/428 = 50.017$ kHz.

Note, for this example, that in moving from the theoretical minimum rate to a rate which includes the guard-bands, a total of $428 - 356 = 72$ separate ranges of disallowed sampling rates have been traversed. This illustrates the pitfall of arbitrarily increasing the sampling rate from its theoretical minimum value for band-pass sampling.

The relative precision of the minimum sampling rate required to avoid aliasing clearly increases with the increasing separation of the band-pass signal and the origin, i.e., increasing n. If an operating point is within the nth wedge, above (i.e., infinitesimally to the left of) the tip of the $(n + 1)$th wedge, then the difference between the minimum and maximum allowed sampling rates is

$$\Delta f_s = \frac{2(f_u - B)}{n - 1} - \frac{2f_u}{n} \tag{29}$$

so that the relative precision required of f_s is

$$\frac{\Delta f_s}{2B} = \frac{1}{n(n-1)}\left(\frac{f_u}{B} - n\right) \approx 0\left(\frac{1}{n^2}\right) \tag{30}$$

i.e., related to the inverse square of the separation of the band from the origin. (For the above-mentioned points, $f_u/B - n = 1$.) For the example with $n = 428$, the relative frequency precision required of the sampling rate is $1/(428 * 427) \approx 5$ ppm. This sensitivity corresponds to the vertical locus within the 428th wedge; above the tip of the 429th wedge. Moving to lower order wedges (smaller n), while remaining above the tip of the 428th wedge (same f_u/B) will relax the sampling oscillator precision requirements and increase the guard-band, but of course requires higher sampling rates.

The relative precision required of the sampling oscillator is depicted in Fig. 6 as the line $1/(n(n - 1))$. This line gives the relative precision required of the sampling rate when operating with $f_u/B - n = 1$ as a function of n. The positions of a typical crystal and RC oscillator are indicated to give a feel for practical requirements. With the inclusion of larger guard-bands by operating in the various nth wedges, the line interpolating the various $\Delta f_s/2B = 1/n'(n' - 1)(f_u/B - n')$ runs above the $f_u/B - n = 1$ line. For example, for $n' = n + 10$, the line would be an order of magnitude in $\Delta f_s/2B$ higher than and parallel to the $f_u B - n = 1$ line in Fig. 6. Varying an operating point within a given wedge corresponds to a positive slope locus in Fig. 6 which steepens as f_u/B decreases. After intersecting the $f_u/B - n = 1$ line, the locus approaches a vertical line at $f_u/B = n$ as Δf_s approaches zero at the tip of the wedge.

The $f_u/B - n = 1$ line represents operating with the minimum allowed set of sampling rates and the maximum available guard-band. The guard-band size decreases with increasing n, and the line is $\sim 0(1/n^2)$. If the relative guard-band is now held at a constant size, then the relative precision of the sampling rate becomes, from (24) and (25), $\Delta f_s/2B \approx B_{GT}/B \cdot 1/n'$. This demonstrates the tradeoff between the precision required of the sampling rate and the guard-band size. For practical situations, a modest increase in the sampling rate permits a

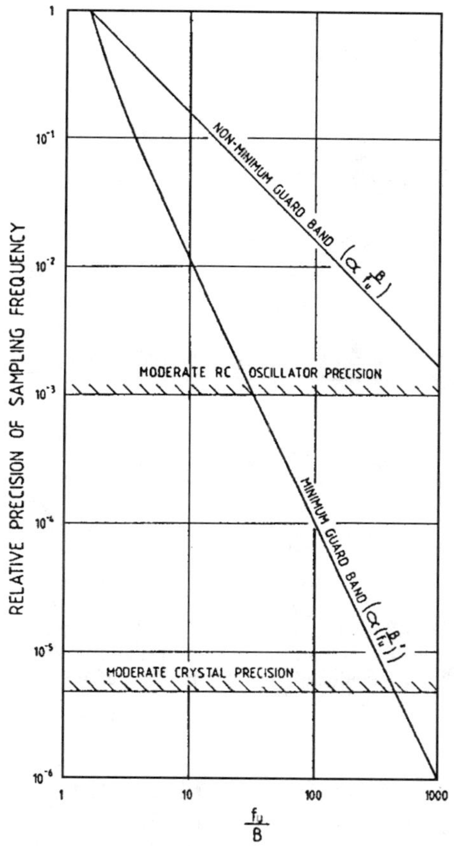

Fig. 6. The relative precision of the set of minimum allowed uniform sampling rates. The line is an interpolation of the operating points corresponding to $f_u/B - n = 1$. In terms of Fig. 4, these operating points are within the nth wedge at a position above the tip of the $(n+1)$th wedge.

realizable precision requirement, as shown in the example above.

In addressing sampling rate variation, it must be borne in mind that so far the discussion has been limited to avoiding aliasing in the sampling process. Exact (re)construction of the bandpass analog waveform requires exact implementation of the sampling parameters used in the interpolant computation, a point addressed within the following section.

III. Interpolation in Second-Order Sampling

Kohlenberg's second-order sampling result gives a theoretical minimum *average* sampling rate which would correspond to the straight line $f_s = 2B$ on Fig. 4. The interest in second-order sampling has been not only from the theoretical viewpoint of maintaining a constant minimum average sampling rate independent of the band position, but also from its practical application in the simplified form of quadrature sampling a bandpass signal.

This section concerns aspects of bandpass signal construction using various k; specifically, the required range (number of bits in the coefficients) and length (number of coefficients) for realizing the interpolant as a finite impulse response filter, and the sampling rate sensitivity. The application is the generation of a band-pass signal from second-order samples.

For second-order sampling, the spectra resulting from each of the two interleaved sampling sequences applied to the bandpass signal are [21]

$$F_A(f) = F(f) * \sum_p B\,\delta(f - pB) \tag{31}$$

$$F_B(f) = F(f) * \sum_p B e^{-j2\pi Bkp}\,\delta(f - pB) \tag{32}$$

where $F(f)$ is the analog spectrum, $*$ denotes convolution, and p is a spectral position index. From here on, the explicit frequency dependence of F_A and F_B is dropped. Equations (31) and (32) show that the sampled spectrum is a series of analog spectra located B Hz apart and scaled by B. For F_B, there is a progressive phase change of $2\pi Bk$ imposed on $F(f)$ at successive spectral locations, caused by the sampling stream being delayed by k seconds.

Fig. 7, which is an elaboration of Linden's [21] Fig. 3, illustrates the geometric aliasing relations. There are two regions in the band in which the aliasing arises from different aliases of the negative frequencies from the analog spectrum. These regions correspond to where the $S_0(t)$ and $S_1(t)$ terms of the interpolant (see (2)) are applied, and are

$$R_0: \quad f_L \le f \le rB - f_L \tag{33}$$

and

$$R_1: \quad rB - f_L \le f \le f_L + B. \tag{34}$$

Note that from the choice of definition for r, R_0 cannot encompass the band whereas R_1 can encompass the band, whence $S(t) = S_1(t)$, i.e., integer or half-integer band positioning.

The spectra resulting from the two sampling sequences are

$$R_0: \quad F_A = F(f) + F(-f_0') \tag{35}$$

$$F_B = F(f) + e^{j2\theta_0}F(-f_0') \tag{36}$$

and

$$R_1: \quad F_A = F(f) + F(-f_1') \tag{37}$$

$$F_B = F(f) + e^{-j2\theta_1}F(-f_1') \tag{38}$$

where

$$R_0: \quad \theta_0 = \pi Bkr \tag{39}$$

$$R_1: \quad \theta_1 = \pi Bk(r+1) \tag{40}$$

and

$$R_0: \quad f_0' = f + 2f_L - rB \tag{41}$$

$$R_1: \quad f_1' = f + 2f_L - (r+1)B. \tag{42}$$

The scaling factor B in (31) and (32) is omitted in (35)–(38) and hereon.

The aliased contribution (the term containing $F(-f')$) is the same for F_A and F_B, except for a phase shift which is constant throughout each region. The phase shifts are, however, different for each of the R_0 and each of the R_1 in the aliased spectra.

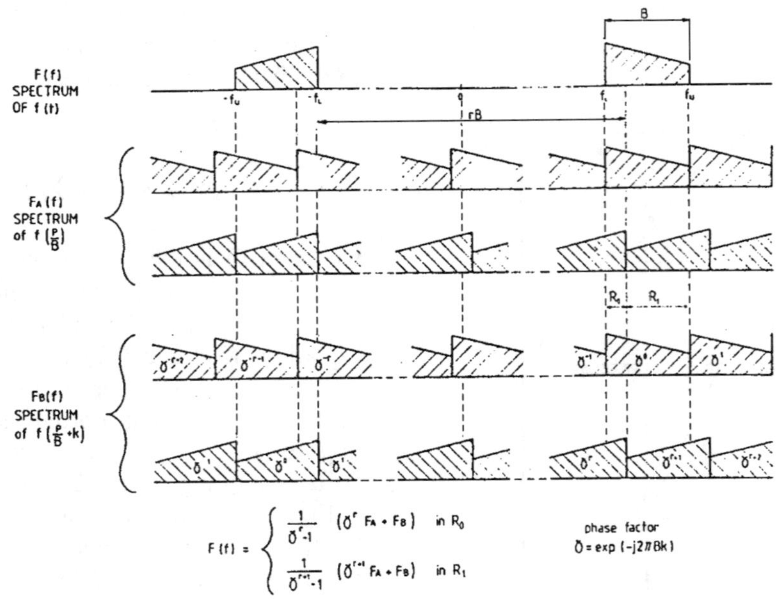

Fig. 7. The geometric relations for aliasing in second-order bandpass sampling and general band position. $F(f)$ is the analog spectrum, and $F_A(f)$ and $F_B(f)$ are the spectra from each of the two interleaved (by k s) uniform sample sequences. The aliased bands are progressively phase shifted relative to the analog band as indicated by the notation $\gamma^r = e^{-j2\theta_0}$. For integer and half-integer band positioning, the region R_0 becomes zero.

The bandpass signal construction is (cf. (1))

$$R_0: \quad F(f) = \frac{1}{e^{-j2\theta_0} - 1}\,(e^{-j2\theta_0}F_A - F_B) \tag{43}$$

$$R_1: \quad F(f) = \frac{1}{e^{-j2\theta_1} - 1}\,(e^{j2\theta_1}F_A - F_B) \tag{44}$$

so the disallowed values for k are

$$R_0: \quad k = \frac{m}{rB}; \qquad m = 0, \pm 1, \pm 2, \cdots \tag{45}$$

$$R_1: \quad k = \frac{m}{(r+1)B}; \qquad m = 0, \pm 1, \pm 2, \cdots \tag{46}$$

which are unique except when $k = 0, 1/B, 2/B, \cdots$; whence the sampling sequences coincide.

In Fig. 7, the phase factor notation

$$\gamma^r = e^{-j2\pi Bkr} = e^{-j2\theta_0} \tag{47}$$

is used for brevity. Considerable insight into the interpolant can be gained by expressing it in terms of the frequencies in Fig. 7. Using the angular frequencies $\omega_u = 2\pi f_u$, $\omega_L = 2\pi f_L$, $\omega_B = 2\pi B$ and $\omega_i = 2\pi(rB - f_L)$ which is the common border of R_0 and R_1 in the analog spectrum, and noting that R_0: $\arg(\gamma^r) = -2\theta_0$, R_1: $\arg(\gamma^{r+1}) = -2\theta_1$:

$$S_0(t) = \frac{\cos(\omega_i t - \theta_0) - \cos(\omega_L t - \theta_0)}{\omega_B t \sin\theta_0} \tag{48}$$

$$S_1(t) = \frac{-\cos(\omega_i t - \theta_1) + \cos(\omega_u t - \theta_1)}{\omega_B t \sin\theta_1} \tag{49}$$

which shows the composition of the interpolant in the same form as the low-pass case of (8).

As already noted for integer and half-integer band positioning, $2f_L/B = r$ and R_0 becomes zero. Further simplification in the signal construction occurs if $e^{-j2\theta_1} = -1$, i.e., for quadrature sampling, in which case

$$F(f) = \tfrac{1}{2}(F_A + F_B). \tag{50}$$

This occurs when

$$j2\pi kB(r+1) = \pm j(2m+1)\pi \tag{51}$$

i.e.,

$$k = \frac{2m+1}{2B(r+1)} \tag{52}$$

$$= \frac{1}{4f_c} + \frac{m}{2f_c} \tag{53}$$

since $f_c = (r + 1/2)B$, so that these values of k indeed correspond to quadrature samples.

As k approaches its disallowed values, construction of the bandpass signal becomes more difficult to implement since the interpolant becomes larger. Both the dynamic range (number of bits required for the filter coefficients) and the length (number of filter taps) required for the interpolant contribute to the cost of an implementation. The general interpolant form (2) predominantly resembles a modulated (by f_c) sinc function, so the dynamic range and the length are related, although the relation is not simple. The dynamic range is a minimum for quadrature sampling when the interpolant assumes its maximum value at $t = 0$, as evident in (10). In this case, the filter length is also clearly defined for a given coefficient size as long as $f_c \gg B$, i.e., for the narrow-band situation. When k deviates from the values corresponding to quadrature sampling, both the dynamic range and the length of the interpolant increase.

The low-pass case illustrates the increase in interpolant length with a poor choice of k. Assuming integer band positioning, the low-pass interpolant can be written, from (8), in the form

$$S_{LP}(t) = -\cot \pi Bk \frac{\sin^2 \pi Bt}{\pi Bt} + \frac{\sin 2\pi Bt}{2\pi Bt}. \quad (54)$$

The first term decays as $1/\pi Bt$ whereas the second term decays as $1/2\pi Bt$. For a given dynamic range and coefficient word size, the magnitude of the first term should be minimized, or at least kept small with respect to the second term, for minimum filter length. Consequently, uniform sampling where $k = 1/2B$ (corresponding to quadrature sampling since $f_c = B/2$), is optimum in this case. For baseband interpolation of a sampled bandpass signal, configuring the system such that the signal is integer positioned and applying uniform sampling is the simplest arrangement. In an implementation, it should be borne in mind that half-integer band positions alias to be spectrally inverted at the low-pass position.

In a nonuniformly sampled low-pass system, the alternative to applying the interpolant of (54) directly is to generate uniform samples and interpolate with sinc $2\pi Bt$. The required resampling is a shift of one of the sample sequences using the sinc $2\pi Bt$ interpolant on alternate samples.

Further expense in the implementation of bandpass signal construction may arise because of the need to include "guard bits" in the sampled signal and the interpolant coefficients. This is because errors in the measurement or generation process of F_A and F_B, such as those due to quantization, may in turn create errors in the reconstructed signal. The effect of the choice of k on the resulting uncertainty in the constructed signal is of interest, and is estimated here using a sensitivity analysis in the frequency domain. The specific aim is to get an idea of the cost of a poor choice of k.

Progress can be made by assuming that the variations of corresponding (same analog frequency) components in F_A and F_B are uncorrelated. The variance of a component in F can then be written in terms of its magnitude:

$$\sigma_F^2 = \left|\frac{\partial F}{\partial F_A}\right|^2 \sigma_{F_A}^2 + \left|\frac{\partial F}{\partial F_B}\right|^2 \sigma_{F_B}^2 \quad (55)$$

where $\sigma_{F_A}^2$ and $\sigma_{F_B}^2$ are the variances of the components of F_A and F_B. For correlated variations, caused for example by k becoming very small, σ_F^2 will increase, so (55) represents a best case. The units of σ_F^2 are watts/hertz and the noise power in the constructed signal is found by integrating over the band of F.

Letting $\sigma^2 = \sigma_{F_A}^2 = \sigma_{F_B}^2$, and using (43) and (44) in (55) and integrating, the noise power in F is

$$\sigma_F^2 2B = \frac{\sigma^2 B}{2}\left(\frac{\Delta r}{\sin^2 \theta_0} + \frac{1 - \Delta r}{\sin^2 \theta_1}\right) \quad (56)$$

where

$$\Delta r = r - \frac{2f_L}{B} \quad (57)$$

denotes the extent of the R_0 portion of the band and $0 \leq \Delta r < 1$. The minimum noise power is $\sigma^2 B/2$, and from the choice of definition of r, occurs for $\Delta r = 0$ and $\theta_1 = \pi/4 + m\pi$, $m = 0, \pm 1, \pm 2, \cdots$; i.e., for integer or half-integer band positioning and quadrature sampling.

For general values of Δr and θ_0, θ_1; the noise power relative to its minimum value can be expressed as the number of bits of resolution irrecoverably lost in calculating F; viz.,

$$N_b = \log_2 \left(\frac{\sigma_F^2 2B}{\sigma^2 B/2}\right)^{1/2} \quad (58)$$

$$= \log_2 \left(\frac{\Delta r}{\sin^2 \theta_0} + \frac{1 - \Delta r}{\sin^2 \theta_1}\right)^{1/2}. \quad (59)$$

In terms of an implementation, (59) indicates for poor choices of k, how the errors in the sampled data magnify during the construction of an analog bandpass signal. N_b is plotted against k normalized by B in Fig. 8(a) for the example where $r = 9$ and $\Delta r = 0$ (half-integer band positioning); and in Fig. 8(b) for $r = 10$ and $\Delta r = 0.5$. The functions are symmetric about $k = 0$ and $k = 1/2B$. The asymptotes correspond to the disallowed values if k; viz., $k = m/10B$, $m = 0$ to 10, for Fig. 8(a); and both $k = m/10B$ and $k = m/11B$ for Fig. 8(b). For the half-integer band positioning case, the quadrature sampling values

$$k = \frac{1}{4\left(\frac{9+1}{2}\right)B} + \frac{m}{2\left(\frac{9+1}{2}\right)B} = \frac{1}{20B}, \frac{3}{20B} \cdots;$$

are located at the function minima of Fig. 8(a). The uniform sampling case is given by $k = 1/2B$ and, as already noted, it is only possible to apply this in the special case of integer band positioning (r even and $\Delta r = 0$) and is clearly not viable (i.e., $k = 1/2B$ is a disallowed value) in Fig. 8(a) and (b). The interpolant is simplified and is the same for each of the quadrature sampling values (*NB*: (10) is independent of k). This is evident in Fig. 8(a) where any choice of quadrature sampling gives the optimum k with respect to sensitivity of the signal construction to uncertainty in the sampled data. Since the quadrature sampling locations are at the minima, small perturbations (relative to $1/r$) of k from the quadrature values do not result in a large penalty in the uncertainty caused by constructing an analog bandpass signal. This is particularly clear for the low-pass case, where there is only one quadrature sampling location which is at $k = 1/2B$.

In Fig. 8(b), the quadrature sampling points are at $k = 1/21B, 3/21B, \cdots$; and these do not occur at the function minima. For constructing a bandpass signal whose band is not integer or half-integer positioned, it is unwise to use an arbitrary choice of quadrature sampling since the construction process could well give rise to an error

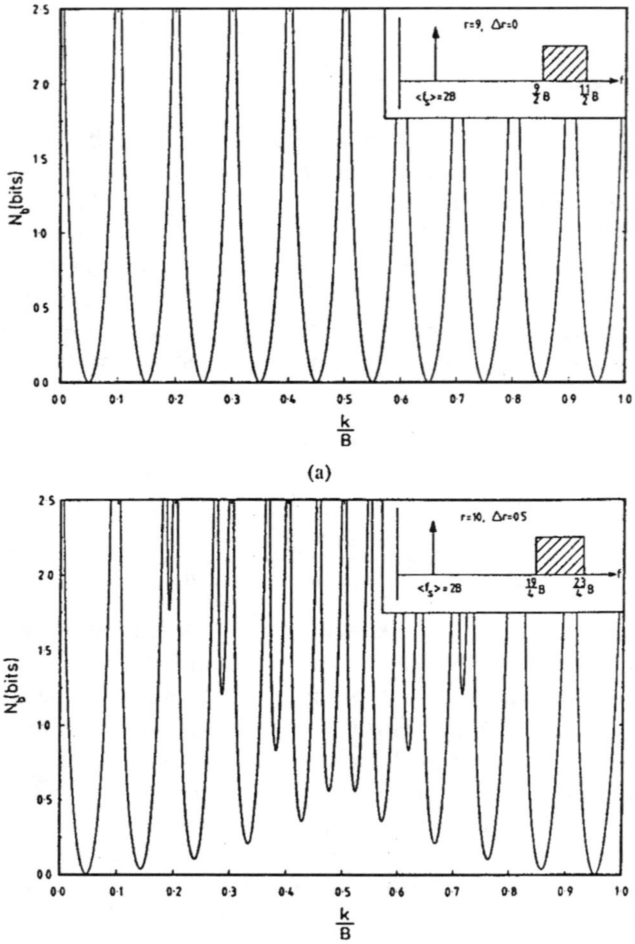

(a)

(b)

Fig. 8. An estimate of the equivalent number of bits of resolution irrecoverably lost in interpolating a bandpass signal using uncertain baseband rate samples, as a function of k. The graphs represent a cost of a poor choice of k in the second-order bandpass sampling process. The average sample rate is $\langle f_s \rangle = 2B$. (a) $r = 9$, $\Delta r = 0$ (half-integer band positioning). The minima correspond to cases of quadrature sampling. (b) $r = 10$, $\Delta r = 0.5$ (neither integer nor half-integer band positioning). The minima do not, in general, correspond to quadrature sampling.

in the signal. However, the first quadrature sampling value is close to the first minimum of the function, which is the global minimum. In this case of general band positioning, optimum quadrature sampling requires the minimum k, i.e., $k = 1/4f_c$. The best general value of k occurs at the first function minimum (or the last minimum, which amounts to the same sampling sequence interleaving), and so the configuration of the optimum general sampling system requires finding this minimum. This disadvantage, as well as the need here for a complicated interpolant, make it worthwhile arranging the band to be integer or half-integer positioned.

The accurate construction of the bandpass signal from its samples also depends on the accuracy of the sampling rate and k. If the sample sequences are interleaved by k_a and the interpolant implementation features $k = k_a + \Delta k$, then the constructed spectrum of interest $\hat{F}(f)$ and its alias $\hat{F}(-f')$ in terms of the wanted spectrum is, expressed here for R_1 only,

$$\begin{bmatrix} \hat{F}(f) \\ \hat{F}(-f'_1) \end{bmatrix} = \frac{1}{e^{-j2\theta_1} - 1} \begin{bmatrix} e^{-j2\theta_1} & -1 \\ -1 & 1 \end{bmatrix} \cdot \begin{bmatrix} 1 & 1 \\ 1 & e^{-j2\theta_{1a}} \end{bmatrix} \begin{bmatrix} F(f) \\ F(-f'_1) \end{bmatrix} \tag{60}$$

where $\theta_{1a} = -\pi k_a(r + 1)B$. For small Δk

$$\hat{F}(f) \approx F(f) + \frac{\pi(r + 1)B\Delta k}{e^{j\theta_1} \sin 2\theta_1} F(-f'_1) \tag{61}$$

which approximates how the degree to which the alias is suppressed depends on the difference of k and k_a. The relative difference between the constructed spectrum and the wanted spectrum can be expressed as

$$\Delta_F = \left| \frac{\hat{F}(f) - F(f)}{F(f)} \right| \approx \pi B(r + 1)\Delta k \tag{62}$$

a result which is also readily obtained from time domain considerations.

As r increases, the timing error Δk becomes increasingly critical. For large values of r pertinent to communications or radar, etc., the demands on Δk become considerable. For the previous example (in Section II, using guard-bands) with $r = 356$ and $B = 30$ kHz, and $f_L = 10.7$ MHz, and an alias suppression of 40 dB, the required accuracy of the sample sequence interleaving is

$$\Delta k \approx \frac{1}{357} \cdot \frac{10^{-2}}{\pi 30(10)^3}$$

i.e., $\Delta k \sim 0.3$ ns. For lower intermediate frequencies in communications, such as 455 kHz, the requirement is correspondingly simpler, viz., $\Delta k \sim 7$ ns ($r = 15$).

IV. Noise in Bandpass Sampled Signals

Noise considerations are important not only where the signal-to-noise ratio is important, such as in communication receivers, but also in the measurement of noise itself, such as in noise power measurements. In a sampling system, the periodicity of the spectrum means that wideband noise, such as the thermal noise introduced by the relevant hardware, is all combined into each of the $fs/2$ bands.

In applying bandpass sampling to relocate a bandpass signal to a low-pass position, the resulting signal-to-noise ratio will be poorer than that from an equivalent analog system (i.e., an ideal image-rejecting mixer), in which the signal-to-noise ratio is preserved.

Consider a system with a bandpass signal of spectral power density S, in-band noise power density of N_p, and out-of-band noise power density N_o. The analog signal-to-noise ratio is thus S/N_p. The signal-to-noise ratio for the sampled signal becomes degraded by at least the noise aliased from the bands between dc and the passband, and is thus

$$\mathrm{SNR}_s \approx \frac{S}{N_p + (n - 1)N_o}. \tag{63}$$

Often, $N_p >> N_o$ and the signal-to-noise ratio is established before sampling. However, if $N_p = N_o$, and assuming $n >> 1$, then the degradation of the signal-to-noise ratio in decibels is simply

$$D_{\rm SNR} \approx 10 \log n. \tag{64}$$

Equivalently, the effective noise temperature is increased by at least n. Note that this analysis assumes a flat analog noise spectrum, ideal filters, and an infinitesimally small sampling aperture.

For the example with $B = 30$ kHz located at the integer position $n = r/2 = 30(f_u \sim 455$ kHz), or $n = 356(f_u \sim 10.7$ MHz), then the increase in the noise power is at least 15 and 25 dB, respectively.

In the case where the noise spectrum is not uniform, a conservative estimate of the noise power density in each of the bands in the sampled spectrum can be obtained by equating the noise power before and after sampling. Denoting $N_{\rm EA}$ as the equivalent noise spectral density (ENSD) and $B_{\rm EA}$ as the equivalent noise bandwidth (ENBW) of the analog signal; and $N_{\rm ES}$ as the ENSD of the sampled signal with $2B$ the sampling rate (it is known [22] that the ENBW of a sampled system cannot exceed B), then

$$N_{\rm EA} B_{\rm EA} = N_{\rm ES} B. \tag{65}$$

The degradation in a signal-to-noise ratio is $N_{\rm ES}/N_{\rm EA}$ which, in decibels, is then at least

$$D_{\rm SNR} \approx 10 \log \left(\frac{B_{\rm EA}}{B}\right). \tag{66}$$

A simple experiment using a noise source and a sampling oscilloscope demonstrates the effect. The test signal was a 10.95-kHz carrier amplitude-modulated by a 200-Hz sine wave. A Bruel and Kjaer 0–20 kHz noise source signal was added to the test signal. The resultant signal was sampled at 125 kHz, windowed using a minimum 3-term Blackman–Harris function, and the spectra calculated by FFT on a LeCroy oscilloscope. The average of 10 power spectra is shown in Fig. 9(a). The signal-to-noise ratio is about 38 dB. The signal was then sampled at 1250 Hz giving a Nyquist bandwidth of 625 Hz, and $n = 20\,000/625 \approx 32$. The average spectra is shown in Fig. 9(b), with the signal-to-noise ratio now being ~21 dB. The observed degradation is thus about 17 dB.

The minimum degradation given by (64) is 10 log 32, i.e., about 15 dB. Recall that this minimum is given by considering only the aliased noise from the noise source, and assumes that $B_{\rm EA}$ is 20 kHz, while in practice it is a little greater.

In general, the potential for signal-to-noise degradation in bandpass sampling systems is considerable, even when the analog spectrum contains only thermal noise. In most communications systems, for example, the out-of-band power (adjacent band signals, etc.) has similar, or even higher levels, than that in the band of interest, so an antialiasing filter is necessary.

To get a feel for the filter requirements, the signal distortion ratio (SDR) [24] is a useful metric. It is the ratio of the mean-square aliasing error and the total signal power passed through the filter and can be interpreted as a signal-to-noise ratio. It is here defined for the bandpass case [23] as

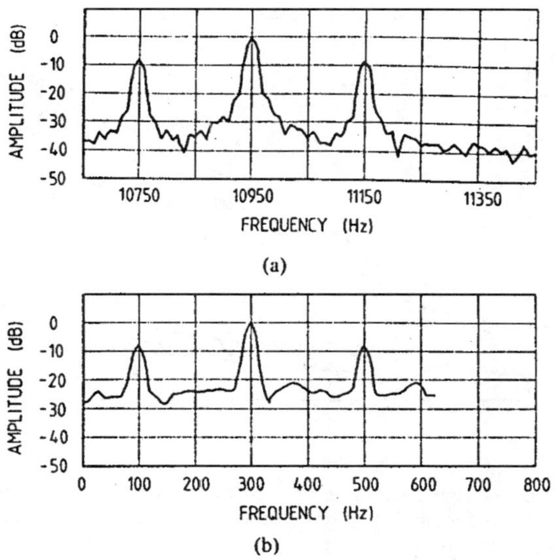

Fig. 9. Averaged sampled spectra of a 10.9-kHz carrier amplitude modulated by a 200-Hz sinusoid in the presence of white noise. The sampling rate is (a) 125 kHz, giving a sampled spectra SNR of about 38 dB, and (b) 1250 Hz, giving a sampled spectra SNR of about 21 dB.

$$\mathrm{SDR} = \frac{\displaystyle\int_0^{\infty} |H(\omega)|^2 \, d\omega}{\displaystyle\int_0^{2\pi f_c - (\pi/2) f_s} |H(\omega)|^2 \, d\omega + \int_{2\pi f_c + (\pi/2) f_s}^{\infty} |H(\omega)|^2 \, d\omega} \tag{67}$$

where $H(\omega)$ is the filter transfer function and f_c is the arithmetic center frequency of the upper and lower 3-dB points. The limits are chosen so that the wanted signal band occupies a Nyquist bandwidth. In practice, this requires selecting f_s such that the band is integer or half-integer positioned.

This signal distortion ratio offers only a rough guide for trading off the filter requirements against sampling rate and is only one of many factors contributing to the effective signal-to-noise ratio [23]. This is because of the assumption in (67) that the signal power at the filter input is constant across the spectrum. In practice, the presence of guard-bands will normally ensure this will not be the case at the edges of the filter. On the other hand, adjacent channel power levels may be tens of decibels above the power level in the channel of interest. In radio communications cases, the filter $H(\omega)$ not only represents the antialiasing filter, but also the receiver front end, including the antenna. Nevertheless, the antialiasing filter will normally be the dominant effect on the signal distortion ratio.

Fig. 10 [23] shows the signal distortion ratio for two filter types against the sampling rate normalized to the

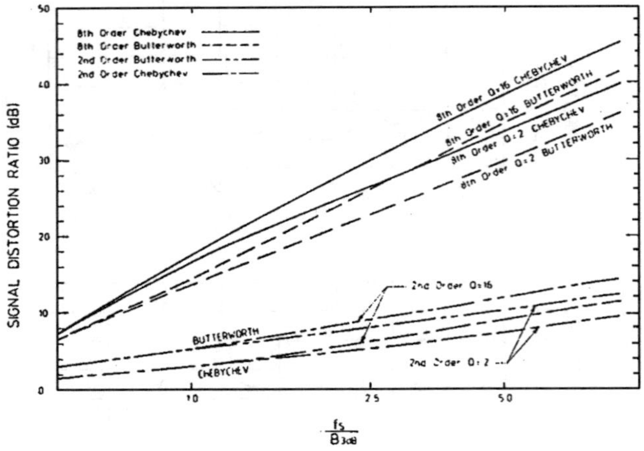

Fig. 10. The signal distortion ratio (SDR) as defined for the bandpass case for second-order and eighth-order Butterworth and Chebychev filters. The envelope edges for each curve represent $Q = 1$ and $Q = 200$ [23].

3-dB bandwidth of the filter $B_{3\text{dB}}$. Because of the selection of f_s mentioned above, there are only discrete points on the curves which are valid. The envelopes enclose the curves for filters with Q values from 1 to 200. The parameter $Q = f_c/B_{3\text{dB}}$ is related to the band position parameter $r \approx 2f_L/B$ (for integer band positioning) by $Q = \frac{1}{2}(r + 1)$. In the example with a 25-kHz band at 455 kHz, i.e., $r \approx 36$, a signal distortion ratio of 37 dB requires a sampling rate of $f_s/B_{3\text{dB}} \approx 2$ (i.e., $f_s \approx 50$ ksamples/s) with a filter equivalent to an eighth-order Chebychev with $Q = 16$.

The definition of the signal distortion ratio in (67) can perhaps be more easily related to a signal power to "distortion power" ratio if the integration in the numerator is over the 3-dB bandwidth only, i.e., the denominator corresponds to the signal power only rather than total power. For such a definition, however, the curves of Fig. 10 change by less than a couple of decibels. Strictly speaking, only interleaved portions of the filter "stop band" contribute to the aliasing distortion, but again, this definition does not appreciably shift the curves of Fig. 10. This matter is discussed in [23]. In applying Fig. 10, it should be borne in mind that the signal distortion ratio is only an approximation to the effective signal-to-noise ratio caused by the aliasing.

V. Conclusion

For practical bandpass sampling, there is a need to sample at above the theoretical minimum rate. Care must be taken when increasing the sampling rate from its minimum value because there are bands of rates for which aliasing will occur. Operating at a nonminimum sampling rate is equivalent to the introduction of guard-bands or a corresponding tolerance to variations in the sampling rate. The relations between the sampling rate tolerance, the guard-band size, and the various allowed nonminimum sampling rates give practical design choices.

For sampling an analog bandpass signal in order to process digitally the signal, the simplest system is to configure the band to be integer positioned and apply uniform (which is also quadrature) sampling.

In the sampling process, the signal becomes degraded owing to aliasing of the noise (introduced after any bandpass antialiasing filter) between dc and at least the original bandpass spectral position. This degradation is unavoidable and a simple estimate for the minimum degradation is available in terms of the band position.

For constructing an analog bandpass signal from its second-order samples, it is wise to configure the system such that the band is integer or half-integer positioned. The advantages are a simplified interpolant and the potential to apply quadrature sampling with the flexibility to use several values for the sequence spacing. Quadrature sampling represents the optimum sample sequence interleaving in the sense that it requires the shortest filter length (least number of taps) for interpolant implementation using a given coefficient word size, and displays the least sensitivity to error propagation in the construction of a bandpass signal from its samples.

For cases where the band is not integer of half-integer positioned, the optimum sampling sequence interleaving for bandpass signal construction is not trivial to find. Here, quadrature sampling spacing should not be applied in general, owing to possible error propagation from imperfect samples, although the minimum quadrature spacing $k = 1/4f_c$ is usually close to the optimum.

References

[1] J. D. Gaskell, *Linear Systems, Fourier Transforms, and Optics*. New York: Wiley, 1978.

[2] M. C. Jackson and P. Matthewson, "Digital processing of bandpass signals," *GEC J. Res.*, vol. 4, no. 1, 1986.

[3] O. D. Grace and S. P. Pitt, "Quadrature sampling of high frequency waveforms," *J. Acoust. Soc. Amer.*, vol. 44, pp. 1432–1436, 1968.

[4] W. M. Waters and B. R. Jarrett, "Bandpass signal sampling and coherent detection," *IEEE Trans. Aerosp. Electron. Syst.*, vol. AES-18, no. 4, Nov. 1982.

[5] D. Del Re, "Bandpass signal filtering and reconstruction through minimum-sampling-rate digital processing," *Alta Frequenza*, vol. XLVII, no. 9, Sept. 1978.

[6] F. J. J. Clarke and J. R. Stockton, "Principles and theory of wattmeters operating on the basis of regularly sampled pairs," *J. Phys. E. Sci. Ints.*, vol. 15, 1982.

[7] H. S. Black, *Modulation Theory*. New York: Van Nostrand, 1953.

[8] H. Nyquist, "Certain topics in telegraph transmission theory," *AIEE Trans.*, vol. 47, 1928.

[9] D. Gabor, "Theory of communication," *J. Inst. Elec. Eng.*, vol. 93, 1946.

[10] A. Kohlenberg, "Exact interpolation of band-limited functions," *J. Appl. Phys.*, vol. 24, no. 12, Dec. 1953.

[11] S. Goldman, *Information Theory* (Prentice-Hall Electrical Engineering Series). London: Constable and Co. Ltd., 1953.

[12] D. Middleton, *An Introduction to Statistical Communication Theory*. New York: McGraw-Hill, 1960.

[13] P. F. Panter, *Modulation, Noise and Spectral Analysis*. New York: McGraw-Hill, 1965.

[14] S. Haykin, *Communication Systems*. New York: Wiley, 1978.

[15] K. S. Shanmugam, *Digital and Analogue Communications Systems*. New York: Wiley, 1979.

[16] C. B. Feldman and W. R. Bennett, "Bandwidth and transmission performance," *Bell Syst. Tech. J.*, vol. 28, pp. 490–595, 1949.

[17] W. D. Gregg, *Analog and Digital Communications Systems*. New York: Wiley, 1977.

[18] T. Kida and T. Kuroda, "Relations between the possibility of restoration of bandpass-type band-limited waves by interpolation and arrangement of sampling points," *Electron. Commun. Japan*, pt. 1, vol. 69, pp. 1-10, 1986.

[19] D. W. Rice and K. H. Wu, "Quadrature sampling with high dynamic range," *IEEE Trans. Aerosp. Electron. Syst.*, vol. AES-18, no. 4, pp. 736-739, Nov. 1982.

[20] J. L. Brown, Jr., "On uniform sampling of amplitude modulated signals," *IEEE Trans. Aerosp. Electron. Syst.*, vol. AES-19, no. 4, July 1983.

[21] D. A. Linden, "A discussion of sampling theorem," *Proc. IRE*, vol. 47, pp. 1219-1226, 1959.

[22] D. R. White, "The noise bandwidth of sampled data systems," *IEEE Trans. Instrum. Meas.*, vol. 38, no. 6, pp. 1036-1043, 1989.

[23] N. L. Scott and R. G. Vaughan, "The signal distortion ratio of filters due to aliasing," *Proc. Int. Symp. Signal Processing and its Applications (ISSPA'90)* (Gold Coast, Australia), Aug. 1990.

[23] N. L. Scott and R. G. Vaughan, "The signal distortion ratio of filters due to aliasing," in *Proc. Int. Symp. Signal Processing and its Applications (ISSPA'90)* (Gold Coast, Australia), Aug. 1990.

[24] R. M. Gagliardi, *Introduction to Communications Engineering*, 2nd ed. New York: Wiley (Series in Telecommunications), 1988 (1st ed. 1978).

RSFQ Front-end for a Software Radio Receiver

Erland B. Wikborg
Ericsson Components AB, Stockholm, Sweden

Vasili K. Semenov and Konstantin K. Likharev
Department of Physics and Astronomy, SUNY at Stony Brook, USA

Abstract - We present results of a preliminary analysis of possible front-ends of a software defined radio base station receiver containing superconductor Rapid Single-Flux-Quantum (RSFQ) components. The main component of such a front-end, an oversampling analog-to-digital converter (ADC) with expected internal clock rate up to 20 GHz, has already been developed and fabricated with commercially available 3.5-μm, 1-kA/cm² niobium-trilayer technology. Simulations show that, even with such crude feature size, the superconductor ADC components can outperform their semiconductor counterparts.

I. INTRODUCTION

The air is rapidly becoming filled with wireless communication systems. As the limited frequency spectrum gets ever more crowded, the next-generation mobile radio communication systems will have to manage the radio spectrum more effectively. A new radio technology, usually referred to as software defined radio (SDR), achieves this by computer-controlled access to RF bands, beam shapes, waveforms and protocols [1, 2].

The driver of this technology is the development, during the last 10-20 years, of faster digital signal processors (DSP), capable of handling data streams at the rates close to, and even beyond 50 MHz. This gave a motivation to move the analog-to-conversion stage as far up the receiver chain towards the antenna as possible (Fig. 1) This would allow all functions of the radio to be controlled with programmable DSP chips.

Unfortunately, the ultimate SDR "dream system", shown in Fig. 1b, would require a direct analog-to-digital converter for GHz frequency carriers with signal band-widths above 60 MHz and an accuracy of 14 to 16 effective bits. This level is far beyond the limits of the present-day semiconductor ADCs technology, which is progressing very slowly (Fig. 2).

This situation creates an interesting opportunity for Rapid Single-Flux-Quantum (RSFQ) technology [3], [4]. This technology, based on the quantum properties of superconductor Josephson junctions, not only enables ultrafast digital signal processing, but also features a natural, uniquely accurate measure of analog signals, the single quantum of magnetic flux. This combination allows creation of unique analog-to-digital converters [6]-[10], which have already reached the performance frontiers (Fig. 2). This has been achieved in spite of the use of the commercially available fabrication technology with very crude, 3.5-μm minimum, feature size [11]. Analysis and simulations show that certain straightforward improvements of these devices, especially complemented with a transfer to submicron fabrication technology, will bring RSFQ ADCs far beyond the reach of semiconductor competition.

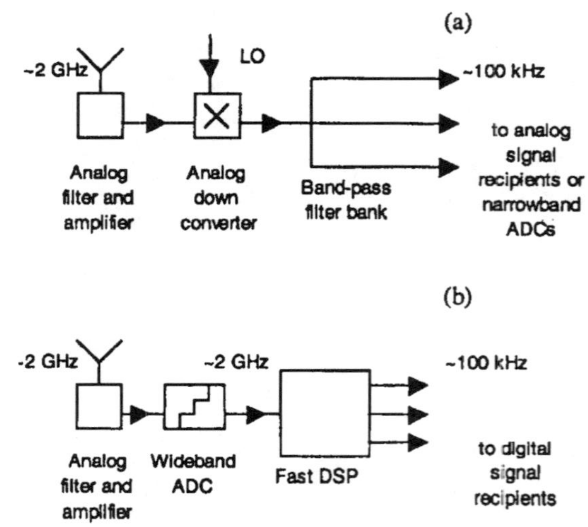

Fig. 1. (a) Traditional base station receiver and (b) Ideal SDR receiver

Recently, Ericsson and SUNY at Stony Brook started a joint effort to develop advanced front ends for SDR base station receivers, using advantages of RSFQ technology. The goal of this paper is to summarize the findings of the initial stage of our work. In Section II examine options which are practicably feasible presently, given the restrictions of the existing RSFQ fabrication technology and semiconductor ADCs. We start with a conventional, "realistic" architecture containing an analog down-converter and an RSFQ ADC as the only superconductor component of the system. As an extension of this architecture, we then discuss the possibility of digital superconductor post-processing, which may simplify the

Manuscript received September 15, 1998
This work was supported in part by Ericsson Components AB and Ericsson Radio Systems AB

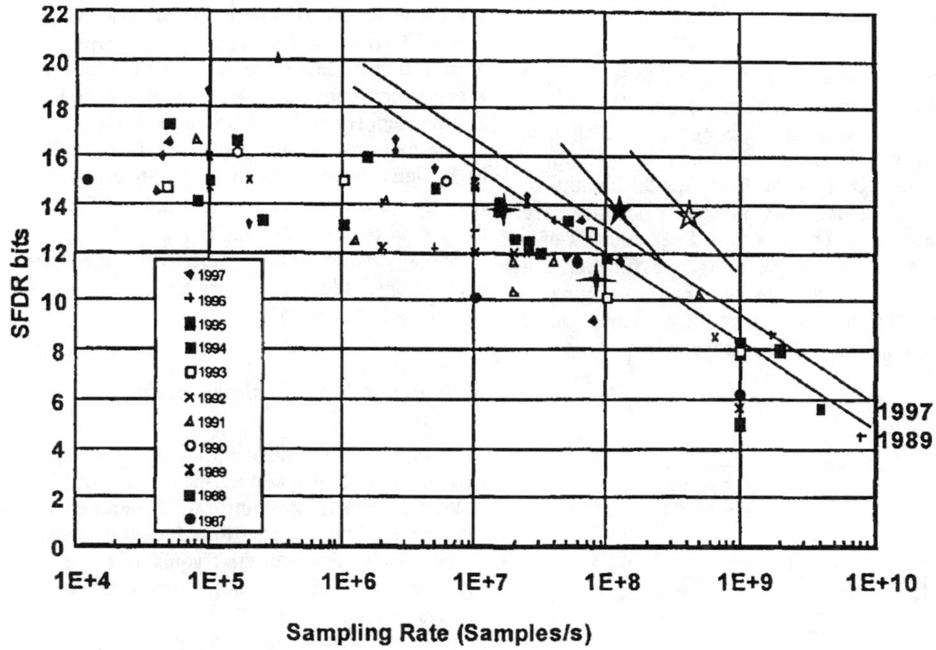

Fig. 2. Performance (SFDR) of semiconductor and superconductor ADCs. The horizontal axis shows the minimum Nyquist frequency of the signal, i e twice its analog bandwidth. Semiconductor data, as collected by R. Walden [3], show a very slow improvement (~1 effective bit over 8 years). Superconductor data are shown by 4- and 5- pointed stars. Solid stars show performance achievable using commercially availabl3.5-μm niobium trilayer technology [11]. Four-pointed stars show experimental data obtained at HYPRES, Inc. in collaboration with SUNY Stony Brook [9]. The solid five-pointed star shows the expected performance of an ADC developed in the SUNY-Ericsson collaboration [10]. The open five-pointed star shows the expected performance of ADCs of similar architecture, fabricated using prospective 0.8-μm technology. Notice that it is SFDR rather than other figures-of-merit (e.g., SNR) which is the most adequate characterization of noise properties of ADCs used for this particular application. (In SDR, each individual recipient recieves noise in proportion to its allocated bandwidth, which is much smaller than the total bandwidth of the ADC. On the other hand, there are important transmodulation components, which in practice limit the SFDR within each narrow band, provided that the signal intensity is high enough.)

interface to the semiconductor DSP stages. Finally, in Section III, we will discuss a possible implementation of a fully superconductive RF front-end, combining a superconductor ADC and a digital down-converter.

II. FRONT END WITH RSFQ ADC

Fig. 3 shows the most straightforward incorporation of an RSFQ-ADC in an SDR front-end: A conventional (analog, semiconductor) down-converter feeds an IF signal with frequency of the order of 100 MHz, via a low-pass filter, to the RSFQ-ADC, which in turn outputs a 200 Msamples/s bitstream to semiconductor DSPs. This architecture allows us to employ conventional ADCs with active signal bandwidth extending from zero up to an upper cutoff frequency f_{max} characterizing the initial signal bandwidth. For SDR applications it should be at least 60 MHz, at 14 to 16-bit SFDR.

The best RSFQ ADCs implemented earlier [6]-[9] could not reach this performance, mainly due to insufficient negative feedback depth. This is why we have developed a new version of such device, which we call "oversampling delta ADC".

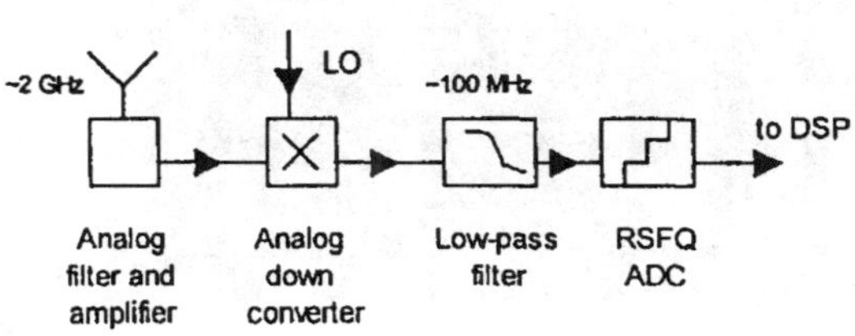

Fig. 3. First option for an RSFQ front-end. Only one component, the ADC, is superconducting.

A. Oversampling Delta ADC.

The main new feature of our RSFQ ADC is a digital integrator in the negative feedback loop. This is an important point: we use a digital integrator, which puts magnetic flux quanta into or withdraws them from a superconductor loop. Due to the fundamental identity of the magnetic flux quanta, this integrator should have unprecedented accuracy. The low-pass filter consists of a passive (L/R) circuit, which performs two functions: it increases accuracy by making an additional averaging of the passing noise (by adding "a half integration"), and it provides necessary stability of the feedback loop.

(a)

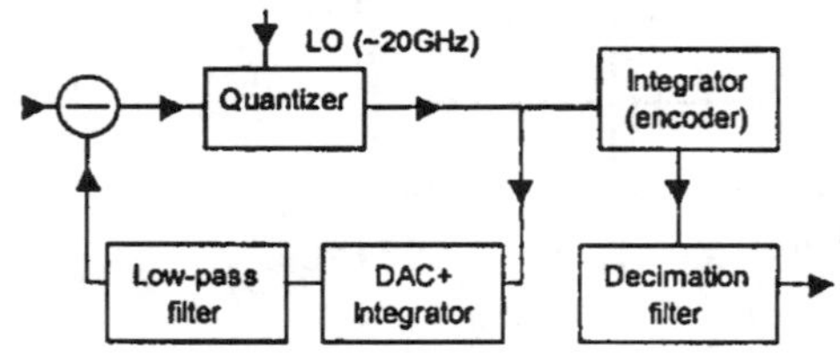

(b)

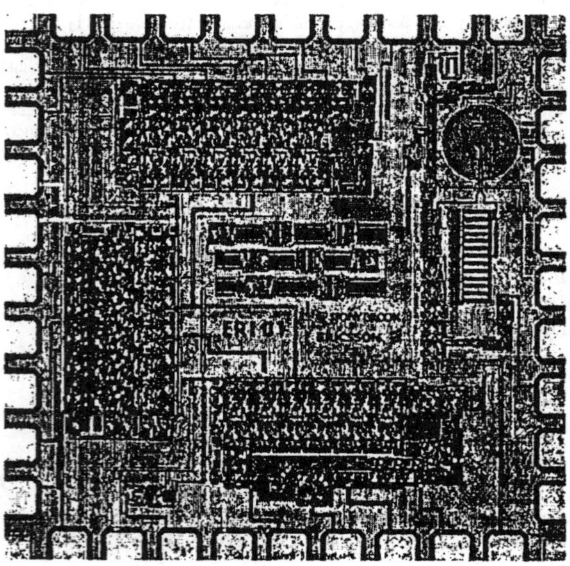

Fig. 4. Block diagram (a) and photograph (b) of an oversampling delta ADC prototype. The chip size is 5x5 mm^2 of which about 3x3 mm^2 is actively used. The chip, containing over 2,000 Josephson junctions, was fabricated at HYPRES, Inc.

A prototype of such an ADC with 20 MHz bandwidth has been designed and fabricated (Fig. 4b) and is now under test. It houses two digital circuits, the integrator and the decimation filter, together containing over 2,000 Josephson junctions. These circuits are expected to operate at over 10 GHz clock frequency and perform encoded differential presentation of data from the quantizer into usual binary form, average high-frequency noise, and decimate the sampling rate to a convenient Nyquist level of 10 to 100 MHz.

The expected accuracy (SFDR) of a similar ADC with 60 MHz signal band-width, is about 86 dB (or 14 effective bits), even with the presently available 3.5-μm fabrication technology, as marked by the solid star in Fig. 2. Using the 0.8-μm RSFQ technology, which is being planned for the Petaflops computing applications, at least two extra bits should be gained, as marked by the open star in Fig. 2.

B. Superconductor post-processing.

If, as we do, consider an analog input signal with a carrier frequency of 2 GHz and an analog bandwidth of 60 MHz, then at least 120 MHz sampling rate is required to transmit the 60 MHz signal. But, in order to reduce linear distortion caused by discreteness of the data, the Nyquist rate should be increased to, say, 200 MHz. Such sampling rates could, however, be too high for available DSPs, so it will be necessary to split data into several, say 4, streams. (We will actually restrict ourselves to approximately 50 MHz data rate, which is what commercially available semiconductor DSPs can handle.) This can be done by a set of signal splitters (or down-converters in conventional terminology), built from two key components: fixed point multipliers (with throughputs of 200 Msamples/sec) and filters with a decimating ratio equal to 4. Both components are "immediately available": the decimation filter is similar to (and in fact simpler than) the one inside the ADC, and suitable ("low-speed", hardware saving) serial multipliers have been designed and successfully tested at HYPRES, Inc [12].

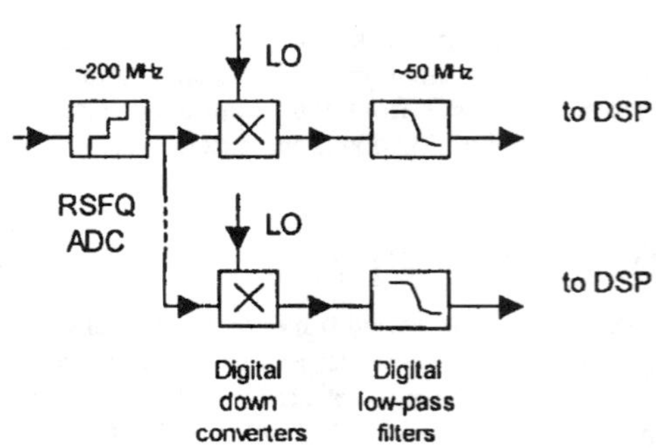

Fig. 5. Structure of a possible RSFQ post-processor for matching the required (200 MHz) sampling rate to commmercially available DSP chips (50 MHz).

III. Towards a Fully Digital Front End

An obvious route towards an "ultimate" receiver, as shown in Fig.1b, is to integrate the down-converter and the ADC into a single digital unit. Such "direct conversion" approaches exist in the semiconductor domain and are known under the names of undersampling, band-pass or sub-harmonic sampling techniques [13]. The noise figures of these direct conversion schemes at RF are, however, inferior to those of conventional analog down-conversion ahead of the ADC. The very high speed of the RSFQ components, on the other hand, allows such high oversampling rates that a front-end with direct sampling at the input RF signal and a digital down- converter is clearly feasible.

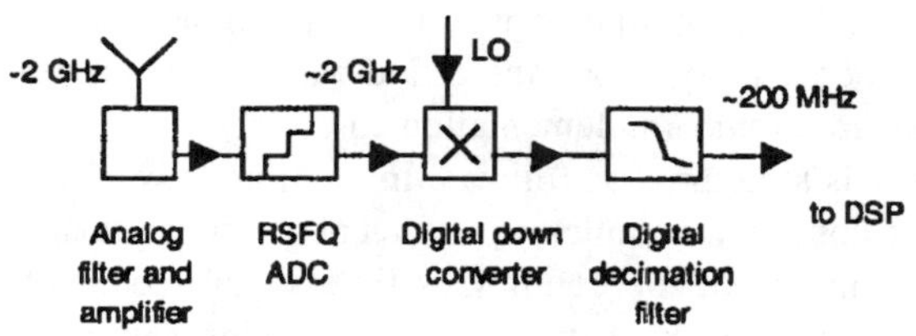

Fig. 6. Block diagram of a fully digital SDR front-end.

We believe such an "ac"ADC with a relatively narrow bandwidth, $\Delta f << f_{max}$ (rather than the usual "dc"ADC with $\Delta f = f_{max}$), can have a structure similar to that shown in Fig. 4a, but with a feedback loop having a frequency response repeating that of a measured signal.

Since the output digital signal of such an ac ADC will be close to a sine-wave with the carrier frequency, it may be down-converted digitally (Fig. 6). The implementation of a digital mixer for a 2 GHz carrier frequency is, in our opinion, not an impossible task. To down-convert a digital signal we multiply it in real time with a sine-wave, the frequency of which is a little below the carrier frequency. A multiplier of two 1-bit numbers can be built from one single XOR cell - a few RSFQ cells. The structure of a serial multiplier based on such elements was suggested in Ref. [4] and has since then been revised and successfully implemented [14].

IV. Conclusion

Superconductor ADCs, of which RSFQ represents the "third generation", can fill the performance gap between what is required for advanced SDR concepts and presently available semiconductor ADCs. The RSFQ ADCs are, even in their present non-optimized stage and 3.5-μm fabrication technology, superior to semiconductor ADCs. They require a much simpler (3-layer) fabrication technology and are clearly on a much steeper improvement curve, given a relatively small investment in submicron technology.

However, although such RSFQ-ADC chips can be readily fabricated, quite a few engineering problems, mostly connected with the optimization of their interface to the DSP, remain to be solved before an implementation of a real SDR front-end system will take place. The cooling requirement is usually seen as the main obstacle to the implementation of superconductor electronics: the low temperature of operation (of about 4 K) can only be reached with complex and rather expensive refrigerators. The "cooling penalty" can, in our opinion, be considerably reduced, if all the ultra-high-frequency components of the receiver (ADC, down-converters, decimation filters, etc.) are placed in the refrigerator, and the room-temperature processors are left to deal only with low data rates.

Acknowledgments

We thank Ericsson Radio Systems for support and Bob Walden of HRL for the kind permission to use his semiconductor ADC data.

References

[1] J. Mitola, "The software radio architecture", *IEEE Communications Magazine*, vol. 33, pp. 26-38, May 1995 (Special issue on " Software Radios").

[2] R.M. Lober, "Base-station technology takes Software-Definable Approach", *Wireless Systems Design*, Feb. 1998, pp.19-30.

[3] R. Walden (HRL), personal communication, Aug. 1998.

[4] K. Likharev and V. Semenov, "RSFQ logic/memory family: a new Josephson junction technology for sub-terahertz clock frequency digital systems," *IEEE Trans. Appl. Supercond.* vol. 1, pp. 3-28, March 1991.

[5] K. Likharev, "Superconductors speed up computation," *Phys. World*, vol. 10, pp. 39-43, May 1997.

[6] J.C. Lin, V.K. Semenov, and K.K. Likharev, "Design of SFQ-counting analog-to-digital converter", *IEEE Trans. on Appl. Supercond.*, vol. 5, pp. 2252-2259, June 1995.

[7] V. Semenov, Yu. Polyakov, and D. Schneider, "Preliminary results on the analog-to-digital converter based on RSFQ logic", in CPEM'96 Conference Digest. Braunschweig, Germany: PTB, 1996, Suppl. II, pp. 15-16.

[8] S. Rylov, L.A. Bunz, D.V. Gaidarenko, M.A. Fisher, R.P. Robertazzi, and O.A. Mukhanov, "High resolution ADC system", *IEEE Trans. on Appl. Supercond.* vol. 7, pp. 2649-2652, June 1997.

[9] S.V. Rylov, D.K. Brock, D.V. Gaidarenko, A.F. Kirichenko, V.K. Semenov, and J.M. Vogt, "High resolution ADC using phase modulation - demodulation architecture", in this volume.

[10] V.K. Semenov, Yu.A. Polyakov, and T.V. Filippov, "Analog-to-Digital converters with on-chip digital low-pass filters", in this volume.

[11] *Niobium Design Rules, Revision 017*, Elmsford, NY: HYPRES, Inc., Jan. 1997, see Web site http://hypres.com/designrule/rules.html.

[12] A.F. Kirichenko (HYPRES, Inc.), personal communication (Aug. 1998).

[13] R. Groshong and S. Ruscak: "Undersampling techniques simplify digital radio", *Electronic Design*, vol. 39, p. 67, May 1991.

[14] O. Mukhanov and A. Kirichenko, "Implementation of a FFT radix 2 butterfly using serial RSFQ multiplier-adders", *IEEE Trans. on Appl. Supercond.* vol. 5, pp. 2461-2464, June 1995.

On Sampling Rate, Analog Prefiltering, and Sufficient Statistics for Digital Receivers

Heinrich Meyr, *Fellow, IEEE,* Martin Oerder, and Andreas Polydoros, *Senior Member, IEEE*

Abstract— **In this paper we consider the joint sequence estimation, timing and phase recovery for linear modulation. The paper differs from the classical ones in the sense that time-discrete algorithms suitable for fully digital receivers are discussed. Sufficient conditions are given such that the signal samples represent sufficient statistics. These conditions involve signal bandwidth, sampling/symbol rate and the analog prefilter characteristics. It is shown that the sampling rate need not be an exact multiple of the symbol rate, i.e., the samples can be taken from a free-running oscillator. All subsequent signal processing operations in the receiver then operate with the clock of this free-running oscillator. Timing recovery is then performed by a time-variant linear digital interpolator and a decimator. Carrier recovery and sequence estimation are performed at an average rate of one symbol per sample. The digital matched filter for this case is derived for an arbitrary colored noise spectrum.**

I. INTRODUCTION

THE PROBLEM of joint sequence estimation, timing and phase recovery in a class of linear modulation data has been treated in the classical paper by Kobayashi [1] and subsequently by others [2]–[4] in a thorough and excellent manner. Although not explicitly stated these papers assume an analog implementation as manifested by the fact that the theoretical results are stated in the continuous-time domain. With the advent of VLSI technology, fully digital receivers operating on samples of the received signal became feasible. In order to match the strength of digital technology, these receivers have an entirely different structure than their analog counterparts [5], [6].

The goal of the paper is to present the theoretical basis for the synthesis of digital receiver structures. While the receiver structures are clearly nonfeasible for implementation reasons, they provide the basis (being an optimal bound) to systematically devise approximations which can be implemented in hardware. A "derived structure" approach (as opposed to an approach where a given structure is found *ad-hoc*) is necessary if the communication engineer wants to truly explore the design space comprised of power/bandwidth and signal processing complexity. In view of the advanced modulation/coding techniques approaching the Shannon limit, it becomes increasingly important to find hardware implementations with minimum implementation loss.

Paper approved by W. R. Braun, the Editor for Transmission Systems of the IEEE Communications Society. Manuscript received January 15, 1991; revised March 15, 1992 and January 15, 1993.

H. Meyr is with Aachen University of Technology, 52056 Aachen, Germany.

M. Oerder is with Philips GmbH Research Laboratories, 52066, Aachen, Germany.

A. Polydoros is with the Department of Electrical Engineering Systems, Communications Sciences Institute, University of Southern California, Los Angeles, CA 90089 USA.

IEEE Log Number 9406153.

The paper is structured as follows. In Section II we discuss under what condition samples of the received signal contain all the information of the continuous time-signal required to jointly perform the task of data detection, timing and carrier recovery. The conditions specify the analog prefilter $F(\omega)$ with the signal bandwidth B, symbol rate $1/T$ and sampling rate $1/T_s$ as parameters. Contrary to widespread belief the analog prefilter does not need to be the matched filter. Rather, the designer can make a trade-off between a simple analog prefilter and higher sampling rate $1/T_s$. Also, it is shown that the sampling rate $1/T_s$ doesn't need to be an exact multiple of the symbol rate. This has important consequences for timing recovery, since it basically implies that timing recovery can be achieved from samples of a free-running receiver clock.

In Section III the structure of the optimal digital ML receiver is derived. The derivation assumes bandlimited signal. The noise process is also assumed bandlimited with an arbitrary power density spectrum. The received samples are first processed in a digital matched filter followed by a time-variant interpolator and a variable-rate decimator which outputs a subset of samples processed at rate $1/T_s$ such that, on the average, exactly one sample per symbol duration T is generated. These samples which form an irregular pattern on the T_s-axis are then further processed in a phase rotator and sequence estimator. By construction, the digital receiver is a possible realization of the optimal ML-receiver. Thus, its performance equals that of its analog counterpart, demonstrating the equivalence of analog and digital signal processing.

II. SUFFICIENT STATISTICS FOR BANDLIMITED SIGNALS

We assume the following model. The received signal $r(t)$ is the sum of a useful signal $s(t,\alpha)$ plus additive noise $w(t)$

$$r(t) = s(t,\alpha) + w(t)$$

where α is a parameter vector. The useful signal $s(t,\alpha)$ depends on the modulation format, the data sequence $\mathbf{a}$, the unknown fractional timing delay $\mathbf{e}$ and the carrier phase Θ. It is assumed to be strictly bandlimited to B. The additive noise is modeled as colored Gaussian.

Since $s(t,\alpha)$ is assumed to be strictly bandlimited, the received signal can be decomposed into

$$r(t) = \underbrace{s(t,\alpha) + n(t)}_{y(t)} + \underbrace{[w(t) - n(t)]}_{\tilde{n}(t)} \tag{1}$$

Reprinted from *IEEE Transactions on Communications,* Vol. 42, No. 12, pp. 3208–3213, December 1994.

where $n(t)$ is the part of the total noise which falls into the bandwidth of the useful signal, while $\tilde{n}(t)$ is the out-of-band noise. Since $\tilde{n}(t)$ is statistically independent of the signal $s(t, \alpha)$ and uncorrelated with $n(t)$, by the Gaussian assumption we conclude that $\tilde{n}(t)$ is irrelevant to the decision of which message was transmitted or to the estimation of the synchronization parameters.

Since $y(t)$ is bandlimited to B, it can be expanded into a series

$$y(t) = \sum_{n=-\infty}^{\infty} y\left(\frac{n}{2B}\right) si\left[2\pi B\left(t - \frac{n}{2B}\right)\right] \tag{2}$$

with orthogonal base functions

$$si\left[2\pi B\left(t - \frac{n}{2B}\right)\right] = \frac{\sin\left[2\pi B\left(t - \frac{n}{2B}\right)\right]}{2\pi B\left(t - \frac{n}{2B}\right)}. \tag{3}$$

Thus, a sufficient statistic can be obtained by prefiltering $r(t)$ with an ideal analog low-pass filter

$$H_L(\omega) = \begin{cases} 1 & \left|\frac{\omega}{2\pi}\right| < B \\ 0 & \text{elsewhere} \end{cases} \tag{4}$$

and subsequent sampling at Nyquist rate $2B$. Involving an ideal low-pass filter is of course a solution of no practical interest. Fortunately, it is possible to generate a sufficient statistic employing a realizable analog filter $F(\omega)$.

Proposition: The samples $\{r_f(kT_s)\}$ represent sufficient statistics for the estimation of the parameter α if the following conditions on the analog prefilter $F(\omega)$ and sampling rate $1/T_s$ are satisfied

$$F(\omega) = \begin{cases} \text{arbitrary, but unequal to zero} & \left|\frac{\omega}{2\pi}\right| < B \\ \text{arbitrary,} & B \le \left|\frac{\omega}{2\pi}\right| < B_F \\ 0, & B_F < \left|\frac{\mu}{2\pi}\right| < \frac{1}{T_s} - B \end{cases}. \tag{5}$$

To prove the assertion we use the concept of reversibility which states that if any preliminary processing employed is reversible it can be made to have no effect on the performance of the system.

In Fig. 2 the input $r(t)$ is processed in the upper path by an ideal low-pass filter $H_L(\omega)$. From the previous discussion we recall that the samples $y(n/2B)$ represent sufficient statistics. In the lower signal path the same signal $r(t)$ is first processed by the analog filter $F(\omega)$ and subsequently sampled at rate $1/T_s$. Due to the conditions on $F(\omega)$ there always exists a digital filter $H^{-1}(e^{j\omega T_s})$ which reverses the effect of $F(\omega)$ in the passband of the useful signal. The samples are then processed in an ideal digital lowpass filter $H_L(e^{j\omega T_s})$ which eliminates all out-of-band noise. Subsequently, the analog signal

$$y_d(t) = \sum y_d(kT_s) si\left[\frac{\pi}{T_s}(t - kT_s)\right] \tag{6}$$

is constructed from the samples by an ideal interpolator $H_I(\omega)$. Using well-known techniques [7, p. 316] it can be shown that

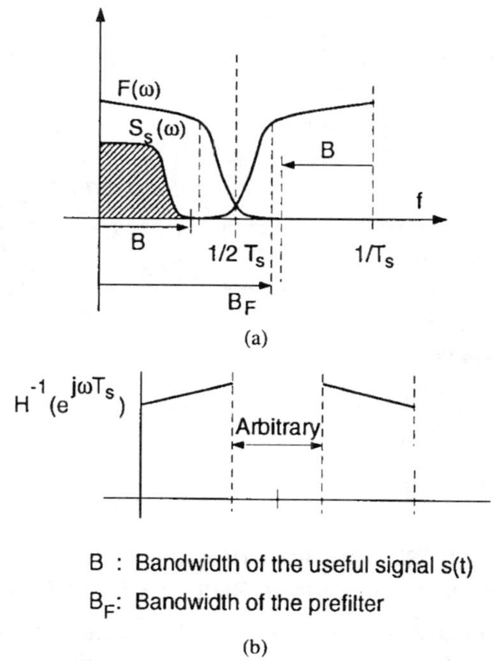

Fig. 1. (a) Conditions on the analog prefilter $F(\omega)$. (b) Conditions on the inverse digital filter $H^{-1}(e^{j\omega T_s})$.

the signals $y(t)$ and $y_d(t)$ are equal in the mean square sense

$$E[|y(t) - y_d(t)|^2] = 0 \qquad \text{for all } t. \tag{7}$$

But since the samples $\{y_d(kT_s)\}$ are obtained from $\{r_f(kT_s)\}$, it immediately follows from the principle of reversibility that the samples $\{r_f(kT_s)\}$ also represent sufficient statistics.

There are several interesting aspects in this result which deserve a discussion. It is tempting to conclude that $r_f(t)$ has a series representation with base function $si\left[\frac{\pi}{T_s}(t - nT_s)\right]$. However, this is faulty reasoning as can be seen by inspection of Fig. 1. The bandwidth B_F of the noise, in general, is larger than half the sampling rate. Thus the sampling theorem is not applicable to $r_f(t)$. It is, however applicable to the useful signal $s_f(t, \alpha)$. Stated differently, the conditions on the prefilter $F(\omega)$ demand that the sampling theorem applies to the useful signal $s(t, \alpha)$ but tolerate an undersampling of the noise process as long as there is no aliasing of the noise into the passband $|\omega/2\pi| < B$.

For bandwidth efficient communication the signal bandwidth lies within

$$B = \frac{1}{2T}(1 + \beta), \qquad 0 < \beta < 1 \tag{8}$$

where β is the excess bandwidth. We recognize from (8) that the sampling rate $1/T_s$ can be chosen independently from B, or equivalently from the symbol rate $1/T$, as long as conditions (5) are satisfied. But since $\{r_f(kT_s)\}$ represent sufficient statistics for the parameter set $\alpha = (\theta, \epsilon, \boldsymbol{a})$ this implies that timing recovery must be possible from the samples of a free-running oscillator taken at kT_s. Put differently, the receiver clock $(1/T_s)$ does *not* necessarily need to be synchronized to the transmitter clock $(1/T)$.

Notice that the entire discussion on sufficient statistics is true irrespective of any decision/estimation criterion chosen.

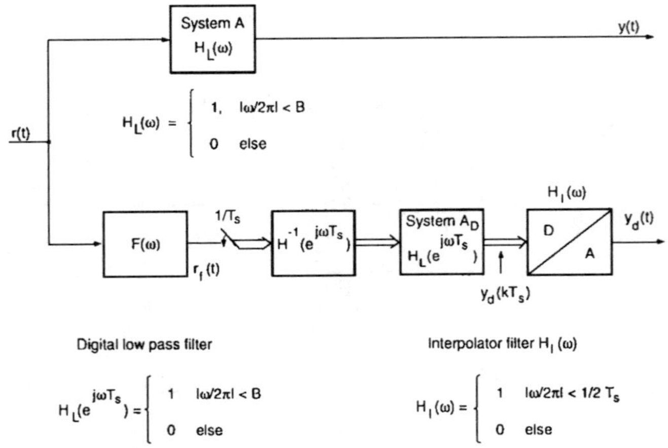

Fig. 2. Reversibility.

III. The ML-Function and Optimal (Noncausal) Digital Receiver Structures

To exemplify the significance of the theoretical results we confine the discussion to linear modulation and the maximum likelihood (ML) optimality criterion. The received signal can then be written in the form

$$r_f(t) = \underbrace{\sum_{n=0}^{N-1} a_n g(t - nT - \epsilon T)e^{j\theta}}_{s_f(t,\alpha)} + n(t) \tag{9}$$

where we define $g(t)$ as the received pulse after the analog prefilter $F(\omega)$. The signal bandwidth B lies within

$$B = \frac{1}{2T}(1+\beta); \quad 0 < \beta < 1.$$

The covariance matrix Λ of the noise samples has the elements

$$\lambda(k, l) = [\Lambda]_{kl}. \tag{10}$$

Using well-known techniques [8, p. 274], the log-likelihood function for the joint estimation of the data sequence and synchronization parameters (θ, ϵ) can be shown to be (see Appendix)[1]

$$L(\boldsymbol{a}, \epsilon, \theta) = \boldsymbol{a}^H e^{j\theta}\boldsymbol{z} + \boldsymbol{z}^H \boldsymbol{a} e^{-j\theta} - \boldsymbol{a}^H \boldsymbol{H}\boldsymbol{a}. \tag{11}$$

The nth element a_n equals the nth data symbol transmitted in a sequence of length N. The element z_n

$$z_n = z(nT + \epsilon T) = \sum_{k=-\infty}^{\infty} r_f(kT_s) g_{MF}(-kT_s + nT + \epsilon T) \tag{12}$$

is the output of a digital matched filter $g_{MF}(kT_s)$ defined by

$$g^*_{MF}(-kT_s) = \sum_{l=-\infty}^{\infty} q[(k-l)T_s]g(lT_s). \tag{13}$$

The pulse $g(t)$ is that of the received signal at $t = lT_s$ and $q(iT_s)$ is the element of the (infinite dimensional) inverse covariance matrix

$$q(iT_s) = [\boldsymbol{Q}]_{kl}$$
$$\text{with } i = k - l.$$

The matrix $\boldsymbol{H}$ has the elements $h(m, n)$ defined by

$$h(m, n) = [\boldsymbol{H}]_{mn} = \sum_{k=-\infty}^{\infty} g(kT_s - mT)g_{MF}(-kT_s + nT) \tag{14}$$

which is the output of the digital matched filter at time nT with input $g(kT_s - mT)$. Notice that the matrix $\boldsymbol{H}$ is independent of the synchronization parameters.

The ML estimates $(\hat{\boldsymbol{a}}, \hat{\theta}, \hat{\epsilon})$ are those elements which maximize $L(\boldsymbol{a}, \epsilon, \theta)$. Equation (11) is the *exact* form of the maximum likelihood function. There is no approximation involved in computing a correlation type integral. This is a consequence of the Parseval theorem employing the base function $si(x)$

$$z(nT + \epsilon T) = \frac{1}{T_s}\int_{-\infty}^{\infty} r_f(t)g_{MF}(-t + nT + \epsilon T)dt = \sum_{k=-\infty}^{\infty} r_f(kT_s)g_{MF}(-kT_s + nT + \epsilon T). \tag{15}$$

The right-hand side of the first line of (15) shows how to obtain $z(nT + \epsilon T)$ by means of an analog matched filter. The second line demonstrates that there exists an equivalent digital matched filter. The digital matched filter is time-variant unless T/T_s is an integer. However, (15) can always be brought into a form such that the digital matched filter becomes time invariant by cascading it with a digital interpolator as will be presently shown.

The role of the interpolator (and its associated control) become most clearly visible by emphasizing that the only time scale available at the receiver is defined by units of T_s and that, therefore, the transmitter time scale defined by units of T must be expressed by units of T_s. We write for the argument of the samples

$$nT + \epsilon T + kT_s = T_s\left[n\frac{T}{T_s} + \epsilon\frac{T}{T_s}\right] + kT_s. \tag{16}$$

The key step is to write the first term of the previous equation as the sum

$$n\frac{T}{T_s} + \epsilon\frac{T}{T_s} = L_{\text{INT}}\left[n\frac{T}{T_s} + \epsilon\frac{T}{T_s}\right] + \mu_n = m_n + \mu_n \tag{17}$$

where $m_n = L_{\text{INT}}(\cdot)$ means the largest integer that is less than or equal to the real number in the argument and μ_n is the remainder. We then obtain for (15)

$$z(nT + \epsilon T) = \sum_{k=-\infty}^{\infty} r_f(kT_s)g_{MF}(-kT_s + m_nT_s - \mu_nT_s). \tag{18}$$

[1] $\boldsymbol{a}$, $\boldsymbol{z}$ are N-dimensional vectors, and $\boldsymbol{H}$ is a $N \times N$ matrix, $\boldsymbol{a}^H\boldsymbol{z}$ equals the inner product of the two vectors, ^{H} denotes Hermitian of vectors.

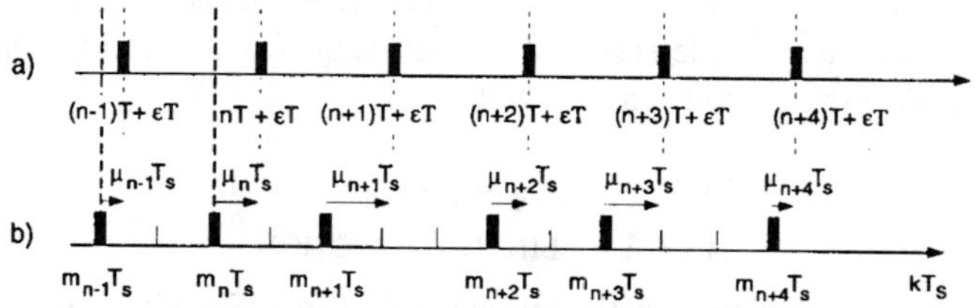

Fig. 3. (a) Transmitter time scale (nT). (b) Receiver time scale (kT_s).

The samples of the impulse response shifted by $(\mu_n T_s)$ follow from the representation of the bandlimited signal [7, p. 142]

$$g_{MF}(t+\mu_n T_s) = \sum_{i=-\infty}^{\infty} g_{MF}(iT_s+\mu_n T_s) si\left[\frac{\pi}{T_s}(t-iT_s)\right] = \sum_{i=-\infty}^{\infty} g_{MF}(iT_s) si\left[\frac{\pi}{T_s}(t+\mu_n T_s-iT_s)\right]. \quad (19)$$

For $t = lT_s$ we obtain from the previous equation

$$g_{MF}(lT_s+\mu_n T_s) = \sum_{i=-\infty}^{\infty} g_{MF}(iT_s) si\left[\frac{\pi}{T_s}(lT_s+\mu_n T_s-iT_s)\right]. \quad (20)$$

Equation (20) has the form of a convolution sum. We can thus view the sequence $\{g_{MF}(lT_s+\mu_n T_s)\}$ as being generated by a digital interpolator filter with impulse response

$$h_I(lT_s;\mu_n) = si\left[\frac{\pi}{T_s}(lT_s+\mu_n T_s)\right] \quad (21)$$

and input $g_{MF}(iT_s)$. The interpolator filter is a linear, time-variant filter (due to μ_n) with output $y(kT_s)$. Notice that the matched filter and interpolator run at rate $1/T_s$.

Unlike conventional digital filters where the input rate equals the output rate, the right-hand side of (18) defines a time-variable decimation process since only outputs $y(m_n T_s)$ at times $m_n T_s$ are taken

$$y(m_n T_s) = \sum_{k=-\infty}^{\infty} r_f(kT_s) g_{MF}(-kT_s+m_n T_s+\mu_n T_s). \quad (22)$$

By comparison of (22) and the definition of $z(nT+\epsilon T)$, (18), one concludes that

$$z(nT+\epsilon T) = y(m_n T_s).$$

Since T/T_s is irrational the time instances $m_n T_s$ form a completely irregular (though deterministic) pattern on the time axis kT_s. (see Fig. 3) This irregular pattern is required in order to obtain an average time between an output $z(nT+\epsilon T)$ of exactly T (given a time quantization of T_s). Finally, the samples $\{y(m_n T_s)\} = \{z_n\}$ obtained at average rate $1/T$ are then phase rotated by $exp(-j\theta)$ before they enter the sequence estimator. Notice that (on the average) *one* sample per symbol is sufficient for ML phase and sequence estimation, (assumed ϵ is known).

The optimal receiver structure employing digital signal processing is depicted in Fig. 4. The received samples $\{r(kT_s)\}$

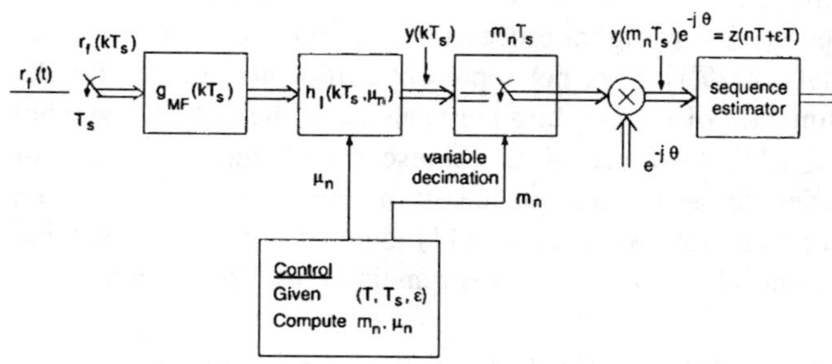

Fig. 4. Optimal (ML) receiver structure.

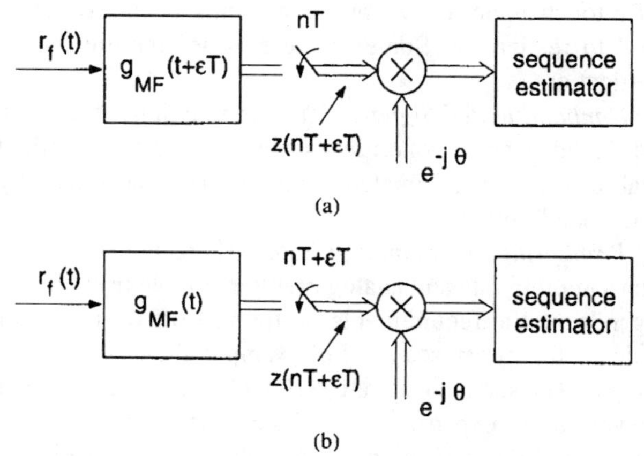

Fig. 5. Analog realization of the matched filter.

are first processed in a digital matched filter followed by the time-variant interpolator, both running at rate $1/T_s$. Since the filters are linear their position may be interchanged. Notice that only the results $y(m_n T_s)$ are required for data detection and phase synchronization, thus only these values need to be computed. As an interesting detail we remark that in the case where the interpolator precedes the matched filter the output of the interpolator does not equal $r_f(kT_s+\mu_n T)$ since $r_f(t)$ is, in general, not bandlimited to $B_F \leq 1/2T_s$, (see Fig. 1).

The structure of the receiver of Fig. 4 is nonrealizable since it contains noncausal operations. However, it serves as basis from which realizable structures can be derived (see following section).

Before we discuss realizable structures we briefly return to the analog realization of the matched filter. Because of the equivalence of analog and digital signal processing as expressed by the Parseval theorem (15) we obtain the well-known analog structure of Fig. 5.

We deduce from Fig. 5, (5), and $B > 1/2T$ that the samples $z(nT+\epsilon T)$ are *not* sufficient for the estimation of the timing parameter ϵ as frequently stated. From Fig. 5 we learn that we have to convolve the incoming signal $r_f(t)$ with $g_{MF}(t+\epsilon T)$ which contains the timing parameter. But this implies that a bank of filters $g_{MF}(t+\epsilon_n T)$, ϵ_n: trial parameters, must be implemented and sampled at symbol rate $1/T$ to obtain the ML-estimate by a maximum search. Alternatively, one may use a single filter with $g_{MF}(t)$ and control the sampling instants to occur at $t = nT+\epsilon T$. In both cases of Fig. 5 the sampling rate is *synchronized* to the symbol timing nT.

Since it is impossible to obtain $z(nT+\epsilon T)$ in the first case from $z(nT)$ by digital processing (interpolation), the sequence $\{z(nT+\epsilon T)\}$ does not represent sufficient statistics for the estimation of ϵ. Sampling the analog matched filter at symbol rate $1/T$, however, does represent sufficient statistics for carrier recovery and sequence estimation as previously pointed out and as follows from eq. (11) together with (12), assuming ϵ is known and only θ and $\boldsymbol{a}$ remain as trial parameters.

IV. Realizable Digital Receiver Structures

The nonrealizable receiver structure of Section III serves as basis for any suboptimal, but realizable structure. We can only briefly touch upon a few important items in this paper with regard to realization. Some of these issues require thorough investigation.

1) Nonbandlimited Signals: The assumption of a strictly bandlimited signal is noncritical since any (nearly) bandlimited signal can be approximated with arbitrary accuracy by a strictly bandlimited one.

2) Prefiltering and Sampling Rate: There is a tradeoff between sampling rate and analog prefilter complexity. Oversampling relaxes the requirements on the roll-off thus leading to simple analog filters, see (5). For example, the analog matched filter $g_{MF}(t)$ sampled at rate $2B$ yields sufficient statistics, however, at the expense of a complex analog filter.

3) Clock Synchronization: The optimal ML receiver requires joint maximization of the log-likelihood function $L(\boldsymbol{a}, \epsilon, \theta)$ (11) which is too complex to be practical. In a practical receiver one assumes that estimates of ϵ and θ are available. These estimates then are used for a detection $L(\boldsymbol{a}, \hat{\epsilon}, \hat{\theta})$ as if they were true values. From Fig. 4 one deduces the important fact that clock synchronization is done by means of an adaptive digital interpolator circuit. To our knowledge, this idea was first presented in [9] and implemented in digital receivers in [5].

4) Interpolation Filter and Digital Matched Filter: The ideal interpolator of (20) can be approximated with negligible performance degradation by simple structures [10]. Also, there is no need to exactly realize the matched filter, rather, a joint optimization of interpolator and matched filter subject to minimal noise power output leads to surprisingly simple solutions [6].

5) Sampling Rate Conversion: A particularly interesting problem (theoretical and practical) poses the rate conversion from $1/T_s$ to symbol rate $1/T$. In the case when the symbol duration T is a multiple of T_s the rate conversion simply requires taking every $M = T/T_s$, samples of the digital matched filter output. Assuming an integer ratio T/T_s however, oversimplifies the problem. In a practical receiver T and T_s are in general incommensurate.

V. Conclusion

We have shown how to systematically derive digital receiver structures based on estimation theoretic concepts. We have demonstrated that the concept of sufficient statistics is not a merely theoretical tool but directly provides guidelines related to analog prefilter design and sampling rate such that the samples of the received signal are indeed all that is required for the simultaneous detection of the data sequence and estimation of the synchronization parameters.

Appendix
Log-Likelihood Function

The derivation directly follows Van Trees [8]. The samples

$$\boldsymbol{r}_{f,2M+1}^T = \{r_f(-MT_s), \cdots, r_f(0), r_f(T_s), \cdots, r_f(MT_s)\}$$
$$\boldsymbol{s}_{f,2M+1}^T = \{s_f(-MT_s), \ldots, s_f(0), s_f(T_s), \ldots, s_f(MT_s)\}$$

define a $(2M+1)$ coefficient approximation

$$p(\boldsymbol{r}_{f,2M+1}|\boldsymbol{a}, \epsilon, \theta) = \text{const} \cdot \exp\left\{-\frac{1}{2}[\boldsymbol{r}_{f,2M+1} - \boldsymbol{s}_{f,2M+1}]^H \boldsymbol{Q} \cdot \boldsymbol{r}_{f,2M+1} - \boldsymbol{s}_{f,2M+1}]\right\} \tag{A.1}$$

$\lambda(k, l)$ are the elements of the covariance matrix Λ, while $\boldsymbol{Q}$ is the inverse covariance matrix. The log-likelihood function $L(\boldsymbol{a}, \epsilon, \theta)$ follows from (A.1) by dividing by $p(\boldsymbol{a}_{f,2M+1}|\boldsymbol{a} = 0)$, taking the logarithm and letting $M \to \infty$

$$L(\boldsymbol{a}, \epsilon, \theta) = \boldsymbol{r}_f^H \boldsymbol{Q} \boldsymbol{s}_f + \boldsymbol{s}_f^H \boldsymbol{Q} \boldsymbol{r}_f - \boldsymbol{s}_f^H \boldsymbol{Q} \boldsymbol{s}_f. \tag{A.2}$$

Inserting for $s_f(kT_s) = \sum_{n=0}^{N-1} a_n g(kT_s - nT - \epsilon T) e^{j\theta}$ yields after some lengthy, but straightforward algebraic manipulations.

$$\boldsymbol{s}_f^H \boldsymbol{Q} \boldsymbol{r}_f = \sum_{n=0}^{N-1} a_n^* e^{-j\theta} \sum_{k=-\infty}^{\infty} r_f(kT_s) \cdot g_{MF}^*(nT + \epsilon T - kT_s) \quad (= \boldsymbol{a}^H z e^{-j\theta}) \tag{A.3}$$

with the matched filter impulse response

$$g_{MF}^*(-kT_s) := \sum_{l=-\infty}^{\infty} q[(k-l)T_s] g(lT_s)$$
$$q(iT_s) = [\boldsymbol{Q}]_{kl} \quad \text{with} \quad k - l = i \tag{A.4}$$

element of the inverse covariance matrix $\boldsymbol{Q}$.

The definition of $g_{MF}^*(-kT_s)$ assumes that $\boldsymbol{Q}$ is a Toeplitz matrix (to be discussed later).

The third term in (A.2) is found to be

$$\boldsymbol{s}_f^H \boldsymbol{Q} \boldsymbol{s}_f = \sum_{n=0}^{N-1} a_n^* \sum_{m=0}^{N-1} a_m \left[\sum_{k=-\infty}^{\infty} g^*(kT_s - mT - \epsilon T) \cdot g_{MF}^*(-kT_s + nT + \epsilon T)\right]$$
$$= \boldsymbol{a}^H \boldsymbol{H} \boldsymbol{a} \qquad \text{(independent of } \epsilon, \theta) \tag{A.5}$$

with

$$h(m, n) = [\boldsymbol{H}]_{mn} = \sum_{k=-\infty}^{\infty} g(kT_s - mT - \epsilon T) \cdot g_{MF}(-kT_s + nT + \epsilon T)$$
$$= \sum_{k=-\infty}^{\infty} g(kT_s - mT) g_{MF}(-kT_s + nT). \tag{A.6}$$

The independence of the matrix elements $h(m,n)$ of ϵ follows by employing the Parseval theorem

$$\begin{aligned}&\sum_{k=-\infty}^{\infty} g(kT_s - mT - \epsilon T)g_{MF}(-kT_s + nT + \epsilon T)\\&= \frac{1}{T_s}\int_{-\infty}^{\infty} g(t - mT - \epsilon T)g_{MF}(-t + nT + \epsilon T)dt\\&= \frac{1}{T_s}\int_{-\infty}^{\infty} g(t - mT)g_{MF}(-t + nT)dt. \qquad (A.7)\end{aligned}$$

The last line of (A.6) is a consequence of (A.7).

Toeplitz Structure of Q: We define

$\lambda(i,j) = [\boldsymbol{\Lambda}]_{ij}$: element of the covariance matrix

$q(l,m) = [\boldsymbol{Q}]_{lm}$: element of the inverse covariance matrix

The matrix Λ is a $(2M+1) \times (2M+1)$ positive definite Hermitian matrix. Furthermore, $\boldsymbol{\Lambda}$ is a Toeplitz matrix. The matrix $\boldsymbol{Q}$ can also be shown to be Hermitian, but has not, in general a Toeplitz form. However, asymptotically, $\boldsymbol{Q}$ becomes a Toeplitz matrix.

By definition we have $\boldsymbol{\Lambda Q} = I$, where $b(k, m)$ is an element of the unity matrix, $b(k, m) = 1$ for $k = m$ and zero else. We have

$$b(k, m) = \sum_l \lambda(k, l)q(l, m), \quad |l| \le M \qquad (A.8)$$

using the Toeplitz form of $\boldsymbol{\Lambda}$ we obtain

$$b(k, m) = \sum_l \lambda(k - l)q(l, m), \quad |l| \le M. \qquad (A.9)$$

The asymptotic relationship of $\boldsymbol{Q}$ may be best understood by applying the z-transform method. The z-transform of $b(k,m)$ (m fixed) is related to (A.9) by

$$\begin{aligned}B(z,m) &= \sum_k b(k, m)z^{-k}\\&= \sum_k z^{-k}\sum_1 \lambda(k - l)q(l, m), \quad |l|, |k| \le M.\end{aligned} \qquad (A.10)$$

Assuming an infinite horizon $M \to \infty$ we obtain by interchanging the order of summation

$$B(z, m) = \sum_{l=-\infty}^{\infty} q(l, m)z^{-l} \sum_{k=-\infty}^{\infty} \lambda(k - l)z^{-(k-l)} \qquad (A.11)$$

since k runs from $(-\infty, +\infty)$ the second factor is independent of l. The sequence $b(k,m)$ is nonzero for $k = m$ only, hence $B(z, m) = z^{-m}$. From (A.11) we obtain in obvious notation

$$q(z,m) = \frac{z^{-m}}{\lambda(z)}. \qquad (A.12)$$

From (A.12) follows that the column vectors, $q(l,m), l = (-\infty, +\infty)$ are identical, apart from a shift, i.e.

$$q(l, k_1) = q(l + k; k_1 + k)$$

$\boldsymbol{Q}$ has an asymptotically Toeplitz form.

From (A.4) follows that g_{MF} is the convolution of q and g. Hence, the frequency response of the matched filter equals

$$G_{MF}(e^{j\omega T_s}) = G^*(e^{j\omega T_s})Q(e^{j\omega T_s}).$$

Replacing $\boldsymbol{Q}$ by the right-hand side of (A.12) yields

$$G_{MF}(e^{j\omega T_s}) = \frac{G^*(e^{j\omega T_s})}{\lambda(e^{j\omega T_s})} = \frac{G^*(e^{j\omega T_s})}{\frac{1}{T_s}\sum_k S_n\left[\omega - \frac{2\pi k}{T_s}\right]} \qquad (A.13)$$

where

$$S_n(\omega) = |F(\omega)|^2 S_w(\omega)$$

is the power spectrum of the filters noise $n(t)$ (see Fig. 2).

Notice in the baseband

$$G_{MF}(e^{j\omega T_s}) = 0, \quad \left|\frac{\omega}{2\pi}\right| > B. \qquad (A.14)$$

Inserting for $\lambda(e^{j\omega T_s})$ the denominator of (A.13) indicates that the covariance matrix $\boldsymbol{Q}$ becomes singular if $B_F < \frac{1}{2T_s}$. (strict oversampling). However, for the definition of the matched filter (A.13) only the frequency response of $F(\omega)$ in the range $|\frac{\omega}{2\pi}| < B$ is relevant. Hence, $F(\omega)$ can be replaced by any filter which coincides with it in the above frequency range and which is arbitrary, but nonzero for

$$B \le \left|\frac{\omega}{2\pi}\right| \le \frac{1}{2T_s}.$$

ACKNOWLEDGMENT

We would like to thank the anonymous reviewers for their constructive criticism.

REFERENCES

[1] H. Kobayashi, "Simultaneous adaptive estimation and detection algorithm for carrier modulated data transmission systems," *IEEE Trans. Commun.*, vol. COM-19, June 1971.

[2] G. D. Forney, "Maximum-likelihood sequence estimation of digital sequences in the presence of intersymbol interference," *IEEE Trans. Inform. Theory*, vol. IT-18, May 1972.

[3] G. Ungerboeck, "Adaptive maximum-likelihood receiver for carrier-modulated data-transmission systems," *IEEE Trans. Commun.*, vol. COM-22, May 1974.

[4] D. D. Falconer, "Optimal reception of digital data over the Gaussian channel with unknown delay and phase jitter," *IEEE Trans. Inform. Theory*, vol. IT-23, Jan. 1977.

[5] G. Ascheid, M. Oerder, J. Stahl, and H. Meyr, "An all digital receiver for bandwidth efficient transmission at high data rates," *IEEE COM*, pp. 804–813, Aug. 1989.

[6] O. J. Joeressen, M. Oerder, R. Serra, and H. Meyr, "DIRECS: System design of a 100 Mbit/s digital receiver," in *IEE Proc. G*, vol. 139, Apr. 1992, pp. 222–230.

[7] A. Papoulis, *Signal Analysis*. New York: McGraw-Hill, 1977.

[8] H. L. van Trees, *Detection, Estimation, and Modulation Theory*. New York: Wiley, 1968.

[9] M. Oerder, G. Ascheid, R. Haeb, and H. Meyr, "An all digital implementation of a receiver for bandwidth efficient communications," *Signal Processing III (EUSIPCO 1986)*, I. T. Young *et al.*, Eds. New York: Elsevier, 1986, pp. 1091–1094,

[10] F. M. Gardner, "Timing adjustment via interpolation in digital demodulators," Final Rep. ESTEC Contract No. 8022/88/NL/DG, Palo Alto, CA: Gardner Res. Co., Aug. 1988.

Digital IF Filter Technology for 3G Systems: An Introduction

David B. Chester, Harris Semiconductor

ABSTRACT Contemporary digital communication systems such as those being developed for the deployment of third-generation cellular require ever increasing performance levels in their signal processing chains to extract higher data rates and to provide decreasing price/performance ratios. Additionally, communication systems like 3G that must support multimode flexibility, such as the software radio, must be able to reconfigure their signal processing chains while keeping circuit complexity to a minimum. Given these constraints, DSP is the only viable alternative for baseband processing and digital IF processing. DSP is in many cases the only viable alternative to analog IF processing. Digital IF affords greater flexibility and higher performance in terms of attenuation and selectivity. It also offers better time and environment stability and lower equipment production costs than traditional analog techniques.

Software radio and digital intermediate frequency (IF) technology are enabling the economic development of third-generation cellular base stations by providing the required flexibility and fidelity at reasonable price points. Digital signal processing (DSP) affords greater flexibility and higher performance (in terms of attenuation and selectivity) than traditional analog techniques. It also offers better stability over time and environmental parameters with lower equipment production costs. Additionally, components available today let DSP extend from baseband to intermediate frequencies. This makes DSP useful for tuning, signal selection, and frequency up- and downconversion. The confluence of DSP's properties with the technologies that allow it to be economically implemented further up in the signal processing path of communication systems is now enabling the concept of software radio to become a product reality.

The term *software radio* has come to stand for a software-reconfigurable radio that derives its flexibility from programmable radio frequency (RF), IF, and baseband signal processing stages. They may be implemented either in programmable DSPs or reconfigurable dedicated function signal processors. The baseband processing section of a software radio performs data modulation and demodulation. The IF processing section translates the information-bearing signal(s) between baseband and IF. The IF stage of a software radio typically consists of a data conversion stage, a digital rate change stage, and a numerically controlled oscillator and modulator stage. In some cases, the rate change and modulation stages can be combined into a single functional block. Each of the signal processing stages are the subject of many publications. This article focuses on the basic concepts of digital IF.

This article begins with an overview of the concept of digital IF. The two basic types of digital filters, infinite impulse response (IIR) and finite impulse response (FIR), are then introduced and contrasted. An explanation is given as to why an FIR filter is usually preferred in the IF and baseband stages of a data communication system.

Some of the basic concepts of multirate digital filters are then introduced. Multirate digital signal processing is used in most digital IF implementations. The article then discusses the three basic types of digital IF processing: narrowband single-channel, wideband single-channel, and transmultiplexing. The article concludes with an overview of digital IF implementation, including a survey of commercially available digital IF products.

This article introduces digital IF. Those interested in further depth may consider the introductory DSP texts *Discrete-Time Signal Processing* [1] and *Theory and Application of Digital Signal Processing* [2], which is unfortunately now out of print. Another text introduces most of the basic concepts used in contemporary digital IF, *Multirate Digital Signal Processing* [3]. Finally, the *Handbook for Digital Signal Processing* [4] is an excellent general reference for both introductory and advanced applications.

An Overview of Software Radio IF Processing Architectures

The top-level block diagram of a generic software radio is shown in Fig. 1. The diagram shows a traditional analog architecture converting the signal of interest from RF to a convertible IF in the downconversion side and from IF to RF in the upconversion side. The automatic gain control (AGC), antialiasing filter, and analog-to-digital converter (ADC) constraints on the downconversion side of the radio are highly dependent on the overall radio architecture. One of the primary considerations is whether the digital downconverter is operating on a wideband signal, a narrowband signal, or multiple narrowband signals.

A wideband signal is defined as a signal with a bandwidth out of the digital downconverter that is greater than 10 percent of the input signal. This defines constraints for the signal processing components. A narrowband output signal has significantly narrower bandwidth than the input signal, implying a reasonable amount of processing gain [5]. Multiple narrowband signals with similar characteristics place constraints on the signal processing components similar to the single narrowband signal. The trend in digital IF processing requirements for evolving standards are summarized in Table 1. In the table W_a is the input bandwidth, W_c is the signal-of-interest bandwidth, and N_c is the number of signals simultaneously processed.

Another consideration is whether the ADC performs baseband or passband sampling [6]. Analagous considerations are relevant to upconversion.

A digital downconverter typically consists of a digital local oscillator, a digital mixer, and a decimation filter [3]. These operations can be performed in a distributed manner when operating on multiple parallel channels with the use of a transmultiplexer [3]. A digital upconverter typically consists of a digital local oscillator, a digital mixer, and an interpolation filter [3]. Again, a transmultiplexer can upconvert multiple channels in parallel [3]. In addition, the direct conversion receiver converts RF directly to baseband without digital or analog IF. This architecture is attractive for handset applications, as described in the companion article by Tsurumi and Suzuki in this issue. Direct conversion is effective for single-

Reprinted from *IEEE Communications Magazine,* Vol. 37, No. 2, pp. 102–107, February 1999.

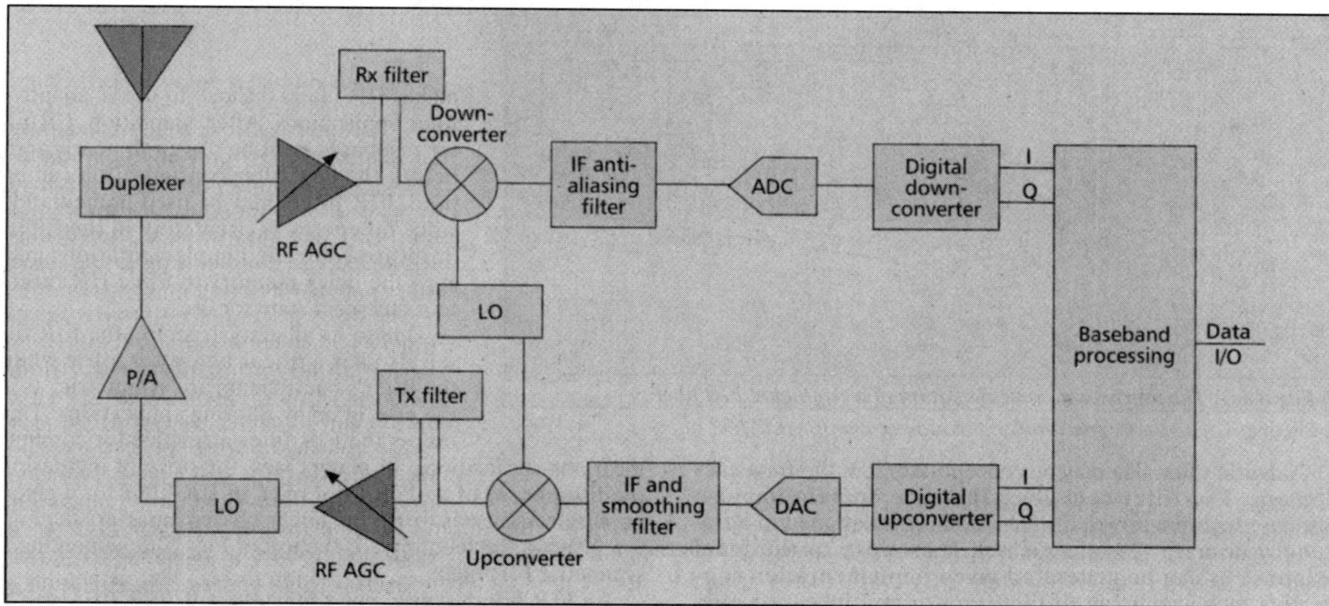

■ **Figure 1.** *The top-level block diagram of a generic software radio configuration.*

channel applications, but multiple simultaneous channels have fewer spurs and better signal-to-noise ratio in near-far conditions through digital IF architectures.

Standard	W_a (MHz)	W_c (kHz)	N_c	Digital IF
First-generation	12.5 or 25	25 or 30	416–1000	Narrowband
GSM	25	200	125	Narrowband
W-CDMA	25–60	5000	5-12	Wideband

■ **Table 1.** *Evolving digital IF requirements.*

DIGITAL FILTERING FOR DIGITAL IF

A digital filter is a discrete-time discrete-amplitude convolver. The linear convolution of two sequences in the time domain is the same as multiplication of two corresponding spectral sequences in the frequency domain. Filtering is in essence the multiplication of the signal spectrum by the Fourier transform of the impulse response of the filter. For an ideal low-pass filter, the passband part of the signal spectrum is multiplied by one and the stopband part of the signal by zero.

The algorithmic structure of a digital filter is shown in Fig. 2. This Nth-order digital filter has both a feedforward and a feedback path. The input to the filter is a discrete-time discrete-amplitude series $x(n)$. The output of the filter is a discrete-time discrete-amplitude series $y(n)$. The series of delay registers shown in the figure is a discrete-time discrete-amplitude tapped delay line. The z-domain variable z^{-1} represents a delay of one clock. The feedforward coefficients are represented by a_0, a_1, ..., a_N. The feedforward section of the architecture shown in Fig. 2 is a simple transversal structure. The feedback coefficients in the figure are represented by b_1, b_2, ..., b_N.

A filter's response can be time-varying. In this case, the coefficients will also be a function of n.

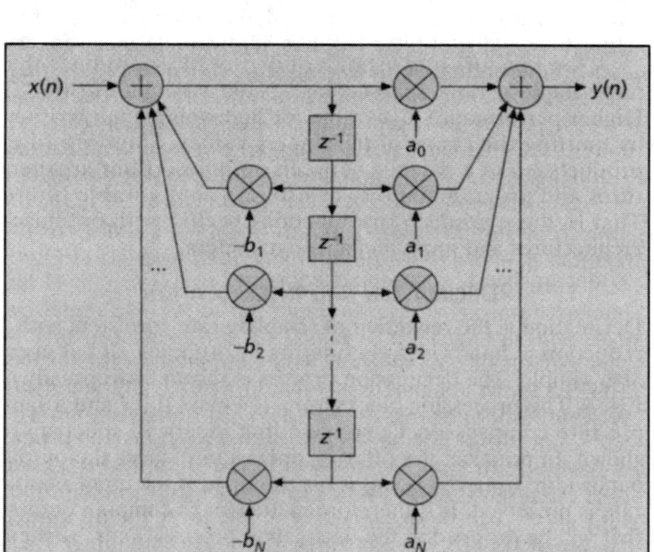

■ **Figure 2.** *Direct form 2 or canonical form architecture of a digital filter.*

ANALOG FILTERS, SOFTWARE-BASED AND HARD-WIRED DIGITAL FILTERS

Analog filters are inherently smaller and more power-efficient than their digital counterparts. Analog filters are more sensitive to component parameters than digital filters, however. As signal frequency increases, the disparity in efficiency increases. Digital filters are an effective alternative to analog filters when the performance required of the filter is beyond the physical limits of the analog devices, and the frequency range is within the limits of DSP implementations.

Digital filters are more size- and power-efficient than analog filters in applications requiring extremely linear phase, very high stop band attenuation, or very low pass band ripple. Also, if the filter's response must be programmable or adaptive, digital filters are more appropriate. A digital filter's shape factor is the ratio of the filter's pass bandwidth plus the filter's transition bandwidth to the filter's pass bandwidth. Digital filters also accommodate very low shape factors.

In the time domain, the operation of an RC low-pass filter is easily envisioned as a capacitor charging and discharging through a resistor. Likewise, a negative-feedback active filter shifts phase as a function of frequency, which is also a time-domain operation.

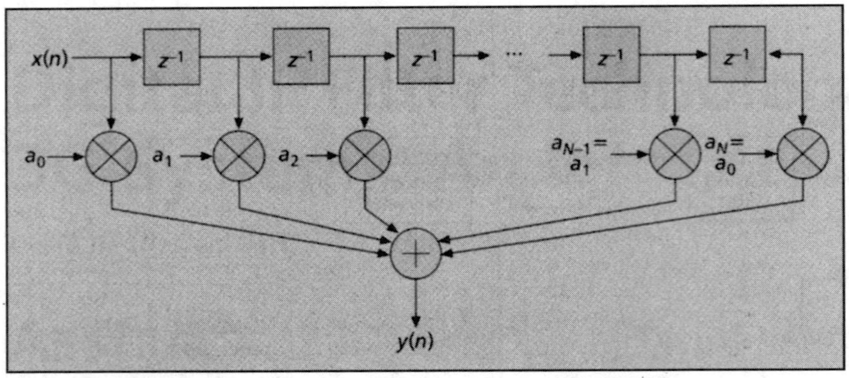

■ **Figure 3.** *The transversal implementation of a symmetric FIR filter.*

A digital filter also may be conceptualized in the frequency domain. The filter convolves the time domain impulse response with the sampled signal. A filter is designed to a frequency domain response which is as close to the ideal response as can be generated given implementation constraints such as number of filter taps or multiplication rate. The frequency domain impulse response is then transformed into a time domain impulse response, which is converted to the coefficients of the filter with a specified number of taps and multipliers.

The characteristics of an analog filter are directly attributable to the physics of the device. In contrast, the characteristics of a digital filter are only indirectly attributable to the physics of the device (e.g., through clock speed and chip area).

A digital filter's passband ripple, shape factor, stopband attenuation, and phase characteristics are functions of the order and type of polynomial used to approximate the ideal impulse response. The number of bits of arithmetic precision also determines the response because of implementation errors. Frequencies have meaning in a digital filter in their relation to the sampling frequency. This is because the impulse response is generated as a function of z^{-1}, the sample interval (time between samples). To achieve a smaller shape factor, the order of the filter (number of taps) and the arithmetic precision can be increased, subject only to the physical limits on the amount of arithmetic processing that can be integrated on a device given the filter order and sampling rate.

Digital Filter Types

The two basic types of digital filters are finite impulse response (FIR) and infinite impulse response (IIR). The form of the digital filter difference equation is

$$y(n) = \sum_{i=0}^{N} a_i \ x(n-i) - \sum_{i=1}^{N} b_i \ y(n-i)$$

where $y(n)$ is the current filter output, $y(n-i)$ is the previous filter output, $x(n-i)$ is the current or previous filter input, a_i is the filter's feedforward coefficients corresponding to the zeros of the filter, b_i is the filter's feedback coefficients corresponding to the poles of the filter, and N is the filter's order.

IIR filters have one or more nonzero feedback coefficients. As a result of the feedback term the filter has one or more poles, and once the filter has been excited with an impulse there is always an output. FIR filters have no nonzero feedback coefficient. That is, the filter has only zeros, and once it has been excited with an impulse the output is present for only a finite, $N + 1$, number of computational cycles.

Because an IIR filter uses both a feedforward polynomial (zeros as the roots) and a feedback polynomial (poles as the roots), it has a much sharper transition characteristic for a given filter order. Like analog filters with poles, an IIR filter has nonlinear phase characteristics. Also, the feedback loop makes IIR filters difficult to use in adaptive filter applications. After adaptation, IIR filter responses must be driven to nearly zero before the new filter configuration can be used. FIR filters may be used immediately since no energy is circulating in the filter. Software-controlled implementations may zero the delay memory to force IIR filters to a quiescent state for use.

Due to its all-zero structure, the FIR filter also has a linear phase response when the filter's coefficients are symmetric, as is the case in many filtering applications. This makes the FIR filter attractive for communications applications. In many cases, the price of increasing the filter order of an FIR filter over an IIR filter for a given impulse response is worth the linear phase characteristic. Figure 3 shows the feedforward section of Fig. 2 implementing a symmetric FIR filter.

An FIR filter's typical noise floor is $(-6.02\,B + 6.02 \log_2(N + 1))$ dB if no intermediate truncation is used. B is the number of bits in the filter's coefficient quantization arithmetic, and N is again the filter order. FIR filters with very narrow transition bands sometimes exhibit a slightly higher noise floor. This noise characteristic explains why some commercially available filter integrated circuits have more coefficient bits than data bits.

FIR filters guarantee stability. An IIR filter's poles may be close to or outside the unit circle in the Z plane. Such a filter is more likely to have stability problems, especially through quantization-induced errors. An FIR filter, on the other hand, is always stable because there is no feedback path. They lend themselves to computationally efficient architectures in decimating and interpolating applications which are used extensively in software radio applications.

While digital filters can be implemented in general-purpose DSPs, now commodity devices, these processors can implement moderately complex digital filters economically in sub-megahertz clock ranges. This is because the architectures of DSPs are usually optimized to perform a sum-of-products calculation with data from RAM or ROM. They are not optimized, for example, for filtering with 121 taps using 11-bit arithmetic precision. To get extended sampling rate performance from a digital filter requires hardware designed for the intended filter at the desired sampling frequencies.

A few vendors offer standard digital filter products that now sample at rates approaching 75–100 Msamples/s (MSPS). Higher performance is possible for high-volume applications by limiting the range of parameter values. Standard filter products strike a balance between optimized filter architectures and programmability by offering configurable filters. That is, these products are function-specific, with optimized architectures and programmable parameters.

Decimation and Interpolation

Decimation is the reduction of sampling rate consistent with a reduction in bandwidth via filtering by dropping all but every Mth sample. The decimation process is shown conceptually in Fig. 4. The processing blocks are a low-pass filter and a sample rate compressor. Corresponding idealized spectra are shown. In practice, the filtering operation reduces the output bandwidth; thus, the signal is oversampled if the input sample rate is preserved. It is therefore inefficient to compute outputs that will be dropped in the decimation process. Thus, efficient FIR implementations compute only the desired output samples. This yields a one-to-one gain in computational efficiency for each increase in decimation rate. This potential computational efficiency cannot be fully realized in an IIR filter because the feedback path must be computed for every input cycle.

Interpolation is the process of increasing the sampling rate while preserving the signal's spectral content. It is most often used when a narrowband signal will be combined with a signal of wider bandwidth or different carrier frequency. The first step in interpolation is to insert $L - 1$ zero-valued samples between each valid input sample, expanding the sampling rate by L. This causes the original signal spectrum to be repeated $L - 1$ times. This process is referred to as *sample rate expansion*. To complete the interpolation, the zero-valued input samples are converted to spectrally accurate approximations of the narrowband signal. This is equivalent to preserving the original signal spectrum. The zero-stuffed input stream may be filtered by a low-pass filter with a passband at the original spectrum location. This filters out all of the repeated spectra. This process is shown in Fig. 5.

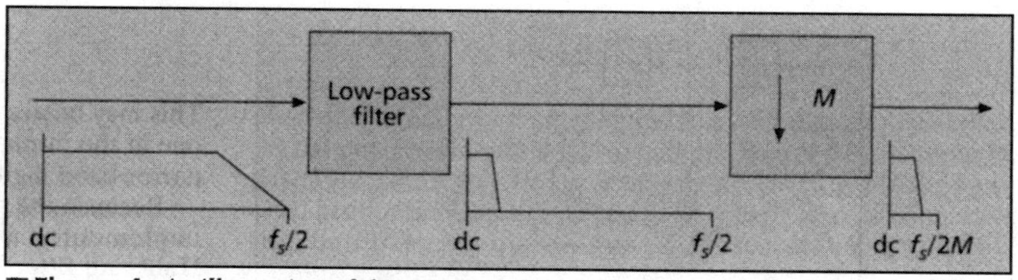

■ **Figure 4.** *An illustration of the conceptual decimation process.*

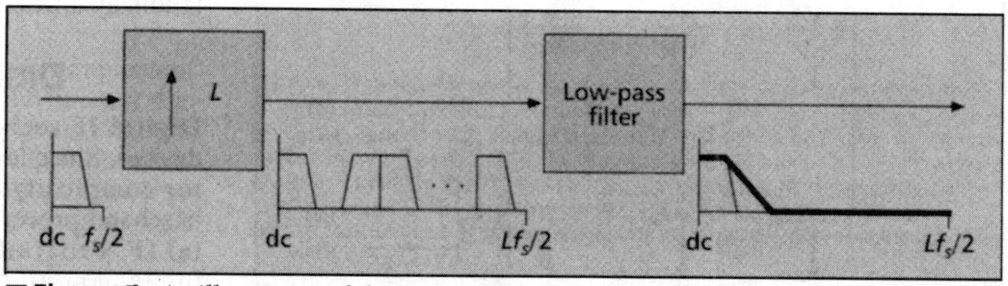

■ **Figure 5.** *An illustration of the conceptual interpolation process.*

In FIR filters, interpolation computational efficiency is possible because only every Lth data value is nonzero, requiring a multiply-add operation to compute an output sample. This approach also yields a one-to-one correspondence between interpolation rate and gain in computational efficiency. Again, the feedback loop in an IIR filter precludes such gains in computational efficiency. Interpolation filters are often implemented as polyphase structures [3].

Digital IF Implementations

This discussion addresses digital IF implementations in receivers, that is, digital downconverters (DDCs). DDCs translate signals from IF to baseband with low-pass decimation filtering. Digital upconversion is the network transpose of digital downconversion (i.e., interpolation filtering and frequency translation from baseband to IF).

Narrowband Single-Channel Digital IF

Narrowband single-channel digital IF is used in software radios implementations where a small number of signals of interest are being extracted from a wideband input. Narrowband digital IF is the architecture of choice in applications where:

- A large dynamic range is desired.
- Accurate selectivity is required.
- There are multiple signals of interest with differing bandwidths.

Narrowband digital IF requires a high degree of decimation. The classical implementation in this case is to sample a real IF signal and to mix it digitally with a quadrature local oscillator (LO) set to the signal of interest's center frequency. This process moves the center of the signal of interest to dc. Contemporary digital LOs in single-chip DDCs can easily achieve 100 dB of spurious free dynamic range (SFDR) with maximum center frequencies approaching 50 MHz. Another advantage of digital LOs is that their center frequencies and phase offsets can be changed to anywhere in the frequency range within one clock cycle. Frequency resolutions for a 100 MSPS LO is typically either 15 Hz for a 16-bit frequency word or 0.0002 Hz for a 32-bit frequency word. A detailed explanation of the performance of digital LOs is found in [7, 8].

The complex signal which results from mixing the real IF signal with the quadrature LO is then low-pass filtered and decimated to yield a narrowband quadrature baseband signal with a baseband sampling rate commensurate with its bandwidth. This is accomplished with two identical real filter chains, one in the in-phase path and one in the quadrature path.

High decimation filtering is most efficient using multiple stages [3]. The first stage decimates at a high rate. The subsequent stages shape the band with lower decimation rates. The aggregate decimation is the product of the decimation rate of each stage.

The architecture of choice for the high decimation stage in a narrowband digital IF is the Hogenauer filter [3, 9]. The Hogenauer filter is most efficient for decimation rates above 10. The author has contributed to the development of Hogenauer filters with programmable decimation rates on the order of 100,000.

The Hogenauer filter is usually followed by one or two filter stages that compensate for the passband droop induced by the Hogenauer filter and then do final shaping. In a two-stage implementation, these functions are combined into one FIR filter. In a three-stage implementation, the functions are separated. Most current implementations use the two-stage approach. However, recent research indicates that in many applications the three-stage approach can offer implementation efficiency [10]. Software radio base stations may employ parallel single-channel digital IF chips to recover a large number of subscriber signals. Handset size, weight, and power constraints generally drive one toward direct conversion in lieu of a single-channel digital IF.

Narrowband Multichannel Digital IF

Narrowband multichannel digital IF is most commonly used where multiple channels of the same bandwidth and equal spacing are of interest. One of the most common examples of this is the cellular base station. A common digital IF in these applications is a transmultiplexer (tmux) based on the discrete Fourier transform (DFT) filter bank [3]. The tmux receiver translates one wideband IF sampled signal stream into multiple quadrature baseband streams, one per subscriber channel. Emerging tmux architectures may be computationally more efficient [11].

Using single-channel filters for the multichannel application, M digital LOs would be set at equally spaced frequencies for M dual real mixers followed by M identical dual real decimating filters. A tmux DFT replaces the LOs and mixers. The

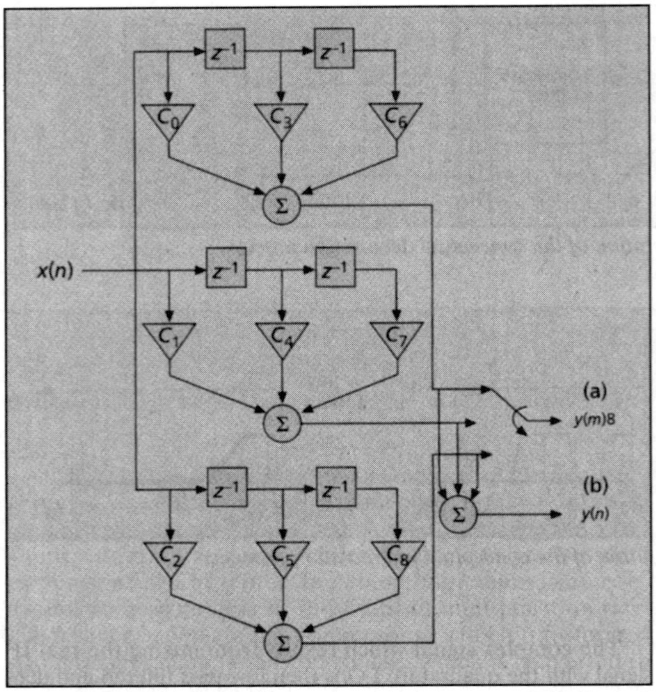

■ **Figure 6.** *Polyphase implementation of a nine-tap FIR filter: a) in an interpolate by three configuration; b) in a non-rate-change configuration (proper delays between phases not shown).*

tmux often relies on a fast Fourier transform (FFT) implementation of the DFT. The FFT acts as M parallel quadrature LOs and mixers operating on the same input data. The decimating filters may then be implemented as one polyphase structure [3] with M phases at the input of the DFT. The computational demand of the tmux is much less than that of M narrowband downconverters. Numerous tmux patents are based on the basic architecture found in [3].

A polyphase (order N with k phases) FIR filter reconfigures the standard $N + 1$-tap filter structure into k phases of an N-element delay line. An eighth-order FIR filter with $k = 3$ phases has three sets of three coefficients. The coefficients for the eighth-order filter are $C_0, C_1, \ldots, C_8$. The first polyphase taps the delay line at the input to the filter (C_0), after the third delay (C_3), and after the sixth delay (C_6). The second phase taps C_1, C_4, and C_7. The third phase taps C_2, C_5, and C_8. The output of the original filter is the sum of the three phases for a noninterpolating filter. The output is a commutated output of the three phases for an interpolating filter. Figure 6 illustrates the polyphase implementation of the nine-tap filter. Phase combining in the tmux is inherent in the DFT.

Wideband Single-Channel Digital IF

Wideband single-channel digital IF is applicable where the ratio of the bandwidth between the IF and baseband is less than 10:1 and the advantages of digital IF are desirable. The advantages include:

- A large dynamic range
- Accurate frequency selection
- Accurate filter characteristics
- Flexibility in the capture of multiple signals with different bandwidths

Wideband digital IF entails decimation rates of less than 10:1. The classical implementation translates a real IF signal to baseband by mixing it digitally with a quadrature LO at the signal of interest's center frequency, moving the center frequency to dc. The resulting complex signal is low-pass filtered and decimated to yield a quadrature baseband signal using a baseband sampling rate commensurate with its bandwidth. This may be accomplished with two identical real filter chains, one in the in-phase path and one in the quadrature path, as in narrowband digital IF.

Because the required decimation is low, the filters may be implemented as one- or two-stage decimating FIR filters. Halfband filters [3] convert to baseband quadrature without much bandwidth reduction. These filters have frequency domain symmetries that make their implementations computationally efficient.

Digital IF Implementations

Digital IF technology is implemented in semiconductor devices. Since levels of integration, data rates, and demodulator complexity are constantly increasing, more and more baseband processing functions are being integrated with digital IF. Programmable DSPs are not always adequate to implement all baseband processing functions. Dedicated function DSP blocks may perform required algorithms that cannot be cost-effectively implemented in DSP software. As a result, many of the recent commercial digital IF devices integrate baseband functions.

Commercial Digital IF Products

The two major suppliers of commercial digital IF products are Gray Chip (www.graychip.com) and Harris Semiconductor (www.harris.com). These two companies pioneered the digital IF standard products business. Other companies are beginning to explore the market.

Gray Chip's products include digital downconverters, upconverters, digital filters, and high-speed quadrature amplitude modulation (QAM) modem chip sets. Their user-friendly Web site lists the GC4014 62.5 MSPS Quad Digital Receiver, a four-channel wideband-narrowband digital IF chip. The GC4114 62.5 MSPS Quad Digital Transmitter supplies the companion upconversion process. The GC1011A 70 MSPS Narrowband Receiver and the GC1012A 80 MSPS Wideband Receiver are designed for second- and third-generation cellular base stations, respectively. The GC3021 80 MSPS Mixer/Carrier Remover has limited filtering. The CG2011 80 MSPS Digital Filter and the GC3011 80 MSPS Digital Resampler are general-purpose DSP parts.

Harris Semiconductor's standard digital IF products are reconfigurable, function-specific signal processing blocks. Examples include the HSP50016 Digital Down Converter, a narrowband digital IF chip. The HSP50110 Digital Quadrature Tuner is a precision digital mixer. The HSP50210 Digital Costas Loop provides stable, high-speed carrier recovery. The HSP50214 Programmable Down Converter has decimation rates capable of narrowband and wideband operation. The HSP50215 Digital Up Converter is the companion part for transmission. The HSP43168 Dual FIR Filter operates standalone or with a digital up- or downconverter. The HSP43216 Halfband Filter is optimized for symmetric wideband conversion. The HSP43220 Decimating Digital Filter and the HSP45116 Numerically Controlled Oscillator/Modulator were the initial product family. Digital IF products soon to be offered from Harris include a wideband multimode modulator, a high-speed digital tuner, a quad programmable digital downconverter, and a family of direct digital synthesizers. Harris is also moving toward more application-specific designs, including designs with ADCs and DACs on board for high-volume applications.

The Gray Chip and Harris products are representative of the diversity of productization for the DSP, digital radio, and emerging software radio systems, but other manufacturers offer similar products.

Application-Specific Implementations

Software radio standard products are reconfigurable function-specific building blocks. While providing viable solutions to specific applications in moderate quantities, they may not be cost effective in high volumes. This is where application-specific architectures play a role. At first glance, there are two viable candidate technologies for such applications: field programmable gate arrays (FPGAs) and custom or semi-custom application-specific integrated circuits (ASICs).

FPGAs allow end equipment providers to easily implement their own custom designs in IC technology. Unfortunately, the gate count for a given function implemented in an FPGA is often 10 times the gate count for the same function in a custom ASIC. Therefore, FPGAs in software radios appear in prototypes and in low-gate-count functions such as glue-logic interfaces between building blocks.

ASICs provide the economies of scale for high-volume commercial software radios. Many ASIC vendors have the digital design capability required for high-gate-count designs. However, software radio ASICs do not just entail digital design problems. More important, they present communication system and DSP architecture design challenges. Only a few ASIC vendors have digital design, communications system, and DSP firmware capabilities. Thus, "soft core" intellectual property is emerging in the semiconductor industry to cope with the related ASIC integration challenges. Soft cores are circuit designs in high-level languages that are transferable between foundry processes. Cores can be interconnected to form complex systems on a chip. Soft DSP, microprocessor, and dedicated function (e.g., filter) cores are becoming increasingly available. Thus, it is becoming easier for organizations with communications systems expertise to develop highly integrated software radio ASICs with reasonable levels of effort. This should accelerate the growth of software radio technology.

Conclusion

Digital IF technology provides many advantages over conventional analog implementations. Flexibility to handle multiple signal types, a high degree of signal processing accuracy, and the ability to efficiently process a large number of similar signals is key. These advantages may meet essential requirements in 3G base stations, for example. Success in digital IF requires communications systems engineering, novel DSP architecture, multirate signal processing, digital local oscillators, and other digital design disciplines. Fortunately, digital IF technology has matured, largely because of commercially available standard products. With higher levels of integration, the increasing viability of multivendor ASIC technology, and the increasing availability of soft core standard products, digital IF is becoming a mainstream technology in software radio base station applications.

References

[1] A. V. Oppenheim and R. W. Schafer, *Discrete-Time Signal Processing*, Englewood Cliffs, NJ: Prentice Hall, 1989.

[2] L. R. Rabiner and B. Gold, *Theory and Application of Digital Signal Processing*, Englewood Cliffs, NJ: Prentice Hall, 1975.

[3] R. E. Crochiere and L. R. Rabiner, *Multirate Digital Signal Processing*, Englewood Cliffs, NJ: Prentice Hall, 1983.

[4] S. K. Mitra and J. F. Kaiser, *Handbook For Digital Signal Processing*, New York: Wiley, 1993.

[5] D. B Chester, D. H. Damerow, and C. Olmstead, "Analog to Digital Converter Requirements and Implementations for Narrow band Channelization Applications," *Proc. ICASSP '92*, vol. 4, San Francisco, CA, Mar. 1992.

[6] D. B. Chester and G. Phillips, "Single Chip Digital Down Converter Simplifies RF DSP Applications," *RF Des.*, Nov. 1992.

[7] F. A. B. Cercas, M. Tomlinson, and A. A. Albuquerque, "Designing With Direct Digital Frequency Synthesizers," *Proc. RF Expo East*, Orlando, FLNov. 13–15, 1990, pp. 625–34.

[8] D. B. Chester and G. Phillips, "DSP Chips for IF Processing, Electronic Products," Apr. 1994.

[9] E. B. Hogenauer, "An Economical Class of Digital Filters for Decimation and Interpolation," *IEEE Trans. ASSP*, vol. ASSP-29, no. 2, Apr. 1981, pp. 155–62.

[10] H. J. Oh *et al.*, "On the Use of Interpolated Second Order Polynomials for Efficient Filter Design in Programmable Downconversion," *IEEE JSAC*, 1998.

[11] Zangi and Koilpillai, "Software Radio Issues in Cellular Base Stations," *IEEE JSAC*, Feb. 1998.

The DSP Bottleneck

The stumbling blocks in development of a commercially usable software radio have been the availability of hardware, in particular the fast DSP required. As high performance ADCs have become available commercially, with the sample rates and SFDRs required, hybrid techniques using specialized digital hardware, operating under software control, have become more common.

Rupert Baines

RUPERT BAINES works in the Communications Division of Analog Devices.

The attractions of the software radio — flexibility and ease of adaptation — are dependent on the manner in which not only coefficients of a filter, but virtually any aspect of a program that implements radio functions, can so easily be changed. This is in contrast with their conventional hard-wired equivalents. Of course, this software requires hardware (in the form of dedicated or programmable processors) to realize these advantages. And it has been here that until very recently, ideas have outpaced the available technology: somewhat paradoxically, the constraint on the software radio has been the hardware. But suddenly that is changing: a new generation of devices is becoming available, processing power is escalating, and implementations are fast enough to be realistically usable.

For simplicity, this discussion will concentrate on the receive side of the radio. The techniques are just as applicable on the transmitter side, but then the challenges lie in subtly different areas. Rather than the DSP-intensive task of trying to recover corrupted/distorted signals from noise, the designer has the somewhat easier synthesis task, but then has the tricky problem of providing a broadband linear power amplifier if the software radio is to be heard by anyone.

The software radio approach employing wideband A/D converters, DSP and general purpose CPUs is more expensive than the traditional hardware intensive technology. For a single channel, software radios now cost significantly more than dedicated hardware. Hard-wired receivers are cheap and perfectly adequate for commodity applications such as home tuners, cellular handsets and the vast majority of radio systems. However, the two advantages of the software approach are flexibility and concentration, and for some systems these are irresistible. Flexibility — being able to switch channels, change modulation schemes or receive different types of signals — has always been valuable to the military, and explains their involvement in much of the early development of technology. But now flexibility has an added attraction for radio designers faced with a plethora of new standards (e.g., different cellular protocols) or varying conditions. For example, PCS base stations can adapt to new protocols by downloading software, or can optimize performance by dynamically assigning channels and capacity between several sites. An optimal receiver realization does not require matched filter hardware, which would be specific to an air interface/protocol, but rather a simple analog (generic) prefilter and sufficiently high sample rate. That rate need not be specific, either, (i.e., it need not be an exact multiple of a symbol rate as long as it is sufficiently high, and sufficient processing power exists) so a single sample rate can be used for channels of differing bandwidths and symbol rates.

The second benefit comes from concentration: multiple channels share the front end radio stage, while having cheap digital processing for each. While the software radio is more expensive than a single traditional receiver, the cost per channel drops rapidly. (At present, the breakeven point lies between four and ten channels.) This is extremely attractive in cellular and PCS applications, where a typical base station will contain 20 or more radio receivers (Fig. 1a and b).

DSP Requirements for Contemporary Applications

In narrowband digital radios, baseband signals are typically digitized and processed with a channel bandwidth of tens or hundreds of kHz (30kHz for AMPS or IS54/IS136, 200kHz for GSM, 1.25 MHz for CDMA, etc.). Figure 2 shows the block diagram of such a narrowband receiver, as used in digital cellular. The radio stages are analog, tuning into a particular frequency slot and using conventional filters to exclude other signals. The software radio requirement is to digitize the entire band (perhaps 25MHz) and to perform IF processing, baseband, bitstream and other functions in software. In the United States, operators have a 12.5MHz band, suggesting a 30MSPS rate minimum (2.5x to allow for real filters, etc.). However, over-sampling this signal is useful to

Reprinted from *IEEE Communications Magazine,* Vol. 33, No. 5, pp. 46–54, May 1995.

shift aliases out of band and simplify filtering, so faster sampling rates, or narrower bandwidths, are used. For example the AirNet system operates on a 5MHz band, requiring a rate of at least 12.5MSPS, which increases to more than 25 MSPS with oversampling. Several currently commercially available U.S. systems employ 30.72MSPS x 12 bits per sample, corresponding to a convenient map of one FFT slot to a frequency bin in an FFT which the system uses to manage the environment. (See related textbox on this page.)

The DSP must be fast enough to perform isochronous operations on this rapid flow of data. As discussed in the companion article, this results in IF processing requirements on the order of 500 MIPS/MFLOPS to upwards of 10 GFLOPS, baseband processing on the order of 10 to 100 MIPS/MFLOPS per channel, and additional processing burdens for bitstream, overhead, and control that put extreme pressure on the DSP architecture.

Tackling the Front End: RF/IF Functions

A modern programmable DSP typically offers up to 200 MIPS or 50 MFLOPS. (Table 1 shows some commercially available DSPs.) As even the most rudimentary demodulation or tuning requires 10 operations per sample, this would limit it to filtering signals with a carrier of a few hundred kHz. (All modern DSPs will do a single-cycle multiply-accumulate with simultaneous internal transfers, so a FIR filter requires just one cycle per tap.) Obviously, more complex modulation, analysis, estimation, and processing require more DSP horsepower, so any practical system will be more seriously limited. In addition, in reality, data fetches and architectural constraints will also play a significant role in determining the effectiveness of those MIPS, but it is clear that they are limited to operations on fairly low bandwidths. A good FIR/IIR channel selection filter could require about 100 operations per sample at 30 Msa/s, or 3000 MIPS. Using a naive brute-force approach, we would require 15 to 60 DSPs cooperating for this section alone, repeated for every channel. As a result, even with faster devices, software on DSPs still cannot be used for the down conversion itself, but must still essentially operate at baseband (albeit a much wider baseband up to a few MHz). Does this mean we are forced to abandon our aim? Not at all; we merely abandon general purpose hardware at IF. Instead, dedicated but highly programmable filtering hardware is used at the earliest stages, doing much of the filtering and processing in fast digital logic, reducing the processing load that must be done to a level that can be handled in software. As long as this specialized hardware is versatile and is controllable from software, this hybrid architecture still meets our requirements!

One of the most common implementations of such a hybrid architecture is a time-domain approach, using a digital down-converter, such as the Harris HSP50016. This chip contains a synthesizer, a quadrature pair of digital multipliers (which act as mixers in the digital domain), and some clever filters that implement both low-pass and decimation (Haugenauer filters). The device extracts a narrow band baseband signal from a wideband digital input, and decimates it to a reduced data rate, proportional to the bandwidth of interest, that a DSP can cope with. Figure 3a contains a block diagram of the DDC, while Fig. 3b shows the fre-

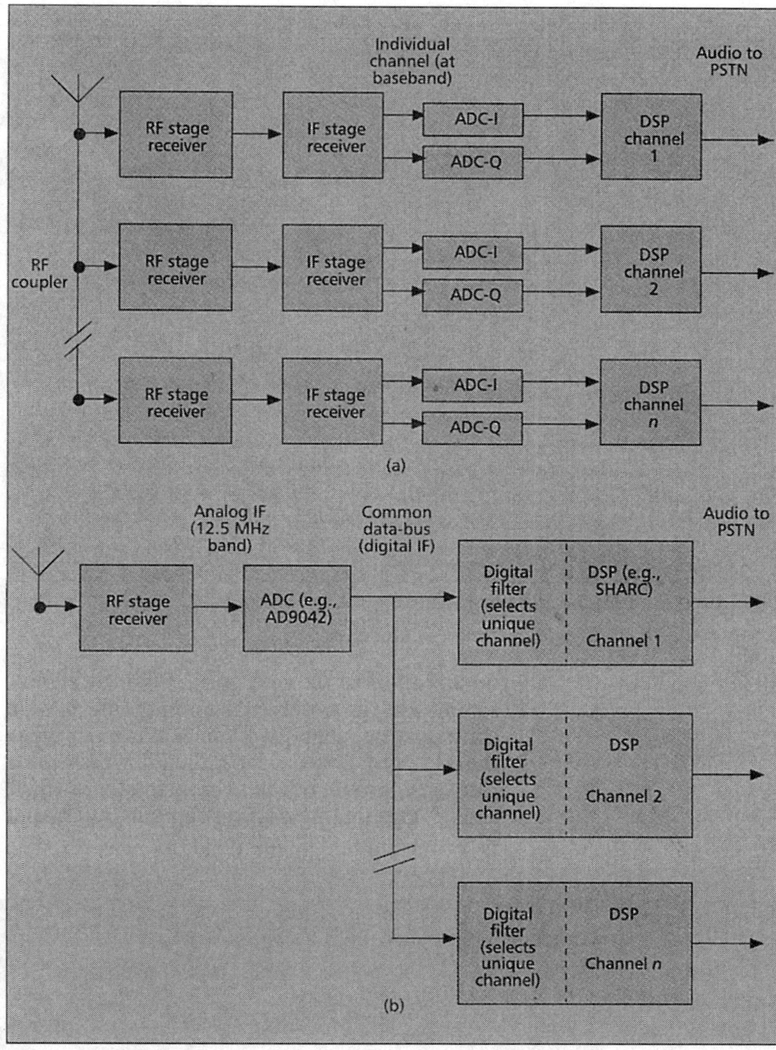

■ **Figure 1.** *Where there are many receivers in one place (e.g., cellular basestations) the wideband approach has attractions of concentration and economy.*

"Magic Numbers" Rapidly Emerging for Cellular and PCS Applications

Applications need anywhere from 5-20 MHz of spectrum with SFDR of at least 80 dB.

GSM is probably the most demanding: blocking requirements include a full signal at –13 dBm, with a simultaneous weak one at –104 dBm; or requirement to distinguish adjacent channels, both at the same strength of –43dB. This > 90 dB dynamic range is still very tough to achieve and few commercial systems are up to it.

However, the U.S. standards are easier. The European PCN system (DCS1800) and its American sibling (TAG5/PCS1900), although a GSM variant, has a 10 dB less demanding radio specification; several manufacturers have systems that should meet the requirements.

Other common (minimum) sample rates are:
- 30.72 MSPS (U.S. analog cellular — 1024 times 30 kHz channel for FFT).
- 38.88 MSPS (IS-54/IS136 digital cellular).
- 40.96 MSPS (military receivers).
- 39 MSPS (GSM, from 13 MHz system clock).
- 50 MSPS (4x over-sampling of a cellular 12.5 MHz band).

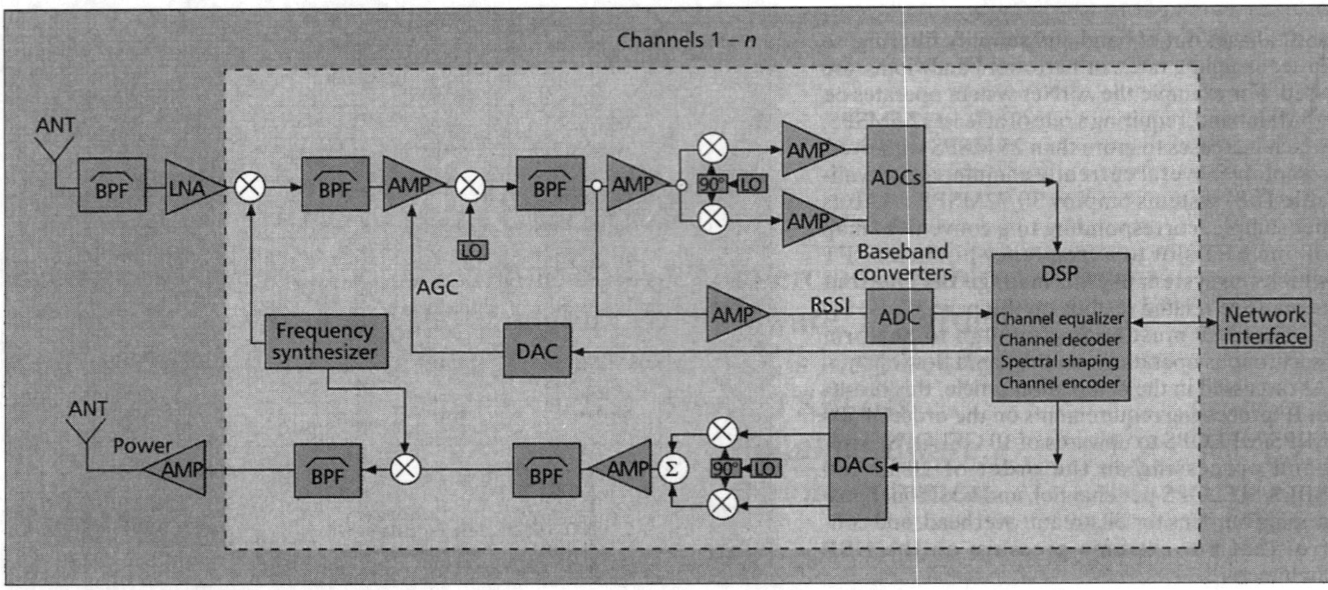

Figure 2. *A traditional narrowband receiver.*

quency plots of the operation. First the signal is sampled at the wideband sampling rate f_s, which shifts it in frequency by f_c (since it is a real signal, the spectrum is two sided); then it is quadrature filtered to extract the band of interest; and finally, the programmable decimator changes the data rate to f_n — giving a slower (more manageable) data stream containing only the desired information. The DDC has a maximum input rate of 52 MSPS, and can operate with 16 b words to give spurious free dynamic range (SFDR) of more than 100 dB. Depending on the clock rates and

Part	Manufacturer	Speed and clock	On-chip memory	Data width	Dedicated I/O	Price*	Comments
ADSP21062 SHARC	ADI	40 MFLOP 120 MIPS 40 MHZ	32 K words RAM	32 b float	Fast I/O port 240 Mbyte/s Six link ports 40 Mbyte/s each Two 40 MHz bidirectional serial ports	$120	Second-generation multiprocessing. Memory and I/O allow consistently peak performance without bottlenecks.
TMS320C32	TI	40 MFLOP 20 MIPS 40 MHz	512 words RAM	32 b float	Two 10 MHz serial ports	$25	Lowest cost; broad commercial base.
TMS 320C40	TI	50 MFLOP 25 MIPS 50 MHz	2 K words RAM	32 b fixed	Six link ports (20 Mbytes/s)	$150	First-generation multiprocessing.
TMS 320C80	TI	Theoretical peak 2 GOPS	12 K words	32 b	Crossbar I/O 400 Mbytes/s	$400	On-chip MIMD multiprocessing (4 DSPS); very powerful. Integrated RISC master processor.
DSP96002	Motorola	60 MFLOP 20 MIPS 40 MHz	2 K words program, 2 * 512 words data RAM	32 b float	None specific (two external bus ports, used for data and program as well as I/O)	$100	Multiprocessing interconnect limited; determind by external circuits.
DSP32C	AT&T	25 MFLOP 12.5 MIPS 50 MHz	1.5 K words RAM	32 b	Serial port, general-purpose buses		

*Approximate price in volume. This chart only discusses "conventional" DSPs, and even then is guilty of mixing apples and oranges. There are other extremely powerful processors that some would argue should be included. The Intel i860 is a fast numeric engine, while the newest DEC Alpha, the 21164, is the "fastest microprocessor on the planet" (> 450 Floating point SPECmarks with a 300 MHz clock). However, these devices are intended for the multifaceted world of general-purpose computing. While some can compete with DSPs they are not intended nor optimized for the single-minded I/O intensive operations that the devices listed above excel at.

Table 1. *Commercial DSPs.*

control parameters, the DDC can be configured in a variety of ways (i.e., real or quadrature outputs) and for a variety of frequency/bandwidth combinations. Given a programmable clock of some description, it is essentially under software control.

Figure 4 shows how the DDC reduces the gigabit per second data rate from the IF A/D converter to digital baseband data rates compatible with follow-on DSPs. The RF conversion stage contains two mixers and associated filters necessary to create a well-filtered wideband IF. The first analog filter rejects out of band components that are to be completely rejected, before the signal is passed to the first mixer. This operates to select the access band of interest and to reject out of band interference. Another mixer then moves this to the second IF. Normally, this would finally be mixed down to baseband, where it would be sampled — perhaps at 25 MSPS for a 10 MHz bandwidth. (A suitable candidate would be the AD9042 — a cost-effective 12-bit ADC that offers 80 dB SFDR and 41 MSPS rate). An attractive alternative to baseband sampling is IF sampling — digitizing the signal in the second or third Nyquist zone, and the desired signals will be aliassed inband by the sampling. This requires a wider analog input bandwidth on the ADC but greatly eases the requirements on the amplifier and filter, as the harmonics will be generated far from the pass-band.

This can be combined with oversampling, which significantly improves the SNR; every time the sampling rate is doubled, the effective noise floor drops by 3dB, as the total noise is spread over more frequencies. When it comes to channelise to narrow band the gain is dramatic — a 30 kHz channel sampled at 40.96 MSPS has a processing gain of some 28 dB!

The programmable clock input rate of the DDC selects a given channel; the filters reject other components (once more, the coefficients are controllable); and the decimator gives an output at a rate usable to follow-on DSPs. A number of DDCs can be used in parallel, with offset clocks, to select a number of channels from the wideband system.

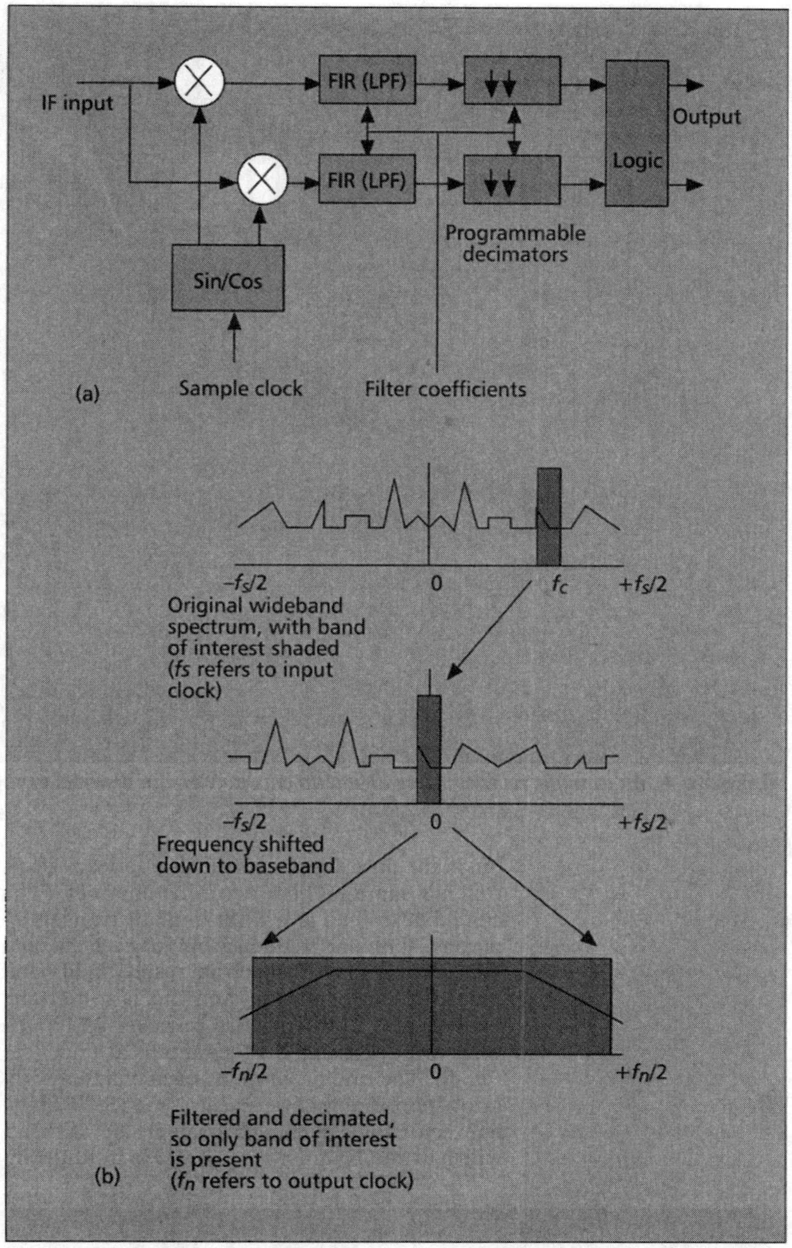

Figure 3. *a) Block diagram of the Harris HSP50016 DDC; b) operation of digital downconverter.*

Spectrum Monitoring Using DSPs Alone

Purely software-based spectrum monitoring has been implemented using the fast Fourier transform (FFT). The FFT extracts frequency domain information from a series of time domain samples, to resolve the signal into a set of frequency bins. With no oversampling, each frequency bin corresponds to a channel. The FFT can then be used to compute a power spectrum that reveals the signal level of that channel during the period of data acquisition (the FFT epoch or window). A series of such FFTs then creates a sliding window filter bank in the time domain. In practical systems, the power per channel yields key information for well managed channel assignments. Such an FFT is generated in newer PCS base stations in dense urban areas, for dynamic spectrum allocation. The base station selects which channels to use so as to optimize efficiency and minimize noise. In military and emerging PCS applications, such spectrum monitoring may be used to help the base station determine which protocols are being used, and how best to configure its channels and operations among the different systems. A completely nonstructured system is not required; for most commercial applications the receiver will know which type of signal it wants to receive (or rather, which selection of types). A given digital standard will have predictable channel bandwidth, modulation scheme, spectral density and so on, and most include training sequences and preambles to provide even more help in the equalization and recovery. However, a base station will need to perform an overview to decide which signals it is receiving, and which protocols they are using, checking bandwidth and spectral properties so that it can select the appropriate software to tune and recover that signal from all the others.

The time domain trace of a single frequency

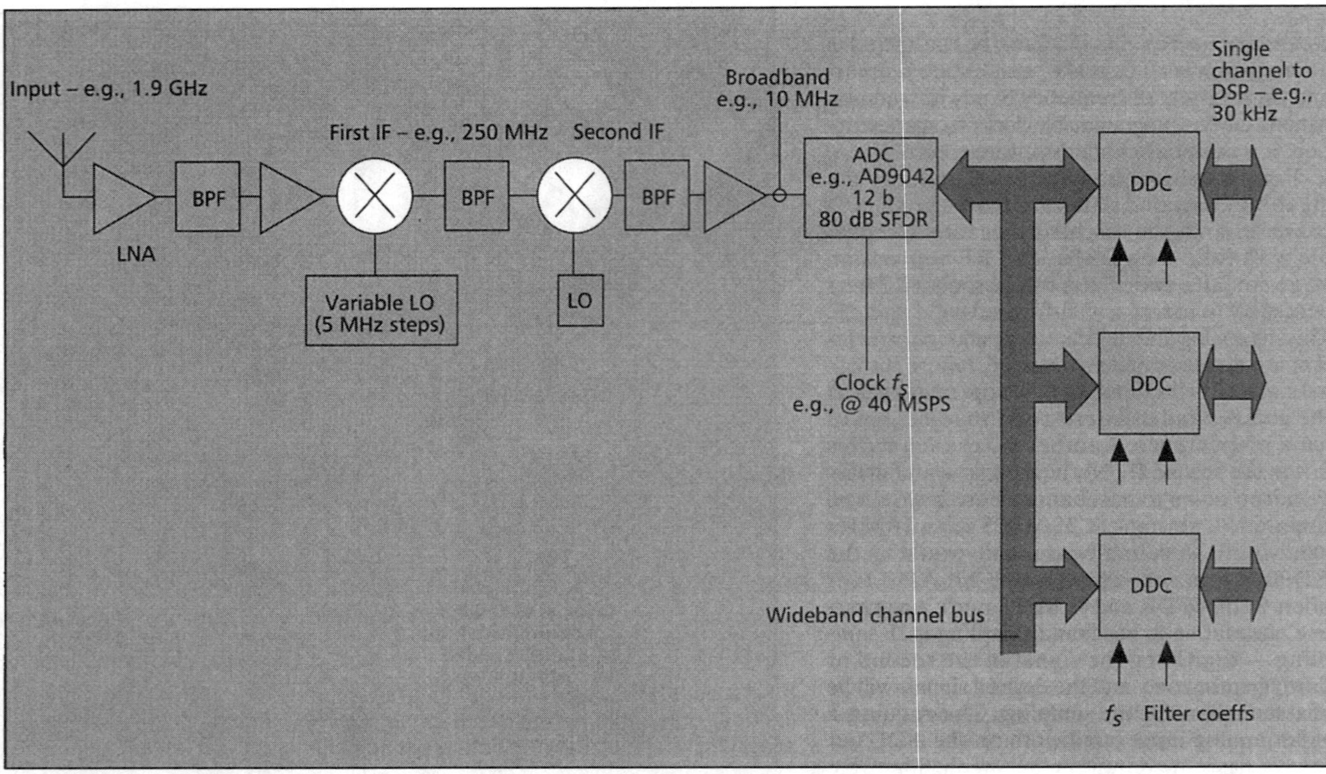

■ **Figure 4.** *An example receiver using digital downconverters for channel tuning and selection.*

bin in the power spectrum yields a series of levels that is a sampled history of that channel. (This can be described as a filter bank or transmultiplexer). With one frequency bin for each channel, the successive power spectrum results yield signal amplitude information, and this is sometimes used as a straightforward parallel AM radio receiver, operating on *N* channels at once (see Fig. 5). The rate of sampling each channel is the rate of performing power spectra, so each FFT must be executed quickly enough to sample the AM bandwidth at the Nyquist rate. A 4kHz modulating bandwidth would require at least 8kHz sampling — or a FFT/power spectrum at 1.25ms intervals. The SHARC performs a 1024 point complex FFT in just 0.46 ms, so that the power spectrum is computed in less than the required 1.25 ms. Rather than using a series of filters or downconverters, one per channel, the FFT performs the required operations in parallel. The parameters of the FFT can be easily changed, reconfiguring the receiver. Other modulation schemes can be recovered in a similar way. For example, an elementary FSK (frequency shift keying) receiver can be implemented

Function / Radio class	RF receiver	IF stage, tuning	Processing 1, baseband	Processing 2 source	Example
Classical (commodity)	Analog-LNA, first.	Analog, IF mixers, PLLs, mixer, etc.	Analog hardware filters.	Analog hardware (depends on modulation).	Analog cellular.
Digital	"	"	Baseband digitization. Digital equalization (carrier, timing recovery) and channel decoding.	Speech decoding, error correction done in DSP.	Current and future digital cellular, PCS handsets (e.g., GSM, IS95). Current digital cellular base stations.
Hybrid	"	IF sampling or wideband ADC.	Digital hardware downconverters, Haugenauer filters.	"	Next-generation commerical basestations (for both digital and analog systems).
Software	"	"	Software-based. Filterbank, Costas loop, integrate and dump, etc. Move to non-structured design.	"	Research. Military. Emerging commercial

■ **Table 2.** *A comparison of different radio architectures.*

directly with two frequency bins per channel, one for the low tone and one for the higher. This is a highly simplified example. In practical implementations, bit synchronization, Costas loops, integrate and dump filters, etc., increase the performance requirements of the DSP. Sophisticated algorithms to transform the signal to a more usable form often increase the DSP requirements to multiple processors. As the versatility increases, the amount of processing power required is not insignificant. And this leads us inevitably to consider multiprocessing.

The DSP Bottleneck: Multiprocessing Then and Now

Software radios invariably require the use of parallel and sequential partitioning of algorithms to get required processing power. DSPs are getting ever faster, but it will be a while before we can use a single "ultimate chip" to do everything. Instead, the idea of using multiprocessing to share the effort seems attractive. Traditionally, real-time multiprocessing with DSPs has been fiendishly tricky. Fortunately, recent products have addressed the issue, and it is becoming much simpler. One approach is to integrate a general purpose CPU and a dedicated DSP core on the same die. Motorola, for example, has released a "Communications Processor," the 68356, which includes a general purpose micro-controller with a 56000 series DSP. For those applications that required both general purpose operations and a DSP, this is an attractive approach. For IF and baseband processing, however, which require significant DSP with minimal supervisory functions, it is less relevant. A different approach has been taken by AT&T, who integrated a dedicated Viterbi decoder as co-processor on some of their 16xx series DSPs. This drastically eases the processing load for systems requiring an equalizer (e.g., GSM handsets). It is very likely that more DSPs will

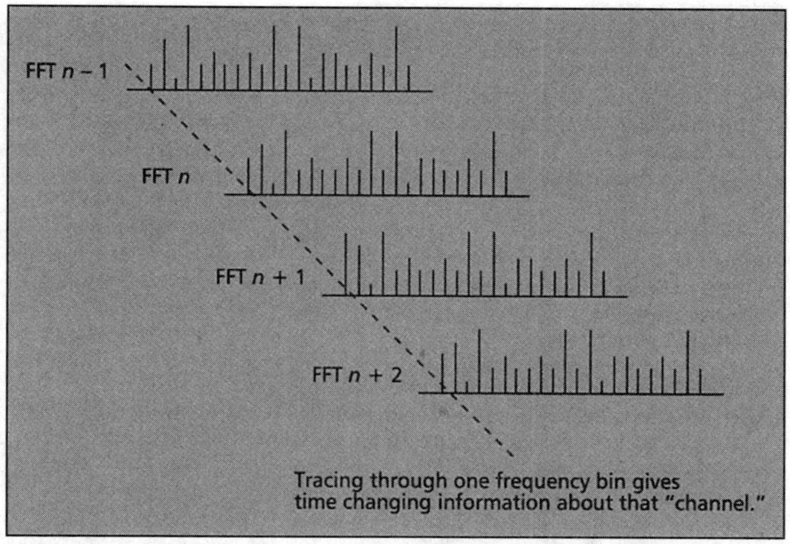

■ **Figure 5.** *A filter bank (such as FFT implementation) can be a way of tuning many channels in parallel.*

support such features in the future. (See text box on next page.)

If the system is to operate in real time (which is usually the requirement), then the data must be able to get in and out of the DSP, which can pose I/O bottleneck problems. For fast data access, fast memory is required, but this is expensive. If several processors are to be used, there must be efficient methods of linking them and coordinating their actions. Traditional DSP architectures were not well-suited to multiprocessing. It is a characteristic of a DSP (as contrasted to a conventional microprocessor) that it must operate on a continuous flow of data. A typical system might require the equivalent of four bus operation each cycle, with an instruction read, two operand fetches (e.g., latest sample of data from the ADC, and filter coefficient), and an I/O transfer (e.g., output of result). Unfortunately, most DSP implementations do not support this many data buses. As a result, the effective processing rate

Integration level	Example	Issues
Within-processor	Superscalar units within MIPS R4000 or DEC Alpha	Very design intensive, requires sophisticated compilers, only feasible for volume applications. However, may be most efficient and cost-effective. Within chip I/O is easy. Can have closely-coupled memory and cache.
Within-chip	Co-processor (e.g., 68356) Multiprocessor (e.g., TI MVP)	Smaller system. Can be cheaper. Can use specialized co-processor (e.g., Viterbi engine). Can integrate DSP and μP. Potential I/O bottleneck to off-chip parts which may be expensive. Less flexibility — processor mix is predefined.
Board	Multiple DSPs on board (e.g., several C40s or SHARCs)	Cost-effective, versatile. May be hard to program. On-chip memory very valuable. System analysis to avoid I/O bottlenecks. Need high I/O bandwidth and interprocessor support to achieve true potential.
System	Multi-VME rack	Custom or semi-custom system. Tradionally hard to program/support. Expensive. However, can give any level of performance required. Can mix different processors.

■ **Table 3.** *Degrees of parallelism.*

Second-Generation DSP Multiprocessing

On-chip bus arbitration:

- Fixed and rotating priority.
- Bus locking and timed release.
- Single cycle bus mastership latency.

These allow the processors to work together smoothly, without requiring the designer to worry about conflicts or the processors to waste time.

Unified address space:

- Direct interprocessor memory read/write/interrupt.
- Broadcast writes.
- Reflective semaphore.
- Read-Modify-Write locking.
- Global memory sharing/arbitration.

Allow DSPs to share and partition memory without conflicts. This is essential for "parallel" architectures, but does simplify pipeline or dataflow systems, too.

Transparent DMA:

- Dedicated I/O processor.
- Dual port memory allows:
  - –DMA chaining/routing.
  - –Multiple DMA channels.

Allows transfers to be coincident with computation, while dual porting allows simulataneous reads and writes — avoiding processor stalls.

On-chip memory:

- Sufficiently large ultra-fast on-chip memory.

Avoids bus bottlenecks. Memory can be faster and cheaper than if external (e.g., 4 Mb)

High-speed I/O

This is crucial.

Current devices (e.g., SHARC):

- Up to 240 Mbyte/s through external port.
- 40 Mbyte/s through each link port.
- 160 Mbyte/s max through six link ports.
- Two general purpose 40 MHz bidirectional serial ports.

decreases, as operands must be passed through a keyhole — the bus bottleneck (Fig. 6a).

On and Off Board Memory

The situation gets much worse if reasonable quantities of memory are required (as they are for any realistic radio, with demodulators or channel decoders, etc.). Commodity memories (SRAMS) are designed for the Von-Neuman processor architectures, where only one access per cycle is required. Using such a memory leaves the DSP designer with substantial bus bottleneck problems, and the need to perform heroic data-flow analysis to try to determine system feasibility. Furthermore, the need for speed means that very fast (i.e., very expensive) memories are required to reduce the impact of this bottleneck. However, the on-chip memory aspect deserves a little bit of explanation. As shown in Fig. 6b, the ideal DSP system integrates memory on-chip, which allows the program sequencer, the external I/O, and all parts of the computational unit to have simultaneous single-cycle access. (In fact, two accesses per cycle are possible under some circumstances.) Additionally, the system designer doesn't have to provide board space for memory ICs, or include cost and power budgets for fast external SRAM. Buses within the chip are faster and less power demanding than the long and capacitatively loaded burden of IC to IC connection. Similarly, the I/O peripherals, which were on-chip all along, no longer require interrupts to go off-chip, through the keyhole of a single bus, to the memory and back to the chip. Instead, a dedicated bus can be used within the IC for direct access whenever desired. Finally, there are commercial implications: the designer no longer needs to do a complex data-flow analysis to assess true performance, and time-to-market is reduced. Some estimate that the performance penalty for using off-chip memory in a real system is that single-cycle operations take an average of 2.31 cycles, equivalent to operating at 43 percent of rated speed (Table 4).

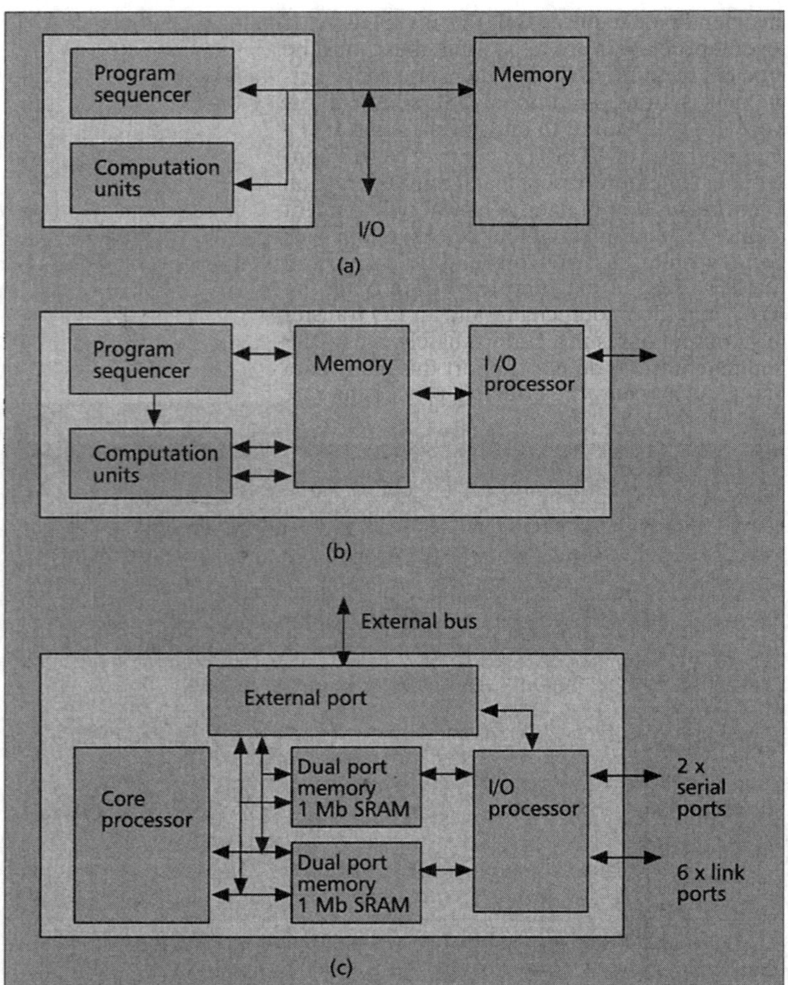

■ **Figure 6.** *a) Traditional DSP architecture. Note the bus-bottleneck; b) Multiple buses and internal memory improve DSP architecture; c) a real world implementation.*

I/O Issues in Multiprocessing

As discussed earlier, the amount of I/O data flow in such a system can be staggering. The ADC could easily produc 60 Msa/s, each 12 or more bits in resolution, requiring the processor system to be reading data at rates of at least 120 Mbytes/s. While there will be redundancy, filtering, and removal of unwanted channels, it is likely that the output data-flow will be roughly comparable, giving an upper bound of some 240 Mbytes/s, just for the

Instruction execution	Relative Frequency	Cycles
Using core registers	30%	1
Using 1 memory location	30%	2
Dual operand fetch	40%	3

Concurrent I/O occurs 10 percent of the time.
Therefore, a typical single cycle instruction requires 2.31 cycles on average:
2.31 = [(0.4*3) + (0.3*2) + (0.3*1) [all times 1.1 for the concurrent I/O penalty].
Or, to put it differently, the DSP only operates at 43 percent of rated processing power if using off-chip memory.
Since each of these off-chip fetches requires power for chip-to-chip drive circuitry, similar savings are possible in power consumption.

■ **Table 4.** *The penalty of off-chip memory access.*

data I/O. In addition, in a real system it is likely that intermediate results, program information, and the like will need to be shared between multiprocessors, necessitating dedicated high-speed interprocessor links. This emphasizes the essential importance of the processor communications architecture. Conventional designs have relatively few data paths, often funneling through bottlenecks. Instead, software radio applications require several parallel routes for information: bulk data I/O, processor to processor links, and access to both global and local memory structures. Ideally, these would be all simultaneous with dedicated DMA or I/O processors.

But most DSPs were not designed to be used as multiprocessing systems. (The TMS320C40 is the notable exception.) They did not support arbitration or prioritization, forcing the programmer to spend both development time and precious processing bandwidth in software solutions. There were no hardware systems for global memory addressing/sharing, nor for sempahores or signaling. And so on... This is somewhat surprising, given the prevalence of multiprocessor DSP systems and the fact that these techniques, and their implementation in silicon, had been seen with Inmos' Transputer (well engineered if not a staggering commercial success). This changed with the release of Texas Instrument's deservedly popular TMS320C40, which corrected several of these gaps with a DSP that was suited to multiprocessing. Indeed, it is fair to describe the C40 as the seminal DSP multiprocessor. Figure 7a shows a typical implementation of a C40 multiprocessor. The second off-chip bus makes it easier to access local memory and helps solve the bus bottleneck, while the dedicated processor-processor links allow for dedicated data-flow paths. In other words, the processor's electrical layout and the algorithms logical layout are mapped, drastically improving the transfer efficiency between processors (and, not incidentally, simplifying the design analysis). A more recent processor from TI goes even further: the C80, or MVP, includes four DSP cores with a RISC supervisor. However, while the MVP is undeniably powerful, it is also undeniably expensive, and does not seem to have been used widely, in contrast to the C40.

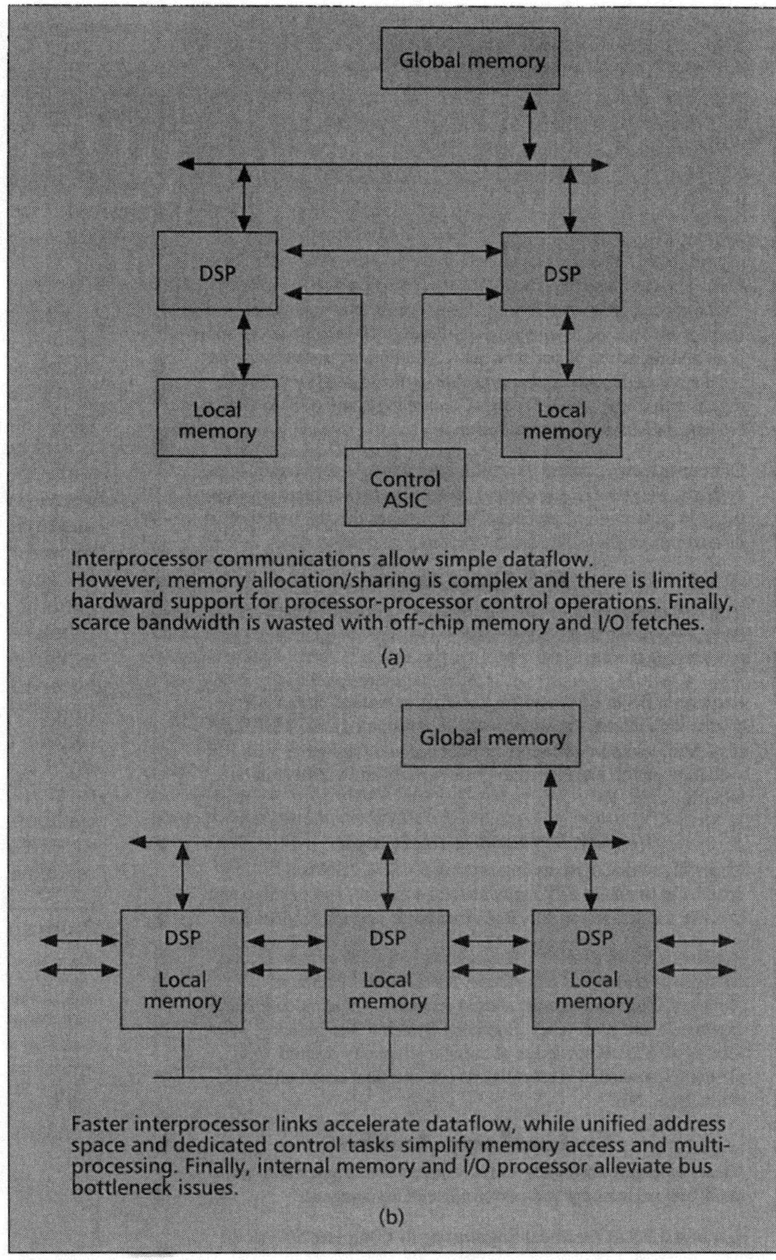

■ **Figure 7.** *a) First-generation DSP multiprocessing; b) second-generation DSP multiprocessing.*

Just as innovation in the ADC realm has suddenly made wideband receivers and IF sampling with excellent SFDR feasible, the processor market has advanced. Second generation DSP multiprocessing is now a reality, learning from the experience of the C40. While that had some of the features required, it has limits in how deep the multiprocessing ran. In particular, while there were the processor-processor links, there were few other hardware tools provided, e.g., for arbitration, semaphores, etc. (The text box on "Second-Generation DSP Multiprocessing" summarizes some of these features.) The SHARC is the first processor to meet the requirements of true multiprocessor DSP as shown in the box. Of course there will be subsequent devices that also implement these features, and no doubt will add new ones. Figure 7b shows a cluster/dataflow multiprocessor architecture based on several SHARCs. In contrast to the earlier case, this system has much greater effective I/O bandwidth and dramatically simplified coordination. The need for fast I/O and a dedicated parallel processor to manage it is hardly surprising; if a single computation core is too slow, one can add more DSPs. But if they cannot read and write fast enough, radio applications are not easy and may not be possible.

Summary

The software radio concept is critically dependent on high performance general purpose processor hardware architectures. The stumbling blocks in development of a commercially usable software radio have been the availability of hardware, in particular the fast DSP required. As high performance ADCs have become available commercially, with the sample rates and SFDRs required, hybrid techniques using specialized digital hardware, operating under software control (e.g., digital downconverters) have become more common. These techniques use digital mixers (multipliers) to select a narrowband channel from the fast, wideband data coming out of the ADC, then filter and decimate it before passing a manageable flow to the general purpose DSPs. However, there are more elegant algorithms—for example, nonstructured receivers, or parallel approaches that recover and tune an entire band at once. These require faster calculation, implying dedicated hardware or, better yet, multiprocessing DSPs. Recent developments in multiprocessing architectures (starting with the TI C40) make this much easier. However, until the SHARC, DSPs were still seriously constrained in their I/O bandwidth, level of multiprocessing support or, crucially, on-chip memory. Now that these issues have been addressed, we can expect to see many more developments in this arena reaching the marketplace.

VLSI Design and Implementation Fuels the Signal-Processing Revolution

The Design and Implementation of Signal-Processing Systems Technical Committee

Edited by
Jan M. Rabaey, with contributions from W. Gass, R. Brodersen, and T. Nishitani

Anniversary Guest Editor's Message

In this issue we are glad to present the contribution of the Design and Implementation of Signal Processing Systems Technical Committee (DISPS-TC), which is the fourth article in our 50th Anniversary series.

Some may think of signal-processing research as mainly algorithm-oriented. But an examination of the history of signal processing shows that new algorithms in signal processing result from the need to improve the efficiency and reduce the cost of implementation. On the other hand, new design and implementation techniques, such as the systolic array, are often triggered by signal-processing algorithms that tend to be very computation-intensive. Therefore, in the field of signal processing, the development of algorithms and the progress of implementation are closely tied to each other.

It goes without saying that signal-processing researchers have always been leading the technology of systems design and software/hardware implementation. In this article, members of the DISPS-TC provide a comprehensive overview of the history of how signal-processing researchers have been effectively transforming signal-processing algorithms into efficient implementations. Starting from the early days of analog circuits for signal processing, to digital signal processors (DSPs), to application-specific DSPs and programmable DSPs, and to the more recent trend of integrating a complete system on a single chip, this article provides a thorough coverage of the past and the present of design and implementation technology for signal-processing systems. Moreover, it presents the exciting challenges faced by the study of the design and implementation of current and future signal-processing applications.

Now, I welcome you to enjoy this article and review how design and implementation technology has been fueling the signal-processing community!

Tsuhan Chen
Guest Editor

Reprinted from *IEEE Signal Processing Magazine,* Vol. 15, No. 1, pp. 22–37, January 1998.

In the last decade, the term *signal processing* has evolved from being an arcane word, only known to a few specialists working in the field, to a household term. The growth of the signal-processing components in consumer, communications, networking, and computing devices has been spectacular, and it is fair to state that signal processing is now one of the main driving forces behind the expansion of the semiconductor industry. In fact, the digital signal processor (DSP) market has grown 40% per year since 1988, which is well over the growth rate of the overall semiconductor market. It is projected [1] that the market for DSPs together with the related mixed-signal devices will reach about $50 billion over the next 10 years (Fig. 1).

This spectacular expansion in applications and capabilities clearly has its roots in the exponential growth in complexity offered by the integrated circuit technology (commonly known as Moore's law). Over the last three decades, integration density of integrated circuits has been increasing at a rate of 50% every year, hence quadrupling every 3.5 years. At the same time, the peak clock speed of the circuits has doubled every three years, yielding ever more performance and computational power. It is expected that this trend will continue without interruption in the next decade to come. Figure 2 projects of the capabilities of the CMOS technology by the year 2010 for a variety of technologies [2]. The potentials offered by this technology are mind-boggling and will open the door for a wide range of new signal-processing applications, some of which have not been invented yet or live only in the mind of some algorithm researcher. What currently can only be implemented on a rack of printed circuit boards will soon fit on a single die.

Yet, technology evolution on its own is not a sufficient argument to explain the rapid developments in the world of signal-processing applications. One of the main properties of a signal-processing application is its ferocious appetite for computational cycles, which typically supersedes the capacity offered by the state-of-the-art microprocessor at that time. The embedded nature of most of these applications has also led to stringent constraints on cost, size, and power dissipation. As a result, signal-processing designers have traditionally led the way in the development of special-purpose architectures, application-specific integrated circuit (ASIC) design methodologies, and high-level design automation. Architectures such as the DSP and the systolic array have truly enabled the revolution in multimedia and communications that we are currently witnessing. The quest to streamline the process of translating a complex algorithm into a functional piece of silicon that meets stringent performance and costs constraints has transformed the electronic design-automation world and has led to concepts such as silicon compilation and behavioral synthesis. Signal processing has had a major impact on the introduction of higher abstraction levels in the design automation process. Ideas such as data-flow modeling, while originating in the general-purpose computer-science community, have found their fruition in the signal-processing community and are having a profound influence on the newest system-design environments that are currently under development in the design-automation industry.

The study of how to effectively translate a signal-processing application into an efficient implementation is exactly the purpose of the Design and Implementation of Signal-Processing Systems (DISPS) Technical Commit-

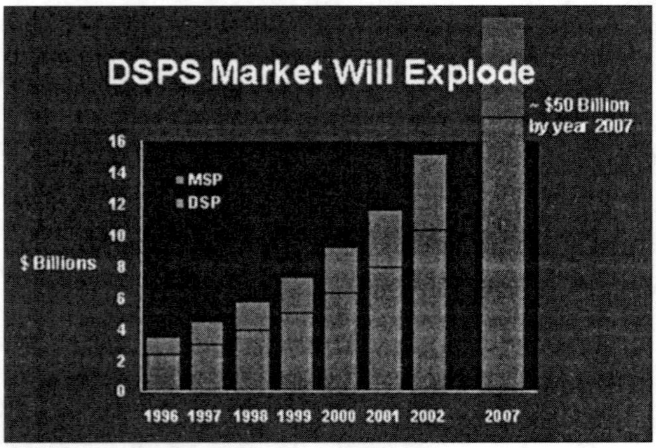

▲ *1. Projected growth in the DSP market.*

List of Current DISPS TC Members

John Allen, *Massachusetts Institute of Technology*
Magdy Bayoumi, *University of Southwestern Louisiana*
Jeff Bier, *BDTI*
Robert Brodersen, *U.C. at Berkeley*
Wayne Burleson, *University of Massachusetts, Amherst*
Chaitali Chakrabarti, *Arizona State University*
Anantha Chandrakasan, *Massachusetts Institute of Technology*
Ed F. Deprettere, *Delft University of Technology*
Wai-Chi Fang, *NASA Jet Propulsion Laboratory*
Wanda Gass, *Texas Instruments* (chair)
Mohammad Ibrahim, *De Montford University*
Vijay Jain, *University of South Florida*
Konstantinos Konstantinides, *Streammachine*
Jit Kumar, *AT&T Bell Labs*
Elias Manolakos, *Northeastern University*
John V. McCanny, *The Queen's University of Belfast*
Teresa Huai-Ying Meng, *Stanford University*
Robert M. Owens, *The Pennsylvania State University*
Keshab Parhi, *Univ. of Minnesota*
V.K. Prasanna, *University of Southern California*
Jan M. Rabaey, *U.C. Berkeley (past chair)*
Bing J. Sheu, *University of Southern California*
Steve Tjiang, *Synopsys*
Ingrid Verbauwhede, *Atmel*
Takao Yamazaki, *Sony Corporation*
Kung Yao, *UCLA*

Die Area: 2.5x2.5cm
Voltage: 0.6 V
Technology: 0.07 μm

	Density ($Gbits/cm^2$)	Access Time (ns)
DRAM	8.5	10
DRAM (Logic)	2.5	10
SRAM (Cache)	0.3	1.5

	Density ($Mgates/cm^2$)	Max. Ave. Power (W/cm^2)	Clock Rate (GHz)
Custom	25	54	3
Std. Cell	10	27	1.5
Gate Array	5	18	1
Single-Mask GA	2.5	12.5	0.7
FPGA	0.4	4.5	0.25

▲ *2. Anticipated capabilities of silicon in 2010 (from [2]).*

tee of the IEEE Signal Processing Society. The mission statement of the committee reads as follows [3]:

The purpose of the committee is to promote and support activities of the Signal Processing Society in the areas of
▲ *design, development, and implementation of signal-processing systems*
▲ *design of algorithms with implementation in mind*
▲ *development of software tools and methodologies to support the design of signal-processing systems*

The committee was founded in the early 1980s as the VLSI Signal Processing Committee and grew out of the then rapidly emerging systolic-array community. The most visible activity of the committee has been the organization of a series of very successful workshops, held annually in different locations all over the world. The latest was held last November in Leicester, England, with a guiding theme of "Multimedia Systems on-a-Chip." These workshops have resulted in 10 edited "VLSI Signal Processing" books, published by IEEE, that present a chronicle of the evolutions in the world of signal-processing implementations over the last two decades. In the early 1990s, the committee was re-christened into the "Design and Implementation of Signal Processing Systems" committee to more precisely reflect the wider range of activities covered by the committee and the focus on

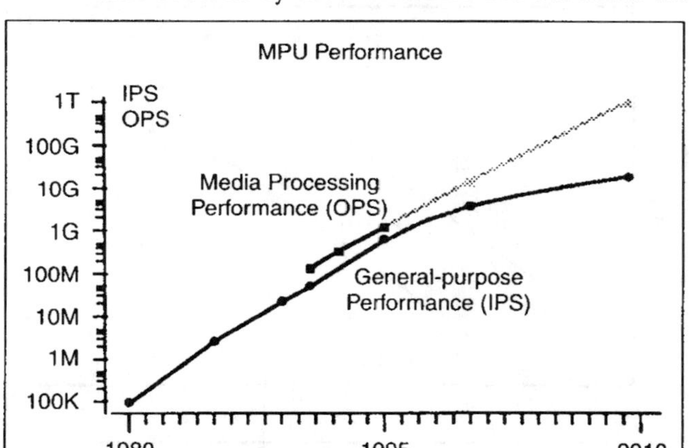

▲ *3. Trends in performance of programmable processors (from [4]).*

system implementation in its totality, including embedded software and design methodologies.

One might wonder if the continuing trends in semiconductor integration with its exponential increases in density and performance will not make research in architectures for signal processing obsolete. Will the Pentium of the future not be capable of easily solving in real time all the signal-processing tasks thrown at it? While this is an often-stated assumption, the adverse actually seems to be true. When evaluating trends in the world of integrated systems, a number of important tendencies can be observed:

▲ Signal processing is currently having a profound impact on the way general-purpose processors are being conceived. Most manufactures of processors have realized that the majority of the compute cycles in current- and next-generation processors are being devoted to either multimedia or communication tasks that have characteristics that are substantially different from general-purpose computations in terms of parallelism, locality, and regularity. This is leading to novel architectures such as multimedia-enhanced processors (MMX, UltraSparc), VLIW processors (Trimedia, Chromatics, TI), DSP super-scalars, etc. In his ISSCC keynote address in 1996 [4], H. Sasaki from NEC Corporation argued that the introduction of multimedia features is the only way the exponential growth in processing performance can be maintained in the next decade (Fig. 3).

▲ The continuing trend of further integration leads to the integration of complete systems on a single die, including the data acquisition and output periphery (sensors, network and wireless interfaces), advanced signal and data processing, and general-purpose computing [5]. Power and performance constraints typically force these systems to be implemented on a heterogeneous architectural platform, combining microprocessors, DSPs, and ASIC accelerator functions. As an example, Fig. 4 shows how the different functions needed in an integrated GSM phone require a combination of implementation options. The study of these heterogeneous architectures and the methodologies to design and program them in a timely fashion is just in its infancy and will probably lead to both new architectural constructions (such as reconfigurable structures) and methodologies [6].

▲ Finally, history has proven that the implementation cycle of signal-processing functions goes through cycles, as shown in the model of Fig. 5. While conceived initially in software, an advanced signal-processing function typically requires dedicated hardware for its first real-time implementation. Later in the cycle as technology progresses, software implementations become desirable again. At that time however, newer and more aggressive algorithms have been devised, causing the cycle to repeat itself (just think about the JPEG -MPEG1 - MPEG2 - MPEG4 - MPEG7 - ... evolution).

These arguments demonstrate that the study of the design and implementation of signal-processing applications and the related methodologies is facing a wide range of

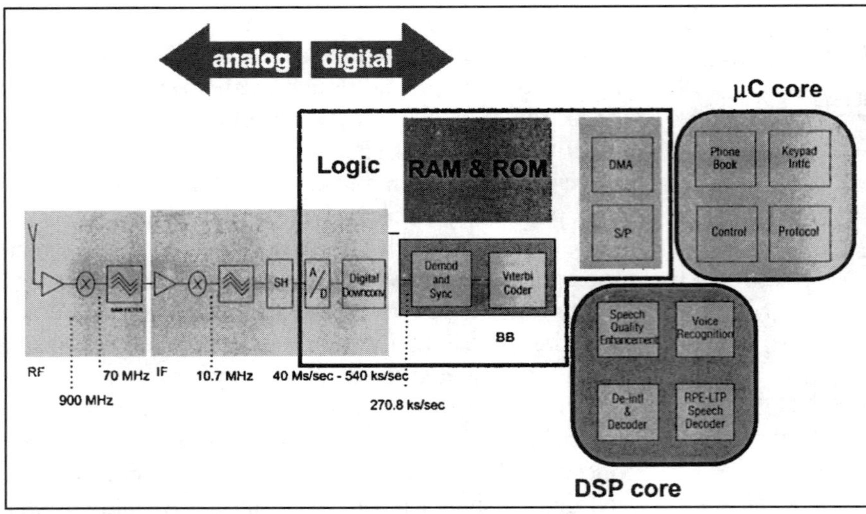

▲ *4. Integrated GSM phone functions and architectural mapping.*

daunting and exciting challenges in the years ahead. It is often stated that complex system design is one of the most important challenges facing the semiconductor industry in the next decade. It is the intention of the committee to ensure that the Signal Processing Society remains a leader in this field. The multidisciplinary nature of this area requires, however, that this is done in concert and cooperation with other IEEE societies such as the Solid-State Circuits Society, the Circuits-and-Systems Society, the Communications Society, and the Computer Society.

The rest of this article revisits different aspects of the past and future of signal-processing implementations, as seen by some of the leaders in the field. Topics covered include milestones in signal-processing integrated circuits, the past and future of the signal processor, signal processing in consumer applications, and design automation for signal processing.

Crucial Developments in the Implementation of Signal-Processing Systems

To get a detailed picture of the evolution of signal-processing circuits, it suffices to study the *Proceedings of the Integrated Solid-State Circuits Conference* (ISSCC) over the years. This section presents a chronology and retrospective on this evolution by doing just that. This helps us to understand the different trends and the underlying motivations.

The Early Days

Signal processing in all its various forms has been around as an important application for circuit design since the very inception of integrated circuit technology. Radio circuits, including amplifiers, oscillators and filters that were implemented using analog circuits, were the primary focus in the early years. In fact, the first paper in the first ISSCC conference (the premier conference in the implementation of integrated circuits) was on signal processing of radio signals in "Transistor Super-regenerative Detection" by W.F. Chow [7]. Though this is a long way from the digital signal-processing circuits that are now being implemented, the progression from analog implementations to digital ones has been followed in virtually all the signal-processing areas, making for many spirited discussions as this technology evolution has unfolded.

Probably of most interest has been the progression of filtering circuits, the most ubiquitous of electronic signal-processing functions. As just mentioned, analog filtering for radio applications was present almost from the inception of integrated circuit technology with tuned circuit techniques discussed in 1957 [8, 9]. 1959 saw the introduction of ceramic filters, which even today remain a serious competitor to electronic techniques at the highest frequencies [10]. Of particular interest in this early period was the presentation of a "Solid State Sampled Data Bandpass Filter" by L.E. Franks and F.J. Witt [11]. Time quantization of the signal (i.e. sampled data) is now the base of all digital techniques, as well as the most common analog approaches. The design of what we now call active RC filters emerged for the first time in 1966 as embodied in the publication of "Design of Integratable Desensitized Frequency Selective Amplifiers" by A.A. Gash, R.S. Pepper, and D.O. Pederson [12].

An important limitation of all these techniques was that it was not possible to fully integrate the filter, either because of their size or requirements on accuracy. A search was therefore underway to find a strategy to fully integrate the complete filtering function. A digital approach to implementing the filtering function was clearly one attractive approach and one of the first to be reported was in the mid 1970s by G.P. Edwards, P.J. Jennings, and T. Preston, titled "A MOS LSI Double Second Order Digital Filter Circuit" [13]. Though this by no means

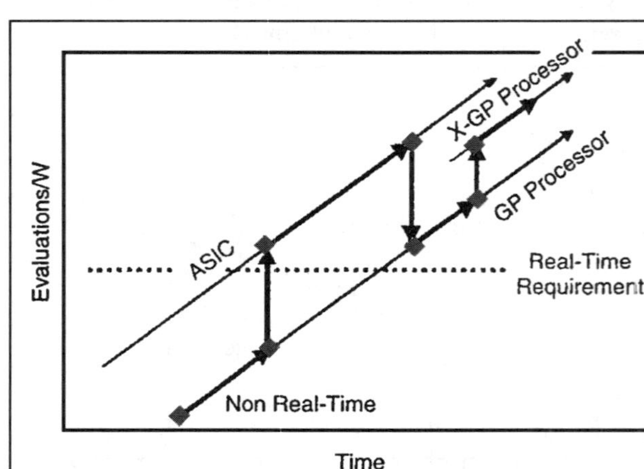

▲ *5. Typical implementation cycle of signal-processing applications [2].*

caused an immediate conversion to digital techniques, it indicated that digital technology had progressed to the point that, at least for audio rates, filtering could be performed entirely in the digital domain. Although there were great advantages to be gained in processing the signals digitally, this required the signal to be sampled in time, requiring an analog anti-alias prefilter followed by an A/D conversion. After the digital processing, D/A conversion and signal reconstruction with another analog filter was necessary. This overhead clearly limited the usefulness of the digital approach if the filter was to be used in a standalone application in which the input and output signals are analog, continuous time waveforms. This overhead made pure digital signal-processing not a viable solution and opened the door for the sampled-data approaches of the 1970s and early 1980s.

Discrete-Time Signal Processors

Another technology that arrived on the scene in the early 1970s held the promise of full integration without the overhead of a digital solution. It was based on the use of charge packets as the representation of the value of the signal, instead of voltage and current. This representation made it possible to realize elegant, compact analog delay lines using charge transfer techniques [14]. The charge-packet representation required a quantization in time (thus still requiring anti-alias and reconstruction filters), but at least the need for A/D and D/A converters was avoided. If it is possible to delay a signal and to scale its size and perform sums (or subtractions), a frequency-selective filter can be realized. Scaling and summing are also relatively easy to do in the charge domain by charge redistribution across various sized capacitors with the use of op-amps. The first fully integrated filters were published and demonstrated in 1977. One approach used switches, capacitors, and op-amps to implement the delay line [15], while others used charge-coupled devices [16, 17, 18]. Though these techniques were able to place all the critical components on the chip, they suffered from various performance limitations that were mostly related to their restriction to delay-line topologies.

Active RC filter structures that were based on doubly terminated LC filters were well known to have the best performance in terms of noise and insensitivity to component variation, while making modest demands on the operational amplifiers. Unfortunately, the size of the RC time constants and their accuracy did not allow direct realization of these filters in a monolithic form. It was discovered, however, that by rapidly switching the plates of a capacitor between voltage sources an approximation to a resistor could be realized. It was this approach that was used in 1978 in what has become known as the switched-capacitor filter technique [19]. Since the approximation gets better as the sample rate is increased, it was found that very simple pre- and postfilters would suffice for anti-aliasing and reconstruction, thus allowing for an efficient, complete filtering solution.

Also in 1978, the first fully integrated solution to the important telephone pulse-code modulation (PCM) codec was presented, which made use of charge-coupled device filters [20]. It was followed the next year by switched-capacitor solutions from Intel [21] and A.M.I. [22]. These circuits demonstrated the manufacturability of that approach for high-volume applications and began a dominance of switched-capacitor techniques for integrated audio band filtering applications that continued into the early 1990s. Higher levels of analog integration were later shown possible in a touch-tone receiver design that integrated over 30 poles of switched-capacitor filtering [23]. It is interesting to note that a plot of that chip was exhibited in the New York Museum of Modern Art for a number of years.

The success of the charge-domain, sampled-data approach in fixed-frequency filtering encouraged others to extend its application to even more complex signal-processing applications. Examples of these are adaptive echo cancelers and equalizers [24, 25] and a complete 1200b/s QPSK Duplex modem from Texas Instruments [26]. Though these designs are a tour-de-force in complex analog signal processing, they mark the end point in pushing that approach to higher levels of complexity. The difficulties in design, primarily due to a lack of adequate simulation capability, and the problems of manufacturability due to difficult testing and parametric control resulted in an ever-increased focus on digital solutions.

The Emergence of Digital Signal Processors

The era of the digital approach to the integrated filtering problem was officially entered when it became possible to integrate A/D and D/A filters on a microprocessor chip, augmented to support the computations required in digital filters. The first design of such nature, by M.E. Hoff and M. Townsend, titled "An Analog Input/Output Microprocessor for Signal Processing" [27] made it possible to clearly identify the tradeoffs between the analog and digital approaches, and it inspired a series of energetic debates in the design community.

During this same time period, a number of other microprocessor-based solutions were being pursued, primarily targeted toward telecommunications applications. Probably the first that focused on support for digital signal processing was a design from Bell Laboratories [28] that attempted to increase the computational throughput of a standard microprocessor through small modifications of the architecture. It was hampered by the small word size (8 bits), the lack of a hardware multiplier, and the use of a relatively standard internal bus structure. In 1980, advances in technology (3 micron NMOS) allowed the inclusion of a full 16-bit array multiplier, a more sophisticated bus structure, and an internal fixed-point 16-bit word size in a processor from NEC [29]. A competing design from Bell Laboratories used an internal word-width of 20 bits and a 16 x 20 bit array multiplier [30]. The focus on filtering applications is obvious as the circuits are benchmarked by the number of 2nd-order filter sections that can be implemented at an 8 kHz sample rate (39 for the Bell chip and 55 for the NEC).

The next round of DSP circuits appeared in 1982 with contributions from Texas Instruments [31] and from Matsushita [32]. This latter design could implement a complete 9600 bps modem and included special I/O support for external A/D and D/A circuitry. The Texas Instruments design was the first in their family of DSP chips that has come to dominate the marketplace today. This early chip made use of the so called "Harvard Architecture," which has separate program and data busses, and had multiple 32-bit internal busses designed to support a continuous data-flow to the 16 x 16 bit array multiplier. It was reported that this circuit had sufficient on-chip memory to be able to do a 64-point FFT, thus setting a new benchmark for comparing these ever more complex designs.

Clearly, these DSP circuits could do much more than filters and, as the applications became more complex, the importance of software support and ease of programming became increasingly important. This argued for the use of floating-point arithmetic to avoid the complex scaling needed in the fixed-point processors when attempting sophisticated applications. Improved technology (1.5 micron NMOS) finally allowed floating-point support in a design by a team from Bell Laboratories in 1985 [33]. This chip was benchmarked to have the capability to implement 125 2nd-order sections at 8 kHz and be able to do a 1024-point FFT in 12 ms. These early DSPs have seen a wide variety of successors in the following decade(s). Due to the importance of the programmable processors and their impact on the signal-processing community; a detailed discussion of their evolution is presented in a following section.

Domain- and Application-Specific DSPs

While programmable solutions have had a major impact on the introduction of digital signal processing into the mainstream, they tend to be too expensive or don't meet the performance and energy constraints for many commercial applications. This has led to a wide range of so-called application- or domain-specific integrated circuits, some of which are discussed below.

The Fourier transform is one of the most common functions performed in advanced signal-processing applications and has led to a number of circuits to accomplish this task in a monolithic form. In 1975, during the heyday of charge-transfer devices, a Fourier transform chip was presented that used analog sampled-data techniques. It employed the Chirp Z Transform algorithm [34], which transforms the primary transform computation into a convolution that could be efficiently performed in a charge-coupled device transversal filter. This circuit could be used to implement a 512-point transform in 256 microseconds. However, the accuracy of the transform was limited because of the complex analog computations. In 1982, TRW presented a "2 micron CMOS/LSI 32-Point Fast Fourier Transform Processor" [35]. This chip, which used an FFT algorithm, could perform a 32-point transform in 47 microseconds and a 1024-point transform would take 188 microseconds. Still yet another approach was presented in 1983 in a chip that directly calculated a 256-point DFT. It was argued that even though the DFT required additional computations, the reduction in silicon area and control overhead was ultimately beneficial [36].

Another area that has had a rich implementation history is the domain of circuits for speech processing, which includes both synthesis and recognition. In 1982, three chips were introduced that indicated that technology had progressed to the point that speech synthesis could be performed on a single chip [37, 38, 39]. In that same year, a systolic array circuit for speech recognition was also presented [40]. This circuit marked one of the few examples of systolic-array circuits presented at the ISSCC conference and was useful in benchmarking different architectural approaches. Two more conventional dedicated circuits were presented the following year [41,42] that demonstrated that the systolic implementations suffered from a considerable overhead.

The improvement in technology also allowed the implementation of circuits that can directly process image data. Image processing has two characteristics that differentiate it from audio or telecommunications applications: the data is two-dimensional and has a sample rate that is about 3 orders of magnitude higher. This requires an architecture that is quite different from standard microprocessors as there is not adequate time available to do much time multiplexing. A first design by researchers at Hitachi [43] was primarily designed to do 4 x 4 convolutions, which it could do over a 256 x 256 image in 11.8 ms if 4 chips were used in parallel. Faster image processors were not introduced until three years later (1986). The microprogrammable real-time image processor from Matsushita [44] used an advanced bipolar process to obtain a 50 MHz cycle time, which allowed a 3 x 3 convolution to be performed on one chip in 11.8 ms. On the other hand, a number of dedicated image processors were presented [45] that could perform 3 x 3 convolutions, sorting, and logical operations on 512 x 512 images in 30 ms using standard NMOS technology. At that time, video processing (which adds the third dimension of time) became feasible as well.

Dedicated digital architectures for coding and decoding color TV signals at real-time rates were introduced for the first time in 1986 [46, 47]. Since then, image, graphics, and video applications have captured a dominant role in the signal-processing implementation arena. The complexity of the circuits currently being complemented supersedes by far the earlier designs. For instance, an MPEG-2 encoder chip (three of which were presented at the 1997 ISSCC conference), requires approximately 3 million transistors and combines an application-specific signal processor with dedicated accelerator blocks for functions such as DCT/IDCT and motion estimation.

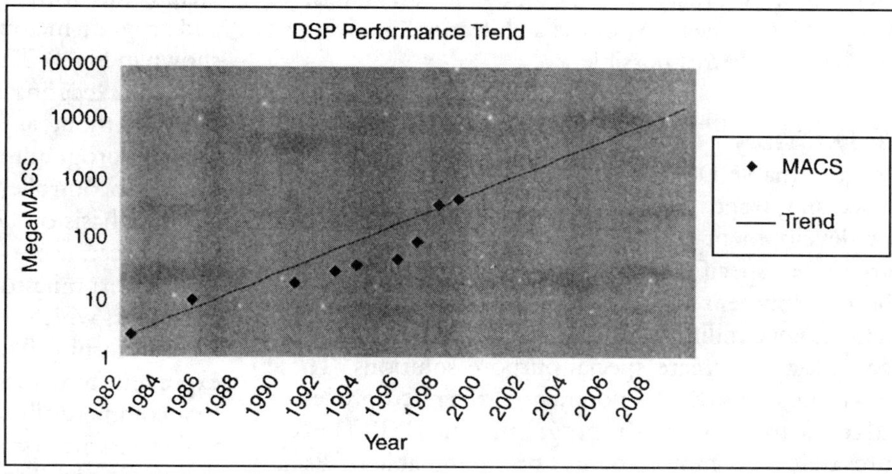

▲ 6. *Performance trends in DSPs.*

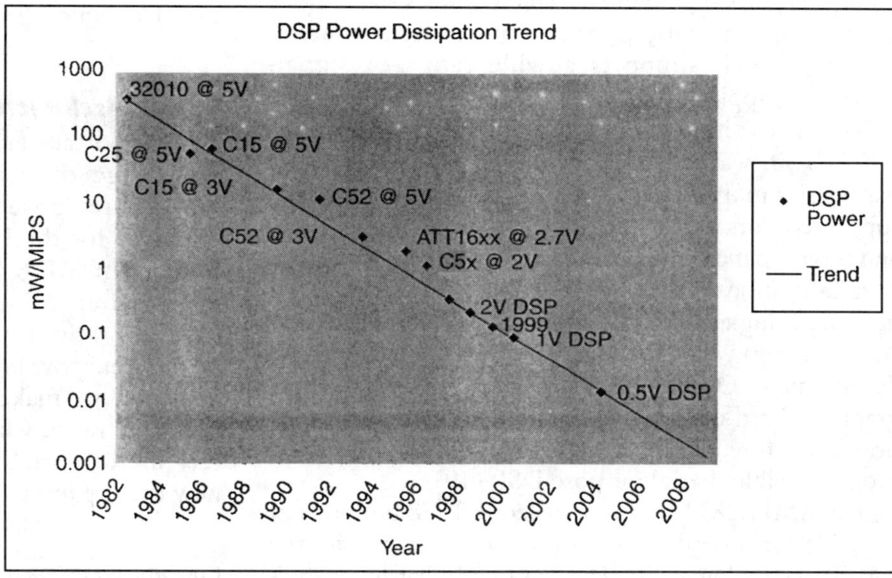

▲ 7. *Energy trends in DSPs.*

The New Frontiers

Signal processing has come a long way since the early days of integrated circuits. As mentioned earlier, it has become one of the dominant drivers of the semiconductor and system industries. This is clearly illustrated by the fact that more than 40% of the papers published in the 1997 conference were related to signal-processing and communication applications, and that even a wide range of microprocessor publications carry the term "multimedia" in their title.

In coming full circle in this review of signal-processing circuits, it is interesting to point out that digital technology is now impacting even radio circuits. The extremely low-cost and high-speed operation required of radio circuits had kept digital solutions from becoming competitive until recent. Now, however, the demands of cellular phones and more advanced radio services has required that digital signal-processing techniques be increasingly applied to radio receivers and transmitters. One of the first circuits demonstrating these capabilities was published in 1991, implementing a 200 MHz all-digital QAM modulator in a 1.2 micron CMOS technology [48]. As the simple radio becomes ever more sophisticated, there is no doubt that the implementation of ever more complex algorithms for processing signals will continue and will yield a rich area for novel circuit architectures and design techniques.

Programmable DSPs

As mentioned earlier, programmable DSPs had a tremendous impact on the popularization of digital signal processing and its penetration into the commercial market. It helped to transform digital signal processing from an academic study area into a viable commercial business for consumer products, which require low-cost, real-time performance, and upgradable solutions. They also caused an important shift in implementation paradigm toward software-oriented solutions. This trend is bound to continue in the next decade.

To understand the evolution of the signal processor, it is worthwhile to study the trends in signal-processor performance as embodied by consecutive generations of Texas Instruments signal processors (Fig. 6). It can be observed that, on the average, signal-processor performance almost doubles every two years. While performance is an important issue, in the embedded world of signal processing other metrics have to be taken into account such as energy per operation. Figure 7 plots the average energy of a number of signal processors (measured in mW/MIP) over time. While technology has a crucial impact on the improvement of both the energy and performance metrics, even more leverage is obtained by the adaptation of the architecture to the specific requirements of the signal-processing applications at hand. This explains why signal processors outperform traditional microprocessors in both computational performance and energy by a wide margin. In the following paragraphs, we will present an overview of the architectural innovations that have made this possible.

Early History

Programmable DSPs evolved out of two digital signal-processing trends in the late 1970s. The first trend was the development of dedicated single-chip DSP devices targeted at specific applications. The second trend was the development of multichip DSP systems for high-performance military applications, which used bit-slice technology to create special-purpose solutions. These two trends plus the development of the microprocessor led to the invention of the programmable DSP. The features that are common to most programmable DSPs included a data path with an integrated hardware multiplier; on-chip memory, which enables stand-alone system implementations; and an instruction-set architecture, which supports a wide range of signal-processing applications.

The first commercially available DSP processor, the NEC 7720, was introduced in 1980. The 7720 had a hardware multiplier and a Harvard architecture, which supported concurrent memory access to data memory and program memory, but the instruction set did not support saturation arithmetic. The 7720 was more like a dedicated single-chip DSP rather than a microprocessor. An emulation version of the chip was used for software development. Once the software was validated, the program was hard-coded in on-chip ROM for mass production. Therefore, the circuit rather represented a one-time programmable special-purpose DSP [49].

The AMI S2811 was announced in 1978 but was not available for several years because of problems with the VMOS technology [50]. The 2811 had a hardware multiplier and concurrent memory accesses like the NEC 7720. It could either execute out of on-chip program ROM, or it could execute instructions that were provided on the I/O pins one at a time. It was designed to be a peripheral part to a microprocessor such as the Motorola 6800 and therefore could not fetch and execute instructions out of program RAM.

Intel introduced the 2920 in 1980 [27]. It was aimed at signal-processing applications because it had A/D and D/A converters on-chip. But its performance for compute-intensive tasks was not impressive because it had no hardware multiplier. In addition, it was hard to get parameters into the chip because it lacked a digital interface.

In 1980, AT&T developed the DSP1, which was used for internal product development but was never marketed outside of AT&T. It had a hardware multiplier, but was a traditional Von Neumann architecture. Like the NEC 7720, program memory was confined to execute out of on-chip ROM and, hence, the DSP1 could not be reprogrammed once it was fabricated [51].

Texas Instruments introduced the TMS32010 in 1982 [31]. The 32010 had a hardware multiplier and Harvard architecture with separate on-chip buses for data memory and program memory. A chip photograph of the 32010 is shown in Fig. 8. This was the first programmable DSP to support executing instructions from off-chip program RAM without any performance penalty. This feature brought programmable DSPs closer to the microprocessor/microcontroller programming model. In addition, TI's emphasis on development tools and libraries led to wide-spread use.

The next milestone in DSPs was the adoption of floating point, which was first introduced in the HD61810 from Hitachi (1982) with a 12-bit mantissa and a 4-bit exponent format [52]. In 1984, AT&T introduced their first commercially available DSP to the external market, which was the first 32-bit floating-point DSP with 24-bit mantissa and 8-bit exponent [33]. Through the end of the 1980s, AT&T provided the fastest DSPs with the DSP16A running at 30 MHz in 1988.

Recent Architecture Advancements

Two important factors have given DSPs a significant advantage over embedded microcontrollers and general-purpose microprocessors: MIPS/$ and MIPS/W for the DSP application domain. While the number of MIPS for general-purpose processing has increased at a faster rate than MIPS for DSPs, the MIPS/$ ratio for DSPs is still much better than general-purpose processing. Another area where DSPs continue to make significant improvements is the MIPS/W ratio, which is critical to extending the battery life of portable devices while continuing to add new the capabilities over time (Fig. 7).

Customized DSPs

While the instruction sets of DSPs were general enough to support a wide variety of applications, it was common to have different program/data memory configurations and I/O peripherals within a product family. This allowed the same set of development tools to cover several products and still allow the device to be customized for certain applications. There were DSPs that allowed customers to modify the internal data path, but the disadvantage of less-sophisticated development tools generally outweighed any performance advantage these cores had over standard DSPs.

As advances in process technology continue to increase the amount of logic that can be integrated onto a single chip, the ability to combine ASIC logic with a DSP becomes feasible. This adds a new level of customization to digital signal processing for high-volume applications. This capability is used mostly to integrate interface circuits with a DSP core. For space-sensitive applications, such as hard disk drives, this is a real advantage. Finally, the ability to specify the memory configuration adds another level of customization that makes the programmable DSP solution more competitive in

▲ *8. Die photograph of the TI 32010.*

price to dedicated DSP solutions. The availability of DSP cores as general building blocks is bound to have an important impact on the way next-generation DSP systems are designed.

Low-Power DSPs

Battery-powered products, such as multimedia notebook PCs and digital cellular phones, are driving the demand for lower-power DSP processors. Power reduction in DSPs is achieved using three design techniques: low voltage, gated clocks, and sleep modes.

Several 1V DSPs have been reported recently [53, 54] that demonstrate the advantage of lowering supply voltage. While lowering the supply voltage causes the delay to increase, performance can be regained by adding more parallelism. Since power is proportional to V^2, the effect of lowering the power is greater than the effect of increasing the gate count.

Due to the highly parallel nature of DSPs, gated clocks can be used to only enable the segment of the data-path and the section of memory required to execute each instruction. Therefore, the more parallelism that is available in a DSP, the greater advantage there is in using gated clocks. This was demonstrated in the low-power version of the TMS320C548 [55], a processor primarily targeting the speech-coder market. A chip photograph of the device is shown in Fig. 9, demonstrating the systems-oriented integrated nature of the processor (e.g., the many embedded memories).

The third technique used to extend the battery life of portable devices is to provide multiple sleep modes. As many as three different sleep modes are available on state-of-the-art DSPs to provide better power management alternatives at the system level. The sleep modes vary from disabling just the CPU to disabling the entire processor, including peripherals and digital PLL. The system must optimize between the power saved during the different sleep modes and the amount of time it takes to resume execution of instructions.

VLIW

Very-long-instruction-word (VLIW) architectures, which were originally developed to address the supercomputer market, have become widely adopted by a new set of media processors. In 1995, the Trimedia and MPACT 3000 architectures were introduced by Philips and Chromatics, respectively [56]. Then in 1997, Texas Instruments announced the TMS320C62xx [57]. VLIW architectures are well suited to DSP applications with their inherent parallelism because they support multiple execution units that can operate concurrently. The execution units can be specialized to support special-purpose operations such as motion estimation or variable-length decoding, which are difficult tasks for most processors. The architecture of the TMS32062xx is shown in Fig. 10. Observe the parallel data path structure that is capable of executing 8 32-bit instructions per cycle (at a clock rate of 200 MHz). The processor further supports large on-chip memories and multiple input-output ports.

The VLIW architecture is similar to superscalar architectures in the sense that multiple instructions are executed per cycle. Yet, they are different in how they handle the scheduling of concurrent operations. VLIW exposes the parallelism to the compiler so that scheduling can be done at compile time. Superscalar hides the concurrent execution of instructions from the programmer and schedules the operations at run time. To that purpose, superscalar adds an amount of hardware overhead support to the CPU that is not needed for VLIW (where the burden is placed on the compiler).

Future Architecture Trends

Heterogeneous Processors

As the complexity of systems increases and integrated-circuits tend to combine multiple processes and functions, the trend exists to integrate multiple processors of a different nature onto a single die [5, 6]. In addition, the systems continue to become more memory-intensive and the availability of high-bandwidth I/O interfaces becomes a critical factor in system performance. Single-chip solutions will increasingly have both a hardware component as well as a software component. Hardware and software interfaces (APIs) for these components will emerge in order to allow the system designer to move to a higher level of abstraction. Optimizations will take place at a higher level such that gate level and assembly level optimizations will be less critical. For some applications, memory will likely be the dominant factor in driving price and performance.

FPGAs

In an attempt to keep the flexibility of a programmable solution and the efficiency of a dedicated solution, field-programmable gate arrays (FPGAs) are beginning to emerge as a viable option [6]. FPGAs give the manufacturer the ability to upgrade in the field, much like a programmable DSP. FPGAs often provide a more efficient solution in terms of area, delay, and power as compared to a programmable DSP solution, especially for applications with either small or mixed word-lengths and operation granularity. It is projected that microprocessors and/or microcontrollers will be combined with FPGAs to provide configurable accelerators that can be configured to handle multiple functions per block of data. Also, it is likely the DSP cores will have on-chip FPGA sections for low volume and emerging applications, or for the implementation of adaptive or agile interfaces.

Fusion of GP and Digital Signal Processing

The most unpredictable of the future trends is the outcome of the much-heralded merger of general-purpose (GP) architectures and digital signal-processing architectures. Both seem to be fast approaching a common ground. GP architectures are adding features that give them similar performance as low-end DSPs, while DSPs are moving to architectures that are more supportive of high-level language programming. The architectures could merge to the point that the two approaches are indistinguishable. Alternatively, cost, performance, energy, and design productivity requirements might cause the market to fragment into segments of domain-specific processors, where each architecture retains its own set of features that makes them unique from each other.

DSP Applications

As became obvious in previous sections and as projected in Fig. 5, the DSP implementation community continually oscillates between programmable and more dedicated implementation platforms. The choice of implementation fabric and methodology depends upon a large number of factors such as cost, volume, performance, time-to-market, and design methodology. Quantifying the impact of each of these factors is currently a dark science. Coming up with more quantifiable metrics is an important challenge for the design community and will gain importance in the next decade with the cost of integrated solutions dramatically increasing. This section gives a number of examples of how evolving technology and applications impact the choice of implementation style.

Initially, programmable DSPs were used for a variety of audio-rate applications. The early DSPs became competitive with the dedicated DSPs and found early success in applications such as voice-band modems, echo cancelers, voice and audio coding, voice synthesis, speech recognition, digital audio equalization, and music synthesis. Initially, these markets were best served by DSPs from the systems houses (Motorola, NEC, AT&T, etc.)

▲ *9. Die photograph of the TI 320C54.*

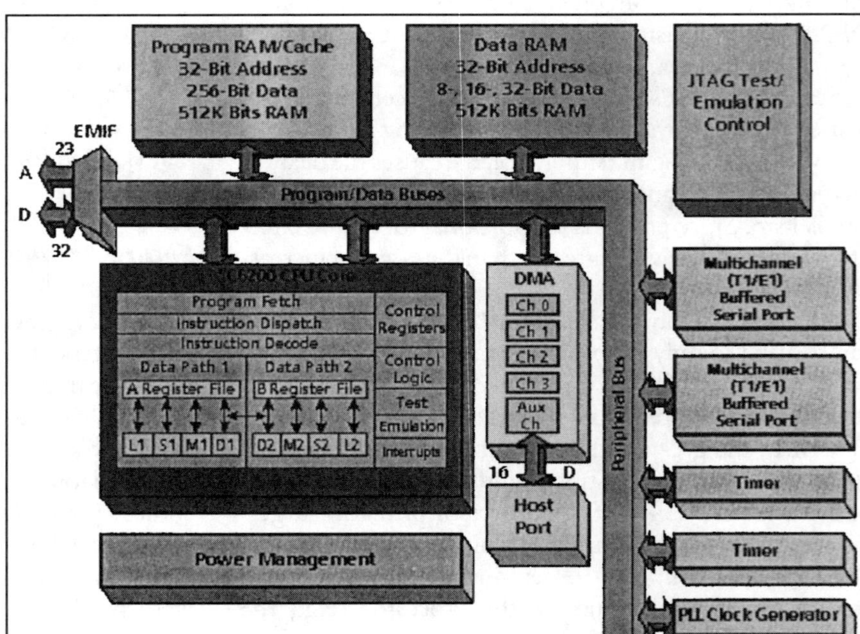

▲ *10. Architecture of the VLIW TI 320C62xx.*

▲ *11. Prototypes of future "silicon terminals" (SiliconAudio and SiliconView).*

where the knowledge of voice, audio, and telecommunication applications existed inside the company. Companies that were primarily semiconductor manufacturers, such as TI and AMD, supported a customer base that required a more programmable solution and were more successful in new DSP markets such as hard disk drives and automotive applications. By the early 1990s programmable DSP markets had reached the point where it had become a viable business for several companies.

Some Case Studies

An interesting example to demonstrate this trade-off was presented by the Julie doll from Worlds of Wonder, which was introduced in 1987. Up until that time, the toy market exclusively used dedicated DSPs such as speech synthesis chips. The Julie doll was a historic event because it marked the first time that programmable DSPs hit a price point that made it feasible to be in a toy.

The modem market presents another worthwhile case study. It was dominated by dedicated DSPs for many years. To get to the market quickly while meeting timing requirements, modem developers would rely on application-specific signal processors, optimized for modem functionality. Only recently have programmable DSPs overtaken the dedicated solutions for early market penetration. When in 1996 the newest modem standard was finalized (56 kBits), the programmable solution allowed manufacturers to upgrade modems in the field.

Wireless communications is another area that continues to grow and change over time. The challenge for implementations in this area is that the solutions have to keep up with evolving features and communication standards. At the same time, the solution is subject to very tight cost and energy requirements (standby/talk time is one of the most important specifications of the cellular phone). It is interesting to observe that the majority of the U.S. and European manufacturers almost immediately opted for a programmable solution for the GSM voice coder (to reduce time-to-market in a time of evolving or ill-defined standards), albeit adding some dedicated hardware support for time- or energy-critical functions (e.g., the TMS320C540 adds hardware support for the Viterbi coder). The idea was to migrate to more dedicated solutions in the 2nd generation. With some exceptions, this never really happened, and most GSM cellular phones are based on a programmable platform. The ability to integrate other functions on-chip with the DSP core, the quick transitions to new technology nodes, and availability of low power DSP cores seem to make the programmable solution the best option for today's phones.

This picture is bound to change again with the introduction of multimodal and software radios, where digital signal processing is bound to take over a large portion of the radio processing and where agile and adaptive solutions are desirable. Lower power requirements as well as higher demand for signal processing will tend to push implementations to a less-programmable solution. Yet the agility and flexibility requirements will increase the demand for programmable solutions. It will be interesting to see what architectures can provide the best solution.

A Perspective from the Consumer Market

The single-chip programmable DSP has been gradually introduced into consumer products from the day it appeared. Unlike the microprocessor, which employs external main memory chips, the single-chip DSP has supported the all-in-one concept for simple applications from its birth. While most cost-sensitive consumer applications make use of ASICs or ASSPs (application-specific signal processors), programmable DSPs have found their way into consumer applications in a number of ways. Very often, parts that originally developed for the telecommunications market (such as voice coders, synthesizers, and recognizers, and data modems) have found their way in consumer applications different than those they were originally intended for. For example, telecommunication processors are currently used as the speech-input module for an automobile navigation system. Besides these derived applications, some other consumer applications of DSPs deserve attention:

DSPs as Accelerators

Low-end DSP chips are heavily employed as assists to embedded CPUs in the video-game area. Both Nintendo's super-family computer and Sega's arcade game machine use DSP chips for the support of computer graphics characters. These examples indicate that the DSP's all-in-one concept works very well for cost-effective implementations in this area and that programmable DSP chips have already played an important role in the multimedia consumer market. More recent implementations replace the DSPs with media-processors to support MPEG decoding on top of the computer graphics generation.

DSPs as Prototyping Tools

Programmable DSP chips are employed in many experimental and first-generation systems to provide a flexible system-analysis and optimization platform under real-world conditions. The CD player is a good example of this. Initial implementations employed up to 4 DSP chips. These were replaced by a single signal-processing ASSP in later generations for cost-reduction reasons. A similar approach was employed for the ghost-canceler in TV sets.

An interesting example of how programmable solutions can help to explore novel product directions and test them with real customers is offered by the following case study. With the dramatic progress in high-quality compression techniques and the continuous increase in density of nonvolatile memories (flash memory), it will soon become economical to provide audio and video playback that does not require any mechanical or moving parts and is completely implemented in silicon. NEC had dubbed these devices "Portable Silicon Terminals." The availability of DSP-based MPEG audio and video decoders, combined with PCMCIA memory cards that hold the compressed data sequences, makes it possible to develop early prototypes that can then be used to improve user interfaces and system quality, and to get customer feedback. A picture of such SiliconAudio and SiliconView terminals is shown in Figure 11. Options and extensions that are currently being experimented with are the addition of a wireless interface and recording capabilities (SiliconCam).

DSPs in Standards Development

Finally, an often-underrated application of DSP chips is their role in establishing world standards. A first example of this was the establishment of the ITU-T32kb/s ADPCM speech and audio bandwidth compression coding standard. The Bell Labs DSP-16 and the NEC7720 were extensively used for the bread-boarding and the fine-tuning of this standard in the early 1980s. A similar strategy was used for the establishment of the ITU-T16kb/s Low-Delay CELP coder, where Bell Labs used their own line of DSP chips and PKI used the NEC SPX. In the late 1980s, the MPEG-1 audio codec competition was carried out with all competitors bringing their own real-time codec. A wide variety of processors was used ranging from the DSP-32, the transputer, and TMS and NEC-77220 DSPs.

Signal-Processing Design Methodologies and Techniques

The performance-driven and computation-oriented nature of digital signal processing has long made it an ideal driver for the development of automated high-level and system-design tools. In the last two decades, digital signal processing has spurred innovation in areas such as algorithm synthesis and optimization, behavioral specification, architectural synthesis and optimization, silicon compilation, and co-design and verification. This resulted in a number of success stories that indicated that automated design starting from a high-level description is a reality and is here to stay. In fact, the techniques and lessons learned from the earlier digital signal-processing implementation tools have a tremendous impact on the way system-design tools are currently being conceived in both industry and academia, an impact that is felt on a much wider range than pure digital signal processing. This section highlights just some of the digital signal-processing design-automation efforts in the past and offers a brief perspective on future developments.

Since the earlier efforts in digital signal-processing implementation were uniquely oriented toward filters, it is no surprise that the earlier design aids were oriented in that direction as well. Actually, one of the most popular packages of design aids for digital signal processing was delivered in the landmark IEEE reprint volume *Programs for Digital Signal Processing*, published by the IEEE Acoustics, Speech, and Signal Processing Society in 1979 [58]. The book was accompanied by a set of FORTRAN programs for the synthesis of IIR and FIR filters. This package has been used extensively in both academia and industry and still forms the core of commercial filter synthesis tools (such as the 1980's Filsyn and Comdisco, and currently in Matlab).

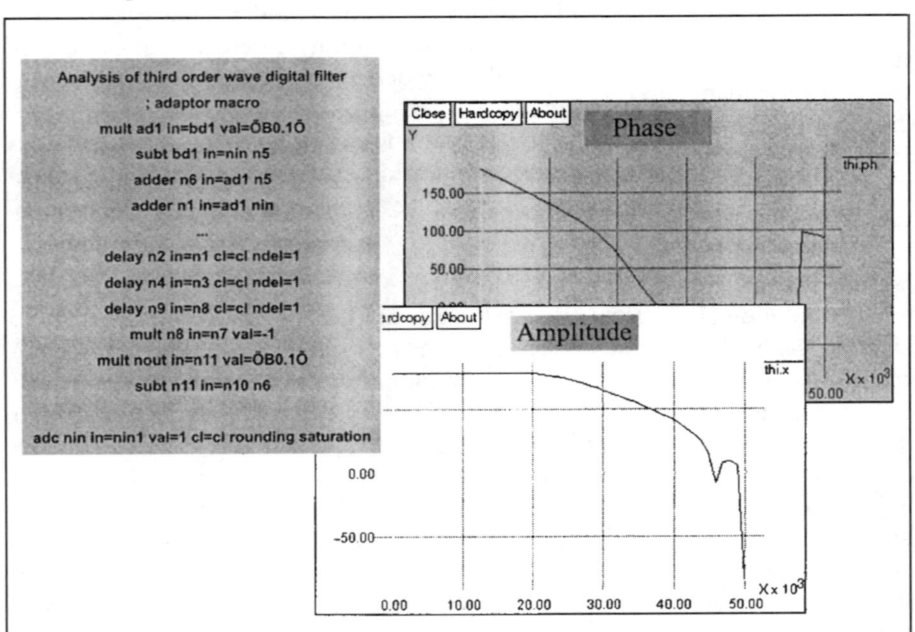

▲ *12. Description and analysis of 3rd-order wave-digital filter in DIGEST.*

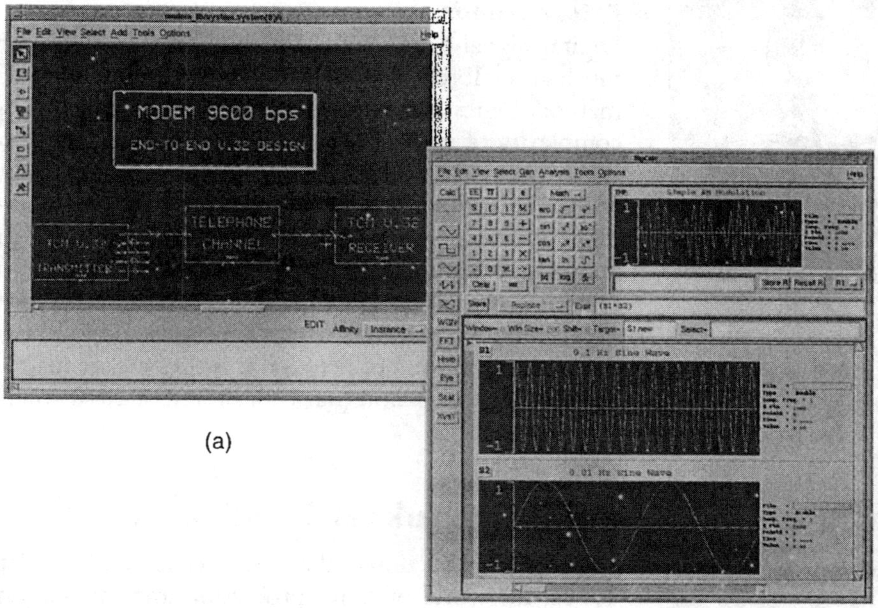

(a)

(b)

▲ *13. SPW block-editor (a) and signal analysis (b) environment.*

Enabled by both integrated circuit technology and the emergence of programmable DSPs, filter implementation, analysis, and optimization tools started to emerge in the early 1980s.

DSP Specification, Modeling, and Analysis

DSP specification differed quite dramatically from the traditional algorithm specification and description languages that were available in the early 1980s (FORTRAN being the premier choice). Major differentiators were the requirements for accurate data representation caused by the finite-precision arithmetic, the concept of stream-based processing (reflected in the delay operator), and the necessity of not precluding parallel execution.

Efforts at the Katholieke Universiteit Leuven, Belgium, resulted in the DIGEST tools [59] that allowed the analysis of the impact of finite precision arithmetic on S/N ratio, and occurrence of small- and large scale limit cycles. The tool used a description model similar to the SPICE circuit analysis program and used as basic operators adders, multipliers, shifters, delay elements, and A/D and D/A converters. An example of a 3rd-order wave-digital filter described and analyzed in DIGEST is shown in Fig. 12.

A major event in the mid-80s was the realization that most digital signal-processing applications are better described by schematic block diagrams. The first generally available tool in that direction was the SPW design entry and analysis environment from Comdisco (Fig. 13). While very effective, these earlier tools lacked a precise definition of the underlying computational semantics. The systolic-array community at the same time was working toward a precise mathematical underpinning that would aid in the mapping of algebraic problems into arrays of tightly-synchronized processing elements [60]. These combined efforts gradually led to the development of the data-flow model for signal processing specification. This model was underpinning both textual specification environments as embodied by the Silage language [61], and the graphical modeling approach of which the Berkeley Ptolemy system [62] became the prime example. Today, data-flow is the accepted specification environment for digital signal-processing applications and is embraced by commercial tools offered by companies such as Cadence, Synopsys, and Mentor Graphics.

Digital Signal-Processing Architectural Optimization

From the early days, digital signal-processing implementations with their tightly limited time and area budgets required that the algorithm be optimized to the maximum extent with respect to architecture or implementation efficiency. An excellent example of such an optimization is the Canonical Signed Digit (CSD) representation. With the cost of parallel multipliers at a premium in the pioneering days of digital signal-processing ASICs, techniques to avoid them altogether were desirable. This was definitely an option in the case of fixed-coefficient filters, where multiplies can be replaced by sequences of add/shift operations. The CSD notation minimizes the number of 1 bits in these coefficients, hence reducing the number of add operations that have to be implemented. The CANDI program [59] took this one step further and performed a global optimization over all coefficients of a digital filter such that overall number of add/shifts is minimized while still staying within the frequency-domain specification of rejection and ripple.

Since then, the digital signal-processing community has been at the core of the developments in the world of algorithmic transformations and optimizations. Techniques such as arithmetic restructuring, retiming, and loop pipelining and folding have either been pioneered or been brought to fruition in the world of signal-processing implementation. The look-ahead transformation (pioneered by Parhi and Messerschmitt [63]) exploits the fact that digital signal-processing algorithms are stream-based to provide virtually infinite speed-ups for linearly-recursive algorithms (at the cost of area, of course). The HYPER synthesis environment of UC Berkeley [64] provided the first integrated environment to combine a wide variety of optimizing transformations and the means to analyze their impact on the eventual implementation cost. These techniques are now

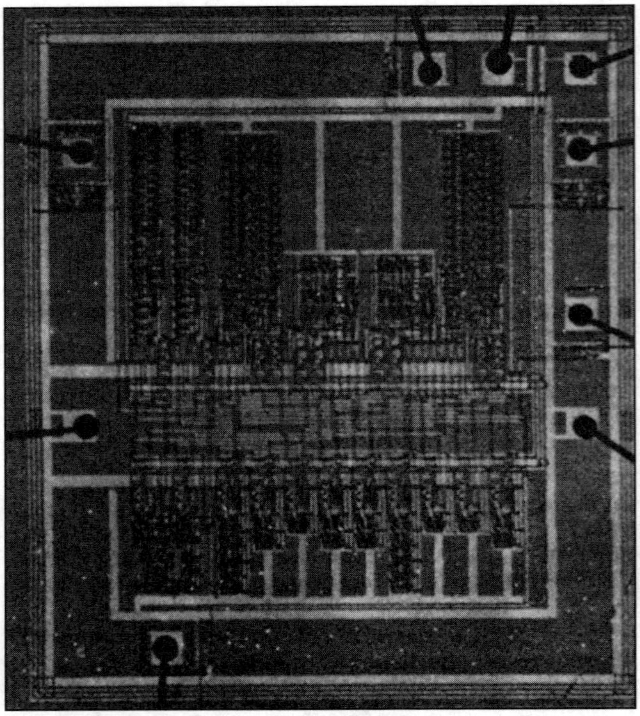

▲ *14. Bit-serial design generated by FIRST silicon compiler.*

finding their way into the worlds of ASIC behavioral synthesis and microprocessor architectural optimization.

Silicon Compilation and Architectural Synthesis

In 1980, Mead and Conway launched the concept of structured VLSI design [65]. This inspired the concept of the "silicon compiler," which would automatically generate a silicon implementation starting from a high-level description (in analogy with the software compiler). In 1981, the FIRST silicon compiler from the University of Edinburgh [66] provided a first glimpse into this prospect. FIRST targeted bit-serial architectures since bit-parallel implementations were considered to be too expensive at that time. An example of such an early-day implementation, generated from a schematic netlist, is shown in Fig. 14.

This led to complete new fields of research and development. Design automation tools were developed for bit-serial, bit-parallel, and systolic architectures. Important milestones in that respect were the CATHEDRAL tool-sets from IMEC (addressing, respectively, bit-serial, application-specific processor, and bit-parallel architectures) [67], and the LAGER silicon compiler from UC Berkeley [68]. The latter demonstrated how a collection of point tools, connected by a well-thought out design flow, could automatically translate a parameterized description of a VLSI architecture into a silicon artifact.

Silicon compilation and behavioral synthesis are now common household words in the design-automation world. Many of the behavioral and architectural synthesis tools currently available from major design automation vendors have their direct roots in the digital signal-processing community.

Future Trends

Digital signal-processing implementation is still very much at the forefront of the development of new design methodologies for integrated circuits. The increasing complexity of signal-processing applications is now focusing the attention on the heterogeneous aspects of system design: combining multiple models of computation, different architectural styles, and various implementation approaches (hardware and software) into a single design. Dealing with complexity is the ultimate challenge, which requires both higher levels of abstractions and increased design reuse. If this problem is not addressed adequately, the cost of the 1 billion transistor design of 2010 will be astronomical.

Closing Remarks and Dedication

Since the introduction of the integrated circuit in the late 1950s, the world of signal-processing implementations has been a constant roller coaster. It is our belief that this thrill ride will continue and even accelerate in the decades to come. New architectures, implementation styles, and design methodologies are desperately needed if we want to track the exponential increase in complexity that semiconductor technology is promising us. The editor would like to thank all the members of the committee for their contributions to both this overview article and to this great and exciting community.

This 50th anniversary celebratory article is dedicated to Robert Michael Owens (Pennsylvania State University), one of the pioneers of our committee who passed away just this September while on sabbatical in Italy. Robert was a major voice in the world of signal-processing implementation, a dedicated teacher and researcher, and a wonderful colleague. We will miss him.

Jan M. Rabaey and *Robert Brodersen* are with the University of California, Berkeley. *Wanda Gass* is with Texas Instruments and *Takao Nishitani* is with NEC Corporation.

References

1. T. Engibous, "Focus on Digital Signal Processing Solutions," Bernstein Strategic Decisions Conference, HYPERLINK "http://www.ti.com/corp/docs/library/bern/bern.html"
2. R. Newton et al., "Silicon After 2010," Final Report 1997 ISAT Study Group.
3. Design and Implementation of Signal Processing Systems Home Page, \HYPERLINK http://infopad.eecs.berkeley.edu/~jan/vsp.
4. H. Sasaki, "Multimedia Complex on-a-Chip," Proceedings 1996 ISSCC Conference, pp. 16-19, San Francisco, February 1996.
5. J. Borel, "Technologies for Multimedia Systems on a Chip," *Proceedings 1997 ISSCC Conference*, pp. 18-21, San Francisco, February 1997.
6. J. Rabaey, A. Abnous, M. Wan, Y. Ichikawa, K. Seno, "Heterogeneous Reconfigurable Systems," in *Proceedings 1997 Workshop on Signal Processing Systems*, pp.24-33, Leicester, November 1997.

7. W.F. Chow, "Transistor Superregenerative Detection," IRE-AIEE, University of Pennsylvania, Conference on Transistor Circuits, Philadelphia, PA, p. 3, February 1955.

8. W.F. Chow and Donald A. Paynter, "Series Tuned Methods in Transistor Radio Circuitry," IRE-AIEE-University of Pennsylvania 1957 Transistor and Solid State Circuits Conference, Philadelphia, PA, pp. 20-22, February 1957.

9. A. Proudfit, K.M. St. John, C.R. Wilhelmsen, and R.J. Farber, "A New Approach to Transistor Receiver Design," RE-AIEE-University of Pennsylvania 1957 Transistor and Solid State Circuits Conference, Philadelphia, PA, pp. 22-23, February 14-15, 1957.

10. S.W. Tehon, "Miniaturized Ceramic Filters, " *ISSCC Digest of Technical Papers*, pp. 34-35, February 1959.

11. L.E. Franks and F.J. Witt, "Solid State Sampled-Data Bandpass Filter," *ISSCC Digest of Technical Papers*, pp. 70-71, February 1960.

12. A.A. Gash, R.S. Pepper, and D.O. Pederson, "Design of Integrable Desensitized Frequency Selective Amplifiers," *ISSCC Digest of Technical Papers*, pp. 34-35, February 1966

13. G.P. Edwards, P.J. Jennings, and T. Preston "A MOS LSI Double Second Order Digital Filter," *ISSCC Digest of Technical Papers*, pp. 20-21, February 1975.

14. L. Boonstra, F.L.J. Sangster, "Progress on Bucket-Brigade Charge Transfer Devices," *ISSCC Digest of Technical Papers*, pp. 140-141, February 1972.

15. I.A. Young, D.A. Hodges, P.R. Gray, "Analog NMOS Sampled-Data Recursive Filter," ISSCC Digest of Technical Papers, pp. 156-157, February 1977.

16. C.H. Sequin, M.F. Tompsett, D.A. Sealer, P.I. Suciu, P.M. Ryan, "Sensing Technique for Self-Contained Charge-Coupled Split-Electrode Filters," *ISSCC Digest of Technical Papers*, pp. 150-151, February 1977.

17. T.G. Foxall, A.A. Ibranim, G. Hupe, L.P. Sellars, "Minimum Phase CCD Transversal Filters," *ISSCC Digest of Technical Papers*, pp. 152-153, February 1977.

18. W.E. Engeler, J.J. Tiemann, R.D. Baertsch, H.S. Goldberg, "A Charge-Transfer Recursive Filter," *ISSCC Digest of Technical Papers*, pp. 154-155, February 1977.

19. D.J. Allstot, R.W. Brodersen, P.R. Gray, "Fully Integrated High Order NMOS Sampled-Data Ladder Filters," *ISSCC Digest of Technical Papers*, pp. 82-83, February 1978.

20. J.T. Caves, C.H. Chan, S.D. Rosenbaum, L. Sellars, J. Terry, "A PCM Voice CODEC with On-Chip Filters," *ISSCC Digest of Technical Papers*, pp. 182-183, February 1978.

21. P.R. Gray, D. Senderowicz, H. O'Hara, B.M. Warren, "A Single-Chip NMOS Dual-Channel Filter for PCM Telephony Applications," *ISSCC Digest of Technical Papers*, pp. 26-27, February 1979.

22. R. Gregoriran, Y.A. Hague, R. Mao, R. Blasco, W.E. Nicholson, Jr., "CMOS Switched-Capacitor Filter for a Two-Chip PCM CODEC," *ISSCC Digest of Technical Papers*, pp. 28-29, February 1979.

23. B.J. White, G.M. Jacobs, G.F. Landsburg, "A Monolithic Dual-Tone Multifrequency Receiver," *ISSCC Digest of Technical Papers*, pp. 36-37, February 1979.

24. E.J. Swanson, R.J. Starke, G.F. Cross, K.H. Olson, C.J. Waldron, "A Fully-Adaptive Transversal Canceler and Equalizer Chip," ISSCC Digest of Technical Papers, pp. 20-21, February 1983.

25. T. Suzuki, et. al. "A CMOS Switched Capacitor Variable Line Equalizer," *ISSCC Digest of Technical Papers*, pp. 74-75, February 1983.

26. K. Hanson, et. al., "1200 b/s QPSK Duplex MODEM," *ISSCC Digest of Technical Papers*, pp. 194-195, February 1984.

27. M.E. Hoff, Jr., M. Townsend, "An Analog Input/Output Microprocessor for Signal Processing," *ISSCC Digest of Technical Papers*, pp. 220-221, February 1979.

28. J.A. Cooper, J.A. Copeland, R.H. Krampeck, D.C. Stanzione, L.C. Thomas, "A CMOS Microprocessor for Telecommunications Applications," *ISSCC Digest of Technical Papers*, pp. 138-139, February 1977.

29. Y. Kawakami, T. Nishitani, E. Sugimoto, E. Yamauchi, M. Suzuki, "A Single Chip Digital Signal Processor for Voiceband Applications," *ISSCC Digest of Technical Papers*, pp. 40-41, February 1980.

30. J.R. Boddie, G.T. Daryanani, I.I. Eldumiati, R.N. Gadenz, J.S. Thompson, S.M. Walters, R.A. Pedersen, "A Digital Signal Processor for Telecommunications Applications," *ISSCC Digest of Technical Papers*, pp. 44-45, February 1980.

31. S. Magar, A. Leigh, E.R. Caudel, "A Microcomputer with Digital Signal Processing Capability," *ISSCC Digest of Technical Papers*, pp. 32-33, February 1982.

32. T. Nakamura, M. Yoshida, T. Uno, Y. Ichikawa, M. Mizutani, K. Sawada, "A Digital Signal Processing LSI," *ISSCC Digest of Technical Papers*, pp. 30-31, February 1982.

33. R.N. Kershaw, L.E. Bays, R.L Freyman, J.J. Klinikowski, C.R. Miller, et al., "A Programmable Digital Signal Processor with 32b Floating Point Arithmetic," *ISSCC Digest of Technical Papers*, pp. 92-93, February 1985.

34. R.W. Brodersen, H.S. Fu, R.C. Frye, D.D. Buss, "A 500-Point Fourier Transform Using Charge-Coupled Devices, *ISSCC Digest of Technical Papers*, pp. 144-145, February 1975.

35. B.L. Troutman, G.W. McIver, W.E. Kingsley, B.H. Whalen, "A 2u CMOS/LSI 32-Point Fast Fourier Transform Processor," *ISSCC Digest of Technical Papers*, pp. 26-27, February 1982.

36. J.L. Van Meerbergen, F.J. van Wyk, "A 2u NMOS 256-Point Discrete Fourier Transform Processor," *ISSCC Digest of Technical Papers*, pp. 124-125, February 1983.

37. F. Tanaka, et al, "C2MOS Speech Synthesis Systems," *ISSCC Digest of Technical Papers*, pp. 266-267, February 1982.

38. T. Oura, T. Isozaki, S. Toufuku, H. Igarashi, "A Single-Chip Sound Synthesis Microcomputer," *ISSCC Digest of Technical Papers*, pp. 268-269, February 1982.

39. L.T. Lin, H.F. Tseng, D.B. Cox, R. Runge, D. Conrad, "A Monolithic Audio Spectrum Analyzer for Speech Recognition Systems," *ISSCC Digest of Technical Papers*, pp. 172-173, February 1982.

40. N.H.E. Weste, D.J. Burr, B.D. Ackland, "A Systolic Processing Element for Speech Recognition," *ISSCC Digest of Technical Papers*, pp. 274-275, February 1982.

41. M. Lowy, H. Murveit, D.M. Mintz, R.W. Brodersen, "An Architecture for a Speech Recognition System," *ISSCC Digest of Technical Papers*, pp. 118-119, February 1983.

42. T. Iwata, H. Ishizuka, M. Watari, T. Hoshi, Y. Kawakami, M. Mizuno, "A Speech Recognition Processor," *ISSCC Digest of Technical Papers*, pp. 120-121, February 1983.

43. T. Fukushima, Y. Kobayashi, K. Hirasawa, T. Bandoh, E. Ejiri, H. Kuwahara, "An Image Signal Processor," *ISSCC Digest of Technical Papers*, pp. 258-259, February 1983.

44. T. Mori, K. Aono, H,.Sakai, K. Hasegawa, H. Yamada, T. Takemoto, "A Micro-Programmable Real-Time Image Processor," *ISSCC Digest of Technical Papers*, pp. 144-145, February 1986.

45. P.A. Ruetz, R.W. Brodersen, "A Realtime Image Processing Chip Set," Digest of Technical Papers," *ISSCC Digest of Technical Papers*, pp. 148-149, February 1986.

46. M. Nagatani, H. Yoshimura, T. Tsuchiya, Y. Suzuki, "Digital Signal Processors for Encoding/Decoding Color TV Signals," *ISSCC Digest of Technical Papers*, pp. 150-151, February 1986.

47. M. Yoshimoto, S.I. Nakagawa, K. Murakumi, S. Asai, Y. Akasaka, Y. Nakajima, Y. Horiba, "A Digital Processor for Decoding of Composite TV Signals Using Adaptive Filtering," *ISSCC Digest of Technical Papers*, pp. 152-153, February 1986.

48. B. Wong, H. Samueli, "A 200MHz All-Digital QAM Modulator in 1.2um CMOS for Digital Radio Applications," *ISSCC Digest of Technical Papers*, pp. 126-127, February 1991.

49. T. Nishitani, R. Maruta, Y. Kawakami, and H. Goto, "A Single-Chip Digital Signal Processor for Telecommunications Applications," *IEEE Journal of Solid-State Circuits*, SC-16, pp. 372-376, 1981.

50. W.E. Nicholson, R.W. Blasco, and K.R. Reddy, "The S2811 Signal Processing Peripheral," *Proceedings of WESCON*, pp. 1-12, 1978.

51. J.R. Boddie, G.T. Daryanani, I.I. Eldumiati, R.N. Gadenz, J.S. Thompson, S.M. Walters, "Digital Signal Processor: Architecture and Performance," *Bell Systems Technical Journal*, 60 pp.1449-1462, 1981.

52. Y. Hagiwara, Y. Kita, T. Miyamoto, Y. Toba, H. Hara, and T. Akazawa, "A Single Chip Digital Signal Processor and its Application to Real-Time Speech Analysis," *IEEE Transactions on ASSP*, ASSP-31(1), 1983.

53. M. Iaumikawa, et al, "A 0.9V 100MHz 4mW 2mm2 16b DSP core," in *ISSCC Digest of Technical Papers*, pp. 84-85, February 1995.

54. S. Mutoh, et al., "A 1V multi-threshold voltage CMOS DSP with an efficient power management technique for mobile phone applications," *ISSCC Digest of Technical Papers*, pp. 168-169, February 1996.

55. W. Lee, et al, "A 1V DSP for wireless communications," *ISSCC Digest of Technical Papers*, pp. 92-93, February 1997.

56. R. Owen, "An enhanced DSP Architecture for the seven multimedia functions: The Chromatic Research Mpact2 Media Processor," in *Proceedings 1997 Workshop on Signal Processing Systems*, pp.76-85, Leicester, November 1997.

57. R. Simar, "DSP Architectures, Algorithms and Code Generation: Fission or Fusion," in *Proceedings 1997 Workshop on Signal Processing Systems*, pp. 50-59, Leicester, November 1997.

58. Digital Signal Processing Committee, *Programs for Digital Signal Processing*, IEEE Press, 1979.

59. R. Jain et al., "Custom Design of a VLSI PCM-FDM Transmultiplexer from System Specification to Circuit Layout using a Computer Aided Design System," *IEEE Journal of Solid-State Circuits*, vol. SC-21, pp. 73-85, February 1986.

60. S.Y. Kung, *VLSI Array Processors*, Prentice Hall, 1988.

61. P. Hilfinger, "SILAGE: A High Level Language and Silicon Compiler for Digital Signal Processing," *Proceedings IEEE CICC Conference*, May 1995.

62. J. Buch, S. Ha, E. Lee, and D. Messerschmitt, "Ptolem: A Platform for Heterogeneous Simulation and Prototyping," *Proc. 1991 European Simulation Conference*, Copenhagen, June 1991.

63. K. Parhi and D. Messerschmitt, "Pipeline interleaving and parallelism in recursive digital filters - I: Pipelining using scattered look-ahead and decomposition," *IEEE Transactions on ASSP*, pp. 1099-1117, July 1989.

64. J. Rabaey, C. Chu, P. Hoang, and M. Potkonjak, "Fast Prototyping of Data-Path Intensive Architectures," *IEEE Design and Test of Computers*, pp. 40-51, June 1991.

65. C. Mead and L. Conway, *Introduction to VLSI Design*, Addison-Wesley, 1980.

66. P. Denyer, "A Silicon Compiler for VLSI Signal Processing," *Proceedings ESSCIRC Conference*, pp. 215-218, Brussels, 1982.

67. J. Rabaey and H. De Man, "Computer-aided Design of Signal Processing Systems," *Proceedings IEEE ICCD Conference*, pp. 134-137, Port Chester NY, October 1987.

68. C. Shung et. al., "An Integrated CAD System for Algorithmic-Specific IC Design," *IEEE Journal on Computer Aided Design*, Vol. 10, No 4, pp. 447-463, April 1991.

ABSTRACT

Contemporary wireless communications are based on digital communications technologies. The recent commercial success of mobile cellular communications has been enabled in part by successful designs of digital signal processors with appropriate on-chip memories and specialized accelerators for digital transceiver operations. This article provides an overview of fixed point digital signal processors and ways in which they are used in cellular communications. Directions for future wireless-focused DSP technology developments are discussed.

Digital Signal Processors in Cellular Radio Communications

Zoran Kostic, AT&T Laboratories-Research

Selvaraj Seetharaman, GlobeSpan Technologies Inc.

Current trends suggest that the wireless communications industry will continue to grow at an impressive rate. Today, the dominant application of wireless technologies is speech communications. A number of cordless, cellular, and personal communications services (PCS) standards have been developed and at least half a dozen of them are successful such as Global System for Mobile Communications (GSM), IS-136, IS-95, PDC, Digital European Cellular Telecommunications (DECT), and Personal Handy Phone System (PHS). The work on other commercial radio applications, such as wireless local loops, wireless local area networks (WLANs), wireless asynchronous transfer mode (W-ATM), and satellite communications, is being pursued at full speed in laboratories at many companies and universities. Tables 1 and 2 show various wireless applications and standards. Although different wireless applications have many commonalities, each is sufficiently special to require different architectures and implementations. This article takes the example of second-generation mobile wireless systems and discusses the use of digital signal processors (DSPs) in their realization. Also, the article presents a view on the future directions in the design and application of DSPs for third-generation digital wireless systems.

Type of application	Wide area (high mobility)	Local area (low mobility)
Voice communications	Cellular phones Satellite systems	Cordless telephones Wireless local loop systems Wireless PBXs
Data communications	Cellular telephones Pagers Specialized packet data (RAM, ARDIS, CDPD) Satellite systems	Cordless telephones Wireless LANs

■ Table 1. *Wireless applications.*

Worldwide deployment of cellular mobile systems and provision of services to large numbers of consumers requires that cost, size, and power consumption of end-user terminals and base stations meet the aggressive expectations of service providers and users. Advances in integrated circuit (IC) technology are among the crucial contributions which enabled the widespread commercialization of wireless communications. Key aspects of very large-scale integration (VLSI) technological development that concern wireless system designers are fundamental advances in semiconductor processes, digital signal processing, and radio frequency circuits.

Second-generation cellular mobile systems are based on digital communications concepts. The number of functions which have to be implemented in digital transceivers and their complexity require that a programmable VLSI device be available. This role has been filled by digital signal processors (DSPs), a special variant of microcomputers. The use of DSPs in wireless has grown many-fold with the start of deployment of GSM, IS-136 and other cellular systems in the early 1990s. Today, tens of millions of DSPs are in use in mobile phones and base stations worldwide.

The development of digital wireless technologies has taken a multiversion/multistandard approach. Opportunities to realize business and intellectual property issues motivate companies to work on proprietary and/or different air interfaces. Although the European-developed GSM is the most successful cellular system, it has recently seen strong competition in emerging markets. The specification of third-generation wireless systems is being done under the umbrella of the International Telecommunication Union (ITU) and is named IMT-2000 (for International Mobile Telecommunication by the year 2000). Although the initial goal of the ITU work was to come up with a single worldwide standard for as many components of the system as possible, it seems that commercial forces are going to modify this goal, and several "compatible" air interface wireless standards will be adopted. The Japanese standardization body Association of Radio Indstries and Businesses (ARIB), European Telecommunications Standards Institute (ETSI), and the U.S. Telecommunications Industry Association (TIA) are soon to receive numerous for-

Reprinted from *IEEE Communications Magazine,* Vol. 37, No. 12, pp. 22–35, December 1997.

mal contributions for future technologies from participating members. Regardless of these developments, the role of digital signal processors in the future of wireless systems is not in doubt.

This article is divided into seven sections. The second section is devoted to DSPs. The discussion is focused on fixed-point devices. DSP architectures and development tools are reviewed. A table of contemporary DSPs most suitable for wireless applications is presented. The third section is an overview of general issues in wireless systems. The fourth section treats the subject of wireless transceiver design. We focus on physical-layer functionality, that is, DSP implementations of transmit and receive algorithms as well as software/hardware partitioning issues. Digital speech coding is at the heart of second generation wireless systems and we devote the fifth section to this topic. The sixth section discusses future trends in DSP design and applications in wireless communications. The article ends with a summary.

	Digital cellular	Digital cordless	Paging
North America	IS-95 (800, 1900) IS-54/136(800, 1900) PCS1900	Unlicensed 900 MHz ISM PACS	FLEX™ ReFLEX™ InFLEXion™ PACT
Europe	GSM DCS1800	DECT	FLEX ReFLEX InFLEXion Ermes
Japan	PDC PDC1500	PHS	FLEX ReFLEX InFLEXion
Asia-Pacific	GSM IS-95 IS-54/136	DECT PHS	FLEX ReFLEX InFLEXion

■ **Table 2.** *Worldwide digital wireless standards for cellular, cordless, and paging. Courtesy of Texas Instruments; reprinted with permission; http://www.ti.com/sc/docs/wireless/97/standards.html*

Digital Signal Processors

Programmable DSPs are specialized microcomputers, VLSI devices designed for implementation of extensive arithmetic computation and digital signal processing functions through downloadable or resident software/firmware [3]. Their hardware and instruction sets usually support real-time application constraints. Classical examples of signal processing functions are finite impulse response (FIR) filters and fast Fourier transforms (FFTs). DSPs are meant for performance, not for extensive functionality or programmer convenience.

Hardware and instruction sets of conventional microcontrollers and microprocessors do not include structures specially designed to execute the previously mentioned functions efficiently. Contemporary high-end microprocessors are expensive devices (hundreds of dollars) with significant power consumption (watts), most of which are computer designs targeted at multimedia computing. Microcontrollers target applications such as automobile electronics, and industrial and home appliance control. On the other hand, DSPs suitable for wireless applications distinguish themselves particularly in low power consumption, low cost, and high number-crunching capabilities.

The following subsections summarize DSP architectures, memory structures, DSP software and development tools, and interfaces to DSPs. Detailed presentations on DSPs can be found in [3, 4, 6, 15, 20, 21]. A list of fixed-point DSPs suitable for wireless applications is given in Table 3.

DSP Architectures

The most basic architectural component of a DSP is the fast multiplier/accumulator (MAC) integrated into the data path, rather than placed on a coprocessor of a microcontroller. This makes possible an instruction cycle time equal to the cycle time of the arithmetic hardware. The multiply-accumulate instruction can be expressed as

$$a = a + b * c.$$

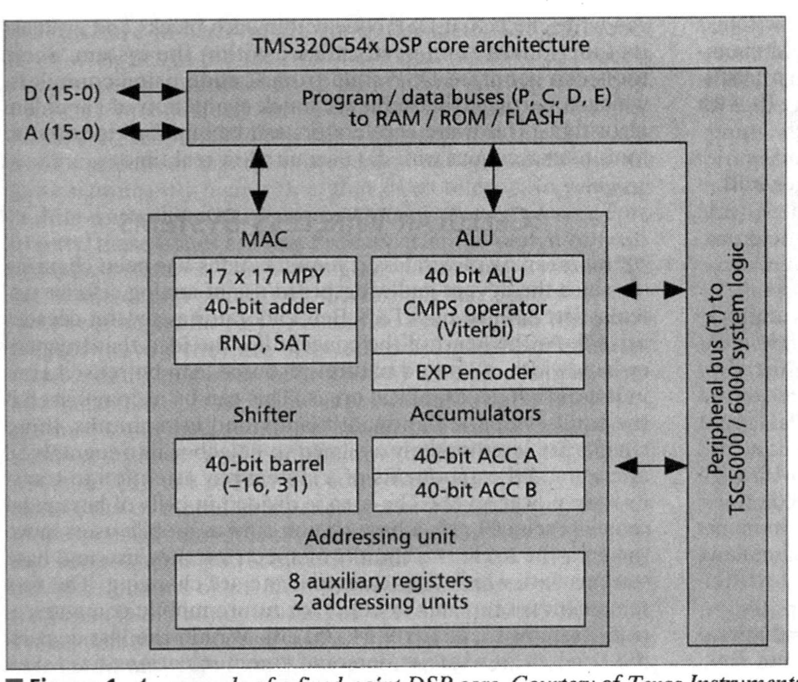

■ **Figure 1.** *An example of a fixed-point DSP core. Courtesy of Texas Instruments, reprinted with permission; http://www.ti.com/sc/docs/wireless/97/core.htm.*

The instruction cycle time of contemporary DSPs is on the order of tens of nanoseconds. One of the benchmarks expressing the power of a particular DSP is the MAC time, which also represents the time to execute a single tap instruction in a FIR filter. The optimization of the MAC unit in DSPs leads to the fact that a typical DSP performs nonoptimally for functions such as conditional branching or a sequence of control instructions.

DSPs use extensive pipelining,[1] several independent on-chip memories, parallel function units, and a hardwired (rather than microprogrammed) design. DSPs utilize instructions such as MAC, bit reverse addressing for FFTs, and those that support circular buffering. Hardware-wise, DSPs contain phase-locked loop (PLL) oscillators, several 8- or 16-bit busses, clock interrupt circuits, onboard ROM and RAM, and sometimes an onboard CPU. DSPs typically process buffers of data, and often these buffers are reused on successive frames of data.

Programmable DSPs are either floating-point

[1] *Pipelining refers to a special form of parallel execution of several operations, where downstream operations depend on results of upstream operations.*

or fixed-point devices. The overwhelming virtue of fixed-point DSPs is that they can be very fast and inexpensive. Wireless applications primarily use fixed-point processors, and this article will focus on their applications. The most often used word size in fixed-point DSPs is 16 bits. Fixed-point arithmetic requires programmers to pay attention to precision, scaling, and overflows which come from quantization errors of analog-to-digital converters and finite-length multiplication. Multiplying two 16-bit numbers yields a 32-bit number, which in 16-bit DSPs needs to be truncated down to a 16-bit number again. Larger word sizes for accumulators, shifters, scaling and saturation hardware are used to simplify the programmer's task.

The dominating architecture for digital signal processor designs is the so-called Harvard architecture and its variants [3]. The major feature of this architecture is that the program and data buses are separate. In all Harvard architectures an instruction is fetched at the same time that operands for a previously fetched instruction are. This suggests that a three-stage pipeline consisting of the instruction fetch, operand fetch, and instruction execution is at the heart of this architecture. An example of a DSP core based on the Harvard architecture is given in Fig. 1. Whereas traditional microprocessor architectures achieve high performance with register-to-register instructions, DSPs have sufficiently high memory bandwidth and use more memory-to-memory instructions. This is done by accessing several memory allocations in each instruction cycle, which is possible due to rich addressing modes and parallel memory banks. The basic Harvard architecture contains two memories and two buses, and more sophisticated Harvard architectures may have four data memories and two program memories. DSPs are microcomputers rather than microcontrollers (since they have memory on-chip). Numerous memory banks are hard to implement off-chip due to pin number limitations and the speed penalty of off-chip busing. Often, additional memory is used off the DSP and internal buses are multiplexed to the outside. Register-indirect addressing is the method used to accomplish multiple-address specification in software. Modulo-mode addressing helps with filtering-type operations where circular access to data may be needed.

Storing data without power is a requirement for wireless portable devices. To this end technologies such as flash memory and ferroelectric memory are being included in DSPs, although the cost remains an issue at this time. DSPs with flash memory help designers by:

- Cutting down on ROM code turnaround time
- Facilitating field trial and last-minute debugging and code fixing
- Making external DSP memory boards unnecessary

DSP Software and Development Tools

Three major approaches to DSP programming exist: interlocking, time stationary, and data stationary coding. These techniques are related to a particular use of the underlying pipeline-capable DSP hardware [3]. Different DSP hardware manufacturers and product generations follow different approaches to this issue. The use of any of these approaches requires skilled and experienced programmers. Even with DSP speeds seen today, much DSP code often needs to be tuned by hand to accomplish the tight real-time requirements of wireless communications. Optimizing compilers are making strides but are not capable of entirely replacing hand-written assembly code.

Signal processing system designers have at their disposal development tools to simulate, edit, compile, assemble, link, load, and debug DSP software programs [6, 7]. Simulators, hardware emulators, and development boards let developers work in non-real time or in real time on the development of their code. Debuggers often have limited capabilities in real-time testing due to limited ability to set breakpoints.

Some system-level tools that deal with code development for DSPs are N!Power by Signal Technology Inc., Hypersignal Windows by Hyperception, DADiSP by DSP Development Corp., Signal Processing Worksystem by Alta, COSSAP by Synopsys, SystemView by Elanix, Omnisys by Hewlett Packard, and DSPStation by Mentor Graphics.

A hierarchical block diagram approach to representing both high- and low-level functions, which can be integrated into large models and systems, is more and more popular [9]. Examples of development tools that take this approach are Signal Processing Worksystem and COSSAP. These tools have optional libraries that contain blocks (representing functions) covering a large number of signal processing and digital communications functions. The suite of software by Hewlett Packard has extensive support for mixed analog/digital communications system design. Some specific blocks included in almost all packages are Viterbi trellis search implementations, quadrature amplitude modulation (QAM) modulators and demodulators, wireless channel models, equalizers, and synchronization blocks.

The first generation of simulation tools dealt with "system" level designs emphasizing behavioral (waveform) models unrelated to the actual implementation. Recent generation of communication system simulation tools is integrated with suites of tools that address VLSI/ASIC (application-specific integrated circuit) implementations, most often through an intermediate step of very high-speed integrated circuits hardware description lagnuage (VHDL)/Verilog code generation. The underlying methodologies of simulation techniques belong to one of the following classes: time-driven, event-driven, data- or stream-driven, or mixed-mode.

Parameters of blocks in block-diagram-based tools are modifiable, and functions can be either floating- or fixed-point. A user can create his own blocks with arbitrary functionality by writing C or other code and embedding it into a system through wrappers and block diagrams. There exist blocks which model DSPs and their functionality. A developer can write the actual DSP code within such blocks and evaluate its functionality and performance within the system. Some tools can generate DSP code from C code using compilers, which is particularly useful for quick evaluation of candidate algorithms. Hardware accelerators can be attached to software tools to experiment with data acquired in real time.

Cellular Wireless Systems

The meaning of wireless communications has been changing since the advent and widespread use of analog cellular systems, first defined by AT&T Bell Laboratories several decades ago [1]. At the heart of the concept lies the idea that frequency bandwidth, a limited natural resource, can be reused even in nondistant geographical areas. This can be accomplished if the total available bandwidth is divided into chunks, those chunks are appropriately assigned to neighboring geographical areas, and all components of a system pay attention to transmission power levels. The area is divided in cells of hexagonal shapes; each cell gets a base station, and as mobile users move through the area, the chunks of spectrum they use and base stations with which they communicate are changing. The very same concept remains in effect for future mobile communications systems [1, 2, 8, 10, 24, 25, 27]. Within the last decade, the concept of wireless personal communications has taken off. In many respects, it is a modification of the cellular concept (at the air interface level) with inclusion of low cost and

Analog devices ADSP-21xx family	16-bit fixed point, enhanced Harvard architecture 10.2–33 MIPS range Program RAM: 0, 1, 2, 16 kbyte; data RAM: 512, 1, 2, 4, 16 kbyte, ROM: 0, 4, 8 12 kbyte 5V or 3.3V operation, power-down modes, host interface ports, DMA port, 16 bit codec http://www.analog.com/products/selection_trees/dsp/dsp.html
ATMEL LODE DSP core	16-bit fixed point, two single-cycle MACS, 40-bit ALLU, four 40-bit accumulators, 64K program and 64K data memory address space, 16-level hardware stack, nestable loops http://www.atmel.com/atmel/products.tcsi/lcore.html
DSP Group: Pine, Oak	25 ns cycle, 40 MIPS, 3.3 V, DMA mode, power managment, expandable RAM/ROM, core-based http://www.lsil.com/products/unit5_5e.html
Lucent Technologies DSP1610/1611 DSP1616 DSP1617/1618 DSP1620 DSP1627/1628/1629 SABRE DSP 16000	 Base stations, RAM up to 12K, ROM up to 2K, 26 MIPS at 3.3 V, 50 MIPS at 4.75 V RAM = 2K, ROM = 12K, 26 MIPS at 3.3 V, 50 MIPS at 4.75 V RAM = 4K, ROM = 24K, Viterbi accelarator, 26 MIPS at 3.3 V, 50 MIPS at 4.75 V Base stations, RAM = 32K, ROM = 4K, Viterbi accelerator, 80 MIPS at 3.3 V, 120 MIPS at 4.75 V RAM up to 16K, ROM up to 48K, Viterbi accelerator, 78 MIPS at 2.7 V Two MACs, 8 40-bit accumulators, three-operand adder, 62 K x 16 dual-port RAM, 31-word cache, 200 MIPS http://www.lucent.com/micro/wam/docs/OT97215.pdf
Motorola DSK56000 family DSP56100 family DSP56300 family DSP5600 family	 24-bit fixed point 16-bit fixed point, 30 MIPS/33 ns/60 MHz, up to 8K RAM and 4K data ROM, 256b RAM and 8K program ROM 24-bit fixed point, 80–100 MIPS, 3-3.6V, up to 24K program RAM, 7K data ROM and RAM 16/24 bit low power for cellular, 60 MIPS at 2.7V, 24K x 24 program ROM, 4K x 16 X-data RAM, 6K x 16 X-data ROM, 4K x 16 Y-data RAM, 6K x 16 Y-data ROM http://www.mot.com/pub/SPS/DSP/LIBRARY/56100/FM_REV0/1.PDF
Texas Instruments TM320C1x TM320C2x TM320C5x TM3206x	 16-bit fixed point, 5 MIPS, 200 ns multiply, up to 256 b RAM and 1.5K ROM, 5 V 16-bit fixed point, 10-12 MIPS, 80 ns multiply, up to 544 b RAM, 4K ROM, 5 V 16-bit fixed point, 40–50 MIPS, 20–50 ns, up to 10K RAM, up to 32K ROM, 3 and 5 V, $15–60 0.25 μ technology, 200 MHz, 2000 MIPS, very long instruciton word architecture, RISC-like instructions, eight 32-bit instrucitions per cycle, 512 b program, 512 b data, 2.5 V CPU with 3.3V peripherals http://www.ti.com/sc/docs/dsps/products.htm

■ **Table 3.** *List of some fixed-point DSP devices suitable for wireless communications applications.*

the anytime/anywhere communications notion [8, 10]. From the point of view of DSP use, the two do not differ much.

In cellular systems, the requirements for base stations and mobile stations are different. At the physical layer, of concern to us, one can say that the functions which base stations and mobiles have to perform are about the same. The difference most often lies in the fact that base stations have to process signals going to and coming from many mobile stations, whereas a mobile deals with a signal of one base station only. There are exceptions to this rule (e.g., soft handoff in IS-95), but it is safe to say that the processing power of base stations has to be significantly higher than that of mobiles. Also, whereas mobile units have hard power consumption and size constraints, these issues are less prominent in the design of traditional base stations. This is changing with the more widespread use of micro- and picocells of increasingly small size.

Wireless systems can be classified according to the power emission levels they are allowed to use. One can consider those systems that operate at 0.5–5 W as high-power systems and those that operate below as low-power systems [11]. Since the most significant amount of power in a radio transceiver is taken by a power amplifier, the way the DSPs are used is not affected. The use of DSPs is going to be determined by the complexity of modulation, coding, specifics of the necessary receiver design, and other similar factors. Examples that prove this point are low-power systems such CT-2 and DECT cordless and versions of cordless in the industrial, scientific, and medical (ISM) band, as well as WLANs and the personal access communications system (PACS) (wireless local loop standard) [11]. Although they require low-power transmission, their physical layers are of significant complexity suitable for DSP applications.

The most flexible development strategy is always the first choice of a seasoned engineer, assuming that all other things are of equal significance. From this point of view, using DSPs is the most desirable option. Algorithms can change conceptually, and the DSP program change is easy to execute. This is not simple when hardwiring is the implementation strategy, as is the case for ASICs. Wireless environments are notorious for undesirable effects on transmitted signals. Simulation of these channels to come up with system parameters is helpful, but only experiments with real-time platforms bring designers to ultimate product solutions. From this point of view DSPs are particularly suitable for wireless. If DSPs are designed making sure that power consumption is low, that they have appropriate amounts of on-chip memory, that they are easy to interface to and very low-cost, it is very compelling to use programmable DSPs rather than other implementation approaches.

Wireless Transceivers

Due to the nature of radio communications, physical-layer design problems fall into one of two categories: those concerned with radio frequency/intermediate frequency (RF/IF) subsystems and those with baseband subsystems. We discuss baseband issues. An overview of design problems in the domain of RF/IF is given in [5].

Radio Architectures

Radio communications require that a desired signal's original bandwidth be up-converted to the range of frequencies suitable for radio transmission, and that the complementary operation be done at the receiver. There are several different approaches to accomplishing the frequency conversion, all of

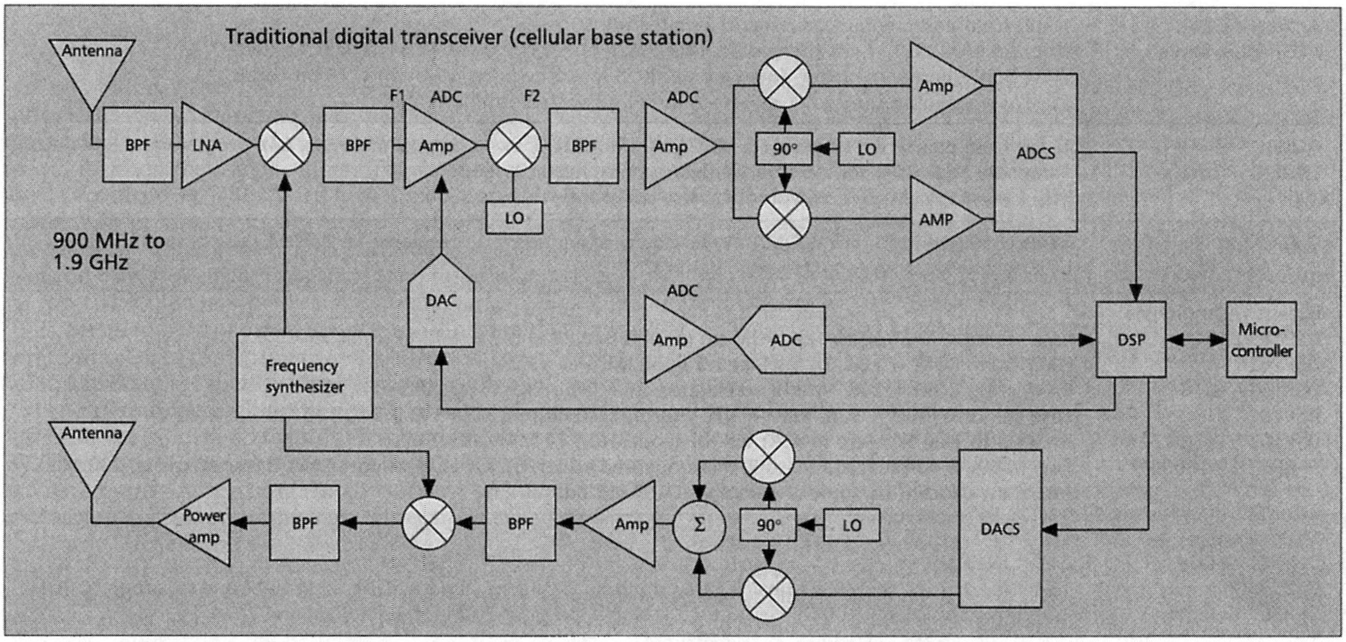

■ Figure 2. *Superheterodyne radio architecture. Courtesy of Analog Devices, reprinted with permission; http://www.analog.com/products/signal_chains/trad_dig_trans/trad_dig_trans.html.*

which have baseband portions amenable to DSP implementations. In [12], four principle radio transceiver architectures are defined: superheterodyne, digital IF receivers, direct conversion receivers, and I/Q-less analog-to-digital (A/D)-less digital receivers. An example of the most often used radio receiver architecture is given in Fig. 2. This figure illustrates the superheterodyne radio architecture where up-/down-conversion is done in two mixing steps. For this architecture, all baseband functionality is allocated to a DSP.

Direct conversion architectures remain vulnerable to problems of high-gain low-noise mixer, high dynamic range, extremely high sensitivity, I and Q phase balancing, rapid large DC offset cancellation, antenna isolation, and high-selectivity filters. Digital IF receivers use superheterodyne principles but move the digitization of the signal higher into the IF frequency band. They do not require analog devices which individually handle quadrature components of the signal, but leave I and Q separation and processing to digital functions suitable for DSPs. The preeminent problem in using this architecture in cellular communications has to do with power-hungry high-speed A/D converters.

In recent times the concept of software radio has been given significant attention in some communications system development circles. The primary goal of the software radio is to replace as many analog components (those dealing with IF/RF sections) and hardwired digital VLSI devices of a transceiver with programmable devices (such as DSPs). This implicitly involves the use of wideband A/D and digital-to-analog (D/A) converters [13]. Several tutorial articles have been published in this magazine on the topic of software radio [14, 15]. This idea is conceptually very appealing, and very friendly to the use of DSPs. The proliferation of different air interface

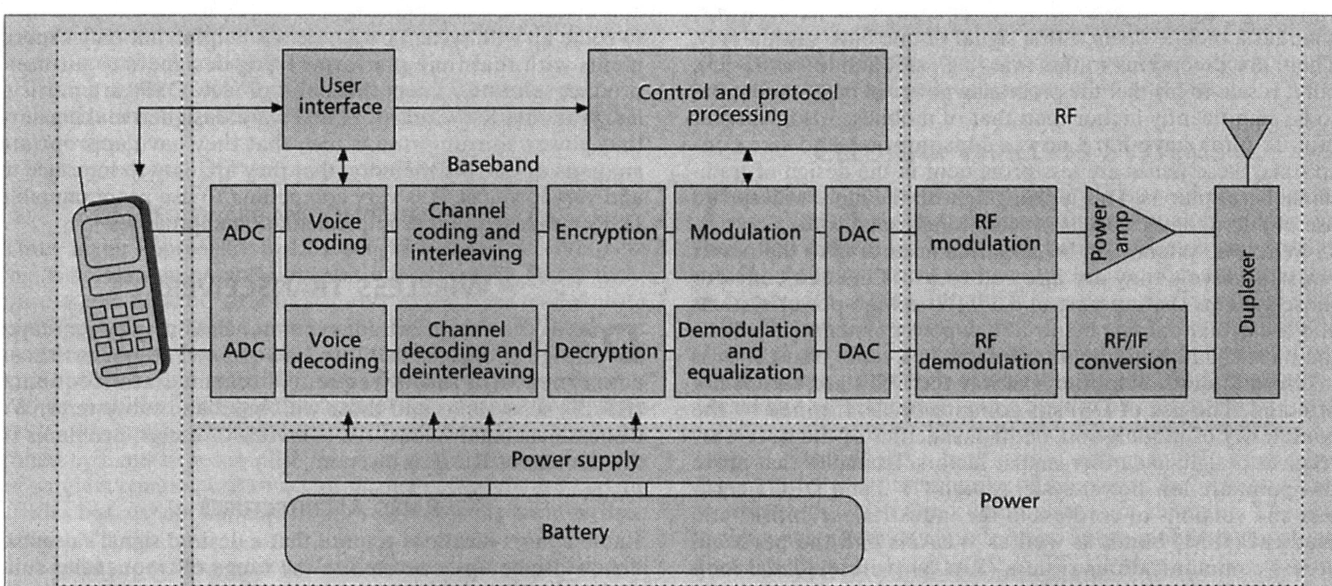

■ Figure 3. *Block diagram of a generic transceiver. Courtesy of Texas Instruments, reprinted with permission; http://www.ti.com/sc/graphics/wireless/97/enable/rfgraph.gif.*

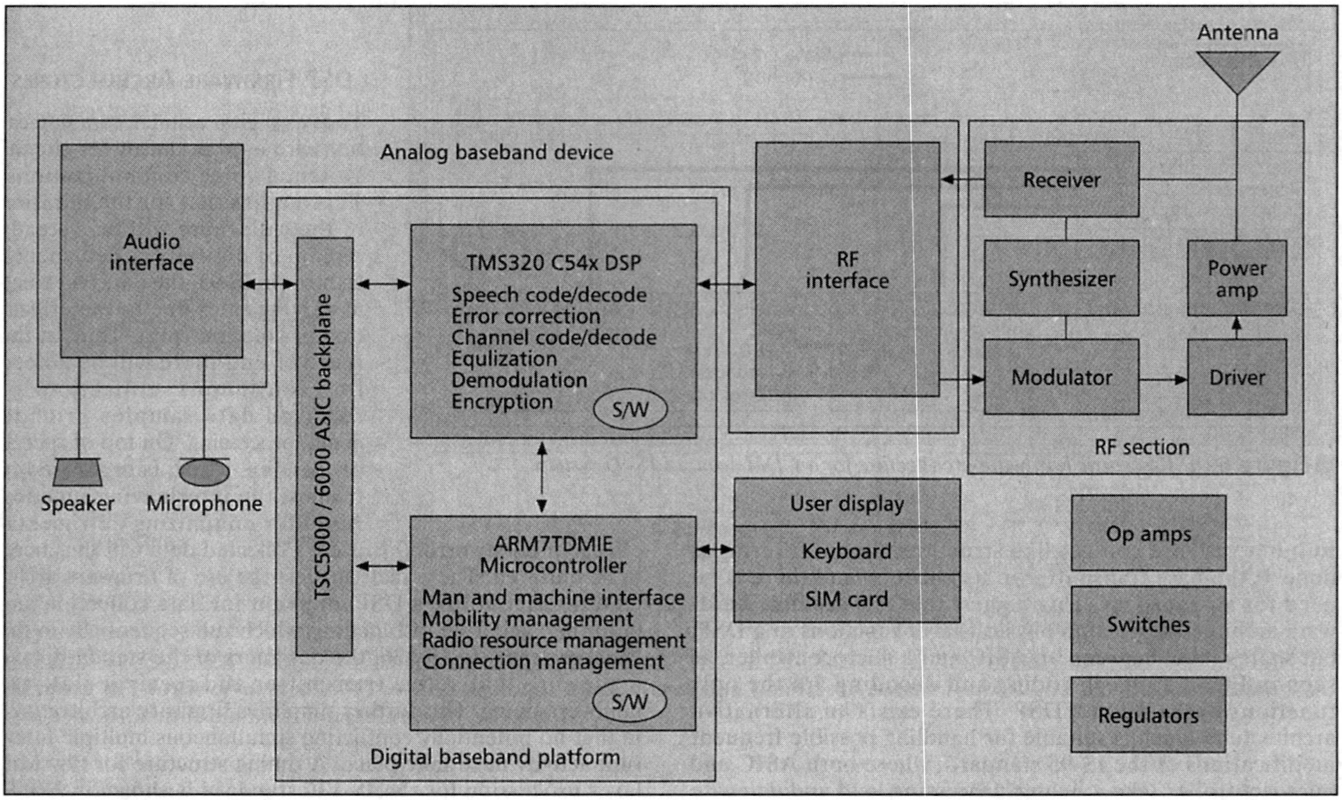

■ **Figure 4.** *An example of partitioning of transceiver functions. Courtesy of Texas Instruments, reprinted with permission; http://www.ti.com/sc/graphics/wireless/97/baseband/fig3.gif.*

standards may provide a boost to multiple-standard-device designs which would benefit from software radio work. However, the time has not yet come for low-cost low-power communications devices, and we shall not discuss this any further.

In this article we focus on the use of DSPs within traditional superheterodyne architectures.

Baseband Hardware Architectures

In designing baseband portions of portable digital communications devices, systems engineers can mix and match microcontrollers, DSPs, and ASICs. The choice can be driven by development time constraints, cost, size, level of comfort with functions that need to be implemented, challenges data rates impose on hardware, and other system requirements. Different wireless standards have different optimal hardware designs. A block diagram illustrating a generic wireless transceiver is shown in Fig. 3. In the following we look at examples of two American digital cellular standards: IS-136 and IS-95.

IS-136 — IS-136 is a time-division multiple access (TDMA)-based standard with channel data rates of 24.3 kbaud with $\pi/4$-DQPSK (differential quadrature phase shift keying) modulation. For heterodyne receiver architectures, this rate creates no extravagant challenges to A/D and D/A converters. Physical-layer receiver functions needed to accommodate this standard belong to the class of compute-intensive signal processing operations suitable for DSPs. This is particularly true for digital filtering, equalization, and channel decoding. Speech coding/decoding is based on code excited linear prediction (CELP) principles and requires vector searches and filtering, both appropriate for DSP implementations. Since cellular standards require numerous call processing operations, the existence of a microcontroller is required. In Figs. 4 and 5 we show examples of a heavily exploited hardware/software architecture for IS-136.

IS-95 — IS-95 is a code-division multiple access (CDMA)-based standard with channel data rates of 1.2288 Mchip/s and 19.2/28.8 kbaud. The chip rate mentioned above is not the rate at which digital samples can be processed in today's general-purpose programmable DSPs. The implication is that IS-95 mobile hardware architecture cannot follow the same path as IS-136 architectures. A considerable portion of the baseband functionality related to signal despreading has to be implemented in an ASIC. The rest of the receiver physical layer does not have an extensive signal processing component

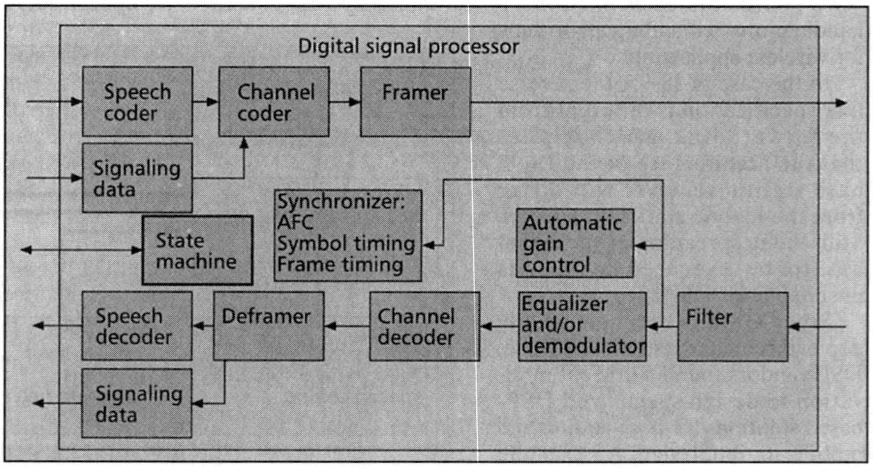

■ **Figure 5.** *Partitioning of IS-136 functions into DSP firmware.*

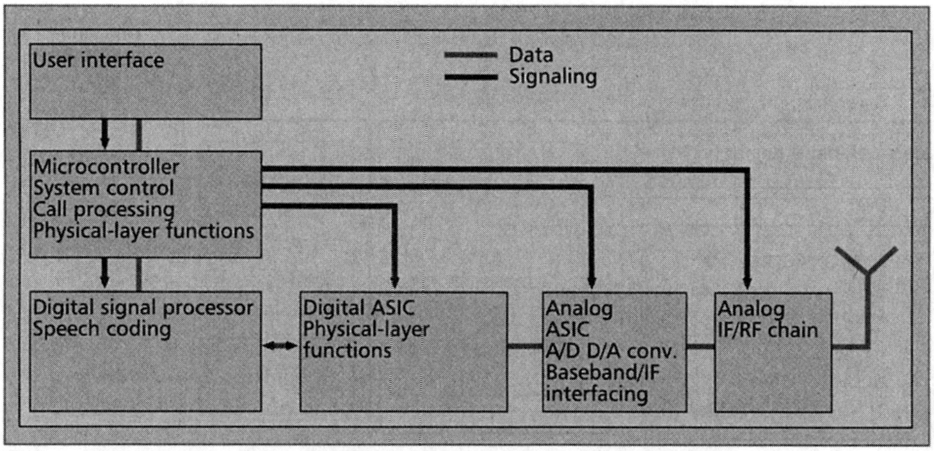

■ **Figure 6.** *ASIC-centric hardware architecture for a CDMA-based IS-95 system.*

to it but rather a control-like structure: digital filtering is done at the base transmitter in its entirety, and there is no need for an equalizer. This suggest that one credible hardware architecture puts no physical-layer functions in a DSP, but shares them between an ASIC and a microcontroller, as seen in Fig. 6 . Speech coding and decoding are the only functions executed in a DSP. There exists an alternative architecture which is suitable for handling possible frequent modifications of the IS-95 standard, where both ASIC and microcontroller take a lighter processing load and delegate whatever is possible of the physical layer to a DSP. This is illustrated in Fig. 7.

Base Stations — In the case of IS-136, for each communications link (mobile user) the functions a base station needs to execute are of the same complexity as those a mobile needs to execute. The difference lies in the fact that base stations need to handle many mobiles. This necessitates that careful thought be given to optimizing DSPs such that they can accommodate this requirement. Indeed, DSP vendors are paying attention to this segment of the wireless market. An illustration of the use of Lucent's DSP 1620 in base station design is shown in Fig. 8. Texas Instruments has recently come up with a family of fixed-point DSPs, TMS320C6x, with 2000 MIPS performance claimed. This DSP uses the VelociTI™ Very Long Instruction Word (VLIW) architecture. It consists of multiple execution units running in parallel to perform multiple instructions during a single clock cycle. In early September 1997, Lucent announced a DSP16000 core family of DSPs using Sabre architecture. This architecture has around 200 MIPS performance, two parallel MACs, a bank of eight 40-bit accumulators, a 31-word instruction cache, and low power consumption. It can handle more demanding algorithms for wireless applications.

In the case of IS-95, the reverse link specification is different from the forward link, which implies that the architecture of the IS-95 base station receiver will differ from the mobile station receiver's. Still, the despreading of the signal has to be executed at speeds unsuitable for DSPs.

Since GSM is the most widely deployed cellular system worldwide, DSP vendors found sufficient motivation to design specialized DSP-based solutions for base station and mobile station design. An example is Lucent's Sceptre chip set [33].

DSP Firmware Architectures

Today's digital cellular communications are used primarily for circuit-switched voice communications. This implies that for the duration of the call, there will be a steady stream of digitally coded speech frames (in IS-95, the data will come at varying rates due to the utilization of voice activity). Thus, at the receiver end there will be a need for continuous collection of received data samples prior to actual processing. On top of speech processing, which is organized in frames, the interleaving function (used for minimizing detrimental effects of bursty errors) is used. Collected data will thus need to be buffered. These facts suggest the use of firmware architectures that utilize a DSP interrupt for data collection and buffering, and a set of functions which run sequentially in the "background." In IS-136, the designers of the standard have made sure that active transmitter and receiver slots are nonoverlapping. This further simplifies firmware architectures in that no potentially conflicting simultaneous multiple interrupt activity need take place. A timing structure for physical-layer processing for the IS-136 standard is shown in Fig. 9, together with a list of DSP-implemented functions which need to be executed in the background. The same structure makes it possible to eliminate duplexers from the phone design (since IS-136 is a dual-mode standard, the analog mode would still require a duplexer in 800 kHz band cellular frequencies).

In a typical phone design, it is usual to dedicate master and control functionality to a general-purpose microcontroller. Microcontrollers are much more suitable than DSPs for branching functions which are dominant in system control software. The DSP is thus typically controlled by the microcontroller, and it runs physical-layer functions in parallel with the microcontroller running system control functions.

A critical issue in the design of a firmware architecture is the type of interface between the microcontroller and DSP. Several hardware options exist: parallel ports, serial ports, and dedicated hardwired first-in first-out buffers (FIFOs). There is usually a mismatch between fast DSP and slow microcontroller clocks. If the data is communicated over a parallel input-output (PIO) or serial input-output (SIO) port from the DSP to the microcontroller with the intention of generating microcontroller interrupts, the flow of the microcontroller software may be undesirably broken by frequent interrupt service routines.

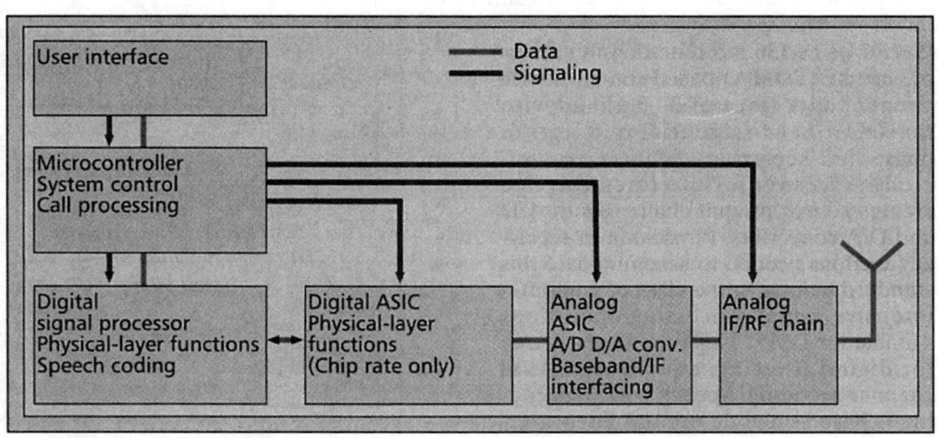

■ **Figure 7.** *DSP-centric hardware architecture for a CDMA-based IS-95 system.*

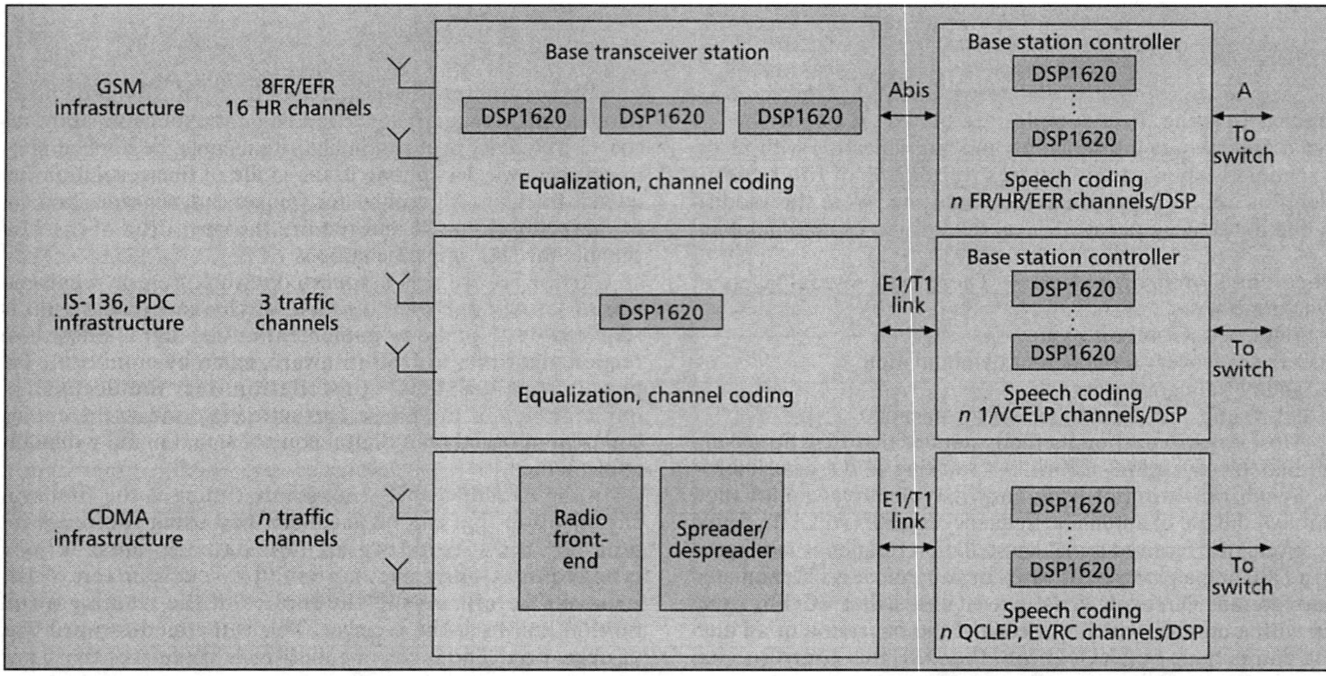

■ **Figure 8.** *The use of multiple DSPs in the base station transceiver design. Courtesy of Lucent Technologies, reprinted with permission; http://www.lucent.com/micro/wam/docs/OT97199.pdf.*

Special-purpose FIFOs are more desirable from this point of view, although they require some additional hardware.

Prudent programming practices such as writing modular code and breaking software into small, easily readable and modifiable functions are most often not applicable to writing real-time-sensitive wireless applications. Programmers have to aggressively optimize the execution time, size, and memory use of the code. It is a norm that the code for some function has to be rewritten several times before the real-time constraints of the DSP are met and all three factors are harmonized. Sixteen-bit digital signal processors have some number of 32-bit or longer registers. In cases when actual numbers involved in computations do not require more than 16-bit results, a programmer may choose to utilize unused portions of a 32-bit-long register in unexpected ways such as a temporary register/memory location.

Although one may buy DSP operating system kernels, DSP programmers have not opted to use them. Again with focus on optimization of firmware from every aspect, custom and minimal state machines are used for wireless applications.

It can easily be forgotten that wireless software has to be field tested in devices running in mobile environments. Large amounts of data may need to be collected and analyzed in real time. This requires that a DSP software architecture supports rapid data movement to devices outside the DSP which are dedicated to massive data processing. An excellent example of this would be the so-called mobile station monitor required for testing of IS-95 physical-layer functionality. This monitor needs to continuously analyze and display multipath profiles critical to the operation of rake receiver fingers.

Algorithms

To illustrate the involvement of DSPs in the wireless transceiver design, we shall briefly describe some issues in the implementation of physical-layer algorithms.

Filtering — Filtering is one of the operations cited to be ideally suited for DSP applications. FIR filtering is a repetitive process performed by multiplying the set of input signal samples with a fixed set of known filter coefficients. In the example of IS-136, pulse shaping is done at a transmitter with a square-root raised cosine filter, and appropriate matched filtering has to be done at a receiver. Although straightforward, there are instances when care needs to be exercised to make sure that filtering is executed within the smallest possible number of machine cycles. In Fig. 10 we provide an example which illustrates required modifications to the DSP code and array indexing when there is a mismatch between sampling rates of filter coefficients and the signal to be filtered. The rate of the signal before passing through the pulse shaping filter is

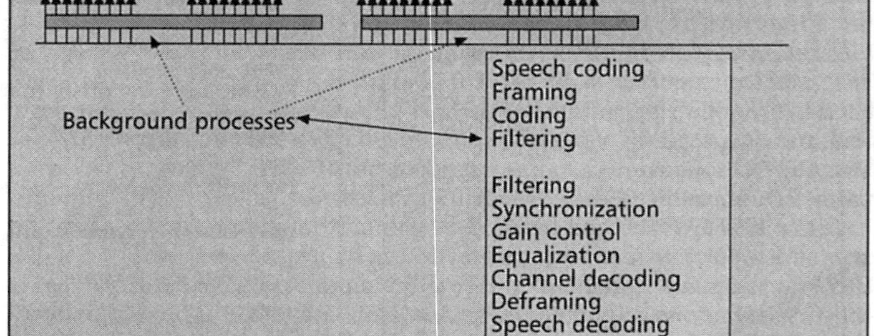

■ **Figure 9.** *The timing structure for physical-layer processes in IS-136.*

$1/T$. If one uses a filter with 4 x oversampling, the requirement is that the filter sample rate be $4/T$. If one wishes to avoid unnecessary interpolation and multiplication with zeros, a scheme is shown that requires reordering of filter coefficients as well as appropriate loop indexing. Were this modification not done, a penalty in real time would be considerable.

Receiver Synchronization — There exist several layers of synchronization:

- Frequency synchronization
- Carrier recovery for coherent demodulation
- Symbol timing recovery
- Slot, frame, and superframe synchronization

Most synchronization methods require that they be accomplished through initial acquisition, tracking, and reacqusition.

Synchronization of frequency is typically accomplished through the use of automatic frequency control (AFC). In digital receivers, the received signal constellation rotation is monitored in a DSP and a phase error based measurement is differentiated and filtered. This error signal is used in a digital VCO to come up with a number which will correct the operation of an analog component (VCXO). From the DSP this control signal can be sent through the dedicated D/A to a digitally controlled device or converted to a pulse-width modulation form suitable for transmission as an analog signal. Components of wireless products are specified to be inexpensive and thus coarse frequency acquisition algorithms must be applied prior to getting a wireless phone in the mode of fine-resolution frequency tracking. Algorithms for frequency synchronization are often feedback-based and require the operation of the PLL suitable for DSP implementations.

Carrier recovery is associated with coherent receivers, where knowledge of the phase of the received signal is required. It is simple to implement a fully digital phase correction algorithm in DSP firmware, again by monitoring the phase error in a signal constellation. It is the decision of implementers if the phase correction is done in the analog component based on a digital control signal or fully digitally implemented.

In the example of IS-136, frame timing is the first synchronization that can be accomplished using training symbols at the beginning of all data frames. Frame synchronization is accomplished by correlation of the received waveform with the replica of the training waveform(s) known to the receiver. This is the feed-forward type of operation. The receiver repetitively correlates the signal until it identifies a peak in the correlation function, and based on the peak's location in time adjust its timing. It is

```
/*--------------------------------------------
FIR filter coefficients
square root raised cosinealpha = .33
Note permuted indices and 4 segments -
done for filtering efficiency in a DSP
--------------------------------------------*/
   /*3*/  coeff(0) = -2.57910E-03;
   /*7*/  coeff(1) = 8.57456E-03;
   /*11*/ coeff(2) = -2.44425E-02;
   /*15*/ coeff(3) = 0254584;
   /*19*/ coeff(4) = 3.33408E-02;
   /*23*/ coeff(5) = -1.10911E-02;
   /*27*/ coeff(6) = 3.63476E-03;
   /*31*/ coeff(7) = -5.64713E-04;

   /*2*/  coeff(8) = -4.19223E-03;
   /*6*/  coeff(9) = 1.58502E-02;
   /*10*/ coeff(10)  = -4.47768E-02;
   /*14*/ coeff(11)  = 0.201956;
   /*18*/ coeff(12)  = 0.117725;
   /*22*/ coeff(13)  = -3.45362E-02;
   /*26*/ coeff(14)  = 1.19383E-02;
   /*30*/ coeff(15)  = -259598E-03;

   /*1*/  coeff(16)  = -2.59298E-03;
   /*5*/  coeff(17)  = 1.19383E-02;
   /*9*/  coeff(18)  = -3.45362E-02;
   /*13*/ coeff(19)  = 0.117725;
   /*17*/ coeff(20)  = 0.201956;
   /*21*/ coeff(21)  = -4.47768E-02;
   /*25*/ coeff(22)  = 1.58502E-02;
   /*29*/ coeff(23)  = -4.19223E-03;

   /*0*/  coeff(24)  = -5.64713E-04;
   /*4*/  coeff(25)  = 3.63476E-03;
   /*8*/  coeff(26)  = -1.10911E-02;
   /*12*/ coeff(27)  = 3.33408E-02;
   /*16*/ coeff(28)  = 0.254584;
   /*20*/ coeff(29)  = -2.44425E-02;
   /*24*/ coeff(30)  = 8.57456E-03;
   /*28*/ coeff(31)  = -2.57910E-03;
```

```
/*****************************************************************
*Function pulse_shape_one_half_frameT4
*Description:
*-  Does filtering for T/4 samples per symbol
*-  Filter coefficients are contained in coeff[32]
*   Note that indexing is customized for efficient real-time
*   excution
*-  Shaping is done at a rate of 4 samples per symbol, this
*   requires interpolation with 0 between every pair of symbols
*   feom I_aymvol_out and Q_symbol_out

*****************************************************************/
void pulse_shape_one_half-framT4(void)
{
/*---------------------------------------------
*Local Variable Definitions
*---------------------------------------------*/
int    i,j,k;
float  sum:
int    fir_segment_offset;
staticfloat   *curr_I_start_pt;
staticfloat   *curr_Q_start_pt;

/*---------------------------------------------
*Pulse Shaping For I channel
*---------------------------------------------*/
/* Loop selecting input symbols */
for (i=0;
  i<NUMBER_OF_SLOTS_PER_HALF_FRAME*NUMBER_OF_SYMBOLS_PER_SLOT;
  i++)
{
  /*Loop selecting one of (4) samples per symbol*/
  for (j=0; j<NUMBER_OF_SAMPLES_PER_SYMBOL;j++)
  {
    /*Computing the output sample index */
    /*out_sample_no = NUMBER_OF_SAMPLES_PER_SYMBOL*i+j;*/
    /*Computing the offset to the correct segment of filter coefficients */
    fir_segment_offset = j*8;
    /*FIR filtering (only 32/4 multiplications because of interpolation */
    sum = 0;
    for (k=0; k<8; K++)
    }
    sum=sum + I_symbol_out[I+k] * coeff ([fir_segment_offset + k];
    }
    *I_fir_out_P++=sum;
  }
}/*End for (i=0; ... */
```

■ **Figure 10.** *Optimized C code for clock cycle minimization in a rate-mismatched pulse-shaping filter implementation.*

		Function	Instruction cycles	Number of repetitions/ frame of 20 ms	MIPS
modem_main	forward channel	voice_decoding	50,000	1	2.5
		analog_codec_isr	50	160	0.4
		deinterleaver	8000	1	0.4
		channel_decoder	10,000	1	0.5
		AFC	30	160	0.24
		AGC	80	160	0.64
		symbol_timing_recovery	80	160	0.64
		frame_timing	800	14	0.56
		channel_delay_detection	30	160	0.24
		doppler_detection	30	160	1.2
		slicing	200	160	1.6
		differential_detector	150	160	1.2
		deframing	150	1	0.0075
		messages_to_uc	1000	1	0.05
		receive_data_isr	200	160	1.6
		equalizer*	150,000	1	7.5
	reverse_channel	voice_coding	40,000	1	20
		interleaving	80,00	1	0.4
		channel_coding	2500	1	0.125
		framing	2000	1	0.1
				Total MIPS* (no equalizer case)	31.3425

■ **Table 4.** *An example of MIPS requirements for some IS-136 algorithms.*

important to note that in cases where frequency offset exists the correlation will fail when it is done only against the ideal original training waveforms. For good performance in realistic environments it is also necessary to correlate the received signal against frequency-shifted original waveforms. A typical DSP is ideal for correlation operations. Tracking of the frame timing can be accomplished by the same operation, except that the span of the received signal which needs to be correlated is significantly smaller since we are close to the actual timing. Here, though, one has the choice of implementing better resolution.

Symbol timing recovery makes sure that a received signal waveform is sampled as close as possible to the optimal sampling point for detection. Since it is desirable to have A/D converters operate at the slowest possible rate, it is required to be able to finely change the sampling position. Indeed, one can choose to adjust the sampling phase of an A/D explicitly. However, more and more often in digital receivers, the preferred choice is to let the converter keep sampling at an arbitrary phase and to use digital interpolation to find the value of the signal at the optimal sampling point from two or more neighboring samples of an arbitrary phase.

Equalization — The fact that wireless channels have an associated delay spread which causes intersymbol interference requires in some instances that this be compensated for. For a 24.3 kb/s IS-136 system, the delay spread which causes trouble and requires an equalizer is on the order of 10 μs. Two principal techniques are used for equalization: decision-feedback equalizers (DFEs) and maximum-likelihood sequence estimators (MLSEs). Equalization is one of the most MIPS-intensive functions in cellular phone receivers. Although equalizers are not always needed, since channels often have smaller delay spreads receivers/DSPs have to be designed to be able to handle equalization. DFEs consist of two FIR filters and are amenable to DSP implementations. MLSE equalization requires clever memory addressing approaches which DSPs support.

Channel Coding/Decoding — Channel coding is almost always applied in cellular communications systems. Exceptions to the rule are systems where fading rates are expected to be very slow, and thus no amount of coding can recover the signal when in a deep fade for significant amounts of time. The operation of coding is always simpler than decoding. Both block-code and convolutional-code decoding can be demanding in terms of the number of cycles required. DSP vendors are paying particular attention to efficient software implementations and/or building specialized hardware for trellis search techniques which are effective for various decoding schemes. These accelerators are probably the first of a number of accelerators that will deal with speeding up the operation of DSPs. Constraint length 7 rate 1/2 coding used for IS-136 can be dealt with in a DSP, but decoding constraint length 9 rate 1/2 codes for IS-95 are more comfortably done in a specialized

Speech coder		kb/s	MIPS	ROM	RAM
VSELP - IS54	1st TDMA IS-54	7.95	24.5	9k	<2 k
ACELP - IS641	2nd TDMA IS-136	7.4	25.0	11.3k	4.5 k
JVSELP - PDC	Full rate PDC	6.7	24.7	10k	2 k
PSI-CELP-PDC	Half rate PDC	3.45	38	23k	4 k
RPE-LTP-GSM	Full rate GSM	13.0	8	4k	1 k
ACELP-GSM	2nd full rate GSM	12.2	25	15k	5 k
VSELP - GSM	Half rate GSM	5.6	28	20k	6 k
QCELP - 8k	1st CDMA	8.5	23	7k	2.4 k
QCELP - 13k	2nd CDMA	13.2	27.5	13.5k	4 k
EVRC- IS127	3rd CDMA	8.5	32	15k	5.5 k

■ **Table 5.** *Summary of various speech coders and resource requirements.*

VLSI accelerator. It is interesting to note that Viterbi MLSE equalization techniques can sometimes share trellis searching structures with channel code decoders. This is most obvious in GSM, where modulation is binary.

Automatic Gain Control — While propagating through a wireless channel, a signal can experience dramatic changes in power levels. Standard deviation of a signal due to shadowing is on the order of 8–12 dB, whereas Rayleigh fading can cause as much as 30–40 dB of rapid signal power fluctuations. It is not always desirable to get rid of all Rayleigh fading fluctuations (especially when they occur rapidly), but shadowing needs to be compensated for. Automatic gain control schemes in modern receivers collect and process data in the digital domain, and then send control information to analog components which adjust signal power levels prior to A/D converters. It is not usual to have overdesigned A/Ds (in terms of the number of bits), which would let DSPs cover most of the dynamic range of radio signals.

DSP State Machines — DSPs used in wireless terminals are usually controlled by microcontrollers. They operate according to microcontroller orders, and, rather than using sophisticated operating systems, follow states driven by state machine flow. Bearing in mind that DSPs are not good in branching operations, it is of some importance to carefully implement them.

To illustrate processing loads with which a DSP might have to deal, we provide an example of approximate MIPS requirements for some IS-136 algorithms in Table 4.

Speech Processing

Speech processing for digital wireless communications refers to the process of reducing the bit rate to represent digitized speech while maintaining good quality of the synthesized speech. Many speech-coding standards have been developed in the last decade because of the advances in DSP technology and the increase in demand for new speech coding techniques [34, 35]. Standards bodies such as ITU, TIA, ETSI, and Radio Communications Research (RCR) have been responsible for defining a set of speech coding standards for digital cellular applications.

Speech coders can be characterized by the following attributes: bandwidth, channel-error sensitivity, bit rate, quality, complexity, and delay. Depending on the application, trade-offs can be made between these attributes to come up with a specific coder. A speech signal for digital cellular applications is band-limited to between 300 Hz and 3400 Hz and is sampled at 8 kHz. Cellular telephone standards use speech coders ranging from 3.45 kb/s to 13.2 kb/s [29, 30, 31, 36, 38–41], as shown in Table 5.

Algorithms

Most of the speech coders used in cellular standards are based on linear prediction based on analysis by synthesis (LPAS) [35]. Figure 11 shows the basic principles involved in LPAS. A speech signal in a speech coder is represented by a set of parameters such as filter coefficients, gain coefficients, and codebook indices.

The basic principle of LPAS coding is that parameters to generate the excitation signal are determined using a decoding structure similar to that used at the decoding end of the channel. The long-term prediction (LTP) filter models the long-term correlations (fine structures in speech spectrum), while the short-term prediction (STP) filter models the short-term correlations of the speech signal. Almost all the coders used in cellular applications use a 10th-order all-pole filter model for STP, whereas a 1st–3rd-order filter is used for LTP. These coders operate on a block of speech samples, usually of 20–40 ms duration, to extract the various parameters. The decoded speech is produced by passing the output from the excitation generator through the filters. The parameters to generate the excitation are determined by minimizing the error between the original input speech and the synthesized speech.

Once the speech parameters are determined, they have to be quantized. The coder used in first-generation IS-54 TDMA, a vector sum excited linear prediction (VSELP) coder, uses the scalar quantizer to represent the STP filters [29]. This coder has two fixed codebooks and one adaptive codebook. The codebook indices are used for transmission. The gains for these code vectors are vector quantized. While there was a bit-exact specification [35] for the first-generation GSM coder, there was no such requirement for the IS-54 VSELP. While the coders for TDMA and GSM cellular standards operate at a fixed rate, the CDMA standards use variable-rate speech coders [34]. The idea behind variable-rate coding is, that depending on the speech activity in the speech signal, the number of bits used to represent the input speech signal varies. Full rate is used to code active speech, while half and quarter rates can be used for noise-like speech and silent background signal. Qualcomm's QCELP 8.5 kb/s coder is used in the IS-95 standard [34].

The quality of the synthesized speech depends on the bit rate. It can be said that, in general, the higher the bit rate the better the synthesized speech quality. Also, for a given bit rate, better quality can be achieved with more complex speech coders. This means that complexities of the computational requirements increase with lower bit rates and with higher quality. The personal digital cellular (PDC) system used in Japan uses a VSELP coder similar to the one used in IS-54 systems in the United States [29]. This coder uses one fixed codebook twice the size of that used in IS-54. Thus, the quality of speech is slightly lower than that of the 8 kb/s VSELP. The half-rate PDC system uses a more complex coder, the pitch synchronous innovation CELP (PSI-CELP) coder at 3.45kb/s. The quality of this coder is as good as the full-rate PDC coder and even better under error-free conditions [35, 37]. The IS-95 CDMA system uses a 13.2 kb/s QCELP coder in order to provide good speech quality [40].

In order to get better speech quality, different vector quantization methods to represent filter parameters as well as gain parameters have been used. By using fractional lag values to represent the LTP (adaptive codebook), the quality can be improved . Based on the reactions and customer satisfaction, the GSM and North American cellular standards committees have adopted new speech coders for the various cellular standards [31, 32, 39–41]. The IS-136 TDMA systems used in 800 MHz and 1900 MHz in the United States utilize the algebraic code excited linear prediction (ACELP) coder [31]. The coder still uses adaptive codebook to represent the LTP effects, but uses fractional pitch values with 1/3 resolution. The fixed codebook is constructed from a structure based on interleaved single-pulse permutation design. The filter parameters are vector quantized using line spectral frequency (LSF) representation. The gains are vector quantized. The resulting quality of this coder is very good [30]. The GSM full-rate standard has adopted a new coder , at 12.2 kb/s, based on ACELP principles [39]. It uses 1/6 fractional pitch representation, and uses two sets of LSF parameters, vector quantized, for every frame to represent the synthesis filter. Similarly, the IS-95 CDMA standards committee has adopted a new coder, the enhanced variable rate coder (EVRC), for the CDMA system [41]. The EVRC operates at 8.5 kb/ and is also a variable-rate coder like QCELP. The EVRC attempts to match the time-warped version of the original residual signal instead of trying to match the original residual signal like the CELP coders. This reduces the number of bits needed to represent the pitch; as a result, more bits are available to represent the excitation. The EVRC also uses the ACELP codebook for the fixed codebook representation. Different pulse structures are used for different rates depending on speech activity. The quality of the EVRC is very good, and it maintains good quality under error conditions [41]. Even though the half-rate GSM coder was defined in 1995 [35], not many half-rate GSM systems have been installed.

IMPLEMENTATIONS

All described speech coders are computationally very demanding. They require powerful DSPs to do the coding of speech in real time. The advances in VLSI technology made it possible to have DSPs which can run at 120 MIPs with 16 kbytes of on-chip RAM and 48 kb of on-chip ROM to store the program. During implementation of these speech coders one can use various software and hardware development tools available from leading DSP suppliers. All these coders are first implemented in floating-point C code. The VSELP, JVSELP, and QCELP coders were only specified at the algorithm level. The standards committee provided the reference coder implemented in floating point using the C language. There were no fixed-point C code specifications, which made the task of evaluating and comparing various implementations of the same coder by different people using different DSPs difficult.

The recent coders have been defined to be bit-exact standards. The basic operations involved in implementing the speech coder are defined using 16 -bit and 32-bit operations. These operations are defined very precisely using C code. The standard also provides a set of test vectors and the corresponding test results so that one can test and compare his/her own DSP code to make sure it meets the bit-exact specification. For example, the IS-686 standard specifies that handsets meet these bit-exact requirements, whereas the base stations can deviate from this requirement as long as they meet other requirements [32]. Most currently available DSPs do not have single assembly instructions to directly implement some of these basic operations specified in the standards. However, there are efforts to incorporate newer instruction sets in future DSPs. Table 5 gives a summary of various speech coders used in cellular standards.

It is often assumed that once you have a floating-point C code, it is easier to use a compiler to generate fixed-point DSP assembly code. There have been efforts to come up with methods for doing this [42], but the resulting DSP assembly code is

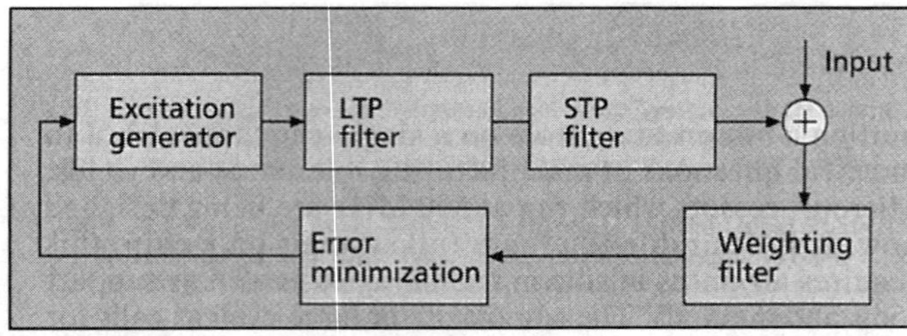

Figure 11. *Linear prediction-based analysis by synthesis (LPAS) speech coding.*

very inefficient. Under these conditions, it is left up to the individual to come up with an efficient DSP code on a given DSP for a specific coder. One of the most important aspects of implementing speech coders is to keep RAM usage to a minimum. There is nothing one can do about the RAM needed for storing the permanent variables. However, one can use various techniques to reduce the temporary RAM required. The first step is to identify the temporary variables and partition them into various sections on frame and subframe levels. Then, by carefully overlaying different sections, one can reduce the RAM requirements. Sometimes one can precalculate some of the intermediate variables and store them in ROM to save RAM. Some DSPs do not save the register contents while calling other subroutines. In those circumstances extra care must be taken in the beginning, and a systematic way of handling the call and return must be defined and strictly followed during implementation. One should structure the modules such that other modules can reuse some parts of the code. Extra care needs to be taken while writing interrupt service routines to handle various interrupts in the system. Even though the speech samples are coming in at 125 μs intervals, there are other real-time requirements in any given cellular system. Some DSPs (e.g., Lucent's 1629) have hardware cache to speed up the execution and at the same time reduce power consumption. In that case one has to pay attention to the interrupt latencies involved. With the availability of flash DSPs from Lucent, one can speed up development of DSP code.

ARCHITECTURES

With the definition of new and better speech coders, the equipment makers are faced with the challenges of supporting all the speech coders on a handset or base station in order to be compatible with other vendors. This has posed a difficult problem because the ROM space to store the coders and RAM usage are such that some combinations of coders cannot be supported easily on a single programmable DSP. One approach in that situation is to integrate multiple DSP cores and memory banks on a single chip. Another is to design a special ASIC hardware implementation of different speech coders. One can add hardware accelerators to current DSP cores, thereby increasing performance and reducing code size. There are some efforts to define DSPs with multiple MAC units in order to meet the ever-increasing demand for more MIPs. By increasing the number of registers used to access the RAM, DSPs can provide more MIPS with less code. Designing more complex addressing techniques, and increasing the on-chip RAM and ROM, one can accommodate increasing requirements for future applications.

FUTURE TRENDS

IC TECHNOLOGIES

The electronics industry has an appetite for integrating more and larger functions in ever smaller chips and devices [16]. No end appears to be in sight to transistor downsizing for the growth in IC functionality [4]. Technologies are being developed to shrink the transistor to 0.1 μ. Major challenges in putting a billion transistors on a single chip are related to practical questions of manufacturing tolerances and yields. Microprocessors which run at 500 MHz are being designed now. It is desirable to create full systems on a chip; this requires advances in silicon technologies as well as support tools and methods. The complexity of large designs calls for the paradigm based on reusable high-level building blocks, or macroblocks. This paradigm requires a careful approach to dealing with block integration and testing [17]. The growth in the number of logic gates put on a chip is far outpacing the growth of device I/O, which bodes ill for controllability and observability. Chip-embedded control and testing (e.g., built-in self-test, BIST) is one of the strategies for dealing with these problems.

Low power consumption is a critical factor determining how widespread cellular communications are and will be. Future opportunities for low-power high-level integration will be governed by fundamental limits, materials, devices, circuits, and systems. There is no consensus on numbers achievable in practical settings. Some projections are shown in [19]. Pessimistically, in practical environments one cannot go below 0.25 μ technologies; optimistically, we shall reach 0.0625 μ sizes. Maximum chip dimensions may be anywhere in the area between 25 mm and 50 mm square.

WIRELESS APPLICATIONS OF DSPS

There exists sufficient difference between analog RF/IF and baseband processing that the integration of the two is not easily achievable. Sound technical solutions to related problems do not exist [22]. Most integration problems are related to the noise generated by high-speed digital circuitry and self-induced codec noise due to PLL operation. The boundary between analog and digital is moving, and low-power fast A/Ds and D/As and high-speed DSPs are shifting it toward higher and higher frequencies.

Whereas current generation of cellular products mostly use off-the shelf DSPs which are standalone, the trend toward integrating DSPs with microcontrollers, special-function hardware, and RF/IF interfaces is picking up steam. Some vendors are offering DSPs as cores through licensing agreements. The first mass market chip for the IS-95 standard is an ASIC which contains an embedded DSP for speech processing. Special-purpose DSPs based on DSP cores and built for large product runs are the likely direction in which the evolution of the DSP will occur. An example of a wireless-specialized device design based on a RISC microcontroller and a DSP can be found in [23].

Functions which are often used in wireless applications and are not supported by traditional DSPs are: complex arithmetic, division (such as needed for automatic gain control), coordinate rotation (useful for constellation-based algorithms in a receiver), and square root. Done on a traditional DSP architecture, a combined division and square root operation requires $5N + 12$ cycles for N-bit operands [20]. It has been shown that coordinate rotation digital computer (CORDIC)-based approaches to these functions may provide significant speedup of execution.

The example of noninteger-related clocks for physical-layer and speech coding algorithms is an excellent motivator for pursuing single-chip DSP designs with multiple and independent arithmetic units [20, 21]. Due to close functional coupling between speech coding/decoding and physical layer, optimized memory-sharing approaches have to be designed. It is desirable to investigate options for independent interruptability of operation sequences executed on individual arithmetic logic units (ALUs). Another direction for improvement of DSPs is the functional improvement of ALUs.

Recent announcements of DSP vendors indicate that DSP architectures with multiple MACs will appear on the market in 1998. For instance, Lucent's DSP16000 core with two MACs can support the computation of the squares of differences between two sets of numbers with accumulation. A special three-operand adder has been included in the architecture for this purpose. This is a function useful for Viterbi decoding, one of the most widely used algorithms on the wireless receiver side. Atmel has announced a general-purpose 16-bit DSP core named LODE, with two 16 x 16 MACs, particularly targeted at wireless applications.

Measuring and comparing the performance of DSPs is not easy, particularly because different vendors have different approaches to the measurement. Lucent has recently suggested an "applications cube" approach to assess any DSP chip with respect to an application. The new unit presents power, performance, and cost as a measure of a function that is implemented in the processor. Consequently, power will be represented as milliamps per function at a particular voltage, performance as MIPS/function and cost by code-size/function.

An issue of hardware-software codesign for embedded systems is being actively pursued [18]. An example of how complex it is to design hardware and software that cooperate well is the problem of timing and synchronization for the IS-95 CDMA-based digital cellular standard. From setting parameters to control the length of correlation to updating the system time for a mobile receiver, every task requires coordinated hardware/software architectures.

Methodologies and tools for dealing with codesign are not widely known and used. Embedded design can be divided into four major tasks: partitioning the function into smaller pieces, allocating partitions to hardware, scheduling the times at which the functions are executed, and mapping a generic functional description into an implementation on a particular set of components as either software or the logic. In future, it is of considerable interest to automate these processes with the motivation of simplifying the designer's job as well as reducing the design error probability.

Reducing voltage supply levels is a desirable goal, and the trend is to go below 2 VDC. At these voltages, analog circuits start failing, and the number of available MIPS is reduced.

Future wide-area wireless systems will require higher data rates and thus impose tougher requirements on signal processing components [24,25]. Data rate evolution envisioned by system developers targets 144 kb/s, 384 kb/s, and 2 Mb/s for third-generation systems and as much as 32 Mb/s for the mobile broadband communications (MBS) project in Europe. Both low-tier and broadband wireless local loop systems [26,27] share many of the problems with cellular systems discussed in this article.

In the realm of speech coding, new speech coders will emerge as more progress is made in i) understanding the structure of speech signal, ii) understanding of how human hearing and perception work, iii) better quantization techniques to represent speech parameters, and iv) much faster DSPs with more advanced addressing and instruction sets.

CONCLUSIONS

Digital signal processors have a prominent role in the design of cellular communications systems. They offer development flexibility and are used primarily for number-crunching operations in signal processing algorithms. DSPs are evolving by being integrated with mirocontrollers, getting new functions and specialized accelerators. Their future in the world of advanced wireless communications is guaranteed.

ACKNOWLEDGMENTS

The authors are grateful to Analog Devices, Lucent Technologies, and Texas Instruments for allowing the use of figures from their Web pages.

REFERENCES

[1] D. Goodman, "Trends in Cellular and Cordless Communications," *IEEE Commun. Mag.*, June 1991, pp. 31–40.
[2] P. W. Baier, "Taking the Challenge of Multiple Access for Third Generation Cellular Mobile Radio Systems – A European View," *IEEE Commun. Mag.*, Feb. 1996, pp. 83–89.
[3] E. A. Lee, "Programmable DSP Architectures," Parts 1 and 2, *IEEE ASSP*, Oct. 1988 and Jan. 1989.
[4] P. M. Grant, "Signal Processing Hardware and Software," *IEEE Signal Processing Mag.*, Jan. 1996, pp. 86–88.
[4] L. Geppert, "Solid State," Annual Technology Overview, *IEEE Spectrum*, Jan. 1997, pp. 55–59.
[5] B. Rezavi, "Challenges in Portable Transceiver Design," *IEEE Circuits and Devices*, Sept. 1996, pp. 12–25.
[6] B. C. Mather, "Embedding DSP," *IEEE Spectrum*, Nov. 1991, pp. 52–55.
[7] J. Snyder *et al.*, "Tools for Real-Time Signal-Processing Research," *IEEE Commun. Mag.*, Nov. 1993, pp. 64–74.
[8] D. Cox, "Wireless Personal Communications, What Is It," *IEEE Pers. Commun.*, Apr. 1995, pp. 20–35.
[9] K. S. Shanmugan, "Simulation and Implementation Tools for Signal Processing and Communication Systems," *IEEE Commun. Mag.*, July 1994, pp. 36–40.
[10] Special Issue, "PCS—The Second Generation," *IEEE Commun. Mag.*, Dec. 1992.
[11] D. Cox *et al.*, "Low-Power Digital Radio as a Ubiquitous Subscriber Loop," *IEEE Commun. Mag.*, Mar. 1991, pp. 92–95.
[12] H. Meyr and R. Subramanian, "Advanced Digital Receiver Principles and Technologies for PCS," *IEEE Commun. Mag.*, Jan. 1995, pp. 68–78.
[13] J. Wepman, "Analog to Digital Converters and Their Applications in Radio Receivers," *IEEE Commun. Mag.*, May 1995, pp. 39–45.
[14] J. Mitola, "The Software Radio Architecture," *IEEE Commun. Mag.*, May 1995, pp. 26–38.
[15] R. Baines, "The DSP Bottleneck," *IEEE Commun. Mag.*, May 1995, pp. 46–54.
[16] M. Hunt and J. Rowson, "Blocking in a System on a Chip," *IEEE Spectrum*, Nov. 1996, pp. 35–41.
[17] R. Chandramouli and S. Pateras, "Testing Systems on a Chip," *IEEE Spectrum*, Nov. 1996, pp. 42–47.
[18] W. Wolf, " Hardware-Software Co-Design of Embedded Systems," *IEEE Proc.*, vol. 82, no. 7, July 1994, pp. 967–89.
[19] J. Meindl, "Low Power Microelectronics: Retrospect and Prospect," *IEEE Proc.*, vol. 83, no. 4, April 1995, pp. 619–34.
[20] H. M. Ahmed, "Directions in DSP Processors," *IEEE JSAC*, vol. 8, no. 8, Oct. 1990.
[21] H. Ahmed and R. Kline, "Recent Advances in DSP Systems," *IEEE Commun. Mag.*, May 1991, pp. 32–45.
[22] R. Franzo and M. Diamondstein, "System Focus is Key to Designing Optimized Wireless DSPs," *Wireless Sys. Design*, Aug. 1997, pp. 35–38.
[23] D. Walsh, "Integrated DSP/RISC Devices Target Cellular/PCS Phones," *Wireless Sys. Design*, July 1997, pp. 50–54.
[24] M. Callender, "Future Public Land Mobile Telecommunications Systems," *IEEE Pers. Commun.*, 4th qtr. 1994, pp. 18–22.
[25] Special Issue, "The European Path toward UMTS," *IEEE Pers. Commun.*, Feb. 1995.
[26] C. Yu *et al.*, "Low-Tier Wireless Local Loop Systems – Part 1 and 2: Comparison of Systems," *IEEE Commun. Mag.*, Mar. 1997.
[27] W. Honcharenko *et al.*, "Broadband Wireless Access," *IEEE Commun. Mag.*, Jan. 1997, pp. 20–26.
[28] I. A. Gerson and M. A. Jasiuk, "Vector Sum Excited Linear Prediction (VSELP) Speech Coding at 8kbs," *Proc. ICASSP*, 1990, pp. 461–64.
[29] Gerson, "Vector Sum Excited Linear Prediction (VSELP) Speech Coding for Japan Digital Cellular, *Proc. IEICE*, Nov. 15, 1990, pp. 35–40.
[30] R. Salami *et al.*, "A Toll Quality 8kb/s Speech Codec for Personal Commun. Systems," *IEEE Trans. Vehic. Tech.*, vol. 43, no. 3, Aug. 1994, pp. 808–16.
[31] TIA/EIA/IS-641, "TDMA Cellular/PCS- Radio Interface — Enhanced Full-Rate (EFR) Speech Coder," June 1996.
[32] TIA/EIA/IS-686, "TDMA Cellular/PCS — Radio Interface — Minimum Performance Standards for IS-641 Full-Rate Voice Coder," Dec. 1996.
[33] AT&T Wireless Products, *Data Book*, AT&T Microelectronics, Sept. 1995.
[34] *Speech and Audio Coding for Wireless and Network Applications*, B. S. Atal, V. Cuperman, and A. Gersho, eds., Kluwer, 1993.
[35] *Speech Coding and Synthesis*, W. B. Kleijn, and K. K. Paliwal, Eds., Elsevier, 1995.
[36] GSM 06.20 (ETS 300 581): "European Digital Cellular Telecommunications Systems: Half Rate Speech Transcoding," Jan. 1995.
[37] K. Mano *et al.*, "Design of a Pitch Synchronous Innovation CELP Coder for Mobile Communications," *IEEE JSAC*, vol. 13, no. 1, Jan 1995, pp. 31–41.
[38] R. Cox, and P. Kroon, "Low Bit-Rate Speech Coders for Multimedia Communication," *IEEE Commun. Mag.*, Dec. 1996, pp. 34–41.
[39] GSM 06.60 (ETS 300 726), "Digital Cellular Telecommunications Systems: Enhanced Full Rate (EFR) Speech Transcoding," ETSI, Nov. 1996.
[40] EIA/TIA, "High Rate Speech Service Options for Wideband Spread Spectrum Communications Systems," IS96B, Feb. 1995.
[41] EIA/TIA/IS-127, "Enhanced Variable Rate Codec, Speech Service Option 3 for Wideband Spread Spectrum Digital Systems," Sept. 1996.
[42] S. Kim, and W. Sung, "A Floating Point to Fixed Point Assembly Program Translator for the TMS 320C25," *IEEE Trans. CAS-II*, vol. 41, no. 11, Nov. 1994, pp. 730–39.

Recent Developments in Enabling Technologies for Software Defined Radio

Dimitrios Efstathiou, Jose Fridman, and Zoran Zvonar

Analog Devices Inc.

ABSTRACT In response to the technical challenges surrounding the realization of software defined radio, high-performance digital signal processors and related software have emerged as key enabling technologies. This article focuses on high-performance digital signal processing architectures and devices. The partitioning of the transceiver functions between DSPs and programmable dedicated hardware is examined.

Software defined radio (SDR) technical implementation issues vary according to the form factor and the application. A number of enabling technologies have allowed the development of commercially viable platforms providing an efficient bridge to third-generation wireless systems [1, 2]. While analog-to-digital converters (ADCs) are still the limiting performance factor [3], recent advances in digital signal processing (DSP) technology and programmable dedicated hardware solutions have significantly changed the landscape of software radio techniques. As a result of DSP architectural innovation, process technology improvements, and system integration, systems that as little as two or three years ago required several DSP boards have now become feasible in fewer DSP boards, and in some cases even in a single device.

SOFTWARE DEFINED RADIO PARTITIONING

The structure of the SDR terminal has evolved from processing source coding and baseband modem functions by software to a complete digital intermediate frequency (IF) receiver that can support adaptation to multiple air interface standards, as depicted in Fig. 1. In addition to wideband radio frequency (RF) front and wideband high-resolution converters, critical additions are programmable digital hardware elements, including digital down- and upconverters (also referred to as digital receiver/transmitter signal processors) [4, 5].

The SDR is essentially a reconfigurable system that can be implemented as either a complete software solution in DSP and/or microcontroller cores, or special-purpose reconfigurable hardware complemented by software. A generic SDR architecture based on functional modules allows the flexibility to exchange any functional modules by maintaining a fixed standard interface, and, as a result, leaving module implementation and partitioning to the manufacturer.

DSP cores are more suitable to handling algorithms characterized by irregular throughput rates, as well as frequent branching and irregular control flow. SDR operations commonly executed in DSPs are frame timing, amplitude gain estimation, carrier frequency and phase recovery, and symbol timing recovery. Dedicated hardware, application-specific integrated circuits (ASICs), or field programmable gate arrays (FPGAs) can be optimized for functions such as matched filtering and correlators, fast Fourrier transform, and Viterbi decoding. Digital downconverters provide the capability to share execution of baseband matched filtering, as well as high computational processes for frequency offset, symbol timing, and amplitude gain estimation.

DSP TECHNOLOGY

In this section state-of-the-art DSP technology and recent trends are reviewed. The DSP industry is currently undergoing a generational change driven by three factors: architectural innovation, process technology, and system integration. Architecturally, DSPs have improved performance by means of higher levels of parallelism, incorporating multiple data paths and execution units, and techniques found in mainstream general-purpose computing [6]. Process technology, which remains the primary driver of performance by increasing clock rates and transistor counts, while at the same time decreasing power consumption, has continued to improve. DSPs with drawn geometries of 0.35 μ in the 1995–1996 timeframe are now being delivered in 0.25 μ, and expected in late 1999 in 0.18 μ. This improvement alone directly translates to a 100 percent logic transistor budget increase, and a nearly similar clock rate improvement. DSPs, which have traditionally integrated a relatively large number of system peripherals on-chip, have continued along this path.

Table 1 lists the major DSP manufacturers and main product families. Although the most often quoted performance measure is the million instructions per second (MIPS) rating, it is also the most unreliable, because different manufacturers count instructions according to different criteria. Instead, a more reliable performance measure for the algorithmic workload in an SDR is the number of peak *million multiply-accumulates per second* (MMACS), shown in the fourth column of Table 1. The MAC rate, expressed in MMACS, is the product of the clock speed times the number of MACs per cycle of which a DSP is capable. Examples of MAC-intensive functional modules for the SDR are finite impulse response (FIR) filters, correlators, and equalizers (discussed later).

Peak MAC rates alone do not fully characterize performance. It is also necessary to determine the extent to which this peak rate can be achieved given all the support infrastructure of a DSP chip. Table 1 lists the main supporting hardware: internal and external buses for input/output (I/O), and on-chip memory (most commonly fast static dynamic random access memory, DRAM). Since the DSP is a block in a larger data flow system, one has to determine if there are enough resources to maintain a sufficiently large data flow into and out of the chip to match the achievable MAC rate.

The 1997–1998 timeframe represents a generational break

Reprinted from *IEEE Communications Magazine,* Vol. 37, No. 8, pp. 112–117, August 1999.

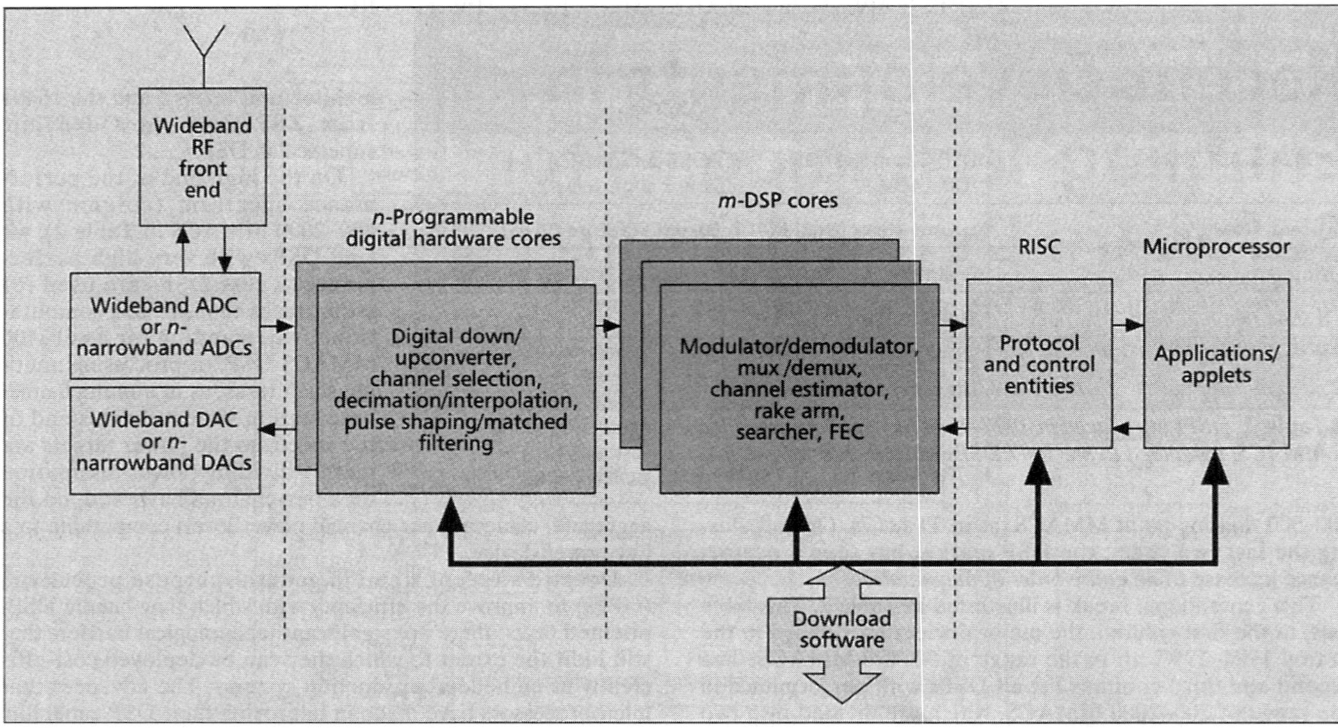

■ **Figure 1.** *DSP partitioning of software defined radio.*

for high-end DSPs. In the 16-bit domain, most high-end devices in the market prior to 1997 had performance in the range of 50–100 MMACS, and had fairly conventional dispatch and data path mechanisms, for example, TI's c54x and Lucent's 16xx series. However, device introductions during 1997 and 1998 have been in the range of 200–400 MMACS, and have incorporated more sophisticated architectures with peak rates up to 2000 MMACS.

A similar transition has occurred in the floating-point domain, where the market has evolved from devices in the range of 50 floating-point MMACS prior to 1998, as in ADI's 21060 and Motorola's 96xxx, to devices in the range of

Manufacturer and part number (a)	Clock (MHz)	Parallel MAC units	MAC rate (MMACS)	Data formats (b)	Instruction issue rate (MIPS) (c)	Internal bus width	External interfaces (PP, SP, LP) (d)	Internal SRAM (kbytes)
ADI 21xx	50	1	50	16	50	16/24	16, 2, 0	40–160
ADI 210xx	60	1	60	float/32	60	32/48	32, 2, 6	512
ADI TigerSHARC	250	2 float 8 16-bit	500 float 2000 16-bit	float/32/16	1000	256	–, –,–	–
Lucent 16xx	120	1	120	16	120	16	16,2,0	64
Lucent 16xxx	100	2	200	16	100	32	16,2,0	240
Lucent StarCore	300	4	1200	16	3000	128	–, –, –	–
Motorola 96xxx	40	1	40	float/32	40	32	32,0,0	8
Motorola 563xx	100	1	100	24	100	24	32,2,0	24–100
TI c54x	100	1	100	16	100	16	16,2,0	20–128
TI c40	30	1	30	float/32	275	32	32,0,6	8
TI c62x	200	2	400	16	1600	64	32, 1, 0	128
TI c67x	167	2	334	float/32	1600	128	32, 1, 0	128

Notes
(a) The entries listed are global part numbers that identify a family of devices, rather than a particular device. The figures shown in the table are for a range of representative devices, and are used to characterize an entire family.
(b) There are four data types listed: single precision floating point; and 32-bit, 24-bit, and 16-bit fixed point.
(c) Instruction issue rates are as reported by manufacturer. In general, MIPS are an inaccurate performance measure, since different manufacturers count instructions according to different criteria.
(d) PP: parallel port, number listed shows port width. SP: serial port, number listed shows number of ports. LP: link port, number of ports.

■ **Table 1.** *Major DSP manufacturers and product families.*

1994–1997: DSPs with performance in the 50–120 MMAC range	1997–1998: DSPs with performance in the 200–2000 MMACS range	
	200–400 MMACS	400–2000 MMACS
ADI 21xx ADI 210xx	DSP Group PalmDSP (Dual MAC)	ADI TigerSHARC (static superscalar)
Lucent 16xx	Lucent 16xxx (Dual MAC)	Lucent StarCore (VLES)
Motorola 96xxx	Philips R.E.A.L. (Dual MAC)	TI c62x, c67x (VLIW dual MAC)
TI c54x	Siemens Carmel (CLIW)	
TI c40	ZSP 16401 (dynamic superscalar)	

■ **Table 2.** *DSPs introduced in 1997–1998 have performance in the range of 200–2000 MMACS, compared to pre-1997 DSPs with 50–120 MMACS.*

200–500 floating-point MMACS, as in TI's c67x. Overall, during the last two years, the DSP market has seen a performance increase of an *entire order of magnitude*.

This generational break is illustrated in Table 2. This table lists, in the first column, the major devices introduced in the period 1994–1997, all in the range of 50–120 MMACS. The second and third columns list all DSPs with performance in the range of 200–2000 MMACS, which are divided into two categories: those in the range of 200–400 MMACS, and those in the range of 400–2000 MMACS.

Along with higher performance, there are two discernible trends that have recently emerged: dual-MAC and low-power architectures. Dual-MAC architectures achieve higher performance by replicating the primary computation engine. Although in some applications this technique can result in a 2x improvement of the peak MAC rate, in practice only a fraction of this factor is realized. An improvement factor of 1.6–1.8 for a dual-MAC DSP on SDR-type algorithms is a typical realizable goal.

Low total power dissipation is also an important and well established trend in all post-1997 device introductions. All DSPs in the middle range of the performance spectrum (shown in Table 2 under the column with 200–400 MMACS) have as a primary target very low overall power consumption. These DSPs are intended for handheld and mobile applications, where the total power dissipation for the complete DSP must be kept below a threshold. During 1998 there were four new entries into this category: DSP Group's PalmDSPCore; Philips R.E.A.L DSP; Siemens' Carmel DSP; and ZSP Corp.'s 16401 processor. All of these processors have dual MAC. The Carmel DSP from Siemens is based on a variation of VLIW, called CLIW (see box with architectural terms), and the 16401 from ZSP Corp. is a dynamic superscalar DSP.

On the high end of the performance spectrum (column with 400–2000 MMACS in Table 2), we find DSPs with very high performance. These DSPs are used for either tasks that are too computationally demanding for a sub-400 MMACS DSP, or processing multiple small tasks, as in a multichannel base station system. In this end of the spectrum the power targets are met by dividing power dissipation on a per-channel basis and, on the aggregate, achieving per-channel power levels comparable to a low-power device.

Despite a recent trend in general-purpose processors (GPPs) to improve the efficiency with which they handle DSP-oriented tasks, there are significant technological barriers that still limit the extent to which they can be deployed cost-efficiently as embedded production systems. The advances that microprocessors have made in improving their DSP capability have been primarily the addition of MAC hardware and parallel operations, and in general low-level hardware feature additions. Most of the media extensions, such as Intel's MMX [7], SUN's VIS [8], as well as HP's MAX-2 and Motorola's Altivec, only provide parallel computational capability.

Even with high computational rates, GPPs have not adequately addressed the larger and more fundamental requirement underlying embedded DSP systems: real-time deterministic execution. In addition, GPPs lack mechanisms for sustaining sufficiently high I/O data rates, and suffer from unpredictable cache and memory management second order effects. Even if some of the real-time problems are mitigated, the system costs associated with GPP platforms remain comparatively high relative to specialized DSPs. Complete system costs include all the components of a chip set, as well as all the peripheral device support needed to provide adequate I/O.

SDR Transceiver Architectures

Two software radio base station architectures are presented that enable the coexistence of older analog wireless systems along with mature digital wireless systems and seamless migration toward third-generation systems.

Current cellular radio base stations typically require one

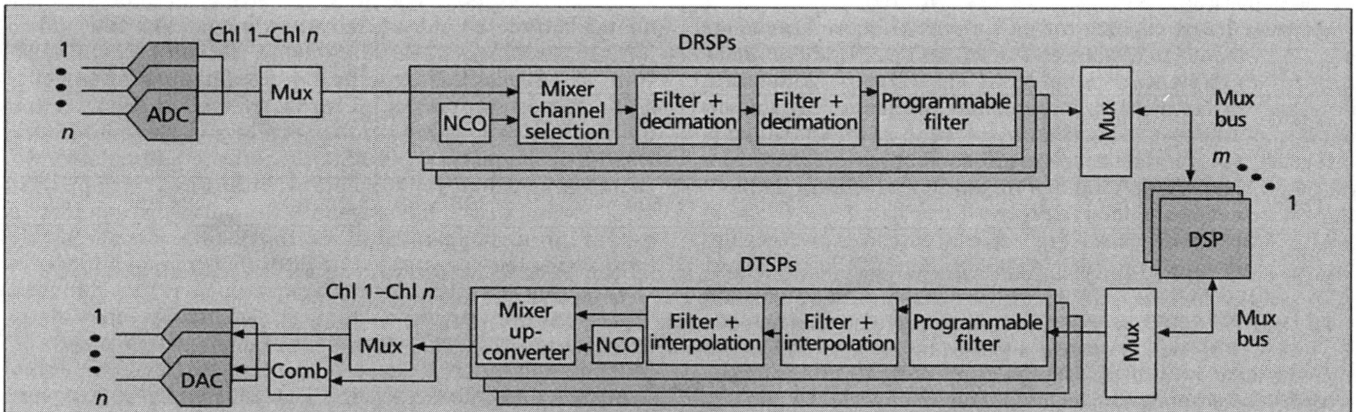

■ **Figure 2.** *Narrowband base station architecture.*

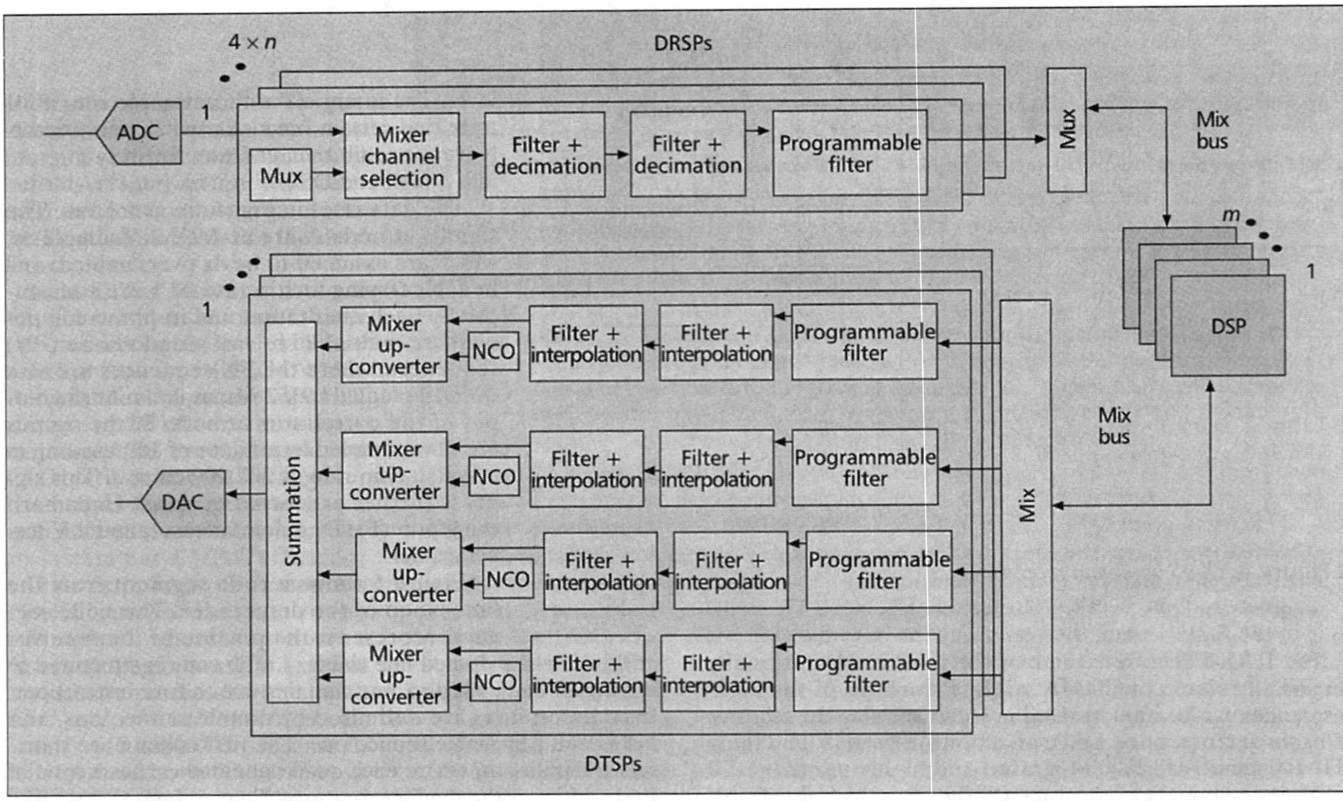

■ **Figure 3.** *Wideband base station architecture.*

radio transceiver for each carrier frequency assigned to the base station. Figure 2 shows a Global System for Mobile Communications (GSM) narrowband base station transceiver. This is an example of IF sampling in which an individual transceiver operates on each RF carrier. In this example, a digital receiver signal processor (DRSP) operates in dual-channel real mode, with 11-bit sampling of a diversity antenna on the channel at 13 Mb/s (48 x 270,833 samples/s). The 26 MHz 2x clock rate of the ADC is used as the processing clock of the DRSP.

On the transmitter side there is a digital transmitter signal processor (DTSP) for each RF carrier. The offline calculated Gaussian minimum shift keying (GMSK) pulse shaped values are stored in a RAM memory, providing an interpolation rate of 4 samples/symbol. The output of the pulse shaping filter is routed to the cascaded interpolation filters, where the GSM samples are interpolated by a factor of 48, resulting in a sample rate of 52 Msamples/s. The interpolated GSM signal is upconverted to an IF frequency of 18 MHz by a sine/cosine sequence generated by the numerically controlled oscillator (NCO).

New-generation base station systems are based on wideband transceiver architectures, shown in Fig. 3. In these architectures, multiple carriers are simultaneously downconverted in a single radio, digitized and separated in the digital domain using digital downconversion techniques. In the transmit direction, the multiple carriers are combined digitally at the output of the NCOs. A number of wideband receiver architectures have been examined in [9].

The GSM multicarrier base station transceiver uses a wideband 12-bit 65 Msamples/s ADC instead of a narrowband ADC. Several DRSPs are pipelined after a single ADC, and used to tune in and isolate separate channels from the digitized wideband data. A DRSP provides a great deal of selectivity by mixing down the channel of interest and filtering the out-of-band noise and adjacent channels. The spurious free dynamic range (SFDR) of an ADC is usually the limiting factor of the wideband system. In addition, the decimation of the DRSP is a more critical factor for wideband than for narrowband systems, because the digital data rate must be lowered to allow subsequent processing by the DSP.

The SDR architecture shown in Fig. 3 can support a wideband base station system capable of processing several IS-95 code-division multiple access (CDMA) channels. The wideband 12-bit 65,000,000 samples/s ADC is used in this architecture, followed by a bank of DRSPs. The input rate to the DRSP is 58.9824 Msamples/s (48 x 1.2288 Mchips/s). The NCO in each DRSP bank will shift the desired signal down to baseband. Rejection of strong narrowband channels outside the operating multichannel band is required in order to decrease the operating channels. Digital matched filtering is implemented in the RAM programmable coefficient filter of the DRSPs. At a rate of 4 samples/symbol, each DRSP bank (group of four DRSPs) can calculate 48 coefficients (12 coefficients per DRSP). At the output of the DRSP bank, despreading is applied to the signal in order to recover the original message bandwidth.

At the transmitter side, four RF carriers are processed in a quad DTSP. Each channel provides independent serial data inputs that are connected directly to the serial port of DSP chips. A programmable coefficient FIR filter is used to pulse-shape filter quadrature phase shift keyed (QPSK) linear inputs. The FIR filter impulse response is 12 symbols long with 48 coefficients. The QPSK samples are interpolated by a factor of 4, resulting in a sample rate of 4.9152 Msamples/s (4 x 1.2288 Mchips/s). The output of the pulse-shaping filter is routed through a gain adjust multiplier to a phase equalizer infinite impulse response (IIR) filter that is unique to a base station implementation complying with the IS-95 standard. The outputs of the phase equalization filter are interpolated by a factor 12 (58.9824 Msamples/s) using a high-speed cascaded integrator comb filter. The interpolated CDMA signal is upconverted to an IF of 19.66 MHz by a sine/cosine sequence, which is generated by each NCO. The outputs of the four upconverters are summed and scaled on-chip.

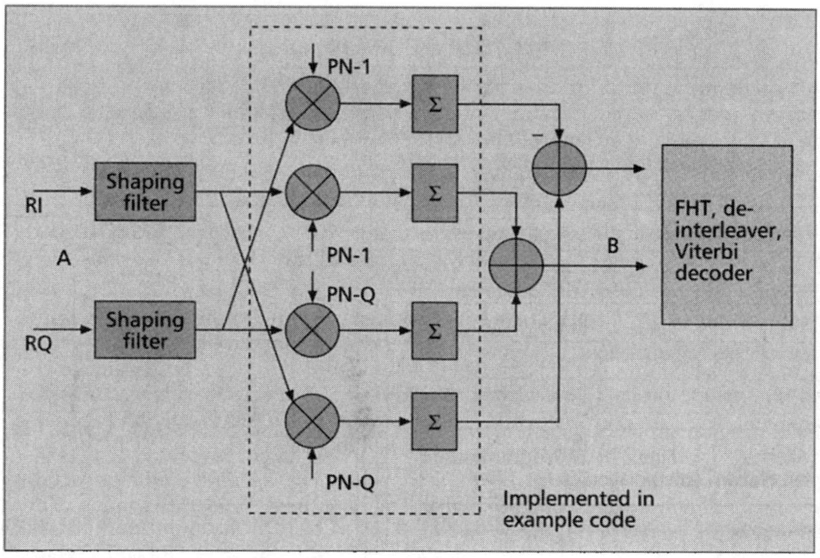

■ **Figure 4.** *IS-95 baseband receiver.*

The RAKE receiver is one of the most computationally demanding blocks in CDMA systems. Portions of the IS-95 despreader can be implemented in software using the features of high-performance DSP, an example being the TigerSHARC family, highly integrated and highly parallel DSP with native support for both floating-point and 16-bit fixed-point data types. The peak MAC rates of the TigerSHARC are 500 MMACS in floating point and 2000 MMACS of 16-bit arithmetic, and the sustained internal data bandwidth is 12 Gbytes/s, at a clock rate of 250 MHz. The most significant aspects of TigerSHARC are:

- The load-store static superscalar dispatch mechanism, with instruction parallelism determined prior to runtime under compiler and programmer control
- Support for multiple data types, including IEEE single-precision floating-point, 16-bit and 8-bit fixed-point
- Parallel arithmetic operations for two floating-point MACs or eight 16-bit MACs per cycle, with an optional single-instruction multiple data (SIMD) execution mechanism in two computation blocks (see box with architectural terms)
- Highly parallel *short-vector*-oriented memory architecture
- 128 architecturally visible and fully interlocked registers
- Eight-stage fully interruptible pipeline, with a regular two-cycle delay on all arithmetic and load/store operations, and a 128-entry, four-way set-associative branch target buffer

Figure 4 shows the block diagram for an IS-95 baseband receiver. In this example we show that the quadrature despreader of the CDMA receiver can be implemented entirely in software. Note, however, that despite the fact that this despreader can be implemented entirely in a DSP, there are many other factors that determine the economic feasibility of CDMA system partition, and trade-offs must be made at the implementation level. The intention of this example is to illustrate that certain functions, which until recently were prohibitive, can now entirely migrate into a high-end DSP.

The data rates in Fig. 4 are as follows. The signals at node A are at 4.9152 Msamples/s, which are assumed to be 4x oversampled, and hence carrying a chip rate of 1.2288 Msamples/s. Each quadrature and in-phase component are multiplied by two pseudo-noise (PN) sequences, where the PN sequences are also oversampled at 4.9152 Msamples/s. At the output of the correlators at node B, the signals are downsampled by a factor of 16, resulting in a Walsh chip rate of 307,200 chips/s. This signal is further processed by a fast Hadamard transform (FHT), a deinterleaver, and a Viterbi decoder.

Figure 5 shows a code segment from the inner loop of the despreader. This code segment represents the portion of the receiver enclosed in the dashed line in Fig. 4. The code is structured in *instruction lines*, where a line contains one to four instructions. Instruction lines are delimited by double semicolons, and instructions by single semicolons. The first column are short-vector data loads, where each quad data load carries a total of four 32-bit words into four registers. To economize on memory, the PN sequences are stored as bytes, and prior to multiplying by I and Q are expanded into 16-bit short words. This is accomplished by the byte-to-short EXPAND instructions in the second column.

The MAC operations are performed by the instructions in the third column, where the partial multiplication results are added to the current contents of the accumulator registers labeled MR. There are two levels of parallelism in MAC instructions. At one level, a MAC instruction is issued to the two computation blocks of the processor, called CBX and CBY, using the SIMD mechanism. In addition, each computation block executes four packed multi-

```
__11:
yr19 : 16 = q[j4+=4] ;                                     MR1:0+=r23:22*r5:4(I);     [k8+=1]=yR31;;
xr19 : 16 = q[j4+=4] ;                                     MR1:0+=r23:22*r7:6(I);     [k8+=1]=yR31;;
yr3 : 0   = q[j0+=4] ;                                     MR1:0+=r25:24*r9:8(I);;
yr7 : 4   = q[j0+=4] ;                                     MR1:0+=r25:24*r11:10(I);;
yr11 : 8  = q[j0+=4] ;                                     MR1:0+=r27:26*r13:12(I);;
yr15 : 12 = q[j0+=4] ;                                     MR1:0+=r27:26*r15:14(I);;
xr3 : 0   = q[j0+=4] ; Sr23:20=EXPAND  Br17:16 (I);            R31:30 = MR1:0;;
xr7 : 4   = q[j0+=4] ; Sr27:24=EXPAND  Br19:18 (I);;
xr11 : 8  = q[j0+=4] ;                                     MR1:0+=r21:20*r1:0(CI);    SR31=R31 + R30;;
xr15 : 12 = q[j0+=4] ;                                     MR1:0+=r21:20*r3:2(I);     [k12+=1]=yR31;;
                                                           MR1:0+=r23:22*r5:4(I);     [k12+=1]=xR31;;
                                                           MR1:0+=r23:22*r7:6(I);;
yr19 : 16 = q[j5+=4] ;                                     MR1:0+=r25:24*r9:8(I);;
xr19 : 16 = q[j5+=4] ;                                     MR1:0+=r25:24*r11:10(I);;
                                                           MR1:0+=r27:26*r13:12(I);;
                       Sr23:20=EXPAND  Br17:16 (I);        MR1:0+=r27:26*r15:16(I);;
                       Sr27:24=EXPAND  Br19:18 (I);;           R31:30 = MR1:0;;
                                                           MR1:0+=r21:20*r1:0(CI);;
if nlc0e,  jump __11;                                      MR1:0+=r21:20*r3:2(I);     SR31=R31+R30;;
```

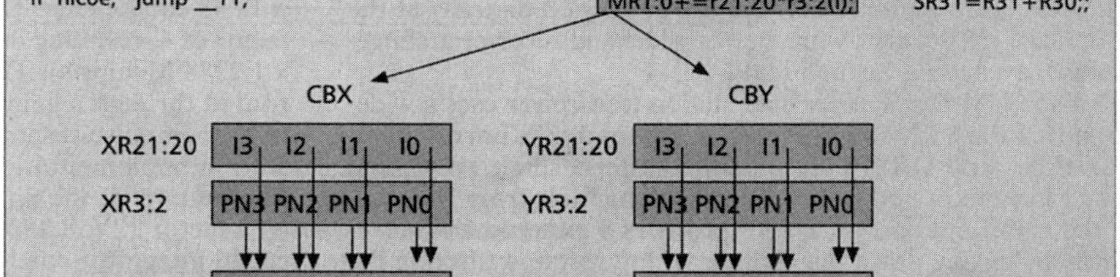

■ **Figure 5.** *Innerloop code segment for the despreader.*

plications in parallel. This is shown in Fig. 5, where a total of eight multiplications are issued by each MAC instruction, four executed in each computation block. The fourth column shows the data stores.

Simulations show that a block of 1,024 samples of I and Q data is processed in 620 cycles. Since there are a total of 4096 MACs in the correlators (ignoring the small number of MACs in the downsamplers), the overall sustained MAC rate is 6.6 MACs/cycle. Given that the processor has a peak potential of 8 MACs/cycle, this system has a MAC efficiency of 82 percent. Some of the sources of inefficiency are evident in the code sample, where we see some lost scheduling opportunities. This shows that a processor with a peak MAC rate of 2000 MMACS can sustain 1640 MMACS. In general, efficiencies above 70 percent in highly parallel machines represent a reasonable performance goal. With a clock cycle of 250 MHz and a block interleaver frame rate of 20 ms/frame, the processing time for this portion of the despreader represents only less than 2 percent loading factor on the processor.

Architectural Terms

A clear trend in the recent wave of DSPs has been the acquisition of several architectural techniques from general-purpose computing. This has brought about a proliferation of terms, some of rather obscure extraction, into the DSP lexical mainstream.

Customizable long instruction word (CLIW) — A variation of VLIW, where a programmer can store customized instructions in internal microcode to reduce code size.

Dual MAC — A technique used to increase the performance of a DSP by replicating the MAC functional unit. Very successful due to the high levels of parallelism present in DSP code. Very likely to remain a future trend, perhaps leading to multi-MAC architectures.

Dynamic superscalar — Traditional superscalar sequencing, where the processor determines at runtime, rather than at compile time, the schedule of instructions to be issued in parallel. It has an enforced programming model that is independent of particular implementations, and is generally simple to program and compile for. Superscalar architectures are also commonly register-based load/store, where all instructions operate only on data resident in the register file, and must be explicitly moved to and from memory.

Single-instruction multiple data (SIMD) — Sequencing mechanism where one instruction controls the execution of more than one functional unit, hence performing the same task on more than one data element.

Static superscalar — Enforces a consistent and functionally well defined programming model, but the schedule is determined prior to run time. It retains many superscalar and RISC attributes (like register-based load/store, interlocked register files, and independence from the latencies of a particular implementation), while incorporating the static scheduling techniques found in VLIW.

Very long instruction word (VLIW) — Sequencing mechanism that relies on an instruction format where every single execution unit in the chip is under direct programmer or compiler control, and has little or no hardware support for maintaining the integrity of data dependencies or scheduling hazards. In VLIW, all operation latencies in a particular implementation are fully exposed to software. VLIW attempts to reduce cost and increase execution speed by reducing hardware complexity.

Variable-length execution sets (VLES) — Statically scheduled, but so-called execution sets (groups of instructions executing as a unit) have prefix instructions with additional parallelism information. Prefixes determine attributes such as set length, predication flags, and additional arithmetic behavior of instructions in the set.

Summary

Major technological advances and trends in high-performance digital signal processing architectures and devices are outlined. Current technology enables SDR configurations based on digital IF, covering multiple modes and enabling migration to third-generation systems.

References

[1] *IEEE Commun. Mag.*, Globalization of Software Radio, Feb. 1999.

[2] *IEEE JSAC*, Special Issue on Software Radio, Apr. 1999.

[3] R. Arnott *et al.*, "Advanced Base Station Technology," *IEEE Commun. Mag.*, vol. 36, no. 2, Feb. 1998, pp. 96–102.

[4] Y. Shinagawa, "Software Radio Technologies from the view point of terminal manufacturer," *Proc. 1st Int'l. Software Radio Wksp.*, Rhodes, Greece, June 1998, pp. 25–35.

[5] D. Chester, "Digital IF Filter Technology for 3G Systems: An Introduction," *IEEE Commun. Mag.*, vol. 37, no. 2, Feb. 1999, pp. 102–7.

[6] J. Eyre and J. Bier, "DSP Processors Hit The Mainstream," *IEEE Comp.*, vol. 31, no.8, Aug. 1998.

[7] A. Peleg, S. Wilkie, and U. Weiser, "Intel MMX for Multimedia PCs," *Commun. ACM*, vol. 40, no. 1, Jan. 1997.

[8] M. Tremblay *et al.*, "VIS Speeds New MediaProcessing," *IEEE Micro*, Aug. 1996, pp. 10–20.

[9] R. Karimi and B. Friedrichs "Wideband Digital Receivers for Multi-Standard Software Radios," *ACTS Mobile Commun. Summit 1997*, Aalborg, Denmark, pp. 392–98.

FPGA in the Software Radio

Mark Cummings , Cadence Wireless Design Services
Shinichiro Haruyama, Keio University

ABSTRACT As new radio standards are deployed without substantially supplanting existing ones, the need for multimode multiband handsets and infrastructure increases. This article describes how emerging FPGA technology's unique combination of size and power efficiency plus field programmability offers a transition of FPGAs from ASIC prototyping to embedded products. Software-defined receiver examples suggest an enlarged role for FPGAs in pragmatic paths toward the productization of software radio technology.

As information tools move off the desktop and into peoples' pockets, purses, and briefcases, the rate of evolution of wireless communications is accelerating. At the same time, business, politics, culture, education, and so on are becoming global. For example, we are designing what some call third-generation systems, while we are rolling out second-generation systems and using widely deployed first-generation systems [1]. In 1998 there will be 140 million cellular and personal communications services (PCS) handsets sold. A few years ago, we had cellular/PCS penetrations on the order of magnitude of 10 percent. Now we are driving toward a global 50 percent penetration. Against this background, some argue that it is possible to create a new single standard which will displace all the previous ones. History tells us that each time this has happened, the new standard just became one more without displacing any of the others. For more detailed information on this problem consult MMITS TR1.2 [2] available through the MMITS web site at www.mmits.forum.org.

What is needed are wireless system components that can be reconfigured to support a range of standards. Multimode multiband handsets and mobile equipment are needed to allow end users to move between domains and footprints and maintain serviceability. Reconfigurable infrastructure equipment (base stations, etc.) is needed to allow those responsible for managing the infrastructure (common carriers, service providers, governmental agencies, enterprises, and consumers) to upgrade infrastructure components without traveling to them and physically manipulating them. Silicon providers and equipment manufacturers need to be able to reduce the number of unique designs they must create and support.

Other Technology Solutions and Their Problems

The old digital design paradigm for baseband and controller functions is based on single-stream instruction set processors optimized for high-speed arithmetic, called digital signal processors (DSPs), hard-coded logic, called application-specific integrated circuits (ASICs), and single-stream instruction set processors optimized for message processing, called microcontrollers. This mix of architectures is referred to here as the *past paradigm*. Currently, the high-speed signal processing requirements for multimode multiband are the most challenging. There are four major categories of *new paradigm* potential technical solutions [3]. Each is summarized in Table 1 and discussed below.

High-speed instruction set processors are increasing in speed, decreasing in feature size, and improving low-power operation. As these new generations of chips begin to appear and feature size continues to decline, there are proportionate increases in speed, power efficiency, and heat dissipation. However, for handheld systems the high-speed processing requirements are still daunting. For example, if a particular algorithm requires 100 instructions to execute and the signal of interest is in the megahertz range, with memory access wait times and processor speed requirements, if performed entirely on a past paradigm DSP, are in the gigahertz range. Gigahertz-range clock speeds are likely to require power consumption and heat dissipation beyond the reach of handsets for many years to come.

An approach that is an easy migration from the past paradigm and has been widely employed is to use an architecture which relies on a DSP complemented by multiple ASICs, one for each service. Although this approach seems simple and direct and appears to solve the performance and power consumption problems, it has serious inherent issues. As more modes and bands are supported by a single handset, the number of ASICs required increases linearly. This results in large silicon area, associated cost, and power consumption problems. Also inherent in this approach is limited ability to add or modify air interfaces once devices are in the field. It demands significant resources for development and support. Each system represents a particular permutation and combination of modes and bands (standards). The full set of permutations and combinations can be quite large. For example, Nokia has stated publicly that it has 1000 different products in its handset product line. Part of the problem comes from the fact that ASICs which have been developed for different services have been developed to complement different DSP implementations, so in most cases new or modified ASIC designs are needed to support multimode multiband systems (silicon evolution issues).

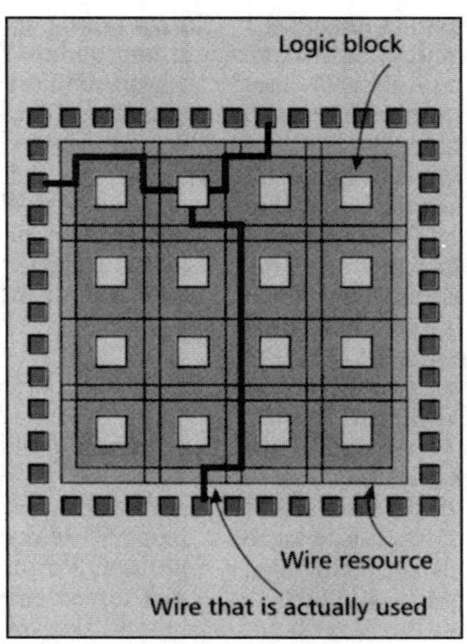

■ **Figure 1.** *A typical FPGA architecture.*

Parameterized hardware comes in several different forms. It can be an ASIC designed for a certain subtask, such as an FIR filter with the ability to change the number of taps by supplying a parameter, or a processor with switch-

Reprinted from *IEEE Communications Magazine,* Vol. 37, No. 2, pp. 108–112, February 1999.

	Power consumption	Size	Cost	Field upgradable	Silicon evolution	Tools
High-speed DSPs	Very high	Modest	Moderate/ high	High	Easy	Some
Multiple ASICs	Moderate	Large	High	None	Difficult	Available
Parameterized hardware	Moderate	Moderate	Moderate	Some	Moderate	Some
Reconfigurable logic	Low	Low	Moderate/ low	High	Easy	Unavai- able

■ **Table 1.** *Comparison of technology solutions.*

able microcode. These approaches can lead to very long instruction word (VLIW) processors. The companies that have tried to implement these approaches have found out that it takes an order of magnitude more effort to create the parameterized ASIC than they expected. This has led to time-to-market (TTM) problems. There are also inherent problems with flexibility. Implementations are typically keyed to specific combinations of standards and are not able to handle additional upgrades in the field.

The use of reconfigurable logic such as field programmable gate arrays (FPGAs) for production implementations is the greatest departure from the past paradigm. In the past, FPGAs were used for fast ASIC prototyping. ASIC designs were tested by being prototyped in FPGAs and, when close to meeting the requirements, transferred to tape and tested in small fab run ASIC's. This testing is done with iterative tapeouts and fab runs as necessary. By designing systems to use the full capability of dynamically reconfigurable FPGAs, it is possible to create systems where silicon area is no longer a function of the number of air interfaces supported, while providing high flexibility in the field. Because the tapeout and small fab run/test iteration time is taken out, TTM can be significantly reduced. What emerges, then, is the possibility of having a single or relatively small number of platforms which will support all the combinations and permutations of standards. Because the tools are not designed specifically to support this design flow, these implementations require knowledgeable staff, which currently are in short supply.

WHAT IS AN FPGA?

An FPGA is an array of gates with programmable interconnect and logic functions that can be redefined after manufacture. Field programmable devices can be grouped into the following two categories:

- Programmable logic device (PLD)
- Field programmable gate array (FPGA)

PLDs usually implement sums of products of input signals. Even though PLDs implement any signals represented as sums of products, it is usually very inefficient to represent all the signals in sum of products form, thus wasting precious silicon area. On the other hand, FPGAs usually consist of an array of blocks, each of which contains logic blocks and an interconnection resource to connect logic blocks. The example of a typical FPGA architecture is shown in Fig. 1.

The difference between PLDs and FPGAs is that the size of an FPGA logic block is usually much smaller than that of PLDs, such that FPGAs can have better resource utilization. An FPGA logic block usually contains lookup tables (LUTs) of n inputs, where n is usually between 3 and 6, and flip-flops to store data. Half of the logic block of Lucent Technologies' ORCA3 is shown in Fig. 2 as an example. Inputs to the logic block are connected to either LUT input ports or flip-flop input ports. Outputs from LUTs are either connected to output ports of the logic block or connected to flip-flop input ports. By using multiplexers (muxes), various combinations of input signals can be chosen.

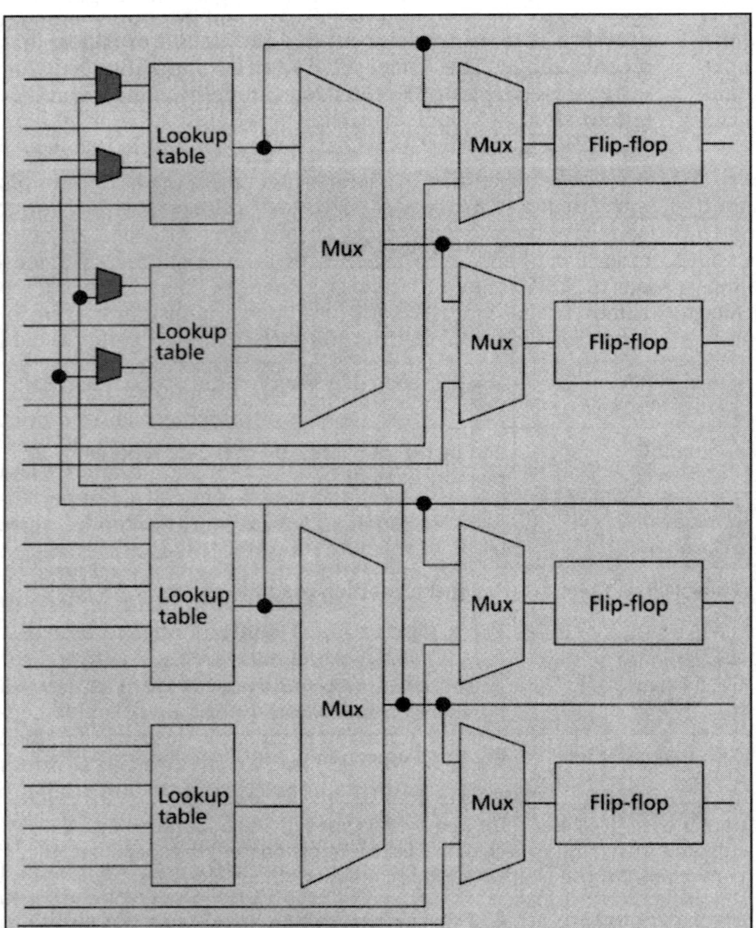

■ **Figure 2.** *An example of FPGA logic block structure (half of Lucent Technologies' ORCA3 logic block).*

LUTs are usually realized by using static RAM where address lines are input signals and data output lines are output signals. Thus, by correctly setting data contents at all the addresses of static RAM, LUTs can implement any Boolean functions of n inputs by feeding address signals to get data output signals. Flip-flops in each logic block are used to store data such as state information of finite state machines. By interconnecting LUTs and flip-flops using programmable routing resources, not only arbitrary combinatorial functions but also arbitrary sequential functions can be realized. The routing resources can usually be grouped into two types: clock routing resources and signal routing resources. Clock routing

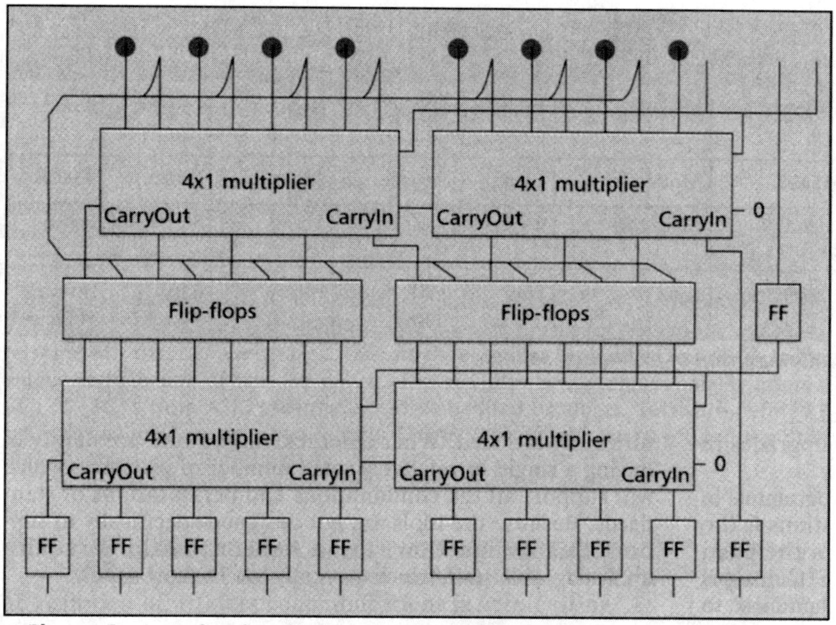

■ **Figure 3.** *A pipelined multiplier using an ORCA2 FPGA.*

resources are used to feed fast clock signals to clock ports of flip-flops with a small delay and a clock skew. Signal routing resources are used to route nonclock signals to inputs and outputs of logic blocks. The routing resources typically have a hierarchy of several wire types: short wires are used to connect adjacent logic blocks, medium-length wires are used to connect logic blocks that are several blocks apart, and long wires are used to send signals all the way across a chip.

Circuit designers usually use hardware description languages such as VHDL [4] or Verilog [5]. Even though the architecture is fairly complex, they usually do not have to worry about internal structures of FPGAs. Instead, vendors' tools perform automatic mapping, placement, and routing of users' circuits onto FPGAs [6]. However, to achieve high performance and low power consumption, "hand wiring" with knowledge of the underlying structures is often needed with today's generation of FPGA tools. Xilinx, Altera, and Lucent are major manufacturers of FPGAs.

There is another category of chip: DSP. DSP chips have been in the market for two decades, and they are important components of signal processing systems. Table 2 shows the comparison between FPGA and DSP chips.

Reconfigurability of FPGAs

Configuration of FPGAs is typically performed when the system power is turned on. During operation the FPGA configuration is usually fixed, so the FPGAs do a fixed operation until the system power is turned off. However, recent FPGAs allow dynamic reconfiguration, where a portion or the entire chip of an FPGA is reconfigured on the fly while it is doing signal processing. For example, when an FPGA-based system works as a transmitter and receiver at different times, FPGAs can perform data compression in a transmitter mode, and decompression in a receiver mode. The Atmel AT40K series and Xilinx XC6200 series are examples of FPGAs that can do dynamic reconfiguration.

	FPGA chip	DSP chip
Programming language	VHDL, Verilog	C, Assembly Language
Ease of software programming	Fairly easy. However, a programmer needs to understand the hardware architecture before programming.	Easy
Performance	Can be very fast if an appropriate architecture is . designed.	Speed is limited by the clock speed of a DSP chip.
Reconfigurability	SRAM-type FPGAs can be reconfigurable infinite times.	Can be reconfigurable by changing program memory content.
Reconfiguration method	Reconfiguration is done by downloading configuration data to a chip electronically.	Reconfiguration is done by simply reading a program at a different memory address.
Areas where FGPAs can outperform DSPs, or vice versa	FIR filter, IIR filter, correlator, convolver, FFT, etc.	A signal processing program of sequential nature.
Power consumption	Can be minimized if the circuit is designed to save power, or the power is dynamically controlled.	Even if program A is larger than program B, power consumption does not change as long as the number of memory chips is the same.
Implementation method of MAC	Parallel multiplier/adder or distributed arithmetic.	Repeated operation of MAC function.
Speed of MAC	Can be fast if a parallel algorithm is used. If a filter is implemented using distributed arithmetic, the speed does not depend on the number of taps.	Limited by the speed of MAC operation of a DSP chip. If a filter is implemented, the speed becomes slower if the number of taps increases.
Parallelism	Can be parallelized to achieve high performance.	DSP chip programming is usually sequential and cannot be parallelized.

■ **Table 2.** *A comparison of FPGA and DSP chips.*

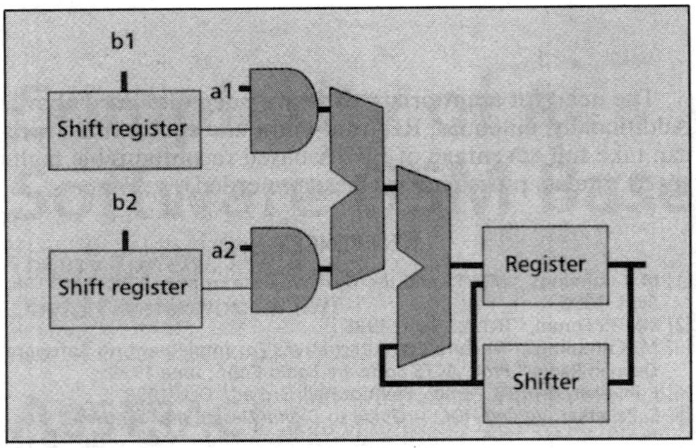

■ **Figure 4.** *Distributed arithmetic MAC.*

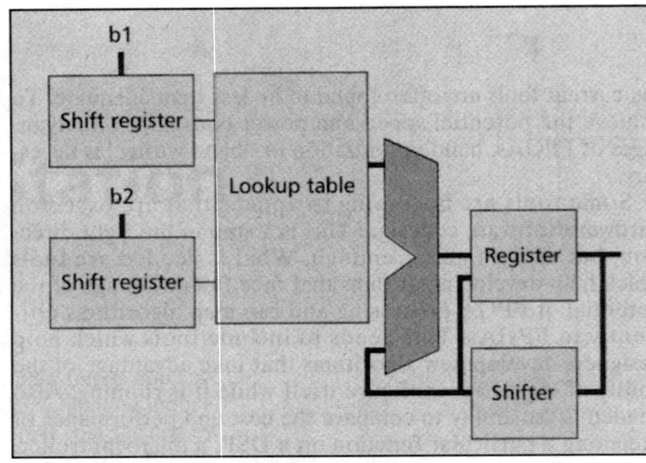

■ **Figure 5.** *Distributed arithmetic MAC using an LUT.*

THE FPGA SOLUTION

FPGAs implement DSP functions such as multiply and accumulate (MAC) efficiently (e.g., for FIR filters). There are several methods to implement MAC operation using FPGAs. One method is array multiplication, where rows of adders are placed in parallel and the decision to add partial products is determined by the corresponding bit of a multiplicand. By inserting flip-flops between rows of adders, a pipeline structure can be constructed, resulting in fast throughput. One of this article's authors, Haruyama, implemented a parameterized data path synthesis tool called Synthesis Compiler for User Programmable Arrays (SCUBA) for Lucent Technologies' ORCA FPGA [7]. One of the data path modules that can be synthesized by SCUBA is an array multiplier. Figure 3 shows an 8 × 2 pipelined multiplier synthesized by SCUBA using ORCA2's 4 × 1 multiplier mode.

The array multiplication method is suitable when fast multiplication is required. However, it needs a large number of FPGA logic blocks, often using up most of them just for one multiplier.

The second method is a method called *distributed arithmetic* [8, 9]. Distributed arithmetic differs from conventional arithmetic only in the order in which it performs operations. Following is an equation of the MAC function.

$$P = a1b1 + a2b2 + \ldots + ambm$$

In conventional arithmetic, when a sum of products P needs to be calculated, multiplication of two values have to be calculated followed by addition of these multiplied values. In the example of an FIR, ai is a tap coefficient and bj is a sampled input. $Aibj$ can be represented as follows by breaking ai into each bit:

$$aibj = (ai0bj) + (ai1bj)s(1) + \ldots + (ain - 1bj)s(n - 1)$$

where aik is the kth bit of ai, and $s(k)$ means shift left operation for k bits.

On the other hand, in distributed arithmetic this equation is rewritten as follows:

$$aibj = (ai0bj) + ((ai1bj)s(1) + (((ai2bj)s(1) + \ldots + (\ldots(((ain - 1\,bj)s(1))))\ldots)s(1)$$

Thus, the addition of all m bits is performed first, followed by multiplication operations (shift and add). Figure 4 shows an example of parallel distributed arithmetic when $m = 2$.

In applications such as FIR, ai, a tap coefficient, is usually a fixed value, and bj, a sampled input, is a changing value. Since ai is a constant, the AND and ADD portion of the distributed arithmetic circuit can be replaced by an LUT as shown in Fig. 5.

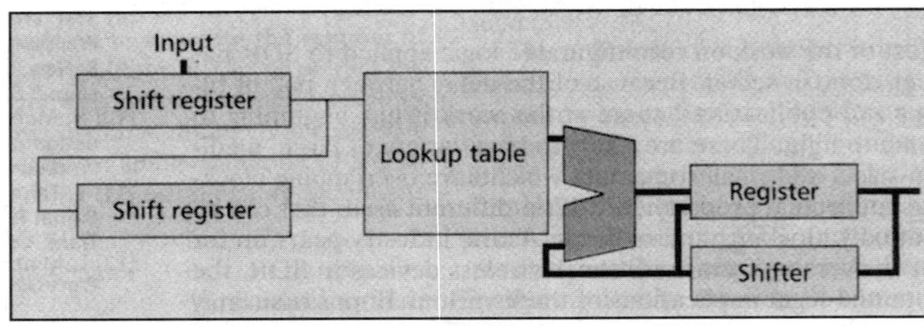

■ **Figure 6.** *A distributed arithmetic FIR filter using an LUT.*

The distributed arithmetic method uses these LUTs for fast calculation, which makes LUT-based FPGAs very suitable. An FIR filter can be implemented by serially shifting input data which are fed into the distributed arithmetic circuit as shown in Fig. 6.

Parallelism is key to achieving high performance using FPGA-based systems, and does not have to be done by distributed arithmetic only. Other appropriate architecture can be designed and implemented in FPGAs to achieve very high performance. The examples of achieving high performance by applying parallelism using FPGAs are an FPGA Viterbi decoder [10] and the use of FPGAs in a digital beamforming antenna [11]. In the future, FPGA flexibility and high performance will find wider application in wireless communication functions such as modulation, demodulation, and fast Fourier transform (FFT) [12].

TOOLS LIMITATIONS

Most of the existing FPGA system development tools have been created for fast ASIC prototyping. They are designed to be very efficient in the use of engineering time, not in the use of FPGA resources. When designing a new wireless system, spending a few thousand dollars on a handful of FPGAs that are going to be used to save six to eight months of time for a large engineering team makes economic sense. If the cost of the FPGA resources doubles, it has little effect on the overall development cost. Power consumption costs are not even measured.

When planning to implement production systems with FPGAs, the economic equation changes dramatically. Now, days and months of time are spent by the engineering team to optimize the bill of materials (BOM) and power consumption. When this economic equation is applied to the design flow, the current tools are often found to be less than adequate. To achieve the potential speed and power consumption advantages of FPGAs, hand optimization or "hand wiring" is necessary.

Some tools are beginning to appear that are based on hardware/software codesign. This is a step in the right direction, but doesn't go far enough. What is needed are tools which help develop algorithms that take full advantage of the potential of FPGA processing and can map algorithms efficiently to FPGAs. This needs to include tools which help designers develop new algorithms that take advantage of the ability of logic to reconfigure itself while it is running. Also needed is the ability to compare the cost and performance of executing a particular function on a DSP, a microcontroller, an ASIC, or an FPGA.

INTELLECTUAL PROPERTY ASPECTS

Most of the work on reconfigurable logic applied to SDR has been done in secret. Because of the delay between patent filings and publications, some of this work is just beginning to come to light. There are likely to be a variety of large, medium-sized, and small companies which have overlapping blocking intellectual property (IP), IP in different areas that can be mutually blocking, and so forth. As the industry gears up for the conversion from traditional wireless devices to SDR, the potential legal implications of these various IP positions may be daunting.

One model to speed the rollout of SDR technology was developed in the Global System for Mobile Communications (GSM). Some observers credit the rapid development of the global GSM industry segment to the creation of the GSM patent pool. Similar techniques may be appropriate in the area of reconfigurable logic for SDR.

DIRECTIONS FOR FUTURE WORK

Today's FPGA chip architectures are heavily influenced by the fast ASIC prototyping application. The pricing and production models are heavily influenced by the volumes inherent in the fast ASIC business model. The industry needs to focus chip architectures, business, and production models on these multimode multiband wireless communications applications and their scale of volume. For example, in 1998 it is projected that 140 million cellular/PCS handsets were sold.

The need for appropriate tools has been discussed above. Additionally, antennas, RF front-ends, and controllers which can take full advantage of FPGA-based reconfigurable high-speed wireless processing engines are needed.

REFERENCES

[1] M. Cummings, "MMITS and the Next Wireless Frontier," *Proc. PCS '96*, Sept. 1996.

[2] MMITS Forum, "TR1.2," Sept. 1998.

[3] M. Cummings, "Technology Alternatives For Implementing Software Defined Radio," *Proc. ACTS Software Radio Conf.*, June 1998.

[4] J. Bhasker, *A VHDL Primer*, Prentice Hall, 3rd ed., Oct. 1998.

[5] S. Palnitkar, *Verilog HDL: A Guide to Digital Design and Synthesis*, Prentice Hall, Mar. 1996.

[6] R. Murgai, R. Brayton, and A. Sangiovanni-Vincentelli, *Logic Synthesis for Field-Programmable Gate Arrays*, Kluwer, 1995.

[7] S. Mohanty *et al.*, "SCUBA: An HDL Data-Path/Memory Module Generator for FPGAs," *IEEE VHDL Int'l. Users Forum 1997*, Arlington, VA, Oct. 1997, pp. 135–42.

[8] S. A. White, "Applications of Distributed Arithmetic to Digital Signal Processing: A Tutorial Review," *IEEE ASSP Mag.*, July 1989, pp. 4–19.

[9] B. New, "A distributed arithmetic approach to designing scalable DSP chips," *EDN*, Aug. 17, 1995, pp. 107–14.

[10] N. Mohd and S. M. Rezaul Hasan, "Application of FPGA device in the design of a Systolic Array Viterbi Decoder," *Proc. 5th Int'l. Conf. Signal Processing Apps. and Tech.*, Oct. 1994, pp. 1459–64.

[11] T. Tanaka, R. Miura, and Y. Karasawa, "Implementation of a Digital Signal Processor in a DBF Self-Beam-Steering Array Antenna," *IEICE Trans. Commun*, vol. E80-B, no. 1, Jan. 1997, pp. 166–75.

[12] L. Mintzer, "The FPGA as FFT Processor," *Proc. 6th Int'l. Conf. Signal Processing Apps. and Tech.*, Oct. 1995, pp. 657–61.

John Villasenor and Brad Hutchings

The Flexibility of CONFIGURABLE COMPUTING

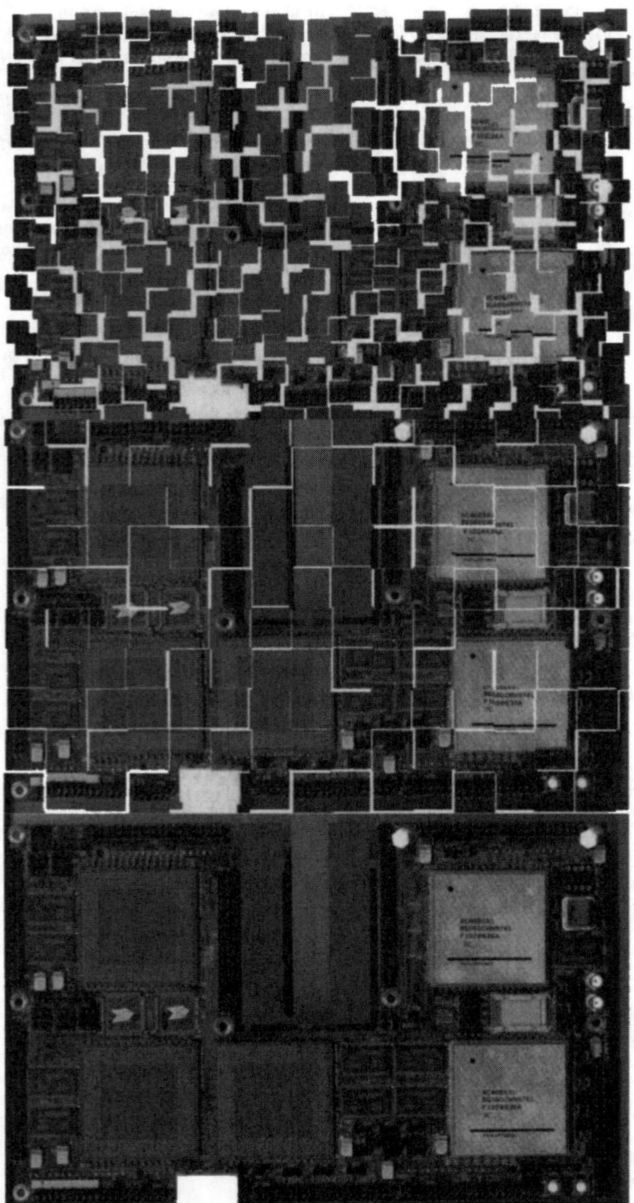

Board courtesy of Hea Joung Kim, J. Villasenor, and B. Hutchings (DARPA funding)

Providing the Hardware for Data-Intensive Real-Time Processing

One of the most fundamental tradeoffs in the design of computing devices involves the balance between flexibility and efficiency. At one end of the spectrum are application-specific integrated circuits (ASICs), which contain dedicated circuitry optimized to a particular set of tasks. ASICs have the advantages of low power dissipation and high clock speeds, but suffer the drawback of being able to perform only the tasks for which they were designed, and thus become obsolete if the task is changed. Because ASICs are highly specialized to a given application, they typically achieve the highest possible performance at the lowest silicon cost. However, this specialization comes at the cost of flexibility and ASICs are only useful for the one task for which they were designed. At the other end of the spectrum are programmable processors such as microprocessors or DSPs. Programmable processors implement a limited and *fixed* set of arithmetic and control operations that can be organized and sequenced to implement any arbitrary computation. However, in cases where the native processor operations are not well suited to the task at hand, or in cases where massive amounts of parallelism can be exploited, programmable processors are inefficient and deliver relatively poor performance.

There has been growing recent interest in configurable computing, which can be viewed as a hybrid between ASICs and programmable processors. Configurable computing machines are implemented with *programmable logic*: flexible hardware that can be programmed or, more correctly, structured to fit the natural organization and dataflow of a computation. Unlike programmable processors where computations must be implemented as some sequence of available operations, configurable com-

Reprinted from *IEEE Signal Processing Magazine,* Vol. 15, No. 5, pp. 67–84, September 1998.

puting machines allow a much more direct and natural approach in which the hardware can be structured to directly implement the native operations required by the *application* and also organized to exploit the concurrency inherent in the computation. These systems can "track" the computational requirements imposed by changes in the processing task or data.

The enabling device for configurable computing is the field-programmable gate array (FPGA). FPGAs have typically been used for prototyping of ASIC designs and as low nonrecurring engineering (NRE) cost parts in applications where the cost of designing and fabricating an ASIC is not justified. FPGAs have not traditionally been viewed as effective computing devices in their own right, mostly because of their higher power requirements, low clock speed, and relatively long (tens of milliseconds) programming time. However, the combination of technology advances and some recent demonstrations of FPGA-based computing machines has led to a re-evaluation of FPGAs as efficient computing devices. It is now clear that for applications characterized by deeply pipelined, highly parallel, and integer arithmetic processing, configurable computing machines can outperform alternative solutions by up to an order of magnitude for metrics such as cost and computation speed. Computational tasks having some or all of these characteristics include pattern matching, image processing, target recognition, cryptography, and some database tasks. Because there are many other applications that require computations that FPGAs do not perform as well, configurable computing today is a niche solution that appears unlikely to make significant inroads into the general-purpose computing applications currently dominated by microprocessors. However, the advantages of configurable computing for these niche applications are quite compelling. The combination in a single device of dedicated hardware and rapid, submillisecond-scale reprogrammability constitutes an exciting and promising development whose implications are only just beginning to be explored.

In this article we begin with a brief tutorial on FPGAs that describes the most common FPGA architectures and how these architectures are used to support computation, memory access, and data flow. We then present FPGAs as computing machines and focus on devices that are reconfigured during run time. Ongoing research involving FPGAs as well as future directions are also discussed.

Field-Programmable Gate Arrays

Architecture

FPGAs share with ASICS the capability to support application-specific circuitry, with the key difference that FPGA circuits are programmed by means of a bitstream that completely specifies the logical functions and connectivity to be implemented. Because FPGAs are static-random-access-memory (SRAM) devices, they can be reprogrammed as often as desired, thereby allowing the silicon resources in a single device to be time shared across a wide range of functions. The flexibility of programmable hardware inevitably comes at a cost in efficiency relative to an ASIC. Therefore, while there is some truth to the statement that FPGAs combine the advantages of ASICs with the flexibility of software-programmable processors, the caveat must be added that an FPGA never achieves the power, clock rate, or die size that could be realized in a full custom chip optimized for a particular task.

Although there is a wide range of architectural approaches that are used in FPGAs, all FPGAs consist of an array of logical units distributed across a grid of programmable interconnect, or routing. For a given total die size, the basic design issues facing an FPGA manufacturer are much the same as those involved in the design of any other computing device. These issues include the amount of resources to devote to routing as opposed to computation; the amount, granularity, and distribution of on-chip memory; and the quantity and accessibility of I/O.

Broadly speaking, FPGAs can be classified as either "coarse-grained," meaning that they possess a smaller number of relatively powerful logical units, or "fine-grained," which refers to a larger number of very elementary logic units. Most FPGAs in current use are coarse-grained, and have logical units that are based on look-up tables (LUTs). For example, the Xilinx 4000 series family of FPGAs has logical units (referred to by Xilinx as "configurable logic blocks") that each contain two 4-input, 1-output LUTs, one 3-input, 1-output LUT, two flip flops, and several multiplexors to select from among the lookup table and flip-flop inputs and outputs. One Xilinx 4000 series configurable logic block has the capacity to implement a two-bit adder or a nine-bit parity checker. By contrast, the Xilinx 6200 series provides an example of a device that lies at the other end of the granularity spectrum. Each logical unit in the 6200 consists of two 4-to-1 multiplexors (though not all of the inputs are independent), three 2-to-1 multiplexors and a flip-flop, and can implement any one- or two-input function as well as a single bit of storage.

To describe an FPGA in terms of its gate count can be broadly useful but also deceptive.

Coarse-grained FPGA architectures also provide some form of distributed memory in addition to logic resources. For example, each LUT in a Xilinx 4K-series FPGA can be used to implement a 16 × 1 RAM/ROM. This makes it possible to trade logic for memory and vice versa. FPGAs manufactured by Altera provide dedicated RAM blocks on the 10K-series that can each function as either a 2048 × 1, 1024 × 2, 512 × 4, or 256 × 8 RAM. Altera RAMs are larger than the Xilinx LUT-based RAMs and are more efficient for implementing larger memories but may go unused in designs that don't need them. It is anticipated that vendors of fine-grained FPGAs will eventually introduce devices that provide RAM blocks, but there are no current fine-grained FPGAs that provide distributed RAM resources (though there is storage capacity in the flip flops that are part of each logical unit in a fine-grained FPGA).

LUTs, memory blocks, and flip-flops are all interconnected via programmable routing. Programmable routing consists of fixed-length metal segments that are interconnected with programmable switches, e.g., a single MOSFET. Metal segments are of various lengths that support nearest-neighbor interconnect, global interconnect, and several lengths in between. Connections are made by programming the switches that interconnect the metal segments between the desired pins. The additional resistance and capacitance introduced by the programmable routing network is one of the primary performance bottlenecks in FPGAs. Current-generation FPGAs are typically clocked at between 10 and 40 MHz, though certain types of algorithms can be optimized to run as fast as 250 MHz [23]. Figures 1 and 2 illustrate the routing and logical unit of a coarse-grained and fine-grained FPGA, respectively.

Logic Capacity

FPGA capacity is often expressed in terms of logic gates, where one logic gate is understood to be a two-input NAND gate. As in other types of computing, in which the pressure placed on manufacturers to characterize performance using a single number has led to oversimplification, to describe an FPGA in terms of its gate count can be broadly useful but also deceptive. As is evident from the

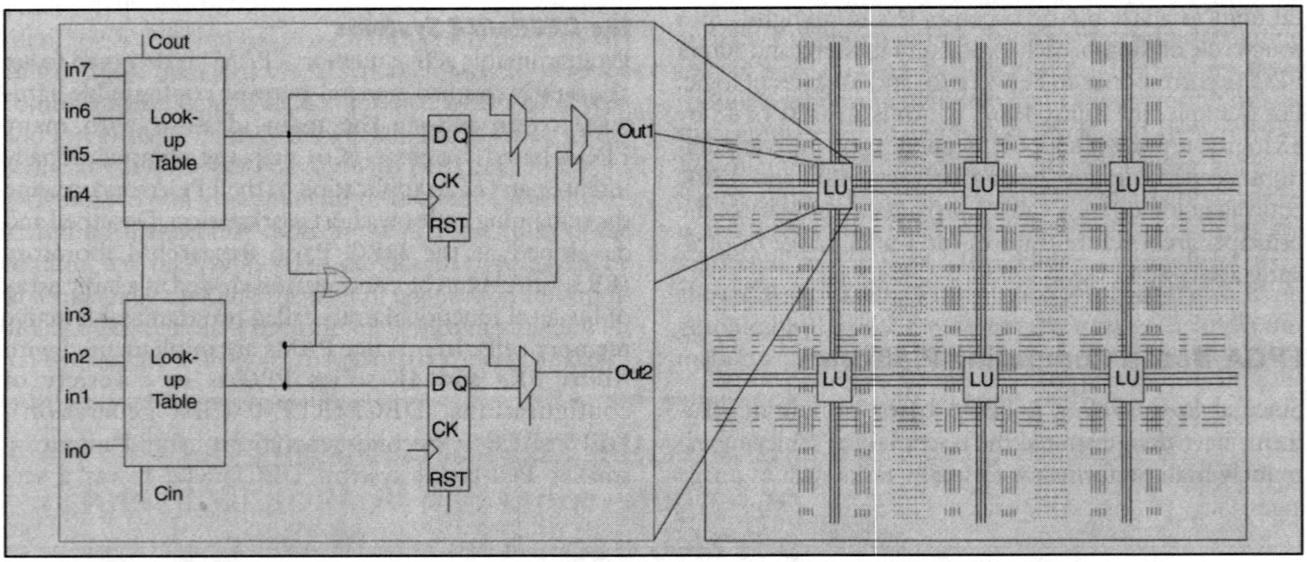

▲ *1. Example of the logical unit and routing in a coarse-grained FPGA. The logical unit typically contains several look-up tables, multiplexors, and flip-flops. Connectivity between logical units is hierarchical, including routing between adjacent logical units (solid lines), among logical units in a local neighborhood (dashed lines), and routing to connect widely separated logical units (dotted lines).*

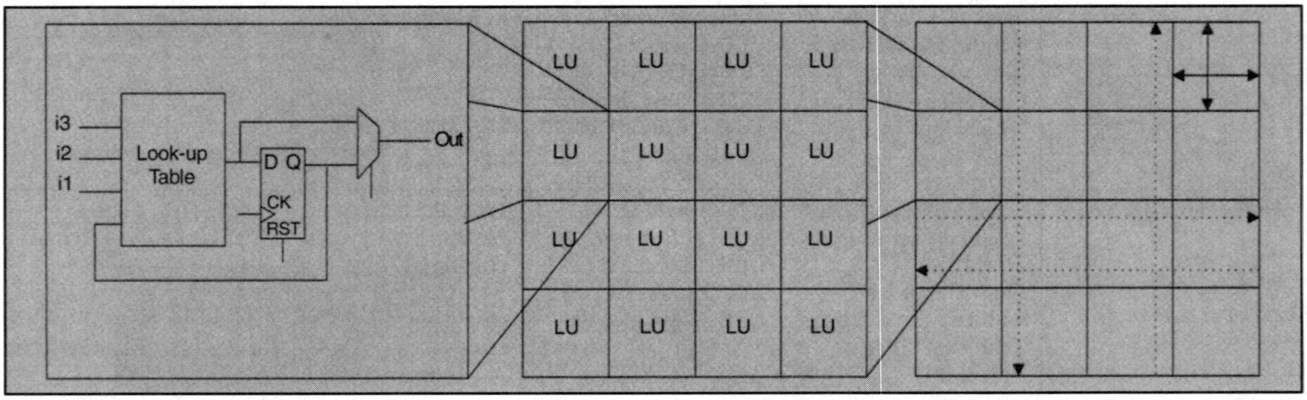

▲ *2. Example of a fine-grained FPGA. The logical unit is much smaller than in a coarse-grained FPGA, typically containing the equivalent of a single, small look-up table and a memory element. Routing permits a high degree of connectivity among logical units in a local neighborhood as well as among widely separated logical units.*

contents of the logical units as described above, the logical functions that are realized in a given FPGA configuration will be highly dependent on computational demands and algorithm-mapping strategies associated with the application. In addition, within any given configuration, some logical units will be extremely efficiently utilized while others will receive little or no utilization. Another complication is that for LUT-based FPGAs, logical elements can be used for logic, memory, or a combination of both, thereby making it necessary to distinguish between logic gates and gates dedicated to storage.

For experienced users of FPGAs, a more meaningful way to characterize capacity is to use the number of vendor-specific logical units. Since vendors tend to scale parts by increasing the array size as opposed to increasing the complexity of the logical units that are elements in the array, a designer with experience using a part with 1,024 logical units will readily understand what it means if the vendor announces an FPGA with 1,600 logical units. In the absence of experience with FPGA designs, using logical units as a size metric becomes less meaningful. As a rough rule of thumb, each logical unit in Xilinx and Altera FPGAs corresponds to between 20 and 50 gates of logic. For example, the Xilinx 4062, which is an array of 48 by 48 logical units, will support, depending on the computations needed, designs having between 40K and 130K equivalent logic gates. Table 1 provides a summary of the principal architecture features for a wide range of presently available FPGAs.

FPGA-Based Computing Platforms

Since at least 1989 [7], FPGA-based computing platforms have demonstrated the potential for achieving extremely high performance for many tasks such as image filtering [36], convolution [5,7], morphology [13], feature extraction [1], and object tracking [32,39]. The potential is real; prototype systems have been constructed and applications developed that achieve performance that is an order of magnitude faster than conventional approaches [3].

Although a wide variety of FPGA-based computing systems have been constructed and reported in the literature, we focus here on two platforms that stand out as good examples of what can be achieved with FPGA-based computing platforms: DECPeRLE, which was built by researchers at Digital Equipment Corporation (DEC), and SPLASH-2, which was built at the Supercomputer Research Center/Institute for Defense Analysis. These two platforms are successful examples of FPGA-based systems: each had clearly defined architectural goals, integrated design and debug tools, and achieved significant speedup for several important applications.

Programmable Active Memory and the DECPeRLE Systems

Programmable active memory (PAM) is designed to act as a tightly coupled, general-purpose configurable hardware co-processor. The main idea—as with many FPGA-based systems—is to map the computationally intense parts of an application to the FPGAs and execute the remaining code on a host workstation. Designed and developed at the DEC Paris Research Laboratory (PRL), PAM can be conceptually viewed as a large array of bit-level functional units called programmable active memory cells. In practice PAMs are implemented with Xilinx 3K- and 4K-series FPGAs in a variety of configurations: DECPeRLE-0 (first generation), DECPeRLE-1 (second-generation), and Pamette, a smaller PCI-based system. DECPeRLe-1 was a sec-

Table 1. Comparison of resources in common FPGA architectures.

FPGA	Logical Unit Name	Routing Hierarchy	Memory	Largest Part (available)
Xilinx 4000 Series	Configurable logic block (CLB) Two 4-input LUTs with fast- carry chain; One 3-input LUT; 2 Flip-flops	5-level routing: single (adjacent CLBs), double (2 CLBs), QUAD (4 CLBs), long (entire length of array) and global lines	Each CLB can be used as 32-bits of RAM	40125XV: 64 x 64 CLB-array, 448 User I/O
Altera Flex 10K Series	Logic element (LE), 8 LEs make up 1 logic array block (LAB); One 4-input LUT with fast-carry chain; 1 flip-flop OR gate	3-level routing Local (LE in one LAB), fast-track row and Column (between LABs), and global	Embedded array blocks (EABs), each contain 2K bits. From 3 to 12 EABs per chip depending on size	EPF10K250, has 1520 LABs, 470 max user I/O

ond-generation system and was the most successful of the DEC PAM systems; to the best of our knowledge more applications have been reported for DECPeRLE-1 than for any other single platform.

DECPeRLe-1 consists of 23 Xilinx 3090 FPGAs: 16 in a tightly coupled array of 4×4 FPGAs with the remaining FPGAs providing interfaces to RAM and a host machine, which is typically a DEC Alpha workstation [7]. In addition, a programmable clock generator and additional static logic is provided to manage the host bus interface and download process. Figure 3 shows the architecture of the DECPeRLe-1. As shown, the core array is connected to four additional 3090s, one on each side of the array, which are, in turn, connected to 1 MB (32-bits wide) of static RAM. Three additional FPGAs are used to interface the core to two FIFOs, which are connected to the host through a turbo-channel interface.

Although PAMs can be programmed with commercial CAD software, the PAM group developed custom design software for use with DECPeRLe. Referred to as PeRLe-1, the design tool is based on C++ and is a structural/physical design tool for capturing synchronous circuit designs. Standard C++ syntax is used to describe combinational logic. This is combined with a *NET* type and a register primitive, thus making it possible to describe any synchronous circuit with PeRLe-1.

Designers can also annotate their circuits with placement information that controls the final floorplan of the resulting circuit. Application development proceeds by first manually dividing the overall application into individual FPGA circuits. Each circuit is then described with a single PeRLe-1 program that, when compiled, linked to a technology library, and executed, generates a native netlist (Xilinx Netlist Format (XNF)) that can then be processed by Xilinx back-end software to route the design and to finally create the downloadable bitstream.

PeRLE-1 is a relatively low-level design tool that allows designers to finely craft high-performance designs. Unlike verilog-hardware-decription-language (VHDL) synthesis environments where structural implementations can be inferred from higher-level behavioral descriptions, PeRLe-1 requires the designer to describe a circuit strictly in structural terms and designers using PeRLe-1 often manually place significant portions of the circuitry to achieve the highest possible performance. The designers of PeRLe-1 consciously chose this design path because the synthesis tools available at the time when PeRLE-1 was under development were often inefficient for high-performance designs. This was a reasonable tradeoff since one of the goals of the PAM project was to outperform supercomputers.

DECPeRLE-1 also provides a rich run-time library and set of debugging tools that allows design debugging to take place on the actual hardware. The run-time environment supports single-step and various tracing modes that simulate for a single clock cycle and then retrieve FPGA internal flip-flop state data for further analysis. Signal values are accessible via the variable names that denoted them in the PeRLe-1 program as long as they are registered in flip-flops. In addition, a graphical tool, **showRB**, is available for analyzing readback traces. It can graphically playback the stored state traces as a simple animation at about 100 Hz [45].

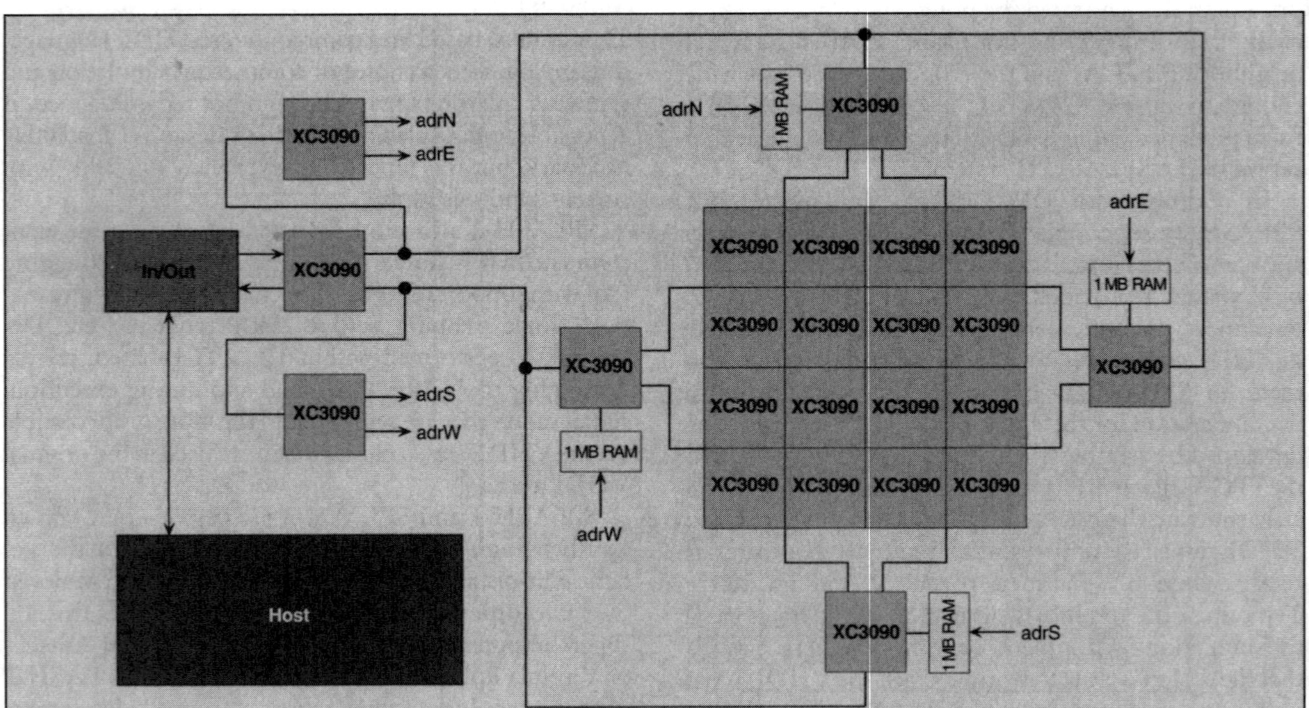

▲ *3. Architecture of the DECPeRLe-1.*

DECPeRLe-1 achieved significant speedup for a wide range of application areas including RSA cryptography, molecular biology, heat and Laplace equations, neural networks, and image classification [45]. In most cases, reported applications were the fastest known implementations at that time (including ASICs and supercomputers) and typically achieved a 10x speedup over other existing solutions.

SPLASH-2

SPLASH-2 is another example of a well-known FPGA-based computing platform. Created by researchers at the Supercomputer Research Center (SRC), SPLASH-2 was specifically designed to support high-performance linear systolic applications. SPLASH-2 was also a second-generation system that was constructed to overcome I/O limitations and the lack of inter-FPGA communication that limited the general application of the SPLASH-1, which preceded it [4].

The SPLASH-2 system connects Xilinx 4010 FPGAs in a linear systolic array. In addition to the linear systolic data-path, a global broadcast data-path and user-defined crossbar data path are available for custom communication networks. The system scales well for linear-systolic applications because the array can be easily extended simply by adding additional boards. Each of the FPGAs is connected to a local 256K 16-bit memory that is also mapped into the address space of the host processor. FIFOs are placed at the input and output sides of the systolic pipeline to maintain high throughput and buffer data to the host processor, which is typically a Sun SPARC-II. Connections between neighboring FPGAs and the crossbar are 36-bits wide, enough to support 32-bit data and 4-bits of tag, and useful in pipelined applications. Figure 4 shows the organization of the SPLASH II system.

In contrast with DECPeRLe-1 and SPLASH-1, SPLASH-2 uses commercial VHDL synthesis tools for application development. A VHDL model of the complete system is provided that includes all board-level interconnect, memories, and the host interface. This body of VHDL code serves not only as the simulation environment for SPLASH-2 designs but also as the definitive documentation for the entire platform. Applications are developed by creating a single VHDL design for each of the FPGAs that will be used in the system. Designers typically simulate these designs in the context of the system VHDL model to verify that the circuitry will operate correctly when downloaded to the actual hardware. Typically, only the initial simulations are performed in the simulation environment; once the designer is satisfied that the design is close to being correct, the VHDL is typically synthesized and downloaded for further verification via the run-time debugging tools. In practice, the hardware almost always operates as predicted by simulations. When differences occur, it is typically caused by a bug in the synthesis tool or by the use of signal initializers, which are ignored by the synthesis tool.

A SPLASH-2 application is developed by first manually partitioning the design into FPGA-sized pieces and then coding the VHDL that describes the circuit that goes into each FPGA. This code is verified by simulation in the SPLASH-2 environment with all of the other user FPGA designs. Each VHDL file is then synthesized separately, placed and routed with the Xilinx backend software, and finally downloaded to the SPLASH-2 hardware. Breaking the design into FPGA-sized pieces is one of the challenges faced when designing for SPLASH-2 because it is impossible to know if a design will fit in an FPGA until it is synthesized and placed and routed. Thus, partitioning is typically an iterative affair with designers attempting 2-3 times to converge on a design that meets the needs of the application and fits within SPLASH-2 resources.

Reconfiguration costs time because the logic being reconfigured is unavailable for processing while the new bitstream is being downloaded.

Although the commercial CAD tools used with SPLASH-2 do not make as efficient use of silicon as the PeRLe-1 tools, they still allow design to proceed rapidly. They are also based on a commonly-used HDL language, thus giving users a choice of commercial simulation and synthesis environments. Although not recommended, it is possible to floorplan postsynthesis designs if absolutely necessary, but this tends to be extremely unwieldy with current synthesis tools.

SPLASH-2 also provides an excellent run-time environment that is similar to a typical software debugging tool with support for reading back flip-flop state information, single stepping, setting clock frequency, etc. Debugging is performed within T2, a TCL-based, textual debugging tool. After download and during execution, signal values that are registered in flip-flops are accessible via the VHDL signal names that were used in the original VHDL text.

SPLASH-1 and SPLASH-2 are perhaps best known for their high-performance implementation of the genetic edit-distance application. This application achieved over two orders of magnitude of speedup over existing supercomputer implementations [24, 28]. Peter Athanas of Virginia Polytechnic University modified SPLASH-2 so that it would accept data directly from a video camera and reported significant speedups for a wide variety of image-processing algorithms [1, 5]. In addition, Graham and Nelson of Brigham Young reported significant speedup of a genetic-programming algorithm [19, 20].

Configurable Computing: Run-Time Reconfiguration

The key attribute distinguishing many of the recent and ongoing efforts in FPGA-based computing from early systems like SPLASH and DECPeRLE is the use of FPGA reconfiguration during computation as a means to further boost performance. While the early systems achieved high performance by focusing on efficient mapping of algorithms into FPGAs, they were capable of only limited reconfiguration during execution. Recent systems specifically target run-time reconfiguration (RTR) as a means to enhance performance. This enables silicon to be time shared across multiple tasks.

Reconfiguration is, of course, not free, and one of the most interesting questions in configurable computing concerns the costs and benefits of RTR for different algorithms, FPGAs, and FPGA platforms. The basic tradeoffs in RTR are relatively straightforward. Reconfiguration costs time because the logic being reconfigured is unavailable for processing while the new bitstream is being downloaded. It also costs power since configuration bitstreams tend to be large (approximately a megabit for the largest currently available FPGAs) and involve a correspondingly large number of pin transitions when loaded onto an FPGA. A few devices, most notably the Xilinx 6200 series and the CLAy chip from National Semiconductor Corporation, permit partial reconfiguration in which a selected region of the logic can be configured without altering the state of the remainder of the FPGA. More often, however, FPGAs allow only a full reconfiguration.

It is only in the last five years or so that FPGAs have supported reconfiguration at speeds fast enough to permit RTR without intolerable overheads. Most FPGAs on the market today can be configured (depending on the device and the size) in between 1 and 50 milliseconds. There have been several proof-of-concept demonstrations of RTR in recent years, and this has stirred interest in the computing community as well as awareness by FPGA vendors of the benefits of fast reconfiguration. FPGA manufacturers, in turn, have made reconfiguration speed a design issue in the newer and next-generation devices, with the result that configuration times of a few milliseconds or faster are likely to become common within a few years.

A full evaluation and understanding of RTR will have to wait several years until the infrastructure of design tools, devices, and machines matures to the point where its potential can be fully realized, and even then it will remain somewhat of a moving target as technology continues to evolve. To date, there have been very few attempts to quantify the tradeoffs in run-time reconfigured systems. It is relatively easy to build a system in which an FPGA is reconfigured multiple times during computation. It is much harder to perform a quantitative evalua-

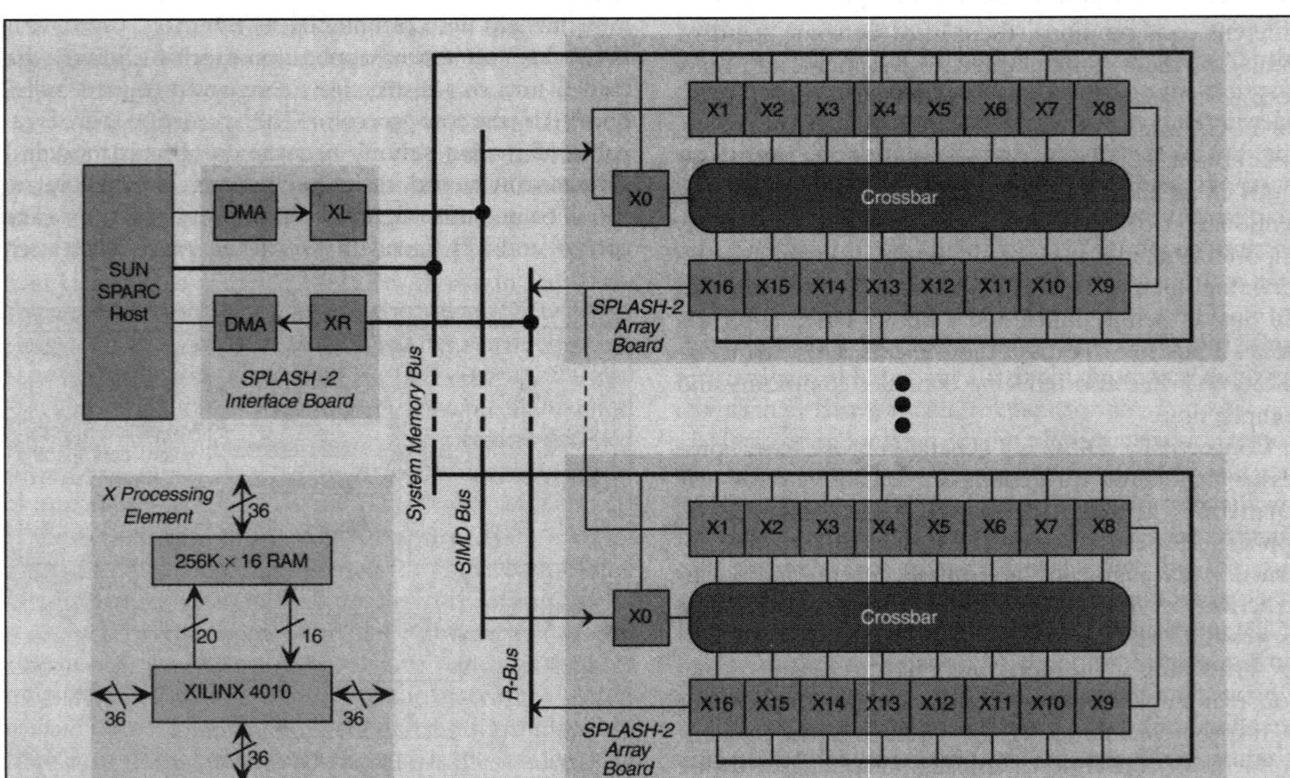

▲ 4. *SPLASH II computing platform.*

tion of the advantages, or lack thereof, that this enables relative to systems that perform the same task using other computational approaches. One of us (Hutchings) [49] has performed the first quantitative comparison of RTR for a range of applications. A few other researchers are also studying RTR, including Lysaght [30], Brebner [8], Luk [40], Singh [9], and Schmitt [37]. We expect that many more such studies will become available in the published literature in the next few years. To illustrate the nature and tradeoffs involved in RTR in more detail, we present two examples of systems below.

System Example 1: DISC

The first system example is DISC (Dynamic Instruction Set Computer), which is a research project started in 1994 at the Configurable Computing laboratory at Brigham Young University [46. 47]. In this ongoing experiment, researchers at BYU are experimenting with the following:

▲ Systematic approaches that ease the design and implementation of RTR applications

▲ New architecture configurations that integrate conventional processors and configurable-computing devices such as FPGAs

▲ Library-based techniques that promote the reuse of previously designed circuit modules to achieve both rapid application development and fast compilation

These three important research areas were motivated by BYU's past experiences with RTR applications. A systematic design framework is essential for RTR applications because these applications typically consist of many different configurations; there must be some standard means for these configurations to communicate results both with each other and to the outside world. Processor-FPGA interfaces are important because many of the basic housekeeping tasks that are unwieldy for FPGAs are much easier and efficient to implement on simple, conventional processors. Finally, a library-based approach is important because it makes it possible to *reuse* past efforts: high-performance, FPGA circuits require substantial human design effort, and a library-based approach makes it possible to exploit these efforts across many applications while also reducing compiler complexity and compile time.

Conceptually, DISC is a general-purpose "core" processor augmented with configurable hardware for performing application-specific tasks. Each application-specific function is treated as a custom instruction that augments the standard, general-purpose instruction set found in conventional programmable processors. For example, a DISC designed for image-processing applications might have a custom instruction that implements application specific circuitry for high-speed convolution. DISC overcomes the inherent problems of conventional application-specific hardware with a novel approach made possible by configurable logic: treat application-specific hardware (instructions) as distinct circuit modules that are stored as FPGA configurations in off-chip memory; at run-time, load and execute these circuit modules only as demanded by the application. Stored off-chip, these circuit modules only occupy valuable silicon resources while performing useful work and as soon as they are no longer necessary, they can be "swapped out" and replaced with other application-specific instruction modules. In addition, this library of application-specific circuit modules can be extended and individual circuit modules can be modified and redesigned over time to suit the needs of ongoing applications.

DISC System Architecture and Operation

The DISC architecture consists of two basic sections: a *static* core processor and a *dynamic* application-specific instruction space, which is slaved to the core processor. The basic organization is shown in Fig. 5. The main purpose of the core processor is control: it manages the application-specific instruction space and is responsible for loading circuit modules, initiating the execution of application-specific instructions, and managing I/O between the instruction space and the core processor. In most cases the core processor is static, i.e., implemented with circuitry that remains structurally unchanged throughout operation, and can be implemented with either a microprocessor or a statically configured FPGA—the FPGA in turn implementing either a simple state machine or programmable processor.

The application-specific instruction space is an undedicated array of reconfigurable logic; this is typically implemented with reconfigurable FPGA(s). Configurations that implement application-specific circuitry are loaded into this instruction space and executed as requested by the core processor. The instruction space is capable of storing several instructions (circuit modules) simultaneously and these instructions individually respond to initiation signals from the core processor. The control and I/O signals that form the interface between

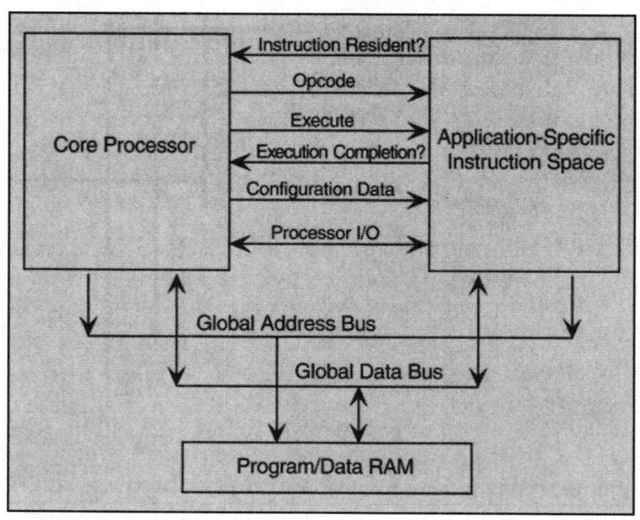

▲ *5. DISC system organization.*

the core and the instruction space are also shown in Fig. 5. As shown in the figure, the core processor and instruction space share a common address and data bus. This particular organization requires mutually exclusive operation: either the core processor is actively fetching instructions or data from memory, or the core processor is idle, waiting for some application-specific instruction to complete. However, this is not a requirement but rather a limitation of the current implementation.

In current implementations the core processor is a stripped-down programmable processor that can be programmed via a modified version of the lcc compiler [10,16]. A DISC program is then just a linear list of instructions that is assembled into machine code and stored in program memory for later fetching by the processor. The core processor fetches instructions from the program memory in a conventional manner and then processes them as shown in Fig. 6. Once the instruction is fetched, it is decoded and the opcode is latched onto the opcode pins (see Fig. 5 for these signal connections). The instruction-space reads the opcode value and quickly searches to see if the required circuit module is currently resident. If the required instruction is not resident, the instruction space requests the core to load the correct configuration. Once the load is complete (or if the circuit module was already resident), the core initiates the execution of the instruction module. The core processor then remains idle until the custom instruction has completed execution.

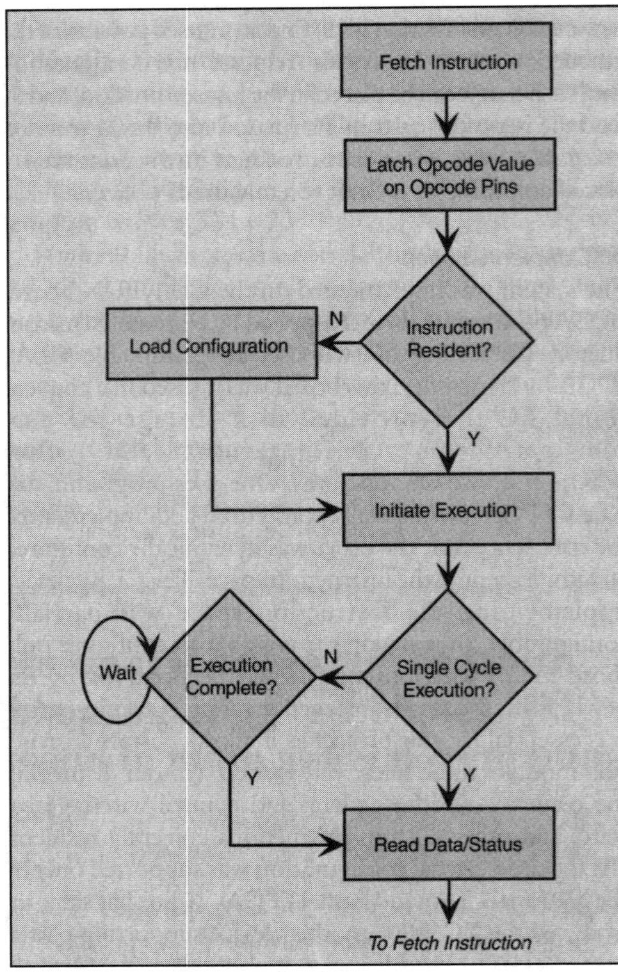

▲ *6. DISC operation.*

Hardware Relocation and Internal Architecture of Instruction Space

The internal architecture of the instruction space is organized to support a concept referred to as relocatable hardware. This organization provides the necessary circuitry that supports the execution of any custom instruction of any size at any physical location within the instruction space. The basic organization consists of several static elements: a global controller and three bundles of wires (control, memory address, memory data) that are collectively referred to as the "context," as shown in Fig. 7(a). The context provides a means for custom instruction modules to communicate with external memory and perhaps other devices, and to the global controller—independent of the physical location they occupy while executing. Memory address and data signals are used by custom instructions to directly access the system memory as shown in Fig. 5.

Execution and control of custom instructions is coordinated by the global controller via the global control signals. The opcode (supplied by the core processor) is broadcast to all resident instruction modules via the global control signals; each custom instruction is then responsible for decoding this signal and asserting a global *resident* signal (another one of the global control signals). If no custom-instruction module asserts the resident signal, the global controller requests a configuration load from the core processor. Once the custom instruction is resident, and if the execution signal is asserted, the custom instruction begins execution and asserts the completion signal when finished. Supporting variable-size instructions is important: custom instructions vary greatly in their hardware needs and allowing their size to vary as dictated by their hardware requirements ensures efficient use of the instruction space. In Fig. 7(b), the instruction space is shown with two instructions resident: one general-purpose (add) and one custom (histogram). Note that while both instructions span the entire width of the instruction space, they vary in size in the vertical direction due to their hardware requirements.

Hardware relocation refers to dynamic physical relocation of circuit modules during system operation and is an important part of the DISC system that dramatically improves efficiency. Because circuit modules are dynamically loaded into the instruction space by the core processor as dictated by the user program; it is not possible to know a priori where a custom instruction will be loaded during operation. Moreover, as custom instructions may be loaded, executed, removed, and reloaded over time, commonly used custom instructions will typically need to be loaded into different physical locations at different times, depending upon where free space is available in the instruction space. The process for loading a new custom instruction is: 1) find an unused portion of the instruction space, 2) process (relocate) the configuration data such that it can be placed in the unused portion, and 3) load the relocated custom instruction into the instruction space. If no free space exists, resident instructions are replaced according to a "least recently used" policy.

DISC Implementation

The system was implemented on the CLAyFUN board, an ISA prototyping board supplied by National Semiconductor. Consisting of two user-programmable CLAy FPGAs and memory, this board was housed in a conventional PC that provided disk storage for custom-instruction configurations and other general-purpose OS functions. One user-programmable CLAy FPGA was statically configured and implemented the core processor; the other was dynamically configured and implemented the instruction space. The CLAy device implementing the instruction space was partially configurable, thus making it possible to configure only those FPGA locations that would be occupied by the newly loaded custom instruction. Partial configuration was essential for this project as it allows custom instruction modules to be selectively loaded without disturbing the context of address, data, and control wires, global state, and other custom instructions currently resident. (At the time, partial configuration was supported only by the National CLAy and Atmel FPGAs. Xilinx has since introduced the XC6200, another device supporting partial configuration.) In addition, partial configuration directly reduces configuration time by reducing the number of configuration bits that are loaded into the FPGA.

Because FPGA resources were extremely limited, only a bare-bones processor with a single 16-bit accumulator was possible. Because of the core processor's limited performance, the actual process of bitstream relocation was handled by the 486 CPU in the PC enclosure; however, all of the functions that directly contributed to data processing were performed by the core processor. Although limited, the core processor was completely programmable in "C" using a modified version of lcc [16] and a custom assembler developed for this project [10]. In the assembler, application-specific instructions are accessed just as any other native machine instruction. In "C" code, these instructions are accessed like inline function calls simply by prepending a special prefix, **native-**, to the custom instruction identifier. The modified "C" compiler processes these calls in a special manner, replacing these custom function calls with native custom instructions in the generated assembly code.

A library of general-purpose and application-specific instructions are available for application development. The general-purpose instructions are a subset of those found on conventional microprocessors: add, subtract, shift, compare, etc. Custom instructions focused on image-processing applications and consisted of: mean filter, threshold, image subtract, morphological operations, etc. Early in the project the decision was made to implement nearly *all* instructions as circuit modules that are executed in the instruction space. Thus, general-purpose instructions compete for space in the instruction space and get swapped out when necessary to provide room for other instructions, either general-purpose or application-specific. Although this approach to general-purpose instructions actually *reduces* performance for general-purpose operations, it met the BYU research goals at the time.

To test the efficacy of the application-specific approach, a line-thinning (skeletonization) application was developed that used many of the custom im-

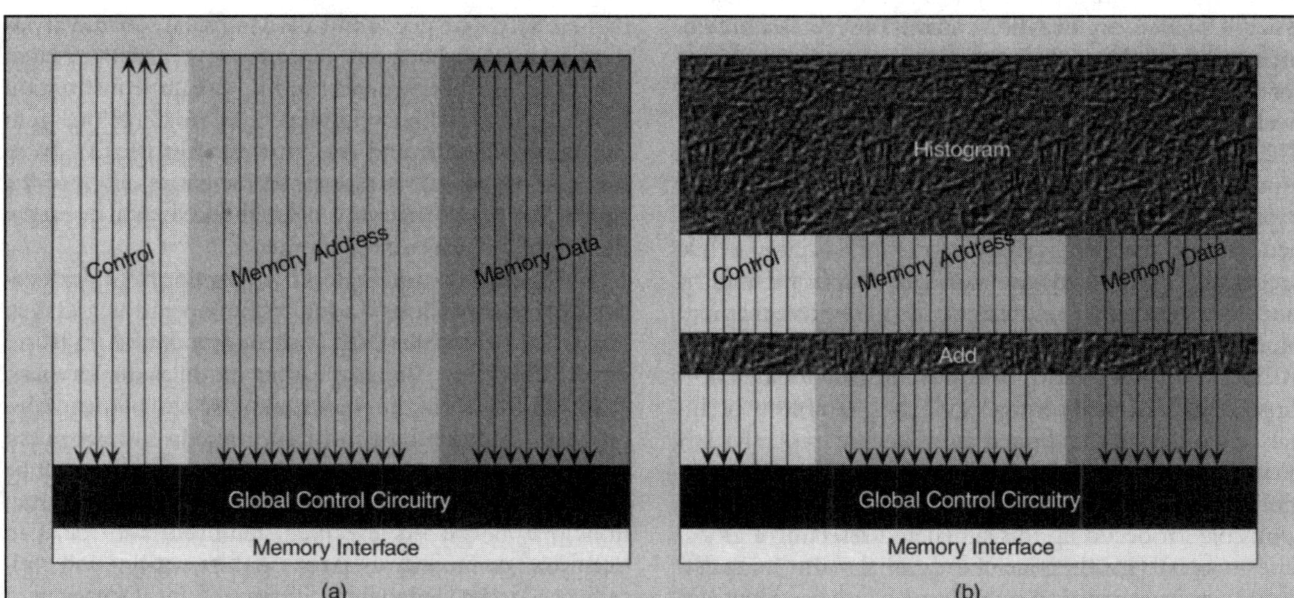

age-processing instructions. Although high-performance was not the focus of this early project, the laboratory prototype still achieved good performance for the line-thinning application: a speedup of nearly an order of magnitude over a software implementation running on a 66 MHz DX-2 [48]. Note that these performance numbers include all configuration overhead and are for a substantially suboptimal system. The application required active paging of both the custom and general-purpose instructions—many of the general-purpose instructions were loaded more than once and to more than one location in the instruction space. In addition to its performance advantages, DISC also demonstrated significant resource savings. The accumulated FPGA resource requirements for all of the executed instructions would require three FPGA devices if statically configured. With DISC, idle hardware could be immediately reclaimed by writing over idle circuitry with new custom instructions as requested by the core processor.

System Example 2: Template-Based Automatic Target Recognition

The second system example is from the area of automatic target recognition (ATR). ATR is among the most demanding real-time computational problems, and one which maps particularly well to configurable computing platforms. ATR is the application focus of the DARPA-sponsored UCLA Mojave project in configurable computing. The challenge addressed by an ATR system is conceptually simple—to analyze a digitally represented input image or video sequence in order to automatically locate and identify all objects within the scene of interest to the observer. Since there are many types of imaging devices and many algorithmic choices available to a designer, there are clearly a large number of possible ways to implement an ATR system. The Mojave project uses ATR algorithms developed by Sandia National Labs for the U.S. Department of Defense Joint STARS airborne radar-imaging platform.

The processing used in ATR is illustrated in simplified synthetic aperture radar (SAR) images consisting of 8-bit pixels and measuring several thousand pixels on a side, and they are generated in real time by a radar imager. Images are input to a focus-of-attention processor that identifies a set of regions of interest, each of which contains a potential target. These regions of interest, referred to here as candidate images, must then be correlated with a very large number of target templates. Target templates are binary; e.g., each pixel is represented using one bit. The correlation results are output to a peak detector that identifies the template and relative offset at which the peak correlation value occurs. The correlation of candidate images with templates is an important computational bottleneck in the system, involving data rates and computational requirements that exceed by several orders of magnitude the processing load most of the other steps in the algorithm (though it should be noted that the focus-of-attention step shown in Fig. 8 is also very computationally intensive). While the precise system parameters vary with implementation, the work under way at UCLA uses candidate image sizes of 128 by 128 and template sizes of 32 by 32.

Figure 9 illustrates the correlation operation targeted for FPGA implementation in more detail. Target templates are binary (e.g., 1 bit per pixel) and occur in pairs, one member of which is called the bright template and contains pixels from which a strong radar return is expected, and the other member of which is the surround template and identifies pixels where strong radar absorption is expected. The templates tend to be sparsely populated, with only a relatively small percentage of the pixels set to 1. This property is important in obtaining high performance in FPGA implementations. The first step of the correlation is known as a shapesum calculation, in which the 8-bit candidate image is correlated with the bright template, providing for every pixel in the image a number that is used for local gain control. The second step is the actual correlation, which is performed on a binary version of the image that is obtained via thresholding. The shapesum value is used to select which threshold to apply to generate the binary image.

What is less clear is the extent to which configurable computing techniques will become useful in more general computing environments.

Run-time reconfigured FPGAs offer an extremely attractive solution to the ATR correlation problem. First, the operations being performed occur directly at the bit level and are dominated by shifts and adds, making them easy to map into the hardware provided by the FPGA. In addition, the sparse nature of the templates can be leveraged to achieve a far more efficient implementation in the FPGA than could be realized in a general purpose correlator. This can be illustrated using the example of the simple template shown in Fig. 10.

In this example template, only five of the 24 pixels are "on." At any given relative offset between the template and image, the correlation output is the sum of the five binary pixels in the image that lie immediately above the "on" pixels in the template. The template can therefore be implemented in the FPGA as a simple adder tree as shown in Fig. 10. The image pixel values can be stored in flip-flops and are shifted to the right by one flip flop with each clock cycle. Though correlation of a large image with a small mask is often understood conceptually in terms of the mask being scanned across the image, in this case the opposite is occurring—the template is hard-wired into the FPGA while the image pixels are clocked past it.

Another important opportunity for increased efficiency lies in the potential to combine multiple templates on a single FPGA. The simplest way to do this is to spatially partition the FPGA into several smaller blocks, each of which handles the logic for a single template. Alternatively, one can seek to identify templates having some topological commonality, and which can therefore share parts of adder trees. This is illustrated in Fig. 11, which shows two templates that share several pixels in common, and which can be mapped using a set of adder trees that leverage this overlap. The advantage of using FPGAs is that FPGAs can be dynamically optimized at the gate level to exploit template characteristics. A general-purpose correlator would have to provide large general-purpose adder trees to handle the summing of all possible template bits. The FPGA, however, exploits the sparse nature of the templates, and only constructs the small adder trees required. FPGAs can exploit other factors such as collapsing adder trees with common ele-

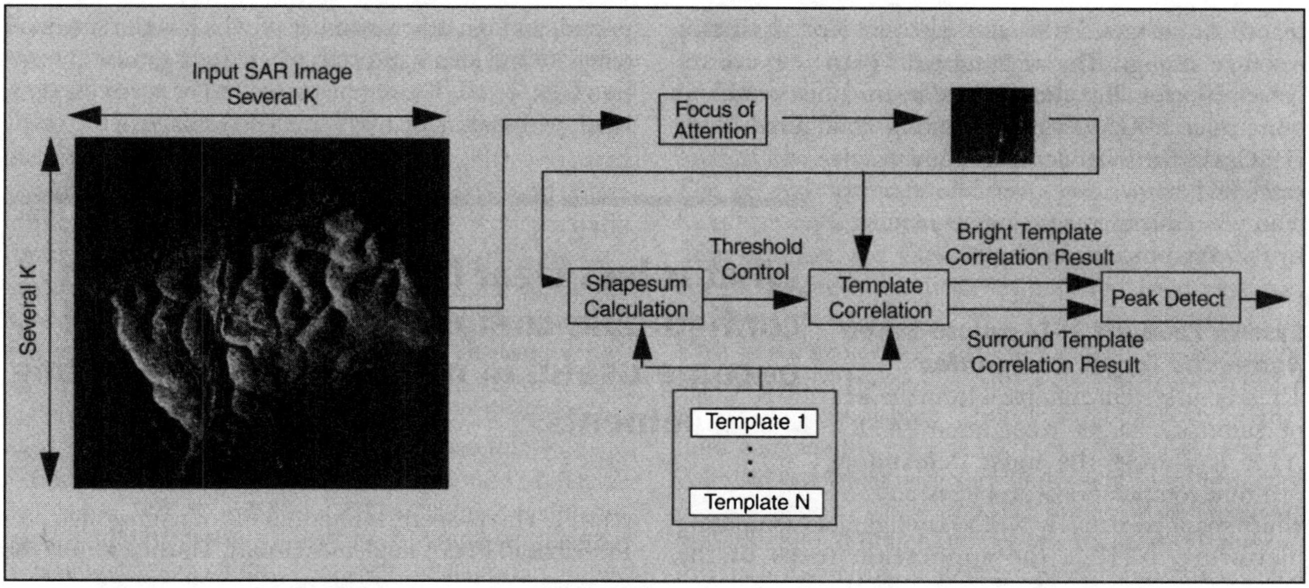

▲ *8. High-level block diagram for ATR processing. The focus of attention algorithm identifies regions of interest in SAR images. Pixels in regions of interest are correlated against a series of binary target template pairs, with each pair containing a bright template (identifying pixels of strong expected radar return) and a surround template (strong radar absorption). Templates with the highest correlation are selected in the peak detection step.*

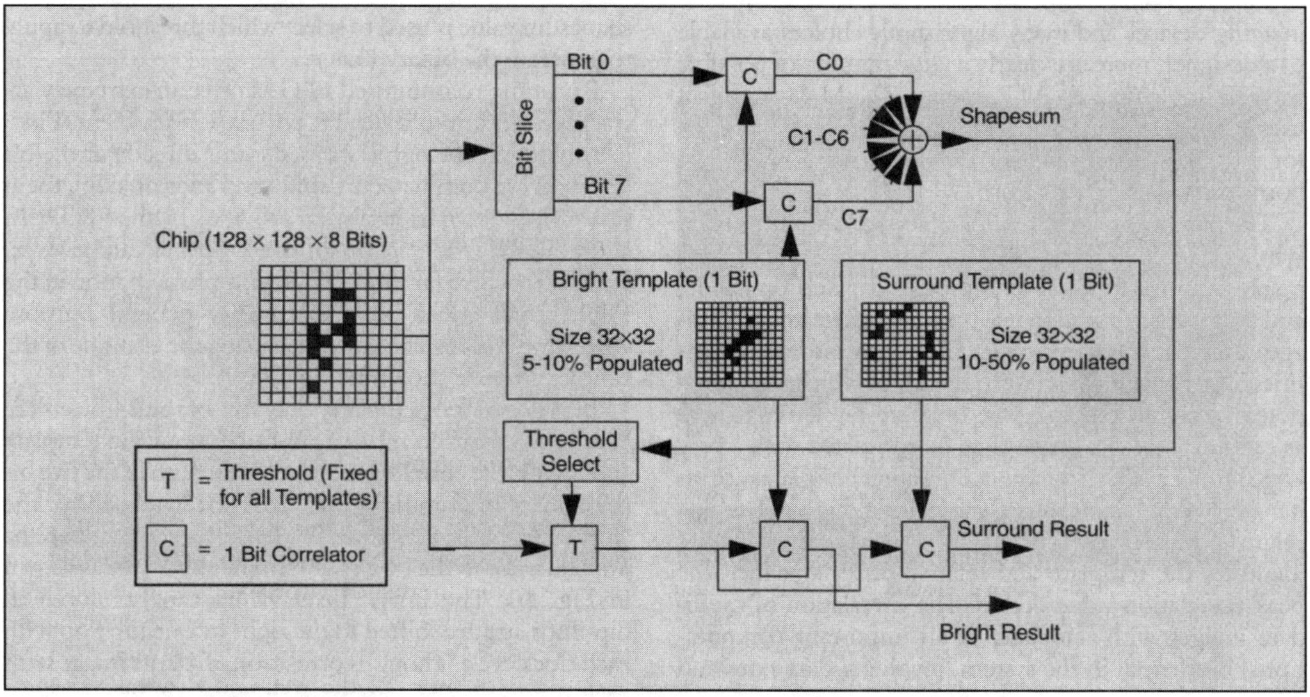

▲ *9. Correlation operation showing shapesum calculation (top) and thresholding/correlation (bottom).*

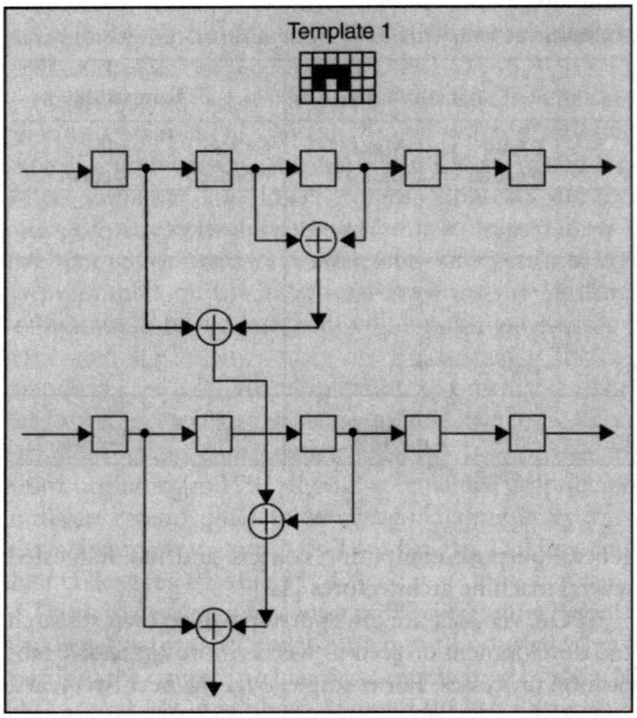

▲ *10. Example binary template with five "on" pixels (top) and corresponding adder tree (bottom).*

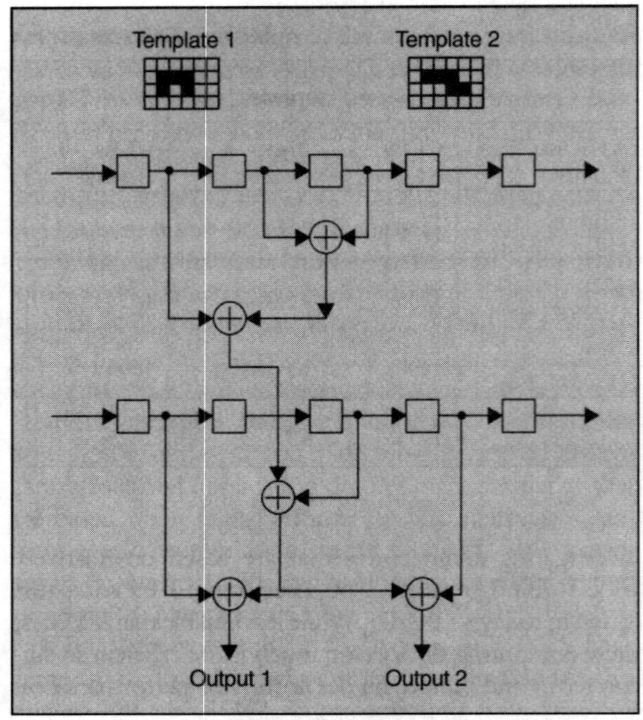

▲ *11. Template commonalities are exploited to reduce hardware requirements for computing multiple correlations.*

ments, and storing pixels that are not needed by the adder trees using RAM-based shift registers.

Table 2 illustrates the FPGA resource trade-offs involved in template mapping. The table gives the FPGA utilization for the Xilinx 4062 when four through seven template pairs are simultaneously mapped into the FPGA using the approach described above. Each template pair consists of two 32×32 binary images and is represented in the hardware using two template-specific adder trees. The number of templates per second that can be evaluated using this approach is a function of many factors including the clock rate, the FPGA configuration time, the number of templates per configuration, the candidate image and target sizes, the number of clock cycles needed to evaluate the templates at each relative image/template offset, and on I/O considerations. The performance can be upper bounded by assuming that the I/O is fully efficient; i.e., that the FPGA is always either computing correlations or being reconfigured. Assuming efficient I/O is fairly reasonable in the prototype systems we have constructed, we have been able to avoid letting the FPGA be idle by using scaled down versions of the templates. When all of these factors are considered together, we find that configuration can consume more time than computation; i.e. there is a significant performance penalty due to reconfiguration. This overhead will diminish to 10% or less when partially reconfigurable FPGAs become more widely available. However, for parts that are not partially reconfigurable, the benefits of increased computation power offered by larger FPGAs are to some extent mitigated by the larger configuration bitstreams and longer reconfiguration times that these parts require.

Figure 12 shows a configurable computing board that was constructed at UCLA as a prototype for the template-matching problem. The board contains a "dynamic" FPGA that is used for template correlations and is run-time reconfigured, a "static" FPGA for control, SRAM for storage of pixels and results, EPROM for configuration bitstream storage, and an interface to an i960 embedded processor for more advanced configuration control.

Ongoing Research

Configurable computing has grown from a field with a handful of researchers in 1989 [45] to one that now receives the attention of hundreds of researchers and engineers in academia, industry, defense, and a rapidly increasing number of start-up companies. In this section we identify some of the open issues in this field and describe selected recent and ongoing projects that aim to address them.

One of the most interesting questions in configurable computing concerns the extent to which current FPGA device and machine architectures should be altered to better support computing as opposed to the prototyping that drove much of the early evolution of FPGAs. Academic researchers pursuing this question face the obvious challenge of being unable to fully exploit the existing infrastructure of commercial FPGAs and design tools, and typically design custom FPGAs to validate their architecture proposals. Various projects are underway, each attacking one or more of the well-known weaknesses of commercial FPGAs. For example, some researchers are

Table 2. CLB resource requirements for Sandia ATR.

	4 Templates		5 Templates		6 Templates		7 Templates	
XC40000XL Family	**CLBs Needed**	**Max. Delay**	**CLBs Needed**	**Max. Delay**	**CLBs Needed**	**Max. Delay**	**CLBs Needed**	**Max. Delay**
4028 (1024 CLBs)	1196							
4028 (1296 CLBs)	1296	86.6 ns	1421					
4028 (1600 CLBs)	1600	63.6 ns	1600	104 ns	1608			
4028 (2304 CLBs)	1734	99.9 ns	2025	94.9 ns	2304	94 ns	2304	127 ns

investigating architectures that are based on relatively wide 16-bit datapaths (as opposed to the 1-bit datapaths found in today's FPGAs). While less flexible than FPGAs, these computing devices are much more efficient in silicon terms and achieve higher arithmetic performance on 16-bit integer data. Other researchers are investigating novel configuration approaches that either reduce configuration time through context-switching or that distribute configuration data with data to be processed. Still other researchers are merging general-purpose processors and FPGA resources on the same die in an attempt to combine the best features of both technologies.

Peter Athanas' group at Virginia Tech is exploring 16-bit computing devices based on the "wormhole" technique: a computing approach that distributes configuration data with the data to be processed [33]. Consisting of a single multiplier and a 4 × 4 array of 16-bit arithmetic logic units (ALUs) interconnected by a crossbar, their COLT device combines configuration data and data into a single packet. Resembling dataflow computing in many aspects, configuration data in the packet are used to route data through the array and to configure ALUs for subsequent processing. COLT has been fabricated and is currently being tested.

Carl Ebeling's group at the University of Washington is working on RAPID, another device based on 16-bit datapaths [14]. A RAPID array consists of a mostly linear array of RAPID cells, each cell consisting of an integer multiplier, three integer ALUs, six registers, and three small memories. RAPID is primarily statically configured but uses limited dynamic control to provide run-time flexibility.

Matrix, developed by Andre DeHon and others at MIT, is based on a cell that can serve as an instruction store, a memory element, or a computational element. All datapaths are 8-bit and these cells are interconnected with multilevel interconnect that can be used both for data and instruction distribution. Matrix is currently undergoing commercial development by a new startup company, Silicon Spice. DeHon has also conducted an in-depth study that sets FPGA-based computing in a general-purpose computing context and has suggested several machine architectures [12].

FPGA vendors are also pursuing their own research and development projects as well as more aggressive fabrication processes. For example, over the next two years, devices using supply voltages of 2.5 Volts and below will become common. In addition, the technology lag of FPGAs with respect to ASICS in terms of feature size and number of metal layers is rapidly shrinking, with 3-5-layer FPGAs fabricated using sub .3 micron technology expected to become common. The vendors are also likely to both introduce and adopt architectural innovations that have shown promise in academic research. Since future FPGAs will track ASIC technology more closely and will benefit from a richer set of architectural features, they are likely to compare more favorably with ASICs for many applications than that of today.

The BRASS project at U.C. Berkeley under John Wawrzynek [22] is developing a single chip (Garp) that incorporates a MIPS-II processor and an FPGA core whose elements roughly correspond to those found in the Xilinx 4000 series. The BRASS researchers have modified the MIPS-II processor, replacing the floating-point unit with an FPGA core of their own design, and have augmented the instruction set to include operations that manage the FPGA resources. Their goal is to execute data-intensive operations on the FPGA core and leave general-purpose operations on the processor. A related effort at National Semiconductor Corporation is building an FPGA that will combine a programmable processor and a fine-grained FPGA on the same chip [18].

Other researchers are investigating solutions to mixing configurable computing elements with more traditional processors. For example, Jan Rabaey of U.C. Berkeley has examined the allocation of tasks in typical digital signal processing and has proposed a multigranularity architecture that allows computations to be directed to the hardware that best supports them [34]. Rabaey is also investigating strategies for low-power FPGAs. Though some power reduction will occur automatically due to technology changes, there is substantial opportunity to redesign the logical units in FPGAs with power as a principal constraint.

Several groups are looking at FPGAs that have multiple configurations, or contexts, stored on-chip simultaneously. At any given time one context is active and the others are stored in lower planes. Contexts can be swapped extremely quickly—requiring from one to several hundred clock cycles to complete—potentially eliminating much of the overhead involved in loading configuration bitstreams from off-chip. Of course, context switching involves other overheads such as the resources needed to hold multiple contexts on-chip, and the hardware and tools to manage context-switching. The earliest work on context-switched FPGAs was done at Xilinx beginning in 1991, though it remained proprietary until very recently [42]. In the academic community context switching was studied by Tom Knight, DeHon, and their colleagues at MIT [11, 41].

Work to develop new configurable computing devices also benefits from an understanding of how algorithms map into the range of architectures represented by today's FPGAs and FPGA systems. Some of the most extensive algorithm mapping work has been performed by the BYU group led by Brad Hutchings, which has experimented with most commercially available (and noncommercially available) FPGAs as well as prototype systems such as the HP Teramac [2] and Splash-2 [4]. BYU has demonstrated applications in the following areas: neural networks [15], morphology [48], ATR [35] and genetic algorithms [19]. BYU has also developed a variety of design and implementation strategies [25] and provides tutorials for many different FPGA platforms via their web site: http://splish.ee.byu.edu. A large bibliography of related papers is also available at this site.

BYU's early research agenda was twofold: one, determine what characteristics make an application a good candidate for implementation on an FPGA-based computing platform, and two, research and understand the strengths and weaknesses of current devices, system organizations, and tools. Following up on this basic research, BYU is now in the process of developing new system organizations and application-development strategies that are based upon high-performance circuit libraries, domain-specific compilation, and RTR. BYU also continues to experiment with applications in an effort to find additional applications that can exploit this technology.

John Villasenor and his colleagues at UCLA have demonstrated a video communications system in which a single 5000-gate FPGA was reconfigured four times per image frame to allow compression and transmission of an image [26]. The Mojave project at UCLA, led by John Villasenor and Bill Mangione-Smith, has resulted in several generations of boards and domain-specific design libraries for the ATR application described previously. These boards included an interface to an embedded processor that performed on-the-fly analysis of results and modified the FPGA configuration sequence accordingly [43, 44].

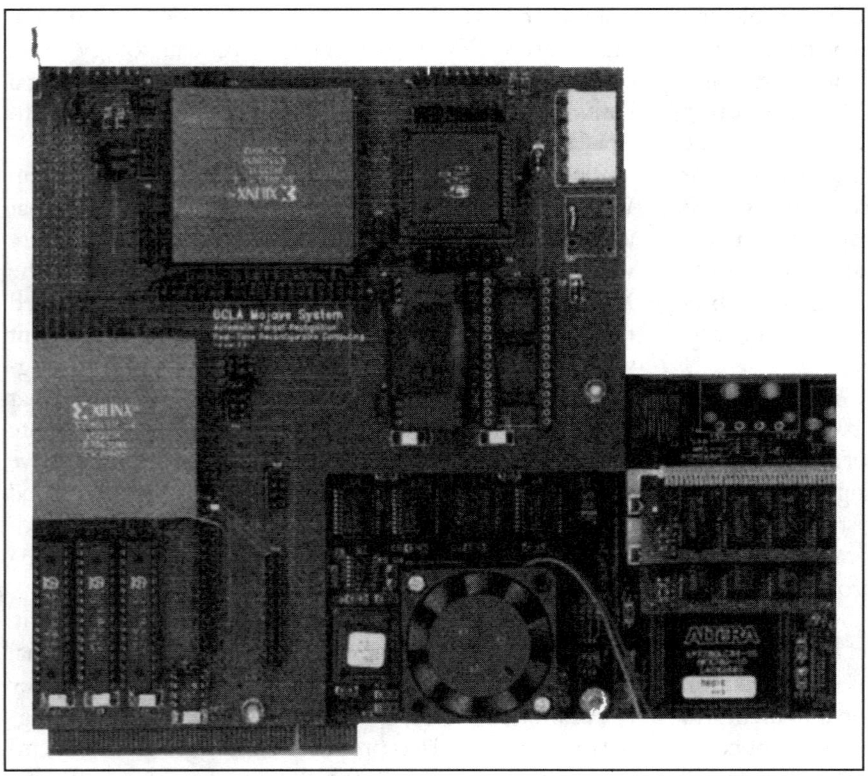

▲ 12. A configurable computing board for ATR built at UCLA. The board includes a "dynamic" FPGA that implements template correlations and is rapidly reconfigured during execution, a "static" FPGA for control, SRAM for image data storage, and an EPROM for configuration bitstream storage. The board resides in a host PC and receives images across a PCI bus.

Researchers including Mohammad Shajaan and John Sorensen of the Technical University of Denmark [38] have examined architectures for performing digital filtering using FPGAs. Because today's FPGAs perform multiplications poorly, much of the attention in filtering using FPGAs has focused on multiply-free implementations. In the future, it is also likely that adaptive filtering algorithms will find application in FPGAs that are partially reconfigured as the filter coefficients evolve.

Another area of research focus is in compilers and tools for configurable computing platforms. Ian Page of Oxford University has developed Handel, a programming language that allows programmers to simultaneously develop the FPGA circuit descriptions and processor software with a single description language based on OCCAM [31]. Reiner Hartenstein of the University of

Kaiserslautern has developed a machine-level abstraction called the Xputer [21] that also derives the target machine description and its program from the same description. Wayne Luk at Imperial College is investigating formal approaches to FPGA design based on the language RUBY [29]. Transmogrifier-C [17], developed by researchers at the University of Toronto, is a programming approach targeted at Toronto's TM-2 custom platform, which is currently under development [27]. Anant Agarwal and his colleagues at MIT are working on automated programming approaches for very large configurable-computing platforms [6]. In addition, HP developed a very easy-to-use compiler for their Teramac system that automatically partitioned, placed, and routed a netlist of 1-million gates into the nearly 1000 custom FPGAs that formed Teramac [2].

Configurable computing is likely to benefit from architectural innovations both in FPGAs and in the hardware to interface to them.

Configurable computing is represented by a growing presence in the commercial world. In addition to FPGA vendors including Xilinx and Altera there is a rapidly growing list of start-up companies with products that are based on configurable computing. These including Annapolis Microsystems of Annapolis Maryland, which commercialized the SPLASH-II architecture; Virtual Computing Corporation of Reseda, California; Morphologic of Nashua, New Hampshire; and Giga Operations of Berkeley.

Conclusions and Future Directions

It is now clear that for applications requiring deeply pipelined, highly parallel, bit-level operations including cryptography, target recognition, and some types of image processing, configurable computing machines offer compelling speed and cost advantages over alternative implementations. For these types of applications, configurable computing machines are likely to become solutions of choice. What is less clear is the extent to which configurable computing techniques will become useful in more general computing environments, in particular for applications that involve high arithmetic complexity. Given the dominance and ever-increasing capabilities of microprocessors for general-purpose computing, it seems highly unlikely that any other computing model, including that offered by configurable computing, will make significant inroads against microprocessors in the foreseeable future. Widespread adoption of configurable computing is also hampered by the lack of exactly what microprocessors possess in abundance: a set of relatively easy to use, widely known software programming languages and associated compilers or interpreters that allow a user with little or no knowledge of the underlying hardware to instruct a computing platform to perform a desired task.

The lack of a sufficiently general high-level software programming model is of course a well-known problem among researchers performing work in configurable computing, and there are many ongoing efforts in which creation of a design tool infrastructure is a goal. Even if such languages can be developed, tested, and adopted, there remains the problem of the "compiler," which in the domain of FPGAs means the tool chain that translates a functional or structural description of the task into a configuration bitstream that fully describes the circuit in the FPGA. FPGA place and route tools have always benefited from place and route techniques used in ASIC design, which involves many of the same challenges and tradeoffs in terms of clock speed, design complexity, etc. However, the several hours needed by current-generation commercial FPGA tools to synthesize, place, and route a design on an FPGA, while fast when viewed in the context of ASIC design, are unacceptably slow when compared to software compilers. To make configurable computing practical will require that FPGA place and route tools be made faster by several orders of magnitude, most likely at the cost of highly suboptimal mappings of tasks into hardware. One exciting, but as yet unproven, approach that has been advocated by William Mangione-Smith of UCLA is dynamic compilation, in which small units of precompiled FPGA configuration bitstreams can be combined extremely quickly at run time to constitute a full FPGA configuration bitstream. There are many challenges in dynamic compilation, not the least of which is the proprietary nature of configuration bitstreams for most commercial FPGAs.

As configurable computing advances it is also important to distinguish techniques that are truly new, such as large-scale run-time hardware reconfiguration, from techniques that have existed in computing for many years. Many of the "new" approaches in configurable computing are in fact existing computing concepts that are being implemented in a new domain. For example, the ATR algorithm described previously gains its efficiency from RTR, which can legitimately be claimed as an innovation due to configurable computing, and from mapping target templates into template-specific adder trees, which is an example of the years-old technique of partial evaluation.

In addition to the obvious trend toward larger devices, configurable computing is likely to benefit from architectural innovations both in FPGAs and in the hardware to interface to them. Configurable computing is a young field with enormous potential to grow as FPGAs, their derivatives, and the tools to use them advance. The FPGAs that will be emerging in the next few years will be in excess of half a million equivalent gates, which is large enough to support a very diverse range of applications. In addition, the state of the art in architectures for configurable computing devices will be significantly enriched by the many ongoing research efforts studying architecture issues. Existing and perhaps new vendors of configurable computing devices, who are now well aware of the potential of configurable computing, can be expected produce devices and the associated tools that will make FPGAs of today look primitive.

References

1. A.L. Abbott, P.M. Athanas, L. Chen, and R.L. Elliott, "Finding lines and building pyramids with Splash 2". In D.A. Buell and K. L. Pocek, editors, *Proceedings of IEEE Workshop on FPGAs for Custom Computing Machines*, pp. 155-161, Napa, CA, April 1994.

2. R. Amerson, R. Carter, B. Culbertson, P. Kuekes, and G. Snider, "Teramac - configurable custom computing." In D.A. Buell and K. L. Pocek, editors, *Proceedings of IEEE Workshop on FPGAs for Custom Computing Machines*, pp. 32—38, Napa, CA, April 1995.

3. J.M. Arnold, "The Splash 2 software environment." In D.A. Buell and K.L. Pocek, editors, *Proceedings of IEEE Workshop on FPGAs for Custom Computing Machines*, pp. 88-93, Napa, CA, April 1993.

4. J.M. Arnold, D.A. Buell, and E.G. Davis, "Splash 2." In *Proceedings of the 4th Annual ACM Symposium on Parallel Algorithms and Architectures*, pp. 316-324, June 1992.

5. P.M. Athanas and A.L. Abbott, "Real-time image processing on a custom computing platform." *IEEE Computer*, 28(2):16-24, February 1995.

6. J. Babb, M. Frank, E. Waingold, and R. Baruna, "The RAW benchmark suite: Computation structures for general purpose computing." In J.M. Arnold and K.L. Pocek, editors, *Proceedings of IEEE Workshop on {FPGA}s for Custom Computing Machines*, Napa, CA, April 1997, to be published.

7. P. Bertin, D. Roncin, and J. Vuillemin, "Introduction to programmable active memories." In J. McCanny, J. McWhirther, and E. Swartslander Jr., editors, *Systolic Array Processors*, pp. 300-309. Prentice Hall, 1989.

8. G. Brebner, "The swappable logic unit: a paradigm for virtual hardware." In J.M. Arnold and K.L. Pocek, editors, *Proceedings of IEEE Workshop on FPGAs for Custom Computing Machines*, Napa, CA, April 1997, to be published.

9. J. Burns, A. Donlin, J. Hogg, S. Singh, and M. de Wit, "A dynamic reconfiguration run-time system." In J.M. Arnold and K.L. Pocek, editors, *Proceedings of IEEE Workshop on FPGAs for Custom Computing Machines*, Napa, CA, April 1997, to be published.

10. D.A. Clark and B.L. Hutchings, "Supporting FPGA microprocessors through retargetable software tools." In J. Arnold and K.L. Pocek, editors, *Proceedings of IEEE Workshop on FPGAs for Custom Computing Machines*, pp. 195-203, Napa, CA, April 1996.

11. A. DeHon, "DPGA-coupled microprocessors: Commodity ICs for the early 21st century." In D.A. Buell and K.L. Pocek, editors, *Proceedings of IEEE Workshop on FPGAs for Custom Computing Machines*, pp. 31-39, Napa, CA, April 1994.

12. A. DeHon, Reconfigurable Architectures for General-Purpose Computing. PhD thesis, Massachusetts Institute of Technology, September 1996.

13. T. Drayer, W. King, J. Tront, and R. Conners, "MORRPH: A Modular and reprogrammble real-time processing hardware." In D.A. Buell and K.L. Pocek, editors, *Proceedings of IEEE Workshop on FPGAs for Custom Computing Machines*, pp. 11-19, Napa, CA, April 1995.

14. C. Ebeling, D.C. Cronquist, and P. Franklin, "RaPiD - reconfigurable pipelined datapath." In J.M. Arnold and K.L. Pocek, editors, *International Workshop on Field-Progammable Logic, FPL'96*, pp. 126-135, Darmstadt, Germany, September 1996.

15. J.G. Eldredge and B.L. Hutchings, "Run-time reconfiguration: A method for enhancing the functional density of SRAM-based FPGAs." *Journal of VLSI Signal Processing*, vol. 12, pp. 67-86, 1996.

16. C.W. Fraser and D. Hansom, *A Retargetable C Compiler*. Benjamin/Cummings Company, 1995. ISBN 0-8053-1670-1.

17. D. Galloway, "The transmogrifier C hardware description language and compiler for FPGAs." In D.A. Buell and K.L. Pocek, editors, Proceedings of IEEE Workshop on FPGAs for Custom Computing Machines, pp. 136-144, Napa, CA, April 1995.

18. M. Gokhale and E. Gomersall, "High-level compilation for fine grained FPGAs". In J.M. Arnold and K.L. Pocek, editors, *Proceedings of IEEE Workshop on FPGAs for Custom Computing Machines*, Napa, CA, April 1997, to be published.

19. P. Graham and B. Nelson, "A hardware genetic algorithm for the travelling salesman problem on SPLASH 2." In W. Moore and W. Luk, editors, *Field-Programmable Logic and Applications*, pp. 352-361, Springer, Oxford, England, August 1995.

20. P. Graham and B. Nelson, "Genetic algorithms in software and in hardware - A performance analysis of workstation and custom computing machine implementations." In J. Arnold and K. Pocek, editors, *Proceedings of IEEE Workshop on FPGAs for Custom Computing Machines*, pp. 216-225, Napa, CA, April 1996.

21. R.W. Hartenstein, A.G. Hirschbiel, M. Riedmuller, K. Schmidt, and M. Weber, "A novel ASIC design approach based on a new machine paradigm." *IEEE Journal of Solid-State Circuits*, vol. 26, no. 7, pp. 975-989, July 1991.

22. J.R. Hauser and J. Wawrzynek, "Garp: A processor with a reconfigurable coprocessor." In J.M. Arnold and K.L. Pocek, editors, *Proceedings of IEEE Workshop on FPGAs for Custom Computing Machines*, Napa, CA, April 1997, to be published.

23. Brian Von Herzen, "Signal processing at 250 MHz using high-performance FPGAs." In *ACM/SIGDA International Symposium on Field Programmable Gate Arrays*, pp. 62-68, Monterey, CA, February 1997.

24. D.T. Hoang, "Searching genetic databases on Splash 2." In D. A. Buell and K.L. Pocek, editors, *Proceedings of IEEE Workshop on FPGAs for Custom Computing Machines*, pp. 185-191, Napa, CA, April 1993.

25. B.L. Hutchings and M.J. Wirthlin, "Implementation approaches for reconfigurable logic applications." In W. Moore and W. Luk, editors, *Field-Programmable Logic and Applications*, pp. 419-428, Springer, Oxford, England, August 1995.

26. B. Schoner J. Villasenor and C. Jones, "Video communications using rapidly reconfigurable hardware." *IEEE Trans. on Circuits and Systems for Video Technology*, pp. 565-567, December 1997.

27. D.M. Lewis, D.R. Galloway, M. van Ierssel, J. Rose, and P. Chow, "The transmogrifier-2: A 1 million gate rapid prototyping system." In *ACM/SIGDA International Symposium on Field Programmable Gate Arrays*, pp. 53-61, Monterey, CA, February 1997.

28. D.P. Lopresti, "Rapid implementation of a genetic sequence comparator using field-programmable gate arrays." In C. Sequin, editor, *Advanced Research in VLSI: Proceedings of the 1991 University of California/Santa Cruz Conference*, pp. 138-152, Santa Cruz, CA, March 1991.

29. W. Luk, "A declarative approach to incremental custom computing." In D.A. Buell and K.L. Pocek, editors, *Proceedings of IEEE Workshop on FPGAs for Custom Computing Machines*, pp. 164-172, Napa, CA, April 1995.

30. P. Lysaght and J. Stockwood, "A simulation tool for dynamically reconfigurable field programmable gate arrays." *IEEE Transactions on Very Large Scale Integration (VLSI) Systems*, vol. 4, no. 3, pp. 381-390, September 1996.

31. I. Page and W. Luk, "Compiling occam into FPGAs." In *FPGAs, International Workshop on Field Programmable Logic and Applications*, pp. 271-283, Oxford, UK, September 1991.

32. G.M. Quenot, I. Kraljic, J. Serot, and B. Zavidovique, "A reconfigurable compute engine for real-time vision automata prototyping." In D.A. Buell and K.L. Pocek, editors, *Proceedings of IEEE Workshop on FPGAs for Custom Computing Machines*, pp. 91-100, Napa, CA, April 1994.

33. R. Bittner Jr. and P.M. Athanas, "Computing kernels implemented with a wormhole RTRCCM." In J.M. Arnold and K.L. Pocek, editors, *Proceedings of IEEE Workshop on FPGAs for Custom Computing Machines*, Napa, CA, April 1997, to be published.

34. J.M. Rabaey, "Reconfigurable processing: The solution to low-power programmable DSP." In *Proceedings of ICASSP'97*, Munich, Germany, April 1997, to be published.

35. M. Rencher and B.L. Hutchings, "Automated target recognition on SPLASH-2." In J.M. Arnold and K.L. Pocek, editors, *Proceedings of IEEE Workshop on FPGAs for Custom Computing Machines*, Napa, CA, April 1997, to be published.

36. D. Ross, O. Vellacott, and M. Turner, "An FPGA-based hardware accelerator for image processing." In W. Moore and W. Luk, editors, *More FPGAs: Proceedings of the 1993 International workshop on field-programmable logic and applications*, pp. 299-306, Oxford, England, September 1993.

37. H. Schmitt, "Incremental reconfiguration for pipelined applications." In J.M. Arnold and K.L. Pocek, editors, *Proceedings of IEEE Workshop on FPGAs for Custom Computing Machines*, Napa, CA, April 1997, to appear.

38. M. Shajaan, K. Nielsen, and J.A Sorensen, "Time-area efficient multiplier-free filter architectures for FPGA implementation." In *Proceedings of 1995 International Conference on Acoustics, Speech, and Signal Processing Conference*, pp. 3251-3254, 1995.

39. M. Shand, "Flexible image acquisition using reconfigurable hardware." In P.M. Athanas and K.L. Pocek, editors, *IEEE Workshop on FPGAs for Custom Computing Machines*, pp. 125-134, Napa, CA, April 1995.

40. N. Shirazi, W. Luk, and P. Cheung, "Compilation tools for run-time reconfigurable designs." In J.M. Arnold and K.L. Pocek, editors, *Proceedings of IEEE Workshop on FPGAs for Custom Computing Machines*, Napa, CA, April 1997, to be published.

41. E. Tau, D. Chen, I. Eslick, J. Brown, and A. DeHon, "A first generation DPGA implementation." *In FPD'94 - Third Canadian Workshop of Field-Programmable Devices*, pp. 138-143, May 1995.

42. S. Trimberger, D. Carberry, A. Johnson, and J. Wong, "A time-multiplexed FPGA." In J.M. Arnold and K.L. Pocek, editors, Proceedings of IEEE Workshop on FPGAs for Custom Computing Machines, Napa, CA, April 1997.

43. J. Villasenor and W. Mangione-Smith, Configurable computing. Scientific American pp. 66-71, June 1997.

44. J. Villasenor, B. Schoner, K.N. Chia, C. Zapata, H.J. Kim, C. Jones, S. Lansing, and B. Mangione-Smith, "Configurable computing solutions for automatic target recognition." In J. Arnold and K. L. Pocek, editors, *Proceedings of IEEE Workshop on FPGAs for Custom Computing Machines*, pp. 70-79, Napa, CA, April 1996.

45. J. Vuillemin, P. Bertin, D. Roncin, M. Shand, H. Touati, and P. Boucard, "Programmable active memories: Reconfigurable systems come of age." *IEEE Transactions on VLSI Systems*, vol. 4 no. 1, pp. 56-69, 1996.

47. M.J. Wirthlin and B.L. Hutchings, "DISC: The dynamic instruction set computer." In J. Schewel, editor, *Proceedings of the International Society of Optical Engineering (SPIE). Field Programmable Gate Arrays {FPGAs} for Fast Board Development and Reconfigurable Computing*, vol. 2607, pp. 92-103, Philadelphia, PA, October 1995.

47. M.J. Wirthlin and B.L. Hutchings, A dynamic instruction set computer." In P. Athanas and K.L. Pocek, editors, *Proceedings of IEEE Workshop on {FPGA}s for Custom Computing Machines*, pp. 99-107, Napa, CA, April 1995.

48. M.J. Wirthlin and B.L. Hutchings, "Sequencing run-time reconfigured hardware with software." In *ACM/SIGDA International Symposium on Field Programmable Gate Arrays*, pp. 122-128, Monterey, CA, February 1996.

49. M.J. Wirthlin and B.L. Hutchings, "Improving functional density through run-time constant propagation." In *ACM/SIGDA International Symposium on Field Programmable Gate Arrays*, pp. 86-92, Monterey, CA, February 1997.

Chapter 3

Systems and Architectures

THIS chapter is about building useful software-defined radios (SDRs). These include cutting-edge products and technology pathfinders. Prill and Antonesco were among the first to describe an SDR precursor product, the PDR. This paper is included for historical perspective. Differentiating the SDR from PDRs and other classes of radio system became important following the publication of this paper. The notion of a baseband, single-channel radio with a dedicated DSP path per channel is fairly easy to grasp. Early PDRs replicated this approach for each band and mode using modular hardware "slices" to obtain multimode capability. Lackey and Upmall's "military software radio" SPEAKeasy I generalized this approach by providing a wideband RF (28–400 MHz), tunable medium-bandwidth ADCs, and programmable digital IF conversion to augment the baseband DSP. With this radio architecture, one can implement the simpler modes of the Single Channel Ground Air Radio System (SINCGARS) in the 28–88 MHz LVHF band and the simpler modes of HAVE QUICK in the 225–400 MHz band with the same general-purpose RF hardware. Today, we might call this a solid-center SDR but not an ideal software radio. In particular, the frequency hopped modes of both waveforms require analog local oscillators because the bandwidths of the ADCs and DACS were not adequate to digitally hop and dehop the FH modes. The enabling technology of Hypress, the 200 MHz ADC with 80 dB of dynamic range, permits one to at least design wideband ADC access needed for digital dehopping. Taken with wideband DACs, this would be a step toward the ideal software radio. The ideal software radio for the military would access 2 MHz to 2 GHz in one very linear RF band, would convert to digital using a 5 or 6 GHz ADC/DAC front end, and would transfer this digital RF via fiber optics from the antenna/conversion module to the processor module, where a pool of general-purpose processors would perform all aspects of all radio modes. This ideal software radio might also be called the extreme software radio because such extreme flexibility is so expensive that it may not be ideal for most applications. Nevertheless, the progress of wideband RF, ADCs, DACs, and processors continues to move implementations in the direction of the extreme software radio. Thus, the design trade-offs are constantly changing.

Subsequently, Cook and Bonser describe SPEAKeasy II with its PCI-based design. The two SPEAKeasy papers show the substantial contributions of technology pathfinders to the formulation of an SDR-based acquisition strategy in the U.S. DoD. Meyer and Subramanian and Tsurumi and Suzuki emphasize PCS applications, and Sevenhans et al. focus on the homodyne receiver. These topics are more central to the trade-offs in commercial applications. Thus, we have attempted to balance the technology development goals of the military with the affordable mass-production orientation of the commercial sector. Arnott provides a cutting-edge tutorial on commercial base stations that rounds out the discussion of prior generation air interfaces. Erben et al. show the way ahead for 3G.

Papers by Srikanteswara, Zangi, Gunn, and Gatherer address practical processing platforms. The Virginia Tech soft radio, first of all, is a unique integration of packet-driven software with reconfigurable FPGA hardware. This architecture facilitates the abstraction of software radio functions without losing the computational efficiency of the FPGA. Zangi's discussion of programmable processors shows how to consolidate the parts count by integrating programmable digital filters in a way that is optimized for specific air interfaces. Gunn explores the partitioning of processing functions between ASICs and a DSP core. He also describes techniques for minimizing power in handsets, with projections for power reduction in CMOS. Finally, Gatherer summarizes DSP architectures for mobile applications.

The final three papers of this chapter stretch the envelope of implementation technologies. Bose's Virtual Radio is a research construct for describing a radio in the SpectrumWare scripting language "out of band" which is interpreted in real-time "in band." The Guppi interface between the RF/ADC front end and the DEC Alpha's UNIX operating system shows how an ideal narrowband software radio can be implemented on essentially unmodified general-purpose processing hardware and operating system. Schuh emphasize the networking aspects of SDR, and Seskar and Mandayam address interference cancellation. Multichannel software radios have a unique capability to generate self-interference as an air interface configured into one channel can interfere with a different air interface configured on a different channel and multiple-access interference mitigation will be of great importance.

PROGRAMMABLE CHANNELIZED DIGITAL RADIO

ROBERT S. PRILL
MICHEL ANTONESCO

GEC-MARCONI ELECTRONIC SYSTEMS CORPORATION
15 JACKSON ROAD, TOTOWA, NJ 07512

ABSTRACT

This paper describes ongoing work by GEC-Marconi Electronic Systems Division in the development of a multifunction, Common Module, Programmable Digital Radio/Modem. The radio system handles both conventional narrow-band and spread spectrum waveforms. The architecture developed is a programmable channelized approach. In the channelized approach the modules of one channel are identical to those of the next, and as such are ideally suited to channel reusability during different parts of a mission or during failure modes of higher priority functional systems. The hardware implementation exploits latest RF, Digital-RF, and Digital Signal Processing component technologies. A PC based brassboard was developed and used to demonstrate real-time programmable AM, FM, and IFF reception.

BACKGROUND

The integrated radio architecture and common module set that is evolving out of JIAWG is based on the Pave Pillar architecture. It is predicated on centralized processing nodes of RF, Signal Processing, Control Processing, Data Processing as depicted by the processing nodes (cubes) in Figure 1a.

The Pave Pillar/ICNIA architecture is designed to handle efficiently multiple simultaneous systems. The common module set includes many module types within each processing node. All/many modules within a node usually participate in processing the signal making it less efficient to apply the architecture to applications where only one to a few system functions are to be integrated within the constraints of existing LRU volumes. Similar conclusions were reached in a recent F-16 MASA (Modular Avionics System Architecture) study funded under an AF contract.

The channelized concept, which addresses the integration of a few system functions, is an open ended architecture that employs one system's worth of processing hardware using a few common but programmable module types; i.e., RF receiver, Signal Preprocessor, Signal Processor and Crypto Processor. Figure 1b illustrates that architecture.

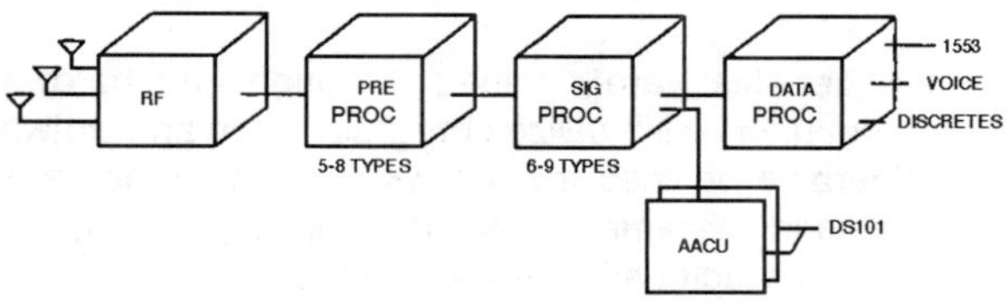

1a

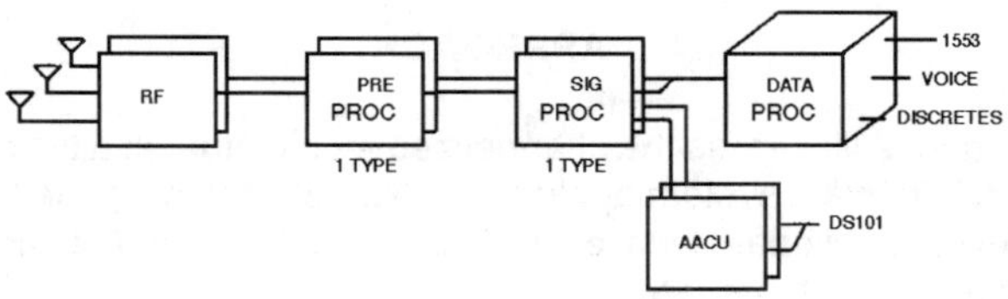

1b

Figure 1 - a) Pave Pillar architecture b) Channelized architecture.

It is an approach implemented with a module set that can be replicated for each channel. Each channel is programmed differently for each function to be implemented. The architecture supports channel reusability during different segments of a mission, channel reconfigurability during fault conditions of a higher priority system and fault tolerant scheduled maintenance concepts. GEC has defined and demonstrated key portions of such an architecture.

ARCHITECTURE OVERVIEW

The attributes of the GEC "common Module" digital radio/modem architecture are:

- It is easily expandable to multifunction simultaneous system operation, by channel replication.
- It minimizes multisystem integration because of the relative independence of channels.
- It simplifies system qualification/multilevel security testing if/when new systems (channels) are added, i.e., channel independence down to time base and crypto functions.

Reprinted from *IEEE National Telesystems Conference,* pp. 21–24, 1992.

- It supports channel reusability/reconfigurability, ie., for time sharing a channel during different mission segment, for priority based system survivability and for scheduled maintenance concepts.

- It is implementable on a small number of SEM-E modules such that a multifunction system could be configured via module replication without module count becoming excessive.

- It will handle narrow-band/wide-band and spread spectrum waveforms, all with a single programmable architecture.

These benefits are gained through a hardware implementation which utilizes programmable and software reconfigurable devices that allows the host computer via mission load parameters to dramatically change the functional characteristics of the channel.

APPROACH

Figure 2 illustrates the channelized architecture in a typical multifunction system application example depicting the top level functional/physical partitioning using TACAN, IFF, landing systems, etc.

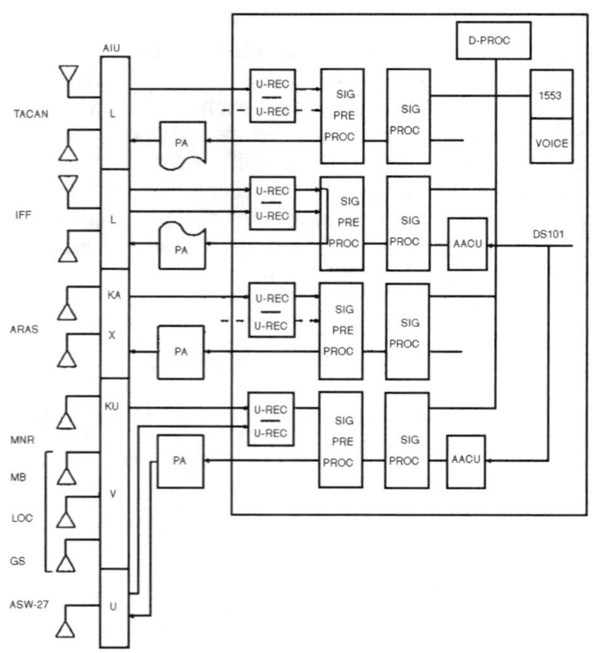

Figure 2 - Channelized Architecture (CNI Systems).

GEC has incorporated a programmable universal modular radio/modem channel by modifying the distribution of RF, IF, baseband, signal message, and data processing to that in the Pave Pillar/ICNIA architecture. In particular the digital radio incorporates:

- Block up/down conversion to UHF band in the AIU

- IF rather than baseband receiver outputs

- Direct point-to-point conflict resolved control of channel RF resource(s)

- Direct point-to-point access to crypto functions

- Dedicated point-to-point low latency interconnect between signal preprocessor and signal processor

The architecture maintains the high level message traffic between the data processor and individual channels on a JIAWG specified PI bus. The data processor, host interface, and power supply modules are also JIAWG defined modules.

BRASSBOARD HARDWARE

GEC-Marconi has developed a brassboard demonstration unit. Figure 3 illustrates the functional/physical partitioning of that brassboard to the functional and physical partitioning previously described for a production SEM-E version. As illustrated each module was constructed on PC compatible cards and plugged into the PC backplane.

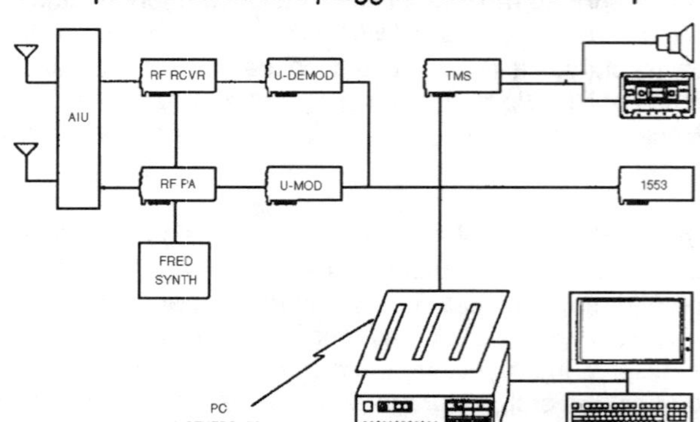

Figure 3 - PC Based Digital Radio Brassboard.

The brassboard hardware consists of an RF receiver module, Universal-Demodulator module, Universal-Modulator module and a commercial TMS digital signal processor module, (for development purposes the preprocessor function, the U-Demod and U-Mod, were partitioned on two cards to expedite development). The PC and all peripheral equipment used were commercial equipment.

Universal Receiver

Some of the key features of the universal receiver are the following:

- All frequency dehopping is performed in the UHF band using digital RF components such as Direct Digital Synthesizers (DDS) and PLL techniques.
- The programmable RF receiver implements linear/AGC, Log, and saturated dynamic range characteristics.
- Wideband programmable filters are employed in the RF/IF area
- The IF frequency plan was chosen to handle up to 250 MHz of bandwidth.

Universal Demodulator

The universal demodulator performs the following functions:

- IF to digital I/Q baseband conversion.
- Programmable sample rate decimation.
- Digital FIR filtering (channel selectivity).
- Cartesian to polar coordinate transformation (I/Q to amplitude and phase).
- Programmable digital demodulation (AM, FM, pulse decoding, correlation).
- AGC feedback/feedforward processing.

The demodulation can support a wide variety of CNI and Landing system functions.

Universal Modulator

The universal modulator performs the following functions:

- Converts Real samples to Inphase and Quadrature (I/Q) digital samples (for USSB, LSSB, ISSB generation).
- Digital baseband FIR filtering to control transmit spectrum.
- Polyphase interpolating FIR filtering running at the final sampling rate (patent pending).
- Digital-to-analog conversion and reconstruction.
- Direct up-conversion to RF.

The universal modulator also provides internal realtime control, time base management and reference frequency generation.

DIGITAL RADIO BRASSBOARD TEST RESULTS

In order to provide initial validation of the architecture and its ability to perform real-time reconfigurability, two system functions were demonstrated. These were: 1) transmit/receive loopback function and 2) over-the-air reception. The over-the-air demo consisted of the reception of local radio stations and commercial airline IFF replies. The Army's Airborne Target Handoff System (ATHS) waveform was chosen for the transmit/receive function.

The brassboard is user-friendly in that all of the hardware/software elements are housed within an PC. Figure 4 depicts GEC's first generation signal preprocessor/signal processor hardware. Also shown is the PC used to hose the cards and a commercial multiband tuner.

ATHS Test Results

To demonstrate a transmit/receive capability the ATHS waveform with a AM RF carrier/CPFSK baseband modulated waveform was generated and looped back through the system at IF as depicted in the Figure. The waveforms at various points are also illustrated in the Figure. The ATHS waveform was generated at 455 KHz for a AM carrier and at 10.7 MHz for a FM carrier. A DSP simulation of the baseband demodulation process of the CPFSK waveform is also shown. Here successive 64-point FFTs are performed on a moving sample window to identify symbol boundaries prior to CPFSK demodulation.

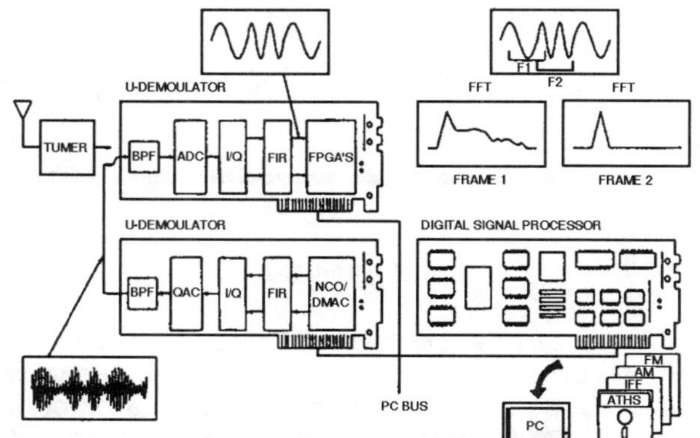

Figure 4 - Over-The-Air AM/FM reception and ATHS loopback function.

AM and FM Test Results

For the over-the-air demo (receive only) the demodulator card was programmed to receive local AM radio broadcasts and then reprogrammed to receive FM broadcasts. The voice quality of the channel was comparable to that of a conventional radio. Some of the parameters that could be varied were channel selection, FIR filter bandwidth, sampling rates, etc.

IFF Test Results

The IFF demo consisted of downloading new configuration data to the preprocessor to characterize it as a pulse decoder rather than an AM or FM demodulator. Figures 5 and 6 illustrate two separate detected IFF transpond replies (mode 3 A/C) and the decoded information received (Table 1). The IFF decoded data was formatted into a histogram of air traffic in the area. The fact that the histogram and PRF were consistent, no "fruit" bins, sufficed for a proof of concept demo.

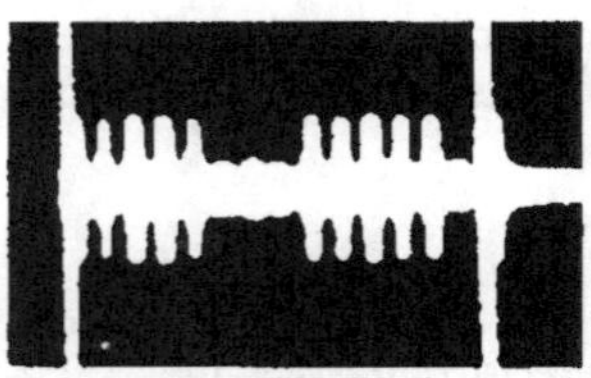

Figure 5 - IFF Reply 1.

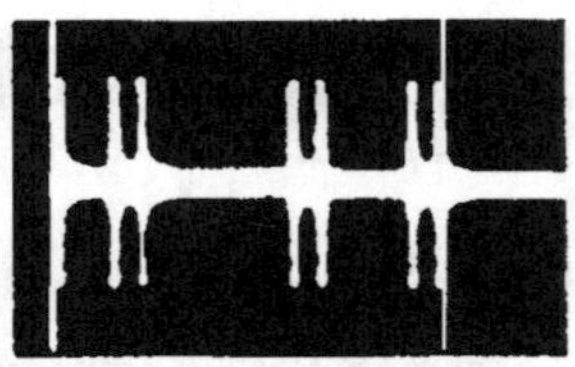

Figure 6 - IFF Reply 2.

Table I - IFF Transpond Replies Decoding.

	F1	C1	A1	C2	A2	C4	A4	X	B1	D1	B2	D2	B4	D4	F2
R1	1	1	1	1	1	0	0	-	1	1	1	1	1	0	1
R2	1	0	1	1	0	0	0	-	1	1	0	0	0	1	1
	A4	A2	A1		B4	B2	B1		C4	C2	C1		D4	D2	D1
R1		3				7				3				3	
R2		1				1				2				5	

MECHANICAL

Figure 7 shows the mechanical outline of a two system resource. Expansion is obtained by replicating an identical module set as required. The modules are based of the standard SEM-E format. The RF Power amplifiers (PA) and Antenna Interface Units (AIU) are system dependent and are therefore housed external to the Radio/Modem cluster.

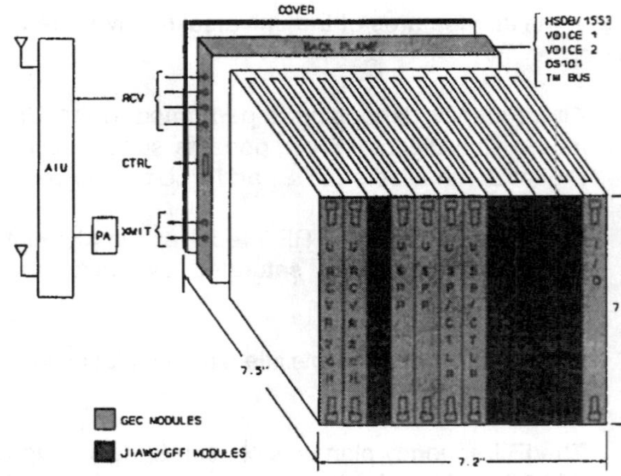

Figure 7 - Two System (expandable) Radio/Modem Cluster

CONCLUSION

The channelized architecture offers significant advantages over the fully integrated approach when only a few number of systems are to be integrated. Also advantages are gained over the Federated (a system at a time) approach when channel programmability/reusability is desired. The hardware implementation incorporating all digital modulation/demodulation techniques, coupled with the advent of software programmable devices maintains the full flexibility of the architecture.

REFERENCES

[1] Pearce T.H., Jay M.J., GEC-Marconi Research Lab "Performance Assessment for Digitally Implemented Radio Receivers", IEE Conference "Radio Receivers and Associated Systems", July 23rd, 1990.

Speakeasy: The Military Software Radio

The Speakeasy Phase II radio will use programmable processing to emulate more than 15 existing military radios. Speakeasy is a challenge, even with recent advancements in DSP technologies. The benefits, however, make the challenge highly worthwhile.

Raymond J. Lackey and Donald W. Upmal

RAYMOND J. LACKEY is a section head with Hazeltine Corporation.

DONALD W. UPMAL is the Army Project Leader on the joint service Speakeasy Multiband Multimode Radio Program.

Speakeasy is sponsored by the U.S. Air Force Rome Laboratory Contract F30602-90-C-0115.

In the past, a military radio was developed for a 30-year lifetime. It performed a single function, and was optimized for a particular field application. This was primarily caused by the slowly evolving technology and the difficulty of fitting the military user's needs into the package space available. Today, commercial applications are driving technology so that the half-life of a component is down to 19 months, that is, the time from product release to the use of its next generation replacement in new designs.

The dilemma facing the Department of Defense is: How can the military ensure communication with its latest allies and global support structure, deny interception by our current enemies, take advantage of the rapid technology changes, and control costs of military spending? Hazeltine, as the prime contractor, and TRW, as waveform software subcontractor, have faced these concerns with the Government on the Speakeasy Multiband Multimode Radio (MBMMR) Program. The Speakeasy Program is a multi-phase, tech base research and development program for the next-generation military radio, managed by Air Force Rome Laboratory (contract number F30602-90-C-0115) with triservice and ARPA funding. The Speakeasy solution is a combination of strategies, the key to which is waveform reconfigurability in modular, programmable signal processors through software implementation of waveforms and associated processing.

As Digital Signal Processors (DSPs) become faster, the Analog-to-Digital (A/D) conversion is moving closer to the antenna. The DSPs then take on additional signal processing tasks from typically analog circuits in radio implementations. The flexibility of the programmable implementation of these functions allows rapid changes to waveform modulation not possible in analog implementations. This flexibility is an advantage in the military environment where communication at different ranges and to different command levels may require different RF bands, waveform modulation types, band- width-efficient voice encoding algorithms, and encryption structures. Speakeasy Phase I proved the concept and demonstrated the capability of a military software radio.

The Speakeasy Phase II radio is planned to utilize programmable processing to emulate more than 15 existing military radios. Many radios have unique modulation forms and thus specific software modules. Other radios have modes with common waveforms, such as amplitude modulation (AM), which can use generic waveform software modules with a custom set of parameters passed as variables. The waveform software can be a standard set, a set loaded for a specific mission, or even a custom waveform designed from existing software modules for a special-operation mission. Software downloading can be accomplished either locally, by authorized user entry, or via over-the-air (OTA) data distribution.

Background

Speakeasy is a multi-phase, joint service technology program to prove the concept of a programmable waveform, multiband, multimode radio. The MBMMR modules and software will be applicable to a number of platform configurations. This will allow the smaller quantity platforms to take advantage of the mass produced modules from the greatest production base. Figure 1 illustrates the Speakeasy modular concept for implementing multiple waveform capability in common programmable modules to be used in a number of field configurations.

The MBMMR's envisioned use is on joint service platforms, could potentially perform the functions of more than ten existing radios, and should pay for itself solely from the savings accrued by the elimination of parallel maintenance and logistics tails. It is to be a modular design with a fixed, defined, open architectural interface that will allow constant upgrades as technology advances. The

Reprinted from *IEEE Communications Magazine,* Vol. 33, No. 5, pp. 56–61, May 1995.

modular design will also allow expansion for enhanced processing, including increased waveform simultaneity, fault tolerance, and survivability.

The Speakeasy-derived MBMMR will operate in bands from 2 to 2000 MHz in its basic configuration with any existing or future waveform. It has programmable digital signal processing capacity on the order of one billion 16-bit integer operations per second (OPS) and 200 million 32-bit floating-point OPS (FLOPS). This will have to be reduced to a small, lightweight package for tactical use, including the other analog circuits, RF modules, and battery. Up to four simultaneous waveforms will be active in receive or transmit. It will truly be a laptop supercomputer with antenna ports for voice and data networking.

Although every voice radio (transmit and receive unit) has a group of functions in common, there are differences that prevent interoperation between different types of units. These differences include the RF carrier frequency, modulation technique, and waveform structure. Digitally encoding the voice information further complicates the compatibility of communication systems.

Physical interaction of the RF frequency and the propagation medium also affect the applicability of a frequency band to a communication system requirement. HF signals propagate well for long distances but require large antennas, tend to have communication outages, and are difficult to use for large data bandwidths. Microwave frequencies are very good for high data rates and ECCM waveforms, but are restricted to line-of-sight situations. The typical military unit has varying communication requirements and thus has various radios for its necessary communications links.

The incompatibility of the communication equipment of even the United States' military branches was demonstrated in the Grenada conflict. The communication complexity was greatly compounded in Operation Desert Storm as a large number of allies and coalition forces performed joint operations and needed timely, efficient communication. Again, extra radios were necessary to guarantee an effective communication link.

It is evident that this proliferation of radios can become a significant logistic, as well as physical, burden to the military personnel. Speakeasy seeks to alleviate this burden by implementing the common functions in an all-digital, software-programmable, baseband signal processor, a multiband (2 to 2000 MHz) low-power RF transceiver, and interchangeable high-power RF amplifiers, and antenna subsystems. Figure 2 shows a possible futuristic manpack MBMMR, with an interchangeable RF module in back of the unit. The MBMMR will be a software radio requiring specialty hardware modules only for high power transmissions and further band extensions.

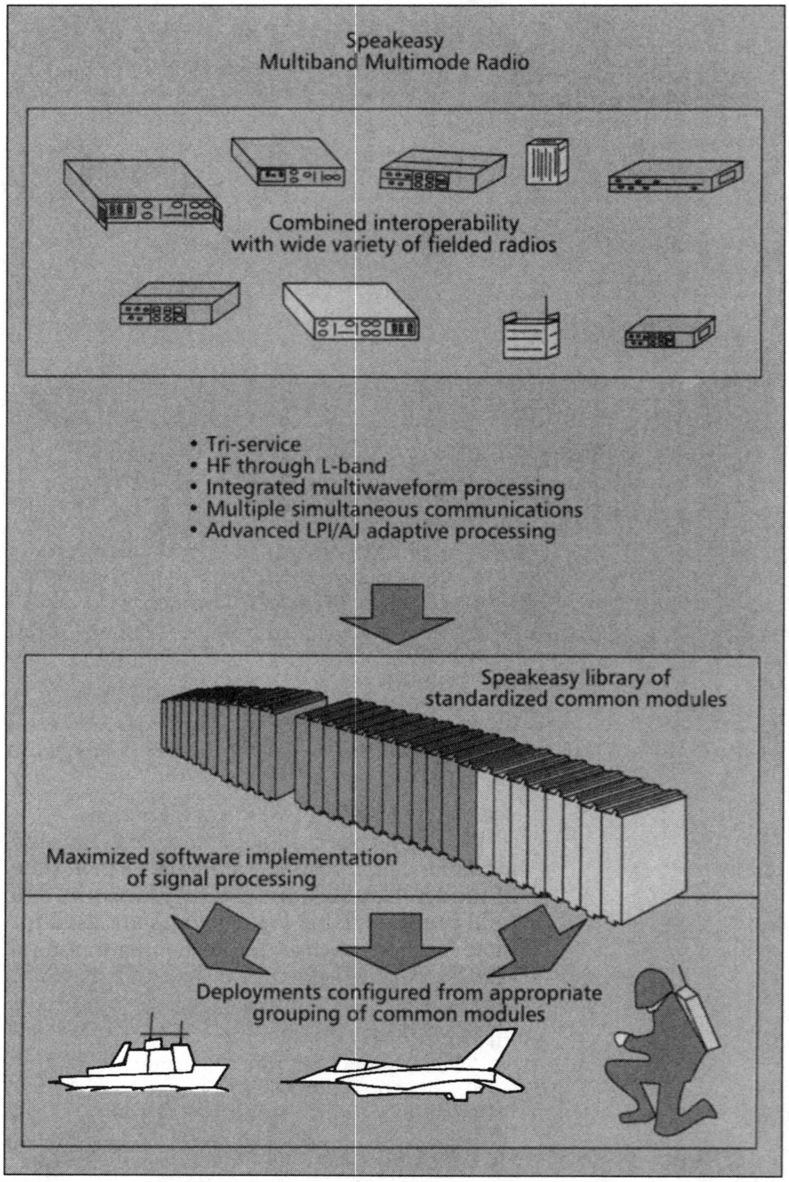

■ **Figure 1.** *The Speakeasy concept implements multiple waveform capability in interchangeable, software-programmable modules.*

■ **Figure 2.** *This futuristic MBMMR has a software-programmable baseband processing section and interchangeable RF modules.*

Approach

Flexibility is achieved by providing required processing functions in a set of processing modules. The modules are interconnected through a fault tolerant, high-speed control and data bus to maximize flexibility and reliability. This multiprocessor architecture allows the major functions to be organized in any required processing sequence dictated by the waveform structure. The Speakeasy Phase I program developed an architecture, mapped it onto a VME backplane, and developed software for the executive, a number of waveforms,

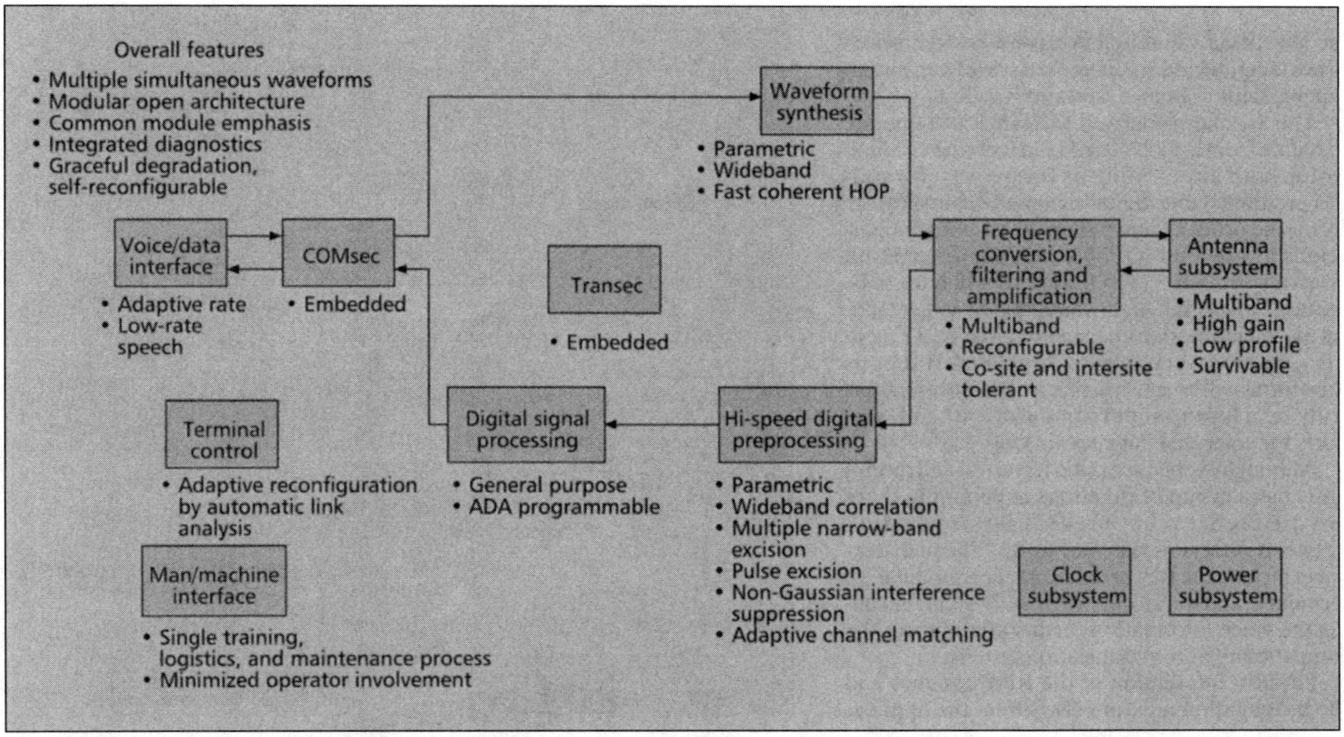

■ **Figure 3.** *The Speakeasy functional block diagram illustrates common functional processing flow of a software-based digital radio.*

and a number of INFOSEC key algorithms.

Programmable, general-purpose DSPs are used for all functions possible within the current state of the art. Application-specific processors and field programmable logic devices are used for those functions, such as the fast Fourier transform (FFT) and spread spectrum matched filters, which are too complex for efficient implementation in today's general-purpose programmable DSPs.

Packaging of the final system will be modular with open standards for both physical and electrical interface specifications. This will ensure a long service life for the radio by allowing the replacement of individual modules as technology advances, much as today's personal computers are continually upgraded. The open standards will allow contractors to repackage commercial technology and demonstrate the resulting performance improvement without major development costs or risks. The VME bus was selected for the Phase I demonstration because of the availability of commercial processor boards.

Power consumption is a major concern for the Speakeasy-derived MBMMR. Current consumer products are moving more functions to digital signal processing because of its production cost savings and enhanced performance. However, nothing in this world is free; there is a price to be paid, for example, higher power requirements. A recently introduced digital video recorder uses twice the power of its analog predecessor. Speakeasy provides enhanced performance in flexibility to emulate a large number of different radios with a wide range of modulations. The likely tradeoff, however, is higher power consumption for an individual waveform than could be achieved with a special-purpose radio designed to handle only that waveform.

Architecture

The functional flow of processing in any radio conforms to a general standard with a few variations in methods of implementation. This functional flow also exists in the digital processing within Speakeasy, but additional algorithm options are made possible. Figure 3 is the block diagram of signal flow within the Speakeasy radio. This could be the block diagram of any radio if "digital" were removed. The addition of that same "digital," and the flexibility carried with it, makes it the block diagram of *all* radios, hence, Speakeasy.

The use of a high-speed bus to interconnect the processing elements allows an implementation block diagram to vary greatly from the functional block diagram. Figure 4 illustrates the Speakeasy Phase I block diagram. All processing is digital except the necessary analog interfaces at each end, the voice I/O with the user, and the RF signals interfacing to the antenna and atmosphere. Most of the digital processing is implemented in programmable subsystems.

The Speakeasy Phase I advanced development model hardware is based upon a VME configuration. The VME bus provides command, control, and low data rate transfers. A segmented, high-speed backplane bus was implemented to provide waveform-bandwidth data transfers. The number and types of DSP, FFT, and other modules can be adjusted to accommodate waveform types, simultaneity, fault tolerance, and unit reliability.

The DSP engine identified for Speakeasy Phase I was Texas Instrument's TMS320C40 processor. This processor is one of an evolutionary family that has wide commercial acceptance. A multichip module (MCM) with substrate measuring approximately two inches square was developed with four 'C40 processors and 5 Mbytes of memory, 1 Mbyte of local memory for each processor and 1 Mbyte of shared global memory. The photo in Fig. 5 shows the interior of this Quad-C40 MCM. The memory chips are stacked in blocks of ten

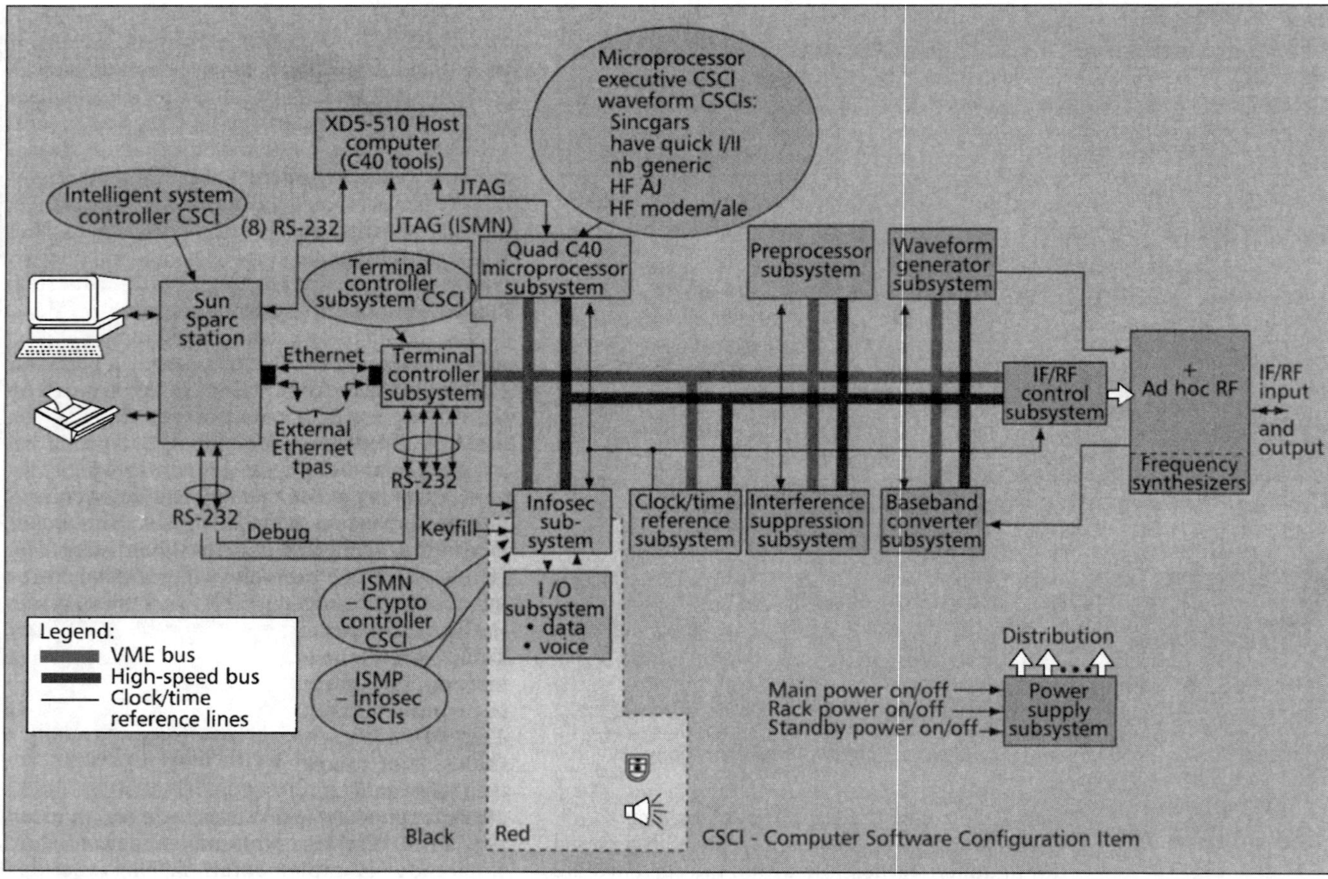

■ **Figure 4.** *The Speakeasy Phase I block diagram shows the main processing blocks and their interconnections.*

and mounted on their sides. The center chip is a programmable logic device programmed for bus arbitration and interrupt handling.

The Quad-C40 MCM is designed for clock speeds of 50 MHz, although those built under the Speakeasy Phase I program had 40 MHz devices. The performance capabilities for a 50 MHz MCM will include 200 million floating point operations per second (MFLOPS), 1100 million integer operations per second, and 300 million bytes per second I/O. All this from a module with a power consumption of less than 10 watts in today's technology.

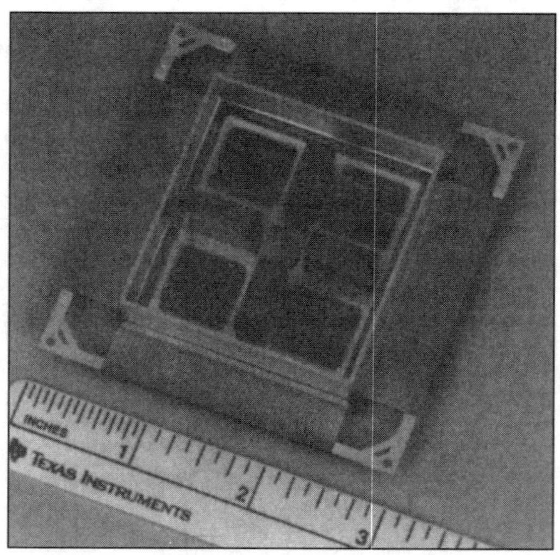

■ **Figure 5.** *The Quad-C40 MCM provides significan t programmable signal processing in a small package.*

Software

Software reprogrammability of waveform processing is the key to Speakeasy's success. Some flexibility is gained by reordering the existing processing functions. However, only limited waveform variability is achievable with the reordering of processing functions. Adding new functions will be required to develop new waveforms. The advantages conceived for Speakeasy will not be realized unless these waveforms can be added without opening the radio. Software, in programmable DSPs, is the best way to accomplish this goal.

Software lifetime is a major cost driver, more important than the platform lifetime, because of the rapid evolution of the hardware technology and the stability of the common waveforms. The higher order language of Ada was used for all waveform coding during the Speakeasy Phase I. However, interrupt service routines in the software executive had to be replaced with hand-coded assembly language modules where today's processor and compiled code failed to meet time constraints. Also, security-critical cryptological code in the programmable INFOSEC module was assembly language coded to meet NSA evaluation requirements.

The Speakeasy-derived MBMMR will utilize programmable processing to emulate more than 15 existing military radios. Many radios have modes with common waveforms, such as amplitude modulation (AM), which can use generic waveform software modules with a custom set of parameters passed

Waveform	RF band	Synopsis
STAJ	HF	Frequency hopping data and voice communications.
HF modem (MIL-STD-188-110A)	HF	Multirate data communications.
Automatic Link Establishment (MIL-STD-188-141A, App. A)	HF	Automatic setup and monitoring of HF links, including channel scanning, automatic answering, and link quality analysis.
PACER BOUNCE	HF	Classical analog HF communications.
SINCGARS	VHF	Frequency hopping voice and data communications.
HAVE QUICK I/II	UHF	Frequency hopping voice and data communications.

■ **Table 1.** *Baseline waveform compatibility.*

Waveform: Narrowband Generics

Modulation/demodulation of:

Amplitude modulation
- (a) AM (DSB) with modulation from 50 Hz to 20 kHz (hopped/non-hopped)
- (b) USB-SC with modulation from 50 Hz to 20 kHz (hopped/non-hopped)
- (c) LSB-SC with modulation from 50 Hz to 20 kHz (hopped/non-hopped)
- (d) ASK (DSB-OOK) (non-hopped)
- (e) CW (non-hopped)

Frequency modulation (hopped/non-hopped)
- (a) FM with modulation from 50 Hz to 20 kHz
- (b) FM (FSK) with tones form 50 Hz to 10 kHz (M = 2 to 8 tones)
- (c) Minimum-shift keying (MSK)
- (d) CPFSK

Phase modulation
- (a) MPSK (M = 2 to 8) (non-hopped)
- (b) DPSK, QDPSK (hopped/non-hopped)
- (c) OQPSK (non-hopped)

Phase/amplitude modulation
- (a) 4, 16, 64 and 256 QAM (non-hopped)

Frequency hopping with above modulations with variable hop dwell and dead times. A predefined acquisition and frequency selection algorithm shall be used.

Data rates up to 20 kb/s (consistent with Ada execution speed)

Error deteciton and correction for digital modulations using Reed Solomon (16, 7) and (31, 15) and Convolutional K = 7, R = 1/2, T = 133, 171.

Waveform	RF Band	Synopsis
Wireless network access (WNA)	L-band	Packet data waveform
GPS	L-band	Precise position and time COTS module
Enhanced EPLRS	UHF	Position location and increased data rates
SINCGARS SIP	VHF	Increased data rates
Low probability of intercept (LPI)	—	Classified
UHF SATCOM DAMA	UHF	5 and 25 kHz DAMA
T1 Data waveform	TBD	Transmit T1 rate data
Cellular phone	Commercial	COTS module

■ **Table 2.** *Planned phase II waveform expansion.*

as variables. The narrowband generic waveforms identified for Speakeasy are shown in the accompanying box on this page. Other radios have unique acquisition protocols and thus specific software modules. Those waveforms in the Phase I baseline Speakeasy program requiring unique modules are shown in Table 1. Unique waveforms planned for expansion in Phase II are shown in Table 2. Physical layer functions are more easily implemented generically with increasing difficulty in the link and network layers of existing waveforms.

Emulation of an existing radio includes much more than just being able to generate a particular RF modulation. All functions of the radio, from the microphone (or data port) to the antenna, must be supported. The high-speed waveform processing at the physical layer, as well as the voice encoding at the host application layer, must also be performed in the MBMMR. The accompanying box on the next page shows the listing of I/O formats of the existing radios with which Speakeasy must be interoperable. Other levels of protocol within the 'radio' function affect interoperability. Table 3 shows the International Standards Organization (ISO) levels for one mode of one waveform to be implemented within Speakeasy Phase II. This variability, and the flexibility to meet it, is thus a major concern for the host processors and their software. More diverse ISO layers will be included as networked protocols are added. An example is the Wireless Network Access (WNA) protocol, which supports real-time, on-the-move (OTM) networking.

The waveform software can be a standard set, a set loaded for a specific mission, or even a custom waveform designed from existing software modules for a special-operation mission. Maximum flexibility is achieved if every radio holds the code for all waveforms. The storage of all of this software can become a burden. It is more conceivable to believe that the radio will be provided to the user with a waveform set common to his command segment or mission. The need for reassignment of a resource after it has been assigned must also be considered.

Maximum flexibility of the MBMMR will require software loading through a secure front panel, over-the-air download, or authorized user entry. This flexibility can also become its own problem because of the sensitivity of military communications. Managing software versions, authenticity checking of change attempts, and protecting from unauthorized snooping will be major security issues to be dealt with in the operation of a radio that has changed all of the existing paradigms.

Programmable INFOSEC

Another feature of the Speakeasy MBMMR is programmable INFOSEC, which includes both communications security (COMSEC) and transmission security (TRANSEC). COMSEC is the encryption of message data. TRANSEC is the modulation overlaid on the transmitted signal, i.e., center frequency of transmission for frequency hopping waveforms and carrier modulation such as a phase modulation for a spread spectrum coded signals. Typically, every radio has its own INFOSEC device. Speakeasy seeks to overcome this logistic nightmare by using a programmable

Baseline I/O formats

Voice

- Analog
- 16 kb/s CVSD
- 2400 b/s LPC-10
- 4800 b/s CELP
- 9600 b/s MRELP

Digital data

- Narrowband: 50 to 38,400 b/s
- Wideband: up to 2 Mb/s

INFOSEC processor, the CrYPtographic Reduced Instruction Set (CYPRIS) processor, and software implementation of the cryptographic algorithms, in the multichannel Speakeasy INFOSEC Module (SIM), also developed in Phase I. Key management has also been implemented in software, allowing BLACK, RED, and over-the-air-rekeying (OTAR) keyfill modes. The Phase I INFOSEC module replaces five individual INFOSEC devices. Additional device capability will be added on Phase II.

Results

The Speakeasy Phase I Advanced Development Models (ADM) were demonstrated to Government representatives in August 1994. Units were operated in links with government furnished standard radios including HAVE QUCIK, HF Modem (MIL-STD-188-110A), Automatic Link Establishment (MS-188-141A), and SINCGARS. Units also demonstrated simultaneity by simultaneously transmitting on both HAVE QUICK and SINCGARS frequency hopping nets. Bridging was demonstrated by linking two nets with different waveforms, i.e., a HAVE QUICK net was bridged to a SINCGARS net so that all users communicated, with Speakeasy acting as a bridge/gateway between the two independent nets, transparent to the users. Demonstrations included both voice and data. A computer-generated video image was transferred through speakeasy on a HF Modem waveform. Programmability was demonstrated by modifying the SINCGARS waveform on two Speakeasy units and showing operation.

ISO Layer	UHF SATCOM/DAMA - 25 kHz
3. Network	Centralized protocol Circuit and message services Pre-assigned and demand assigned services
2. Data link	Frame/slot format/deformat TDMA DAMA protocol Link quality measurement CW followed by repeated 110 I-CH (Inverted for Q-CH) and legendre framing word preamble fields Address fields and frame numbering PT and CT orderwire messages/KGV-11 CRC: 16-bit for PT or CT orderwire messages
1. Physical	FEC: Convolutional R = 1/2, K = 7 @ 75 to 4800 b/s : Convolutional R = 3/4, K = 9 @ 300 to 16 k/s Interleave/deinterleave Mod/demod: 9600 and 19200 SPS BPSK 32000 SPS DEQPSK Transmit/receive: Uplink - SHF and UHF Downlink - UHF

Table 3. *Radio emulation requires more than just waveform modulation compatibility.*

Conclusion

Speakeasy is a challenge, even with recent advances in DSP technologies. The benefits, however, make the challenge highly worthwhile. Today's cooperative efforts in military engagements, as demonstrated in Desert Storm, requires flexible communication systems to interconnect with all allies. At the same time, defense budgets are being reduced and systems must last longer and cost less to maintain. Speakeasy holds the promise of being such a system. Software reprogrammability of waveform processing is the key to Speakeasy's success. This capability has been demonstrated on Speakeasy Phase I. Software lifetime is more important than the host hardware lifetime because of the rapid evolution of the hardware technology. The higher order language of Ada was used for all waveform coding during the Speakeasy Phase I development. Phase II of the Speakeasy program will be awarded in 1995 to continue the advances already made and to finalize the architecture of the objective "Military Software Radio" of the future.

Advanced Digital Receiver Principles and Technologies for PCS

The synergy between digital radio communications and VLSI signal processing is revolutionizing the design of wireless terminals. Driving this synergy are certain fundamental paradigms in modern communication theory, digital signal processing, and VLSI design.

Heinrich Meyr and Ravi Subramanian

The explosive growth of wireless communications has given rise to a variety of portable voice, data, and messaging systems. In many cases, PCS is defined by some combination of these various systems. Future PCS terminals for example, can be expected to work in 900 MHz terrestrial cellular systems, in the 1.9 GHz PCS band, as well as in 1.6 GHz satellite-mobile systems. The defining feature of PCS is that of coverage. PCS terminals must enable users of PCS services to truly be accessible anywhere and anytime.

Advances in, and new applications for, each of these radio systems are being spawned at an unprecedented rate. The complexity of radio communication systems is significantly increasing with the application of more sophisticated multiple-access and digital modulation techniques in order to accommodate the tremendous growth in the number of subscribers. Employing advanced digital communications principles is the norm. Furthermore, the requirements for frequency-agile terminals are also growing as frequency-hopping techniques find their way into commercial systems.

Rapid advances in solid-state integrated circuit (IC) technology are fueling the growth in commercial wireless communications systems. New technologies are being spurred on by the desire to produce high-performance, low-power, small-size, low-cost, and high-efficiency components. Today's IC and process designers are teaming up with communications engineers to define the new directions in which to push semiconductor processes for PCS. Digital technologies have seen a breakthrough in both performance and size through the use of complementary metal-oxide semiconductor (CMOS) device scaling. These technologies can be expected to scale rapidly owing to the demand for denser and faster memory chips as well as digital microprocessors. Unlike digital logic, which seems to gain speed with increased packing density, analog performance is often compromised with higher levels of integration.

The increasing use of integrated circuits in radio has resulted in significant improvements in the reliability and performance of receivers. The number of parts that require tuning or whose characteristics could significantly change over time or temperature has dropped markedly. Manufacturing costs have also fallen due to improved procedures in assembly and testing. Rapid advances in the packaging of integrated circuits have contributed significantly to compact designs of wireless terminals. Technologies, such as multi-chip module (MCM) allow designers to integrate many individual chip die into a single substrate structure. This new structure reduces the size of the inter-chip capacitances and minimizes the power consumed to drive signals from one chip to another. This drives down the total power consumed by the system since most of the power is now consumed for processing rather than for interfacing signals from chip to chip.

These trends are giving rise to the increasing use of silicon-based signal processing techniques, which include discrete-time analog signal processing techniques (such as switched-capacitor systems and delta-sigma modulators) as well as advanced digital signal processing (DSP) techniques. Traditionally, DSP techniques were used for pre-modulation and post-detection functions in radio communications receivers, such as speech and channel codecs. Over the last ten years, DSP techniques have been used extensively for advanced digital communications transceiver designs, finding their way into detection, equalization, demodulation, frequency synthesis, and channel filtering. In addition, a totally new and broad range of digital algorithms are emerging for which there exists no analog counterpart.

The Communications System Design Space

Design Space and the Role of DSP

Traditionally, the communications engineer has been concerned with a two-dimensional design space comprising power and bandwidth. Employing advanced very large-scale integration (VLSI) technology, the implementation of complex digital algorithms has

HEINRICH MEYR is a professor of electrical engineering at the Aachen University of Technology (RWTH Aachen).

RAVI SUBRAMANIAN is with AT&T Bell Laboratories in Crawford Hill, New Jersey.

Reprinted from *IEEE Communications Magazine,* Vol. 33, No. 1, pp. 68–78, January 1995.

become feasible and economical. As a consequence, the communications engineer nowadays is faced with a three-dimensional design space comprising power, bandwidth, and complexity. This has enabled the design engineer to trade the physical figures of merit (bandwidth and power) against signal processing complexity. Also, the capability of realizing highly complex algorithms has made it possible to approach the theoretic performance limits imposed by Shannon theory. Advanced source and channel coding algorithms are already used in many commonplace products such as CD players and satellite receivers. The success of the personal communications revolution critically depends on the ability to efficiently transmit information over the scarce physical resource bandwidth.

When it comes to realizing digital receivers, system modeling and algorithm design closely interact with the architectural choices for implementation. As a consequence, the more traditional approach — presently pursued by many companies and research institutions — where algorithm design (found in "theory" groups) is well-separated from the implementation (found in "VLSI" or "DSP" groups) should be abandoned in favor of an interdisciplinary "center-of-gravity view" where model/algorithm, architecture, and tools are closely interacting centers of gravity, as illustrated in Fig. 1.

For the purposes of solving a given detection or estimation problem in digital radio modem design, it is not sufficient to have a good circuit architecture. It is of paramount importance to use a good algorithm that is reliable and well-conditioned. There is little point in trying to exploit a desirable architecture if the associated algorithm is not stable. A better algorithm (even the best), on the other hand, may not correspond to a suitable computational architecture. In order to avoid this type of conflict, it is essential to develop algorithms and architectures in parallel. In this way, it is often possible to achieve the best of both worlds for a given application.

The task of *interactively exploring the design space* therefore calls for the "tall thin" engineer rather than the engineer specializing in algorithms, software, hardware, or tools alone. This is not to say that specialists are unimportant. Rather, for this particular task, a specialist is not the best-prepared engineer. Moreover, in order to meet the stringent time-to-market requirements, the use of proper DSP Design Automation (DSPDA) CAD tools is of crucial importance.

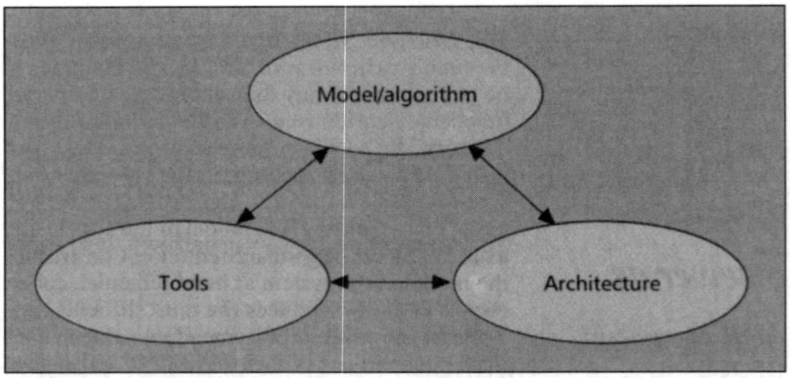

Figure 1. *"Centers of gravity" view of communication systems design.*

Communication System Model

In his seminal paper [41], Viterbi summarizes the three basic lessons learned from information theory:

- Separate source and channel encoding.
- Never discard information prematurely.
- Make the channel look like an AWGN channel.

These lessons immediately translate into important considerations for practical transmission system and receiver design. The basic structure of the information-theoretic system model according to these lessons is depicted in Fig. 2.

Lesson 3 states that the channel capacity is maximized if the decoder faces a symbol sequence a that is corrupted by Gaussian noise n only. Consequently, any receiver designed according to the three lessons above can be decomposed into two fundamental parts as shown in Fig. 3. The sole task of the inner receiver is to (ideally) produce a sequence of symbols a that is disturbed by additive Gaussian noise. Stated differently, the inner transmission system has to provide a "good" channel for the outer receiver, i.e., the source and (possibly) channel decoding system. To accomplish this, the inner transmission system must be adapted to the channel characteristics.

Principles for Deriving Receiver Structures

From the phrase "the *sole* task of the inner receiver..." one must not conclude that the design of the inner receiver is a trivial task that can be accomplished easily and with little complexity. More often than not, quite the contrary is true. Most of the design effort of a digital receiver as well as most of the hardware and software resources are usually spent on the inner receiver and not on the design of codes.

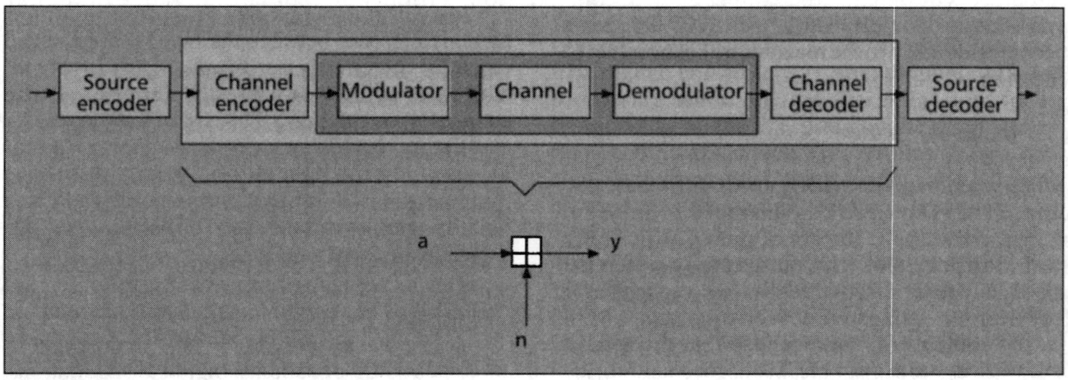

Figure 2. *Information theoretic system model.*

Most of the design effort of a digital receiver as well as most of the hardware and software resources are usually spent on the inner receiver and not on the design of codes.

The Derived Structure Approach — Detection and synchronization are the key functions to be performed by any digital receiver. For both functions, algorithms can be systematically derived. The basic idea behind this derived structure approach is that no a priori guess about the receiver structure is made. Rather, the algorithms are derived based on a mathematical model of the system and a suitably chosen performance criterion. Constructing the model of the system at hand (channel, coder, etc.) is in many instances the most difficult task. This task, however, is of tremendous value in itself since it forces the designer to organize and analyze the problem more thoroughly. It also allows for systematically incorporating a priori knowledge that may be available.

The key features of the derived structure approach are the following:

- Algorithms can be systematically derived for a given modulation/coding scheme and channel. These algorithms are optimal with respect to the chosen performance criterion.
- If, for complexity reasons, the optimal algorithms are not feasible, they form the basis from which suboptimal yet realizable solutions can be derived.
- It is possible to derive lower (not necessarily best-case) performance bounds. This is of great importance since it is often difficult to assess the performance of highly complex systems analytically.

Parameter Estimation and Detection — To illustrate the concept of systematic algorithm derivation, consider the simple transmission model

$$r(t)=\sum_{k=0}^{N-1} a_k g(t-kT-\varepsilon T)e^{j\phi}+n(t)$$

with transmitter pulse forming filter $g(t)$ and additive noise $n(t)$. From the received signal $r(t)$, the data symbol sequence $\boldsymbol{a} = \{a_0, a_1, \ldots, a_{N-1}\}$ and the synchronization parameters $\theta = (\varepsilon, \phi)$ with (relative) timing offset ε and phase ϕ have to be jointly estimated.

The performance criterion leading to minimum probability of error is the maximum a posteriori (MAP) criterion [20]

$$P\{(\boldsymbol{a}, \theta \mid r\} \rightarrow max$$

which maximizes the conditional probability of the desired quantities, given the observed received signal $r(t)$. If all possible data sequences $\boldsymbol{a}$ and sync parameters θ are equally likely (i.e. no prior knowledge is available), the MAP criterion becomes identical to the maximum likelihood (ML) criterion:

$$P\{r \mid (\boldsymbol{a}, \theta)\} \rightarrow max$$

which maximizes the likelihood of the observation, given the desired parameters. In other words, we look for the set of parameters $(\boldsymbol{a}, \theta)$, now interpreted as trial parameters, which has most likely generated the received signal. In general, however, an exhaustive search over all $(\boldsymbol{a}, \theta)$, is far too complex to implement so that suboptimal solutions must be derived [23].

For symbol detection, a lower bound is given by Shannon's channel capacity formula. For parameter estimation, an equally important lower bound exists, namely the Cramer-Rao bound (CRB) on estimator accuracy $var\{\hat{\theta} \mid \theta\}$. If the observation r is linearly dependent on the parameters θ and the noise is additive Gaussian, the ML estimate $\hat{\theta}_{ML}$ satisfies the CRB with equality, i.e. $\hat{\theta}_{ML}$ is an efficient estimate [20]. Note that in our simple transmission model, this is not the case.

PCS Examples — To illustrate the principles described above, we begin with a simple example, and then move on to more complicated scenarios. AWGN Channels. In the simplest case of an AWGN channel (e.g., satellite), the inner receiver consists of a downconverter, a sampler, and a synchronization unit where symbol-timing, carrier-phase, and possibly frequency recovery are performed. Synchronization can be interpreted as estimation and correction of the (slowly) varying channel parameters, here $\theta = (\varepsilon, \phi, \Omega)$ where Ω is the (relative) frequency shift [23]. The estimator unit produces values $\hat{\theta} = (\hat{\varepsilon}, \hat{\phi}, \hat{\Omega})$ that are subsequently used in correction units as if they were the true parameters; this is the widely applied principle of synchronized detection. In a well-designed receiver, the synchronization errors and implementation losses should manifest themselves solely in a slight degradation of the required minimum signal-to-noise ratio (SNR) to achieve a specified performance target. This is indeed achieved in practice.

While the bit error rate (BER) is the appropriate performance measure for the outer receiver, the performance measure relevant for the inner receiver (for AWGN channels) is the error variance of the estimated parameters $\hat{\theta}$. Notice that it is inappropriate to design the inner receiver for optimal BER (as is frequently done), since it violates Shannon's lesson 3: the inner receiver must provide for a "good" channel for the outer coding system, and this is achieved by minimizing the error in the synchronization parameters. As mentioned above, for both kinds of receivers (inner and outer) there exist fundamental performance bounds [20, 29]

- Channel Capacity

$$C=\frac{1}{2}\times\log_2\left(1+\frac{E_b}{N_o}\right)$$

- Cramer-Rao Bound on Estimator Accuracy

$$var\{\hat{\theta} \mid \theta\} \geq J^{-1}$$

where J is the Fisher information matrix. Both bounds are logarithmic measures, but note the fundamental difference between the two: for the channel capacity the energy per bit, E_b/N_o, is involved, while the estimator variance depends on the energy per symbol, E_s/N_o. This is seen, for example, by evaluating the Fisher information matrix J that arises when the parameters $\theta = (\varepsilon, \phi, \Omega)$ are jointly estimated; for this problem, the following bounds on error variances can be established [24]:

- Relative Frequency Offset

$$var\left(\hat{\Omega}\middle|\Omega\right)=\frac{12}{N(N-1)^2}\times\frac{1}{2}\times\left(\frac{E_s}{N_o}\right)^{-1}$$

- Phase Offset

$$var(\hat{\theta}|\theta) = \frac{1}{N} \times \frac{1}{2} \times \left(\frac{E_s}{N_o}\right)^{-1}$$

- Relative Timing Offset

$$var(\hat{\varepsilon}|\varepsilon) = \frac{1}{N} \times \frac{\int |G(\omega)|^2 d\omega}{\int \omega^2 |G(\omega)|^2 d\omega} \times \frac{1}{2} \times \left(\frac{E_s}{N_o}\right)^{-1}$$

where $G(\omega)$ is the transmitter pulse shaping transfer function, and N denotes the estimation window length (in symbol intervals). The energy per symbol (relevant for synchronization) is linked with the energy per bit (relevant for detection) via the code rate R_c (number of source bits per channel symbol) via $E_s/N_o = R_c(E_b/N_o)$. The minimum E_b/N_o tolerated by the detector can be made small by using powerful codes with low code rates R_c, but then E_s/N_o may become very small. On the other hand, any practical synchronizer requires a minimum E_s/N_o to be operational which, in turn, sets a lower limit on the code rate that can be achieved in practice; e.g., see Massey's instructive papers [18, 19] for more information. The systematic design of digital synchronization algorithms by invoking principles of estimation theory for timing, phase and frequency recovery is treated in detail in [24].

Fading Channels — In the case of fading channels, the synchronization task is far more complex since, instead of the static (or slowly varying) parameters of the Gaussian channel, randomly varying channel parameters have to be continuously estimated and compensated for. Furthermore, in order to comply with Shannon's lesson 3, the transmission system should provide for and make use of various kinds of diversity, e.g., time, space, or multipath diversity. Exploiting as many diversity mechanisms as possible is effective in bridging deep fades and leveling out wide variations in the information-bearing signal. As an example, Fig. 4 shows the structure of an inner receiver for linear modulation formats and frequency-selective fading channels [8]. The sampled front-end output signals $r_d(k)$ ($d=1,\ldots,D$ antennas, D RF front ends) are organized in frames where the stream of useful data is interspersed with blocks of training signals.

In the synchronization path (lower part of Fig. 4), the training blocks are extracted and used for feedforward estimation of time-varying channel parameters such as the start-of-frame time instant k, frequency offset Ω and, most importantly, the (sampled) diversity channel impulse response vectors h_d. The receiver structure is therefore based on the so-called *separation principle* that makes it possible to break the two-fold randomness of random data and randomly varying channel(s) which, if channel estimation and random data detection were to be performed simultaneously, would lead to an explosion in complexity. The separation of detection and synchronization paths enables the channel(s) to be identified from known symbols alone; this process is not disturbed by random data. Conversely, the channel estimates so obtained can be assumed to be known during random data detection.

From the estimated channel(s), coefficient sets of various receiver components are computed. As

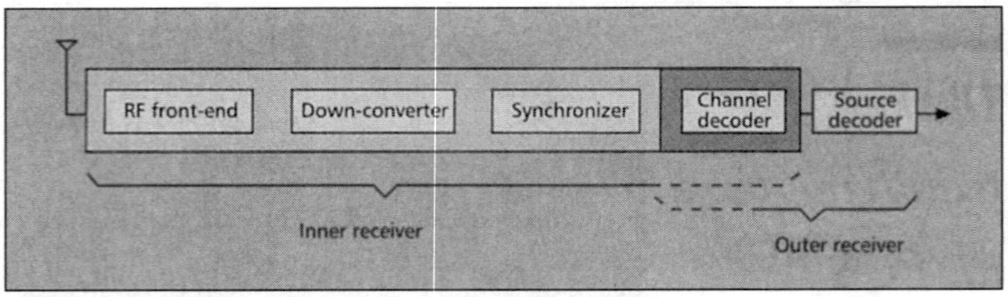

Figure 3. *Fundamental parts of a digital receiver.*

illustrated by Fig. 4, these may encompass sets of prefilter (whitening matched filter) coefficients w_d, combiner weights p, and equalizer taps f. Notice that with feedforward synchronization, this computation is just a mathematical mapping from estimated channel parameters to the coefficient sets. As opposed to recursive-type decision-feedback synchronizers, no lag error is introduced by this mapping, nor has error propagation a chance to thwart continuous receiver operation.

In the detection path (upper part of Fig. 4), the received signals are processed using the appropriate coefficient sets. In this receiver design example, antenna diversity, time diversity via interleaved channel coding, and multipath diversity via energy equalization and detection are made use of simultaneously. On fading channels, it is particularly important to avoid discarding relevant information prematurely (Shannon's lesson 2). This is achieved by transferring soft decisions between the functional blocks, e.g., equalizer, channel decoder, and source decoder.

Spread Spectrum — The design of advanced spread-spectrum multiple access systems is another example of how the key receiver design principles described earlier can be applied to produce a high-performance digital communications link. The advances in DSP techniques, A/D technology, and the speed and density of digital circuits have made these digital receivers realizable with reasonable complexity and cost.

Direct-sequence spread-spectrum systems operating in environments with multipath diversity can exploit multipath through the use of RAKE receivers. These receivers use despreading to extract the symbol energies spread over several signal paths, and use diversity combining to combine these energies to maximize the received symbol-to-noise ratio. These operations are all founded on the principle of 1) making the channel appear as an AWGN channel and 2) ensuring that the receiver uses all the information available to it at reasonable cost and complexity, including multipath versions of the transmitted signal.

Digital spread-spectrum receiver designers exploit the fact that there are actually three regimes of DSP going on in such a receiver [36]. If digitization of the received signal occurs at IF, then the first regime encountered is that of demodulation processing, where the receiver operations focus on capturing all the PN (pseudonoise) chip energy into one or two samples. This is followed by a PN-chip-rate processing regime, where code synchronization and despreading operations take place. The symbol energy is captured at the output of the despreader. Finally, the receiver operates in a symbol-rate regime, where symbol energy is maximized, estimators are provided for various parameters necessary for detection, and final data-detection takes place. Subsequent operations are performed on a symbol-by-symbol or block-of-symbols basis.

The key requirements for PCS portable terminals are performance, cost, power consumption, and size.

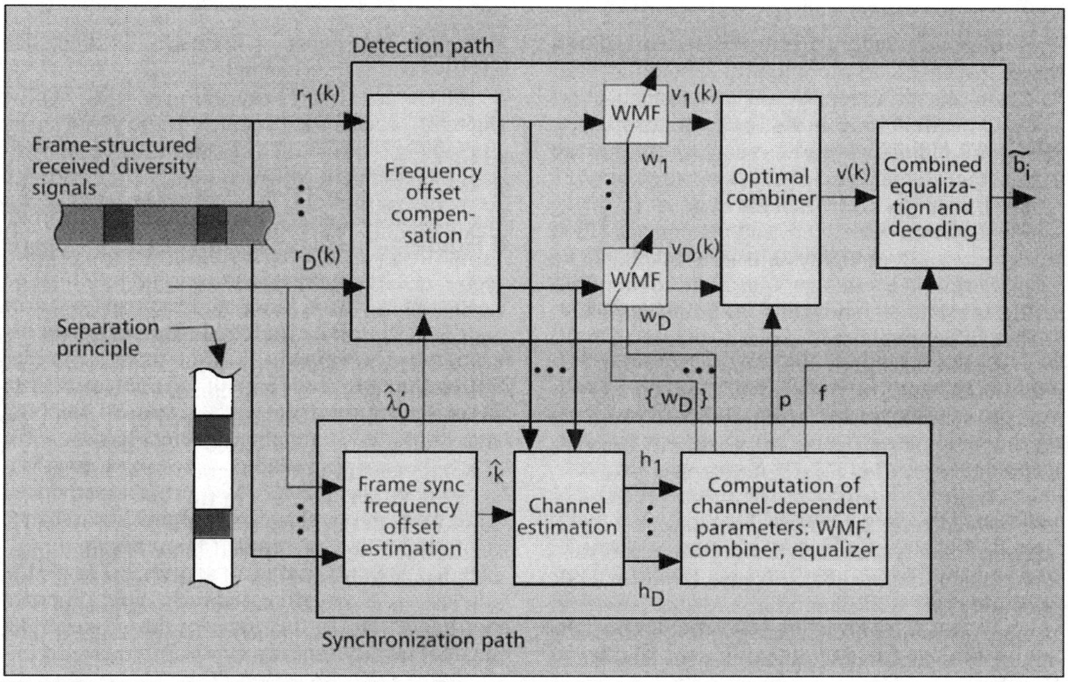

■ **Figure 4.** *The inner receiver decomposed into detection and estimation paths.*

Accurate code synchronization is essential for reliable performance of these links. Without accurate and reliable code synchronization, the receiver cannot guarantee performance to any degree.

The code synchronization operation typically consists of two stages of processing. When a receiver is first turned on, the proper code phase for the local replica of the spreading code needs to be guessed. The receiver must operate at several times the PN-chip rate in order to complete the acquisition operations. Once a reliable first guess is made, the particular code phase is tracked in a code-tracking loop. Here, variations of the classical delay-locked loop dominate most solutions. In systems with short spreading codes, however, scanning-window tracking techniques based on capturing despreader outputs in FIFOs can result in simpler implementation and control complexity [36].

PCS Radio Transceiver Architectures

Portable Radio System Requirements

The key requirements for PCS portable terminals are performance, cost, power consumption, and size. The bill of materials (BOM) of most cellular and cordless transceivers today is dominated by the radio signal processing components. These account for more than 50 percent of the BOM in most consumer phones, and they include all components related to RF/IF filtering, mixing, demodulation, and frequency synthesis. The target terminal size is typically on the order of 50 to 150 cc. This invariably requires using advanced integration techniques coupled with an intelligent partitioning of the radio subsystems.

Achieving low power consumption is done through technology and system-level trade-offs. The actual metrics are the standby-time and talk-time of the terminal between charging. Depending on which mobile communication system the terminal is designed for, the functions required to be performed by the mobile during standby widely vary. Designers need to optimize the current consumed by all the RF components, as well as the baseband DSPs, digital application-specific integrated circuits (ASICs), A/Ds, and any mixed-signal devices. At the system level, terminal designers need to understand the system operation requirements such as time-division-multiplex (TDM) or voice-activity (VA)-based operation. Optimizing power consumption here mainly involves firmware engineering for microcontrollers and DSPs as well as relegating certain firmware functions to customized hardware accelerators. Many digital processors, for example, feature power-down modes that allow for "shutting down" peripherals and certain computational units. But they do not always allow for fast ramp-ups, so that care should be exercised in their use.

Radio Receiver Principles

A radio receiver front-end is defined by an architecture chosen in order to satisfy a set of requirements. These include sensitivity, dynamic range, selectivity, signal blocking, intermodulation performance, and, of course, manufacturability, and cost. The tools of mixing, filtering, and gain control are at the designer's disposal. The typical approach used in the design of portable radio receivers considers two main desirable properties of the equipment: selectivity and sensitivity. Limitations on transmitter power and poor antenna orientation require a receiver to perform well at input levels down to 1 microvolt. The noise figure of the front-end now becomes a significant factor in defining the receiver sensitivity. Moreover, because the receiver may have to work with very low input signal levels in the presence of very large levels of unwanted signal, the linearity of the early stages of the receiver are of key importance.

The Superheterodyne Principle — The key functional requirements for a receiver are that its front-end structure must 1) accurately translate the desired RF signal to a frequency band where the processing required to extract the message is relatively cheap and 2) condition the signal (i.e., filter and gain control) to remove adjacent channel interference while maintaining receiver sensitivity. The great majority of radio receivers employ the superheterodyne principle because the fixed IFs can lead to high selectivity. That is, a first IF (high) is used for image-rejection, and a second IF (low) is used for ease of highly-selective filtering. The primary reason for this popularity is the predominance of certain standard IF frequencies, which makes components of IF strips available at a reasonable cost through economies of scale. The superheterodyne receiver offers significant benefits. Chief among them is the ability to carefully control gain, noise figure, and selectivity through different stages of radio signal processing.

Digital IF Receivers and DSP Techniques — Analog hardware has dominated mobile radio subsystems since the early days of radio. However, DSP techniques are well established today in baseband modem functions, such as speech coding and decoding, interleaving, channel coding and decoding, detection, and equalization, as well as in frequency synthesis. These techniques are now finding their way into IF processing sections of radio receivers.

The digital conversion (or Digital IF) receiver is characterized by the principle of using the inherent aliasing property of sampling to realize a demodulator. The digital IF receiver typically uses a superheterodyne architecture, where analog frequency translation is used to move the received RF spectrum to a low IF, and digitization occurs at the low IF. Thereafter, DSP techniques are used to extract the signal. The DSP techniques used include direct digital frequency synthesis (DDS), digital downconversion, high-speed digital filtering, and multirate techniques such as decimation and interpolation.

Figure 5 shows a typical configuration for a digital IF receiver. Configurations may vary depending on whether a single on double conversion superhet structure is used. The digital conversion principle exploits the fact that the lowpass I and Q components are expressible in terms of bandpass samples. The receiver is then taking samples of a passband signal, and numerically demodulating to producing I and Q samples serially. In fact, it is easy to show that to sample and reconstruct a passband signal centered at f_{IF}, the sampling rate f_s should meet the criterion

$$f_S = \frac{4f_{IF}}{2k-1}$$

where k is the image number. Recall that there are folded spectra present when the bandlimited passband signal is sampled, as shown in Fig. 5.

The simplest implementation occurs when the $k=1$ image is selected, and the IF signal is sampled at $4f_{IF}$. From an implementation point of view, a designer can explore how to trade-off a lower sampling rate for more complex DSP. Any IF sampling procedure will yield I and Q samples which are not coincident in time. As the sampling rate is dropped (i.e., k is increased) from $4f_{IF}$, these samples will be further apart in time. Thus, any detection procedure based on these samples will suffer from this time distortion. As a result, digital processing techniques to reconstruct pulse shapes will be necessary in order to compensate for the resulting distortions in I/Q samples.

There are several factors driving the use of digital IF processing. They include the advances and availability of low-cost, high-speed A/D converters, ever-increasing levels of integration, higher price/performance value, lower manufacturing costs, greater flexibility, and increased robustness and maintainability (i.e., little or no calibration). These characteristics are already making digital IF the technology of choice in non-mobile applications like cellular and PCS base-stations. One of the biggest advantages of the digital conversion receiver is the possibility it opens for multimode terminals which share a similar RF/IF strip, possibly with switchable filters, and a DSP-based programmable processor to yield savings in integration and testing. The main problem standing in the way of its implementation in portable terminals centers on the availability of very high-speed, large-dynamic range, wide-bandwidth sample-and hold circuits and high-performance A/D converters at reasonable cost. One should always note, however, that analog functions, where feasible, are often smaller, cheaper, and less power hungry. These benefits may offset the fact that the analog solution may suffer some increased manufacturing cost.

Digital signal processing techniques are well established in baseband modem functions, as well as in frequency synthesis.

Direct-Conversion Receivers and DSP Techniques — The direct-conversion receiver structure is shown in Fig. 6. The received RF signal is filtered in a duplexer, and passed through a low-noise amplifier. After bandpass filtering, the signal is split, and fed to a quadrature mixer. The mixer is fed with the local oscillator (LO), and a 90-degree phase-shifted version of the LO, which is in fact at the RF carrier frequency. The RF passband is thus translated to baseband, and the receiver must now process the full RF spectrum at baseband. This requires extremely high dynamic range as well as extremely high selectivity, together with low noise and proper amplitude and phase balancing between the I and Q arms. Figure 6 shows the processing involved in one signal path. D_1 denotes the desired signal, and I_i denotes interferers.

The greatest advantage of the direct conversion architecture is its low component count. Its simplicity, however, is deceptive. Few commercial receivers employ this architecture. The much acclaimed benefits of a wide tuning range and high selectivity are limited by the dynamic range of the receiver. Among the challenges in its realization are the need for a very high-gain low-noise mixer (to combat $1/f$ amplifier noise at baseband) as well as rapid large-DC-offset cancellation techniques. Almost all the gain required to achieve the desired sensitivity in this architecture must come from the front-end low-noise amplifier. Furthermore, because of the LO signal path, the antenna needs to be isolated from the mixer. Building in reverse isolation of up to 80 dB on some RF IC components is no easy task. Furthermore, since all the selectivity in this receiver comes from the baseband channel filters, one requires very high-dynamic range, low-noise, highly-selective filters. It was not until the late '80s that commercial solutions were announced for simple FSK paging receivers at 450 MHz and, more recently, at 930 MHz. Recently, some 900-MHz cellular and wireless LAN products have emerged using this architecture.

General-purpose programmable digital signal processor chips continue to climb up the performance curve, driven by market needs, technological advances, and architectural innovations.

Achieving a competitive direct conversion solution for commercial cellular mobile systems is a still a big challenge. Some researchers have focused on designing specific components in the RF chain to optimize performance, oftentimes going to more expensive IC processes to achieve the desired results. Others have taken the route of using more conventional IC processes, and trading-off analog performance requirements at the expense of increased DSP. In fact, the direct conversion receiver is a good example of an architecture where a successful competitive solution requires close interaction between analog high-frequency design, analog baseband design, and signal conversion technology.

Novel Low-cost Digital Architectures — Recently, some receiver structures that do not employ *I/Q* demodulation and A/D conversion when going from the IF regime to the baseband regime have appeared commercially. Many of these structures focus on distorting the IF signal (typically by limiting) and passing the output into a high-speed digital circuit, which produces digital information about the phase of the IF signal at symbol-rate intervals. This, together with the digital version of the filtered received signal strength information (RSSI), typically produced by a limiting amplifier followed by an A/D converter, produces a digital representation of the amplitude and phase of the received signal sample, allowing digital reconstruction in the form of $a_k e^{j\theta_k}$. Thus, only one A/D converter is required per received signal branch. Some receivers throw away the amplitude information, and operate solely with the phase information, thereby suffering a penalty in performance. Whether this penalty is deemed to be significant or not depends on the system under consideration and the minimum performance requirements. Techniques such as antenna diversity can offer significant performance benefits to offset the degradation of suboptimum detection schemes. Several researchers have realized such systems on dedicated low-power custom integrated circuits for cellular and cordless systems using these techniques [14, 16].

Key Digital Signal Processing Technology Issues for PCS

DSP Technology Trends

General-purpose, programmable, digital signal processor chips continue to climb up the performance curve, driven by market needs, technological advances, and architectural innovations. DSP chips have recently been the fastest growing segment of the semiconductor industry, with an annual growth rate of more than 35 percent, and will continue that growth pattern for the next several years, according to market research firm Forward Concepts Inc. This high growth for DSPs is fueled by expanding applications in modems, disk drives, digital cellular radio, digital cordless telephony, and emerging PCS applications.

Both fixed-and floating-point DSP technologies are rapidly evolving to meet the PCS market's shifting requirements. Today's DSP market breaks down to 80 percent fixed-point and 20 percent floating point (Forward Concepts). Broadly speaking, a floating point DSP has characteristically taken three to four times as much current per instruction as a fixed point processor (e.g., TI TMS320LC31). More DSP manufacturers are unveiling true 3V versions of both processors to address the low-power needs of battery-operated portables. New tiers of power-down modes allow careful control of the current consumed in the application being executed. New core architectures are taking shape slowly. And the demand for application-specific digital signal processors based on the DSP-core methodology is growing. This is bringing chip vendors, ASIC houses, and microcontroller suppliers into the DSP arena. Finally, complementing the obvious values of DSP's raw computing power, the availability of an ever-growing spectrum of silicon-level DSP-system integration and customization options has helped DSPs proliferate rapidly into vertical applications.

Today, DSP instruction sets are becoming more efficient in the execution of complex signal processing tasks. In many instances, they are implemented in a single machine cycle through the use of hardware accelerators or enhanced cores. Clocks are now distributed only to the required switching circuits within the DSP. Designs are stripped of all dynamic logic. The most advanced DSPs are manufactured today on 0.6- or 0.5-micron CMOS processes. This has significantly reduced the internal load capacitances, which in turn offer reductions in current consumption. Moreover, the reduction in geometry leads directly to increased performance from the same design as it can now sustain a faster cycle time and more functionality. DSPs are fully reaping the benefits of the advances in silicon-CMOS technologies.

The Role of DSPs, ASICs, and Microcontrollers

There are typically three digital "building blocks" available to the system designer: programmable general-purpose DSPs, microprocessors, and ASICs.

A general-purpose DSP is a software-programmable integrated circuit used for speech coding, modulation, security and integrity of information (through signal coding and scrambling), detection, equalization, and associated modem tasks such as frequency, symbol timing recovery, and gain control. Moreover, the DSP is also used for flexibility in application, and the ability to add new features with minimum redesign and re-engineering.

Microprocessors are usually used to implement protocol stacks, system software, and interface software. Most of the cellular systems require an 8- or 16-b microprocessor dedicated to these functions to address the software size and complexity. Microprocessors are better suited to perform the non-repetitive, control-oriented, input/output operations, as well as all housekeeping chores.

ASICs are used for various purposes in PCS. They are utilized for high-throughput tasks in the area of digital filtering, synchronization, equalization, and channel coding or decoding. The ASIC is often likely to also provide the glue logic to interface components. In some systems, the complete digital receiver is "poured" into silicon as an ASIC, coupled with a microcontroller, without a general-purpose DSP. ASICs have historically been

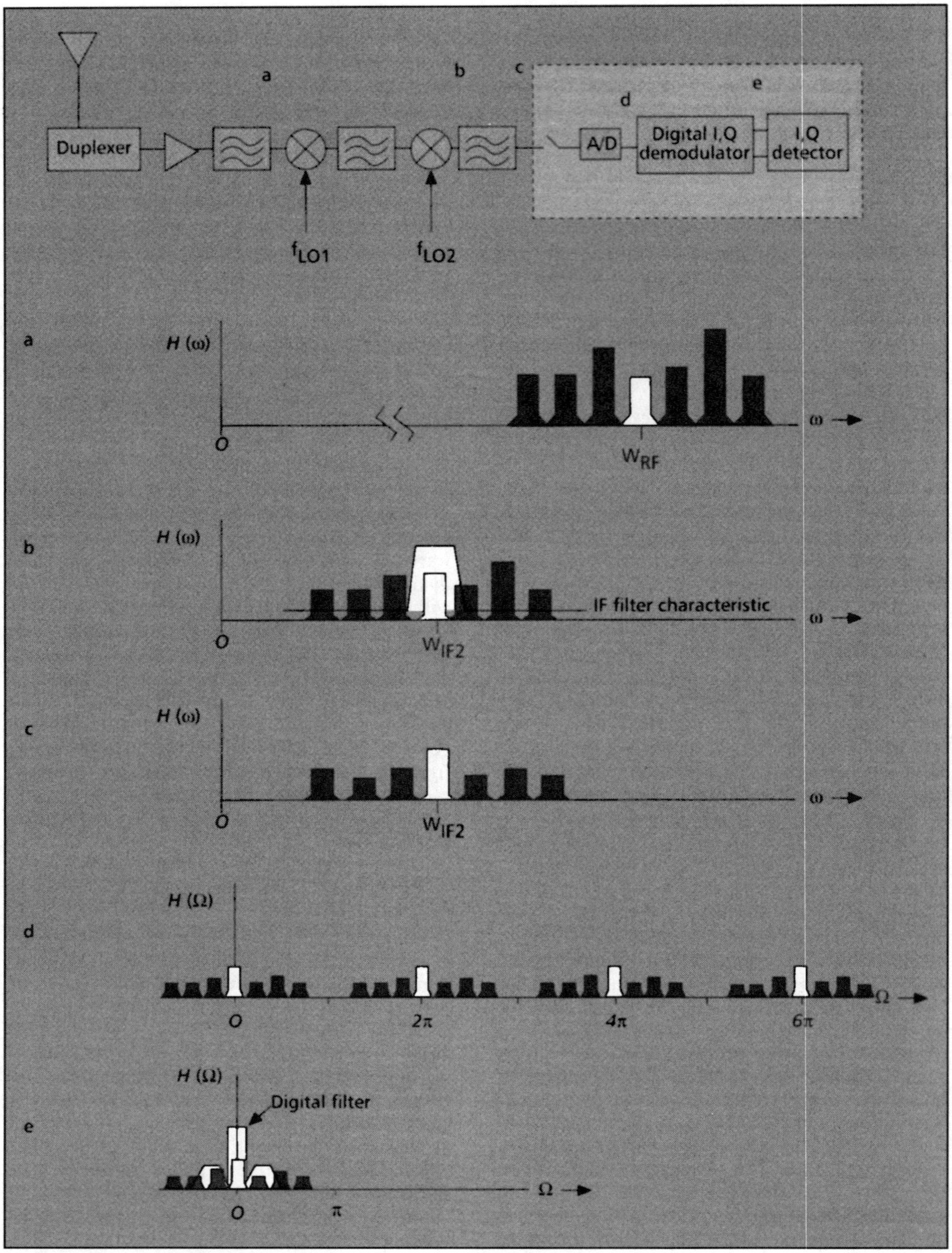

■ **Figure 5.** *Digital IF receiver principles.*

There are *typically three digital "building blocks" available to the system designer: programmable general-purpose DSPs, microprocessors, and ASICs.*

used because of their lower power dissipation per function. In certain applications like spread-spectrum systems, digital receiver designs require ASIC solutions in order to execute the wideband processing functions such as despreading and code synchronization. This is primarily because the chip-rate processing steps cannot be supported by current general purpose DSPs.

Over the last few years, as cellular and cordless manufacturers have brought out first-and second-generation digital cellular and cordless solutions, programmable general-purpose digital signal processors are slowly being transformed into accelerator-based DSP-microcontroller hybrids. This transformation is a result of the severe pressures being put on reducing power consumption. As firmware solutions become finalized, cycle-hungry portions of algorithms (e.g., equalizers) are being "poured into silicon," using various VLSI architectural ideas. This has given rise, for example, to new DSPs with hardware accelerators for Viterbi decoding, vectorized processing, and specialized phase-domain functions. They have been driven by applications in digital cellular telephones, such as GSM and NADC(IS-54), and cellular data (e.g., CDPD).

In the PCS arena, design costs and timescales will become more important that chip processing cost. In many markets, product life-cycles will be very short. To compete successfully, companies

In many markets, product life-cycles will be very short. To compete successfully, companies will need to be able to turn systems concepts into silicon VLSI very rapidly.

will need to be able to turn systems concepts into silicon VLSI very rapidly. This puts a premium on high-quality computer-aided design, and not least on system-level tools that enable algorithm and architecture exploration.

Tools

Design Automation Tools for analysis and modeling in the algorithm and architecture space are critical for efficient simulation and realization of DSP systems. These tools are characterized by a particular design methodology, which, for digital communications and signal processing systems, must be structured to cope with the time scale explosion problem faced by digital system designers. This time scale explosion problem in modeling can best be thought of as requiring a toolset to be able to handle time scales from 1 s (channel correlation time) to 1 ns (circuit gate delay). Allowing different levels of abstraction at different points in the time scale lead to efficient and powerful tools. This is typically done, for example, by partitioning modeling levels into functional, structural, and physical domains.

Exploring the Algorithm Space

Algorithms in DSP are comprehensively specified using block diagram descriptions. The blocks represent signal processing operations or functions, such as filters, whereas the interconnecting signals represent the data flow. Such block diagram specifications are used as input specification in state-of-the-art design environments for digital communication systems [3, 4, 15]. Preferably, these block diagrams are specified graphically in a hierarchical manner. Besides comprehensiveness, block diagram specifications allow for very structured simulation programs with a maximum reusability of components. For reusability purposes and for fast exploration of the algorithm space, blocks are parameterizable. Parameterizable simulations, in turn, allow for automatically evaluating different algorithmic variants based on the same block diagram structure.

As for any type of computation, a fundamental feature of block-diagram oriented simulation is the underlying processing paradigm. According to the nature of digital communication systems, multirate and variable-rate processing must be supported. Data-flow-oriented and time-driven approaches are applied in practice. The market is dominated by the design tools SPW (from Cadence) and COSSAP (from Synopsys), which mainly differ in the way they represent systems with multiple sampling rates and in the optimization strategies performed at the system level [3, 31]. While SPW uses a time driven approach, COSSAP is based on data flow semantics. See [33] for a lucid discussion of the merits of both tools. Recent work [4, 27] has demonstrated that heterogeneous specification languages are very powerful for that purpose. These may include various types of data flow and state diagram descriptions.

Exploring the Architecture Space

As discussed above, algorithm and architecture of digital communication systems are strongly interdependent. Therefore, joint optimization of both aspects is highly desirable. In the past, system design on the one hand and implementation on the other hand have been separate tasks. In general, different development teams have been in charge of these tasks. The system specifications had to be translated manually into implementations, which is an error prone process. To allow for a smooth interaction between the exploration of both the algorithm and implementation spaces, a tool supporting this two-fold exploration should be based on the same input specification, i.e., a block diagram description. This also offers a safe and rapid transition from algorithm design to architecture implementation.

Software-Based Architectures — As mentioned above, a wide range of functionality can be implemented on custom DSP processors. Once the choice of an appropriate target DSP processor has been made, the key optimization criteria for this type of implementations are throughput and RAM/ROM memory consumption.

Hence, design tools for software-based architectures must support the selection of a target platform and allow trade-offs between the above criteria. According to the requirement that both algorithms and architectures are based on identical input specifications, the task of such tools is to generate optimized real-time code from generic block diagram specifications [31]. The main issue is to exploit the algorithm-specific features of digital signal processors. Vectorization plays a very important role in this context. Most optimizations improve one of the implementation parameters at the cost of at least one other parameter. Therefore, optimization strategies must not always be applied. Instead, the user must be able to select the ones that lead to the prescribed implementation goals.

While memory consumption of an application can be determined exactly off-line, throughput cannot. Therefore, the generated program must be run on the actual hardware or on a hardware simulator to extract this parameter. Many signal processing algorithms, e.g., synchronization based on preamble structures, are in need of realistic input data. Consequently, the DSP hardware must be coupled to the simulation environment. This allows, for example, generation of the transmitted data and simulation of the channel distortion on that data on the simulation computer, while running the receiver implementation on the actual hardware. Exact comparisons between simulation model and DSP implementation can be carried out. If the DSP hardware is unavailable or not accessible via interprocessor communication to the simulation host, a DSP hardware simulator may also be coupled to the system simulation [43]. This coupling is powerful, as users can develop, debug, and functionally test DSP firmware within a system simulation environment.

Hardware-Based Architectures — In the case of applications requiring a very high data throughput or very large production volumes, special-purpose hardware in the form of ASICs is the preferred implementation choice. Many different implementation technologies such as full-custom layout, standard-cell-based design, and gate arrays in different feature sizes are available. Here one has to select based on production volume, required throughput, time to market, testability, power consumption, etc. The technology chosen for a cer-

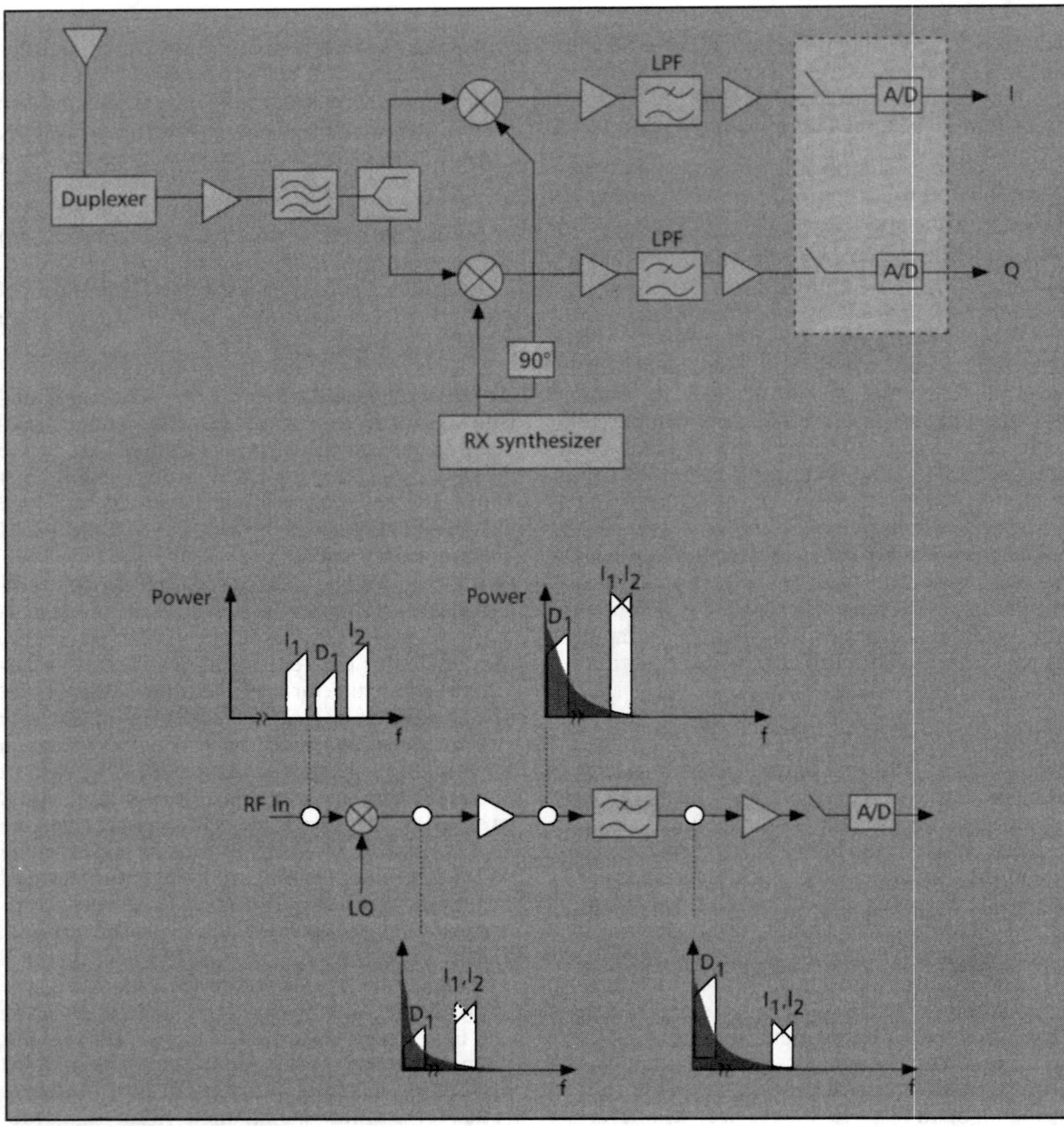

■ **Figure 6.** *Direct conversion receiver principles.*

In the case of applications requiring a very high data throughput or very large production volumes, special-purpose hardware in form of ASICs is the preferred implementation choice.

tain product may change during its life cycle (e.g., prototype realized as several field programmable gate arrays, consumer product realized as one ASIC).

To minimize nonrecurrent engineering cost and to maintain the advantages of using the high level block diagram tools, the communication system designer should use a standardized intermediate format as an interface to industry-standard hardware design tools supporting the path to hardware, such as logic synthesis [7], time-based discrete event simulation, and physical design. Hardware description languages (e.g., VHDL [11] or Verilog) serve best for this purpose. Today all major electronic design automation suppliers offer VHDL and Verilog based logic synthesis tools [2, 5, 21, 38, 40].

Today's algorithm design environments offer a fairly sophisticated way to generate synthesizable VHDL descriptions from the algorithm specification. There are many options for this transformation process. The resulting architecture may be processor-like or lowly multiplexed, consisting of several processing units and memories. A time-driven synthesis environment has a straightforward link to VHDL generation [17, 32, 39], but the algorithm design capabilities are restricted by the required configuration style. Data-flow-driven algorithm specification offers the possibility to generate hardware descriptions for different area and throughput requirements [44]. A library of reusable components which encompasses not only the behavior but also different implementation choices [12] in terms of area, throughput, and power consumption, is of great advantage in achieving a shorter time to market. Recently, design automation software suppliers have announced behavioral synthesis tools which allow a higher level of abstraction in the VHDL input than the current logic synthesis tools. Instead of register transfer level (RTL) code describing every storage cell and every operation, behavioral coding style allows an design description which is closer to that of software-based implementations. Special purpose synthesis tools for signal processing applications based on the programming language Silage [10] have been developed by IMEC, Belgium [28,26] and are in-part also available commercially [22].

Future Directions in Modeling and Tools

Whatever the route, a hierarchical approach to "pouring an algorithm into silicon" is giving rise to new

Algorithm-specific design know-how can be incorporated anywhere from the abstract algorithm description down to the actual physical architecture implementation.

tools and methodologies that give system designers the power to explore the communications, DSP, and VLSI design parameter space to make design trade-offs. These tools exploit the fundamental design concepts of hierarchy, modularity, regularity, and locality. In a nutshell, these concepts give rise to well-defined interfaces in a toolset that provide the appropriate level of abstraction to the designer at different levels of the design hierarchy. This is essential to master the ever-increasing complexity of the upcoming generations of personal communication systems. The CAD tools for system design integrating algorithm and architecture will further move toward abstraction, while offering the means for user interaction at any level. In particular, algorithm-specific design know-how can be incorporated anywhere from the abstract algorithm description down to the actual physical architecture implementation.

Conclusion

Our focus in this article has been on the new principles underlying the design and implementation of modern digital wireless terminals for PCS. We introduced the modern centers-of-gravity model, which we believe is emerging as the basis for the successful design and implementation of advanced digital communication systems. Central to this model are design principles that enable engineers to systematically derive digital receiver structures and explore algorithm and architecture trade-offs using sophisticated tools. DSP technology is critical in the implementation of these digital receiver structures efficiently. Finally, CAD tools for digital communications system design and design space exploration will be of crucial importance in the efficient execution of these designs. The tools and the technology are advancing at a phenomenal pace today, and it is this growth that will clearly be the enabler of universal low-cost wireless digital communications.

References

[1] A. Aghamohammadi, "Adaptive Phase Synchronization and Automatic Gain Control of Linearly-Modulated Signals on Frequency-Flat Fading Channels," Ph.D. Dissertation, Aachen University of Technology, 1989.

[2] "Booledozer User Manual," Altium, An IBM Company.

[3] B. Barrera and E. A. Lee, "Multirate Signal Processing in Comdisco's SPW," Proceedings of ICASSP, Toronto, April, 1991.

[4] J. Buck *et al.*, "Ptolemy: A Platform for Heterogenous Simulation and Prototyping," Proc. 1991 European Simulation Conf., Copenhagen, Denmark, June, 1991.

[5] "Synergy User Manual," Cadence Design Systems,555 River Oaks Parkway, San Jose, CA 95134, USA.

[6] "COSSAP User's Manual," Synopsys Inc., 700 E. Middlefield Rd., Mountain View, CA 94043, USA.

[7] A. J. de Geus, "Logic synthesis speeds ASIC design," *IEEE Spectrum*, 1989, no. 8, pp. 27-31.

[8] S. A. Fechtel and H. Meyr "Combined Equalization, Decoding and Antenna Diversity Combining for Mobile / Personal Digital Radio Transmission using Feedforward Synchronization," Proc. IEEE Int. Conf. Vehicul. Technol., VTC'93, Secaucus, NJ, USA, pp. 633-636, May 1993.

[9] S. A. Fechtel "Verfahren und Algorithmen der robusten Synchronisation f\"ur die Daten\"ubertragung \"uber dispersive Schwundkan\"ale (in German)," Ph.D. Thesis, Aachen Univ. Tech. (RWTH), Wissenschaftsverlag Mainz, Aachen, 1993.

[10] D. Genin *et al.*, "DSP Specification Using the Silage Language," Proc. ICASSP'90, 1990, pp. 1057-1060.

[11] "IEEE Standard VHDL Language Reference Manual, IEEE Std 1076-1987," (IEEE, March, 1988).

[12] R. Jain *et al.*, "Computer-Aided Design of a BPSK Spread-Spectrum Chip Set," IEEE J. of Solid State Circuits, Jan. 1992, vol. 27, no. 1, pp. 44-57.

[13] R. Jain *et al.*, "Computer-Aided Design of a BPSK Spread-Spectrum Chipset," IEEE Journal on Selected Areas in Communications, Special Issue on Computer-Aided Modeling, Analysis and Design of Communication Links, vol. 11, no. 3, pp. 348-358," April 1993.

[14] S. Kato *et al.*, "A Very-Low-Power High-Quality Signal Transmission Baseband LSIC for PCS: Advanced Signal Processing and One-Chip Integration," to appear in Proc. GLOBECOM 1994.

[15] J. Kunkel, "COSSAP: A Stream Driven Simulator," Proc. IEEE International Workshop on Microelectronics in Communications, Interlaken, Switzerland, March 1991.

[16] C. P. LaRosa and M. Carney, "A Fully Digital Hardware Detector for /4-QPSK," Proc. VTC 1992, pp. 293-297.

[17] L. Lundberg, "Generating VHDL for Simulation and Synthesis from a High-Level DSP Design Tool," EURO-VHDL'91 Proceedings 1991, pp.321-327, Institute for Microelectronics, Stockholm, Sweden, Sept. 1991.

[18] J. L. Massey, "Coding and Modulation for Code-Division Multiple Accessing," Proc. 3rd Int. Workshop on DSP Techniques Applied to Space Communications, ESTEC, Noordwijk, The Netherlands, pp.. 3.1-3.17, Sep. 1992.

[19] J. L. Massey, "Deep-Space Communications and Coding: A Marriage Made in Heaven," Advanced Methods for Satellite and Deep Space Communications, Bonn, Germany, pp. 1-17, Sept. 1992.

[20] J. L. Melsa and D. L. Cohn, "Decision and Estimation Theory," (McGraw-Hill, 1978).

[21] "Autologic User Manual," Mentor Graphics Corporation, 1001 Ridder Park Drive, San Jose, CA 95131, USA.

[22] "DSP Station User's Manual," Mentor Graphics, 1001 Ridder Park Drive, San Jose, CA 95131, USA.

[23] H. Meyr and G. Ascheid, "Synchronization in Digital Communications-vol. 1," John Wiley, 1990.

[24] H. Meyr *et al.*, "Synchronization in Digital Communications-vol. 2," [John Wiley (in preparation)].

[25] M. L. Moher and J. H. Lodge, "TCMP-A Modulation and Coding Strategy for Rician Fading Channels," *IEEE JSAC*, vol. 7, no. 9, pp. 1347-1355, Dec. 1989.

[26] S. Note *et al.*, "Cathedral-III: Architecture-Driven High-Level Synthesis for High Throughput (DSP) Applications," in Proc. of the 28th ACM/IEEE Design Automation Conference, 1991, pp. 597-602.

[27] M. Pankert *et al.*, "Dynamic Data Flow and Control Flow in High Level DSP Code Synthesis," Proc. of ICASSP, Adelaide, South Australia, April, 1994.

[28] L. Philips *et al.*,"Silicon Integration of Digital User-end Mobile Communication Systems," in Proc. ICC '93, May 1993, pp. 212-216.

[29] J. G. Proakis, Digital Communications, (McGraw-Hill, 1989).

[30] D. Pucknell and K. Eshragian, "Basic VLSI Design: Systems and Circuits," (Prentice Hall, Australia, 1988).

[31] S. Ritz *et al.*, "High Level Software Synthesis for the Design of Communication Systems," *IEEE JSAC*, Special Issue on Computer-Aided Modeling, Analysis and Design of Communication Links, Vol. 11, No. 3, pp. 348-358," April, 1993.

[32] "SPW User's Manual," Cadence Design Systems, 919 E. Hillsdale Blvd., Foster City, CA 94404, USA.

[33] K. S. Shanmugam, "Simulation and Implementation Tools for Signal Processing and Communication Systems, " *IEEE Commun. Mag.*, July 1994, vol. 32, no. 7, pp. 36-40.

[34] I. Shimizu *et al.*, "New Digital Mobile Radio Technologies," *NTT Review*, Jan. 1992, pp. 64-70.

[35] R. Subramanian, "Synchronization Systems for Spread-Spectrum Receivers," Ph.D. Thesis, University of California, Berkeley, Dec. 1991.

[36] R. Subramanian, "Digital Synchronization Subsystems for Spread-Spectrum Receivers," Proc. IEEE Workshop on VLSI in Communications, Sept. 15-171, 1993, Lake Tahoe, CA, pp. 22-34.

[37] "Behavioral Compiler User Guide," Synopsys Inc., 700 E. Middlefield Rd., Mountain View, CA 94043, USA.

[38] "Design Compiler Reference Manual," Synopsys Inc., 700 E. Middlefield Rd., Mountain View, CA 94043, USA.

[39] P. B. Tjahjadi, *et. al.*, "Vanda -a CAD System for Communication Signal Processing Circuits Design," in VLSI Signal Processing IV, eds. H. Moscovitz, K. Yao, and R. Jain (IEEE Press, 1990).

[40] "SilcSyn User Manual," Viewlogic Systems, Inc., 293 Boston Post Road West, Marlboro, MA 01752-4615, USA.

[41] A. J. Viterbi, "Wireless Digital Communication: A View Based on Three Lessons Learned," *IEEE Commun. Mag.*, vol. 29, no. 9, Sept. 1991, pp. 33-36.

[42] N. Weste, and K. Eshraghian, "Principles of CMOS VLSI Design: A Systems Perspective," 2nd Ed., (Addison Wesley, 1993).

[43] P. Wilhelmsen *et al.*, "Simulating a Programmable DSP at the System Level," DSP Applications, pp. 6-18, Nov. 1993.

[44] P. Zepter and T. Grotker "Generating Synchronous Timed Descriptions of Digital Receivers from Dynamic Data Flow System Level Configurations," Proc. of European Design And Test Conference (poster session), Feb. 1994.

Broadband RF Stage Architecture for Software-Defined Radio in Handheld Terminal Applications

Hiroshi Tsurumi and Yasuo Suzuki, IEICE

ABSTRACT Broadband RF is a general-purpose common RF stage for every standard within a set of RF bands. The RF stage architecture presented in this article is suitable for software-defined radios.

Various mobile communication standards will continue to exist even in the future, so it is important to develop a handheld terminal that can be used as a multimode transceiver. Such a terminal not only enables regional and global roaming, it also reduces the costs of introducing new technology over antiquated legacy technology. Software-defined radio (SDR) has the flexibility to be a multistandard connection with replacement of the application program, such as modem, equalizer, channel codec, synchronization, and so on. The multimode SDR hardware should consist of a reconfigurable baseband digital signal processor (DSP) or field programmable gate array (FPGA) and a broadband analog radio frequency (RF) front-end. However, it is difficult to realize the broadband analog RF front-end that is fit for the ideal SDR.

In terms of the RF section, the conventional heterodyne transceiver architecture requires some analog filters fitted to the carrier frequency and channel bandwidth of each of the communication standards. The disadvantage of this architecture is that it is difficult to make it broadband, in part because these filters are fixed narrowband passive components not fit for a broadband system with multimode operation.

The direct conversion approach is one of the candidates for such broadband requirements since this architecture has no image response, and thus the frequency-fixed image rejection filters can be eliminated. Furthermore, in principle, an anti-aliasing baseband filter with a frequency adjustment function can be embedded in a large scale integrated (LSI) circuit chip. Therefore, the direct conversion architecture is suitable for a broadband transceiver architecture, and is also a promising candidate for realization of a single-chip flexible SDR.

This article first defines the conceptual scheme of a handheld SDR terminal, and then describes how the direct conversion principle is applied to the SDR. The discussion focuses on receiver hardware implementation and systematic control.

A Conceptual Scheme of a Handheld SDR Terminal

Figure 1 shows a conceptual scheme of the handheld SDR terminal including the broadband RF stage. The key subjects considered for putting the SDR to practical use as a handheld terminal are the applications program, radio function library, software specification language, digital radio processor, and broadband RF stage.

Application Program

The AP is written in some software specification language that describes the radio architecture and performance using a radio function library. APs are prepared for a specific radio standard, for example, one for GSM, one for cdma-One, an AP for IMT-2000, and so on. The performance of the handheld SDR is easily reconfigured by rewriting the AP to change the use of the library. In Fig. 1, the AP is downloaded over the air from a central or base station to the SDR terminal.

Radio Function Library

This is a software set that expresses basic radio function. The examples of this library are control programs for hardware such as analog-to-digital converter (ADC) and digital-to-analog converter (DAC). Software programs for modem function embedded in DSP are another example. In this case, the changeable input parameters include resolution and sampling rate corresponding to modulation scheme, bandwidth, and multiplexing.

Software Specification Language[1]

This will be a high-level language simply to express radio characteristics and specifications for each mobile communication standard. Using this language, the AP programmer will write the physical-layer specification such as transmission power, data rate, modulation scheme, channel codec, and speech codec.

Digital Radio Processor

The DRP is a general-purpose hardware functional module or LSI configuration, which can be used for several standards. The DRP typically includes ADC, DAC, and FPGA and/or DSP hardware. FPGAs and/or DSPs express digital down/up converters, digital filters, and modem functions. Modulation scheme, bandwidth, and transmission rate can be controlled by changing DRP parameters. In general, two architectures will be available where one is intermediate frequency (IF) input configuration and the other baseband input. The configuration should be selectable according to the modulation scheme.

The receiver specifications should be changed by changing parameters. It is preferable to be able to change ADC resolution for each standard, for example.

Broadband RF Stage

A general-purpose common RF stage is needed for standards with different RF specifications such as carrier frequency, bandwidth, modulation scheme, and transmission power. In the following, the architecture of the broadband RF stage for handheld SDR is discussed.

Reprinted from *IEEE Communications Magazine,* Vol. 37, No. 2, pp. 90–95, February 1999.

RF Stage Architecture for SDR

Figure 2 shows the schematic view of the multimode radio using a conventional multihardware receiver and SDR in terms of channel selection.

For multistandard applications, the multichannel receiver in Fig. 2a is equipped with one receiver chain for each standard. It selects the channel with different carrier frequency and different channel bandwidth using fixed analog-defined channel selection filters.

On the other hand, in the case of SDR, the received multichannels are fed to the single wideband analog stage (Fig. 2b). Then all channels are converted to digital by one wideband high-speed ADC. The desired channels are then selected from the digitized multichannels by the software-defined channel selection filters of the DSP. This latter architecture has flexibility in that the performance of filter can be selected by changing some parameters in the DSP stage.

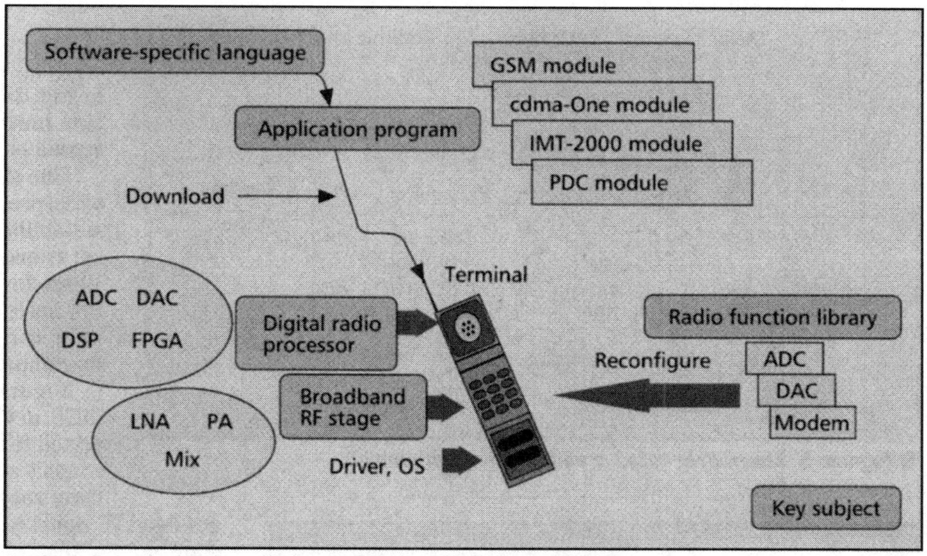

■ **Figure 1.** *A conceptual scheme of a handheld SDR terminal.*

The SDR in Fig. 2b requires the broadband RF stage which ideally feeds the signals over the entire system bandwidth to the ADC. The relevant system bandwidths of significant mobile communication standards are shown in Table 1.

Figure 3 shows an example of a receiver chain of the heterodyne architecture. This architecture requires frequency-dependent passive components, including dielectric filters in the RF stage and surface acoustic wave (SAW) filters in the first IF stage for image suppression. A ceramic or crystal filter is also needed in the second IF stage for channel selection with another crystal for the second local oscillator.

In general, the center frequency and bandwidth of these filters cannot be variable, but are fixed to the radio standard specification. Furthermore, the bandwidth of the passive filter is restricted by the quality factor and usually narrower than the entire system bandwidth. Therefore, the heterodyne approach is not attractive for the wideband RF configuration. It is possible to overcome the functional limitation on wide bandwidth and programmability using switched capacitor filter banks and precision direct digital synthesis. However, the size and weight penalties incurred may be acceptable for base stations, but are unacceptable for handsets. Similar restrictions

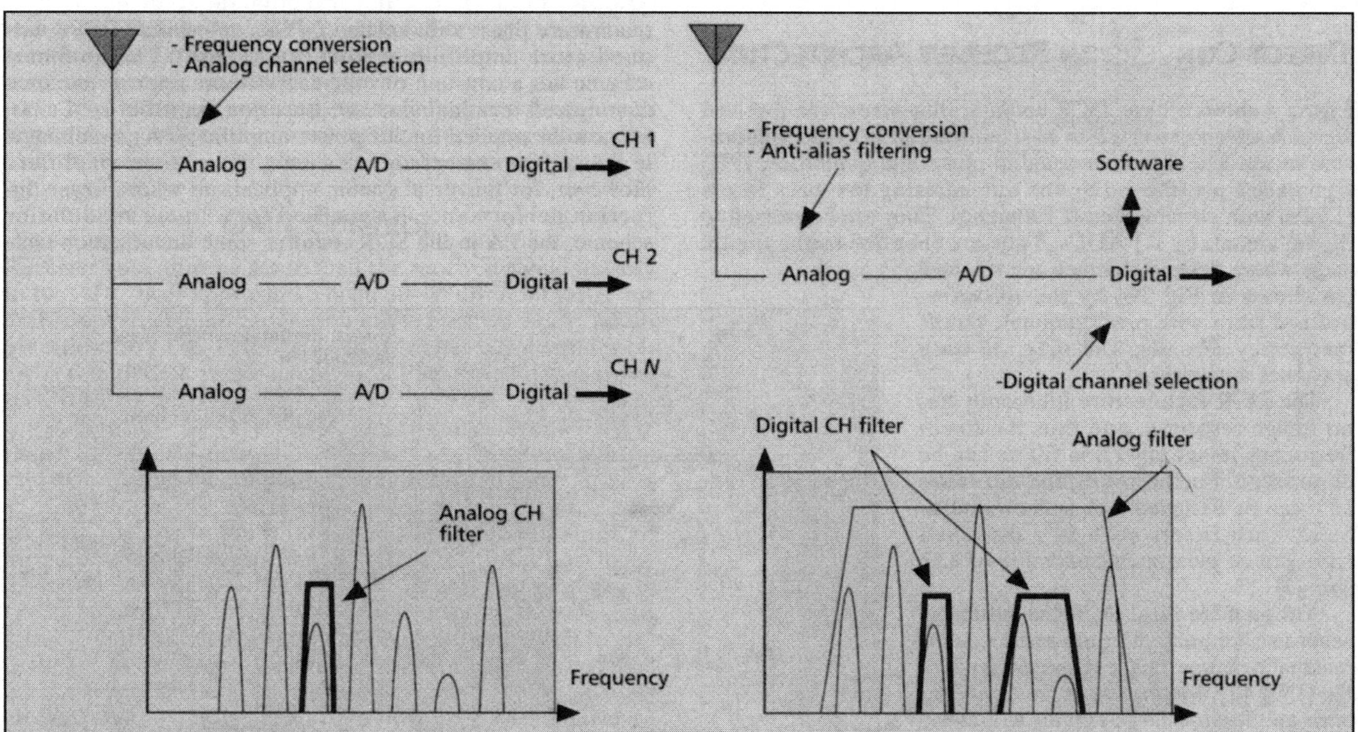

■ **Figure 2.** *Comparison of channel selection between conventional multihardware radio and SDR.*

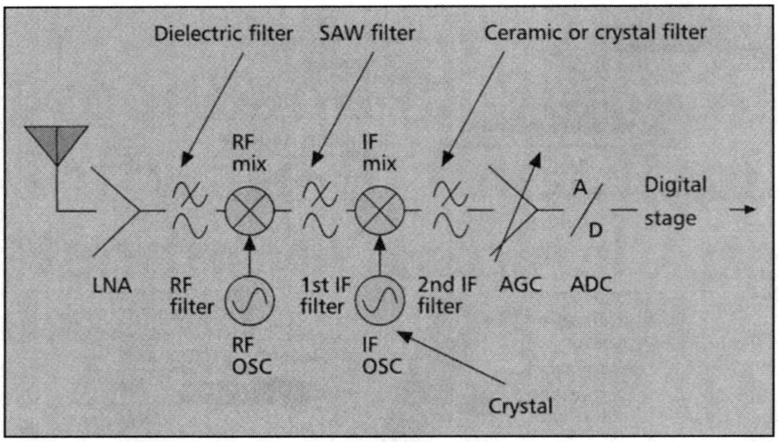

■ **Figure 3.** *Heterodyne receiver with passive components.*

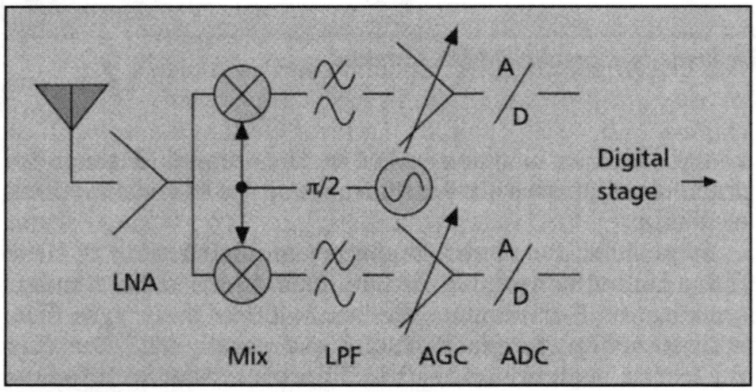

■ **Figure 4.** *A basic DCR configuration.*

exist in the transmission chain in the heterodyne architecture with fixed frequency IF and RF filters.

One of the candidates to realize such a wideband RF section for SDR is the "direct conversion" or "zero-IF" receiver [2].

Direct Conversion Receiver Architecture

Figure 4 shows a basic DCR configuration where the received signal is downconverted to baseband directly by the quadrature mixer. The downconverted in-phase and quadrature (IQ) signals are prefiltered by the anti-aliasing low-pass filters (LPFs) with variable cutoff frequency. They are converted to digital signals by IQ ADCs. They are then fed to the digital stage where desired channels are selected (as shown in Fig. 2b) by the software-defined filter with programmable cutoff frequency. Finally, the original data sequence is recovered.

The DCR architecture inherently has no image response, and thus the fixed-frequency image rejection filters can be eliminated. Furthermore, the anti-alias LPF can be designed with active variable-bandwidth filters such as a switched capacitor or gyrator embedded in an LSI chip [2].

Although an ideal DCR has no image response, imperfect components cause residual or latent image responses at zero Hz (DC) [2]. Furthermore, in-band linearly modulated large signals will cause second-order nonlinear distortion around DC in the downconverter mixer, especially for multichannel applications as shown in Fig. 2b [3]. However, a system-level compensation method for the above problems will be presented later.

The direct conversion principle has attracted widespread attention in recent years because of its suitability for achieving a miniature low-cost handset transceiver, since bulky and high-cost passive filters for both transmitter and receiver chains are not necessary. Recently, the direct conversion principle is in the spotlight again for another reason: its compatibility with multiple standards [4].

Figure 5 shows another configuration of the DCR in which the received signal is first direct-converted to baseband, where the entire system band is roughly selected by LPF for anti-alias purposes. It is then reconverted to the programmable IF. The resultant IF signal is band-limited. Therefore, it may be directly digitized and then digitally downconverted to the baseband to select the desired channels. In this configuration, the digital downconverter by Harris [5] may be well suited to the digital quadrature demodulator (QDEMOD) and digital filter section. In Fig. 5, a desirable frequency can be selected for the IF, where an ADC of suitable power consumption for portable terminals can be operated. The digitized IF signal can also be demodulated using an IF demodulator such as a pulse-count detector, IF differential detector, or IF digital discriminator. In the case of requency shift keying (FSK) reception, the ADC may be replaced by a hard-limiter amplifier to reduce power consumption.

A Direct Conversion Receiver for SDR Applications

A multimode SDR terminal will be used under both nonlinear modulation schemes such as Gaussian minimum shift keying (GMSK) and FSK, and linear modulation schemes such as quadrature phase shift keying (QPSK), π/4-shifted QPSK, and quadrature amplitude modulation (QAM.) The nonlinear scheme has a constant envelope. It is more appropriate for a customized terminal since a saturation amplifier or C-class bias can be applied for the power amplifier (PA), resulting in higher power conversion efficiency than linear amplifiers. However, for universal system applications where lower-distortion performance is specified for a linear modulation scheme, the PA in the SDR requires some linearization tech-

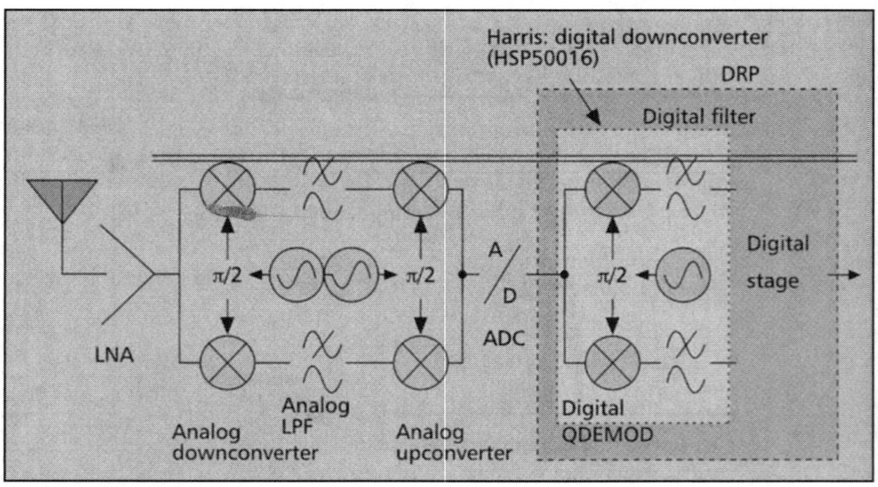

■ **Figure 5.** *An upconversion DCR configuration.*

nique. The feed-forward or Cartesian loop feedback approach [4] linearizes the amplifier. Alternatively, one may use a linearizer-ON/OFF switch or function that selects the most appropriate linearizer (i.e., reconfiguration).

A similar technique must also be applied in a receiver chain, including a low noise amplifier (LNA), mixer, and baseband LPF that also suffer from nonlinear distortion. In the receiver chain, the interference input distorts the desired signal even in the active LPF. Therefore, compensation will be required.

The following presents the compensation method in an adaptive DCR for a universal system, focusing on the receiver chain. The items that should be compensated for in a multimode SDR are as follows:

- Analog stage (LNA, mixer, LPF)
  - Signal saturation (i.e., gain control)
  - Nonlinear distortion
  - DC offset
- Baseband or digital stage
  - IQ gain mismatch
  - IQ phase mismatch
  - DC offset

Baseband stage compensation is achieved generally by DSP algorithms using received signal sequences stored in memory [7].

In contrast, RF analog stage performance depends on circuit linearity, which defies the laws of physics. A device-level approach results in excessive demands on the circuit in order to make it meet every standard. This increases terminal cost and size. Therefore, system-level compensation techniques are desirable.

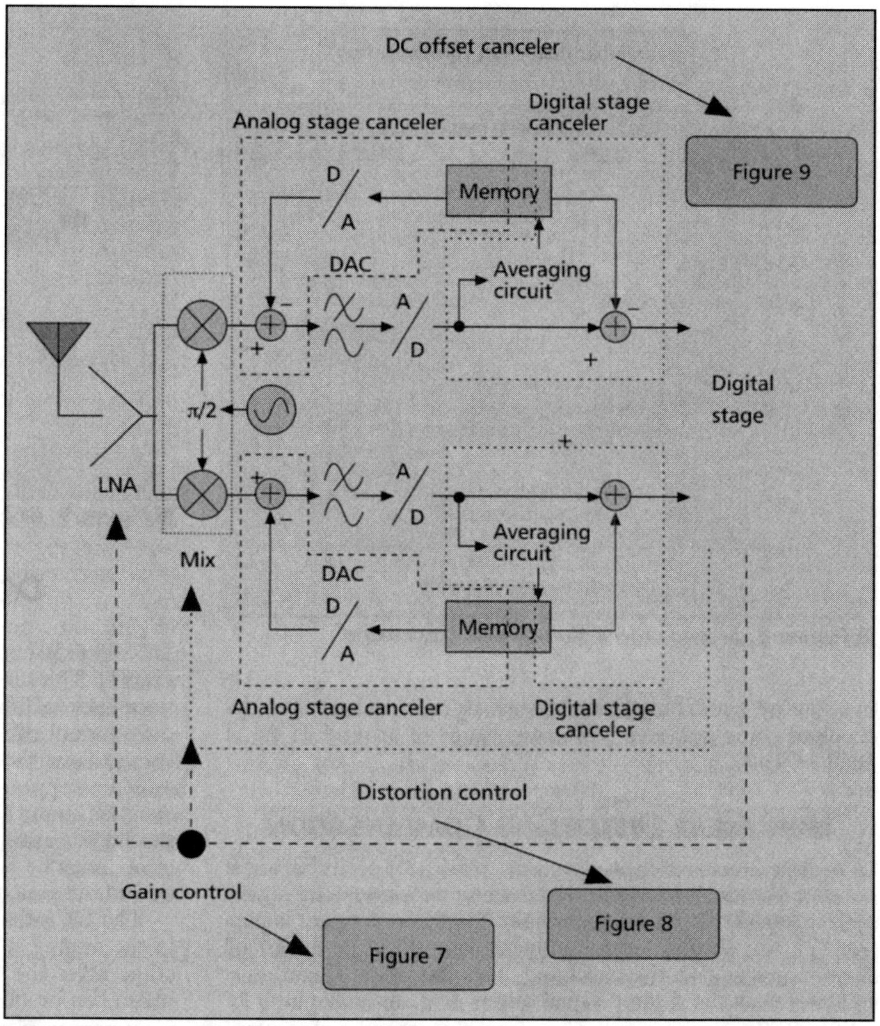

■ **Figure 6.** *System-level compensation for the RF stage.*

In the following, a system-level compensation approach that overcomes distortion or saturation is presented, using the basic DCR configuration shown in Fig. 4 as an example.

GAIN CONTROL

Receiver gain must be set to feed the received signal linearly to the ADC over approximately an 80–90 dB input range for practical applications. Saturation in the baseband LPF, which precedes the AGC as shown in Fig. 4, seriously degrades bit error rate (BER) performance for a linear modulated signal such as QPSK or QAM.

For example, a received dynamic range over 90 dB is required for PHS terminals, so RF stage "gain control" is valuable to avoid the saturation in the baseband stage [8].

As shown in Fig. 6, the gain-control function is distributed in the LNA, in the mixer, and the baseband stage to increase receiver dynamic range. Each circuit gain is changed discretely according to the received signal power detected in the baseband digital received signal strength indicator (RSSI). Both the LNA and mixer operate in high-gain and low-gain mode to produce four different gain modes. The optimum gain mode is selected automatically to realize sufficient channel-to-noise ratio (C/N) over the entire receiver input range.

The measured BER for the Personal Handyphone System (PHS) terminal is shown in Fig. 7. The optimum gain mode

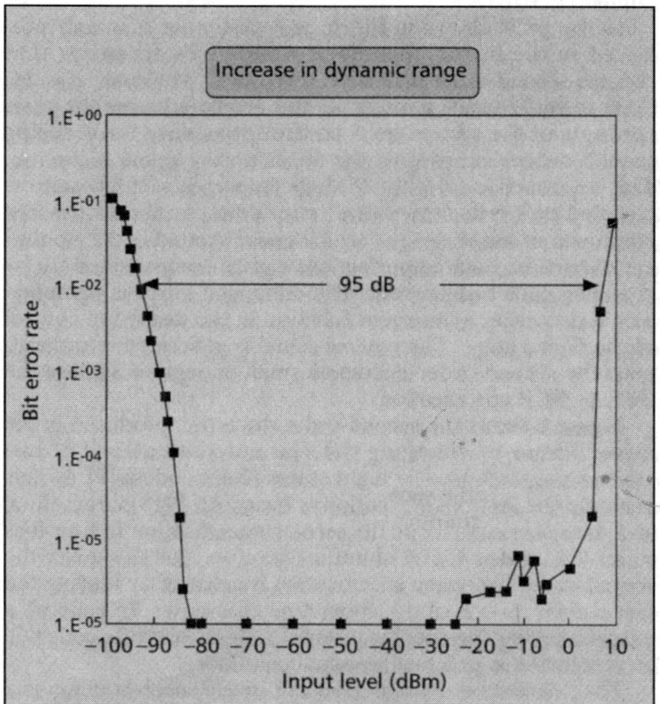

■ **Figure 7.** *BER performance with RF gain control.*

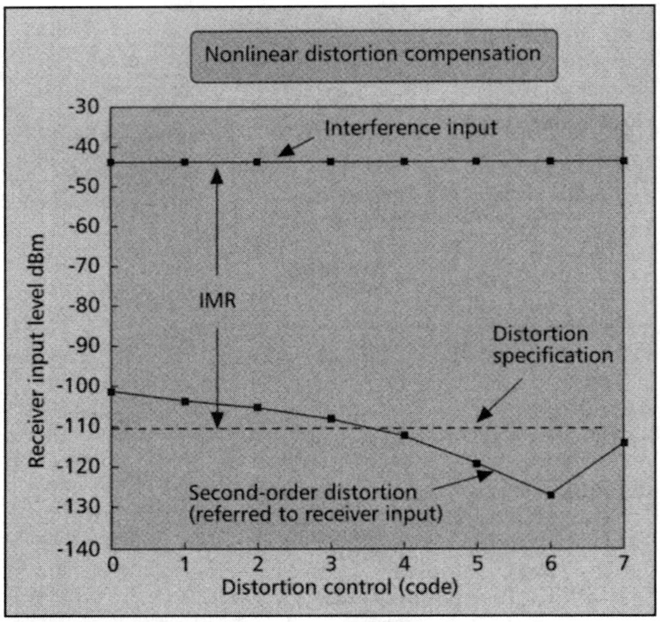

■ **Figure 8.** *Second order distortion with bias control.*

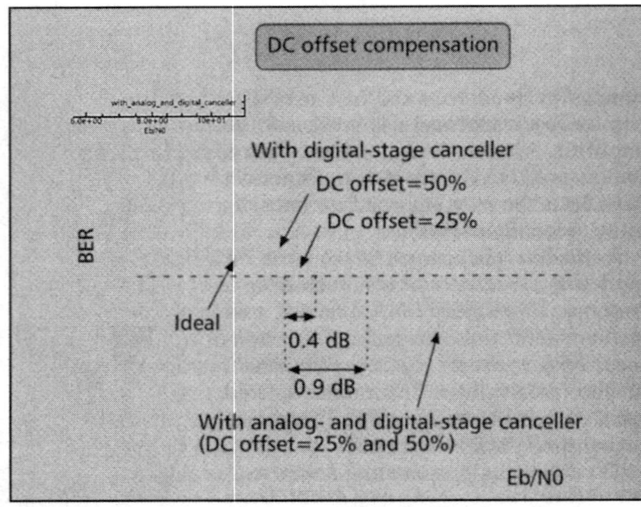

■ **Figure 9.** *BER performance with DC offset canceller.*

is selected according to the input signal level so that the receiver chain achieves a dynamic range of around 95 dB at BER of 10^{-2}.

Nonlinear Distortion Compensation

In mobile communication systems, the radio terminal must receive the desired signal even when an undesired signal approximately 60 dB stronger than the desired signal is present [3]. Any second-order nonlinearity produces an undesired distortion signal in the baseband. This distortion signal must be lower than the desired signal output (e.g., approximately 15 dB for QPSK modulation). Therefore, the second-order intermodulation rejection ratio (IMR) for the mixer must be 75 dB in linear modulation systems such as PHS or personal digital cellular (PDC).

In the DCR shown in Fig. 6, such distortion is mainly produced in the mixer; therefore, a highly linear mixer that reduces second-order distortion is required. However, it is difficult to realize such a mixer, so this article presents the basic concepts of the system-level control procedure for reducing second-order distortion in the mixer during signal reception. This approach is tolerant of slight imperfections in emitter-coupled pair transistors and other semiconductor devices employed in mixer designs which cause second-order nonlinear distortion. Such imperfections can be compensated for by changing each bias current. The enhanced mixer is equipped with a distortion adjustment function in the distortion control of the digital stage. The control signal is generated in order to keep the second-order distortion small enough to achieve the IMR or BER specification.

Figure 8 shows the second-order distortion produced in the mixer circuit by changing the bias current balance of two emitter-coupled pairs in eight states (3-bit codes) [9]. In this example the IMR specification is based on PHS parameters, and the mixer satisfies the distortion specification in four bias states (i.e., codes 4, 5, 6, and 7). Therefore, for this mixer the second-order distortion specification is satisfied by leading the bias current to one of the above four bias states. In general, a mixer satisfying the specification for at least one bias code will be acceptable in practical terminal conditions.

The "distortion control" concept in the receiver chain is a powerful means for commercialization of SDR with the DCR architecture.

DC Offset Compensation

DC offset is mainly caused by "self-mixing" in the mixer circuit [7]. This circuit is divided into two components: the time-invariant offset inherent to the mixer circuit and the time-variant offset caused by local-leakage reflection. The latter varies with changes in the environment. The DC offset must be approximately 30 dB lower than the desired signal at the ADC input for QPSK differential detection. In the case of the DCR configuration shown in Fig. 6, DC offset from the mixer must be reduced to several microvolts. This is unrealistic without cancellation.

The DC offset canceler removes the time-invariant offset in the analog stage and the time-variant offset in the digital stage after the ADC. When a small signal is received, DC offset can be eliminated by using only a digital-stage canceler. However, if a large desired signal is received, the DC offset will saturate the ADC. In this case, if the receiver gain is set low to avoid overflow, the receiver total noise figure (NF) is degraded. The analog-stage canceler allows the receiver gain to be kept high by removing the inherent DC offset produced in the RF stage, thus not causing NF degradation. A DC offset canceler, shown in Fig. 6, consists of analog-stage and digital-stage cancelers with memories for storing the detected DC offset data. The analog-stage canceler consists of analog subtractors at the input port of the LPFs and DACs to supply DC data to the analog subtractors. The digital-stage canceler consists of subtractors following the ADCs. Averaging circuits detect the DC offset in the output of the ADCs.

The analog-stage canceler functions as follows. Before TDMA burst reception, the time-invariant DC offset in the respective RF gain mode is estimated with LNA input termination and stored in memory tables. During TDMA burst reception the stored DC value is read out with the corresponding gain mode and fed back to the analog subtractor through the DAC to compensate for the time-invariant DC offset.

The digital-stage canceler is activated during TDMA burst reception to compensate the time-variant DC offset. Digitized signals, including DC offset, are accumulated during a period of one or two time slots. The accumulated value is then divided by the time period to obtain the mean value. For example, in the case of QPSK modulation the mean value can be regarded as the DC offset, because the mean value for the pure QPSK signal approaches zero with sufficiently long time averaging. The DC value is held in the latch circuit and then subtracted from the input signal during the following receiving slot, while the averaging circuit is activated again to detect the DC offset in the slot.

Figure 9 shows the BER simulation results with DC offsets of 25 and 50 percent relative to the full scale of the ADC. Employing only the digital-stage canceler, the Eb/No degradation from the ideal at a BER of 10-2 is 0.6 dB for 25 percent DC offset and 0.9 dB for 50 percent DC offset. On the other hand, employing both the digital- and analog-stage cancelers, the Eb/No degradation is only 0.4 dB.

NF design in the DCR tends to be strict, so even slight improvement in the required Eb/No is quite effective with regard to total receiver design.

Conclusion

This article describes the basic conceptual scheme of a handheld SDR terminal and broadband RF stage using direct conversion architecture. The focus is on receiver hardware implementation with system-level control to overcome signal saturation, nonlinear distortion, and DC offset.

The direct conversion architecture is suitable not only for broadband transceiver applications but also for single-chip receivers because it does not need passive filters. The direct conversion principle minimizes area and cost of the front-end in the portable SDR terminal for the near future.

System-level compensation presented in this article is well-suited to the software-controlled SDR concept. It is tolerant of slight imperfections in analog circuits and semi-conductor devices, which is necessary for mass production and commercial success of the SDR.

References

[1] B. Robinson, "Software radio: The standards perspective," Software Radio Wksp., Brussels, Belgium, May 1997.

[2] A. A. Abidi, "Low-power radio-frequency ICs for portable communications," *Proc. IEEE*, vol. 83, no. 4, 1995, pp. 544–69.

[3] C. Takahashi *et al.*, "A 1.9 GHz Si direct conversion receiver IC for QPSK modulation systems," *TEICE Trans. Elect.*, vol. E79-C, no. 5, 1996, pp. 644–49.

[4] A. Fernandez-Duran *et al.*, "Zero-IF receiver architecture for multistandard compatible radio systems, Girafe Project," *Proc. 46th IEEE Veh. Tech. Conf.*, vol. 2, 1996, pp.1052–1056.

[5] Harris Semiconductor, "HSP50016 Digital down converter," data sheet, 1996.

[6] D. W. Bennett, J. R. MacLeod, and J. P. McGeehan, "Broadband high dynamic-range RF transmitter technology for flexible multi-standard radios," *Proc. ACTS Mobile Commun. Summit '98*, Rhodes, Greece, June 1998, pp. 127–32.

[7] J. K. Cavers and M. Liao, "Adaptive compensation for imbalance and offset losses in direct conversion transceivers," *Proc. 41st IEEE VTC*, 1991, pp. 578–83.

[8] H. Tsurumi *et al.*, "Design study on RF stage for miniature PHS terminal," *TEICE Trans. Elect.*, vol.E79-C, no. 5, 1996, pp. 629–35.

[9] T. Yamaji, H. Tanimoto, and H. Kokatsu, "An I/Q active balanced harmonic mixer with IM2 cancelers and a 45° phase shifter," *IEEE ISSCC '98 Dig. Tech.*, Paper SP 23.3, 1998.

Additional Reading

[1] J. A. Wepman, "Analog-to-digital converters and their applications in radio receivers," *IEEE Commun. Mag.*, vol. 33, no. 5, May 1995, pp. 39–45.

Trends in Silicon Radio Large Scale Integration: Zero IF Receiver! Zero I & Q Transmitter! Zero Discrete Passives!

Jan Sevenhans, Bart Verstraeten, and Silvio Taraborrelli, Alcatel

ABSTRACT

After 10 years of advances in silicon RF integration, what used to be an art is becoming a "normal practice." Historically, RF-design was the art of s-parameters, shielding, impedance matching, and standing wave ratios. Modern silicon RFICs are designed using the same SPICE-like tools as in low-frequency analog ICs with the addition of important software for system-level simulation and mixer circuit noise analysis. New RF design practices away from the 50 ohm culture, novel chip architectures, and powerful technological advances will drive radio integration toward the ultimate single-chip phone. The obstacles in this quest are high system requirements on noise figure, substrate cross talk, and parasitic coupling, not the silicon IC technology.

RFIC BEGINNINGS

Recently, a new and exciting RF circuit capability came to light as radio front-end integrated circuits have been developed in silicon bipolar and BiCMOS technologies. Such chips are used widely in cellular phones, Global Positioning System devices, and similar "low-GHz-band" products. Although the necessity of having many components on a board is not likely to vanish in the near future, highly integrated radio parts represent an important new niche for ICs.

The original source for the creation of this new RFIC niche can be traced to two events that took place about 10 years ago. First, the progress in silicon bipolar technology employing polisilicon emitters yielded transistors with approximately 10 GHz transition frequency. Since these devices were readily fabricated in low-cost silicon production facilities, the IC industry pushed the development of a new market for this technology. Second, the worldwide telecommunication industry became strongly motivated to spread the use of the mobile digital phone. This device became a consumer fashion item as the Walkman or the Discman did in previous years. For some people the cellular phone was just a gadget, for others a status symbol, but for most it became a useful tool in daily work and life. Starting with these conditions several synergetic technological and economic forces were put in motion to rapidly create the modern advanced products with which we are familiar today.

In 1989 we already had the "Groupe Special Mobile" and ETSI specifications for the GSM system. Years of work in system definition done by telecommunication engineers, the European commission, and national authorities had produced the dictionary and grammar of the GSM language. In parallel, the semiconductor industry was working on sub-micron bipolar, CMOS and BiCMOS technologies. In the late eighties, when we started work on the first CMOS baseband analog and digital chips for the first generation GSM handset, 1.2 μm was the minimum dimension of the CMOS transistors we had available. However, the technology road map for the following five years was showing sub-micron silicon technologies moving from solid state research laboratories to reliable engineering and production. The silicon technology, the GSM system engineering, the ASIC design, the system software, and the commercial market for cellular phones were growing in a harmonic environment orchestrated by the European commission and the telecommunications industry.

RF-design for applications below 2 GHz moved from the printed circuit board and discrete components to large-scale integration in silicon technology. The time was ripe for the creation of a new generation of engineers: RF ASIC or "silicon MMIC" designers. In the early eighties, the GHz spectrum had been GaAs territory. By the end of the decade, the silicon technology conquered the cellular market for the transmit/receive mixers while GaAs technology kept its presence in low-noise amplifiers (LNA) and power amplifiers (PA). The above 10 GHz, high-level applications for military and space remained primarily in GaAs technology. However, at the horizon, one could see how the silicon bipolar RF technology with self aligned emitters and selectively implanted collectors was moving up to 30 GHz transition frequencies and above while the use of the silicon germanium (SiGe) base could push that figure of merit to 100 GHz [1].

The production volumes for cellular phones had to grow massively to match the demand illustrated in Fig. 1. Today, the major successful companies for telecommunications consumer products are or soon will be manufacturing twenty million or more cellular phones per year. In Europe and other places, the number of digi-

Series Editor: M. Banu.

Reprinted from *IEEE Communications Magazine,* Vol. 38, No. 1, pp. 142–147, January 2000.

tal cellular phones has reached a level that rivals the number of plain old wired telephones. The GSM success story has played a major role in this matter. Meanwhile, similar capabilities are being developed in Japan and the U.S. The worldwide compatibility of the Universal Mobile Telecommunication System (UMTS) will refuel this dramatic expansion.

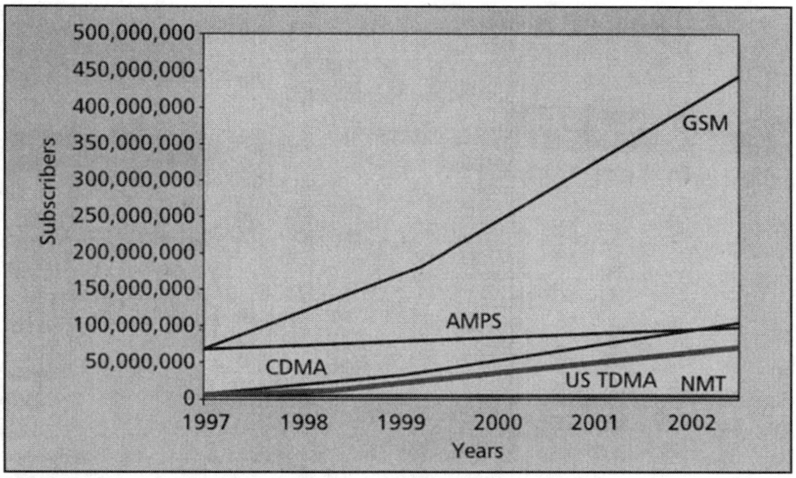

Figure 1. *World subscriber evolution per wireless standard.*

New Directions

The pioneering days for the silicon bipolar RFICs are over but the quest for an even more advanced radio remains. The single-chip cellular phone, the multi-standard phone, the software radio, and the UMTS mobile broadband communication phone are familiar new challenges. We look at these targets with optimism because, just like 10 years ago, the same economic and technological factors are present. The mobile telecommunications application and the constant improvements in silicon technology are once again changing the scene of radio design. This is not unlike the replacement of analog speech processing with digital processing in old telephone systems. For example, the united performance of SiGe bipolar transistors (Figs. 2 and 3) with the massive digital integration of CMOS is opening the horizon for single chip radio. Mobile broadband communication will likely be enabled by new IC technologies such as Silicon-on-Insulator (SOI) CMOS. This technology improves digital power consumption and RF performance compared to regular CMOS. In addition, the principal issues for RFICs, parasitic coupling and cross talk, are simplified since all active devices are isolated naturally by silicon oxide. Similarly, in bipolar technology, the buried collector layer placed under the CMOS logic is a natural shield against substrate-coupled noise injection into the RF/analog blocks. Silicon-carbide bipolar devices will be even faster than SiGe transistors, and copper metalization together with low dielectric-constant insulators on silicon technology will further reduce the interconnect resistivity and the parasitic capacitance, respectively.

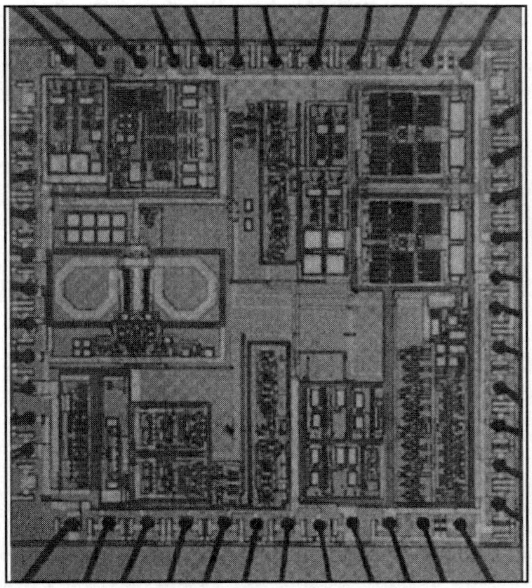

Figure 2. *Photomicrograph of a multiband SiGe transceiver including zero-IF receiver and transmitter, I/Q mixers, prescaler, and 3.6 GHz VCO, for GSM/DCS/PCS.*

Despite a general optimistic landscape, we do have many uncertainties. The SiGe and SOI technologies will be available but, in the next few years, will their cost match the requirements of BiCMOS RF-analog-digital radio integration? Are these or other technologies the right ones indeed? What are the appropriate new transceiver architectures? Are the current wireless standards suitable? Despite the absence of absolute answers, we will still discuss several issues related to these questions.

The RFIC Technology Dilemma

Today, SiGe technology is available for the same wafer cost as silicon RF-bipolar technology. The long promised superior performance of SOI will become cost effective in "smart cut unibond wafers" and "thin oxide SIMOX" technologies, competing in the low-power, low-voltage CMOS market. The SOI wafer cost was an obstacle in the past for large volume, low cost applications, but, as we migrate to 1V-supply voltages, the situation is changing. The major power saving in

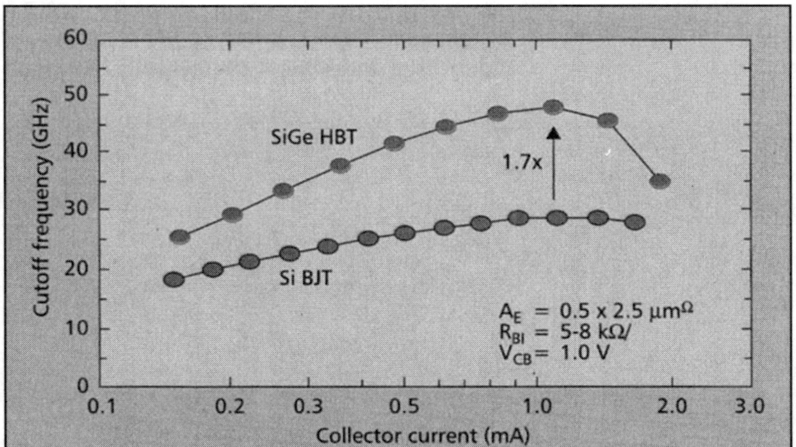

Figure 3. *The transition frequency of a Si and SiGe transistor with comparable geometry.*

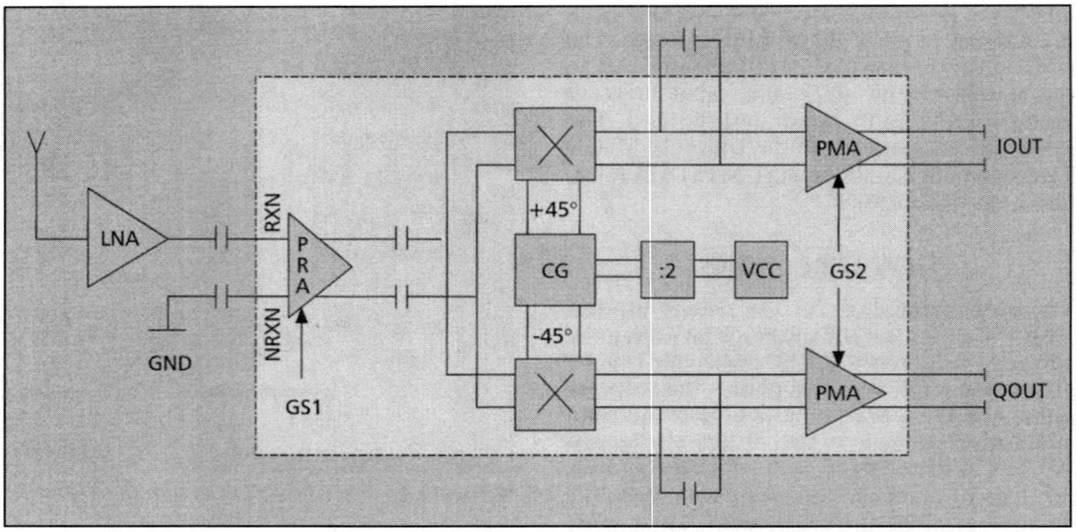

■ **Figure 4.** *Zero-IF radio receiver in full production for GSM 900 since 1991 and for DCS1800 for several years.*

the reduced source and drain junction capacitance is now relevant compared to the loss in metal interconnect for large digital system integration. As mentioned previously, the metal interconnect itself will be improved by the introduction of copper for low resistivity and low dielectric constant insulators for reduced parasitic capacitors.

The plurality of technology innovations is creating a dilemma for present day R&D engineers and managers. Which ASIC technology should they consider? During a single design cycle the technology scene is changing so rapidly that a conservative approach can create a major strategic mistake for the next product generation. Concurrent engineering on the product level, architecture, and technology requires clear roadmaps on each level to make sure that each of the research and engineering teams stay in line to meet the common goal in the next product generation.

Transceiver Architecture

Several options are open for transceiver architectures and each manufacturer is likely to pick the one that fits the expertise of its own RF design team. Superheterodyne I/Q receivers are widely used and some manufacturers have built up expertise in homodyne receivers. Direct conversion digital phones [2] have been in full production since 1991 [3] despite the opinion of many conservative radio designers (Fig. 4). They never believed that GSM homodyne radios would work due to the baseband offsets produced in the front-end mixers through self-mixing of various signals. This offset problem can be handled successfully with digital averaging algorithms. In order to handle the dynamic offsets from high blocking levels, two-dimensional algorithms are needed which calculate jumping offsets in a signal burst with a constant envelope modulation. This is possible because three noncoincidental symbols are enough to calculate the offset as the center point of the I/Q circle. The algorithm searches for maximum I & Q signals to fit the GMSK circle in a square and adjusts the position of the square to the origin of the I/Q plane.

The previous know-how is helping the popularity of zero-IF receivers, the cornerstone of full radio integration. RF-to-baseband direct conversion avoids the need for discrete surface acoustic wave (SAW) filters as IF components. All receiver-filtering requirements for adjacent channels, blockers, and anti-aliasing filtering before sampling for digitization is now performed by lowpass silicon analog integrated circuits that have been applied routinely in speech processing for decades.

In parallel with the Cartesian I/Q modulators traditional for GMSK (Fig. 5), some manufacturers use polar modulators [4], popular in FSK systems for many years. Such a direct VCO modulation method avoids the spurious components in the oscillator output spectrum. This relaxes the linearity requirements of the PA and allows a design compromise in favor of the power-added efficiency [5]. In addition, the harmonics of the mixers in the traditional I/Q TX modulator and the intermodulation requirements of the PA dictate the use of class B or AB output stages. The resulting power efficiency is low and limits the talk time. As the cellular system operators reduce the cost of conversations, talk time

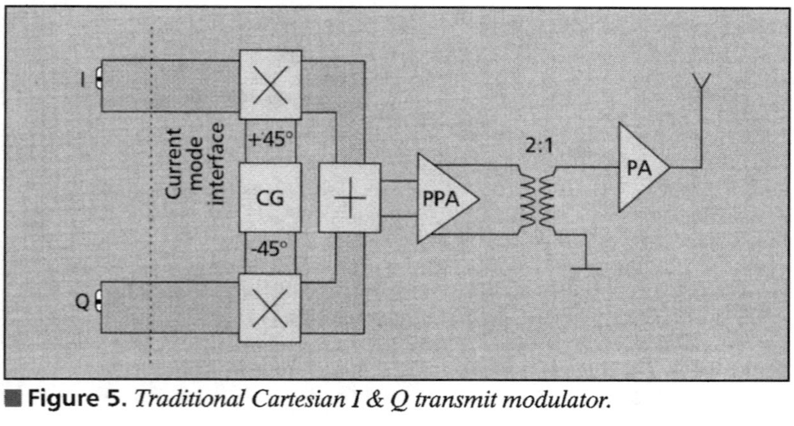

■ **Figure 5.** *Traditional Cartesian I & Q transmit modulator.*

will be the next strategic priority. GSM was conciliated for class C in the first place with the constant envelope GMSK phase modulation.

The new trend in cellular transmitters is modulation in the synthesizer loop (Fig. 6). The digital implementation of the channel stepper and the phase modulation in a fractional N loop are moves toward "software radio." Such architecture is introduced to reduce the analog circuits and move functionality to VHDL code. It is a key element for further "automatic" improvements since new digital IC technologies will continue to reduce the chip size and cost, boosting the overall performance in sensitivity and battery discharge time.

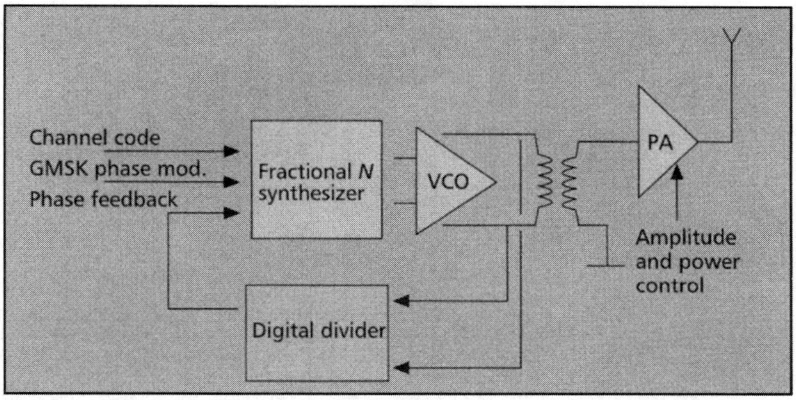

■ **Figure 6.** *Maximum digital "Zero I & Q" modulator.*

CIRCUIT INTERFACES

An important aspect of large-scale radio integration rests in the circuit interfaces. Traditional radio designers have created a "50 ohm culture" with impedance-matching networks, standing wave ratios, s-parameters, power gains, noise figures, and signal levels measured in dB and dBm. With the exception of the s-parameter theory extremely useful in the modeling of RF transistors, the other microwave concepts are not easily applicable to integrated circuit design. For example, Gummel-Poon models for bipolar transistors and state of the art MM9, BSIM, or EKV models for regular or SOI CMOS transistors work very well in RF-simulations. S-parameters will remain the preferred way to measure the input impedance of the receiver and the output impedance of the transmitter. However, in fully integrated radio systems, these two quantities will be the only interfaces with the outside world. The rest of the radio will be an internal silicon affair. It is quite difficult to measure RF signals on sub-micron lines on a chip. Therefore, the major skill of traditional radio engineers, RF measurements, will be mapped to another skill, circuit simulations in SPICE-like simulators.

Many GaAs designers have continued the linear AC analysis culture from the printed circuit board RF-design days. The plurality of bias conditions applied to the integrated RF transistors requires the flexibility of a SPICE-like simulator. For many years, RF semiconductor foundries have perfected tools and software for transistor RF modeling, parameter prediction, and parameter extraction. This methodology is applied even for a new technology under development. The devices are measured on wafers and well known de-embedding procedures are applied to compensate for the parasitic of bonding pads, probe stations, and measurement equipment.

The new silicon RF-analog-ASIC designers speak a different language. They use simple words such as milli-volts and amperes for signal levels and noise, and ohms for interface impedance, because this is what the SPICE-like simulator gives them on the screen of the design terminal. The 50-ohm interfaces no longer exist between integrated building blocks like RF-mixers and RF-amplifiers on a single chip. It was a good impedance choice for discrete components with pico-farad parasitic pin capacitance, but on IC chips we have femto-farad parasitic capacitors. Likewise, the pins of low-cost plastic and ceramic packages are also well below one pico-farad. Even the 50-ohm standard for SAW filters and antenna will be questionable in the near future. The impedance transformer of an antenna changing the 300-ohm air-interface to 50-ohm SAW filter input impedance is just an additional cost for the product. The optimum noise impedance of integrated silicon or SiGe LNAs is closer to 300 ohm than 50 ohm. We can save several discrete components on the radio receiver printed circuit board if 300 ohm is used for SAW filter and antenna impedance.

The transmitter tradeoffs are different. With power supply voltages below 3V it is difficult to generate 1-2 W power levels even into 50 ohms. A 300-ohm antenna impedance will require a bigger transformer ratio to couple the output signal. However, the point remains that the current 50-ohm standard is not a good match for either the receiver or the transmitter. This strongly suggests that work needs to be done in this area to discover new tradeoffs and optimizations.

In conclusion, the external interfaces will remain 50 ohm at least for a while since measurement equipment, discrete filters, antenna production, and tradition will tend to maintain the current standard. Inside the RF-chip the 50-ohm interface does not make sense since it requires too much power dissipation and even jeopardizes the signal-to-noise ratio due to the fact that voltage noise is dominant in the silicon technology at this low impedance level.

DESIGNING FOR MINIMUM POWER DISSIPATION

The interconnect between RF-building blocks remains an issue. Although on silicon we count the capacitance in femto-farads compared to pico-farads on a printed circuit board, the bandwidth on chip is limited by the interconnect capacitance. This comes from raising the impedance level to reduce the power consumption. In the discrete component radio culture this problem was handled by selective LC tuning. But the cellular phone market requires multi-standard and multi-band operation. Therefore, in order to wideband the circuits we cut the interconnect length to the minimum possible by optimum layout floor planning. Auto-routing tools are not always applicable. As the integration

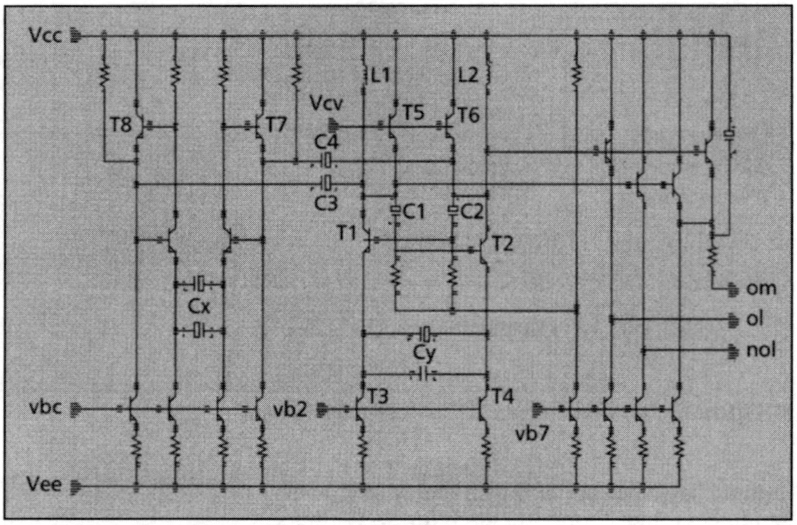

■ **Figure 7.** *A typical VCO circuit with integrated inductors.*

level increases, careful hand crafted layout generation is a must in the RF sections of the chip.

Integrated Inductors

The integration of spiral inductors in silicon RFIC technology has been the subject of many recent studies and developments. The best reported quality (Q) factors are between 10 and 20, depending on the technology. For example, in standard silicon bipolar and CMOS a Q factor of 10 is the usual limit, given by the metal resistivity and by substrate Eddy current losses. Many technology variations can be applied to improve the quality factor, such as thick oxide under the inductor, trench isolation employed to cut the Eddy currents, gold or copper interconnect layer, and thick aluminum layer. Some manufacturers place the inductor on top of the chip passivation. This minimizes the substrate parasitic capacitance and allows the deposition of a thick copper or gold layer through electroplating. A major drawback of on-chip inductors is the fact that they are an order of magnitude bigger in silicon area than transistors. In addition, their modeling is difficult for better than 10 percent accuracy. However, compared to discrete inductors, their reproducibility is a lot better in production. Two to three design iterations are usually enough to meet the desired values. Passive RF RLC circuits are becoming feasible in silicon technology, but the integration of duplex quality filters for cellular phones still requires Q-factors an order of magnitude higher than available today.

In addition to filters, fully integrated voltage controlled oscillators (VCO) use on-chip inductors whose Q factor determines directly the VCO phase noise. A typical bipolar circuit example is shown in Fig. 7. For a GSM handset the performance of such a VCO is marginal (Fig. 8), but for a GSM base station the standard silicon inductors are still far too noisy. Inductorless VCOs resonating on the natural poles of two active elements may be a possibility when SiGe transistors are available with base-to-collector current gains in excess of 200.

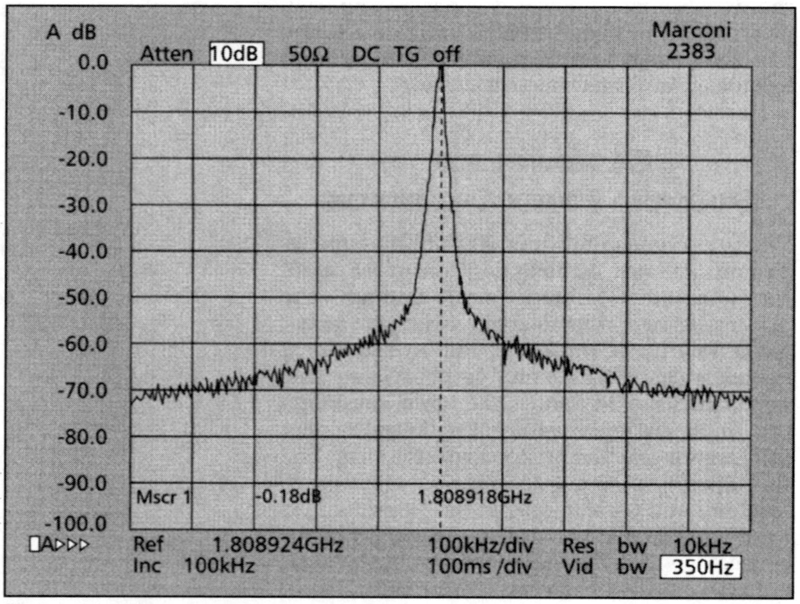

■ **Figure 8.** *Typical phase noise for an integrated VCO with spiral inductors.*

Modeling and Simulations

Over the past 10 years, the integration level of RFICs was limited by modeling and simulation time constraints, but today this situation is improving. For example, some suppliers provide accurate package models, and soon, flip-chip technology will reduce the chip interface parasitic components to a bare minimum. Still, complex phenomena such as the VCO pulling by a 2 W PA on the same chip are very difficult to simulate accurately. For such problems, the solutions are more likely to come from novel architectures rather than from design tools or technology advancements. For example, operating the VCO and the PA in different bands of the spectrum is a good option. This approach has the added advantage of lowering the VCO phase noise. If we double or quadruple the frequency of the resonator and we use subsequent frequency division to obtain the cellular carrier frequency, we get a dual benefit. First, the inductor Q factor is increased as the reactive impedance goes up relative to the metal line resistivity; second, frequency division reduces phase noise by 6 dB for each division by two.

The Role of Intellectual Property

An important aspect of any ASIC design in general is IP transfer. Exchange and cross licensing of intellectual property is the key to speeding up the process of full radio integration. In the area of RFICs we have a win-win situation. The advanced silicon technology is mature and available for full radio integration to merge digital, analog, and RF on a single chip, but the limited availability of RF-silicon and ASIC designers worldwide is the bottleneck. Exchange of building blocks and design effort sharing are beneficial for all cooperative partners.

CONCLUSION

Despite 10 years of pioneering work and much advancement, the RF ASIC field will continue to expand driven by new silicon IC technology capabilities and challenging product requirements. While radio large-scale integration in silicon technology has been the area of main interest in the past ten years, the next crucial topic is mixed RF-analog-digital integration. The main hurdles in this development will come from the isolation between building blocks, the on-chip interference, and cross talk via the supply lines and parasitic elements. This problem is similar to mixed analog-digital designs. New tools for system-level and circuit-level analysis/synthesis will be employed, but architecture and system innovations will be the driving force for further growth in silicon integrated radio.

Despite 10 years of pioneering work and much advancement, the RF ASIC field will continue to expand driven by new silicon IC technology capabilities and challenging product requirements.

REFERENCES

[1] J. D. Cressler, "Silicon Germanium Heterojunction Bipolar Technology: The Next Leap in Silicon?," *ISSCC Digest of Technical Papers*, Feb. 1994, pp. 24–27.

[2] A. Abidi, "Direct Conversion Radio Transceivers for Digital Communications," *ISSCC Digest of Technical Papers*, Feb., 1995, pp. 186–87, 363–64.

[3] J. Sevenhans *et al.*, "An Integrated Si Bipolar RF Transceiver for a Zero If 900 MHz GSM Digital Mobile Radio Front-End of a Hand Portable Phone," *IEEE '91 CICC*.

[4] N. M. Filiol *et al.*, "An Agile ISM Band Frequency Synthesizer with Built-In GMSK Data Modulation," *IEEE JSSC*, vol. 23, July 1998, pp. 998–1008.

[5] N. O. Sokal and A. D. Sokal, "Classe-E: A New Class of High Efficiency Tuned Single-Ended Switching Power Amplifiers," *IEEE JSSC*, vol. SC-10, June 1975, pp. 168–76.

Advanced Base Station Technology

Robert Arnott, ERA Technology Ltd.
Seshaiah Ponnekanti, ERA Technology Ltd.
Carl Taylor, ERA Technology Ltd.
Heinz Chaloupka, University of Wuppertal

ABSTRACT The authors present an overview of advances in three areas including software radio, adaptive antenna technology, and high-temperature superconductivity as currently envisioned for use in advanced cellular base stations. The conclusion broadly drawn is that the advancement in DSP and ASIC technologies will make the major contribution in making these technologies technically and commercially realistic.

Advances in new applications coupled with the complexity of radio communication systems paved the way for software radio to have the capability to operate not only with Universal Mobile Telecommunications Service (UMTS) standards but also with multiple other radio standards. The drive for better grades of service and the need to serve the exponentially growing number of users demand synergy between multiple access systems and the spatial filtering technology. Fading-resistant beamforming requires better understanding of the channel models as well as the development of robust adaptive algorithms. Exploiting the scientific and technology base in superconductivity to better identify the components for supporting advanced base stations may result in better packaging density and low power dissipation.

A Proposed Advanced Base Transceiver Station

The designer of an advanced base station is confronted with a multitude of challenges, and it is essential to identify the subsystems encompassing the fundamental blocks. The situation therefore calls for incorporating the specialized algorithms in the design. The following sections outline the tenets of the approach and a basic exploration of the technology. The scope of an advanced adaptive base station is shown in Fig. 1, which indicates how replacement of existing second-generation services may be accomplished by the use of a UMTS beacon channel. It can be seen that the replacement can be carried out initially by addition of the UMTS beacon channel, followed by the fully functioning adaptive base station.

Requirement for Broadband Downconversion

Current cellular radio base stations typically require one radio transceiver for each carrier frequency assigned to the base station. If many carrier frequencies are used this leads to a high degree of hardware replication. It also necessitates the use of RF power combiners to combine the signals at the antenna. These devices often have poor linearity and efficiency characteristics. If adaptive antenna technology is to be used in the base station, there is a further level of hardware replication, since a radio transceiver is required per array element per carrier frequency. In this case the cabling and interconnectivity between all the transceivers can become very complex. Recent advances in analog-to-digital conversion (ADC) technology and digital processing are pointing the way to a new generation of base station systems based on wideband transceiver architectures. In these schemes multiple carriers are simultaneously downconverted in a single radio and digitized. The carriers are then separated in the digital domain using digital downconversion (DDC) techniques. In the transmit direction the multiple carriers are combined digitally at the baseband and simultaneously upconverted. This avoids the need for radio frequency (RF) power combiners.

When combined with software radio techniques the wideband transceiver provides an extremely flexible base station solution. Carrier allocations, bandwidth allocations, and even modulation schemes may be modified adaptively under software control according to operating conditions.

Choice of Baseband Technology

Software radio is computationally powerful, given sufficiently high-speed digital signaling processors (DSPs). It offers high MIPS performance and easy direct memory access (DMA), suitable for parallel processing at the expense of high power consumption. Present technology requires hardware accelerators alongside the DSPs. The integration density is potentially far greater than that offered by conventional dual-mode (stacked radio) design, and the derivation of the software from the network offers a degree of reconfigurability for software debugging and, to a certain degree, future-proofing. Its weakness is that its computational efficiency is low, implying high power consumption, but this is not a very important factor in base station design. It requires an array of DSPs, ideally in an integrated circuit (IC) for commercial implementation; however, the size of the IC or DSP card is again unimportant for base station applications and could be extended to a small-package low-voltage version for mobile use. The freedom to use large high-power-consumption devices makes software radio base stations possible using today's technology, but flexibility clearly exacts a price due to technical complexity.

This software radio approach is markedly different from hardware-configurable design employing field programmable gate arrays (FPGAs). FPGAs offer lower power consumption and present an easy route to application-specific ICs (ASICs), but designs are physically large due to requirement for multiple FPGAs. They have a longer reconfiguration time and reduced MIPs than the DSP approach with present technology, and DMA is more difficult.

RF Techniques Required for a Base Station Supporting both TDD and FDD Access Schemes

For an advanced base station to be truly adaptive, capable of supporting second- and third-generation access schemes, both

Reprinted from *IEEE Communications Magazine,* Vol. 36, No. 2, pp. 96–102, February 1998.

time-domain duplex (TDD) and frequency-domain duplex (FDD) RF architectures must be supported. In principle, a perfect circulator (highly linear and broadband) is required. Such a component does not exist, but research is ongoing into the area of frequency-elimination diplexers, which employ active techniques to cancel unwanted products and provide Tx/Rx isolation over at least the 200 MHz proposed for UMTS by the ACTS FRAMES project. This assumes 2 x 5 MHz of FMA2 wideband code-division multiple access (CDMA), uplink/downlink symmetry, and 190 MHz uplink/downlink frequency separation. A frequency-elimination diplexer would reside immediately behind the antenna, as shown in Fig. 2.

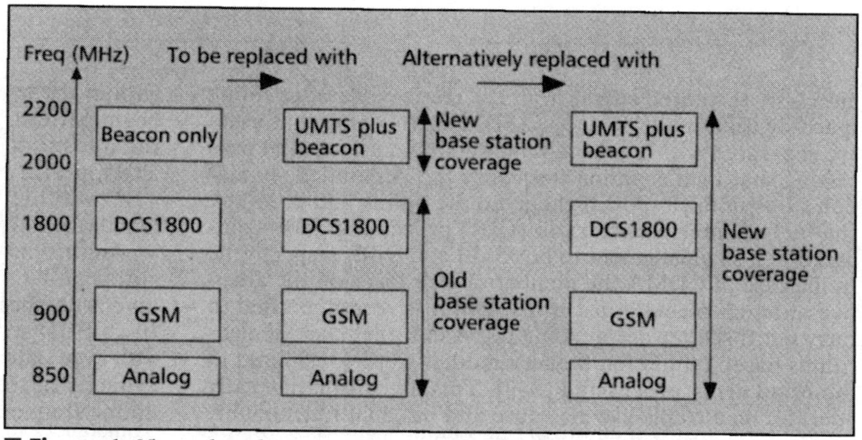

■ **Figure 1.** *Network architecture required for early IMT deployment.*

ADC and DAC Technology Roadmap

The advances in DSP and ADC/digital-to-analog conversion (DAC) technologies experienced in recent years has made the concept of the "software radio" technically and (increasingly) commercially realistic. However, it is still ADC which is the limiting performance factor. Over the last 10 years, flash ADC performance has improved at the rate of about 1 b/3 yr for a fixed sampling rate. If this trend is followed, 18-bit, 60MSPS devices may be routinely available by 2002, for the beginning of UMTS. The alternative method to obtain high dynamic range at baseband is to compress analog signals logarithmically before digitization, followed by digital anti-log processing. The problems with this approach are that it is extremely expensive, and care must be taken not to introduce intermodulation products in the analog signals due to the use of a nonlinear (logarithmic) function.

A current benchmark flash device use d by the ACTS FIRST project is the AD9042. This is a 41MSPS, 12-bit monolithic device which runs from 5 V supplies, and it has its track-and-hold amplifiers and reference voltage on-chip. The device works internally on a two-stage subrange architecture. The input signal drives a first track-and-hold amplifier and a 6-bit coarse analog-to-digital (A/D) converter samples this signal. A 6-bit digital-to-analog (D/A) converter with 12 bits of accuracy then reconverts the coarse ADC output for subtraction from a delayed version of the input. This "residue" signal is then passed to another track-and-hold and then a 7-bit ADC. The two ADC outputs are then added and passed to digital error correction logic to generate the final 12-bit output. It offers 80 dB spurious free dynamic range and a SINAD signal-to-noise ratio figure of 64 dB minimum. Ultimately, in order to provide a large instantaneous dynamic range, sigma-delta A/D converters may be developed with a high enough sampling rate to sample data at intermediate (IF) and RF frequencies.

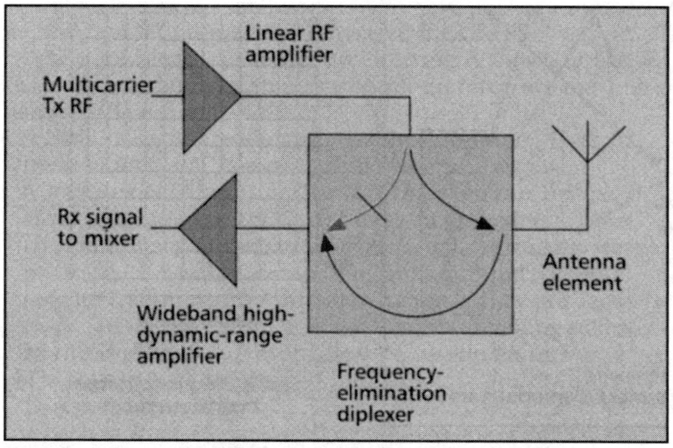

■ **Figure 2.** *Analog components around the frequency-elimination diplexer.*

Adaptive Antenna Algorithms

The key requirement for an advanced base station antenna is to employ robust adaptive algorithms to ensure reliable operation of the adaptive antenna. The explosive growth of interference communications systems has given rise to a variety of algorithms in different frameworks. The crux of the problem is to devise algorithms suitable for implementation when there is no a priori knowledge of the multipath environment. Employing advanced blind-weight updating principles is the norm. Temporal structure methods span from Constant Modulus (CM) [1–2] and Higher-Order-Statistics (HOS) methods [3]. The CM algorithm tries to alleviate the structural damage of the signal introduced by the channel and the interference. This is accomplished by maintaining the constant output at the array while preserving the phase of the array. Special properties of the received signal, including the cyclostationarity and modulation format, are exploited in these formulations. CM algorithms can in principle employ a steepest-descent or least-squares-based approach to minimize the CM cost function. When an adaptive solution is sought, techniques to improve the convergence speed without falling into the local minima and the tracking ability are of paramount importance. Provision of a reference signal is crucial for the training algorithms. In the case of a decision-directed algorithm, after the demodulation of the array output a binary decision is made which is fed back to provide a reference signal. A temporal reference signal is also needed on the uplink to enable efficient estimation of spatiotemporal channel estimation or 2D impulse response in time-division multiple access (TDMA) systems. However, suitable long training sequences with desirable correlation properties are needed to help the adaptive antenna to distinguish the mobiles within the vicinity of the base station. The length depends on the number of antenna elements employed within the adaptive antenna and the type of spatiotemporal processing applied. On the other hand, in CDMA systems user-specific spreading codes satisfy this requirement. The length of the codes is also significant in relation to identifying different paths and for more complex interference cancellation systems. Another closely related topic is the channel estimation technique based on the Space-Alternating Generalized Expectation-Maximization (SAGE) algorithm [4]. Performance studies for the real channels were also reported in the literature [5]. Adaptive antenna architectures operate by estimating the direction of wanted signal from the received data, enabling space-division multiple access (SDMA), implying that users are separated within one cell by predominantly spatial processing sharing a common frequency [6]. A detailed simulation study in this report highlighted the uplink and downlink channel-to-interference ratio (CIR) processing gains compared to a single element to be 15 dB and 6 dB, respectively. In the case of CDMA the number of user loads on the adaptive antenna places constraints, and it is recommended to carry out this after despreading [7]. A different class of algorithms based on hidden Markov models were considered in the blind array processing [8–9]. This algorithm basically operates on a trellis-like structure and has dual functionality through initialization and then blind mode using reestimation procedures. Effective beamforming for both uplink and downlink is crucial to utilize the technology in the future base stations [10–13]. It may be noted that, for the third-generation systems where CDMA may be a possible candidate, joint mitigation of intersymbol interference (ISI) and multiple access interference (MAI) is a requirement to attain maximum spectral efficiency. Toward this objective, multi-user detection [14] was proposed as a strong candidate to relax the stringent requirements of fast power control, soft handover,and macrodiversity. Both inter- and intracell interference suppression can be detected by this formulation.

It is envisaged that adaptive antenna of a future base station will necessarily employ a great variety of algorithms, to offer best performance in all situations. A number of optimum combining techniques show promise for providing excellent gain in the uplink. TDD must be implemented if the same beam pattern and similar performance is to be obtained on the downlink. However it should be noted that even with TDD, performance will be lost if the mobile is moving. So TDD will help improve the performance for a proportion mobiles and therefore should be an option.

Additionally, implementing a TDD solution where time slots can be swapped from up- to downlink to cope with up/down asymmetries maximizes the capacity of the network. In an FDD scenario where a base station is handling traffic with even data rate requirements on uplink and downlink, the capacity of the system is dominated by the downlink. The uplink frequency band is left underutilized. In the same circumstances in a TDD system, unused time slots on the uplink can be allocated to the downlink, and more traffic can be installed.

BASEBAND TRANSMULTIPLEXING AND BEAMFORMING

Introduction — In a cellular radio system each base station may be required to process the signals on more than one carrier simultaneously. Multicarrier reception and transmission require FDMA demultiplexing and multiplexing operations, respectively, and it is important to consider how such operations are to be integrated with the beamforming operation.

The operations of FDMA demodulation and beamforming may be combined in the digital domain by using a digital filter bank, or transmultiplexer (Tmux), to separate all of the carriers at each base station prior to beamforming and demodulation.

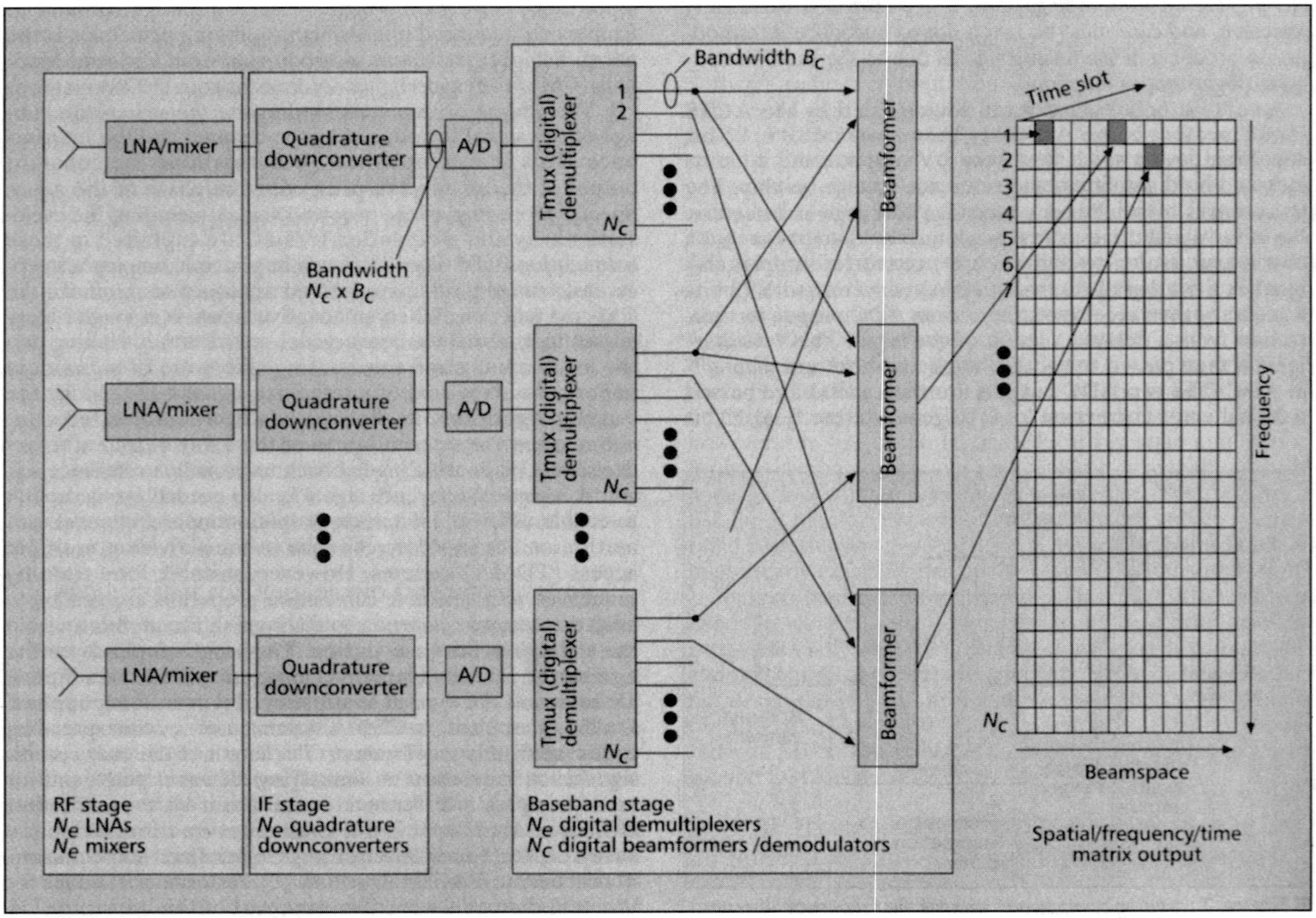

■ **Figure 3.** *Transmultiplexer (TMUX) receiver architecture.*

At first sight such an approach may seem inefficient, since not all of the channels provided by the demultiplexer will be used. However, this scheme provides a very high degree of flexibility, easily supporting dynamic frequency allocation, for example. By integrating the beamforming and frequency routing functions entirely within the digital domain using efficient DSP methods, an efficient beam-frequency processor architecture may be realized.

A possible receiver architecture based on digital Tmuxes is shown in Fig. 3. The key features of the architecture of Fig. 3 are as follows:

- An RF stage consisting of N_e LNAs and N_e mixers, where N_e is the number of antenna elements in the array.
- A bank of N_e quadrature downconverters which translate the whole band of interest to zero IF. N_e dual A/D converters are used to digitize the wideband element signals.
- A bank of N_e digital Tmuxes. Each performs a frequency demultipexing operation, separating the wideband digital input signals into N_c zero-IF components, each of bandwidth B_c.
- A bank of N_c digital beamformers and demodulators, each serving a single carrier signal. Note that more than one beam may be formed at each carrier frequency, as implied by Fig. 3. This feature of the architecture would be required for SDMA operation. Some form of programmable switch would be used to route the Tmux output channels to the beamforming modules. The switch could rapidly be reconfigured to support dynamic channel allocation (DCA) functions.

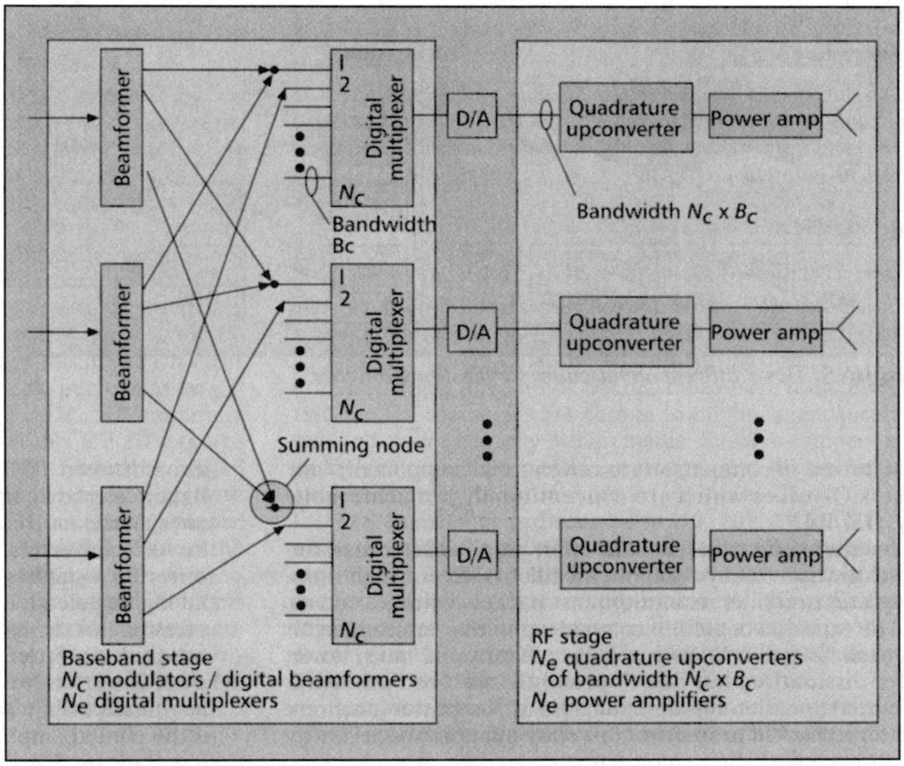

■ **Figure 4.** *Transmultiplexer transmitter architecture.*

A key advantage of this approach is the transfer of the element-to-beam connection problem from the analog domain to the digital one. This increases the inherent flexibility of the architecture, allowing any received carrier to be rapidly rerouted to a beamformer processor under software control. Another advantage of the Tmux scheme is the possibility of implementing mixed levels of frequency demultiplexing in order to offer a variety of different bandwidth channels, corresponding to the different bit rate services envisaged for UMTS.

The equivalent transmitter architecture is shown in Fig. 4. The main attributes of the transmitter architecture are summarized below.

- A bank of digital modulators and beamformers. Each of these produces N_e complex digital baseband signals.
- A bank of N_e digital multiplexers. Each multiplexer stacks the incoming digital data in frequency, thus translating each baseband signal to the desired carrier position. Note that more than one modulated signal may be transmitted on the same carrier if they are separated spatially by SDMA. This requires summing nodes at the inputs to the multiplexers, as shown in Fig. 4. The outputs of the digital multiplexers are fed to N_e D/A converters.
- A bank of N_e quadrature upconverters and power amplifiers.

By combining the transmitted signals digitally at baseband, the need for RF power combiners is eliminated. Digital combining is lossless and offers greatly increased linearity and flexibility.

The Tmux approach transfers the operations of quadrature downconversion and upconversion from the analog domain to the digital domain. This reduces the number of analog components required, but the performance requirements of the analog components is increased because they must operate at increased bandwidth and linearity.

In the Tmux receiver automatic gain control cannot be performed in the analog domain on a per-carrier basis. This implies that the A/D resolution be such that the full dynamic range of the carrier signals can be supported. The practical bandwidth of the Tmux architecture is limited by the sample rate at which A/D technology can support the required dynamic range.

Intermodulation distortion between carriers in the analog part of the receiver and transmitter is also a significant problem. The degree of linearity required from these components necessitates the use of advanced adaptive linearization techniques in the receiver and transmitter chains.

Impact of High-Temperature Superconductivity on BTS Technology

Potential of HTS Technology for Communication Systems

High-temperature superconductivity (HTS) offers, at operation temperatures below about 80 K, a variety of physical effects which, in principle, can all be utilized to realize novel and/or superior functions in the analog or digital part of the base station.

The surface resistance of HTS thin films at RF frequencies is three orders of magnitude lower than the surface resistance of normal conductors like copper. This allows the realization of planar HTS components which are significantly

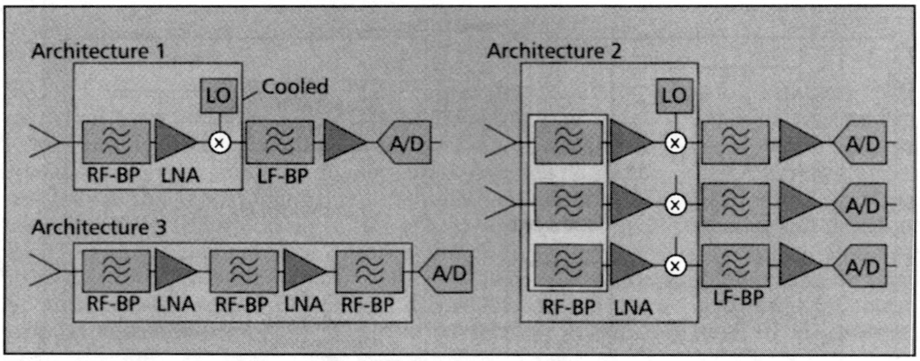

■ **Figure 5.** *Three different architectures for the downconverter.*

miniaturized in comparison to conventional components and possess Q-values which are conventionally not achievable (e.g., [15, 16].

Josephson effects in junctions offer, in principle, a basis for the realization of key components like ADCs for high sampling rates and high resolution, fast packet switches, and so on. The expected benefit in comparison to the semiconductor approach is a much higher packaging density and much lower power dissipation, but due to presently unsolved problems with the fabrication of large numbers of Josephson junctions with reproducible properties, this class of applications has to be considered as a long-term potential.

The SUCOMS project intends to exploit the scientific and technology base in HTS to demonstrate new approaches at the component and subsystem level as required for the UMTS third-generation mobile telecommunication system.

THE BENEFIT OF LOW CONDUCTOR LOSSES ON THE COMPONENT LEVEL

The achievable frequency selectivity of passive components for analog RF signal processing (e.g., bandpass and bandstop filters) is limited by dissipative losses (finite unloaded quality factor). In conventional technology Q-values only up to about 20,000 can be achieved and only at the cost of large geometric size. With HTS miniaturized planar circuits with Q-values up to 200,000 at 60 K have already been demonstrated [17]. The low losses transform into higher achievable selectivity and lower passband insertion loss. These properties together with the small size offer a new degree of freedom to the system designer, since high-performance RF signal processing functions become available which up to now were unrealizable.

THE FRONT END OF A MULTICARRIER ARRAY RECEIVER

The possible impact of HTS technology is seen in accordance with current receiver evolution. Besides the requirements for sufficient sensitivity, selectivity, and dynamic range, an ultimate goal in receiver design is to implement as many receiver functions as possible in programmable digital processors [18]. Consequently, A/D converter will move much closer to the antenna and hence the partitioning between the analog and digital parts will become an important issue which may, under certain circumstances and depending on the chosen architecture, be influenced by the availability of high-performance HTS components.

Architecture 1: Wideband Downconverter and Digitization at IF — As shown in the upper part of Fig. 5, a preselect filter for entire receive band of width Δf is followed by LNA, mixer, IF bandpass filter, IF amplifier, and an A/D converter. Bandpass sampling can be employed to reduce required sampling rate to about 2.5 Δf. In this architecture the expected benefit from using an HTS preselect filter together with a cooled LNA can be summarized as follows:

- Based on the increased Q, higher-order filters with higher selectivity become available. The significantly higher out-of-band rejection provided increased freedom in network planning, allowing the coexistence of systems with large power level differences (e.g., co-site interference, near-far problem, interaction between TDD and FDD).
- Higher selectivity in the image-reject function allows a lower maximum IF and hence a lower input bandwidth of the A/D converter. Instead of using a multistage downconverter, a single-stage version can be employed, where the higher selectivity of the RF filter partially compensates for the selectivity usually obtained in the second and third converter stages.
- Lower insertion loss and lower physical temperature of the preselect filter together with the reduced noise figure of the cooled amplifier lead to a reduced system noise temperature. This in turn transforms into an increased area coverage in noise-limited rural areas or into a lowered required transmitter power in interference-limited regions. The expected improvement of 2 dB in the system noise figure provides an increase in area coverage by about 25 percent.

By decreasing the phase noise of local oscillators by means of a superconductive resonator, the effect of reciprocal mixing is reduced, and this in turn transforms into a lower conversion of adjacent channel interferers.

Architecture 2: Parallel Subband Downconversion and Digitization at IF — Especially with receivers for umbrella cell base stations, a very high dynamic range is required to cope with different signal strengths in the macro- and microcells. In multichannel receivers for smart antennas, a high dynamic range is required if nulling of cochannel interferers for increased capacity is employed. In wideband amplification and downconversion, intermodulation and desensitization effects occur due to the nonlinearities of the LNA and mixer. This causes an upper limit of the dynamic range. Assuming the ideal case of a homogeneous signal spectrum, it can be shown that the dynamic range of the analog part of a downconverter increases inversely proportional to the bandwidth. Furthermore, it is well known that for a given ADC technology the product of sampling rate and resolution can be considered constant. A reduced sampling rate therefore transforms into an improved resolution. For these reasons it can be advantageous, instead of downconverting the entire receive band (e.g., with Δf = 75 MHz), to split the band by means of a filter bank into a number of subbands (e.g., of 10 MHz width) which are separately amplified, downconverted, and digitized. In this approach the full programmability of signal processing is partially lost. Hence, dynamic range is gained by cost of flexibility. For this approach a relative large number of filters with high selectivity and low dissipative insertion loss is required, which makes HTS technology especially well suited to reduce the width of guard bands between adjacent subbands.

Architecture 3: Direct Digitization at RF — By using bandpass sampling with a sampling rate on the order of 2.5

Δf directly at RF frequencies, the full information in the receive band is retained and the process of bandpass sampling can be considered as incorporating the function of downconversion. The advantage of this approach is that no mixer and local oscillator (LO) are required, and hence problems connected with temperature effects and aging, for example, of the LO are overcome. At present, this approach is not utilized because of the required high input bandwidth of the ADC and the problems with the short aperture time (e.g., aperture jitter). Assuming that ADCs for this purpose will be available in the future, the realizability of this approach turns out to depend on the availability of high-performance RF bandpass filters. Additionally to the RF ADC, RF amplifiers with a total gain up to 60 dB and some RF filters are required. To avoid out-of-band signals (including noise) being folded into the processed band, a bandpass filter with steep skirts and high stopband attenuation must be placed between the last amplifier stage and the ADC. Due to the required high RF gain, dynamic range problems due to intermodulation and desensitization occur. The latter problems are overcome by placing additional high-performance bandpass filters in front of the first amplifier stage and possibly between the amplifier stages. As in architecture 2, subband filtering with separate digitization of the subband signals increases the dynamic range.

In case of all three architectures, increased selectivity and reduced insertion loss of HTS filters transforms into significant system benefit. In case of array receivers for smart antennas, a large number of high-performance filters are needed. Miniaturization via HTS allows a tower mounting of the front end which otherwise may be difficult to be realized.

Base Station Transmitter

In order to feed the signals of the different carriers into a common antenna, a "signal combiner" must be included. With respect to the approach used for signal combining, different transmitter architectures can be distinguished. In all architectures the digital signals allocated to the different carriers are separately digitally preprocessed (e.g., filtered).

In accordance with the idea of a fully software-controllable base station, these individual signals should be digitally combined to a multicarrier composite signal which is, by means of a wideband DAC, converted into a analog signal and than upconverted into the final frequency. Before it is fed into the antenna this composite multicarrier signal must be amplified in a power amplifier (HPA). Because of the strongly varying envelope of the composite signal, very stringent linearity requirements are imposed on the amplifier. With current technology these requirements can only be met if the number of simultaneous carriers is relatively low (e.g., four) and only by cost of low power efficiency on the order of 5 percent.

For this reason, a base station for a relatively large number of carriers will use separate DACs for the different signals and will, after upconversion to the final frequency, also separately amplify the carriers. For power-efficient combining of the RF signals, a transmitter combiner composed of single high-Q resonators and a manifold is used. To allow for adaptive frequency allocation, the frequencies of the resonators must be electrically tunable. If, instead of conventional bulky resonators with Q values below about 20,000, miniaturized HTS resonators with Q up to 200,000 at 60 K are implemented, the insertion loss per channel can typically be reduced from about 1.5 to 0.2 dB. This allows significant reduction of the power load of the HPAs, by about 50 percent. Furthermore, the smaller resonator size may help to implement the transmitter on the masthead, if the compressor of the cooling system is placed on the ground. This expected impact justifies the development of an electrically tunable HTS transmitter combiner within the framework of the SUCOMS project.

New Technology for HTS RF Circuits, Encapsulation, and Cooling

Since a detailed description of the technology is beyond the scope of this article, only some important issues are summarized.

Design challenges of the RF, encapsulation, and cooling are inseparably linked. For the masthead base transceiver station the RF topologies are chosen to minimize encapsulation size and cooler capacity requirements, the cryo-temperature is as high as possible for efficient design. and the unit must be environmentally robust.

HTS technology is discussed in [19, 20]. All HTS RF circuits, including the preselect bandpass filters and stopband filters, as well as the resonators for oscillator stabilization and for the transmitter combiner incorporate YBaCuO or GdBaCuO thin films epitaxially grown on compatible single-crystal substrates ($LaAlO_3$, MgO) up to 2 in diameter. The filters employ distributed and lumped "microstrip-like" structures.

Encapsulations featuring novel designs and proven vacuum manufacturing techniques yield reduced thermal loads compared to conventional microwave designs. Modular construction allows the assembly of stacked HTS functional building blocks which can be tested *in situ* prior to final closure. The basic cooler is compact. Adding additional RF functionality does not significantly increase the size.

For a BTS the cooler has to meet technical performance and cost criteria. To meet the challenges of UMTS a new cooler incorporating new features and materials, together with proven technology and production methods, is used. Refrigeration is provided by a special design of cold head and balanced Stirling compressor. This cooler has many advantages over the GM cooler (e.g., smaller, more efficient, no water cooling, long maintenance life cycle, low audio output and vibration).

Conclusion

In conclusion, advanced research is required in the area of RF carrier combination, which may ultimately be superseded by baseband Tx combination techniques. However, these techniques would require very linear transmitters. In the immediate future, advanced techniques such as flexible software radio architectures, adaptive antennas, and high-temperature superconductors are likely to offer the following advantages: a higher level of integration, a degree of "future-proofing" due to upgradable software, extension of coverage, increased capacity due to frequency reuse, increased interference tolerance, lower receiver noise floor, less stringent linearity requirements, and improved filter performance. Many of these techniques depend on high-speed DSP processing, the improvement of which is clearly a key enabling mechanism.

Acknowledgments

The authors wish to thank CEC for funding the TSUNAMI, FIRST, and SUCOMS projects. The authors from ERA would like to thank all the partners in TSUNAMI (Motorola ECID, Orange PCS Ltd., CASA, Bosch Telecom GmbH, France Telecom CNET, WSI Ltd., University of Aalborg, University Politecnica de Catalunya, and University of Bristol) for their contributions to the work.

The author from FIRST would like to thank all the partners in FIRST (CSEM, Dassault Electronique, Motorola ECID, Orange PCS Ltd., Motorola Semiconductor Products Sector, Orange, Robert Bosch GmbH, Thomson CSF Communications, University of Bristol, and University of Southampton).

The author from SUCOMS would like to thank all the partners from GEC-Marconi Materials Technology, Bergische Universtat Gesamthochshule Wuppertal, Thomson-CSF Communications, Thomson-CSF Laboratories Central de Recherches, and University of Birmingham.

REFERENCES

[1] B. G. Agee, "Blind Separation and Capture of Communication Signals Using Multitarget Constant Modulus Beamformer," *Proc. IEEE MILCOM '89*, 1989, May 1989, pp. 304–46.

[2] A. Van der Veen and A. Paulraj, "An Analytical Constant Modulus Algorithm," *IEEE Trans. Sig. Processing*, vol. 44, no. 5, May 1996.

[3] J. F. Cardoso and A. Souloumiac, "Blind Beamfoming for Non-Gaussian Signals," *Proc. IEE-F Radar and Sig. Processing*, vol. 140, no.6, 1993, pp. 362–70.

[4] B. Fleury *et al.*, "Wideband angle of arrival estimation using the SAGE algorithm," *Proc. IEEE 4th Int'l. Symp. Spread Spectrum Techniques and Applications (ISSSTA '96)*, Mainz, Germany, Sept. 1996, pp. 79–85.

[5] P. E. Mogensen *et al.*, "TSUNAMI (II)Stand Alone Testbed," *Proc. ACTS Mobile Telecommun. Summit '96*, Granada, Spain, Nov. 1996, pp. 517–27.

[6] "Algorithms and Antenna Array Recommendations — Part 1," TSUNAMI II del. AC020/AUC/A1.2/DR/P/005/b1.

[7] T. S. Rapport, J. H. Reed, and B. D. Woerner, "Position Location Using Wireless Communications on Highways of the Future," *IEEE Commun. Mag.*, Oct. 1996, pp.33–41.

[8] J. A. R. Fonollosa and J. Vidal, "Application of Hidden Markov Models to Blindchannel Characterisation and Data Detection," *Proc. IEEE Int'l. Conf. Acoust. Speech and Sig. Processing*, Special Issue on Signal Processing for Advanced Communications, Jan. 1997.

[9] J. D. Laster and J. H. Reed, "Interference Rejection in Digital Wireless Communications," *IEEE Sig. Processing*, May 1997, pp. 37–62.

[10] D. Gerlach and A. Paulraj, "Adaptive Transmitting Antenna Arrays with Feedback," *IEEE Sig. Processing Lett.*, vol. 1, Oct. 1994, pp. 150–52.

[13] J.H. Winters, "Increased Data Rate for Communication Systems with Adaptive Antennas," *Proc. IEEE Int'l. Conf. Commun.*, June 1982.

[12] M. Wales *et al.*, "The Performance Enhancement of Multibeam Adaptive Base Station Antennas for Cellular Land Mobile Radio Systems," *IEEE Trans. Vehic. Tech.*, vol. 39, Feb. 1990, pp. 56–57.

[13] Barrett and R. Arnott, "Adaptive Antennas for Mobile Communications," *Elect. Commun. Eng. J.*, vol. 6, no.4, Aug. 1994, pp. 203–14.

[14] P. Jung, F. Berens, and J. Plechinger, "Joint Detection for Multicarrier CDMA Mobile Radio Systems," *Proc. IEEE Int'l. Symp. Spread Spectrum Techniques and Apps.*, vol.3, Sept. 1996, pp. 991–95.

[15] N. Newman and W. G. Lyons, "High Temperature Superconducting Microwave Devices," *J. Superconductivity*, vol. 6, 1993, pp. 119–60.

[16] M. J. Lancester, *Passive Microwave Device Applications of High-Temperature Superconductors*, Cambridge Univ. Press, 1997.

[17] H. Chaloupka *et al.*, "Superconducting Planar Disk Resonators and Filters with High Power Handling Capability," *Elect. Lett.*, vol.32, 1996, pp. 1735–37.

[18] J. Mitola, "The Software Radio Architecture," *IEEE Commun. Mag.*, vol. 33, May 1995, pp. 26–38.

[19] M. Nisenoff and W. J. Meyers, eds.,Special Issue on the Microwave and Millimeter Wave Applications of High Temperature Superconductivity," *IEEE Trans. Microwave Theory Tech.*, vol. 44, July 1996.

[20] H. Chaloupka, M. Hein, and G. Müller, "HTS Microwave Applications in Europe," *Proc. SPIE-/LASE '94*, Los Angeles, CA: Int'l. Soc. for Optical Eng., 1994, pp. 36–54.

ADVANCED SOFTWARE RADIO ARCHITECTURE FOR 3RD GENERATION MOBILE SYSTEMS

Heiko Erben, Kyriacos Sabatakakis
CSEM - Centre Suisse d'Electronique et de Microtechnique S.A.,
Jaquet-Droz 1, CH-2007 Neuchâtel, Switzerland
heiko.erben@csemne.ch

Abstract: *This contribution presents a architecture model for integrating flexible and adaptable radio interface features into a 3rd generation terminal and system. This model is based on describing the exchangeable features as functional entities with their corresponding interactions rather than using a layer structure such as the OSI model but aiming on supporting any transmission scheme, user defined as well as standardised 3rd and 2nd Generation systems. Exemplary, the main issues related to flexible and dynamic use of radio resource are addressed in particular for the Physical Layer.*

I. INTRODUCTION

Traditionally, common air interface (CAI) standards have been rigorously defined in great detail to ensure optimised operation for very specific applications, e.g. voice and telephony, and interoperability between the equipment of different manufactures. Examples of such standards include in particular 2nd Generation systems such as GSM, DECT, TETRA, IS95, IS54/136, PHS, PDC, etc. Contrary to the application specific mobile system design for 2nd Generation systems, the European 3rd Generation system (UMTS) aims towards the provision of highly flexible and adaptive transmission schemes to provide optimised transmission for each user with a wide range of supported services up to 2Mbit/s. In this context, software radio technology as one possible means of flexible communication will facilitate the implementation of adaptive and flexible transmission schemes, protocols as well as applications.

In general, the term software radio is not very precise defined and often used in different context. Trying to find clear definitions and solutions for software radio matters, it is necessary to refer to one well known definition, which is to implement all aspects of the air interface in software, hosted in general purpose programmable processors (DSPs and MPs) and using wideband ADCs/DACs for sampling high intermediate frequencies (IF) or even radio frequencies (RF). This approach is currently under investigation in the MMITS Forum and the European projects FIRST and SLATS for 3rd Generation UMTS operation.

In this paper, the term software radio or software radio technology should be defined as any specific hardware and software co-design to support highly flexible terminal (HW & SW) operation and providing means of highly adaptive and optimised data transmission by keeping low costs and power consumption in general. Obviously, regarding this definition, the software radio implementation issue is based on a software controlling/operating system together with dedicated HW (ASSP, ASIP, ASIC, etc.) rather than pure SW implementation. Either case, this kind of software radio implementation has to be independent on used technology and transmission techniques.

Considering these both complementary software radio definitions, the software radio architecture on the other side becomes an issue of providing implementation and manufacturer independent flexibility and functionality which needs to be reflected in the standard to ensure operation for virtually unlimited types of terminals and transmission modes.

Within the 3rd Generation UMTS standardisation process of ETSI, the objectives of software radio are discussed in the context of adaptive and flexible terminals which are supposed to support different air interfaces and modes to provide functionality for various telecommunication services. Recently, ETSI has made the decision in favour of TD-CDMA (FRAMES Multiple Access Mode 1 -FMA1) and W-CDMA (FRAMES Multiple Access Mode 2 - FMA2) as supported modes within UMTS.

This contribution presents an architecture model investigated in the context of the FIRST & SLATS project for integrating flexible/adaptable radio interface features into a UMTS terminal and system. This architecture defines a set of functional entities that are only rigorously defined on their interface part leaving a free choice of implementation. This definition will permit the manufacturer to define optimal implementation for any kind of terminal in terms of complexity, costs and maximum reuse of resources in functional entities.

II. Scope of adaptive Terminals

The choice of an air interface parameter set corresponding to a multiple access scheme seems to be a critical issue as

Reprinted from *Vehicular Technology Conference*, pp. 825–829, September 1998.

far as spectral efficiency is concerned. There is much research in progress to enhance the different access techniques TDMA, FDMA, CDMA and OFDMA with the aim to define the best access scheme for all conditions. Within the European Programmes RACE (projects CODIT and ATDMA [4,5,6]), the European ACTS Programme (projects FRAMES, MEDIAN, WAND [8,9,11,12,13]), COST 231 [7] as well as in the framework of other research activities, assessments of several different schemes have been carried out, each claiming advantages over the other but for particular transmission environments. It is thus clear, that without further tests and investigation it is difficult to declare a clear winner, baring in mind that the possible advantages and/or disadvantages of each of the proposed schemes depends on a number of circumstances such as channel characteristics, Quality of Services (QoS) supported, traffic requirements, etc.

One possible solution seems to be the realisation of flexible and adaptable software radio technology, supporting virtually any parameter set of transmission scheme and being capable for adaptation to different parameter set while transmitting (changing slot length, burst structure, bandwidth, spreading, modulation, interleaving, coding, etc.). In this way, advantages from each transmission scheme can be integrated and the varying requirements on channel capacity, traffic load and transmission quality in different overlaying hierarchical cells can be met.

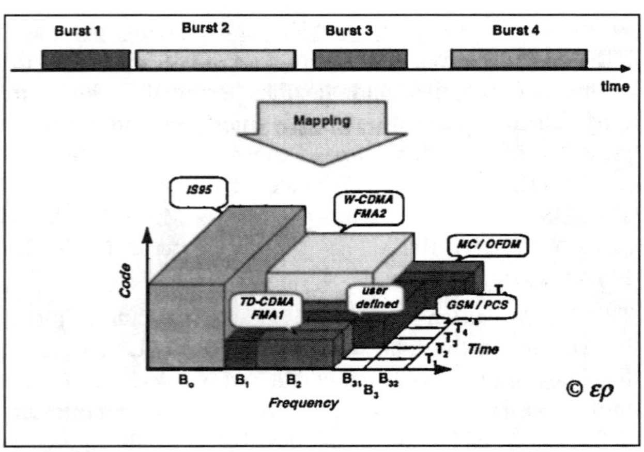

Figure II.1: Flexible multiple mode operation

A conceivable configuration in the code-frequency-time domain is schematically depicted in Figure II.1. Bursty traffic is mapped on the radio channel using the most appropriate physical transmission parameter within a user defined slot. The slot used for transmission can be selected either from predefined or user defined slots (see Figure II.2). In this way, each user is able to define its own optimised physical channel keeping the capability to change the transmission parameter during transmission, for example on frame basis. Changing transmission parameter applies primarily for the modulation index [2], varying slot length, spreading option, coding and interleaving.

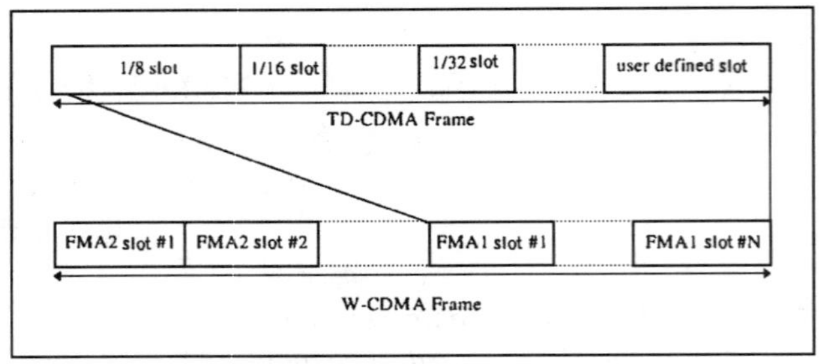

Figure II.2: Flexible frame structure

Furthermore, already defined or envisaged transmission schemes such as the W-CDMA (FMA2), TD-CDMA (FMA1), as well as 2nd Generation standards (GSM, PDC, IS95, etc.) are able to coexist in the same operator frequency band (see Figure II.1). Dynamic sharing between such well defined standards or modes could be performed in the frequency and time domain during transmission as well as on fixed predefined sharing method. The code domain is mainly reserved for multiple user sharing. Hence, any transmission mode can be applied on a operator defined sharing mechanism and network visitors are able to be supported within the limits of their terminals.

In either case, a lookup table needs to be defined maintaining the user defined transmission schemes as well as the predefined transmission modes for each user. This information needs to be known completely on the network side and partially on the terminal side and has to specify the capabilities of the terminal to enable the network to support terminals with different features, 3rd generation specific as well as traditional 2nd generation ones. During call set-up, the base station negotiates the terminal capability and adapt the transmission scheme within the terminal capability. A outlined feature description table is defined in Table II.1.

Table II.1: Outlined feature description table

Bw	bandwidth supported
Nc	number of carriers supported
Dm	duplex mode supported
Cs	carrier spacing
Ss	slot length and structure supported
Ms	modulation schemes supported
Sc	spreading code supported
Mc	channel codes supported
Um	user defined mode supported

In a similar way, the definition of optimised modes for certain transmission condition might be subject for standardising further predefined operation modes within the UMTS standardisation process.

Obviously, this flexible operation is more complex, but with respect to recently investigated hybrid schemes [7] the realisation of such an adaptive air interface is not so far away. Advantages can be seen in realising the combination of any kind of radio transmission as well as flexibility and adaptivity. Furthermore, such an adaptive terminal and basestation, including intelligent medium access and dynamic channel assignment routines realising service and channel specific adaptation functions, will lead to a significant increase in capacity and service quality. Defining parameter sets for different cells, traffic loads, QoS, etc., within a common UMTS standard can be seen as logical choice for a better radio resource management giving "bandwidth or traffic capacity on demand".

Disadvantages of such an flexible realisation are the high complexity, novelty and difficulties in achieving unknown simplified low power and low cost transceiver design as well as efficient flexibility management in several hierarchical cell structures (HCS).

III. Architecture Model

Applying adaptive system features for the physical transmission part as described in the previous chapter are primarily an issue of applied adaptive signal processing. Dynamic protocol and routing issues have to be solved as well but they are not addressed in this paper. On the other side, implementing flexibility and adaptivity becomes an issue of interface definition to ensure conformity with any manufacturer design by keeping proprietary realisations for competitive design. Hence, appropriate models and specifications have to be used to standardise all interactions as detailed as necessary with maximised freedom for changes, flexibility and proprietary solutions.

Modelling features such as flexibility and adaptivity for the physical transmission part implies strong reconsideration of existing approaches. The most common models are the layered OSI and ATM model. The OSI model does not allow any clear separation between user (transport), control and management data and message flow and restricts several functions to reside in corresponding layers independent on suitability. Hence, this model doesn't seems to be very efficient to be used for modelling flexibility and adaptivity issues. On the other side, the ATM model provides already the possibility to separate data and message flow via the transport/user, control and management planes but still restricts functions to the layer approach. Even when the layered ATM model relaxes already some deficiency, still it does not fulfil all requirements to model flexibility and adaptivity issues.

Therefore, a novel unlayered model based on functional entities is applied in this proposal to define functional entities for the physical transport of data and control messages. A functional entity in this context is defined as follows (see also Figure III.1):

A functional entity performs one or more functions, may contain one or more than one functional subentities, is able to communicate with any other functional entity (receive, request and respond to messages) and performs corresponding actions to given commands. Note: Functional entities will be grouped in this approach by functional groups such as for example analogue RF, digital baseband, burst assembly, etc.

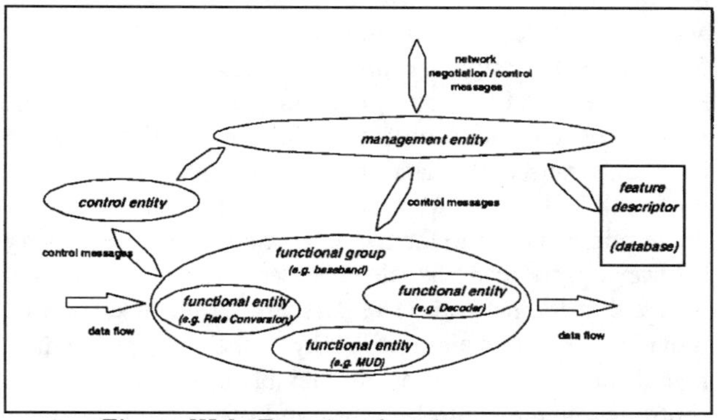

Figure III.1: Functional architecture model

The advantage of this functional entity model can be seen in the possibility to exchange any functional entity with any other one performing similar functions with different parameter, complexity and performance but keeping the interface between the functional entities fixed as specified. Various demodulator, equaliser, interference cancellation algorithm, etc. could be mentioned as example.

Based on this exchangeable functional entities, it is possible to define a managed generic architecture which is able to replace any functional entity in any type of terminal or basestation by keeping the algorithm and implementation issue subject to manufacturer dependent solution. Note, this model intends to specify the interfaces between the functional entities on a high level approach and keeps even the physical implementation and realisation of these interfaces subject for processor or hardware dependent solutions but depends on an implemented management system, e.g. controlling or operating system, handling and co-ordinating flexible operation. A suitable implementation model is shown in Figure III.2.

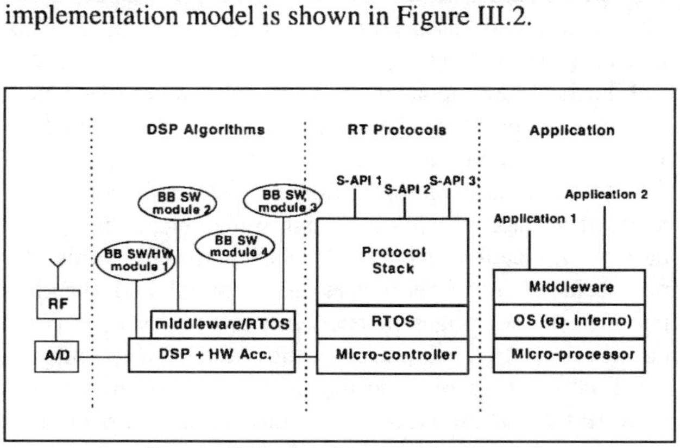

Figure III.2: Implementation model

In this context, this way of specifying flexibility and adaptivity might be taken over and explored further by the standardisation bodies to enable multiple access and transmission independent operation of a variety of different terminals and basestations.

Further remaining questions are how to replace several functional entities as well as how to handle various implementations of functional entities. The first issue can be supported by either hardware implementation of generic algorithm which are able to be configured for various functions with similar arithmetic structure (e.g. equaliser, RAKE and FIR filter) or by changing the software content in case of software implementation. A possible generic baseband transceiver data flow architecture is shown in Figure III.3. In the later case, the software can remain in the terminal subject to the available memory or the missing software parts can be downloaded from a dedicated database in the network. The database service itself can be provide and maintained by the corresponding manufacturer. In any cases, the flexible operation has to be controlled by the control entities and managed in terms of data processing architecture and handshaking with the network by an management entity.

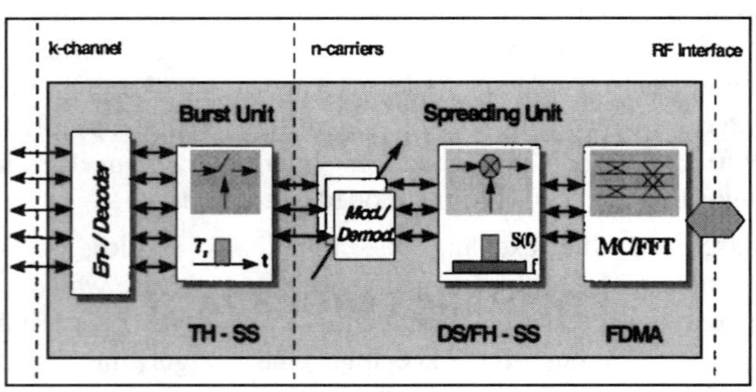

Figure III.3: Generic baseband data flow architecture

Concerning the second issue, the handling of various functional entity configuration has to be done by the management entity. For this purpose, the management entity has to negotiate or indicate during call establishment possible transmission modes by clarifying the capability of the terminal and basestation or wireless access point. This will be supported by a global feature descriptor database and leads to an updated copy of the feature descriptor database for the established call (see step 1 in Figure III.4). The feature descriptor database itself contains information for several transmission modes such as bandwidth, modulation, slot length, data burst, midamble, coding, etc., but with indication of parameter variation. Afterwards, the management entity configures the terminal and specifies parameters and ranges of operation to the several control entities (step 2). The control entities otherwise reconfigure the functional entities during transmission without any network interaction within the specified range of parameter settings and are furthermore able to exchange dedicated functional entities (step 3). The whole date flow will be controlled in this way by the responsible control entities taking further into account direct command with associated data exchange between functional entities as well as the corresponding control entity which initiates all necessary actions (step 4). Reconfiguration is mainly aimed towards setting parameter in dedicated functional entities. Exchanging functional entities otherwise, substitutes one entity with the other one supported by the toolbox approach, downloadability or flexible HW component feature (step 5). Furthermore, one of these control entity also performs actions concerning dynamic channel allocation and MAC data PDU queuing and handling. In case of changing complete the transmission mode, e.g. between different standards, the control entity communicates to the management entity and gets information directly from the updated feature descriptor database for changing the transmission mode (step 6).

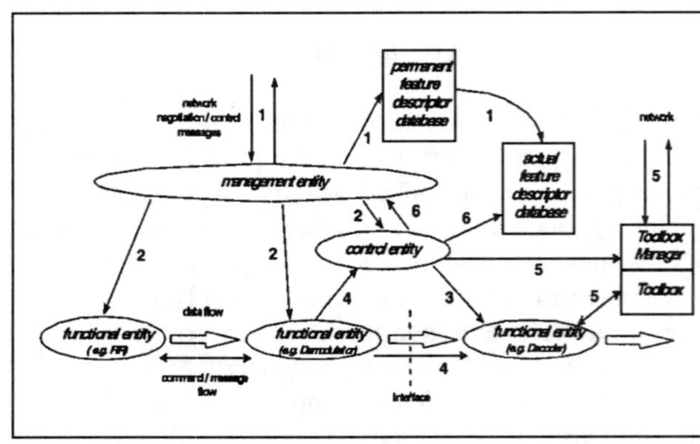

Figure III.4: Control and exchange of functional entities

Applying such an functional architecture definition provides all means for flexible and adaptive operation. Well defined transmission modes or standards can be applied as well as user defined operation without interfering each other. Adaptive operation with unknown future transmission schemes are possible to introduced even after having defined the UMTS standard subject to standardisation of such an interface definition.

IV. Conclusions - Terminal survey

In this article, the terminology adaptive terminal or software radio addresses mainly the realisation of adaptive and flexible physical layer operation. That means mainly hardware and software co-design which is capable to be flexible configured for optimised data transmission by all means of software operating or controlling systems.

There is obviously a trade-off between flexibility and optimised high speed data transmission. Specialised hardware (ASIC/VLSI) is more capable to support high speed transmission by low power consumption but only for very few specific system configurations which are based on a single chip or hardware design. This finally leads to a number of implementations, each of them optimised for specific conditions. In particular in mobile communication systems it becomes a critical issue in terms of various incompatible systems and terminals on the market. On the other side, sophisticated system design is paid by high complexity and power consumption limiting the market penetration even in case of optimised transmission capabilities and lower costs.

One aim should be therefore for such software radio technology to lower complexity based on generic architecture and arithmetic algorithm design. This can be achieved by providing a higher degree on flexibility based on software controlled hardware configurations incorporating generic signal processing algorithm to lower multiple chip systems in terms of hardware complexity.

On the other side, software radio technology is supposed to support any terminal operation. Based on the concept of negotiating terminal capabilities for adaptive operation, a wide variety of terminals with limited functionality can be supported as well. Implementing this concept of flexible operation in the basestation would not only allow more spectrum efficient operation for adaptive terminals, furthermore, it would permit very simple terminals with minimised functionality to be handled by the basestation which gets adapted to the terminal operation. In this context, a wide market for any kind of terminals will be opened by standardising a flexible architecture and corresponding interfaces between functional entities.

V. ACKNOWLEDGEMENTS

The authors wish to express their appreciation to their colleagues in the ACTS and IT projects, in particular the FIRST, SLATS, FRAMES, RAINBOW, etc., who spare no efforts to contribute, through their creative ideas, to the definition of the advanced Third Generation Mobile and Personal Communications.

VI. References

[1] H.Erben, J.S.Dasilva, B.Arroyo, B.Barani, D.Ikonomou: *European R&D towards Third Generation Mobile Communication Systems.* International Conference on Telecommunication ICT'96, Istanbul, Turkey, 14-17 April, 1996.

[2] J.M.Torrance, L.Hanzo, and T.Keller: *Interference resilience of burst-by-burst adaptive modems.* ACTS Summit 1997, Aalborg, 7.-10. Oct. 1997.

[3] Proceedings of the *ACTS Mobile Telecommunications Summit.* European Commission, Granada, Spain, November 27-29th, 1996.

[4] Proceedings of the *RACE Mobile Telecommunications Summit.* European Commission, Cascais, Portugal, 22-24 November 1995.

[5] Mobile Communications: *UMTS Radio Aspects. IBC Common Functional Specifications, Issue E1*, Specification RACE D734, May 1995.

[6] Mobile Communication Networks: *General Aspects and Evolution. IBC Common Functional Specifications, Issue E1*, Specification RACE D731, May 1995.

[7] Mobile and Personal Communications. Proceedings of the 2nd Joint COST 227/231 Workshop on Mobile and Personal Communications, Florence, Italy, 20-21 April, 1995.

[8] J. S. Dasilva, D. Ikonomou and H.Erben: *European R&D programs on Third-Generation Mobile Communication Systems.* IEEE Personal Communications, February 1997. Vol.4,No.1

[9] Erben H.: *UMTS - A Vision of 3rd Generation of Mobile Communication - A Service Perspective.* International Conference on Telecommunication, Bali, 3rd-5th April, 1995.

[10] Urie A., Streeton M., Mourot C.: *An advanced TDMA mobile access system for UMTS.* IEEE Comms. Mag., pp38-47, Feb 1995.

[11] Steiner B., Jung P.: *Optimum and suboptimum channel estimation for the uplink of CDMA radio systems with joint detection.* European Transactions on Telecommunications and Related Techniques, vol. 5, pp. 39-50, 1994.

[12] Klein A., Pirhonen R., Sköld J., Suoranta R.: *FRAMES Multiple Access Mode 1 - Wideband TDMA with and without spreading.* Proceedings of the 1997 International Symposium on Personal, Indoor and Mobile Radio Communications, PIMRC'97, pp. 37-41, 1997.

[13] Klein A., Pirhonen R., Sköld J., Suoranta R.: *FRAMES Multiple Access Mode 2 - Wideband CDMA.* Proceedings of the 1997 International Symposium on Personal, Indoor and Mobile Radio Communications, PIMRC'97, pp. 42-, 1997.

A Soft Radio Architecture for Reconfigurable Platforms

Srikathyayani Srikanteswara, Jeffrey H. Reed, Peter Athanas, and Robert Boyle,
Virginia Tech

ABSTRACT

While many soft/software radio architectures have been suggested and implemented, there remains a lack of a formal design methodology that can be used to design and implement these radios. This article presents a unified architecture for the design of soft radios on a reconfigurable platform called the layered radio architecture. The layered architecture makes it possible to incorporate all of the features of a software radio while minimizing complexity issues. The layered architecture also enables a methodology for incorporating changes and updates into the system. An example implementation of the layered architecture on actual hardware is presented in the article.

INTRODUCTION TO SOFT RADIOS

Software radios are evolving as flexible all-purpose radios that can implement new and different standards or protocols through reprogramming [1–3]. Software radios give rise to the possibility of having multimode terminals without the "Velcro approach" of including separate silicon for each possible standard. Software radios can reduce the cost of manufacturing and testing, while providing a quick way to upgrade the product to take advantage of newer signal processing techniques and new wireless phone applications. The ideal software radio has a set of features that are not yet realizable in commercial systems due to limitations in current technology and cost considerations. Specifically, digital signal processing (DSP) microprocessors are not fast enough to implement all radio functions, and thus some dedicated application-specific integrated circuits (ASICs) are required. This is a consequence of processing speed being traded for flexibility and efficiency.

In this article the term *soft radio* denotes a completely configurable radio that can be programmed in software to reconfigure physical hardware. In other words, the same piece of hardware can be modified to perform different functions at different times. This allows the hardware to be specifically tailored to the application at hand, resulting in greatly increased speed and silicon efficiency, while maintaining a high degree of flexibility.

Although most researchers agree on the advantages of the soft radio, there is a lack of a general design methodology for soft radio architectures. Most practical implementations are based on ad hoc approaches that can only be used for a few specific systems. In this article a formal architecture for soft radios, called the *layered radio architecture*, is developed for reconfigurable platforms.

DESIGN ISSUES IN A SOFT RADIO

A number of papers describe various software radio architectures and design considerations [1–4]. The key design issues include design of fast and efficient analog–digital (A/D) converters, flexibility at the radio frequency (RF) front-end, effective data management procedures, resource allocation, and smooth reconfigurability of the hardware. The A/D converter and flexible RF front-end are extremely challenging to design, particularly for a handset, and set the limitations of dynamic range and bandwidth. The design issues involving the A/D converter and RF design are beyond the scope of this article but are covered in [5].

A soft radio is a complex entity that needs to handle very high data rates efficiently. To process data efficiently entails good computing resource allocation. There are very important differences in the resource allocation of a system implemented in dedicated hardware and the same system implemented in a soft radio. While both approaches strive for compactness and power efficiency, a soft radio design requires additional flexibility and reusability, which impacts the partitioning of a process into its hardware components. This resource allocation is also closely related to the data flow properties of the system since data has to be processed with minimal delays and overheads.

Another important design issue is the smooth reconfiguration of the radio, including the ability of runtime reconfiguration and over-the-air updates. Smooth reconfiguration is necessary to minimize delays, and make the system robust and easy to handle. The reconfigurable feature of a soft radio gives rise to some secondary design issues such as reusability of hardware, scalability, and processing power. For mobile communication systems, power consumption is an important issue. Reusability of hardware supporting multiple communication standards is what makes a soft radio efficient in terms of silicon utilization.

A soft radio should be inherently scalable, at

Series Editors:
J. Mitola and Z. Zvonar.

Reprinted from *IEEE Communications Magazine*, Vol. 38, No. 2, pp. 140–147, February 2000.

both the system and hardware element levels. Hardware scalability supports the easy addition of more computational hardware, ideally as easily as adding additional memory to a conventional processor. At the system level, the radio has to be conducive to *replication* (i.e., it should be easy to add channels to support more users). Finally, the hardware platform must provide high-rate computational processing.

SOFT RADIO RESEARCH AND COMMERCIALIZATION

DARPA's Adaptive Computing Systems Project — Development of commercial off-the-shelf (COTS) hardware components, design, programming, and runtime environments to enable an application to reach through to the hardware layer and directly manipulate the data-path-level architecture at runtime to optimize the application-level performance; exploration of new techniques that result in rapid realization of algorithm-specific hardware architectures on a low-cost COTS technology base; http://www.darpa.mil/ito/research/acs/index.html

Virginia Tech — Research in developing algorithms and architectures for soft radios under the sponsorship of DARPA's GloMo I & II projects; implementation of a multi-user receiver based on reconfigurable computing under GloMo I [6]; development and implementation of a generic soft radio architecture for reconfigurable hardware under GloMo II; http://www.mprg.ee.vt.edu/research/glomo/index.html

University of California at Berkeley — Work on Pleiades, part of the Adaptive Computing Systems project to achieve ultra-low-power high-performance multimedia computing through the reconfiguration of heterogeneous system modules; achievement of high power efficiency by providing programmability at the right granularity and exploiting other energy-reducing techniques, such as parallelism, pipelining, and dynamic voltage scaling; http://bwrc.eecs.berkeley.edu/Research/Configurable_Architectures

Brigham Young University — Development of JHDL, Java-based Hardware Description Language, to facilitate hardware synthesis in reconfigurable processors; http://jhdl.ee.byu.edu/docs/jhdlDocs.html

Chameleon Systems Inc. — Development of reconfigurable processor platform architecture to reduce the trade-off between performance and flexibility needed in wireless base stations; http://www.chameleonsystems.com/newspg.htm

Morphic Inc. — Programmable hardware reconfigurable core using dynamically reconfigurable logic (DRL) to replace multiple hardwired logic cores in wireless baseband processors; http://www.morphics.com/html/tech.html

Quicksilver Technology Inc. — The Wireless Universal 'Ngine (WunChip); supports a variety of baseband algorithms that are downloaded from software; http://www.quicksilvertech.com

Sirius Inc. — Software-reconfigurable code-division multiple access (CDMAx) test chip to support mobile station and base station configurations; can also be reconfigured as a Global Positioning System (GPS) navigation baseband receiver and a satellite Universal Mobile Telecommunications Service (UMTS) transceiver core; http://www.siriuscomm.com

While most researchers agree on the usefulness of the soft/software radio and most of its desirable features, there is a lack of general architecture applicable to reconfigurable systems. Most designs are based on ad hoc approaches which are only appropriate for the problem at hand. There is a lack of a design methodology needed for the systematic realization of soft radios. Although much work is being done in the software and network layers of the soft radio, there has been little progress in the development of a generic hardware architecture for a soft radio.

In this article we develop an architecture leading to a design methodology for soft radios. We show that the architecture can handle complex data processing with efficient resource allocation, while maintaining hardware reusability, flexibility, and scalability. The architecture is illustrated with an example that has been implemented in hardware.

THE LAYERED RADIO ARCHITECTURE

AN OVERVIEW OF THE LAYERED RADIO ARCHITECTURE

The approach to designing a reconfigurable radio using *hardware paging* is formalized in the layered radio architecture. Hardware paging refers to hardware modules being paged in and out of the system in a manner similar to software paging performed with the use of virtual memory. The layered architecture leverages on stream-based processing, where a common bus is used for data as well as programming information. A few bits within the bus are reserved for the header, which indicates the nature of the stream packet. The use of stream-based processing simplifies the interface between modules, making it easy to replace one module with another or add additional modules. The use of a reconfigurable hardware platform enables hardware paging. The architecture can handle complex data processing with efficient resource allocation, while maintaining hardware reusability, flexibility, and scalability.

The functionality of the radio is divided into layers, where each layer attaches/modifies the header and passes the information to the next layer. Once the processing is complete, the information is sent back through the layers in the same way. The layered radio architecture defines three layers (Fig. 1): the soft radio interface (SRI) layer, the configuration layer, and the processing layer. The application software residing above all these layers consists of high-level graphics and software that interface the radio with the user. The three layers of the soft radio are implemented using stream-based processing. The SRI layer is responsible for interfacing the radio hardware with the external world, coordi-

Hardware paging refers to hardware modules being paged in and out of the system in a manner similar to software paging performed with the use of virtual memory. The layered architecture leverages on stream-based processing where a common bus is used for data as well as programming information.

Stream-based processing provides a means to exploit the processing power attainable through deep pipelining, while still maintaining some degree of flexibility.

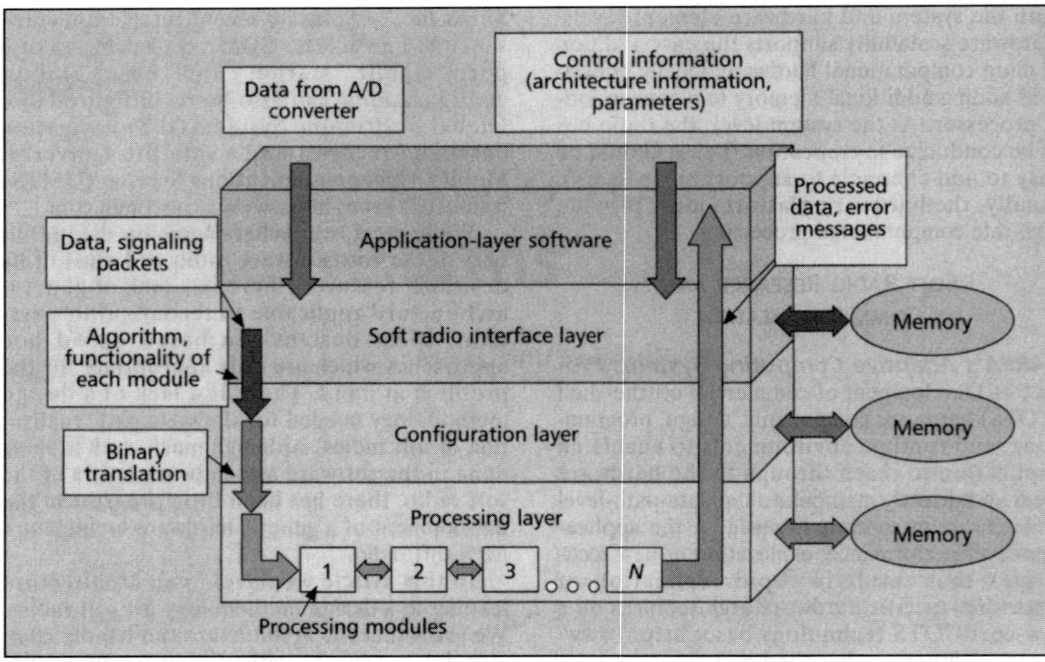

■ **Figure 1.** *The layered architecture.*

nating the various sources of information coming into and going out of the radio, prioritizing the requests made to the radio, and packetizing the incoming information before sending it to the lower layers. The information coming into the radio can be either data, programming information requesting a new setup of the radio, or a modification to the existing setup. The setup of a new system is handled as follows.

The SRI layer contains the system-level description of various radio configurations in its local memory. Each system-level description contains a list of all the algorithms in the appropriate order needed to implement the system. It is important to note that the SRI layer does not contain the bits needed to configure the hardware. Rather, it contains the codes for the algorithms that are to be used and their sequence of interconnection.

The configuration layer has in its memory the actual bits needed to configure the processing layer hardware. The configuration layer extracts the configuration bits from its memory based on the programming packets sent by the SRI layer. The processing layer consists of a series of reconfigurable modules, called *processing modules*, that perform the actual operations on the data to implement the functionality of the radio (Fig. 2). Each processing module has static and reconfigurable sections; the static section configures the reconfigurable section based on the programming packets sent by the configuration layer. Once the radio is configured, data is packetized by the SRI layer and sent to the lower layers using the same bus used for the programming packets. After completion of processing, data is sent back through the configuration layer to the SRI layer, which in turn depacketizes the data and outputs it to the host PC.

The Use of Stream-Based Processing in the Layered Architecture

The soft radio architecture relies on a concept called stream-based processing. A stream is a packet of known length containing either programming (configuration) information or the data to be processed. The stream packets are first formed at the SRI layer. These stream packets are then interpreted by the lower layers of the radio based on the header information.

Stream-based processing provides a means to exploit the processing power attainable through deep pipelining, while still maintaining some

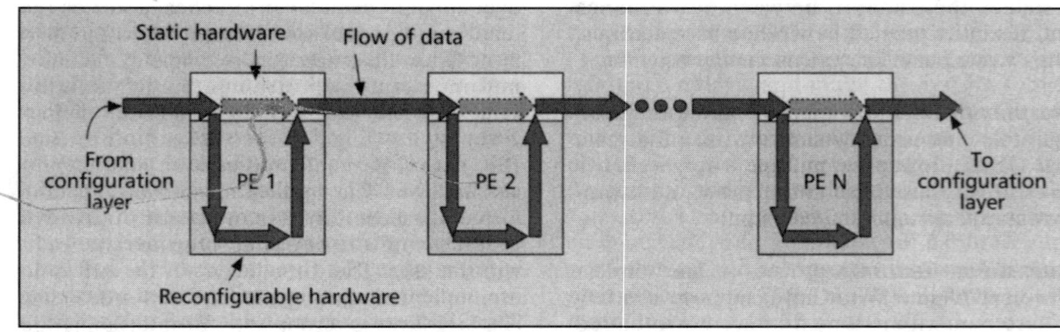

■ **Figure 2.** *Processing elements in the processing layer.*

■ **Figure 3.** *Stream-based processing.*

degree of flexibility [7]. The algorithm to be implemented is first represented as a *data flow graph*. The data flow graph is then decomposed into smaller computational primitives called processing modules. Each processing module performs a unique subset of the overall processing on the data and then passes the data and control information to the next module. An analogy can be drawn to an assembly line process where each module performs a specific task as the component moves forward in the assembly line, as shown in Fig. 3.

Implementation of the Layered Approach

The Soft Radio Interface Layer

The SRI layer receives the incoming digitized RF data and control information from the host PC that contains the desired configuration and settings of the radio. The SRI layer transmits processed data to the host PC, performs initialization of the system on power-up, and performs the first step in configuring the radio. This layer has the system-level description of the desired radio configuration in its local memory, which contains the algorithm codes for each processing module needed to implement the system. This system-level description is sent to the configuration layer, which in turn configures the processing layer modules.

The incoming data to the SRI layer is either from the A/D converter (which always provides data) or control and status requests from the host PC. Once the user decides on the algorithms to be used, the SRI layer inserts the corresponding codes and forwards it to the configuration layer. Data that follows will be processed using these preset algorithms until the user decides to change the configuration of the system. Control packets containing over-the-air updates are used to change the configuration of the library. The requests from the PC (control information), as well as data from the A/D, are buffered and prioritized by the SRI layer and sent to the configuration layer.

The stream packets received by the SRI layer from the configuration layer have more information in them. The configuration layer sends the processed data back to the SRI layer along with other status and error messages. First, the SRI layer checks to see if the data contained in the packet is valid. If it is valid and the packet contains data, it is sent to the host PC, along with error messages if any. When the SRI layer sends program or control information that was requested by the host PC to the configuration layer, the packets from the configuration layer will contain acknowledgments that indicate if the operation was successful. Either these messages can be sent back to the host PC, or the SRI layer can retransmit the control packets.

The Configuration Layer

The information from the SRI layer indicates whether the packet is a control or data packet. If it is the former, the packet contains the codes that represent the algorithms of the desired configuration. For example, the control packet may request a differential quadrature phase shift keying (DQPSK) demodulator followed by a Viterbi decoder. The appropriate configuration code is extracted from the local memory and sent to the processing layer with the address/addresses of the processing layer module/modules. The local memory contains a table that has the addresses of the stored configuration code for each operation. The configuration layer also maintains a status list of the processing layer modules. When the configuration layer receives a data packet, the configuration layer headers are attached to the data and sent to the processing layer. Data coming in from the processing layer is sent back to the SRI layer for appropriate action.

The Processing Layer

The processing layer forms the core of the soft radio that performs the actual processing of data. This layer is responsible for processing data sent from the configuration layer and sending the processed data back to the configuration layer. It consists of a set of linearly connected processing modules, as shown in Fig. 2. Each processing module has the capability to reconfigure itself to perform the desired function without disrupting synchronization in the main pipeline. The main flow of data is in the forward direction from the first to last module, as shown in Fig. 2. Each processing element maintains the continuity of this stream pipeline at all times. Each processing element has both static and reconfigurable sections. The static sections have the capability of configuring the reconfigurable section using the control packets sent by the configuration layer. Data can thus either bypass each processing element entirely or be routed through the reconfigurable section for processing. The structure and operation of the processing element is described in the next section.

Many communication systems have feedback loops. To accommodate such systems, the stream pipeline is designed to be bidirectional in the processing layer. The input/output ports of the processing element can either be implemented as bidirectional ports or have a separate bus in the reverse direction. The reverse slots/reverse buses are used to implement feedback loops. If the ports are bidirectional, the main clock cycle is divided into forward and

The incoming data to the SRI layer is either from the A/D converter (which always provides data) or control and status requests from the host PC. Once the user decides on the algorithms to be used, the SRI layer inserts the corresponding codes and forwards it to the configuration layer.

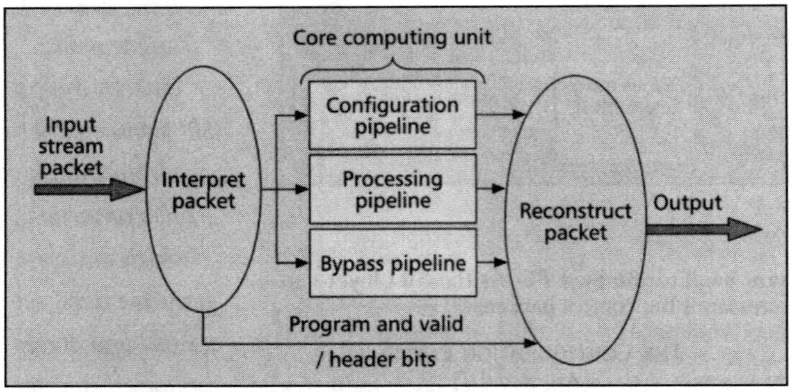

■ **Figure 4.** *A functional description of a processing element.*

reverse slots, and the directions of the interconnections are reversed in the reverse slots. If a separate reverse bus is used, then with every clock cycle, a module both accepts and outputs a forward and reverse packet to the neighboring module. Reverse packets are not needed in the absence of feedback loops.

When valid data enters an appropriate module, it is processed and sent to the next module. The address field is updated to indicate the next module that needs to act on the data. Similarly, if there is control information in the packet for the module, the corresponding code is executed to change the configuration of the reconfigurable section. If there is any latency while the module is being reconfigured, the static portion of the module buffers the subsequent incoming data and sends out idle packets. However, if it gets control packets intended for modules further down the chain, those packets are bypassed and sent to the next module. Thus, the continuity of the stream is maintained at all times.

Data in the stream packet is sent to the processing pipeline. At the end of the processing pipeline, the packet is reconstructed, and error or status messages if any are added to the header. When the processing is completed, data is sent from the output module back to the configuration layer, and finally to the SRI layer that delivers the output data. If the packet contains program information, the module checks to see if the address on the header matches the module address. If so, the code present in the body of the packet is executed in the configuration pipeline, thereby changing the configuration of the processing pipeline. It is also possible to have special error packets to handle more serious errors. If an error occurs in the execution of the code, the module sends a signaling packet indicating the location and nature of the error to the configuration layer. The address field is set to a special value so that none of the succeeding processing modules intercept the packet containing the error message.

The Structure of the Processing Element — The functional description of each processing module is shown in Fig. 4. Running on each processing element are three sets of pipelines and a state machine that interprets how the packet is channeled through the pipelines. The processing pipeline processes the data, that is, performs an operation on the data as part of the demodulation process. The configuration pipeline controls the hardware configuration of the processing pipeline. The ability of the configuration pipeline to execute programming information makes it possible to modify the low-level parameters and functionality through high-level software. The bypass pipeline allows the packet to bypass the module if the packet is not intended for the module. The bypass pipeline is present to ensure that the module does not corrupt data on which it is not supposed to act. It is important that each of these pipelines have the same amount of delay so that the packets are synchronized with the main clock cycle. At the end of the pipeline, the stream packet is reconstructed with the updated header and routed to the next module.

Figure 4 gives a functional description of the processing module, which varies slightly when implemented in hardware. The static sections are identical in all processing modules. The static section of the processing module consists of a packet interpreter, configuration pipeline, and packet reconstructor. The packet interpreter is essentially a switch. By looking at the header bits, all control packets are sent to the configuration pipeline and all data and signaling packets are sent to the processing pipeline. The function of each of these packets was described earlier. The configuration pipeline accepts all the control packets and has the capability to configure the processing pipeline accordingly. The modules maintain the configuration to which they are set and act accordingly on valid data until the configuration is changed. The packet reconstructor accepts data from all the pipelines, buffers them if needed, and sends them out to the next module. The processing pipeline is located in the reconfigurable section of the processing module. The reconfigurable section of the processing module also contains a data interpreter and data reconstructor that takes care of the byte ordering for the data going into and coming out of the processing pipeline.

When a stream packet enters a processing module, the module interprets the packet and performs the necessary action. The processing element examines the contents of a packet only if the valid bit is set. The valid bit is introduced to maintain synchronization between the modules. Every clock cycle, each processing element accepts a packet and sends out a packet. However, the module acts on the packet only if the valid bit is set. Similarly, a module sends out an invalid packet instead of inserting wait states.

Implementation of an Adaptive Single-User CDMA Receiver

At Virginia Tech, the layered architecture was used to develop a single-user CDMA receiver with a least mean square (LMS) equalizer and differential binary PSK (BPSK) demodulation. The receiver was designed using adaptive filtering which performs both despreading and equalization. The parameters that can be varied include the equalizer algorithm, filter length, spreading code, and step size. The receiver was implemented on the

Giga OPS G900 Board by Spectrum, which holds the XC4028EX-series of Xilinx FPGAs, operating at a modest 1.25 MHz. The RF front-end filtered, amplified, and mixed the received signal down to an intermediate frequency (IF) of 68 MHz. The IF was fed into the Harris digital downconverter on the Sigtek evaluation board, ST-114, where it was sampled by the onboard A/D converter. The signal was then digitally downconverted to baseband, and the chip rate was decimated to 1 MHz. A picture of the receiver is shown in Fig. 5 which shows the transmitter, the ST-114 Sigtek evaluation board, and the G900 Board by Spectrum. The output of the ST-114 was the complex envelope of the received signal. It consisted of two 16-bit buses, which represent the in-phase and quadrature channels.

The SRI and configuration layers were implemented in one unit, called the *input module*. The processing modules consisted of the adaptive filter and decision module, and the acquisition and tracking module. Due to the limited number of buses available on the board, a common bus was used for forward as well as reverse flow of data. The main clock was divided into eight slots: four forward slots, two reverse slots, and two guard slots. The guard slots were present to prevent two modules from outputting data to the same bus simultaneously when the direction of data flow was reversed. The forward slots were used to transmit data and programming packets. The reverse packets were used for feedback, including synchronization and timing information. The processing modules were implemented using the structure shown in Fig. 4. The length of the adaptive filter, convergence factor of the LMS algorithm, and user spreading code could be varied in the receiver system.

The processing modules need to have a unique address, and the address initializations were performed in the following way. On power-up, the address bits of all the processing modules were set to 0. Subsequently address initializations were performed using programming packets sent by the input module. When the first processing module encountered a valid programming packet, the module assigned itself the new address and set the invalid bit in the stream before sending the packet to the next module. This ensured that the other modules did not get the same address. When the next programming packet with a new address was sent by the input module, the first module did not act on it, since its address was now nonzero. The next uninitialized module assigns itself the new address, and the process continues until all the modules have a unique nonzero address. The next set of programming packets perform other initializations, including setting the user spreading code, filter length, and so on. Data is sent through the system after completion of all programming.

The implementation on hardware consumed about 300,000 gates. The implementation also indicated the need for more sophisticated hardware with runtime reconfiguration and partial reconfiguration capabilities for successfully implementing the layered radio architecture. The hardware requirements are further elaborated later.

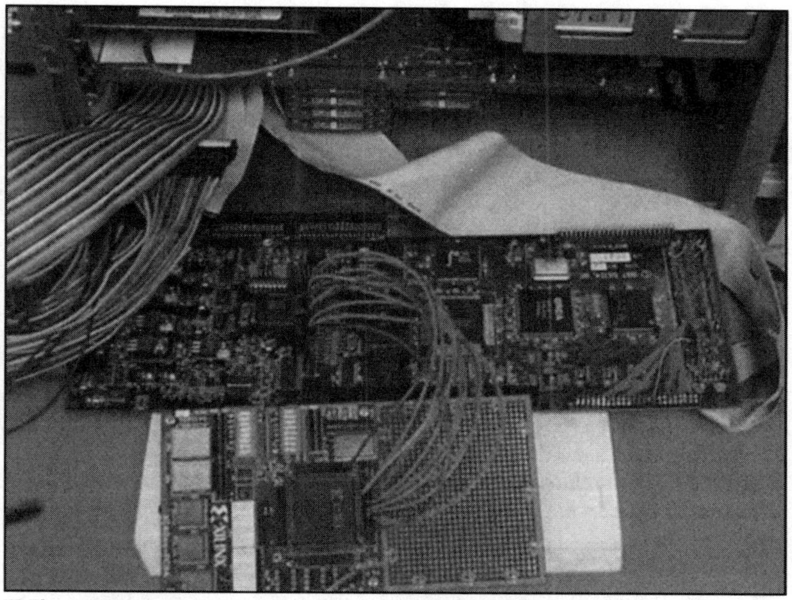

■ **Figure 5.** *A single-user CDMA receiver with adaptive filtering.*

ADVANTAGES AND TRADE-OFFS IN THE LAYERED ARCHITECTURE

A number of wireless standards have emerged in recent years. However, when third-generation (3G) systems are launched, they will have to accommodate second- and perhaps even first-generation wireless systems [8]. The proposed 3G systems differ from existing standards in their modulation formats, channel coding, multiple access, and other features. Furthermore, the 3G standards also place additional demands on the computing power through complex signal processing operations like multi-user receivers that cannot currently be handled by traditional DSPs. Virginia Tech successfully implemented a very computation-intensive multi-user receiver using stream-based processing to evaluate the stream-based approach in reconfigurable systems [9]. The receiver pages information into the system to support each of the different users. The layered radio architecture defines the methodology to design multimode radios using hardware paging, without making any assumptions on the actual algorithms used in the standards. The layered radio architecture can thus be used as a general methodology to design any multimode radios.

The layered architecture provides the framework for building a flexible soft radio at the expense of the overhead for packetizing data. It is possible to implement radically different radios on the same hardware platform through the use of this architecture. Additionally, the architecture is conducive to the implementation of adaptive modulation schemes, achieved by reconfiguring the hardware in real time.

One of the main advantages of the architecture is its excellent hardware reusability. For example, core units like BPSK and QPSK demodulators can be used in a wide variety of standards when built in hardware with the right parameters. Thus, it is possible to build libraries of hardware functions much like building libraries of software functions.

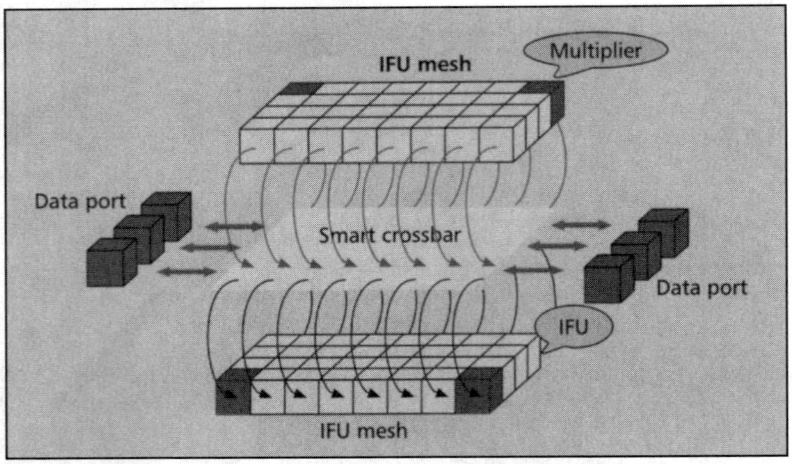

■ **Figure 6.** *The architecture of the Stallion reconfigurable processor.*

The layered architecture has good data flow properties and a simple interface between the processing layer modules, which make it scalable at the system level as well as the hardware element level. The use of stream-based processing ensures that each processing module has one input and output bus that is common for control information and data bits. This simplifies the interface between two processing modules and makes it easy to add additional hardware modules to support higher computational complexity, or integrate designs from different engineers.

Insight into Hardware Development for the Processing Layer

The processing layer is composed of processing modules that perform the actual computations. The layered radio architecture makes certain assumptions about the processing layer in addition to the hardware being reconfigurable. The configuration layer should be able to reconfigure the processing modules, independent of other modules. This inherently assumes partial reconfigurability of the hardware, along with a very high reconfiguration speed. Virginia Tech has developed an FPGA specifically suited to flexible, high-throughput, low-power computations, called the *Stallion*, based on wormhole reconfigurable computing [10]. The architecture of Stallion is shown in Fig. 6. It consists of six bidirectional data ports, two 8 x 4 meshes of interconnected functional units, a crossbar, and four hardware multipliers. The functional units are programmed to process data, while the crossbar aids in routing data through Stallion. The use of Stallion for the processing modules makes it possible to implement the full functionality of the radio, which cannot be achieved with the use of commercial FPGAs. With the use of Stallion, the processing modules can be partially configured as well as configured during runtime as required by the architecture. The Stallion also interfaces easily with other Stallion chips, thus maintaining scalability in hardware. The preferred mode to connect these chips is by using a smart crossbar.

The essence of wormhole runtime reconfiguration is that independent, self-steering streams of programming information and operand data interact within the architecture to perform the computational problem at hand. The stream is self-steering, and as it propagates through the system, configuration information is stripped from the front of the header and used to program the unit at the head of the stream. Since the streams independently guide themselves through the system using the information contained in the stream header, the configuration process is inherently distributed. Multiple independent streams can wind their way through the chip simultaneously. Part of the system can be used to process data operands while any set of neighboring units are accepting configuration data from the header of one or more streams, thus allowing overlap of the configuration and processing operations, as well as partial reconfigurability.

Conclusions

The layered architecture presents an open yet formal and unified structure for implementing soft radios on reconfigurable platforms. The layered approach lends itself easily to standardization of soft radio systems and also allows code reuse. The processing modules and header information can also easily be standardized, allowing support from a variety of vendors. One of the main advantages of the structure of the processing modules is that the interface between two modules is very simple. Each module is able to accept a stream packet and send out a stream packet in the forward and reverse directions in each cycle. The layered architecture is scalable and completely flexible.

The layered architecture is suited for today's FPGAs that support partial reconfiguration and for tomorrow's configurable computing platforms. Virginia Tech's *Stallion* is one such architecture that can support the layered architecture in an efficient manner. Flexibility, of course, is gained at the cost of additional overhead in dealing with packetized information. Current research at Virginia Tech focuses on the development and testing of a soft radio based on the layered architecture and building a library of soft radio modules.

In this article we have presented a scheme whereby the processing modules are connected in a sequential manner. Connecting the modules in a sequential manner is a first step in designing soft radios, but a more sophisticated and powerful approach is to network these modules in the processing layer. For example, creating a mesh of processing modules with 4-connectivity can increase the processing power tremendously. This, of course, opens up a whole new area of research where the stream packets no longer travel in just one (or two) directions. This design approach would have to consider routing aspects, packet collision and prevention, and packet ordering, which are the same issues encountered in packet radio communication systems. However, this change can be viewed as just an extension of the layered architecture. A new layer, a *network layer*, can be inserted between the configuration and processing layers to take care of the packet routing and collating of information obtained from the processing layer. Other aspects of the architecture can remain unchanged.

Acknowledgments

Support to this project was provided by the Defense Advanced Research Projects Agency (DARPA) GloMo program, the Navy Collaborative Integrated Information Technology Initiative program, and Conexant Systems, Inc.

References

[1] J. Mitola III, "The Software Radio Architecture," *IEEE Commun. Mag.*, May 1995, pp. 26–38.

[2] M. Butler, J. Providakes, and G. Blythe, "The Layered Radio," *MILCOM*, New York, NY, 1998, pp. 174–79.

[3] J. Mitola III, "Software Radio Architecture: A Mathematical Perspective," *IEEE JSAC*, vol. 17, no. 4, Apr. 1999.

[4] P. M. Athanas, J. H. Reed, and W. H. Tranter, "A Prototype Software Radio Based on Configurable Computing," *Advancing Microelect.*, 1998, pp. 34–39.

[5] H. Tsurumi and Y. Suzuki, "Broadband RF Stage Architecture for Software-Defined Radio in Handheld Terminal Applications," *IEEE Commun. Mag.*, Feb. 1999, pp. 90–95.

[6] P. M. Athanas *et al.*, "A High Capacity Adaptive Wireless Receiver Implemented with a Reconfigurable Computer Architecture," *ARPA GloMo Principle Investigators Conf.*, San Diego, CA, Nov. 1995.

[7] A. L. DeCegama, *Parallel Processing Architectures and VLSI Hardware, Vol. 1*, Englewood Cliffs, NJ: Prentice Hall, pp. 157–60.

[8] P. Leppanen *et al.*, "Software Radio — An Alternative for the Future in Wireless Personal and Multimedia Communications," *IEEE Int'l. Conf. Pers. Wireless Commun.*, Jaipur, India, Feb. 1999, pp. 364–68.

[9] S. Srikanteswara *et al.*, "Configurable Computing for Communication Systems," *Proc. Wireless Commun. Conf., IMAPS*, 1998, pp. 180–85.

[10] R. Bittner, "Wormhole Run-Time Reconfiguration: Conceptualization and VLSI Design of a High Performance Computing System," Ph.D. dissertation, Dept. of Elec. and Comp. Eng., Virginia Tech, 1997.

The layered architecture presents an open yet formal and unified structure for implementing soft radios on reconfigurable platforms. The layered approach lends itself easily to standardization of soft radio systems and also allows code reuse

Software Radio Issues in Cellular Base Stations

Kambiz C. Zangi, *Member, IEEE*, and R. David Koilpillai, *Member, IEEE*

Abstract—**The use of the "software radio" concept in cellular applications is a topic of widespread interest. Two key issues in the implementation of software radios are the development of optimal receivers that require the minimum number of bits in the wide-band analog-to-digital converter (ADC) and efficient channelizers that extract individual channels from the digitized wide-band signal. In this paper, both of these issues are studied in detail for cellular base stations.**

A computationally efficient wide-band channelizer is presented in this paper. This channelizer is closely related to the discrete Fourier transform filter bank used in transmultiplexers. It is shown that the complexity of the proposed channelizer is significantly less (2–50×) than the complexity of conventional channelizers. An optimal receiver that explicitly takes into account the effect of the quantization noise of the wide-band ADC is also derived in this paper. The analysis of the ADC noise provides guidelines for specifying wide-band ADC for use in cellular applications. The development of the channelizer and the optimal receiver yield important insights into the implementation of cellular software radios.

All of the key results of this paper are applied to a detailed example based on the Digital Advanced Mobile Phone System (D-AMPS, IS-54/IS-136) cellular standard. The bit-error rate (BER) performance simulations of a D-AMPS wide band receiver is presented as a part of this example.

Index Terms—**Cellular base stations, filter banks, quantization, radio communication equipment, software radio, transmultiplexers, wideband receiver.**

I. Introduction

ORIGINALLY conceived for military applications, the concept of software radio has attracted much attention lately in the context of commercial applications [1]–[3]. The drivers for these two applications are somewhat different. The main driver for military software radio is the need for a single radio which can communicate with the many types of military radios that use different RF bands and different modulation schemes [1].

In the commercial arena, software radio applications can be divided into user terminals and radio base stations. The computational load of software radios is on the order of billions of operations per second [2]; hence, given today's state of the art in digital signal processing (DSP) hardware, it is extremely challenging to implement a software radio in a battery-powered terminal. Today, each cellular operator chooses one cellular standard. There does not seem to be a large need for base stations that simultaneously support multiple cellular standards. However, with the rapid growth in demand for cellular services and with the proliferation of microcells and picocells, there is tremendous interest in reducing the cost, complexity, and size of radio base stations.

Manuscript received December 1997; revised January 22, 1998, June 17, 1998, and August 24, 1998.

The authors are with Ericsson Inc., Research Triangle Park, NC 27709 USA.

Publisher Item Identifier S 0733-8716(99)02984-4.

A. Motivation

Wide-band receivers in software radios can significantly reduce the cost and complexity of today's cellular base stations. Two of the challenging issues in the implementation of wide-band receivers are: 1) digitization of a large block of the cellular spectrum using a single wide-band ADC, and 2) extraction of individual channels from the wide-band digitized signal. This paper focuses on both of these issues.

Currently, cellular base stations employ a bank of narrow-band receivers, each having its own complete receiver chain as depicted in Fig. 1. The cost/complexity of these base stations grows *linearly* with the number of received channels, and this cost/complexity is dominated by analog components. Today's base stations use three-sector antennas, requiring one receiver per carrier per antenna element in each sector. With the introduction of smart antenna arrays into cellular systems, the number of channels received by each base station will increase even further. Hence, it becomes even more important to avoid receiver architectures whose complexity grows linearly with the number of received channels. In a base station with a wide-band receiver, a single analog front end can be used to receive all channels; hence, the complexity of the analog part of the resulting receiver remains *constant*, i.e., independent of the number of received channels.

B. Efficient Channelizer

In the context of software radio architectures, wide-band receivers have been discussed by a number of authors [2], [3]. A first-order estimate of resources needed to implement a wide-band receiver is presented in [2], and these estimates show that the extraction of individual radio channels from the output of the ADC (e.g., 30 kHz channels in D-AMPS [4]) is the most computationally intensive part of a wide-band receiver. This part of the receiver is typically referred to as the "channelizer." As indicated in [2], details regarding the implementation of wide-band channelizers are not available in the literature.

Digital downconverter chips for implementing the channelizer are available from several manufacturers (e.g., [5]). Since these chips extract the received channels individually, the complexity of the wide-band receiver using these chips grows linearly with the number of received channels.

The function of the channelizer in a wide-band digital cellular receiver is similar to the function of a traditional frequency-division multiplexed (FDM) ↔ time-division multiplexed (TDM) transmultiplexer [6]. Both systems extract

Reprinted from *IEEE Journal on Selected Areas of Communications,* Vol. 17, No. 4, pp. 561–573, April 1999.

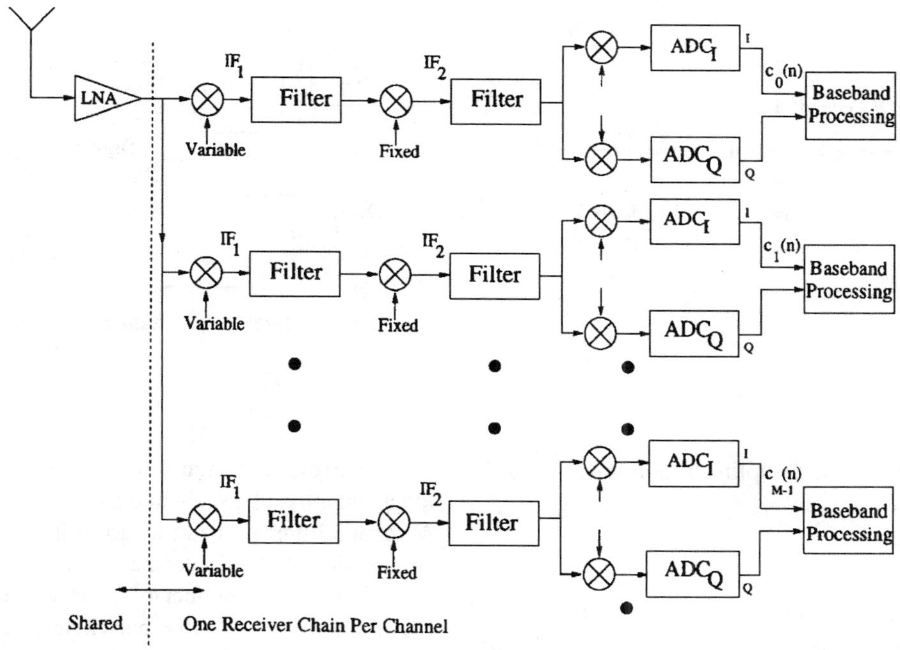

Fig. 1. Conventional radio base station receivers.

narrow-band channels from a wide-band digitized signal. For telephony applications, computationally efficient algorithms for implementing transmultiplexers were developed in the 1970's based on discrete Fourier transform (DFT) filter banks [6], [7]. Each channel in the telephony application contains digitized voice, while each channel in the digital cellular application contains modulated digital symbols. For digital satellite broadcast, several authors have proposed using the DFT filter bank to extract individual narrow-band channels from a wide-band radio signal; however, these proposals require additional processing after the DFT filter bank to account for incompatibility between the baud rate and the channel spacing [8]–[10].

This paper derives a new DFT filter bank specifically for wide-band receivers that operate in cellular environments. This new channelizer is called the "opened filter bank channelizer (OFB channelizer)." The complexity of this channelizer is independent of the number of received channels. For D-AMPS, the per-channel complexity of the OFB channelizer is $2\times$ less than the per-channel complexity of the traditional DFT filter bank, and is $50\times$ less than the per-channel complexity of a channelizer that extracts each channel independently.

C. Optimal Receiver

Several authors [11]–[13] have discussed wide-band ADC's for software radio receivers. Cellular receivers must simultaneously receive a weak carrier and a strong carrier. For example, GSM requires the simultaneous reception of a -104 dBm carrier and a -13 dBm carrier for a dynamic range of 91 dB. By contrast, all carriers in the FDM transmultiplexing application have approximately equal strength.

The spurs generated by nonlinearities of the ADC can block the reception of the weak carrier in the persence of a strong carrier [11]–[13]. Wide-band receivers must reduce the impact of these spurs on the performance of the receiver. Consider a nonlinear ADC with an input signal that is the sum of a strong carrier at frequency "A" and a weak carrier at frequency "B." At the output of the ADC, there could be a spur at frequency "B" due to the strong carrier at frequency "A." This large spur can make it impossible to demodulate the weak carrier. Several authors have proposed "dithering" on the input to the ADC to reduce these nonlinearities [14].

The wide dynamic range in cellular context requires that the quantization steps of the wide-band ADC be set relatively large so that the ADC will not saturate; however, large quantization steps lead to large quantization noise at the output of the ADC. Intuitively, this quantization noise can be viewed as additional thermal noise that reduces the signal-to-noise ratio at the output of the ADC. This additional noise can be sufficient to increase the bit-error rate of the weak carrier above acceptable levels, i.e., the reception of the weak carrier can be "blocked." An important question in choosing ADC's for a wide-band receiver is the number of quantization steps needed so that the bit error is not substantially degraded due to the quantization noise.

In Section III, the impact of quantization noise on the performance of wide-band receivers is analyzed. Note that this analysis is only valid for an ideal linear ADC, i.e., the spurs/nonlinearities of the ADC are ignored.

For a wide-band radio system, the optimal receiver (minimizing the probability of bit error) in the presence of thermal noise *and* quantization noise in a Rayleigh fading channel is derived. For this receiver, new analytical expressions relating the bit-error rate of the receiver to the parameters of the wide-band ADC are given. Using these expressions and the specifications of a particular cellular standard, one can determine the performance requirements for wide-band ADC's that must be met so that the base stations complies with the given cellular standard. For example, a 13 or 14 bit ADC is

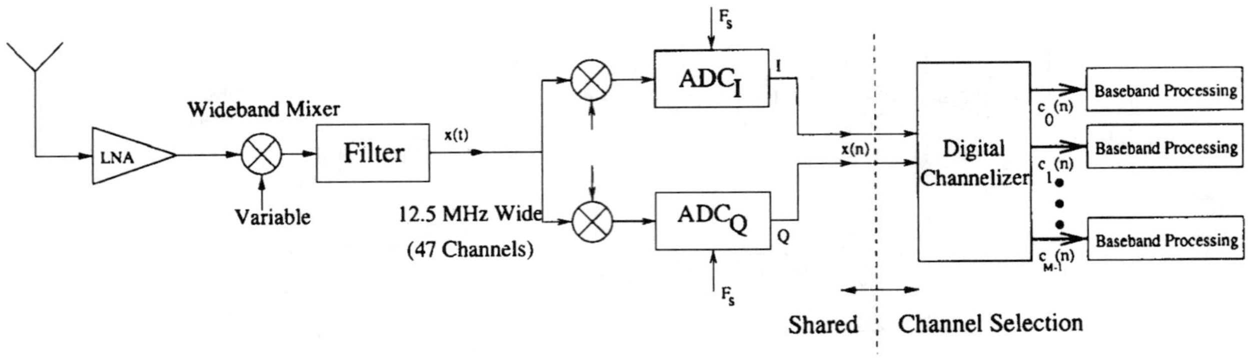

Fig. 2. Generic wide-band receiver.

sufficient to meet the D-AMPS uplink requirements (10–12 MHz bandwidth).

D. Outline

The paper is organized as follows. In Section II, the classical DFT filter bank is presented, and its shortcomings for use in a wide-band cellular receiver are pointed out. Starting from this DFT filter bank, an OFB channelizer is derived that is particularly well suited for use in wide-band cellular receivers. In Section II-C, expressions for the complexity of the OFB channelizer are derived in terms of the number of multiplications per second. Section III derives the optimal receiver (minimizing the probability of bit error) that takes into account the thermal noise *and* quantization noise of the ADC. Performance results (in terms of BER) for a wide-band D-AMPS receiver with several wide-band ADC's are presented in Section III-A.

II. Filter Bank Channelizers

A generic wide-band receiver is depicted in Fig. 2. In this receiver, the entire frequency band of interest (e.g., 12 MHz) is digitized using a pair of wide-band ADC's. In the wide-band receiver of Fig. 2, the final channel selection is performed by the digital channelizer. Each output of the channelizer is fed to a baseband processing block. The sampling rate of the input to the baseband processing block must be an integer multiple of the baud rate (typically between four to eight times the baud rate to enable choosing the optimum sampling point). The channelizer performs bandpass filtering, followed by decimation from the very high sampling rate of the ADC's to the relatively low sampling rate of the baseband processing units.

The complexity of the wide-band receiver is dominated by the complexity of the channelizer since the channelizer must operate at the highest sampling rate in the system (the rate of the ADC's). Computationally efficient algorithms and structures are essential for cost-effective implementation of the channelizer.

The per-channel approach for the channelizer is most widely discussed in the software radio literature [3], [2]. Specifically, many authors have mentioned using digital downconverter chips to implement the channelizer [5]. A per-channel wide-band receiver with M received channels essentially consists of M narrowband digital filters, each extracting one channel independently. This channelizer is the digital equivalent of M parallel narrow-band analog filters. More importantly, the complexity of the digital part of the resulting receiver grows *linearly* with the number of received channels.

An alternative to the per-channel approach is the filter bank channelizer. Using multirate signal processing theory, the filter bank channelizer can extract every channel between $[(-F_s/2), (F_s/2)]$, where F_s is the sampling frequency of the wide-band ADC's. Using the fast Fourier transform (FFT), computationally efficient algorithms for implementing the filter bank channelizer have been developed [15], [16], [6], [7]. The complexity of a filter bank channelizer remains *constant*, i.e., independent of the number of received channels. Hence, when a large number of channels is needed in a receiver, the filter bank channelizer is computationally attractive.

Functionally, the channelizer is a bank of complex-valued, digital bandpass filters. The output of each filter is followed by a mixer, decimator, and a sample rate converter. As shown in Fig. 3, each bandpass filter $H_i(w)$ is centered at the carrier frequency of a particular radio channel. Each mixer in the channelizer converts the bandpass output of the corresponding filter to baseband. Since the bandwidth of the output is much less than the bandwidth of the digitized input $x[n]$, one may decimate the output of each filter by N.

The baseband processing block needs an input sampling rate that is an *integer* multiple of the baud rate. In general, the decimated sampling rate is not an integer multiple of the baud rate. Hence, a sample rate conversion is needed before baseband processing.

A. Classical Filter Bank Channelizer

Provided that the $H_i(w)$'s satisfy certain conditions, the FFT can be used to efficiently implement the channelizer of Fig. 3. This implementation has shortcomings that are pointed out in this section.

Let $h_0[n]$ be a real, causal, low-pass filter with discrete-time Fourier transform $H_0(w)$ and with z transform $H_0(z)$. Assume that there are M equally spaced channels between $[-(F_s/2), +(F_s/2)]$, and let f_{cs} denote the frequency separation between two consecutive channels, i.e.,

$$\boxed{F_s = M \times f_{cs}.} \tag{1}$$

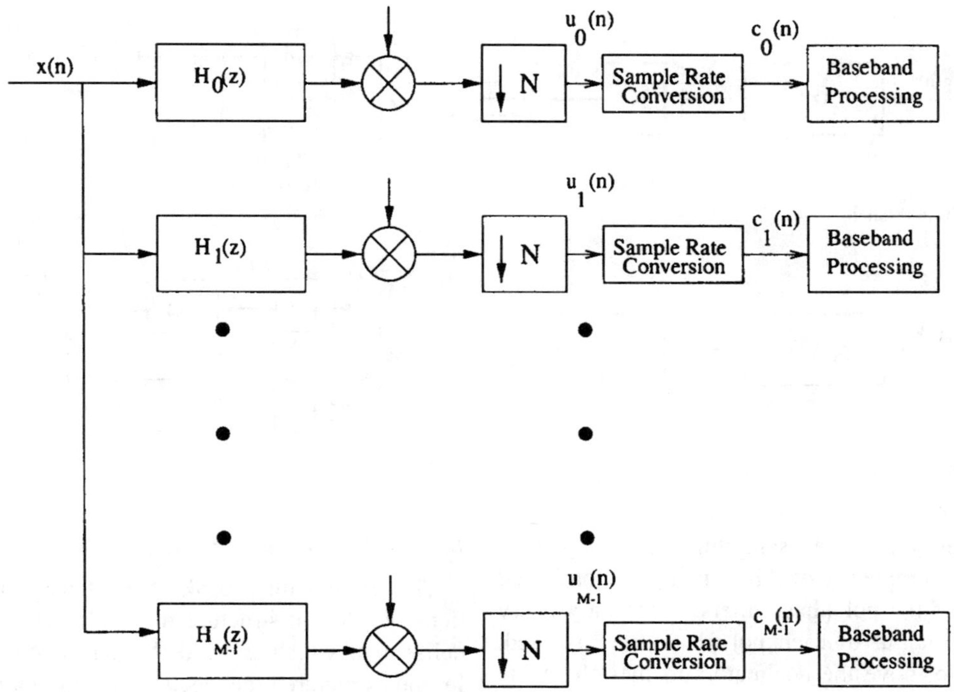

Fig. 3. Filter bank view of a wide-band channelizer.

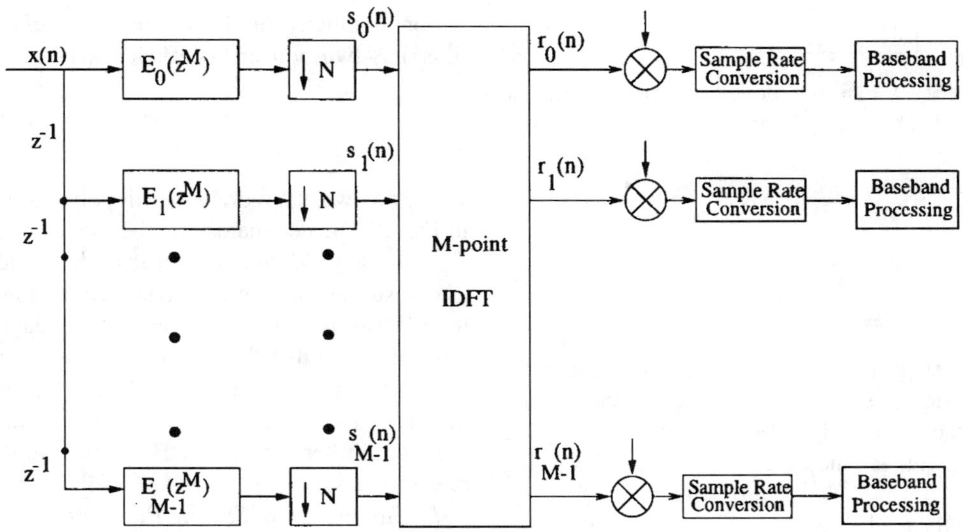

Fig. 4. Classical DFT filter bank channelizer.

Finally, assume that each bandpass filter in the filter bank channelizer $H_i(w)$ is a modulated version of $H_0(w)$, i.e.,

$$H_i(w) = H_0\left(w - \frac{2\pi}{M}i\right), \qquad 0 \le i \le M-1. \tag{2}$$

Note that the center frequency of the ith filter $H_i(w)$ is $(2\pi/M)i$ (or $(F_s/M)i$ in continuous time). The single input/M-output system with input $x[n]$ and outputs $\{r_i[n]\}_{i=0}^{M-1}$ is called a DFT filter bank [15], [16]. Polyphase decomposition of $H_0(z)$ implements this DFT filter bank efficiently [15], [16]. To this end, express $H_0(z)$ in polyphase form as

$$H_0(z) = \sum_{l=0}^{M-1} z^{-l} E_l(z^M) \tag{3}$$

where

$$e_l(n) = h_0(nM + l), \qquad 0 \le l \le M-1 \tag{4}$$

and $E_l(z)$ is the z transform of $e_l[n]$. Using the polyphase filters $E_l(z)$'s, it can be shown [15], [16] that the structure of Fig. 4 is equivalent to the filter bank channelizer of Fig. 3 as long as constraint (1) is met.

In the structure of Fig. 4, the polyphase filters operate at the highest sampling rate in the system (F_s), and the inverse

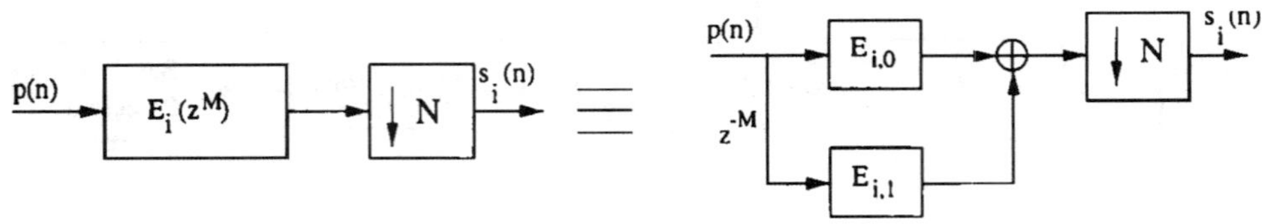

Fig. 5. The ith polyphase branch.

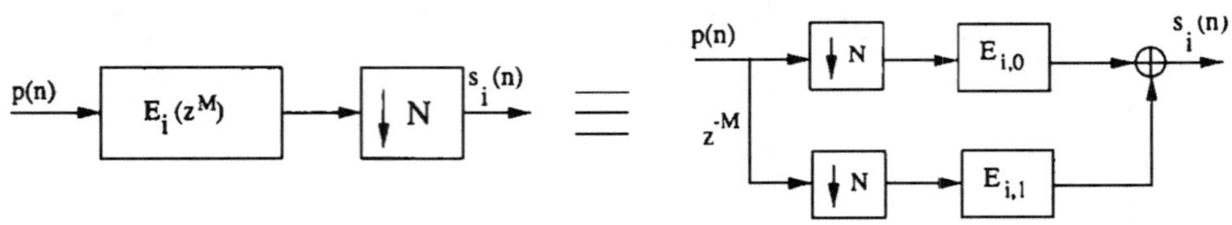

Fig. 6. The ith polyphase branch.

DFT (IDFT) operates at the lowest sampling rate in the system (F_s/N). Hence, the complexity of this structure is dominated by the complexity of the polyphase filters. Since only every Nth sample of the output of each polyphase filter is used, it is natural to try to move the decimators to the left of the polyphase filters in Fig. 4, so that the polyphase filters will operate at the lowest sampling rate (F_s/N). It is well known that if M is an integer multiple of N, i.e.,

$$\boxed{M = K \times N} \tag{5}$$

the decimators in Fig. 4 can be moved to the left of the polyphase filters [15], [16]. Let f_s denote the sampling rate of $r_i[n]$: then

$$f_s = \frac{F_s}{N} \tag{6}$$

$$= K \times \frac{F_s}{M} \tag{7}$$

$$= K \times f_{cs}. \tag{8}$$

Equation (7) is derived from (6) using (5), and (8) is derived from (7) using (1). Equation (8) implies that efficient implementation of the structure in Fig. 4 requires that the sampling rate of $r_i[n]$'s be an integer multiple of the channel spacing f_{cs}.

To interface directly to the baseband processing block, the sampling rate of $r_i[n]$'s must be a small multiple of the baud rate f_b. The smallest f_s that is both a multiple of f_{cs} and a multiple of f_b is the least common multiple of (f_{cs}, f_b), denoted by $\mathrm{LCM}(f_{cs}, f_b)$. In many communication standards, the ratio of the baud rate to the channel spacing is the ratio of two relatively large numbers (e.g., (81/100) in D-AMPS). Hence, typically, $\mathrm{LCM}(f_b, f_{cs})$ is a very large multiple of the baud rate. In D-AMPS, $f_b = 24.3$ kHz and $f_{cs} = 30$ kHz; hence, the smallest allowable f_s is 2.43 MHz, which is 100 times the baud rate. To efficiently implement the DFT channelizer, the parameters of the DFT channelizer are chosen so that f_s is *not* a multiple of the baud rate, and a sample rate conversion block is placed between each $r_i[n]$ and the corresponding basedband processing unit [8], [9]. Sample rate conversion is computationally intensive since the ratio of the baud rate to the channel spacing is usually the ratio of two relatively large numbers.

B. The Opened Filter Bank Channelizer

The opened filter bank channelizer (OFB) is a computationally efficient structure for implementing filter bank channelizers in which M and N can be chosen independently. In this structure, the decimators are moved to the left of the polyphase filters for any M and N. Furthermore, the OFB channelizer eliminates sample rate converters before the baseband processing blocks.

For simplicity of derivation, assume that each polyphase filter has two nonzero coefficients, i.e.,

$$E_i(z^M) = E_{i,0} + E_{i,1}z^{-M}. \tag{9}$$

Using (9), we can redraw the ith polyphase branch as depicted in Fig. 5. The decimators can be moved to the left of $E_{i,0}$ and $E_{i,1}$ for any M and N because $E_{i,0}$ and $E_{i,1}$ are scalars. The resulting structure is depicted in Fig. 6. Finally, using the identity in Fig. 6 and some block diagram manipulations, one can show that the structure in Fig. 7 is equivalent to the filter bank channelizer of Fig. 3, provided that (1) holds. Note that this structure is computationally efficient because both the polyphase filters and the IDFT operate at the lowest sampling rate in the system [i.e., at (F_s/N)].

1) Elimination of Sample Rate Converters: With a specific choice of sampling rate F_s, the OFB channelizer in Fig. 7 can be implemented without sampling rate conversion. The sampling rate of $r_i[n]$ is denoted by f_s and is related to F_s according to

$$\boxed{F_s = N \times f_s} \tag{10}$$

and from (1)

$$\boxed{F_s = M \times f_{cs}.} \tag{11}$$

Combining (10) and (11) shows that, for f_s to be a multiple of the baud rate, F_s must be divisible by the baud rate f_b and be divisible by the channel spacing f_{cs}. The smallest F_s that satisfies these conditions is the LCM (f_b, f_{cs}).

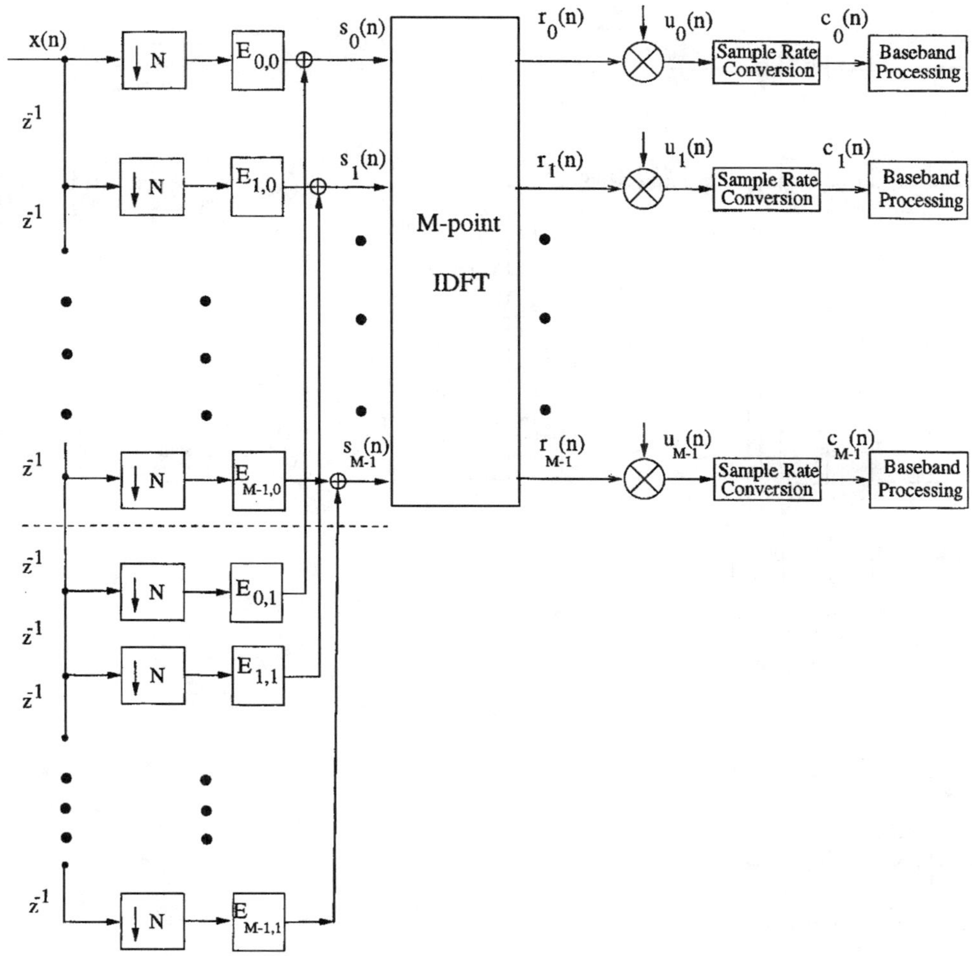

Fig. 7. Opened filter bank channelizer.

Example: Consider the D-AMPS standard in which the baud rate is $f_b = 24.3$ ksymbols/s and the channel spacing is $f_{cs} = 30$ kHz. Then, $\text{LCM}(f_s, f_b) = 2.43$ MHz. Hence, if the sampling rate of the ADC is chosen as a multiple of 2.43 MHz, the filter bank channelizer of Fig. 7 does not require sample rate converters. One possibility is to choose $F_s = 34.02$ MHz, which yields $M = 1134$ and $N = 350$. The sampling rate of the outputs of the decimators in this case will be $f_s = 4 \times 24.3$ kHz, exactly four times the baud rate. With $F_s = 34.02$ MHz, the cellular spectrum allocated to one DAMPS operator (12.5 MHz) can be digitized by one wide-band ADC.

2) Filter Bank Channelizer Matched to Frequency Reuse Pattern: Most cellular systems conform to a predetermined frequency reuse pattern which restricts the set of channels received by each base station. This frequency reuse pattern can be exploited to reduce the complexity of the OFB channelizer.

For cellular radio services, frequency reuse is central to realizing multiple access for a large number of users scattered over many, often contiguous, areas [17]. In D-AMPS, a 12.5 MHz allocated spectrum with 30 kHz channels yields a total of approximately 400 channels. To avoid cochannel interference, the same 30 kHz channel is not reused in contiguous cells, but may be used in a noncontiguous cell. As a consequence, if each cell has $(L-1)$ neighbors, each cell will be allocated $(1/L)$th of the total number of channels. In one popular configuration, the shape of each cell is a hexagonal (with six neighbors), which results in reuse factor $L = 7$, i.e., each cell only uses every seventh channel [17].

Assume that the filter bank channelizer extracts only every Lth channel between $[-(F_s/2), +(F_s/2)]$. In this case, the M-point IDFT in the OFB channelizer is replaced with an (M/L)-point IDFT, resulting in a channelizer with lower overall complexity. For an (M/L)-point IDFT to be defined, M must be an integer multiple of L

$$M = Q \times L. \tag{12}$$

If only every Lth output channel is needed, the OFB channelizer needs to compute only

$$\{r_{iL}(n)\}_{i=0}^{Q-1} = \{r_0(n), r_L(n), r_{2L}(n), \cdots, r_{M-L}(n)\}. \tag{13}$$

Since $\{r_i(n)\}_{i=0}^{M-1}$ are the IDFT coefficients of $\{s_i(n)\}_{i=0}^{M-1}$, the sequence in (13) is obtained by decimating, by a factor of L, the IDFT coefficients $\{r_i(n)\}_{i=0}^{M-1}$. The subsampled IDFT coefficients in (13) are computed by taking a Q-point IDFT instead of an M-point IDFT. The key step is to form a Q-point sequence $\{z_i(n)\}_{i=0}^{Q-1}$ from the M-point sequence $\{s_i(n)\}_{i=0}^{M-1}$ in such a way that the Q-point IDFT of $\{z_i(n)\}_{i=0}^{Q-1}$ is the sequence in (13).

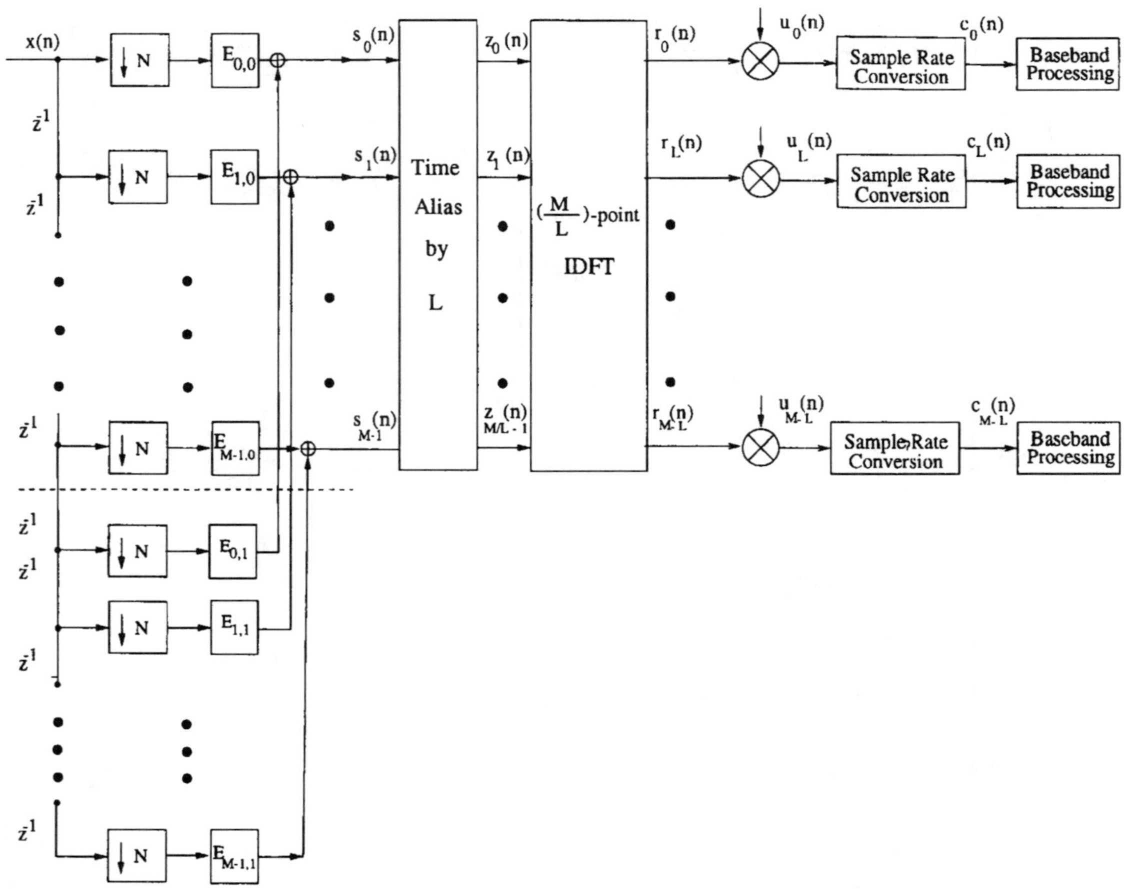

Fig. 8. Subsampled filter bank channelizer.

The desired sequence $\{z_i(n)\}_{i=0}^{Q-1}$ is obtained from $\{s_i(n)\}_{i=0}^{M-1}$ according to

$$z_i(n) = \sum_{r=0}^{L-1} s_{i+Qr}(n) \qquad 0 \le i \le Q-1. \tag{14}$$

Proof of this property of the IDFT is given in the Appendix. The sequence $\{z_i(n)\}_{i=0}^{Q-1}$ is typically referred to as the "time-aliased" version of $s_i(n)$ [18]. The block diagram of the resulting subsampled DFT filter bank is depicted in Fig. 8. To visualize the construction of the sequence $\{z_i(n)\}_{i=0}^{Q-1}$ from the sequence $\{s_i(n)\}_{i=0}^{M-1}$, think of $\{s_i(n)\}_{i=0}^{M-1}$ as a vector of length M. Break this vector into L vectors of length Q. The vector $\{z_i(n)\}_{i=0}^{Q-1}$ is the vector sum of these L vectors.

For a filter bank channelizer with reduced IDFT size, the following restrictions must be met from (1) and (12):

$$F_s = M \times f_{cs} \tag{15}$$

$$M = Q \times L. \tag{16}$$

Combining (15) and (16) yields

$$\boxed{F_s = L \times Q \times f_{cs}.} \tag{17}$$

The smallest F_s that satisfies (17) is $\text{LCM}(L, f_{cs})$. Note that an F_s that satisfies (17) might still result in a filter bank channelizer that requires sample rate conversion since the sampling rate of the outputs of the filter bank is not necessarily an integer multiple of the baud rate. For the sampling rate of the outputs of the channelizer to be an integer multiple of the baud rate, the following additional constraint must be met:

$$F_s = N \times f_b. \tag{18}$$

The smallest F_s that satisfies (17) and (18) is $\text{LCM}(L, f_{cs}, f_b)$.

Consider D-AMPS with a frequency reuse factor $L = 7$. Recall that the baud rate is $f_b = 24.3$ ksymbols/s and the channel spacing is $f_{cs} = 30$ kHz. The smallest allowable F_s is $\text{LCM}(7, 30\times10^3, 24.3\times10^3) = 17.01$ MHz. To digitize the entire cellular band, F_s can be chosen to be 34.02 MHz, which yields $N = 350$ and $M = 1134$. The size of the reduced IDFT in the subsampled filter bank channelizer of Fig. 8 will be 162 since $Q = (M/L) = 162$. The sampling rate of the output of the decimators in this case will be $f_s = 4 \times 24.3$ kHz, i.e., exactly four times the baud rate.

C. Computational Complexity of Filter Bank Channelizers

Expressions for the computational complexity of the filter bank channelizers in Figs. 7 and 8 are derived next. The number of real multiplies per second is a useful measure of complexity. Each n requires the compution of $\{s_i[n]\}_{i=0}^{M-1}$ and the computation of an M-point IDFT or an (M/L)-point IDFT. Let $\Psi(P)$ denote the number of real multiplications needed to compute $\{s_i[n]\}_{i=0}^{M-1}$ for each n, where P is the number of real nonzero coefficients of the low-pass prototype filter $h_0(n)$. Similarly, let $\Phi(X)$ denote the number of real

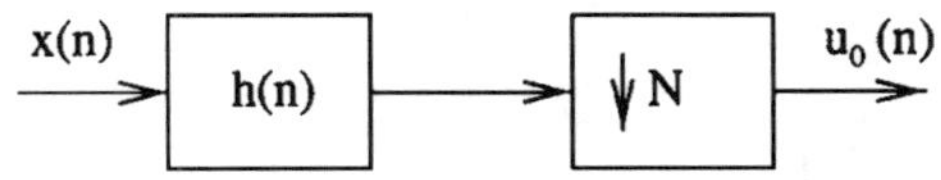

Fig. 9. Extracting one channel individually.

multiplications of an X-point IDFT. Since $x[n]$ is complex-valued and each $E_{i,j}$ is real-valued

$$\Psi(P) = 2 \times P. \tag{19}$$

These computations repeat every (N/F_s) seconds [i.e., at the decimated rate of (F_s/N)]. The total computational demand of the OFB channelizer with a full-size IDFT is

$$\mu(M,P) = [\Psi(P) + \Phi(M)] \times \frac{F_s}{N}. \tag{20}$$

The total computational demand of a decimated filter bank channelizer with reuse factor L is

$$\mu(M,P,L) = \left[\Psi(P) + \Phi\left(\frac{M}{L}\right)\right] \times \frac{F_s}{N}. \tag{21}$$

To put (20) in perspective, consider the computational complexity of extracting just one channel from $x(n)$. For simplicity, consider extracting the channel centered at 0 Hz (dc). To extract this channel, filter $x(n)$ by a low-pass filter centered at dc and decimate the result by N (as depicted in Fig. 9). For a fair comparison to the OFB channelizer in Fig. 9, let $h(n) = h_0(n)$, where $h_0(n)$ is the low-pass prototype filter of the OFB channelizer. As mentioned above, $h_0(n)$ has P nonzero coefficients. It is well known [15], [16] that polyphase decomposition of $h_0(n)$ efficiently computes $u_0(n)$ in Fig. 9 at a cost of

$$2 \times P \times \frac{F_s}{N} = \Psi(P) \times \frac{F_s}{N} \quad \text{real multiplications/s.} \tag{22}$$

Comparing (22) to (20) shows that the complexity of the OFB channelizer is equal to the sum of the complexity of one low-pass digital filter and the complexity of one M-point IDFT. The OFB channelizer simultaneously extracts *every* channel between $[-(F_s/2), (F_s/2)]$. Hence, at an additional cost of one M-point IDFT, the OFB channelizer simultaneously extracts every radio channel between $[-(F_s/2), (F_s/2)]$. Similarly, the subsampled filter bank channelizer simultaneously extracts *every* Lth radio channel at the additional cost of one (M/L)-point IDFT. Since these channelizers extract all channels simultaneously, the complexity of these channelizers remains *constant* as a function of the number of extracted channels.

Given computationally efficient algorithms for the IDFT, the additional complexity due to the IDFT's is quite low. If M [or (M/L)] is a highly composite number, efficient algorithms for computing the M-point [or (M/L)-point] IDFT exist [19]–[21]. For DAMPS base stations, the parameters of the OFB channelizer can be chosen so that the resulting IDFT's are computed efficiently using the FFT.

1) Order of the Low-Pass Prototype Filter: The low-pass prototype filter is the digital equivalent of the analog low-pass filter at the second IF stage in Fig. 1. Recall that the purpose of the analog filter is to ensure against aliasing at the output of the ADC's. The sampling rate at the output of the decimators in the filter bank channelizer, F_s/N, is the same as the sampling rate of the ADC's. The output of the decimators and the output of these ADC's are fed to the same baseband processing block. The passband edge of each analog low-pass filter is $f_{cs}/2$, where f_{cs} is the channel spacing. The stopband edge of each analog low-pass filter is typically set at $(F_{\text{ADC}} - (f_{cs}/2))$, where F_{ADC} is the sampling rate of the ADC's in Fig. 1. However, $F_{\text{ADC}} = F_s/N$. Therefore, the stopband edge of each analog low-pass filter is at $((F_s/N) - (f_{cs}/2))$. Translating these analog specifications to the digital domain yields the passband edge and the stopband edge of the digital low-pass prototype filter $H_0(w)$. Given a peak passband ripple specification δ_1 and a peak stopband ripple specification δ_2, the minimum order linear-phase equiripple FIR filter that meets these specifications [22] has length

$$P = \frac{2\log_{10}\left(\frac{1}{10\delta_1\delta_2}\right)}{3\Delta_f} \tag{23}$$

where $\Delta_f = (\text{stopband edge} - \text{passband edge})/(2 \times \pi)$.

Example: Again consider D-AMPS with $F_s = 34.02$ MHz, $N = 350$, and $M = 1134$. Using typical values of the peak passband ripple and peak stopband ripple in (23) shows that the order of the digital low-pass filter prototype must be at least 1800. If the order of the low-pass prototype filter is $P = 1800$, the complexity of a filter bank channelizer with a full-size IDFT (i.e., 1134-point IDFT) is [from using (20)]

$$\mu(M,P) = [3600 + 4900] \times 97.2 \times 10^3 \tag{24}$$

$$= 826 \times 10^6 \quad \text{(real multiplications/second).} \tag{25}$$

In this case, $\Psi(P) = 3600$ and $\Phi(M) = 4900$. Recall that $\Phi(1134)$ is the number of real multiplications of a 1134-point IDFT. Since $1134 = 7 \times 2 \times 3^4$ is a highly composite number, the Cooley–Tukey FFT algorithm can be used to compute this IDFT very efficiently [19]–[21].

For the same example, consider a system in which only every seventh channel is needed, i.e., $L = 7$. In this case, the size of the IDFT will be $162 = 1134/7$. The complexity of a subsampled filter bank channelizer with this IDFT from (21) is

$$\mu(M,P,L) = [3600 + 700] \times 97.2 \times 10^3 \tag{26}$$

$$= 410 \times 10^6 \quad \text{(real multiplications/second).} \tag{27}$$

In this case, $\Psi(P) = 3600$ and $\Phi(M/L) = 700$. Again, a Cooley–Tukey FFT algorithm is assumed for efficiently computing the 162-point IDFT. Comparing (25) to (27), shows that the complexity of the filter bank channelizer with reduced IDFT is half the complexity of the filter bank channelizer with a full IDFT.

Referring to Fig. 9, recall that $\Psi(P) \times F_s/N$ is also the complexity of implementing one digital filter that extracts one channel from the wide-band output of the ADC. A per-channel

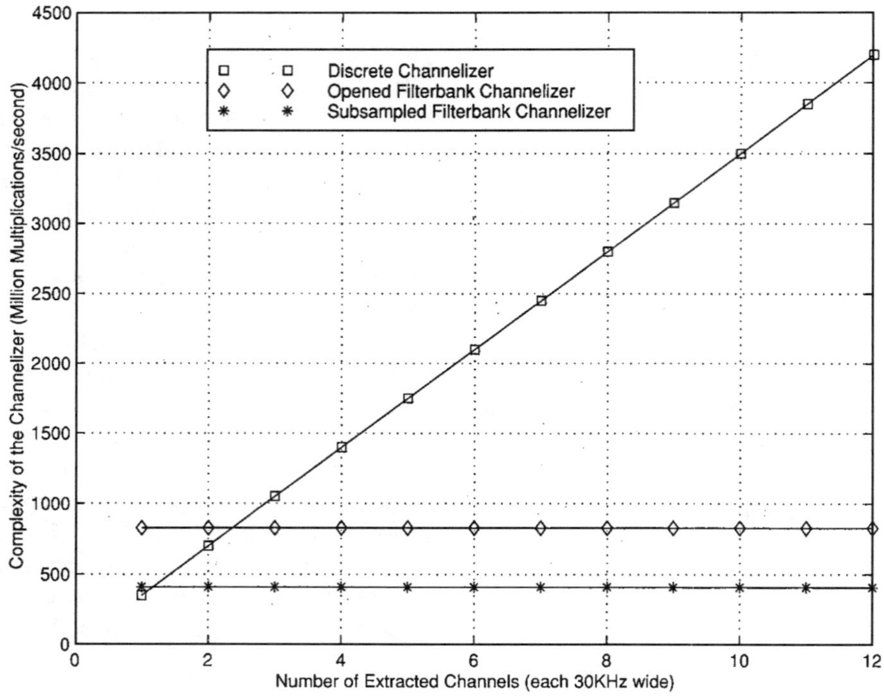

Fig. 10. Complexity of various channelizers.

channelizer with "m" output channels uses "m" separate filters folowed by "m" decimaters to extract these "m" channels. This per-channel channelizer requires $m \times \Psi(P) \times F_s/N$ multiplications per second. For this example, Fig. 10 depicts the complexity of various channelizers as a function of the number of extracted channels. From this figure, it is clear that if two or more channels need to be extracted, it is more efficient to use a filter bank channelizer that extracts *every* channel than to use two or more digital filters to extract these channels individually. In other words, the breakeven point between the complexity of the filter bank channelizer and the complexity of the per-channel channelizer is just two channels.

III. Detection with ADC Quantization Noise

So far, the wide-band ADC's are assumed to be ideal. However, nonideal converters can significantly impact the performance of wide-band receivers [11], [12]. In a traditional narrow-band receiver, automatic gain control (AGC) in front of each ADC ensures that the signal corresponding to a weak carrier is amplified sufficiently for the full-scale dynamic range of the ADC. With such an AGC, the signal-to-quantization-noise ratio at the output of each ADC is relatively high. Hence, the quantization noise of the ADC does not significantly reduce the performance of the narrow-band receiver. In a wide-band receiver, the input to the ADC contains multiple channels. In this case, the input to the ADC can contain the sum of a strong carrier and a weak carrier simultaneously. In this case, AGC is not effective in front of the wide-band ADC since the dynamic range of the input to the ADC is already close to full scale of the ADC. When a weak carrier and a sufficiently strong carrier are received simultaneously, the quantization noise of the ADC significantly degrades the reception of the weak carrier. The net result is that the weak channel is "blocked." Rayleigh fading, shadowing, and the physical size of each cell make it likely that a weak channel and a strong channel simultaneously arrive at the radio base station, typically in 20% or more the time.

In this section, the minimum probability of error receiver is derived taking into account the quantization noise of the ADC and the thermal noise. For this receiver, analytical expressions relate the parameters of the ADC to the bit-error rate of the receiver. These expressions are derived for a frequency nonselective Rayleigh fading channel.

Fig. 11 is a simplified block diagram of a wide-band receiver that uses a wide-band ADC. Assuming BPSK modulation in Rayleigh fading yields the following model for the received signal at baseband:

$$r(t) = \left[A\sqrt{E_b}p(t)\right]\alpha(t) + n(t) \tag{28}$$

where $n(t)$ is the real-valued Gaussian thermal noise with power spectral density $\Phi_{nn}(f) = N_0, E_b$ is the energy per bit, $p(t)$ is the real-valued pulse shape normalized so that $\int |p(t)|^2\,dt = 1$, "A" is the transmitted bit which can be either $+1$ or -1 (with equal probability), and $\alpha(t)$ is the real-valued fading coefficient normalized to have average magnitude of unity, i.e., $E(|\alpha(t)|) = 1$.

In this receiver, $r(t)$ is first band limited by a low-pass filter with continuous-time Fourier transform $B(f)$. The output $x(t)$ of this low-pass filter is then digitized by the ADC. Assume that $B(f)$ is an ideal low-pass filter

$$B(f) = \begin{cases} 1, & |f| < \dfrac{F_s}{2} \\ 0, & \text{otherwise} \end{cases} \tag{29}$$

where F_s is the sampling frequency of the wide-band ADC. This implies that $X(f) = 0$ for $|f| > F_s/2$. The output of the

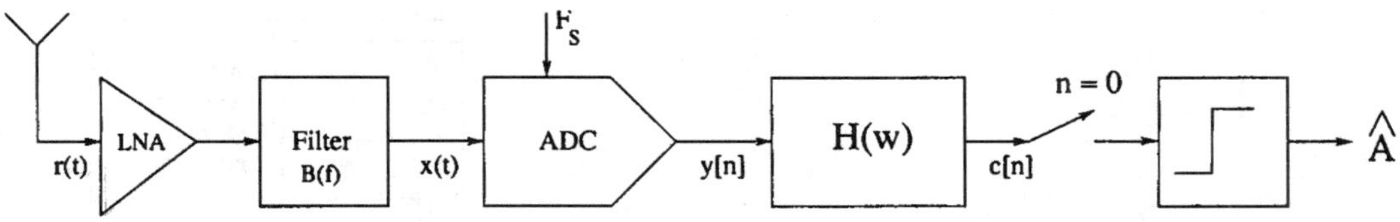

Fig. 11. Simplified block diagram of a wide-band receiver.

ADC is

$$y[n] = \left[A\sqrt{E_b}p[n]\right]\alpha[n] + n_c[n] + n_q[n]. \tag{30}$$

The first term on the right-hand side of (30) corresponds to samples of $A\sqrt{E_b}p(t)\alpha(t)$, the second term is the contribution from the thermal noise, and the third term is the contribution of the quantization noise of the ADC. The noise term $n_c[n]$ is a zero-mean Gaussian white noise sequence with variance $\sigma_c^2 = N_0 F_s$. The quantization noise [23] of a linear ADC can be modeled as a zero-mean white noise sequence uniformly distributed between $[-(R/2^{B_q}), +(R/2^{B_q})]$, where (B_q+1) is the number of bits in the quantizer, and the dynamic range of the quantizer is $[-R, +R]$. The variance of $n_q[n]$ is $\sigma_q^2 = (1/12)(R^2/2^{2B_q})$.

Grouping $n_c[n] + n_q[n]$ together shows that the noise part of $y[n]$ is a white sequence. Assume that the fading coefficient does not vary over a symbol period, and that its value α can be approximated without error at the receiver. Using a Karhunen–Loeve [24] expansion of $y[n]$ with the first basis function equal to $p[n]$, it can be shown that the weight associated with the first basis function is a *sufficient statistic* for detection of "A" when α is known. To compute this weight, convolve $x[n]$ with a filter whose impulse response $h[n]$ is given by

$$h[n] = \frac{p\left(-n\frac{1}{F_s}\right)}{\sum\limits_{m=\infty}^{+\infty} p\left(m\frac{1}{F_s}\right)}. \tag{31}$$

The output of this filter at time zero is the desired sufficient statistic. The normalization in the denominator of (31) ensures that $H(0) = 1$, where $H(w)$ is the discrete-time Fourier transform of $h[n]$. Referring to Fig. 11, the desired sufficient statistic is $c[0]$.[1]

Typically, $h[n]$ is a low-pass filter whose bandwidth is matched to the channel bandwidth. Hence, this filter extracts one channel from the output of the ADC. Essentially, $h[n]$ is a digital matched filter.

To find the receiver with minimum probability of error, find the pdf of the decision statistic $c[0]$ conditioned on the value of "A" being transmitted. To this end, $c[n]$ can be expressed as

$$c[n] = \alpha A\sqrt{E_b}p[n] * h[n] + n_c[n] * h[n] + n_q[n] * h[n]. \tag{32}$$

[1] To show that $c[0]$ is the sufficient statistics for the detection of "A," it suffices to show that the coefficients of the Karhunen–Loeve expansion of $(n_c[n] + n_q[n])$ can be approximated as jointly Gaussian given "A." Then, since $(n_c[n] + n_q[n])$ is white, these coefficients are uncorrelated, and therefore independent of each other. Only the first coefficient in this expansion contains anv statisticallv relevant information about "A."

At time zero, the first term in (32) is

$$\alpha A\sqrt{E_b}\frac{\sum\limits_n p^2[n]}{\sum\limits_m p[m]}. \tag{33}$$

If we approximate $p[n]$ by an ideal low-pass filter with bandwidth F_{BW}, the expression in (33) reduces to $\alpha A\sqrt{E_b F_{BW}}$.

At time zero, the second term in (32) is a zero-mean Gaussian random variable with variance approximately equal to $(F_{BW}/F_s) \times \sigma_c^2$. This approximation is based on modeling the matched filter as an ideal low-pass filter with bandwidth F_{BW} Hz. Since $\sigma_c^2 = N_0 F_s$, the variance of this Gaussian random variable is $N_0 F_{BW}$.

At time zero, the third term in (32) may be modeled as a Gaussian random variable because $n_q[n] * h[n]|_{n=0}$ is a weighted sum of P independent identically distributed random variables, where P is the number of nonzero coefficients in $h[n]$. Using the same reasoning as in the previous paragraph (for thermal noise component), the variance of this Gaussian random variable (quantization noise component) can be approximated by

$$\frac{R^2}{12}\left(\frac{1}{2^{2B_q}}\right)\frac{F_{BW}}{F_s}. \tag{34}$$

Therefore, the third term in (32) at time zero is a zero-mean Gaussian random variable with variance given in (34).

Given α and "A," the first term in (32) is a constant, and the second and third terms are independent zero-mean Gaussian random variables. Therefore, the pdf of $c[0]$ conditioned on knowing "A" and α is a Gaussian with mean $\alpha A\sqrt{E_b F_{BW}}$ and variance

$$\sigma^2 = N_0 F_{BW} + \frac{R^2}{12}\left(\frac{1}{2^{2B_q}}\right)\frac{F_{BW}}{F_s}. \tag{35}$$

Based on the above Gaussian conditional pdf for $c[0]$, the minimum probability of error detection rule is [25]

$$\begin{aligned}&\text{if}\quad \operatorname{sign}(\alpha)c[0] > 0 \rightarrow A = +1 \text{ was transmitted}\\ &\text{if}\quad \operatorname{sign}(\alpha)c[0] < 0 \rightarrow A = -1 \text{ was transmitted.}\end{aligned} \tag{36}$$

The probability of bit error associated with the above decision rule is [25]

$$P_e = \frac{1}{2}\left(1 - \sqrt{\frac{\gamma}{2+\gamma}}\right) \tag{37}$$

where

$$\gamma = \frac{E_b F_{BW}}{N_0 F_{BW} + \left(\frac{R^2}{12}\right)\frac{1}{2^{2B_q}}\left(\frac{F_{BW}}{F_s}\right)}. \tag{38}$$

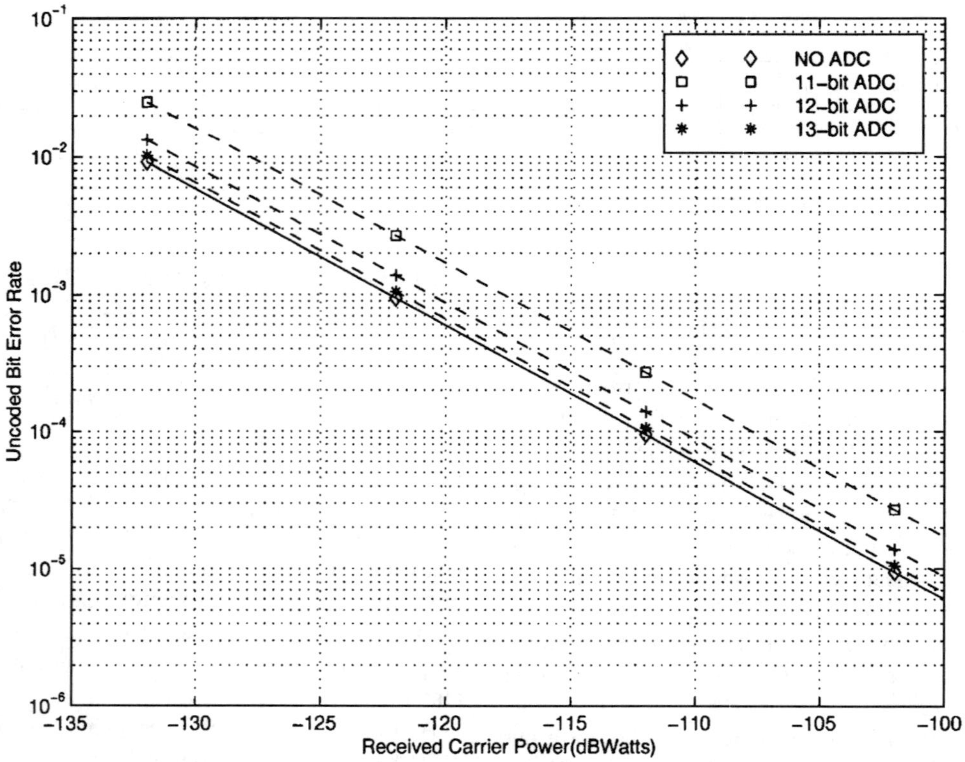

Fig. 12. Analytical BER for a wide-band receiver with quantization noise.

Note that the probability of bit error in (37) is valid only for ideal coherent demodulation where the sign of the fading coefficient α is estimated at the receiver without error.

An interesting feature of (38) is the ratio F_{BW}/F_s multiplying the variance of the quantization noise in the denominator. In a wide-band receiver, this ratio is typically much less than 1, and as this ratio gets smaller, the probability of a bit error decreases. This oversampling ratio is sometimes referred to as the "processing gain" of a wide-band ADC [12], [11], and is one of the inherent benefits of the wide-band architecture. Further insight regarding processing gain is obtained by writing (38) in a slightly different form. As an example, assume that $F_{BW}/F_s = 1024 = 2^{10}$. Then (38) can be rewritten as

$$\gamma = \frac{E_b F_{BW}}{N_0 F_{BW} + \left(\frac{R^2}{12}\left(\frac{1}{2^{2(B_q+5)}}\right)\right)}. \tag{39}$$

With processing gain of $(F_{BW}/F_s) = 2^{10}$, (39) shows that the effect of processing gain is equivalent to having an ADC with five additional bits. The reason for this improvement is that only a small portion of the quantization noise falls within the passband of the matched filter. More specifically, only F_{BW}/F_s of the white quantization noise of the ADC falls within the passband of the matched filter. Equation (39) quantifies the impact of processing gain on the bit-error rate.

A. *Performance*

In this section, the analytical results of the previous section are used to quantify the impact of the wide-band ADC on the bit-error rate of a wide-band receiver for D-AMPS in a frequency-nonselective Rayleigh fading environment. Monte Carlo simulations are also presented which verify the analytical results.

The parameters of the wide-band receiver are as follows. Assume that the strongest signal at the input to the ADC is -50 dBW, and the full-scale dynamic range of the ADC is set 3 dB larger than the standard deviation of this largest signal (i.e., full-scale range $= [-0.0045\,\text{V}, +0.0045\,\text{V}]$). Further assume that this strongest carrier is being received simultaneously with a second carrier. The strength of the second carrier will be plotted on the x axis, and the bit-error rate of this carrier will be plotted on the y axis. For all curves, the sampling rate of the ADC is 34.02 MHz, the bandwidth of each carrier is 30 kHz, and the symbol rate is 48.6 kHz. Assume BPSK modulation with $N_0 = 1.832 \times 10^{-20}$ W/Hz, corresponding to a noise temperature of 293 K and a 6.5 dB noise figure in RF conversion. The average magnitude of the fading coefficient is normalized to 1. The D-AMPS standard requires that a -133 dBW carrier must achieve a BER of 3% or less in Rayleigh fading, i.e., the sensitivity of the receiver must be -133 dBW.

The solid line in Fig. 12 represents the performance of an ideal wide-band receiver with no quantization error. Each dotted line in Fig. 12 depicts the BER predicted by (37) of a wide-band receiver with different ADC resolution. This figure shows that the performance of the wide-band receiver with a 12 bit wide-band ADC is approximately 2 dB worse than the performance of the ideal wide-band receiver. The performance of the wide-band receiver with a 13-bit wide-band ADC is within 0.1 dB of the performance of the ideal receiver.

The wide-band receiver with a 13-bit ADC was also simulated. Fig. 13 shows the simulated BER and the analytically

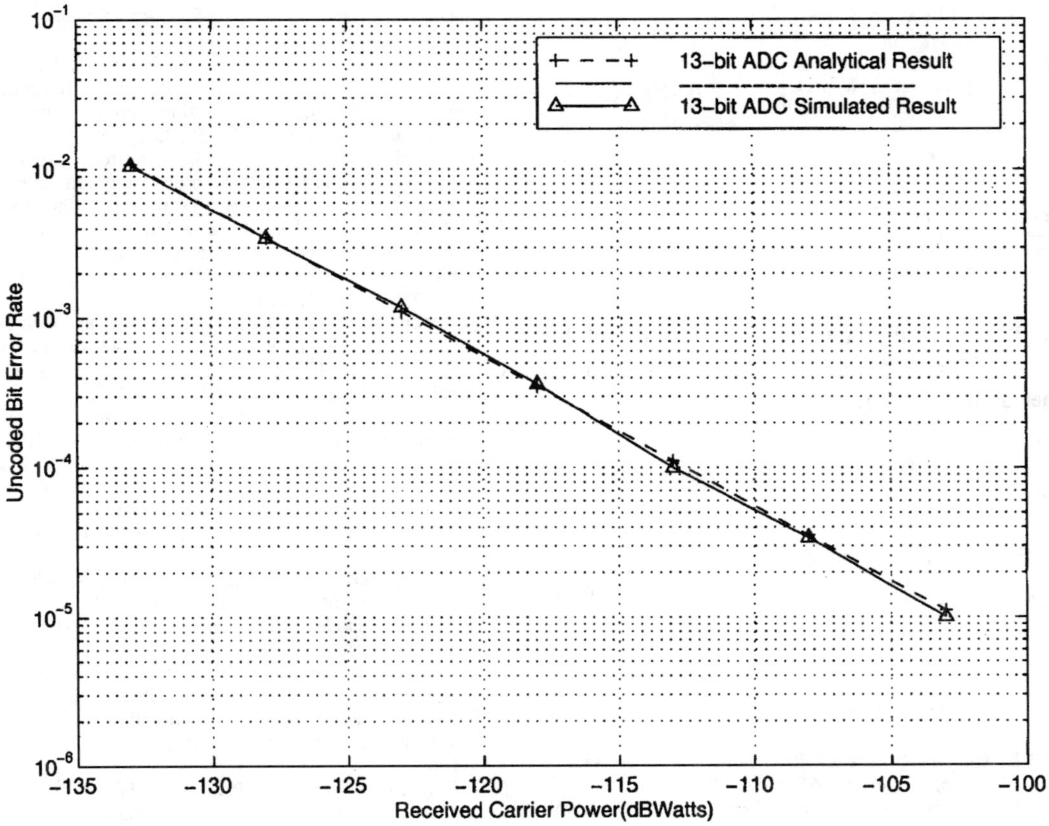

Fig. 13. Analytical/simulated BER for a wide-band receiver with quantization noise.

estimated BER from (37). The simulations used a square-root raised-cosine pulse with excess bandwidth of 35%. Fig. 13 shows that the analytical results agree well with the simulation.

IV. Conclusion

Two key issues in the implementation of wide-band cellular base stations were addressed in this paper: a computationally efficient channelizer, and an optimal receiver that takes into account the quantization noise.

A computationally efficient algorithm for extracting narrow-band signals from a wide-band cellular signal was presented. The algorithm was derived from the conventional DFT filter bank used in FDM ↔ TDM transmultiplexers. The resulting channelizer can be directly interfaced to the baseband processing block of the wide-band receiver. The sampling rate at the output of the OFB channelizer can be made an exact multiple of the baud rate. For a wide-band receiver operating in a cellular environment with a regular frequency reuse pattern, the size of the IFFT in the OFB channelizer can be further reduced, resulting in a more computationally efficient channelizer. The DFT filter bank channelizer is designed specifically for the requirements of digital cellular receivers. For D-AMPS, the OFB channelizer is two times less complex than a traditional DFT filter bank channelizer and 50 times less complex than a per-channel channelizer.

In this paper, the minimum probability of error receiver was derived that takes into account the thermal noise *and* quantization noise of the ADC for a frequency-nonselective Rayleigh fading channel. Analytical expressions relating the parameters of the ADC to the bit-error rate of the receiver were derived. Based on these expressions, one can gain insight into the number of bits needed for the wide-band ADC. According to these expressions, a 13 or 14 bit ADC will meet the uplink requirements for the D-AMPS cellular standard. Monte Carlo simulations validated the analytical results derived.

Appendix
Subsampled IDFT Property

Lemma: Let the M-point sequence $\{r_k\}_{k=0}^{M-1}$ be the IDFT of the sequence $\{s_m\}_{m=0}^{M-1}$, and assume that $M = Q \times L$ for some integers Q and L. The subsampled sequence $\{r_{kL}\}_{k=0}^{Q-1}$ is obtained by taking the Q-point IDFT of the sequence $\{z_i\}_{k=0}^{Q-1}$ defined as

$$z_i = \sum_{r=0}^{L-1} s_{i+Qr}, \qquad 0 \le i \le Q-1. \tag{40}$$

Proof: From the definition of IDFT

$$r_k = \sum_{m=0}^{M-1} s_m e^{j(2\pi/M)km}, \qquad 0 \le k \le M-1. \tag{41}$$

Following [21], define a two-dimensional index map for "m" as

$$m = i + Qr, \qquad \begin{cases} 0 \le i \le Q-1 \\ 0 \le r \le L-1 \end{cases} \tag{42}$$

and subsitute (42) in (41) with $k \leftarrow kL$ to get

$$r_{kL} = \sum_{i=0}^{Q-1} \sum_{r=0}^{L-1} s_{i+Qr} e^{j((2\pi/M)kL)(i+Qr)}, \qquad 0 \le k \le Q-1 \tag{43}$$

$$= \sum_{i=0}^{Q-1} \left(\sum_{r=0}^{L-1} s_{i+Qr} \right) e^{j((2\pi/Q)k)i}, \qquad 0 \le k \le Q-1. \tag{44}$$

In going from (43) to (44), the fact that $M = Q \times L$ and that $e^{j(2\pi/M)kLQr} = e^{j2\pi kr} = 1$ is used. From the definition of z_i in (40), the inner sum in (44) is just z_i; hence,

$$r_{kL} = \sum_{i=0}^{Q-1} z_i e^{j((2\pi/Q)k)i}, \qquad 0 \le k \le Q-1. \tag{45}$$

The right-hand side of (45) is the definition of the Q-point IDFT of the sequence $\{z_i\}_{i=0}^{Q-1}$, i.e., $\{r_{kL}\}_{k=0}^{Q-1} = Q$-point IDFT of $\{z_i\}_{i=0}^{Q-1}$.

References

[1] R. Lackey and D. Upmal, "Speakeasy: The military software radio," *IEEE Commun. Mag.*, vol. 33, pp. 56–61, May 1995.

[2] J. Mitola, "The software radio architecture," *IEEE Commun. Mag.*, vol. 33, pp. 26–38, May 1995.

[3] R. Baines, "The DSP bottleneck," *IEEE Commun. Mag.*, vol. 33, pp. 46–54, May 1995.

[4] EIA/TIA Interim Standard, *Cellular System Dual Mode Mobile Station-Base Station Compatibility Standard IS-54*, May 1990.

[5] Harris Semiconductor Corporation, *Digital Signal Processing Data Book*, 1994.

[6] H. Scheuermann and H. Göckler, "A comprehensive survey of digital transmultiplexing methods," *Proc. IEEE*, vol. 69, pp. 1419–1450, Nov. 1981.

[7] M. Bellanger and J. Daguet, "TDM-FDM transmultiplexer: Digital polyphase and FFT," *IEEE Trans. Commun.*, vol. COM-22, pp. 1199–1204, Sept. 1974.

[8] F. Takahata, M. Yashunaga, Y. Hirata, T. Oshawa, and J. Namiki, "A PSK group modem based on digital signal processing: Algorithm, hardware design, implementation, and performance," *Int. J. Satellite Commun.*, vol. 6, pp. 253–266, July 1988.

[9] I. Corden and R. Carrasco, "Fast transform based complex transmultiplexer algorithm for multiband quadrature digital modulation schemes," *Proc. Inst. Elect. Eng.*, vol. 137, pt. I, pp. 408–416, Dec. 1990.

[10] B. Evans, F. Coakley, H. El-Amin, S. Lu, and C. Wong, "Baseband switches and transmultiplexers for use in on-board processing mobile/business satellite system," *Proc. Inst. Elect. Eng.*, vol. 133, pt. F, pp. 356–363, July 1986.

[11] J. Wepman, "Analog-to-digital converters and their applications in radio receivers," *IEEE Commun. Mag.*, vol. 33, pp. 39–45, May 1995.

[12] B. Brannon, "Using wide dynamic range converters for wideband radios," *RF Design*, pp. 50–65, May 1995.

[13] ——, "Wide-dynamic-range A/D converters pave the way for wideband digital-radio receivers," *EDN*, pp. 187–205, Nov. 1996.

[14] ——, "Overcoming converter nonlinearities with dither," Analog Devices Appl. Note AN-410.

[15] P. P. Vaidyanathan, "Multirate digital filters, filter banks, polyphase networks and applications: A tutorial," *Proc. IEEE*, vol. 78, pp. 56–93, Jan. 1990.

[16] R. Crochiere and L. Rabiner, *Multirate Digital Signal Processing*. Englewood Cliffs, NJ: Prentice-Hall, 1983.

[17] V. H. MacDonald, "The cellular concept," *Bell Syst. Tech. J.*, vol. 58, pp. 15–41, Jan. 1979.

[18] A. V. Oppenheim and R. W. Schafer, *Discrete-Time Signal Processing*. Englewood Cliffs, NJ: Prentice-Hall, 1989, pp. 548–549.

[19] J. Cooley and J. Tukey, "An algorithm for the machine computation of complex Fourier series," *Math. Comput.*, vol. 19, pp. 297–301, Apr. 1965.

[20] C. S. Burrus and T. W. Parks, *DFT/FFT and Convolution Algorithms: Theory and Implementation*. New York: Wiley, 1985.

[21] C. Burrus, "Index mapping for multidimensional formulation of the DFT and convolution," *IEEE Trans. Acoust., Speech, Signal Processing*, vol. ASSP-25, pp. 239–242, June 1977.

[22] M. Bellanger, *Digital Processing of Signals: Theory and Practice*. New York: Wiley, 1989.

[23] D. H. Sheingold, *Analog-Digital Conversion Handbook*. Englewood Cliffs, NJ: Prentice-Hall, 1986.

[24] H. L. VanTrees, *Detection, Estimation, and Modulation Theory, Part I*. New York: Wiley, 1968.

[25] J. G. Proakis, *Digital Communications*. New York: McGraw-Hill, 1989.

A Low-Power DSP Core-Based Software Radio Architecture

James E. Gunn, *Senior Member, IEEE,* Kenneth S. Barron, and William Ruczczyk

Abstract— **This paper describes an approach to developing a low-power digital signal processor (DSP) subsystem architecture for advanced software radio platforms. The architecture is intended to support next-generation wide-band spread-spectrum military waveforms. The methodology illustrates how a next-generation programmable DSP core forms the basis for an application-specific integrated circuit (ASIC). It also shows how semiconductor technologies can be integrated into such chips to achieve algorithm performance while minimizing subsystem power consumption. The ASIC is run-time configurable to maintain high flexibility. The range of RF channel modulation ("waveforms") and air interfaces is intended to include both wide-band and traditional narrow-band waveforms. Estimated gate counts and power-consumption estimates are presented. DSP circuit-design and power-management strategies necessary to achieve low-power operation are presented. While the architecture discussion focuses on military waveforms, the approach is also applicable to commercial waveforms.**

Index Terms—**Communication, configurable ASIC, DSP architecture, low power, programmable DSP, software radio, wireless.**

I. Introduction

SOFTWARE radios migrate the traditional hard-wired radio platforms to flexible software radio platforms that can support multiple modulation waveforms and multiple air interface standards. This approach should allow the graceful evolution of the technology over time. Thus, software radio hardware platforms can serve a range of applications including: analog cellular; digital cellular/personal communications services (PCS); advanced wide-band, spread-spectrum military waveforms; legacy narrow-band military waveforms; navigation waveforms (e.g., the global positioning system); and emergency preparedness, public safety, and other waveforms. Depending on the waveform(s), architecture, and implementation, a single software radio platform could have the flexibility to support a broad range of such waveforms.

Three general classes of programmable digital and software radio architectures are emerging: base-station, mobile, and battery-powered handheld units. Typically, base-stations support large numbers of channels and users with only a few types of waveforms (e.g., cellular). They have less sever power and space constraints than the other classes. Conversely, battery-powered handheld units typically support a single user, have aggressive power and space constraints, and typically support one or two services. Users of handheld units have traditionally carried multiple devices to access multiple services (e.g., cellular telephone and a pager). Users seem to want expanded, single-platform, multiservice flexibility, encompassing multiple services (e.g., cellular/PCS, paging, data networks, private dispatch, military networks, and perhaps limited video). Mobile units fall between these extremes, mounted on vehicles or easily transported. Mobile units typically address military and dispatch applications (e.g., police, taxi, fire, military vehicles, etc.).

This paper focuses on the DSP subsystem for emerging handheld software radio devices. This technology enables the support of multiple waveforms on a programmable, configurable hardware platform that consumes low power.

Section II summarizes communication waveforms and software radio concepts applicable to the design of the low-power DSP platform. It provides an overview of illustrative communication waveforms, emphasizing the characteristics that define DSP processing requirements. Section III provides a brief overview of relevant RF technologies, their function, and the related DSP subsystem integration concepts. Section IV gives an overview of data-acquisition requirements and the important technical parameters from the DSP subsystem perspective. Section V discusses low-power design and power-management techniques that support handheld, low-power implementations. In Section VI, a run-time configurable ASIC architecture is presented that processes throughput-intensive algorithms in support of a standard programmable DSP core. Section VII summarizes gate-count and power estimates. Section VIII summarizes the design methodology, and Section IX draws conclusions.

II. Communication Waveforms

The software radio uses RF circuits, DSP's, and microprocessors to implement wireless-communication-system functions, typically with the algorithmic structure of Fig. 1. Software radios have been canonically partitioned [1] into antenna, radio frequency (RF), intermediate frequency (IF) processing, baseband (BB), bitstream, and source segments. As RF, data acquisition, and digital signal processing (DSP) technologies have advanced, the trend has been to move the analog-to-digital converter (ADC) closer to the antenna and to do

Manuscript received February 11, 1998; revised June 23, 1998 and July 9, 1998. This work was supported in part by the U.S. Army Communications and Electronics Command and the Defense Advanced Research Projects Agency under Contract DAAB07-97-D016.

J. E. Gunn is at Box 833157, Richardson, TX 75083-3157 USA.

K. S. Barron and W. Ruczczyk are with Raytheon Systems Co., Dallas, TX USA.

Publisher Item Identifier S 0733-8716(99)02971-6.

Reprinted from *IEEE Journal on Selected Areas of Communications,* Vol. 17, No. 4, pp. 574–590, April 1999.

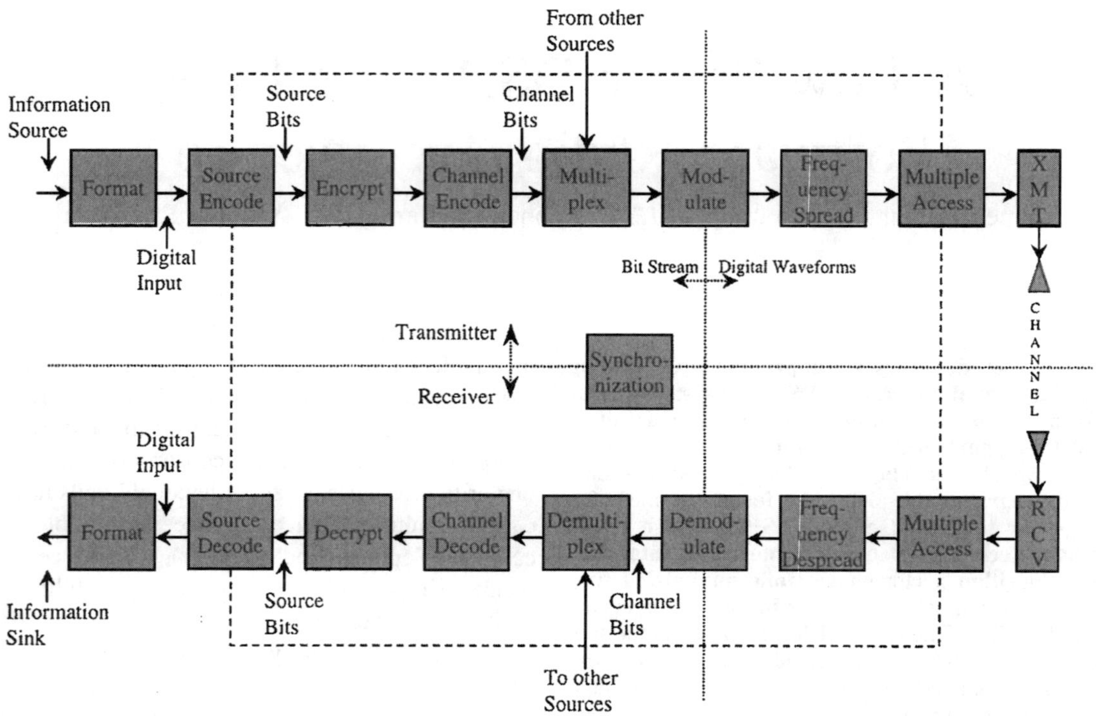

Fig. 1. Generic software radio algorithm block diagram (adapted from [2]).

TABLE I
REPRESENTATIVE DoD AND COMMERCIAL WAVEFORMS

WAVEFORM FAMILY	FREQUENCY RANGE	CHANNEL BANDWIDTH	MODULATION
FH HF	2 to 20 MHz	3 kHz	AM/FM
SINCGARS	30 t0 88 MHz	25 kHz	FH (100 h/s)
Have Quick	225 to 400 MHz	25 kHz	FH
EPLRS	420 to 450 MHz	5 MHz	DSSS/FH
JTIDS	960 to 1310 MHz	6 MHz	DSSS/FH
GPS	1.5 GHz	10 MHz	DSSS
Speakeasy	All of the above		
SATCOM	7 GHz	variable	variable
IS-95	800 MHz, 1.9 GHz	1.25 MHz	DSSS/CDMA
IS-54/136	800 MHz	30 kHz	TDMA
ITU 3G	1.8 TO 2.1 GHz	5, 10, 15 MHz	DSSS/CDMA

more functions digitally. Therefore, the ability of a software radio architecture to support a communication waveform is predominantly determined by

- the largest instantaneous signal bandwidth (W);
- the frequency range and bandwidth of the RF;
- the ADC sampling rate (greater than 2 W);
- the maximum dynamic range;
- DSP throughput requirements including translation of IF to baseband, modulation, demodulation, coding, and decoding.

Table I lists the frequency band, bandwidth, and modulation format of some current DoD and commercial waveforms. Table II presents similar descriptions of waveforms developed for Phase 1 of DARPA's small unit operations (SUO) project. These waveforms represent the state of the art in direct-sequence spread-spectrum (DSSS) and frequency-hopped spread-spectrum (FHSS) concepts. The architecture concepts presented in this paper address the DSP requirements implicit in these current and advanced waveforms. The emerging cellular/PCS wide-band, CDMA waveforms [3] currently being defined in international standards bodies should generally be addressed by this architecture as well.

The processing capacity required to modulate and code a waveform in the transmitter and to demodulate and decode the waveform in the receiver defines the DSP processing requirement of the modem. Fig. 2 graphically presents esti-

TABLE II
PROPOSED SMALL UNIT OPERATIONS WAVEFORMS [4]

Wave form	FREQUENCY RANGE	BAND-WIDTH	CHIP RATE	FH BAND-WIDTH	FH RATE	MODULATION FORMAT
1	30 to 88 MHz 225 to 400 MHz 0.8 to 2.45 GHz	Variable to 10 MHz	0.32 to 16 Mcps			Quasi Bandlimited MSK
2	77 to 88 MHz 225 to 400 MHz 1.8 to 2.0 GHz	Variable to 20 MHz	N/A	>= 100 MHz	100 hps	Wavelet (featureless, good side lobe suppression)
3	20 to 2000 MHz	Variable to 12 MHz	20 Mcps	200 MHz	400 hps	MSK
4	6 to 2000 MHz	Variable to 26 MHz	N/A	N/A	N/A	Transform Domain DSPN
5	10 to 2000 MHz	1.6 MHz	680 Kcps I & Q	Maximum 60 MHz or 20% Carrier	1,200 hps	non-LPI: MSK LPI: Filtered DS

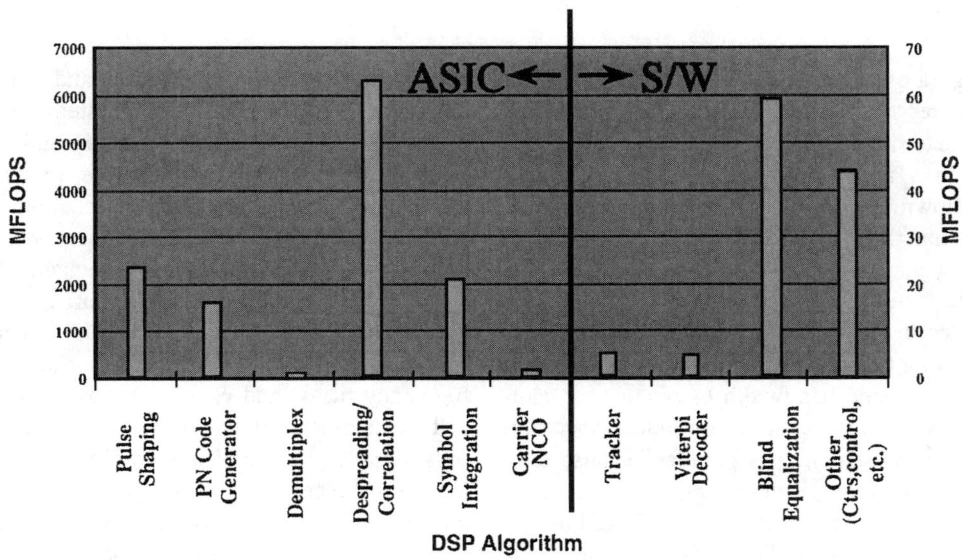

Fig. 2. Software radio receiver DSSS processing capacity estimates for 10 Mchips/s, 40 Msample/s (ADC), 9600 bits/s, 5 μs delay spread; total: 12.78 GFLOPS, ASIC: 12.67 GFLOPS, S/W: 0.11 GFLOPS.

mated processing capacity for the key functions required to demodulate a proposed wide-band DSSS military waveform. This representative waveform has a DSSS chip rate of approximately 10 MCPS and a bandwidth of approximately 10–20 MHz (including frequency-domain sidelobes). The capacity requirements for spreading and despreading are not achievable on a programmable DSP. Thus, the DSP subsystem requires a communication preprocessor. This paper considers a design based on emerging DSP core and ASIC semiconductor technologies [5].

III. EMERGING RF SEGMENT ALTERNATIVES

Fig. 3 shows a typical system-level block diagram of a software radio transceiver. This superheterodyne receiver translates the RF to IF, digitizing the IF bandpass signal. This architecture is well suited for traditional hardwired transceiver implementations that address both limited frequency range and fixed, typically narrow-band (e.g., 30 kHz cellular) channelization plans. Extended frequency range and variable channelization plans (narrow band and wide band) require multiple RF and IF filters and RF switches. Thus, flexibility is not easily implemented by using traditional design methods, and is not well suited for low-cost or low-power applications.

Fig. 4 shows three common RF receiver configurations. Fig. 4(a) shows the analog in-phase and quadrature (I, Q) superheterodyne receiver. The RF frequency is translated to baseband through one or more IF stages. The (I, Q) baseband signals are digitized and demodulated digitally in the DSP. The sampling rates of the ADC's need be at least the highest baseband signal frequency, whereas real sampling requires

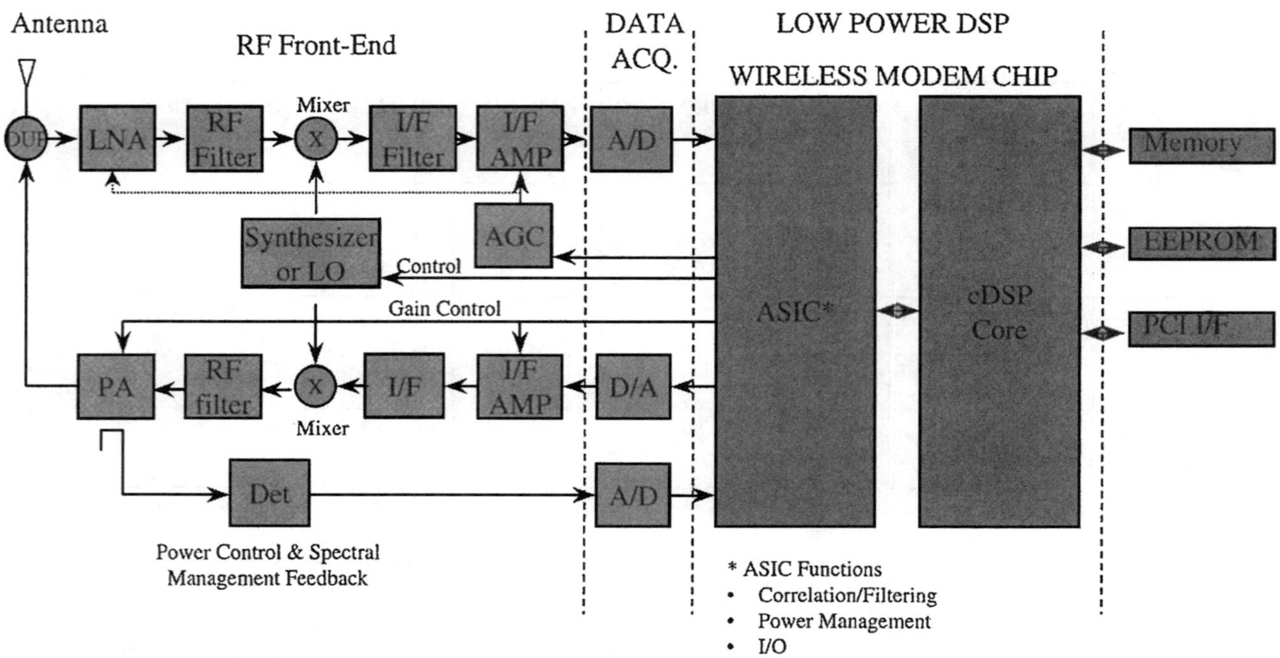

Fig. 3. Example system-level software radio transceiver block diagram [4], [6].

twice this sampling rate to meet the Nyquist criterion for unambiguous signal reconstruction. A second configuration is a passband superheterodyne receiver that translates in one or more IF stages to a final passband IF frequency where it is digitized, as shown in Fig. 4(b). The sampling rate of the passband signal must be at least twice the bandwidth of the upper cutoff frequency of the passband signal. Since the lower passband cutoff frequency will be greater than zero, typical sampling rates are two to three times the bandpass bandwidth. The sampling rate must be sufficiently greater than two times the signal passband bandwidth to ensure that sampled images of the positive and negative continuous spectrum [7] do not alias. Passband sampling is popular because only one ADC converter is required, simplifying the component configuration. For moderate bandwidths (e.g., 25 MHz), one can configure passband superheterodyne receivers using off-the-shelf ADC's (e.g., the Analog Devices 70 MHz ADC).

Fig. 4(c) shows a direct conversion (or homodyne or zero IF) receiver where the RF signal is translated directly to baseband. This requires that the numerically controlled oscillator (NCO) or synthesizer be locked to the carrier. The advantages of the direct conversion receiver include fewer signal translation steps, and the ability to use simpler analog filters cascaded with low-pass baseband digital filters in the DSP. This creates a more flexible (wider) tuning range and potentially greater channel bandwidths. The disadvantages include leakage from high-gain low-noise mixers, requirements for very high dynamic range analog components, the requirement for higher sensitivity than a comparable superheterodyne receiver, the need for precise I and Q phase balancing, dc offset cancellation, antenna isolation, and high-selectivity filters [8]. As a result, homodyne receivers are extremely challenging to implement.

Traditionally, tuning and translation to/from IF and baseband have been accomplished in the analog RF and IF segments [9], while (de)modulation and (de)coding have been done in the baseband DSP subsystem. With emerging commercial and military application for single platforms covering 2–3000 MHz having configurable modulation bandwidths, advanced RF conversion is more frequently based on passband superheterodyne or direct conversion receiver designs. Ongoing research on micro electromechanical systems (MEMS) [10], [11] offers significant miniaturization and power reduction potential for the filters, oscillators, and switches required for configurable RF IF segments that support the desired frequency bands and channel bandwidths.

The superheterodyne receiver with real, (I, Q), and bandpass sampling is applicable to low power software radios. Homodyne receivers are also applicable, with the advantage of reduced parts count, but at the risk of introducing artifacts into the baseband. A flexible ASIC-DSP core entails the flexibility to accommodate any of these RF approaches.

IV. Signal Conversion

Signal conversion refers to analog-to-digital conversion (ADC) and its inverse, digital-to-analog conversion (DAC). The theory of sampling and quantization (i.e., digitization) of analog signals has a rich history [12]. From an implementation perspective, the ADC is more demanding than the DAC. In particular, ADC's have generally consumed considerable power, posing technical challenges for low-power software radios.

A technical discussion of ADC's for software radios is provided in a companion paper in this issue [13]. This includes an analysis of sample-and-hold and conversion circuits, aperture uncertainty, jitter, number of bits of resolution, precision, and related issues. The design issue of primary concern in this paper is power dissipation. Due to the need to rapidly change voltage levels in such a way that the power in the

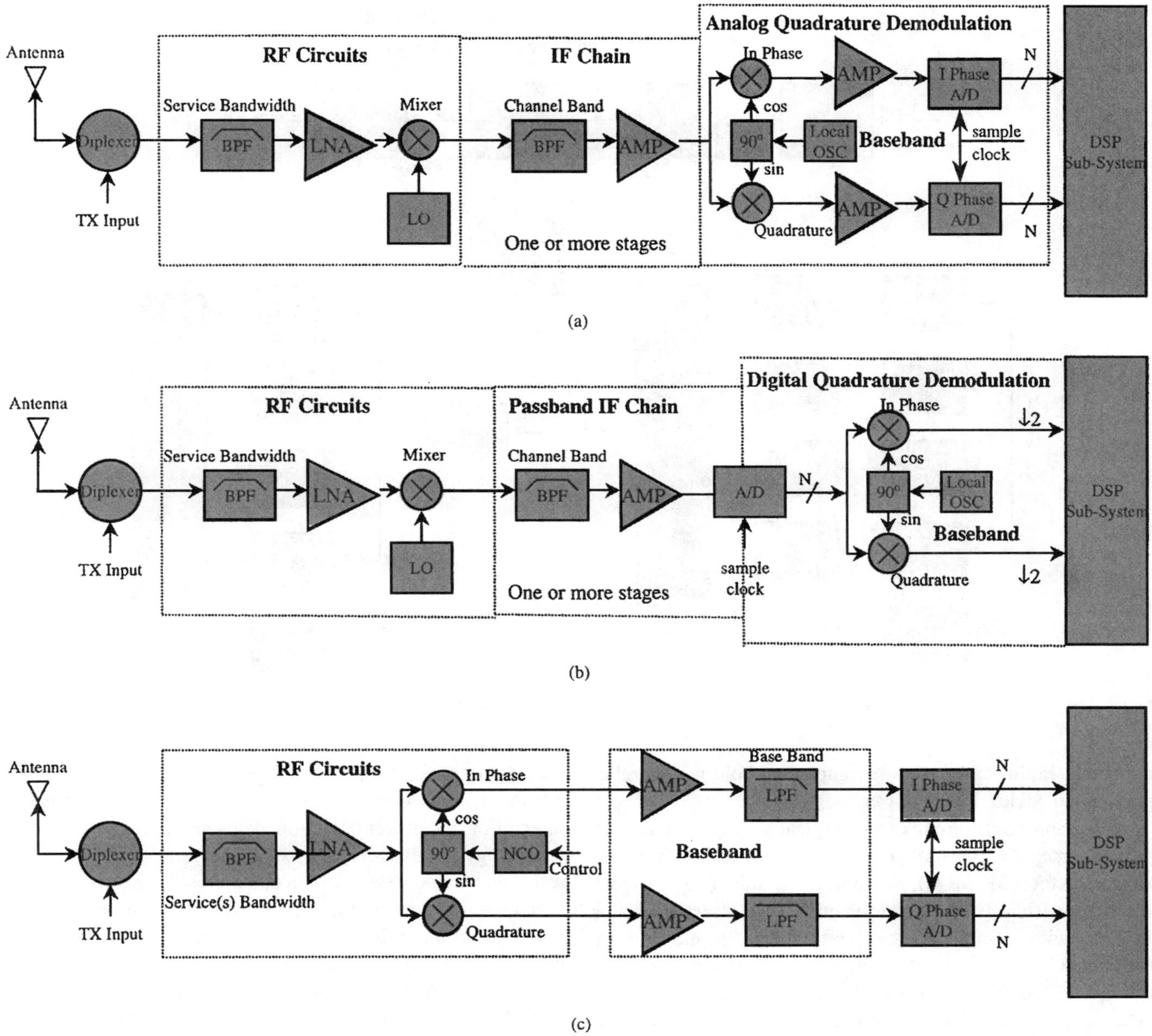

Fig. 4. Common RF receiver configurations for software radios. (a) I, Q superheterodyne (with passband alternatives, (b) passband superheterodyne, and (c) direct conversion (zero IF or homodyne).

least significant bit is greater than ambient noise, sample-and-hold circuits dissipate significant power. Fig. 5 shows the log–log relationship between increased resolution and dissipated power.

It is possible to reduce ADC power by interleaving the sample-and-hold and quantizer circuits. Interleaved ADC's repeat functional blocks of the serial ADC [Fig. 6(a)] to distribute functions over parallel sample-and-hold circuits (Fig. 6(b), left) or over parallel quantizers (Fig. 6(b), right). The interleaved circuits can be integrated on monolithic chips. In many cases, greater performance is achieved at lower total power due to the lower circuit frequencies per parallel path. This is achieved at the expense of larger chip area. Parallel circuits may not be the best low-power option for passband ADC's in which the highest IF frequency is much greater than the signal

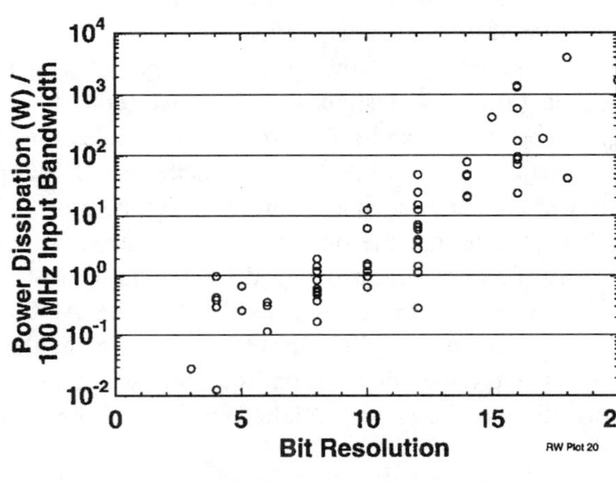

Fig. 5 ADC power dissipation versus resolution [13], [21].

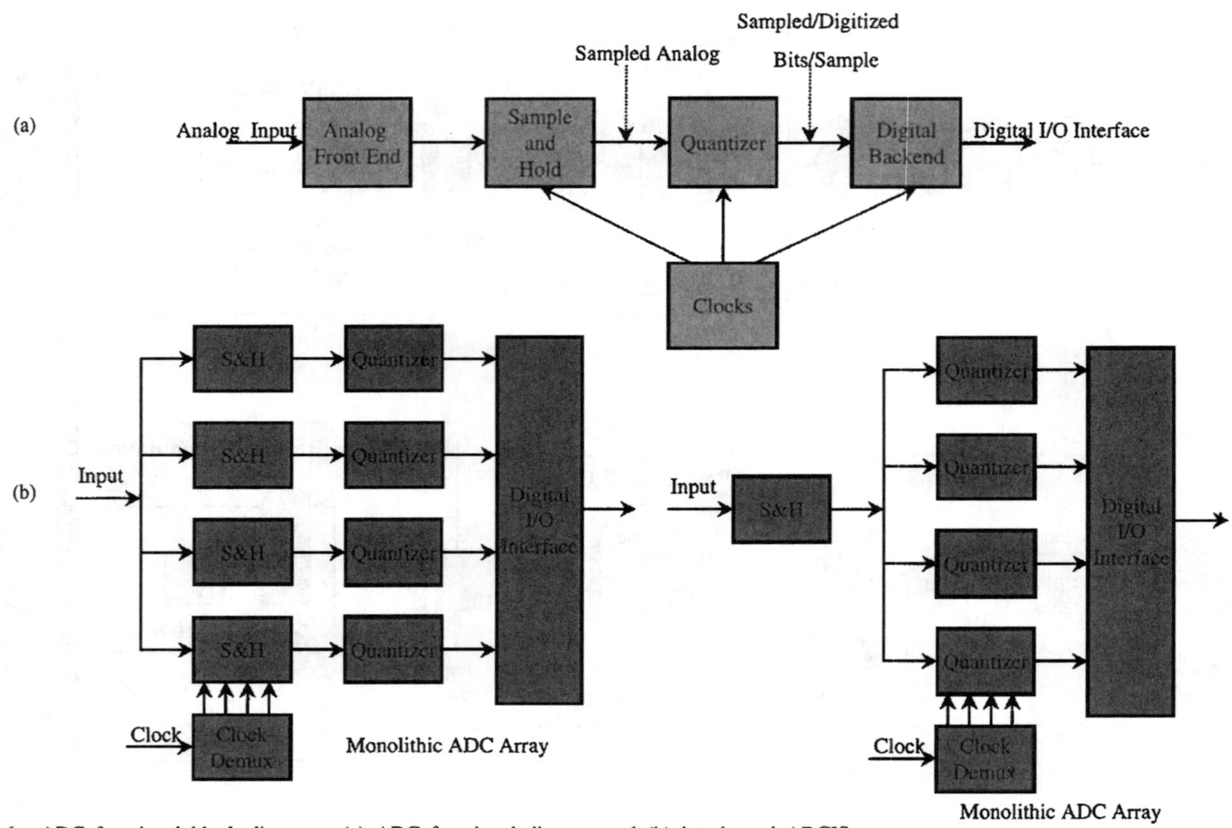

Fig. 6. ADC functional block diagrams. (a) ADC functional diagram and (b) interleaved ADCIS.

bandwidth. In this case, a single sample and hold with parallel quantization yields lower power.

Sample-and-hold circuits are typically implemented in gallium arsenide (GaAs), bi complementary metal-oxide semiconductor (BiCMOS), indium phosphide (InP), hybrid bipolar transistor (HBT), heterojunction field-effect transistor (HFET), and complementary metal-oxide semiconductor (CMOS). GaAs and InP provide high performance with high power dissipation. HBT and HFET circuits have the highest upper cutoff frequencies reported in the literature, and therefore may yield the highest performance ADC's ultimately, but the technology is not yet mature. CMOS traditionally has a lower performance (measured as the product of sampling frequency times the number of bits) than the others, but it dissipates the least power for a given combination of sampling rate and quantization accuracy.

Another approach to low power is the oversampling delta–sigma (also called sigma–delta) converter [14] shown in Fig. 7. The sample-and-hold circuit oversamples the input analog signal by a factor of N. This allows the use of a simpler, lower power converter such as a threshold comparator (a 1 bit converter). In addition, the quality of the antialiasing analog filter (measured as the ratio of -3 dB bandwidth to -23 dB bandwidth) need not be as high as the Nyquist ADC. The cascade of the decimator and digital filter with the antialiasing filter yields a product filter, each stage of which need not be as effective as the single antialiasing filter of the Nyquist ADC.

The delta–sigma and Nyquist converters differ in dissipated power, number of components, and pace of development of the underlying technology. Since digital technology traditionally advances faster than RF and analog technology in terms of integration and power reduction, delta–sigma converters have a technology advantage. Delta–sigma converters with 2–4 bits of resolution are now deployed in commercial low-power, spread-spectrum wireless terminals. However, increased dynamic range is required for multiple-user interference mitigation, near/far performance, and jammer suppression [15].

In analyzing options for the low-power applications such as SUO, a reasonable aggressive compromise envisions 8–12 bits of dynamic range, 20–30 MHz sampling rate, and 150–300 mW of power dissipation. By the year 2000–2001, the sampling rate is likely to advance to 50–60 MHz, with the same accuracy and dissipated power. It is anticipated that the sample-and-hold circuits would be implemented in CMOS, InP, or HFET with CMOS digital circuits.

V. Low-Power DSP Design Techniques

This section discusses semiconductor power-dissipation trends, sources of CMOS power dissipation, and low-power design techniques.

A. Semiconductor Technology Trends

Fig. 8 shows recent trends in ASIC gate density and power consumption. As the feature size decreases linearly, the number of gates per unit area increases quadratically. As the operating voltage is reduced, power dissipation decreases in proportion to the square of the voltage. At a given operating frequency, the change from 5 to 3.3 V results in a power

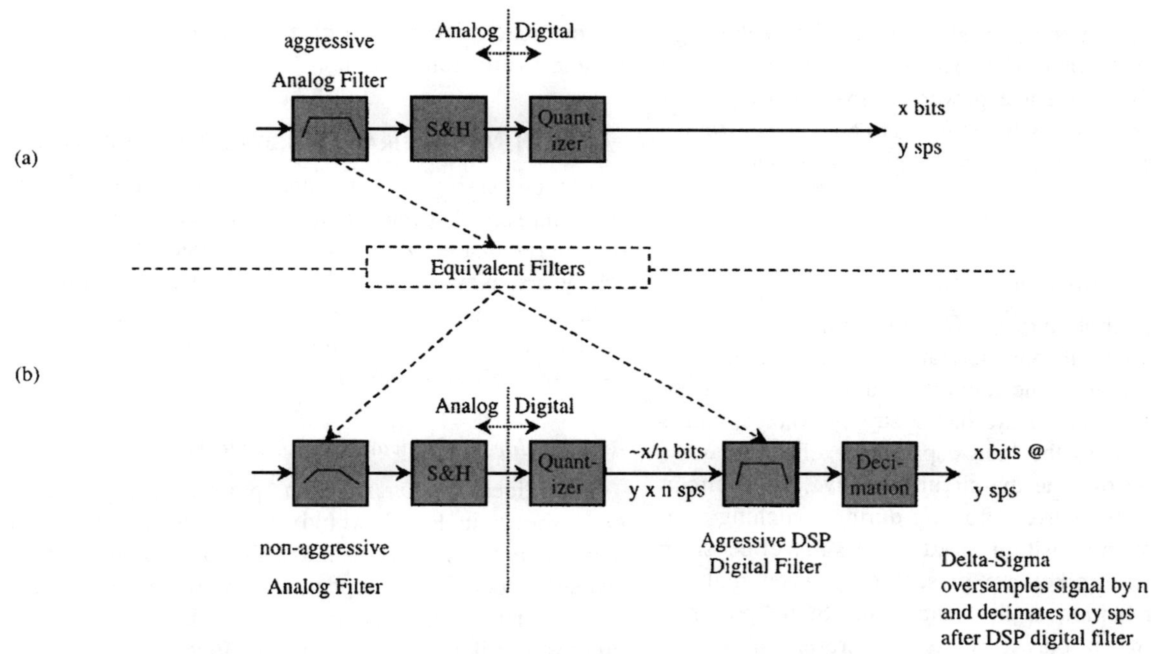

Fig. 7. Nyquist (traditional) versus delta–sigma ADC. (a) Traditional ADC and (b) delta–sigma ADC.

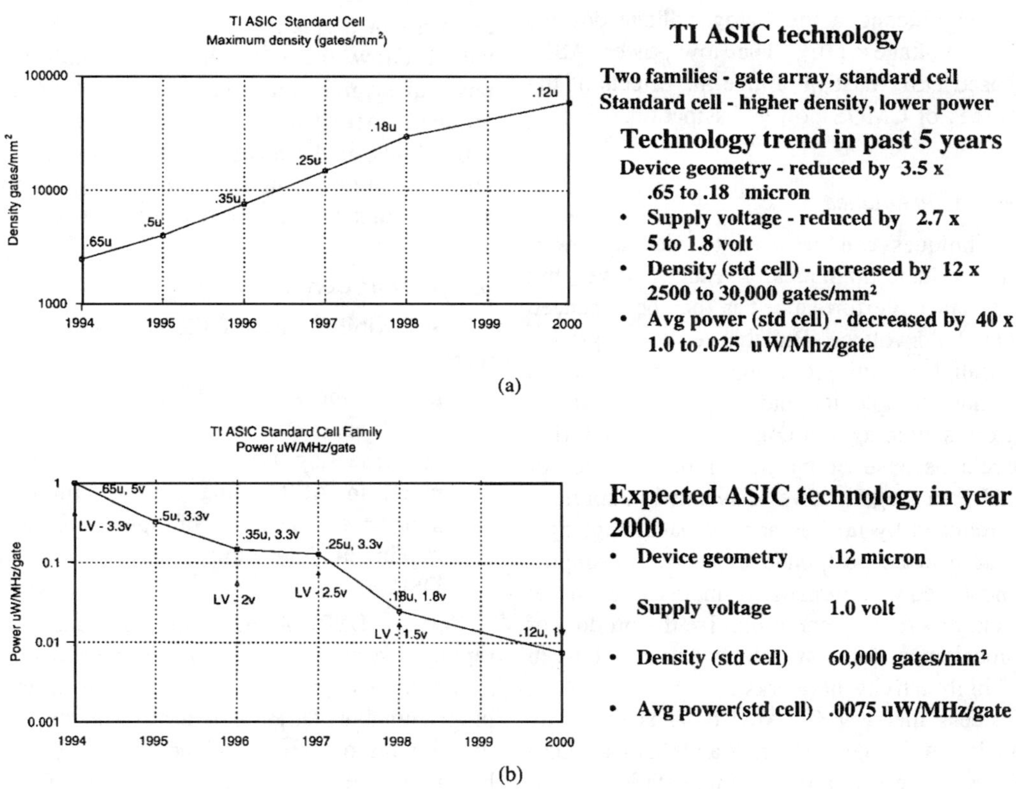

Fig. 8. Texas Instruments ASIC technology trends.

decrease by a factor of 2.5. The change from 3.3 V to lower voltages is already in progress, and will provide an even larger power reduction. A 1.8 V power supply will further reduce power dissipation by a factor of 3.5. And the progression to 1 V technology will reduce power by a factor of about 11 compared with 3.3 V technology.

The standard cell ASIC family shown has higher density and lower power than the gate array family. The most recent member of the Texas Instruments (TI) ASIC family [5], the TSC6000, has a 0.18-μm geometry, and operates at 1.8 V. Its density is 30 000 gates/mm^2, with an average power dissipation of 0.025 μW/MHz/gate. In comparison to the 5 V devices of about five years ago, the gate density has increased by a factor of 12. The average power per gate has decreased by a factor of 40. This reduction is considerably more than the difference resulting from the square of the supply voltage.

Projecting the expected power dissipation of 1.0 V technology to the year 2000 as the ratio of the square of the supply voltage (1.8–1.0 V) yields an average power dissipation per gate of 0.0075 μW/MHz/gate. Density increases to 60 000 gates/mm^2. Low-power designs of several million gates will therefore be feasible.

B. CMOS Power Dissipation

The power consumption of CMOS circuits includes switching power, short-circuit power, static power, and leakage power. Switching power, the dominant component, is essentially proportional to the square of the supply voltage, and is linearly proportional to the load capacitance, the frequency, and the percentage of time the circuit is active. Short-circuit power depends on the current flowing during switching transients. An increase in transition speed reduces this effect since short-circuit current spikes are present only when multiple circuits are in transition at the same time. Static power is dissipated only when bias currents are present in analog circuits. Leakage power is static power that is unintentionally dissipated by the current in the device during the "off" state. This can become a significant factor in low-voltage designs using low threshold voltages [16]. The low-power ASIC designs for core-based DSP must minimize the effects of the entire range of sources of CMOS power dissipation.

C. Low-Power Design Techniques

A number of techniques can be used to reduce power. Lowering the supply voltage dramatically reduces power, but also degrades performance. Performance can then be enhanced by reducing the threshold voltage. But this results in greater leakage current. Parallel circuits yield higher performance at the expense of greater complexity and chip area. Spurious transitions occur from switching a circuit several times during the same clock cycle because of multiple input changes or differences in signal path length. Power dissipated in spurious transitions can be reduced by latches and by balancing logic paths. Switching can also be postponed by reordering inputs to introduce the most frequently changing inputs later in the logic path. Placement and routing can minimize the product of interconnection capacitance times switching activity through the localization of high-activity networks.

Static random access memory (SRAM) designs that only activate a small portion of the memory array and that use latch-style sense amplifiers will essentially eliminate static current. Power-supply switching transistors that turn off internal power to circuits not in use reduce standby power. For example, high-threshold transistors can turn off power to circuits using low-threshold transistors.

Power management techniques, such as "sleep" and "standby" modes, minimize power during times when only a portion of the circuit is needed. Other portions can then be selectively turned on as necessary, and active circuits can be selected in software or hardware. For example, clock gating can turn off the clock to inactive circuits, based on a sleep bit set by software. Alternatively, clock gating could be based on an instruction decode that clocks only those circuits required for the instruction.

VI. Configurable Digital Preprocessors

The processing capacity needed for wide-band, high-bit-rate, and spread-spectrum waveforms exceeds the capabilities of programmable DSP technology projected during the next five years. Nevertheless, the flexibility needed for software radio implementations can be achieved using configurable digital ASIC's with a microprocessor or DSP that is tailored for software radio applications.

A. The "Ideal" Software Radio Concept

In a "ideal" software radio, practically all of the functions shown in Fig. 1 would be implemented in a general-purpose processor. Transmitter functions ranging from the source encoder to the upconversion of the baseband signal to the final carrier frequency would be performed by this processor. Likewise, the converse functions in the receiver would also be accomplished by the processor, including carrier phase recovery and symbol or pseudonoise (PN) code timing recovery in a spread-spectrum application. In principle, this would allow the same hardware platform to support *any* physical layer imaginable as well as the higher layers of the protocol stack. This ideal radio would only be limited by the capabilities of the analog components (e.g., ADC's, DAC's, power amplifiers, low-noise amplifiers, antialiasing filters, and antenna subsystems), and by the capacity of the processor.

B. The DSP Core and ASIC Approach

The flexibility goal of the software radio can be approximated by:

- moving the ADC and DAC as close to the antenna as possible;
- implementing functions with very high processing demand in ASIC's that can be run-time configured to support a wide range of signal structures;
- maximizing the number of functions performed by the DSP.

Current DSP, ASIC, and semiconductor technology is rapidly evolving to support DSP core macro cells via an ASIC library [5]. The DSP core is a microprocessor-like programmable DSP that has been designed for efficient gate count, die area, and performance. The DSP core can be augmented with a customized ASIC on a single chip. For software radio applications, the ASIC includes high-throughput modulation functions such as correlators and high-speed filters. It also includes power management and input/output functions. The combination of DSP core and other ASIC functions can be closely tailored to the application. This offers greater potential throughput, lower gate count, smaller die area, and lower power consumption than other semicustom approaches, such as field-programmable gate arrays (FPGA's). Of course, the nonrecurring engineering of the ASIC will be greater than that of an equivalent design using FPGA's and standalone DSP chips.

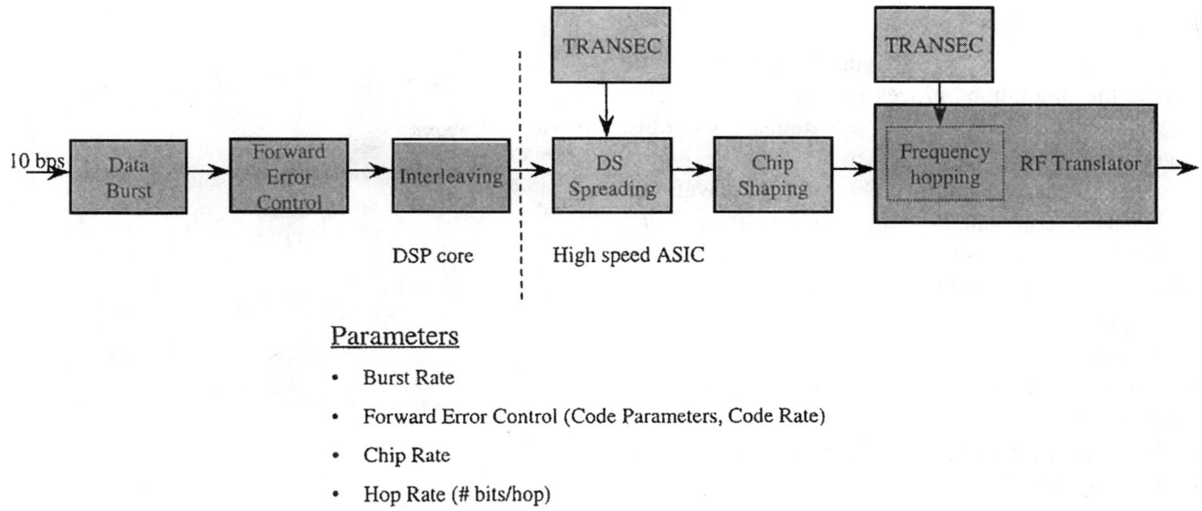

Fig. 9. DSSS transmitter block diagram [17].

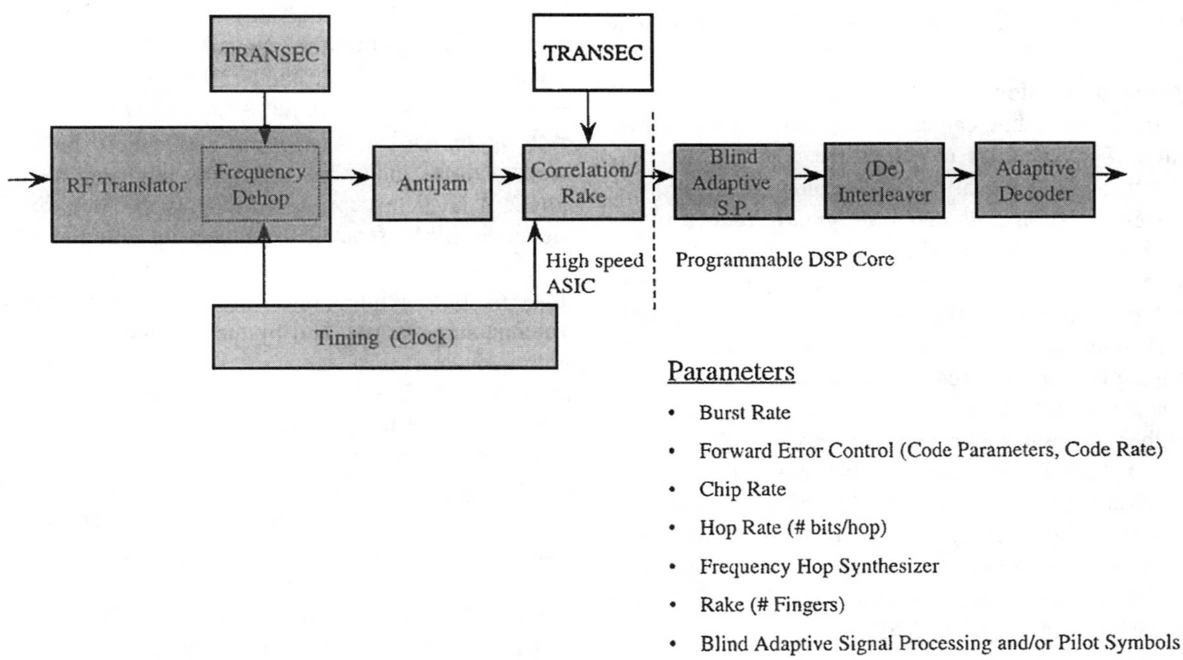

Fig. 10. DSSS receiver algorithm block diagram [4], [6].

To some, it seems inappropriate to use the term "flexible" when describing an ASIC because the logic of an *application-specific* chip cannot be modified once the part has been fabricated. However, the generic functions shown in Fig. 1 are common to a broad spectrum of wireless waveforms. Thus, one may be able to design an ASIC that accommodates many waveforms.

A waveform may be characterized by a parameter set defining modulation type, symbol rate, chip rate, pulse shape, constellation, etc. If implemented in an ASIC with programmable parameters, the ASIC can support a variety of waveforms. Consider the pulse-shaping function of the modulate block shown in Fig. 9. Suppose this function is implemented in an ASIC using a finite impulse response filter. If the filter coefficients are stored in RAM, and if this RAM is accessible to the DSP, the DSP can change the pulse shape by changing the contents of the RAM. A reasonable ASIC design would also allow the number of taps in the filter to be programmable. As long as the number of taps required does not exceed the maximum number available on the ASIC, an arbitrary pulse shape can be supported.

Similar analysis applies to the receiver ASIC of Fig. 10. Timing recovery, transmission security (TRANSEC), and a multifingered correlation receiver that is programmable probably consumes more chip area and power than its waveform-specific counterpart. It is therefore clear that significant effort will be expended in the design of such ASIC's. The implementation of parameterized ASIC's might not be as efficient with respect to gate count and power consumption as that of a point design. Thus, waveform flexibility requires tradeoffs that include waveform types supported, gate count, die area, power consumption, and cost.

C. Waveform Supportability

For handheld radios, size, power, and weight constraints generally limit the amount of processing capacity available for hosting the functions of Fig. 1 in software. This implies that the computationally intensive functions must be implemented in ASIC's. A partitioning of DSSS/CDMA waveforms onto ASIC and DSP components is presented in Figs. 9 and 10. The most crucial parameters in determining waveform supportability by the DSP subsystem include:

- ADC sampling rate;
- dynamic range;
- processing capacity requirement for the translation of digital IF signals to baseband;
- processing capacity requirements of modulation and demodulation algorithms;
- processing capacity requirements of error coding and decoding algorithms (especially, e.g., Viterbi decoders);
- processing capacity of synchronization algorithms—especially the demanding burst mode and satellite applications.

Partitioning a function for ASIC's generally emphasizes increased throughput, but can also emphasize lower power consumption. The precision of arithmetic operations can be customized through the analysis of dynamic range requirements. Lower precision arithmetic can greatly reduce power consumption in the DSSS correlator, for example.

In addition to the physical layer considerations, protocols can limit the flexibility of the proposed software radio. In particular, it may be difficult or impossible to support burst protocols that place arduous requirements on carrier-phase and symbol (or PN code) timing recovery. Such protocols may place a higher processing demand on the system than can be provided in a DSP architecture. Satellite signal tracking has similar requirements. These timing constraints may be met using dedicated ASIC tracking circuits. Thus, a software radio may support the desired range of waveforms and protocols if RF conversion and digital ASIC's with the requisite flexibility are teamed with an appropriate DSP.

D. A DSP Core-Based Software Radio ASIC

To illustrate these concepts, an archetypal software radio is described. The design consists of a single ASIC that includes wireless modem functions, advanced power management, a Texas Instruments TMS320C6xx programmable DSP core [5], [16], and input/output interface logic. This ASIC (Fig. 11) may be called the multipurpose modem chip (MMC).

The spread-spectrum modem logic within the MMC may be programmed by the DSP core for a wide range of functions, including:

- narrow-band jammer suppression;
- digital-to-digital (D/D) conversion;
- demodulation;
- modulation;
- transmit power management;
- military transmission security (TRANSEC) features;
- power management.

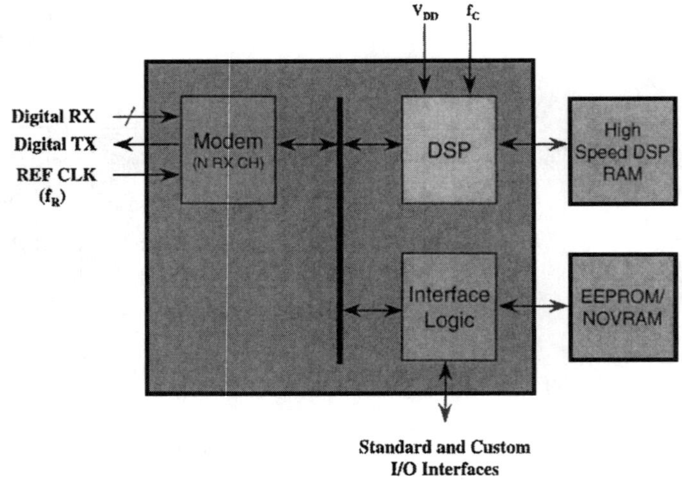

Fig. 11. Multipurpose modem chip [4], [6].

The MMC also generates and demodulates conventional, nonspread-spectrum signals by turning off the PN code generator and mixers. Figs. 9 and 10 show the allocation of the above functions between the ASIC and DSP core. Included in each figure is a list of key parameters that determine whether functions are implemented in the core or elsewhere on the ASIC. With the exception of the military TRANSEC features, the above MMC functions are now described in more detail.

1) Jammer Suppression: Fig. 12 illustrates the main functions of the digital front end of the ASIC including the placement of the jammer suppression block. Mitigation algorithms for this part of the ASIC [15] are chosen to address the characteristics of expected jammers and interference.

The adaptive transversal filter (ATF), for example, excises narrow-band noise and continuous-wave (CW) jammers. The processing for this filter is described in [17] and illustrated in Fig. 13. A typical two-sided ATF design for the MMC calls for 12 bits of precision, 33 taps, and the Widrow–Hoff least mean-square (LMS) algorithm [18], [19] to determine the weight values. The tap weights are maintained in 16 read-only registers within the ASIC, and are accessible to the DSP core. Since the weights are accessible to the DSP core, it computes the ATF transfer function, jammer frequency, and bandwidth in parallel while the ATF is operating.

As shown by the multiplexer (MUX) block after the ATF in Fig. 12, the filter can be bypassed when there is no need for jammer mitigation. Under these circumstances, the low-power ASIC design allows the ATF reference clock to be gated off, thereby conserving power.

2) Digital-to-Digital Conversion: The digital-to-digital converter reduces the number of bits of precision from the jammer mitigation block to the minimum required to demodulate the signal. This substantially reduces the overall gate count and power consumption required for the computationally intensive front-end filtering and correlation functions. Existing DSSS receivers have used lower precision since the dynamic range can be recovered through integration in despreading.

3) The General Demodulator: The number of independent demodulator channels is determined by the application. In terrestrial systems, at least four channels are needed to im-

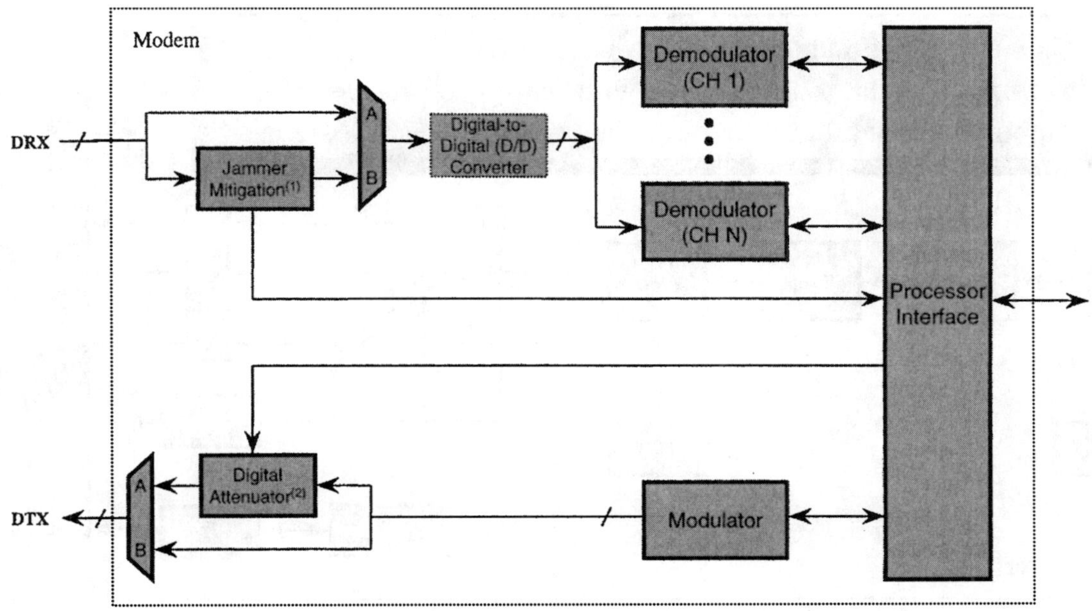

(1) Jammer mitigation includes space-time adaptive antenna processing, an adaptive transversal filter or a combination
(2) Digital attenuator required for precise transmit power management

Fig. 12. Front-end processing [4], [6].

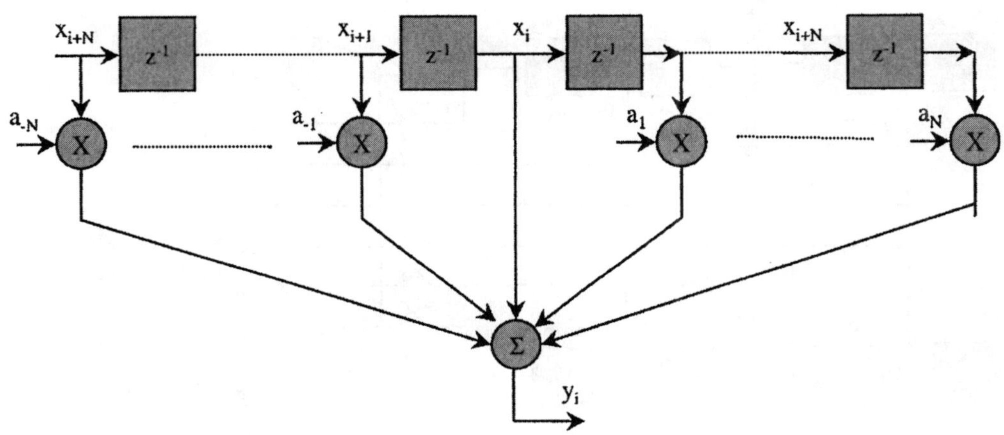

Fig. 13. Adaptive transversal filter.

plement a traditional rake receiver because of nature of the channel. Reception of global positioning satellite (GPS) signals requires as few as four and as many as 12 independent channels. Depending on the number of active channels required, power management provisions in the design allow the DSP to disable the reference clock to unused channels to conserve power.

A more detailed description of the programmable general demodulator is shown in Fig. 14. Conventional digital modulation formats supported by the general demodulator in the MMC design include M-ary phase-shift keying (M-PSK), offset quaternary phase-shift keying (OQPSK), M-ary pulse amplitude modulation (M-PAM), and M-ary quadrature amplitude modulation (M-QAM). Spectral shaping may be used with any of these waveforms. Trellis-coded modulation (TCM) may be generated and demodulated in the DSP core. Further design tradeoffs include the possibility of a hardware Viterbi decoder in the ASIC. Minimum shift keying (MSK) and Gaussian MSK may also be demodulated primarily using the DSP core.

The first operation performed by each demodulator block of the general demodulator is final quadrature downconversion of the passband signal to baseband. Doppler and reference-oscillator drift are compensated by a numerically controlled oscillator (NCO) having a mean phase and frequency precision of 2^{-16} degrees and 2^{-32} Hz, respectively. The phase and frequency of the NCO are set by the DSP via a memory-mapped I/O (MMIO) interface. The complex baseband samples are then decimated so that matched filter processing uses the lowest possible frequency. The decimator includes a complex

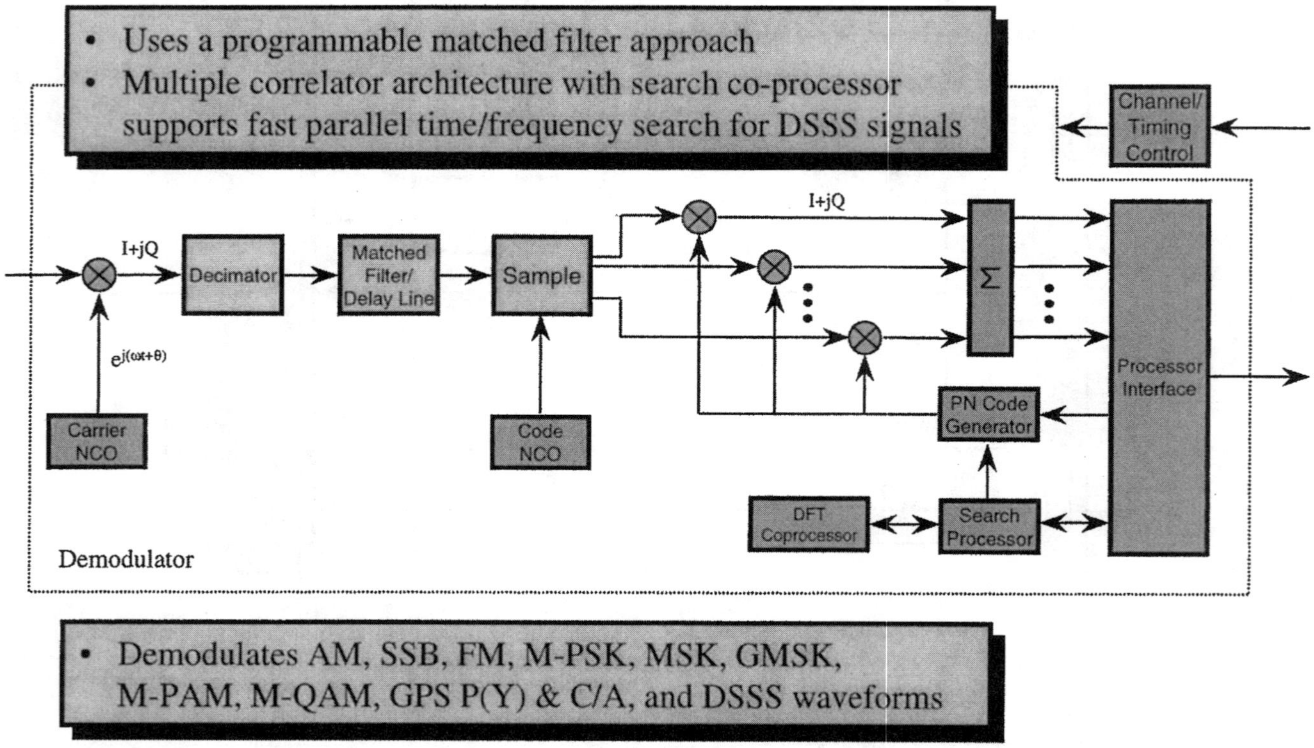

Fig. 14. General demodulator architecture [4], [6].

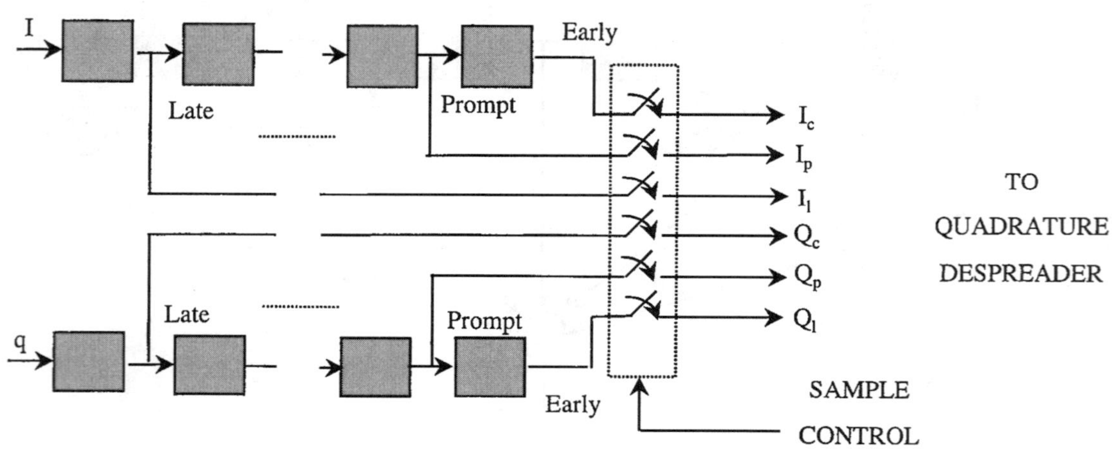

Fig. 15. Delay line and sample functions.

finite impulse response (FIR) filter followed by a decimator. The filter coefficients and decimation rate are controlled by the DSP core.

Following decimation, the complex signal is applied to a matched filter, the coefficients of which are specified by the DSP via the MMIO. The output of the matched filter is sent to the delay line and sample functions, which are shown in greater detail in Fig. 15. The PN chip separation associated with each output in the delay line is a function of the final sample frequency at the output of the decimator. In nonspread applications, the delay line can be used to recover symbol timing. The output of the code NCO determines when the switches in Fig. 15 are closed. The outputs of the delay line are then mixed with the PN code generator output and accumulated during one symbol period or less. At least three code mixers are needed to track the PN code phase using a delay-locked loop (DLL). As many as 32 code mixers could be used in an alternative ASIC design to reduce PN code acquisition time.

The demodulator design allows the DSP to specify precisely when the accumulation period begins, as well as how long the period will be. This flexibility allows the DSP to track the phase of the data edges in satellite communications systems in which propagation path distances are changing rapidly. It can also track Doppler that measurably lengthens and shortens the symbol period, particularly at low data rates. In nonspread-spectrum applications, the PN code is inhibited, allowing the same demodulator structure to accommodate

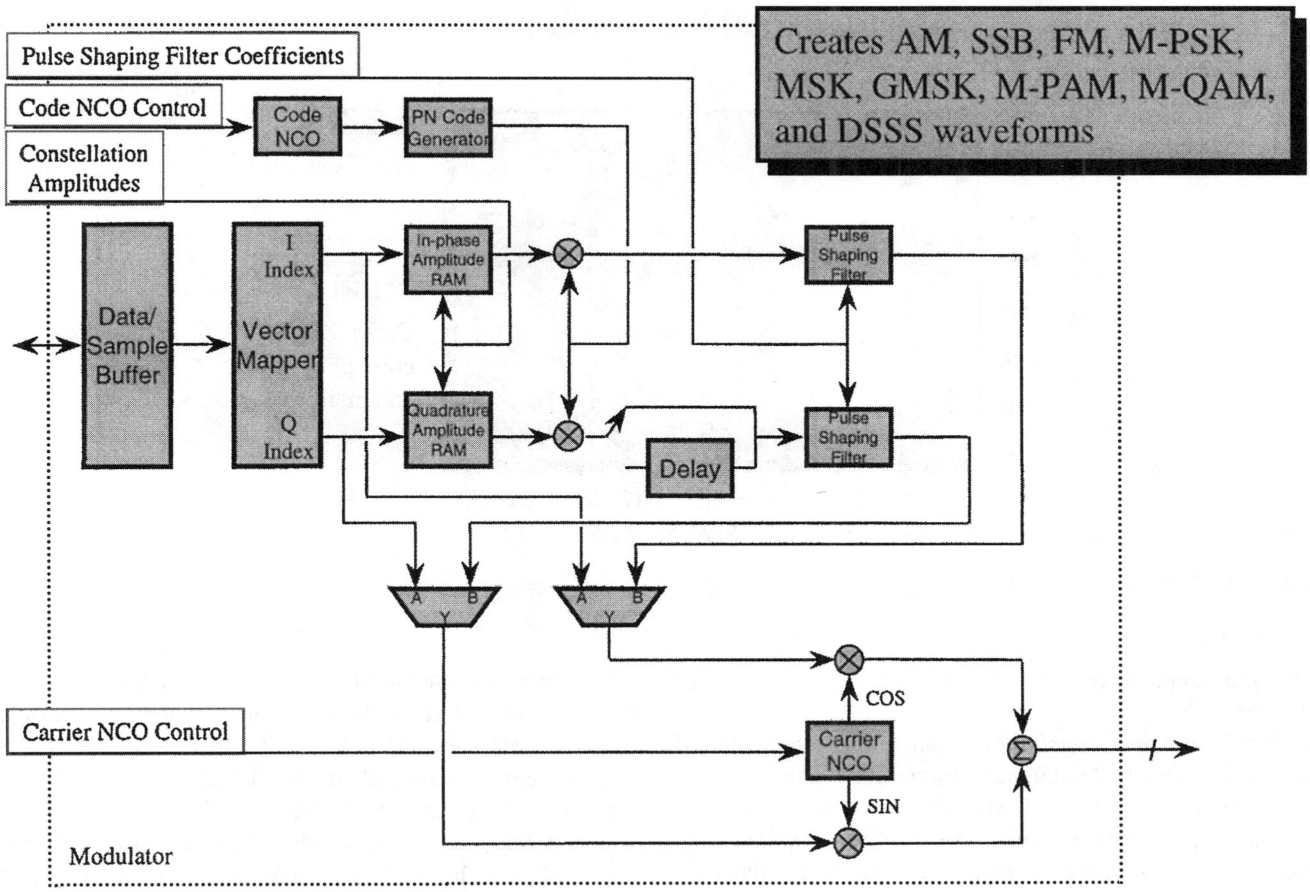

Fig. 16. Programmable general modulator architecture [4], [6].

other waveforms. After accumulation, the I and Q samples are stored in ASIC RAM with an interrupt to the DSP. The DSP reads the I and Q samples via the MMIO interface to complete the demodulation process in software. The code mixer and accumulator (or code correlator) design is such that the DSP can disable the reference clock to unused correlators in order to conserve power. To simplify the logic and conserve gates, code correlators are turned off in groups of four or eight.

Finally, the demodulator contains a search processor and discrete Fourier transform (DFT) coprocessor that accelerate the PN code-timing acquisition process. The search processor forms a time–frequency array of data by computing the DFT of the I and Q samples from each correlator. A sequential probability ratio test is then performed on the time–frequency data to rapidly determine the location of the signal in both time and frequency independently of the DSP core. To initiate a search, the DSP specifies the number of PN chips to be searched and the sample integration period, the inverse of which yields the desired DFT frequency coverage. Then, the search processor signals the DSP with a flag indicating whether the signal was found, and with the PN chip offset and frequency bin of the signal if appropriate. As with other parts of the chip, the search processor may be powered down by the DSP when not in use.

4) The General Modulator: The general modulator function of the MMC is illustrated in Fig. 16. It is designed to generate the following baseband signals: amplitude modulation (AM), single sideband (SSB), frequency modulation (FM), M-PSK, MSK, GMSK, M-PAM, and M-QAM. Others may be programmed using the MMC and DSP core.

The modulation process begins when the DSP writes the coded bits to be transmitted into an ASIC data buffer via the MMIO interface. The vector mapper arranges these bits into an index, which selects the appropriate constellation point from the constellation lookup table. These constellation points have been previously initialized by the DSP core. The I and Q samples associated with the selected constellation point are mixed with the applicable PN code sequence. They are then interpolated to the transmit clock frequency, exciting two pulse-shaping filters. These filter coefficients are also initialized by the DSP. A programmable delay can be inserted into the quadrature signal path as needed to generate offset QPSK, MSK, or GMSK.

The outputs of the pulse-shaping filters are then fed through a multiplexer to a quadrature upconverter consisting of the carrier NCO and two mixers. The quadrature upconverter operates like the quadrature downconverter in the demodulator. If the sample buffer contains complex samples of an analog waveform, the digital waveform logic is bypassed, feeding the vector mapper output directly to the quadrature upconverter. The DSP core programs the frequency and phase of the quadrature upconverter and the PN code. This allows the radio

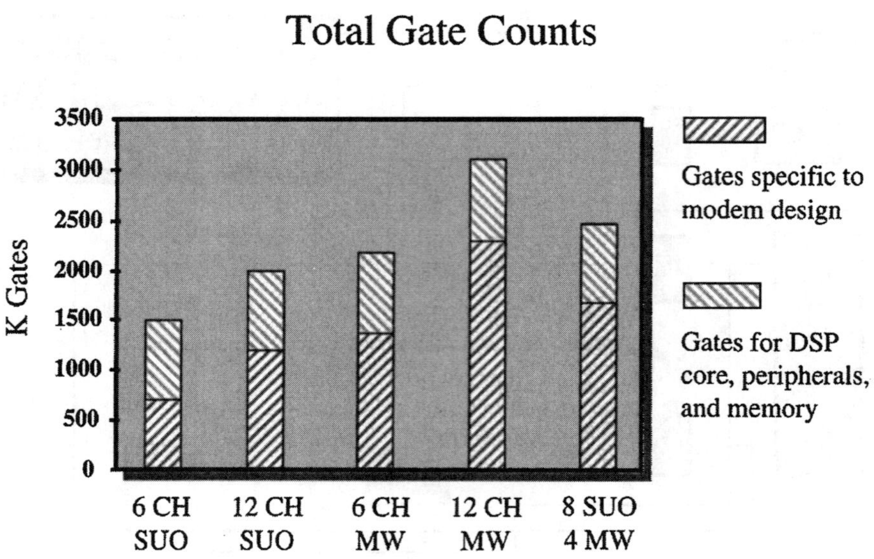

Fig. 17. Total MMC gate count [4].

to perform Doppler precorrection and to correct reference clock drift.

5) Precision Programmable Attenuator: The attenuator of Fig. 12 scales the modulator output amplitude using a fixed-point multiplier, the value of which is set by the DSP core. This assures precise power management when required. The numerical precision of the attenuator depends on the dynamic range over which the output power must be controlled. If the dynamic range is large, the precision of the attenuator output could exceed that of the modulator, distorting the waveform. The attenuator block may be bypassed, sending the output of the modulator directly to the DAC.

VII. Parts Count and Power Consumption

A gate-count estimate was generated for a single MMC ASIC incorporating a high-performance DSP core. The receiver complexity drives gate count, which is nearly linear in the number of parallel receiver channels. In addition, the "SUO" configurations include the complex DSSS waveform and a GPS receiver. Multiple-waveform ("MW") configurations accommodate multiple conventional waveforms. Device configurations comparing gate count therefore varied the number of receive channels and the mix of SUO and MW waveforms, as shown in Fig. 17. The corresponding gate counts show the fixed gates required for the DSP core and the variable gates required for the ASIC portions of the MMC design. The number of gates required for 6 MW channels is about the same as the number required for 12 SUO channels. Twelve MW channels requires more than 3 million gates. The mixed configuration of eight SUO channels and four MW channels reduces the gate count to about 2.5 million while providing much of the capability of the 12 MW configuration.

The gate count of the DSP core is based on the C6xxx series high-performance DSP chip [20]. This core has an instruction cache and internal data memory with minimal local memory and DMA interfaces.

The matched filter and increased numerical precision of the correlators increase the gate count of the MW channels. Demodulators were sized with 32 correlators each. For the SUO waveform, the correlators are based on an existing design consisting of separate accumulators for I and Q. This consists of a two-stage accumulator with 7-bit precision in the first stage, followed by a 23-bit accumulator. To save power, the 23-bit accumulator operates at a lower rate, using carries from the first stage accumulator. The precision of the MW first correlator stage was increased to 12 bits.

Fig. 18 shows the estimated power for the demodulator implemented in 1 V technology using a 40 MHz clock. All channels and all 32 correlators per channel are active in the acquisition mode. In the track mode, four channels are used, and only four of the correlators are used per channel.

The six-channel SUO device operates communications and GPS as separate modes, picking one or the other. In each case, all six channels are used for acquisition and four channels are used for tracking. The dissipated power is therefore the same for either GPS or communication mode. The 12-channel MW device may operate the GPS and communication modes simultaneously. The eight SUO channel device could use eight channels for GPS acquisition and four for tracking. At the same time, the four MW channels could be used for communications. In this case, four channels are used for both acquisition and tracking. When all 12 channels are active for acquisition and eight are active for tracking, the device dissipates the power shown in the figure. For the 12-channel MW device, simultaneous operation is also shown, with six channels each for acquisition and four each for tracking. The 12 channel MW demodulator dissipates over 700 mW in the acquisition mode. Track mode power shown in the figure is reasonably low for all of the configurations since the unused channels and correlators are inactive, consuming essentially zero power. These power estimates are based on scaling the power per gate from 1.8 to 1.0 V. There would be some

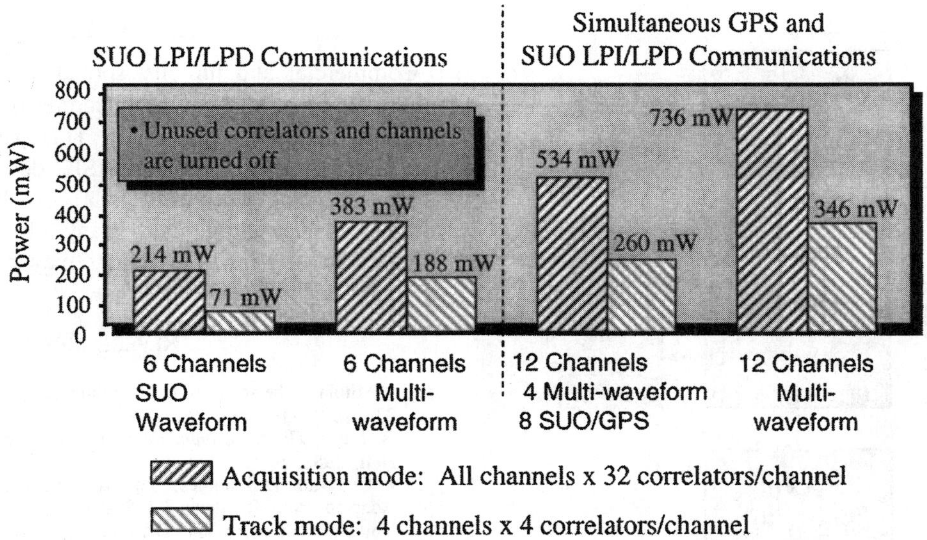

Fig. 18. 1 V demodulator power consumption [4], [6].

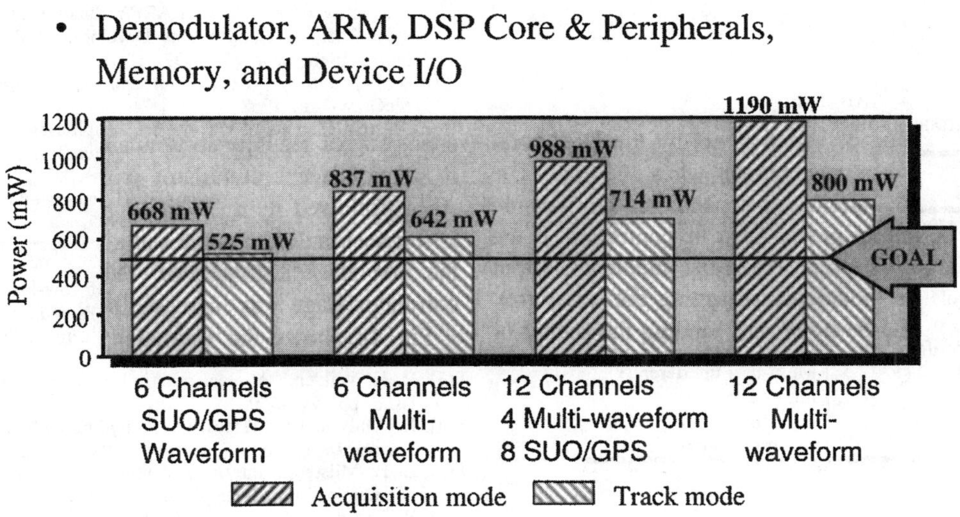

Fig. 19. Receive mode power consumption [4], [6].

additional benefit to reducing feature size from 0.18 to 0.13 or 0.12 μm.

Fig. 19 shows the estimated power for the entire device in the receive mode. The acquisition and track modes are the same as described above. The DSP core operates at 160 MHz and dissipates 440 mW. Its advanced RISC machine (ARM) core operates at 80 MHz to dissipate 12 mW.

VIII. Design Methodology

The design of low-power DSP-based ASIC's for high-performance applications requires appropriately structured tradeoffs. Advanced wide-band waveforms such as those contemplated for DARPA's SUO program present large processing demand. Satisfying this demand at low power requires attention to power dissipation throughout the design of ASIC's like the MMC. The design methods used to develop this architecture are shown in Fig. 20. Significant tradeoffs consider algorithms that are optimum from a performance perspective but that dissipate excessive power versus suboptimum algorithms that dissipate less power. Programmable DSP cores yield less throughput or consume more power than configurable ASIC hardware, but at greater cost. In each design, power consumption must be minimized per gate, function, and subsystem, in the context of overall power management strategies such as clock control, sleep, and off modes.

The architecture presented in this paper is motivated mainly by military requirements that include wide-band spread-spectrum waveforms, as well as legacy military waveforms. But the approaches used to develop this architecture are also relevant to commercial applications. However, the specific device designs might not be cost effective for commercial applications. Table III compares military and commercial communication goals. Military requirements

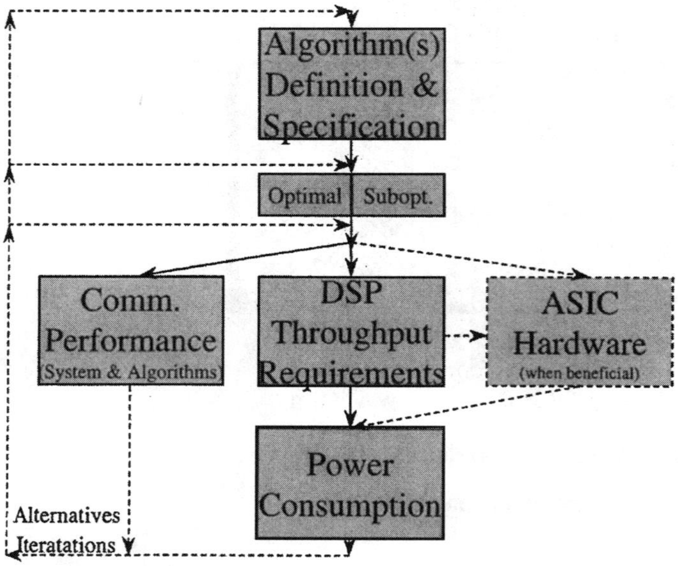

Fig. 20. Low power ASIC development methodology [4], [6].

TABLE III
COMPARISON OF MILITARY AND COMMERCIAL COMMUNICATION GOALS

MILITARY	COMMERCIAL
Peer-to-peer (no base station) operation mode	Basestation/mobile operation/control
Basestation/mobile operation/control	Fixed network infrastructure
Low probability of intercept	Maximum channel capacity
Low probability of detect	Multi-user interference mitigation
Special PN code requirements	Pilot channel for synchronization
Jammer suppression	

for security force hardware commitment to PN codes and channel modulations with features such as low probability of intercept and detection that may be inappropriate to commercial applications. Jammer suppression may also be less appropriate for the commercial sector. In addition, this paper has not addressed emerging multicarrier [22] and wavelet-based [23] communication waveforms. The architecture can support such emerging waveforms provided necessary ASIC circuits are provided. Fast Fourier transform (FFT) circuits may be necessary for multicarrier waveforms. Wavelet waveforms may require filter bank hardware.

IX. CONCLUSION

Commercial and military spread-spectrum waveforms require DSP core and ASIC architectures incorporating advanced power-management techniques. This paper has provided an overview of DSP core-based ASIC designs using the MMC as an example. The result is a complex system-on-a-chip component that implements many of the principles of the software radio in an ASIC that may be implemented with contemporary technology.

REFERENCES

[1] J. Mitola, "The software radio architecture," *IEEE Commun. Mag.*, vol. 33, pp. 26–38, May 1995.
[2] B. Sklar, *Digital Communications.* Englewood Cliffs, NJ: Prentice-Hall, 1988, p. 1.
[3] P. Clark and D. Lammes, "The world phone," *EE Times,* Dec. 8, 1997.
[4] Adapted from Raytheon TI System (RTIS), Purdue University, and University of Michigan; DARPA Small Unit Operation (SUO) Software Radio and Algorithms Development Program: Open Review 2 Presentation, Nov. 18, 1997.
[5] Texas Instruments Web site at http://www.ti.com/sc/docs/news/1996/96037a.htm#support.
[6] Adapted from Raytheon TI System (RTIS), Purdue University, and University of Michigan; DARPA Small Unit Operation (SUO) Software Radio and Algorithms Development Program: Open Review 1 Presentation, Sept. 17, 1997.
[7] A. Papoulis, *The Fourier Integral and Its Applications.* New York: McGraw-Hill, 1962.
[8] A. Kostic and S. Seetharaman, "Digital signal processors in cellular radio communication," *IEEE Commun. Mag.,* vol. 35, pp. 22–35, Dec. 1997.
[9] B. Razavi, "Recent advances in RF integrated circuits," *IEEE Commun. Mag.,* vol. 35, pp. 36–43, Dec. 1997.
[10] G. Stix, "Beam it up: Micromechanical beams may enable a radio on a chip," *Sci. Amer.,* Sept. 1997.
[11] C. M. Goldsmith, "Performance of low-loss RF-MEMS capacitive switches," *IEEE Microwave Guided Wave Lett.*, vol. 8, Aug. 1998.
[12] J. G. Proakis and D. G. Manolakis, *Digital Signal Processing.* New York: Macmillan, 1992, ch. 6.
[13] R. H. Walden, "Analog-to-digital converter survey and analysis," this issue, pp. 539–550.
[14] J. Candy and G. Temes, *Oversampling Delta-Sigma Data Converters: Theory, Design, and Simulation.* New York: IEEE Press, 1992.
[15] L. B. Milstein, "Interference rejection techniques in spread spectrum communication," *Proc. IEEE,* vol. 76, June 1988.
[16] W. Lee *et al.*, "A 1 V DSP for wireless communication," in *ISSCC Dig. Tech. Papers*, 1997, pp. 92–92.
[17] L.-M. Li and L. Milstein, "Rejection of narrow-band interference in PN spread spectrum systems using transversal filters," *IEEE Trans. Commun.*, vol. COM-30, May 1982.
[18] J. G. Proakis, "Adaptive filters," in *Proc. NATO Adv. Study Inst. Commun., Random Process Theory,* 1977, pp. 661–678.
[19] J. M. McCool and B. Widrow, "Principles and applications of adaptive filters: A tutorial review," in *Proc. IEEE Int. Symp. Circuits Syst.*, 1980, pp. 1143–1157.
[20] Texas Instruments Web site at http://www.ti.com/sc/docs/dsps/products/c6000/index.htm.
[21] A. C. Seabaugh, Raytheon TI Systems Nanoelectronics Technol. Dept., private communication.
[22] S. Hara and R. Prasad, "Overview of multicarrier CDMA," *IEEE Commun. Mag.*, pp. 126–133, Dec. 1997.
[23] A. M. Akansu *et al.*, "Wavelet and subband transforms: Fundamentals and communication applications," *IEEE Commun. Mag.*, pp. 104–115, Dec. 1997.

DSP-Based Architectures for Mobile Communications: Past, Present and Future

Alan Gatherer, Trudy Stetzler, Mike McMahan, and Edgar Auslander, Texas Instruments

ABSTRACT

Programmable DSPs are pervasive in the wireless handset market for digital cellular telephony. We present the argument that DSPs will continue to play a dominant, and in fact increasing, role in wireless communications devices by looking at the history of DSP use in digital telephony, examining the DSP-based solution options for today's standards, and looking at future trends in low-power DSPs.

INTRODUCTION

Programmable digital signal processors (DSPs) are pervasive in the wireless handset market for digital cellular telephony. There have historically been two distinct approaches to the implementation of cellular handsets and infrastructure equipment. One approach emphasizes programmable DSPs for flexibility in the face of changing standards, while the other approach focuses on hard-wired application-specific integrated circuit (ASIC) techniques to improve implementation efficiency. The right answer has been and will likely continue to be some combination of both approaches.

In this article we present the argument from a pro-DSP perspective by looking at the history of DSP use in digital telephony, examining the DSP-based solution options for today's standards, and looking at future trends in low-power DSPs. We show that some very compelling arguments in favor of the unsuitability of DSPs for second-generation digital telephony turned out to be spectacularly wrong and that, if history is to teach us anything, it is that DSP use increases as a wireless communications standard matures. We summarize some of the up and coming applications for the new third-generation wireless personal assistants to show that, if anything, flexibility is becoming more of an issue, and therefore the programmability offered by DSPs is even more desirable. Since power is the greatest potential roadblock to increased DSP use, we summarize trends in power consumption and millions of instructions per second (MIPS), and we also describe the use of coprocessors as a complementary technology to today's DSP, allowing flexible solutions to third-generation wireless.

A HISTORICAL PERSPECTIVE ON WIRELESS HANDSET ARCHITECTURES FOR GSM

As we mentioned in the introduction, there is a continuing debate over the role of DSPs in wireless communications. To provide a historical basis for our arguments, in this section we examine the case of Global System for Mobile Communications (GSM) evolution. The assumption is, of course, that third-generation cellular (3G) products will evolve in a similar manner to GSM, which is in itself debatable, but we believe that history does have some good points to make with respect to 3G.

A common functional block diagram of a GSM system is given in Fig. 1. We recognize a classical digital communication model with signal compression, error correction, encryption, modulation, and equalization. In the early days of GSM it was assumed that the low power requirement would mean that most of the phone would be implemented in ASIC. In what follows we show that the power difference between DSP and ASIC was not significant enough compared to other factors that drove GSM phone evolution.

The coder used in GSM phase 1 compressed the speech signal at 13 kb/s using the rectangular pulse excited linear predictive coding with long-term prediction (RPE-LTP) technique as per GSM 06-10 specification. The voice coder is the part of the architecture that most engineers agree should be done on a DSP, and in early designs the DSP was included mainly to do the vocoding. However, once the DSP was included a certain amount of "mission creep" started to occur. As DSPs became more powerful, they started to take on other physical layer 1 tasks until all the functions in the "DSP functions" box in Fig. 1 were included.

Flexibility was also important in the evolving standard. GSM phase 2 saw the introduction of half rate (HR) and enhanced full rate (EFR). HR was supposed to achieve further compression at a rate of 5.6 kb/s for the same subjective quality, but at the expense of an increased complexity, and EFR had to provide better audio qualities and better tandeming performance,

Reprinted from *IEEE Communications Magazine,* Vol. 38, No. 1, pp. 84–90, January 2000.

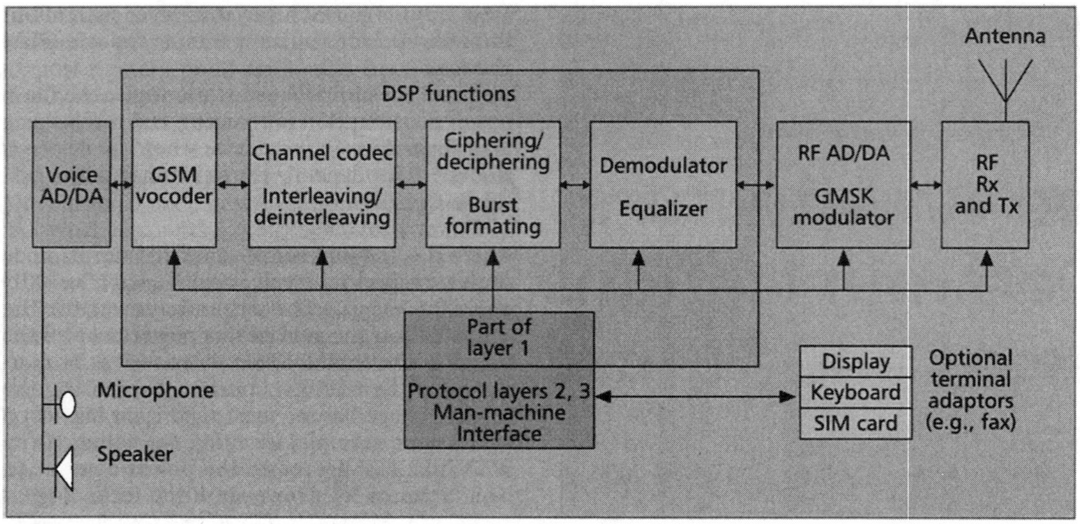

■ **Figure 1.** *A functional block diagram of a GSM phone.*

also at the expense of higher complexity, using an enhanced Vector-Sum Excited Linear Prediction (VSELP) algorithm. Along with these changes came changes in the implementation of the physical layer as better performance, cost, and power savings combinations were found. As a result, each generation of phone had a slightly different physical layer from the previous one, and upgrades to ASIC-based solutions became costly and difficult. Because DSPs were now being designed with low-power wireless applications in mind, the power savings to be had from ASIC implementation of DSP functions was not significant enough that system designers were willing to live with the lack of flexibility. After 1994, a single DSP was powerful enough to do all the DSP functions, making the argument for a DSP-only solution for the baseband even more compelling. To improve system power consumption and board space, several DSP such as the Motorola 56652 [1] and TI's Digital Baseband Platform [2] integrate a RISC microcontroller to handle the protocol and man-machine interface tasks to free the DSP for communication algorithm tasks. The presently most popular partitioning of GSM is as shown in Fig. 2.

It is also true that as GSM phones have evolved, they have gradually moved beyond the simple phone function, and this has led to an increase in the fraction of the DSP MIPS used by something other than physical layer 1. This evolution is shown in Fig. 3. With the advent of wireless data applications and the increased bandwidth of 3G, we expect this trend to accelerate. We discuss this more later.

To summarize, despite the initial belief that a full ASIC implementation would be the norm for GSM, this turned out to be far from the case for the following reasons:

- Making an ASIC vocoder was like replicating available commercial DSP architectures. This provided the initial hook to get the DSP into the system.
- The product life cycle shortened from 2.5 years to 1 year thanks to technology and understanding, but also thanks to the phone becoming a personal fashion statement. DSPs have a significant advantage over ASIC when the product life cycle is short.
- Different worldwide standards related to GSM and the need for product families addressing different market segments called for a platform-based architecture so that original equipment manufacturers (OEMs) could spin different products quickly.
- A DSP-based baseband approach can cope better with different radio frequency (RF) and mixed-signal offerings which occur due to technology improvements and market changes; for example, automatic gain control (AGC) and AFC will change with different front-ends.
- Spare DSP MIPS come for free and enable product differentiation (echo cancellation, speech recognition, noise cancellation, better equalizers).

Trends in Low-Power DSPs

The digital baseband section is critical to the success of wireless handsets and, as mentioned previously, programmable DSPs are essential to

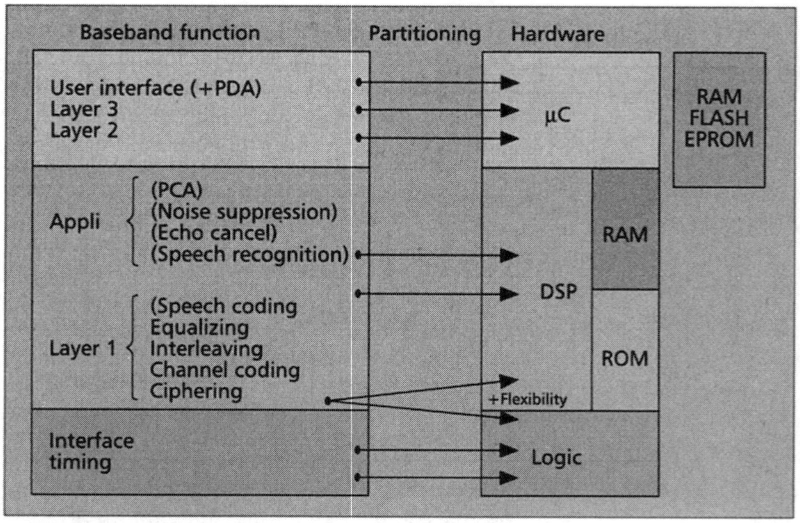

■ **Figure 2.** *GSM function partitioning.*

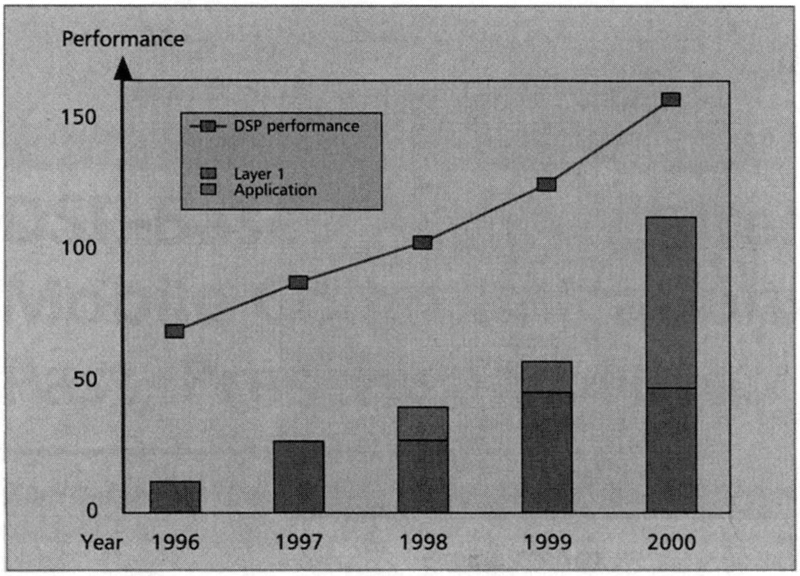

■ **Figure 3.** *Layer 1 and application MIPS with time.*

provide a cost-effective, flexible upgrade path for the variety of evolving standards. Architecture, design, and process enhancements are producing new generations of processors that provide high performance while maintaining the low power dissipation necessary for battery-powered applications. Many communications algorithms are multiply-accumulate (MuAcc) intensive. Therefore, we evaluate DSP power dissipation using milliwatts per million MuAccs (MMuAcc), where a MuAcc consists of fetching two operands from memory, performing a multiply-accumulate, and storing the result back in memory. As shown in Fig. 4, DSP power dissipation is following a trend of halving the power every 18 months [3]. As the industry shifts from second-generation to 3G wireless we see the percentage of the physical layer MIPS that reside in the DSP going from essentially 100 percent in today's technology for GSM to about 10 percent for wideband code-division multiple access (WCDMA). However, the trend shown in Fig. 4, along with more efficient architectures and enhanced instruction sets, implies that the DSP of three years from now will be able to implement a full WCDMA physical layer with about the same power consumption as today's GSM phones.

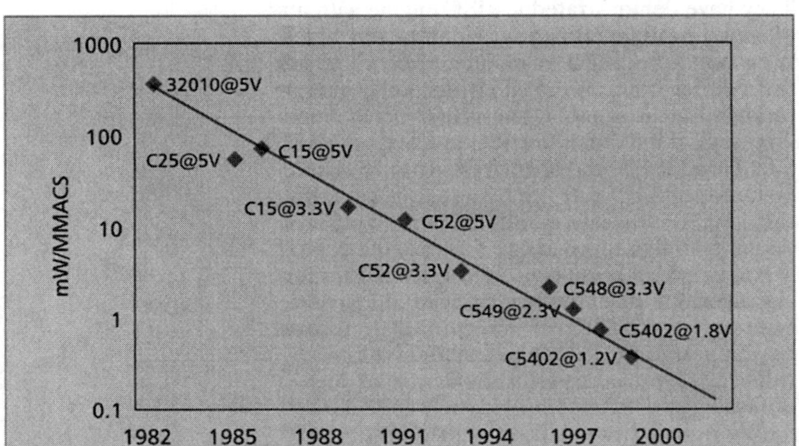

■ **Figure 4.** *Power dissipation trends in DSP.*

Since these DSPs use static logic, the main power consumption is charging and discharging load capacitors on the device when the device is clocked. This dynamic (or switching) power dissipation is given by:

$$\text{Power} = \alpha C^* V_{swing} V_{supply}{}^* f$$

where α is the number of times an internal node cycles each clock cycle, and V_{swing} is usually equal to V_{supply}. The dynamic power for the whole chip is the sum of this power over all the nodes in the circuit. Since this power is proportional to the voltage squared, decreasing the supply voltage has the most significant impact on power. For example, lowering the voltage from 3.3 V to 1.8 V decreases the power dissipation by a factor of 3.4. However, if the technology is constant, lowering the supply voltage also decreases performance. Therefore, technology scaling (which decreases capacitance) and power supply scaling are combined to improve performance while decreasing the total power consumption of the DSP, as shown in Fig. 5 [3]. The current TMS320C54x is capable of 120 MMuAccs at 1.8 V and 0.72 mW/MMuAcc in 0.18 μm CMOS technology.

In what follows we use the TI C54x [4] as an example of an evolving DSP that is optimized for wireless applications. However, the reader should note that because of the growing importance of the wireless market (more than 400 million units projected for 2000 [5]), there are now several DSPs on the market that have been designed with wireless applications in mind, for instance, the Lucent 16000 series [6] and the ADI21xx series. This level of effort by several companies is a sign that the collective wisdom of the marketplace has chosen to bet on a programmable DSP future for wireless technology. We should also note that, although designed for wireless applications, these DSPs are finding major markets in other low-power applications such as telephony modems, digital still cameras, and solid-state audio players.

As mentioned earlier, the power difference between DSP and ASIC solutions was significantly reduced by designing the DSP for low-power wireless applications. Several power saving features are built into the C54x architecture and instruction set to reduce the code size and processor cycles required. The core uses a modified Harvard architecture that incorporates three data memory buses (two read, one write) and one program memory bus coupled with two data address generators and one program address generator. This architecture leads to high memory bandwidth and enables multiple operand operations, resulting in fewer cycles to complete the same function.

Another strategy used by DSP designers is to add instructions that, although fairly generic in themselves, allow efficient implementation of algorithms important to wireless applications. For instance, in the C54x, one of the arithmetic logic unit (ALU) inputs can be taken from a 40-bit barrel shifter, allowing the processor to perform numerical scaling, bit extraction, extended arithmetic, and overflow prevention. The shifter and exponent detector enable single-cycle normalization of values and exponential encoding to support floating-point arithmetic for voice coding. A compare-select-store unit contains an accelerator that, for channel decoding, reduces the Viterbi "butterfly update" to four cycles. This unit generally provides acceleration for any convolutional code based on a single shift register, which accounts for all the codes commonly in use in wireless applications today.

The C54x instruction set also contains several dedicated instructions, including single and block repeat, block memory move, conditional instructions, Euclidean distance calculation, finite impulse response (FIR), and least mean square (LMS) filtering operations.

All these customizations permit power efficient implementations of 7.4 mW for IS-54/136 VSELP and 1.3 mW for the GSM speech coder. The trend toward more specialized instructions will continue increasing as the cost of supporting these instructions goes down. Other instruction enhancements for bit manipulation, which is traditionally done much more efficiently in ASIC, will occur in the near future.

Another trend in DSP evolution is toward VLIW processors to support a compiler-based programmer-friendly environment. Examples of this include TI's TMS320C6x [7], ADI's TigerSHARC [8], and Lucent and Motorola's Star*Core [9]. These VLIW processors use explicitly parallel instruction computing (EPIC) with predication and speculation to aid the compilers. The processors are also statically scheduled, multiple-issue implementations to exploit the instruction-level parallelism inherent in many DSP applications. Although the application of this to physical layer processing in the handset is not apparent so far, these devices allow very efficient compilation of higher-level code, thus reducing the need for DSP-specific assembly-level coding of algorithms. As explained later, the trend of wireless toward an open applications-driven system will make this kind of DSP much more compelling as a multimedia processor in the handset.

Power management is very important in a low-power DSP, and the C54x utilizes a hybrid power management strategy with automatic local clock gating and three user-controlled idle modes to shut down the CPU only, CPU and peripherals, or a complete device shutdown preserving the memory states. A flexible digital phase-locked loop (PLL)-based clock generator and multiplier allows the user to optimize the frequency and power for their application. In general, these techniques allow a DSP not designed for a specific function to optimize its power usage for that function, bringing its power level closer to that of a dedicated ASIC design.

COPROCESSORS

In this section we discuss how coprocessors can complement the function of programmable DSPs in the implementation of a flexible 3G platform. We will use the ARIB/European Telecommunications Standards Institute (ETSI) WCDMA standard [10] to illustrate our arguments. However, the points we make are quite general for

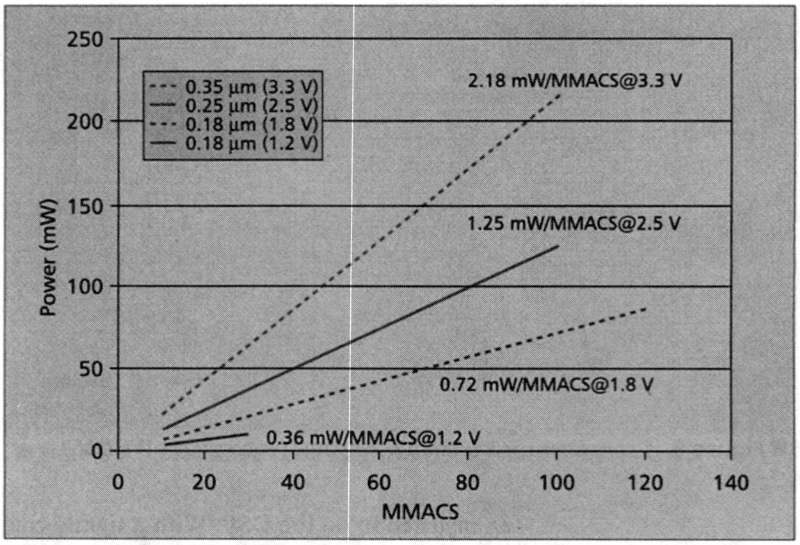

■ **Figure 5.** *C54x MMACS vs. power*

communications physical layer design. Alhough the second section implies that, for a given wireless standard, the amount of the physical layer performed by a DSP will increase to close to 100 percent as the standard matures, we are still faced with the problem of implementing a new standard with today's DSPs. Generally, because standards are driven by what is possible for ASIC implementation at a given power and cost point, a newly defined standard cannot be implemented in a DSP alone. For a WCDMA voice rate terminal, if we make a rough count of the "operations" required, only 10 percent are suitable for implementation on a current DSP, these being essentially the functions operating on data at the symbol rate as opposed to the chip rate. But a fixed function solution would be a high-risk option due to a lack of flexibility, especially in a new standard. Therefore, the system designer is faced with the problem of balancing the power and flexibility requirements. If we assume a long-term trend to increased use of more powerful DSPs, the designer also requires a roadmap for his/her design to migrate toward these devices.

One appealing solution to this problem is a coprocessor-based architecture with a single programmable device at its core. An example of such an approach is the Pleades project [11]. They have demonstrated a RISC engine with an attached configurable array of multipliers which is capable of doing a 16-point complex radix 2 fast Fourier transform (FFT) with a delay energy product for the function that is 0.02 percent and 4 percent of the same function on a Strong ARM and TMS320C54x, respectively. In a complete system the gains may not be as significant, but this nevertheless shows that there is significant potential in this approach. In this case the coprocessor provided significant performance gain for any algorithm that could be mapped to its multiplier array architecture. The gain is therefore only apparent for a domain-specific set of algorithms. But within this set it allows a good degree of flexibility.

We divide the world of coprocessors into loosely coupled and tightly coupled, which are defined relative to the average time to complete

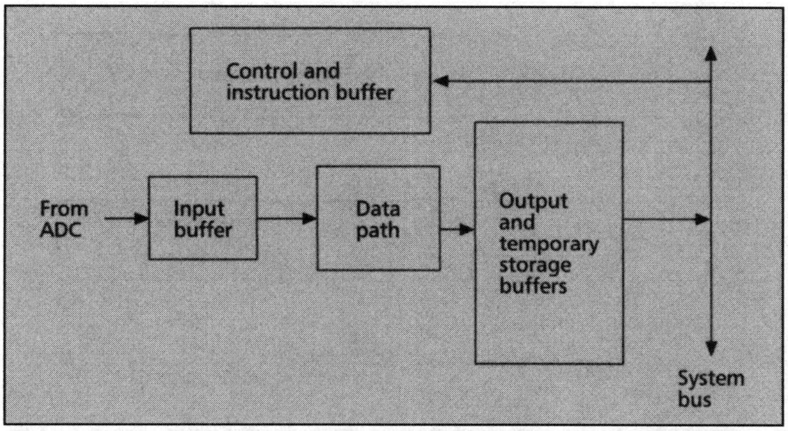

■ **Figure 6.** *A correlator coprocessor for chip rate processing in WCDMA.*

an instruction on the DSP. With a tightly coupled coprocessor (TCC) the DSP will initiate a task on the coprocessor that completes in a few instruction cycles. A task initiated on a loosely coupled coprocessor (LCC) will run for many instruction cycles before it requires more interaction with the DSP.

The TCC will generally have a specific interface to the DSP core and will have access to some registers within that core. Since each task only takes a few cycles, it will naturally only involve a small amount of data. Also, parallel scheduling of tasks on the DSP and TCC will be very difficult, since the DSP will interrupt its task after a few cycles to service the TCC. Therefore, the DSP will generally freeze during the operation of the TCC. The TCC is therefore a user-definable instruction set enhancement that provides power and speed improvements for small tasks where there is no data bottleneck through the DSP. A TCC also may have a very specific task and be relatively small compared to the DSP. With time, the function of the TCC may be absorbed into the DSP by either replacing it with code in a faster, lower-power DSP, or absorbing the function of the TCC into the core of the DSP and giving it a specific instruction. An example of this sort of function would be a Galois arithmetic unit for coding purposes or a bit manipulation coprocessor providing data to symbol mappings that are not presently efficiently implemented in the DSP instruction set.

LCCs are more analogous to a subroutine call than an instruction. As they perform many operations without further DSP intervention, they will generally operate on large data sets. Unlike the TCC, the LCC will have to run in parallel with the DSP if it is to achieve its full benefit. This means that the programmer will have to be more careful with the scheduling of LCC instructions. However, since the LCC has minimal contact with the DSP, this should not be a problem. The main advantage of the LCC is that it solves the serious problem of bus bandwidth that can occur when either the raw input data rate to the system is very high, or else the number of times data is reused in calculations is very high. In either case, the bus bandwidth becomes the bottleneck to performing the computation because the data is stored at the other end of the bus from the computational units. An LCC removes this bottleneck by having the computational units local to the data and arranged specifically for the data access required for a class of computations. In time the DSP will evolve to a point where its bus bandwidth and computational power are sufficient for the LCC's task and the pseudo subroutine implemented by the LCC will become a real subroutine.

The LCC concept applies easily at the chip rate to symbol rate boundary of a CDMA system. In the WCDMA physical layer the DSP would still perform many of the symbol rate processing tasks such as the timing recovery, frequency and channel estimation, and finger allocation. The chip rate processing tasks such as despreading, path delay estimation, and acquisition would be farmed out to a coprocessor that is designed to perform such tasks efficiently. One possibility is a correlator coprocessor, which performs the common despreading tasks for fingers and path delay estimation. The coprocessor can also perform some simple but high-MIPS tasks that occur directly at the chip to symbol boundary. Examples of these are coherent and noncoherent averaging for channel estimation. However, the DSP still chooses the type of averaging that should occur and how to post process the data to produce the final channel estimate. In effect, the system is fully programmable within the domain of CDMA chip rate processing. The DSP also has control of how the despreading MIPS are allocated. For instance, the DSP may choose to allocate a portion of the MIPS to one user with 6-multipath. Alternatively, it may reallocate these same MIPS to several users with less multipath to despread. Apart from allowing different users to differentiate and improve their WCDMA solutions completely in software, this allows the same system to be reprogrammed to perform WCDMA, GSM, CDMA2000, IS95, Global Postitioning System (GPS), and other CDMA-based demodulation systems. It also provides a common platform for both low-cost voice-only terminals and high-end multimedia terminals. A simplified block diagram of this coprocessor is shown in Fig. 6. Note that the coprocessor is connected directly to the frontend to remove the chip rate data completely from the bus. Another important feature is that the instruction and output buffers are memory mapped to allow flexible access to the coprocessor by the DSP. Comparisons to fixed function designs show that, with careful design of the coprocessor logic, there is no significant power penalty to be paid for the flexibility. This is essentially because the data flow dominates the power budget, and this is independent of the flexibility of the design. The size of the coprocessor is somewhat larger than a dedicated design, but not significantly so within the complete system budget.

Decoding is another area which can benefit from the application of LCCs. Voice rate Viterbi decoding is easily performed on today's DSPs, but the higher data rate requirements in 3G make decoding hard to do programmably. Nevertheless, it is possible to find a DSP/coprocessor partition that maintains the flexibility required along with a reasonable MIPS level on the DSP. As an example, for Viterbi decoding in the base station, the DSP could perform all the data processing up to the branch metric generation, and a coprocessor could perform the remaining high-MIPS tasks of state metric update and traceback. This allows the DSP to programmably define a decoder for any code based on a single shift register, including puncturing to other rates. We estimate that to support 240 3G voice channels (with constraint length $K = 9$) would require a 45K gate coprocessor and only 6 MHz of DSP processing on a C6x.

Applications and Architectures for Future Wireless Devices

There have been several trends which have dominated cellular communications since its commercial debut in 1983. First is the trend from analog to digital. This has been a boon for cellular carriers because it has allowed them to triple their subscriber capacity with little to no increase in cost. However, the biggest impact of digital technology is yet to come. The transition to digital technology is also driving a transition from voice-only cell phones to an environment which can deliver a variety of data services to a broad range of mobile devices. Bits, after all, are just bits, and they can be used to transmit data as well as voice.

The key to this transition is the fact that the evolution of cellular technology from first to second and soon to third generation technology is mirroring the evolution of wireline systems by delivering more and more bandwidth to mobile subscribers. This bandwidth lags wireline technology by several years, but the year 2000 should represent a watershed for wireless technology as available bandwidths exceed 100 kb/s and packet-switched networking technology is introduced. Although voice communications will probably always be the dominant mobile service, the next century will begin to see a proliferation of targeted data services for mobile subscribers.

Examples of such services could include:

- Imaging services: Such capabilities would allow a businessperson to transmit the image of a document he/she is discussing. It would allow a real estate agent to show a client the image of a new house or a grandmother to receive the latest snapshots of a new baby via e-mail.
- Location-based services would allow travelers to locate and navigate to new destinations. It would allow advertising to be delivered to mobile subscribers in a specific area. It would even allow parents to keep track of their children.
- Audio and visual entertainment could be delivered wirelessly to mobile subscribers. You could even imagine mobile subscriber-to-subscriber gaming environments.

The best of these applications are yet to be conceived. Those applications will come when low-cost high-bandwidth packet networks are generally available. A common theme in all of these services will be an insatiable thirst for signal processing power. Figure 7 illustrates that even as communication processing needs become more demanding, applications requirements will also drive a need for more powerful DSPs.

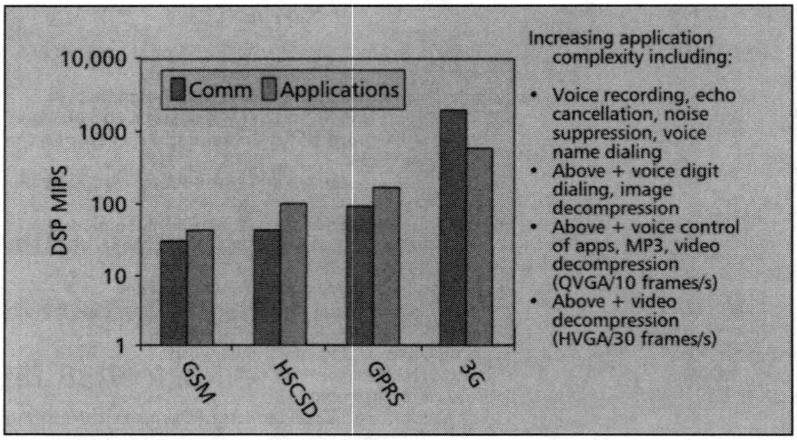

Figure 7. *Applications drive DSP MIPS.*

Traditional cell phones use a dual processor architecture to implement what is in effect a simple modem. Let's call that the communications engine. Data-centric terminals will have new challenges and potentially new architecture. Top-of-the line phones developed in 1999 are starting to integrate personal organizer and information exchange features [12, 13], and this will lead to wireless phones with open operating systems onto which a user can download third-party applications. Mobile devices which deliver data services will have to add processing resources to support an increasingly complex user interface, more sophisticated operating environments, and support for added applications beyond voice. Let's call those resources the applications engine.

Starting with a phone that implements the communications engine using a dual-core RISC + DSP architecture, there are several options for adding resources to support data services. Most current approaches incorporate a single RISC core in addition to the communications engine. Other options include multiple DSPs or even hardware coprocessors to accelerate the heavy additional media processing required to implement the applications engine.

In summary, support for future data-centric mobile devices will require a new approach to the architectures that implement these devices. Care must be taken to consider application and user interface performance, system-level power consumption, hardware costs, software complexity, and time to market. The winning system will likely incorporate multiple processors tuned to fit the tasks for which they were designed. You will see RISCs, which implement protocol stacks, user interfaces, and high-end operating systems. You will also see programmable DSPs and coprocessors, which can provide power-efficient media processing, and support needed application flexibility and upgradeability. The biggest open issue is the extent to which the applications and communication functions can be effectively combined in one programming environment.

Support for future data-centric mobile devices will require a new approach to the architectures that implement these devices.

References

[1] http://www.mot.com/SPS/WIRELESS/products/DSP56652.html

[2] http://www.ti.com/sc/docs/wireless/97/digbase.htm

[3] P. Gelabert and T. Stetzler, "Industry's Lowest Power General Purpose DSP," Embedded Processor Forum, May 3–4, 1999.

[4] http://www.ti.com/sc/docs/products/dsp/c5000/index.htm

[5] Dataquest "Mobile Communications Semiconductor Applications Markets, 1997–2002," Apr. 12, 1999, WSAM-WW-MT-9901.

[6] http://www.lucent.com/micro/wireless.html

[7] http://www.ti.com/sc/docs/products/dsp/c6000/index.htm

[8] http://www.analog.com/industry/dsp

[9] http://www.starcore-dsp.com

[10] http://ftp.3gpp.org

[11] http://infopad.EECS.Berkeley.EDU/research/reconfigurable

[12] http://www.qualcomm.com/pdQ

[13] http://www.symbian.com

Virtual Radios

Vanu Bose, Michael Ismert, Matt Welborn, and John Guttag

***Abstract*— Conventional software radios take advantage of vastly improved analog to digital converters (ADC's) and digital signal processing (DSP) hardware. Our approach, which we refer to as virtual radios, also depends upon high performance ADC's. However, rather than use DSP's, we have chosen to ride the curve of rapidly improving workstation hardware. We use wideband digitization and then perform all of the digital signal processing in user space on a general purpose workstation. This approach allows us to experiment with new approaches to signal processing that exploit the hardware and software resources of the workstation. Furthermore, it allows us to experiment with different ways of structuring systems in which the radio component of communication devices is integrated with higher-level applications.**

This paper describes the design and performance of an environment we have constructed that facilitates building virtual radios and of two applications built using that environment. The environment consists of an input/output (I/O) subsystem that provides high bandwidth low latency user-level access to digitized signals and a programming environment that provides an infrastructure for building applications. The applications, which exemplify some of the benefits of virtual radios, are a software cellular receiver and a novel wireless network interface.

***Index Terms*— Digital down conversion, software radio, software signal processing.**

I. Introduction

A virtual radio is a communications device that: does all its digital signal processing in user space on an off-the-shelf workstation (a PC in our case).

These devices are very different from programmable digital radios using application specific digital hardware or digital signal processors (DSP's) under software control [4], [11]. They represent, to our knowledge, the first implementation of a software radio which digitizes a wideband (e.g., 10 MHz) of the RF spectrum and performs all of the signal processing in application level software using a general purpose processor [12].

The SpectrumWare project is devoted to building infrastructure to support the construction of virtual radios and to building virtual radios that take advantage of the resources available on the workstation to either provide distinctive functionality or to implement traditional functionality in a distinctive way.

Of course, using a general purpose workstation to do signal processing can impose a significant cost. Not only are most DSP's considerably cheaper than workstations but they are also considerably smaller and use less power. However, exploiting the resources available on the workstation offers many potential advantages.

- *Experimentation:* There is a large and growing gap between the capabilities and programming environments available on general purpose workstations and those available on specialized DSP hardware. Implementing the signal processing on the workstation makes it much easier to experiment with new algorithms and protocols.
- *Rapid Deployment:* Users can easily deploy new devices or enhancements to existing devices in the same manner in which they currently upgrade device drivers and software installed in their workstations, for example, as a self-extracting archive downloaded from an ftp site.
- *Integration with Other Applications:* It is often the case that functionality provided by a radio device is only a small part of a larger application. The functionality provided by a wireless modem, for example, is only a part of what is needed for a network interface. Virtual radios allows one to blur the line between the radio and the rest of the application, thus allowing improved functionality and end-to-end efficiency.
- *Multipurpose Devices:* Increasingly, people are dealing with multiple wireless communication devices. They have both a cellular modem and a wireless Ethernet for their notebook computer (not to mention more mundane devices such as garage door openers). While multipurpose DSP devices, e.g., FAX modems, are becoming increasingly available, they perform a relatively small number of predefined functions. In our approach, one adds devices merely by downloading software.
- *Reduced Cost (for Special Purpose Devices):* While general purpose computers are never likely to be less expensive than specialized hardware for mass market items (e.g., cell phones), they are often less expensive than the hardware devices for small markets. Furthermore, the virtual radio system architecture encourages sharing of hardware resources across devices. These devices need not even be what one conventionally thinks of as radios. Technology of the virtual radios has been used to implement the signal processing for medical ultrasound machines with the idea of allowing a number of ultrasound front ends to share the same backend workstation [18].
- *Improved Functionality:* Performing all of the signal processing in modular software permits not only the

Manuscript received September 21, 1997; revised January 11, 1998 and July 2, 1998. This work was supported by the Advanced Research Projects Agency under Contract DABT-6395-C-0060 (monitored by U.S. Army, Fort Huachuca) and by equipment grants from Intel Corporation.

The authors are with the Software Devices and Systems Group, Laboratory for Computer Science, Massachusetts Institute of Technology, Cambridge, MA 02139 USA.

Publisher Item Identifier S 0733-8716(99)02972-8.

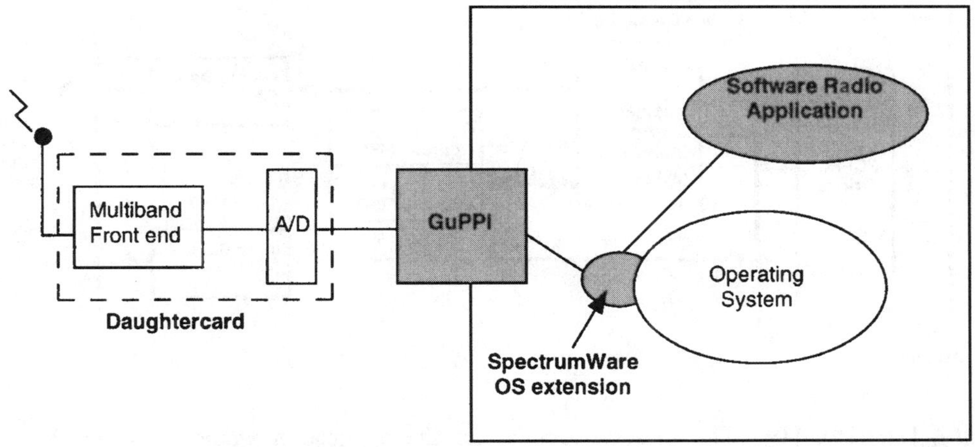

Fig. 1. Block diagram of the virtual radio. The hardware is used only to convert the desired RF band down to the IF frequency, digitize it, and to stream the samples into host memory. All subsequent processing of this digital wide-band IF signal is performed in user-level software.

dynamic assignment of channel locations and widths, but also of the modulation and coding used on each channel. This allows for the construction of heterogeneous systems which employ different communications standards on different channels. This mechanism can be used to accelerate the migration to new standards.

- *Improved Computational Efficiency:* This is a bit surprising. Why should it be possible to get better performance running on a general purpose processor than on a specialized DSP? A partial answer is that market factors are driving rapid improvements in workstation performance. However, there are three other key factors involved
  - The temporal decoupling afforded by the workstation's memory and the flexibility of the workstation's processor allows one to implement novel algorithms, which can be more efficient than the constrained DSP algorithms.
  - The ability to integrate the signal processing with higher-level applications affords the opportunity for system-wide optimization.
  - The signal processing functions can be dynamically modified based on measurements of changing channel or system characteristics to improve performance. This information could come from background tasks that perform functions such as channel estimation when spare cycles are available.

Section II of this paper describes an environment for building virtual radios that run on off-the-shelf personal computers. This environment facilitates the construction of devices that enjoy many of the potential benefits listed above. It has two key parts: an input/output (I/O) subsystem that provides high bandwidth, low latency, user-level access to digitized signals and a programming environment that provides an infrastructure for building applications. The I/O subsystem is discussed in some detail, and the key properties of the software environment are outlined.

Section III describes the design and reports on the performance of two applications built using this environment. The applications are relatively simple, yet they illustrate many of the benefits of virtual radios. Section III-A describes a software cellular receiver and Section III-B a software wireless network interface.

The paper concludes by relating the specific results reported in this paper to the more general assumptions behind the on virtual radios.

II. Enabling Technology

The virtual radio research goal is to move the analog/digital boundary as close to the antenna as possible and to move the software/hardware boundary right up to the ADC. Since current ADC technology and available processors will not support the direct sampling of wide RF bands, the approach, as illustrated in Fig. 1, is to use a multiband hardware front end to convert the desired RF band to the IF frequency, directly sample the wideband IF waveform and then transfer these samples into host memory. All subsequent processing is performed in user level software.

We have used several front ends for the system, but the most flexible is the AR5000,[1] which provides the ability to tune in any 10 MHz wide band located between 100 kHz and 2.6 GHz. The AR5000 offers no transmit capability, however, and for this we generally use a transmitter dedicated to a particular band. Several vendors are working on wideband receive/transmit units that provide better dynamic range and increased bandwidth.[2] These units should be available in the coming year.

Given the high rate digital samples, one is faced with the somewhat daunting task of transporting the samples into host memory. Traditionally, I/O devices have a device driver resident in the kernel which receives and processes the data, and then copies it into user space. However, the performance problems associated with using the default Unix I/O system to move data across the kernel/user boundary are well known, and the extra time required is unacceptable for real-time signal processing applications. To avoid the expense of performing the data copy, previous research efforts have relied on different schemes using virtual memory manipulation and/or shared

[1] AOR Ltd., Tokyo, Japan.

[2] Such receivers are currently under development by Rockwell/Collins and Hughes Electronics.

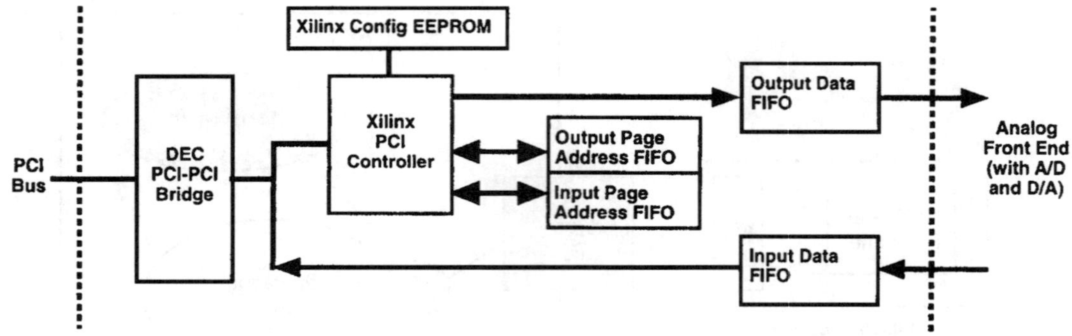

Fig. 2. GuPPI block diagram.

memory [1], [3], [6], [7], [15], [19]. The solution to this problem was to use background direct memory access (DMA) to stream the samples directly into special buffers in the kernel; these buffers are mapped into the user's address space using virtual memory manipulations with very little overhead. The general purpose peripheral component interconnect (PCI) I/O system, described below [9], supports this functionality. This interface utilizes the gigabit capacity of the PCI bus, and supports continuous bidirectional I/O streams of up to 32 megasamples/second (ms/s) with 16 bits/sample.

The virtual radio includes a programming environment for constructing applications. This environment consists of a library of portable (across platforms) signal processing routines designed to support a data-pull style of programming and various kinds of dynamic adaptation.

The next section describes the design of the I/O system and reports on its performance. The following section provides a quick overview of the architecture of the programming environment.

A. The I/O System

In standard signal processing systems based on dedicated digital hardware or DSP's, the incoming samples arrive at a constant rate and are processed with a fixed delay between when a sample enters the system and when the output based on that sample leaves the system. The processing happens in lockstep with the I/O, so the DSP is guaranteed that it will have a constant stream of regularly spaced sample.

In a general-purpose workstation, however, such simple guarantees do not exist. Virtual memory, multiple levels of caching, and competition for the I/O and memory buses add jitter to the expected amount of time required for a sample to travel from an I/O device to the processor. In addition, using a multitasking operating system ensures that the signal processing application will not always be the active process, which adds jitter to the rate at which samples are processed. Smoothing out these sources of jitter is one requirement addressed by the I/O system.

The other major requirement which must be addressed is the need for high throughput between the application and the ADC. Consider, for example, a software cellular receiver. The A-side cellular telephony band (reverse link) is 12.5 MHz wide. If digitized at 25.6 MHz with a sample size of 16 bits, the data rate necessary to transfer this stream of samples to the application would be 409.6 Mb/s. This throughput need exposes two bottlenecks in the existing workstation architecture. First, workstations lack a high-throughput port capable of supporting the required samples stream, creating the need to develop custom hardware. Second, the path between a device driver and the application is inefficient, requiring modifications to the operating system. For comparison, the VuSystem [5] reported sustained throughput of only 100 Mb/s to the application with an unmodified Digital Unix operating system.

There are two main components in the architecture of the I/O system: the GuPPI (for General Purpose PCI I/O), which physically connects the analog front end to the workstation's I/O bus; and the operating system additions, which provide the means for the application to access the sample streams.

1) I/O System Architecture: The GuPPI provides the system's external interface to the analog front end. It bursts data between the front end and main memory at near the maximum I/O bus rate. In addition, the GuPPI decouples the timing between the fixed rate domain of the analog front end and the variable rate I/O bus without losing any samples. This absorbs any jitter caused by the bursty access to the I/O bus. These functions are performed without significant intervention from the processor; the required processing overhead per sample is less than half a cycle.

A block diagram of the GuPPI is shown in Fig. 2. The GuPPI has a simple, generic daughter card interface to which analog front end-specific daughter cards are designed. This interface is directly connected to a set of first-in–first-out (FIFO) buffers in both the input and output directions. These FIFO's provide buffering to absorb jitter caused by the bursty access to the PCI bus.

The GuPPI implements a new variant of scatter/gather DMA which we have named *page-streaming*. The GuPPI has two page address FIFO's, one each for input and output, which hold the physical page addresses associated with buffers in virtual memory. At the end of a page transfer, the GuPPI reads the next page address from the head of the appropriate page address FIFO and begins transferring data to/from it. The GuPPI triggers an interrupt when the supply of page addresses runs low, and the page addresses are replenished by the interrupt handler in the device driver. Page-streaming is unique in two ways. First, the physical page addresses are stored in the GuPPI rather than in a table in memory. Since the processor replenishes the addresses, the GuPPI uses its

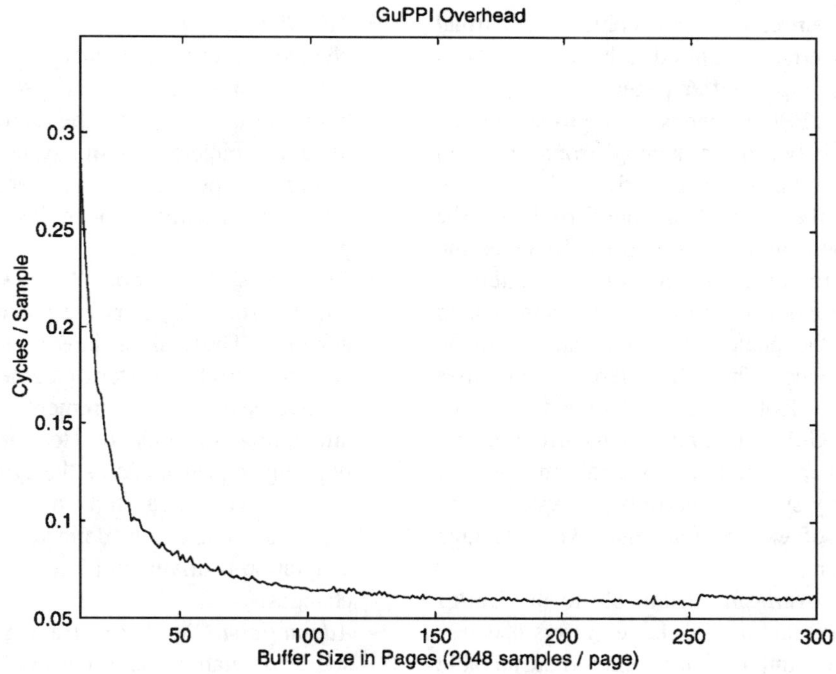

Fig. 3. GuPPI processing overhead (input).

bus grants only to transfer data; this simplified the design of the GuPPI and resulted in more efficient use of the PCI bus. Second, with page-streaming only complete pages are transferred; this is made possible by the constant flow of samples and automatically results in page-aligned, integral page length buffers which are easy for the operating system to manipulate.

The operating system components are responsible for ensuring that the flow of data between the GuPPI and kernel buffers is continuous. This includes facilities for absorbing the jitter due to scheduling and interrupts. The operating system support consists of a device driver for the GuPPI and several small additions to the virtual memory system, all for the Linux kernel, version used is 2.0.30. The total size of the code is just under 1000 lines, with the virtual memory system additions representing just 200 of those. The additions do not affect the performance or functionality of any part of the system not related to the GuPPI; all other applications run completely undisturbed.

The virtual memory additions provide the low-overhead, high-bandwidth transfer of data between the application and the device driver. These components also provide the external interface to the application. To the application, the GuPPI appears to be a standard Unix device with copy semantics. However, virtual memory manipulations are used to make the `read` and `write` system calls to the GuPPI copy-free. It is the responsibility of the *application* to provide real-time guarantees. If the application wishes to run in real time, then it must process the data at or above the average rate at which the system will transfer data. If the application does not run in real time, then the system will eventually begin to drop input samples or produce gaps in the output stream.

2) Performance: The measurements of the GuPPI and I/O modifications were taken on a 200 MHz Pentium Pro system running Linux with a 33 MHz, 32 bit wide PCI bus. All cycle values were gathered using the Pentium Pro cycle counter.

TABLE I
RAW GuPPI THROUGHPUT (Mbits/s)

Input	Output
933	790

The maximum rate at which an application using the GuPPI driver can maintain a continuous flow of input samples from the GuPPI is 512 Mb/s. This number was determined using an application that only accessed enough samples per input buffer to verify data continuity. This an upper bound on the possible throughput that an application can achieve using the GuPPI. At rates above this point there is insufficient depth in the input data FIFO on the GuPPI to absorb jitter due to the PCI bus, however, the next revision of the GuPPI will have deeper data FIFO's, increasing the maximum continuous throughput.

To provide this high continuous throughput, the GuPPI must have higher raw throughput. Table I shows the maximum input and output burst performance of the GuPPI. These numbers reflect the amount of time required for the GuPPI to complete the DMA of approximately 1.2 Mb of data. The measurements include only the time required to DMA the data, not the time required to write page addresses into the appropriate page address FIFO. The maximum PCI throughput available is 1056 Mb/s, so the GuPPI is coming reasonably close to saturating the workstation's PCI bus. The lower maximum throughput for output is due to the latency incurred when reading values from main memory.

Fig. 3 shows the average processing overhead imposed on the workstation by using the GuPPI to generate input. This measurement reflects the number of cycles required per input sample and takes into account both the overhead required to perform the `read` system call (which includes the virtual memory swap) and the overhead required to handle interrupts to replenish the supply of input buffer pages.

3) Discussion: The GuPPI provides a high-bandwidth, low-latency connection between an analog front end and the I/O bus of a PC. The software drives the GuPPI, smoothes jitter, and makes the data transferred by the hardware accessible to applications in user space. Together, the hardware and software provide an application-level interface that simplifies the construction of virtual radios and related applications by making the analog front end appear to be a conventional Unix device. This characteristic simplifies experimentation and rapid deployment by hiding the details of the operation of the GuPPI behind a standard interface while still providing high performance. In addition, the PCI bus, a widely used industry standard, permits the system to be portable to a wide range of workstations and take advantage of newer, faster processors.

The average processor overhead attributable to the GuPPI is less than half a cycle per sample. For large buffers it is less than a tenth of a cycle per sample. The maximum sustainable receive data rate visible to applications is 512 Mb/s. This is more than enough for most of the applications for which the system was designed. The two applications, discussed subsequently, are not I/O-limited but processor-limited instead.

B. Programming Environment

SPECtRA is a signal processing environment for continuous real-time applications. The system embodies a set of abstractions and design rules that allow for simple implementation of signal processing functions and enable significant software reuse. The system has three basic components:

- a library of signal processing modules;
- a set of objects to connect the processing modules;
- a scripting language used to define the topology of the system and the interactions between modules.

The design of SPECTRA was heavily influenced by the experience obtained through building and using the VuSystem [10]. The VuSystem, which was designed to support multimedia applications running over a desk area network, provides considerable support for composing signal processing modules. This made it an excellent environment in which to conduct preliminary experiments with virtual radios. However, as we came to understand better the opportunities that virtual radios afford for adaptive signal processing, we realized that the adaptive signal processing applications could benefit from a different implementation model. The SPECtRA environment thus was designed to support several different notions of adaptive signal processing.

- *Adaptation to the Environment*: Consider a system equalizing for a wireless channel. This is what is most commonly thought of as adaptive signal processing. It requires algorithms to detect environment changes (such as a channel estimation algorithm) and sometimes a mechanism for specifying system modifications based on the detected changes.
- *Adaption to the User*: The user's requirements may change, necessitating changes in the system. A simple example of a conventional system adapting to the user is changing the station on the radio. This causes changes in certain parameters of the system. A more sophisticated example is pushing the AM/FM button, which requires different algorithms to be inserted into the processing path.
- *Functional Adaptation*: Many signal processing techniques, such as a phase-locked loop, are inherently adaptive. These algorithms typically adapt to the signal and may modify system parameters and/or functionality to meet specified requirements. For example, a receiver attempting to lock on to the start frequency of a hopping sequence adapts the system to examine different frequencies until a start code is found. Unlike the two types of adaptation defined previously, this type of adaptation is driven by a specified constraint, not external changes.
- *Adaptation to Resources*: Having the ability to adapt to resource availability can improve system performance. For example, if there are many spare CPU cycles available, then running a more computationally intensive channel estimation algorithm could lead to better overall application performance. Conversely, if the resources suddenly become scarce due to a burst of activity by other processes, the system should be able to adapt, perhaps by running less demanding algorithms and sacrificing some robustness or accuracy. This form of adaptability also enables applications to transparently improve performance as faster processors become available.

A visual description of an application built in SPECtRA is shown in Fig. 4. The system is partitioned into *in-band* and *out-of-band axes*. The in-band axis is where the temporally sensitive, computationally intensive work takes place. The out-of-band axis is used for control and intermodule communication. All of the application specific functionality is contained in the out-of-band section. This partitioning allows for maximal reuse of the computationally intensive in-band signal processing modules, since none of their functionality is specific to a particular application.

The processing modules are objects that contain the code required to perform a particular signal processing task, as well as class variables and accessor functions to allow for configuration and monitoring by the control script. The connectors can be thought of as a wire that carries signals from the output of one processing module to the input of one or more processing modules.

One key innovation in the design of this programming environment is that data is not pushed down the pipe as in most systems, but it is pulled down the pipe by the sink. This allows for the implementation of lazy evaluation along the pipeline, which leads to significant computational savings in applications that use decimation or skip certain portions of the input data. In the data pull case, the downstream module only asks for samples that it needs and can transparently eliminate considerable amounts of computation. The pull architecture also provides performance improvements in systems employ-

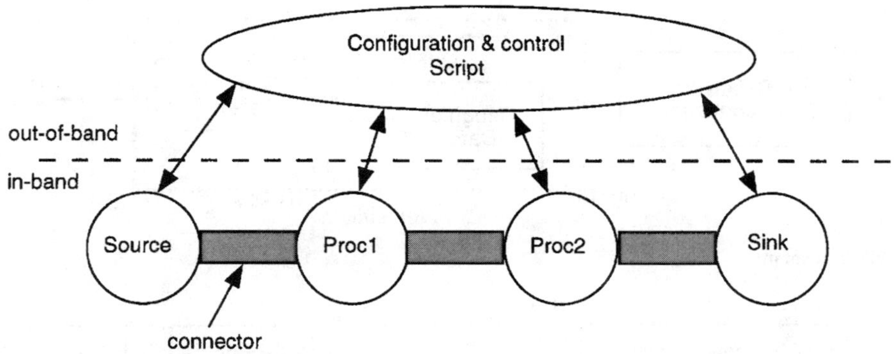

Fig. 4. Graphical description of the programming environment.

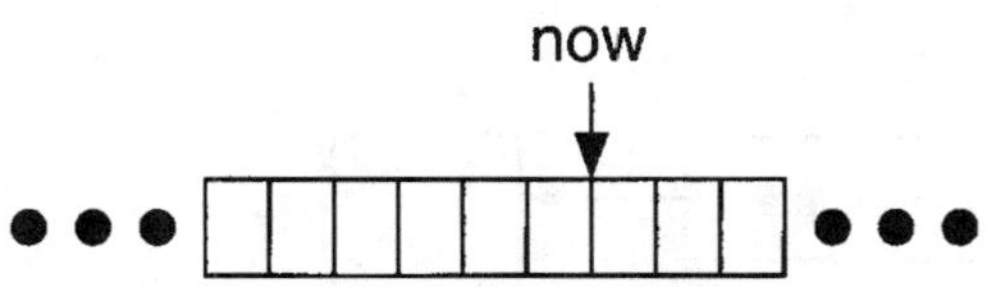

Fig. 5. Representation of streams as infinite buffers.

ing data caching, since the data being operated on is the data most recently produced, and therefore is very likely to be present in the cache.

One of the main motivations behind the development of SPECtRA was to create a programming environment that provides for simple and straightforward implementation of signal processing functions to reduce code development and maintenance overheads. Ideally, there should be little difference between the equations that represent a function, and the code required to implement it.

For many signal processing applications, a stream is a natural I/O model. Functions such as filters operate on a continuous stream of input samples and produce continuous streams of output samples. Many systems use buffers to pass data between one processing module and the next, but this leads to considerable extra code to take care of end conditions and to make the sequence of buffers appear to be a seamless infinite stream of data. Furthermore, since the code required to achieve this effect is dependent on the processing function, this extra code must be incorporated into each processing module.

A stream abstraction was developed to provide a more natural interface to the data samples. A stream serves as the interface between a processing module and a connector, as shown in Fig. 4. Each streams can be thought of as infinite an buffer, with a pointer that indicates the value corresponding to the current instance in time, indicated by the *now* pointer in Fig. 5. The *now* pointer eliminates the need for a absolute time index into the buffer and allows values to be addressed relative to the current time. Eliminating the need for an absolute index is beneficial not only for systems that are designed to run for arbitrarily long times, but also enables straightforward implementation of functions that do not have a consistent relationship between the input and output streams. Consider a detector in a communications system that recognizes the transmitted waveforms corresponding to binary digits. When there is no waveform present, the processing module would continually process input data without writing output data. When a waveforms is present and recognized, only then is the output stream updated.

The system is an efficient mechanism for implementing real-time wireless communications systems. The base system is reasonably compact, consisting of roughly 8000 lines of code. Writing applications is greatly simplified. For example, the AMPS cellular receiver required writing only 600 lines of code. Furthermore, other applications that reused some of the signal processing functions (e.g., the FM demodulation) realized up to 90% code reuse. Section III examines two example applications in detail.

III. Demonstration Applications

Several applications have been built with the virtual radio system, including a radio capable of receiving four half-duplex communications channels simultaneously and a *virtual patch panel* which enables interoperation between two systems using different RF bands, modulation schemes, and channel widths.

In this section we describe two prototype applications: a software cellular receiver and a software wireless network interface. These applications demonstrate both the feasibility as well as some of the advantages of virtual radios.

A. A Software Cellular Receiver

This section describes a wideband digital receiver which operates in the "A-side" of the U.S. cellular band. It continuously monitors 10 MHz of the cellular band and can demodulate FM signals anywhere in that band. Although the control channel [which uses frequency shift keying (FSK)] and supervisory tone were not performed in this implementation, other work using this architecture has demonstrated the implementation of an (FSK) system [16].

A block diagram of the cellular receiver is shown in Fig. 6. The front end is a Tellabs receiver that translates the 825–835 MHz band to a baseband signal and then samples the signal at 25.6 ms/s. The 12-bit sample stream is connected to a GuPPI daughtercard which digitizes and then DMA's the samples into the main memory of the host for processing. All subsequent processing is performed in software.

1) Software Architecture: The sequence of processing steps for the wideband advanced mobile phone system (AMPS) receiver is shown in Fig. 7. Since this wideband receiver

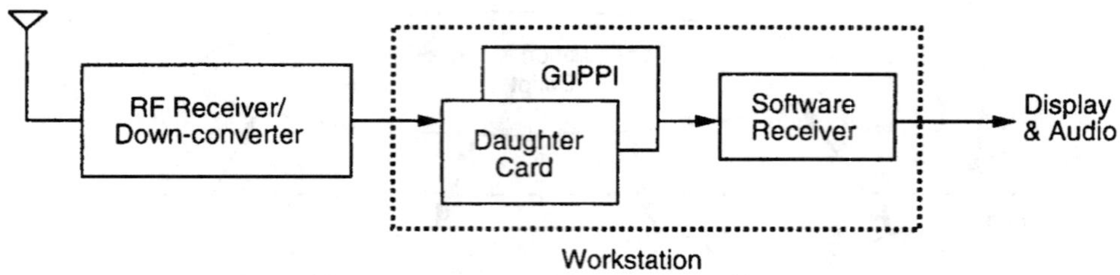

Fig. 6. Cellular receiver block diagram.

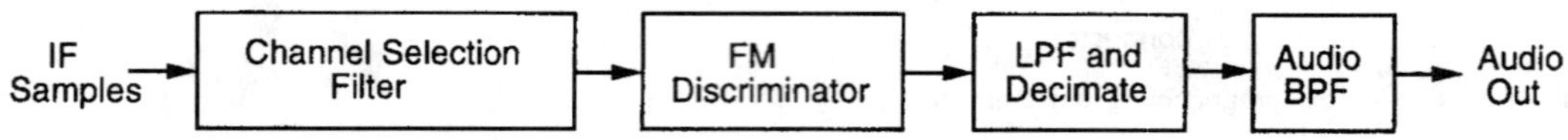

Fig. 7. Software architecture block diagram.

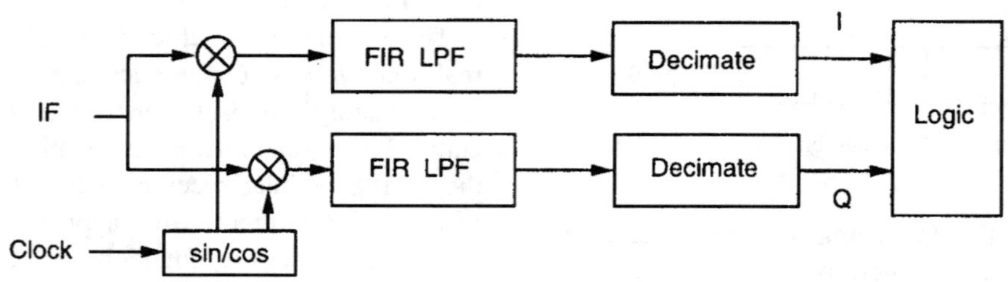

Fig. 8. Dedicated hardware digital downconverter.

demodulates a narrowband signal, the sample rate becomes lower at successive processing stages.

It is worth noting that all of the system parameters, including the number and values of the channel filter coefficients and the sample rates, are under software control. This allows many of these parameters to be easily modified, even while the receiver is operating.

The first processing step is the channel selection filter. This module extracts a 30 kHz FM AMPS channel from a 10-MHz input band. This step uses a novel filter that combines the three conventional steps of translating the signal to baseband, lowpass filtering, and decimating. This step comprises the largest portion of the overall computational load. The details of its design are given in a later section. The channel selection filter consumes the raw samples from the RF front end at $R_S = 25.6$ ms/s and produces a complex baseband signal with a variable intermediate sample rate R_D. The channel selection filter, in this implementation, is initially tuned to the nominal carrier frequency of the desired channel. No tracking of the carrier is required, and any small frequency offset will create only a small dc offset at the output of the discriminator which is then removed by the audio filter.

In a complex FM signal, the desired information (the voice, in this case) is carried by the instantaneous frequency, which is the time derivative of the phase of the complex signal. This signal is therefore demodulated using a simple FM demodulation algorithm that approximates the derivative of the signal phase by the phase difference between successive samples, appropriately scaled.

The two final steps of processing are implemented using finite impulse response (FIR) filters. The first is a lowpass decimating filter that removes high frequency components that would cause aliasing when the sample rate is reduced to the audio rate, typically 8 Ksamples/s. The final step is a bandpass filter which removes out-of-band noise from the voice signal.

The cellular FM demodulator presented here is implemented on a personal computer with a Pentium II/300 MHz microprocessor. The implementation runs in real-time without any glitches or dropped samples and requires approximately 60% of the CPU. Quantitative measurements of the audio output were not made, but the subjective quality is as good or better than a commercial AMPS cellular handset.

Separating narrowband channels in a wideband receiver is a computationally intensive task that is often allocated to special purpose hardware. Fig. 8 depicts a typical digital down-converter (DDC) implemented in dedicated hardware [2]. The wideband signal is translated to a complex baseband signal by the quadrature multiplier and then lowpass filtered to prevent aliasing due to decimation. Special purpose FIR filters for decimation perform this operation very efficiently—which is important with the sample rates for a wideband receiver could be in excess of 30 ms/s. One author [2] estimates that a good channel selection filter requires about 100 operations per input sample for a total of 3000 million operations per second (MOPS).

It is possible to build software that has the same structure as a hardware DDC. However, on current workstation hardware performance would be far too slow. The limiting factor is the high sample rate of a wideband receiver allows time for only a few operations per sample. In hardware DDC's, the separate steps of down-conversion are each done in separate physical devices in the DDC, resulting in a high degree of pipeline parallelism. Furthermore, there is additional fine grained parallelism within the FIR filter itself. Fortunately,

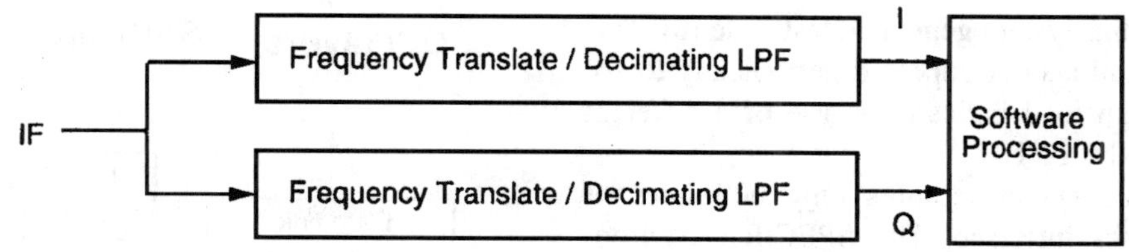

Fig. 9. Proposed software radio downconverter.

there is no compelling reason for our software to mimic the design of the hardware it replaces. In fact, one of the great advantages of virtual radios is the freedom to develop and evaluate unconventional solutions to problems.

Fig. 8 depicts the design of the software DDC. The main consideration is to minimize processing at the high sample rate and reduce the rate as early as possible. The choice of an FIR filter (versus a lower-order IIR filter) means that an output that depends only upon the filter input samples, $x[n]$, and therefore it computes only those output samples, $y[n]$, that are required after decimation [14]. This greatly reduces the computation load by taking advantage of large decimation factors. Furthermore, the precomputed frequency shift factors can be incorporated into a composite FIR filter which then require only a phase correction after each output sample is calculated. This feature makes this a time-varying FIR filter, but all of the variation is combined into a single, time-varying multiplication factor which is applied at the lower sample rate after filtering.

The multiplication for frequency shifting and filtering have been combined into a composite filter. The first step in the selection of a specific channel is to translate the real-valued received signal samples $r[n]$ to baseband by multiplication by a complex exponential

$$\begin{aligned} x[n] &= r[n]e^{-j2\pi f_c n T_s} \\ &= r[n]\{\cos(2\pi f_c n T_s) - j\sin(2\pi f_c n T_s)\} \end{aligned} \tag{1}$$

where f_c is the carrier frequency before translation to baseband and T_s is the sample interval. The result is filtered with the order-M FIR filter $h[m]$

$$\begin{aligned} y[n] &= \sum_{m=0}^{M} h[m]x[n-m] \\ &= \sum_{m=0}^{M} h[m]r[n-m]e^{-j2\pi f_c (n-m)T_s}. \end{aligned} \tag{2}$$

The two steps of frequency translation and filtering can be combined

$$\begin{aligned} y[n] &= e^{-j2\pi f_c n T_s} \sum_{m=0}^{M} h[m]\cdot r[n-m]e^{j2\pi f_c m T_s} \\ &= e^{-j2\pi f_c n T_s} \sum_{m=0}^{M} c[m]r[n-m] \end{aligned} \tag{3}$$

where $c[m] = h[m]e^{j2\pi f_c m T_s}$ are the composite filter coefficients. Not only is the number of computations reduced, but there is no need to compute the unfiltered baseband signal $x[n]$, nor to write the intermediate results to memory only to recall them in the next step. Also, while many techniques are available to design real-valued FIR filters to meet desired specifications with minimum order, (3) shows that the use of complex-valued filter coefficients for $h[m]$ imposes no additional computational cost. For this reason the algorithm can take advantage of recent advances in the design of complex-coefficient FIR filters to reduce the required filter order M of the original low pass filter (LPF), relative to a real-valued $h[m]$, without increasing the required computation load for the final composite filter [13].

This technique also has costs, however. Although the filter requires less real-time computations, it is more complicated to set up and can require more memory to store filter coefficients. Because the frequency translation and filtering are combined, knowledge of the desired carrier frequency is required to compute the filter coefficients, $c[m]$. This also requires re-computation of the coefficients when there is a change in the frequency to which the filter is tuned (or else precompute and store separate filters for any several specific frequencies). Since each set of filter coefficients is used many thousands or millions of times, however, the cost of precomputing or recomputing is well worth the improved efficiency. This technique is similar to using an analytic filter to select only the positive frequency components of the real-valued signal for the passband of interest and decimating. The combination of translation and lowpass filtering in Fig. 9 allows the filter to be quickly redesigned to select a different f_c.

The key improvement over the technique of Fig. 8 is that no operations are performed at the high sample rate. In the case of AMPS, the final rate is on the order of 100 KHz and a complex filter with several hundred taps might require less than 100 MOPS, which is achievable.

Another interesting feature of the software receiver is the ability to allow the system to dynamically optimize its own performance in a variety of ways. A particularly striking example of this is the way in which the system performs frequency-domain analysis. The user may perform frequency-domain analysis of either the entire band or individual signals using fast Fourier transform (FFT's) using a library called FFTW.[3] FFTW evaluates the performance of many different FFT algorithms and selects the best one for the particular processor and data set size. The evaluation of different FFT algorithms is based either on an estimate of relative performance or measured performance of different algorithms in the

[3]The FFTW package was developed at M.I.T. by M. Frigo and S. G. Johnson. Additional information, including source code, is available at http://theory.lcs.mit.edu/~fftw.

current environment using system generated test code [8]. This measurement process can also be repeated periodically to see if changes in the system load indicate the use of a different algorithm to compute the FFT.

2) Discussion: This receiver requires only 600 lines of code, including the user interface. The SPECtRA environment enables rapid implementation of wireless systems. Many applications have been completely implemented in only a few hours. During development, standard software debugging tools were used and modifications required only seconds to recompile and execute—a refreshing change from the hardware design and debugging cycle. The development team took advantage of the ease of modification to experiment with a variety of different filtering algorithms during development.

The use off-the-shelf components facilitated the construction of this receiver (for example, the FFTW package). This not only shortened the development cycle but encouraged us to be more ambitious in adding functionality.

The team also took advantage of the ease with which the receiver could be integrated with application software to provide a user-friendly interface. For example, the output of the FFT calculations is fed to standard plotting software which graphically displays any signals present in the receiver's frequency band.

In designing this receiver, a conscious effort was made to take advantage of the available computational resources to experiment with new filtering algorithms. The channel selection filter uses optimized data structures that take advantage of the large memory and computational flexibility of the workstation. The novel channel filter achieves excellent performance and can be redesigned on-the-fly. The time required to recompute the filter tap weights for a different filter size or carrier frequency is not noticeable during operation.

Currently work includes experimenting with even more radical filter designs. In particular, exploring the possibility of using randomized algorithms as an even more efficient way to prevent aliasing during the decimation process.

B. A Software Wireless Network Interface "Card"

The recent rapid growth in wireless network technology has greatly expanded the capabilities of mobile computing devices. However, the multitude of wireless network standards hinders seamless interoperability by requiring different physical devices to interoperate with different networks. Not only do wireless LAN's operate in different RF bands, but even those using the same band employ different coding, modulation and network protocols. The implementation of network interface cards (NIC's) in dedicated hardware limits the flexibility of these devices. One approach to this problem is to implement as much of the processing as possible in software, allowing the wireless network interface to be dynamically modified. The SPECtRA software environment provides all of the processing needed to convert between wideband IF signals and network packets.

This section presents a software wireless network interface designed to be compatible with a commercial frequency hopping radio operating in the 2.4 GHz ISM band employing

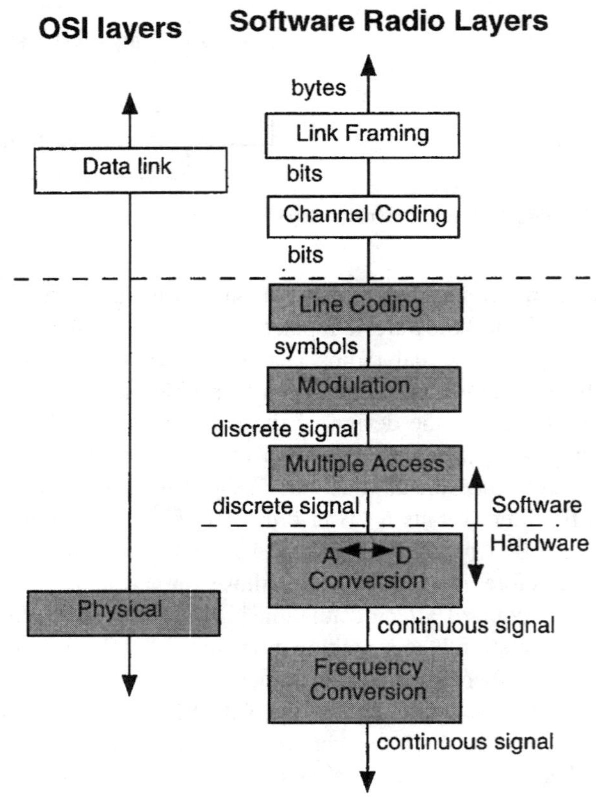

Fig. 10. Software network layering model.

FSK modulation [16].[4] Parameters such as the FSK frequency deviation and the spacing of the hopping channels can be dynamically modified in software. The only constraints imposed by the hardware are the width of the IF band and the center of the RF band. The ability to dynamically modify the channel width, channel spacing, and the hopping sequence allows the system to adapt to its environment and provide better noise rejection and immunity from hostile jamming attacks.

1) Architecture: The software network interface architecture is a refinement of the OSI layering model [17], which subdivides the *Link* and *Physical* layers as shown in Fig. 10. The signal processing performed in these layers can be naturally subdivided into a finer-grained model, but has traditionally been lumped into one layer because of its implementation in dedicated hardware. For the purposes of a virtual radio, however, this is too coarse. To interoperate with different networks, it may only be necessary to change small parts of the existing layers. For example, two different systems may employ the same modulation and coding, but use different multiple access protocols. In the future, we envision flexible networks, where the type of coding or access protocol may be altered dynamically to adapt to changing conditions and/or user requirements. To facilitate this flexibility, we would like to create new network interfaces by simply combining existing functional modules, rather than by writing a new piece of software for the link and physical layer functions of each interface.

[4]The wireless network interface was designed to be compatible with the 2.4 GHz frequency hopping radio from GEC Plessey, model DE6003.

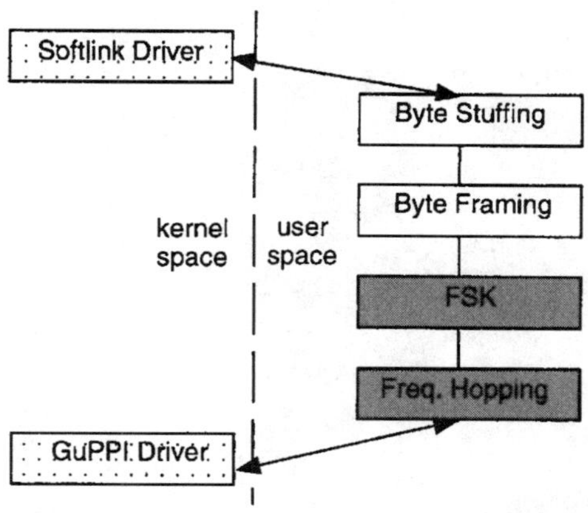

Fig. 11. Software components of the transmission application.

2) Transmission: The sequence of processing modules for transmission is shown in Fig. 11. The system interfaces with the host at the IP layer, through the *SoftLink* device driver. This appears to the kernel as a normal network device driver. However, instead of handing the packets off to a hardware device, this driver hands them up to user space, where the virtual radio processing is performed. The first level of processing is the network framing. For this example, the packets are framed by inserting a start code and byte stuffing the data. A length code, indicating the total length of the packet including the stuffed values, was also inserted after the start code. The next module takes the sequence of bits output by the network framing layer and performs byte framing, inserting start, stop and parity bits.

The conversion of each bit into a discrete signal is performed by the FSK module. The frequency hopping module then assigns this waveform to the appropriate frequency. All of the possible transmission waveforms are known *a priori*. There are two possible waveforms, corresponding to 1 or 0, for each hop frequency. All of these waveforms can be precomputed and stored at startup, significantly reducing the computation required to produce these waveforms. On a 180-MHz PentiumPro the IF waveform corresponding to a single bit required 2.2 μs to produce. This corresponds to a maximum transmit data rate of $\approx$450 kbps.

The generation of continuous phase waveforms is fairly straightforward in software. The precomputed waveforms are oversampled, and only a subsampled set, corresponding to the output sampling rate, are copied into the output buffer. The oversampling allows for indexing into the buffer to match the phase. The pattern is treated as a circular buffer, allowing the generation of waveforms for any bit period. After copying the samples to the output buffer, the phase value is updated and used as the index for the waveform corresponding to the next bit. In a similar manner, continuous phase is maintained between hops, even when the hop occurs in the middle of the bit.

3) Reception: Reception is considerably more complex than transmission even though the sequence of processing modules is essentially the reverse of the transmission system shown in Fig. 11. The receiver detects the presence of a valid transmission and synchronizes to it. It also reverses the function of each of the transmission layers. Combining the parameters of the frequency hopping and the FSK demodulation, the receiver to looks for one of the two valid waveforms at a given hop frequency. Separate functions track the hopping sequences and lock onto and demodulate the bits. These bits are deframed, and then the IP packets is extracted. The driver then hands the packet off to the host IP layer for processing.

TABLE II
AVERAGE TIME REQUIRED FOR EACH RECEPTION FUNCTION ON A 180 MHz PENTIUMPRO

Function	Time
Frequency Hopping	0.5 μs / hop
FSK lock	66.3 μs / bit
FSK Demodulation	4.5 μs / bit
De-Byte Framing	5.7 μs / bit
De-Packet Framing	4.6 μs / bit

4) Performance: The current implementation uses a 4.8 MHz wide IF sampled at 10 MSPS with 12-bit resolution and an RF band centered at 2.45 GHz. The transmission system generates continuous phase waveforms at a sustainable data rate of 320 kbps while hopping 1000 times per second. The receiver sustains a rate of 64 kbps and supports the same hopping rate. The amount of time to perform each reception function is given in Table II.

Rather than insuring real-time performance with the tight synchronous control over the processing that is typical of many DSP and digital hardware designs, the virtual radio employs an approach that is statistical in nature. It is required that, on average, there are enough cycles available to perform the processing, given that the actual number of cycles available in a given period of time varies due to other demands on the system.

A hard real-time system imposes a hard deadline for each task and provides a mechanism to insure that this deadline is met. To quantify the performance of the system, we introduce the notion of *statistical real-time performance*. The system is characterized by defining a probability that the work will be completed within the specified time limit and specifying the action that is to be taken when the deadline is not met.

The probability is determined by profiling the algorithm, as shown in Fig. 12, under the expected conditions on the work station. In this case the expected load is a Linux workstation running an X server, an NFS file system, and the usual network daemons (e.g., sendmail, inetd, etc.). In other cases, the expected load might involve other signal processing tasks, or significant user activity.

The action to be taken when the deadline is not met is application dependent. Possible action could include dropping the data, continuing processing for an additional period of time, or saving the data to be processed at a time when resources are available. Dropping the data would be appropriate in applications such as a full-duplex real-time voice system, where missing small amount of data may be more tolerable than increased latency. On the other hand, a packet data application may be able to deal with jitter, so a more graceful failure mode could be chosen.

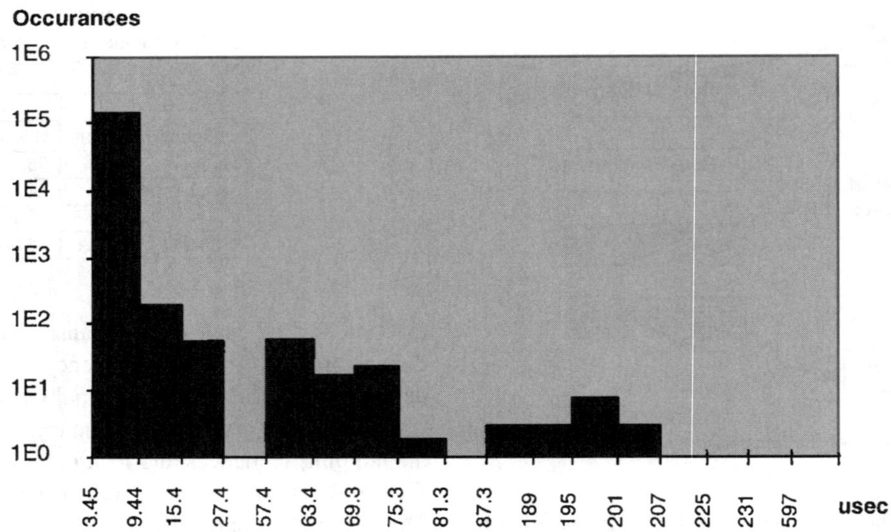

Fig. 12. Histogram of the time required (per bit) to perform the FSK demodulation function on a 180 MHz PentiumPro. The histogram is comprised of data from 170 000 trials, the vertical axis is the number of trials occurring for a given time bin, plotted on a log scale. The probability is less than 0.003 that more than 10 μs were required for processing for a given bit.

Fig. 12 shows a sample distribution of the time required to extract one bit using quadrature demodulation. The probability that more than 10 μs is required is less than 0.003. If the processing is stopped after 10 μs and an arbitrary decision as to the value of the bit is made, this would correspond to an increase in the probability of a bit error by only 0.0015.

5) Discussion: The frequency-hopping wireless network interface presented in this section demonstrates the feasibility of using application level software to perform the real-time signal processing. More importantly, it is illustrative of several of the advantages associated with the virtual radio approach to building virtual wireless devices.

- The network interface was built in a straightforward way using conventional software development tools. The entire implementation is only 520 lines of C++.
- During development the network interface was debugged in simulation using the same code that was later used to implement the actual radio. No porting from a simulation environment was necessary, but the full benefits of off-line simulation are retained.
- The design of the software network interface refines the OSI layering model in a way that would not be practical with a conventional hardware/software boundary.

IV. Conclusion

The last several years have seen dramatic changes in the hardware used to build software radios. Conventional software radios take advantage of vastly improved ADC's and DSP hardware. The virtual radio approach also depends upon high performance ADC's. However, rather than use DSP's, virtual radios ride the curve of rapidly improving workstation hardware.

This choice was dictated by three key assumptions.

1) Today's off-the-shelf workstations have enough processing power to support real-time signal processing applications.
2) Easy access to the computational resources and development environments available on conventional workstations will lead to new and potentially better approaches to signal processing.
3) Increasingly, there are advantages to be gained by coupling the "radio" component of communication devices with applications.

The work described in this paper provides strong evidence of the validity of these assumptions.

When this project began, the I/O bandwidth was the bottleneck. With the completion of the I/O system discussed here, the bottleneck moved to processing. Measurements show that a 200 MHz Pentium class machine is (just) fast enough to keep up with the demands of the kinds of applications described above. As processor speeds improve and caches grow it will become even easier to implement these kinds of applications and a variety of new applications will become tractable.

The two applications presented above provide evidence about the validity of the second two assumptions. Each required a relatively small amount of code and was developed quickly using standard programming tools. The implementation of the cellular receiver took particular advantage of the ease of incorporating existing software that had been built without signal processing applications in mind. Furthermore, its novel channel selection filter is suggestive of the opportunities for algorithmic improvement afforded by virtual radios. Each application also demonstrates the advantages of close coupling of the communications devices with applications. This was particularly important in the design of the software network interface, where refining the OSI layering model led to a flexible interface that facilitates interoperation with a variety of standards.

While there has been considerable progress thus far, this work still has a long way to go. We believe that we have demonstrated that virtual radios provide new and useful ways to build familiar communications devices using algorithms that are not very different from existing algorithms. In the next stage of our work, we plan to use the flexibility and power of virtual radios to experiment with radically different algorithms and with new kinds of functionality.

ACKNOWLEDGMENT

The authors would like to thank the many individuals who have influenced and contributed to the development of this project. In particular they thank D. Tennenhouse for his guidance and vision, A. Shah, A. Chiu, and J. Lin for contributions to the testbed, and D. Wetherall and A. Joseph for their comments on the manuscript.

REFERENCES

[1] E. Anderson, "Container shipping: A uniform interface for fast, efficient, high-bandwidth I/O," Ph.D. dissertation, Univ. California, San Diego, 1995.

[2] R. Baines, "The DSP bottleneck," *IEEE Commun. Mag.*, vol. 33, pp. 46–54, May 1995.

[3] J. C. Brustoloni and P. Steenkiste, "Effects of buffering semantics on I/O performance," in *Proc. 2nd Symp. Operating Syst. Design and Implementation (OSDI'96)*, Seattle, WA, Oct. 1996, USENIX, pp. 277–291.

[4] A. Brown and B. Wolt, "Digital L-band receiver architecture with direct RF sampling," in *IEEE Position Location and Navigation Symp.*, Apr. 1994, pp. 209–216.

[5] D. D. Clark, H. H. Houh, and D. L. Tennenhouse, "Aurora at MIT," Project Aurora Final Rep., Oct. 1995.

[6] C. D. Cranor and G. M. Parulkar, "Universal continuous media I/O: Design and implementation," Dep. Comput. Sci., Washington Univ., Tech. Rep. TR 94-34, Dec. 1994.

[7] P. Druschel and L. Peterson, "Fbufs: A high bandwidth cross-domain transfer facility," in *Proc. 14th Symp. Operating Syst. Principles,* Asheville, NC, Dec. 1993, pp. 189–202.

[8] M. Frigo and S. G. Johnson, "The fastest Fourier transform in the west," Lab. Comput. Sci., Massachusettes Inst. Technol., Tech. Rep. MIT-LCS-TR-728, 1997.

[9] M. Ismert, "The GuPPI: Hardware and software I/O support for PC-based real-time signal processing," in *Proc. IEEE INFOCOM'98,* 1998.

[10] C. J. Lindblad and D. L. Tennenhouse, "The VuSystem: A programming system for compute-intensive multimedia," *IEEE J. Select. Areas Commun.*, 1996.

[11] R. J. Lackey and D. W. Upmal, "Speakeasy: The military software radio," *IEEE Commun. Mag.*, vol. 33, pp. 56–61, May 1995.

[12] J. Mitola, "The software defined radio: Technology and marketplace," in *IEEE Western Electron. Conf. (WESCON),* Oct. 1996.

[13] H. Ochi and N. Kambayashi, "Design of complex coefficient FIR digital filters using weighted approximation," in *Proc. IEEE Int. Symp. Circuits Syst.*, 1988, pp. 43–46.

[14] A. V. Oppenheim and R. W. Schafer, *Discrete-Time Signal Processing.* Englewood Cliffs, NJ: Prentice-Hall, 1989.

[15] V. S. Pai, "IO-lite: A copy-free UNIX I/O system," Master's thesis, Rice Univ., Jan. 1997.

[16] A. B. Shah, "Software-based implementation of a frequency hopping two-way radio," Master's thesis, M.I.T., May 1997.

[17] A. S. Tanenbaum, *Computer Networks,* 2nd ed. Englewood Cliffs, NJ: Prentice-Hall, 1988.

[18] S. R. Thadani, "Software-based ultrasound system for medical diagnosis," Master's thesis, M.I.T., May 1997.

[19] T. von Eicken, A. Basu, V. Buch, and W. Vogels, "U-net: A user-level network interface for parallel and distributed computing," in *Proc. 15th ACM Symp. Operating Syst. Principles,* Copper Mountain, CO, Dec. 1995.

Architectural Overview of the SPEAKeasy System

Peter G. Cook and Wayne Bonser

Abstract— **SPEAKeasy is a successful implementation of a software-defined radio (SDR) for military applications. It permits general-purpose digital hardware to communicate over a wide range of frequencies, modulation techniques, data encoding methods, cryptographic types, and other communication parameters. The background of SDR's for military and commercial needs is discussed, and the SPEAKeasy architecture is defined.**

Index Terms—**Analog–digital conversion, CYPRIS, digital signal processors, joint tactical radio system, SDR Forum, software-defined radio (SDR), SPEAKeasy, waveform applications.**

I. Introduction

THIS paper describes the background of the SPEAKeasy system, and provides an overview of its architecture.

A. Background

Wireless communication using radio frequency links is a key element of increased mobility in military applications. Emerging technology has permitted increased reliability and functionality with reduced size, weight, and power.

With current levels of embedded computational power, the concept of digital signal processor (DSP)-based modems with software-defined capability has become a reality. Analog-to-digital converters (ADC's) are moving closer and closer to the antenna as conversion rates and the dynamic range increase. Similar advances in cryptography, multimedia processing, networking, and operator interfaces improve other aspects of the system. As digital processing elements take on more and more tasks previously accomplished in less capable programmable and single-purpose hardware devices, system flexibility is enormously increased. The potential of this technology was first described in the open literature as a result of the SPEAKeasy I program [1]. Software-defined radios (SDR's) offer a wide range of capabilities defined in software running on common hardware. This yields advantages over the less flexible implementations of previous technology. They allow for improvements or enhancements without altering the circuitry. SDR capabilities enable customers to acquire common hardware, and to satisfy individual requirements with software that fits their specific application, as is done in the personal computer marketplace [2].

In the late 1970's, the U.S. Air Force began work on a system known as the Integrated Communications, Navigation, Identification, and Avionics system (ICNIA). This was one of the first systems to use DSP-based programmable modem and control to obtain a fully integrated capability for airborne platforms rather. Prior technology required a set of independent federated hardware modules. ICNIA was based on three devices: a 25-MHz 6800-series low-latency processing element, a general-purpose processing element, and an enhanced security unit. The enhanced security unit interfaced to an external federated communications security (COMSEC) module. This technology laid the foundation for systems like SPEAKeasy and the Advanced Tactical Fighter's (F22) Communications, Navigation, and Identification system that evolved over the next two decades.

B. SPEAKeasy Phase-1

Building on ICNIA technology, in 1989, the Air Force initiated the Tactical Anti-Jam Programmable Signal Processor (TAJPSP) with the objective of developing a modular, reprogrammable modem. The military projected that it would need robust electronic counter-counter measures (ECCM) and low probability of intercept (LPI) waveforms that would necessarily evolve. An adaptable, software-defined platform, in the form of a programmable modem, appeared to be key to fielding successive waveforms as they became mature. The TAJPSP captured the attention of others to become the triservice programmable radio program called SPEAKeasy.

The primary objective of SPEAKeasy was to develop a modular, reprogrammable, modem with an open architecture. It was to demonstrate this SDR core on the air. A secondary objective was to develop a generic software architecture to facilitate the addition of new waveforms. SPEAKeasy exploited the ICNIA technology, but also adopted a more tactically acceptable security architecture and improved DSP—the TMS320-C40.[1] Technology examined under the initial program[2] was crucial to eventual military products. This included advanced designs for:

- fast fourier transforms (FFT's);
- ADC's with high-speed sampling and large dynamic range;
- a quad-DSP module (four TMS320-C40's);

Manuscript received November 10, 1997; revised March 12, 1998, June 3, 1998, and August 11, 1998. This work was supported by the U.S. Air Force Rome Laboratories under Contract F30602-95-C-0115.

P. G. Cook is with the Systems Solutions Group, Motorola, Inc., Scottsdale, AZ 85257 USA.

W. Bonser is with the Air Force Laboratories, USAF, Rome, NY 13441-4505 USA.

Publisher Item Identifier S 0733-8716(99)02977-7.

[1] The TMS320-C40 (Texas Instruments: www.ti.com), 1998.

[2] The initial SPEAKeasy project has been reported in the literature as SPEAKeasy I; its formal designation at this time is SPEAKeasy Phase 1.

Reprinted from *IEEE Journal on Selected Areas of Communications*, Vol. 17, No. 4, pp. 650–661, April 1999.

- RF up- and downconversion with wide instantaneous bandwidth;
- a programmable information security (INFOSEC) module architecture based on a 40 MHz reduced instruction set computer (RISC) known as CYPRIS.

The Phase-1 program conducted from 1992 to 1995 was to demonstrate a four-channel, wide-band architecture for high-speed frequency-hopped and pseudorandom spread-spectrum waveforms. It was to operate over the military HF, VHF, and UHF frequency bands. RF studies established that the RF operating range from 2 MHz to 2 GHz could not be implemented in a single wide-band RF channel. The RF section was therefore based on available technology using three RF bands. The low band ranged from 2 to 30 MHz; the midband ranged from 30 to 400 MHz, and the high band ranged from 400 MHz to 2 GHz. Only the midband was implemented in the Phase-1 feasibility demonstration. The hardware included a VME-chassis in a 6 ft rack with the VME bus for control and a unique high-speed ring bus for data. The user interface included a Sun SPARC workstation to download script files into a terminal controller unit. The Phase-1 equipment did not have sufficient FFT-based processing capacity to demonstrate the wide-band waveforms. The system was demonstrated on the air in June 1995 at Hanscom Air Force Base, MA, during the Joint Warrior Interoperability Demonstration (JWID-95). The primary objective, the development of the modem, was achieved. In addition, the Phase-1 software performed well enough that it was released on request to other agencies. Unfortunately, the modem software, user interface, and waveform development environment did not achieve the objective of ease of use.

C. Current Deficiencies and Mission Needs

During the early 1990's, the Air Force evolved its approach to command, control, communications, and computer (C4) systems to support operations that integrate multiple military services (e.g., Army, Navy, and Air Force, called "joint" operations[3]). These documents outline the basic military needs that the SPEAKeasy program addresses.

The military employs hundreds of communications systems that were developed when the services executed missions more or less autonomously. Many of these systems were never meant to work together ("interoperate"). Thus, they currently preclude the military from achieving the seamless communications it needs for an integrated battlefield. Law enforcement, public service, and emergency service organizations also are disadvantaged due to the proliferation of noninteroperable radios at federal, state, and local levels [4]. Interoperability can be increased by SDR's capable of bridging these incompatible systems. SDR's also permit the incremental deployment of new capabilities.

The dramatic pace of advances in communications technology, coupled with the military's traditionally long systems acquisition cycles have resulted in the technological obsolescence of new systems before they are fielded. Costs have prohibited retrofitting old systems with improved capability, resulting in reduced military readiness. Current radio systems cannot be technologically updated cost effectively. Software radios provide the opportunity for "future proofing" via preplanned-product improvement.

The U.S. Defense Department's emphasis on reducing life-cycle costs and the shift toward an integrated commercial and military national industrial base motivated technical change [5], [6]. For the military, postdeployment costs are by far the largest portion of a system's total cost [7]. These concerns are not unique to the military, and have resulted in the formation of the Software Defined Radio (SDR) Forum, formerly the Modular Multifunction Information Transfer System (MMITS) Forum. The SDR Forum includes among its members the SPEAKeasy team, The MITRE Corporation, the Federal Aviation Administration (FAA), and BellSouth [8]. It was created to facilitate the adoption of radio standards for software-defined communications systems with open system architectures. Such systems are required by military communicators and commercial wireless service providers [9]. Today, the SDR Forum is an international not-for-profit corporation that includes members from academia, the military, component manufacturers, wireless service providers, and regulatory bodies. Together, they are addressing critical issues such as software downloading, hardware and software module interfaces, and protocols [10], [11].

Single-purpose radios have created size, weight, and power problems for command and control centers that require interoperability across multiple military services. Platforms that require increased radio resources include the Airborne Battlefield Command and Control Center (ABCCC),[4] the Joint Surveillance and Target Acquisition Radar System (Joint-STARS),[5] the Airborne Warning and Control System (AWACS),[6] Army mobile command centers, and Navy ships. Increased numbers of radios have been used to ensure that all potentially required waveforms and channels are available, even if the number of radios simultaneously used is only a small fraction of the number carried. A reconfigurable radio ameliorates size, weight, and power requirements without compromising the availability of radio channels. It also reduces the number of types of spare parts.

Military doctrine[3] states that C4 systems should be "interoperable, flexible, responsive, mobile, disciplined, survivable, and sustainable." Commonality and standardization are the core mechanisms that support these attributes. Equipment and systems are common when:

1) they are compatible;
2) each can be operated by personnel trained on the others without additional specialized training;
3) spare parts are interchangeable;
4) consumable items are interchangeable.

In the military, standardization is the process by which commonality is achieved. One standardization approach consolidates procurements across military departments. If

[3] Joint Pub. 6-0 (Washington, DC: U.S. Department of Defense), 1996.

[4] See http://www.dm.af.mil/ec130e.htm.

[5] See http://www.af.mil/news/factsheets/E_8C_Joint_Stars.html.

[6] See http://www.af.mil/news/factsheets/E_3_Sentry_AWACS_.html.

SPEAKeasy technology were deployed through such a consolidated procurement, the costs of unique equipment training, spare parts, unique maintenance, and unique documentation would be reduced. This SPEAKeasy vision is embraced by U.S. military communications organizations. For example, the U.S. Office of the Secretary of Defense's (OSD) Programmable Modular Communications System Integrated Product Team (PMCS-IPT) issued a Guidance Document [11] that embodies this vision. This Guidance Document follows the functional partitioning of SPEAKeasy. It is also the basis for the current Joint Tactical Radio System (JTRS) procurement. All systems developed or procured after October 1, 1999 are to be JTRS compliant. SPEAKeasy hardware and software documents have been used by both the JTRS joint program office and the SDR Forum in support of standards processes.

D. SPEAKeasy Phase-2

The on-going SPEAKeasy Phase-2 program was initiated in 1995. Its objective was to expand beyond the modem to encompass an open, modular, reprogrammable architecture for the entire radio system, from user input and output (I/O) to RF. To lower the system life-cycle costs, the Phase-2 design emphasized commercial off-the-shelf (COTS) modules and commercial standards [3]. Its capabilities were to include several different types of I/O, internetworking, reprogrammable security services, programmable wide-band modems, and continuous RF coverage at a power output of 2 W from 2 MHz to 2 GHz.

Motorola, the prime contractor, designed a special wide-band RF transceiver unit for Phase-2. It reduced intermodulation distortion of intermediate frequency (IF) conversion through its homodyne-based design. It employs TMS320-C40 DSP's augmented with field-programmable gate arrays (FPGA's). The program initially continued the development of the CYPRIS INFOSEC module. Advanced waveforms, however, cannot tolerate the long context-switching delays of CYPRIS. High-speed switching among data encryption, generation of hop sequences for transmission security (TRANSEC), and generation of pseudorandom codes for spectrum spreading is necessary. This requires rapid context switching among encryption and TRANSEC states. The programmable Advanced INFOSEC Module (AIM)[7] consists of three distinct 32-bit data RISC machines with 100 MHz clocks that provide 1200 million instructions per second (MIPS) of processing power in a 3 V package. AIM is expected to redress the context-switching problems identified using CYPRIS. As built, the current Phase-2 units are comprised of about 70% COTS modules. They employ the industry standard PCI bus. A commercial palm-top computer with Windows 95 provides the user interface. They operate in three bands from 20–400 MHz (HF, VHF and UHF).

SPEAKeasy Phase-2 was to be a four-year research and development program consisting of four sequential "model-year" development cycles. Model-year-3 and model-year-4 units were to be sufficiently robust to be tested in the field. However, model-year-1 units were enhanced for field demonstrations in the Army's Task Force-XXI Advanced Warfighting Experiment (TF-XXI-AWE) at Ft. Irwin, CA, in March 1997, approximately 15 months into the program. The Air Force Tactical Air Control Parties controlled supporting aircraft, bridging Air Force HAVE QUICK UHF to Army Single Channel Ground Air Radio System (SINCGARS) VHF radios. SPEAKeasy also bridged voice channels between aircraft using HAVE QUICK and soldiers using a commercial hand-held land mobile radio (LMR). The LMR waveform was developed in less than two weeks, and was downloaded to the SPEAKeasy units in California from the laboratory in Arizona by telephone. This extended SPEAKeasy's capability during the exercise. One unit was repaired in the field using low cost-parts from a commercial computer. This experience somewhat validates the expected low cost of replacement parts for COTS-based implementations. It also raises questions about long-term availability when the commercial sector no longer stocks the parts.

Due to the success of SPEAKeasy at TF-XXI-AWE, a decision was made to immediately enter production rather than invest in further research and development. Consequently, Phase-2 had no opportunity to implement AIM and the full RF range of 2 MHz–2 GHz. The model-year-1 units also did not include wide-band waveforms, networking, or data–gateways. These units were limited in RF to the range from 4–400 MHz. The only waveforms were AM and FM voice. A small development program plans to demonstrate secure voice and data modes. The program demonstrated the benefits of an open, modular, COTS-based, reprogrammable system architecture.

E. Challenges

SPEAKeasy addresses multiple technical challenges. Open architecture, modularity, and reprogrammability objectives affect the technical aspects of each module of the system. For example, SPEAKeasy developed new methods to control a tactical radio. With the assistance of military operators, a COTS-based user interface was developed. Operation uses a touch-screen graphical user interface and the Windows 95 operating system. The granularity of the architecture had to be adjusted so that capabilities could be extended and units could be repaired in the field. INFOSEC is a fundamental military design issue not fully addressed by the commercial sector. Therefore, the use of COTS-based products in a secure ECCM-capable radio was a challenge. An INFOSEC module that would handle multiple encryption and TRANSEC algorithms simultaneously, context switching between these at various rates depending on the waveform, is still a challenge. The wide RF spectrum coverage (2 MHz–2 GHz) in a linear front end was also a critical technology driver. The exciter must deliver continuous linear output to synthesize spectrally pure waveforms defined by software and digital-to-analog converters (DAC's). Spread-spectrum (ECCM and LPI) waveforms typically require such high linearity.

[7] The Advanced INFOSEC Module (AIM) (Scottsdale, AZ: Motorola Corp.), 1998.

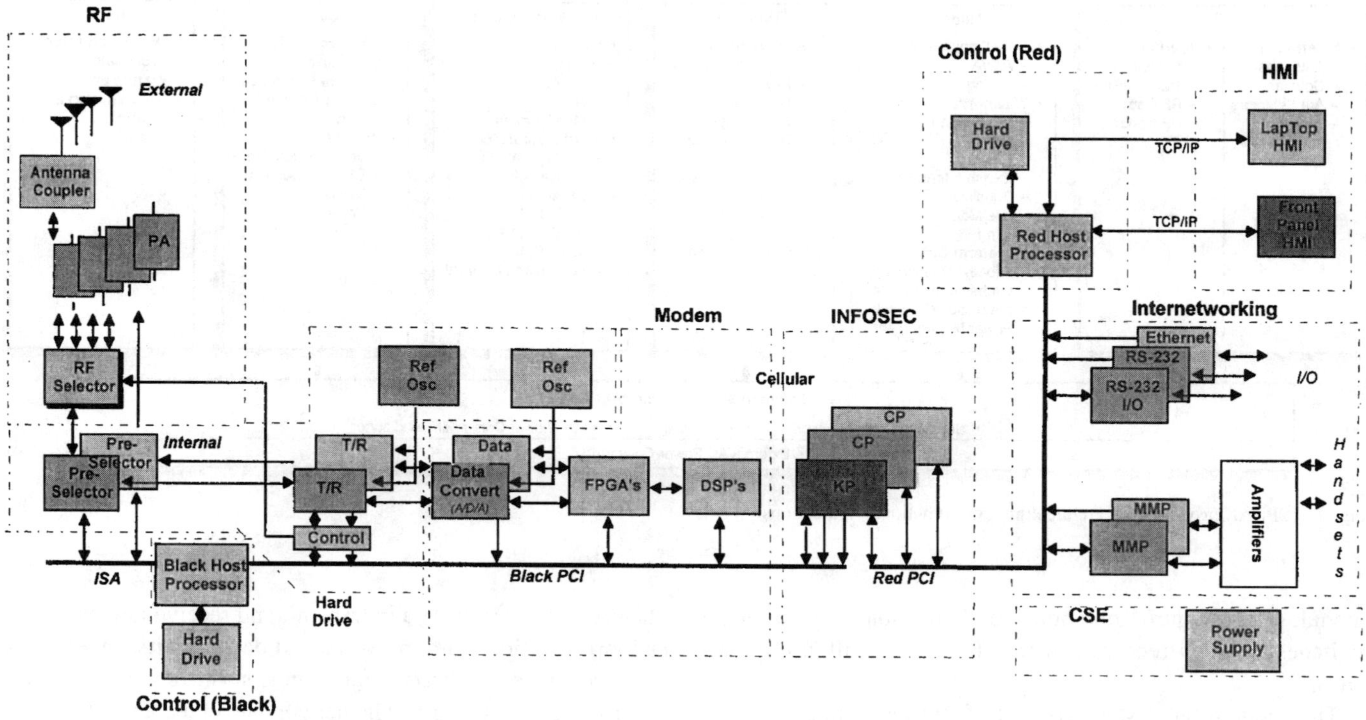

Fig. 1. SPEAKeasy block diagram.

F. The Future

Future proofing is the inclusion in current systems of facilities for the incremental addition of additional capability and new technology to avoid obsolesce. Software modularity and associated applications programmer interfaces (API's) are critical to future proofing. As an SDR, its multiprocessor architecture enables multiple RF channels, each of which employs a waveform determined by dynamically loaded application software. The SDR Forum provides a venue for technical exchange on these issues. This consortium of government, industry, and academia has the objective of defining standards to achieve the potential of SDR's, avoiding the proliferation of competing proprietary standards.

SPEAKeasy demonstrated that the technology of SDR's is feasible. Future military and commercial requirements will be met through SDR technology brought to the marketplace as commercial off-the-shelf (COTS) equipment. Special mission requirements may necessitate nonstandard hardware configurations, but open software architecture will permit both the substantial reuse of software and the introduction of mission-specific application software.

II. Architecture

The following sections describe the hardware and software architecture of SPEAKeasy Phase-2.

A. Hardware Subsystems

Fig. 1 is the hardware block diagram of SPEAKeasy Phase-2. System resources are attached to either the red (unencrypted) or the black (encrypted) peripheral component interconnect (PCI) bus. System resources are initially in a pool of available elements. They are logically assigned to virtual channels as they are instantiated. Each virtual channel is independent, executing any waveform for which there is software in the system archive. The two buses are implemented separately, with INFOSEC services and interbus communication provided by the key processor (KP) and crypto processors (CP). A CP is required for each virtual channel, but information flow does not require KP intervention once the channel is set up.

The external RF section, to the left in the figure, consists of antennas, power amplifiers, couplers, and selectors external to the unit. Modulation and conversion of digital data into analog RF takes place in the chain of transmit/receive (T/R) modules, data converters, FPGA's, and DSP's. This chain operates in half-duplex mode, reversing the flow of information between transmit and receive modes. Full-duplex operation uses two channels, one receiving and the other transmitting. To the right in the figure are facilities for voice operation, network interconnection, and user interface.

Fig. 2 shows the corresponding subsystem organization with the tasks allocated to each element. A description of the function of each subsystem follows.

The RF subsystem transmits and receives RF signals, converting signals from the analog modem into propagating RF waves, and conversely. RF subsystem functions include upconversion, downconversion, filtering, amplification, preamplification, diversity combining, T/R switching, cosite interference cancellation, antenna coupling, multicouplers, and antennas. The RF subsystem transmits and receives multiple channels simultaneously in the frequency range from 2 MHz to 2 GHz. It includes the antennas, couplers, cosite filters, power amplifiers, transmit/receive switching system, T/R

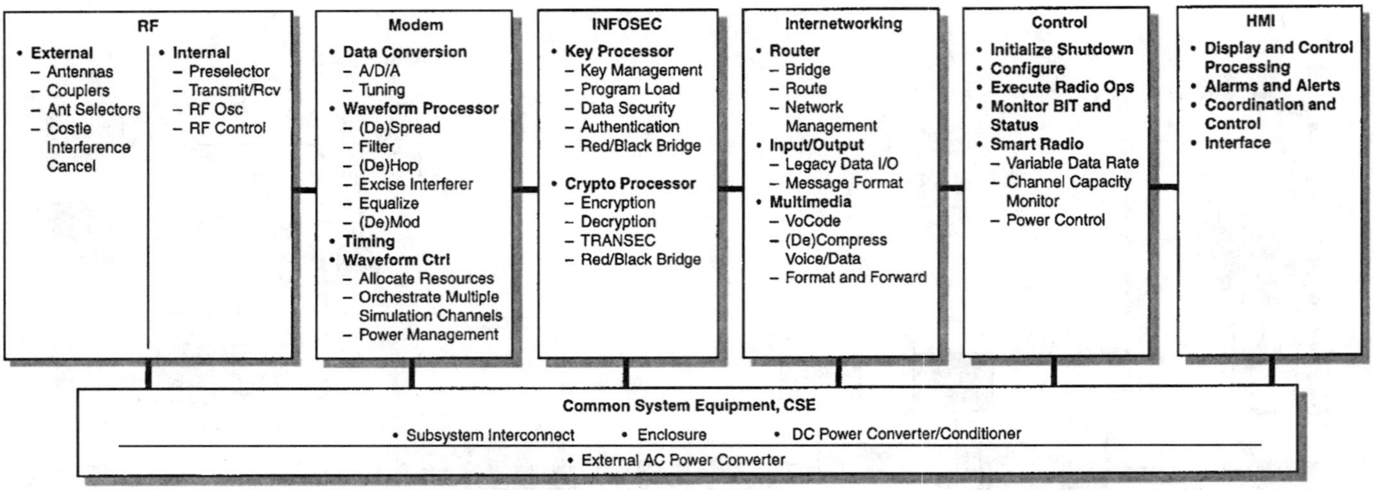

Fig. 2. SPEAKeasy system architecture and functional allocations.

modules, and control functions for half-duplex operation. It switches between receive and transmit states within milliseconds.

The modem subsystem converts between analog and digital format, modulating and demodulating the waveforms. The subsystem can transmit and receive multiple channels simultaneously, receiving analog signals and producing demodulated digital information during receive mode. During transmit mode, the input digital information is modulated and converted to output analog signals to excite power amplifiers at the specified transmitter power level. Frequency-hopping software in the modem generates frequency changes from hop sets maintained by the INFOSEC module. The modem subsystem includes signal processing functions, such as ADC's and DAC's, modulation and demodulation, forward error correction, interference suppression, carrier tracking, bit interleaving, and data framing. Modem resources may be allocated to any of the installed channels based on the waveform selected when the channel is assigned. Simple, low-speed waveforms place less demand on available resources than wide-band, complex waveforms. Changing parameters and reallocating channels may be performed without interrupting the operation of concurrently established channels.

The INFOSEC subsystem provides cryptographic security and TRANSEC. It contains one or more crypto-processor modules to encrypt baseband data and to create TRANSEC bitstreams. The INFOSEC subsystem also manages keying material to support these functions. The COMSEC and TRANSEC services enable the system to interoperate with legacy communications equipment and to provide communications at the same security level as that equipment.

The internetworking subsystem provides routes Internet protocol (IP) packets, accesses I/O ports, and processes channelized voice using multimedia processors (MMP's). These processors compress and restore voice, bridge voice and data from one RF mode to another, condition the signals, and deliver voice to the network or to the user. The subsystem simultaneously supports multiple independent half-duplex radio channels of voice or data in transmit and receive modes. It also supports multiple wireline voice and data circuits. In addition, it interconnects arbitrary data channels via IP routing to any other IP host or network. The network-level features efficiently route packet-switched traffic to the intended destinations. The router function has primary responsibility for routing data packets between appropriate links, and for determining which of the available paths in an attached network is used. It also provides I/O data bridging services, gateway services, and data-gram services.

The control subsystem generates commands that initialize, operate, and shut down the system. Initialization includes configuration of the radio, router, and I/O; downloading of waveforms and algorithms; built-in test (BIT); and diagnostics. It manages networks, subsystem configurations, files, and databases for the other subsystems. It receives commands from the user interface. It also provides long-term storage, including waveform software, cryptographic keys, status, logs, and other data. Operational functions include channel and waveform management; default system configuration controls, user authorization and levels of control, smart radio, and internetworking functions. Smart radio functions use software to improved performance through variable data rates, channel monitoring, and power control. The shutdown function releases resources, updates persistent storage, and prepares for reactivation.

The human–machine interface (HMI) subsystem provides the system-level user interface. An embedded palm-top computer provides the primary user interface for the local operator using the Windows 95 operating system. The HMI also includes remote control on a data circuit through a communications port.

The common systems equipment (CSE) subsystem houses the system, interconnects the subsystems, and supplies power, thermal management, and environmental protection. The enclosure addresses physical security, electromagnetic interference (EMI), and radio frequency interference (RFI) shielding and isolation from the environment (temperature, shock, and vibration). The degree of protection allows commercial com-

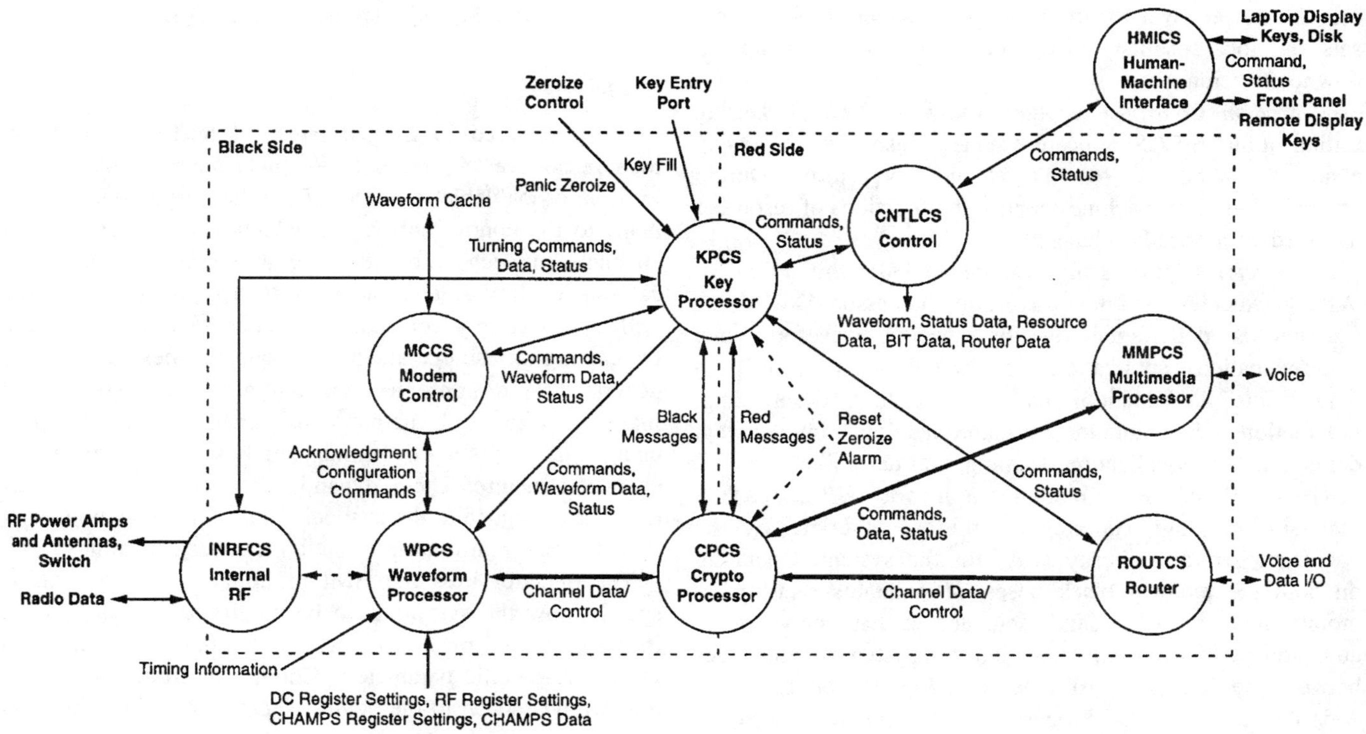

Fig. 3. SPEAKeasy control software (CS) and data flows.

ponents to operate at design capabilities in spite of field conditions.

Hardware was selected through extensive trade studies. Alternate approaches for each of functions to be performed were rated according to the following:

- initial cost
- life-cycle cost
- size
- power
- weight
- risk
- network compatibility
- interface robustness
- availability of alternate form factors
- commercial applications
- preplanned product improvement (P3I) potential
- performance.

The primary acceptance criterion was the composite score for system cost. Procurement cost was weighted by scores assigned to the factors listed above in the composite score.

III. Software Structure

This section provides an overview of the software structure. Fig. 3 indicates the names of the primary system modules and relevant data flows.

Table I indicates the functionality of each software module with its size in executable lines of code.

TABLE I
SOFTWARE MODULES

Module Name	Functions	Language	Lines of Code
INRFCS	RF API RF Control T/R Control	C	2500
WPCS	Common SW SINCGARS SIP UHF SATCOM HaveQuick I & II HF ALE (141A)	C C C C C	4400 3685 3150 3500 8820
MCCS	Allocate Resources Manage Channels Manage Power	C++	2340
KPCS	Key Management Control Red-Black Bridge Program Load SBA Authentication	Ada	33853
CPCS	Encryption Decryption TRANSEC	Assembly	23737
CNTLCS	Startup/Shutdown Waveform Instantiate BIT Management Status Management System Data Management	C++	12000
MMPCS	Vocode (De)Compress VVD Format and Forward	C	3500
ROUTCS	Router I/O MMP Proxy	C	62500
HMICS	Operator Command Process Display process	C/C++	12000

1) Internal RF Control Software (INRFCS): INRFCS is the Internal RF control module. Because the RF stage conditions analog signals, its functionality is limited to control functions. These include interconnection of hardware including antennas, power amplifiers, preselectors, and interference cancellers. It relays commands to those components that have programmable elements. It reconfigures for transmit/receive, sets the power level for the high-speed power amplifier, and sets the local oscillator frequency when used for up- or downconversion.

2) Modem Control Computer Software (MCCS): During initialization, MCCS allocates modem resources to logical channels as needed for the channel waveform. During operation, its state machines control the functions of resources assigned to a specific channel.

3) Waveform Processing Computer Software (WPCS): WPCS executes on the waveform processor DSP's. Its function is determined by the specific waveform and communications application task modules that have been loaded into it. Functions include spreading, despreading, modulation, demodulation, synchronization, interleaving, deinterleaving, equalization, framing, and deframing.

4) Key Processing Computer Software (KPCS): KPCS maintains cryptographic key control as authorized by the cognizant central authority (CA) for the system. Operating on both the red and black buses, it maintains security by monitoring control messages sent across that boundary for authenticity. It is the only entry into the system for messages between the system control module and other modules.

5) Cryptographic Processor Computer Software (CPCS): CPCS is assigned to a specific channel by the KPCS. It provides a virtual connection between the red bus and black bus, and performs simple bus-to-bus operation in plaintext mode. It supplies real-time encryption and decryption to data streams that need that service.

6) Multimedia Processing Computer Software (MMPCS): MMPCS serves as a vocoder to interface analog voice facilities with the digital data carried on an application channel. It also compresses, decompresses, and bridges among voice channels channels.

7) Control Computer Software (CNTLCS): CNTLCS starts the system and shuts it down. Under operator, script, or remote control it controls application instantiation and channel setup. It is the system data manager, and maintains system status. It is also the conduit for information passage between the system and the user interface software.

8) Routing Computer Software (ROUTCS): ROUTCS routes traffic and interconnects protocols to interconnect landline ports. It serves as a networkwide connectivity function for SPEAKeasy. It routes data packets between appropriate links, and determines which of the available paths are used within communications subnetworks. Services provided by the router include I/O data bridging, gateway, and datagram routing. It services external router requests. As the local communications control processor, it supports RS232, Ethernet, FDDI, and MIL-STD-1553. It also routes IP packets and terminates IP protocol stacks.

9) Human–Machine Interface Computer Software (HMICS): HMICS employs a display, keyboard, and other I/O devices for local or remote operator control of the system. Its control panels replicate functions of legacy waveforms in operation. Multiple channels can be in operation simultaneously, and the HMI permits the operator to switch between their control displays.

IV. SPEAKeasy System Operation

A. Overview

When powered up, the red and black host processors scan their respective PCI buses to establish memory maps of the location of available resources. They then communicate those maps to the control subsystem, which sets up static control channels to each resource using a parent–child hierarchical relationship. Agents are also created in the subsystems with responsibility for managing subsystem resources and responding to resource allocation requests. These requests are generated in response to user commands through the user interface software. Dynamic virtual channels are established using control messages communicated over the static control-channel structure. These dynamic channels directly connect resources assigned to the channel at the communications layer level to meet real-time information transfer constraints.

The operator uses the HMI to initiate operation, and to specify how the system is to be configured. A menu specifies the waveform to be used, the radio frequency, and waveform-specific parameters. Commands from that interaction are passed on to the control subsystem, which translates them into specific instructions to each of the other subsystems. These subsystems allocate resources, initialize, and instantiate channel application tasks to communicate on the specified channel in the desired configuration. Channel instantiation includes the following steps:

allocate devices that comprise a channel
instantiate channel application tasks
interconnect the dynamic information paths
set waveform-dependent parameters
activate the channel.

Channel deactivation reverses this process. Resources freed by channel deactivation may be reassigned to another channel. Resource pools increase reliability through redundancy, with graceful degradation in the event of equipment failure.

B. Communication Protocol Model

Communication among software modules employs messages transmitted over the system PCI bus. A three-layer software model forms a stack of logical layers of abstraction that build on the hardware of the physical layer as shown in Fig. 4. The three software layers are

link layer
communications layer
application layer.

1) Link Layer: The link layer moves data on a byte-by-byte basis between two devices over the physical bus. The link layer API function calls are the following.

Function Call	*Parameters*
`lnkStart`	`(pErrorNotify)`
`lnkStop`	`( )`
`lnkRead`	`(pLocal, Remote, Len_In_Bytes)`
`lnkWrite`	`(Remote, pLocal, Len_In_Bytes)`
`ErrorNotify`	`(ErrorNumber)`

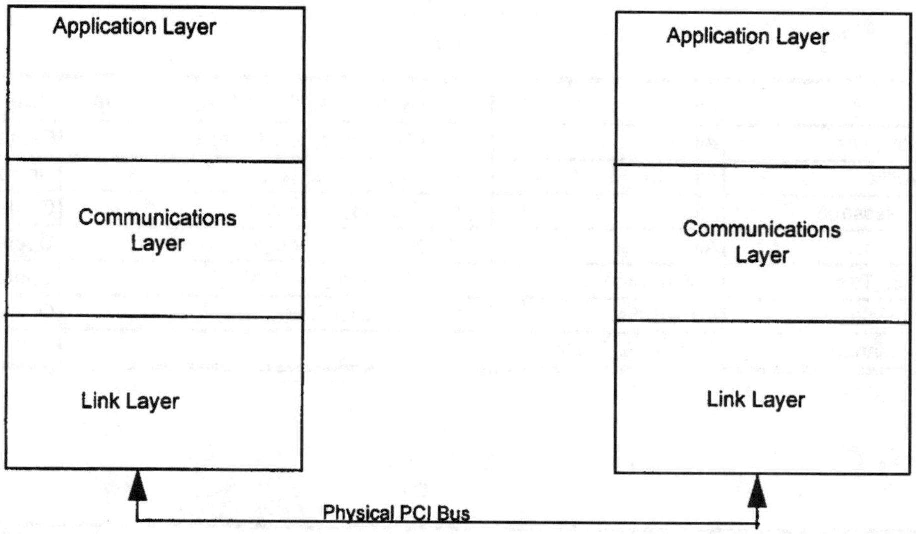

Fig. 4. SPEAKeasy internal protocol stack.

TABLE II
RADIO FREQUENCY CONTROL

No	Name	Phase	No	Name	Phase
1	ACK	All	18	Antenna_Select	Params_&_Mode
2	Buffer_Complete	All	19	DeAllocate_Resources	Params_&_Mode
3	Buffer_Notify	All	20	Define_Child	Params_&_Mode
4	Forward_Message	All	21	RF_Direction	Params_&_Mode
5	NACK	All	22	RX_Calibration	Params_&_Mode
6	BIT_Request	Power_up	23	Software_Version_Request	Params_&_Mode
7	Define_Remote_Child	Power_up	24	Hop_Strobe	Operation
8	Define_Remote_Parent	Power_up	25	Initiate_TX_Calibration	Operation
9	Allocate_Resources	Instantiation	26	Receive_Mode_Ind	Operation
10	Connection_Test	Instantiation	27	RF_Frequency	Operation
11	Define_Remote_Child	Instantiation	28	Set_Gain	Operation
12	End_Download	Instantiation	29	Standard_Data_Msg	Operation
13	File_Download_Complete	Instantiation	30	T/R_Transmit	Operation
14	File_Download_Start	Instantiation	31	Transmit_Mode_Ind	Operation
15	Initiate_Download	Instantiation	32	Destroy_Agent	Teardown
16	New_Agent	Instantiation	33	PA_Power	Teardown
17	Standard_Data_Msg	Instantiation	34	Reset_to_Boot	Teardown

These functions require a hardware bus address to transmit data. The virtual channel address to bus address mapping is handled at the higher level communications layer.

2) Communications Layer: The term "communication layer" is consistent with SPEAKeasy documentation. The term "infrastructure layer" may be used interchangeably. This layer employs link layer facilities to create a message-passing facility with an API used by waveform applications. It establishes links among the installed resources, and automatically identifies the devices on the bus when powered up. It performs the buffering and queueing of data internally, prioritized as necessary to achieve system performance requirements. The communications layer API's are the following.

Function Call	*Parameters*
`CommStart`	`(ErrorNotify)`
`commStop`	`( )`
`commCreateAgent`	`(Agent, SendComplete-Notify, ReceiveNotify, ReceiveCompleteNotify)`
`commDestroyAgent`	`(TAgent_Number_Type)`
`commListen`	`(Listen_Desc)`
`commConnectChildren`	`(Agent, ChildDesc, ChildPCIInterfaceDesc, ParentHint)`

TABLE III
MODEM

1	ACK	All	9	Adjust_RX_Calibration_Response	Params_&_Mode
2	Buffer_Complete	All	10	TX_Calibration_Complete	Params_&_Mode
3	Buffer_Notify	All	11	Crypto_Status	Operation
4	Forward_Message	All	12	Pacing_Indication	Operation
5	NACK	All	13	Receive_Mode_Ind	Operation
6	Connection_Test	Instantiotion	14	Standard_Data_Msg	Operation
7	Standard_Data_Msg	Instantiotion	15	Transmit_Mode_Ind	Operation
8	Activate_Channel	Params_&_Mode			

TABLE IV
INFOSEC

No	Name	Phase	No	Name	Phase
1	ACK	All	20	Standard_Data_Msg	Instantiotion
2	Buffer_Complete	All	21	Activate_Channel	Params_&_Mode
3	Buffer_Notify	All	22	Crypto_Algorithm	Params_&_Mode
4	Error_Status	All	23	Define_Child	Params_&_Mode
5	Forward_Message	All	24	Key_Tag	Params_&_Mode
6	NACK	All	25	Load_Key	Params_&_Mode
7	BIT_Request	Power_up	26	Mode_Req	Params_&_Mode
8	Define_Remote_Child	Power_up	27	Plaintext_Confirm_Response	Params_&_Mode
9	Define_Remote_Parent	Power_up	28	Software_Version_Request	Params_&_Mode
10	Agent_Description	Instantiotion	29	Suspend_Channel	Params_&_Mode
11	Connection_Test	Instantiotion	30	Channel_Status	Operation
12	Connect_Children	Instantiotion	31	Deactivate_Channel	Operation
13	Define_Red_Channel	Instantiotion	32	Pacing_Indication	Operation
14	Define_Remote_Child	Instantiotion	33	Receive_Mode_Ind	Operation
15	End_Download	Instantiotion	34	Standard_Data_Msg	Operation
16	File_Download_Complete	Instantiotion	35	Start_Traffic	Operation
17	File_Download_Start	Instantiotion	36	Transmit_Mode_Ind	Operation
18	Initiate_Download	Instantiotion	37	Disconnect_Children	Teardown

Function Call *Parameters*

```
CommDisconnectChildren
                      (Connection)
commSend              (Agent,
                       Buffer_Addr, Size)
commReceive           ( )
commBufferComplete    (Mailbox)
commLengthToSize (Length, Size)
commGetUserDataPtr    (Agent, Addr)
commSetUserDataPtr    (Agent, Addr)
```

3) Application Layer: This layer consists of waveform-specific software. All of the software modules in Fig. 3 reside in this layer, communicating through exchange of messages dispatched through the communications layer. The messages categories include data and control (which includes commands, status, and parameters). Control messages are virtual function calls.

C. Message Passing Architecture

There is no centralized operating system in SPEAKeasy. All modules communicate by cooperatively passing messages across the bus. In general, message passing is asynchronous and concurrent. Multiple communication channels and associated control functions are active concurrently. The messages are shown in Tables II–VII. These tables are organized according to the system module in which the implicit function call is executed. Only the common messages are identified; each waveform has additional messages. Each table indicates the phase of operation to which it applies.

D. Phases of Operation

1) Power Up: During the power-up phase, agents and their child–parent relationships are established. Static control channels are established, and built-in-test (BIT) is performed.

TABLE V
INTERNETWORKING

No	Name	Phase	No	Name	Phase
1	ACK	All	21	Default_Voice_Rate	Params_&_Mode
2	Buffer_Complete	All	22	Default_Voice_Type	Params_&_Mode
3	Buffer_Notify	All	23	Define_Child	Params_&_Mode
4	Forward_Message	All	24	External_Access_Tone	Params_&_Mode
5	NACK	All	25	Loopback_Test_Mode	Params_&_Mode
6	BIT_Request	Power_up	26	Mode_Req	Params_&_Mode
7	Define_Remote_Child	Power_up	27	Mode_Resp	Params_&_Mode
8	Define_Remote_Parent	Power_up	28	PTT_Priority	Params_&_Mode
9	Connection_Test	Instantiation	29	Sidetones	Params_&_Mode
10	Define_Remote_Child	Instantiation	30	Software_Version_Request	Params_&_Mode
11	End_Download	Instantiation	31	Suspend_Channel	Params_&_Mode
12	File_Download_Complete	Instantiation	32	Test_Mode	Params_&_Mode
13	File_Download_Start	Instantiation	33	User_Tone	Params_&_Mode
14	Initiate_Download	Instantiation	34	Channel_Status	Operation
15	New_Agent	Instantiation	35	Deactivate_Channel	Operation
16	Standard_Data_Msg	Instantiation	36	Standard_Data_Msg	Operation
17	Activate_Channel	Params_&_Mode	37	Start_Traffic	Operation
18	Alert_Tones	Params_&_Mode	38	Volume_Level	Operation
19	Bridging	Params_&_Mode	39	Destroy_Agent	Teardown

TABLE VI
CONTROL

No	Name	Phase	No	Name	Phase
1	ACK	All	27	External_Access_Tone	Params_&_Mode
2	Buffer_Complete	All	28	Key_Tag	Params_&_Mode
3	Buffer_Notify	All	29	Loopback_Test_Mode	Params_&_Mode
4	End_Ext_Channel_Control	All	30	Modem_AGC_Loop_Bandwidth	Params_&_Mode
5	Error_Status	All	31	Modem_TX_Power	Params_&_Mode
6	Ext_Channel_Control	All	32	Mode_Req	Params_&_Mode
7	Forward_Message	All	33	Mode_Resp	Params_&_Mode
8	NACK	All	34	Plaintext_Confirm_Request	Params_&_Mode
9	Add_Agent	Power_up	35	Plaintext_Confirm_Response	Params_&_Mode
10	BIT_Request	Power_up	36	PTT_Priority	Params_&_Mode
11	BIT_Response	Power_up	37	Sidetones	Params_&_Mode
12	Define_Remote_Parent	Power_up	38	Software_Version_Request	Params_&_Mode
13	Connection_Test	Instantiation	39	Software_Version_Response	Params_&_Mode
14	Download_Complete	Instantiation	40	Test_Mode	Params_&_Mode
15	Download_Files	Instantiation	41	User_Tone	Params_&_Mode
16	Instantiate_Waveform	Instantiation	42	Channel_Status	Operation
17	Resources_ Allocated	Instantiation	43	Crypto_Status	Operation
18	Security_Level	Instantiation	44	Deactivate_Channel	Operation
19	Standard_Data_Msg	Instantiation	45	Receive_Mode_Ind	Operation
20	Activate_Channel	Params_&_Mode	46	Signal_Strength	Operation
21	Alert_Tones	Params_&_Mode	47	Squelch_Level	Operation
22	Antenna_Select	Params_&_Mode	48	Standard_Data_Msg	Operation
23	Bridging	Params_&_Mode	49	Transmit_Mode_Ind	Operation
24	Channel_Direction	Params_&_Mode	50	Volume_Level	Operation
25	Crypto_Algorithm	Params_&_Mode	51	PA_Power	Teardown

TABLE VII
HUMAN–MACHINE INTERFACE

No	Name	Phase	No	Name	Phase
1	ACK	All	19	Loopback_Test_Mode	Params_&_Mode
2	Buffer_Complete	All	20	Modem_AGC_Loop_Bandwidth	Params_&_Mode
3	Buffer_Notify	All	21	Modem_TX_Power	Params_&_Mode
4	End_Ext_Channel_Status	All	22	Plaintext_Confirm_Request	Params_&_Mode
5	Error_Status	All	23	PTT_Priority	Params_&_Mode
6	Ext_Channel_Status	All	24	Sidetones	Params_&_Mode
7	Forward_Message	All	25	Software_Version_Response	Params_&_Mode
8	NACK	All	26	Test_Mode	Params_&_Mode
9	BIT_Response	Power_up	27	User_Tone	Params_&_Mode
10	Connection_Test	Instantiation	28	Channel_Status	Operation
11	Security_Level	Instantiation	29	Crypto_Status	Operation
12	Standard_Data_Msg	Instantiation	30	Receive_Mode_Ind	Operation
13	Alert_Tones	Params_&_Mode	31	Signal_Strength	Operation
14	Antenna_Select	Params_&_Mode	32	Squelch_Level	Operation
15	Channel_Direction	Params_&_Mode	33	Standard_Data_Msg	Operation
16	Crypto_Algorithm	Params_&_Mode	34	Transmit_Mode_Ind	Operation
17	External_Access_Tone	Params_&_Mode	35	Volume_Level	Operation

2) Instantiation: Communication occurs between agents within the protocol stack. Links between agents are children to a broker agent that instantiates the connection. Once the child agents exchange information, they test the connection to assure correct instantiation of the channel. Operating software uses these channels to load the subsystem processors. Once the load is complete, the agents deliver messages among modules.

3) Parameters and Mode: Waveform parameters are communicated to individual subsystems for proper operation. Typical parameters include receive/transmit state from the handset's push-to-talk switch, transmit power, and antenna selection.

4) Operation: Normal operation of the channel transfers information from source to destination through the internal protocol stack. RF channels and network interfaces may be either sources or destinations, providing transmit, receive, RF channel bridge, and landline switching modes of operation.

5) Teardown: Upon completion of operation, agents are deactivated, channels disconnected, resources returned to the pool, secondary (persistent) storage updated, and the system made ready for reactivation.

V. CONCLUSION

This paper has summarized the background and military needs that led to SPEAKeasy. It has also described the upper levels of the SPEAKeasy architecture. Using current technology, SPEAKeasy has demonstrated the vision of an information transfer system that implements many waveforms using common hardware. It also holds the promise of the economies of scale that are possible with commercial components in volume production.

The SDR Forum is working to satisfy a major need for future development: the adoption, by the industry, of standards comparable to those in effect in the personal-computer market. Such standards could assure vendors of a larger, consolidated market of diverse platforms manufactured in high volumes. The adoption of such standards will also permit the independent development of waveform applications to meet existing and new applications.

REFERENCES

[1] R. J. Lackey and D. W. Upmal, "Speakeasy: The military software radio," *IEEE Commun. Mag.*, vol. 33, pp. 56–61, May 1995.

[2] J. Mitola, "The software radio architecture," *IEEE Commun. Mag.*, vol. 33, pp. 26–38, 1995.

[3] A. McAuliffe, "Technology drives development of multiband, multimode radio," *Mil. Aerosp. Electron.*, vol. 6, Oct. 1995.

[4] R. E. Hundt, Chairman of the Federal Communications Commission, prepared remarks before the Assoc. of Public Safety Commun. Officers (APCO) Annu. Conf., Aug. 17, 1995.

[5] Under Secretary of Defense for Acquisition and Technology, Honorable P. G. Kaminski, Prepared Remarks before the Res. and Develop. Subcommittee of the House Committee on Nat. Security, Mar. 28, 1995.

[6] ——, Prepared Remarks before the Subcommittee on Defense Technol., Acquisition and Industrial Base of the Senate Committee on Armed Services on Dual Use Technol., May 17, 1995.

[7] F. J. Schrenk *et al.*, "SPEAKEASY—The universal radio for the 21st century," *MILCOM'95 Classified J.*, Nov. 1995.

[8] S. M. Blust, "Software defined radio: Industry request for information," BellSouth Wireless, Inc., Dec. 1995.

[9] D. M. Climek and E. J. Newman, "The SPEAKeasy software defined radio: An opportunity for transparent wireless telecommunications standards in the Asia-Pacific region," in *Proc. Pacific Telecommun. Council (PTC) Conf.*, Honolulu, HI, Jan. 1997.

[10] J. A. Hoffmeyer and W. Bonser, "Standards requirements and recommendations development for multiband, multimode radio systems," *MILCOM'97 Unclassified J.*, Nov. 1997.

[11] SDR Forum Tech. Rep. 1.1, SDR Forum, Jan. 1998.

AN ARCHITECTURE FOR RADIO-INDEPENDENT WIRELESS ACCESS NETWORKS

R. E. Schuh, C. Schuler, M. Mateescu

GMD FOKUS, Kaiserin-Augusta-Allee 31
10589 Berlin, Germany
PH: +49 30 34 637180 FAX: +49 30 34 637180
e-mail: schuh@fokus.gmd.de, schuler@fokus.gmd.de, mateescu@fokus.gmd.de

Abstract - This paper discusses a novel architecture for a Hybrid Fiber Radio (HFR) access network using Software Defined Radio (SDR) in a combined Base Station (BS) headend unit. While most of the current SDR approaches consider software radio to be used in the portable terminals to allow multi-mode operation, we propose a SDR-HFR based access network architecture. The combination of HFR and SDR allows smooth migration from current to future radio standards and realization of multi-standard networks by using the existing fiber infrastructure. In the first part of the paper trends in HFR and SDR as the two enabling technologies for radio-independence are theoretically investigated. The key components of a new multi-standard wireless access network are identified and some practical realization concepts are introduced. The focus of this paper is put on the availability and usability of SDR-HFR concepts from the network operator and the equipment manufacturer viewpoint.

I. INTRODUCTION

In the future wireless fixed or mobile broadband access for supplying multimedia services to the end user will be essential for public and private network providers (1-3). Some general network operators wishes are:

- ability to support a broad range of services
- preserving competition by flexibility
- protection of previous and future investments
- reliability and low maintenance costs
- seamless upgrade of the existing access network

Broadband services (bit-rate ≥ 2 Mbit/s) are supported by the future radio standards like UMTS / IMT-2000, MBS or ETSI BRAN. They will be based on different air interfaces for various telecommunication services. An overview on the occupied frequency spectrum, which ranges from ~1 to above 60 GHz, is given in Figure 1. These standards not only use different carrier frequencies but also different modulation formats, e.g. QPSK, nQAM or GMSK. Access techniques like FDMA and TDMA are combined with CSMA or wideband DS-CDMA. Future base stations and mobile terminals should allow:

- flexible control of the radio interface, adaptation to the different modulation and access formats
- easy adoption of new services and standards
- seamless roaming between different networks with different radio interfaces
- simultaneous support of several different radio standards

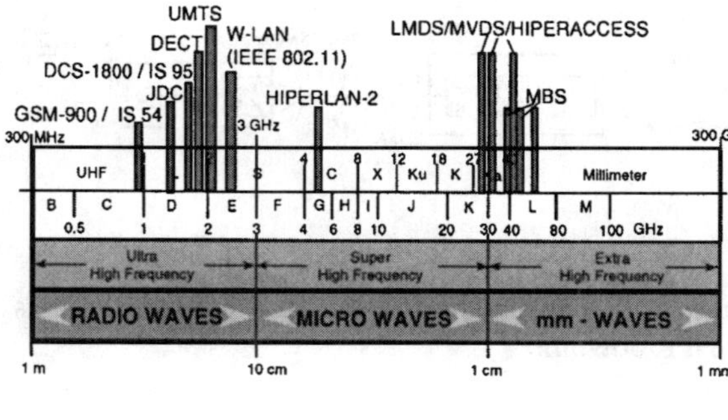

Figure 1 Radio spectrum used by current and future digital radio systems. The HIPERACCESS spectrum will be somewhere in the 3 GHz to 60 GHz range, e.g. in the DAVIC LMDS band

Reprinted from *IEEE Vehicular Technology Conference,* pp. 1227–1231, 1999.

II. ENABLING TECHNOLOGIES FOR RADIO-INDEPENDENCE

Hybrid Fiber Radio:

Microwave radio-frequency transport over fiber (1-2, 4-5) is a novel approach which allows the radio functionality of several base stations to be integrated in a centralized headend unit. Moreover, it offers fixed and mobile wireless broadband access with a radio-independent fiber access network. Different radio feeder concepts as Intermediate Frequency (IF) over fiber with electrical frequency up/down-conversion at the Remote Antenna Unit (RAU) or direct Radio Frequency (RF) transport are possible, as shown in Figure 2.

Radio frequency over fiber for GSM, W-LAN and video distribution systems have already been demonstrated (1-2, 4-5). There are different feeder concepts for HFR depending on the used radio frequency f_{RF}. For microwave transmission systems, e.g. the Local Multipoint Distribution System (LMDS) with 20 GHz < f_{RF} < 60 GHz, the most appealing solution is external modulation with single sideband or double sideband with suppressed carrier modulation, as chromatic dispersion penalties otherwise limit the usable fiber length (4).

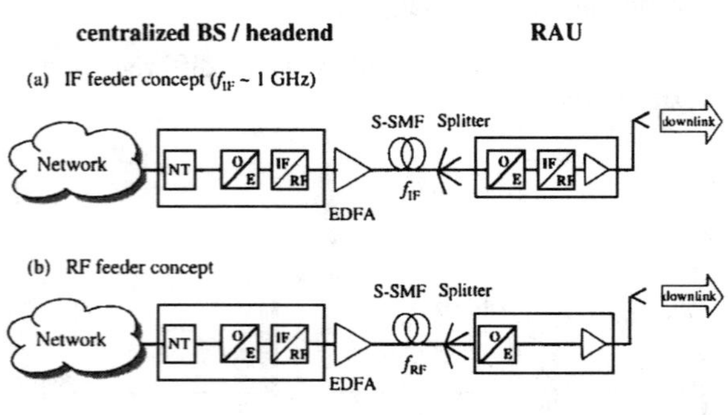

Figure 2 Block diagrams of optical feeder concepts for HFR downlink

Software Defined Radio:

The SDR concept allows the development of hardware independent radio transceivers by extensive use of digital signal processing. Ideal SDR assumes direct wideband ADC/DAC of a radio signal, which offers a programmable multi-band - multi-mode radio architecture (6). Current limitations in processors capabilities, memory capacity and ADC/DAC speed require hybrid solutions, where only a part of the radio modem is implemented in software, as shown in Figure 3. Nevertheless the advances in semiconductor technology, namely the progress in processing power, may enable pure SDR realizations in some years (7). SDR offers the following advantages:

- re-usability of radio hardware for new standards through reconfiguration by downloading signal processing software in a single radio / modem architecture
- reuse of the same base station design for different products – advantageous from a manufacturing viewpoint.
- reduced size, weight and power due to fewer radio units
- efficient usage of available radio spectrum by introducing new channel access modes into bands where existing modes are supposed to co-exist for a number of years

The maximum sampling frequency f_S and the Spurious-Free Dynamic Range (SFDR) may not be sufficient for some air interfaces in the BS. For example the DCS-1800 interface utilizes a bandwidth of 75 MHz for up or downlink, and requires a dynamic range of > 70 dB for different indoor and outdoor environmental conditions, see Reference (2). Possible realization approaches in order to soften the requirements on software radio are:

- parallel operation of multiple ADC/DAC and DSP
- automatic gain control in the analog receiver part to reduce the required dynamic range of the ADC
- integration of programmable hardware concepts

Although software radio is interesting for the mobile terminal and the network side, in this paper we are mainly interested in SDR base stations. This eases at least the restriction on size and power consumption compared to the mobile terminal.

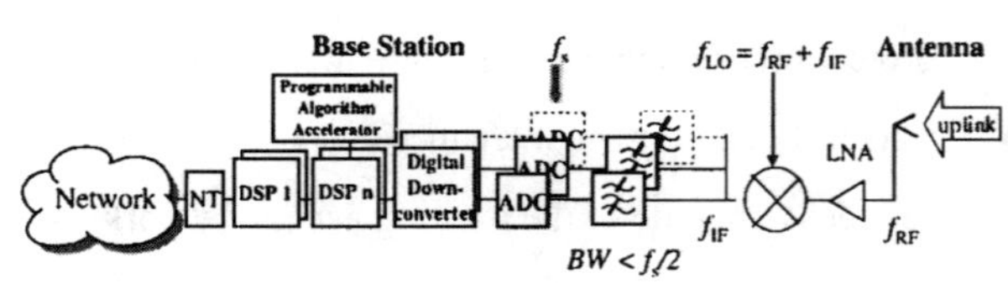

Figure 3 Sketch of a hybrid-SDR architecture

III. ARCHITECTURES FOR MULTI-STANDARD BASE STATIONS

Our proposal is to use HFR technology in combination with hybrid-SDR as a wireless multi-standard access method. Figure 4 and Figure 5 show the concepts for the combined HFR hybrid-SDR architecture for the downlink and uplink respectively. In the following paragraphs the technical requirements and different concepts for realization of key components of the proposed architecture are analyzed.

Digital Signal Processing Unit: The DSP unit is required on the receiver and the transmitter side. Typical tasks are:

- filtering
- carrier and timing recovery
- modulation / demodulation and signal conditioning
- encoding and decoding

If using multiple ADC converters to extend the bandwidth, chip synchronization has to be performed. An interesting new approach is the integration of programmable hardware circuits like Field Programmable Gate Arrays (FPGAs). Many FPGA architectures are in system programmable and can therefore be reconfigured with new functionality during regular operation. The optimum partitioning between hardware and software components allows a reduction of the number of MIPS required without sacrificing flexibility (8).

Programmable Up/Down Conversion: In the transmitter digital up-conversion of the baseband (BB) signal before transmission over the optical fiber may be either to IF or RF, as indicated in Figure 4. For the receiver frequency down-conversion using bandpass sampling is limited by ADC sampling speed. Multiple band sampling over e.g. the frequency band from 1.7 GHz to 2.2 GHz, which includes DCS-1800, IS-95, DECT and UMTS, may be possible in the future if ADCs for an IF input > 500 MHz and SFDR > 70 dB are available. Sampling and conversion can be performed by using parallel operation of multiple ADCs with corresponding bandpass filters. Specially designed commercial software radio chip sets for up- and down-conversion are available and can be used to relieve the multi-purpose DSP from the most computing intensive tasks. Examples are from Analog Devices the AD6622/AD6624 or from Harris Semiconductor the HSP50215/HSP50214B. These programmable devices can for example perform the tuning of the baseband signal to the desired digitized IF and vice versa.

ADC/DAC Converter: An attractive ADC solution for the receiver side is bandpass sampling with $f_S \geq 2\times BW$, where BW is the signal bandwidth. The converter is used to alias down the signal to baseband during the sampling process. The baseband signal is then fed into the DSP unit for further processing. Typical specifications are 12 to 14 bit ADC with IF input from 70 to 200 MHz, $f_S < 80$ MSPS with an in-band SFDR of > 80 dB over a digitized spectrum < 25 MHz. For the transmitter high linearity 14 bit DAC with conversion speeds up to 125 MSPS (SFDR > 70 dB for $f_{out} < 20$ MHz) are available. It should be mentioned that the current bottleneck in SDR design are the digital up- / down-converters, which requires especially for wideband modulation formats (W-CDMA) the use of programmable hardware.

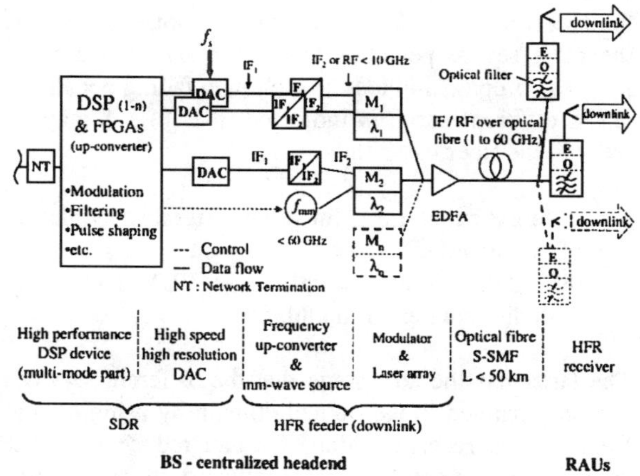

Figure 4 Downlink architecture for HFR and hybrid - SDR

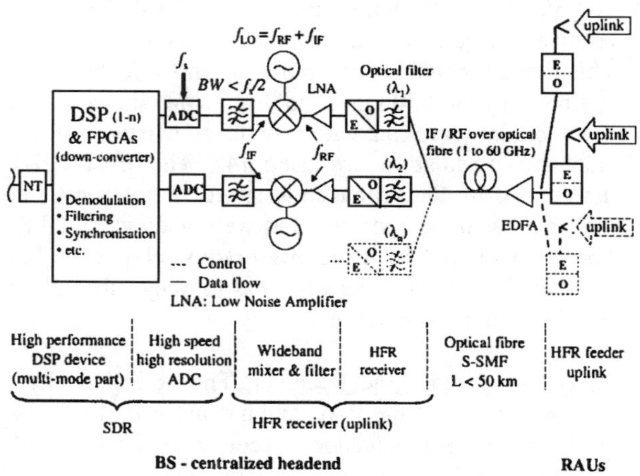

Figure 5 Uplink architecture for HFR and hybrid - SDR

Headend HFR Feeder and Receiver: Optical receivers up to frequencies of ~ 60 GHz and with responsivity > 0.1 A/W are available. External modulation of narrowband lasers up to 50 GHz, in order to generate the radio carrier wave, is state of the art. For higher modulation frequencies double sideband with suppressed carrier modulation may be used, doubling the modulation frequency (4). The magnitude of the dominant noise and non-linearity penalties - in addition to the those arising from the free space radio link - depend on the optical feeder concept and modulation format used. Of these, the most problematical are laser phase noise, non-linearities caused by the transmitter modulation, optical amplifier noise such as Erbium Doped Fiber Amplifier (EDFA) signal-amplified spontaneous emission, fiber dispersion causing destructive interference in the generated radio wave at the receiver end and receiver thermal noise. However, the most severe penalties are in the free space radio link, e.g. propagation loss or multipath fading penalties.

The different services indicated in Figure 1 may be fed into the fiber by using:

- different subcarriers, for direct/external modulation of the same DFB laser
- different laser wavelengths, e.g. WDM laser array with direct/external modulation

The latter has the advantage that the different services can be separated in the optical domain by using optical filters at the receiver side. This can reduce the fiber count but adds extra cost and complexity to the system.

Optical Fiber: Standard-Single Mode Fiber (S-SMF) is the choice as there are millions of kilometers installed worldwide. S-SMF designed for 1.3 μm operation has at 1.55 μm an attenuation of ~0.25 to 0.3 dB/km but a relatively large chromatic dispersion of ~ 17 ps/nm/km. This limits the radio wave transmission for f_{RF} > 5 GHz to fiber lengths < 50 km, if no dispersion compensation or a modulation format with suppressed carrier or single sideband modulation are used (4). The fiber delay determined by the propagation speed (2×10^8 m/s) in the fiber has to be considered as well when using HFR. For GSM the maximum allowed radio delay is ~ 220 μs, therefore the HFR network has to obey: fiber delay + radio path delay < 220 μs.

RAU HFR Feeder and Receiver: The optical feeder / receiver located at the RAU site has basically the same functionality as the feeder / receiver at the headend. Transmitting IF over fiber requires analog frequency up-conversion at the downlink side. For the return link (uplink) the radio-wave carrier may be extracted from the received signal, saving a local oscillator (LO) at the RAU. For strong asymmetric traffic, like in video distribution systems, a direct modulated laser at the RAU side may be used for the uplink. Down and uplink can be physically separated by using two fibers or using a single fiber with WDM technology, e.g. downlink at 1.55 μm and uplink at 1.3 μm.

A novel design for the RAU transceiver has been demonstrated in Reference (5) by using an electro-absorption-modulator for direct electrical-to-optical and optical-to-electrical conversion. However, the radio range, due to low conversion efficiency is < 50 meters, for a DCS-1800 system, if not using some electrical amplification (1).

IV. CONCLUDING REMARKS AND FUTURE WORK

We have shown a novel architecture combining state of the art HFR concepts with hybrid-SDR. Possible applications for the proposed architecture are telecom mobile and access networks as well as in-building wireless LAN infrastructures. There are several advantages for network operators, customers and equipment manufacturers.

HFR using IF / RF over fiber can reduce complexity at the remote antenna side and allows the concentration of the radio related functions of several radio cells in a single centralized BS. The huge bandwidth of the optical fiber offers the possibility of transmission of multiple radio standards simultaneously, over the same feeder network. The generation of the different radio signals, modulation/demodulation, filtering, etc, could be done in general by conventional analogue techniques. However, the centralized BS concept of the proposed HFR – hybrid-SDR architecture allows flexible upgrade approaches, which are essential for the rollout of new mobile network services like UMTS/IMT-2000. By shifting radio functionality to a centralized location at the headend equipment cost and investment risks can be reduced. For the realization of the proposed architecture future research is required, e.g.:

- analysis of the power budget, dispersion penalties and overall noise accumulation
- nonlinearity penalties (SFDR)
- development of fast DSP algorithms for the multi-mode radio transceivers
- implementation and integration of efficient multi-band SDR architectures
- cost analysis of the RAU and of the centralized BS - headend side due to the use of HFR and SDR

V. ACKNOWLEDGMENT

The authors wish also to acknowledge H. Schmuck from Alcatel SEL, Germany for very useful discussions and comments.

REFERENCES

(1) P816 EURESCOM study group, "Implementation frameworks for integrated wireless - optical access networks, ". http://www.eurescom.de/.

(2) Schuh, R.E., Wake, D., Verri, B., and Mateescu, M., "Hybrid Fibre Radio Access: A Network Operators Approach and Requirements," Accepted by the 10th Microcoll Conference, Microcoll'99, Budapest, Hungary, 21 - 25 March, 1999

(3) Schuh, R.E., and Brandt, H., "Simple Mobile-ATM," Proceedings of 5th Int. Workshop on Mobile Multimedia Communication MoMuC'98, pp. 345-348, Berlin, Germany, Oct. 12 - 14, 1998.

(4) Schmuck, H., and Heidemann, R., "Hybrid fibre-radio field experiment at 60 GHz," ECOC'96., Oslo, Sept. 15-19, 1996.

(5) Wake, D., and Moodie, D., "Passive picocell- A future integrated wireless communications network, " British Telecommunications Engineering, Vol. 16, pp. 106-112, July 1997.

(6) Mitola, J., "The software radio architecture, " IEEE Comm. Magazine, vol. 33, pp. 26-38, 1995.

(7) Drew, N.J., Tottle, P., "IC Technologies and Architectures to Support the Implementation of Software De-fined Radio Terminals, " ACTS Mobile Summit 98, Rhodes Greece, pp. 38-44, June 8-11, 1998.

(8) Taylor, C., "A DSP System for a Software Radio Testbed", ACTS Mobile Summit 98, Rhodes Greece, pp. 32-37, June 8-11, 1998.

Abstract

Third-generation wireless systems call for strategies that can improve achievable performance and data rates while providing flexibility and affordability. Software-defined radio technology is promising to provide the required flexibility in radio frequency, intermediate frequency, and baseband signal processing stages. Baseband signal processing techniques such as interference cancellation for direct-sequence code-division multiple access systems have the potential to provide the higher performance requirements of 3G systems. With advances in digital signal processor technology, the gap between the complexity of interference cancellation algorithms and available processing speeds is narrowing. However, 3G system requirements are ever pushing the envelope of signal processing algorithms required, including multi-user detection schemes, that are considered to be attractive and viable candidates. WINLAB has been working on an attractive SDR solution, to provide flexibility and handle the processing speeds required of 3G radio receivers. The approach is based on SDR architectures that partition radio receiver processing into two core technologies (field programmable gate arrays and DSP devices). We present a summary of the SDR work at WINLAB that is based on using this mixed signal processing approach for implementation of nonlinear interference cancellation and linear multi-user detection algorithms.

Software-Defined Radio Architectures for Interference Cancellation in DS-CDMA Systems

IVAN SESKAR AND NARAYAN B. MANDAYAM
WIRELESS INFORMATION NETWORK LABORATORY

Third-generation (3G) wireless systems are being driven by the desire to support innovative broadband multimedia services. While second-generation systems such as Global System for Mobile Communications (GSM), IS-136, and IS-95 are designed to carry speech and low-bit-rate data, 3G systems are expected to provide seamless wireless access to multimedia services, some of which are not yet defined. The common theme to which all variants of 3G systems adhere is the need to provide high bit rates. For example, it is desired that these systems will offer at least 144 kb/s (possibly higher) for high-mobility users with wide-area coverage and 2 Mb/s for low-mobility users with local coverage. Additionally, multimedia traffic presents coexistence of diverse services with varied traffic characteristics. The diversity of services presented results in the need for provisioning variable quality of service (QoS) in these networks. Air interfaces in these systems will need to be able to support services with different bit-error rates, delay, and rate requirements. Direct-sequence code-division multiple access (DS-CDMA) technology is being promoted as an air interface that holds great promise for 3G systems. Specifically, in Japan and Europe the common air interface that has been consolidated[1] is now commonly referred to as *wideband CDMA* (WCDMA). In the United States, the Telecommunication Industry Association (TIA) has chosen a slightly different CDMA air interface referred to as *cdma2000*. Both air interfaces, though, require backward compatibility with existing second-generation interfaces.

The above standardization activities, the evolution of new services, and varied QoS requirements all call for increasing the capacity of CDMA systems that are primarily limited by multiple access interference (MAI). This has prompted interest in signal processing algorithms that can mitigate interference and achieve increased capacity. Furthermore, recent results from information theory considerations also show that successive interference cancellation with single-user decoding can achieve all points in the capacity region of an additive white Gaussian noise (AWGN) channel [3]. Another approach to interference mitigation is multi-user detection [4], where signals of all users in the system are jointly decoded. The significant capacity gains achieved by such interference cancellation and multi-user detection algorithms makes them very attractive candidates for baseband receiver processing in 3G systems. These improved receivers that have been proposed over the conventional receiver have often been accompanied with an increase in complexity that has raised questions about the feasibility of these techniques for practical implementation. There is a wide range of performance/complexity combinations possible [4–6 and references therein].

The main impediment to proving the feasibility of multi-user detection in real systems has been the disparity in processing speeds afforded by current digital signal processor (DSP) technology and that required by the detection/estimation algorithms. While the processing speeds of DSP technology are increasing quite significantly, the requirements of 3G systems are ever pushing the envelope of signal processing

This work is supported in part by ONR under Grant No. 313/96/0211 and by the NSF under Grant No. EEC-9714727.

[1] *Japan's standardization body, the Association for Radio Industry and Business (ARIB), and the European Telecommunications Standards Institute (ETSI) each agreed on the common air interface in 1998 [1, 2]).*

Reprinted from *IEEE Personal Communications Magazine,* Vol. 6, No. 4, pp. 26–34, August 1999.

capabilities required. These include tasks such as (but not necessarily limited to) acquisition of high-data-rate signals, channel estimation for frequency-selective fading channels, fast signal quality estimation algorithms to be used in power control, and optimum combining of signals for diversity gains in space and time. Customized ASIC designs for interference canceling receivers that seem like a feasible option in the near future encounter problems of flexibility [7]. Flexibility in 3G wireless systems is desirable not only for compatibility with other systems/standards, but also in terms of QoS requirements. While radio resource management algorithms (e.g., power control and call admission control) can provide certain levels of variability in QoS, an alternate approach is to use reconfigurable radio architectures to provide diverse QoS guarantees. Ideally, a combination of receiver flexibility and integrated radio resource management may offer the widest range of QoS guarantees. At WINLAB, we have been addressing the issues of both signal processing complexity and reconfigurability through a software-defined radio (SDR) testbed for interference cancellation.

SDRs were historically considered candidates for military applications where the focus was on the flexibility to communicate over several different radio frequency (RF) bands and possibly different air interfaces. With the recent interest in commercialization of software radios for 3G systems [8], the term SDR has now come to stand for more than just that. While compatibility between standards is still attractive, SDRs are shaping up to be software and hardware reconfigurable radios in the RF, intermediate frequency (IF), as well as baseband processing stages [9–12]. The wide range of wireless system components that require this flexibility ranges from multimode multiband handsets to reconfigurable base stations. Our work on SDR at WINLAB to date has been primarily on baseband signal processing architectures for use at the base station. WINLAB has been developing implementation proofs for nonlinear interference cancellation and linear multi-user detection via a signal processing dichotomy into two core technologies, FPGA and DSP devices [12–15]. The attractiveness of this approach is that it simultaneously handles the complexity issue of signal processing algorithms as well as the dynamic reconfiguration of receivers to achieve variable QoS requirements.

Interference Mitigation Algorithms

While it may seem that the partitioning of functionality leads to rigid architectures, it is possible to achieve dynamic reconfigurability among a wide class of interference cancellation schemes [12, 14, 15]. We consider the implementation of the following receivers using the mixed FPGA and DSP approach:

- The matched filter receiver (MF)
- The successive interference canceler (SIC) [16]
- The approximate decorrelator (AD) [17]
- The exact decorrelator(DC) [18]
- The minimum mean squared error detector (MMSE) [19]

Using a "modified" matched filter approach, dynamic reconfigurability among these receiver structures can be achieved via a software radio implementation.

For simplicity in illustration, consider the case of synchronous reception of K users in a system. In this case, the output of the matched filter for the kth user can be written as

$$y_k = A_k b_k + \sum_{j \neq k} A_j b_j \rho_{jk} + n_k, \qquad k = 1, \cdots, K, \tag{1}$$

where A_k, and $b_k \in \{-1, 1\}$ are the amplitude and bit of the kth user, respectively, and $n_k = \sigma \int_0^T n(t) s_k(t) dt$, with $s_k(t)$ being the signature waveform of the kth user which is assumed to have unit energy. The cross-correlation between the signature waveforms is defined as

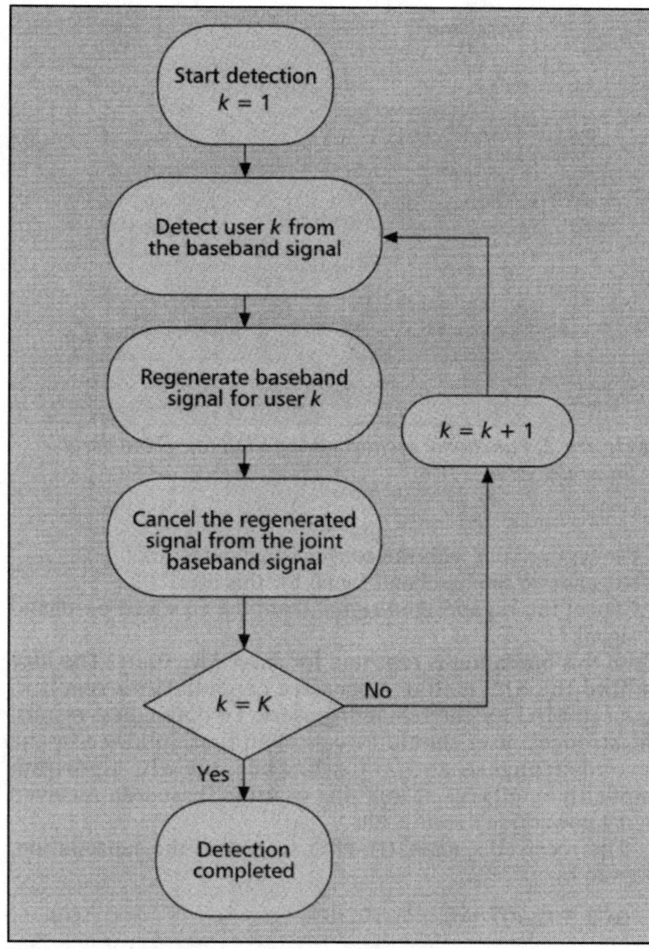

Figure 1. *The flow chart of an SIC scheme.*

$$\rho_{jk} = \int_0^T s_j(t) s_k(t) dt. \tag{2}$$

Let $\mathbf{R}$, the matrix of crosscorrelations among the user pairs, be such that its ijth entry is ρ_{ij}, and let $\mathbf{A} = \text{diag}\{A_1 \ldots A_K\}$ be a diagonal matrix of the users' amplitudes. In this case, linear detectors corresponding to MF, DC, AD, and MMSE can easily be realized for the kth user by correlating the incoming signal with a "modified" filter $h_k(t)$ [12, 14]. Specifically, the decorrelator is realized as

$$h_k(t) = \sum_{j=1}^{K} \mathbf{R}^{+}_{\mathbf{kj}} s_j(t), \tag{3}$$

where $\mathbf{R}^{+}_{\mathbf{kj}}$ denotes $(\mathbf{R}^{-1})_{kj}$. The approximate decorrelator can be realized as

$$h_k(t) = s_k(t) - \sum_{j \neq k} \rho_{jk} s_j(t) \tag{4}$$

and the MMSE detector as

$$h_k(t) = \sum_{j=1}^{K} (\mathbf{R} + \sigma^2 \mathbf{W}^{-1})^{+}_{\mathbf{kj}} s_j(t), \tag{5}$$

where $(\mathbf{R} + \sigma^2\mathbf{W}^{-1})^{+}_{\mathbf{kj}}$ denotes $[(\mathbf{R} + \sigma^2\mathbf{W}^{-1})^{-1}]_{kj}$, with $\mathbf{W} = \mathbf{A}^T\mathbf{A}$. Note that for the matched filter $h_k(t) = s_k(t)$.

The SIC scheme is based on the algorithm [16] illustrated in Fig. 1. The basic operations of the SIC algorithm are:

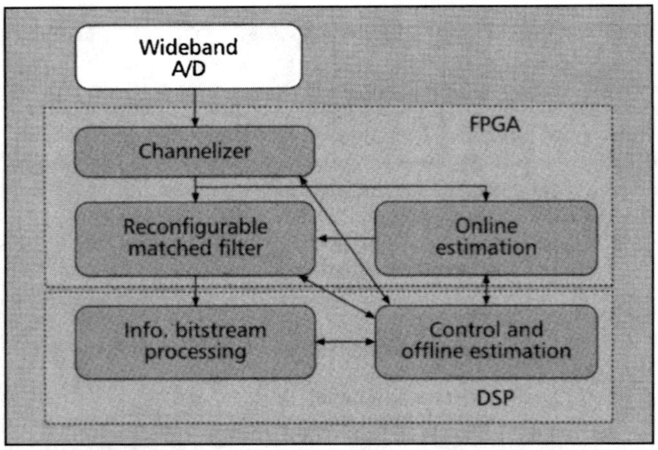

■ **Figure 2.** *Functional architecture of SDR for reconfigurable linear detection.*

- Detect one user with the conventional detector.
- Regenerate the baseband signal for this user.
- Cancel the regenerated signal from the received baseband signal.

Then this operation is repeated for successive users. The idea behind the SIC is that successive cancellations result in reduced MAI for the remaining users. To obtain best results, the strongest user should be cancelled first, followed by the second strongest, and so forth. Thus, the SIC algorithm implicitly requires a reordering of users (based on received signal powers) for best results.

The received signal $r_k(t)$, after stage k of the cancellation, is given by

$$r_k(t) = r_{k-1}(t) - A_{k-1}\hat{b}_{k-1}s_{k-1}(t), \tag{6}$$

where $r_{k-1}(t)$ is the received signal at stage k–1 and $\hat{b}_{k-1}$ is the corresponding estimate of the bit decision. Using the above recursion, it can easily be shown that the SIC algorithm can be written as matched filtering through a bank of K receivers whose outputs are

$$\begin{aligned} y_1^{SIC} &= y_1 \\ y_2^{SIC} &= y_2 - A_1\hat{b}_1\rho_{1,2} \\ y_3^{SIC} &= y_3 - A_1\hat{b}_1\rho_{1,3} - A_2\hat{b}_2\rho_{2,3} \\ y_K^{SIC} &= y_K - \sum_{i=1}^{K-1} A_i\hat{b}_i\rho_{i,K}. \end{aligned} \tag{7}$$

The above implementation of the SIC algorithm is nonlinear in that it uses hard decisions in successive stages, and we refer to this as *nonlinear successive interference cancellation* (NSIC). It is possible to realize a modified matched filter structure for the SIC [15] by replacing the bit estimates in the above set of equations with the true bit values. This results in a *linear successive interference cancellation* (LSIC) scheme where the set of equations in Eq. 7 in vector form are

$$\mathbf{y}^{\mathbf{SIC}} = (\mathbf{R^L A})^{-1}\mathbf{y}, \tag{8}$$

where $\mathbf{y}^{\mathbf{SIC}} = [y_1^{SIC} \cdots y_K^{SIC}]^T$, and $\mathbf{R^L}$ is a lower triangular cross-correlation matrix obtained from $\mathbf{R}$ by setting the elements above its main diagonal to zero. Then it follows that the SIC can be realized for the kth user (in the cancellation order) as a modified matched filter given as

$$h_k(t) = \sum_{j=1}^{K} (\mathbf{R^L A})^+_{\mathbf{kj}} s_j(t), \tag{9}$$

where $(\mathbf{R^L A})^+_{\mathbf{kj}}$ denotes $((\mathbf{R^L A})^{-1})_{kj}$. The reason for implementing the LSIC algorithm is that it facilitates reconfigurability [15]. In other words, the MF, AD, DC, MMSE, and LSIC receivers can all be implemented as appropriate modified matched filters or single-user receivers for multi-user systems.

SDR Architeture for Interference Cancellation

Figures 2 and 3 show the canonical SDR architecture for reconfigurable linear detection and nonlinear interference cancellation, respectively. The architecture includes channel processing (e.g., translation from IF to baseband), environment processing (e.g., estimation of signal and interference parameters and correlations), matched filtering, and information bitstream processing (forward error correction, FEC, or convolutional decoding, soft decisions, etc.). These functionalities are partitioned into FPGA and DSP devices according to processing speed requirements. The SDR architecture for linear detection is designed to dynamically reconfigure the (modified) matched filter according to the desired QoS (corresponding to one of the appropriate linear detectors) [12]. The SDR architecture for NSIC is designed to streamline the operation of successive cancellations among users [13].

The core of the hardware part of the SDR testbed at WINLAB (Fig. 4) is based on an APTIX MP3 board [20] with up to 12 FPGA components (up to 432,000 programmable gates) for sample-level processing tasks with rates of up to 30 MHz. Information-rate tasks (Viterbi decoding, deinterleaving, etc.) are performed by the VMEbus-based Pentek 4270 Quad TMS320C40 floating point DSP processor board [21]. For fixed point implementations, a Pentek 4290 Quad TMS320C6201 fixed point DSP processor board is used. As a baseband front-end, the testbed has two sets of dual (I and Q) analog-to-digital converters for input (Analog Devices AD9762XR 12-bit 41 MHz converters [22] and PENTEK 6472 10-bit 70 MHz converters [23] and one set of digital-to-analog converters (Analog Devices AD9042ST 12-bit 100 MHz [24]). Analog Devices converters are connected to

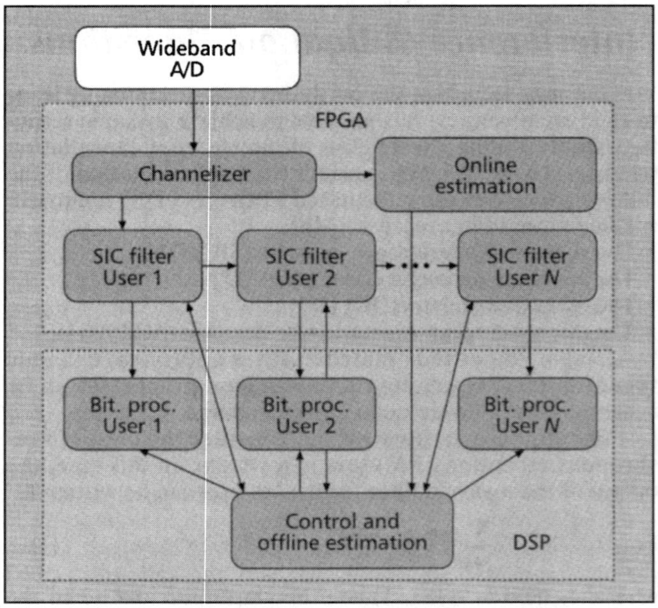

■ **Figure 3.** *Functional archictecture of SDR for nonlinear interference cancellation.*

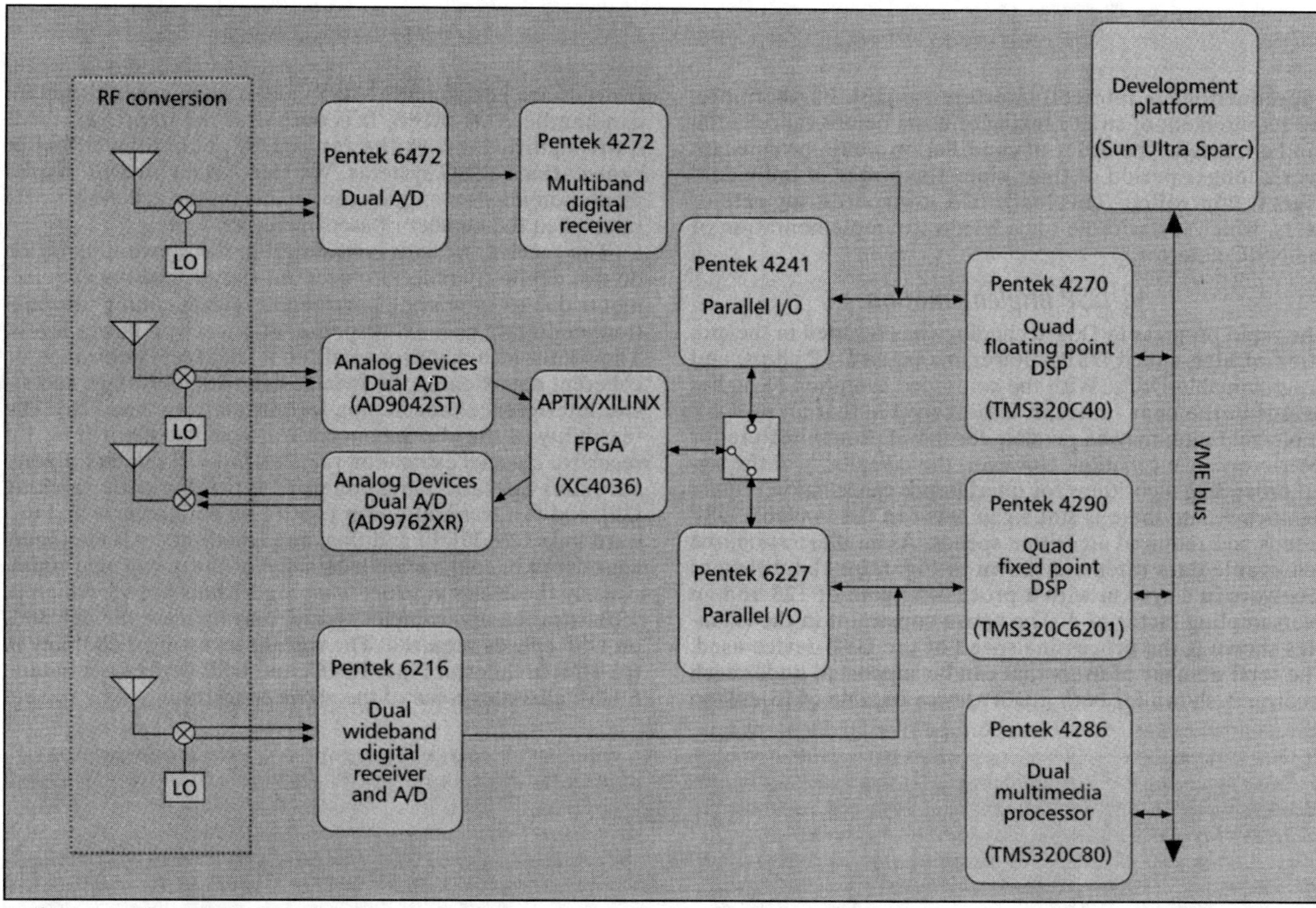

■ **Figure 4.** *The testbed block diagram.*

the FPGA board through the custom adapters, while PENTEK converters are connected to the DSP via a Multiband Digital Receiver (PENTEK 4272) [25] acting as a *channelizer* in Figs. 2 and 3. In addition, the fixed point DSP board has a separate dual wideband digital receiver and A/D converter board (PENTEK 6216). The channelizer performs frequency downconversion, low-pass filtering, and decimation of the sampled baseband signal. It is used for selection of the service bandwidth (i.e., the *tuning band*) from among those available in the sampled signal. The multiband digital receiver used as a channelizer has two narrowband receivers with dynamic range of 1 kHz–1 MHz and one wideband receiver with dynamic range of 2 MHz–35 MHz. Thus, it is capable of supporting a wide range of output sample rates. The wideband digital receiver features two channels with input bandwidths of up to 30 MHz.

Logical Partitioning of the Architecture

Figure 5 shows the necessary operations required by each type of detector among the MF, LSIC, NSIC, AD, DC, and MMSE receivers. All of the above receivers require two common generic operations, namely, estimation of path delays of the users and generation of pseudo-noise (PN) sequences $\{s_k(t)\}$ of the users in the system. The MF is probably the simplest in that for any user k, it just uses the information from these two generic operations in determining the timing offset of the PN sequence $s_k(t)$ for the specific user. For the AD, the complexity is slightly higher than that of the MF in that the modified matched filter taps are adjusted according to the formulation in Eq. 4. As seen in Fig. 5, the additional functionality required here is the cross-correlation values $\{\rho_{kj}\}$ and the signature sequences of all the users $\{s_k(t)\}$. For the LSIC receiver, the complexity involves ordering users according to received power strengths followed by a computation of the lower triangular matrix $\mathbf{R_L}$ and then subsequent inversion of the matrix $\mathbf{R^L A}$ (Eq. 9). From Eq. 3 we see that the functional operations required for the DC are additionally the computation of the inverse matrix of crosscorrelations $\mathbf{R}^{-1}$ and the column vector corresponding to the kth user (i.e., $\mathbf{R}_{kj}^{+}$). The MMSE receiver of Eq. 5 incurs additional complexity over the DC in that it also requires estimates of the received signal powers of the users in the system.

For the NSIC receiver, the relevant operations include estimation of path delays of the users, generation of PN sequences, and amplitude (power) estimates. This information is needed in order to regenerate the baseband signal before cancellation in successive stages. The amplitude estimates could be obtained from the linear correlator output of the conventional detector. This has been shown to be a sufficiently good estimate, even though better estimates can be obtained by using more advanced signal processing algorithms or even simple averaging over symbols. We do not explicitly address the issue of estimating the path delays or received signal powers of the users in the system. However, the SDR architecture presented here is versatile in that it can easily allow a variety of signal processing algorithms to be implemented for accomplishing the required estimation. The detection and cancellation procedure is repeated until as many users as necessary are cancelled using the SIC scheme. The remaining users are then demodulated by a bank of conventional detectors. The users need to be ranked in order of received powers before initiating cancellation. Power control can be integrated with NSIC to obtain potential reductions in intercell interference [13]. Furthermore, the requirement of strict ranking of users before cancellation can be relaxed. The order of cancellation would be constant over a longer period of time, since the power of individual users is controlled. This leads to a low reordering rate of users, which is desirable for a hardware implementation of the NSIC detector.

All-DSP Implementation

The rapid progress in DSP technology has resulted in the creation of high-speed DSPs, power-managed DSP chips, and programmable DSPs. With the continued evolution to higher speeds in the near future, it is conceivable that an all-DSP implementation may be possible for the SDR architecture for interference cancellation. However, the complexity of the signal processing algorithms for interference cancellation is quite significant, and there is still a gap between the available DSP speeds and required algorithm speeds. As an illustration, the achievable data rates are shown in Fig. 6 for the different receivers in a system with a processing gain of 128 and an oversampling factor of 4. The active constraint in the examples shown is the processing speed of the DSP device used. The total number of users that can be supported under each receiver is shown for both a DSP device capable of 25 million instructions per second (MIPS) and a more recent one that can handle 2000 MIPS. It is seen that the improvement is quite significant with the faster DSP, but still well below expectations of 3G systems. We see that an all-DSP implementation imposes serious constraints on the achievable data rates when the number of users increases.

The receiver designs considered in the above illustration do not explicitly take into account the processing requirements due to estimating parameters such as timing information, multipath channel response, and received power levels. The solutions being proposed for WCDMA systems include coherent detection on the reverse link with multistage successive interference cancellation for multirate systems [26]. The feasibility of the above concept requires fast algorithms for recursive channel estimation [27, 28], fast cell search for asynchronous operation and coherent spreading-code tracking [29], and fast transmit power control on both reverse and forward links [26, 30]. In addition, antenna diversity is also being considered in conjunction with some of the above algorithms. Clearly these signal processing algorithms are very high in sophistication and complexity, and only increase the demands on DSP speeds required. The signal processing dichotomy in the SDR architecture into FPGA and DSP devices is a natural fit that alleviates some of the above concerns.

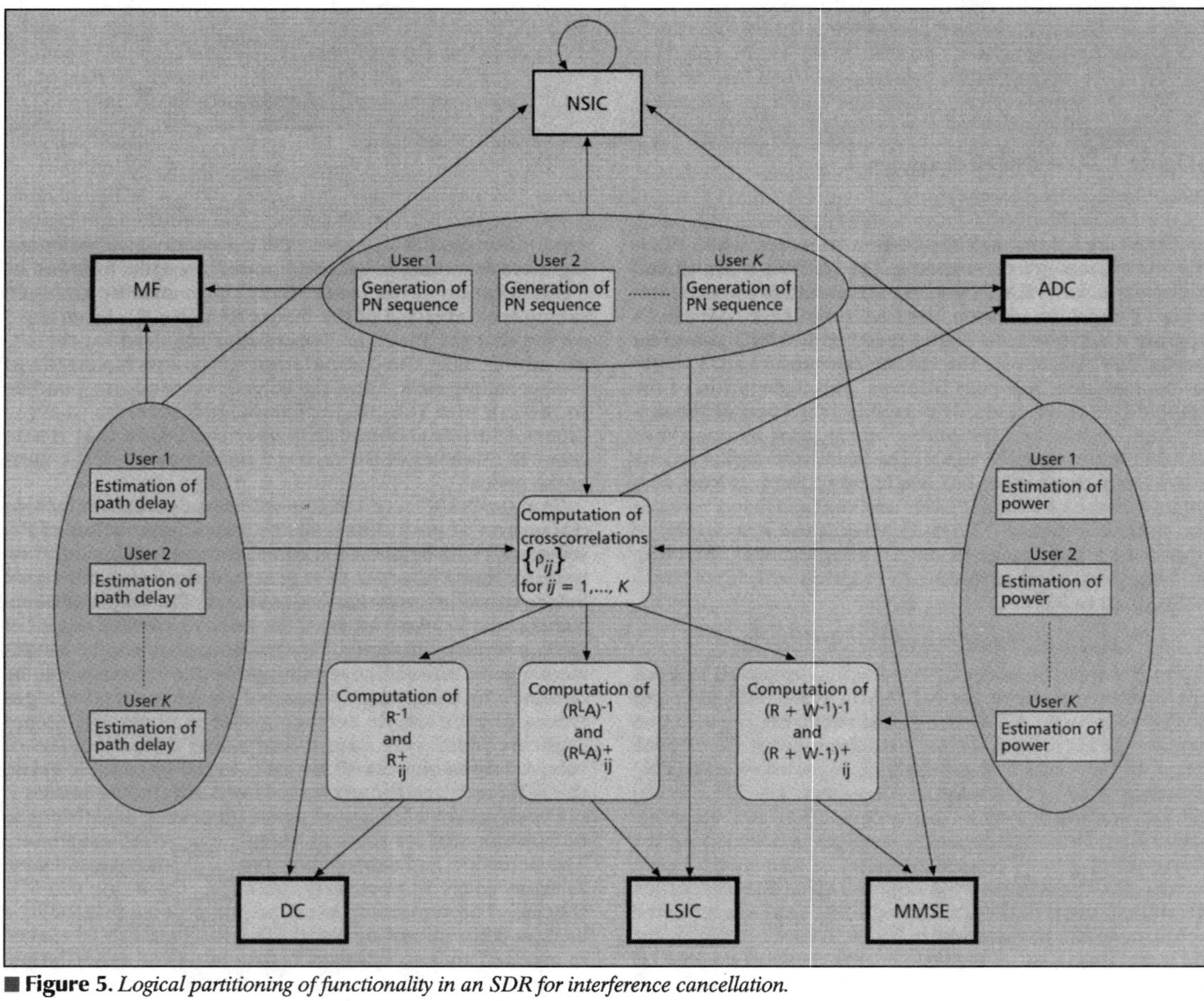

■ **Figure 5.** *Logical partitioning of functionality in an SDR for interference cancellation.*

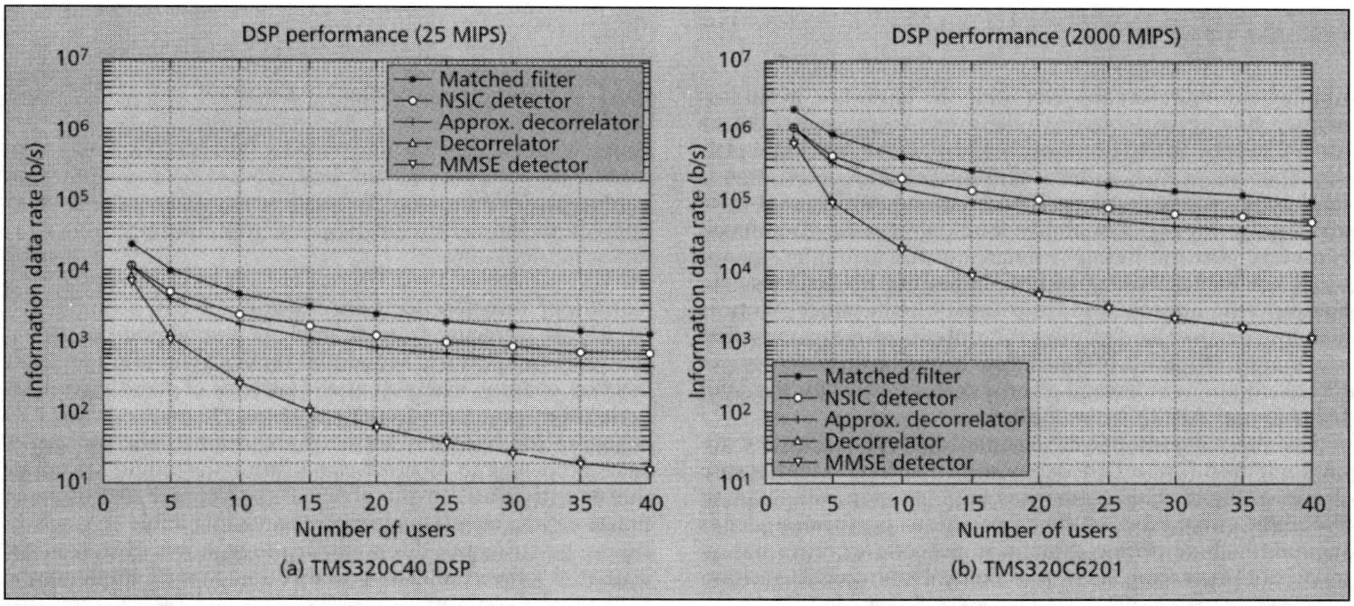

■ **Figure 6.** *Achievable rates vs. number of users.*

Resource Partitioning of the Architecture

Reconfigurable Linear Detection – The reconfigurability of linear multi-user receivers allows for the integration of multimedia services over wireless channels with variable QoS requirements. While radio resource management algorithms can provide certain levels of variability in QoS guarantees, reconfigurable radio architectures can also provide diverse QoS guarantees ranging over several orders of magnitudes in terms of BER requirements [12].

The SDR architecture for reconfigurable linear detection is implemented by partitioning the resources between FPGAs and DSPs, as shown in Fig. 7. The FPGA segment of the architecture includes the PN sequence generators, two reconfigurable blocks (which determine the filter taps of the appropriate linear receivers), an online estimator module, and the correlator. The motivation for the particular selection of constituents making up the FPGA segment is that the functionality provided by each constituent here can easily be handled by the processing speeds of the FPGA hardware. The two reconfigurable blocks contain the core of the sample level processing for each of the detectors and are actually implemented in separate FPGA components to facilitate on-the-fly reconfigurability.[2] This particular partitioning enables us to reconfigure one of the blocks while the other is

[2] *The above duplicity may be avoided in the XC62xx XILINX FPGA family, which has in-system reprogrammability, allowing sections of the device to be reprogrammed without disturbing circuits running in other sections.*

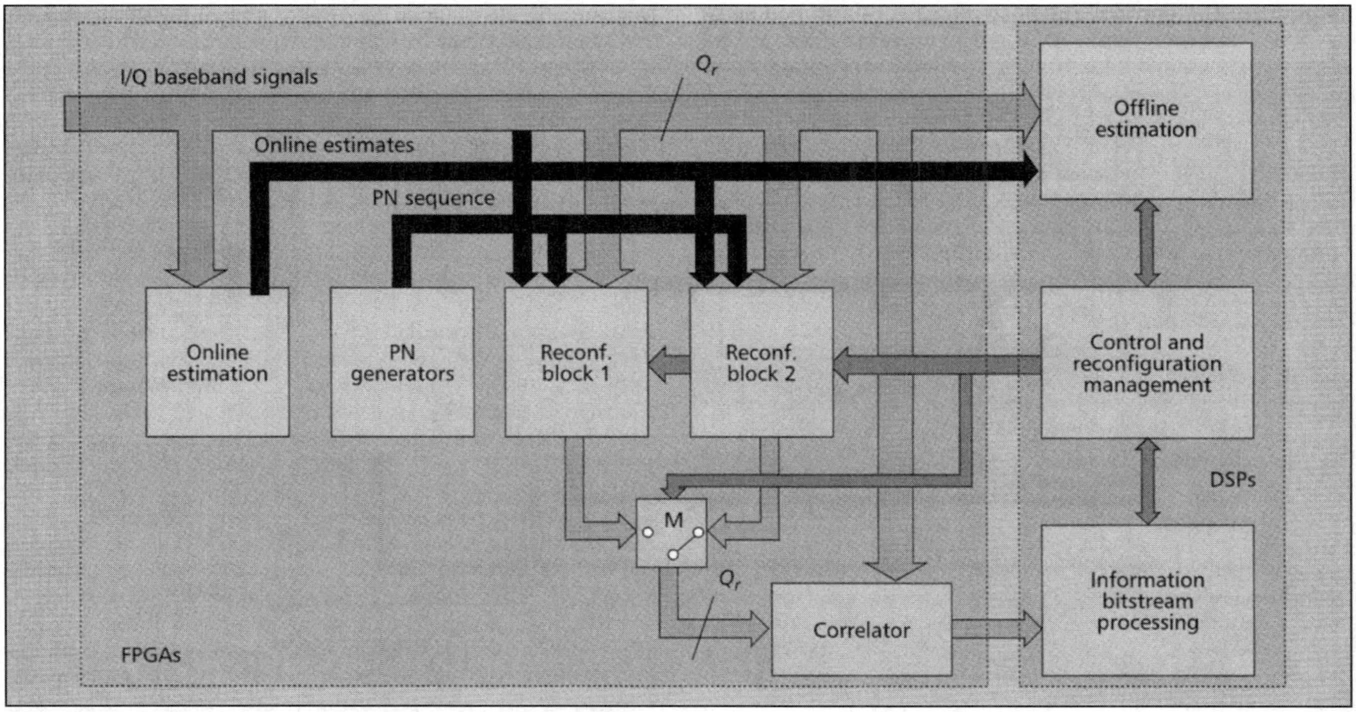

■ **Figure 7.** *SDR implementation for recongifurable linear detection.*

running and therefore, by oversizing the hardware, avoid loss of data during switchover to different receiver structures [14]. The rest of the sample level processing logic (PN sequence generators, control multiplexer, and correlator) is implemented in a separate FPGA component since it does not require reprogrammability. The online estimation block is used to perform timing estimates of the incoming signals, which are then fed along with the reference PN sequences to compute the appropriate filter taps. Furthermore, there is also a provision for refining the online estimates by using more sophisticated offline algorithms. Additionally, an FPGA component is used as an interface between the DSP segment and APTIX board [20].

The more algorithmically complex operations required are implemented in the DSP segment of the SDR architecture shown in Fig. 7. This segment is implemented using a quad TMS320C40 DSP board. The operations performed in this segment include offline estimation procedures, information bitstream processing, as well as control and reconfiguration management. The offline estimation block includes estimation of powers of the received signals (for the MMSE detector), and computation of the inverse of the matrix of cross-correlations(in the case of both the decorrelator and MMSE detector). Furthermore, it can also refine estimates of the online estimation block in the FPGA segment. The information bitstream processing functions performed vary from error control techniques, such as FEC or convolutional decoding, to soft decision decoding.

The control and reconfiguration management block in the DSP segment of the architecture basically determines the variable QoS that can be achieved by dynamic reconfiguration of the receiver structures in the FPGA segment. The reconfiguration of the blocks in the FPGA segment is directly controlled by the control and reconfiguration management block. The subsequent switching among receivers is achieved by a control multiplexer M (shown in Fig. 7) that is also controlled by the control and reconfiguration management block. Thus, the DSP segment of the device acts to achieve the variable QoS requirements of the specific type of service. Implicit in the control and reconfiguration management block is also the notion of sensing the QoS demand of a specific data stream as well as the actions that encompass environment processing such as sensing interference levels. The reader should note that the DSP devices have access to the configuration bitstreams for each of the receiver structures and can download the particular configuration bitstream as and when required.

Nonlinear Interference Cancellation – NSIC schemes [16, 31, 32] are attractive from an implementation point of view in that they can perform sequential replica generation and subtraction of MAI. The potential drawback of these algorithms is that they require extremely accurate channel estimates for regeneration and subtraction. Furthermore, the ranking of users according to received signal strengths requires accurate power estimates. In the absence of accurate channel and power estimates, these algorithms are susceptible to degradation in performance due to error propagation. However, the potential gains that can be had from accurate implementations [3, 33] make these worthwhile candidates for 3G systems [34].

The SDR architecture for NSIC [13] is implemented by partitioning the resources between FPGAs and DSPs, as shown in Fig. 8. Although the NSIC gain is nontrivial and worth pursuing, a requirement for very low complexity still exists. The approach taken in [13] was to view the NSIC scheme as an extension to the existing conventional detector, thus minimizing the number of modifications. Due to the issue of backward compatibility to second-generation systems, it is desirable for the NSIC implementation that it be possible (if necessary) to support the same bit rate and modulation scheme as the conventional detector. The choice of implementation in [13] was a *pipelined structure* shown in Fig. 8 for implementing the NSIC scheme.

The structure consists of a chain of SIC filters. Each of these filters adaptively detects and cancels a specific user from its input signal. The filter is matched to a user corresponding to the applied PN sequences (PNI and PNQ in Fig. 8) and estimated path delays (TAU in Fig. 8). The output of the SIC filter is essentially the same as the input signal, but with the current

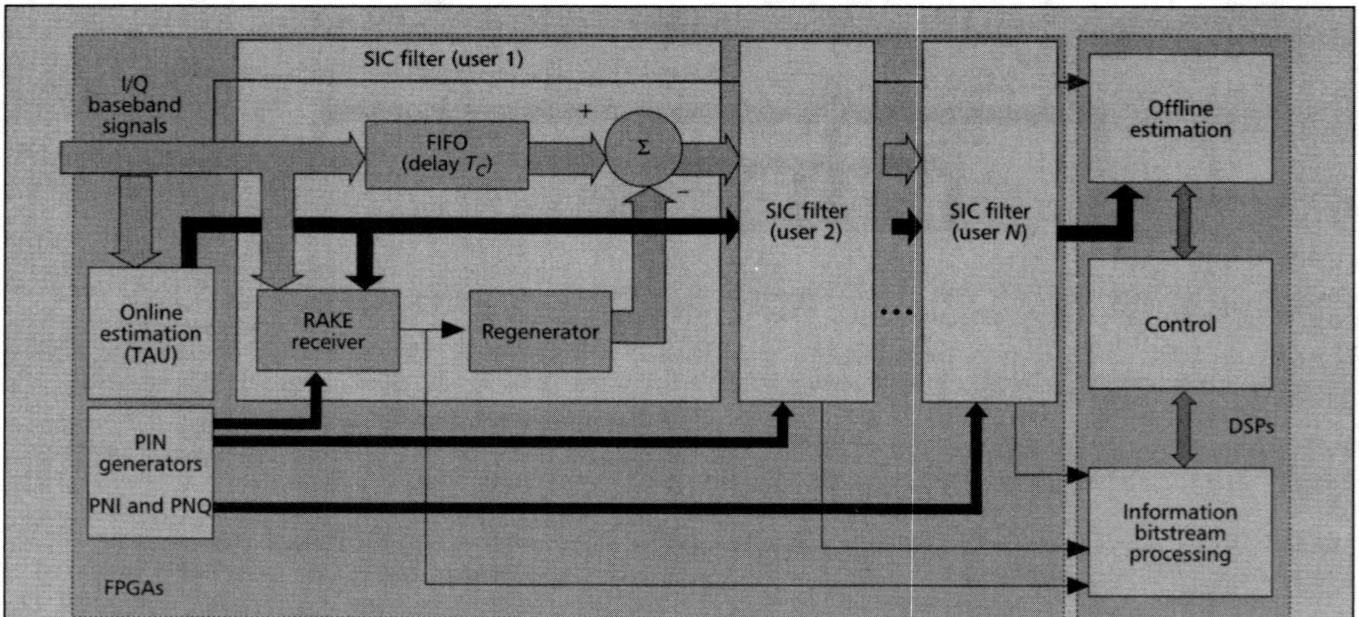

■ **Figure 8.** *SDR implementation for nonlinear interference cancellation.*

Quantization	MF	NSIC	AD	LSIC, DC, and MMSE
$Q_r = 4$ bits	25	36	$31 + 13K$	$37 + 3K(K-1)$
$Q_r = 5$ bits	27	44	$40 + 14K$	$47 + \lceil 13/4 \rceil K(K-1)$
$Q_r = 6$ bits	29	57	$57 + 15K$	$64 + \lceil 7/2 \rceil K(K-1)$

Table 1. *The hardware complexity of various cancellation schemes.*

user cancelled. It is seen that each SIC filter operates on both an I and Q baseband branch due to noncoherent reception (as in IS-95). From the description of the NSIC scheme it should be recalled that the first SIC filter should be matched to the strongest user and so forth.

Because the NSIC scheme with symbol averaging is based on hard decisions, a symbol estimate is necessary before the corresponding user can be cancelled. This implies that each SIC filter must complete its correlation before it can initiate cancellation. This introduces a delay through each SIC filter, denoted T_c. This delay corresponds to the symbol time plus the multipath delay spread because all multipath signals must be received before a reliable estimate can be made. It is seen that the I and Q input signals are passed to a RAKE receiver. The RAKE receiver is responsible for performing multipath combining and estimating the transmitted symbol. The symbol estimate and the correlator output specifying the amplitude of the user are next used by the *regenerator* to recreate the I and Q baseband signals of the current user. The detection of a symbol has a duration of T_c before sample-by-sample subtraction can be initiated with the regenerated signals. For delaying the input signals a first-in first-out (FIFO) buffer is used as shown in Fig. 8. From a complexity point of view it is convenient to have a fixed delay in the FIFO buffers since variable delays are more difficult to implement in hardware. A fixed delay is obtained by setting T_c equal to the symbol time plus the maximum expected delay spread of the channel.

From considering the pipelined structure it is seen that each SIC filter operates independent of the others, and both synchronous and asynchronous transmission are supported by the structure. All filters operate simultaneously, leading to full utilization of the hardware. By introducing the detection delay, the processing time requirements to the RAKE receivers remain unchanged. With the regenerator implemented in real time, the proposed structure can support the same bit rate as the conventional detector. However, the increased detection delay may set a limit on how many cancellations can be performed. As the symbol rate increases this potential limitation is relaxed.

A disadvantage of the pipelined structure appears when considering the dynamic features. As mentioned earlier, the performance of the SIC scheme is based on canceling the strongest users first. However, with the pipelined structure it is difficult to reorder users since the SIC filters are operating on samples with different delays. To illustrate this, consider what happens if the weakest user suddenly becomes the strongest user. This user should then be moved up in front of the filter chain. If this is done instantaneously all the samples stored in the SIC filters located in between would be lost. By using a power control scheme in conjunction with cancellation [13], the power ranking is greatly stabilized and the reordering rate reduced. The regeneration process contains a set of parallel multiplications and additions. The desired simplicity and static behavior of the regenerator further suggest an implementation in dedicated hardware.

Hardware Complexity

In Table 1 we show the relative hardware complexity[3] of the different cancellation schemes in terms of the required number of configurable logic blocks (CLBs). A CLB, in a XILINX 4000 series FPGA, comprises a pair of flip-flops and two independent (Boolean) logical four-input function generators [35]. As can be seen, the complexity for each detector increases with both increasing number of users K and increasing precision in quantization (Q_r denotes the number of bits used in quantization). As expected, the matched filter complexity remains invariant to the number of users in the system and depends only on the precision of quantization.

Even though the decorrelator (and MMSE) achieves better QoS, it comes at the expense of an exponential increase in complexity (with increasing number of users) over the SIC[4] and AD detectors. This directly maps to an increase in processing power requirements of the FPGA segment of the SDR architecture. Switching to lower-order receivers when the QoS requirements are moderate has been considered [14]. The complexity of operations required to be performed in the DSP segment of the architecture involve matrix inversion and offline estimation of amplitudes. The complexity trade-offs for linear multi-user schemes in terms of add-multiply floating point operations can be found in [12, 26].

Conclusions and Future Directions

Provisioning of services in third-generation wireless systems require flexible and versatile air interfaces. SDR architectures promise to bring this capability for 3G systems such as WCDMA by making possible implementation of sophisticated interference mitigation algorithms. Our studies at WINLAB on the feasibility of such a concept show that it is possible to achieve the flexibility and versatility required of air interfaces by using a combination of DSP with a reconfigurable hardware processing subsystem. The key idea is to enable the offload into hardware of processing (speed) intensive operations. Achieving the appropriate choice of digital architecture is a trade-off between power consumption, flexibility in performance, and cost. Experimental results on the performance and reconfigurability of the SDR architecture at WINLAB show that diverse services can be simultaneously supported by flexible radio interface design. The success of nonlinear interference cancellation schemes requires accurate channel estimation and ordering of users according to relative received signal strengths. Future work will include studying the feasibility and constraints of implementing sophisticated estimation algorithms for use in conjunction with detectors. Also of great interest is the implementation study of blind adaptive multiuser detectors, such as for use in unlicensed bands.

References

[1] F. Adachi, M. Sawahashi, and H. Suda, "Wideband DS-CDMA for Next-Generation Mobile Communication Systems," *IEEE Commun. Mag.*, vol. 36, no. 9, Sept. 1998, pp. 56–68.

[2] T. Ojanpera and R. Prasad, "An Overview of Third-Generation Wireless Personal Communications: A European Perspective," *IEEE Pers. Commun.*, vol. 5, no. 6, Dec. 1998, pp. 59–65.

[3] *The complexity estimates for NSIC given in Table 1 are based on the implementation described in [13], which has much lower complexity than LSIC, the modified-matched-filter-based implementation (which has the same complexity as DC and MMSE detectors).*

[4] *Note that the SIC receiver requires explicit ordering of users according to their relative received powers. This is not accounted for in Table 1.*

[3] B. Rimoldi and R. Urbanke, "A Rate-Splitting Approach to the Gaussian Multiple-Access Channels," *IEEE Trans. Info. Theory*, 42, no. 2, Mar. 1996, pp. 364–75.
[4] S. Verdu, *Multi-user Detection*, New York: Cambridge Univ. Press, 1998.
[5] A. Duel-Hallen, J. Holtzman, and Z. Zvonar, "Multi-User Detection for CDMA Systems ," *IEEE Pers. Commun.*, Apr. 1995, pp. 46–58.
[6] S. Moshavi, "Multi-User Detection for DS-CDMA Communications," *IEEE Commun. Mag.*, Oct. 1996, pp. 124-36.
[7] M. Cummings and S. Haruyama, "FPGA in the Software Radio," *IEEE Commun. Mag.*, vol. 37, no. 2, Feb. 1999, pp. 108–12.
[8] J. Mitola, "Technical Challenges in the Globalization of the Software Radio," *IEEE Commun. Mag.*, vol. 37, no. 2, Feb. 1999, pp. 84–89.
[9] H. Tsurumi and Y. Suzuki, "Broadband RF Stage Architecture for Software-Defined Radio in Handheld Terminal Applications," *IEEE Commun. Mag.*, vol. 37, no. 2, Feb. 1999, pp. 90–95.
[10] R. H. Walden, "Performance Trends of Analog-to-Digital Converters," *IEEE Commun. Mag.*, vol. 37, no. 2, Feb. 1999, pp. 96–101.
[11] D. B. Chester, "Digital IF Filter technology for 3G Systems: An Introduction," *IEEE Commun. Mag.*, vol. 37, no. 2, Feb. 1999, pp. 102–7.
[12] I. Seskar and N. B. Mandayam, "A Software Radio Architecture for Linear Multi-user Detection," *IEEE JSAC* Special Issue on Multimedia Network Radios, May 1999, pp. 814–23.
[13] K. Pedersen *et al.*, "Practical Implementation of Successive Interference Cancellation in DS/CDMA Systems," *Proc. ICUPC '96*, Cambridge, MA, Sept. 1996, pp. 321325.
[14] I. Seskar and N. B. Mandayam, "A Software Radio Architecture for Linear Multi-User Detection," *Proc. CISS '98*, Princeton, NJ, Mar. 1998.
[15] I. Seskar, N. B. Mandayam, and P. Hatrack, "Reconfigurable Testbed for Interference Cancellation in DS-CDMA Systems," *Int'l. J. Wireless Networks*, to appear.
[16] P. Patel and J. M. Holtzman, "Analysis of a Simple Successive Interference Cancellation Scheme in DS/CDMA System," *IEEE JSAC*, vol. 12, no. 5, June 1994, pp. 796–807.
[17] N. B. Mandayam and S. Verdu, "Analysis of an Approximate Decorrelating Detector," *J. Wireless Pers. Commun.*, vol. 6, Jan. 1998, pp. 97–111.
[18] L. Lupas and S. Verdu, "Linear Multi-User Detectors for Synchronous Code-Division Multiple Access Channels," *IEEE Trans. Info. Theory*, vol. 35, no. 1, Jan. 1989, pp. 123–36.
[19] U. Madhow and M. Honig, "MMSE Interference Suppression of Direct-Sequence Spread-Spectrum CDMA," *IEEE Trans. Commun.*, vol. 42, no. 12, Dec. 1994, pp. 3178–88.
[20] Aptix Corporation, FPCB Development System documentation, 1993.
[21] PENTEK, Inc. *Operating Manual: MODEL 4270 Quad TMS520C40 DSP Board*, 1995.
[22] Analog Devices, Inc., Data Sheet: AD9762* 12-bit, 100 MAPS TODAY D/A Converter, 1996.
[23] PENTEK, Inc. *Operating Manual: MODEL 6472 Dual Channel, 70 MHz, 10-bit, A/D Converter*, 1996.
[24] Analog Devices, Inc., Data Sheet: AD9042 12-bit, 41 MSPS Monolithic A/D Converter, 1996.
[25] PENTEK, Inc. *Operating Manual: MODEL 4272 Multiband Digital Receiver*, 1996.
[26] F. Adachi and M. Sawahashi, "Wideband Wireless Access based on DS-CDMA," *IEICE Trans. Commun.*, vol. E81-B, no. 7, July 1998, pp. 1305–16.
[27] M. Sawahashi *et al.*, "Pilot Symbol-Assisted Coherent Multistage Interference Canceller Using Recursive Channel Estimation for DS-CDMA Mobile Radio," *IEICE Trans.*, vol. E79-B, no. 9, Sept. 1996, pp. 1262–70.
[28] M. Sawahashi, H. Andoh, and K. Higuchi, "Interference Rejection Weight Control for Pilot Symbol-Assisted Coherent Multistage Interference Canceller Using Recursive Channel Estimation for DS-CDMA Mobile Radio," *IEICE Trans. Fund.*, vol. E81-A, no. 52, May 1998, pp. 957–97.
[29] K. Higuchi, M. Sawahashi, and F. Adachi, "Fast Cell Search Algorithm in Inter-Cell Asynchronous DS-CDMA Mobile Radio," *IEICE Trans. Commun.*, vol. E81-B, no. 7, July 1998, pp. 1527–34.
[30] S. Seo, T. Dohi, and F. Adachi, "SIR-Based Transmit Power Control of Reverse Link for Coherent DS-CDMA Mobile Radio," *IEICE Trans. Commun.*, vol. E81-B, no. 7, Duly 1998, pp. 1508–16.
[31] M. K. Varanasi and B. Aazhang, "Multistage Detection in Asynchronous Code-Division MultipleAccess Communications," *IEEE Trans. Commun.*, vol. 38, no. 4, Apr. 1990, pp. 509–19.
[32] Y. C. Yoon, R. Kohno, and H. Imai, "A Spread-Spectrum Multi-Access System with Co-Channel Interference Cancellation for Multipath Fading Channels," *IEEE JSAC*, vol. 11, no. 7, Sept. 1993, pp. 1067–75.
[33] M. K. Varanasi and B. Aazhang, "Near-Optimum Detection in Synchronous Code-Divisions Multiple-Access Systems," *IEEE Trans. Commun.*, vol. 39, pp. 725–36, May 1991.
[34] M. Sawahashi *et al.*, "Experiments on Coherent Multistage Interference Canceller for DS-CDMA Mobile Radio," *Proc. PIMRC*, Boston, MA, Sept. 1998.
[35] XILINX, *XA CT Reference Guide*, 1994.
[36] R. M. Buehrer, S. N. Correal, and B. D. Woerner, "A comparison of multi-user receivers for cellular cdma," *Proc. 1996 IEEE GLOBECOM*, London, U.K, Nov. 1996, pp. 1571–77.

Chapter 4

Software-Defined Radio Emerging Technologies

BECAUSE software radio itself is an emerging technology, this chapter samples the new technologies related to software radio that represent either different directions or particularly challenging areas. This chapter does not cover all challenging areas—wideband highly linear RF is notably absent. But it does cover important emerging areas in smart antennas, networking, digital preprocessing, and software.

Smart antennas digitally combine antenna channels to adaptively form beams and point nulls and equalize the received signal. Kennedy and Sullivan provide a tutorial for those with little background in this area, and Razavilar et al. extend the tutorial to algorithms and traffic engineering aspects. Evans applies beamforming principles to the transmitter, mitigating the asymmetry that smart antennas can introduce.

Hentschel and Fetweiss address the critical question of providing clock references for multiple air interfaces needed for multiple SDR software personalities. Given the accuracy requirements for diverse air interfaces, the authors show that deriving multiple clocks from a single master clock has more to do with antialiasing and preserving frequency-domain properties than with time-domain interpolation. This treatment transforms the relatively well understood domains of interpolation and clock conversion into an emerging area for software radio.

Papers by Munro and Shepherd address emerging aspects of software. Shepherd sets the software issues in a deployment context. Munro critically examines the emerging needs for middleware that insulates radio applications from the rapidly evolving radio hardware platforms. His example of a Bluetooth headset controlling a UMTS/GSM core is revealing. Although CORBA and Java/JINI are the emerging middleware standards of the information technology (IT) community, the intelligent network (IN) with Signaling System 7, TINA, and IP each provide many of the capabilities now being standardized in IT, including, for example, naming service, downloads, and QoS monitoring. This emerging area will no doubt be the subject of substantial research, engineering, and international debate as these diverse middleware approaches are on a collision course in the future PDA.

The final paper by Mitola looks somewhat further ahead toward the integration of machine learning into such high-performance PDAs. Location-aware SDR PDAs capable of learning, it is shown, could rent spectrum from each other and from equally adaptive "cognitive" networks. Spectrum rental, it is argued, would substantially reduce spectrum congestion, albeit at the cost of an increase in the complexity of the handsets and infrastructure. This paper and the companion doctoral dissertation point to an interesting area of convergence in teleinformatics—the pervasive integration of communications, speech recognition, and machine learning.

Direction Finding and "Smart Antennas" Using Software Radio Architectures

Software radio architectures, originally developed for military applications, are now becoming economically viable in commercial products because of the rapid advance of DSP technology.

Joseph Kennedy and Mark C. Sullivan

Operational tests and demonstrations of systems based on software radios are currently being performed in the 800 MHz mobile cellular radio band. These field trials are for mobile unit geolocation systems and adaptive phased array "smart antenna" appliqués. The geolocation system trials are in response to the demand for high confidence geolocation of mobile units for enhanced emergency 911 service and for use in the U.S. Department of Transportation's Intelligent Vehicle Highway System (IVHS) initiative. The smart antenna array appliqués address the cellular service providers' need for more user channel capacity and/or geographic coverage from existing base station installations. Software radio architectures were selected because of their ability to provide superior performance at low life cycle cost. These systems use 4 to 8 wideband coherent channels and fully characterize the arriving RF energy to either geolocate the emitter or to maximize the carrier-to-interference (C/I) ratio.

Software Radio Background

The application of software radio techniques to military communications was conceived in the 1970s. The initial goal was to provide higher fidelity in the HF band, where over-the-horizon communications via ionospheric propagation were important. The HF band is ideally suited for testing software radio principles for two reasons. First, advanced signal processing is necessary in the HF band because of the highly dynamic nature of the channel due to ionospheric propagation and also because of the high density of signals. Second, the HF band spans only a few tens of megahertz, and generally only about 10 to 20 percent of the band will propagate over the horizon. Software radios capable of processing even hundreds of kilohertz of bandwidth may thus be effective in operating over RF bands of potential interest.

Software radio architectural concepts were reduced to practice in about 1980 in the first operational military digital HF communications system. The hardware required to implement the system filled many racks and consumed a great deal of power. The system provided simultaneous coverage of only 750 kHz of the HF band with a dynamic range of less than 60 dB. Still, the ability to employ sophisticated signal processing algorithms, even in this modest hardware configuration, proved the worth of software radio techniques.

Exponential increases in DSP technology performance coupled with reductions in size and power (Moore's Law[1]) over the last 20 years have opened up applications of software radios to line-of-sight RF bands. Throughout the '80s and '90s, these systems have been upgraded through five generations.

The first few generations of software radios were developed to provide enhanced fidelity and flexibility unavailable from analog receivers. Changing needs could be met quickly and inexpensively with software upgrades. This ability to adapt to difficult problems makes the software radio attractive in commercial telecommunications applications, where infrastructure equipment might otherwise quickly become obsolete due to cumulative double digit growth in markets, rapidly evolving air standards, and requirements for new services. The action of Moore's Law has driven the cost of the software radio down at a rate greater than other competitive technologies, providing the ultimate motivation for their use in the commercial telecommunications industry.

Application to Cellular/PCS Infrastructure

The current generation of commercial wireless products employs readily available DSP technology to digitize the RF spectrum with great fidelity and then perform all of the base station functions using software radio architectures. Most of these functions are performed by executing software on a general-purpose DSP processor, but the first stage of frequency conversion and filtering is performed by a specialized decimating

JOSPEH KENNEDY is the director of the Advanced Communications Systems group at Engineering Research Associates.

MARK C. SULLIVAN is a scientist with Engineering Research Associates.

[1] In 1979, Intel co-founder Gordon Moore noted that the achievable density of transistors on chips appeared to double every 18 months.

Reprinted from *IEEE Communications Magazine,* Vol. 33, No. 5, pp. 62–68, May 1995.

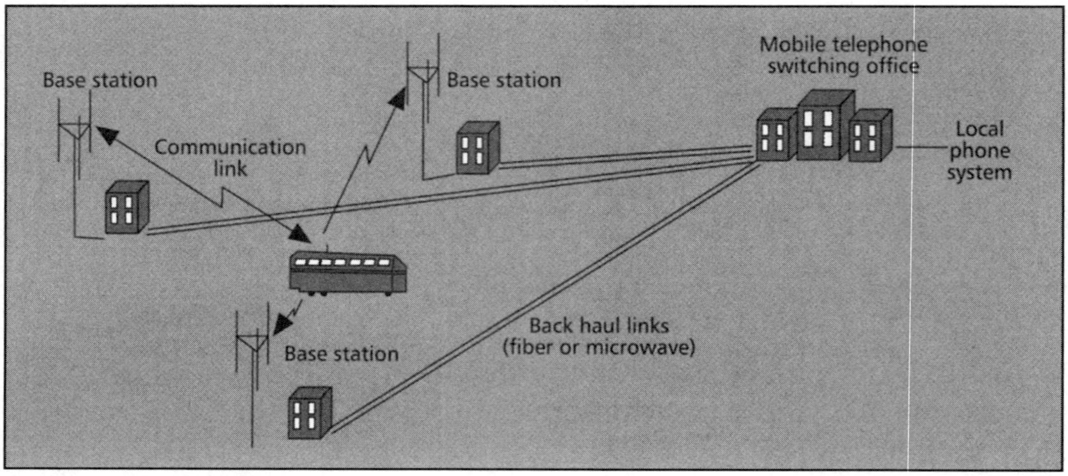

■ **Figure 1.** *Typical system layout.*

downconverter chip produced by Harris Semiconductor and others. This circuit implements a cascaded integrator-comb filter followed by a conventional transversal filter to provide computationally efficient sample rate conversion from the tens of megahertz at the A/D converter to the tens of kilohertz occupied by the signal [2].

Using software radio techniques, a common hardware platform may provide all of the current base station functions and yet retain enough flexibility to meet future requirements through alternative software application programs. The products are configured from a set of hardware and software modules, interconnected using industry standard mechanical, electrical and software interfaces. These modules may be combined to provide user-configurable equipment suites including complete Cellular/PCS macro- and microcell base stations and "Smart Antenna" appliqués. Software selectable options include domestic and international air standards, base station controller-to-switch interfaces, fixed omni/sector or "smart antennas," and new service features such as fraud detection and mobile terminal geolocation.

CAPITAL Direction Finding Operational Demonstration

One example of commercial equipment using software radio architectures is the Cellular Applied to IVHS Tracking & Location (CAPITAL) project, currently undergoing field trials in the Washington, D.C. suburbs. The concept behind CAPITAL is that automobile traffic congestion may be monitored by tracking the progress of vehicles equipped with cellular phones. This comprehensive program demonstrates all of the capabilities of such a system, beginning with the collection and processing of cellular data and proceeding to the dissemination of traffic information to remote users and in-vehicle equipment. The operational test makes extensive use of the in-place cellular infrastructure for both wide area automobile traffic congestion monitoring and communications. Geolocation equipment has been colocated at selected towers to collect cellular phone usage statistics and to geolocate phones on designated roadways.

Cellular-based wide area surveillance techniques are being evaluated for use in automobile traffic monitoring applications because of their comparative advantage over magnetic sensors buried in the roadway, video cameras, and other conventional techniques. These advantages include:

- Area coverage based on square miles, not vehicle count or road miles.
- No disruption of road service for installation or repair.
- Orders of magnitude lower in cost than loop-based approaches.
- Very high reliability with low maintenance costs.
- Other uses include fleet management and emergency assistance.
- Geolocation equipment could be owned and operated by cellular service providers.

CAPITAL geolocates mobile users of cellular radios through the reception and processing of RF signals emitted by ordinary mobile cellular telephones. In this application the content of the signal is of no interest and the privacy of the user is protected by not demodulating the audio portion of the signal and by replacing the electronic serial number with a randomly chosen identifier. Geolocation is achieved using either the Reverse Voice Channel (RVC) or the Reverse Control Channel (RECC) transmitted by the mobile "probe" vehicle. The system geolocates the emitter to within ~100 meters using a combination of techniques. The line of bearing to the emitter is estimated at each receiving site using the Alternating Projection (AP) maximum likelihood direction-finding algorithm [4, 7]. Time of arrival for each signal is also measured with respect to a GPS time reference so that time-difference of arrival (TDOA) techniques may also be employed to fix the location. If the RVC is being processed, then the frequency of the Supervisory Audio Tone (SAT) is used to determine if the received signals are from a common mobile unit, or from interfering mobile transmitters.

RF signals are received at multiple receiving sites, characterized by angle of arrival, time stamped, and then passed to a common site where the geolocation is determined. A typical layout is shown in Fig. 1.

Geolocation equipment based on software radios is located at selected towers and interconnected

The action of Moore's Law has driven the cost of the software radio down at a rate greater than other competitive technologies.

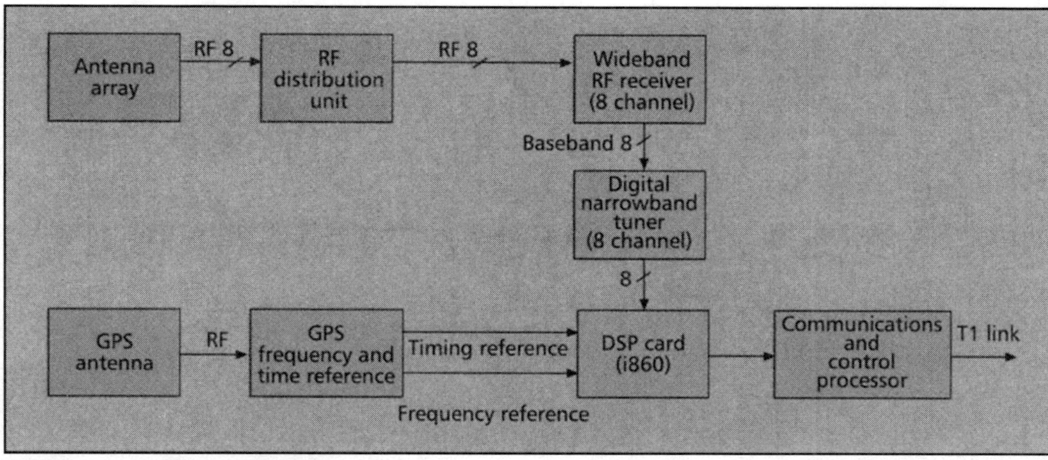

■ **Figure 2.** *Hardware block diagram.*

through existing back haul links to a control subsystem at the Mobile Telephone Switching Office (MTSO). In Fig. 1, a transit bus is shown communicating through a base station via a cellular telephone. Geolocation equipment located at one or two other sites are receiving the mobile cellular transmission and estimating a position. The algorithms employed by the system to combat multipath and interference would be all but impossible to implement without the use of software radio technology. Geolocation accuracy is enhanced by the following features:

- GPS-derived clocks provide less than 100 ns differential timing error from site to site.
- GPS-based tower siting removes any siting-based bias errors.
- Coherent multichannel time delay allows DSP-based spatial filtering to be used to combat multipath and interference effects.
- Phase is calibrated to hundredths of a degree through DSP.

A block diagram of the CAPITAL hardware is shown in Fig. 2. Signals from the eight-element antenna array are block downconverted to an IF before the entire band of interest is digitized. Eight digital tuner ASICs coherently downconvert each narrowband channel being monitored to a zero IF complex baseband format. The resulting narrowband data is then processed by Intel i860 chips in the Signal Processing Unit.

Software implemented in the "C" language enumerates and isolates the components of the received signal using the AP algorithm and spatial filtering. The SAT frequency of each component is then measured to distinguish between components of the desired signal, which will all exhibit the same frequency, and those of co-channel interference, which will generally be assigned a different frequency. The isolated common SAT signals are then cross correlated to identify the component of the desired signal that arrives earliest. The components arriving later are deemed to have resulted from multipath propagation and are discarded. The line of bearing and time of arrival of the direct-ray component are sent to a central site over a T1 link to complete the geolocation of the emitter.

The major impediment to reliable geolocation of mobile units in the cellular telephone band is severe multipath propagation. The effects of this phenomena must be mitigated in order to achieve accurate position estimates with low variance. Typical examples of the multipath encountered in the suburban Washington, D.C. operating environment were collected using instrumentation in the geolocation equipment. Specular multipath can be characterized in terms of power, delay, and angle-of-arrival relative to the direct ray.

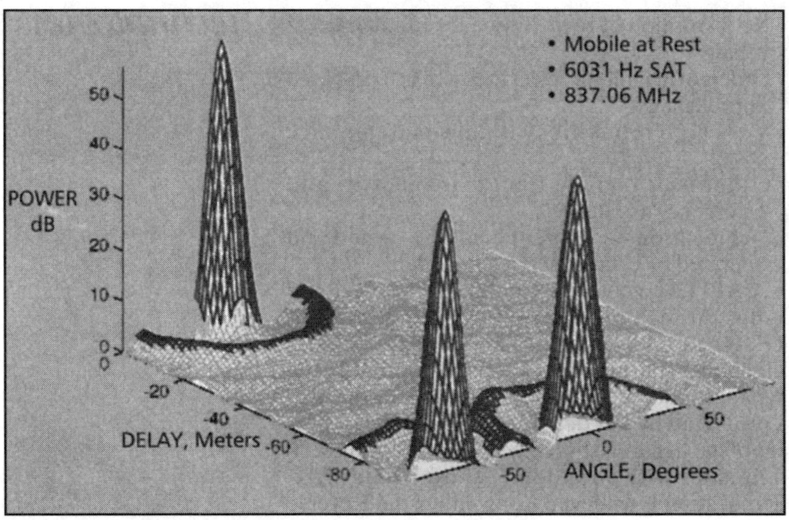

■ **Figure 3.** *PDAgram showing mulitpath.*

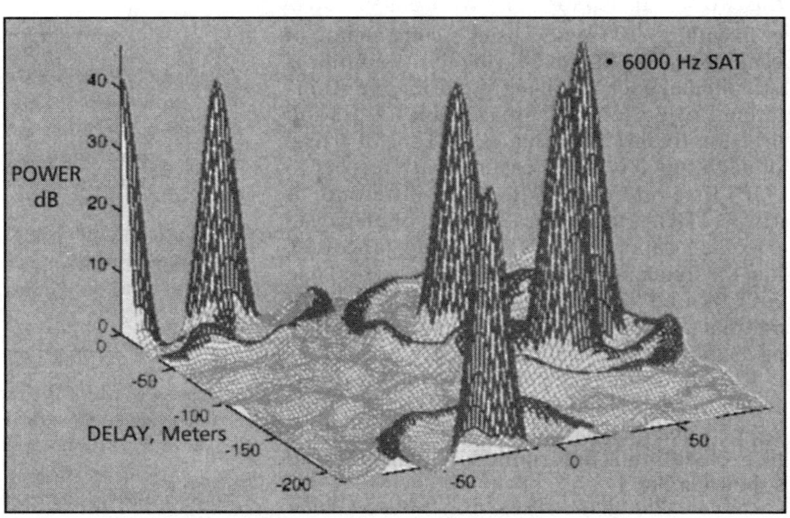

■ **Figure 4.** *PDAgram showing multipath and interference.*

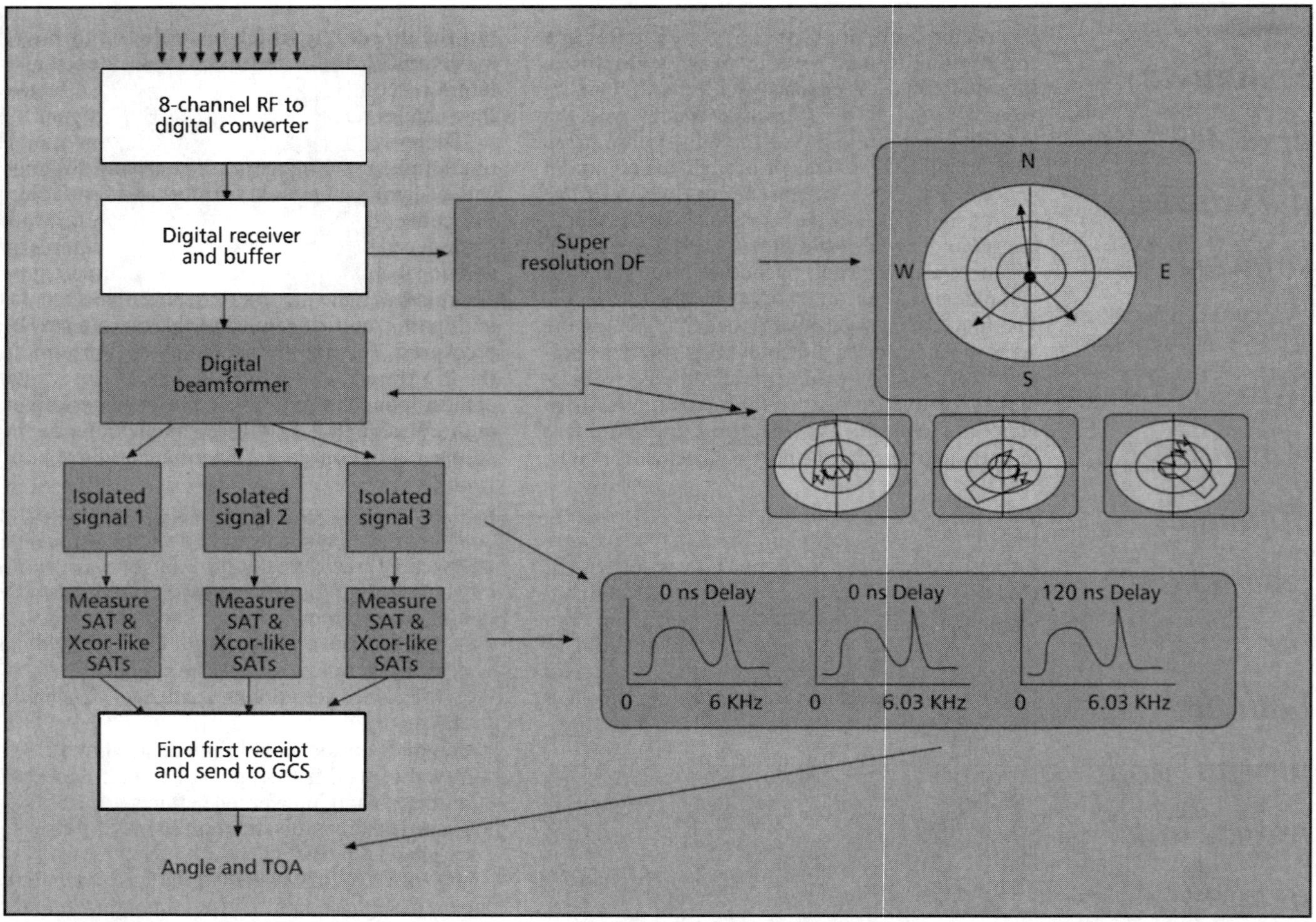

■ **Figure 5.** *PDAgram signal isolation process.*

The ideal tool to measure these parameters is the software radio, with its inherent fidelity and digital storage capability. The signal parameters may be displayed in a "PDAgram" (Power-Delay-Angle diagram) such as the example shown in Fig. 3. This PDAgram was generated from data collected by the CAPITAL system and shows three signal components, the direct ray shown on the "0" delay axis and two multipath components with differential path length delays of about 80 meters. Figure 4 shows another PDAgram obtained from a signal data record exhibiting multipath and interference. One signal exhibits a SAT frequency of 6000 Hz and the other, interfering, signal originates from a cell using a 6030 Hz SAT frequency assignment. Note that the interferer also has a multipath component present.

This record was processed to determine direction of arrival for the desired signal. The desired mobile signal multipath components, the interfering mobile (shown in a different color), and interfering mobile multipath component had to be removed to determine the correct angle and time of arrival for the "correct" direct ray. Once the signal components have been isolated, the software examines the SAT to determine which components came from interfering cellular users or from multipath reflections. The process used to isolate the signals is shown in Fig. 5. This process is employed in the DF equipment to raise the confidence in the geolocation estimates by being sure the "right" signal is being geolocated. The result is accurate geolocation needed for Emergency 911 service.

"Smart Antenna" Arrays

Another commercial application of software radio technology is the "Smart Antenna" appliqué for existing base stations. A field trial at a service provider's network this Spring will be used to evaluate the ability of the "Smart Antenna" to increase C/I for an Advanced Mobile Phone Service (AMPS) system in an area with a very high density of users.

Increased user capacity and expanded geographic coverage from existing installations are the two strong economic motivations voiced by service providers as the impetus for these developments. The need for increased area coverage is of primary concern in rural environments and for prospective PCS providers. Increases in user capacity are desired by existing service providers for high-use areas where they are faced with loss of potential revenue, customer dissatisfaction, and reduction in spectrum available for AMPS created by the introduction of digital voice services.

Focusing on the capacity challenge, the real need is to make more voice channels available in a given geographic area in the finite spectrum allocated. When translating this operational need to a technical requirement for an antenna system, the need is not for a steerable "higher gain" array, but instead an array which can increase the carrier to interference ratio (C/I) at both the base station and the mobile terminal. If this can be achieved, then more RF channels can be added to the base stations in the area without driving the C/I below an

Preliminary analyses indicate that software-radio-based "Smart Antenna" appliqués can enhance the C/I in both the forward and reverse link to allow frequency reuse of 3 to 1.

acceptable limit. Suppose that the channels in a conventional system without an array have been allocated to maximize capacity subject to a minimum acceptable C/I. If the smart array is employed, the C/I will increase, allowing the addition of more channels until the C/I is again brought to a minimum acceptable level. By increasing the C/I of the desired signal for both the forward and reverse links, the smart array provides more channel capacity.

There are three approaches which can be used when designing a smart array to maximize C/I. The first and most obvious is to create gain on the array in the intended direction using antenna aperture. This is done by using physically directive elements, or by combining the outputs of more elements to create aperture, and thus gain. It is important to note that it is not the gain that is of value, but the directivity which has value in trying to maximize C/I. The "lack of gain" in the direction of interference, ideally a null, may be more effective than gain in the direction of the desired signal.

The second approach is the mitigation of multipath fading. C/I budgets are generally designed with a significant margin for slow and fast fading. For conventional mobile radio systems, fast fading induced by multipath propagation requires an additional link margin of 8 dB [3]. If the smart array can minimize the effects of destructive specular multipath combining, most of this margin may be recovered. The multipath components present in the PDAgrams shown in Figs. 3 and 4 are typical of those found in a long-term experiment performed in the Washington, D. C. area to characterize the mobile radio propagation environment. Preventing

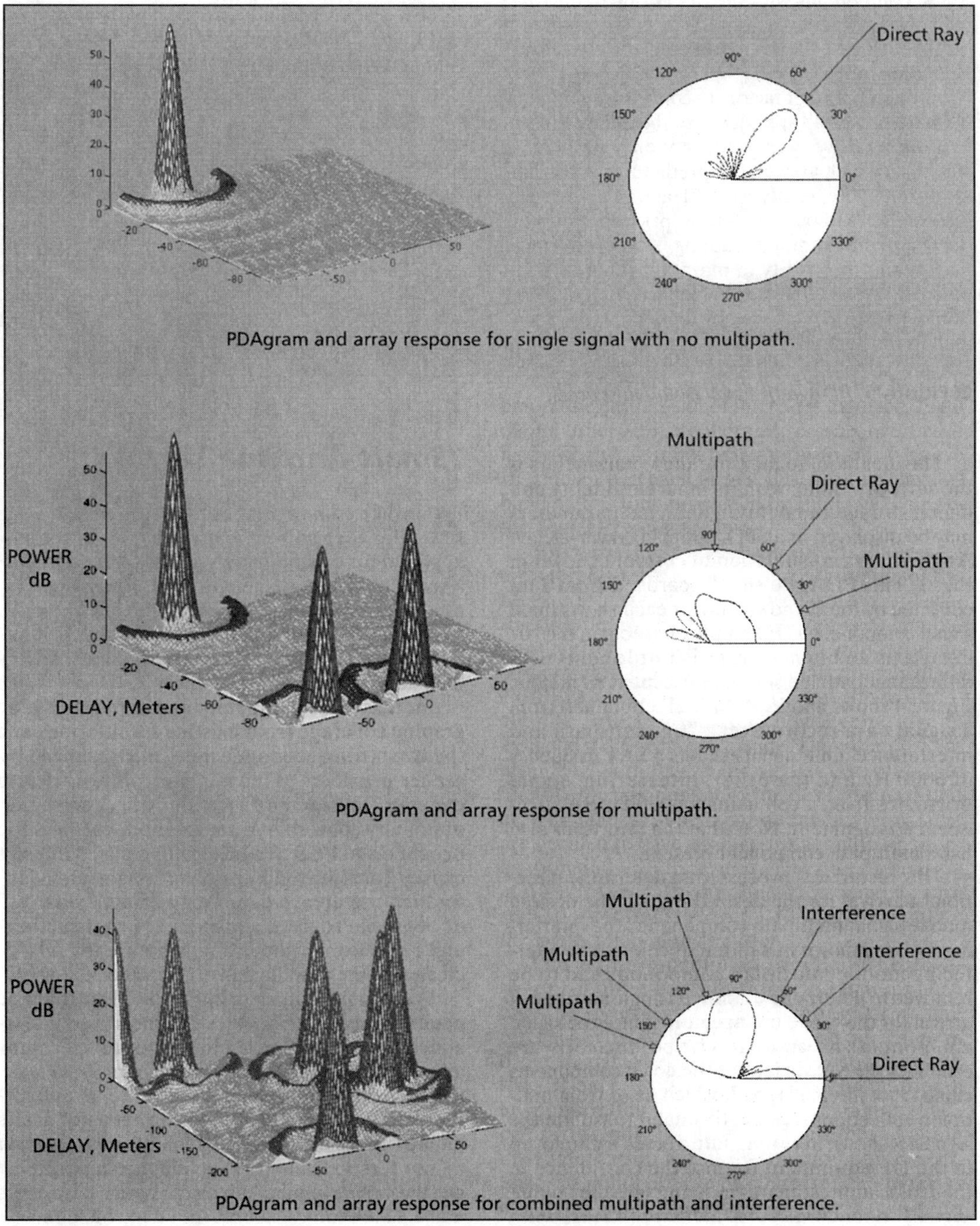

PDAgram and array response for single signal with no multipath.

PDAgram and array response for multipath.

PDAgram and array response for combined multipath and interference.

■ **Figure 6.** *PDAgrams and array responses of the "Smart Antenna."*

the significant losses in C/I induced by multipath propagation is one very effective use of a smart antenna array.

The final approach used to maximize C/I is the identification and nulling of interferers. When designing an antenna element combiner, it is generally much easier to form sharp nulls than beams for a given number of elements in the array. It is generally easier to place a 10 dB null on an interferer than to create a beam with 10 dB of gain in the direction of the desired signal. The operation of the "Smart Antenna" is illustrated in Fig. 6. In the first pair of graphs, the PDAgram shows a single component and the corresponding array response shows a single beam. In the second pair of graphs, the PDAgram shows two multipath components. The array response now includes gain with phase shifts to coherently combine these components with the direct ray. In the third pair of graphs a source of interference is present with multipath and the array response shows corresponding nulls.

Other important factors to consider in designing "Smart Antenna" appliqués are the ability to form beams on the reverse control channel access requests which arrive at unpredictable directions and times, the ability to interface to an existing base station without violating proprietary data interface agreements and disrupting hand-off management, and the ability to maintain the necessary linearity in the equipment with increased dynamic range (> 80 dB) imposed by higher antenna gains and more channels per base station.

Software radio architectures support the implementation of all three approaches to maximize C/I and also address the factors mentioned above. The architecture for the N-channel beamformer is shown in Fig. 7. Each antenna element is followed by a wideband RF to baseband converter, where the entire operating RF band is moved to baseband at very high dynamic range. The digital hardware is a combination of custom boards and commercial "off-the-shelf" boards. Digital tuner ASICs coherently downconvert each narrowband channel to a zero IF complex baseband format. The digital beamformer is implemented in assembly language software executed by a Texas Instruments TMS320C40 DSP chip. Assembly language was chosen after a "C" language implementation failed to execute in real time. Each of the resulting narrowband signals is then upconverted digitally and summed to produce a single wideband signal that is then converted back to an analog format using a high-speed D/A converter. An RF block upconverter completes the process by translating the wideband IF back to the original cellular RF band.

The signal processing algorithms provide a fully adaptive receive beam and a switched transmit beam with four azimuthal beams per 120 degree sector. The net aperture gain is in excess of 20 dBi. The adaptive receive beams are created by a specialized variant of the constant modulus algorithm (CMA) [1, 6, 7] which executes in software and maximizes C/I through a combination of spatial beam steering, interferer null steering, and maximal ratio combining of multipath.

The beamformer software stores the incoming signals in random access memory (RAM) while the beam steering coefficients are being calculated.

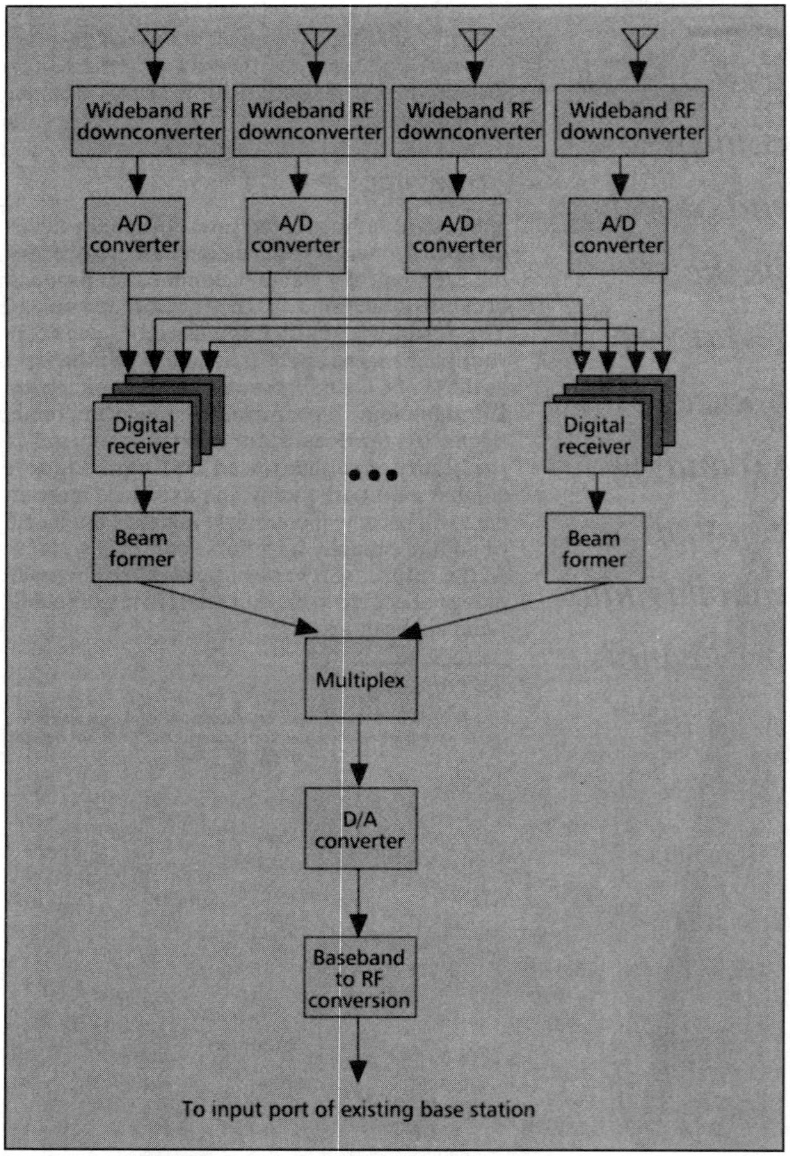

■ **Figure 7.** *Wideband digital beamformer architecture.*

Once they are calculated, the data is read out of RAM through the beamformer. Signals with unpredictable arrival times and directions of arrival may thus be accommodated using the RAM delay buffer to give the beamforming algorithm enough time to initialize the steering vector. The algorithm also separates and identifies signals by SAT frequency so that the beamformer response has gain in the directions of desired signal components and nulls in the directions of interfering components.

This architecture also uses newly developed techniques to implement "software transmission." Once the spatial filtering has been accomplished, the signal is reconstituted digitally and converted with high fidelity to RF so that the beamformed signals can be "received" by a standard base station through the omni or sector antenna port.

In current cellular systems with omnidirectional or sectored antennas, interference limits the choice of frequencies in each cell to one out of every seven. Preliminary analyses indicate that software-radio-based "Smart Antenna" appliqués can enhance the C/I in both the forward and reverse link to allow frequency reuse of 3 to 1. Use of this

In the future, software radios will be the technology of choice in an expanding array of mobile radio applications.

type of a smart array represents a greater than 2 to 1 increase in capacity and supports a 3 to 1 increase in call arrival rate for the same probability of blocked calls.

Summary

Software radio architectures, originally developed for military applications, are now becoming economically viable in commercial products because of the rapid advance of DSP technology. The advantages of digital techniques over conventional analog approaches are much the same as those of CD audio over the now-obsolete vinyl LP technology: higher dynamic range and better fidelity. As the demand for radio spectrum grows, the ability of sophisticated DSP algorithms to combat multipath fading and to reduce interference will become increasingly valuable as a means of adding capacity to cellular and PCS systems. In the future, software radios will be the technology of choice in an expanding array of mobile radio applications.

References

[1] D. N. Godard, "Self-recovering equalization and carrier tracking in two-dimensional data communication systems," *IEEE Trans. on Commun.*, vol. COM-28, no. 11, Nov. 1980, pp. 1867-1875.

[2] E. B. Hogenauer, "An economical class of digital filters for decimation and interpolation," *IEEE Trans. on Acoustics, Speech, and Signal Processing*, vol. ASSP-29, no. 2, April 1981, pp. 155-162.

[3] W. C. Y. Lee, Mobile Cellular Telecommunications, 2nd Ed., (New York: McGraw-Hill, 1995).

[4] S. K. Oh and C. K. Un, "Simple computational methods of the AP algorithm for maximum likelihood localization of multiple radiating sources," *IEEE Trans. on Signal Processing*, vol. 40, no. 11, Nov. 1992, pp. 2848-2854.

[5] J. O. Smith and J. S. Abel, "Closed-form least-squares source location estimation from range-difference measurements," *IEEE Trans. on Acoustics, Speech, and Signal Processing*, vol. ASSP-35, no. 12, Dec. 1987, pp. 1661-1669.

[6] B. J. Sublett, R. P. Gooch, and S. H. Goldberg, "Separation and bearing estimation of co-channel signals," Proceedings of the 1989 IEEE Military Communications Conference, Oct. 1989.

[7] I. Ziskind and M. Wax, "Maximum likelihood localization of multiple sources by alternating projection," *IEEE Trans. on Acoustics, Speech, and Signal Processing*, vol. 36, no. 10, pp. 1553-1560, Oct. 1988.

Software Radio Architecture with Smart Antennas: A Tutorial on Algorithms and Complexity

Javad Razavilar, *Member, IEEE*, Farrokh Rashid-Farrokhi, *Member, IEEE*, and K. J. Ray Liu, *Senior Member, IEEE*

Abstract— **Recently, there has been considerable interest in using antenna arrays in wireless communication networks to increase the capacity and decrease the cochannel interference. Adaptive beamforming with smart antennas at the receiver increases the carrier-to-interference ratio (CIR) in a wireless link. This paper considers a wireless network with beamforming capabilities at the receiver which allows two or more transmitters to share the same channel to communicate with the base station. The concrete computational complexity and algorithm structure of a base station are considered in terms of a software radio system model, initially with an omnidirectional antenna. The software radio computational model is then expanded to characterize a network with smart antennas. The application of the software radio smart antenna is demonstrated through two examples. First, traffic improvement in a network with a smart antenna is considered, and the implementation of a hand-off algorithm in the software radio is presented. The blocking probabilities of the calls and total carried traffic in the system under different traffic policies are derived. The analytical and numerical results show that adaptive beamforming at the receiver reduces the probability of blocking and forced termination of the calls and increases the total carried traffic in the system. Then, a joint beamforming and power control algorithm is implemented in a software radio smart antenna in a CDMA network. This shows that, by using smart antennas, each user can transmit with much lower power, and therefore the system capacity increases significantly.**

Index Terms—**Adaptive beamforming, handoff, power control, smart antennas, software radio.**

I. Introduction

THE software radio is emerging from military applications. Because of recent advances in high-speed digital signal processors (DSP's) and analog-to-digital converters (ADC's), the commercial implementation of software radios has become viable. In software radio systems, the IF signal is digitized using wide-band ADC's, and all of the subsequent processing is implemented in software [1]–[8]. The main advantage of software radio is its great flexibility such that it can be programmed for emerging standards. It also can be dynamically updated with new software without any changes in hardware and infrastructure. Rapid deployment is another important feature of the software radio. In wireless applications where different standards might be deployed, users' roaming can be a big issue in existing platforms. In the software radio system, users just need to download the new air interface upon entering the new territory.

The reallocation of bandwidth is a standards-setting process that takes a considerably long time. On the other hand, due to fixed finite resources, one cannot assume that more physical radio channels will be added to a spectrum allocation to reduce the probability of blocking or forced termination [9]–[13]. The increasing popularity of wireless communication services together with the limited amount of the available radio spectrum calls for highly efficient usage of resources (traffic channels) in the system [14]. The interference reduction capability of antenna arrays has been considered as a means to increase the capacity of wireless systems [15]–[21]. Specifically, using beamforming techniques at the receiver, two or more transmitters can share the same traffic channel to communicate with the base station at the same time. An adaptive antenna array is used at the base station to form several antenna beams simultaneously. Each beam captures one transmitter by automatically pointing its pattern toward that transmitter while nulling other cochannel transmitters and multipath signals [19]–[21]. In this way, the cochannel interference (CCI) is minimized, and therefore the carrier-to-interference ratio (CIR) for the signal of interest is maximized. In urban wireless environments, the signal transmitted by each user is reflected by surrounding buildings and the terrain. Therefore, several copies of the transmitted signal are received at the base station with different delays and different attenuation. This effect is called multipath fading. In space-only diversity, given a cell with L channels, an M-element antenna array that forms $M-1$ distinct array nulls at the receiver allows a maximum of KL users to be served in a cell, where $K = M/\alpha$ and α is the multipath factor representing the number of distinct strong path components received from each mobile station at the beamforming array. In space–time diversity, spatial nulls reject the CCI, while intersymbol interference (ISI) is eliminated by time diversity.

Sectorization is an alternative to smart antennas. Sectorized cell sites employ hardware beamforming array antennas in which each beam is assigned a distinct RF channel set. The direction and gain of sectorized antennas are fixed in the hardware. They cannot place nulls on interference like

Manuscript received November 15, 1997; revised April 2, 1998 and August 17, 1998.

J. Razavilar was with the Department of Electrical Engineering and the Institute for Systems Research, University of Maryland, College Park, MD 20742 USA. He is now with Corporate R&D, 3COM Laboratories, San Diego, CA 92128 USA.

F. Rashid-Farrokhi was with the Department of Electrical Engineering and the Institute for Systems Research, University of Maryland, College Park, MD 20742 USA. He is now with the Wireless Communications Research Department, Bell Laboratories, Lucent Technologies, Holmdel, NJ 07733 USA.

K. J. R. Liu is with the Department of Electrical Engineering and the Institute for Systems Research, University of Maryland, College Park, MD 20742 USA.

Publisher Item Identifier S 0733-8716(99)02978-9.

Reprinted from *IEEE Journal on Selected Areas of Communications,* Vol. 17, No. 4, pp. 662–676, April 1999.

the smart antenna. The cost of a sectorized cell site is an increase in the (analog) hardware complexity of the cell site, but there is virtually no impact on the single-channel cell site software. When the capacity of a sectorized site is exceeded, one may then overlay a smart antenna with its dedicated digital signal processing (DSP) capacity and unique software architecture which accommodates the vastly increased computational demand of the beamforming network. The smart antenna's beams are not fixed, but are dynamic per user, and will place nulls that cancel interference. Another advantage of the smart antenna system is the reduced rate of handoffs because of tracking the user. In this paper, the term "cell" means either a conventional cell or one sector of a sectorized cell.

This paper first presents a tutorial review of beamforming algorithms. It then reviews a software radio architecture, first with omnidirectional antennas, and then with smart antennas. Two applications show the advantages of a system with a smart antenna. First, a wireless network with beamforming capabilities at the receiver allows two or more transmitters to share the same channel to communicate with the base station. Each cell is modeled by a multiuser multiserver service facility. Each server is a beamformed channel formed by the cell's base station. For a set of cochannel transmitters, the probability of successful capture by a separate antenna beam is computed. The success probabilities are taken into account in the queueing model of the system. From this generalized model, the closed-form blocking probabilities of the calls and total carried traffic in the system under different traffic policies are derived. The second application shows how to implement a robust power control algorithm when using a smart antenna at the base station. In CDMA networks, power control is an important issue because of the near–far effect. Therefore, the effective power control algorithm results in a capacity increase in the system.

The paper is organized as follows. In Section II, various adaptive beamforming algorithms are discussed. The software radio architecture in a system with an omnidirectional antenna and also with a smart antenna is presented in Section III. The complexity issue is also discussed in this section. Some applications and advantages of using software radio smart antennas are presented in Section IV, where traffic improvement and capacity enhancement are addressed. Section V includes our conclusions and remarks.

II. Review of Spatial Multiplexing Using Adaptive Beamforming

A smart antenna is a multibeam adaptive array with its gain pattern adjusted dynamically [23]. For a fixed M-beam sectorized antenna and an M-element smart antenna system, the antenna gain and system capacity are increased approximately M-fold. In a fixed beam system, the handoff from one beam to another is the main disadvantage, while in a smart antenna system, each beam tracks its user within the coverage of the cell. Moreover, smart antenna systems provide diversity gain against multipath signals and strong interferers by placing antenna nulls toward those sources. Since the beamforming in

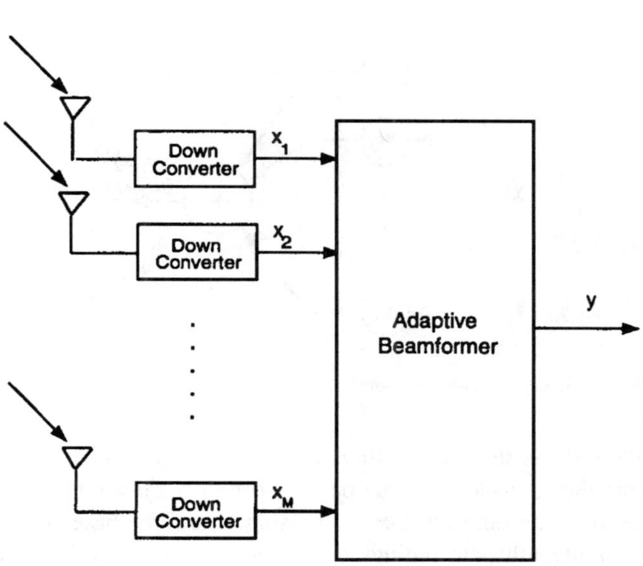

Fig. 1. A diversity combining system.

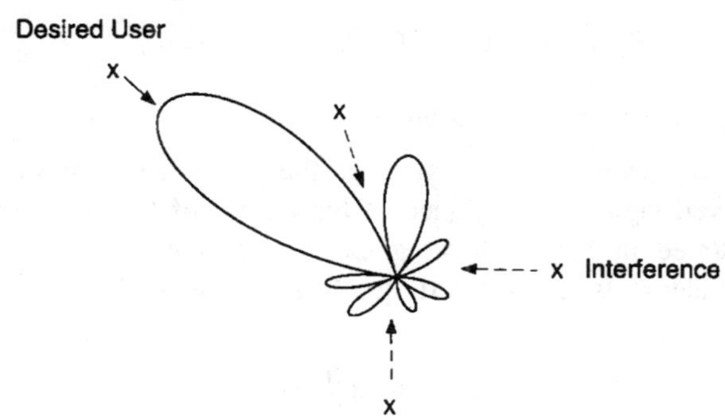

Fig. 2. Sample array pattern.

adaptive arrays is done in IF and baseband, these systems are well suited for software radio implementation.

Space and time diversity reduce the cochannel interference (CCI) and intersymbol interference (ISI). If the desired signal and the cochannel signals have different temporal or spatial signatures, time and space diversity can improve the signal-to-noise ratio (SNR). The block diagram of an adaptive array is depicted in Fig. 1. In general, the objective is to form the main beam toward the desired user and nulls toward the interference sources as illustrated in Fig. 2.

An antenna array consisting of M elements is considered at the receiver. Adaptive beamforming capabilities of antenna arrays may be used to maintain a constant gain for the signal along the direction of interest, and adjusting the nulls so as to reject the cochannel interference. In this way, the interference is minimized and the CIR for the signal of interest is maximized. A set of J cochannel transmitters is considered. Let $s_j(k)$ for $j = 1, 2, \cdots, J$ denote the jth transmitted signal. Assume that at most N multipath signals from each user arrive in the base station with different delays (τ_n). Therefore, the received signal vector due to the $s_i(k)$ is

$$\boldsymbol{x}(k) = \sum_{j=1}^{J} \sqrt{P_j G_j} \sum_{n=1}^{N} \alpha_j^n \boldsymbol{a}(\theta_j^n) s_j(t - \tau_n) + \boldsymbol{n}(k) \quad (1)$$

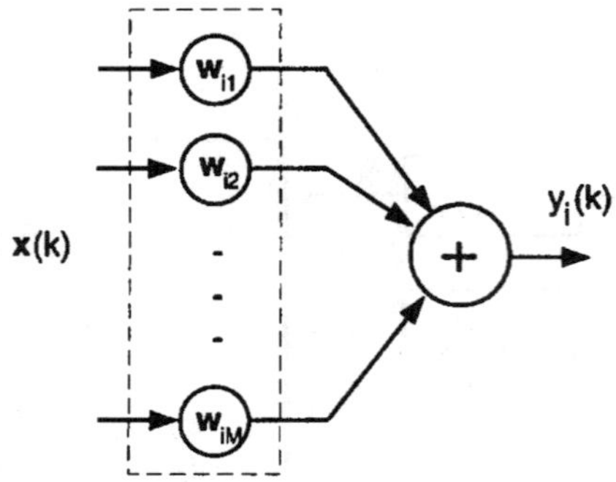

Fig. 3. Space diversity combiner.

where θ_i^n is the arrival direction of the nth multipath signal from the ith user, P_i is the power of the ith transmitter, G_i is the link gain between the ith transmitter and the base station, α_i^n is the nth path fading coefficient, and $\boldsymbol{a}(\theta_i^n)$ is the array response to the multipath signal arriving from direction θ_i^n with

$$\boldsymbol{a}(\theta_i^n) = [a_1(\theta_i^n), a_2(\theta_i^n), \cdots, a_M(\theta_i^n)]^T. \tag{2}$$

A. Space Diversity Combining

In a space diversity system, the weighted sum of the received signal is combined at the output of beamformer as illustrated in Fig. 3. A weighted sum of the outputs of the array elements is generated by a beamformer in the following way:

$$y_i(k) = \boldsymbol{w}_i^H \boldsymbol{x}(k) \tag{3}$$

where $\boldsymbol{w}_i^T = [w_{i1}, \cdots, w_{iM}]$ is the weight and $\boldsymbol{x}^T(k) = [x_1(k), \cdots, x_M(k)]$ is the received signal vector sampled at the output of the downconverters.

Minimum Variance Distortionless Response (MVDR): One can choose the weight vector $\boldsymbol{w}_i$ to steer a beam toward the direction of the signal of interest θ_i and adjust the nulls to reject the interference. This is done by attempting to maintain a distortionless response in the direction of interest and placing the nulls in the directions of other cochannel interferers. The average output power is given by

$$E_i = E\left[y_i(k)y_i^H(k)\right] = \boldsymbol{w}_i^H E\left[\boldsymbol{x}(k)\boldsymbol{x}^H(k)\right]\boldsymbol{w}_i = \boldsymbol{w}_i^H \boldsymbol{\Phi}_i \boldsymbol{w}_i \tag{4}$$

where

$$\boldsymbol{\Phi}_i = P_i G_i \alpha_i^1 \boldsymbol{a}_i(\theta_i^1)\boldsymbol{a}_i^H(\theta_i^1) + \sum_{j=1, j\neq i}^{J} P_j G_j \sum_{n=2}^{N} \alpha_j^n \boldsymbol{a}_j(\theta_j^n)\boldsymbol{a}_j^H(\theta_j^n) + \frac{N_0}{2} I. \tag{5}$$

The correlation matrix due to the interference terms is

$$\boldsymbol{\Phi}_N = \sum_{j=1, j\neq i}^{J} P_j G_j \sum_{n=2}^{N} \alpha_j^n \boldsymbol{a}_j(\theta_j^n)\boldsymbol{a}_j^H(\theta_j^n) + \frac{N_0}{2} I. \tag{6}$$

If the array response in the direction of the desired user is known, the beamformer tries to minimize the output power E_i subject to maintaining a distortionless response in the direction of interest such that $\boldsymbol{w}_i^H \boldsymbol{a}_i(\theta_i^1) = 1$. For this reason, this adaptive beamformer is called an MVDR [15] beamformer. From (4) and (5), with the constraint that $\boldsymbol{w}_i^H \boldsymbol{a}_i(\theta_i^1) = 1$, the received signal power plus the interference power as a function of $\boldsymbol{w}_i$ is

$$E_i = P_i G_i + \sum_{j=1, j\neq i}^{J} P_j G_j \sum_{n=2}^{N} \alpha_j^n \boldsymbol{w}_i^H \boldsymbol{a}_j(\theta_j^n)\boldsymbol{a}_j^H(\theta_j^n)\boldsymbol{w}_i + \frac{N_0}{2}\boldsymbol{w}_i^H \boldsymbol{w}_i. \tag{7}$$

Let I_i denote the total interference plus noise power given by

$$I_i = \sum_{j=1, j\neq i}^{J} P_j G_j \sum_{n=2}^{N} \alpha_j^n \boldsymbol{w}_i^H \boldsymbol{a}_j(\theta_j^n)\boldsymbol{a}_j^H(\theta_j^n)\boldsymbol{w}_i + \frac{N_0}{2}\boldsymbol{w}_i^H \boldsymbol{w}_i. \tag{8}$$

In (7), $P_i G_i$ is the received power from the signal of interest, while I_i given by (8) is the contribution to the output power E_i from the interference and noise.

The *optimum weight vector* minimizes the interference I_i while maintaining a unity gain in the direction of interest by imposing $\boldsymbol{w}_i^H \boldsymbol{a}_i(\theta_i^1) = 1$. The solution of this problem is

$$\hat{\boldsymbol{w}}_i = \frac{\Phi^{-1}\boldsymbol{a}_i(\theta_i^1)}{\boldsymbol{a}_i^H(\theta_i^1)\Phi^{-1}\boldsymbol{a}_i(\theta_i^1)}. \tag{9}$$

Since the desired signal (arriving along θ_i) will not be affected by the beamforming process and only the interference is rejected, the CIR is maximized for the signal in the direction of interest, i.e.,

$$\Gamma_{i,\max} = \frac{P_i G_i}{\sum_{j=1, j\neq i}^{J} P_j G_j \sum_{n=2}^{N} \alpha_j^n \hat{\boldsymbol{w}}_i^H \boldsymbol{a}_j(\theta_j^n)\boldsymbol{a}_j^H(\theta_j^n)\hat{\boldsymbol{w}}_i + \frac{N_0}{2}\hat{\boldsymbol{w}}_i^H \hat{\boldsymbol{w}}_i}. \tag{10}$$

Minimum Mean-Square Error (MMSE): If the array response is not known, one may employ a training sequence and minimize the difference between the training sequence and the output of the beamformer in the mean-square sense

$$\hat{\boldsymbol{w}}_i = \arg\min_{\boldsymbol{w}_i} E\left[|d_i - \boldsymbol{w}_i^H \boldsymbol{a}_i(\theta_i^1)|^2\right]. \tag{11}$$

The solution to this problem is given by [14]

$$\hat{\boldsymbol{w}}_i = \Phi_i^{-1}\boldsymbol{p}_i \tag{12}$$

where Φ_i is defined as before and $\boldsymbol{p}_i$ is the cross correlation between the received vector and the training sequence; $\boldsymbol{p}_i = E[\boldsymbol{x}_i d_i^*]$. The maximum CIR in this case is given by

$$\Gamma_{i,\max} = \frac{P_i G_i \hat{\boldsymbol{w}}_i^H \boldsymbol{a}_i(\theta_i^1)\boldsymbol{a}_i^H(\theta_i^1)\boldsymbol{w}_i}{\sum_{j=1, j\neq i}^{J} P_j G_j \sum_{n=2}^{N} \alpha_j^n \hat{\boldsymbol{w}}_i^H \boldsymbol{a}_j(\theta_j^n)\boldsymbol{a}_j^H(\theta_j^n)\hat{\boldsymbol{w}}_i + \frac{N_0}{2}\hat{\boldsymbol{w}}_i^H \hat{\boldsymbol{w}}_i}. \tag{13}$$

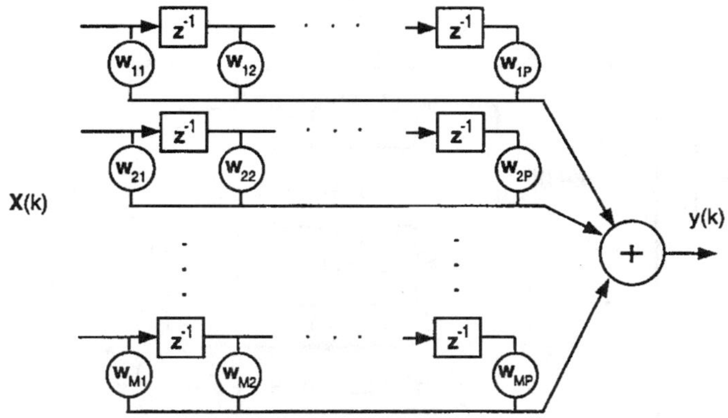

Fig. 4. Space–time diversity combiner.

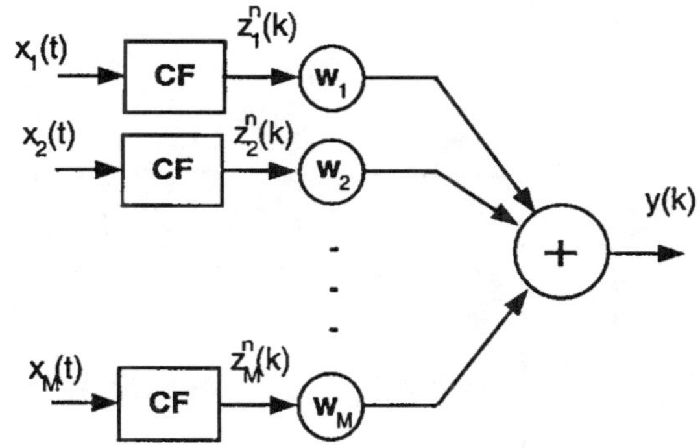

Fig. 5. Space–time diversity combiner in a CDMA system.

B. Space–Time Diversity Combining

In the space diversity combining system of Fig. 3, the interference and multipath signals are rejected by placing nulls at the directions of those signals. In a broadband linear combiner, illustrated in Fig. 4, the desired signal and its multipath are combined at the combiner output to estimate the desired signal.

MMSE Space-Time Combining: The output of a broadband combiner can be expressed as

$$y(k) = \sum_{m=1}^{M} \boldsymbol{w}^H(m)\boldsymbol{x}(m)$$

where $\boldsymbol{w}^T(m) = [w_{m1}, \cdots, w_{mP}]$ is the weighting vector and $\boldsymbol{x}^T(m) = [x_{m1}, \cdots, x_{mP}]$ is the received signal vector at the mth element equalizer.

Let $\boldsymbol{W} = [\boldsymbol{w}^T(1), \cdots, \boldsymbol{w}^T(P)]^T$ and $\boldsymbol{X} = [\boldsymbol{x}(1)^T, \cdots, \boldsymbol{x}^T(P)]^T$. The output of the beamformer can be written as

$$y(k) = \boldsymbol{W}^H \boldsymbol{X}.$$

In this case also, the objective is to minimize the mean-square error between the output of combiner and the desired signal d, i.e.,

$$\hat{\boldsymbol{W}} = \arg \min_{\boldsymbol{W}} |d - \boldsymbol{W}^H \boldsymbol{X}|^2.$$

The optimal beamformer coefficients are similar to space diversity MMSE by

$$\hat{\boldsymbol{W}} = E\{\boldsymbol{X}\boldsymbol{X}^H\}^{-1} E\{\boldsymbol{X}\boldsymbol{d}^*\}.$$

Many adaptive methods that update weight vectors according to the incoming data have been developed [51], such as recursive least square (RLS) [21] and minimum mean-square error (MMSE) [23].

Blind Methods: Using a training sequence consumes some bandwidth. In order to save the bandwidth allocated to the training sequence, one may use blind methods. For *single-input single-output* (SISO) systems, numerous blind identification algorithms [24]–[27] and blind equalization algorithms [28]–[37] have been proposed that exploit the higher order statistics of channel output. Among these algorithms, the *Godard algorithm* (GA) [30], also known as the *constant modulus algorithm* (CMA) [35], [36], is one of the best and simplest adaptive blind equalization algorithms. *Single-input multiple-output* (SIMO) systems may be viewed as fractionally spaced sampled communication systems which receive the distorted versions of one input signal. Fractionally spaced CMA adaptive blind equalizers under symbol timing offsets is considered in [38]. The equalization of multiple-input multiple-output (MIMO) transmission systems is studied in [39] and [40], where the MIMO channel impulse response is known. When the channel parameters of the MIMO systems are unknown, blind identification and equalization techniques must be used to separate and capture signals. The capture properties of the CMA algorithm used in MIMO systems with constant modulus input signals are studied in [36], [41], and [42]. The blind identification of MIMO systems using second-order statistics or higher order statistics are also studied in [43]–[46]. Some subspace algorithms with fast convergence rate are proposed in [47].

C. Space–Time Diversity in CDMA Systems

In code-division multiple-access systems (CDMA), all users are sharing the same channel and each user has a different pseudonoise (PN) sequence. The received signal due to J cochannel users at the receiver is given by

$$\boldsymbol{x}(k) = \sum_{j=1}^{J} P_j G_j \sum_{n_b} s_{n_b} p_j(t - n_b T) + \boldsymbol{n}(k)$$

where $\boldsymbol{x}(k)$ is the received signal at the array, s_{n_b} is the transmitted sequence, and

$$p_j(k) = \sum_{n} \boldsymbol{a}_j(\theta_n)\alpha_j^n c_j(t - \tau_j^n).$$

τ_j^n is the nth path delay associated with the jth user, and c_j is the jth user PN sequence. In a space-diversity combiner, the signals from the main path are despread and combined at the beamformer output, as shown in Fig. 5. Code filters (CF) are matched to the desired user code and the delay is matched to the pth path delay, i.e.,

$$z_j^n(k) = \int_T \boldsymbol{x}(k) c_j(t - kT - \tau_j^n)\, dt$$

and

$$y_j^n(k) = \boldsymbol{w}^H \boldsymbol{z}_j^n(k).$$

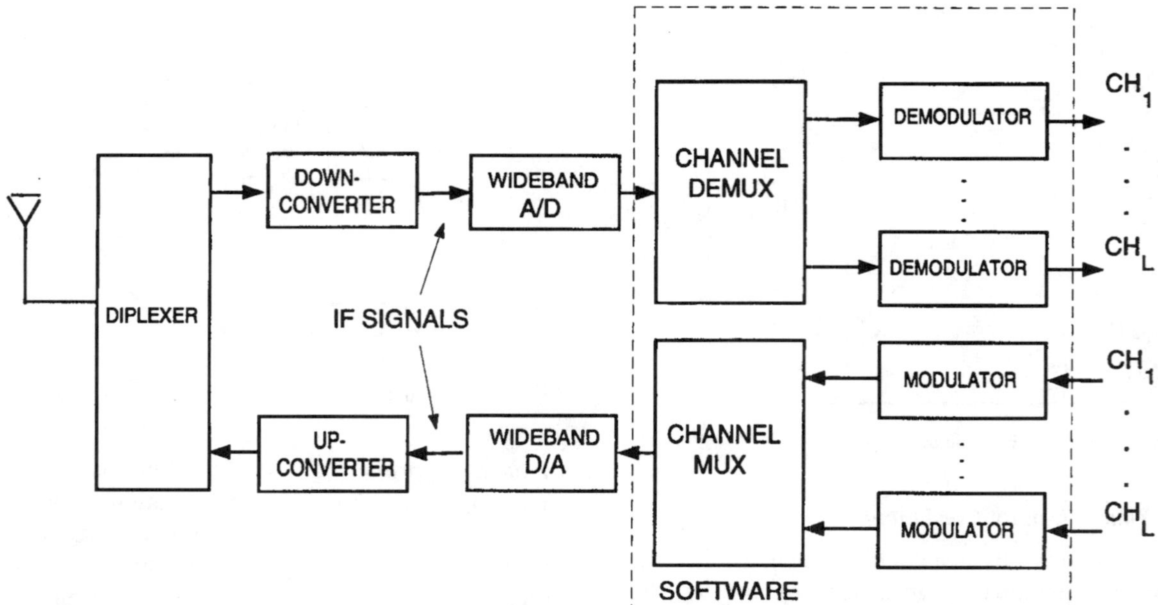

Fig. 6. Functional block diagram of the software radio for a base station with single omnidirectional antenna.

It has been shown [16] that the optimal beamformer for the above architecture is given by

$$\hat{\boldsymbol{w}} = \frac{\Phi_{in}^{-1}\boldsymbol{a}(\theta_j^n)}{\boldsymbol{a}(\theta_j^n)^H\Phi_{in}^{-1}\boldsymbol{a}(\theta_j^n)}$$

where Φ and Φ_{in} are defined as before, and

$$\Phi_{in} = \frac{\rho}{\rho - 1}\left(\Phi - \frac{1}{\rho}\Phi_o\right)$$

where ρ is the processing gain and

$$\Phi_o = \Phi_{in} + P\boldsymbol{a}(\theta_j^n)\boldsymbol{a}(\theta_j^n)^H.$$

If the delay spread is larger than a chip period, a correlator may be employed for the signal received from each path. In signal antenna systems, signals from different correlators may be combined using maximum ratio combining. In those receivers, known as Rake receivers, each correlator is called a finger of the Rake receiver. In a multiple-antenna system, one can design a beamformer for each resolvable path, and then combine the outputs with a standard Rake receiver [48], i.e.,

$$\hat{z}_j(k) = \sum_{n=1}^{N} z_j^n(k)\alpha_j^n$$

where N is the maximum number of paths in each link.

III. Software Radio Architecture with Smart Antenna

The software radio allows one to build flexible, multiband radio systems rapidly. The functional block diagram of the software radio in a system with an omnidirectional antenna is shown in Fig. 6. The architecture of the software radio is described in more detail in [1]. As illustrated in this figure, a wide-band front end downconverts the received signal to the IF where it is sampled and digitized by a high-speed ADC. The rest of the processing is implemented in software [2]. Such a radio may be dynamically updated with new software without changes in hardware and infrastructure [2]. One may achieve better performance by introducing coding on the bitstream using coding gain to improve the bit-error rate (BER). In the case of the omnidirectional antenna, the software of the cell site may either have an IF software radio architecture, in which the waveforms are synthesized and received in software, or they may have a baseband DSP radio architecture, in which the software is limited to bitstream signal processing. There are P MIPS of processing capacity required per server, with P(IF+baseband) $\gg$ P(baseband). A block diagram of a typical digital downconverter is depicted in Fig. 7. The wide-band IF signal is translated to a complex baseband signal by the quadrature multiplier. The numerically controlled oscillator (NCO) block generates the quadrature signals for the multipliers. Then the complex baseband signal is low-pass filtered to prevent aliasing due to decimation [5]. In a software radio architecture, these blocks are implemented in software. The high sampling rate of a wide-band receiver allows time for only a few operations per sample; however, the digital downconverter combines the frequency translation and filtering in one step. The specific signal processing steps show how the multiplication and convolution required to perform the frequency shift and filtering may be combined. The first step is frequency translation of the real-valued received signal samples $r[n]$ to baseband by multiplication with the appropriate complex exponential as follows:

$$\begin{aligned} y[n] &= r[n]e^{-j2\pi f_c n T_s} \\ &= r[n]\{\cos(2\pi f_c n T_s) - j sin(2\pi f_c n T_s)\} \end{aligned} \quad (14)$$

where f_c is the carrier frequency before translation to baseband and T_s is the sample interval. The next step is to filter $y[n]$

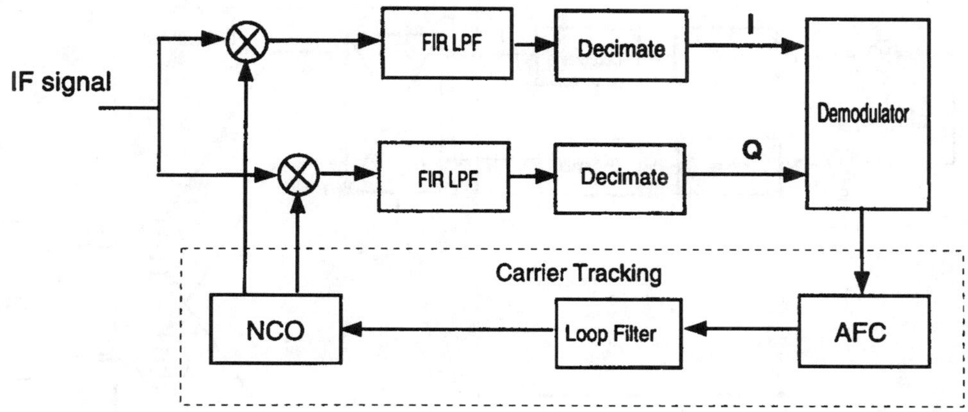

Fig. 7. Block diagram of the down converter and demodulator.

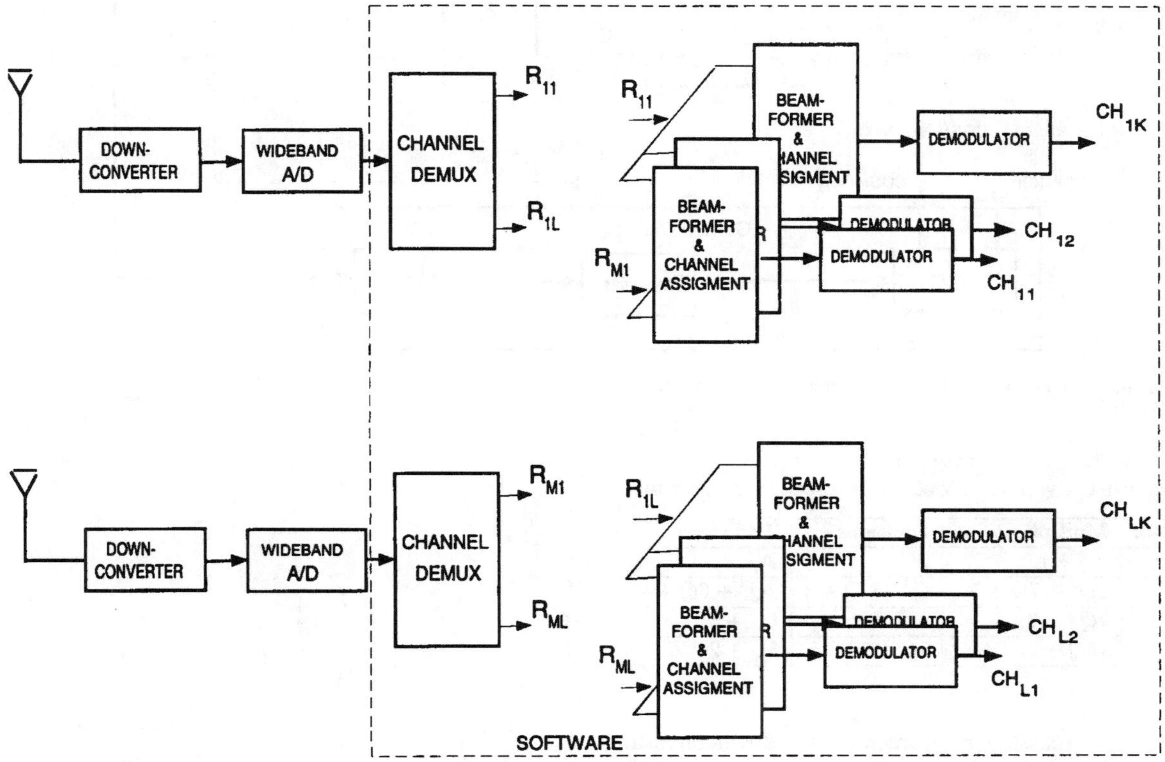

Fig. 8. Functional block diagram of the software radio for a base station with smart antenna.

with an FIR filter with filter coefficients $h[n]$

$$x[n] = \sum_{m=0}^{M} h[m]y[n-m]$$
$$= \sum_{m=0}^{M} h[m]r[n-m]e^{-j2\pi f_c(n-m)T_s}. \quad (15)$$

From (14) and (15), combine these two processes as follows:

$$x[n] = e^{-j2\pi f_c nT_s} \sum_{m=0}^{M} h[m]y[n-m]$$
$$= \sum_{m=0}^{M} h[m]r[n - me^{-j2\pi f_c mT_s}$$
$$= e^{-j2\pi f_c nT_s} \sum_{m=0}^{M} c[m]r[n-m] \quad (16)$$

where $c[m] = h[m]e^{j2\pi f_c nT_s}$. Here, not only is the number of computations reduced, but there is no need to compute the unfiltered baseband signal $y[n]$, which further reduces the number of operations [5].

The functional block diagram of the software radio with smart antennas is shown in Fig. 8. Each antenna element has its own downconverter and ADC. But the subsequent beamforming and demodulation are implemented in software and are shared among all of the elements. Fig. 9 illustrates the software architecture of the beamformer for each channel. The received signal from each antenna element is passed through the same software block as depicted in Fig. 7 to generate I and Q signals. These signals are combined using the "combiner" block as shown in Fig. 9. The received signal from the user would be the output of the demodulator in Fig. 9. Using K similar blocks for each channel one allows up to K users to share the same channel. Since each user generates its carrier

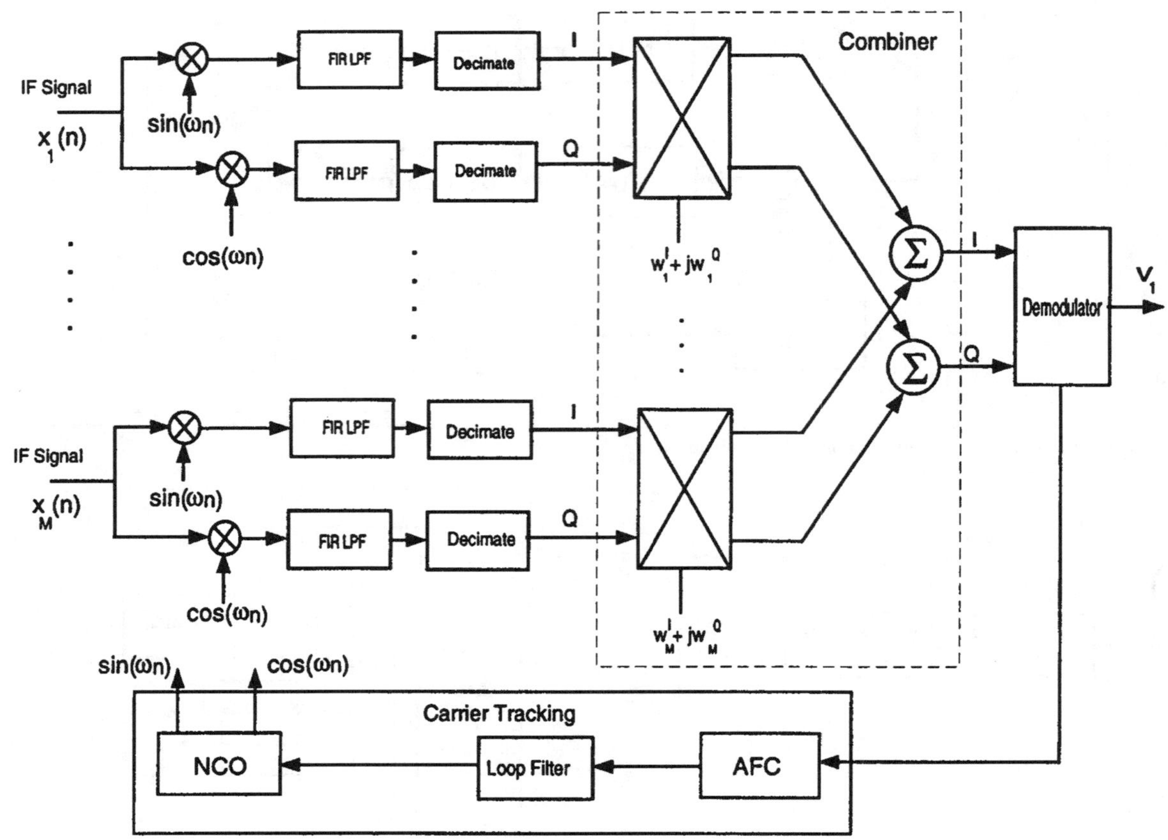

Fig. 9. Block diagram of the software beamformer for each user.

TABLE I
COMPUTATIONAL COMPLEXITY OF VARIOUS BEAMFORMING ALGORITHMS

Algorithm	Multiplications	Divisions	Additions
LMS	$2Q+1$	0	$2Q$
RLS	$2Q^2+7Q+5$	Q^2+4Q+3	$2Q^2+6Q+4$
FTF	$7Q+12$	4	$6Q+3$
LSL	$10Q+3$	$6Q+2$	$8Q+2$

locally, there are separate carrier tracking and downconverting blocks per user. It was mentioned that one can assign K users to each channel by using K beams for each channel. In this manner, a system with L physical channels serves at most KL users. Such a software radio with a smart antenna is scalable. Increasing the number of beams and users is a software process within the constraints of the hardware costs. The proportion of increase in computational resources required for the smart antenna depends on the cell site architecture. With baseband DSP, one converts the smart antenna channels back to analog IF to provide only L channels in the conventional cell site as the maximum capacity of each cell. In principle, with an IF software radio architecture, the smart antenna processing could be part of the larger pooled DSP resources, and one could implement KL or KL/α service channels. In the sequel, a software radio architecture is assumed in which one synthesizes up to KL servers in the integrated DSP architecture of the software radio cell site. As illustrated in Fig. 8, the beamforming process can be implemented on DSP's. Usually, adaptive algorithms are used to update the weight vectors. The

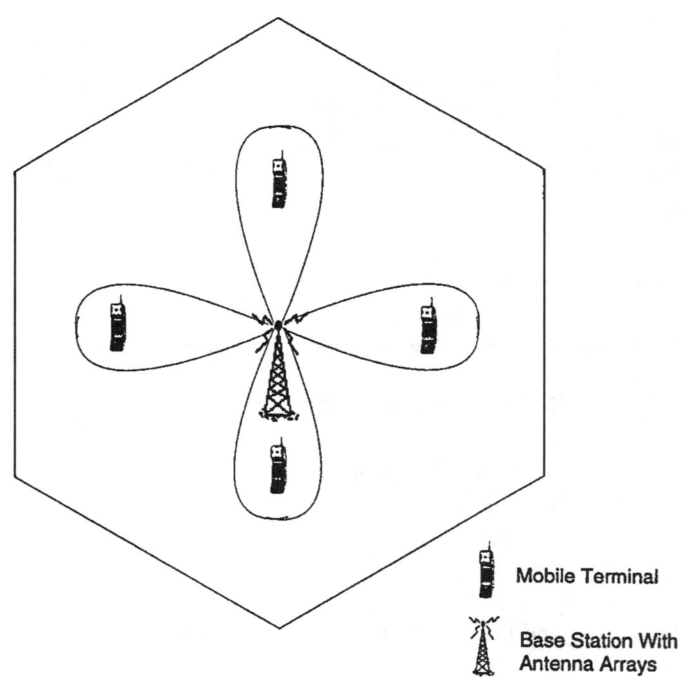

Fig. 10. Capturing four mobile users over the same channel successfully by an adaptive antenna array at the BS.

optimum beamformer weight vector presented in the previous section requires knowledge of second-order statistics. These statistics are usually not known, but with the assumption of ergodicity, they can be estimated. Statistics may also change

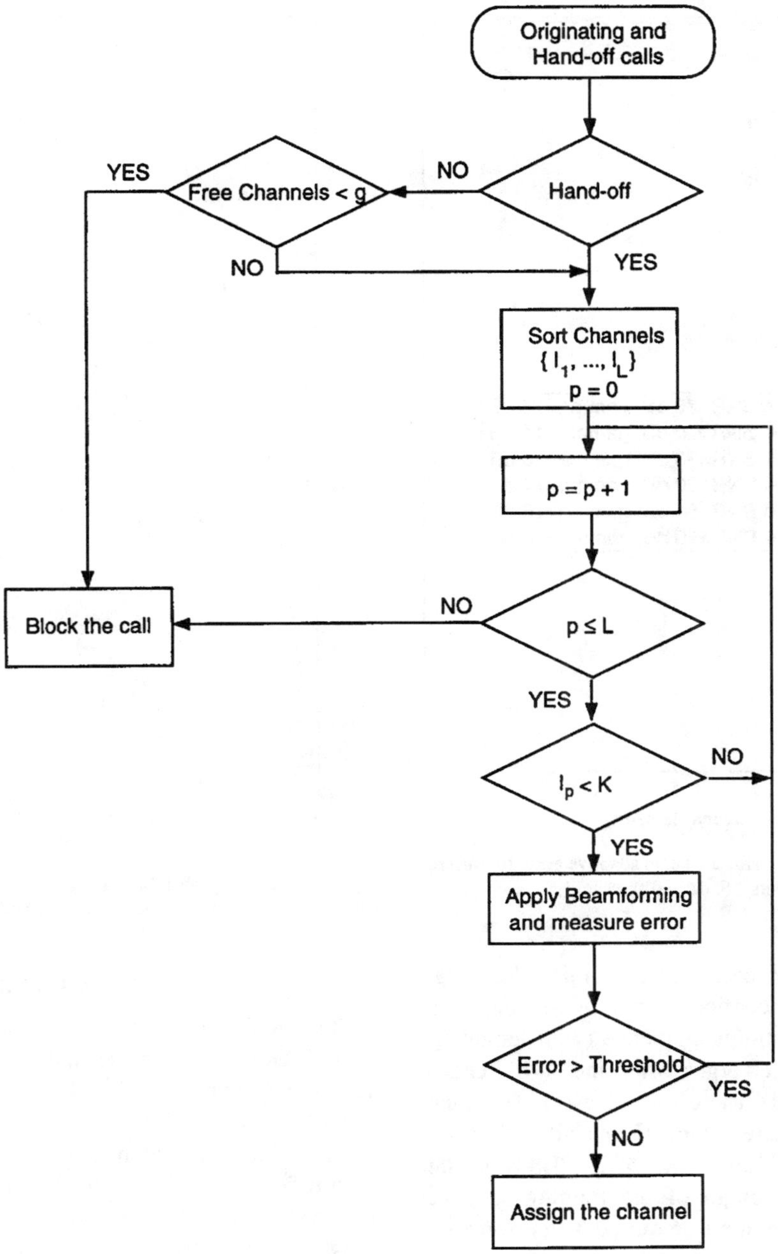

Fig. 11. Flow chart of the channel assignment algorithm in a system with smart antenna.

over time (due to moving interferers). To solve these problems, weight vectors are typically determined adaptively. One may employ training sequences to update the weight vectors. The training sequence is known at the receiver. Let $d_i, i = 1, \cdots, Q$ denote the training sequence. The received training sequence at the receiver is denoted by $X_i, i = 1, \cdots, Q$.

The direct matrix inversion (DMI) is the straightforward method for calculating the weight vectors. DMI minimizes the difference between the training sequence and the output of the beamformer in a mean-square sense

$$\hat{\boldsymbol{w}}_i = \arg\min_{w_i} \sum_{q=1}^{Q} |d_i - \boldsymbol{w}_i^H \boldsymbol{X}_i(q)|^2. \tag{17}$$

The solution to this problem is given by [14]

$$\hat{\boldsymbol{w}}_i = \Phi_i^{-1} \boldsymbol{p}_i \tag{18}$$

where Φ_i is the estimated correlation matrix and $\boldsymbol{p}_i$ is the cross correlation between the received vector and the training sequence given by

$$\Phi_i = \frac{1}{Q} \sum_{q=1}^{Q} \boldsymbol{X}_i(q) \boldsymbol{X}_i^H(q)$$

$$\boldsymbol{p}_i = \frac{1}{Q} \sum_{q=1}^{Q} \boldsymbol{X}_i(q) \boldsymbol{d}_i. \tag{19}$$

The complexity of the DMI method is $O(Q^3)$ where Q is the length of the training sequence. Adaptive algorithms that update the weight vector taps reduce the complexity. In recursive estimation procedure the weights are adjusted iteratively. Such an approach eliminates the estimation of the correlation matrix or cross-correlation vector frequently as in

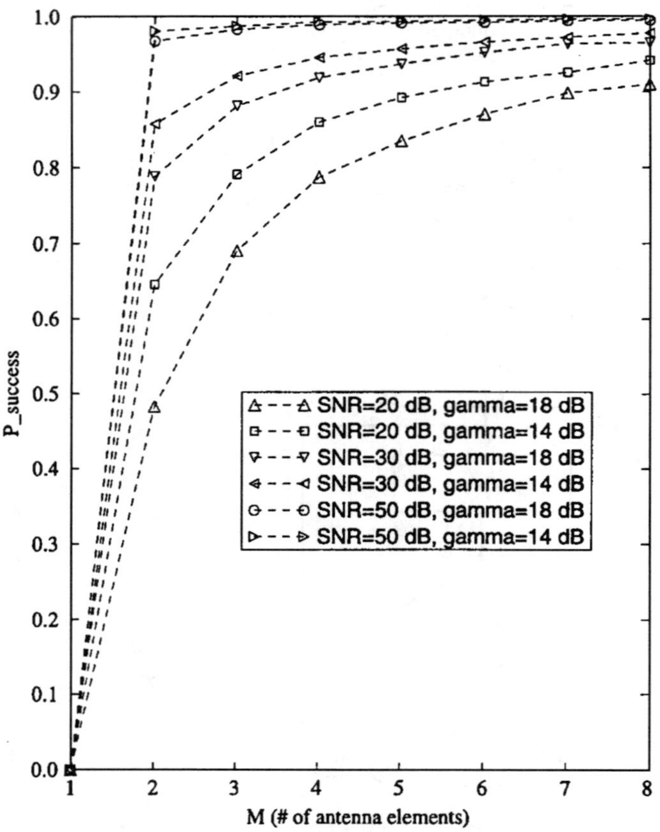

Fig. 12. Success probability $p_{2|1,M}$ for a 2-beam adaptive array for different values of M (# of antenna elements), SNR and threshold γ.

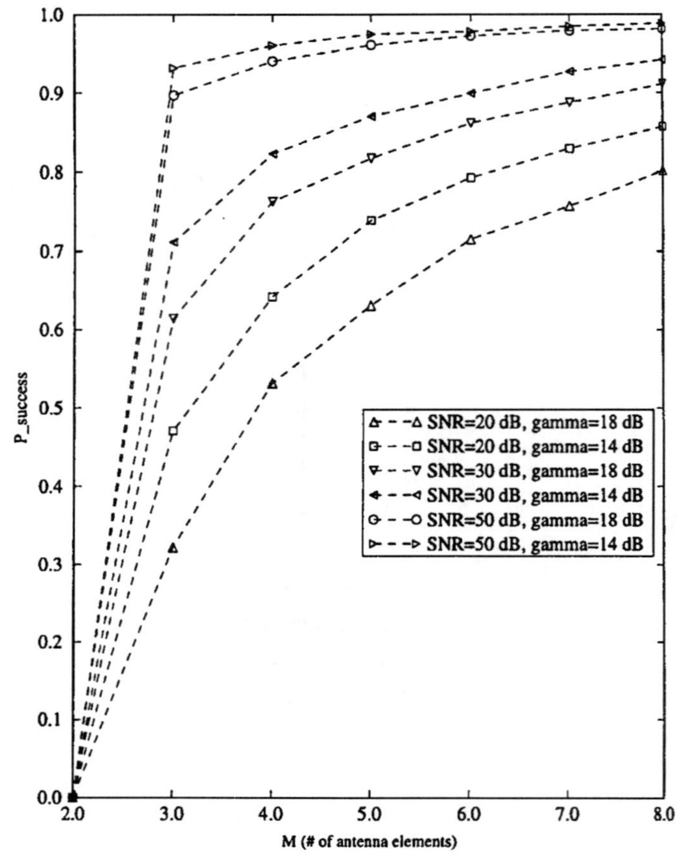

Fig. 13. Success probability $p_{3|2,M}$ for a 3-beam adaptive array for different values of M (# of antenna elements), SNR and threshold γ.

the DMI method. Moreover, each iteration trains the weight vector taps to track the incoming signal more accurately. Finally, at the end of the training sequence (with length Q), the new weight vector $\boldsymbol{w}$ emerges after the Qth iteration satisfying (17). The complexity of the least mean-square (LMS) algorithm, least square lattice (LSL) algorithm, fast transversal filter (FTF) algorithm, and QRD–RLS algorithm increases linearly with the length of the training sequence Q [14]. The number of operations required to perform one iteration of the algorithm is a measure of computational complexity. Operations consist of multiplications, divisions, and additions. Table I compares the computational complexities of four important adaptive algorithms: LMS, RLS, FTF, and LSL. All those operations use complex operands and Q is the length of the training sequence. It is worth mentioning that the numerically stable QRD–RLS algorithm also achieves $O(Q)$ complexity [14].

From Table I, the following observations can be made. The LMS algorithm is the least demanding in computational complexity. But the LMS algorithm has a slow convergence rate. The other algorithms belong to the recursive least squares family with much faster convergence rates. For a software radio smart antenna with L channels and K parallel beamformers for each channel, the overall complexity of the software beamformer is KL times that of the Table I entries. Another way to update the weight vectors is to employ the blind methods which do not need training sequences. But in general, the complexity for a single channel is more than $O(Q^2)$.

IV. Applications

This section presents two applications of smart antennas in a wireless network. In the first example, the smart antenna is deployed to reduce the blocking probability (or forced termination) of the handoff calls. The second example illustrates the use of a smart antenna in a CDMA network to achieve a more efficient power control algorithm. These two applications are computationally demanding, but can be implemented in software efficiently.

A. Traffic and Handoff Improvement with Smart Antennas

This section analyzes the network with adaptive array under traffic policies proposed in [12] and [13]. In this network model [14] it is assumed that when a new call (handoff or originating call) arrives, the adaptive array points one beam toward that user, and assigns one channel out of those L channels to that user. Each channel c_i for $i = 1, 2, \cdots, L$ may be assigned to one of the K users by K separate beams, using the K beamformers in parallel for each channel, as shown in Fig. 10. If the first beam of all channels is occupied, the new call is assigned to another beam. If there are no multipaths in the system, with K beamformers for each channel and L RF frequency channels, at most KL users can be accepted into the system. On the other hand, if there are multipath signals with effective α paths per user, then the number of antennas has to increase to null those multipaths. In this case, the effective number of usable channels would be KL/α. The flowchart

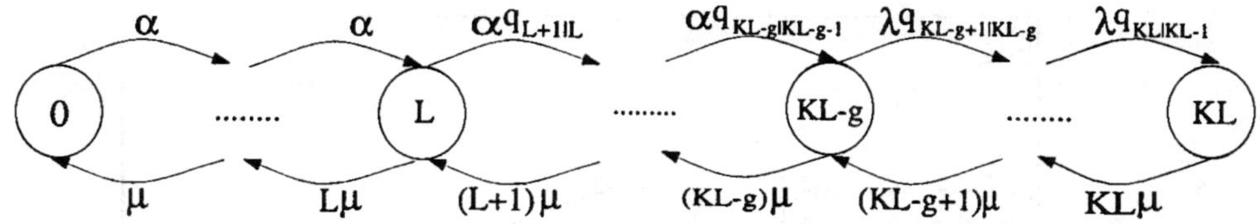

Fig. 14. State transition diagram of discrete Markov chain, the network model of the system with an adaptive array at the base station and guard channels.

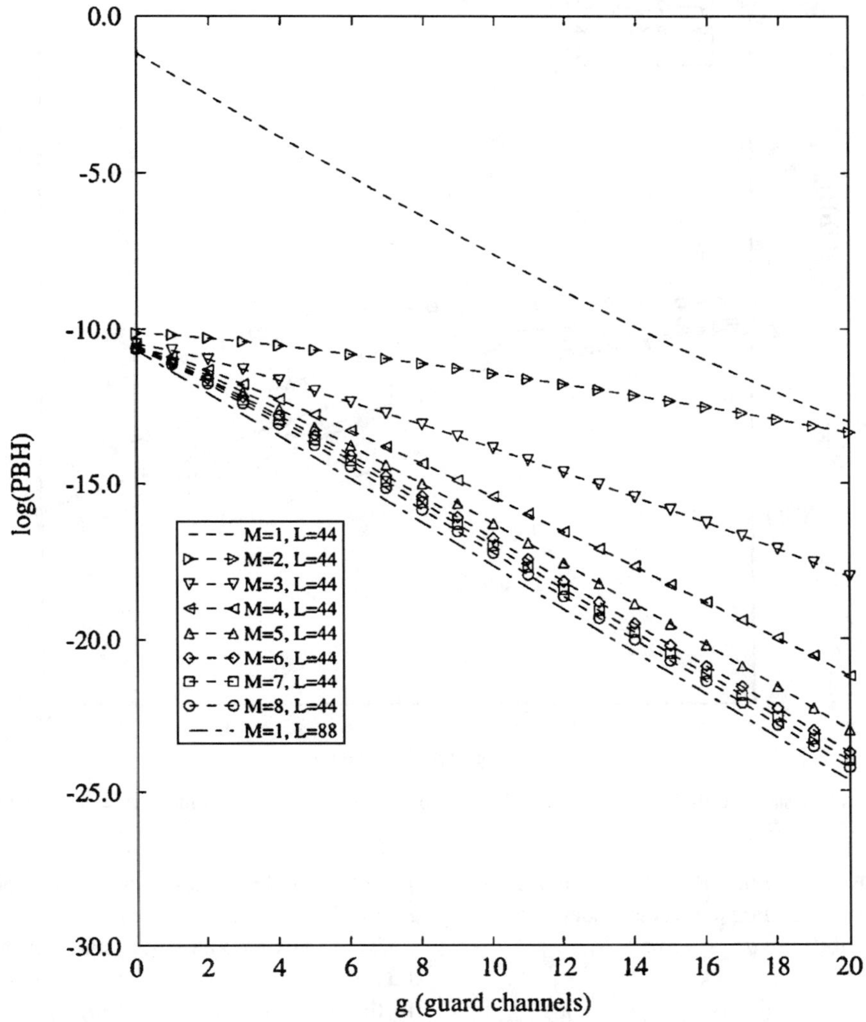

Fig. 15. Blocking probabilities of hand off calls (B_H) for different numbers of antenna elements M, with SNR $= 20, \gamma = 18$ dB, $a = 40, c = 8$ $\beta = 4$, and a two-beam adaptive array systsem.

of the channel assignment with smart antennas is illustrated in Fig. 11.

Assume that $(i - 1)$ cochannel transmitters successfully share the same channel. For acceptable link quality, the newly arrived ith transmitter shares that channel if $\Gamma_i \geq \gamma$, where Γ_i is given by (10) and γ is a system parameter which is dictated by the governing standard. For instance, in the IS-54 standard, γ is 14 dB, while in AMPS, it is 18 dB. The probability of establishing an acceptable link, i.e., $P(\Gamma_i > \gamma)$, is estimated by Monte Carlo simulation. In the network model, the probability density function for the angular position of the transmitters in the system is assumed to be $f_{\Theta_j}(\theta_j)$ for $1 \leq j \leq J$. The Monte Carlo simulations of the success probabilities $p_{2|1,M}$ and $p_{3|2,M}$, for two-beam and three-beam adaptive arrays, respectively (with M antenna elements), are shown in Figs. 12 and 13. A *call admission control* (CAC) is required to avoid degradations in system performance. If $\Gamma_i < \gamma$ or $i > K$, that user is not accepted into the system. The probability of the event $\Gamma_i \geq \gamma$ in the system, would yield the success probability $p_{i|i-1,M}$ that the ith transmitter can share the same channel, given that i transmitters are already using that channel. Clearly, the success probability $p_{i|i-1,M}$ depends on M. Since there are L distinct channels in the system and each channel may be reused up to K times, define the *probability of successful reception of* the $(n+1)$*th user into the system given that there are already* n *users in the system* as

$$q_{N_{t+1}|N_t}(N_{t+1} = n+1|N_t = n)$$

where N_t is the number of users in the system at time t before a new call (user) arrives into the system and t is the time

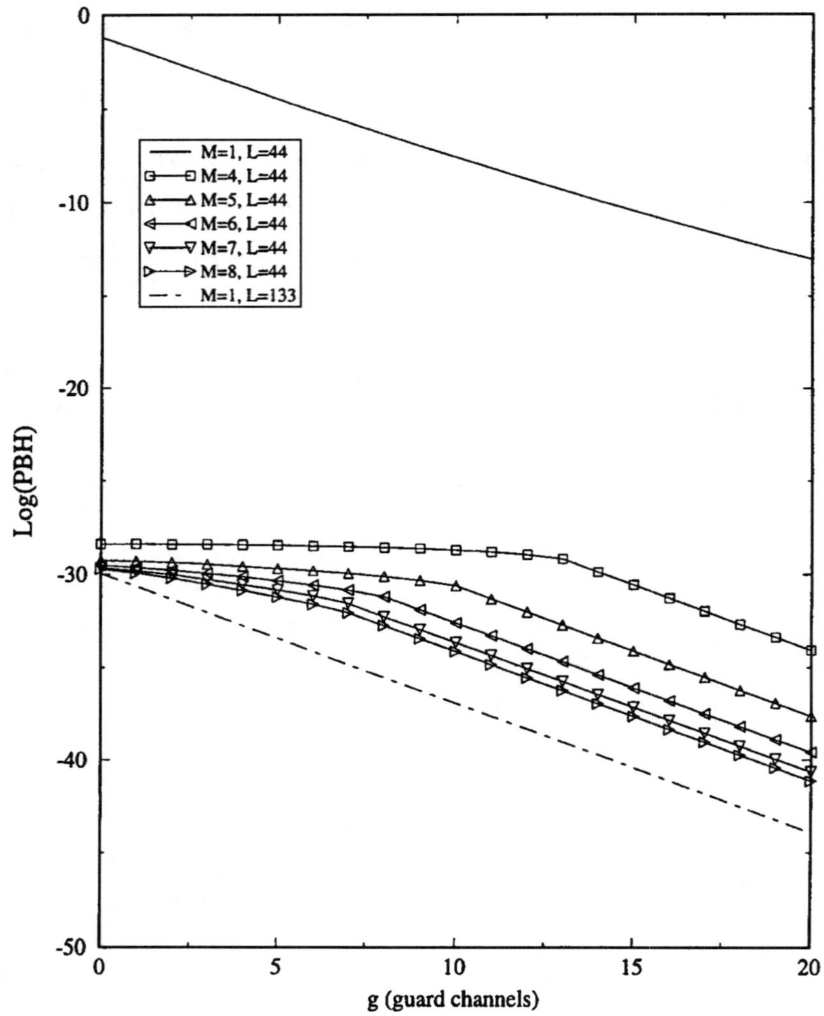

Fig. 16. Blocking probabilities of hand off calls (B_H) for different numbers of antenna elements M, with SNR $= 20, \gamma = 18$ dB, $a = 40, c = 8$ $\beta = 4$, and a three-beam adaptive array system.

index which increases by 1 at each epoch corresponding to a new handoff or originating call. For a K-beam adaptive array system, the success probabilities $q_{n+1|n}$ are [14]

$$q_{n+1|n} = \sum_{\{x_2 \cdots x_K | x_K \le \cdots \le x_2\}} \hat{q}(n+1, x_2, \cdots, x_K) \cdot p[(x_2, \cdots, x_K) | x_2 + \cdots + x_K = n - L] \quad (20)$$

where $\hat{q}(n+1, x_2, \cdots, x_K) = 1 - (1 - p_{2|1,M})^{(L-x_2)} \times (1 - p_{3|2,M})^{(x_2 - x_3)} \times \cdots \times (1 - p_{K|K-1,M})^{(x_{K-1} - x_K)}$, and $(x_2, x_3, \cdots, x_K)$ denotes the vector representing the number of calls assigned to the channels in the second beam up to the Kth beam. Let $\alpha = \lambda + \gamma, \mu = \eta + \nu, a = \alpha/\mu, b = \gamma/\mu, c = \lambda/\mu$, where γ and η are the arrival rate and service rate for originating calls, and λ and ν are the arrival rate and service rate for handoff calls, respectively. In the network with an adaptive array, the effective arrival rate into the cell at state $n, \lambda_{\text{eff}}(n)$ is [14], [52]

$$\lambda_{\text{eff}}(n) = \begin{cases} \alpha q_{n+1|n}, & n = 0, 1, \cdots, KL - g - 1 \\ \lambda q_{n+1|n}, & n = KL - g, \cdots, KL - 1. \end{cases} \quad (21)$$

Similarly, the effective service rate in the system at a given state n, $\mu(n) = n\mu n = 1, 2, \cdots, KL$. Fig. 14 illustrates the queueing model of the system with adaptive arrays under a guard channel traffic policy.

Using this model, the state probabilities $P(n)$, $n = 0, 1, \cdots, KL$ (where n is the number of ongoing calls in the cell) are derived [50], [51], and from there, the blocking probability B_O of originating calls and the blocking probability B_H of handoff calls are given by [14]

$$B_O = \left[\sum_{n=L}^{KL-g-1} (1 - q_{n+1|n}) \frac{a^n}{n!} \prod_{j=0}^{n-1} q_{j+1|j} + a^{KL-g} \sum_{n=KL-g}^{KL} \frac{c^{n-(KL-g)}}{n!} \prod_{j=0}^{n-1} q_{j+1|j} \right] P(0)$$

$$B_H = \left[\sum_{n=L}^{KL-g-1} (1 - q_{n+1|n}) \frac{a^n}{n!} \prod_{j=0}^{n-1} q_{j+1|j} + \sum_{n=KL-g}^{KL-1} (1 - q_{n+1|n}) \frac{a^{KL-g} c^{n-(KL-g)}}{n!} \prod_{j=0}^{n-1} q_{j+1|j} + \frac{a^{KL-g} c^g}{(KL)!} \prod_{j=0}^{KL-1} q_{j+1|j} \right] P(0). \quad (22)$$

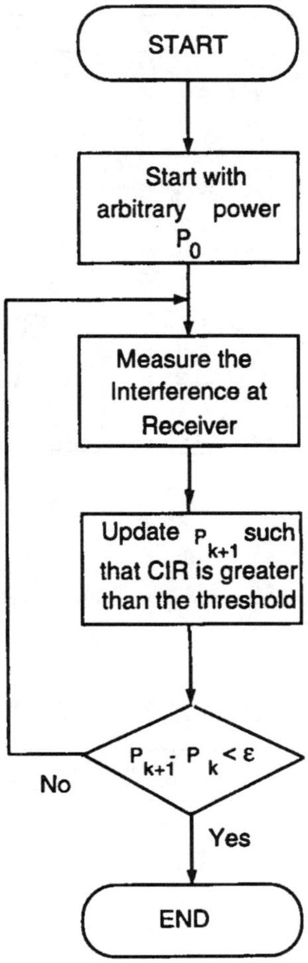

Fig. 17. Power control algorithm.

The performance of the system is evaluated with parameters drawn from [12]. Choose a cell with $L = 44$ channels, with a total offered traffic of $a = 40$ Erlangs (heavy traffic) and a handoff traffic of $c = 8$ Erlangs. Blocking probabilities of handoff and originating calls for different values of SNR, threshold γ, and number of antenna elements M are plotted in Figs. 15 and 16. In these simulations, the path loss exponent β is assumed to be four.

Using (22), the blocking probabilities for handoff calls B_H, for different numbers of antenna elements in a two-beam and three-beam adaptive array system, are plotted in Figs. 15 and 16. These two figures illustrate that for a given SNR and threshold γ, as the number of antenna elements M increases, the B_H decrease. Because of array limitations, the blocking probability B_H in a two-beam system ($K = 2, L = 44$) is slightly higher than the B_H for the system with $K = 1, L = 88$. The same property is observed for the three-beam system compared to the system with $K = 1, L = 3 \times 44$. These results confirm that a system using a two-beam adaptive array has almost the same effect as doubling the number of RF channels in the system. A three-beam array nearly triples the capacity [14].

B. Power Control and Beamforming in CDMA Networks

Power control is necessary to reduce the near–far effect in CDMA [53], to balance the interference in a wireless network,

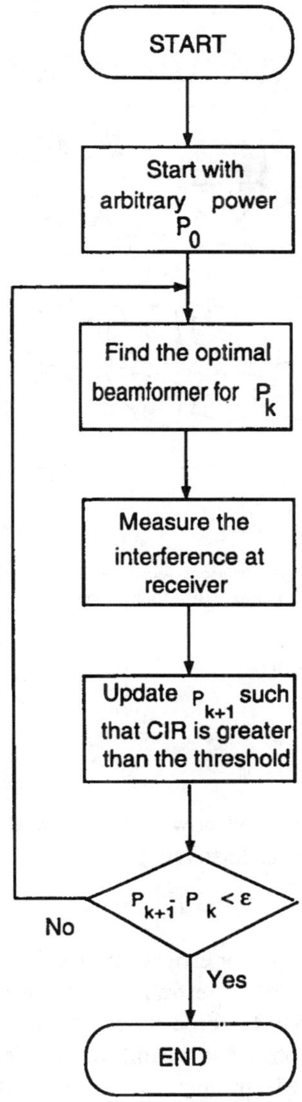

Fig. 18. The joint power control and beamforming algorithm.

and to minimize the transmitted power while the SNR is set to a threshold [54]. Distributed algorithms have been proposed [55] that adjust the power of each transmitter independently. It has been shown that these algorithms converge to the optimal power allocation that minimizes the transmitted power for each mobile while the link quality is guaranteed for each receiver. The nth step of the algorithm is given by

$$P_i^{n+1} = \frac{\gamma}{G_{ii}} \left(\sum_{j \neq i} G_{ji} P_j^n + N_i \right) = \frac{\gamma}{G_{ii}} I_i^n$$

where P_i^n is the transmitted power at the nth iteration, and G_{ij} is the link gain between the ith mobile and the jth base station. N_i is the thermal noise, and I_i^n the interference at the ith receiver. The right-hand side of this equation is a function of the interference at the ith receiver. In order to update the power, the interference I_i is measured at each receiver, and the power is updated by multiplying a constant γ / G_{ii}. The flowchart of the algorithm is shown in Fig. 17.

Each receiver measures the interference power by calculating the average power at the output of the receiver during a

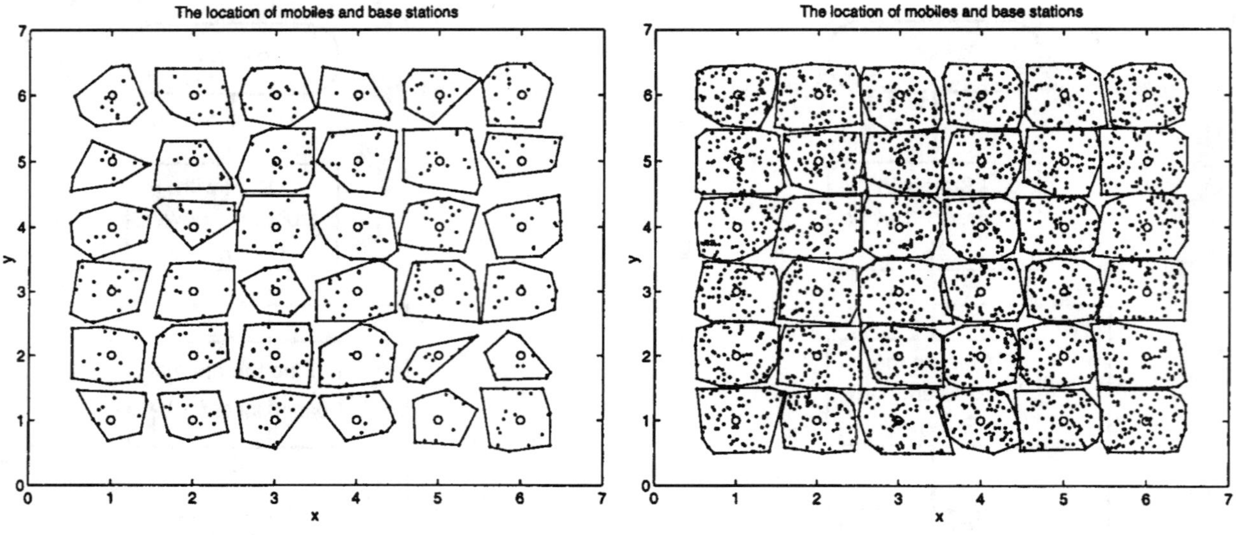

Fig. 19. Mobile and base stations locations for maximum number of users in: (a) power controlled network with single antennas and (b) power controlled network with smart antenna with four elements.

window of length W

$$E_i = \sum_{i=1}^{W} x_i^2.$$

It subtracts the received power due to the desired transmitter power from the total received power

$$I_i^n = E_i - G_{ii} P_i.$$

The complexity of power estimation is $4W + 1$ multiplications and $W + 1$ additions. The transmitted power can be readily updated up to a few hundred times per second [53].

Joint power control and beamforming algorithms are proposed in [56]–[58]. In these algorithms, the beamforming weight vectors and power allocations are updated jointly by a distributed algorithm. The nth iteration of the algorithm is as follows [58].

- The minimum error is calculated by

$$E_i = \sum_k |d(k) - \boldsymbol{w}_i^H \boldsymbol{x}_i(k)|^2.$$

- The power is updated by

$$P_i^{n+1} = P_i^n \frac{\gamma}{\Gamma_i} = \gamma P_i^n \frac{E_i}{1 - E_i}.$$

The weight vectors are updated by the beamforming algorithm only during the training phase, and are constant between the training intervals. The complexity of the beamforming algorithms is as discussed in the previous sections. The complexity of the first step of the power control algorithm is $4M + 1$ multiplications and one addition for the calculation of $E_{\min}$ and one division and addition for updating the power. This algorithm is depicted in Fig. 18.

The above algorithm converges to the jointly optimal beamforming weight vector and power allocation such that the transmitted power is minimized [56]. In order to show the performance of the algorithm, we have simulated a network of mobiles and base stations where mobiles are randomly distributed in the network as illustrated in Fig. 19. The total transmitted power as a function of total number of users in the network is shown in Fig. 20. As the number of users approaches the capacity of the network, the total power is increased significantly. As illustrated in this figure, with smart antennas with four elements at the base station, for the same number of users, one can reduce the transmitted power. It is also possible to increase the maximum number of users in the network while the SNR is above a threshold.

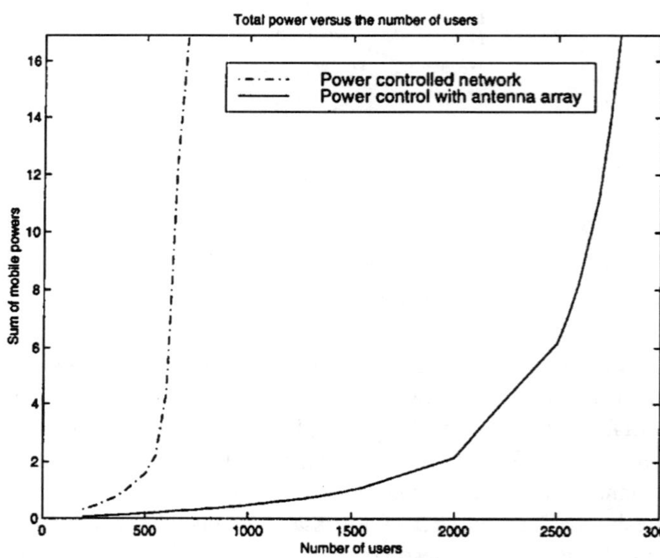

Fig. 20. The total mobile powers versus the number of users.

V. Conclusions

This paper has presented the software radio architecture with smart antennas. The advantages of software radio implementation of wireless communication systems include flexibility for adaptive beamforming with novel algorithms. This paper also has reviewed the commonly employed adaptive algorithms and their complexities. Finally, two examples illustrated the effectiveness of software radio smart antenna systems in increasing the system capacity and enhancing the CDMA power control.

ACKNOWLEDGMENT

The authors wish to express their gratitude to the anonymous reviewers for their valuable suggestions and helpful comments in the preparation of this paper.

REFERENCES

[1] J. Mitola, "The software radio architecture," *IEEE Commun. Mag.*, pp. 26–38, May 1995.

[2] ——, "Software radio technology challenges and opportunities," presented at the 1st European Workshop on Software Radios, May 1997.

[3] W. Tuttlebee, "The impact of software radio," presented at the Software Radio Workshop, Brussels, Belgium, May 1997.

[4] D. L. Tennenhouse and V. G. Bose, "The spectrumWare approach to wireless signal processing," *Wireless Networks*.

[5] V. Bose, M. Ismert, M. Welborn, and J. Guttag, "Virtual radios," this issue, pp. 591–602.

[6] V. G. Bose, A. B. Shah, and M. Ismert, "Software radios for wireless networking," presented at IEEE Infocom'98, San Francisco, CA, Apr. 1998.

[7] T. Turletti and D. Tennenhouse, "Estimating the computational of software GSM base station," in *Proc. IEEE ICC'97*, Montreal, Canada, June 1997.

[8] J. Kennedy and M. C. Sullivan, "Direction finding and smart antennas using software radio architectures," *IEEE Commun. Mag.*, pp. 62–68, May 1995.

[9] D. Everitt, "Traffic engineering of the radio interface for cellular mobile networks," *Proc. IEEE* (*Special issue on Wireless Networks for Mobile and Personal Communications*), pp. 1371–1382, Sept. 1994.

[10] ——, "Combined analysis of transmission and traffic in micro cellular mobile communications system," in *Proc. IEEE VTC*, May 1993, pp. 577–580.

[11] B. Jabbari, G. Colombo, A. Nakajima, and J. Kulkarni, "Network issues for wireless communications," *IEEE Commun. Mag.*, pp. 88–102, Jan. 1995.

[12] E. Posner and R. Guerin, "Traffic policies in cellular radio that minimizes blocking of handoff calls," in *Proc. 11th Teletraffic Cong. (ITC 11)*, Kyoto, Japan, Sept. 1985, pp. 294–298.

[13] R. Guerin, "Queueing-blocking system with two arrivals stream and guard channels," *IEEE Trans. Commun.*, vol 36, pp. 153–163, Feb. 1988.

[14] J. Razavilar, "Signal processing and performance analysis for optimal resource allocation in wireless networks," Ph.D. dissertation, Elect. Eng. Dept., University of Maryland, College Park, December, 1998.

[15] S. S. Haykin, *Adaptive Filter Theory.* Englewood Cliffs, NJ: Prentice–Hall, 1991.

[16] A. Naguib, A. Paulraj, and T. Kailath, "Capacity improvement with base-station antenna arrays in cellular CDMA," *IEEE Trans. Veh. Technol.*, vol. 43, Aug. 1994.

[17] B. Suard, A. Naguib, G. Xu, and A. Paulraj, "Performance analysis of CDMA mobile communication systems using antenna array," in *Proc. ICASP'93*, vol. IV, Apr. 1993.

[18] G. Xu, Y. Cho, A. Paulraj, and T. Kailath, "Maximum likelihood detection of co-channel communication signals via exploitation of spatial diversity," in *Proc. 26th Asilomar Conf. Signals, Syst., Comput.*, CA, Oct. 1992.

[19] A. Klouche-Djedid and M. Fujita, "Adaptive array processing applications for mobile telephone communication," *IEEE Trans. Veh. Technol.*, pp. 405–416, Aug. 1996.

[20] S. Swales, M. Beach, D. Edwards, and J. McGeehan, "The performance enhancement of multibeam adaptive base-station antenna for cellular land mobile radio systems," *IEEE Trans. Veh. Technol.*, pp. 56–67, Feb. 1990.

[21] T. Gebauer and H. Gockler, "Channel-individual adaptive beamforming for mobile satellite communications," *IEEE J. Select. Areas Commun.*, pp. 439–448, Feb. 1995.

[22] S. J. Orfanidis, *Optimal Signal Processing: An Introduction.* New York: Macmillan, 1985.

[23] J. H. Winters, "Smart antennas for wireless systems," *IEEE Personal Commun. Mag.*, vol. 5, pp. 23–27, Feb. 1998.

[24] B. Widrow and S. Stearns, *Adaptive Signal Processing.* Englewood Cliffs, NJ: Prentice-Hall, 1985.

[25] D. Donoho, "On minimum entropy deconvolution," in *Applied Time Series Analysis II.* New York: Academic, 1981, pp. 565–608.

[26] G. B. Giannakis and J. M. Mendel, "Identification of nonminimum phase systems using via higher order statistics," *IEEE Trans. Acoust., Speech, Signal Processing*, vol. 37, pp. 360–377, 1989.

[27] C. L. Nikias, "ARMA bispectrum approach to nonminimum phase system identification," *IEEE Trans. Acoust., Speech, Signal Processing*, vol. 36, pp. 513–525, Apr. 1988.

[28] J. K. Tugnait, "Identification of linear stochastic systems via second and fourth-order cumulant matching," *IEEE Trans. Inform. Theory*, vol. IT-33, pp. 393–407, May 1987.

[29] A. Benveniste, M. Goursat, and G. Ruget, "Robust identification of a nonminimum phase system: Blind adjustment of a linear equalizer in data communications," *IEEE Trans. Automat. Contr.*, vol. AC-25, pp. 385–399, June 1980.

[30] Y. Chen and C. L. Nikias, "Blind equalization with criterion with memory nonlinearity," *Opt. Eng.*, vol. 31, pp. 1200–1210, June 1992.

[31] D. N. Godard, "Self-recovering equalization and carrier tracking in two-dimensional data communication systems," *IEEE Trans. Commun.*, vol. COM-28, pp. 1867–1875, 1980.

[32] F. Gustafsson and B. Wahlberg, "Blind equalization by direct examination of the input sequence," *IEEE Trans. Commun.*, vol. 43, pp. 2213–2222, July 1995.

[33] G. Picchi and G. Prati, "Blind equalization and carrier recovery using a 'stop-and-go' decision-directed algorithm," *IEEE Trans. Commun.*, vol. COM-35, pp. 877–887, 1987.

[34] Y. Sato, "A method of self-recovering equalization for multi-level amplitude modulation," *IEEE Trans. Commun.*, vol. COM-23, pp. 679–682, June 1975.

[35] O. Shalvi and E. Weinstein, "New criteria for blind deconvolution of nonminimum phase systems (channels)," *IEEE Trans. Inform. Theory*, vol. 36, pp. 312–321, Mar. 1990.

[36] J. R. Treichler and B. G. Agee, "A new approach to multipath correction of constant modulus signals," *IEEE Trans. Acoust., Speech, Signal Processing*, vol. ASSP-31, pp. 349–372, 1983.

[37] J. R. Treichler and M. G. Larimore,"New processing techniques based on the constant modulus adaptive algorithm," *IEEE Trans. Acoust., Speech, Signal Processing*, vol. ASSP-33, pp. 420–431, Apr. 1985.

[38] S. Verdu, B. D. O. Anderson, and R. A. Kennedy, "Blind equalization without gain identification," *IEEE Trans. Inform. Theory*, vol. 39, pp. 292–297, Jan. 1993.

[39] J. K. Tugnait, "On fractionally-spaced blind adaptive equalization under symbol timing offsets using Godard and related equalizers," in *Proc. IEEE ICASSP'95*, vol. 3, May 1995, pp. 1976–1979.

[40] B. R. Petersen and D. D. Falconer, "Suppression of adjacent-channel, co-channel, and intersymbol interference by equalizers and linear combiners," *IEEE Trans. Commun.*, vol. 42, pp. 3109–3118, Dec. 1994.

[41] J. Yang and S. Roy, "On joint transmitter and receiver optimization for multiple-input-multiple output (MIMO) transmission systems," *IEEE Trans. Commun.*, vol. 42, pp. 3221–3231, Dec. 1994.

[42] J. R. Treichler and M. G. Larimore, "The tone capture properties of CMA-based interference suppressors," *IEEE Trans. Acoust., Speech, Signal Processing*, vol. ASSP-33, pp. 946–958, Aug. 1985.

[43] S. Mayrargue, "A blind spatio-temporal equalizer for a radio-mobile channel using the constant modulus algorithm (CMA)," in *Proc. IEEE ICASSP'94*, Adelaide, Australia, Apr. 1994, pp. III 317–320.

[44] D. Slock, "Blind joint equalization of multiple synchronous mobile users using oversampling and/or multiple antennas," in *Proc. 28th Asilomar Conf. Signal Syst. Comput.*, Pacific Grove, CA, Oct. 1994, pp. 1154–1158.

[45] E. Weinstein, A. Swami, G. Giannakis, and S. Shamsunder, "Multichannel ARMA processes," *IEEE Trans. Signal Processing*, vol. 42, pp. 898–913, Apr. 1994.

[46] E. Weinstein, A. V. Oppenheim, M. Feder, and J. R. Buck, "Iterative and sequential algorithms for multisensor signal enhancement," *IEEE Trans. Signal Processing*, vol. 42, pp. 846–859, Apr. 1994.

[47] D. Yellin and E. Weinstein, "Criteria for multichannel signal separation," *IEEE Trans. Signal Processing*, vol. 42, pp. 2158–2168, Aug. 1994.

[48] E. Moulines, P. Duhamel, J. Cardoso, and S. Mayrargue, "Subspace methods for the blind identification of multichannel FIR filters," *IEEE Trans. Signal Processing*, vol. 43, pp. 516–525, Feb. 1995.

[49] A. Naguib and A. Paulraj, "Performance of CDMA cellular networks with base-station antenna arrays," in *Proc. Int. Zurich Seminar Digital Commun.*, Zurich, Switzerland, Mar. 1994, pp. 87–100.

[50] R. B. Cooper, *Introduction to Queueing Theory*, 2nd ed., 1981.

[51] L. Kleinrock, *Queueing Systems, Vol. I*, 1985.
[52] R. A. Monzingo and T. W. Miller, *Introduction to Adaptive Arrays*. New York: Wiley, 1980.
[53] J. Razavilar, F. Rashid-Farrokhi, and K. J. R. Liu, "Blocking probability of handoff calls and carried traffic in wireless networks with antenna arrays," in *Proc. 31st Asilomar Conf. Signals, Syst., Comput.*, Pacific Grove, CA, Nov. 1997, pp. 635–639.
[54] "An overview of the application of code division multiple access (CDMA) to digital cellular systems and personal cellular networks," Qualcomm Inc., May 1992.
[55] J. Zander, "Performance of optimum transmitter power control in cellular radio systems," *IEEE Trans. Veh. Technol.*, vol. 41, pp. 57–62, Feb. 1992.
[56] ——, "Distributed cochannel interference control in cellular radio systems," *IEEE Trans. Veh. Technol.*, vol. 41, pp. 305–311, Aug. 1992.
[57] F. Rashid-Farrokhi, L. Tassiulas, and K. J. R. Liu, "Joint optimal power control and beamforming for wireless networks with antenna arrays," in *Proc. IEEE Global Commun. Conf. (GLOBECOM-96)*, London, England, Nov. 1996, pp. I-555–559.
[58] ——, "Joint optimal power control and beamforming in wireless networks with antenna arrays," *IEEE Trans. Commun.*, vol. 46, pp. 1313–1324, Oct. 1998.
[59] F. Rashid-Farrokhi, K. J. R. Liu, and L. Tassiulas, "Transmit and receive diversity and equalization in wireless networks with fading channels," presented at IEEE Globecome'97, Phoenix, AZ, Nov. 1997.

The Rapidly Deployable Radio Network

Joseph B. Evans, *Member, IEEE*, Gary J. Minden, *Member, IEEE*,
K. S. Shanmugan, *Fellow, IEEE*, Glen Prescott, *Member, IEEE*, Victor S. Frost, *Fellow, IEEE*,
Ben Ewy, Ricardo Sanchez, Craig Sparks, *Member, IEEE*, K. Malinimohan,
James Roberts, *Senior Member, IEEE*, Richard Plumb, *Senior Member, IEEE*, and Dave Petr, *Senior Member, IEEE*

***Abstract*—The Rapidly Deployable Radio Network (RDRN) is an architecture and experimental system to develop and evaluate hardware and software components suitable for implementing mobile, rapidly deployable, and adaptive wireless communications systems. The driving application for the RDRN is the need to quickly establish a communications infrastructure following a natural disaster, during a law enforcement activity, or rapid deployment of military force. The RDRN project incorporates digitally controlled antenna beams, programmable radios, adaptive protocols at the link layer, and mobile node management. This paper describes the architecture for the Rapidly Deployable Radio Network and a prototype system built to evaluate key system components.**

***Index Terms*— Adaptive protocols, mobile networking, programmable antenna beamforming, software radios, wireless networks.**

I. Introduction

QUICKLY evolving events, such as natural disasters, law enforcement actions, or military force depoyment require adaptive mechanisms to rapidly establish a communication infrastructure with minimal *a priori* planning and configuration. The Rapidly Deployable Radio Network (RDRN) project created an experimental systems to develop and evaluate architectures, protocols, network control software, and prototype hardware for mobile, rapidly deployable, and adaptive wireless communications systems. The approach used in the RDRN is to implement a radio network based on switches. When deployed, the switches establish a backbone network among themselves over high-capacity radio links and self-configure into an appropriate topology. Users access switches over a "cellular-like" architecture. That is, each switch supports users in a limited cell. Users are mobile, and there will be handoffs within a switch and between switches. Switches can connect to other communication facilities, such as fiber links, wire links, or satellite links.

Rapid deployment and mobility imply limited opportunity for *a priori* radio frequency, topology, and deployment-specific engineering. Hence, adaptive techniques at the link, network, and management levels are required. The RDRN system uses the following adaptive techniques.

1) Digitally beam-steered phased array antennas establish the high-speed point-to-point links and track mobile users.
2) Programmable waveform generation changes modulation, frame lengths, and forward error correction at the data link layer.
3) Asynchronous transfer mode (ATM) virtual circuits follow mobile users through the system.
4) Switch and user locations to anticipate handoffs and required switching resources.

RDRN represents an initial step beyond the conventional view of a software radio as a configuration of radio frequency sections, analog-to-digital converters, digital-to-analog converters, and a digital signal processor. The RDRN system incorporates a software radio in the larger context of programmable antennas, mobile nodes, varying topologies, and networking services.

The RDRN project defined the architecture and system components, implemented and evaluated the components, and built two prototype edge nodes. The remaining sections of this paper discuss the RDRN architecture and system components, the prototype system, and results of experiments.

II. System Overview

When the need arises for a communications infrastructure in a place that has little or no infrastructure or the infrastructure has been destroyed, RDRN is deployed. An RDRN system consists of two node types: 1) *edge nodes* that include network management, switching functions, and interconnection to other networks; and 2) *remote nodes* that are mobile and serve as user access points.

Deployment involves placing edge nodes across the area to be covered. Edge nodes support three radio networks: 1) a low-capacity order-wire network, 2) multiple high-capacity point-to-point radio links, and 3) multiple "cellular-like" radio links. Edge nodes also contain a switch for routing traffic between the radio links, control and management software, and possible interfaces to wired or satellite communication services. When activated, the edge node determines its location via the global positioning system (GPS), and begins broadcasting its location on the low-capacity omnidirectional channel and listens for other systems broadcasting their locations. As other nearby edge nodes are detected, the local node attempts to establish a high-capacity point-to-point radio link to these nearby nodes. See Fig. 1.

Manuscript received September 1997; revised November 2, 1997, March 19, 1988, June 15, 1998, and September 15, 1998. This work was supported by the Defense Advanced Research Projects Agency under Contract J-FBI-94-223, with funding from Sprint Corporation and the State of Kansas.

J. B. Evans, G. J. Minden, K. S. Shanmugan, G. Prescott, V. S. Frost, R. Sanchez, C. Sparks, J. Roberts, R. Plumb, and D. Petr are with the Information and Telecommunication Technology Center, University of Kansas, Lawrence, KS 66045 USA.

B. Ewy is with Sprint Corporation, Overland Park, KS USA.

K. Malinimohan is with Cadence Corporation, Santa Clara, CA USA.

Publisher Item Identifier S 0733-8716(99)02979-0.

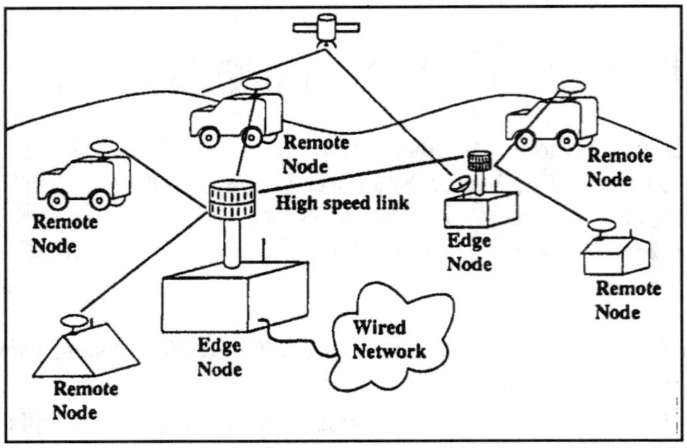

Fig. 1. RDRN system architecture. Edge nodes are deployed covering the area of interest. Edge nodes perform switching, and can interconnect to wired or space-based services. Mobile users access edge nodes via remote nodes that provide a cellular-like service. The RDRN tracks the location of remote nodes and adapts to their current positions.

The three radio networks may each be implemented with multiple radio transceivers. For example, the high-capacity point-to-point links will require a separate transceiver for each beam. Likewise, the "cellular-like" user access network will require a separate transceiver for each beam. The interconnection of transceivers to possibly shared antennas will depend on the frequency bands, the antenna characteristics, and performance issues.

Establishing the link requires consideration of multiple criteria.

1) Available carrier frequencies within the band.
2) Modulation and coding techniques (BPSK, QPSK, spread-spectrum codes).
3) Avoiding interference from other edge nodes in the system that are not directly linked.
4) Ensuring the edge nodes achieve and maintaining a connected topology.
5) Steering an antenna beam toward the nearby nodes and nulls toward interferers.

Link coordination takes place over the low-capacity, omnidirectional, order-wire link, while the local antenna is configured to steer its beams toward nearby edge nodes, establishing the high-capacity links. The backbone topology is based on the consistent labeling technique [1], [2] and is reported in [3]. The consistent labeling problem is a generalization of assigning labels to objects under a set of constraints, and is similar to the problem of assigning colors to map entities, or scheduling jobs on equipment, or packing small boxes in larger crates. While the problem is recognized as NP complete, the approach serves as a solid foundation for the establishment of the backbone topology.

Edge nodes implement a switching function to route traffic between radio links and possible wired or satellite services. The asynchronous transfer mode (ATM) provides the switching function and multiplexing between multiple streams which require independent classes of service. It is easily connected to wired services, and takes advantage of standards in the areas of signaling, quality of service, and flow control, among others. At the relatively low bandwidths of mobile wireless networks, the relatively quick multiplexing of ATM allows time-constrained traffic access to the channel in a timely manner.

TABLE I
NETWORK PROTOCOLS IMPLEMENTED IN THE RDRN PROTOTYPE

Protocol	Management
TCP/UDP	
IP	IP Routing
IP over ATM	Classical IP over ATM
AAL 5	
ATM	Q.2931 and RDRN NCP
Adaptive HDLC	RDRN NCP
Radio Link	RDRN NCP

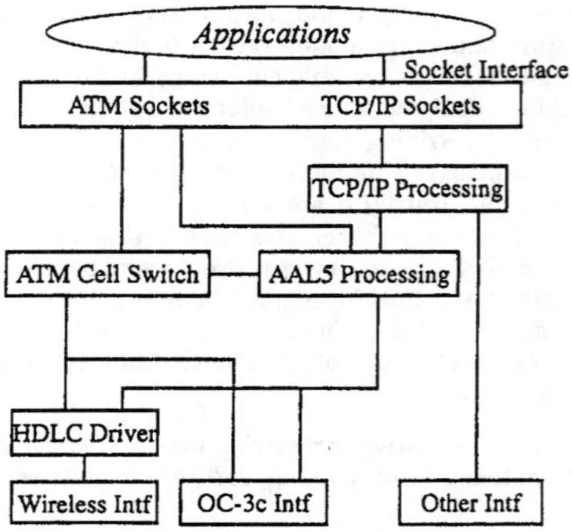

Fig. 2. RDRN protocol stack.

Users access the edge backbone network through remote nodes and the cellular-like radio links around edge nodes. Edge nodes generate multiple digitally formed beams, each of which can be modulated independently. The prototype implements a 1 Mbaud (1 Mbit/s BPSK or 2 Mbit/s QPSK) TDMA protocol to support up to 64 users per beam. TDMA slots may be allocated to users for capacity management purposes.

Remote nodes obtain location information via GPS or other location techniques, and report their location to their connected edge node. Location information is used by the edge node for two purposes: 1) to steer the antenna beams to maximize the minimum signal-to-interference ratio for a set of nodes at known locations [4], and 2) to determine when to assign a remote node to a different beam or different edge node.

III. RDRN NETWORK CONTROL

This section describes the RDRN adaptive protocols and the network control program (NCP) that manages those protocols. Each protocol is first described, followed by the overall control architecture. The protocols are listed in Table I.

A. RDRN Wireless ATM Protocols

The RDRN protocol stack is shown in Fig. 2. This stack is implemented in the edge nodes and the remote nodes; however, radio link management is only implemented in the edge nodes.

RDRN protocols and management components have been implemented in the Linux operating system running on Intel ×86 compatible computers. Linux provides an open development system with many contributors around the world running on a wide variety of computer architectures. The system supports applications running over both native-mode ATM as well as TCP/IP over ATM. Considerable development work was done to support standard ATM on Linux by researchers at the Laboratoire de Reseaux de Communication (LRC) at EPFL in Switzerland [5]. In particular, they developed a BSD-sockets-based application programming interface (API) to support native-mode ATM applications, as well as to support classical IP over ATM [6] for TCP/IP-based ATM applications [7].

The RDRN NCP supports applications through the standard UNIX socket interface to the networking protocols within the operating system kernel. The socket interface provides conventional TCP/IP services and ATM services. Applications can interface to the ATM service either through a framing protocol, such as AAL5, or through raw ATM cells. Within the kernel, the network layer implements the Internet Protocol (IP), whose primary purpose is to provide routing between subnetworks with service for the Transmission Control Protocol (TCP), User Datagram Protocol (UDP), and other transport layers.

Classical IP (CLIP) and the ATM Address Resolution Protocol (ATMARP) [6] are initial standard solutions for carrying IP traffic over the ATM links. However, there are several weak points such as requiring a router to connect separate logical IP subnets (LIS), even when the subnets are directly connected at the ATM level, and requiring an ATMARP server to provide address resolution for a single LIS. The nonbroadcast multiple-access Next Hop Resolution Protocol (NHRP) [8] provides a better solution, but it is still in draft form. The RDRN architecture implements CLIP and supports both permanent virtual circuits and switched virtual circuits via ATMARP.

To this base system there were added four software components in the prototype system: 1) the Bellcore Q.Port ATM signaling system [9], 2) an ATM driver for handling AAL5 protocols, 3) an ATM cell switch, and 4) a driver for RDRN wireless links. The Bellcore Q.Port system provides a reference implementation of the ATM user–network interface (UNI). This software is used for call setup and teardown, virtual circuit management, and interoperability with wired systems. The ATM driver for AAL5 is a custom software component that can interface to conventional ATM interfaces and/or the RDRN wireless interface at the frame level, and handles the encapsulation of IP for the network layer. The ATM driver also distinguishes between virtual circuits (VC's) that carry AAL5 frames and those that carry raw ATM cells. In the case of VC's carrying AAL5 frames, segmentation and reassembly functions are handled in hardware on conventional interfaces and software for the RDRN wireless interface. Otherwise, for nonframed VC's, single ATM cells are handed to the ATM cell switch for switching to other VC's and then passed to the appropriate interface. The ATM cell switch is not present in remote nodes.

RDRN nodes implement an adaptive HDLC link layer over the wireless link. ATM cells are collected into a modified HDLC frame, called the wireless ATM frame or WATM frame for transmission. In the case of edge-node-to-edge-node traffic, ATM cells from any VC can be included in the WATM frame. In the case of edge-node-to-remote-node traffic, ATM cells must be sorted into separate queues, one for each remote node. Only cells destined for a single remote are included in the WATM frame for transmission under the TDMA media access protocol. The WATM frame consists of a start-of-frame flag (1 byte), a link identifier (1 byte), a control byte (1 byte), one–nine ATM cells (53–477 bytes), and a 16-bit CRC field (2 bytes). Multiple ATM cells are included in each WATM frame to reduce the effect of encapsulation overhead necessary for addressing in the TDMA access mechanism and error control. No ATM header compression is used at this time. Likewise, we have not included forward error correction in the WATM frame. The number of ATM cells in a frame is based on resources at each end of the link and the link condition.

Radio link quality at the link level is estimated by counting the number of WATM frames received in error. This information is communicated upstream to the transmitting node. The HDLC link layer adapts to link quality in two ways. As the link quality goes down, fewer ATM cells are included in each WATM frame. Thus, any noise burst or fade over a few tens of microseconds will cause fewer ATM cells to be dropped. The second technique, used for loss-sensitive traffic, is to transmit multiple copies of the WATM frame, for example, two or three copies. Experiments have shown that WATM frame length adjustment with up to two copies of each frame transmitted is sufficient for most situations [10]. The protocol uses a sliding window and go-back-N ARQ scheme.

B. Network Control Program

The RDRN Network Control Program (NCP) is the primary control program for the system. As such, it monitors the backbone topology, locating remote nodes, computes antenna beam patterns, and implements connection handoffs between antenna beams and between edge nodes. The NCP executes on edge nodes. To perform these functions, the NCP monitors remote node locations and link quality to make decisions on configuration changes. The NCP causes antenna beams to be adjusted as remote nodes move, and reroutes ATM virtual circuits as remote nodes move between beams and edge nodes.

1) Maintaining Backbone Topology: The current RDRN implementation assumes that edge nodes remain stationary during the operation of the network. However, edge nodes can enter or exit the system, causing backbone topology changes. As mentioned above, the algorithms for determining backbone topology are based on a consistent labeling approach. In consistent labeling, constraints are expressed as pairwise constraints, e.g., if node A transmits to node B on frequency F, then node B can transmit to node A on any frequency except F. The consistent labeling algorithm considers all pairwise constraints, and determines a consistent labeling of all links.

The consistent labeling problem is a generalization of the graph coloring problem, the cellular frequency assignment problem, and many others [1], [2]. Suppose that one has cellular frequencies $F_1, \cdots, F_7$. Further, suppose that cells A

and B are adjacent, and that cell A is assigned frequency F_1. Then cell B can only use frequencies $F_2, \cdots, F_7$. This constraint can be expressed as a list of 4-tuples

$$\{\langle A, F_1, B, F_2\rangle, \langle A, F_1, B, F_3\rangle, \cdots, \langle A, F_1, B, F_7\rangle\}$$

indicating the allowable combinations of cell A, the frequency assigned to cell A, cell B, and the frequencies assigned to cell B. The consistent labeling algorithm takes such a list describing the allowable combinations for all cell pairs, and determines a consistent assignment of frequency to cells meeting all of the constraints.

The advantage of the consistent labeling approach is that multiple criteria, such as frequency, beam direction, and spread-spectrum code, can easily be included in determining the backbone topology. The disadvantage is that the labeling problem is NP complete. However, due to the limited number of edge nodes, the computational speed of available computers, and heuristics that can be incorporated into the implementation, this approach is expected to be sufficient for an RDRN-type system.

2) Monitoring Locations of Remote Nodes: The NCP at each edge node monitors the location of nearby remote nodes through either the order-wire link or in-band signaling. From this information, the NCP determines whether to adjust an antenna beam to accommodate remote node motion or to switch the remote node to a different beam. In the former case, the edge node considers the locations of all of the remote nodes within a current beam, and computes new antenna beam patterns to maximize the minimum signal-to-interference ratio across all remote nodes [4]. In the latter case, the edge node moves a remote node to a new antenna beam, re-calculates the antenna beams for the "source" and "destination" beams, and reroutes all necessary ATM virtual circuits. In the current system, the beams are re-calculated every 5 seconds.

An RDRN variant predicts the motion of remote nodes. Future positions of remote nodes are predicted from recent positions and a simple velocity vector. Using predicted positions, the RDRN system can preallocate resources to make smoother handoffs between beams or nodes. The system is described in [11].

IV. RDRN Radio Transceiver

The radio transceiver supports the RDRN system requirements for rapid deployment, and responds to a changing radio-frequency environment. The radio system allows multiple transmit beams to be steered to multiple remote node users from the same antenna at the same frequency, thus allowing spatial reuse of the frequency bands. It simultaneously allows the steering angles of the beams and the modulation types used within them to adapt to a changing mobile environment.

Before the specifics of the hardware design are presented, introductory information is given that places perspective on its design. Flexible software-based modulation and spatial reuse of frequencies are the main functions of the hardware. These topics are therefore introduced first.

The details of beamforming are discussed, and various analog and digital implementations are compared and contrasted. They are followed by a description of the hardware implementation that combines the task of beamforming and flexible IF modulation into a single cost-effective solution. Experimental results are then offered in support of the design. Finally, the receiver system and digital beamforming in the receiver are discussed.

A. Software-Radio-Based Flexible Modulation

Recent advances in analog-to-digital converter technology have stimulated an increasing interest in software-defined radio technology. With the capacity to digitize an IF or even an RF signal comes the ability to implement radio receivers that can be reprogrammed to emulate virtually any type of traditional receiver. One of the main goals of the RDRN proof of concept system is to provide a research testbed for the development of adaptive software-defined radios. The system presented in this section is a programmable transmit modulator only. The design is restricted to any type of phase or amplitude modulation. A software-radio-based beamforming receiver is the subject of future research.

B. Spatial Reuse

Spatial reuse of frequencies requires that the same carrier frequency be reused at least twice by the transmitter within what is considered to be a traditional cell. This can be achieved in different ways. Fig. 3 shows a two-user scenario where the positions of the transmitting antenna labeled edge node and both of the remote nodes are all known. The steering angles for the two beams Θ_1 and Θ_2 can therefore be calculated by simple geometry.

A brute force and intuitive approach to spatial reuse of the same carrier frequency would be to use two separate microwave "dishes" both located at the transmitting edge node, and simply point them at the individual users. This approach would work, and it describes well the requirements of a system that employs spatial reuse. However, it also has some serious drawbacks. The major disadvantage is that the individual beams must be steered by physically rotating the dishes to track the moving end users, and the beam patterns are determined at the time of dish manufacture.

The approach used in the RDRN prototype system is an electronically controlled beam-steering system in which multiple beams can be produced by a multielement phased-array antenna. The prototype system implements an equally spaced eight-element linear array.

C. Transmit Beamforming

Several antennas can be arranged in space and interconnected to produce a directional radiation pattern. This configuration is called an antenna array or, simply, an array [12]. When the phase of the signals reaching the individual antenna elements is modified in a useful way, then the antenna is called a phased array. This is the type of antenna being used in the RDRN prototype system to achieve spatial frequency reuse. The radiation pattern of an array antenna is controlled by the type of individual elements used, their orientations, their positions in space, and the amplitude and phase of the

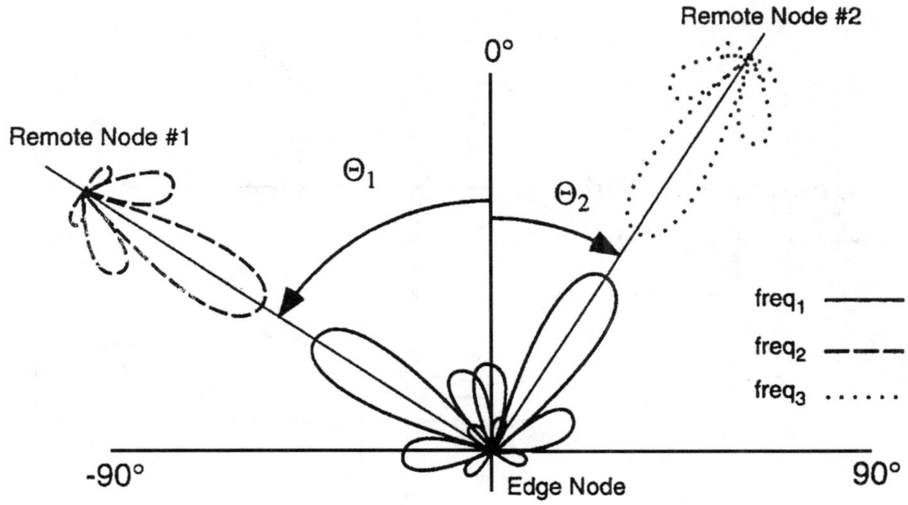

Fig. 3. Spatial frequency reuse by beamforming.

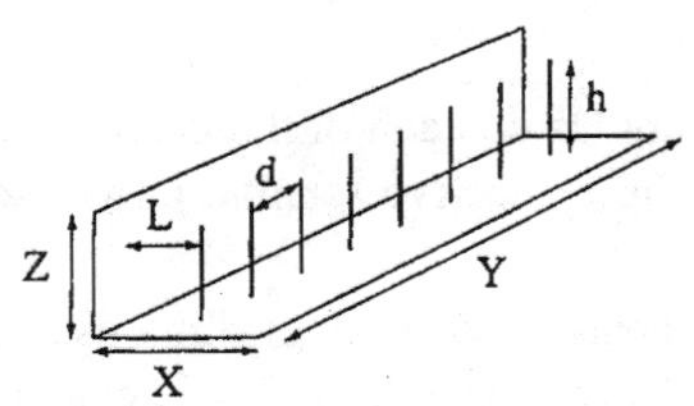

f = 1.25 GHz (λ=24.1 cm)
h = 5.4 cm (0.224 λ)
d = 12.1 cm (0.5 λ)

X = 15.9 cm (0.66 λ)
Y = 100.3 cm (4.16 λ)
Z = 15.9 cm (0.66 λ)
L = 7.6 cm (0.32 λ)

Fig. 4. A linear array antenna.

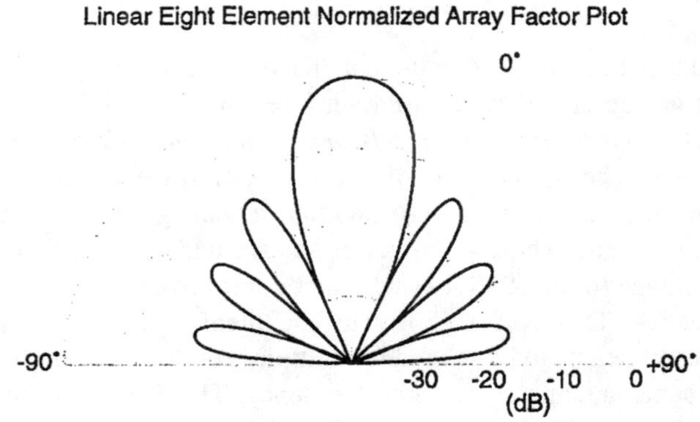

Fig. 5. Polar array factor of an equally excited eight-element linear array.

currents feeding them [12]. When the radiation pattern, called the array factor, of an antenna can be controlled in such a way as to selectively radiate energy in desired directions while simultaneously attenuating energy in others, this is called beamforming. Beamforming is effectively the result of far-field constructive and destructive interference patterns. A discussion of beamforming and how it applies to the RDRN prototype system follows.

1) Beamforming via Multiple-Element Linear-Array Antennas: A uniformly excited, equally spaced linear array provides the simple example. It consists of N active elements in a line spaced at regular intervals that are usually 1/2 wavelength or less of their center operating frequency. Fig. 4 is an example of a linear array of the type used in the RDRN prototype system.

If the eight-element antenna depicted in Fig. 4 were fed eight identical signals of equal phase and amplitude, the theoretical broadside far-field pattern or array factor in Fig. 5 would result.

The normalized array factor appearing in Fig. 5 is given by

$$f(\psi) = \frac{\sin(N\psi/2)}{N\sin(\psi/2)} \approx \frac{\sin(N\psi/2)}{N\psi/2} \tag{1}$$

where N is the number of antenna elements and is equal to eight and Ψ is the angle off broadside [12]. It should be noted that the width of the main lobe decreases with increasing N. The reduction of the main lobe width is what allows for spatial reuse. The cost involved in a large number of active antenna elements and the associated electronics needed to drive them must be weighed against the benefits of such an antenna pattern. The design goal for a multiple-beam system is to find the minimum number of elements that can support the number of beams required at the desired performance or signal-to-interference level. For the RDRN system, these cost and performance issues were balanced. The resultant tradeoff placed the number of antenna elements at eight.

2) Beamsteering by Application of Complex Weights: The broadside far-field antenna pattern shown in Fig. 5 is only as useful as the previously mentioned microwave dish example in that it, too, would have to be physically pointed at the mobile user to be of any use. A linear array antenna becomes more useful when it is used as a linear phased array. By manipulating the relative phase and amplitude of the RF waveform on a per-antenna-element basis, different interference patterns result, and it is possible to rapidly "steer" a beam in a particular direction. Fig. 6 is a block diagram of this process that depicts the application of N complex weights to an N element array.

The weights are applied to the same RF signal on a per-antenna-element basis, and the effect is the ability to spatially direct the beam in a desired direction without the need to physically rotate the array. Determination of the proper complex weights needed to effect the desired amount of

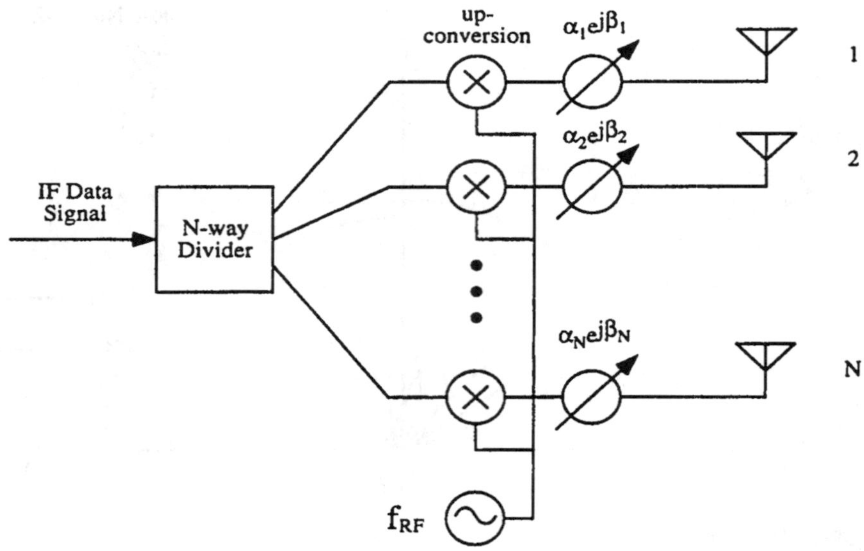

Fig. 6. Beamsteering by the application of complex weights.

beamsteering is the function of the beamforming algorithms that are described in the following sections.

3) Simultaneous Multiple Beams at the Same Carrier Frequency: The ability to rapidly steer a beam by electronic rather than physical means is an important advantage of a phased-array antenna. There is, however, another and more significant advantage to the RDRN system that comes from phased-array antennas. This is the ability to use them to form multiple transmit beams to be steered to multiple mobile end users from the same antenna at the same frequency. This is spatial reuse, and is one of the main goals of the RDRN proof-of-concept system.

The quantity and quality (defined by signal-to-interference ratio) of the multiple beams is controlled by many factors of which the number of antenna elements, their physical topology, and the proper application of complex weights are some of the most important factors. Fig. 7 depicts a way to form two simultaneous beams at the same frequency.

This system is the same as the single beam case appearing in Fig. 6 with the addition of a second set of inputs. The separate inputs from the two different beamforming processes are simply summed before the antenna. The system is linear, and because superposition holds for this electrical summation, the output from the antenna array will be the spatial sum of the individual antenna patterns. This process can be indefinitely expanded to form as many simultaneous beams as desired. There are, of course, some practical limits. As seen in the previous graphs of typical linear array antenna patterns, sidelobes exist in the form of $\sin(Nx)/Nx$ outputs. These sidelobes will cause interference to any other beams that are simultaneously being produced at the same frequency. Ways to reduce the level of interference include increasing the number of antenna elements, steering the beams further away from each other, and the proper application of complex weights to effect nulling patterns in the direction of other beams.

The proper application of the complex weights requires answers to at least the following questions.

1) How many beams need to be formed?
2) At what angles do each of the beams need to be steered?
3) What is the respective relative power contained in each beam?

The current beamforming algorithms used for RDRN were developed by Haas and Petr [13]. The algorithm takes the number of beams and the locations of users surrounding a transmitter, and calculates the steering angles and relative beam power such that a local maximum in the minimum signal-to-interference ratio is achieved.

4) Beamforming for Arbitrary Modulation Schemes: Beamforming is only of use in a communication system when its carrier frequency is modulated with some information-bearing signal. Beamforming in this case can be described for a generic modulation scheme as follows [14]. Let the signal samples be represented by the vector

$$s = (s_{-\infty}, \cdots, s_{-1}, s_0, \cdots, s_t) \tag{2}$$

where s is assumed to include the time index t implicitly. The most general form of modulation can be expressed as

$$x(t) = \mathrm{Re}[\alpha(s) \exp(j2\pi f_{\mathrm{IF}} t + \phi_0 + \Phi(s))] \tag{3}$$

where $\alpha(s)$ represents any form of amplitude modulation, f_{IF} is the IF carrier frequency, ϕ_0 is an initial phase, and $\Phi(s)$ represents any form of phase modulation. In most modulation schemes, $\alpha(s)$ will reduce to $\alpha(s_t)$ where t represents the current time index or, in phase-shift keying, further reduces to a constant α. $\Phi(s)$ can represent continuous-phase frequency-shift keying (CPFSK) schemes, of which minimum shift keying (MSK) is a special case. It can also represent all phase shift keying methods where the dependency reduces to $\Phi(s_t)$, or even coded modulation schemes like trellis-coded modulation.

Beamforming introduces an additional relative phase and amplitude of the signal to each antenna element. Thus, the modulated signal of a single user $x_i(t)$ is modified to

$$x_{ij}(t) = \mathrm{Re}[a_{ij}(t) \exp(j\phi_{ij}(t))\alpha_i(s_i) \exp(j2\pi f_{\mathrm{IF}} t + \Phi(s_i))] \tag{4}$$

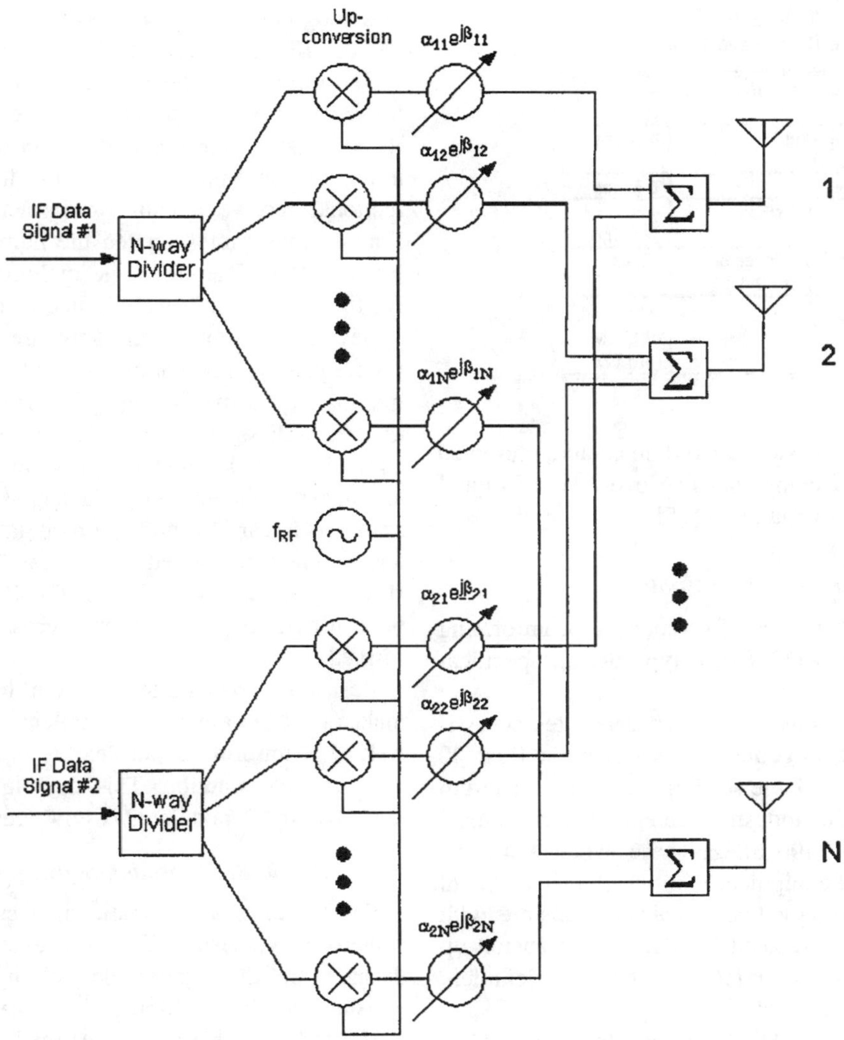

Fig. 7. Multiple beamforming by summing individual beams.

where subscript i refers to a user IF signal index and j refers to a specific element of a phased antenna array. Thus, the composite signal to an element is given by

$$X_j(t) = \sum_i \mathrm{Re}[A_{ij}(t)\exp(j\phi_{ij}(t))\alpha_i(s_i)\exp(j2\pi_{\mathrm{IF}}t + \Phi(s_i))]. \tag{5}$$

Equation (5) is the per-antenna-element summing operation that appeared in Fig. 7. Implementing phase shifts and amplitude control on RF signals can be done at frequencies up to 10 GHz. The devices used to do this are expensive and prone to elaborate calibration requirements in the phased array. Manipulation of the phase and amplitude at RF is, however, unnecessary. This is because relative phase shifts at the intermediate frequency are maintained after upconversion. This makes the task of applying the complex weights much easier and less expensive.

D. Hardware Methods for the Application of Complex Weights

There are various analog and digital techniques for applying complex weights. Digital beamforming (DBF) techniques are the preferred method, and require some type of digital control over the complex weights. Several analog techniques are briefly presented that make the digital design choices clear.

1) Analog Control of Phase and Amplitude: The current predominant technology used to implement attenuators at RF frequencies up to several gigahertz are gallium arsenide (GaAs) field-effect transistors (FET). They are used as voltage variable attenuators where the controlling voltage is applied to the gate of the FET. They are widely available from many manufactures in inexpensive matched 50 Ω surface mount packages. Unfortunately, they impart a phase shift that is a nonlinear function of the level of attenuation. Analog phase shifters suffer similar problems, and both require complex calibration schemes.

2) Digital Control of Phase and Amplitude: Digitally manipulating the phase of data at RF rates is computationally unfeasible, and as previously mentioned, unnecessary. Taking advantage of the fact that phase shifts and relative amplitudes are maintained after conversion to RF, it is possible to cost effectively modify the lower frequency IF signals. Directly generating the IF signals with an arbitrary waveform generator is the most straightforward way to apply the desired complex weights. The technique used for this function in the RDRN

TABLE II
PROTOTYPE DESIGN SPECIFICATIONS

Maximum number of simultaneous beams	4
Maximum transmit rate (only 1 beam active)	8 Mbps
Maximum receive rate	2 Mbps
Minimum number of Tx/Rx frequencies	4 (2 pairs)
Maximum Transmit Distance at a Bit Error Rate (BER) of 10-5	10 km
Modulation types	AM, PM, or QAM (No FM)

proof-of-concept system is summarized next and presented in detail in Section V. A comprehensive overview of digital beamforming techniques is found in [15].

E. Beamforming Transmitter Design Goals

The RDRN implements a proof-of-concept beamforming system. Table II lists the RDRN prototype design specifications.

1) Radio Frequency Options: The high data rates necessitate a large section of contiguous bandwidth of at least 20 MHz. The first publicly available sections of the RF spectrum with this capacity are the industrial scientific and medical (ISM) bands at 915 and 2400 MHz. These bands, however, are crowded by industrial equipment with which it is unlawful to interfere. Equipment in these bands typically transmits with less than 1 W of power. The need for RDRN to transmit up to 10 km requires higher power levels that would definitely cause interference to these bands.

The amateur radio service (ARS) band located at 1.27 GHz was chosen instead. It has a contiguous 60 MHz of bandwidth from 1240–1300 MHz, and a transmit power limit of 2000 W peak envelope power (PEP). The only requirement is that anyone transmitting in the bands be an FCC licensed amateur radio operator, and participants in the RDRN project have met this requirement.

2) Intermediate Frequency: Regardless of whether beamforming is done by analog or digital means, or if it is done at all, some type of IF information-bearing signal must be upconverted to RF. The process of upconversion is detailed later. During the analog mixing process, the sum and difference frequencies are produced in addition to other nonlinear products. Only the signal of interest, usually the sum, is passed on, amplified, and transmitted on the antenna. Extracting the signal of interest is done in the image-reject filter.

In order for the image-reject filter to be physically and cost-effectively feasible, the IF frequency must be sufficiently high as to place the difference frequency $(f_{\mathrm{IF}} - f_{\mathrm{LO}})$ far enough away from the sum frequency $(f_{\mathrm{IF}} + f_{\mathrm{LO}})$ that it can be filtered out. The higher the RF frequency, the higher the IF frequency required. Operation of the RDRN system in the ARS band at 1.27 GHz places the practical minimum IF frequency at 70 MHz, with 150 MHz being more the ideal. Digital beamforming places competing requirements on the IF signal frequency; however, the digital clock limits on digital-to-analog converters (DAC's) constrain how high the IF can be. The IF chosen for the RDRN prototype was 150 MHz.

3) Prototype Implementation Constraints: Due to complexity-driven cost constraints and the prototype nature of the RDRN system, the transmitter section was limited to a maximum of four directional beams. It used a simple omnidirectional receive antenna to implement a full-duplex link. Similar constraints dictated the number of antenna elements implemented. Therefore, the multielement phased array used for the transmit link was limited to eight elements.

Several additional limitations are key to the design. First, no frequency modulation is used so that all data streams modulate the same frequency IF carrier. This constraint means that, regardless of the number of beams, end users, or the complexity of the modulation being used, only the amplitude and phase of the composite IF signal can be changed (i.e., the system is linear). Second, the modulation schemes and steering angles must change at a rate several orders of magnitude slower than the symbol rate. Finally, the symbol rate must be constant and synchronous across all of the beams being formed.

Restricting the RDRN system to the above constraints makes possible a hardware implementation that combines the task of beamforming and flexible IF modulation into a single cost-effective solution. This flexible beamforming software-radio-based IF modulator is now presented in detail.

F. A Flexible Beamforming Software-Based IF Modulator

Fig. 8 shows a schematic of a simple, single-channel IF waveform synthesizer. The components of the waveform synthesizer are a lookup table holding waveform samples, a register file and multiplexer to meet speed requirements, a DAC, and a bandpass filter. At the beginning of a transmitted symbol, the symbol value is used to retrieve signal values for the waveform from the lookup table. There are eight samples per cycle of a sinusoidal waveform, but due to symmetry, only four values need be stored in the lookup table. Retrieved values are stored in a register file to feed to the DAC at appropriate rates. In the prototype system, waveforms for symbols are retrieved every 1 μs. Waveform samples are passed to the DAC every 0.1 μs, resulting in sine waveforms at 10 MHz. These waveforms have multiple harmonics, and we select the harmonic at 150 MHz with the bandpass filter for the IF signal.

The IF waveform synthesizer is replicated for each antenna element. The lookup table for each antenna element is independent, and stores a waveform appropriate for the phase and amplitude assigned to that antenna element to implement a steered beam.

For a single BPSK modulated signal, the lookup table holds two waveforms of four samples each, one for the 0 symbol and one for the 1 symbol. For a single QPSK-modulated signal, the lookup table holds four waveforms of four samples each, one for the 00 symbol, one for the 01 symbol, one for the 10 symbol, and one for the 11 symbol. To form multiple beams, we need to consider whether each beam is on or off and the symbol sent on each active beam. Consider forming two beams, each using BPSK modulation. The lookup table entries are given in Table III.

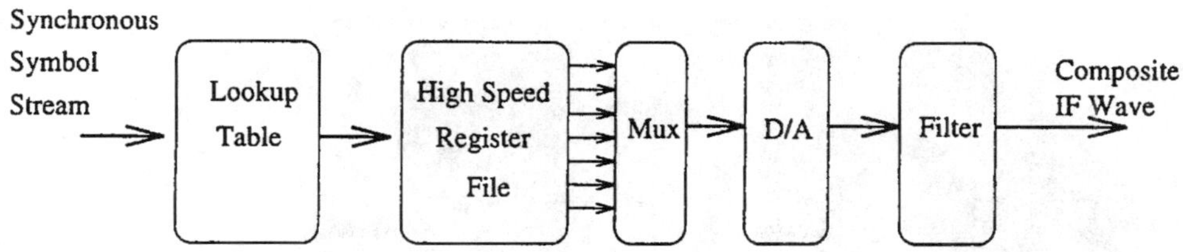

Fig. 8. Schematic of a simple IF modulator.

TABLE III
EXAMPLE LOOKUP TABLE FOR TWO BPSK BEAMS

Beam #1	Data #1	Beam #2	Data #2	Wave-form
Off	X	Off	X	0
Off	X	On	0	$A_2\sin(\omega t)$
Off	X	On	1	$A_2\sin(\omega t+\pi)$
On	0	Off	X	$A_1\sin(\omega t)$
On	1	Off	X	$A_1\sin(\omega t+\pi)$
On	0	On	0	$A_1\sin(\omega t)+A_2\sin(\omega t)$
On	0	On	1	$A_1\sin(\omega t)+A_2\sin(\omega t+\pi)$
On	1	On	0	$A_1\sin(\omega t+\pi)+A_2\sin(\omega t)$
On	1	On	1	$A_1\sin(\omega t+\pi)+A_2\sin(\omega t+\pi)$

The size of the lookup table is determined by the number of beams plus the sum of the number of symbols per beam:

$$TableSize = 2^{(NBeams+\Sigma_{i=1}^{NBeams} NSymbol_i)} \tag{6}$$

where $NBeams$ is the number of formed beams and $NSymbol_i$ is the number of symbols per beam. In the RDRN the number of beams is restricted to four or fewer.

G. IF and RF Generation

As pointed out above, higher intermediate frequencies make rejecting the undesired frequencies easier prior to the final RF amplifier. The RDRN uses an IF of 150 MHz. Generating a 150 MHz sine wave with eight samples per cycle would require a DAC operating at 1.2 Gsample/s, a daunting design task. Options include generating a lower frequency IF, and then implementing two upconversion stages. An alternative method takes advantage of the second image frequency that appears after digital-to-analog reconstruction. By proper choice of the sampling frequency F_s and the fundamental frequency F_c, the second image can be placed at the desired IF, bandpass filtered, and then upconverted directly to RF. In order to repetitiously use the same samples over an entire symbol interval, the desired fundamental frequency F_c must be chosen so that the samples used to represent it repeat as an integer sub-multiple of F_s. The desired IF is then $(2F_s - F_c)$. F_s was chosen to be 80 MHz and F_c to be 10 MHz. The first frequency image appears at $F_s - F_c$ (80–10 MHz = 70 MHz) and the second frequency image appears at $2F_s - F_c$ ($2 \cdot 80$ MHz − 10 MHz = 150 MHz). The image at 150 MHz must be selected by a narrow-bandwidth bandpass filter. The quality of the alias is limited by the bandpass filter, of the DAC number of bits, and rise-time characteristics.

Even though there is a 32 dB loss, the signal-to-distortion ratio can be kept high by picking a converter with the necessary resolution. Most importantly, changes in both the amplitude and phase of the 10 MHz signal are linearly translated into all of the image frequencies. This unconventional procedure eliminates the need for upconversion to a second IF frequency.

H. IF to RF Upconversion

The IF outputs from each of the eight beamform cards are upconverted in tandem so that the relative amplitudes and phase shifts between each of them are maintained all the way to the antenna. The 10 MHz IF signal from the DAC is filtered with a 150 MHz bandpass filter, buffered, and directly input into a double-balanced mixer that is driven by a local oscillator. Each of the eight separate RF mixers uses a local oscillator signal with the same phase. This is achieved by a passive eight-way power divider that distributes the same RF local oscillator to each mixer. The RF LO is a synthesized phase-locked oscillator with a 900–1600 MHz operating range. The LO has a parallel interface allowing its operating frequency to be changed under software control.

The output of the mixer is comprised mainly of the sum and difference frequencies produced by the mixer. It also contains LO feedthrough, in addition to other nonlinear products and sums. The signal of interest is the sum frequency that falls in the transmission band from 1240–1300 MHz. This signal is filtered out by an image reject bandpass filter, preamplified, and fed to the RF power amplifiers, 1240–1300 MHz Mitsubishi M67715 linear amplifier. The proper operation of the beamforming algorithm requires all chains of amplification to be linear. Therefore, the maximum output of the amplifier is limited to avoid compression of the output signal. In the prototype system, each RF amplifier delivered approximately 2 W to the antenna element.

The initial prototype operates between 1240–1300 MHz. However, the system is designed to be capable of operation at up to 2500 MHz by changing the RF LO and a single bandpass filter.

I. Antenna Array

The transmit section uses a linear, equally spaced, eight-element monopole phased array optimized to work at the center of the 23 cm (1240–1300 MHz) amateur band. A ground plane prevents the antenna from radiating in the backward direction. The field of coverage is limited to 180° in the proof-of-concept system.

J. Implementation of RDRN Transmitter

Fig. 9 illustrates the configuration of the completed RDRN transmitter. On the right is the interface to the host computer system. At the bottom is a VME rack holding eight IF cards,

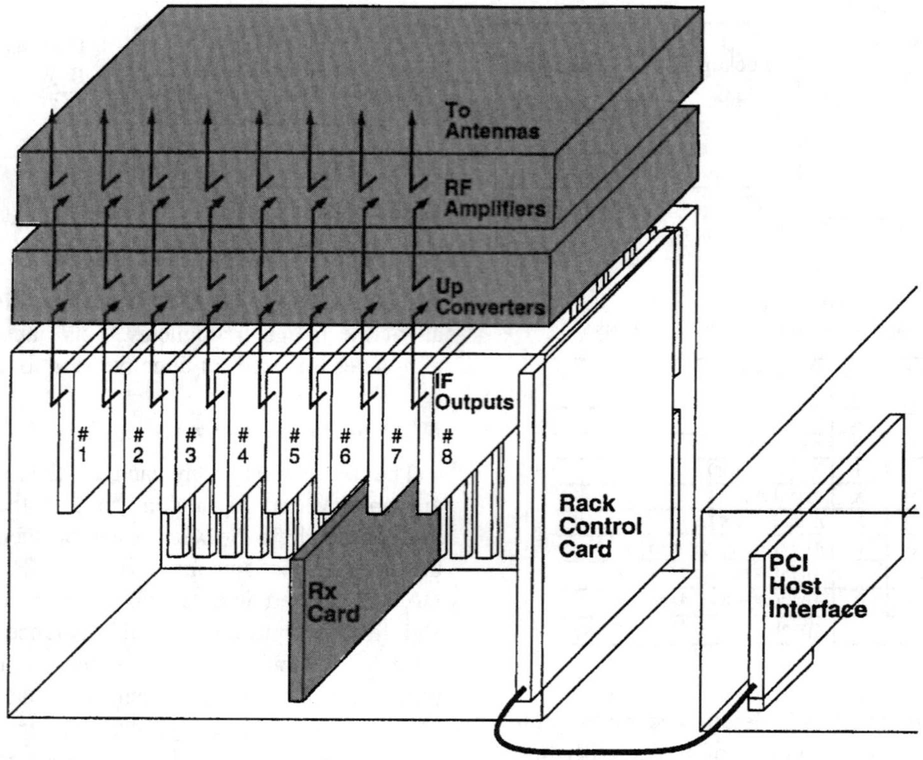

Fig. 9. Component level view of the hardware.

each containing a data buffer, beamforming lookup table, and the DAC for one antenna element. Each IF card feeds a separate upconverter, and each upconverter feeds a separate RF amplifier which, in turn, drives separate antenna elements. The entire stack occupied 20 vertical inches in a 19 in rack.

The results of forming a beam with a synthesized IF waveform are shown in Figs. 10 and 11. In Fig. 10, the element amplitude and phases are set equal to direct the beam toward the boresight. In Fig. 11, the element amplitude and phases are set to direct the beam 15° to the right of boresight.

V. Digital Receiver Implementation

The RDRN system architecture calls for a digital beamformed receiver. However, due to a lack of suitable components at the time of the first prototype system, the team was only able to validate the functionality of a single digital receiver element. This section first outlines the design of a digital beamforming receiver and then presents a prototype implementation for a single element.

A. Design of a Digital Beamforming Receiver

In traditional analog beamforming, the RF signal at each antenna element is downconverted to IF, the phase is changed using analog phase shifters, the phase-shifted signals are summed, the summation signal is downconverted to baseband, and demodulated to a digital symbol stream. Beams formed in this manner are limited in the number of beams formed, exhibit sidelobe level control problems, have high loss, lack individual beam shape control, require complex construction, and are correspondingly heavy.

In a digital beamforming system, receivers digitize the received signals at the earliest possible time on a per-element basis. The element signals are processed by a special-purpose digital processor to form the desired symbol streams. Ideally, one digitizes the RF signal after amplification. However, practical considerations for conversion and processing bandwidth dictate conversion to an IF prior to digitization.

Fig. 12 shows a block diagram of a digital beamforming receiver. The received RF band is amplified, downconverted, and digitized at the IF with one section for each antenna element. A particular signal from the IF band is digitally downconverted to baseband, producing a complex (I and Q) signal. The complex signals are summed, and the sum is passed to a digital receiver for demodulation. One can replicate the digital downconversion to baseband, complex weighting, and digital receiver functions to form additional receiver beams. The RF amplification, downconversion, and analog-to-digital conversion can be shared by all digital receivers within the same band.

In digitizing the analog received signal, one analog-to-digital converter (ADC) is required for each antenna element. The final step in the demodulation process is to translate the sampled IF signal into a complex baseband signal. The complex baseband signal retains the magnitude and phase detail of the original real input signal. Low-pass filtering and decimation of this signal to a lower sample rate will reduce the number of samples needing to be processed, The digital downconversion (DDC) process is illustrated in Fig. 13.

B. Prototype Digital Receiver

The receiver consists of a conventional low-noise RF amplifier as the front end, followed by a downconversion to an

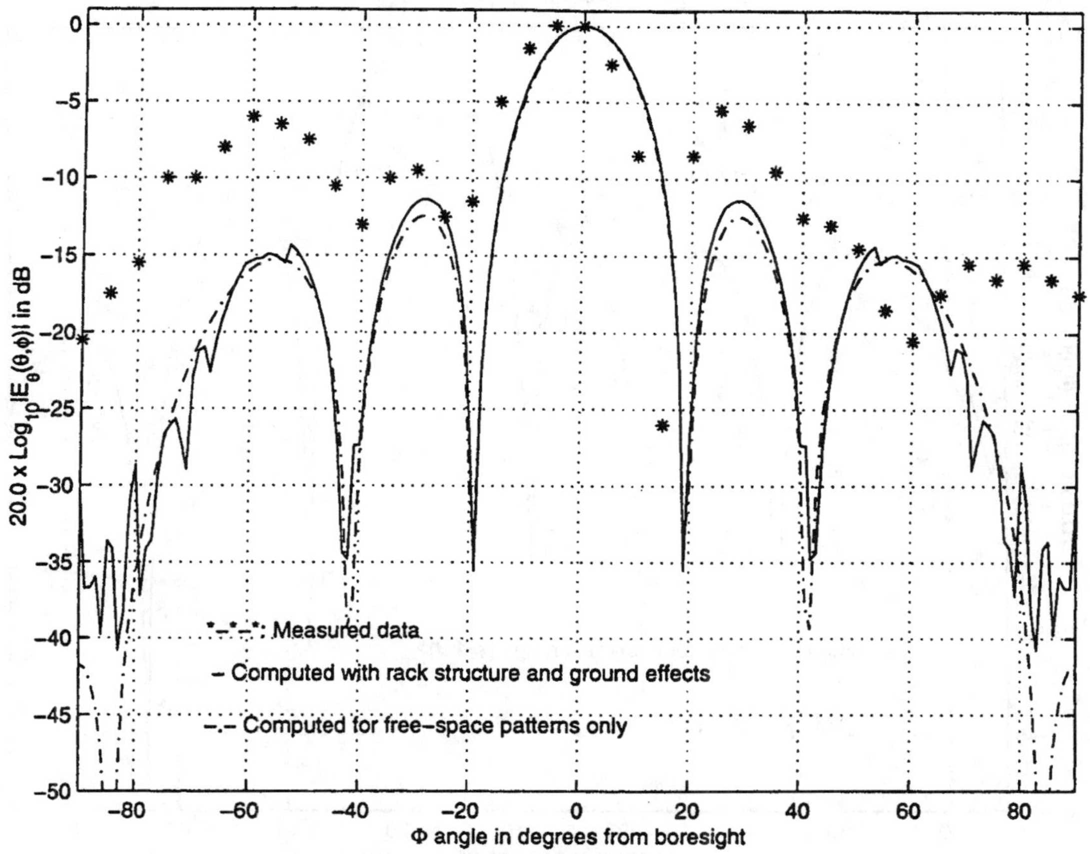

Fig. 10. Comparison of measured and computed antenna patterns at boresight.

IF signal. The IF signal is digitized, and the digital sample stream is fed to a programmable digital receiver. The receiver is capable of demodulating both BPSK and QPSK signals. The digital sections of the receiver are as follows.

The IF signal is centered at 70 MHz with a bandwidth of 5 MHz, allowing for a data rate up to 10 Mbit/s with QPSK modulation. The IF bandpass signal is digitized using bandpass sampling, which consists of sampling the signal at a multiple of the information bandwidth, but at a rate which is much lower than the low-pass Nyquist frequency.

The IF signal is digitized to 10 bits resolution at 25 Msamples/s with a Harris HI5703 ADC.

Following the ADC, the receiver uses a Harris HSP50110 digital quaduature tuner (DQT) and an HSP50210 digital Costas loop (DCL) as shown in Fig. 14. Both of these devices are programmable. The DQT selects a specific signal from the IF bandpass by mixing with a local digital oscillator and low-pass filtering the result. The output is a complex signal at baseband. The final stage of the receiver is a standard QPSK demodulator built with the Harris DCL. The DCL performs the baseband processing tasks required for the demodulation of BPSK, QPSK, and OQPSK waveforms. These tasks include matched filtering, carrier tracking, symbol synchronization, and automatic gain control (AGC). The output of the DCL is a digital bit stream.

The HSP50110 and HSP50220 chips are programmed for the particular operation of the receiver by loading the registers in these chips via the microprocessor interface. The main programmable parameters of the receiver are the following.

1) Local Oscillator (HSP50110): The center frequency and the phase offset of the local carrier can be programmed. The carrier frequency can be adjusted dynamically to account for the carrier phase offset effects.
2) Input Level Detector Threshold (HSP50110): This is used for IF AGC control. The level detector output is externally averaged to set the gain of an amplifier in front of the ADC which closes the AGC loop.
3) Baseband AGC (HSP50110): The level of the mixer output is adjusted by the baseband AGC. The baseband AGC provides coarse gain correction necessary to maintain the output of the HSP50110 at an appropriate signal level to maintain an acceptable dynamic range.
4) Low-Pass Filter Configuration (HSP50110): The low-pass filter can be configured as a one-stage or three-stage CIC filter.
5) Resampler Configuration (HSP50110): The resampler sets the output sample rate by controlling the sample rate of the decimation filters. The output sample rate can be adjusted dynamically to synchronize with baseband waveforms.
6) Matched Filtering (HSP50210): Both the raised root cosine (RRC) and integrate and dump (I/D) filters were used for matched filtering. The number of samples to integrate per symbol is programmable.
7) Baseband AGC (HSP50210): This is used to maximize the dynamic range, adjust for signal-to-noise variations, and maintain an optimal signal level at the input to the soft decision slicer.

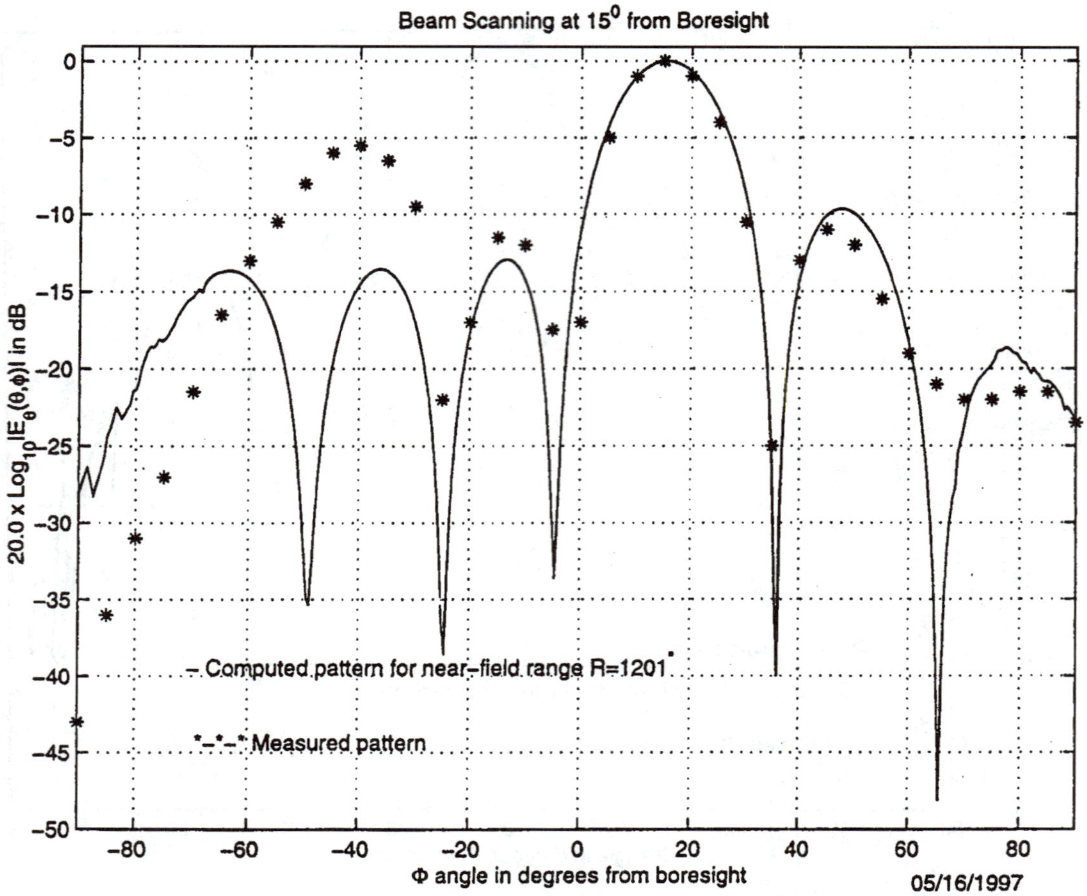

Fig. 11. Comparison of measured and computed antenna patterns 15° to the right of boresight.

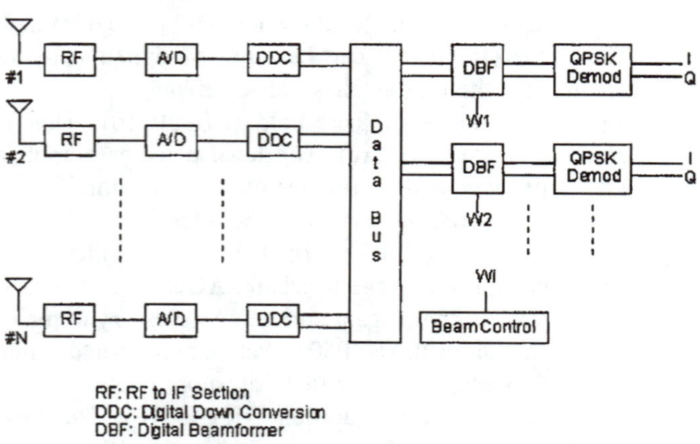

Fig. 12. Block diagram of the digital beamforming receiver.

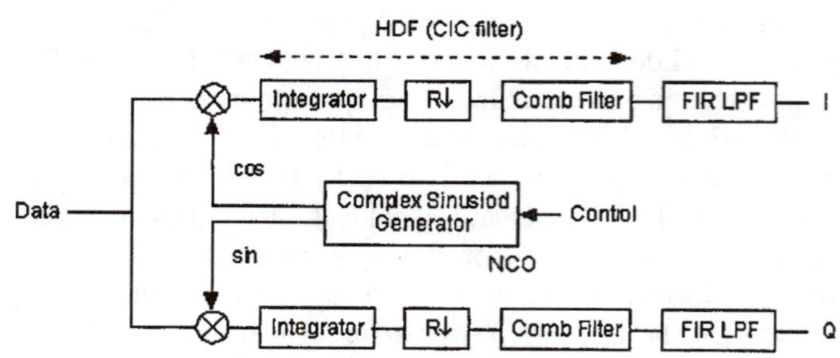

Fig. 13. Digital down conversion block diagram.

VI. Physical Construction Overview

The RDRN prototype system is contained within a mobile 19 in rack mount case. The ruggedized case is suitable for transport. Fig. 15 shows a component level view of the physical setup of a node. The host is at the bottom, and has interfaces to the transmitter and receiver and, for an edge node, interfaces to the wired infrastructure. The transmit cards are contained in a VME chassis along with a host controller card. Above the VME chassis is the transmitter upconverter and the RF amplifiers. The antenna feeds are on the back of the RF amplifier chassis.

The low-noise amplifier and downconverter for the receiver are mounted on the receiver mast to minimize interference and noise effects. The received IF is fed to the digital receiver mounted in the rack.

The transmit antenna array, receiver antenna, GPS antenna, and order-wire antenna are mounted to the 19 in rack, as shown on the right in Fig. 15.

VII. Test Results

The radio transmitter/receiver system developed for the RDRN project represents an initial step in the development and implementation of an end-to-end software radio system that adapts to the deployed environment. The first prototype incorporated beamforming at the transmitter, topology management, and link adaptation. The transmitter used digital

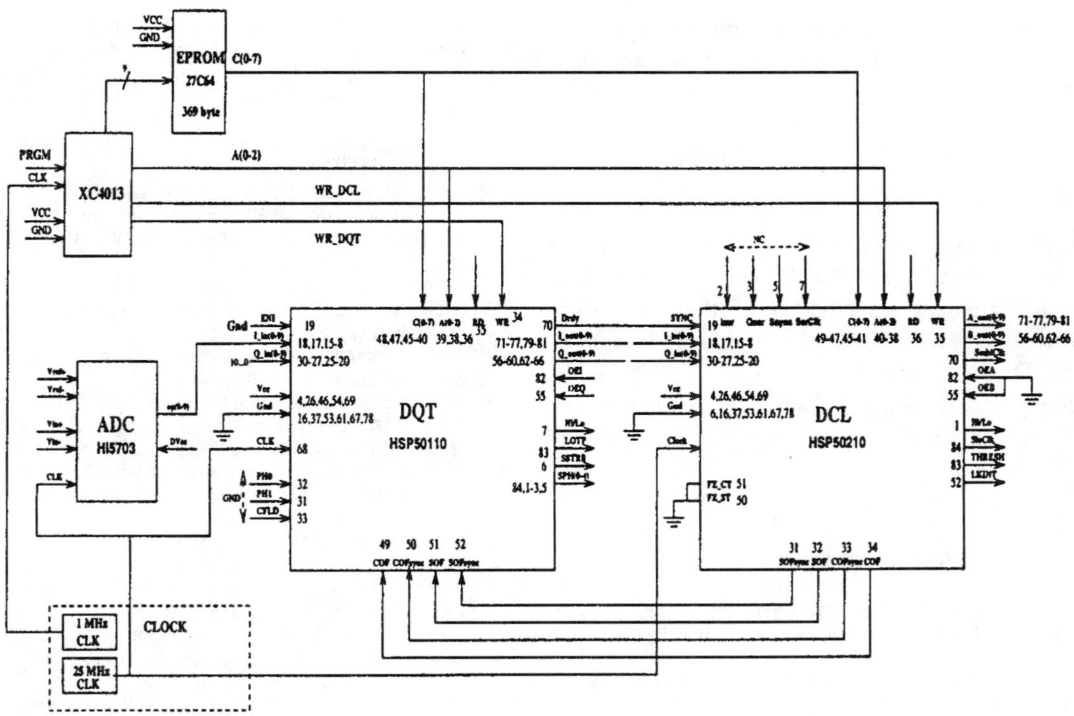

Fig. 14. Reconfigurable digital receiver.

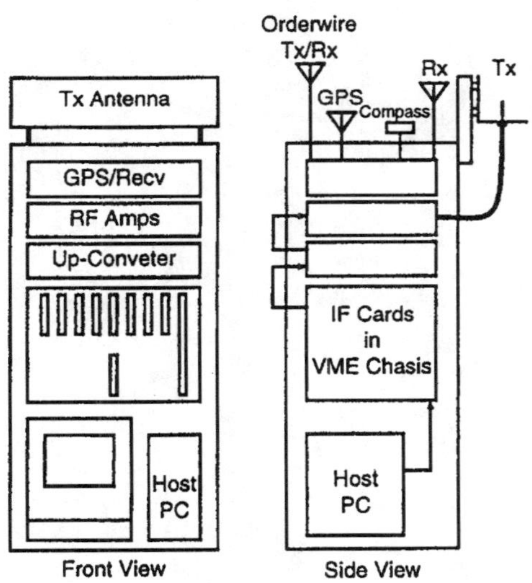

Fig. 15. The antenna and rack mount case.

waveform synthesis at IF to generate separate signals for each antenna element. The transmitter beamforming helped prove the concept of frequency reuse, while gaining experience in prototype fabrication with complex RF and signal processing hardware. The radio receiver system used the latest techniques in digital signal processing, including undersampling at the IF, followed by in-phase and quadrature downconversion to baseband via an application-specific numerical controlled oscillator (NCO). All demodulation and tracking algorithms were implemented digitally using programmable, application-specific integrated circuits.

The system performed well in field tests over a distance 6–7 km. The error rate was consistently within the link quality requirement. We tested the link under the following conditions.

TABLE IV
PREDICTED RECEIVED SIGNAL LEVELS FROM THE RDRN TEST LINK BUDGET

PATH MODEL	PATH LOSS	PREDICTED RSL
Free-space	111.42 dB	-91.31 dBW
Two-ray	113.14 dB	-95.03 dBW
Okumura	129.45 dB	-109.24 dBW

1) The distance between the transmitter and the receiver was approximately 7 km.
2) Four antenna elements were active. The transmitted power in each of the four antenna elements was 2.4, 2.55, 2.65, and 2.65 W, summing to a total of 10.25 W.
3) The received power level was observed to be -67 dBm or -97 dBW (-50 dBm and a -17 dBm receiver gain).
4) The four element transmit antenna gain was 10 dB.
5) From the test link budget system parameters, the received signal level (RSL) can be predicted as

$$\text{RSL} = P_t + G_t - L_s + G_r = 19.11 - L_s. \tag{7}$$

Table IV shows the predicted receiver signal levels using three different path models.

VIII. PLANS

RDRN next steps could include beamforming at the receiver, extending both receiver and transmitter beamforming to cylindrical antennas for 360° coverage, and implementing multiple beams. Techniques for adapting the link to maintain consistently high throughput could be explored through the use of algorithms in the radio processor which adjust power level, coding depth, antenna pointing, modulation type, and data rate.

Acknowledgment

The Rapidly Deployable Radio Network is a large project that could not be completed without the help of many staff, graduate, and undergraduate students. The following worked on the project over the years: S. Haas, S. Bush, S. Jagannath, T. Mai, Y. Lui, D. Chatterjee, V. Paulrajan, S. Gurrapu, H. Samra, P. Rajagopalan, C. Smit, J. Paden, T. Gallagher, and D. DePardo. The authors wish to thank them for their diligent work.

References

[1] R. M. Haralick and L. G. Shapiro, "The consistent labeling problem, Part I," *IEEE Trans. Pattern Anal. Machine Intell.*, vol. PAMI-1, Apr. 1979.

[2] ——, "The consistent labeling problem, Part II," *IEEE Trans. Pattern Anal. Machine Intell.*, vol. PAMI-2, May 1980.

[3] S. M. Haas and D. W. Petr, "A consistent labeling algorithm for the frequency/code assignments in a rapidly deployable radio network (RDRN)," TISL Tech. Rep. TISL-10920-04, Jan. 1995.

[4] ——, "Maximizing the minimum signal-to-interference ratio in a wireless communications network," TISL Tech. Rep. TISL-10920-21, May 1996.

[5] W. Almesberger, *Linus ATM*, Dec. 1995, [Online]. Available WWW: http://lrcwww.epfl.ch/.

[6] M. Laubach, "Classical IP and ARP over ATM," *RFC-1577*, Jan. 1994, [Online]. Available WWW: http://ds.internic.net/rfc/rfc1577.txt.

[7] W. Almesberger, *Linus ATM API*, July 1996, [Online]. Available WWW: http://lrcwww.epfl.ch/.

[8] Internet Engineering Task Force, "NBMA next hop resolution protocol (NHRP)," 1995, [Online]. Available: gopher://ds.internic.net/oo/internet-drafts/draft-ietf-rolc-nhrp-04.txt.

[9] "Q.Port. Portable ATM signaling software. Overview," Bellcore, Licensed Product LP-25, Release 1.3., July 1995.

[10] R. Sanchez, J. B. Evans, G. J. Minden, V. S. Frost, and K. S. Shanmugan, "RDRN: A prototype for a rapidly deployable radio network," *ACM Mobile Computing Commun. Rev.*, vol. 2, Apr. 1998.

[11] S. Bush, "The design and analysis of virtual network configuration for a wireless mobile ATM network," Ph.D. dissertation, Univ. Kansas, Aug. 1997.

[12] W. Stutzman and G. Thiele, *Antenna Theory and Design.* New York: Wiley, 1981.

[13] S. Haas and D. Petr, "Beam Matlab Code," TISL Tech. Rep. TISL-10920-03, Univ. Kansas, Jan. 1995.

[14] B. Ewy, and C. Sparks "A flexible beamforming digitally synthesized IF modulator," *ISCAS '96 Paper Proposal*, TISL Tech. Report TISL-10920-15, Univ. Kansas, Sept. 1995.

[15] H. Steyskal, "Digital beamforming at Rome laboratory," *Microwave J.*, vol. 39, pp. 100–126, Feb. 1996.

Sample Rate Conversion for Software Radio

Tim Hentschel and Gerhard Fettweis, Dresden University of Technology

ABSTRACT

Software radio terminals must be able to process many various communications standards. These standards are generally based on different master clock rates and thus employ different bit/chip rates. The most obvious solution to cope with the diversity of master clock rates in one terminal is to provide a dedicated master clock for each standard of operation. Not only too costly, this kind of solution limits the applicability of a realized terminal. Hence, it is much more elegant to run the terminal on a fixed clock rate, and perform digital sample rate conversion controlled by software.

INTRODUCTION

Software radio is a notion as ambiguous as any object of research can be. Depending on the point of view, different facets can be emphasized (e.g., the network or terminal aspect), where either the hardware or software aspects may be of more interest. From a signal processing point of view, software radio is a means to realize as many as possible signal processing tasks of a communications transceiver by means of software. Hence, there must be a hardware platform on which this software can run. As technology advances, this hardware platform will more and more appear as a kind of compute engine driven by a digital signal processor (DSP) and equipped with certain accelerators. Together with the necessary interfaces — the analog-to digital converter (ADC) and digital-to-analog converter (DAC) — this compute engine is clocked by a master crystal. The interfaces especially require high-quality clocks with very low jitter (Fig. 1).

Since different communications standards are based on different master clock rates, it is mainly necessary to provide these different clock rates. However, due to the strong requirements for clock quality, it is reasonable to assume that only one fixed master clock will be provided in practical software radio applications. A solution to this is to provide the different clock rates virtually by means of digital sample rate conversion (SRC). Hence, with the advent of software radio a new functionality has to be introduced to the signal processing of digital communications transceivers: SRC. Apart from being just a lukewarm rehash of the well-known interpolation problem, SRC is the task of converting the sample rate of a digital signal to another sample rate while a certain amount of information, usually in a limited frequency band, must be preserved. As will be seen in the course of the article, the main concern is not interpolation but anti-aliasing [1].

SRC has to take place at each of the analog-to-digital or digital-to-analog interfaces. Depending on the characteristics of a signal and the amount of information to be preserved, different solutions for SRC exist. Among those, interpolation as suggested in [2] is a solution for a certain application scenario only. The reason interpolation is often favored is a "historical" one. Interpolation is a well-known solution to the mathematical problem of calculating in-between values of tabulated functions. In the context of signal processing the table elements are replaced by signal samples. However, when comparing the requirements for calculating in-between values of tabulated functions and calculating those of discrete-time signals, it becomes apparent that the classical approach of employing interpolation for SRC is not sufficient. The idea of interpolation is to have a smooth curve whose samples are the tabulated values. Thus, interpolation is a time-domain approximation. In signal processing the constraints on calculating in-between values can be different. A smooth course of the signal is generally not the issue; the main concern is the information carried by the signal or parts of it. To understand this it is helpful to shift the point of view from the time to the frequency domain, where usually different frequency bands carry different information. By doing this the requirement for SRC can be formulated as *calculating in-between values of a discrete-time signal such that a certain frequency band of the signal is not distorted.*

Once the signal characteristics (and thus the distribution of the interesting information in the frequency domain) are known, the problem of SRC can be tackled with the idea of *resampling after reconstruction*, where an analog signal is (virtually) reconstructed from the digital signal by means of DAC and filtering. Eventually the reconstructed signal is resampled [3]. From this idea the basic equations can be derived, resulting in an all-digital description of the necessary filtering task. Although this all-digital description enables one to forget the

Reprinted from *IEEE Communications Magazine,* Vol. 38, No. 8, pp. 142–150, August 2000.

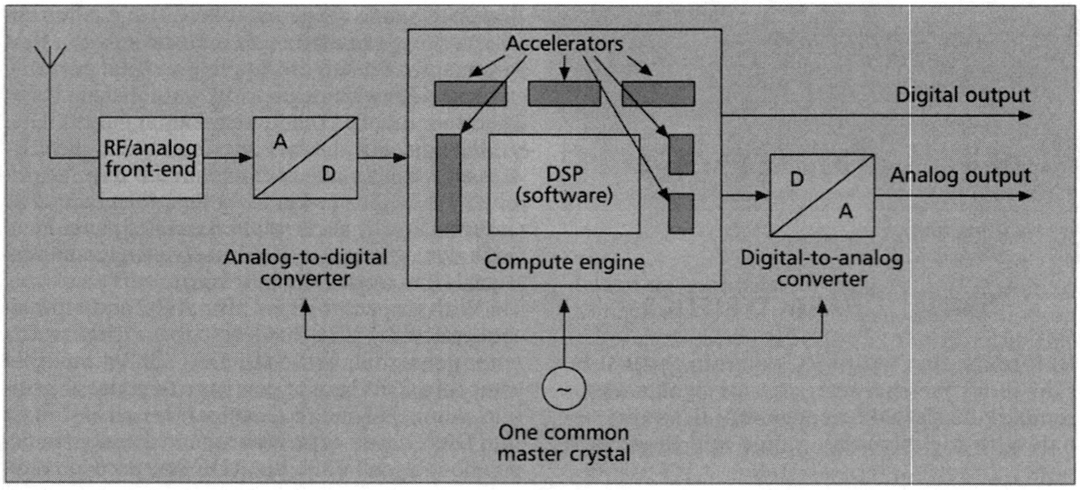

■ **Figure 1.** *The hardware platform of a software radio receiver.*

detour via reconstructing an analog signal and resampling it, it is of paramount importance to always keep in mind that *SRC is a process of resampling*. The fundamental effects of sampling — imaging and aliasing — must be expected to appear with SRC. Hence, SRC turns out to be mainly a problem of designing appropriate (anti-imaging and anti-aliasing) filters and hardware structures implementing these filters. In particular, the underlying hardware structures should not be underestimated. Since power consumption is a major issue in mobile communications systems — at least on the mobile terminal side — hardware structures are sought which implement the necessary filtering tasks efficiently. Moreover, these hardware structures should be adaptive in such a way that SRC can be performed independent of the current standard of operation of the software radio terminal. Different rate change factors, pass-bandwidth, and stop-band attenuation are the main characteristics which must be variable, and call for reconfigurability of the hardware platform for SRC in software radio transceivers. In order to show how demanding this reconfigurable hardware platform is, the article starts with a brief discussion of signal processing issues of software radio.

A Signal Processing Perspective on Software Radio

Approaching the software radio concept from a signal processing perspective means that one tries to implement the different functionalities of a transceiver by means of software. Since the signals at the antenna are analog signals, there will always be the need for a radio frequency (RF) part and an analog front-end. Depending on how many functionalities of the transceiver can be realized in the digital domain, the complexity of the RF part and analog front-end varies. It should be noted that realizing a certain functionality in the digital domain does not necessarily mean that it is fully reflected by software running on a DSP. There are certain DSP algorithms which are too demanding for a (conventional) DSP to achieve a cost-effective solution. A typical example is digital down-conversion (DDC) of a received signal, which is performed directly after ADC, and thus at a relatively high sample rate. DDC is multiplication of a bandpass signal with a rotating complex phasor. It can be realized by two separate multiplications with sine waves having a mutual phase difference of $\pi/2$. One just has to imagine a sample rate of 65 Msamples/s to quickly realize that a conventional DSP with one multiply-accumulate (MAC) unit being clocked at 65 MHz is not capable of handling this task.

Therefore, such "number-crunching" functions are realized by dedicated pieces of hardware named *accelerators*. These accelerators are not programmable. Still, in the context of software radio they should be adaptable to the different modes of operation of the transceiver. This can be reached by making them parameterizable. Due to the very limited functionality of a certain accelerator, it is sufficient to just change its parameters in order to adapt it to a certain task; for example, it is sufficient to change the carrier frequency and phase offset of a DDC.

The different accelerators and DSP form the compute engine of a software radio transceiver. In Fig. 1 is sketched how a compute engine, the ADC, the DAC, and the RF part form the hardware platform of a software radio receiver. By reversing the order of the signal flow, the principal structure of a hardware platform of a software radio transmitter can be obtained.

There are a manageable number of different signal processing tasks which might require acceleration: Viterbi equalization, despreading/spreading, digital filtering (at high sample rates), digital down/upconversion, and so on. At any time some of them are part of every communications transceiver. Hence, it is sensible to merge them with the DSP, yielding a new generation of application-tailored DSPs; for further details on this concept see [4, 5].

As will be seen later in this article, SRC should be implemented on a cascaded multirate architecture where some parts have to run at very high clock rates. Therefore, the necessary filtering tasks overstretch the capabilities of conventional DSPs, as DDC does. Hence, SRC is a candidate to get its own accelerator.

Approaching the software radio concept from a signal processing perspective means that one tries to implement the different functionalities of a transceiver by means of software. Since the signals at the antenna are analog signals, there will always be the need for an RF part and an analog front-end.

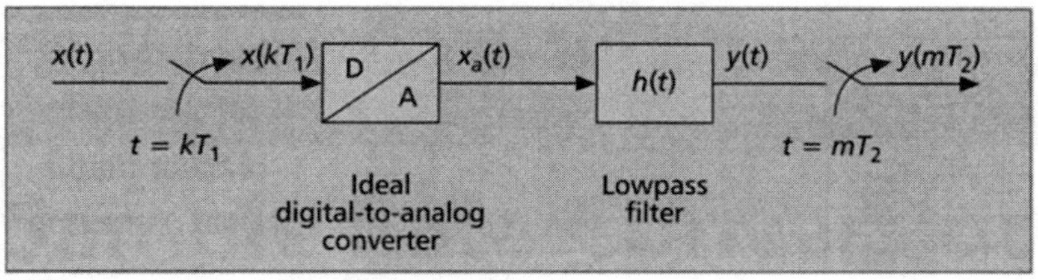

■ **Figure 2.** *Resampling after reconstruction.*

SIGNAL CHARACTERISTICS

Before dealing with SRC something must be said about the characteristics of signals whose sample rate should be converted. Basically, signals with high dynamic range and those with lower dynamic range must be distinguished. In the context of mobile communications the first are mainly a result of sampling a signal covering a bandwidth which comprises several channels of a certain mobile communications standard. Due to the adjacent channel interference characteristics allowing an adjacent channel to have a considerably higher power level than the channel of interest itself, such *multichannel* signals have high dynamic range. Conversely, single-channel signals have lower dynamic range.

With respect to Fig. 1, the ADC and a possible following SRC have to cope with a multichannel signal, while the DAC and a possible preceding SRC might just have to process a single-channel signal (e.g., voice). In a transmitter the DAC has to cope with a multichannel (transmission) signal, while the ADC just needs to digitize a single-channel signal (e.g., an audio signal).

With respect to the definition of SRC in the introduction to this article, it should be noted that in many cases it is sufficient to preserve the information of one single channel of interest of a multichannel signal when performing SRC on a multichannel signal.

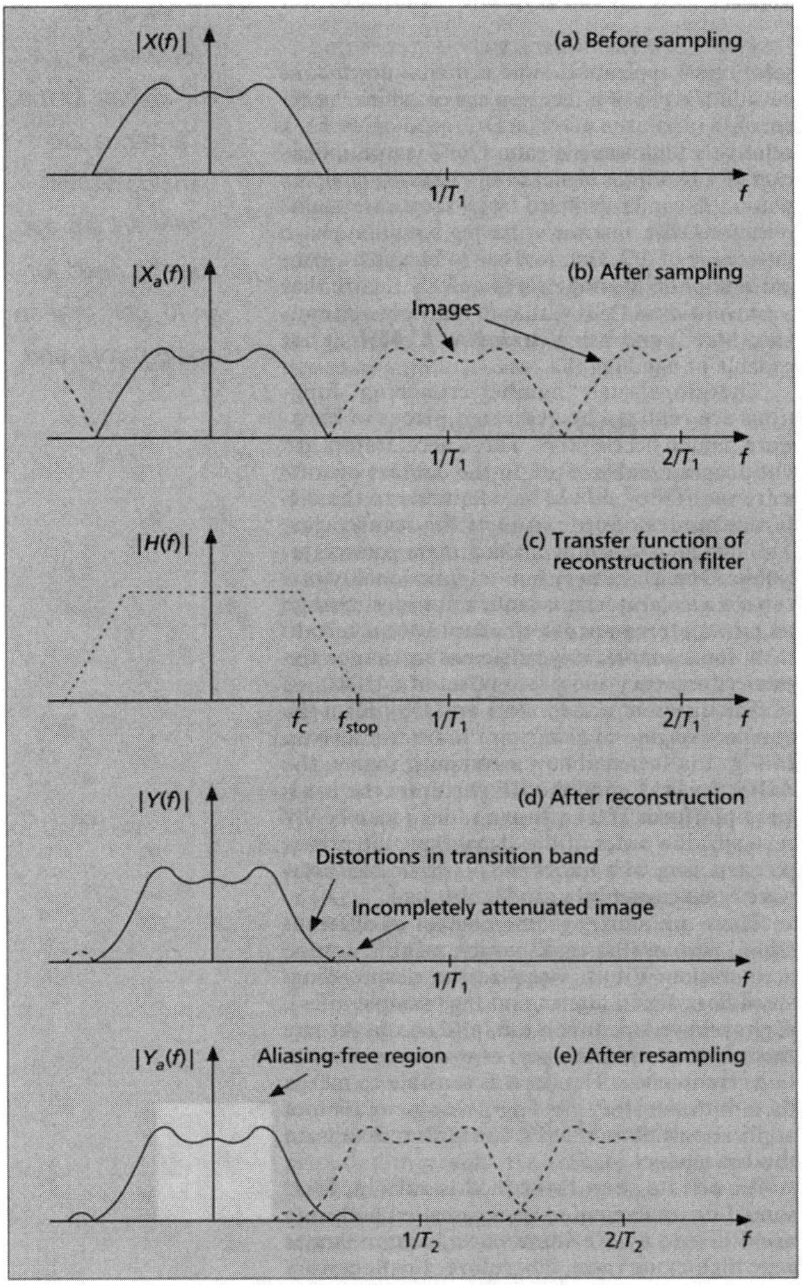

■ **Figure 3.** *Spectral interpretation of SRC (given for the case* $T_1 < T_2$*).* *(*$|Y_a(f)|$ *is the magnitude spectrum of the sampled version of* y(t), *since* $|X_a(f)|$ *is the magnitude spectrum of the sampled version of* x(t).*)*

RESAMPLING AFTER RECONSTRUCTION

A straightforward solution to SRC is to reconstruct the original analog signal from the discrete-time signal and resample the reconstructed signal with the new clock period T_2. This approach is sketched in Fig. 2.

Assuming ideal DAC (ideal Dirac impulses) and a reconstruction filter with impulse response $h(t)$, the discrete-time signal after SRC can be related to the discrete-time signal before SRC as follows:

$$y(mT_2) = \sum_{k=-\infty}^{\infty} x(kT_1)\cdot h(mT_2 - kT_1) = \sum_{n=-\infty}^{\infty} x\left(T_1\left(\left\lfloor \frac{mT_2}{T_1}\right\rfloor - n\right)\right)\cdot h(T_1(n+\mu_m)) \quad (1)$$

with

$$\mu_m = \frac{mT_2}{T_1} - \left\lfloor \frac{mT_2}{T_1}\right\rfloor$$

and

$$n = \left\lfloor \frac{mT_2}{T_1}\right\rfloor - k.$$

The quantity μ_m is limited to the interval $\mu_m \in [0,1)$. It reflects the position of the current sample inside to be calculated inside the sample period T_1, and is commonly called the *intersample position*. Equation 1 is a direct digital representation of SRC. The impulse response $h(t)$ is sampled with the period T_1 of the input signal, still, with different time offsets $\mu_m T_1$ which depend on the position of the output sample with respect to the input samples. This means that for each calculation of an output sample, a different set of samples of the impulse response $h(t)$ is used. Hence, the digital representation of SRC is a time-varying filtering operation. The filter $h(t)$ influences the quality of the SRC process, which itself is determined by the effects *aliasing* and *imaging*. This can be understood from interpreting SRC spectrally, as in Fig. 3.

If the sampling period T_1 is short enough with respect to the bandwidth of the bandlimited signal $x(t)$, the images caused by spectral repetition due to sampling do not overlap and thus do not cause aliasing, as shown in Fig. 3b. Perfect reconstruction would mean that $h(t)$ is an ideal low-pass filter canceling all images. Still, this is neither possible nor necessary in most cases. Depending on the band to be kept free from distortions by SRC, the filter $h(t)$ can employ more or less relaxed constraints, possibly causing pass-band distortions of the original signal and/or incomplete attenuation in the stop-band (e.g., incomplete attenuation of the images). These effects are sketched in Fig. 3c, d. Since any sampling process causes spectral repetition with respect to the sampling period, resampling also does. According to the characteristics of $h(t)$ the images overlap. Consequently, aliasing distortions arise, as shown in Fig. 3e. Thus, the design of the filter is the first task to be solved when tackling the problem of SRC. The main purpose of this filter is to control aliasing. As stated in [1], *anti-aliasing is the most prominent constraint to be obeyed by any sample rate conversion system.*

The power of the aliasing distortions depends on the signal characteristics. It should be remembered that in the context of SRC the aliasing components are a result of the discrete-time nature of the original signal $x(kT_1)$ and the respective spectral repetitions (Fig. 3). In a single-channel signal the aliasing components cannot be stronger than the channel of interest, while in multichannel signals the aliasing components can have a much higher power level than the channel of interest. This occurs if the channel of interest in a multichannel signal has low power while some adjacent channels potentially causing aliasing to the channel of interest have high power. From this short discussion it can be concluded that it is sensible to stick to the frequency domain approach to SRC rather than the time domain approach.

From Fig. 3 it is evident that the smaller the region which should be kept free from aliasing, the smaller the stop-bands of the filter attenuating the aliasing components can be. This enables the application of comb filters with reduced effort.

RATIONAL FACTOR SRC

Rational factor SRC is the case if the sample periods of the input and output signals are related by two positive integers L and M (i.e., $T_1/T_2 = L/M$). Now, the second part of Eq. 1 can be rewritten as

$$y(mT_2) = \sum_{n=-\infty}^{\infty} x\left(T_1\left(\left\lfloor \frac{mM}{L} \right\rfloor - n\right)\right) \cdot h(T_1(n+\mu_m)). \tag{2}$$

The intersample position

$$\mu_m = \frac{(mM)(\mathrm{mod}\, L)}{L}$$

can only take L distinct values. Thus, the number of different sets of samples of $h(t)$ used for the calculation of $y(mT_2)$ is limited to L. Since μ_m is periodic with period L, the filter $h(\cdot)$ can be implemented as a periodically time-varying filter with period LT_2.

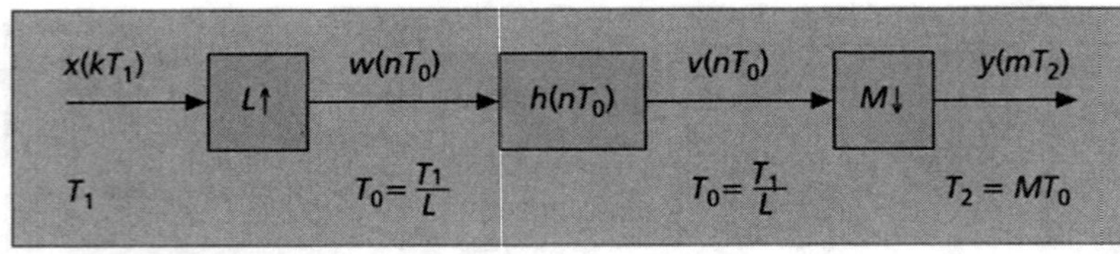

■ **Figure 4.** *Rational factor SRC.*

Introducing the period $T_0 = T_1/L = T_2/M$, and the two signals $v(nT_0)$ and $w(nT_0)$ enables us to rewrite Eq. 1 and derive the well-known block diagram of a system for rational factor SRC [3] sketched in Fig. 4. The signal $w(nT_0)$ is the result of upsampling the signal $x(kT_1)$ by L. It is generated by the $L\uparrow$ block, called an *upsampler* or a *sample-rate expander*. Upsampling by L is realized by inserting $L - 1$ zeros between two consecutive samples of the original signal. In order to obtain the output signal $y(mT_2)$ the signal $v(nT_0)$ is downsampled by M. This is done by the $M\downarrow$ block, called a *downsampler* or a *sample-rate compressor*. Downsampling by M is realized by deleting all but every Mth sample from the signal.

Although very useful for investigations, the structure of Fig. 4 is not applicable in practice. This is due to the possibly very high intermediate sample rate $1/T_0$ at which the filter would have to be clocked.

INTEGER FACTOR SRC

By setting L or M to 1, integer factor SRC as a special case of rational factor SRC can be achieved. All considerations made with respect to rational factor SRC can be applied directly to integer factor SRC.

In integer factor sample rate increase ($L > 1$, $M = 1$), the images caused by the SRC process can be regarded as not overlapping, and thus not causing aliasing. The filter is a pure anti-imaging filter. Often, it is called an *interpolation filter*, which — together with the upsampler — forms an interpolator (i.e., a system for integer factor sample rate increase).

Setting $L = 1$ leads to a reduction of the sample rate by M. Depending on the bandwidth of the signal before SRC the images resulting from resampling might overlap, which means that aliasing occurs. Aliasing can be reduced by proper anti-aliasing filtering with $h(\cdot)$. The combination of the anti-aliasing filter and downsampler is called a *decimator*.

CONCEPTS FOR SRC

Since the sample rate both before and after SRC can be expressed as an integer number of samples per time unit, the rate change (conversion) factor is a rational number. An exception to this is the case where two asynchronous digital systems are interfaced [3]. Besides the conversion factor, another design parameter is important: the ratio between the sample rate of the signal,

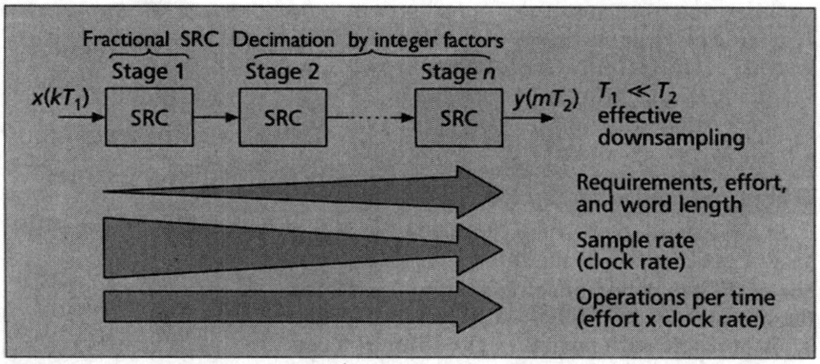

■ **Figure 5.** *Effort vs. sample rate in a cascaded SRC system (with the suggestion of placing fractional SRC at a high sample rate).*

and the bandwidth of the channel of interest, that is, the oversampling ratio (OSR) of the channel of interest. The OSR *after* the SRC process directly determines the relative bandwidth (compared to the sample rate) of potential aliasing components that have to be attenuated by the SRC filter. The higher the OSR, the smaller the passband and the stop-bands of this filter can be. Hence, a high OSR (after SRC) relaxes the design constraints, leading to simpler filter structures. A consequence of this is that SRC is advantageously implemented on a cascaded multirate system (i.e., a cascade of filters with relaxed requirements at high sample rates and strong requirements at low sample rates) [3].

Since the number of coefficients of a filter complying with a given tolerance scheme is heavily dependent on the type of filter, it is not possible to give quantitative estimates that are generally valid. Therefore, it should be sufficient to note that decreasing the sample rate while keeping the bandwidth of the channel of interest constant means a decreasing OSR. Again, this leads to an increase in the necessary number of coefficients of the respective filter stage. Hence, in multirate systems a trade-off between sample rate and hardware effort (number of coefficients) can be made.

The ratio of the output and input sample rates is $f_2/f_1 = L/M$, where L and M are relatively prime (i.e., the greatest common divisor of L and M is 1). It can be distinguished between effective reduction ($L < M$) and effective increase ($L > M$) of the sample rate.

Moreover, it can be distinguished between integer factor and fractional SRC. Although having several commonalities, integer factor and fractional SRC also have differences. Therefore it is sensible to separate them. The rate change factor can be factorized to a fractional part $(L/M)_{\text{frac}}$ and an integer part L_{int} (or M_{int}):

$$\frac{L}{M} = \begin{cases} \left(\frac{L}{M}\right)_{\text{frac}} \cdot L_{\text{int}} & \text{effective increase} \\ \left(\frac{L}{M}\right)_{\text{frac}} \cdot \frac{1}{M_{\text{int}}} & \text{effective reduction} \end{cases}$$

It should be noted that the fractional factor $(L/M)_{\text{frac}}$ is limited to the interval (0.5,2). If the overall conversion is an effective reduction of the sample rate, this factor can be limited further to the interval (0.5,1) or (1,2) in the case of an effective increase of the sample rate. Whether this further limitation is sensible or not depends on the filter algorithms and their implementation (e.g., it might be a problem to temporarily increase the sample rate in a system realizing an overall sample rate reduction).

The order of arranging the fractional and integer parts is an open question. It should be noted that the OSR before fractional SRC is of the same order as after fractional SRC. Therefore, the requirements for integer factor SRC remain the same regardless of its position within the sequence of filtering. Thus, the above mentioned trade-off between sample rate and hardware effort can be made for the fractional part of the SRC system independently.

The sample rate of the filter directly determines the multiplication rate. Still, the multiplication rate and hardware effort are determined by the filter type. For equiripple finite impulse response (FIR) low-pass filters, the number of coefficients N depends on the pass-, transition-, and stop-band width, and thus on the OSR [3]. The relation between N and the OSR is given by the proportionality $N \sim OSR/(OSR - 1)$. Hence, as the OSR approaches 1, the number of coefficients explodes. Although this number cannot be infinite in practice, it can be concluded that placing the fractional part of SRC at a very low sample rate in a cascaded multirate system is not the optimal solution to minimizing the number of coefficients, at least if equiripple FIR low-pass filters are employed.

In Fig. 5 a cascaded multirate structure for SRC is sketched, indicating the increase of hardware effort as the sample rate, and thus the OSR, decreases. This leads to a (theoretically) constant effort in terms of operations per time unit regardless of the position in the cascaded structure.

Different filter types have different advantages and disadvantages. This leads to the fact that placing fractional SRC at a high sample rate in a cascaded multirate system would lead to applying another type of filter compared to the type which would be used in case of placing fractional SRC at a low sample rate in the cascade. For instance, if the OSR is high, comb filters merely suppressing the aliasing components are a very efficient choice. Still, for low OSRs such filters are not sufficient.

Generally, the advantages of placing fractional SRC at a high sample rate are:

- Lower complexity due to relaxed requirements.
- If noise-shaping ADCs with low output word length feed the filter, coefficient multipliers can be realized by simple switches or lookup tables of relatively small size.
- The complete system for SRC has greater flexibility. The fractional part runs on the fixed input sample rate, while the integer part can be made parameterizable with respect to its output rate. If placed at the end of the cascade, the fractional part would have to run on different output sample rates, making it more difficult to find efficient solutions.

The disadvantages are:

- The high clock rate at which the filters have to run
- The required high aliasing attenuation for multichannel signals

From these advantages and disadvantages the pros and cons of placing fractional SRC at a lower sample rate can be derived. Since the requirements on the filters are stronger with lower OSRs, the effort for fractional SRC at a lower sample rate is higher. This is mainly reflected by the number of coefficients of the filter, and thus the number of multipliers. If the sample rate is so low that time-division hardware sharing is possible, several coefficient multipliers can be realized by a small number of MAC units of a DSP. In a field programmable gate array (FPGA) or an application-specific integrated circuit (ASIC) based implementation, each coefficient is usually realized by its own multiplier. Therefore, in this case it is far more advantageous to place fractional SRC at a high sample rate. This enables the application of simple comb filters whose implementation requires only a small number of multipliers.

Systems for Integer Factor SRC

Being a fundamental signal processing task, integer factor SRC is not only a part of a cascade for rational factor SRC but is also a part of any system employing oversampling. In such systems it is necessary to convert between an oversampled rate and a processing rate, which could be the chip or symbol rate of a certain communications standard.

The *direct approach* of realizing integer factor SRC is determined by the fact that downsampling has to be preceded by filtering in order to avoid aliasing, while upsampling can be followed by filtering in order to remove spectral images. Thus, the filter is always placed at the high sample rate. Principally, any kind of filter can be used, insofar as it obeys a given tolerance scheme for magnitude and phase response.

The signal at the input to an interpolation filter comprises $L - 1$ zeros between two consecutive samples of the original signal. At the output of the decimation filter $M - 1$ out of M samples are deleted. Exploiting this, the filters can be combined with the upsampler or downsampler, respectively. This leads to solutions much more efficient than the direct ones. Their main parts are clocked at the lower of the two involved sample rates. These solutions are based on the polyphase representation of both the signal and the impulse response of the filter, and can be derived, for example, from the corresponding block processing structure [6].

Block processing is a very graphic way of analyzing systems for SRC. It is as simple as *constructing a system with* M *inputs and* L *outputs which takes blocks of* M *samples of the input signal and generates blocks of* L *samples of the output signal.* Such systems are generally time-invariant, enabling the use of the large apparatus of time-invariant system theory, which is the great advantage of applying block signal processing in this respect. The relation between a periodically time-varying system (e.g., Fig. 4) and the respective time-invariant block processing system is given by the so-called *lifting isomorphism* or *raising procedure* [1, 7]. It is based on the state-space description of linear systems. For integer factor SRC the well-known structures of the polyphase-decimator and polyphase-interpolator result. The hardware platform of these structures can be made independent from the rate change factor by just foreseeing one polyphase branch and changing the set of coefficients periodically. For details on polyphase structures the reader is referred to [3, 8].

A disadvantage of these solutions is that L sets of coefficients of the impulse response $h(t)$ (sampled with $1/T_1$ and L different offsets $\mu_m T_1$) must be stored. Especially for large values of L a considerable size of coefficient memory results. This can be avoided if the coefficients are not stored but calculated on demand. If the effort for calculating the coefficients is less than the effort for storing them, this approach is sensible. In order to keep the effort low for calculating the coefficients, the impulse response $h(t)$ must be a function (or must be approximated by a function) of relatively low complexity. Polynomials are such functions. Thus, $h(t)$ can be described by piecewise polynomials. Combining polynomial filtering and block processing leads to structures which can be realized very efficiently. One of those is the Farrow structure, resulting from implementing a polynomial impulse response on a polyphase interpolator [9, 2]. Although it is very efficient, there are still several multipliers required in the Farrow structure. The actual number depends on the length of the impulse response of the filter and the degree of the polynomials of which the impulse response consists. Therefore, the Farrow structure and similar implementations of polynomial impulse responses should be realized on a DSP at a low sample rate where the few MAC units of the DSP can realize all necessary multipliers in a time-shared manner.

Another class of simplified structures for integer-factor SRC results from sensible factorizations of the transfer function of certain filter types. These are cascaded comb filters [10] and cascaded integrator comb (CIC) filters [11]. Especially, the latter have enjoyed great success and wide application. They are multiplierless comb-filters with low effort. Due to the small width of their stop-bands they should be applied at high OSRs of the channel of interest. Their simple structure supports an implementation at high sample rates.

Finally, it should be mentioned that the application of block processing also leads to novel filter structures employing time-varying feedback loops [12, 13]. These structures exhibit very low complexity with respect to the state-space (and thus, the number of registers).

Systems for Fractional SRC

Implementing Eq. 2 as shown in Fig. 4 can be regarded as the *direct approach* to realizing fractional SRC. A time-invariant realization of the filter on the high intermediate sample rate $1/T_0$ circumvents additional controlling effort, which is necessary in time-varying systems. Still, very quickly the intermediate sample rate can reach values that are not realizable. As mentioned above, the intermediate sample rate is directly

Being a fundamental signal processing task, integer factor SRC is not only a part of a cascade for rational factor SRC but is also a part of any system employing oversampling.

Time-varying polyphase structures for fractional SRC can be obtained from combining a polyphase interpolator with a downsampler, or an upsampler with a polyphase decimator.

determined by the ratio of the two rates at the input and output of the sample rate converter. Thus, realizations according to Fig. 4 are only feasible in certain applications where the fractional rate-change factor $(L/M)_{frac}$ is given as a ratio of two small positive integer numbers L and M.

If the impulse response of the reconstruction filter can be realized by a comb filter, its implementation as a CIC filter can result in reduced effort, as in the case of integer factor SRC. However, parts of the filter have to be clocked at the high intermediate sample rate, which again limits the application to small numbers L and M.

Directly employing block processing is not a solution either, due to the fact that a system with M inputs and L outputs is hardly realizable if L or M get large. Moreover, the dependence of the number of inputs and outputs on the rate change factor makes such systems unsuitable for arbitrary rate change factors, which is, however, necessary in software radio applications.

The limitations of implementing time-invariant filters (on a high intermediate sample rate, or as block processing filters) for fractional SRC suggest the application of time-varying structures. A straightforward solution would be to realize Eq. 2 directly. This leads to time-varying polyphase structures.

Time-varying polyphase structures for fractional SRC can be obtained from combining a polyphase interpolator with a downsampler, or an upsampler with a polyphase decimator. Hence, it is simply a certain partitioning of Fig. 4 where either the upsampler and filter are realized by a polyphase interpolator, or the filter and downsampler are realized by a polyphase decimator. The great advantage of these structures is that they are exactly the same as the polyphase structures for integer factor SRC (interpolation and decimation). Only the additional downsampling or upsampling processes have to be taken into account. This is done by appropriately selecting the respective polyphase defined by the intersample position μ_m. It is also a simple matter to extend the application of polynomial filters to fractional SRC. Hence, the Farrow structure as an efficient realization of polynomial polyphase filters is also applicable [14], but with the restriction to low sample rates.

Polyphase realizations can also be given for CIC filters. They can be found, for example, by exploiting the block processing idea. The result is time-varying CIC-filters for fractional SRC with arbitrary factors [15, 16] which provide a means to realize fractional SRC at high sample rates.

A Case Study

Although concrete solutions and structures for SRC are beyond the scope of this article, an illustrative example should be given. It is based on references given at the end of the article. They should help the reader fully comprehend the example.

The object of the example is designing a system for fractional SRC which is part of a receiver for the Global System for Mobile Communications (GSM), IS-95, and Universal Mobile Telecommunications System (UMTS) air interfaces. Digitization of the signal at intermediate frequency (IF) with a sample rate of 80 Msamples/s and sufficient dynamic range of the ADC is taken for granted. Furthermore, it is assumed that the analog filtering before digitization is solely for anti-aliasing. Hence, 40 MHz of (real) bandwidth is available, comprising the channel of interest as well as adjacent channel interferers.

Two solutions to SRC should be examined that are principally shown in Fig. 6: time-varying CIC filters [15, 16] placed first in a cascaded structure, and the Farrow structure [14] running at twice the chip or bit rate of the current mode of operation. Both approaches have been subject to investigation in the context of the ACTS project Software Radio Technology supported by the European Commission. Time-varying CIC filters have been chosen for an implementation in a demonstrator of this project.

Placing fractional SRC at the end of a cascaded structure for SRC means that all adjacent channel interferers have been attenuated before. Potential aliasing components can only stem from spectral replications of the channel of interest itself, which can distort the signal. Therefore, an aliasing attenuation of 40 dB is claimed to be sufficient for fractional SRC. For an OSR of 10 this attenuation can be achieved by means of linear interpolation (i.e., a linear interpolator suppresses images of a 10x oversampled signal by 40 dB at minimum).

In order to achieve this OSR the signal must be upsampled by 10. The resulting nine spectral images must be attenuated. In order to keep up with the linear interpolator, an attenuation of 40 dB is required for the anti-imaging filter. This can be achieved with an equiripple low-pass filter with approximately 40 taps.

An implementation of the anti-imaging filter in conjunction with the linear interpolator on the Farrow structure circumvents high intermediate sample rates (e.g. a 10x increased input sample rate) at the output of the equiripple low-pass filter. It results in a time-varying system requiring 2 x 40 coefficients to be implemented. The doubling of the number of coefficients comes from the linear interpolation combined with the filtering. The filter implemented on the Farrow structure has a piecewise linear polynomial impulse response (straight lines connecting the 40 coefficients). Hence, to describe each of the linear polynomial pieces two values are needed.

Since the incoming signal to the system is (virtually) 10x oversampled, only four samples of the impulse response are used at a time. This results in a requirement of eight multipliers for calculating the required samples from the linear polynomial pieces. For each incoming signal sample different polynomial pieces are sampled. Therefore the 8 multipliers must be general purpose and cannot be fixed coefficient multipliers. Additionally, one general purpose multiplier is necessary for the calculation of the linear interpolation itself.

Principally any desired fractional rate-change factor can be implemented on this system without the need to change the number of coefficients. This is due to the fact that the actual SRC with arbitrary rational factors takes place in the linear interpolator having a continuous-

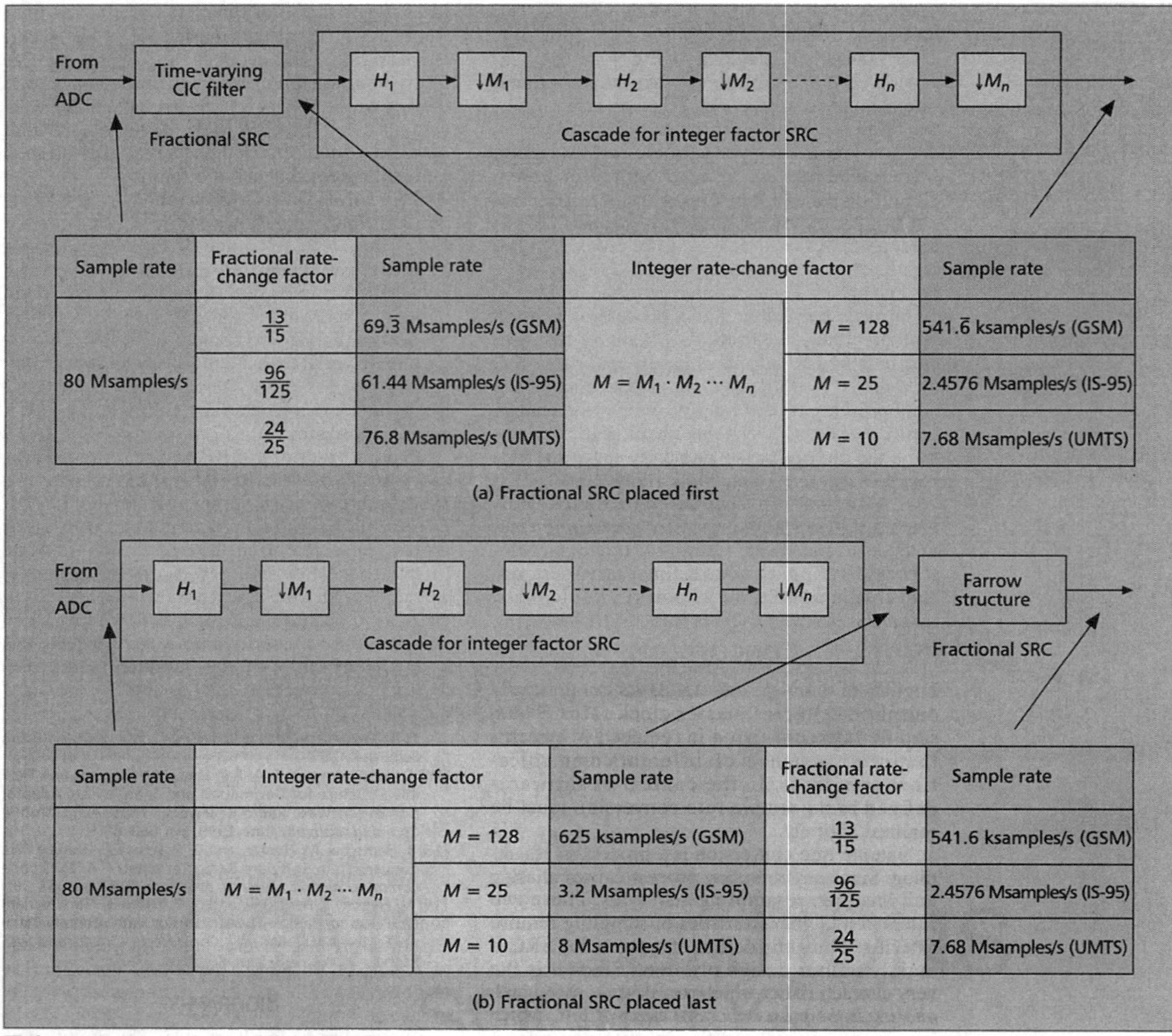

■ **Figure 6.** *A cascaded structure for SRC (down to twice the bit/chip rate of the standard of operation).*

time impulse response. Any sample of this impulse response can be calculated. For further details see [14].

When placing fractional SRC first in the cascade (i.e., directly after the ADC) and digital downconversion of the signal to baseband, all adjacent channel interferers are still part of the signal. Hence, the aliasing attenuation of the filter for SRC must be much higher than in the previous case. This is where the anti-aliasing constraint is important. It states that the required high aliasing attenuation is necessary only in the very frequency bands which contribute to aliasing in the channel of interest. Comb filters suppressing these aliasing components result from this idea. For fractional SRC, time-varying CIC filters should be chosen. They achieve high attenuation in a narrow bandwidth (i.e., for high OSRs) and less attenuation as the bandwidth of the aliasing bands increases (i.e., the OSR decreases). Within the context of this example second-order time-varying CIC filters suffice. Concretely, they achieve 110 dB aliasing attenuation for GSM, 70 dB for IS-95, and 50 dB for UMTS. A second-order time-varying CIC filter requires one general-purpose multiplier and one fixed coefficient multiplier. The latter requires one coefficient to be stored in memory. For further details on time-varying CIC filters see [16].

Assuming a digitization rate of 80 Msamples/s, the effort of the two solutions is compared in Table 1. The Farrow structure is implemented on two times the target rate of the standard of operation (i.e., two times the bit rate of GSM, and two times the chip rate of IS-95 or UMTS, respectively). With 270.83 ksamples/s the bit rate of GSM, 1.2288 Mchips/s the chip rate of IS-95, and 3.84 Mchips/s the chip rate of UMTS, the multiplication rates in million multiplications per second can be calculated.

It can be concluded that the hardware effort for the time-varying CIC filter is much lower,

	Farrow structure (first-order interpolation, 40 taps)	Time-varying CIC filter (second-order)
Number of general-purpose multipliers	9	1
Number of fixed coefficient multipliers	0	1
Number of coefficients to store	80	1
Multiplication rate (millions of multiplications per second)		
for GSM	4.9	160
for IS-95	22	160
for UMTS	69	160

Table 1. *Efforts of the Farrow structure vs. time-varying CIC filter for SRC.*

while the multiplication rate is lower for the Farrow structure. Hence, the Farrow structure is advantageously implemented if hardware effort is not an issue, or time-shared hardware is available (e.g., on a DSP), while time-varying CIC filters have the great advantage of a small occupied chip area in an ASIC or FPGA.

Conclusions

Digital communications standards are generally based on different master clock rates. Thus, sample rate conversion is required in systems that process signals of different communications standards. In the context of software-defined radio sample rate conversion must be realized digitally.

Sample rate conversion is a process of resampling. Since any sampling process causes aliasing and imaging, resampling also does. These two fundamental characteristics of sampling require filtering. Thus, the design of systems for SRC is mainly a filter design problem. Since it is the very characteristics which can destroy the signal, *aliasing is the most important effect of SRC* which must be avoided by proper filter design.

Since the constraints on the filters are most relaxed at high oversampling ratios of the signal of interest, it is reasonable to place the most complex part of SRC, namely fractional SRC, at high sample rates.

Feasible implementations for integer factor SRC can be found by realizing block processing structures, or by properly factorizing the transfer functions of the filters. Among the first are the well-known polyphase decimators and interpolators, while CIC filters belong to the second group.

Based on time-varying realizations of the combination of block filters for integer factor SRC with either an up- or a downsampler, systems for fractional SRC can be derived. These structures are parameterizable, and thus can adopt arbitrary rate change factors, which is required for software radio applications.

Acknowledgment

This work has been partially supported by the European Commission ACTS Software Radio Technology project.

References

[1] T. Hentschel, M. Henker, and G. Fettweis, "The Digital Front-End of Software Radio Terminals," *IEEE Pers. Commun.*, vol. 6, no. 4, Aug. 1999, pp. 40–46.

[2] H. Meyr, M. Moeneclaey, and S. A. Fechtel, *Digital Communication Receivers: Synchronization, Channel Estimation and Signal Processing*, Wiley, 1998.

[3] R. E. Crochiere and L. R. Rabiner, *Multirate Digital Signal Processing*, Prentice Hall, 1983.

[4] G. P. Fettweis *et al.*, "Breaking New Grounds Over 3000 MOPS: A Broadband Mobile Multimedia Modem DSP," *ICSPAT '98*, Toronto, Canada, Sept. 13–16, 1998.

[5] M. Weiss, F. Engel, and G. P. Fettweis, "A New Scalable DSP Architecture for System on Chip (soc) Domains," *ICASSP '99*, Phoenix, AZ, Mar. 15–19, 1999.

[6] T. Hentschel, *Systems for Sample Rate Conversion in Software Radio*, Ph.D. thesis, in preparation, Dresden Univ. of Tech., 2000.

[7] A. Feuer and G. C. Goodwin, *Sampling in Digital Signal Processing and Control*, Birkhauser, 1996.

[8] N. J. Fliege, *Multirate Digital Signal Processing: Multirate Systems, Filter Banks, Wavelets*, Wiley, 1994.

[9] C. W. Farrow, "A Continuously Variable Digital Delay Element," *Proc. IEEE Int'l. Symp. Circuits and Sys.*, Espoo, Finland, June 1988, pp. 2641–45.

[10] M. E. Frerking, *Digital Signal Processing in Communication Systems*, Van Nostrand Reinhold, 1994.

[11] E. B. Hogenauer, "An Economical Class of Digital Filters for Decimation and Interpolation," *IEEE Trans. Acoustics, Speech and Sig. Proc.*, vol. ASSP-29, no. 2, Apr. 1981, pp. 155–62.

[12] T. Hentschel and G. P. Fettweis, "Reduced Complexity Comb-Filters for Decimation and Interpolation in Mobile Communications Terminals," *Proc. 6th IEEE Int'l. Conf. Elect., Circuits and Sys.*, Papfos, Cyprus, Sept. 5–8, 1999, vol. 1, pp. 81–84.

[13] T. Hentschel and G. P. Fettweis, "Time-Varying Recursive Filters for Decimation and Interpolation," *Proc. 10th Euro. Sig. Processing Conf.*, Tampere, Finland, Sept. 5–8, 2000.

[14] L. Lundheim and T. A. Ramstad, "An Efficient and Flexible Structure for Decimation and Sample Rate Adaptation in Software Radio Receivers," *Proc. ACTS Mobile Commun. Summit*, June 1999, pp. 663–68.

[15] T. Hentschel, M. Henker, and G. P. Fettweis, "Sample Rate Conversion in Software Radio Terminals," *ACTS Mobile Commun. Summit*, Sorrento, Italy, June 1999, pp. 733–38.

[16] M. Henker, T. Hentschel, and G. P. Fettweis, "Time-Variant CIC-Filters for Sample-Rate Conversion with Arbitrary Rational Factors," *IEEE 6th Int'l. Conf. Elect., Circuits and Sys.*, Paphos, Cyprus, 5.-8. Sept. 1999, pp. 67–70.

Mobile Middleware for the Reconfigurable Software Radio

Alistair Munro, Bristol University

ABSTRACT

The maturing technologies and concepts for software radio mean that reconfigurability is fast becoming reality. The functions and RF configuration of the mobile terminal, the services it supports, and the applications accessible to it can all be selected, installed, and executed on demand. Mobile middleware provides the glue that matches the requirements of the applications and their users to the capabilities of the terminals and networks. The state of the art in distributed processing systems built on an Internet platform must be integrated with advanced telecommunications functions of the intelligent network and UMTS to provide the environment in which this middleware operates. In this article we explore the issues of integration, the components of mobile middleware, and likely demands placed on such systems when mobile access comes to dominate personal communications.

INTRODUCTION

Mobility has undermined the traditional orderly processes of specifying, implementing, and deploying standardized telecommunications systems. Telephones and computers that were formerly fixed have become nomadic, and their users will expect to be able to connect them anywhere, anytime, to any network. The rapid expansion and globalization of mobile telephony has forced the pace of change in telecommunications to the extent that it is no longer possible to tolerate the relatively slow processes of traditional standardization. Nor is it acceptable to users to have to radically adapt their activities to suit apparently arbitrary local conditions. Consequently, it is now necessary that users, their terminals, the applications they provide, and the supporting networks and service providers cooperate to allow communication on demand.

Visions of a world of untethered access, to any service, anywhere, anytime were discussed in the mid-1990s [1, 2], and, to a large extent, the proposed solutions for the radio platforms have become feasible. A radio interface that is agile in frequency and air interface is a reality; solutions are needed to the problems of exploiting this versatility effectively. In seeking these solutions we can draw on the experience and insights consequent to the penetration of information technology (IT) and computing technologies into the communications domain. Mobile access to information services is now commonplace; again, experiences in this field have encouraged a dynamically reconfigurable, downloadable approach for components that support personalized, localized services on a flexible application platform.

The objective of this article is to examine how this convergence will impact the network processes that support reconfigurable radio systems. With Universal Mobile Telecommunications System (UMTS) and the 3rd Generation Partnership Project (3GPP) [3] being the focus of global standardization efforts for future mobile systems, we begin with a review of some key aspects of UMTS. Following some outline scenarios to motivate the benefit of reconfigurability, we consider some conceptual aspects of distributed processing systems and relate these to real examples. Finally, we summarize the possibilities for solutions emerging from this state of the art.

KEY RECONFIGURABILITY ISSUES IN THE EVOLUTION OF GSM TOWARD UMTS

Starting from the requirement to reconfigure the radio subsystem of a mobile network, there are three areas where specific support for this reconfigurability must be provided: the radio access network (terminal, base station, and air interface), the network core, and the set of subscription management functions that give users permission to use UMTS services. Changing one leads to the need to change other parts, to revalidate and to redeploy resources.

MULTIBAND AND MULTIMODE TERMINALS

The processes of communication tend to stand alone at a level close to the physical reality of the radio medium. For example, synchronization, coding and modulation, equalization, and the distributed coordination needed when connected to shared communications media define a set of rules that devices must obey to be able to use the radio channel at all. However, diver-

Reprinted from *IEEE Communications Magazine,* Vol. 38, No. 8, pp. 152–161, August 2000.

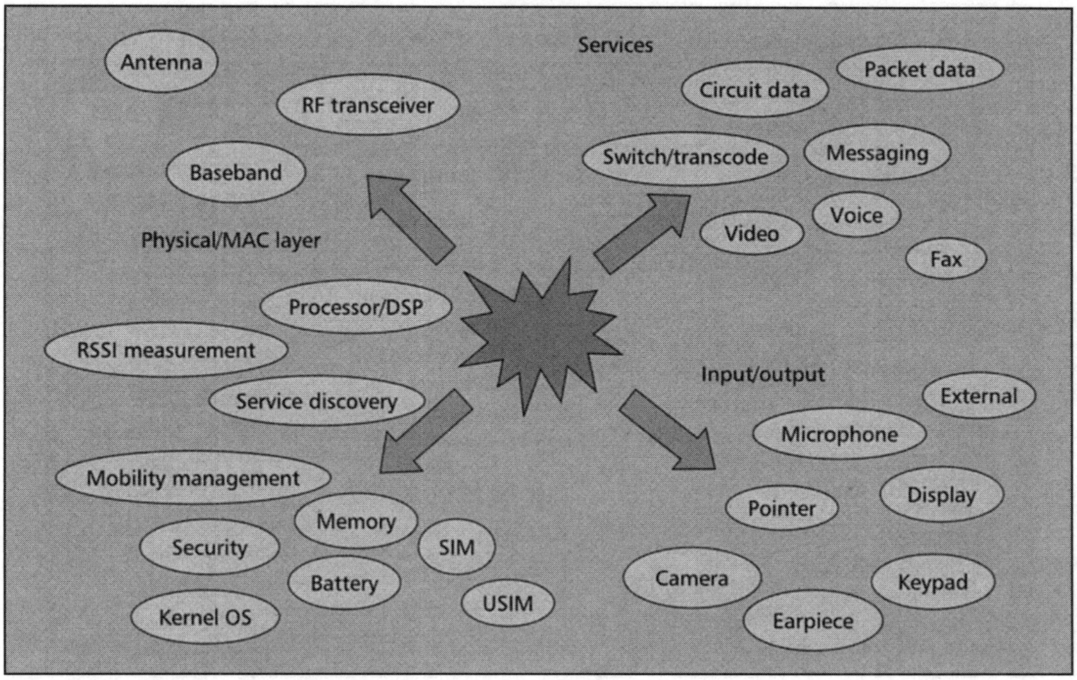

■ **Figure 1.** *Potential objects in the reconfigurable terminal.*

Trends in operating system kernel technologies for terminal platforms have already motivated the industry to seek convergence on common functions and APIs to be used by applications hosted by the terminal, the so-called Mobile Station Application Execution Environment (MExE).

sity at this level is already commonplace. The current state of the art in multiband mobile terminals illustrates that it is already economic to implement at least three physical layers (e.g., GSM 900 and 1800, and PCS 1900) supporting the same layer 2 air interface and upper layer signaling. Common semantics for air interface signaling processes is clearly essential, but there is no technological reason to demand a single air interface; in fact there is no time in which to do this, given user expectations, and no economic need since the components become more powerful and cheaper along with increasing demand.

The form commercial multimode terminals take will depend ultimately on the consolidation of demand for them. By observing the success of individual technologies in some countries but not in others, this has strong local dependency. An established local service must be accommodated as well as a global one. It is therefore very unlikely that there will be one multimode terminal to fit all requirements. There is thus an issue concerning the optimum form and combination this will take (which permutations of GSM, DECT, UMTS, and BRAN, for example, [4]) and the mechanisms by which this will be done (parameter setting, DSP firmware download).

Trends in operating system kernel technologies for terminal platforms have already motivated the industry to seek convergence on common functions and application programming interfaces (APIs) to be used by applications hosted by the terminal, the so-called mobile station application execution environment (MExE) [3]. This presumes that application components will be downloaded and, accordingly, draws heavily on the specifications published by the World Wide Web Consortium (W3C) [5], the Wireless Application Protocol (WAP) Forum [6], and extensions to JAVA [7] such as the JAVA Telephony API (JTAPI) [8]. Protocols for downloading are derived from well established, reliable Internet specifications, thus providing a guarantee that objects will arrive intact; but the main concerns are what these objects are, who created them, and the permissions the user may have to use them. The terminal may survive if malfunctioning information service objects are installed. If the objects are part of the air interface, there must be a high degree of certainty of their integrity.

Comparable specifications have not emerged so far for base stations but, with the increasing separation between service provider and network operator combined with regulatory pressure for open access, similar trends can be envisaged. We can see this complementarity by exploding a reconfigurable terminal into its component objects, such as those shown in Fig. 1.

These objects can be instantiated in various ways and distributed into many possible configurations. Figure 2 shows an example that is already near market in which the effects of splitting a mobile handset into multiple components can be seen: functions are replicated and must communicate to configure themselves into a useful product. This can be contrasted with the base station shown in Fig. 3, where many objects are absent but others are much more prominent.

Reconfigurable Radio Networks

Established standards provide for roaming and handoff between multiple platforms, wireless mobile as well as fixed. The specification of Capability Set 2 (CS2) in the ITU-T intelligent network (IN) recommendations defines procedures for cordless terminal mobility (CTM) and initial phases of Universal Personal Telecommunications (UPT). The European Telecommuni-

The algorithms that determine what is allocated, where and when, depend on data from the user, the terminal and the operational network. This is an open-ended scenario in which new approaches will be needed to ensure system availability.

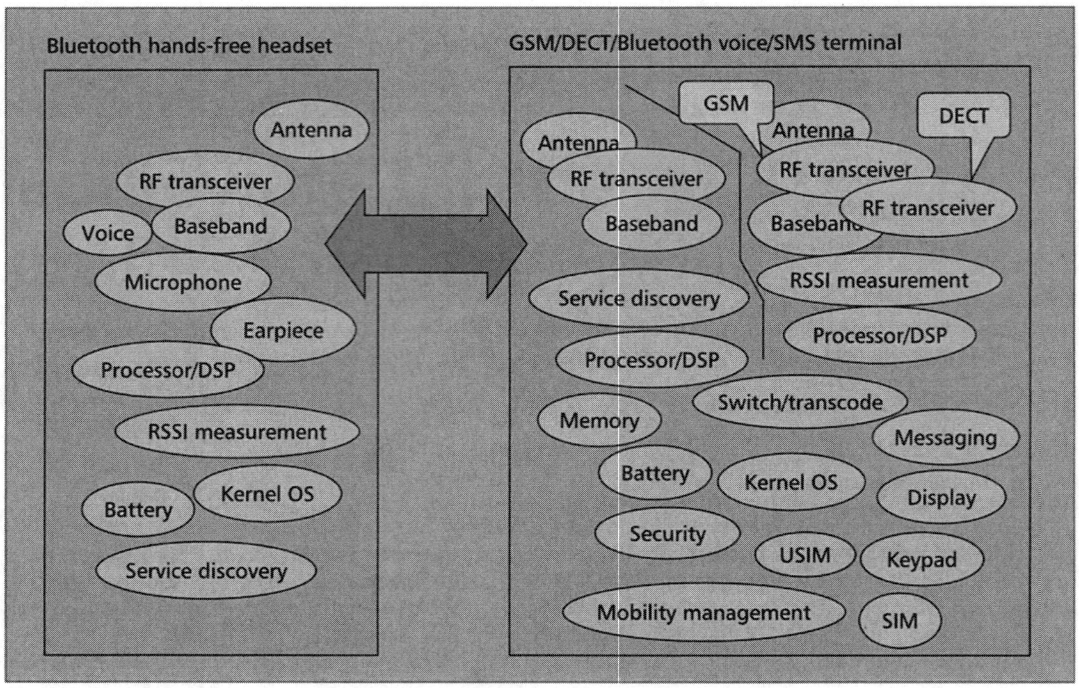

■ **Figure 2.** *Object instantiation in a simple multimode distributed terminal.*

cations Standards Institute's (ETSI's) Customized Applications for Mobile Network Enhanced Logic (CAMEL) builds on IN CS1 to integrate the GSM mobility management system mobile application part (MAP) into an IN framework. Emerging specifications from the 3GPP take care of the UMTS situation. Comparable recommendations are emerging in the Internet domain for mobility management: Mobile IP for v.4 and v.6 is already established, and other proposals have been tabled for wireless IP networks [9], which will become widespread as ETSI HIPERLAN interfaces become available.

Our reconfigurable multimode terminal will have a degree of certainty that it can stay in contact after a handoff from one domain to another, technological or administrative or both. It will also have some expectation that it can maintain an executing application within certain bounds at the level of network connectivity. (Whether it is *allowed* to do so is a separate issue, dealt with below.) These bounds are set by the limit of available resources in the network according to the aggregate resource requirements of the application and the demands of other users. For existing systems, where the granularity of resources is in most cases equivalent to the bit rate of a typical voice codec, and most application instances are person-to-person voice calls, the handoff is likely to succeed. With the anticipated increase in volume of data traffic and diversity of mobile multimedia applications there will be much more uncertainty about the success of a handoff due to doubt about the amount of resources needed and whether the technology of the new domain can support the application.

Traditional radio mobile network planning has much to offer when enhanced with more quantitative data about how users move and new teletraffic models that complement traditional Erlang planning approaches, to yield an initial configuration. However, it is likely that regimes for dynamically allocating and managing resources will be executed in the network to achieve economic utilization in the presence of wide variations in traffic load and user density. The algorithms that determine what is allocated, where, and when depend on data from the user, the terminal, and the operational network. This is an open-ended scenario in which new approaches will be needed to ensure system availability.

Upper Layers: Service Mobility and the Virtual Home Environment

One key function in the management of roaming users, alluded to above, is whether a user arriving in a new network, and associated with a certain terminal and subscription, is permitted to gain access to the resources of that network. This is achieved through consulting the home environment registers for the subscriber/user, and the user's UMTS Subscriber Identity Module (USIM), and matching the information with roaming agreements negotiated between the service providers and operators in the home and visited networks. When the right to use the services in a visited network has been verified, the installation of the necessary components can proceed. At present, where bearer services are generally the same everywhere, most issues are not relevant to reconfigurable radio components. As wireless technologies become more diverse and specialized according to specific services, matching the visiting terminal to its local environment, consistent with the services it expects to receive (its virtual home environment, VHE), will become an important challenge.

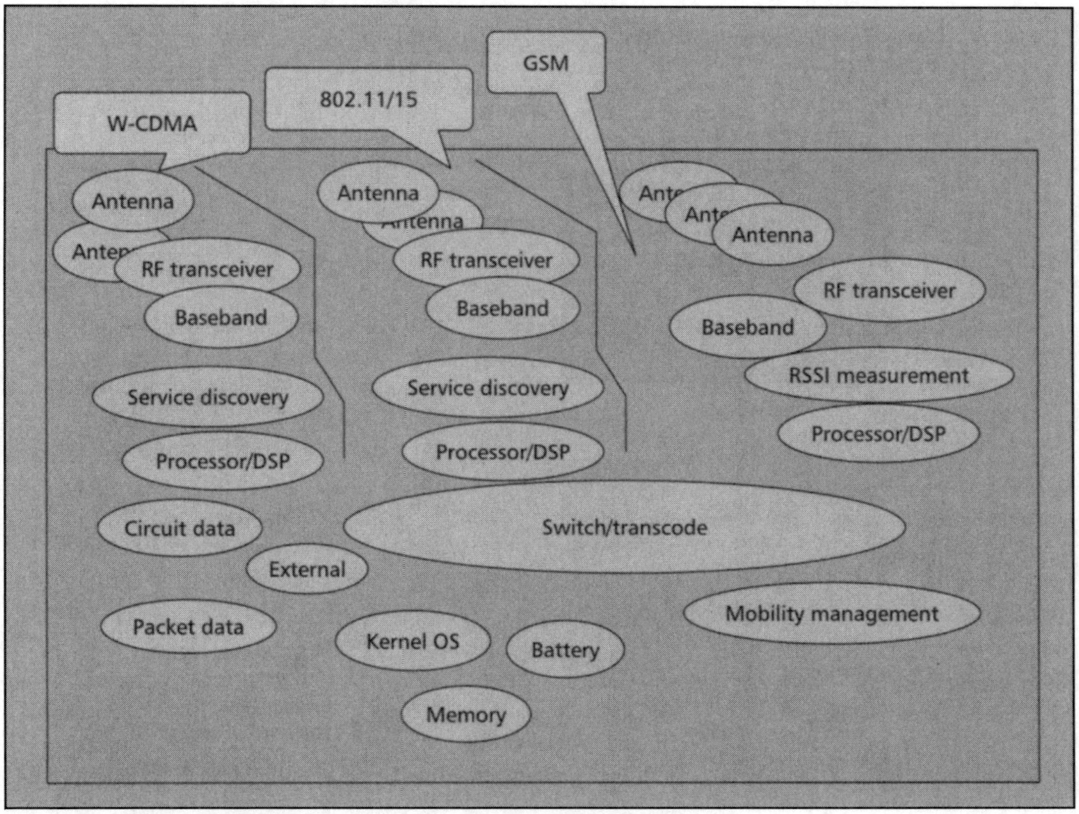

■ **Figure 3.** *Object instantiation in the reconfigurable base station.*

As wireless technologies become more diverse and specialized according to specific services, matching the visiting terminal to its local environment, consistent with the services it expects to receive, will become an important challenge.

MOTIVATING SCENARIOS

Having introduced the key elements where reconfigurability support is needed, it is useful to analyze the typical situations where this support will be invoked. Requirements on the functionality the network must provide can be projected from this. Some scenarios are more complex than others:

- Regular update of parameters or complete firmware components, invoked by the service provider or on user demand. There are several over-the-air mechanisms that can be used to achieve the download; for GSM/UMTS there is messaging, supplementary service data streams, or packet/circuit data. The controlling process could be the kernel operating system of the terminal or an application accessed through a service such as WAP. The activation of the new components must be signaled to the serving network so that coordinated updates to the serving base stations can be implemented along with administrative processes related to the user's USIM and VHE profile.
- On-demand installation or activation of radio system components, according to the user's location, application needs, and subscription to available services. This will arise when the mobile user moves from home to the car or train, arriving in a public space such as a station, airport, or shopping mall, transferring eventually to work or back home. As the travel cycle progresses, the reconfigurable terminal will be switching between air interfaces as subscriptions permit. In parallel, there will be extensive network activity to implement the handoffs between cells, location areas, and visited networks, possibly brokered with several service providers. Visiting a network for the first time will introduce an extended negotiation phase to validate the user's right to access resources. If the user is engaged in a conversation lasting over one or more of these events, this must be maintained, subscriptions and network capabilities permitting. Any resources required to support the conversations will be renegotiated according to the desired quality of service. This, too, will generate infrastructure activity, involving not just the mobile user whose actions initiated the reconfiguration but other parties participating in the session.

  It is quite possible that this traffic channel reconfiguration will fail. The scenario often quoted is the handoff of a low-bit-rate data "call" from the UMTS domain into a high-speed wireless LAN. For information-serving applications (such as Web browsing) with connectionless client/server configurations, the participants will perceive an improved response time in this case. The reconfiguration will be limited to the terminal and the serving base stations. However, reconfiguration of interactive person-to-person calls will involve potentially all elements in the chain. One user trying to take advantage of local high bit rates and higher-quality codecs may not succeed if, in doing so, it becomes impossible to maintain communication with other

Handsets may maintain multiple concurrent registrations with local and wide-area cellular services, as well as with each other in intercom mode. We also see middleware entities emerging that make this arbitrary, ad-hoc, continuously varying system usable.

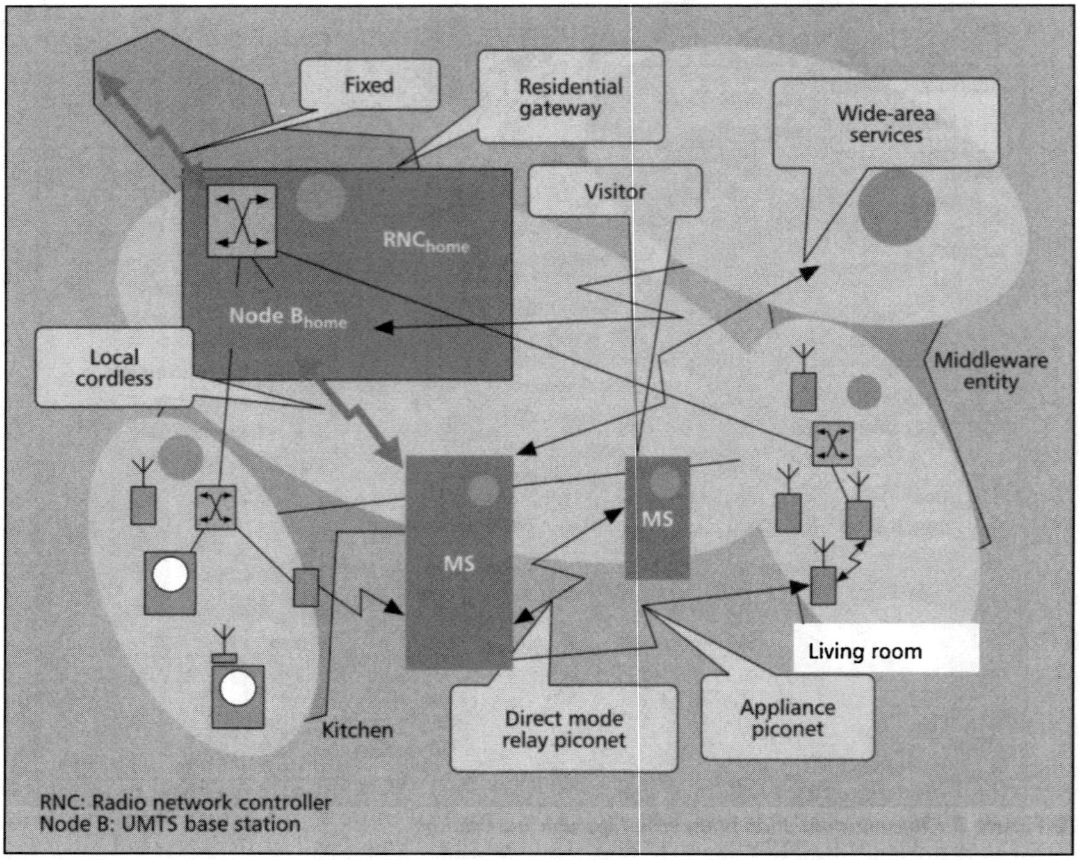

Figure 4. *MAC-layer connectivity and middleware entities in a residential network.*

parties who are tied to lower bit rates. Nevertheless, quality of service may be improved by a partial reconfiguration, giving higher quality on one leg combined with a different transcoding function (for example) to provide a correspondingly higher-quality source for transfer to other legs. In the worst case, the entire reconfiguration will have to be rolled back to restore the original relationships and their parameters.

Figure 4 shows a *simplified* and *small* part of such a system, forming the subset potentially installed in the home. There is a very high density of possible connectivity at layer 2, which will vary as terminals move in range of appliances and then away again. Handsets may maintain multiple concurrent registrations with local and wide-area cellular services, as well as with each other in intercom mode. We also see *middleware* entities emerging that make this arbitrary, ad hoc, continuously varying system usable.

- Applications may request certain configurations of radio subsystem to support specific services, regardless of the current configuration. In the preceding scenario, the service providers have cooperated to provide as far as possible the same or better quality of service. In this case, the choices are made available to the application, which may direct the decisions based on other criteria such as selection of a specific codec or combination of bearers to suit a special need. A recently announced example of this scenario is the installation of a V-series modem in place of a speech codec for text telephony between deaf users.

These scenarios illustrate the wide scope of activity that could arise to support radio system reconfiguration and the extent to which elements of the mobile communications system have to be involved. Figure 5 illustrates the flow of information involved in a switch between one air interface and another.

The middleware platform provides the framework for communicating with and coordinating the flow of information between components below and external to the terminal. Some typical middleware functions concerned with handover are shown. The terminal is registered in an IEEE 802.11 WLAN picocell and has the capability to use Bluetooth and a wide-area cellular service as well. Key features of this scenario include:

- The *QoS Monitoring* module, which has requested notification of radio signal quality from each air interface object. At some stage it decides that the WLAN IP service must be moved to the cellular platform and indicates this to:
- The *Service Mobility* module, which interacts with internal components (SIM, USIM) and external components of the VHE, in the course of which a missing component is detected, so:
- The *Download* module is contacted to acquire and validate this component, installing it into the *Memory* object, and finally:

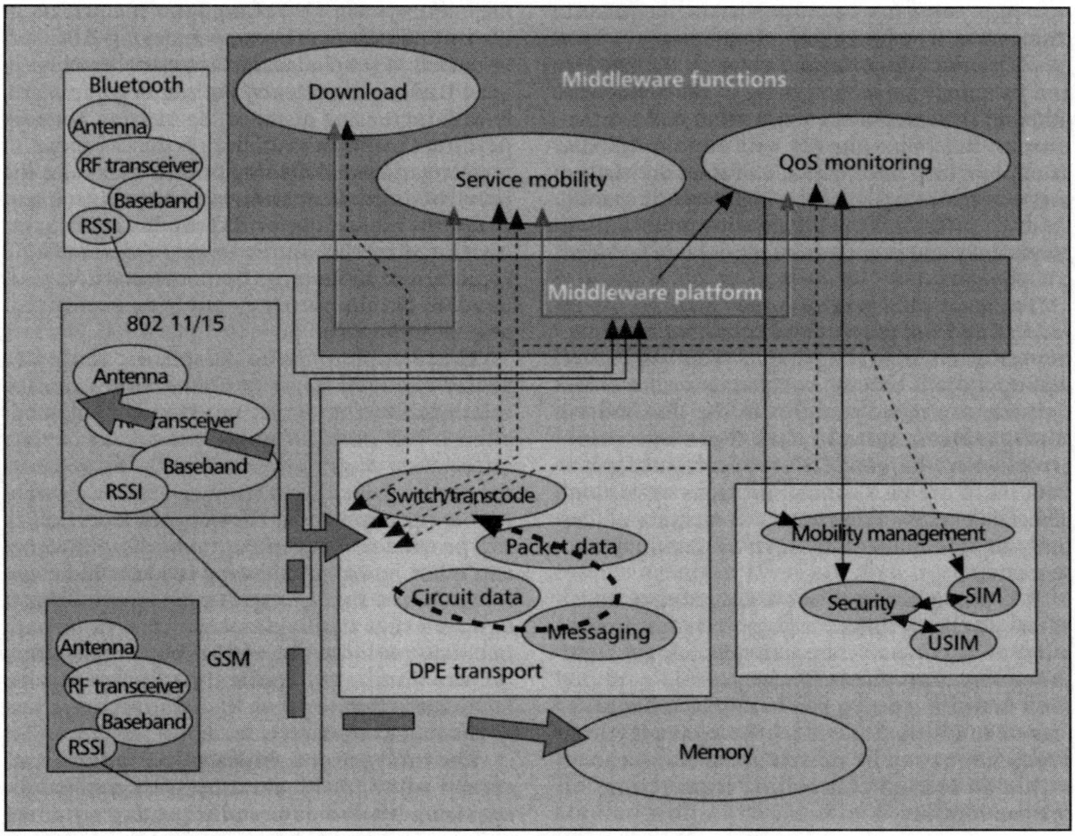

■ **Figure 5.** *Information flows in handover across air interfaces, with download.*

Middleware's role is to make it easy for the components of distributed applications to identify, acquire, and use services via appropriate interfaces. Its primary objective is to achieve transparency — to location, to access method, to implementation platform...

- The *Switch/Transcoder* object is reconfigured to use the GSM packet, or circuit-mode, service objects as appropriate.

This is the simplest path the handover process can take. Because it succeeds, none of the most important aspects of the middleware system are visible, including actions when any one or more of the actions fail. We now examine the essential elements of these middleware functions and how they could be provided within a mobile distributed environment.

Middleware in the Distributed Processing Environment

Middleware functions are built into all distributed processing environments (DPEs). It is useful to categorize these functions in a very general conceptual sense before relating them to specific examples, because apparent differences can often be exaggerated by divergent terminology and domain-specific architectural features.

Middleware's role is to make it easy for the components of distributed applications to identify, acquire, and use services via appropriate interfaces. Its primary objective is to achieve *transparency* — to location, to access method, to implementation platform, and other more subtle transparencies, including so-called transactional transparency, which assists in maintaining consistent state in a group of distributed objects. As well as discussing these issues briefly below, we also give an overview of key functions.

Interfaces: The Naming and Typing of Objects

Object-oriented design has been very successful in mastering the complexities of distributed systems for reasons similar to those that have made it a success in other application execution environments, of which Microsoft Windows is well known. Encapsulation of functionality, accessible via a well-defined external interface, is easily mapped into a standardized abstract syntax and interface definition language (IDL). This is convenient for generating the modules that perform the communications functions, but its role in providing a framework for global naming and typing is much more significant. When an application is split up into separate communicating components, their namespaces diverge: the identifiers they use locally to refer to functions, objects and datatypes, and other services lose the global meaning they had in the centralized application. Any shared object must be named globally and uniquely, including methods of interacting with it, so-called interface definitions; this global naming and strong typing provide the foundation for transparency to location and implementation platform.

Consistency among Distributed Objects

Transactional transparency is required because distribution introduces several discontinuities that are not suffered by centralized systems. The functional separation, which may be logical or physical, geographically large or small, is obvious through the fact of distribution. There is also

Specifications of communications systems use terminology and concepts that are closely linked to the underlying behavior of the entities they define. They become steadily more abstract as functionality is built up layer by layer.

time dislocation and, most important, namespace dislocation. The latter has a fundamental effect on all transactions executed between the parties, and its impact varies according to the configuration and structure of the application and interaction modes. The impact will depend on the extent to which information and state are shared between the components. A high level of sharing leads to potential problems with maintaining consistency and requires closer coupling between components.

Transactional transparency also assists in recovering from failures and coordinating activities across multiple components. This will ensure that operations execute completely or not at all. Failures are more complex in the distributed configuration: not only are there execution errors where parameters are incorrectly specified; there are also communications errors and detectable system failures, not to mention potentially undetectable ones such as deadlock and starvation.

It is possible to achieve transactional transparency in many different ways, including explicit use of distributed transactional dialogues and commitment protocols. These provide a framework in which a dialog can be conducted regardless of individual parts of the conversation. Transactions can be nested as subtransactions within an overall controlling transaction, or chained together.

Communications Paradigms

The base level of DPE communications operates above the end-to-end transport level, using a mixture of connectionless and connection-oriented paradigms with a varying degree of stored state. In the simplest case there is a messaging service that retains no state: commands are transmitted asynchronously between elements, and any relationship between them is held by the elements themselves. A common service built on top of bidirectional messaging is the remote operation (RO): a request is issued by an application component to an RO service element (ROSE) in one location and transmitted to a serving ROSE instance in another location. The initiator may block until the responder replies, or continue with other tasks; when the reply is received, sufficient information comes with it for ROSE to identify the originator. There is also the facility to link ROSE requests so that context can be maintained across multiple elements. One particular feature of linking is callback, where the responder invokes an operation on the initiator within the context of the original request. Remote procedure call (RPC) represents the most tightly coupled case, where the calling party blocks until a response is received from the serving party, in effect transferring control to it.

It is rare for current DPEs to insist on one specific approach, recognizing that there are trade-offs which can be exploited in specific applications. The conceptual distinctions do not preclude mixing one or all of the paradigms: RPC can be built on top of a bidirectional messaging service, or ROSE, or vice versa if some compromises in efficiency are accepted.

In a complementary way, the interaction mechanisms can also compound the effects of discontinuities. Synchronous blocking RPC survives well in simple local interactions (familiar in most LAN environments) but can degrade rapidly with increasing distance, decreasing network performance, host availability, the overhead of scatter/gather marshalling of user data, and the level of logical nesting or chaining. Asynchronous messaging or RO modes incur extra overhead in the application to disentangle replies, and to manage timeouts and retries if needed, but do not block awaiting events that may never happen.

There appears to be no specific reason to prefer connectionless or connection-oriented relationships (but lower layers may implement either). The trade-off is in the overhead of state information stored at various levels. Some benefits can be gained from strong connection-orientation; for example, at the network layer, which can provide early notification of disconnection and other problems, allowing requests to be terminated in a timely way. It may be used if it is believed that a reliable communications path improves performance, although generally other factors dominate; applications must survive many more failures than the relatively rare loss of communications services.

The interactions discussed so far are concerned with individual requests or commands, requiring limited data and expecting a limited response, executed by one application component on another. These are often stateless or very short-lived individually, but they may be executed in the context of a long-term association between application components. Applications will often set up a context, such as a managed transactional dialog, in which individual interactions can be sequenced, synchronized, and recovered or cancelled if failures are detected. They may also involve many other kinds of interaction, including other application objects, such as reliable bulk data transfer.

Directories, Service Lookup, Brokering, and Trading

Specifications of communications systems use terminology and concepts that are closely linked to the underlying behavior of the entities they define. They become steadily more abstract as functionality is built up layer by layer until the implementer can identify the other parties in an application by their network identity (roughly equivalent to the Internet address and TCP/UDP port number), the attributes of a session conducted with them, the languages they speak (presentation contexts), and a collection of object identifiers that allow further discrimination of functions they provide to each other.

This information is simultaneously over-specified — applications are generally much simpler — and insufficiently detailed: how are the values chosen, who is in charge of them, where can they be found? However, the communications API will generally need all the parameters. Additionally, provision of an appropriate service is often all the user requires, not the full specification. Finally, little guidance is given on how to conduct a conversation with the other parties. It is not strictly the role of such specifications to do this, but the gaps become apparent very quickly.

Applications will expect support from the DPE to allow them to discriminate between partners according to specific services, and their attributes and qualities. It is implicit that they can consult a well-known authority who may deliver the partner details directly or a list of possible candidates. The functions of the authority are to act as a directory, delivering just locations; or as a service lookup, delivering the identity of a set of providers; or contact with a set of intermediaries, who might direct the requestor to the requested services, with an implied negotiation between the parties who offer the services (hence the term *broker* or *trader*).

This function, which ultimately returns the network location and service parameters of the target partner(s), is absolutely central to DPEs. It can be realized in many ways and, as the examples below will illustrate, this reflects on the application domain. One common complaint is the vulnerability of broker/trader functions as central points of failure and, because of this centralization, the potential problems of scalability. This can be mitigated by moving the trading activity into a multicast or anycast space, where service providers can respond directly to service requestors.

Notification

A further consequence of distributing functions is that events affecting the ability of the system to operate correctly may not be detectable by the components of the system. One component that detects the event may not know which other components need to be informed or, by the nature of the event itself, may not be able to pass the knowledge on. The simplest example is the loss of a link. The entities using that link will not be able to inform each other directly of its loss, and a separate mechanism is needed.

Notifications are events generated by the middleware and passed to entities that have indicated an interest in being informed about them. The source of the event may be the middleware itself (e.g., when it detects the unexpected termination of an application component), but more generally it will be the application components themselves that register the event. Notification is an important contribution to location transparency — no component need be aware of any other. Implementation of the feature has proved very difficult in practice. A completely general approach suffers from scalability problems, and delivery of the notifications can be slow. The scope of delivery of the notification may therefore have to be limited or, as with trading, executed in a multicast network.

Middleware for Mobile Systems

Recent middleware specifications presume that objects will be mobile. Transparency to location and the inherent location independence of many objects allows this to be achieved with minimal difficulty if the only requirement is to execute a component anywhere. However, objects that move while they are executing introduce quite different needs.

From the point of view of the reconfigurable radio system, we have already presented the scenario of the mobility of terminals across several different radio environments. In GSM networks the middleware to support this is provided by the MAP, which guarantees transparency both geographically to location and to terminal platform. It creates two separate global namespaces, one for subscribers to the services of a particular operator and the other for terminals manufactured by a specific vendor. MAP subnetworks are linked together by roaming agreements. More detailed study of MAP protocols reveals analogies to the basic middleware functions outlined above. For the Internet, the comparable specification at the terminal level is mobile IP, in which middleware functions (home agents, foreign agents in IPv4) are used in conjunction with temporary address assignment and tunnels to enable roaming. Developers of IP mobility drafts are beginning to address terminal mobility in a cellular context with some indication that concepts current in MAP may be adopted. The integration of mobile IP with mobility management in GSM mobile data (General Packet Radio Service, GPRS) and UMTS has also been studied with a view to migrating to an all-IP core network.

As well as provision for mobile objects, specific provisions can be made for the radio environment and the transient connectivity of mobile terminals (which may often be switched off). The initiation of handoff from a poor-quality wide-area service into a better-quality LAN-like environment can be triggered by notifications from objects that measure radio conditions such as signal strength or error rates, for example. Specifications such as WAP make provision for proxy gateways and optimizations of the application protocol to shield the wireless domain from the fixed domain.

Although location transparency is important from the perspective of the distributed application, mobile middleware must provide information about location to allow personalized adaptation to the local situation.

Typical Middleware Architectures and Examples

There are several mainstream middleware platforms that have gained popularity in the telecommunications domain:

- CORBA — the Common Object Resource Broker Architecture, from the Object Management Group [10]. This has its roots in the IT domain and is now used widely in all kinds of distributed system. Implementations support centralized service lookup and a range of communications services based on IP, from asynchronous messaging through RPC. CORBA has been deployed in the implementation of IN and TINA (see below).
- JAVA/JINI — JAVA is a programming language and platform-independent execu-

Developers of IP mobility drafts are beginning to address terminal mobility in a cellular context with some indication that concepts current in MAP may be adopted. The integration of mobile IP with mobility management in GSM mobile data (GPRS) and UMTS has also been studied with a view to migrating to an all-IP core network.

At least for MAP but also for other APs in the functions supporting UMTS, it is likely that the abstract middleware functions will be observable within the signalling functions. However, it is still speculative that more concrete specifications, such as CORBA, will be adopted in the short term.

tion environment. It is familiar to users of the World Wide Web, in which it adds portable secure application objects that can be downloaded to a local terminal and executed, often involving interaction with remote JAVA objects. The interaction is supported by the Remote Method Invocation (RMI) RPC protocol. JAVA has no middleware itself, although RMI requires a name lookup service to locate the partner components. JINI is an initial attempt to fill this gap. JINI [7] provides more sophisticated ways of building applications from services advertised in a local community, either through a lookup service or using local multicast. It has some resilience to local transient availability through a notification service. JAVA has already been mentioned in the context of a software-reconfigurable terminal, and some extensions are being drafted for IN applications.

Both these examples are firmly Internet-based. From the telecommunications domain come two major specifications:

- IN, comprising a family of functional specifications that allow rapid creation of services, location-transparent numbering and number translations, mobility, supplementary services, and many other features, in-call and out-of-call, that extend the capabilities of the public switched telephone network (PSTN). IN is independent of protocol platform in principle, but the standardized specification for its "physical plane" (an application layer in conventional terms) is based on ROSE modules mapped onto a reliable connectionless service. An alternative physical plane that uses an IP stack (SCTP) has already been specified and appears in UMTS draft specifications. IN makes a strict separation between switching, control, database access, and other distributed functions. Initially this allowed vendors to use proprietary technologies, but these are gradually being displaced by standards, in some cases with an Internet base. IN is seen as a key component in implementing the VHE in UMTS using CAMEL.
- TINA [11] builds on the basis of IN and open distributed processing specifications to create an environment in which telecommunications operators can engage simply and quickly in new business relationships, creating new services and making them accessible to service retailers via a fully comprehensive brokering process. Although proposals have been made for business models appropriate to UMTS, such as rapid setup of roaming agreements, the impact in actual UMTS standards is still uncertain.

Finally, there are specifications based exclusively on IP:

ICEBERG, [12] a project sponsored by telecommunications companies at the Berkeley Wireless Research Center, is aiming to implement an integrated fixed/mobile telecommunications service based entirely on Internet protocols. The architecture is strongly reminiscent of IN with objectives comparable to those achieved by telecommunications services in terms of security, charging, and so on, all implemented within an IP framework. Objects in the scope of ICEBERG can be distinguished at a fine level of granularity.

MASE [13], the Mobile Applications Support Environment, is the output of a European Commission ACTS project, ON THE MOVE. MASE provides middleware functions for messaging (for transient connectivity), evaluation of radio environment quality of service including handoff between GSM and an IEEE 802.11 WLAN with seamless continuity of a voice (VoIP) call, and a range of personalization and localization functions. Optimization of access to fixed services is achieved through proxy HTTP servers and specializations of the IP protocols on the air (similar to WAP, to which the MASE made a large contribution).

Reconfigurable Radio Systems: Are There New Middleware Requirements?

We now have to be specific about the functionality needed to achieve radio system reconfigurability, taking into account the scenarios under which it will be required.

The first issue to address is architectural and structural: are any enhancements to existing middleware specifications needed to provide an integrated framework in which UMTS software-defined radio systems will operate? Is there a new style of communication, of notification, of service lookup and brokering needed for rapid changes to the radio processes in the terminal, serving base stations, and core network, for download, or for inter/intrasystem handoff?

As noted above, at least for MAP but also for other APs in the functions supporting UMTS, and remembering the trend toward an all-IP core network, it is likely that the abstract middleware functions will be observable within the signaling functions. However, it is still speculative that more concrete specifications, such as CORBA, will be adopted in the short term, mainly because of the strong IN/SS7 legacy. IN/TCP or IN/UDP platforms can use protocol toolkits based on IDL, but it is not necessary to take up the whole architecture at the levels in the access network where throughput and timeliness are crucial. The work of the SDR Forum [14] gives an indication of probable directions.

The requirements in the core may well be different where business relationships between providers are involved. For the terminal, finally, a clear pointer toward a JAVA framework can be inferred.

Conclusions

In this article we have followed a logical thread encompassing the evolution of GSM toward UMTS, technological motivations for reconfigurable radio systems, possible usage scenarios, and the options for mobile middleware solutions. For the latter we took a deliberately abstract view so that we could characterize typical current examples, and project how middleware solutions might emerge.

Whatever form these solutions take — and it seems likely they will refer strongly to existing standards but with an IP communications platform — certain aspects of the design and implementation of distributed processing systems are still crucial:

- Integrity — The system should do nothing bad, respond correctly, and execute operations verifiably. Interactions that prevent correct operation are already familiar to implementers of IN applications, and similar problems will be encountered when these extend to the management of radio system.
- Security — It must be impossible to deny service to a bona fide user through incorrect reconfiguration or for an intruder to misuse the reconfiguration service.
- Performance — The PSTN is a mature high-throughput distributed application; DPEs and middleware originating in the IT domain are less so.

Whatever emerges eventually in reconfigurable radio systems, these criteria must be applied in assessing its viability and success. The rapid pace of technological development and the move toward multimode multiband agile terminals will focus attention on these issues very rapidly.

The requirements in the core may well be different, where business relationships between providers are involved. For the terminal, a clear pointer toward a JAVA framework can be inferred.

REFERENCES

[1] R. Steele, "The Evolution of Personal Communications," *IEEE Pers. Commun.*, vol. 1, no. 2, 2nd qtr., 1994.
[2] R. Katz, "Adaptation and Mobility in Wireless Information Systems," *IEEE Pers. Commun.*, vol. 1, no. 1, 1st qtr. 1994.
[3] 3GPP, http://www.3gpp.org
[4] ETSI, http://www.etsi.org
[5] W3C, http://www.w3c.org
[6] WAP Forum, http://www.wapforum.com
[7] JAVA, and JINI, http://www.javasoft.com
[8] S. Roberts, *Essential JTAPI*, Prentice Hall, 1999.
[9] IETF, http://www.ietf.org
[10] OMG, http://www.omg.org
[11] TINA-C, http://www.tina-c.com
[12] H. Wang, and B. Raman *et al.*, "ICEBERG: An Internet-Core Network Architecture for Integrated Communications," *IEEE Pers. Commun.*, 1999.
[13] A. Munro *et al.*, "Services and Applications: Requirements and Realizations in the UMTS Era," *IEEE Commun. Mag.*, vol. 36, no. 2, 1998.
[14] SDR Forum, http://www.sdrforum.org

Engineering the Embedded Software Radio

Rob Shepherd, Cambridge Consultants Ltd.

ABSTRACT Software-defined radios allow flexible implementation of different basebands in digital signal processors. For communications transceivers this introduces an engineering problem: how to manage the different protocol stacks that may be required to work above the baseband software. To solve this, we propose a consistent software architectural framework for the dynamic implementation of these different protocols within an embedded environment. We discuss its requirements and present one possible approach based on STREAMS.

The basic concepts behind software radios have been in existence for well over ten years. During the 1980s the advances in signal processing silicon and analog-to-digital converters allowed the basic concepts to be investigated [1, 2]. There is currently a resurgence of interest in software defined radio. This is partly driven by the idea that a truly global commercial communications terminal may be realized as a software radio. In the intervening years we have seen significant advances in key technological areas such as signal processing, semiconductors and battery technology, and the feasibility of a commercially viable product will be determined by the advances in these areas together with the software architecture. In these years we have also seen significant effort put into the development and implementation of complex air interface protocols, most of which are implemented in software. What we see today are the same basic concepts required to implement, for example, the analog air interface functionality in software. However, managing the software required for these functions, the protocols and the applications, is now a significantly larger problem.

While there have been significant advances in software technology during this period, we should be aware that the realization of the software radio could be hampered by a lack of attention to some of what may be considered to be the less glamorous engineering aspects. The subject of this article is one such aspect.

The requirement for an architectural framework for a software radio has already been expressed [3–5]. In this article I intend to present a solution that is achievable today using a widely adopted, understood, and proven methodology. It is not the only applicable methodology. The purpose of presenting it is to stimulate thought in this area.

THE LEGACY OF OSI

The air interface protocols, and those that we may wish to realize over and above the air interface, have generally been influenced by the International Organization for Standardization (ISO) seven-layer reference model (ISORM) for open systems interconnection (OSI). We make the assumption that the protocols we will be dealing with are layered to a certain degree. However, the approach described is applicable to protocols that, for example, merge layers 2 and 3, introduce sublayers (e.g., IEEE 802), as well as those that adhere more closely to the ISORM.

What has emerged during the development of these air interface protocols is the need for close cooperation between layers that are not necessarily adjacent, for example, the physical layer and the radio resource management sublayer, which is part of layer 3 in GSM. We show in this article how to include a layer-to-layer management scheme that retains both the formal structure of the software framework and the layering of the protocol.

A SOFTWARE ARCHITECTURAL FRAMEWORK IN CONTEXT

Before considering the requirements of a software architectural framework, we should indicate where it is to be found within the software radio. In Fig. 1 we show the context of the framework. It is located above the digital signal processing (DSP) implementation of the baseband and below the middleware. The middleware may use Common Object Request Broker Architecture (CORBA), and is outside the scope of our discussion here. The architectural framework we are interested in is to support the real-time communications protocol stacks that are typically implemented on an embedded processor.

We define a software architectural framework as providing the basic services by which individual software modules can perform their individual tasks and communicate with each other. Traditionally this was the sole responsibility of neither the operating system nor the protocol software. It is thus necessary to extend the framework to provide formalized support specifically for communications-based software modules. The inevitable reliance on the services provided by the operating system limited the reusability of the software across other hardware platforms and operating systems. The formalized framework improves that reusability since it can be constructed in a portable manner.

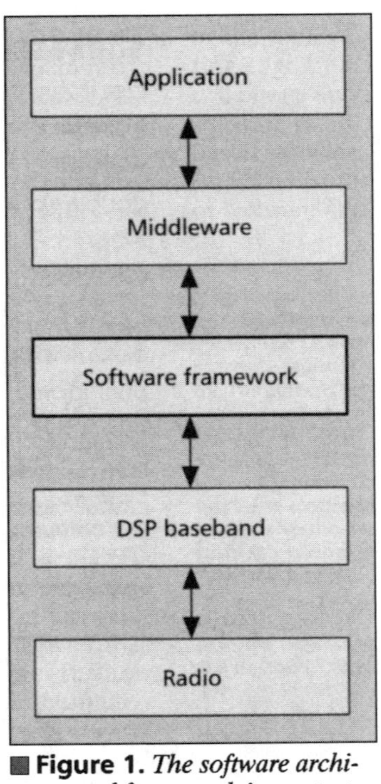

Figure 1. *The software architectural framework in context.*

A STREAMS-BASED FRAMEWORK

One industry standard architectural framework is STREAMS. STREAMS was developed as a set of tools for the development of UNIX system communication services. Figure 3 shows the basic architecture of STREAMS (shown inside Stream Space).

Editorial liaisons: J. Mitola and Z. Zvonar.

Reprinted from *IEEE Communications Magazine,* Vol. 37, No. 11, pp. 70–74, November 1999.

A STREAM consists of a driver that interfaces to the communications hardware, a number of modules above that, and finally a Stream Head that provides a defined application programming interface (API) to the user application. The similarity of this approach to the seven-layer ISORM for OSI is immediately apparent. The STREAM is actually built by the user via a process. Using Fig. 3 to illustrate this, the STREAM is "opened" in a similar manner to a file. At this point only the Stream Head and Driver are present. The opening process establishes the interface between the driver and the Stream Head. The interface is a queue pair. Separate queues exist for the upstream and downstream directions. The next step is to "push" a module, which is then inserted between the Driver and the Stream Head. Subsequent modules can be pushed and these are each inserted immediately below the Stream Head (i.e., between the module shown in Fig. 3 and the Stream Head). This allows the required protocol stack to be built dynamically by the users to match their individual requirements. In an embedded software radio terminal, the user process is a supervisor task that is not visible to the end user of the product.

The STREAM is opened when the communications services are required. In the context of an embedded software radio, this would be when the required communications protocol had been identified and the terminal wanted to communicate, possibly to register, with the system.

Each protocol layer, or sublayer, is implemented as a module. The lower layer interface to the hardware is a driver. A STREAMS module contains the processing functions that make up the specific layer or sublayer being implemented. For example, an OSI transport layer could be implemented as a STREAMS module.

Communication between layers is by means of STREAMS messages. Numerous types are defined, but we will generally concern ourselves with the types defined as M_PROTO messages for protocol-related control and data, and M_IOCTL for STREAMS management. With STREAMS we define a service interface between the different protocol layers (i.e., using primitives). A STREAMS message for such a primitive includes both a control part and a user data part, as illustrated in Fig. 2.

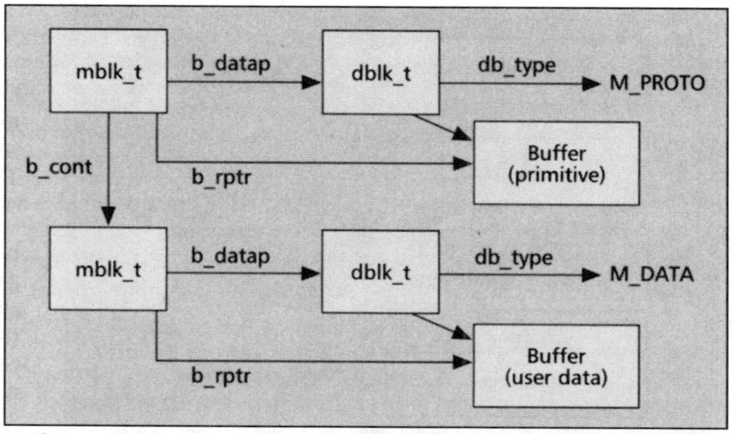

Figure 2. *Message and data blocks associated with an M_PROTO message.*

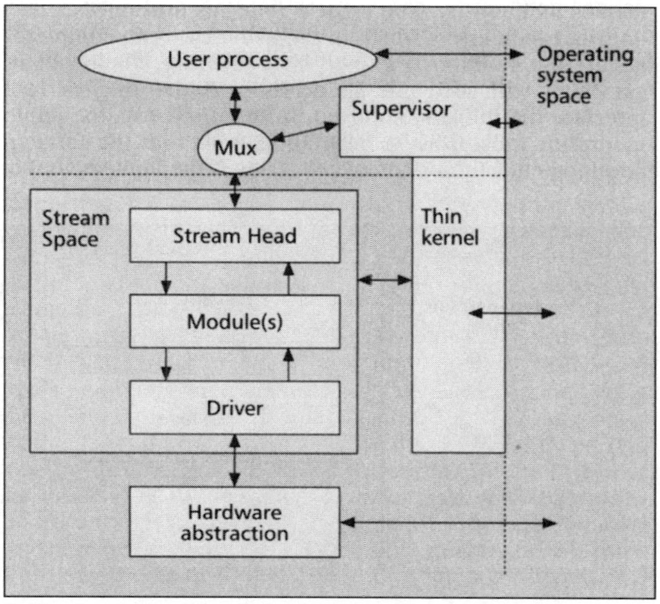

Figure 3. *An embedded software architectural framework based on STREAMS.*

In Fig. 2 the construction of an M_PROTO message is shown. The block represented by mblk_t is a STREAMS-defined structure known as a message block. The block represented by dblk_t is a STREAMS-defined structure known as a data block. The combination of a message block, data block, and a memory block for the buffer is known as a *triple*. The message block structure contains pointers to the next and previous messages (on a queue, the queues are managed as linked lists), any associated message blocks (b_cont), the associated data block (b_datap), and read and write pointers to the buffer. The data block structure contains pointers to the base location of the buffer and to the memory location after the final byte of the buffer, plus the message type. This arrangement allows the user data buffer to be passed between different layers without the need to copy the data; the buffer containing the primitive is used only across the interface for which the service primitive is defined. When the user data is sent to the next layer, the primitive information in the buffer will be different. This can be achieved by linking the M_DATA triple to a new triple that contains the primitive.

STREAMS only passes a reference (pointer) to the data between the modules; thus, the data itself is not copied between layers. This has benefits in terms of both RAM and processing speed. If the data is copied, there is a higher RAM requirement due to the need to store multiple copies of message data (if only when copying), and there is the processor overhead of copying the data, which has been a significant bottleneck on some data communications implementations. The messages are passed between modules by placing the M_PROTO message on the input queue of the next-layer module. This generates an event which is placed on the event queue for the STREAMS scheduler. Once scheduled, the event causes the service routine of the layer module receiving the message to be executed, which then will result in what is now a service data unit (SDU) being processed by that layer.

We can consider the module to be a composite of a STREAMS domain and a protocol domain. The STREAMS domain contains the processing elements required for interface processing; the protocol domain contains the layer protocol processing. The service routine simply detects that an M_PROTO message has been received and passes a pointer to the primitive into a procedure that defines the interface to the protocol domain. This requires further standardization of interface structure over and above that directly provided by the STREAMS message and data block structures. The interface primitives are defined in a consistent manner. Each one is defined as a structure, the first member of that structure being the primitive type (e.g., an Establish Request). Finally, the interface is defined as a union of all the primitive structures.

Once the service routine is called, it obtains from the queue

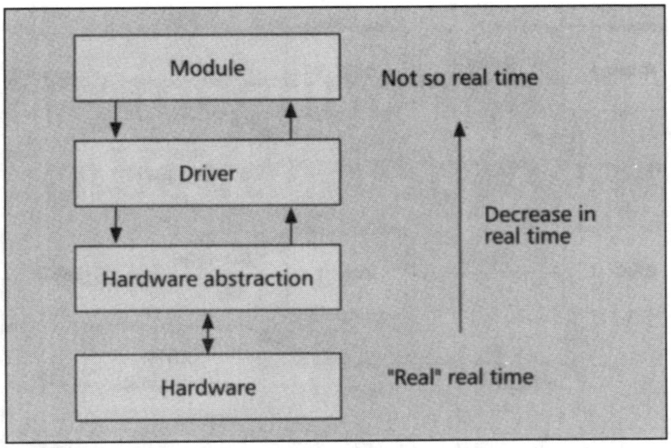

■ **Figure 4.** *How real time decreases up the protocol stack.*

a message pointer (*mp), which points to the first message block. It can then switch on the message type. For the case of M_PROTO the primitive type is then determined by pointing the message pointer at the buffer containing the primitive structure, and extracting the primitive type from the structure. This is consistent with the definition of the structures of the primitives as a union. Once we have reached this step, the module is now in the protocol domain, and should then process the service request according to the protocol rules.

As well as being dynamically built (opened), a STREAM can be closed in a similar manner. This allows the user process to free the resources that have been allocated to the STREAM, which can then be reassigned for, say, a different protocol stack.

The framework itself provides a very useful software development discipline. Since the only means of communicating between layers is via the defined STREAMS interface, the input and output points of the module software are readily identifiable and standardized between modules. The actual modular construction of the modules allows easy portability between STREAMS environments. Not only does the standard manner in which the modules are written promote reusability, but also, because the only means of communicating between modules is via the defined interfaces (the queues), the code produced is both testable and maintainable. For testing purposes, additional modules can be pushed between the protocol layer modules, allowing points of control and observation (PCOs) to be established. Alternatively, a target module can be placed into a test harness, again within a STREAMS environment. By using the STREAMS environment to provide portability this test harness could exist on a development workstation, rather than the target hardware.

STREAMS is not the only software framework available for the provision of communications services. There are a number of commercially available proprietary solutions for embedded systems, such as SoftStax™ [6].

Architectural Aspects

Figure 3 shows the STREAMS basic architecture within the embedded environment.

The STREAMS utilities are realized wholly within Stream Space and are decoupled from the operating system. The Thin Kernel acts as an operating system adaptation layer and allows the STREAMS utilities to use the facilities of the operating system (e.g., timer support) in a consistent manner. By introducing the Thin Kernel we make the STREAMS utilities independent of the actual operating system and thus increase portability across different operating platforms.

Within the user space we introduce the concept of a supervisor entity. This is the *user process*, which is responsible for building the STREAM and is protocol-independent. In addition, the specific protocol being used at that moment in time will require additional protocol-specific management support. The mux entity is shown in the figure, since even with the supervisor entity there is still only one API to the Stream Head.

The hardware abstraction layer is logically a partitioning of the STREAMS Driver. If a protocol stack is to be ported to a different hardware platform, there is normally a need to change the actual driver. In the architecture shown in Fig. 3, the driver is partitioned such that the hardware abstraction contains the code that will need to be modified with a port to new hardware, and the Driver contains the invariant parts.

Real-Time Issues

One of the general implementation problems facing designers is the concept of real time. The problem is not real-time, but *how real-time*?. The real-time requirements of transmitter power control in code-division multiple access (CDMA) are totally different in terms of orders of magnitude regarding what is acceptable as real-time responsiveness at the user interface, for example.

We can make the generalization that there is a hierarchy of real-time requirements within any protocol stack. This is illustrated in Fig. 4.

Inevitably there is the case where the actual real-time requirement of the protocol being implemented is not consistent with this model. To deal with these cases we introduce the concept of delegation. This allows a module (or driver) associated with a specific layer of a protocol to delegate the real-time functional aspects to a lower-layer module, driver, or hardware abstraction layer. The lower layer does not take over the full functionality, it simply provides a real-time service according to a set of rules that the higher layer has provided. Since the higher layer can change these rules dynamically, the actual service can be tailored to suit the requirements of the current state of the higher layer.

Management

The ISORM for OSI introduces a management entity that sits alongside the protocol layers (Fig. 5).

A potential design error for dynamically constructed protocol stacks is to introduce a vertical layer in the form of the management entity. The error is that this introduces a back door between layers which in the worst case can completely negate the modularity gained from layering. The problems associated with such a design decision are that the interlayer interface discipline is removed. In the worst case the implementation may prove to be so untestable that the software development is never completed. Even if the implementation

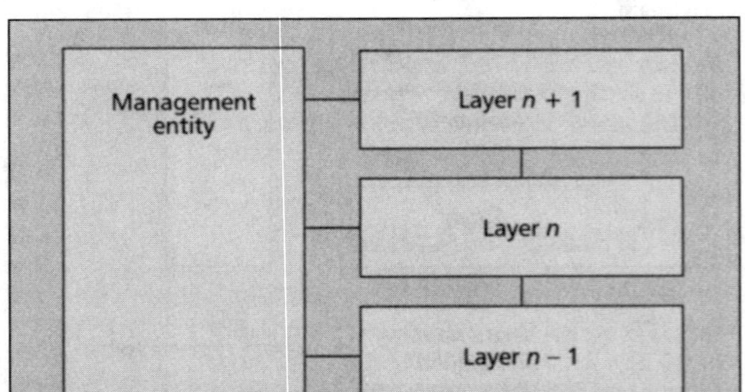

■ **Figure 5.** *The seven-layer ISORM for OSI representation of the management entity.*

is completed successfully, the lack of modularity does not promote easy reuse in subsequent platforms. We are convinced that the protocol software for the software radio will need to be a well-engineered modular design.

Within the architectural framework the implementation of the management entity is as shown in Fig. 6.

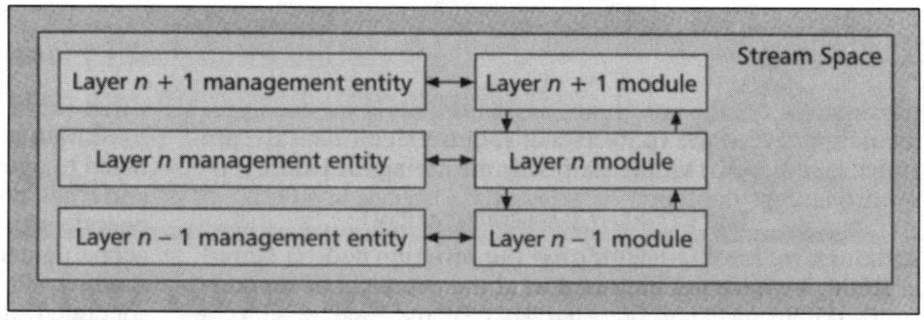

■ **Figure 6.** *Implementation of the management entity within the architectural framework.*

The layer management functionality is partitioned (correctly) with the layer to which it relates. The set of primitives supported using M_PROTO messages is extended to include four basic management types:

- Get — Allows a layer to request information from another layer
- Response — The response to a Get request, containing the information requested
- Set — Allows a layer to set data or a notification request in another layer
- Notify — The notification of an event that was requested

Thus, if a higher layer wished to know that a lower layer had, for example, found a suitable system, it would set a request in the lower layer to notify it when the system had been found. Once the lower layer has found the system, the higher layer is notified and can then start to establish its connection.

Unlike the service access point primitives that are defined between layers n and $n + 1$, the management primitives are valid across all interfaces. This is consistent with the M_IOCTL messages used for STREAMS management, but differentiates the protocol-related management messages from the STREAMS ones.

Considerable care needs to be taken when designing the management strategy for a particular protocol. The management of the protocol stack is driven from the supervisor. This is responsible for the building of the appropriate protocol stack (initialization) and for the configuration of that protocol stack. During initialization the protocol stack is built using STREAMS management, and during configuration using the protocol management.

Allied with the management of the protocol stack are the configuration and tuning aspects. While these can be dealt with at compile time, a more universal software radio implementation would require that they are set up at runtime. This would be achieved using the layer management messages partly when the stack is built, and partly when the stack is running, either to optimize performance or as a result of the subscription. Examples of the types of operation included here are: the setting of options within a protocol stack to reflect what the underlying hardware is actually capable of supporting, the setting of options to reflect what the system may require as a condition of the subscription, or the adaptive tuning of algorithms where the specific layer is unable to adapt the algorithm autonomously or in conjunction with its peer entity.

The supervisor thus has the picture within the radio as to which is the exact protocol stack implemented; individual layers need only have knowledge of the layers immediately above and below them. The management messages communicated from the supervisor thus need to incorporate certain types of information to ensure the successful operation of the total software. This includes information which directs modules to perform exactly the functions that are required (which may be a subset of that module's capabilities), coordination to ensure that the modules act as an integrated entity, and informing modules of what information the supervisor requires to be notified about to maintain the total picture. In addition to this the supervisor is required to act as an arbitrator; one example is the relative processor bandwidth that each layer can be allowed to use for a given protocol.

Interlayer management is particularly important for certain protocol stacks during normal operation. Using this interlayer communications mechanism within the STREAMS framework allows the management communication to coexist with the normal protocol primitives. However, the interlayer management can have real-time constraints in excess of the time constants assumed for the protocol primitives. In this case there are two strategies that can be adopted. The first is the concept of delegation, which we have covered in the section on real-time aspects. The second is that the STREAMS interface also defines an M_PCPROTO message, which can be implemented to have a higher priority than the normal M_PROTO messages. This is useful where the normal primitive messages can be queued, for example, during radio link reestablishment, where the radio resource management sublayer may be directing the physical layer to reestablish the link.

Practical Realization and Issues for Further Investigation

Memory

In an embedded system such as a radio terminal there are usually memory issues to be considered, the normal desire being for more memory than the design or commercial constraints allow. The embedded STREAMS approach stores the program code in read only memory (ROM) — which we take as a generic category including EPROM, Flash, and one time programmable devices. What STREAMS does *not* do is to then copy the program code into RAM when it builds the stack. All that is copied into RAM is a module control block that allows the modules to be linked together. Further scalability is achieved by then allowing in RAM one connection control block per connection (which is usually that layer's meaning of *connection*). Again, we see a distinction between the STREAMS domain and the protocol domain, the module control block being associated directly with STREAMS, the connection control block with the protocol domain. This memory is all dynamic (i.e., it is not statically allocated by the compiler) and thus only appears on the software stack when required. This is attractive for a software radio because it allows both the ready coexistence of different protocol stacks within a single software environment and the later addition of further protocol stacks.

Our realization of the software radio therefore requires sufficient ROM for the different protocols, but only requires a small RAM overhead. This is for the module control blocks and is seen as part of the amount of RAM allocated for the software stack or heap.

The ROM required need not be a simple summation of the ROM required for each protocol supported, but a basic requirement for the core functions, plus a delta for each protocol. However, this in itself will require significant design effort, and may not ultimately justify the savings in ROM.

The Current Implementation

Although we have indicated that the program code is stored in ROM, we have not indicated what the overhead of supporting the STREAMS utilities actually is. What is important here is that the overheads imposed by the architectural framework do not significantly increase the cost of a commercial software radio. In the case of a STREAMS-based framework this requires that the framework itself be engineered for the embedded environment of the terminal. Currently we have implemented an embedded STREAMS framework on an ARM 7TDMI processor where the program code overhead of the framework is only 11 kbytes and the total RAM available for the whole software is only 4 kbytes. It should be noted that the program code overhead includes many functions that would have to be realized in any application. The RAM overhead of using STREAMS for a protocol stack of four modules and one driver is about 1 kbyte. This is a combination of the STREAMS queues and the module control blocks.

The software was designed using object-oriented (OO) methods (Unified Modeling Language) and implemented in C.

The additional power consumption attributable to the use of STREAMS is negligible. The only significant overhead is in the additional processor cycles required to dynamically build the protocol stack, which is negligible. As we have already indicated, there is some saving in that the data is not copied between layers, but this is not an exclusive feature of STREAMS.

An Example Scenario

One possible use of such a software radio is that at power-up the radio is initialized and the supervisor "pushes" a set of modules to operate the terminal in a surveillance mode. The purpose of this is to determine the local systems that are available to the subscriber. The modules specifically would allow the modulation type, data rate, and frequency of operation to be determined, along with one or two key signaling channel characteristics. Once these have been determined, the supervisor compares the gathered information against a locally held database. In our (hypothetical but realizable) example we assume that the database contains information about wideband CDMA (WCDMA), UTRA/time-division duplex (TDD), GSM, and Personal Handyphone System (PHS), and that subscriptions and roaming agreements held mean that any of the above systems are potentially valid for making calls on.

There is no reason why, alternatively, a default communications protocol stack cannot be built at this stage to obtain the necessary software to operate the terminal according to local protocols which are not already loaded.

Let us assume, however, that the result of the surveillance search is that WCDMA and GSM 900 systems have been identified (along with several others). The supervisor will then select with which of these systems it wishes to configure the terminal to operate. We assume initially it is the GSM 900 system, and it will push down the modules to configure the terminal to operate with GSM 900. If the terminal is successful at obtaining access rights, it will continue to operate as a GSM 900 terminal; if not, it will configure for WCDMA . If access rights are rejected here, the terminal reverts to surveillance mode, but updates its local database so that it seeks other systems.

Multiprocessor Enhancements

One lesson learned from the 1980s was that software radios would have to cope with multiprocessor implementations. The approach adopted was to place specific code on specific processing blocks from a common point within the architecture. Logically this is managed from the supervisor within the current architecture. We believe that this is still true today. What is also still true is that the type of code placed on a specific processing block will be determined by that processing block. Thus, DSP code will still be executed on a DSP engine, while protocol state machines will be implemented on more general-purpose processors.

Conclusions

STREAMS can be used as the software architectural framework for the realization of protocol-based communications services in an embedded software radio. The memory overhead of STREAMS in a well designed implementation is not particularly significant. The resulting protocol layers can be managed without introducing vertical layers, and the real-time requirements met by the use of delegation. The STREAMS environment provides a framework in which the protocol layer modules implemented are readily portable between different STREAMS.

References

[1] J. Masterton and P. A. Ramsdale, "Integration Aspects of Digital Radio," *IERE 4th Int'l. Conf. Radio Rec. and Assoc. Sys.*, Bangor, North Wales 1986.

[2] A. P. Cheer and P. Diamond, "A Digitally Implemented Radio Receiver," *IERE 4th Int'l. Conf. Radio Rec. and Assoc. Sys.*, Bangor, North Wales 1986.

[3] MMITS Forum paper presented to Software Radio Workshop, Brussels, Belgium, May 29, 1997.

[4] Lucent Technologies paper presented to Software Radio Workshop, Brussels, Belgium, May 29, 1997

[5] Motorola paper presented to Software Radio Workshop, Brussels, Belgium, May 29, 1997

[6] SoftStax™, http://www.microware.com

Abstract

Software radios are emerging as platforms for multiband multimode personal communications systems. Radio etiquette is the set of RF bands, air interfaces, protocols, and spatial and temporal patterns that moderate the use of the radio spectrum. Cognitive radio extends the software radio with radio-domain model-based reasoning about such etiquettes. Cognitive radio enhances the flexibility of personal services through a Radio Knowledge Representation Language. This language represents knowledge of radio etiquette, devices, software modules, propagation, networks, user needs, and application scenarios in a way that supports automated reasoning about the needs of the user. This empowers software radios to conduct expressive negotiations among peers about the use of radio spectrum across fluents of space, time, and user context. With RKRL, cognitive radio agents may actively manipulate the protocol stack to adapt known etiquettes to better satisfy the user's needs. This transforms radio nodes from blind executors of predefined protocols to radio-domain-aware intelligent agents that search out ways to deliver the services the user wants even if that user does not know how to obtain them. Software radio [1] provides an ideal platform for the realization of cognitive radio.

Cognitive Radio: Making Software Radios More Personal

JOSEPH MITOLA III AND GERALD Q. MAGUIRE, JR.
ROYAL INSTITUTE OF TECHNOLOGY

A Global System for Mobile Communications (GSM) radio's equalizer taps reflect the channel multipath structure. A network might want to ask a handset, "How many distinguishable multipath components are you seeing?" Knowledge of the internal states of the equalizer could be useful because in some reception areas, there may be little or no multipath and 20 dB of extra signal-to-noise ratio (SNR). Software radio processing capacity is wasted running a computationally intensive equalizer algorithm when no equalizer is necessary. That processing capacity could be diverted to better use, or part of the processor might be put to sleep, saving battery life. In addition, the radio and network could agree to put data bits in the superfluous embedded training sequence, enhancing the payload data rate accordingly.[1]

Two problems arise. First, the network has no standard language with which to pose a question about equalizer taps. Second, the handset has the answer in the time-domain structure of its equalizer taps, but cannot access this information. It has no computational description of its own structure. Thus, it does not "know what it knows." Standards-setting bodies have been gradually making such internal data available to networks through specific air interfaces, as the needs of the technology dictate. This labor-intensive process takes years to accomplish. Radio Knowledge Representation Language (RKRL), on the other hand, provides a standard language within which such unanticipated data exchanges can be defined dynamically. Why might the need for such unanticipated exchanges arise? Debugging new software radio downloads might require access to internal software parameters. Creating personal services that differentiate one service provider from another might be enhanced if the provider does not need to expose new ideas to the competition in the standards-setting process. And the time to deploy those personalized services could be reduced.

Cognitive radio, through RKRL, knows that the natural language phrase *equalizer taps* refers to specific parameters of a tapped delay-line structure. This structure may be implemented in an application-specific integrated circuit (ASIC), a field programmable gate array (FPGA), or an algorithm in a software radio. Since a cognitive radio has a model of its own internal structure, it can check the model to find out how the equalizer has been implemented. It then may retrieve the register values from the ASIC (e.g., using a JTAG port) or find the taps in the proper memory location of its software implementation. A radio that knows its own internal structure to this degree does not have to wait for a consortium, forum, or standards body to define a level H33492.x7 radio as one that can access its equalizer taps. The network can pose such an unanticipated question in (a standard) RKRL, and *any* RKRL-capable radio can answer it. To enable such a scenario, cognitive radio has an RKRL model of itself that includes the equalizer's structure and function, as illustrated in Fig. 1.

In this example, the radio hardware consists of the antenna, the radio frequency (RF) conversion module, the modem, and the other modules shown in the hardware part of the figure. The baseband processor includes a baseband modem and a back-end control protocol stack. In addition, this processor contains a cognition engine and a set of computational models. The models consist of RKRL frames that describe the radio itself, including the equalizer, in the context of a comprehensive ontology, also written in RKRL. Using this ontology, the radio can track the user's environment over time and space. Cognitive radio, then, matches its internal models to external observations to understand what it means to commute to and from work, take a business trip to Europe, go on vacation, and so on.

Clearly, significant memory, computational resources, and communications bandwidth are needed for cognitive radio, so this technology might not be deployable for some time. In addition, a collection of cognitive radios may not require human intervention to develop their own protocols. Initially, intervention will be required in order to ensure that networks of such radios remain stable (or that we know who to blame if this is not the case). Networks of such radios are complex

[1] *This raises a host of questions about the control of such complex adaptive agents, network stability, and the like.*

Reprinted from *IEEE Personal Communications Magazine*, Vol. 6, No. 4, pp. 13–18, August 1999.

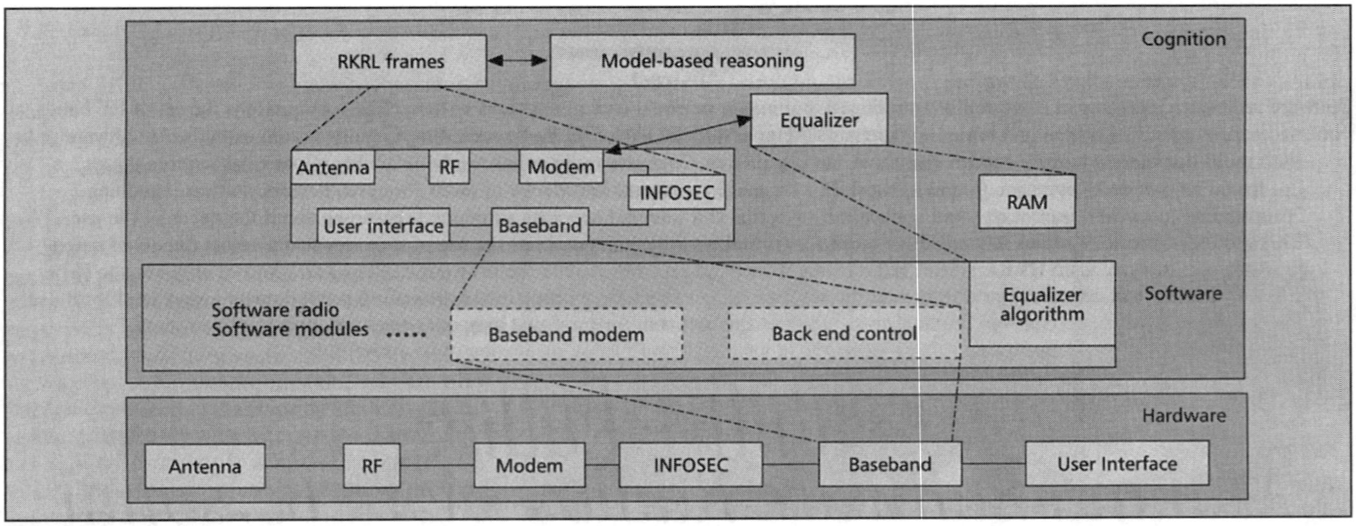

■ **Figure 1.** *Cognitive radio.*

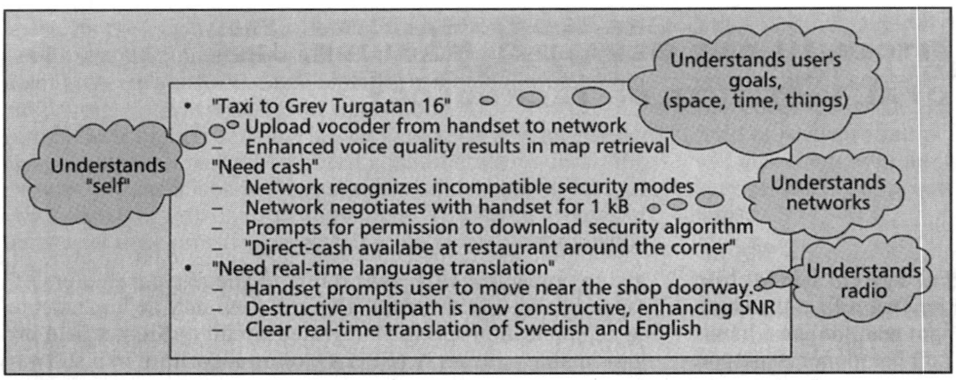

■ **Figure 2.** *More personalized service concepts.*

adaptive systems [2], the study of which is an emerging discipline concerned with the nonlinear behavior of large collections of adaptive entities that have complex interactions. Although there are many technical challenges, the opportunities for enhanced personal services motivate the development of cognitive radio. This article, therefore, outlines the key technical ideas behind cognitive radio, RKRL, and related research at the Royal Institute of Technology (KTH), Sweden.

Personalized Services Scenarios

The services enhancements to be enabled by cognitive radio are motivated by a set of use cases [3] that require the radio to have an advanced degree of "understanding" of topics illustrated in Fig. 2. Next-generation personal communications services (PCS) will know the location of handsets and wireless personal digital assistants (PDAs) to within 125 m for emergency location reporting. Location-aware research [4] is creating technologies for location-aware services, such as flexible directory services [5]. Cognitive radio adds locally sensed recognition of common objects, events, and local RF context. Thus, for example, a cognitive radio can infer the radio-related implications of a request for a taxi to a specific address. It can then tell the network its plan to move from its present location to "Grev Turgatan 16." The network then knows that this user (with high probability) will move across three cell sites into a fourth within the next ten minutes. If this user is headed for a conference center equipped with a local cellphone jammer, it is unlikely to offer the usual load to the network after the taxi ride. Such exchanges could reduce uncertainty about the load offered to a network, potentially enhancing the efficiency of the use of radio resources.

Software radios as presently conceived cannot have such an intelligent conversation with a network because they have no model-based reasoning or planning capability and no language in which to express these things. For example, a software radio from the United States may have the RF access, memory, and processing resources to operate in Sweden. If it lacks compatibility with release level G of the host service provider, it will not work. A software radio cannot "discuss" its internal structure with the network to discover that it can be reconfigured to accept a download of the required software personality. Cognitive radio, however, employs a rich set of internal models useful for a wide range of such dialogs. In addition, the space-time models of the user, network, radio resources, and services personalize and enhance the consumer's experience. The analysis of such use cases yielded a large set of models, conceptual primitives, and reasoning schema necessary for cognitive radio. Which computer languages should be used to express these things?

Radio-Related Languages

In addition to natural language, several computer-based languages are relevant to the expression of radio knowledge (Table 1). The International Telecommunication Union (ITU), for example, adopted the Specification and Description Language (SDL) in its Z.100 Recommendations. SDL readily expresses radio state machines, message sequence charts, and related data dictionaries. The European Telecommunications Standards Institute (ETSI) recently adopted SDL as the normative expression of radio protocols, so one expects SDL modeling of radio to continue to expand. SDL, however, lacks primitives for general ontological knowledge needed, for example, to reason about a travel itinerary.

The Unified Modeling Language (UML) resulted from the unification of diverse object-oriented analysis, modeling, design, and delivery methods. This language readily expresses software objects, including attached procedures (*methods*), use cases, and the packaging of software for delivery. In principle, it can be used to model common-sense knowledge, including

Language	Strengths:	Lacks (or not designed for)
SDL [6]	State machines, message sequence charts, large user base, much encoded knowledge	Plan representation, uncertainty
UML [7]	General ontologies, structure, relationships	Hardware, RF propagation
IDL [8]	Interfaces, object encapsulation	General computing
HDLs [9]	Hardware, electronic devices, interfaces	Geospatial, plans
KQML [10]	General dialogs (ask/tell), tagged semantics	General computing
KIF [11]	Axiomatic treatment of sets, relations, numbers, frames; ontologies	General computing, hardware, RF propagation

■ **Table 1.** *Radio knowledge languages.*

```
(Tell :language RKRL :ontology Stockholm/Europe/Global/Universe/Version 0.1
:Move_Plan (:owner User (:from Kiosk :to "Grev Turgatan 16") :distance 3522 m
(:via (Taxi :probability .9) (Foot :probability 0.03))
(:PCS-needs (:DECT 32kbps) (:GSM GPRS) (:backlog Composing-email)))
```

■ **Figure 3.** *The KQML expression of a plan.*

plans, space, time, relationships, people — just about anything. In practice, it has a strong presence in software design and development, but is weak in the modeling of hardware devices. In addition, although UML can provide a design framework for radio propagation modeling, the target languages are likely to be C or FORTRAN for computational efficiency in tracing tens of thousands of rays of radio waves.

The Common Object Request Broker Architecture (CORBA) defines an Interface Definition Language (IDL) as an implementation-independent syntax for describing object encapsulations. In addition to the 700 companies that comprise the Object Management Group (OMG), IDL is being used by the Software-Defined Radio (SDR) Forum [12] to represent interfaces among the internal components of SDRs. Since this language is specifically designed to declare encapsulations, it lacks the computational power of general languages such as C or Java. IDL excels at architecture integration (e.g., the interface to an equalizer ASIC), but not at expressing the functions and contributions of a component (e.g., the enhancement of bit error rate, BER, at low SNR).

The hardware description languages (HDLs), primarily Verilog HDL and VHDL, readily express the internal structure of ASICs and the personalities of FPGAs. However, cognitive radio does not need the level of detail present in most HDL data sets. Moreover, it needs to know the functions and contributions so that it can make trade-offs, create plans, and reprogram itself. While the documentation package associated with HDL may provide some of this insight, the information is not in a computationally accessible form.

The Knowledge Query and Manipulation Language (KQML), on the other hand, was explicitly designed to facilitate the exchange of such knowledge. Based on *performatives* such as *tell* and *ask*, KQML readily expresses the dialog about equalizer taps and multipath by introducing a few new *tags*. The KQML plan to take a taxi from the information kiosk to Grev Turgatan 16 uses the tell performative to tell the network of the plan, as shown in Fig. 3. In this example, the radio also warns the network that its user is composing some e-mail and so will need either a Digital European Cordless Telecommunications (DECT) data channel or the GSM general packet radio service (GPRS) while in transit.

The ontology performative could invoke an existing ontology or could express the local context, as in this case. It normally would be defaulted unless it changed. The other declarations are self evident, which is one of the strengths of KQML. Like IDL, however, KQML is an interface language. Although, for example, one can express rules from a knowledge base using KQML, one must translate these rules into a convenient internal form (e.g., LISP or PROLOG) in order to use them. In addition, the expression of general spatial knowledge, such as the three-dimensional structure of adjacent city blocks, is better expressed in structured arrays than in KQML. KQML could be used to send changes to such arrays, however.

The knowledge interchange format (KIF) provides an axiomatic framework for general knowledge including sets, relations, time-dependent quantities, units, simple geometry, and other domain-independent concepts. Its main contribution is strong axiomatization. It has a LISP-like structure and, like IDL and KQML, is not specifically designed for internal use, like C or Java.

Finally, most radio knowledge is represented in natural language. It lacks precision, but in some sense has ultimate expressive power, particularly if one includes graphics and multimedia in natural language. Natural language suffers from ambiguities and complexity that at present limit its use as a formal language. RKRL v. 0.1 was created to fill the voids in the expressive power of computer languages while enforcing a modicum of structure on the use of natural language.

Cognitive Radio as a Chess Game

RKRL is supposed to represent the domain of information services that use software radios for mobile connectivity. Since a software radio has a choice of RF bands, air interfaces, data protocols, and prices to be paid, in competition with other users, the

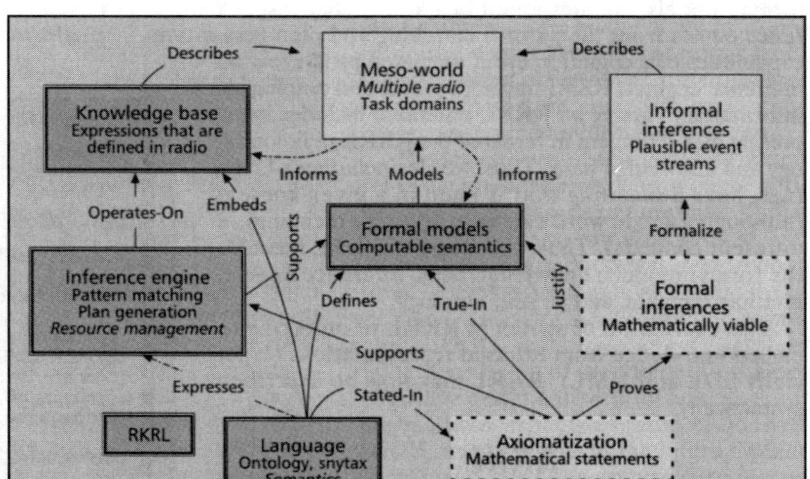

■ **Figure 4.** *Meso-world structure of cognitive radio.*

Handle	Model	Body	Context
Global plane	Contains	South America	Physical world model/universe/RKRL 0.1

■ **Table 2.** *An RKRL frame asserting that the global plane contains South America.*

Handle	Model	Body	Context
Contains	<Process>	SetAccumulate	Models/Universe/RKRL/Version 0.1
SetAccumulate	Excel	<Control><Shift>S	Contains/Models/Universe/RKRL/ Version 0.1
Contains	<Test>	modelVar = "contains"	Models/Universe/RKRL/Version 0.1
Contains	<Domain>	Sets	Models/Universe/RKRL/Version 0.1
Contains	<Range>	Sets	Models/Universe/RKRL/Version 0.1
Contains	Definition	This string-model asserts that the body is a member of the handle.	Models/Universe/RKRL/Version 0.1

■ **Table 3.** *Statements defining the verb* contains *as a formal model.*

domain is analogous to a chess game. The network may orchestrate the game, or in some bands (e.g., the U.S. industrial, scientific, and medical band) cognitive radios will simply compete with each other, hopefully with some radio etiquette. The game board is the radio spectrum with a variety of RF bands, air interfaces, smart antenna patterns, temporal patterns, and spatial locations of infrastructure and mobiles. RKRL provides a consistent way of describing this game board. The future wireless PDAs are the game pieces. Which moves are legal? How will one move impact others in the neighborhood? If a game piece expresses its future needs or plan for use of services to the network, can the network better orchestrate the use of radio resources? And to what degree should the way in which spectrum is used over physical space, code space, power, parameter space, and time be defined by the mobile units themselves? No one knows the answers to these questions because software radios are just emerging, but the need to address them in the future seems clear. RKRL expresses the game board and the legal moves.

Davis [13] defines *micro-worlds* as performance domains for naive physics. Cognitive radio consists of the multiple micro-worlds (the *meso-world*) represented in Fig. 4.

The meso-world of RKRL 0.1 consists of the 41 micro-worlds summarized in Fig. 5. Each is structured according to formal models, and described in a knowledge base. Competence comes from the pattern matching and plan generation capabilities of a cognition cycle, mediated by the related inference engines. RKRL includes syntax and ontological information. Parsing an RKRL statement includes interpreting that statement in terms of the RKRL radio ontology and knowledge base. Thus, words, including KQML tags, have a meaning that is fixed in a given context (although a single word can have different meanings in different contexts). Thus, the scope of RKRL includes the formal models, knowledge base, inference engine, multiple syntaxes, and a radio ontology.

The expression of syntax in RKRL permits one to embed knowledge from external representations (especially SDL and UML). RKRL may now be described syntactically.

RKRL

Instead of attempting to replace SDL, UML, IDL, or KQML, RKRL integrates them through model-based reasoning as follows. RKRL is a parallel frame language. Each RKRL statement is a frame:

<frame> = [<handle>, <model>, <body>, <context>].

The frame expresses a relationship between the handle and body, in a given context. The <model> part defines the exact relationship being expressed. Handles should be thought of as names for things. If a thing contains other things, it can be viewed as an object. If not, it is a terminal constant. Frames are interpreted in parallel, like the cells in a spreadsheet. For example, the following RKRL statement says that South America is part of the global plane in the physical world model of the universe of RKRL 0.1. Additional frames assert Europe and the rest into the global plane. Since the global plane is part of the universe, it can be thought of as an attribute of universe, with a value that is a set of regions of the world. However, it can also be thought of as just a list of the names of countries. These semantics (object, list, etc.) are *not* part of the semantics of RKRL. Instead, the semantics are explicitly declared and coded in RKRL using formal computational models (Table 2).

For example, the word *contains* is a verb from natural language, used in an obvious way. But contains is also a formal model defined in the model's micro-world using the statements in Table 3.

Contains, the model, is defined in terms of a <domain> (the micro-world of sets), a <range>, a <test> for membership, and a <process> for finding members in a local context. The first frame says that Contains has a <process> called SetAccumulate. The fourth line says that Contains is defined over sets (which is a complete micro-world). The items in <brackets> are defined in the meta micro-world. The <test> process consists of a chunk of Excel Visual Basic for Applications (VBA) macro language that will test a frame with a standard binding (modelVar is bound to the model part of the frame). These string models tell the RKRL interpreter (embryonic at present) how to use chunks of code to construct programs that create and manipulate objects, perform model-based reasoning, and otherwise control the software radio platform. To attach an existing VBA macro to an entity, one

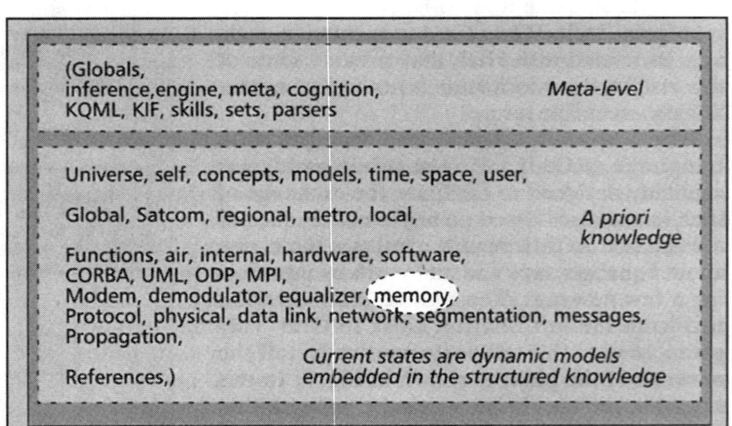

■ **Figure 5.** *RKRL micro-worlds.*

states (line 2, Table 3) that its Excel model is <Control> <Shift>S, the associated keyboard macro. In addition, one may redefine contains in some other context. If it is undefined in a local context, the interpreter searches the micro-worlds ascending the context hierarchy, and then looks across micro-worlds until it either finds a definition or sets a goal of getting a definition. In RKRL, to Get includes to Create (through inferencing) and to Inquire (e.g., of the user or network).

	Plane	Objects	Space	Time	Information sources
1	Global	Regions	10,000 km	1 yr	Travel itinerary
2	Regional	Cities	1000 km	1 week	Weekly planner
3	Metropolitan	Districts	100 km	1 day	Commuting pattern
4	Local	Buildings	1 km	1 hr	GPS, lunch?
5	Immediate	Rooms	100 m	1 s–1 min	Dead reckoning
6	Fine scale	Body parts	1 m	1 μs	Equalizer taps
7	Internal (radio)	HW, SW	.1 m	1 ns	Architecture

■ **Table 4.** *The phusical world inference hierarchy; HW, SW: hard/software.*

Knowledge about Equalizers

For a more radio-oriented example, consider the adaptive equalizer. RKRL knows about the equalizer running on its own platform from the following context: Equalizer/Demodulator/Modem/Internal/Fine Scale/Immediate/Local/Metropolitan/Regional Plane/Global Plane/Physical World Model/ Universe/RKRL/Version 0.1. It also knows about a generic equalizer in Equalizer/Demodulator/Modem/Concepts/RKRL/ Version 0.1. Facts that are known a priori of an equalizer include a frame [Equalizer, Property, Reduces intersymbol interference ...]. The natural language phrase *reduces intersymbol interference* (ISI) is parsed because the concept of reduce has the Excel fragment "handleVar(bodyVar)<bodyVar" and ISI is defined in the physical layer as intersymbol interference with the property that it increases BER. Since all the elements reduce to either Excel, VBA, or a call to an embedded model, the frame is completely interpretable.

The equalizer is defined from the <domain> IF-signal onto the <range> IF-Signal. Its taps are defined using statements such as [Tap-0, Numerical model, 1.2745, Taps/Delay Line/Equalizer/...]. The output is defined in a frame [Weighted-result, Numerical model, *, ...], where * is the Excel expression:

= Weight 3 Model * Tap 3 Model
+ Weight 2 Model * Tap 2 Model
+ Weight 1 Model * Tap 1 Model

If the present RKRL were embedded in a PDA, the a priori model of a three-tap equalizer would be as above, but the internal model would be a dynamic model. Dynamic models contain the values from the current system that they are modeling. A UNIX stream can be a dynamic model, for example. Thus, cognitive radio could tell the network about its equalizer by binding its generic model to the dynamic model stream and reporting the results in KQML to the network. Since such values change as a function of time, RKRL will access (and log) signals as fluents [14] in order to detect regular patterns.

The current version of RKRL is implemented in Visual Basic attached to 41 Excel spreadsheets. Object linking and embedding from Excel allows RKRL to access almost any existing software as an executable model. Thus, instead of writing the large number of subordinate models needed for a comprehensive RKRL, the RKRL framework points to those that exist. In addition, KQML, SDL, IDL, and UML primitives are represented in RKRL. One of the benefits of this approach is an ability to express a given item in more than one standard way. Another benefit is the ability to parse expressions from other languages in order to extract existing knowledge for use in cognitive radio.

Spatial Inference Hierarchy

RKRL embeds a standard spatial inference hierarchy for space and time, as shown in Table 4. Each of the planes consists of objects with associated space-time properties. RKRL also declares ways in which the radio can autonomously obtain information about objects on that level. The global plane, for example, divides the Earth into large regions. The properties of the global plane change in annual cycles (e.g., through annual holiday patterns of travel). RKRL statements at this level define the components of standard annual cycles including seasons of the year, weather, and holidays. The radio can get information about its user's interaction with the global plane by examining the user's travel itinerary. Other planes contain objects appropriate to that level of abstraction, including space-time characteristics and information sources.

The lowest level of this hierarchy represents the physical architecture of the software radio. It describes antennas, digital signal processors, memory (RAM and ROM), user interface devices, and so on in terms of physical capabilities and interconnections. Although there is nothing to preclude RKRL from invoking a complete HDL description of the radio, the goals of cognitive radio concern inference about higher-level aspects of radio etiquettes. RKRL micro-worlds in the internal plane therefore embed the architecture framework, applications programmer interface (API), and IDL of the SDR Forum.

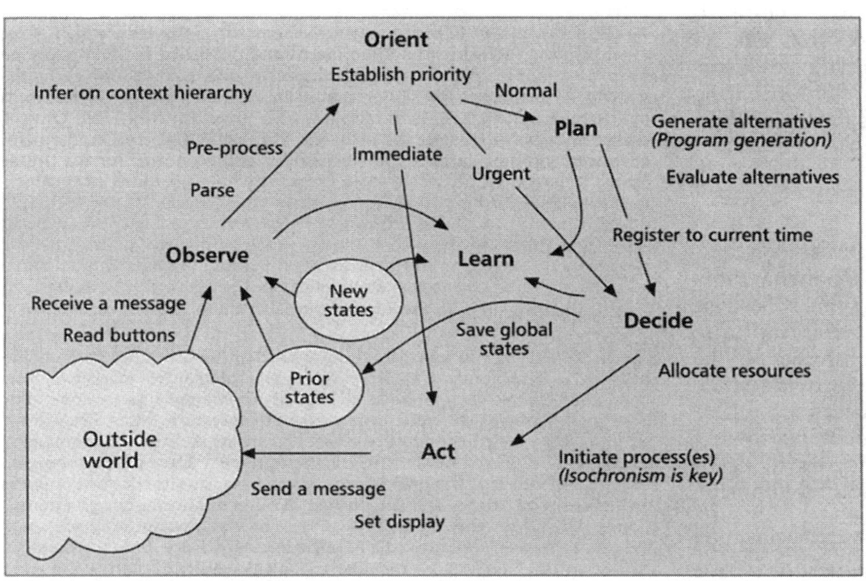

■ **Figure 6.** *The cognition cycle.*

Model Matching

Detailed models of radio functions are embedded in RKRL 0.1 for each micro-world in which competence has been developed. As an example, Table 5 shows an executable model of the segmentation of a message into packets (from the segmentation micro-world).

If a cognitive radio sets the packet number to 0, the payload becomes SAMPL, the first five octets of the outgoing message. To determine the impact of changing the protocol, the system first copies the values of the model to temporary RKRL frames. It then changes the parameters to correspond to its hypothesized protocol (e.g., to a 7-byte payload). Finally, it compares the values of the model to the previously stored values to determine how the change of protocol will change the payload. It can thus independently reason about the correctness of a downloaded packetization module.

Parameter	Value
Message	SAMPLE MESSAGE
Payload size	40
Payload bytes	5
Packet number	1
Payload	E MES

■ **Table 5.** *An Excel model of a message being parsed into packet payloads.*

The Cognition Cycle

RKRL supports the cognition cycle illustrated in Fig. 6. The outside world provides stimuli. Cognitive radio parses these stimuli to extract the available contextual cues necessary for the performance of its assigned tasks. It might analyze GPS coordinates plus light and temperature to determine whether it is inside or outside a building. This type of processing occurs in the observe stage of the cognition cycle. Incoming and outgoing messages are parsed for content, including the content supplied to/by the user. This yields contextual cues necessary to infer the urgency of the communications and related internal tasks. This task is akin to topic spotting in natural language processing. Even relatively high word error rates can result in high probability of detection and low false alarm rate in detecting ordinary events. Thus, the radio "knows" it is going for a taxi ride (with some probability) if the user packets at the wireless information kiosk order a taxi. If the main battery has just been removed, however, the orient stage *immediately* acts to save data necessary for a graceful startup and to shut the system down. Loss of carrier on all available links (e.g., due to entering a building) can result in *urgent* steps to restore connectivity, such as scanning for an in-building PCS or RF LAN. Most other *normal* events might not require such time-sensitive responses, resulting in the plan-decide-act cycle. The act step consists of allocating computational and radio resources to subordinate (conventional radio) software and initiating tasks for specified amounts of time. RKRL also includes some forms of supervised and unsupervised learning.

Conclusion

Software radios provide a vast untapped potential to personalize services, but the contemporary process of modifying radio etiquettes is extremely labor-intensive. In part this is because there is no generally accepted way of representing radio knowledge. This limits the flexibility and responsiveness of the radio to the network and user. RKRL may provide some insights into how to better automate this process. Cognitive radio, built on RKRL, is envisioned as a competence system over the domain of radio resources and protocols. Its agent knowledge and inference mechanisms are under development, as is the initial critical mass definition of RKRL.

Finally, RKRL is designed to be used by software agents that have such a high level of competence, driven in part by a large store of a priori knowledge, that they may accurately be called cognitive. This goal may be very far off, or may emerge from the current research program. Cognitive radio approaches the software radio as a micro-world. But radio engineering is such a large, complex world that it will require much effort to describe it in computationally accessible, useful ways. The present research is therefore offered as a mere baby step in a potentially interesting research direction.

References

[1] J. Mitola, *Software Radio: Wireless Architecture for the 21st Century*, 103255.2507@compuserve.com, Mitola's STATISfaction, ISBN 0-9671233-0-5.

[2] S. Kaufman, *At Home In The Universe*, New York: Oxford Univ. Press, 1995.

[3] J. Mitola, "Cognitive Radio," Licentiate proposal, KTH, Stockholm, Sweden, Dec. 1998.

[4] Location-Aware Computing and Communications Home Pages, KTH, Sweden, and Wollongong University, Australia: www.elec.uow.edu.au/people/staff/beadle/badge/location_aware.html

[5] M. Henning, "Location Aware Mobile Applications Based on Directory Services," Mobicom '97, 1997.

[6] ITU Z.100 Recs., "Specification and Description Language," Geneva, Switzerland, 1991

[7] Eriksson and Penker, *UML Toolkit*, New York: Wiley, 1998.

[8] Mowbray and Zehavi, *The Essential CORBA, Object Management Group*, New York: Wiley, 1995.

[9] Verilog, Object GEODE, Paris, France, 1998.

[10] T. Finin, "KQML – A Language and Protocol for Knowledge and Information Exchange," Int'l. Conf. Building and Sharing of Very Large-Scale Knowledge Bases," Tokyo, Japan, Dec. 1993.

[11] Genesereth and Fikes, "Knowledge Interchange Format Version 3.0 Reference Manual," Logic Group Rep. Logic 92-1, Stanford Univ., Palo Alto, CA, 1992.

[12] SDR Forum Web Site, www.sdrforum.org

[13] E. Davis, "The Naïve Physics Perplex," *AI Mag.*, Menlo Park, CA, Winter 1998.

[14] Cohen *et al.*, "Neo: Learning Conceptual Knowledge by Sensorimotor Interaction with an Environment," *Proc. Autonomous Agents '97*, 1997, www.acm.org

Chapter 5

Software Defined Radio Applications and Economics

SDR applications and economics are inextricable. Cummings and Heath's software-download paper was motivated by the need for service providers to offer differentiated services such as voice coders customized to language families, such as Asian languages. The need for differentiated service, of course, is driven by market economics. Similarly, as GSM fades from cutting edge to relatively easy compared with 3G, the economics of software realizations of GSM handsets and base stations loom large. Turletti's paper characterizes processing requirements in terms of industry-standard benchmarks, substantially simplifying the cost/benefit trade-offs. Posti extends the consideration of GSM to hybrid multicarrier base stations.

Murotake et al., similarly, look toward reconfigurable 3G base stations. Their economics driver is the opportunity to employ standard parallel processing computing platforms for the delivery of such base stations, expanding the production base of the commercial products, and potentially reducing the costs of the base station. Cost of entry into the base station market would also seem to be reduced using their approach.

Smith considers an innovative approach to software radio technology using software-based CDMA as a proximity sensor for a virtual mouse. This paper suggests the pervasive nature of low-cost software-based solutions to accomplishing tasks without wires.

Cummings offers an analysis of the requirements of software radio applications for large-scale airborne applications, such as an airborne command post or surveillance system. The economics issues in this application have been engaged fully by the U.S. DoD's Joint Tactical Radio System (JTRS) Program (*www.jtrs.sarda.mil*). The consolidation of 200 families of discrete radio systems into a dozen or so substantially simplifies logistics and improves the flexibility of such airborne applications.

Mitola and Margulies conclude the text with their analysis of software radio economics. The key finding is that it takes about 2 years and U.S. $25 million to transition a strong single-channel digital radio capability into a multiband, multimode SDR capability with wider bandwidth RF and ADCs and DACs, large-scale software, middleware, and pooled processing resources.

Mode Switching and Software Download for Software Defined Radio: The SDR Forum Approach

Mark Cummings, enVia, Inc.

Steve Heath, Motorola Semiconductors

ABSTRACT The proliferation of air interfaces has created the need for a standard way of switching services supported by a wireless system and downloading new code to systems in the field. Both conventional and software defined systems need this capability. This article describes the work of the SDR Forum in developing such a switcher/downloader standard.

Today, if a user has an IS-95 — dual-mode code-division multiple access (CDMA) and Advanced Mobile Phone System (AMPS) — phone, on power-up, it tries to register in CDMA mode. If it is not within the footprint of an IS-95 base station, it automatically switches to AMPS mode and remains in AMPS mode until it is powered down. If the handset succeeds in registering in CDMA mode and leaves the CDMA footprint (coverage area), it automatically switches to AMPS mode and stays there until it is powered down. Neither the user nor the carrier (and/or service provider) has any ability to switch the handset between modes, to balance traffic loads, improve voice quality, lower cost, and so forth. The same situation occurs today with other commercial (e.g., dual-mode North American Global System for Mobile Communications, GSM, and AMPS), public safety, civil and military systems.

This inability to switch between standards results in loss of efficiency and usefulness. Service providers cannot load balance, use financial incentives to guide subscribers through migrations, or provide consistent service across properties with dissimilar technologies. Users cannot choose service characteristics or take advantage of financial incentives. Switching between capabilities embedded in a system at manufacture is only half the problem; there is also a need to download new code to devices in the field. New code can fix bugs and add new features, (e.g., improved vocoders or complete new air interfaces). As handsets and personal digital assistants (PDAs) merge, there will be additional types of software that could be downloaded in ways similar to Internet plugins. These applets could support new services and service extensions.

The switcher/downloader problem exists for today's systems. With the introduction of third-generation (3G) and other new air interfaces, and software defined radio technology coming online [1–3], this problem will increase in magnitude very quickly.

For both the switcher and the downloader, the technology base exists. These functions are within the capabilities of the microcontroller hardware and software found in wireless devices. The software tools required to develop these tools are readily available. What is missing is a standard to build to.

THE SOLUTION

Problems such as these led to the creation of an industry association named the SDR Forum. The association has defined the architecture and application programming interfaces (APIs) for a switcher and downloader. This is an ongoing process. The switcher and downloader described in this article are documented in SDR Forum TR 2.0 [4] voted out of the SDR Forum in December 1998 and published in April 1999. Work is continuing and will result in a revision voted out every six months.

API FOUNDATION

In order to have an effective switcher/downloader there must be a set of interface standards for the components that are being switched or downloaded. The SDR Forum has taken the approach of standardizing APIs in such a way as to allow maximum opportunity for innovation in implementing the modules inside the APIs. The rules for creating APIs and the messages that implement them are described in detail in SDR Forum TR2.0. A generic architecture is described there. Figure 1, first published in MMITS TR1.1 and later quoted in [1], shows how this generic architecture is configured for handhelds (extreme constraints on size, power consumption, etc.).

The T-shaped component labeled "Switcher between standards" contains the switcher/downloader. It resides in the application software layer, but functions as a monitor. It takes control only when a switching or downloading event is in process, otherwise leaving control to the appropriate controller (Controller Std. *N*). The generic architecture has also been configured for mobile systems (those in vehicles with relatively relaxed size and power consumption constraints), and the SDR Forum is working on similar configurations for base stations and satellite systems.

MESSAGE TYPES TO IMPLEMENT SWITCHING

There are five types of messages that are critical for implementing an SDR Forum switcher: enable/start, reset, capability exchange/configuration change, and place module. The enable/disable and start/stop message pairs provide two levels of operation: enabled but stopped (EBS) and enabled and started (EAS). This is important to allow fine control over the module, and effectively corresponds to online and offline states where changes can be made.

It is important, especially when trying to recover from an error or system failure, to reset a module into a known state. Two reset commands are supported: a reset using the current configuration and a reset to the default. The reset to default allows a different configuration to be used, and thus prevents an infinite loop.

Capability exchange messages are used to exchange capabilities between modules and thus allows a negotiated set of capabilities to be defined before commencing operation. This is achieved

The Software Defined Radio (SDR) Forum was previously called the MMITS Forum.

Reprinted from *IEEE Communications Magazine,* Vol. 37, No. 8, pp. 104–106, August 1999.

by requesting a capability table, which is an extensible data structure that contains information about what the module can do (meta information). This may be different options, different specifications and/or revisions, and so on. Each API would have its own set of definitions for the capability table. In parallel with the capability table is the concept of configuration tables. These data structures are similar to the capability table but contain the actual settings used to configure the module. Multiple tables are supported with table 0 defined as the default. It is possible to set up the detailed configuration information and then to change the configuration by simply selecting the appropriate table. This approach gives several options: configurations can be changed at either a high level through table selection or at a lower level by changing the table contents. An example command for changing configurations is shown in Table 1.

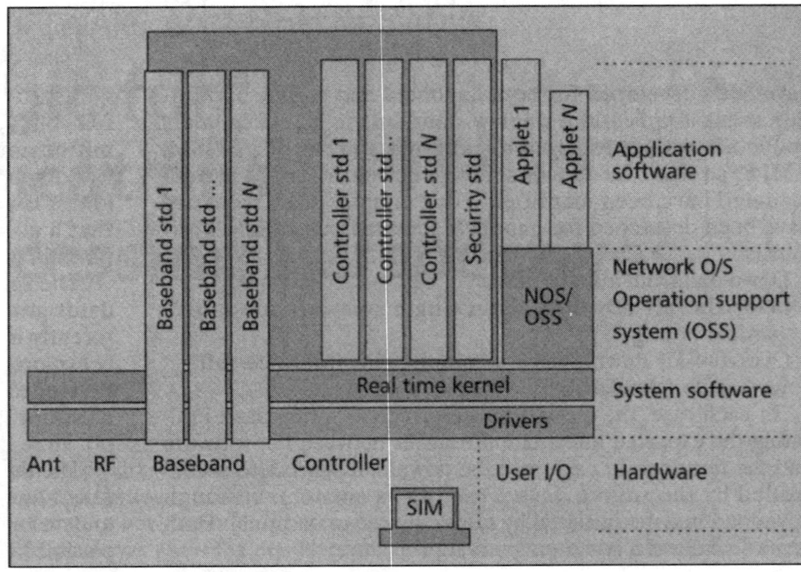

■ **Figure 1.** *MMITS handheld architecture.*

Switching is achieved with the `place_module` command, which allows the current module to be replaced by another locally available one. This simple mechanism can support many physical implementations such as switching between modules, or downloading new software or extensions. By assuming that the modules are local and thus separating the selection process from downloading, many of the testing and security software problems can be avoided.

There are three levels of granularity of modules that can be switched. The regulatory level is the switching of complete standards specified by regulatory bodies (e.g., AMPS in place of GSM). This greatly simplifies the interface and allows virtually any type of lower-level implementation to exist. It also assumes that all other messaging and control information, including configuration, is handled outside the mode switcher API. The next level of granularity, nonregulatory, is the switching of high-level configuration information to enhance the higher-level choices. This may include the choice of voice encoder and other similar information, which can configure the device's capabilities. At the detailed level, low-level configuration information can be changed, thus allowing the complete setup and definition of a waveform, modem coding, and so on. This will need to be extremely detailed to provide the breadth of support that is necessary. Initially, this level of support is of interest to mobile radios (most widely used in government applications).

Migration Support

The SDR Forum has also considered how to bring together legacy and SDR-Forum-compliant systems. While in an ideal world the best approach is to embrace the new set of APIs, this can be difficult. To allow reuse, some form of technology is needed to link existing code to the new API structure. In the SDR Forum environment, this is performed by the wrapper. The wrapper uses three techniques — translation, simulation, and integration — to provide a link between legacy and SDR Forum modules. The translation technique exploits content and/or syntax commonality between the two APIs. It essentially takes each message, looks it up in a dictionary, and translates it so that it can be understood by the other API (e.g., parameter reordering, and bit/byte swapping for little- and big-endian data organizations). The simulation technique will create messages or behavior to support one API without involving the other. This can be used where there are messages or behavior unique to one API. The integration technique is a hybrid of the previous two techniques, combining both simulation and translation to provide a communication path between the two APIs.

set_config (destination, table_config_id, parameters...) ← (status)	
destination	Destination ID of the module to be enabled through the API.
table_config_id	this identifies the configuration table that is used to receive the new parameters. The API supports up to n tables with table ID 0 being the default. The total number of tables is defined in the capability exchange information. The default table 0 can either be fixed and not capable of modification or capable of modification. Again, these characteristics are returned in the capability table.
parameters	these are the parameters that are sent and stored in the configuration table. The actual format will depend on the API definition for each module and will reflect the functionality controlled by the API.
status	This is returned to indicate the success or otherwise of the message and the current status of the module. Included in the status word will be the API module ID to confirm that it was sent to the correct location.
Description	This message allows the module configuration via the API to be changed as required. The changes do not imply or demand any activation and therefore the process of changing the functionality should be seen as a two-stage process: • Set up the new configuration using set_config and one of the available configuration tables. • Select this table using the select_config message to activate it.

■ **Table 1.** *The set-config message.*

Software Download

The SDR Forum has developed a set of scenarios as a means of identifying the architectural, implementation, regulatory, and standardization issues posed by the requirement to support software download. Download scenarios have been developed for both handheld and mobile application areas. Implications on new standards (e.g., 3G cellular) and modifications required to existing standards (e.g., GSM, AMPS) so that over-the-air software download can be accommodated have been identified. For example, three scenarios have been developed for handheld (commercial cellular) terminals:

- Download from a smart card
- Over-the-air download of a single software module (update or bug fix)
- Over-the-air download of a complete air interface software suite (standard)

In each case, the scenario is described as a flowchart indicating the expected information transfer between the software and the terminal. In each case the download procedure is controlled by the source device (acting as master), although download may be initiated by either source or terminal. Each scenario follows a common general procedure:

- *Initiation:* Download source device or terminal initiates a download request.
- *Mutual authentication:* Between source device and terminal.
- *Capability exchange:* To ensure terminal can be configured to accept, install, and successfully run the downloaded code.
- *Download acceptance exchange:* Source provides information on type approval of the code, download procedures and schedules, installation procedures, billing and licensing options, and procedures. Options are selected by the terminal or terminal user, which are validated by the source device.
- *Download/integrity test:* Code is downloaded from source device to a buffer area within the terminal according to the agreed download schedule. In-line integrity testing of received data is performed with retransmissions requested where errors are detected. Only 100 percent error-free reception of code is acceptable.
- *Installation:* Code is appropriately prepared (decompressed, configured, compiled, etc.) and installed within the system. Because there may be a time delay between download of code and installation, a capability exchange internal to the SDR Forum device is required prior to installation to ensure that the configuration of the terminal at install time remains acceptable for correct operation of the downloaded entity.
- *In-situ testing:* This is a provision for testing the downloaded code on the system. Test vectors downloaded with the program code are used to verify the modified operation. In-situ testing is optional.
- *Nonrepudiation exchange:* Response from SDR Forum device to confirm successful installation, permitting required billing to take place.

Implementation Issues

The switcher/downloader has to operate in networks administered by different authorities, consisting of system components from different vendors. Mobile terminals have to be able to roam between different networks. Network operators have to be able to keep their network infrastructure software stable. This stability requires a small number of infrastructure software upgrades that are well preplanned and at fairly infrequent intervals. These requirements lead to the conclusion that switcher/downloader implementations take on the properties of system software in that they must be interoperable across a wide range of platforms and vendors, and come from a relatively small number of sources. On first inspection it appears that there are four alternative paths that could take the industry to such an outcome; however, closer examination reveals that the fourth alternative is just a longer path to the third.

The first alternative is that the SDR Forum can create a full gold standard implementation within the Technical Committee and a set of conformance tests. The SDR Forum is working to carry the switcher/downloader specification to the lowest level of detail possible, but there is a distinct possibility that a gold standard is beyond the economic and political reach of an industry association.

The second path is to work with other industry and standards groups to create a gold standard. The SDR Forum has recently initiated a broad-based liaison activity and is currently exploring the technical benefits of incorporating previously developed output or working with a wide range of other organizations doing work in this area, including MExE and WAP [5]. Here again, however, the task may still be beyond reach.

The third alternative is to form an industry consortium which would fund the creation of a team with the resources necessary to develop and maintain a gold standard implementation. One possible business model for such a consortium is Symbian [6].

The fourth apparent alternative is to allow proprietary implementations to come into the market and compete until there is a shake-out. Because of the nature of the switcher/downloader, this will just result in delay and a forced move to alternative three. When proprietary implementations come into the market, they are likely to support the SDR Forum standard in noninteroperable ways (the IrDA experience is instructive here). When these proprietary solutions come to market, network operators will ask questions about interoperability and roaming. This will lead to long rollout delays. Finally, enough pressure will build up and the market will force the various proprietary solution providers to come together and compromise on a gold standard. The problem with this proprietary approach is the long inherent delay.

Directions for Further Work

The SDR Forum is open to all organizations interested in this area. Contributions can be made by both members and nonmembers. In addition to carrying the switcher/downloader API to lower levels of detail, the Forum is working on the regulatory implications. Meeting and key people contact information can be found on the SDR Forum Web site, http://sdr.forum.org.

References

[1] J. Mitola, "Technical Challenges in the Globalization of Software Radio," *IEEE Commun. Mag.*, Feb. 1999.

[2] M. Cummings and S. Haruyama, "FPGA in the Software Radio," *IEEE Commun. Mag.*, Feb. 1999.

[3] M. Cummings, "Technology Alternatives For Implementing Software Defined Radio," *Proc. ACTS Software Radio Conf.*, June 1998.

[4] SDR Forum, "TR2.0," http://sdr.forum.org, Apr. 1999.

[5] http://wapforum.org

[6] http://www.symbian.com

Toward the Software Realization of a GSM Base Station

Thierry Turletti, *Member, IEEE*, Hans J. Bentzen, and David Tennenhouse, *Member, IEEE*

***Abstract*— Recent advances in processor and analog-to-digital conversion technology have made the software approach an increasingly attractive alternative for implementing radio-based systems. For mobile telephony base stations, the advantages with the new architecture are obvious: great cost savings by using one transceiver per base transceiver station (BTS) instead of one per channel, tremendous flexibility by moving system-specific parameters to the digital part, and allowing the support of a wide range of modulation and coding schemes.**

This paper considers the software implementation of a GSM BTS, and analyzes the performance of each of its radio interface modules. The performance of each software module is evaluated using both a % CPU metric and a processor-independent metric based on SPEC benchmarks. The results can be used to dimension systems, e.g., to estimate the number of software-based GSM channels that can be supported by a given processor configuration, and to predict the impact of future processor enhancements on BTS capacity. Two novel aspects of this work are the portability of the software modules and the platform-independent evaluation of their computational requirements.

***Index Terms*— Base station, digital signal processing, GSM, software complexity, software radio, SPEC.**

I. Introduction

HISTORICALLY, the use of software within radio systems has been restricted to low-bandwidth *out-of-band* tasks, such as signaling, operations, and management. Implementation of the *in-band* radio interface functionality has leveraged analog front ends that partition the band into individual FDMA channels, and digital components that perform the time-division and speech-processing functions. The design of the digital portion is typically based on a compromise between a full application-specific integrated circuit (ASIC) implementation and programmable digital signal processors (DSP's). In particular, past efforts to develop *programmable radios* [9], [11], [21] use some general-purpose processor for embedded control and internetworking functions, but leave the computationally intensive radio functions part to ASIC, field programmable gate array (FPGA), or DSP devices.

In contrast, the *virtual radio* approach [18], [20] involves the reexamination of the overall system from a software designer's perspective, with specific emphasis on the *virtualization* of RF sources. Advances in processor [16] and analog-to-digital conversion technology [7], combined with recent improvements in memory and I/O bandwidth, have enabled a complete software solution. In particular, wide-band analog-to-digital converters (ADC's) can now be used to sample an entire RF band, and the resultant sample stream can be directly deposited into memory for software-based analysis. This eliminates the need for dedicated per-channel frequency partitioning hardware. Furthermore, the temporal decoupling of the sample stream, using large memory buffers, allows conventional operating systems and networks to bridge the gap between the ADC and the software processing modules. Accordingly, both the frequency partitioning and the digital processing (previously performed by ASIC's and DSP's) can be implemented in portable software modules that execute on general-purpose processing platforms. This approach is being pioneered by the SpectrumWare[1] project, whose aim is to extend the reach of application software as close as possible to the analog-to-digital conversion boundary [17], [18].

This paper focuses on the design and performance of a library of software modules that can be used to implement the *in-band* portions of a GSM BTS. GSM, the global system for mobile communications, is a digital cellular telephony system that has gained worldwide acceptance.[2] Besides its widespread adoption, GSM is an ideal experimental vehicle because its complexity is high enough to generate a variety of interesting problems; it is a real system whose implementations are worthy opponents, i.e., they use state-of-the-art analog and DSP technology; and its design is well documented.

The structure of the paper is as follows. Section II briefly describes the hardware and software environment in which we are developing the software BTS. Sections III and IV describe the implementation of the sending side and the receiving side of the BTS, respectively, and evaluate the computational requirements of each of the individual modules. In Section V, we discuss the overall computational requirements of a complete BTS and a potential system configuration for its realization.

Except where specifically noted in the text, little time has been spent optimizing the code of the individual modules, so it is likely that the reported performance can be improved. On this basis, within three years, it will be possible to implement a *minicell* that supports 7–15 speech channels, i.e., one–two FDMA channels, using a commercially available multiprocessor server or a small cluster of workstations.

Manuscript received November 13, 1977; revised April 15, 1998 and August 23, 1998. This work was supported by DARPA under Contract F30602-92-C-0019 (monitored by AFSC, Rome Laboratory) and Contract DABT-6395-C-0060 (monitored by U.S. Army, Fort Huachuca).

T. Turletti is with the High Speed Networking RODEO Group, INRIA, BP 93, 06902 Sophia Antipolis, France.

H. J. Bentzen is with McKinsey and Co., Zurich, Switzerland.

D. Tennenhouse is with the Software Devices and Systems (SDS) Group, LCS, Massachusetts Institute of Technology, Cambridge, MA 02139 USA.

Publisher Item Identifier S 0733-8716(99)02975-3.

[1] See URL http://www.sds.lcs.mit.edu/SpectrumWare/.

[2] By the end of 1997, GSM networks accounted for more than 66 million customers of the world (110 countries/areas)—equivalent to 31% of the world's wireless market; see URL www.gsmworld.com/.

Reprinted from *IEEE Journal on Selected Areas of Communications*, Vol. 17, No. 4, pp. 603–612, April 1999.

Channel partitioning, equalization, and demodulation modules are those most in need of further development and optimization.

Two novel aspects of this work are the portability of the software modules and the platform-independent evaluation of their computational requirements. For each module, one must first determine the fraction of CPU it requires when executing[3] on a range of workstation platforms. One must then normalize these individual results with respect to the known performance of the workstations, using the industry-standard SPEC[4] benchmarks. This gives an estimate of the module's computational requirements in SPECmarks. Although this is far from a precise characterization, it has the benefit of being platform independent. Furthermore, it allows one to predict module performance on future platforms, based on anticipated improvements in the SPECmark ratings of next-generation workstations and servers.

In summary, the software realization of a GSM BTS is becoming a viable alternative to present designs. A software-only approach has the potential of offering many benefits: a software BTS can support several modulation schemes and otherwise incompatible standards (e.g., AMPS, NA-TDMA, CDMA, GSM). Furthermore, the elimination of dedicated hardware reduces the complexity of the BTS, and introduces tremendous flexibility and platform independence into the overall system.

II. System Overview

This section gives a brief description of the minimal hardware required to support a software BTS, and describes the software environment for development of the GSM modules.

A. Simple Wide-Band RF Transceivers

Wide-band ADC technology allows the construction of simple RF transceivers, consisting of only an antenna, a preamplifier, a bandpass filter, and a wide-band ADC in the receive path; and a digital-to-analog converter (DAC) and an amplifier in the transmit path. Today, state-of-the-art ADC's and DAC's are about to comply with GSM performance requirements [6]—the critical issue is the high dynamic range in both receiver and transmitter, and especially the multicarrier power amplifier.

The sampled signal data are directly transferred into the memory of the workstation via a direct memory access (DMA) interface, such as the general-purpose PCI interface (GuPPI) developed at M.I.T. This PCI expansion board provides an efficient means for the continuous streaming of sample data to/from the host memory [2].

B. The Software Environment

The prototype software is based on the VuSystem, a programming system for the dynamic manipulation of temporally sensitive data [10]. This programming system runs on a standard Unix platform not specifically designed for the manipulation of digital media, such as audio and video. The VuSystem implements a two-level strategy, in which programs are partitioned along an *in-band* axis, that supports the flow of temporally sensitive information, and an *out-of-band* axis that supports the event-driven program components, including the user interface and the configuration and control of the in-band processing pipeline (see Fig. 1). In-band processing and out-of-band processing are best handled in separate partitions instead of together because choices of language and architecture can then be made for each partition separately (e.g., C++ for in-band and Tcl scripting language [15] for out-of-band processing). The in-band processing partition is arranged into processing *modules* which logically pass dynamically typed data *payloads* through input and output ports. The payload abstraction is an efficient way to hide the implementation details such as shared-memory data regions, so that the designer of in-band processing modules needs only to know a few simple rules about payload handling.

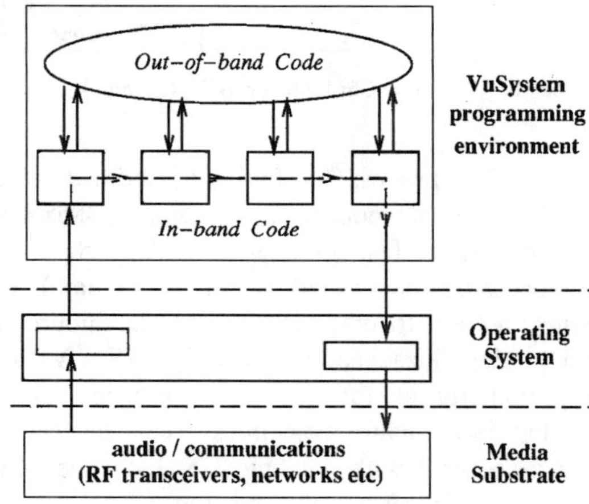

Fig. 1. VuSystem programming environment.

The following sections describe the software implementation of the in-band[5] portion of a GSM BTS; for further detail concerning GSM, see [12]. The in-band modules may be classified into two classes: the downlink modules that correspond to the sending side of the BTS, and the uplink modules corresponding to functions performed in the reverse path. For each module, the section briefly reviews the functionality provided, the algorithm chosen, and the performance obtained.

C. Estimation of Computational Requirements

The functionality and performance of the downlink and uplink modules will be discussed in detail in Sections III and IV, respectively. The following describes the methodology used to measure the performance of the individual modules.

To measure performance, each module is isolated and runs separately in a loop, processing $T = 20$ ms of audio chunk each time through the loop.[6] Consider the speech coder module in detail. To estimate the module's execution time, measure

[3] See details in Section II-C.

[4] SPEC stands for system performance evaluation cooperative; see URL www.specbench.org/spec/.

[5] Currently, we have not implemented any signaling functions.

[6] In GSM, speech is transmitted using groups of 260 bits every 20 ms; see Section III-A.

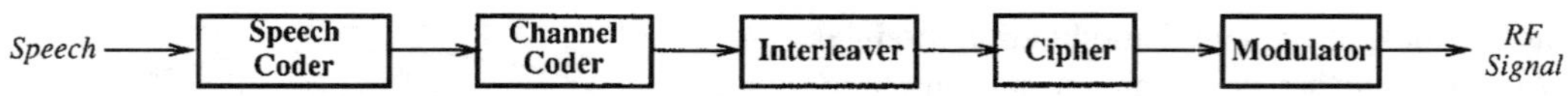

Fig. 2. Block diagram of a GSM sender BTS (downlink).

the time needed to process N chunks of audio data on each platform, allowing the module to sequentially process these N chunks of data as fast as possible (i.e., without pausing between processing the individual chunks of data). Use the c-shell *time* function to measure the overall amount of time needed.[7] Dividing this execution time by NT (N times the required periodicity of the process) gives the fraction of the CPU that the module consumes when performing the processing associated with a single GSM channel. In order to get an accurate measure of time, the N value is chosen high enough[8] to get at least 10 s of processing time on the fastest platform.

To measure performance of the channel coder module, run the speech coder module once to get the 256 input bits, and then run the channel coder module N times over these data. The execution time of the speech coder module is subtracted from the overall execution time measured to give the overall channel coder execution time. Then, the ratio of N divided by T yields the corresponding fraction of CPU used by this module, etc.

The resultant "% CPU/GSM channel" is a good metric to report the performance of each module on a given platform. However, these results will rapidly become outdated as new generations of processors appear. Quantification of computational requirements in a *processor-independent* way is critical for a vision of plug-and-play telecommunications. Accordingly, the SPEC benchmark is adopted as a vehicle for the normalization of our results. SPEC benchmarks have been designed to provide a comparable measure of performance of a system executing a known compute-intensive workload. The benchmarks SPEC92 are included because they still have high popularity in the computer community. Since 1995, SPEC is replacing SPEC92 with an improved SPEC95, with the subcomponents CINT95 (focusing on integer/nonfloating-point compute-intensive activity) and CFP95 (focusing on floating-point compute-intensive activity). The benchmark of the software modules normalize the % CPU results with respect to the SPEC92 and SPEC95 ratings of each platform, as presented in Tables XIV and XV, respectively. This gives an estimate of the module's computational requirements in SPECmarks. Although the mapping to a module's SPECmark requirement is not entirely platform independent, the SPECmark range can provide a basis for performance predictions. This generic metric is not precise, but it is nonetheless an improvement on platform-dependent measures of computational requirements.

III. From Speech to Radio (Downlink)

The sequence of downlink modules is shown in Fig. 2. Basically, the GSM speech coder compresses speech into digital blocks. Channel coding adds redundancy to the blocks, which are then interleaved and spread into pieces called bursts. Finally, after the ciphering operation, the bursts are used to modulate the phase of a carrier.

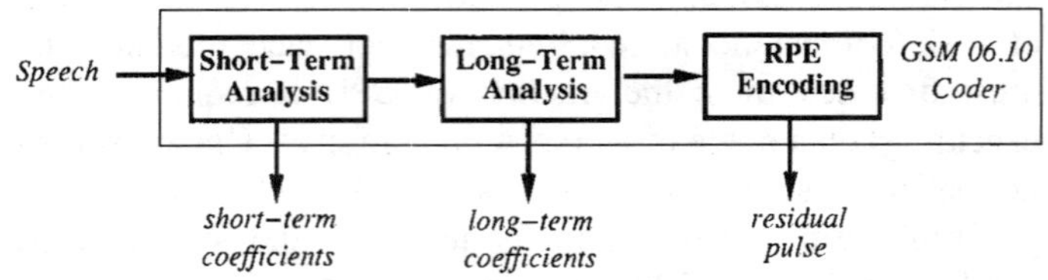

Fig. 3. Block diagram of the GSM speech coder.

TABLE I
GSM Speech Coding Performance

Speech Coder (RPE-LTP 06.10)	Perf. (metric/user-channel) % cpu	SPECint 92	SPECint 95	SPECfp 92	SPECfp 95
SUN Ultra M170	8.1	20	0.45	28	0.73
DEC 3000/800	10	14	–	19	–
Pentium 166	10	20	0.48	–	0.34
Pentium Pro 200	5.9	19	0.48	17	0.40

A. GSM Speech Coding

The GSM full-rate speech coding algorithm is called RPE–LTP, which stands for regular pulse excitation with long-term prediction [5]. The analog voice is first digitized in an ADC to 8000 samples/s, uniformly coded at 13 bits each. Then, the encoder divides the speech into a short-term predictable part, a long-term predictable part, and the remaining residual pulse (see Fig. 3). Finally, it encodes the residual pulse and parameters describing the two predictors. The speech coding algorithm produces a speech block of 260 bits every 20 ms. These 260 bits are classified into three classes (Ia, Ib, and II) according to their importance.

The software implementation of this vocoder was implemented by Degener and Bormann from the Technical University of Berlin.[9] The performance[10] of this vocoder on a number of platforms is shown in Table I.

B. Channel Coding

Class Ia bits are first protected by three CRC bits for error detection. The Class Ib bits are then added to this result. A convolutional code with rate $r = 1/2$ and constraint length $K = 5$ is then applied to this complete class I sequence. The resulting 378 bits are used in conjunction with the 78 unprotected Class II bits, to form a complete coded speech frame of 456 bits (see Fig. 4).

1) Error-Detecting Codes: As indicated in Fig. 4, the most sensitive portion of the speech frame (i.e., the 50 bit category

[7] The c-shell *time* function ensures that the application uses the whole CPU.

[8] For example, we chose $N = 10^4$ for the speech coder module measurement.

[9] The implementation is available in the URL www.cs.tu-berlin.de/~jutta/toast.html.

[10] All C and C++ modules are compiled with gcc v2.7.2 using the −O2 optimization flag.

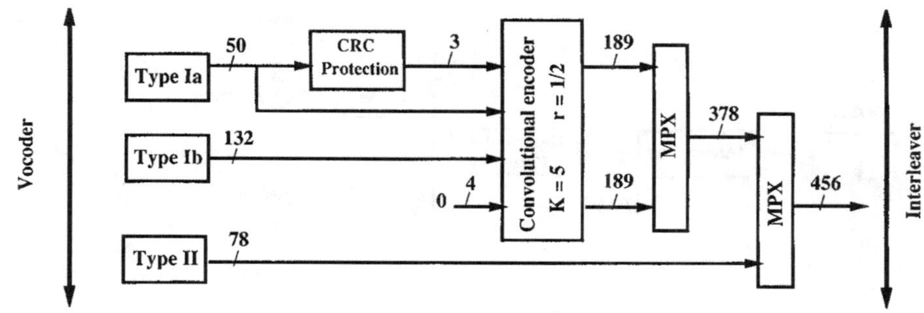

Fig. 4. TCH/FS transmission mode.

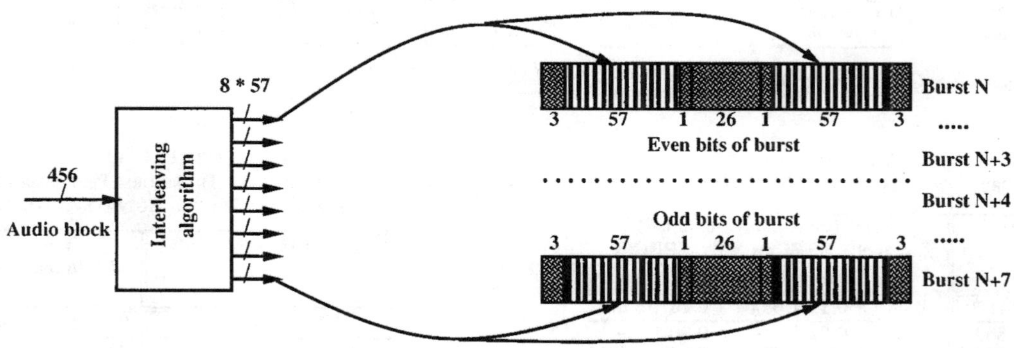

Fig. 5. Interleaving.

TABLE II
CHANNEL CODING PERFORMANCE

Channel Coder	Perf. (metric/user-channel)				
	% cpu	SPECint		SPECfp	
		92	95	92	95
SUN Ultra M170	0.20	0.50	0.011	0.70	0.018
DEC 3000/800	0.45	0.73	–	0.85	–
Pentium 166	0.38	0.76	0.018	–	0.013
Pentium Pro 200	0.24	0.76	0.019	0.68	0.016

TABLE III
INTERLEAVING PERFORMANCE

Interleaver	Perf. (metric/user-channel)				
	% cpu	SPECint		SPECfp	
		92	95	92	95
SUN Ultra M170	0.04	0.11	0.002	0.15	0.004
DEC 3000/800	0.12	0.17	–	0.23	–
Pentium 166	0.15	0.30	0.007	–	0.005
Pentium Pro 200	0.045	0.14	0.004	0.13	0.003

Ia) is protected by three CRC bits used for error detection. These bits are computed using the polynomial generator[11]: $G(X) = X^3 \oplus X \oplus 1$. The 3-bit remainder of this division is transmitting along with input data.

2) Convolutional Coding: The convolutional step involves the padding out of the class I sequence with four extra bits and a convolution using the two following polynomials:

$$P_1(X) = X^4 \oplus X^3 \oplus 1 \quad (1)$$

$$P_2(X) = X^4 \oplus X^3 \oplus X \oplus 1. \quad (2)$$

The result of this step is two 189 bit sequences, which are multiplexed with the (unprotected) type II bits to yield a 456 bit coding block, see Fig. 4. The performance of this channel coding algorithm is summarized in Table II.

C. Interleaving

The aim of interleaving is to decorrelate the relative positions of the coded bits within the code words and in the modulated radio bursts. GSM coding blocks are interleaved on eight bursts: the 456 bits of one block are split into eight groups of 57 bits. Each group of 57 bits is then carried in a different burst (see Fig. 5).

The performance of the software implementation is shown in Table III.

D. Ciphering

Ciphering is achieved by performing an XOR (exclusive OR) operation between a pseudorandom bit sequence and the 114 bits of each burst. The deciphering operation is identical to ciphering. The pseudorandom sequence is derived[12] from the burst number and a session key that itself is determined through signaling when a call is established. The performance of the algorithm is shown in Table IV.

E. Modulation

GSM uses Gaussian modulation shift keying (GMSK) [13] with modulation index $h = 0.5$, BT (filter bandwidth times bit period) equal to 0.3, and a modulation rate of 271 kbaud. There are many ways to implement a GMSK modulator. A block diagram of such a modulator is shown in Fig. 6. A

[11] Note that the "$\oplus$" operation stands for the XOR logical operation.

[12] Note that the algorithm used to generate the pseudorandom sequence is not fully included in the public GSM specification.

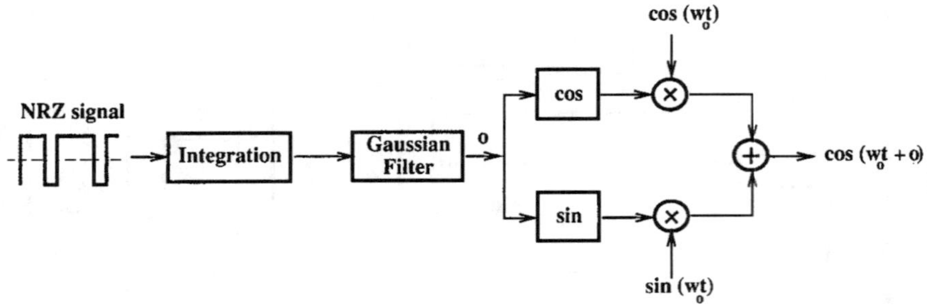

Fig. 6. GMSK modulation block diagram.

Fig. 7. Block diagram of a GSM receiver BTS (uplink).

TABLE IV
CIPHERING AND DECIPHERING PERFORMANCE

De/Cipher	Perf. (metric/user-channel)				
	% cpu	SPECint		SPECfp	
		92	95	92	95
SUN Ultra M170	0.08	0.19	0.004	0.26	0.007
DEC 3000/800	0.16	0.22	-	0.30	-
Pentium 166	0.09	0.17	0.004	-	0.003
Pentium Pro 200	0.05	0.16	0.004	0.14	0.003

TABLE V
GMSK MODULATOR PERFORMANCE

GMSK Modulator	Perf. (metric/user-channel)				
	% cpu	SPECint		SPECfp	
		92	95	92	95
SUN Ultra M170	2.1	5.3	0.12	7.4	0.19
DEC 3000/800	4.5	6.2	-	8.4	-
Pentium 166	3.7	7.3	0.18	-	0.12
Pentium Pro 200	1.7	5.4	0.14	4.8	0.11

previous modulator reported in [20] used an IIR algorithm to process the Gaussian filtering. In the current algorithm, the Gaussian filter is achieved by directly using the phase-shaping response of the Gaussian filter $\Phi(t)$. Basically, $\Phi(t)$ consists of a $\pi/2$ step function, smoothed in order to have a more narrow spectrum than if the step were steeper. The new algorithm precomputes the function $\Phi(t)$, and then obtains the sum over all input bits k_i using (3)

$$\phi(t) = \phi_0 + \sum_{i=0}^{147} k_i \, \Phi(t - iT) \text{ (}\phi_0 \text{ may take any value).} \quad (3)$$

Theoretically, a bit k_i influences the output infinitely long. However, in practice [3], this influence becomes negligible outside a $3T$ period. So, the algorithm can precompute pieces of outputs of bit sequences of length 3, and find the whole output by assembling pieces together. The speedup compared to the IIR-based algorithm is about 9 [3]. Table V reports the performance of the new modulator.

IV. FROM RADIO TO SPEECH (UPLINK)

The sequence of uplink modules is shown in Fig. 7. First, the demodulator reconstitutes the GSM bursts. Then, after the deciphering and the deinterleaving operations, the channel decoder recovers the transmitted blocks of data. Finally, a GSM speech decoder converts them into audio samples.

TABLE VI
POLYPHASE TRANSFORM PERFORMANCE
(200 kHz SPACING BETWEEN CHANNELS)

Polyphase Transform	Perf. (%cpu)			
	# frequency channels			
	1	2	4	8
SUN Ultra M170	20	60	180	620
DEC 3000/800	25	80	220	670
Pentium 166	40	110	340	1200
Pentium Pro 200	25	70	210	720

A. Channel Partitioning

The number of frequency channels used by a GSM BTS depends mainly on the density of the network and the type of cells used (micro cells, urban, rural or road cells). It can be one in a low-density area, and two–four in high-density areas. Typically, adjacent channels are not used within the same cell to avoid spectral recovering, and a GSM BTS uses every second channel (i.e., 200 kHz spacing) to avoid high level interferences. In the GSM terminology, a logical user channel corresponds to a speech channel (defined by its frequency and its time slot number), whereas a physical channel stands for an FDMA channel (which contains eight logical channels).

Multichannel receivers for digitized data can be synthesized using a polyphase transform algorithm [7], [22]. Basically, this algorithm performs the following tasks: 1) frequency translation (each center frequency to baseband), 2) bandwidth reduction of the translated spectrum to match the signal bandwidth, and 3) resampling of the output to match the reduced channel bandwidth.

Table VI reports the percentage of CPU per platform required to partition N GSM frequency channels ($N \in [1, 8]$) at a sampling rate equals to 2.5 times the maximal frequency and with a 200 kHz spacing scheme. The filter used in the implementation has 255 coefficients. The corresponding SPEC performances are shown[13] in Table XII, Section V-A.

[13] Note that the SPECint numbers are not very significant since the partitioning algorithm mainly executes floating-point multiplications.

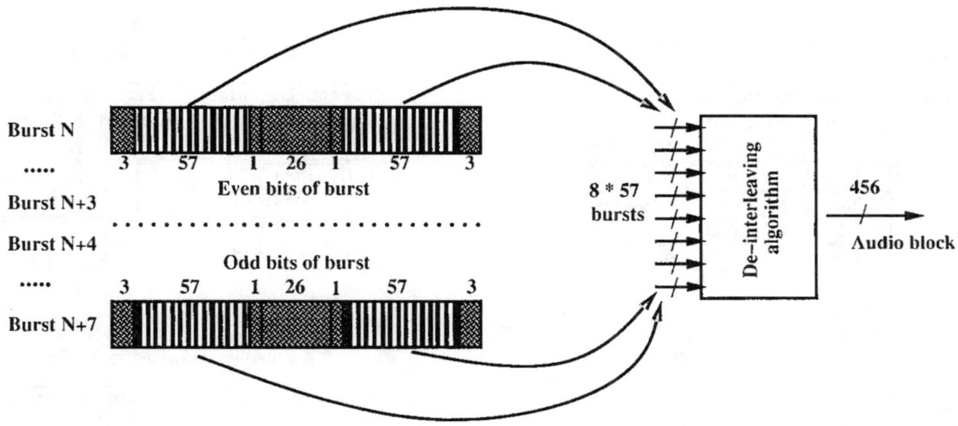

Fig. 8. Deinterleaving.

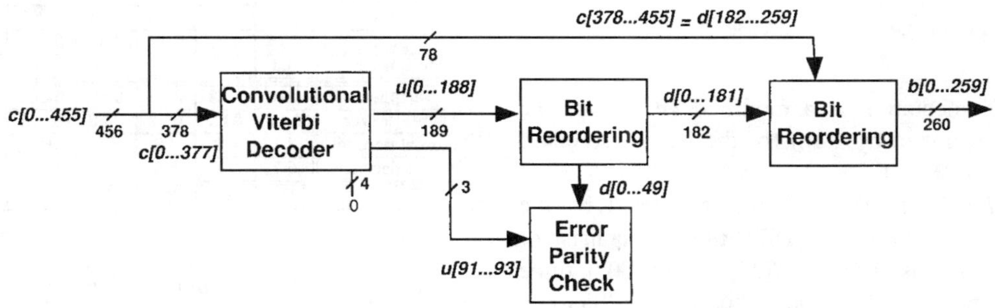

Fig. 9. GSM channel decoding.

The high CPU requirement[14] is due to the high input data rate and the need to compute imaginary and real components of output samples for each frequency channel (requiring multiple floating-point operations).

B. Demodulation and Equalization

The GSM specifications do not impose a particular demodulation algorithm. However, they impose minimal performance criteria, and the algorithm used is expected to cope with two multipaths of equal power received at intervals of up to 16 μs (i.e., more than four symbols). With such a level of intersymbol interference, simple demodulation techniques are ineffective, and an equalizer is required. We are currently implementing an equalizer based on the Viterbi algorithm [4], [14], and this module is expected to be processor intensive[15] [3].

C. Deciphering

Deciphering performs the same operation as the ciphering module described in Section III-D, and hence its performance is summarized by the results in Table IV.

D. Deinterleaving

Deinterleaving performs the inverse operation to the interleaving described in Section III-C. Every 40 ms, eight 114 bit bursts are merged into a 456 bit buffer as shown in Fig. 8.

The performance of this module is shown in Table VII.

[14] The software implementation of this algorithm has been optimized in C.

[15] It has previously been shown that a GSM BTS equalizer can be implemented using a full TMS 320/C40 DSP, i.e., consuming 50 MIPS.

TABLE VII
DEINTERLEAVING PERFORMANCE

De-interleaver	Perf. (metric/user-channel)				
	% cpu	SPECint		SPECfp	
		92	95	92	95
SUN Ultra M170	0.04	0.11	0.003	0.16	0.004
DEC 3000/800	0.17	0.24	–	0.32	–
Pentium 166	0.16	0.32	0.008	–	0.005
Pentium Pro 200	0.04	0.13	0.003	0.12	0.003

E. Channel Decoding

Channel decoding involves the retrieval of the original compressed speech data from the (possibly corrupted) received flow.

Fig. 9 shows the different steps of the channel decoding algorithm. The deinterleaving algorithm generates a 456 bit data buffer from eight GSM bursts. Some of these bits (the 378 bits corresponding to the Type I data bits) are fed into the convolutional decoder, which tries to reconstitute the 189 bits corresponding to the original sequence. After a bit-reordering step, the 50 bits that have the highest priority (Type Ia) are checked using an error control algorithm. If there is no error, a final bit reordering is performed. These steps are further described in the following paragraphs.

1) Convolutional Decoding: Convolutional decoding can be performed using the Viterbi algorithm. The encoder memory is limited to K bits; a Viterbi decoder in steady-state operation takes only 2^{K-1} paths. Its complexity increases exponentially with the constraint length K, which is equal to 5 for the GSM convolutional code; see Section III-B.

TABLE VIII
CHANNEL DECODING PERFORMANCE

Channel Decoder	Perf. (metric/user-channel)				
	% cpu	SPECint		SPECfp	
		92	95	92	95
SUN Ultra M170	1.7	4.2	0.09	5.9	0.15
DEC 3000/800	2.2	3.1	–	4.2	–
Pentium 166	2.1	4.2	0.10	–	0.07
Pentium Pro 200	1.5	4.8	0.12	4.2	0.10

TABLE IX
GSM 06.10 DECODING PERFORMANCE

Speech Decoder (RPE-LTP 06.10)	Perf. (metric/user-channel)				
	% cpu	SPECint		SPECfp	
		92	95	92	95
SUN Ultra M170	4.4	11	0.24	15	0.40
DEC 3000/800	5.2	7.2	–	9.8	–
Pentium 166	3.9	7.7	0.19	–	0.13
Pentium Pro 200	2.1	6.7	0.17	5.9	0.14

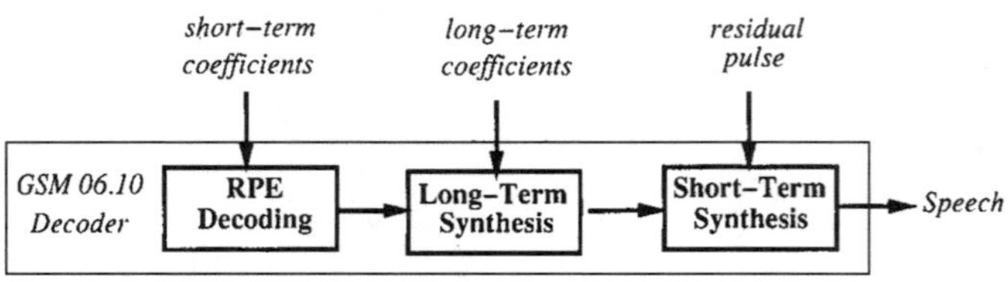

Fig. 10. Block diagram of the GSM speech decoder.

The present implementation is a modified version of the Viterbi decoder[16] developed by Karn for the NASA standard code ($K = 7$, $r = 1/2$) [8].

2) Reordering Algorithm: The 189 bits generated by the Viterbi decoder ($u(0)$, $u(1)$, $\cdots$, $u(187)$, $u(188)$) consist of 182 information bits of class 1 ($d(0)$, $d(1)$, $\cdots$, $d(180)$), three CRC bits, and four tail bits. The relation between u bits and d bits is defined by (4):

$$u(k) = d(2k) \quad \text{and} \quad u(184 - k) = d(2k + 1), \qquad k \in [0, 90]. \tag{4}$$

3) Error-Detecting Codes: The polynomial operation described in Section III-B is applied to the CRC-protected data. If the remainder differs with the received CRC bits, an error is detected, and the audio frame is ignored and eventually discarded.

4) Reordering Algorithm: The last step of the channel decoding regroups the 260 bits of the speech block from the position that corresponds to the order of decreasing importance ($d(0)$, $d(1)$, $\cdots$, $d(259)$) to the position that matches their classification ($b(0)$, $b(1)$, $\cdots$, $b(258)$, $b(259)$) as specified in [5] and [12]. This operation (which requires no effort when implemented in hardware) can be efficiently implemented in software using lookup tables.

The performance of the whole channel decoding algorithm is summarized in Table VIII.

F. GSM Speech Decoding

The decoder reconstructs the speech by passing the residual pulse first through the long-term prediction filter, and then through the short-term predictor (see Fig. 10). The performance of this module is shown in Table IX.

V. PUTTING IT ALL TOGETHER

In this section, we discuss the overall computational requirements of a generic BTS configuration and our efforts to integrate the modules into an experimental platform.

[16] See URL http://people.qualcomm.com/karn/.

TABLE X
AGGREGATE GSM MODULE PERFORMANCE

GSM Modules (for one GSM logical channel)	Performance (metric/user-channel)							
	SPECint92		SPECfp92		SPECint95		SPECfp95	
	min	max	min	max	min	max	min	max
Speech coder	14	20	17	28	0.45	0.48	0.34	0.73
Channel coder	0.50	0.76	0.68	0.85	0.011	0.019	0.013	0.018
Interleaver	0.11	0.30	0.13	0.23	0.002	0.007	0.003	0.005
Cipher	0.16	0.22	0.14	0.30	0.004	0.004	0.003	0.007
Modulator	5.3	7.3	4.8	8.4	0.12	0.18	0.11	0.19
Decipher	0.16	0.22	0.14	0.30	0.004	0.004	0.003	0.007
Deinterleaver	0.11	0.32	0.12	0.32	0.003	0.008	0.003	0.005
Channel decoder	3.1	4.8	4.2	5.9	0.09	0.12	0.07	0.15
Speech decoder	6.7	11	5.9	15	0.17	0.24	0.13	0.40
Total	30	45	33	59	0.85	1.1	0.68	1.51

A. Estimating a BTS's Computational Requirements

Table X summarizes the performance of the modules required to encode and decode one GSM logical (TDM) channel, exclusive of the demodulator (including equalizer) and partitioner modules. We give the minimum and maximum SPEC numbers we have obtained with the four different platforms.

Table XI (based on Table X) shows the computational requirements for the encoding and decoding of N GSM physical FDMA channels ($N \in [1, 8]$), where each physical channel supports eight logical user channels.

Here, again, the entries are exclusive of the partitioner and demodulator/equalizer modules. Furthermore, it is assumed that, for each group of up to four physical channels, one of the derived logical channels is dedicated to signaling, and the remaining logical channels require speech processing. This table suggests that a single Sun UltraSparc M170 or Pentium Pro 200 workstation can just about support the encoding/decoding requirements of one physical GSM channel. Fig. 11 shows the corresponding percentage of CPU required per GSM module on the Pentium Pro 200 platform; a total of 93% of CPU is used.

Table XII reports the SPEC-normalized performance of the partitioning algorithm discussed in Section IV-A. The results suggest that our current Sun UltraSparc M170 platform would be sufficiently powerful to perform the partitioning of two physical FDMA channels (i.e., 16 logical channels).

Table XIII provides a summary of computational demands, including the partitioning algorithm and exclusive of the demodulator/equalizer.

Taken together, the above tables suggest that a *minicell* base station supporting two physical channels, i.e., up to 15 logical channels, could be assembled using one processor to perform the partitioning of incoming samples; one processor to multiplex the outgoing sample streams, two processors to per-

TABLE XI
GSM BTS PERFORMANCE (WITHOUT PARTITIONING AND EQUALIZATION)

BTS	# log. ch	SPECint92		SPECfp92		SPECint95		SPECfp95	
# frequency channels	data/total	min	max	min	max	min	max	min	max
1	7/8	240	360	260	470	6.8	8.8	5.5	12
2	15/16	480	720	530	940	14	18	11	24
4	31/32	960	1440	1060	1890	27	35	22	48
8	62/64	1920	2880	2110	3780	54	70	44	96

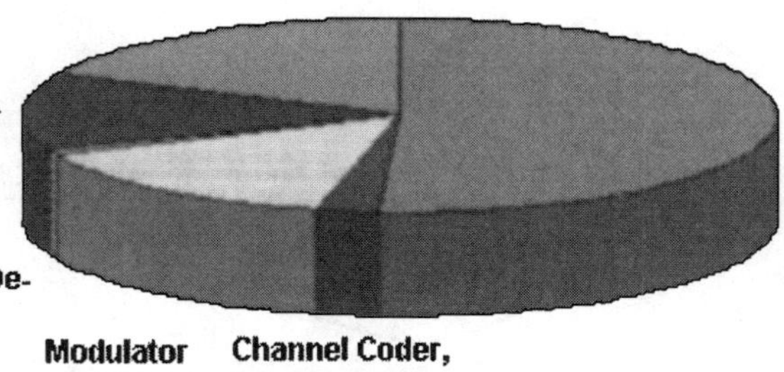

Fig. 11. CPU requirement for eight logical channels on a Pentium Pro 200.

TABLE XII
PARTITIONING PERFORMANCE

BTS	SPECint92		SPECfp92		SPECint95		SPECfp95	
# frequency channels	min	max	min	max	min	max	min	max
1	35	80	47	70	1.1	2.0	1.4	2.0
2	110	220	150	210	3.3	5.6	3.7	5.6
4	300	670	410	630	10	17	11	17
8	930	2400	1300	2300	34	57	40	58

form the encoding/decoding, and some number of additional processors to realize the equalization/demodulation stages. Even after allowing a considerable margin for management and overhead,[17] it will be possible to realize a complete two channel system using a cluster of 8–15 workstations, or possibly, a single multiprocessor server. Although this seems rather expensive today, improvements in processor technology suggest that this level of functionality will be available in a standard server within three years.

B. Module Integration

With the exception of the equalizer, which is currently being implemented, the modules described in this paper were implemented in C and C++. These modules have been incorporated into the VuSystem programming environment, which supports their integration into experimental applications. Fig. 12 shows the flow graph of the VuSystem-based BTS application. The next steps will be to tune the performance of the individual modules, and to experiment with the system as a whole. The project is moving toward *real-world* experiments, initially using an RF channel simulator, such as the HP 11759c. The goal is, eventually, to demonstrate interoperation of the uplink modules with a production handset. Two key questions to be addressed involve: the multitasking and communication overhead associated with the concurrent operation of the entire suite of modules, and the latency characteristics of the software-based system. The latter may be less of an issue here than in other domains, given the built-in latency of the GSM speech coder.

[17]The present analysis is somewhat optimistic concerning the level of CPU utilization that can be attained in a practical system. On the other hand, little effort has been spent tuning the code of the module implementations, which are processor intensive and should be amenable to optimization.

VI. RELATED WORK

The Speakeasy multiband multimode radio is the most aggressive implementation of a programmable radio [9]. That project's objective was to emulate more than 15 existing military radios, operating in several frequency bands (HF, VHF, and UHF). The Speakeasy system was designed to facilitate the addition of new coding and modulation standards. However, the effort put into the DSP processor design, coupled with the low-level programming environment, limit the portability of the system to other processing platforms.

In the commercial sector, one of the first products to implement a programmable radio has been developed by Securor in the U.K. Based on a 50 MHz DSP engine, the radio allows intelligent programmable channeling between 5–25 kHz and adaptive data rate switching using dynamic measurement of bit-error rates [21].

VII. CONCLUSIONS AND FUTURE WORK

This paper has presented the design and implementation of the in-band components of a software-only realization of a GSM base station. The computational requirements of each of the components has been evaluated. Modulation, equalization, and demodulation stages have been identified as the key targets for optimization and tuning.

This paper has introduced the SPEC benchmark metric as a platform-independent means to estimate the requirements of the individual modules. Although this method is not perfect, it provides a very useful way to benchmark the algorithms

TABLE XIII
GSM BTS Performance (Including Partitioning and Without Equalization)

BTS # frequency channels	# log. ch data/total	SPECint92 min	SPECint92 max	SPECfp92 min	SPECfp92 max	SPECint95 min	SPECint95 max	SPECfp95 min	SPECfp95 max
1	7/8	275	440	307	540	7.9	10.8	6.9	14
2	15/16	590	940	680	1150	17.3	23.6	14.7	29.6
4	31/32	1260	2110	1470	2520	37	52	33	65
8	62/64	2850	5280	3410	6080	88	127	84	154

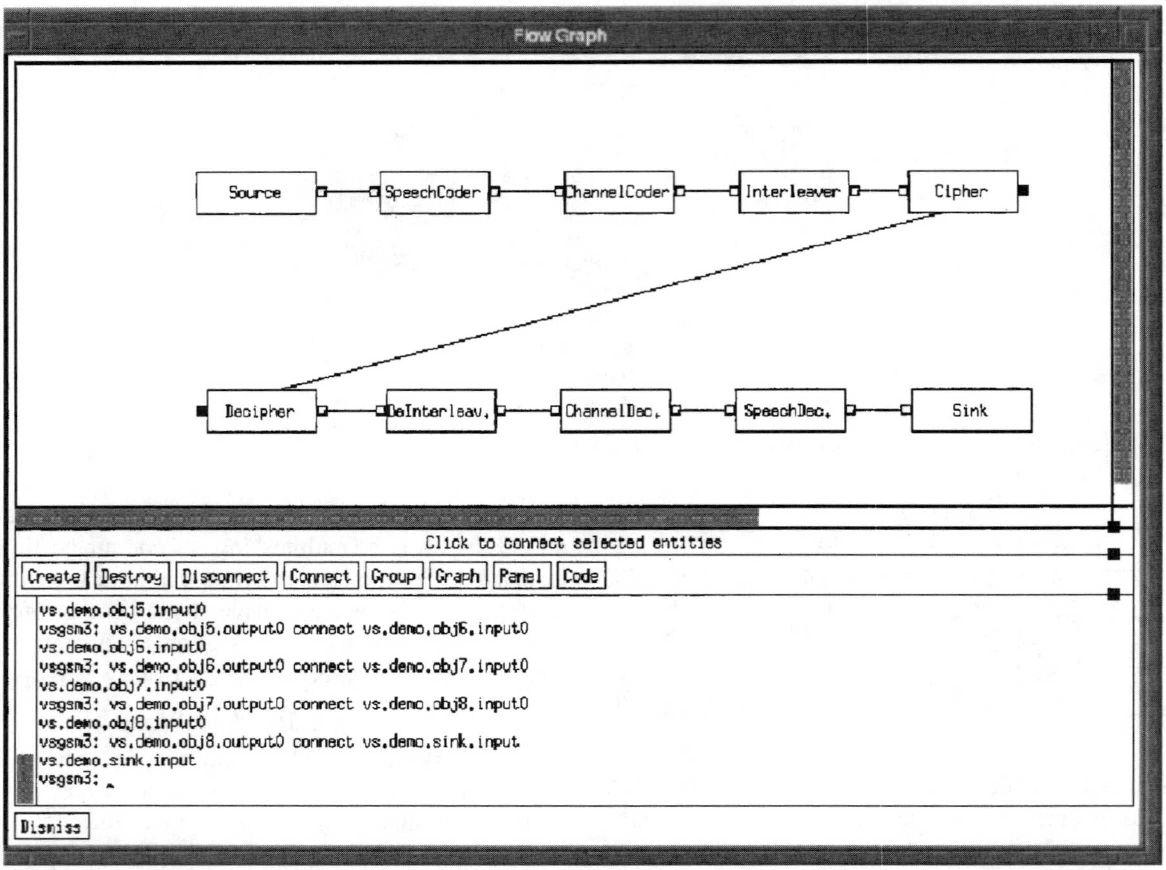

Fig. 12. GSM BTS flow graph.

and to forecast module performance on future platforms. This appears to be the first use of the SPEC metrics as a means of benchmarking communications software to project its performance on future programmable radio platforms.

This approach allows one to project the impact of software-based processing on wireless communications. Within three years, it may be possible to implement a complete BTS on an off-the-shelf server. This suggests that organizations embarking on new design efforts should be cognizant of the software-only approach. As processor and wide-band ADC technology continue to improve, it will be possible to implement increasingly sophisticated and flexible processing on ever wider bands of sampled RF spectrum.

The software approach provides tremendous flexibility for mobile communications systems. For example, changing modulation schemes or frequency assignments only involves the loading and/or configuration of software modules. Similarly, the BTS software can be upgraded to support new modulation schemes without having to upgrade hardware, especially the RF components. At the same time, the scheme preserves the vendor's software investment, as modules can be ported to higher performance platforms, e.g., to support increased cell capacity. Finally, note that the software-only strategy can also be applied to the design of the mobiles; units moving between service providers, or continents, can be dynamically configured to interoperate with local facilities.

The Vusystem programming environment was designed for multimedia applications, and not specifically for virtual radios [10]. A more flexible programming environment supporting adaptive signal processing is currently being designed at M.I.T: the SPECTRA programming environment [2]. SPECTRA will adapt to the environment (e.g., adaptive channel coding), the user requirements (e.g., AM/FM switch), and the available resources (e.g., some robustness or accuracy may be sacrificed if the CPU resources become scarce).

A future SPECTRA environment may include new functionalities to help programmers implement the control-intensive parts of the application. Most applications have both data-intensive and control-intensive handling activities that require different programming techniques. By linking a formal

TABLE XIV
SPEC92 BENCHMARKS

Platform	Clk (MHz) ext/in	Cache ext+I/D	SPECint 92	SPECfp 92	Info date	Source obtained
SUN Ultra M170	83/167	512+16/16	252	351	Nov95	SunIntro
DEC 3000/800	40/200	2M+8/8	138.4	187.6	May94	c.sun.hw
Intel XXpress Pentium	66/166	1M+8/8	197.5	–	Jan96	www.intel
Intel Alder PentiumPro	200	256+8/8	318.4	283.2	Jan96	www.intel

TABLE XV
SPEC95 BENCHMARKS

Platform	Clk (MHz) ext/in	Cache ext+I/D	SPECint 95	SPECfp 95	Info date	Source obtained
SUN Ultra M170	83/167	512+16/16	5.56	9.06	Mar96	c.bmarks
DEC 3000/800	40/200	2M+8/8	–	–	–	–
Intel XXpress Pentium	66/166	1M+8/8	4.76	3.37	Jan96	www.intel
Intel Alder PentiumPro	200	256+8/8	8.09	6.75	Jan96	www.intel

synchronous language such as Esterel [1] to SPECTRA, a large set of GSM signaling functions could be directly implemented from automata described in the GSM specifications.

Finally, one must be prepared to use FPGA's if these devices become available on future general-purpose workstations. They could significantly help with the most CPU-greedy application functions. For example, speedups of close to 50 over strict software implementations have already been achieved for computing the DCT [23].

VIII. SPECmarks for Several Machines

Tables XIV and XV report, respectively, SPEC92 and SPEC95 benchmarks for the four platforms[18] used in our experiments.

Acknowledgment

The authors would like to thank J. Kurose and the Editors for several comments and suggestions that have improved the quality of this paper.

References

[1] G. Berry and G. Gonthier, "The Esterel synchronous programming language: Design, semantics, implementation," *Sci. Comput. Programming,* vol. 19, no. 2, pp. 87–152, 1992.

[2] V. Bose, M. Ismert, M. Welborn, and J. Guttag, "Virtual radios," this issue, pp. 591–602.

[3] H. Bentzen, "A software GSM base station," Master's thesis, ETH Zurich, M.I.T., Cambridge, Mar. 1997.

[4] G. D. Forney, Jr., "The Viterbi algorithm," *Proc. IEEE,* vol. 61, pp. 268–278, Mar. 1973.

[5] "GSM 06.10—European digital cellular telecommunications system (phase 2); Full rate speech transcoding," *ETS 300 580-2, European Telecommunication Standard,* Sept. 1994.

[18] Some SPECmarks are missing in the list of benchmarks available in the public domain; see URL www.specbench.org/osg/cpu95/results/.

[6] B. Hedberg, "Technical challenges in introducing software radio for mobile telephony base stations," in *ACTS Software Radio Workshop'97,* Brussels, Belgium, May 1997.

[7] F. Harris and D. Steinbrecher, "Wireless interface design for digital processors," in *ICUPC'93 Tutorial 5,* Oct. 1993.

[8] P. Karn, "Convolutional decoders for amateur packet radio," in *Proc. ARRL'1995 Digital Commun. Conf.,* 1995.

[9] L. Lackey and U. Upmal, "Speakeasy: The military software radio," *IEEE Commun. Mag.,* vol. 33, pp. 56–61, May 1995.

[10] C. J. Lindblad, D. Wetherall, and D. L. Tennenhouse, "The VuSystem: A programming system for visual processing of digital video," in *Proc. ACM Multimedia 94,* San Francisco, CA, Oct. 1994.

[11] J. Mitola, "Software radios survey, critical evaluation and future directions," in *Proc. IEEE Nat. Telesys. Conf.,* Washington, DC, May 1992.

[12] M. Mouly and M. B. Pautet, *The GSM System for Mobile,* ISBN 2-9507190-0-7, 1992.

[13] K. Murota and K. Hirade, "GMSK modulation for digital radio telephony," *IEEE Trans. Commun.,* vol. COM-29, pp. 1044–1050, July 1981.

[14] S. Ono, H. Hayashi, T. Tanak, and N. Kondoh, "A MLSE receiver for the GSM digital cellular system," in *Proc. 44th IEEE Veh. Technol. Conf.,* June 1994, pp. 230–233.

[15] J. K. Ousterhout, *Tcl and the Tk Toolki* (Addison-Wesley Professional Computing Series). Reading, MA: Addison-Wesley, 1994.

[16] L. C. Steward, A. C. Payne, and T. M. Levergood, "Are DSP chips obsolete?," TR 92/10 DEC Cambridge Res. Lab., Cambridge, MA, Nov. 1992.

[17] D. L. Tennenhouse and V. G. Bose, "SpectrumWare—A software-oriented approach to wireless signal processing," in *Proc. ACM Mobile Computing and Networking 95,* Berkeley, CA, Nov. 1995.

[18] D. L. Tennenhouse, T. Turletti, and V. G. Bose, "The spectrumware tesbed for ATM-based software radios," in *Proc. ICUCP'96,* Boston, MA, Sept. 1996.

[19] T. Turletti, "A brief overview of the GSM radio interface," Tech. Memo., TM 547, MIT, Mar. 1996.

[20] T. Turletti and D. L. Tennenhouse, "Estimating the computational requirements of a software GSM base station," in *Proc. ICC'97,* Montreal, Canada, June 1997, pp. 169–175.

[21] W. Tuttlebee, "The impact of software radio," in *ACTS Software Radio Workshop '97,* Brussels, Belgium, May 1997.

[22] P. P. Vaidyanathan, "Multirate digital filters, filter banks, polyphase networks, and applications: A tutorial," *Proc. IEEE,* vol. 78, Jan. 1990.

[23] R. D. Wittig and P. Chow, "OneChip: An FPGA processor with reconfigurable logic," in *Proc. IEEE Symp. FPGA's for Custom Compute Machines (FCCM'96),* 1996.

Real-Time Implementation of a Reconfigurable IMT-2000 Base Station Channel Modem

David Murotake, John Oates, and Alden Fuchs, Mercury Computer Systems, Inc.

ABSTRACT

Research suggests that joint methods combining smart antennas, RAKE reception, multi-user detection or other adaptive methods may be practically implemented for IMT-2000 channel modems using computationally simplified algorithms. Using software-defined radio methods, these modems can be employed in a new generation of adaptive multimode base stations which permit software reconfiguration from second- to third-generation air interfaces. Practical implementation is made possible by corresponding advances in hardware technology, including new processors and high-bandwidth I/O fabrics which replace traditional computer buses with their inherent limitations in bandwidth and scalability. In this article recent adaptive processing research is reviewed, implementation requirements for second- and third-generation base stations are considered, and the capabilities of selected new monolithic silicon devices are examined. A possible implementation approach for a reconfigurable multimode base station channel modem using SDR design methods is proposed.

INTRODUCTION

International Mobile Telecommunications-2000 (IMT-2000) is the International Telecommunications Union's (ITU's) vision of global wireless (mobile) access in the 21st century. It is scheduled to start service after the year 2000 subject to market considerations, using one of five defined air interfaces. To achieve the required system performance and spectrum efficiency, ITU M.1036 recommends adaptive technologies be used [1]:

> *"6. Spectrum efficiency: that methods should be employed to ensure efficient use of the spectrum, such as the following:*
> *a) radio transceiver technology and access protocols, including access technology, modulation and coding, adaptive interference management, diversity techniques, and smart antenna technology;"*

During the spectrum allocation process for IMT-2000, several national administrations assumed the use of smart antennas and adaptive technologies when calculating the spectrum requirements for IMT-2000 deployment. Thus, third-generation (3G) network operators *may not achieve full network capacity without these methods.* Competitive pressures from growing numbers of 3G users, demanding high bandwidth, high quality of service, and small phones, may further motivate IMT-2000 operators to implement smart antennas, multi-user detection (MUD), 2D RAKE receivers, and other space-time adaptive processing (STAP) algorithms. In addition, because of the gradual implementation of 3G systems around the world, the ability to upgrade base stations from 2G to 3G, using software reconfiguration, is a highly desirable feature. The chip technologies needed to practically implement these new adaptive, software-reconfigurable modems are now becoming available to base station designers.

A *software-defined radio* (SDR) is defined as a radio in which the receive digitization is performed at some stage downstream from the antenna, typically after wideband filtering, low-noise amplification, and downconversion to a lower frequency in subsequent stages. A reverse process occurs for the transmit digitization. Software radios employ one or more reconfigurable processors embedded in a real-time multiprocessing fabric, permitting flexible reprogramming and reconfiguration using different software downloads. SDRs often employ wideband radio frequency (RF) front-end architectures in addition to reconfigurable digital signal processing in their design (Fig. 1). Such designs are well suited for implementing multiband, multimode base stations equipped with smart antennas and adaptive processing [2]. As new services or signal processing algorithms become available, the service provider may upgrade the base station using software downloads, adding new hardware if more processing is required for real-time performance. Another primary consideration of SDR technology application in base stations is reductions in cost [3]. Software base stations also offer cross-standards support and reconfigurability across networks, and may be used to deploy new features and instantiate new air interfaces on demand. Standards and architectures for SDR are under development by the SDR Forum and other groups around the world [4].

SPACE-TIME ADAPTIVE PROCESSING

Application of STAP can be used to enhance the performance of 2G and 3G systems [5]. In a com-

Series Editors:
J. Mitola and Z. Zvonar.

Reprinted from *IEEE Communications Magazine,* Vol. 38, No. 2, pp. 148–152, February 2000.

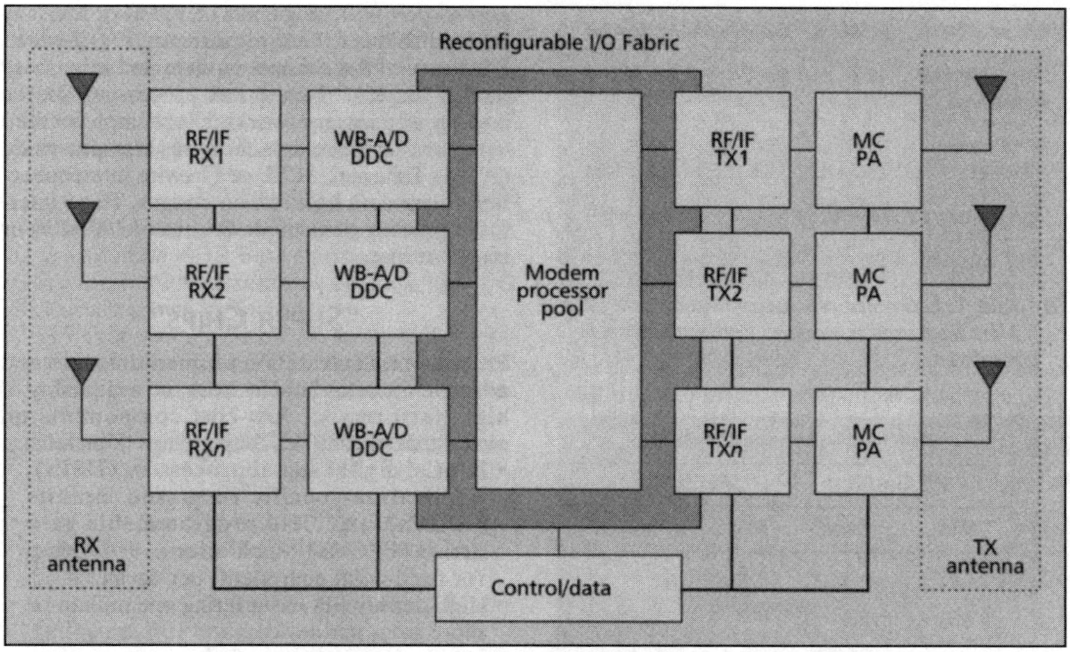

■ **Figure 1.** *In a software-defined base station, a reconfigurable, reprogrammable "pool of processors" including RISCs, DSPs, and ASICs offers unprecendented flexibility in designing wideband modems for IMT-2000 base station front-ends.*

In a commercial GSM field trial conducted by Ericsson and Mannesmann Mobilfunk, smart antennas achieved a 120 percent capacity increase under real-world network conditions.

mercial Global System for Mobile Communications (GSM) field trial conducted by Ericsson and Mannesmann Mobilfunk, smart antennas achieved a 120 percent capacity increase under real-world network conditions [6]. In simulations of IS-95 code-division muliple access (CDMA) networks using smart antennas, "sculpting" of sector coverage to provide load balancing increased forward link capacity by 27 percent [7].

While smart antennas for 3G systems involve greater implementation complexity than systems for GSM or CDMA, similar benefits are achievable [8]. Switched fixed-beam smart antennas are under serious consideration by several wideband CDMA (W-CDMA) base station designers. These antennas offer significant performance benefits over conventional (two-element diversity) base station antenna designs.

Research shows promising combinations of spatial diversity with other methods, giving new meaning to "hybrid vigor." One method, 2D RAKE, combines traditional RAKE receivers with spatial diversity. Separate beamformers are assigned to each finger of the receiver [9]. Space-time array receivers (STARs) may improve antenna performance by 14 dB over conventional adaptive array methods for W-CDMA, and also mitigate co-channel interference (CCI) [10].

Reduced Complexity Methods

The computational complexity of adaptive algorithms has been reduced by use of suboptimal methods. In the example of multi-user detectors (MUDs), which may double or even triple the number of simultaneous users in CDMA networks, optimal detectors for MUD are exponential in complexity with respect to the number of users. Typically several teraflops of processing and several gigabytes per second of interprocess I/O are required to implement optimal detectors for 3G base stations. In contrast, suboptimal detectors such as multistage parallel interference cancellation (PIC) may be implemented with quadratic complexity, and are of great interest to base station designers [11].

With two- and three-branch diversity antennas, use of groupwise successive interference cancellation (GSIC) may increase system capacity by over 200 percent for CDMA networks that support multiple bit rates [12]. Implementation complexity for basic and extended GSIC requires approximately 21.7 Gflops and 43.4 Gflops, respectively. However, when implementing multistage interference cancellation systems, *computational latency must be kept low* to avoid intolerable processing delays in digitized voice and video applications.

Joint space-time processing methods are desirable, not only for their improved performance, but also for their *synergistic reduction in computational complexity over single-discipline methods* when tightly coupled to overall 3G system design [13].

Implementation Requirements

A key design issue is the processing complexity for multimode air interface implementation. Table 1 shows the estimated implementation requirements for GSM, IS-95 CDMA, or W-CDMA using SDR methods on a wideband (5 MHz) channel modem.

A second key consideration is the effect of partitioning. Table 2 shows the effect of partitioning on processing requirements to implement one 200 kHz GSM air interface without equalizers or adaptive processing [14]. By using sufficiently powerful processors, partitioning can be eliminated or minimized. This may reduce implementation complexity by an order of magnitude or more. (Note: the complexity estimates in Table 1 do not consider the effect of parti-

Air interface	GSM	CDMA	W-CDMA
Proc. reqt per RF channel	405 Mflops	1945 Mflops	12,300 Mflops
RF channels allocated	16x 200 kHz	3x 1230 kHz	1x 5000 kHz
Users/channels	$N = 8$	$N = 13$	$N = 16$
Total Gflops	6.6	5.4	12.3

■ **Table 1.** *Estimated processing requirement for a 5 MHz base station modem. Partitioning is not considered.*

Partitioning (for eight GSM logical channels)	Implementation requirement			
	SPECint95		SPECfp95	
	Min	Max	Min	Max
1	1.1	2	1.4	2
2	3.3	5.6	3.7	5.6
4	10	17	11	17
8	34	57	40	58

■ **Table 2.** *Processing requirement for a GSM modem without equalizer, with consideration given to processor partitioning. For best implementation efficiency, partitioning should be kept to a minimum. (Turletti & Tennenhouse, 1999).*

Sample BW	5 MHz	10 MHz	15 MHz	25 MHz
2x oversample	40 Mbytes/s	80 Mbbytes/s	120 Mbytes/s	200 Mbytes/s
4x oversample	80 Mb/s	160 Mb/s	240 Mb/s	400 Mb/s

■ **Table 3.** *Stream bandwidths per RF stream for 16-bit I and Q data, assuming 2x or 4x oversampling, range from 40 to 400 Mbytes/s.*

tioning.) New architectures such as the PowerPC/G4 with integrated 128-bit vector processors permit efficient implementation of software radio algorithms with minimal partitioning.

A third design consideration is the performance of wideband digitizers. To meet GSM and IMT-2000 specifications, analog–digital converters (ADCs) require true 14-bit performance, and support a sustained sample rate of at least 20 megasamples/s with a spur-free dynamic range (SFDR) in excess of 80 dB. Minimum bandwidths of 40 Mbytes/s assuming 4 bytes per complex sample (I and Q) must be supported (Table 3). Maximum stream bandwidths may range as high as 400 Mbytes/s for 25 MHz wideband designs.

A fourth design consideration is the need for high-performance I/O, both between processors on a circuit board and between circuit boards. Numerous adaptive applications like MUD, 2D RAKE, and adaptive antennas are I/O-intensive and sensitive to latency. Industry estimates of interprocessor I/O bandwidths required to implement certain adaptive processing algorithms such as multistage PIC and fully adaptive antennas exceed 1 Gbyte/s. The low-latency, high-bandwidth crossbar fabrics and multiprocessing with port-to-port switching times of 100 ns or less may be used to meet these requirements. Off-board I/O between the channel modem and other base station "layers," such as link processing, source coding, and internetworking, are implemented using one or more asynchronous transfer mode (ATM), Ethernet, SCSI, or Firewire interconnections in modern base station designs. These interfaces typically handle bandwidths of 100 Mb/s or more per line.

"SUPER CHIPS"

Practical implementation of smart antennas and adaptive modems benefit from the availability of high-performance low-cost components on monolithic silicon (i.e., "super chips"), including:

- RISCs, digital signal processors (DSPs), application-specific inegrated circuits (ASICs), and field programmable gate arrays (FPGAs) which offer over 1 Gflop (or fixed-point equivalent) per device
- High-density FPGAs offering one million or more gates per device
- I/O devices capable of digitizing, tuning, and filtering 2G and 3G signal channels
- Highly integrated devices such as the PowerQUICC II or Star*Core combining RISC or DSP cores with communications processors and I/O fabric gateways
- New I/O fabrics using crossbar switches with over 1 Gbyte/s bandwidth per device

PROCESSORS

Low-latency processing is needed to meet air interface synchronization requirements, avoid degradation of beamforming performance, and eliminate intolerable delays in digitized speech and video. High processing bandwidth can eliminate or minimize partitioning, which reduces implementation cost and complexity. When algorithms can be parallelized (as in the case of OFDM, MUD, and STAP), improvements in real-time computing speed may also be obtained by parallel execution of multiple data streams.

Powerful new RISC processors, such as the 3.2 Gflop MPC7400 PowerPC/G4 with 128-bit vector processor, have recently emerged as ideal implementation platforms for software radio. While the Alpha and Pentium III processors may offer comparable fixed-point or floating-point performance to the PowerPC, those devices also consume substantially more power (Table 4). The new RISCs offer numerous implementation advantages over traditional DSPs, such as ease of programmability, cross-platform software portability and FFT performance. Using such a processor, researchers successfully implemented a 19.7 Mb/s OFDM modem using a single 375 MHz PowerPC/G4 [15].

HIGH-DENSITY FPGAS

While ASICs may continue to be the technology of choice for high-volume production implementations, new high-density, high-speed reconfigurable logic devices may aid the designer in prototype development. An example is the Xilinx Virtex, a new high-density FPGA offering devices with up to 1,000,000 system gates (27,648 logic cells) on monolithic silicon. The chip offers over 500 user I/O pins. Using a 2.5-V, 0.22 μ, five-layer complementary metal oxide semiconductor (CMOS) process, this super-FPGA features compile times of 200,000 gates/hr, making "overnight" reprogramming for base stations feasible.

I/O Devices

New I/O devices with high levels of integration have also made SDR base stations more practical to implement. Examples of these new I/O devices are shown in Table 5.

Another useful device class is the highly integrated combination of RISC or DSP cores and I/O peripherals on monolithic silicon. Examples include:

- MPC8260 PowerQUICC II (200 MHz, PowerPC 603e RISC core, ATM, Ethernet, SCSI, PCI bridge)
- MPC8101 Star*Core (300 MHz, 16-bit fixed point DSP core, ATM, Ethernet, SCSI, PCI bridge)
- TMS320-C6205 (200 MHz, 16-bit fixed point DSP core, PCI bridge)

Future examples of these highly integrated devices may also include "glueless" interfaces to high-performance I/O fabrics, further reducing implementation costs.

High-Speed Crossbar Fabrics

An open standard method for implementing high speed fabrics is RACEway (ANSI/VITA 5-1994). RACEway has been used to implement several experimental smart antenna and adaptive processing systems. The high processor-to-processor bandwidth provided by crossbar fabrics is essential for implementing adaptive methods such as MUD and fully adaptive antenna processing. Crossbar fabrics can also be used to reconfigure data flows between processing elements. Such reconfiguration may be necessary to change a smart-antenna-equipped base station from time-division multiple access (TDMA) to CDMA air interfaces (e.g., GSM to W-CDMA conversion).

These "super chip" components may be employed in the design of a *software-reconfigurable, scalable* IMT-2000 channel modem, using the "pool of processors" approach (Fig. 1). A "building block" approach may be economically employed in a range of 2G/3G applications, starting with a single-board "minicell" modem, and scaling up as required for higher-volume macrocell designs, smart antennas, and adaptive processing as additional capacity and features are required. The following configurations, while conceptual, are feasible using available components.

Minicell IMT-2000 Base Station Modem — Features: 560 MIPS, onboard 32-bit RISC processing plus Virtex, implemented on a single 6U-sized card. It can be programmed to support multiple 2G and 3G air interfaces, including IS-136 TDMA, IS-95 CDMA, GSM, and W-CDMA. Can be scaled up to basic configuration (#2) by addition of one quad PPC/G4 circuit card. PowerQUICC II's embedded I/O capability includes up to 2x PCI — for use with PCI mezzanine cards (PMC) or backplane bus — 4x ATM, 2x Ethernet, 2x SCSI, and 2x DUART. Components include:

Processor	MHz	Watts	SPECint95	SPECfp95
PPC G3	400	6	18.8	12.2
PPC G4	400	8	21.0	21.0
Sun Ultra 10	360	20	15.2	19.9
Pentium III	500	28	20.6	14.7
Alpha	600	46	18	27
TI-C62XX	200	1.2	1600 MIPS	—
Star*Core	300	0.5	3600 MIPS	—

■ **Table 4.** *Processor comparison. A balance of good performance with low power, as exhibited by the new MPC-7400 PowerPC G4 and Star*Core, are desirable features when designing software radio modems.*

- 2x MPC8260 PowerQUICC II
- 1x Virtex FPGA
- 1x RACE++ crossbar

Basic IMT-2000 Base Station Modem — Features: 12 Gflops onboard 32-bit RISC processing plus Virtex, implemented on a two 6U-sized circuit cards. It is imilar to the minicell modem, but has increased system capacity. It can be scaled up to the adaptive version (#3) by addition of four quad PPC/G4 circuit cards for additional processing. Components include:

- 2x MPC8260 PowerQUICC II
- 1x Virtex FPGA
- 4x MPC7400 PowerPC/G4
- 2x RACE++ crossbar

Adaptive IMT-2000 Base Station Modem — Features: 60 Gflops onboard processing, implemented on six 6U-sized circuit cards. It supports multiple 2G and 3G air interfaces, including IS-136 TDMA, IS-95 CDMA, GSM, and W-CDMA. It supports adaptive processing algorithms such as multistage PIC or GSIC for MUD, smart antennas, and joint methods. Components include:

- 2x MPC8260 PowerQUICC II
- 1x Virtex FPGA
- 20x MPC7400 PowerPC G4
- 6x RACE++ crossbar

Conclusions and Recommendations

Using software radio techniques and recently developed "super chips," practical implementation of software-reconfigurable IMT-2000 modems featuring smart antennas and other space-time adaptive methods has become feasible. While reducing overall computational complexity, these methods may involve iterative or serialized computations which require parallel multiprocessing to keep processing delays tolerable. While requiring fewer "chips" to implement, high chip-to-chip I/O bandwidth with short transfer latencies should be provided between multiprocessing streams. Future "super chips" may soon include on-chip interfaces to crossbars, satisfying this demand for interprocessor bandwidth while further reducing implementation costs.

Using software radio techniques and recently developed "super chips," practical implementation of software-reconfigurable IMT-2000 modems featuring smart antennas and other space-time adaptive methods has become feasible.

Device type	Model, manufacturer	Required features	Advertised performance
Wideband A/D Converter	AD6644 Analog Devices	20 MSPS 14-bit 80 dB SFDR	65 MSPS 14-bit 90 dB SFDR 100 dB w/dither
Quad Digital Receiver	GC4014 Graychip	14-bit 25 MHz 1 channel	14-bit 62 MHz 4 channels
CDMA Receiver	CSM2000 QualComm	IS-95 CDMA RAKE 1 user	IS-95 CDMA 4-finger RAKE 8 users
Crossbar I/O Fabric	RACE++ Mercury Computer	200 MB/s/port; 100 ms latency	266 MB/sec/port 75 ns latency; 8 ports
Integrated RISC + Comms Processor + PCI	MPC8260 Power-Quicc II Motorola	PCI Bridge ATM Ethernet SCSI Serial RISC Core	PCI-C Bridge 2x ATM 100BT Ethernet SCSI 2x Serial PPC 603e RISC
Integrated DSP + Comms Processor + PCI	MSC8101 Star*Core Motorola	PCI Bridge 2x ATM Ethernet SCSI Serial DSP Core	PCI-X Bridge 2x ATM 100 BT Ethernet SCSI 2x Serial Star*Core DSP
Integrated DSP+ PCI	C6205 TI	PCI Bridge DSP Core	PCI Bridge TI C62 DSP

Table 5. *New I/O devices, including chips which combine communications processors and PCI bridges with DSP or RISC cores, simplify implementation of software-reconfigurable 3G base station modems.*

References

[1] ITU Rec. M.1036, "Spectrum Considerations for Implementation of IMT-200 in the Bands 1885 - 2025 MHz and 2110 - 2200 MHz," Geneva, Switzerland.

[2] J. Mitola III, "Software Radio Architecture and Technology," *Proc. 1998 Int'l. Symp. Adv. Radio Tech.*, Boulder, CO, Sept. 1998.

[3] J. Pereira, "Beyond Software Radio, Towards Reconfigurability Across the Whole System and Across Networks," *Proc. VTC Fall '99*, Amsterdam, Sept. 1999.

[4] SDR Forum, "Section 3.0 — Architecture," Tech. rep. 2.0, http://www.sdrforum.org, Sept. 1998.

[5] R. Kohno, "Spatial and Temporal Communication Theory Using Adaptive Antenna Array," *IEEE Pers. Commun.*, Feb. 1998, pp. 28–35.

[6] H. Dam *et al.*, "Performance Evaluation of Adaptive Antenna Base Station in a Commercial GSM Network," *Proc. VTC Fall '99*, Amsterdam, Sept. 1999.

[7] S. Gordon, M. Feuerstein, and M. Zhao, "Methods for Measuring and Optimizing Capacity in CDMA Networks Using Smart Antennas," *Proc. 9th VA Tech/MPRG Symp.*, Blacksburg, VA.

[8] G. Tsoulos, M. Beach, and S. Swales, "Adaptive Antennas for Third Generation DS-CDMA Cellular Systems," *Proc. 9th ICAP*, Eindhoven, the Netherlands, Apr.1995.

[9] B. Adrian and S. Haggman, "UMTS Radio Network Simulation with Smart Antennas," *Proc. 9th VA Tech/MPRG Symp.*, Blacksburg, VA.

[10] M. Kyeong and D. Im, "Spatial Processing in N-/W-CDMA Mobile Systems," *Proc. 5th Annual Wksp. Smart Antennas in Wireless Mobile Commun.*, Stanford Univ./SARG, Stanford , CA, July 1998.

[11] G. Xue *et al.*, "Adaptive Multistage Parallel Interference Cancellation for CDMA over Multipath Fading Channels," *Proc. IEEE VTC '99*, Houston, TX, May 1999.

[12] C. Wijting *et al.*, "Groupwise Serial Multi-User Detectors for Multirate DS-CDMA," *Proc. IEEE VTC '99*, Houston, TX.

[13] A. Paulraj, "Smart Antennas in Wireless Communications," *Proc. 5th Annual Wksp. Smart Antennas in Wireless Mobile Commun.*, Stanford Univ./SARG, July 1998.

[14] T. Turletti and D. Tennenhouse, "Complexity of Software GSM Base Station," *IEEE Commun. Mag.*, Feb. 1999.

[15] S. Chuprun *et al.*, "High Data Rate OFDM Waveform Capabilities of Emerging Software Defined Radio Platforms," *Proc. IEEE MILCOM '99*, Atlantic City, NJ, Nov. 1999.

Code-Division Multiplexing of a Sensor Channel: A Software Implementation

Joshua R. Smith, Christopher Salthouse, and Neil Gershenfeld, *Member, IEEE*

***Abstract*—This paper demonstrates the use of software radio techniques in the context of sensing, rather than communications. It describes code-division multiplexing (CDMA) and time-division multiplexing (TDMA) of a receiver channel in an electric field sensing system. The only hardware used is a front-end gain stage consisting of two opamps and a microcontroller. The modulation and demodulation operations are implemented entirely in the microcontroller software. Multiple coded waveforms are transmitted simultaneously, and induce a combined signal on a single receive electrode. The combined signal, after passing through a single analog front end terminating in an analog-to-digital converter, is separated into the four original component signals by a software demodulation operation.**

The signal-to-noise ratio (SNR) achieved by the code-division multiplexed system given a fixed measurement time is compared to the SNR achieved by a time-division multiplexed implementation given the same total measurement time. The paper also compares the scaling of TDMA and CDMA performance with the number of transmitted channels and the number of demodulated channels.

***Index Terms*—CDMA, electric field sensing, sensing, software radio, spread spectrum.**

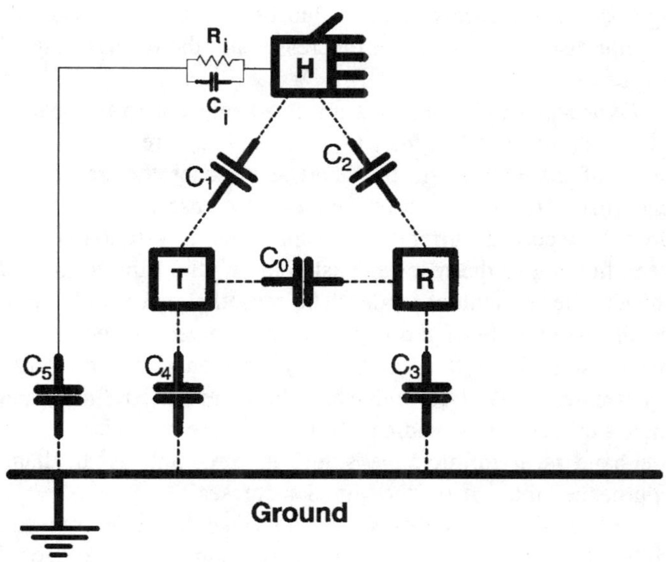

Fig. 1. Lumped circuit model of electric field sensing.

I. Introduction

THIS paper demonstrates the use of software radio techniques in the context of sensing, rather than communications. It describes code-division multiplexing (CDMA) and time-division multiplexing (TDMA) of a receiver channel in an electric field sensing system. The demodulation operation is performed entirely in software. Several sensor signals can simultaneously share the same analog front-end hardware, including ADC, because software performs different processing operations on one set of samples to extract several distinct signals. This means that the number of channels that may simultaneously be received is not rigidly fixed by hardware. Additional receiver channels are extra processing steps on the data already being collected. Additional coded waveforms do appear as noise to the other channels (in the CDMA case), or require additional sensing time (for TDMA), so additional channels are not "free," but the number of channels is not fixed by the hardware. As long as one is able to pay the price in either signal-to-noise ratio (SNR) or measurement time, and also in processing time, additional channels may be added. Furthermore, the software receiver readily changes which particular channels it is demodulating.

The work is set in the context of electric field sensing, which is introduced below, but the same principles could be applied in virtually any sensing or measurement system in which the quantity being sensed is modulated by a carrier. For example, these techniques could be applied to systems of ultrasound or infrared emitters and detectors.

II. Motivation: Electric Field Sensing

The term electric field sensing refers to a family of noncontact methods for measuring the position and orientation of the human body, or parts of the human body such as a hand. Electric field sensing has been used for human–computer interface [1], to make a three-dimensional noncontact mouse [2], for creating new musical instruments (3-D) [3], and in the automotive industry as a solution to the rear-facing infant seat problem.[1]

In a typical implementation of electric field sensing, a low-frequency (from 10–100 kHz) voltage is applied to a transmit electrode (labeled T in Fig. 1), and the displacement current induced at a receiver R is synchronously detected. Fig. 1 shows a lumped circuit model of the electrodes and the body.

Manuscript received September 1997; revised June 19, 1998 and July 31, 1998.

J. R. Smith is with the Escher Laboratories, Cambridge, MA 02142 USA.

C. Salthouse and N. Gershenfeld are with the Physics and Media Group, M.I.T. Media Laboratory, Cambridge, MA 02139 USA.

Publisher Item Identifier S 0733-8716(99)02976-5.

[1] Because the violent inflation of an airbag can injure infants in rear-facing infant seats, it is desirable to sense the orientation of the child and disable the airbag as appropriate [4].

Reprinted from *IEEE Journal on Selected Areas of Communications*, Vol. 17, No. 4, pp. 725–731, April 1999.

The term capacitive sensing ordinarily refers to measuring the change in the loading of the transmitter as the hand approaches, increasing the value of C_1. There is no distinct receiver in this measurement, so C_1 is the only significant capacitance in the problem. This type of measurement is called a *loading mode* measurement.

But the other current pathways in the diagram suggest other measurement techniques. In transmit mode, the transmitter is coupled strongly to the body—C_1 is very large—and the body is essentially at the potential of the transmitter. As the body approaches the receiver, the value of C_2 (and C_0—the two are not distinct in this mode) increase, and the received signal increases.

Shunt mode measurements are most relevant to this paper. In the shunt mode regime, C_0, C_1, and C_2 are of the same order of magnitude. As the hand approaches the transmitter and receiver, C_1 increases and C_0 decreases, leading to a drop in received current: the displacement current that had been flowing to the receiver is shunted by the hand to ground (hence, the term shunt mode). The sensed signal is defined to be the magnitude of the decrease in received current as the hand moves in from infinity. In other words, one measures a baseline received current when the hand is at infinity, and then subtracts later readings from this baseline. Thus, when the hand is at infinity, the signal is zero, and as the hand approaches the sensor, the signal increases.

With n ordinary capacitive sensors (loading mode), one can collect n numbers. These n numbers turn out to be the diagonal of the capacitance matrix for the system of electrodes. In shunt mode, one measures the $n(n-1)$ off-diagonal elements.[2]

As the size of the electrode array grows, it becomes possible to infer more about the geometry of the hand or other body parts—as n gets large (say, above 10), a kind of fast, cheap, low resolution 3-D imaging becomes possible [5]. However, the large amounts of data are not without a cost. If one were to measure sequentially each value of the capacitance matrix, the time required would be $O(n^2)$. Fortunately, it is possible to make all of the receive measurements for a single transmitter simultaneously. This brings the measurement time down to $O(n)$. One might expect the measurement time to be constant rather than linear since all of the measurements are being made simultaneously. However, as will be explained in the section on resource scaling, the simultaneous measurements interfere with one another, leading to linear rather than constant scaling. But before the scaling is considered in more detail, the basics of code-division multiplexing will be reviewed.

III. Direct-Sequence Spread-Spectrum and Code-Division Multiplexing

In direct-sequence spread spectrum [6], [7], the signal is modulated with a pseudorandom carrier, usually generated by a maximum length linear feedback shift register (LFSR). With code-division multiple access (CDMA), multiple users share the same physical channel by choosing different coded waveforms. Another common channel-sharing technique is time-division multiple access, in which transmitters avoid transmitting simultaneously.

The simple implementation of direct sequence spread spectrum for electric field sensing can be understood in terms of linear algebra. Each measurement is made using a single short pseudorandom measurement burst, during which the hand is assumed to be stationary.

If the pseudorandom carrier signal for transmitter i at time t is $\phi_i(t)$, then the signal received on electrode j, as modified by the capacitance matrix to be measured, is

$$R_{ij}(t) = C_{ij}\phi_i(t) + N_{ij}(t)$$

where C_{ij} is a constant because the hand geometry, and therefore the capacitance matrix, are assumed to be static on the time scale of a sensing burst, and $N_{ij}(t)$ is noise. The noise is indexed by the transmitter and receiver because, for each demodulation operation, the signal from the other transmitters appears as noise. Thus, the noise depends not only on the receiver, but also on which transmitter is being demodulated.

In the ideal case, the transmitted waveforms would be orthogonal to one another, so that channels do not interfere

$$\langle \phi_i, \phi_j \rangle = \frac{1}{s}\sum_{t=1}^{s} \phi_i(t)\phi_j(t) = G^2\delta_{ij}$$

where s is the number of chips in a burst and G^2 is the average power per chip. One can view time division multiplexing as a case in which the carriers do not overlap, and thus satisfy this orthogonality condition exactly.

In CDMA systems, there are nonzero cross correlations between different code sequences. This can be a serious problem in a communications scenario in which receivers may spuriously lock onto the wrong code sequence. However, in the sensing application, there is no synchronization problem because the transmitter and receiver are on the same circuit board, so the only problem is a decrease in SNR due to interference between the channels.

It would be best if the basis functions were also orthogonal to the ambient noise N. In reality, they are not completely orthogonal to N; if they were, one could sense using arbitrarily little energy or time. Since the noise is uncorrelated with the carrier, then by the law of large numbers, the fluctuations around zero of the inner product of the carrier and the noise are on the scale of $1/\sqrt{s}$, where s is the number of samples

$$\langle \phi_i, N \rangle = \frac{1}{s}\sum_{t=1}^{s} \phi_i(t)N(t) \approx 0 \pm \frac{1}{\sqrt{s}}.$$

To measure an entry in the capacitance matrix, form the inner product of the received signal R_{ij} with the transmitted signal ϕ_i

$$D_{ij} = \langle R_{ij}, \phi_i \rangle = \frac{1}{s}\sum_{t=1}^{s} R_{ij}(t)\phi_i(t) = G^2C_{ij} \pm \frac{1}{\sqrt{s}}. \quad (1)$$

[2] Because the capacitance matrix is symmetrical, there are ideally only $(1/2)n(n-1)$ distinct values. In practice, apparent deviations from symmetry can be used to calibrate the measurement system.

IV. Resource Scaling

With the code-division sensor multiplexing techniques explored in this paper, one might argue that it should be possible in constant time to collect data that completely characterize the n^2 entries in the capacitance matrix. Each transceiver unit would simultaneously transmit its coded carrier, measure the combined signal induced by the other transmitters, and store its measurement samples in memory. The vector of samples in unit $\hat{\imath}$ contains a complete, but encoded, representation of all of the C_{ij} values.

Of course, the "constant time" figure ignores deviations from orthogonality of the coded waveforms. A simple algebraic argument suggests that the number of samples necessary to maintain orthogonality grows linearly with n. If sinusoids are used instead of pseudorandom sequences, then the length of the measurement vector must be equal to (or greater than) the number of sensor channel amplitudes to be extracted since the DFT matrix is square. Thus, more realistically, one can form a complete but computationally encoded representation of the capacitance matrix using measurement time proportional to n, and storage of n values in each of the n units.

So far, this discussion has ignored the computational operations necessary for unit $\hat{\imath}$ to do signal separation, transforming the raw measurement vector into an explicit representation of the n capacitance values. The naive separation algorithm (for a single rseceiver unit—multiple receivers can process in parallel) requires time mT, where T is the number of samples in the measurement vector and m is the number of channels being demodulated. By the argument from the previous paragraph, T cannot be less than n, and in practice, should be some constant multiple of n. In the case in which one wishes to demodulate all n channels (so that $m = n$), then the naive demodulation algorithm requires n^2 time. (It is natural to wonder whether an $n \log n$ fast spread-spectrum demodulation algorithm exists, analogous to the fast Fourier transform, that could be used with coded rather than sinusoidal carriers. A pseudorandom generator with hidden symmetry properties would probably be needed.) The naive algorithm to extract just a small number m of the possible n sensor values would require time mn.

In the time-division multiplexed (TDMA) case, the transmitted carriers are exactly orthogonal to one another since they do not overlap at all. As in the CDMA case, the total measurement time for one unit is proportional to n. Because the carriers are exactly orthogonal, the time required for a single measurement is constant, rather than proportional to n as in the CDMA case. The amount of computation required to demodulate TDMA is also proportional to n. Thus, in the limit in which all possible measured values are actually demodulated, TDMA appears to have an advantage, requiring n rather than n^2 or $n \log n$ time. In the limit in which just a few of the possible sensor values are extracted, the algorithms scale similarly since TDMA still requires linear measurement time, and CDMA's processing time becomes linear instead of quadratic.

In the practical examples discussed in this paper, there were four transmitters and one receiver (rather than measuring the full 5×5 capacitance matrix of this five-electrode system),

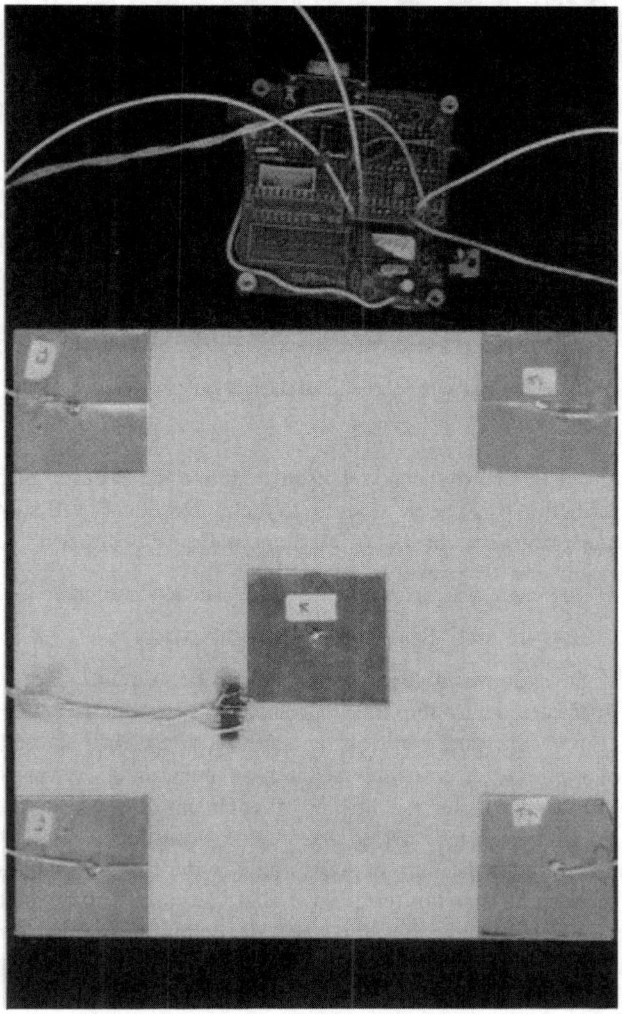

Fig. 2. Hand-wired prototype circuit board used to make measurements, together with electrode array. In the center is a receive electrode, with front-end op-amp attached. In the corners are the transmit electrodes.

so m was four. The theoretical discussion above considers the measurement and processing time required to achieve the same SNR using software TDMA and CDMA. In the examples, the same total measurement and processing time is allotted to the two schemes, so that a meaningful comparison of the resulting SNR may be made.

The next sections describe the hardware and software used to implement the sensing schemes introduced above.

V. Hardware

The analog front end consists only of a MAX 474 dual opamp. It is configured as a transimpedance amplifier with a 1M resistor and 22 pF capacitor in its feedback network, followed by an inverting voltage gain stage with a 10K input resistor and 100K feedback resistor, for a gain of 10. Because the opamp uses a single supply (5 V), but the received signals are bipolar (positive and negative), 2.5 V was used as analog ground. The amplified signals are read by the analog-to-digital converter on the PIC16C71 microcontroller. The PIC transmits data through a Maxim MAX233 RS-232 transceiver to a host computer for display and analysis. Fig. 2 shows the

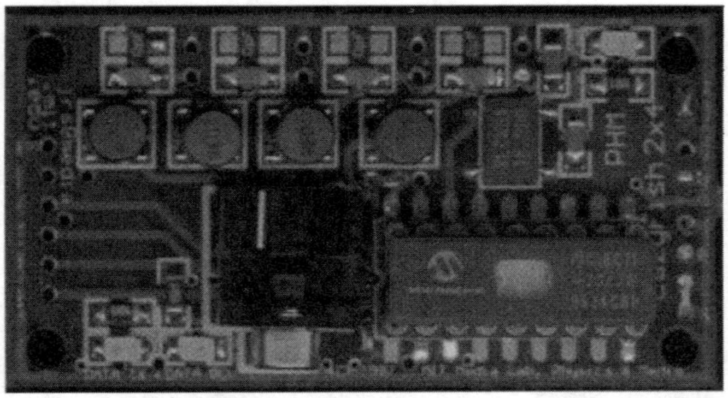

Fig. 3. Later printed circuit board implementation that supports four transmit and two receive channels.

hand-wired prototype board used to make the measurements described in this paper, along with the electrodes. Fig. 3 shows a later printed circuit board implementation that supports four transmit and two receive channels.

VI. Software Demodulation

If the gain variable G from Section III is unity, then the multiplications in the inner product operation described in Section III become additions and subtractions. Performing the demodulation in software requires an accumulation variable in which to store the value of each inner product as it is being calculated. When the first transmitter is high, the processor adds the current ADC value to the first accumulator; if the second transmitter is low, the processor subtracts that same ADC value from the second accumulator. The same ADC value is operated upon differently for each of the demodulation calculations. The addition operation can be thought of as a multiplication by +1 followed by addition to the accumulator; the subtraction is a multiplication by -1 followed by addition to the accumulator. Thus, the operation just described effectively takes the inner product of a vector of samples with a vector representing a transmitted waveform.

An 8-bit LFSR was used, with taps at bits 3, 4, 5, and 7 (counting from zero). This set of taps is known to be maximal [8]. The C code fragment below calculates the four LFSR's, and then sets the states of the transmit pins accordingly.

```
for(k = 0; k < 3; k++) { // k indexes the 4 LFSRs
  low = 0;
  if(lfsr[k]&8) // tap at bit 3
    low++; // each addition performs
           // XOR on low bit of low
  if(lfsr[k]&16) // tap at bit 4 low++;
  if(lfsr[k]&32) // tap at bit 5 low++;
  if(lfsr[k]&128) // tap at bit 7 low++;
  low& = 1; // keep only the low bit
  lfsr[k] <<= 1; // shift register up to make
                 // room for new bit
  lfsr[k]| = low; // or new bit in
}
OUTPUT_BIT(TX0, lfsr[0]&1); // Transmit according
                            // to LFSR states
OUTPUT_BIT(TX1, lfsr[1]&1);
OUTPUT_BIT(TX2, lfsr[2]&1);
OUTPUT_BIT(TX3, lfsr[3]&1);
```

The variable `lfsr` is an array of four bytes. Each byte in the array represents the state of one LFSR. The variable `low` also holds a byte. It is used as temporary storage to calculate the new value of the LFSR's low bit, which must be set to the sum modulo 2 of the current "tap" bits. The value of the least significant bit of `low` contains the appropriate modulo 2 sum after the compare and increment operations above. Its LSB is then masked off and ORed into the low bit of `lfsr`.

The next code fragment performs the demodulation operation. Each value returned by the ADC is operated upon differently for each channel, as specified by the appropriate LFSR.

```
meas = READ_ADC(); // get sample
for(k = 0; k < 3;k++) {
  if(lfsr[k]&1){ // check LFSR state
    next = accuml[k]; // keep backup of low
                      // byte of accum
    accuml[k]+ = meas; // add measured val to accum
    if(accuml[k] < next)accumh[k]++;
    // if overflow, then carry
  }
  else {
    next = accuml[k];
    accuml[k]- = meas;
    // subtract measured valfrom accum
    if(accuml[k] > next)accumh[k]--;
    // carry if necessary
  }
}
```

On the host computer, the final sensor value is found by dividing the accumulated value (`accumh*256 + accuml`) by s, the number of samples taken, as in (1). Because of the division by s, the fluctuations diminish as $1/\sqrt{s}$, but the typical magnitude of the sensed value (that is, the mean of the random variable describing the final sensed value) is independent of s.

VII. Results: Comparison with Time Division

The performance of the CDMA and TDMA techniques was compared by measuring four channels in 105 ms. The ADC on the current PIC16C71 can sample at about 40 kHz. Calculating the four LFSR's, performing the ADC, and doing the four demodulation computations, the PIC's chip time was 136 μs or 7.4 kchip/s. To examine the limit in which the processing time is negligible compared to the ADC time, one compares the performance of the CDMA algorithm with a TDMA algorithm generating a 3.7 kHz square wave (7.4 kchip/s). The total

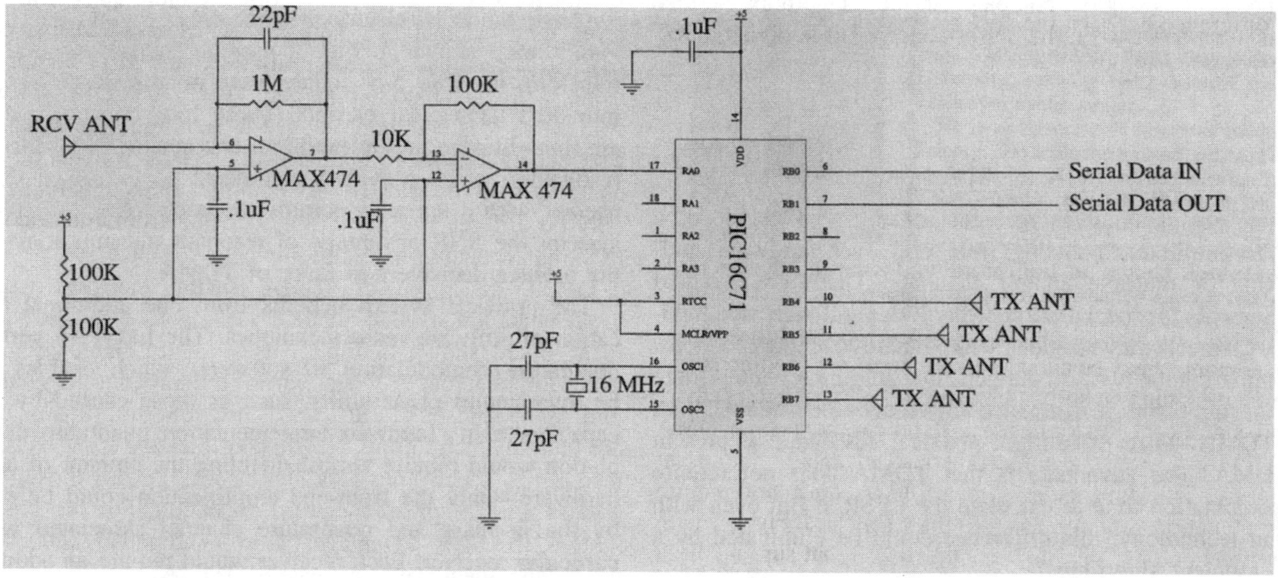

Fig. 4. Analog front end consists of two opamp gain stages. These feed into the ADC built into the PIC. Demodulated values are sent serially to a host computer for display analysis.

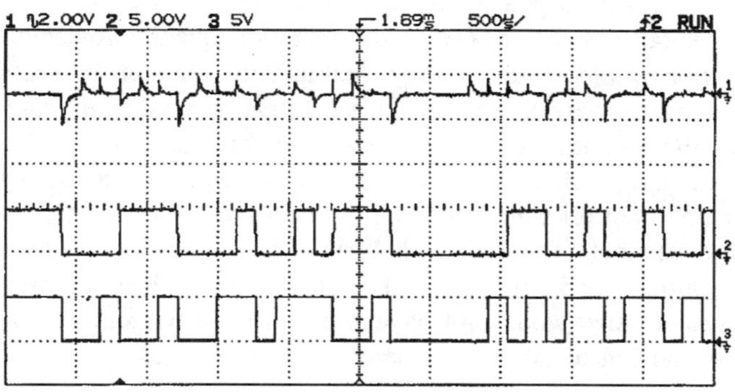

Fig. 5. Top trace: received signal induced by four coded waveforms. Middle and bottom traces: two of the four transmit channels.

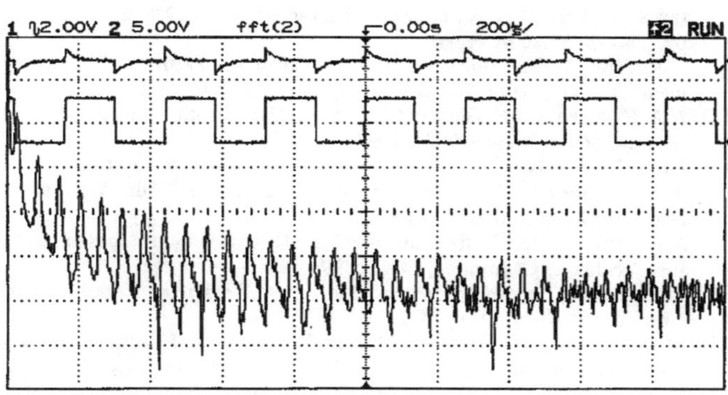

Fig. 6. Top trace: received signal from one transmitted square wave. Middle trace: transmitted square wave. Bottom trace: Fourier transform of transmitted square wave (frequency span: 244.1 kHz; 10 dB/division).

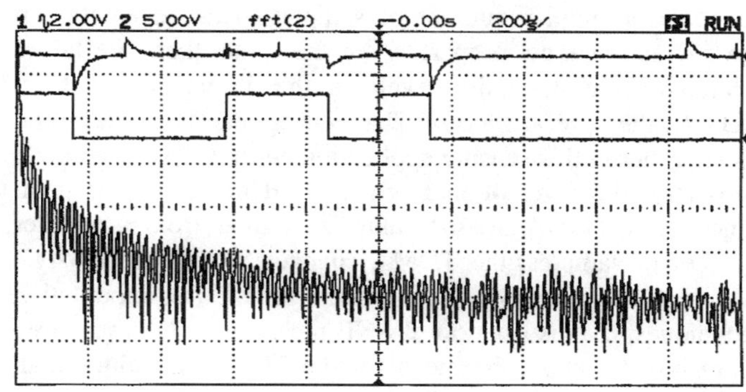

Fig. 7. Top trace: signal received from four coded waveforms. Middle trace: one of the transmit channels. Bottom trace: Fourier transform of one transmit channel (frequency span: 244.1 kHz; 10 dB/division). Notice how much more uniformly the LFSR-generated carrier fills the bandwidth than the square wave of Fig. 6.

measurement time for the four time multiplexed channels was fixed at 105 ms. An additional comparison was made in which the PIC generated a square wave at top speed (given the time required for analog-to-digital conversion and demodulation calculations), which was 13.5 kHz.

To measure the SNR on a particular channel requires calibration of the maximum and minimum observed values as a hand moved from outside the sensor's active region (defined to be zero signal) to the sensor's most sensitive point (maximum signal). With the hand outside the sensitive region, 100 samples were used to calculate the standard deviation. The SNR estimate is an average of five ratios of maximum signal to standard deviation.

When limited to the same chip rate, the SNR for the CDMA and TDMA approaches were as follows: for CDMA, the SNR was 340, or 51 dB, or 8.4 bits; for TDMA, the SNR was 320, or 50 dB, or 8.3 bits. When the TDMA system was operated at the maximum data rate, the SNR increased to 55 dB. This is not surprising. The total output signal power is higher in the higher chip-rate case, and so the SNR should be higher. Thus, in the practical systems of today, the higher computational cost of generating the pseudorandom sequence (which limits the chip rate) diminishes the performance. When the computation time becomes negligible compared to the ADC time, there is little difference between the performance of time-division and code-division multiplexing.

VIII. Conclusion

There are clear advantages for sensing applications of a software-multiplexed analog front end over hardware multiplexing or multiple analog channels. There are minimal requirements for special-purpose hardware and increased flexibility. Channels may be added without hardware modifications.

Comparing the two software channel-sharing schemes, there are no significant performance differences between CDMA and TDMA. There is a slight practical advantage at present to TDMA. The advantage is that TDMA does not require the computation time to calculate the LFSR's. But even with present technology, this difference could be eliminated by a more efficient algorithm.

Although the mean SNR values were the same for CDMA and TDMA, the variance of the SNR's for TDMA was higher. In the TDMA scheme, each single-channel measurement is based on a smaller number of samples. Thus, even though the average channel properties are the same, there are larger fluctuations in the noise levels in the TDMA channel. In CDMA, the several transmit channels appear as noise to one another, but in this sensing application, with its fixed electrode geometry, the level of that self-induced noise is consistent. Since each TDMA sensor value is formed from a smaller number of samples, the TDMA channels are more "bursty." Although the mean SNR is the same for the two modulation techniques, the distribution of SNR measurements is scattered more broadly about the mean for the TDMA technique than for the CDMA technique. For bulk communications purposes, the increased SNR variance does not matter, but for streamed data or sensor values, this is a distinct disadvantage. The performance of a sensor system is only as good as the worst case SNR, measured in short windows of time. During a bulk communications operation like an FTP download, the short time performance of the channel is irrelevant; the relevant figure of merit is the global performance, calculated over the entire time required to complete the operation.

Apart from the subtle SNR variance advantage over TDMA, it is difficult to determine whether the properties of CDMA sensing channels make them significantly more attractive than TDMA channels. The additional flexibility—such as the possibility of determining which sensor channels, or how many, to extract after the raw measurements have been made—might be useful in certain applications. For making multiple measurements with an array of sensors, the more consistent noise levels may make CDMA more attractive than TDMA. In either case, it is useful to process multiple channels of sensor data using a single analog front end, ADC, and DSP software. And while it is common practice to send multiple streams of communications data simultaneously through a single physical channel, it is historically less common (although fundamentally no different) to use a single analog front end and ADC to simultaneously process multiple-sensor channels.

Despite the appeal of CDMA sensing, the most practical implementation of electric field sensing to date, the "LazyFish" board shown in Fig. 3, uses LC resonant circuits to transform the four 5-V square-wave microcontroller outputs into 80-V sinusoidal carriers. These four transmit channels are time-division multiplexed into two analog front ends. The resonators, which provide a significant increase in SNR, are useless with a spread-spectrum excitation. For the practical system, the SNR advantage of resonant transmission tipped the balance decisively in favor of TDMA.

The practical system benefits from one additional application of software radio techniques. The LazyFish performs quadrature demodulation in software, which enables it to be invariant to phase shifts, such as those caused by cable capacitance. In a hardware implementation, quadrature demodulation would require almost doubling the amount of analog hardware—only the front-end amplification could be shared by the in-phase and quadrature channel associated with a particular receiver. Each receiver would require an additional analog multiplier and low-pass filter for the additional quadrature channel. The LazyFish board fits 16 sensing channels (counting quadrature channels) into a 1 in $\times$ 2 in footprint. Its small size would not have been possible without shifting the division of labor from hardware to software.

References

[1] T. G. Zimmerman, J. R. Smith, J. A. Paradiso, D. Allport, and N. Gershenfeld, "Applying electric field sensing to human-computer interfaces," in *CHI'95 Human Factors in Comput. Syst..*, Denver, CO, 1995, pp. 280–287, ACM Press.

[2] J. R. Smith, "Field mice: Extracting hand geometry from electric field measurements," *IBM Syst. J.*, vol. 35, no. 3 and 4, 1996.

[3] J. A. Paradiso and N. Gershenfeld, "Musical applications of electric field sensing," *Comput. Music J.*, 1997.

[4] NEC Automotive Electronics, *Smart Airbags and the Future of Automotive Sensing Technologies from NEC*, World Wide Web page, http://www.nectech.com/auto 1997.

[5] J. R. Smith, "Electric field imaging," Ph.D. dissertation, Massachusetts Inst. Technol., Cambridge, Feb. 1999.

[6] R. C. Dixon, *Spread Spectrum Systems with Commercial Applications*. New York: Wiley, 1994.

[7] M. K. Simon, J. K. Omura, R. A. Scholtz, and B. K. Levitt, *The Spread Spectrum Communications Handbook*. New York: McGraw-Hill, 1994.

[8] P. Horowitz and W. Hill, *The Art of Electronics*. Cambridge, England: Cambridge Univ. Press, 1990.

Software Radios for Airborne Platforms

Jonathan P. Cummings

***Abstract*—Defense contributions to the Programmable Modular Communications System (PMCS) Integrated Product Team (IPT) included designs for an RF module based on software radio configurations useful in airborne systems. Several configurations were examined, analyzes were consolidated, and concepts of operation (CONOPS) were evaluated. Geographic separation of platforms and on-board separation of radio modules have consequences for the remote control of reconfigurable radios. This paper identifies organizational roles in software radio development, characterizes the need for software radios in airborne applications, and highlights those configurations found to be attractive. The methodology, CONOPS, and conclusions are summarized.**

***Index Terms*— Aircraft radio, air interface, antenna control, common data link (CDL), HAVE QUICK, joint tactical radio system (JTRS), multiband ratio, programmable modular communications system (PMCS), radio configuration, radio modules, RF network, reconfigurable data link, reconfigurable radio, SINCGARS, software radio, waveform.**

I. Organizational Roles in Software Radio Development

BRIEFLY, the Defense Airborne Reconnaissance Office (DARO) was established in December 1993 to oversee the architecture of airborne reconnaissance for the United States Department of Defense (DoD). It was absorbed into the Office of the Under Secretary of Defense (OUSD) in 1998. Its oversight role includes DoD program funding for manned and unmanned airborne reconnaissance systems. The OUSD establishes communications architectures, monitors progress, and assists the military services in evolving their communications systems toward greater interoperability, the smooth transfer of information among diverse radio infrastructure. OUSD's communications group works with the military services and with multiservice ("joint") organizations. The Joint Spectrum Center establishes frequency allocations. The Defense Information Systems Agency (DISA) defines messaging systems, tactical data link standards, and joint long-haul ("backbone") services. The National Security Agency (NSA) provides cryptographic systems. This collection of agencies, among others, collaborates on the definition of software radio applications for airborne reconnaissance applications, both explicitly and implicitly.

The acquisition of radios and the migration to new radios differs between fielded and future aircraft fleets. For aircraft yet to be fielded, such as unmanned aerial vehicles (UAV's), the OUSD exercises oversight through the Defense Advanced Research Projects Agency (DARPA) to acquire new radio systems. For aircraft already fielded, such as the Navy's EP-3E, the OUSD exercises oversight for acquisition of supplemental radios and replacements through the military service's manned aircraft programs. In both cases, radios generally need to be backward compatible with existing radio systems, such as the Common Data Link (CDL), Single Channel Ground Air Radio System (SINCGARS), Fleet Satellite Communications System (FLTSATCOM), or satellite transponders leased through DISA. Joint and DoD-level groups work to assure the interoperability of these systems. The military services are responsible for the operations and maintenance (O&M) of radios installed in existing airborne fleets. But the DoD oversees the definition and acquisition of radios to be installed on Advanced Concept Technology Demonstration (ACTD) airframes yet to be fielded. Additionally, the OUSD has architecture oversight, and is responsible for migrating fielded radios toward interoperable, standards-compliant radio networks. Migration is distinct from O&M in that migration to interoperable radios may conflict with the direct replacement of existing, single-function radios. In addition, migration and replacement strategies may compete for the same pool of financial resources. The migration of radios to newer systems such as software radios is often not funded in an O&M approach, unless the replacement of an entire group of unserviceable radios is contemplated. The migration of radios simultaneously across multiple fleets must be a centrally managed, DoD-wide effort because individual services only budget for their individual fleets. The OUSD has the architecture oversight and acquisition authority for DoD-wide central management. For these reasons, DoD-wide communications forums such as the PMCS IPT have benefited from its participation.

II. Scope of Existing Narrow-Band Radio Configurations

The majority of narrow-band radios aboard some aircraft fleets are aging and difficult to support with spare parts. In cases where production lines have ceased, parts may no longer be available. In cases where production lines remained open through several versions of a radio, spare parts for early production models may not be available. Modern radios with multiyear mean time between failure (MTBF) may outlive the availability of parts, particularly with rapidly changing communications technologies. Small acquisitions, with small production runs, may encounter greater problems with parts availability over their life cycle than larger, DoD-wide production runs.

There is a mix of narrow-band radios in airborne fleets. An exact count depends upon the baseline configuration. This

Manuscript received September 4, 1997; revised March 14, 1998, May 19, 1998, and August 1, 1998.

The author is with the Advanced Development Division, Adroit Systems, Inc., Alexandria, VA 22314-1764 USA.

Publisher Item Identifier S 0733-8716(99)02980-7.

Reprinted from *IEEE Journal on Selected Areas of Communications*, Vol. 17, No. 4, pp. 732–747, April 1999.

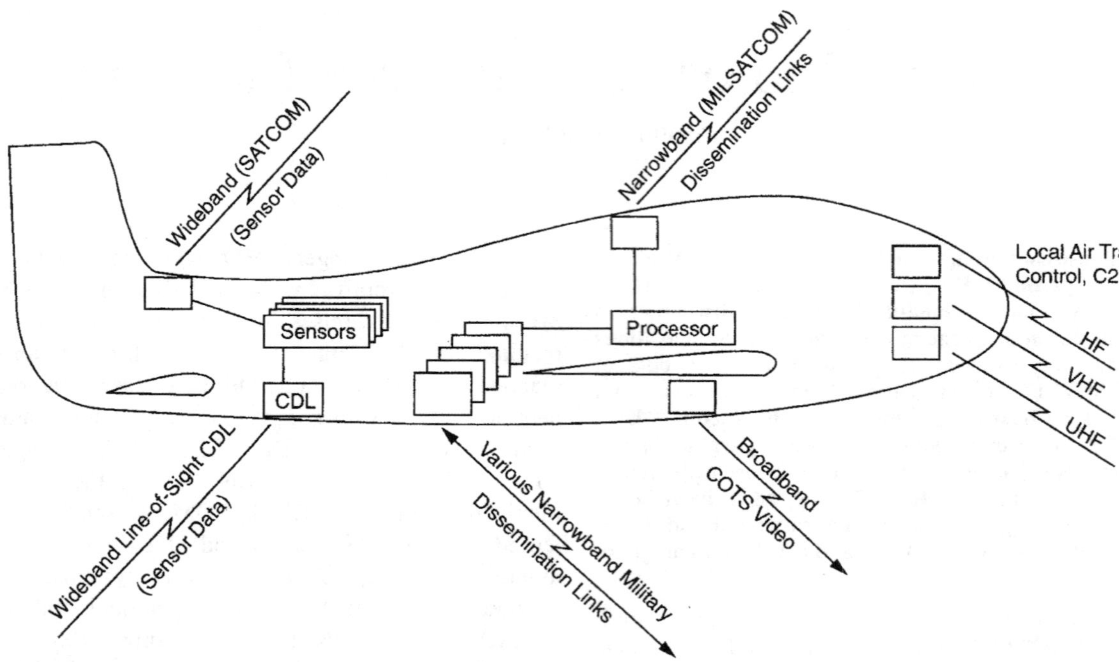

Fig. 1. Radio usage aboard reconnaissance aircraft.

paper used configurations expected to be prevalent in airborne reconnaissance fleets during the next five years. Exact numbers depend on the target year and configuration baseline, but currently, there are five types of HF radios, including the AN/ARC-34, -94, -161, -190, and -718 V-7, with a total of over 300 HF radios installed. There are four types of VHF radios, including the AN/ARC-51A, -182, -186, and -197, with more than 250 VHF radios installed. There are 12 types of UHF radios, including AN/ARC-143, -158, -164, -171, -187, -206, -211, -218, Commanders Tactical Terminal (CTT) variants, Mobile UHF Satellite Terminal (MUST), RT-1273, SIS D-1, and -56 Y-1, with nearly 500 in use. Multiband radios include the dual-band HF/VHF AN/ARC-153A and dual-band VHF/UHF AN-ARC-210. These are becoming more common, with 300 radios of six types in use. Some of these radios have been out of production for several years.

If a decision were made to replace these radios, the production rates would depend on whether the radios were installed in the field or at a depot. Field installation sets may be purchased quickly in quantities sufficient for an entire fleet. Sets for slower depot installation are more often purchased at the same pace at which aircraft enter periodic depot-level maintenance. Depot-set production rates and timing might match these depot-level maintenance schedules. Airframe integration plans also influence radio installation. If the replacement of the radios significantly changes the aircraft's center of gravity or flight parameters, depot-level installation may become necessary. This can be the case for smaller, lightweight aircraft. On the other hand, the replacement of single-channel radios by dual-channel radios (i.e., an AN/ARC-210) may save enough weight to increase the payload or mission duration.

If multiple fleets are served by a radio production batch, then acquisitions might be consolidated. The Air Force U-2 and Army RC-12 fleets might be outfitted with lightweight radios with one of the services serving as executive agent. There are validated requirements for the current mix of over 1300 radios. HF radios facilitate transoceanic deployment, while VHF and UHF meet line-of-sight requirements. UHF demand assigned multiple access (DAMA) satellite communications (SATCOM) is sometimes preferred for transoceanic flights.

Alternatively, radios can be acquired for an entire fleet. The Navy P-3 fleet and the Air Force C-135 fleets, including the reconnaissance derivatives, are candidates for fleet-size radio production batches. In such consolidated acquisitions, the distinction between depot and field installations is even more important for sizing manufacturing lots.

Fig. 1 depicts representative functions and use in a notional reconnaissance aircraft. The flight crew uses radios for take off and landing, taxiway access, runway access, intermediate waypoint traversal, foreign country entry and exit, and similar matters. If the crew operates sensor equipment, they may use radios to disseminate reports. Radios also broadcast video. Wide-band radios relay unprocessed data to ground or to shipboard sites for processing. Airborne imagery sensors require very high data rates to transfer multispectral pixels. Processors at ground sites convert the raw pixel bitstreams into standard image files. Sometimes, satellites relay reports or image files to theater-wide users. They may also relay unprocessed sensor data to remote processing sites. The CDL, for example, can forward unprocessed sensor data to local processing sites at high data rates.

III. Configuration Methodology

Radio frequency ranges and air interface standards ("waveforms" in military jargon) are prescribed by several organizations. The DoD defines the Joint Technical Architecture (JTA) for military waveforms. The International Civil Aviation Organization (ICAO) defines voice waveforms for international

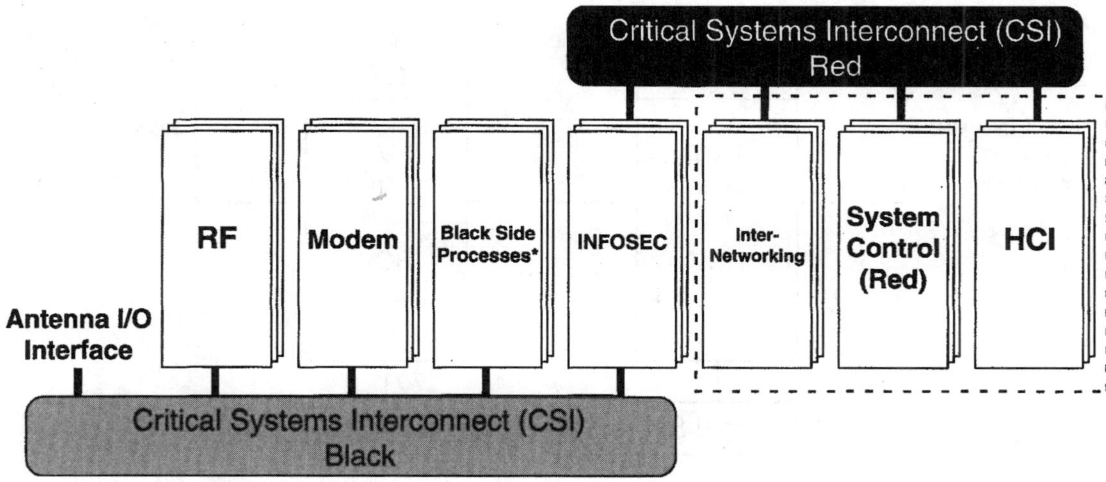

Fig. 2. PMCS architecture entity reference model.

flight. The Federal Aviation Administration (FAA) specifies voice and data waveforms for domestic flight. These requirements and recent DoD C4ISR led to a set of radio configurations, including graphical descriptions and requirements checklists. A probable CONOPS for each configuration was also prepared to establish operating ranges, aircraft altitudes, and associations of waveforms to operations. The CONOPS was used to develop estimates of RF amplifier power and other aspects of RF designs.

From a radio communications perspective, the JTA is both a body of DoD and commercial communications standards, and a mandate to adhere to those standards. Common standards promote interoperability through the acquisition of new and replacement systems. JTA standards are used like "building codes" to constrain the way in which hardware and software are created, while allowing necessary differences such as different shock and vibration limits with related price differences. The services and command organizations define the operational architectures, while developers propose systems architectures. The technical architectures are provided in contract solicitations to implement interoperable commercial standards. The JTA is available at www.itsi.disa.mil/jta/jta.html.

Potential software radio configurations for airborne applications were provided to the PMCS IPT for design analysis, specification analysis, and architecture planning. These configurations were not meant to impose a specific physical model, but to assist in analyzing the impact of the physical distribution of functions. The distribution of functions over space requires a modular approach, with functional and physical modularity. The design of RF modules and other reconfigurable functions were considered in some detail, resulting in benchmark RF designs for the PMCS IPT Guidance Document.

The physical locations of radio modules on an aircraft are not obvious for common modules that are defined in functional terms. Manned aircraft, for example, often require remote control units ("heads") in the cockpit or crew workstations. Remotely operated modems, encryption devices, RF conversion, and power amplifiers may be in nonpressurized parts of the airframe. On unmanned aircraft, the remote control heads remain in ground-based or ship-based ("surface") control stations. Remotely operated radio modules are placed throughout the airframe. Encryption devices for unmanned aircraft may be either onboard the aircraft or in the surface control station at locations that are readily accessible to operators. Considering the constraints of the military services and the FAA assists in partitioning airborne radio configurations into logical modules. Common modules may be used in multiple applications. Finding modules common to many potential applications promotes larger production volumes. The analysis of functions by physical location supports an acquisition approach based on modules common to both the DoD and the FAA.

The modular approach to the functional architecture of reconfigurable radios as defined in the PMCS Guidance Document presents opportunities to examine physical architectures and the impact of functional decisions on physical implementations. Fig. 2 shows the architecture of the Guidance Document. The entity reference model both guided and benefited from the analysis of airborne configurations.

The PMCS architecture is reprinted with simplifications. Each rectangle depicts a function. Each thick black line depicts an interface relationship. The two CSI rectangles depict the two bus or backplane functions. For a more complete description, see the PMCS Guidance Document.[1] One implementation of the PMCS architecture is the Joint Tactical Radio System (JTRS). The JTRS requirements and mission are available on the Internet as well.[2] The interconnects drawn in Fig. 2 indicate either interfaces or functional relationships that should be mandated by specification or standardization in a contract solicitation, rather than depicting data flows. By contrast, IF signals flow between the RF and the modem, but the RF–modem interconnect was not drawn because this interface's standard may be proposed by vendors using best commercial practices. Thus, it is not shown.

Configurations of software radios that appeared to be both feasible and technically attractive for airborne applications served as a basis for the analysis of PMCS options and designs. Subordinate functions like antenna switching and

[1] www.dtic.mil:80/c3i/jtrpage.htm.

[2] www.dtic.mil:80/jcs/j6/jtr_announce.htm.

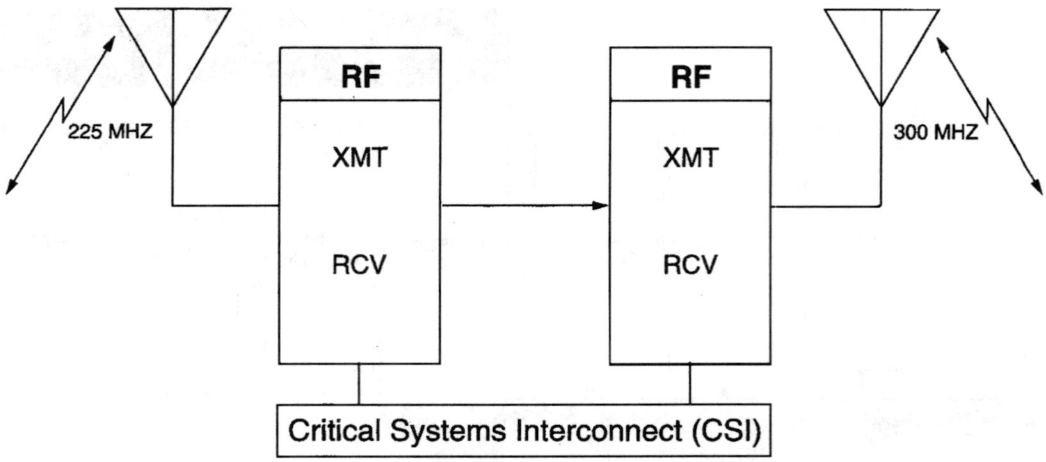

Fig. 3. Duplex airborne transponder.

bandwidth control emerged as common requirements across most configurations. A divided bus consisting of two backplanes was also needed in most configurations to separate encrypted from unencrypted data to meet information security (INFOSEC) constraints. Airborne equipment environments naturally coalesced into three categories. Aircraft skin, the most severe environment, is the domain of antennas. The unprotected cargo bay is the domain of RF components. And the habitable cockpits and ground stations are the domain of encryption equipment and controls. Environments of surface terminal equipment were also examined. High-speed serial data streams were required only in the RF and modem functions of the PMCS architecture, alleviating backplane data rate requirements. A set of control words, status words, and interfaces for RF control also emerged from the analysis. Timing requirements at the interfaces were developed from legacy Tactical Digital Information Link (TADIL) formats, timing signals, and timing standards.

IV. Configurations Examined

Initially, more than 25 configurations of software radios for airborne applications were examined. Eighteen found to have tactical value were included in the follow-on analysis. These configurations included repeaters, transponders, unencrypted relays, encrypted relays, relays for line-of-sight ("tethered") operations, and radios for manned aircraft. UAV's provide excellent platforms for repeater, transponder, and relay services. The absence of a crew and crew compartment aboard a UAV constrains one to remotely operated or preprogrammed radios, as in the configuration diagrams. Manned aircraft need numerous types and quantities of radios. The Airborne Warning and Control System (AWACS) and Joint Surveillance Target Acquisition Radar System (Joint-STARS), for example, require 24 conventional radios. Manned aircraft also accommodate on-board displays and data dissemination. For example, a manned electronics support mission results in a volume of short event reports that is proportional to the activity of the radar environment. This large volume of reports must be disseminated to tactical users via narrow-band channels. Manned surveillance aircraft, like Joint STARS, similarly, can produce large volumes of moving target indicator (MTI), and track reports to be disseminated over the same narrow-band channels. The majority of the manned aircraft processes and analyzes the raw data on board to produce tactical reports. Such on-board processing reduces the amount of raw data to be transmitted to processing sites, but increases the number of product reports to be disseminated in near real time. Weapons controllers aboard manned surveillance aircraft might require several radios each, for their own purposes.

Configurations discarded from the PMCS set include audio-, data-, and most video-broadcast systems, and systems that require steered antennas. The broadcast systems include timing signal broadcasts, beacons, and imagery broadcasts. Imagery broadcasts are to be provided through the separate facilities of the Joint Broadcast Service for Europe and the Global Broadcast Service worldwide. The specialized technical demands of steered antenna systems were judged to be outside the envelope of PMCS-class radios. Finally, video-broadcast systems (particularly for *C*-band) available as commercial-off-the-shelf (COTS) items were judged to be outside the scope of PMCS.

A. Duplex Configurations

Fig. 3 illustrates the building-block approach available with PMCS modules. RF modules may be paired with the PMCS critical systems interconnect (CSI) backplane functions to provide transponder capabilities. Such a device readily transponds analog RF signals, but would not necessarily have the processing gain of modems. Since baseband content is not reconstructed, the processing gain of decoding and reencoding, error detection and correction, and parity checking is unavailable.

Fig. 4 illustrates the promise of the PMCS modular approach. RF modules, modems, and CSI backplanes are connected into an unmanned relay. This is a functional rather than physical representation. Encryption and decryption are not provided. However, the modems provide processing gain. This relay configuration is applicable to unencrypted digital signal systems, such as the improved data modem (IDM).

To relay baseband information from one encrypted network to a different network requires dual INFOSEC. Bridging from

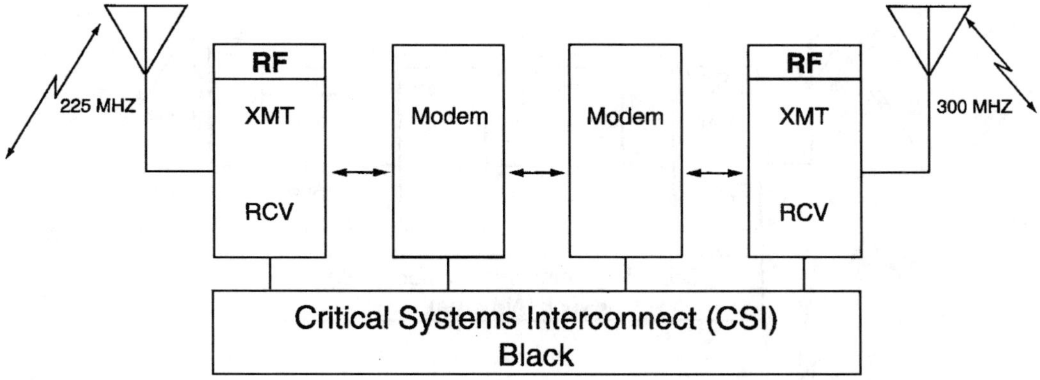

Fig. 4. Duplex airborne relay.

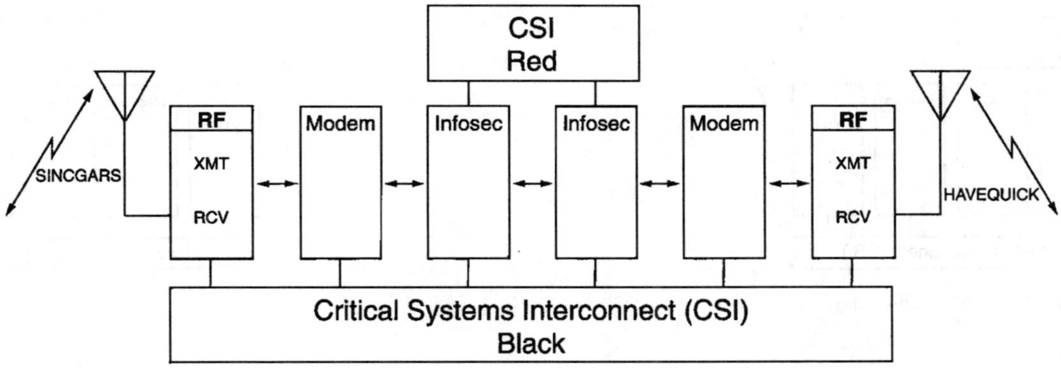

Fig. 5. Duplex multi-INFOSEC radio relay.

SINCGARS to HAVE QUICK, an Air Force frequency-hopped network, for example, requires this capability. Fig. 5 depicts the functions (but not necessarily the physical components) of such internetworking relay services. If PMCS building blocks accommodate multiple channels within each module, then a single RF module, modem, and INFOSEC module accomplish secure bridging. A single network relay configuration (e.g., bridging multiple SINCGARS networks) generally requires only one encryption key. But if the network has multiple security levels, multiple encryption keys would be required. The configuration of Fig. 5 also provides processing gain for digital signals.

The configuration of Fig. 6 supports distributed Tactical Internet networks, geographically separated TADIL networks, and geographically separated wireless LAN's. The figure is simplified to emphasize the operational function rather than the design. A processor needed on the encrypted bus ("black side") is assumed. Black-side processors described in the PMCS Guidance Document are needed for TADIL network relay functions. The INFOSEC modules may also be required to relay from one TADIL network to another.

B. Shared Configurations

Radios aboard manned aircraft can be shared. The cockpit flight crew needs multiple radios for take off, landing, and air traffic control. During the phase of a flight in which the aircraft has attained cruising altitude and assigned flight pattern, the flight crew normally has less need for these radios. The mission crew, however, has a greater need for radios to disseminate reports and to control weapons. Thus, some radio resources can be shared. In this case, a single RF module and modem may service multiple consoles, as Fig. 7 depicts. If cockpit and weapons controller consoles each include individual INFOSEC keys, then multiple security levels are available through the modularity of the architecture.

C. CDL Configurations

Fig. 8 depicts the use of the high-data-rate CDL backbone with PMCS modules. CDL provides a variety of data rates, multiplexing data from multiple sensors, avionics, and airborne radios. The HAVE QUICK and UHF voice basebands are multiplexed through CDL.

Because PMCS RF waveforms are reconfigurable during a mission, a single PMCS unity may be time shared among voice product net (VPN), clear voice, HAVE QUICK, and IDM. The crew could be physically on the ground or aboard ship near the CDL surface terminal, but be virtually in the air with the supported forces. Analogous relay applications for SINCGARS are also possible.[3] The use of SINCGARS or VHF voice in ground operations may require VHF antennas on the underside of the aircraft, to complement UHF antennas installed on the top side of the aircraft for air operations. CDL multiplexing also would have to be reconfigurable to accommodate on-the-fly changes in the PMCS baseband channels, as highlighted in Figs. 9 and 10.

[3] Configuration diagrams used for the analysis are available by e-mail. Request from jcumming@alexandria.adroit.com.

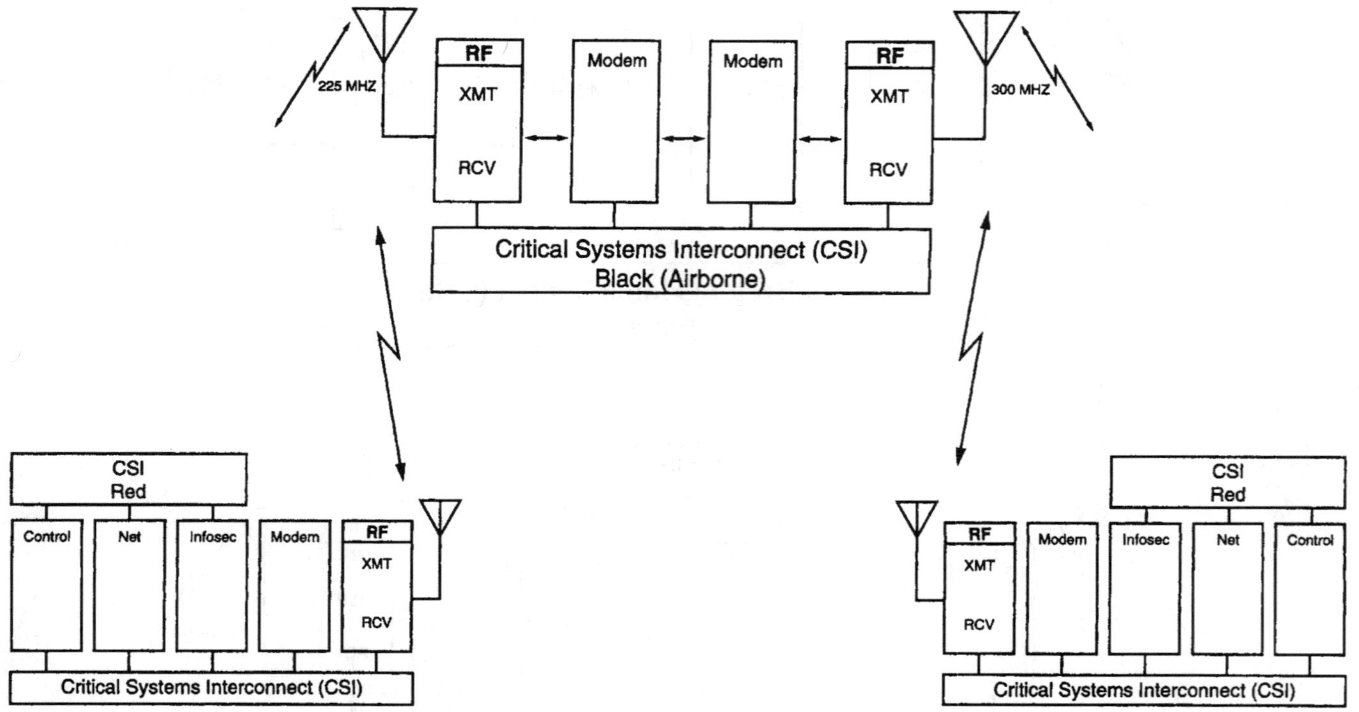

Fig. 6. Duplex LAN-to-LAN radio relay.

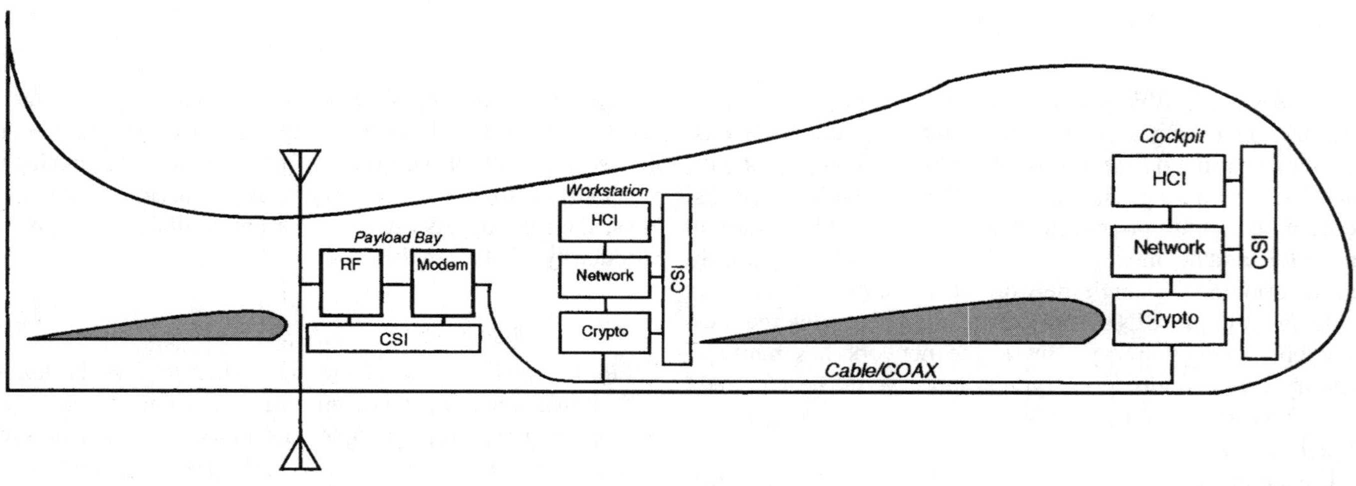

Fig. 7. Shared radios for manned aircraft.

Fig. 9 depicts the use of a four-channel (equivalent to four radios) PMCS configuration. The UHF and VHF waveforms of the Joint Technical Architecture (JTA) are accommodated, with full communications relay. Video and imaging sensors on long-endurance UAV's are useful in CSAR operations. The multiplexing of PMCS baseband channels in the CDL backbone means that UAV's could be used instead of manned aircraft for such operations in high-threat environments. PMCS workstations installed near the CDL surface terminal are also dynamically reconfigurable for weapons controller, communicators, and CSAR staff, adapting to the RF waveform and baseband content as necessary. The CDL backbone multiplexing structure is set in firmware at older CDL terminals. It can be reconfigured through programmable read only memory (PROM) in the field. Erasable PROM or software is necessary for dynamic reconfiguration, however.

Fig. 10 shows how reprogrammable CDL multiplexing supports reconfigurable PMCS radios, for command links, return links, and SATCOM multiplexing. Since the PMCS radios can switch data rates and baseband structure, and since these baseband channels are multiplexed via either CDL or SATCOM, the multiplexing of either the CDL and SATCOM must be reprogrammable. If the CDL and SATCOM multiplexing is implemented in software, then channel structure and data rates of the PMCS baseband channels can be changed as needed for operations. Note also that an encrypted CDL backbone link partially negates the need for PMCS INFOSEC modules aboard the aircraft. Ground-based PMCS INFOSEC

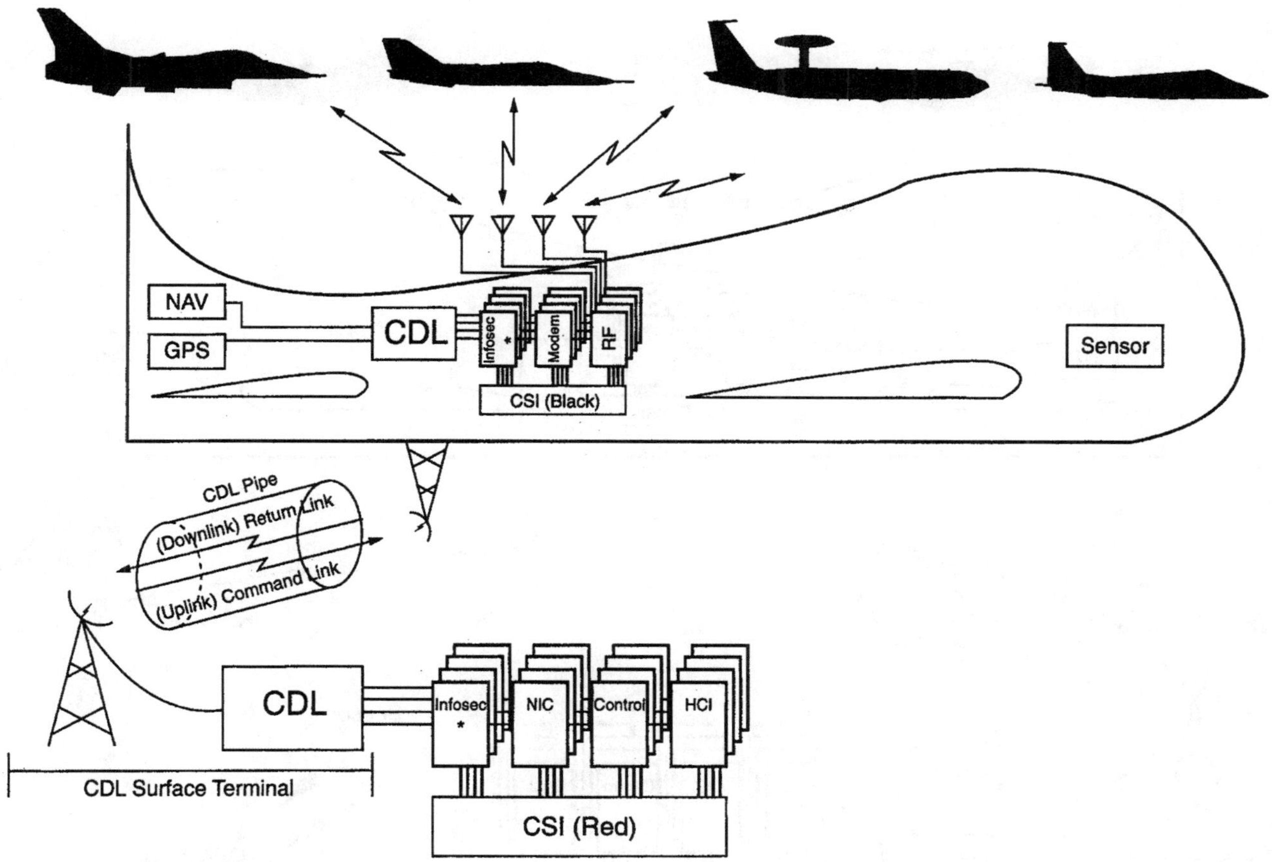

Fig. 8. HAVEQUICK/UHF voice relay.

modules might reduce the aircraft payload burdens. But in cases where INFOSEC modules are required to implement the waveform, PMCS INFOSEC modules are needed aboard the aircraft. If the CDL backbone link is not encrypted when the PMCS radios must support encrypted content, then each baseband communications channel from the surface terminal must be encrypted before transmission to the aircraft. In other words, any lack of encryption on CDL or SATCOM may put a new burden of encryption on each workstation. With the advent of the Defense Information System Network (DISN), and especially the Secret ATM Network, PMCS workstations need not be collocated with the CDL surface terminal. These workstations could be located at any DISN secret ATM network node. The PMCS operators would then be connected to the PMCS RF communications on the aircraft through the DISN secret ATM network and through the CDL or SATCOM backbone link.

V. Subfunctions and Submodules

The concepts subfunction and submodule need to be defined. A subfunction defines a service or capability within a module. Submodules are physical partitions of the top-level modules of the architecture. RF submodules could be physically as large as circuit boards or as small as a few gates in an FPGA. For example, switches might be located in the cockpit in the pilot's remote control panel, but the switching would occur in the remote RF module. Preamplifier selection, antenna selection, transmit/receive, and power amplifier on/off exemplify such distributed subfunctions. Microphones and headphones are located in helmets, but analog submodules of the RF module may accommodate these analog voice circuits.

Telemetry subfunctions may be appropriate for flight tests, range support, and periodic calibration. The most common military telemetry standards are the Inter-Range Instrumentation Group (IRIG) specifications. IRIG submodules are currently required aboard a few manned aircraft. The software radio might include subfunctions which, when combined with communications subfunctions, implement the telemetry functions without additional on-board equipment.

Specialized TADIL submodules might be needed to meet data link timing and processing requirements. Power conditioning submodules might be required to accommodate differences among the voltage available from the airframe, the voltage needed by the PMCS RF module, and power supply ripple specifications. Receiver/exciter or synthesizer submodules could be required for frequency hopping, frequency shift keying, or multichannel relay. Phase detector submodules might be needed for some waveforms as well.

Table I summarizes requirements for subelements of the RF module.[4] Safety-of-flight mandates drive requirements for analog voice for the flight crews of manned aircraft. Differential analog cable from the cockpit to the RF module would bypass bus or backplane for voice. This permits the use of a radio during ground taxiway operations when the backplane is not powered or inoperable. This requirement

[4]PMCS Guidance Document.

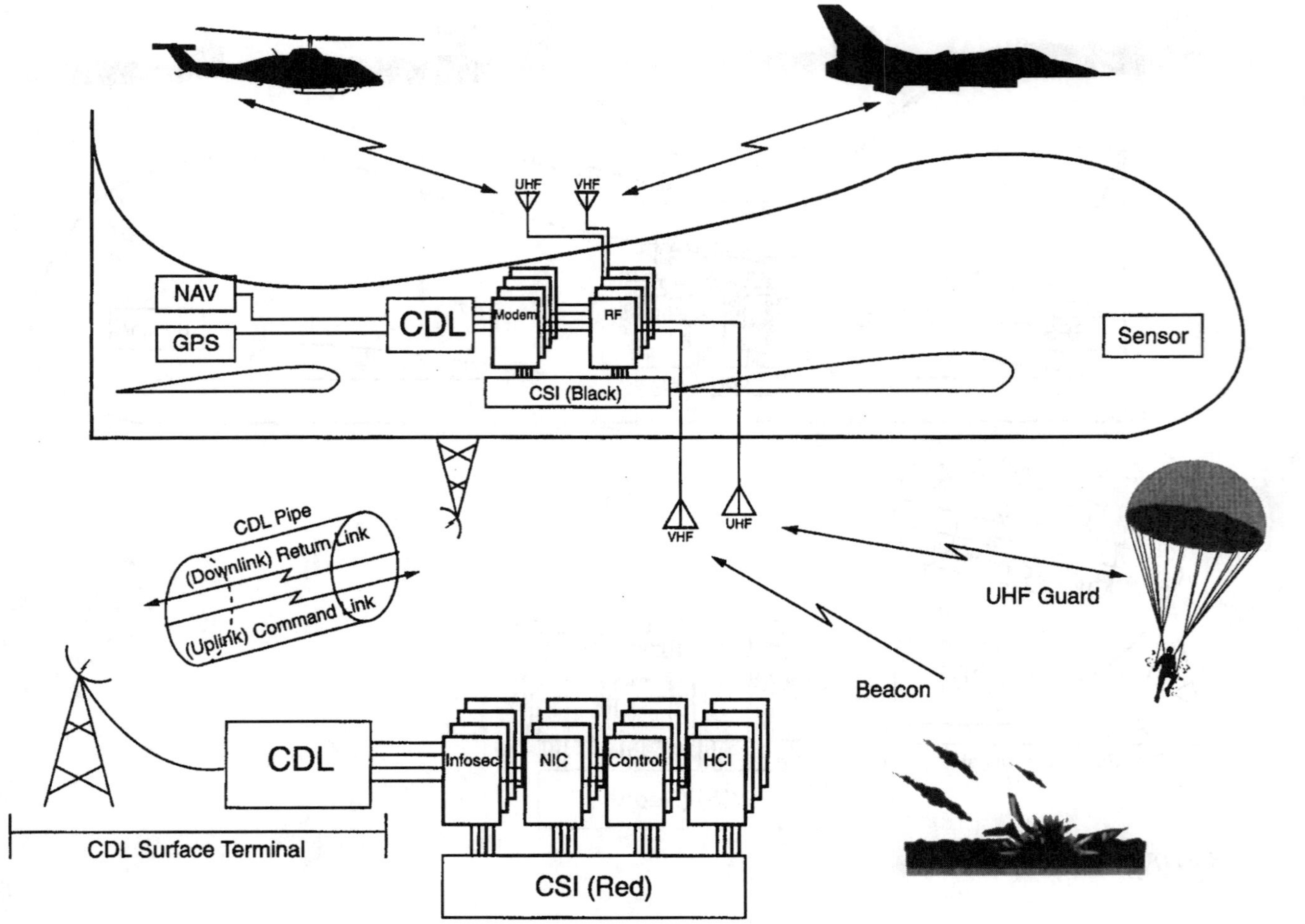

Fig. 9. Combat search and rescue (CSAR) configurations.

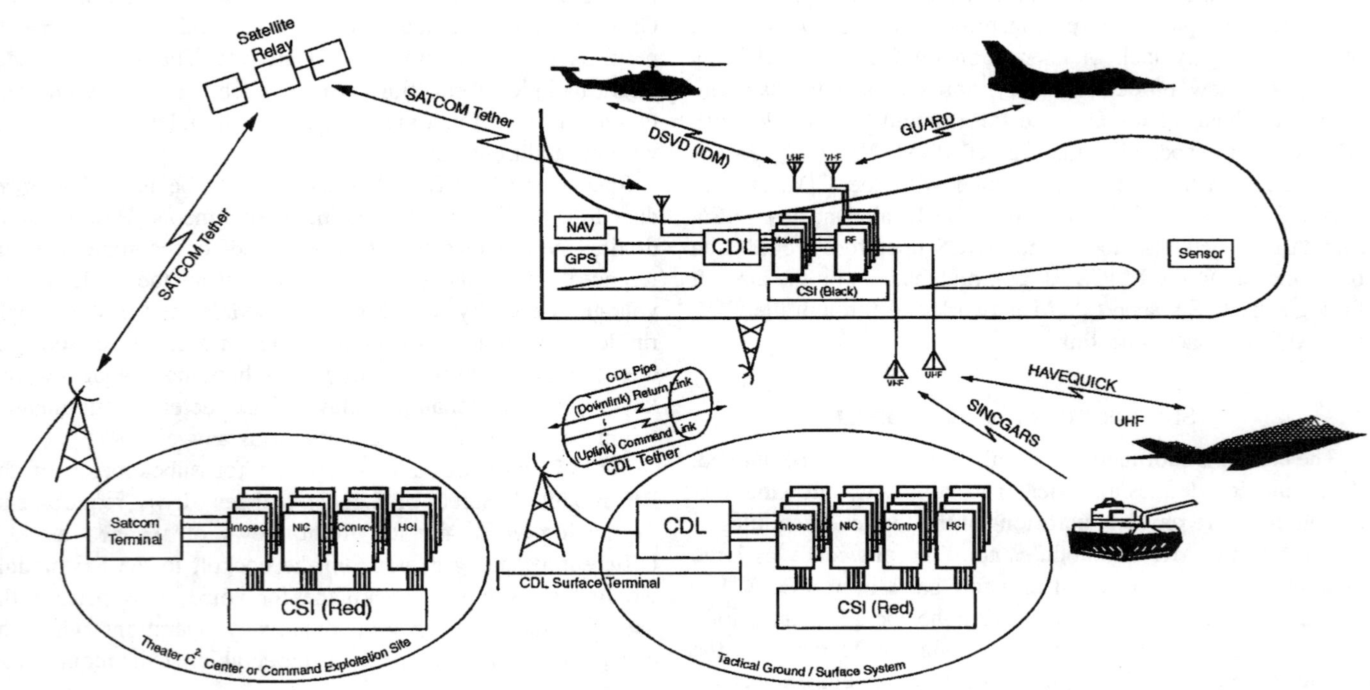

Fig. 10. Programmable tether configuration.

TABLE I
REQUIREMENTS FOR SUBELEMENTS IN THE PMCS RF MODULE

Potential physical submodules required	Repeater with diplexer	Repeater with delay	Transponder, simplex	Transponder, duplex	Relay, simplex	Relay, duplex	Dual crypto relay, simplex	Dual crypto relay, duplex	LAN-to-LAN relay, simplex	LAN-to-LAN relay, duplex	Manned aircraft, front end crew	Manned aircraft, back-end crew	Manned aircraft, shared radio	Tethered HAVEQUICK relay	Tethered SINCGARS relay	Tethered IBS relay	Tethered CSAR relay	Tethered SATCOM relays
Transmit/receive switch	R	R	R	R	R	R	R	R	R	R	R	R	R	R	R	R	R	R
Antenna select switch	R	R	R	R	R	R	R	R	R	R	R	R	R	R	R	R	R	R
Internal RF preamp/amp	R	R	R	R	R	R	R	R	R	R	R	R	R	R	R	R	R	R
External RF power amplifier (add-on PA)							d	d	d	d	R	R	R	R	R	R	R	R
Multiplexer/MUX switch																		
Multi-frequency receivers/synthesizers (down converters) [FH, FSK, multi-channe							R	R	R	R	R	R	R	R	R	d	R	R
Down converter phase detector [for PSK and some FSK waveforms]									d	d								
Multi-frequency exciters/synthesizers (up converters) [FH, FSK, multi-channels]							R	R	R	R	R	R	R	R	R	d	R	R
Up converter phase detector [for PSK and some FSK waveforms]									d	d								
Preselector bandwidth control	R	R	R	R	R	R	R	R	R	R	R	R	R	R	R	R	R	R
Preselector media/mode switch	R	R	R	R	R	R	R	R	R	R	R	R	R	R	R	R	R	R
Legacy analog voice input (PTT microphone input)											R	R	R					
Legacy analog voice output ("pilot's headphones")											R	R	R					
Legacy differential analog (hum-free PTT microphone) input [zvo/zco]											R	R	R					
Legacy differential analog (hum-free headphone) output [zvo/zco]											R	R	R					
Timing processor: GPS data stream processor							R	R	R	R	R	R	R	R	R	R	R	R
Timing processor: IRIG-B data stream processor (IRIG 104-70)												R	R					
Timing processor: 1pps (MIL STD 1553) processor							R	R	R	R	R	R	R	R	R	R	R	R
Timing processor: IRIG 106-93 (telemetry timing)																		
Timing processor: TADIL (data link network timing)									R	R	R	R	R			R	R	R
Timing processor: US Naval Observatory UTC bitstream from GPS												R	R					
Power: module's DC power conditioner/power supply	R	R	R	R	R	R	R	R	R	R	R	R	R	R	R	R	R	R
Power: RF preamp/amp DC power conditioner/power supply	R	R	R	R	R	R	R	R	R	R	R	R	R	R	R	R	R	R

is applicable only to civilian and military manned aircraft using AM, FM, and single sideband (SSB) waveforms. In the interest of examining the potential for an all-digital RF module, this requirement for analog input and output (I/O) was further analyzed as shown in Table IV. Timing subfunctions are stringent for electronic reconnaissance platforms. They are also useful for TADIL, frequency-hopping waveforms, and navigation. Finally, dc power is a required subfunction or submodule.

The configuration requirements list includes all candidate subelements. In addition to identifying the common modules, it can also help determine which to exclude from the architecture. The elimination of subelements leads to cost savings and simplified designs. Complex waveform and antenna requirements, in particular, were examined. For example, steered antennas require embedded microprocessor submodules to steer the antennas. Multichannel waveforms with multiple sidebands have stringent intermodulation distortion requirements. Meeting these requirements can be especially difficult if data streams in different channels have different timing or contain information classified at different security levels with different encryption. Waveforms with separate in-phase and quadrature (*I&Q*) streams also have stringent intermodulation distortion and noise requirements. These are particularly difficult to meet if the *I&Q* components carry data streams with different timing or encryption.

Subfunctions deleted from the RF module include separate *I&Q* bitstream channels and buffers, separate multichannel bitstream preamplifiers and buffers, separate antenna processors, separate multiplexing components, and separate IRIG 106-93 processors. Requirements for multiplexing streams in the RF module were also deleted, simplifying the interface between the RF and modem module. These functions may be accomplished in the modem or elsewhere in the system. Timing requirements were biased toward reconnaissance, with IRIG-B required only for radios supporting mission crews. Requirements for timing messages from the TADIL-J data link may be deleted eventually by the TADIL management organizations if the global positioning system (GPS) can provide adequate timing services for those networks. Phase coherent up- and downconversion were not required based on the lack of *I&Q* waveform requirements [1]. The definition of local area network (LAN) was expanded to construe TADIL networks to be LAN's. In this paper, LAN-to-LAN relays were allowed to use *I&Q* waveforms, as an exception, to examine a wireless LAN relay concept.

VI. INTERFACES

This section summarizes the analysis of interfaces among software radio entities, with emphasis on the RF module.

TABLE II

Platform Interface Analysis of the PMCS RF Module

Potential platform interfaces required	Repeater with diplexer	Repeater with delay	Transponder, simplex	Transponder, duplex	Relay, simplex	Relay, duplex	Dual crypto relay, simplex	Dual crypto relay, duplex	LAN-to-LAN relay, simplex	LAN-to-LAN relay, duplex	Manned aircraft, front end crew	Manned aircraft, back-end crew	Manned aircraft, shared radio	Tethered HAVEQUICK relay	Tethered SINCGARS relay	Tethered IBS relay	Tethered CSAR relay	Tethered SATCOM relays
RF interfaces:																		
- HF antenna coupler with diplexer											R	R	R				d	d
- HF antenna coupler without diplexer (may be "receive only")		d			d	d	d	d	d	d								
- VHF antenna coupler with diplexer	R			R		R		R		R	R	R	R		R		R	R
- VHF antenna coupler without diplexer		R	R		R		R		R									
- UHF antenna coupler with diplexer	R			R		R		R		R	R	R	R	R		R	R	R
- UHF antenna coupler without diplexer		R	R		R		R		R									
- Dummy loads for antennas	R	R	R	R	R	R	R	R	R	R	R	R	R	R	R	R	R	R
Timing bitstream interfaces:																		
- GPS data stream from navigation gear							R	R	R	R	R	R	R	R	R	R	R	R
- IRIG-B data stream from aircraft bus (IRIG 104-70)												R	R					
- 1 pps (MIL STD 1553) data stream							R	R	R	R	R	R	R	R	R	R	R	R
- IRIG-106-93 (telemetry timing) data stream												d	d					
- TADIL (data link) timing messages data stream									R	R	R	R	R			R		R
- US Naval Observatory UTC from GPS data stream												R	R					
Power interface: RF module DC power supply	R	R	R	R	R	R	R	R	R	R	R	R	R	R	R	R	R	R
Power interface: RF power amplifier (external PA) power supply					R	R	d	d	d	d	R	R	R	R	R	R	R	R
EMI/cosite interface: filters at the coupler switches	R	R	R	R	R	R	R	R	R	R	R	R	R	R	R	R	R	R
EMI/cosite interface: filters at the antenna interfaces	R	d	R	R	R	R	R	R	R	R	R	R	R	R	R	R	R	R
EMI/cosite interface: filters at the up and down converters							R	R	R	R	R	R	R	R	R	R	R	R
EMI/cosite interface: filters at the RF preamp/amp	R	d	R	R	R	R	R	R	R	R	R	R	R	R	R	R	R	R

R = Required
d = Desirable

A. Host Platform Interfaces

Requirements for PMCS interfaces to the host platform were also examined, as summarized in Table II. This analysis raised several possible tradeoffs. The interface could be part of the RF module or part of the platform, for example. The imposition of extensive interface requirements on the RF module would increase the cost of PMCS. Host platform interfaces include antennas, timing, navigation, power, and electromagnetic interference (EMI) (also called "cosite interference"). Antenna interfaces for HF, VHF, and UHF include diplexers for both full- and half-duplex operations, although the table indicates that diplexers are not required for half-duplex modes. Diplexers make radios capable of either full- or half-duplex operation. In some cases, diplexers are not necessary (e.g., broadcast), so the cost of diplexers might be avoided. However, the economies of scale of including diplexers into all of the PMCS antenna interfaces reduce these costs. Thus, each of the antenna couplers has a diplexer. Dummy loads are required primarily because the PMCS RF module configuration will change on the fly. If a signal were erroneously routed to an uninstalled antenna with no dummy load, the RF amplifier could burn out. Moreover, the absence of a dummy load in a programmable radio would complicate the software. The fail safe approach would prevent RF emissions until the presence of the antenna is confirmed. Dummy loads also facilitate on-board, on-the-fly diagnostics.

Power requirements include interfaces to the host platform's dc power supply for both the RF module and an (optional) external RF power amplifier. The external RF power amplifier may be needed for extended operational range, power-sensitive waveforms, or frequency management. Finally, EMI drives the need for programmable analog filters and other EMI suppression techniques. EMI on a small airborne platform can result from antennas placed very close to each other. Particularly with HF radios, antennas may be in the near field of each other. Four functional locations are suggested for filters: antenna interfaces, coupler switches, up- and downconverter submodules, and RF amplifier submodules. As with diplexers, there are tradeoffs in filter acquisition strategies. Filters may be acquired in the host platform interface or in the PMCS radio. Finally, other interfaces are outlined in Tables III–V. Interfaces not categorized as "host platform interfaces" include modem interfaces (Table III), legacy interfaces (Table IV), bus or backplane interfaces (Table V), and backbone multiplexing (see the tethered relay figures, for example).

B. Modem Interfaces

Table III presents requirements for the bitstream flows between the modem and the RF modules. The question of whether IF flows on separate RF or fiber-optic cables or the backplane was not clarified by this analysis, primarily because the characterization of IF data rates was out of scope. The

TABLE III

Modem Interfaces of the PMCS RF Module

Potential modem interfaces required	Repeater with diplexer	Repeater with delay	Transponder, simplex	Transponder, duplex	Relay, simplex	Relay, duplex	Dual crypto relay, simplex	Dual crypto relay, duplex	LAN-to-LAN relay, simplex	LAN-to-LAN relay, duplex	Manned aircraft, front end crew	Manned aircraft, back-end crew	Manned aircraft, shared radio	Tethered HAVEQUICK relay	Tethered SINCGARS relay	Tethered IBS relay	Tethered CSAR relay	Tethered SATCOM relays
Interface: High-speed data (serial) I/O [Fiber Channel, etc.]																		
- Data I/O or data bus control (logic or voltage path) or control word									R	R		d	d					
- Data I/O or data bus status (BIT or add-on logic) or status word									R	R		d	d					
Interface: Low-speed data I/O (data modem, file transfer, etc.)																		
- Data I/O or data bus control (logic or voltage path) or control word					R	R	R	R	R	R	R	R	R	R	R	R	R	R
- Data I/O or data bus/backplane status (BIT, add-on logic) or status word					R	R	R	R	R	R	R	R	R	R	R	R	R	R
Interface: 2 to 4 GHz phase-locked loop (PLL) timing signal					R	R	R	R	R	R	R	R	R	R	R	R	R	R
Interface: IF receive [from modem] (programmable data rates)																		
- IF receive control (logic or voltage path) or control word					R	R	R	R	R	R	R	R	R	R	R	R	R	R
- IF receive status (BIT, add-on logic) or status word					R	R	R	R	R	R	R	R	R	R	R	R	R	R
Interfaces: I&Q receive [from modem] (programmable data rates)																		
- I&Q receive control (logic or voltage path) or control word																		
- I&Q receive status (logic or voltage path) or status word																		
Interface: IF transmit [to modem] (programmable data rates)																		
- IF transmit control (logic or voltage path) or control word					R	R	R	R	R	R	R	R	R	R	R	R	R	R
- IF transmit status (logic or voltage path) or status word					R	R	R	R	R	R	R	R	R	R	R	R	R	R
Interfaces: I&Q transmit [to modem] (programmable data rates)																		
- I&Q transmit control (logic or voltage path) or control word																		
- I&Q transmit status (logic or voltage path) or status word																		

TABLE IV

Legacy Interface Analysis of the PMCS RF Module

Potential legacy interfaces required	Repeater with diplexer	Repeater with delay	Transponder, simplex	Transponder, duplex	Relay, simplex	Relay, duplex	Dual crypto relay, simplex	Dual crypto relay, duplex	LAN-to-LAN relay, simplex	LAN-to-LAN relay, duplex	Manned aircraft, front end crew	Manned aircraft, back-end crew	Manned aircraft, shared radio	Tethered HAVEQUICK relay	Tethered SINCGARS relay	Tethered IBS relay	Tethered CSAR relay	Tethered SATCOM relays
Legacy analog voice (microphone) PTT input where mandated by FAA/ICAO "Safety of Flight" rules											R	R	R					
Legacy analog voice (headphone) output where mandated by FAA/ICAO "Safety of Flight" rules											R	R	R					
Differential analog ("hum-free" microphone) PTT voice input											R	R	R					
Differential analog ("hum-free" headphone) voice output (zvo/zco)											R	R	R					
Analog signal ground (ground return path)	R	R	R	R	R	R	R	R	R	R	R	R	R	R	R	R	R	R
Analog power ground (ground return path to power supply)	R	R	R	R	R	R	R	R	R	R	R	R	R	R	R	R	R	R

only firm requirement for a high-speed serial stream (e.g., fiber channel) is associated with tactical LAN-to-LAN relay. Manned electronic surveillance aircraft may need a fast serial stream for reports, but this is unclear, even in the latest *Joint Airborne SIGINT Architecture (JASA) Standards Handbook* [2]. Some military operations aspects are evolving. Low-speed I/O may be implemented using buses or backplanes. It supports I/O for control of IF signals. Requirements for separate *I&Q* interfaces were examined in detail, but for PMCS radios, separate *I&Q* data streams are not required in the IF–modem interface.

Each interface includes control and status data. Control data would be preset for autonomous UAV's, UAV's that fly preprogrammed routes beyond the range of radio control. Control would be remotely set for tethered UAV's. UAV's with SATCOM tethers may accept control via the tether link. UAV's that remain within radio line-of-sight and manned aircraft may reconfigure the modem interfaces as needed.

TABLE V
CSI INTERFACE ANALYSIS OF THE PMCS RF MODULE

Potential CSI interfaces required	Repeater with diplexer	Repeater with delay	Transponder, simplex	Transponder, duplex	Relay, simplex	Relay, duplex	Dual crypto relay, simplex	Dual crypto relay, duplex	LAN-to-LAN relay, simplex	LAN-to-LAN relay, duplex	Manned aircraft, front end crew	Manned aircraft, back-end crew	Manned aircraft, shared radio	Tethered HAVEQUICK relay	Tethered SINCGARS relay	Tethered IBS relay	Tethered CSAR relay	Tethered SATCOM relays
Data interfaces: Low-speed I/O (data modem, file transfer, etc.)																		
- Data I/O/data bus/backplane control (logic or voltage path) or control word			R	R	R	R	R	R	R	R	R	R	R	R	R	R	R	R
- Data I/O/data bus/backplane status (BIT, add-on logic) or status word					R	R	R	R	R	R	R	R	R	R	R	R	R	R
Interface: Bus or backplane functions																		
- System bus/backplane status (BIT, add-on logic) or status word					R	R	R	R	R	R	R	R	R	R	R	R	R	R
- Black side processor					d	d	R	R	R	R	R	R	R	R	R	R	d	d
- T/R switch control (logic or voltage path) or control word	R	R	R	R	R	R	R	R	R	R	R	R	R	R	R	R	R	R
- T/R switch status (BIT or add-on logic) or status word					R	R	R	R	R	R	R	R	R	R	R	R	R	R
- Antenna selector control (logic or voltage) or control word	R	R	R	R	R	R	R	R	R	R	R	R	R	R	R	R	R	R
- Antenna selector status (BIT or add-on logic) or status word					R	R	R	R	R	R	R	R	R	R	R	R	R	R
- Antenna tuning control (logic or voltage) or control word	R	R	R	R	R	R	R	R	R	R	R	R	R	R	R	R	R	R
- Antenna tuning status (BIT or add-on logic) or status word					R	R	R	R	R	R	R	R	R	R	R	R	R	R
- Internal RF preamp/amp control (logic or voltage) or control word	R	R	R	R	R	R	R	R	R	R	R	R	R	R	R	R	R	R
- Internal RF preamp/amp status (BIT, add-on logic) or status word					R	R	R	R	R	R	R	R	R	R	R	R	R	R
- External RF power amp control (logic or voltage) or control word					d	d	d	d	d	d	R	R	R	R	R	R	R	R
- External RF power amp status (BIT, add-on logic) or status word					d	d	d	d	d	d	R	R	R	R	R	R	R	R
- Multiplexer (MUX) switch control (logic or voltage) or control word																		
- Multiplexer (MUX) switch status (BIT, add-on logic) or status word																		
- Downconverter single frequency control (logic or voltage) or control word					R	R	R	R	R	R	R	R	R	R	R	R	R	R
- Downconverter multi-frequency control (logic or voltage) or control word							R	R	R	R	R	R	R	R	R	R	R	R
- Downconverter single frequency status (BIT, add-on logic) or status word					R	R	R	R	R	R	R	R	R	R	R	R	R	R
- Downconverter multi-frequency status (BIT, add-on logic) or status word							R	R	R	R	R	R	R	R	R	R	R	R
- Upconverter single frequency control (logic or voltage) or control word					R	R	R	R	R	R	R	R	R	R	R	R	R	R
- Upconverter multi-frequency control (logic or voltage) or control word							R	R	R	R	R	R	R	R	R	R	R	R
- Upconverter single frequency status (BIT, add-on logic) or status word					R	R	R	R	R	R	R	R	R	R	R	R	R	R
- Upconverter multi-frequency status (BIT, add-on logic) or status word							R	R	R	R	R	R	R	R	R	R	R	R
- Preselector mode control (logic or voltage) or control word	R	R	R	R	R	R	R	R	R	R	R	R	R	R	R	R	R	R
- Preselector mode status (BIT or add-on logic) or status word					R	R	R	R	R	R	R	R	R	R	R	R	R	R
- Preselector single frequency control (logic or voltage) or control word	R	R	R	R	R	R	R	R	R	R	R	R	R	R	R	R	R	R
- Preselector single frequency status (BIT, add-on logic) or status word					R	R	R	R	R	R	R	R	R	R	R	R	R	R
- Preselector multi-frequency control (logic or voltage) or control word							R	R	R	R	R	R	R	R	R	R	R	R
- Preselector multi-frequency status (BIT, add-on logic) or status word							R	R	R	R	R	R	R	R	R	R	R	R
- Preselector bandwidth control (logic or voltage) or control word	R	R	R	R	R	R	R	R	R	R	R	R	R	R	R	R	R	R
- Preselector bandwidth status (BIT, add-on logic) or status word					R	R	R	R	R	R	R	R	R	R	R	R	R	R
- RF module power control (logic or voltage) or control word	R	R	R	R	R	R	R	R	R	R	R	R	R	R	R	R	R	R
- RF module power status (BIT, add-on logic) or status word					R	R	R	R	R	R	R	R	R	R	R	R	R	R
- RF preamp/amp power control (logic or voltage) or control word	R	R	R	R	R	R	R	R	R	R	R	R	R	R	R	R	R	R
- RF preamp/amp power status (BIT, add-on logic) or status word					R	R	R	R	R	R	R	R	R	R	R	R	R	R
System or GPS timing interfaces																		
- System timing control (logic or voltage path) or control word					R	R	R	R	R	R	R	R	R	R	R	R	R	R
- System timing status (BIT or add-on logic) or status word					R	R	R	R	R	R	R	R	R	R	R	R	R	R
System clock interfaces																		
- System clock control (logic or voltage path) or control word					R	R	R	R	R	R	R	R	R	R	R	R	R	R
- System clock status (BIT or add-on logic) or status word					R	R	R	R	R	R	R	R	R	R	R	R	R	R
Timing pulse, bitstream or data stream interfaces:																		
- GPS data stream input from bus							R	R	R	R	R	R	R	R	R	R	R	R
- IRIG-B data stream input from bus (IRIG 104-70)												R	R					
- 1 pps (MIL STD 1553) pulse stream input from bus							R	R	R	R	R	R	R	R	R	R	R	R
- IRIG-106-93 (telemetry timing) data stream input from bus												d	d					
- TADIL (data link network timing message) timing input from RF network									d	d	R	R	R			R		R
- US Naval Observatory UTC (extracted from GPS) input from bus												R	R					
Power interfaces:																		
- Digital DC power source (" + 5V")	R	R	R	R	R	R	R	R	R	R	R	R	R	R	R	R	R	R
- Digital DC power ground	R	R	R	R	R	R	R	R	R	R	R	R	R	R	R	R	R	R
- Analog DC power source (" + 28V")	R	R	R	R	R	R	R	R	R	R	R	R	R	R	R	R	R	R
- Analog DC power ground	R	R	R	R	R	R	R	R	R	R	R	R	R	R	R	R	R	R

The corresponding status words are similarly affected by the presence or absence of an RF tether. Those configurations that do not use a modem (for example, a transponder) set the modem interface to a predefined null value.

The phase-locked loop (PLL) timing signals originate in the RF module. A 2–4 GHz PLL reference signal would facilitate up- and downconversion, frequency hopping, and modem timing. Such a timing signal might be distributed to the modem via an EMI-resistant high-speed serial bitstream such as a low-voltage differential signaling (LVDS) copper wire or fiber channel. The tradeoffs for these decisions are beyond the scope of this paper.

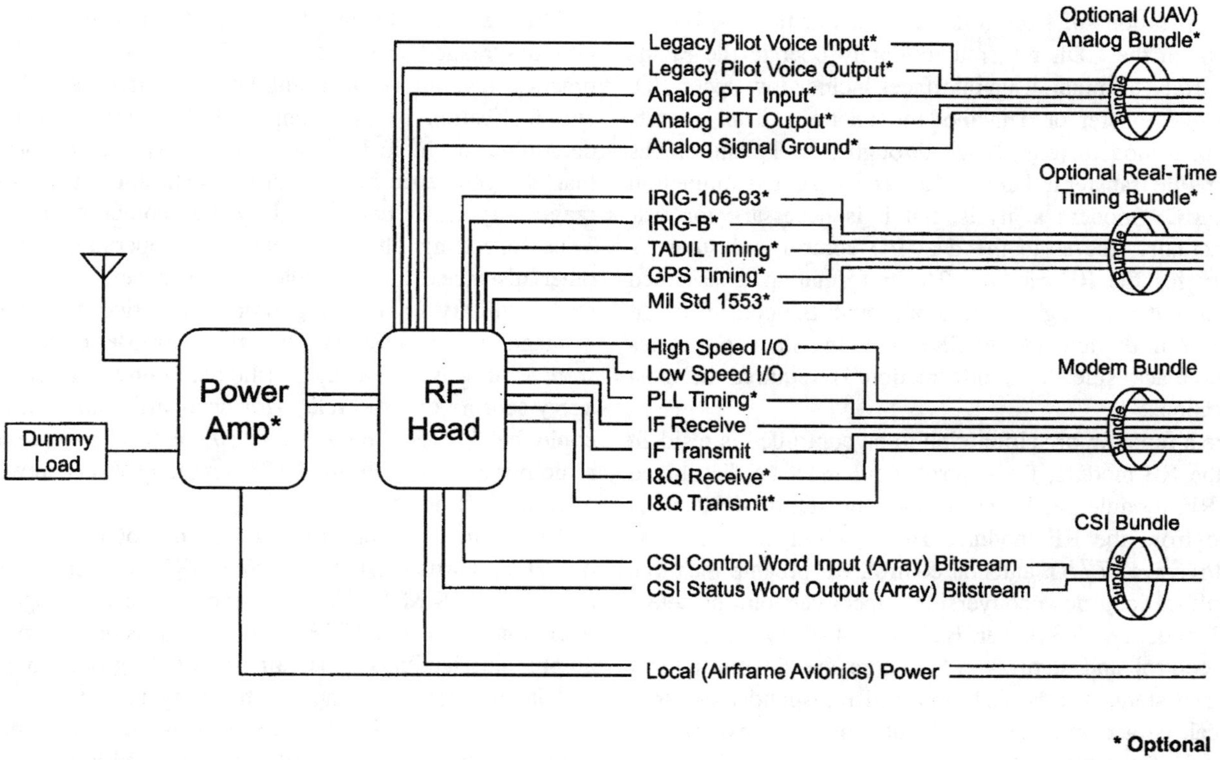

Fig. 11. Example interfaces for a remotely controlled PMCS RF module.

C. Legacy Interfaces

The interfaces seen as absolutely necessary for interoperability with legacy systems include analog voice and analog grounding, as summarized in Table IV. USAF and FAA representatives to the PMCS IPT stressed the need for an analog voice path from the cockpit to the RF module that bypasses all other modules. Such a connection permits the radio operations during ground operations when these modules are commonly powered down. Maintenance, some taxiing, refueling, and emergency operations (e.g., on-board fires, engine loss, and engine restart) typically preclude power to the backplane.

Requirements for analog ground include the signal and power return paths. The trend in the construction of aircraft is toward increased use of composite materials and decreased use of metal. The RF ground plane no longer behaves as an infinite mass in smaller UAV's. This deteriorates grounding paths for signal and power. In aircraft built decades ago, a connection to a bulkhead or cabinet could be designated as a ground path. But increased use of composites results in a need for separate grounding cables. Airborne PMCS radios therefore require two connectors for ground in addition to the chassis.

D. Remote Control Interfaces

Fig. 11 provides a system block diagram of a remotely controlled RF module showing connections necessary for airframe integration. The legacy analog subfunctions outlined in Table IV may be viewed as either a bundle of low-voltage differential cables or as the output of a digital-to-analog converter (DAC) located near the RF module.

The timing bitstreams, optional for many aircraft, could be part of the platform interfaces (Table II), bundled in a shielded cable for noise immunity, or defined in a high-performance backplane (as in Fig. 10). The modem interfaces of Table III could be provided by a backplane if the modem and RF module shared the same chassis as in Figs. 4 and 5. Alternatively, these interfaces may be provided by high-performance cables, like IEEE-STD-1394 FireWire or LVDS. This approach is useful if the RF head is physically remote from the modem.

E. CSI Interfaces

The CSI shown notionally in Fig. 11 as a pair of serial cables might be implemented as a parallel cable, depending upon distance. CSI interface requirements are outlined in Table V. The CSI bundle of Fig. 11 provides the bus or backplane functions of Table V. Remote control of the RF head could be accomplished by passing a parallel array of control words to the RF head, and passing a similar parallel array back from the RF head. Alternatively, one might use serial bitstreams with array parsing and array concatenation within the RF head. The nature of these arrays is summarized in Table V. Finally, interfaces to airframe power are also needed.

CSI is presented in the PMCS Guidance Document as a separate functional entity, but not as a physical entity [3]. For this paper, CSI functions were examined for software (control and status words), hardware (power, clock, etc.), and functional (data, timing) requirements. The categories of CSI interfaces include data interfaces, bus or backplane functions (distribution of control and status), system timing, system clock, timing bitstream, and power.

The bundling of Fig. 11 together with the interface analysis of Table V characterize the data streams needed for the tethered operations of Figs. 8–10. Portions of the CSI bundle pass through the CDL tether to remotely operate the radios from a surface terminal. Data interfaces include low-speed I/O, such as file transfer or data modem functions. This permits the use of a modem (e.g., IDM) through a backplane or bus interface. File transfer did not surface as a separate subfunction in the PMCS modem analysis, but it is necessary for red-side functions. Access to the data I/O control and status is necessary for the RF module. This information is assumed to be distributed through the CSI, although the control might originate from the red side, while responding status originates from the black side. This information is required in most configurations.

Where a function or subfunction to be controlled is resident within the RF module, the control word must be distributed to the RF module, and the responding status word must originate from the RF module. These subfunctions include transmit/receive (T/R), antenna control, RF preamplification, RF amplification, downconversion, upconversion, preselection, RF mode, and RF power. Built-in test (BIT) is supported via the control and status words. There are no multiplexer control and status words. Repeater and transponder configurations can be accommodated without a bus or backplane, as indicated in the figures.

System timing interfaces are required if aggregate system timing is provided by the CSI. The RF module uses the system clock for synchronization of internal functions. By contrast, the RF module might use GPS or another timing source to synchronize to an external SATCOM DAMA or TADIL network. The timing bitstream interfaces are more important for electronic reconnaissance platforms than for other systems. With imaging platforms, the timing stream is processed in the sensor. But with electronic reconnaissance platforms, timing needs to be distributed throughout the system via the PMCS CSI. However, tactical dissemination networks are migrating to TADIL-J, which requires accurate timing for network synchronization. The PMCS Guidance Document also recommends that GPS be available from the black-side processor through the CSI. IRIG-class timing streams would also flow through the CSI. IRIG-B timing has been particularly well characterized as deterministic and stable. The existence of IRIG-B software source code potentially reduces costs.

Power interfaces to the CSI include power for the digital devices (referred to as "+5 V") and power for analog components such as the RF amplifier (referred to notionally as "+28 V"). Separate power ground paths are accommodated by the CSI interface, augmenting the chassis or airframe grounds.

VII. CONOPS

The CONOPS was examined for 18 configurations to assess requirements in detail.

A. CONOPS Parameters

Variables derived from these CONOPS include aircraft altitude, range and other distances, frequency ranges, use of RF waveforms, use of IF and baseband data formats, and the number of simultaneous channels needed. The analysis is summarized in Table VI.

Altitudes ranged from 10 000–65 000 ft. The most common altitudes ranged from 12 000–15 000 ft, with 45 000 ft as a practical maximum. Operating altitudes affect both the design and fabrication of the equipment. Operating altitudes also determine the radio horizon, which is typically more distant than the geometric horizon in the VHF and HF bands. UHF range is typically determined by the geometric line of sight. Thus, operating altitudes determined operating ranges and other distances. For example, the distance from the aircraft to the military forces being supported varied by three orders of magnitude. Small UAV platforms for close-in video support were as little as 5 mi away. For higher altitude aircraft with imagery sensors or radio relays for deep strike missions, UAV's could be 300–500 mi distant. 300 nmi is a representative radio horizon estimate for UHF communications between two aircraft aloft at 45 000 ft.

The frequency range for support to ground forces centers on VHF, particularly the SINCGARS frequency range of roughly 30–88 MHz. The frequency range for support to air operations centers on UHF. Combined arms operations include amphibious landings, close air support to ground operations, and interdiction from the air, resulting in a combination of frequency ranges. HF is used for over-the-horizon communications, and the HF receive-only mode is used in many military operations.

RF waveforms included AM, FM, and SSB in HF. SINCGARS and HAVE QUICK are also ubiquitous. UHF SATCOM is desired, but not universally required. The UHF Follow-On (UFO) satellites provide more UHF channels than FLTSATCOM, reducing the need for UHF broadcast from tactical aircraft. C-band video in the CONOPS was based on COTS products rather than on a PMCS mode.

Baseband formats include analog voice for most relay applications and all manned aircraft. Digital simultaneous voice and data (DSVD) is needed for IDM functionality. Digitized voice, advanced narrow-band digital voice transmission (ANDVT), linear predictive coding (LPC-10), TADIL bitstreams, and digital file formats were also included in the CONOPS.

The number of simultaneous channels or virtual radios varied from one or two up to as many as 36. For example, if one SINCGARS channel is dedicated to each Infantry Company, the number of simultaneous SINCGARS relays depends on the number of companies engaged. The number of companies in the organization that operates the relay platform may also affect the number of relays. Support to combined arms operations with simultaneous support to ground, naval, and air components offers the most demand and communications complexity.

C-band video is included in Table VI, but is not a PMCS configuration. However, C-band COTS video is useful for reconnaissance. If a PMCS radio could receive the C-band video waveforms, then the C-band radios could be eliminated, freeing the payload for additional sensors.

B. CONOPS Observations

Functionally, unmanned and manned aircraft present differing communications planning considerations. Potential software radio configurations aboard unmanned aerial vehicles

TABLE VI
CONOPS Analysis for the PMCS Configurations

Potential CONOPS variables	Repeater with diplexer	Repeater with delay	Transponder, simplex	Transponder, duplex	Relay, simplex	Relay, duplex	Dual crypto relay, simplex	Dual crypto relay, duplex	LAN-to-LAN relay, simplex	LAN-to-LAN relay, duplex	Manned aircraft, front end crew	Manned aircraft, back-end crew	Manned aircraft, shared radio	Tethered HAVEQUICK relay	Tethered SINCGARS relay	Tethered IBS relay	Tethered CSAR relay	Tethered SATCOM relays
Altitude of aircraft foreseen or predicted																		
- minimum altitude for CONOPS in thousands of feet	15	10	10	10	10	10	10	10	10	10	18	18	18	18	12	15	15	18
- maximum altitude for CONOPS in thousands of feet	45	45	45	45	45	45	45	45	45	45	45	45	45	45	45	45	45	65
Ranges and distances:																		
- From aircraft sensor to probable target in nautical miles (NM)																		
-- minimum distance (video sensors, direct overhead)	5	5	5	5	5	5	5	5	5	5	20	20	20	20	5	5	5	5
-- maximum distance (max distance to wavelength horizon)	150	150	150	150	150	150	150	150	150	150	150	200	200	300	100	200	50	200
- Ground forces (from aircraft to supported ground forces) in NM																		
-- minimum distance	5	5	5	5	5	5	5	5	5	5	20	20	20	20	5	5	5	5
-- maximum distance (Direct Action Teams)	100	150	150	100	100	100	100	100	100	100	150	150	150	100	50	100	50	100
- Naval forces (from aircraft to supported naval forces) in NM																		
-- minimum distance (amphibious operations)	5	5	5	5	5	5	5	5	5	5	5	20	20	20	5	5	5	5
-- maximum distance (ASW, ASUW, Patrols)	200	200	200	200	200	200	200	200	200	200	200	300	300	300	50	200	150	300
- Air forces (from aircraft to supported forces aloft) in NM																		
-- minimum distance (Close Air Support, etc.)	5	5	5	5	5	5	5	5	5	5	5	20	20	20	5	5	5	5
-- maximum distance (Deep Strike, Interdiction)	200	200	200	200	300	300	300	300	300	300	300	300	300	300	200	300	300	500
Frequency ranges anticipated																		
- Ground forces (used to support/contact forces on the ground)																		
-- minimum (in MHz) (HF might be "receive only")	30	30	30	30	30	30	30	30	3	3	3	3	3	200	30	225	3	3
-- maximum (in GHz)	0.5	0.5	0.5	0.5	0.5	0.5	0.5	0.5	0.5	0.5	0.5	0.5	0.5	0.5	.09	0.5	0.3	0.5
- Naval forces (used to support/contact naval forces)																		
-- minimum (in MHz) (HF might be "receive only")	30	30	30	30	30	30	30	30	3	3	3	3	3	200	30	225	3	3
-- maximum (in GHz)	0.5	0.5	0.5	0.5	0.5	0.5	0.5	0.5	0.5	0.5	0.5	0.5	0.5	0.5	.09	0.5	0.3	0.5
- Air forces (used to support forces aloft)																		
-- minimum (in MHz) (HF might be "receive only")	30	30	30	30	30	30	30	30	3	3	3	3	3	200	30	225	3	3
-- maximum (in GHz)	0.5	0.5	0.5	0.5	0.5	0.5	0.5	0.5	0.5	0.5	0.5	0.5	0.5	0.5	.09	0.5	0.3	0.5
RF waveforms anticipated																		
- AM	R	R	R	R	R	R	R	R	R	R	R	R	R	d	d		R	R
- SSB (HF SSB might be "receive only")							R	R	R	R	R	R	R				d	R
- FM	R	R	R	R	R	R	R	R	R	R	R	R	R	d	d		R	R
- SINCGARS			d	d	d	d	R	R	R	R	R	R	R		R			R
- HAVEQUICK			d	d	d	d	R	R	R	R	R	R	R	R				R
- UHF SATCOM (RHCP) for surrogate satellite broadcasts			d	d	d	d	d	d	R	R	R	R	R	d	d	R		R
- C-band COTS video																		
Baseband data formats/IF data streams anticipated																		
- Analog voice	R	R	R	R	R	R	R	R	R	R	R	R	R					
- Digitized voice	d	d	R	R	R	R	R	R	R	R	R	R	R	R	R		R	R
- Digital Simultaneous Voice and Data (DSVD) [IDM]	d	d	d	d	d	d	R	R	R	R	R	R	R	d	d		d	R
- Adv Narrowband Digital Voice Transmission (ANDVT) [LPC-10]							R	R	R	R	R	R	R	d	d		d	R
- TADIL bitstreams									d	d	d	R	R			R		R
- Digital file formats	d	d	d	d	d	d	R	R	R	R	R	R	R	d	d	d		R
Number of Simultaneous Channels Anticipated																		
- Minimum number of channels	1	1	3	3	4	4	4	4	4	4	4	4	4	4	4	1	4	4
- Maximum number of channels	1	3	6	6	8	8	8	8	8	8	8	8	36	24	36	4	8	32

(UAV) include remotely operated repeaters, transponders, relays, gateways, and broadcast nodes. Narrow-band radios aboard UAV's may be viewed as remotely operated communications nodes for single-channel and multichannel streams. Despite the absence of a crew, UAV nodes are suited to voice relay services. Manned aircraft configurations include locally operated multimode, multichannel nodes. These platforms need reconfigurable data links (TADIL, IDM, etc.) and dissemination services. The presence of a human crew does not automatically trigger an increased need for voice communications. Instead, the larger flow of processed data intensifies the data traffic.

Airborne HF, VHF, and UHF relays complement SATCOM UHF, SHF, and EHF relays. Relays may be installed on UAV's or in pods mounted on manned aircraft. Low-band relays on airborne platforms and high-band relays on satellites extend the communications architecture and capabilities. In particular, UAV's with reconfigurable communications nodes may be operated remotely via multiplexed satellite relay tethers. Data rate agility complements these operations.

VIII. Conclusions

Modular, reconfigurable, software radios are attractive for airborne communications. The requirements analysis addressed configurations and concepts of operations. The three key airborne equipment environments include the aircraft skin for antennas, unprotected cargo bays for black-side RF and modem components, and habitable crew bays for red-side display and INFOSEC components. Tables summarize required control and status words, submodules, sets of interfaces for the RF module, and operational concepts for such radios in airborne applications. Radios that are remotely software reconfigurable support tether operations where SATCOM or CDL provide the backbone tether communications path.

REFERENCES

[1] E-mail from T. F. Roome, Jr., MITRE Corporation, Bedford, MA, Aug. 15, 1997, "RE: Which waveforms need separate I and Q?"; E-mail from J. Boland, Boltec Corporation, Crystal City, VA, Aug. 16, 1997, "RE: Which waveforms need separate I and Q?".

[2] *Joint Airborne SIGINT Architecture (JASA) Standards Handbook (JSH)*, version 2.0, draft for comment, Appendix C, JSH Standards, National Security Agency, July 29, 1997 [Online]. Available WWW: www.jswg.org/handbook.html.

[3] Programmable Modular Communications System (PMCS) Guidance Document, July 31, 1997, revision 2 [Online]. Available WWW: www.dtic.mil:80/c3i/jtrpage.htm.

Abstract

Wideband multicarrier receivers are an interesting technology for software-defined radios. When combined with suitable digital signal processing hardware and software, they enable flexible adjustment of the number of carriers as well as provide the capability of simultaneous reception of signals with differing bandwidths and modulation formats. Due to technological limitations, the required dynamic range may not be available in the near future, however. For this reason, it has been proposed to install two separate receivers in a base station: a high gain receiver for serving distant terminals and a low gain receiver for nearby terminals. In GSM base stations, it is also necessary to include a narrowband receiver with large dynamic range to enable random access capture. Since the low and high receivers will have only partly overlapping coverage areas, it is interesting to study what the relative traffic through each receiver would be. This will affect the number of carriers to be supported by each receiver. In GSM base stations, it is also important to consider the narrowband receiver, which can serve terminals anywhere within the cell. A series of simulations was performed in order to define the respective traffic capacity requirements in a suburban macrocellular and urban microcellular environment. The number of high input levels in a base station was found to be rather small. Therefore, in base stations of small or moderate capacity, the narrowband receiver can apparently serve those terminals which cannot be supported by the high gain receiver. Consequently, no low gain wideband receiver is needed, and the base station architecture can be simplified. The novel base station architecture brings about the need for a level-based intracell handover. Parameterization of this handover as well as the strategy of allocating channels in the different receivers should be done with care, in order to avoid loss of traffic capacity.

Receiver Dimensioning in a Hybrid Multicarrier GSM Base Station

HARRI POSTI AND RAULI JÄRVELÄ, NOKIA TELECOMMUNICATIONS

PENTTI A. LEPPÄNEN, UNIVERSITY OF OULU

In recent years, the concept of software-defined radio has received much military, commercial, and academic interest. In its extreme form, a software radio would consist of digital hardware connected to the antenna. However, the implementation technology has not yet sufficiently matured for such applications [1], and near-future products will be based on less ambitious architectures.

Important building blocks of a more realizable software radio include a wideband analog–digital (A/D) converter (ADC) with large dynamic range preceded by a wideband analog receiver module [2]. With these technologies, it is possible to build multicarrier receivers, where a number of narrowband radios are replaced by a single piece of wideband equipment [1]. This would be most attractive in cellular base stations, where the number of carriers may be quite large [3, 4]. Wideband receivers also have the capability to process signals with different bandwidths and modulation formats, and enable flexible assignment of traffic capacity between neighboring sites, as dictated by the traffic [5].

Potential benefits of multicarrier receivers are numerous, including reduced size and weight [1, 3, 5, 6], lower power consumption [6], reduced cost [5–7], as well as savings in R&D, production, and logistics [6, 8]. Another category of possible gains is related to improved receiver functionalities. Examples of these include multistandard capability [6], more practical implementation of smart antennas [3, 7], and flexible upgrading of base station capacity and services [9]. For the sake of realism, it appears necessary to rememeber that the actual gains will depend on the base station application [7] and configuration.

Since the receiver specifications vary between cellular standards, the available components may not provide sufficient performance for some systems. Therefore, there is interest in searching architectures, which would circumvent some of the most serious challenges. An example of such a solution is presented in [10] and illustrated in Fig. 1. The principle of the proposed method is providing separate receivers for strong and weak signals. Hence, the receivers need not exhibit the full dynamic range required of the base station.

Global System for Mobile Communications (GSM) base station receiver specifications are considered among the most demanding in cellular standards. Therefore, the presented architecture is particularly attractive for GSM. The purpose of this article is to explore the feasibility of this and to point out specific considerations for application of the architecture. In particular, the target is to find out whether the low gain receiver can be omitted and nearby terminals served by the narrowband receiver. This will take place by first identifying the critical GSM base station receiver specifications and approximating the magnitude of the dynamic range requirement. Next, the technological constraints on the multicarrier receiver dynamic range are presented and viable performance estimated. Following that, the need for common access and signaling channels and their implications are discussed. The expected distribution of input signal levels is estimated by simulations and, based on the results, the relative capacities of the high and low gain receivers determined. Finally, implications of the architecture on radio resource management are given, and conclusions of the study are drawn.

Reprinted from *IEEE Personal Communications Magazine,* Vol. 6, No. 4, pp. 56–64, August 1999.

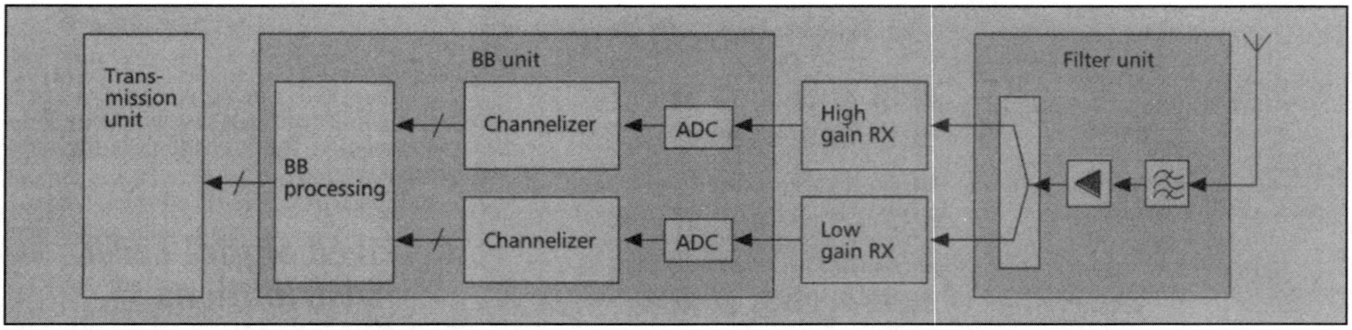

■ **Figure 1.** *The proposed high/low gain architecture.*

Required and Achievable Receiver Dynamic Range

GSM Base Station Receiver Dynamic Range

The dynamic range of a receiver is determined by the expected spread of input signal levels. In GSM base station specifications, this is determined by two measurements: reference sensitivity level and nominal error rates [11].

Reference sensitivity level defines the lowest expected input power, whereas nominal error rate specification determines the highest input level to be supported. The respective values for normal GSM 900 base stations, which constitute the most demanding case, are:

- Reference sensitivity level: useful input signal at –104 dBm
- Nominal error rates: useful input signal level up to –15 dBm

In narrowband single-carrier receivers, it is typical to apply automatic gain control (AGC). This enables positioning of the signal at or close to the optimal level, in terms of noise, spurious, and intermodulation performance. Hence, the requirement on instantaneous dynamic range remains moderate. In multicarrier receivers, all the input signals are contained within the receiver bandwidth. Therefore, application of AGC could drown a weak signal in noise or cause a strong signal to saturate the receiver [12].

Contemporary GSM base stations regularly exhibit sensitivity which surpasses the specified performance. Although the exact figures are not public information, it appears safe to assume –108 dBm reference sensitivity presently and –110 dBm in the nearest future. Hence, the required dynamic range becomes (–15 + 108) = 93 dB or even 95 dB.

Achievable Dynamic Range

According to some references (e.g., [3]), the ADC is the most critical component in multicarrier receivers. Therefore, the obtainable dynamic range is estimated based on the available ADCs. A few further assumptions are necessary in order to translate the dynamic range into ADC signal-to-noise ratio (SNR), namely:

- Desired SNR after A/D conversion of 9 dB
- Equal noise contributions from the ADC and analog receiver chain
- Maximum ADC input signal placed 3 dB below the full scale in order to optimize spurious performance

Based on these figures it is seen that the ADC noise must be 12 dB below the lowest expected ADC input signal. Full-scale input, on the other hand, should be 3 dB above the highest input level. Therefore, the obtainable receiver dynamic range is given by

$$DR = [SNR_{ADC} - (9 + 3 + 3)]dB = (SNR_{ADC} - 15)dB. \quad (1)$$

In a multicarrier receiver, the ADC sampling rate must be equal to or higher than twice the receiver bandwidth, according to the Nyquist criterion. Since the receiver bandwidth is considerably larger than that of a single channel, oversampling takes place. This results in SNR improvement, or processing gain, of [13]

$$G = \left[10 * \log\left(\frac{f_s}{2 * bw}\right)\right] dB, \quad (2)$$

where

f_S = ADC sampling rate

bw = channel bandwidth, 200 kHz

According to preliminary component data [14], a Burr Brown ADS852 ADC will exhibit sampling rate of 65 MHz and yield typical SNR of 75 dB. Applying Eqs. 1 and 2 yields the respective dynamic range of 82 dB. It should be noted that the sampling rate is not quite sufficient for digitizing the full 35 MHz bandwidth of GSM900. Since the ADC performance is dependent on the input signal, it is likely that the worst-case performance is somewhat lower than the reported value. Therefore, receiver dynamic range of 70 dB appears a safe value for the subsequent analysis.

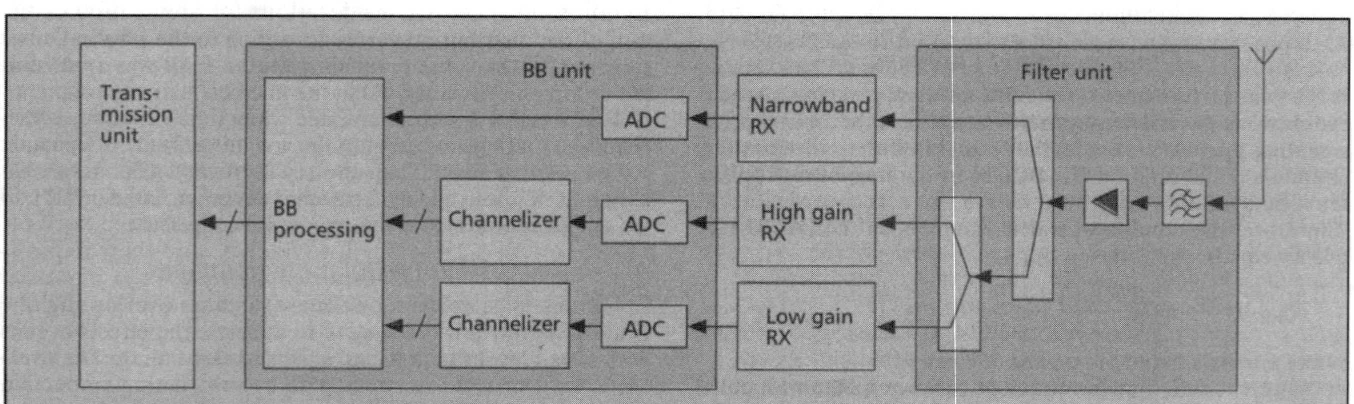

■ **Figure 2.** *High/low gain architecture for GSM base stations.*

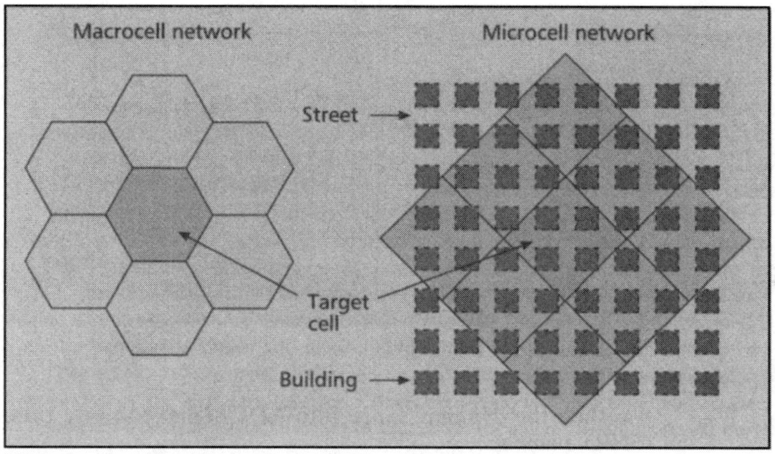

■ **Figure 3.** *Network topologies in the signal level simulations.*

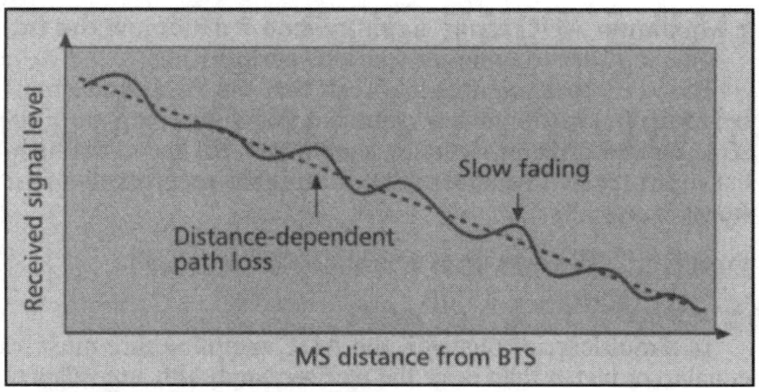

■ **Figure 4.** *Path loss components.*

GSM Signaling Channels

Random Access Channel

The purest implementation of the proposed architecture is not suitable for GSM, because the random access channel (RACH) receiver must be capable of capturing mobile station accesses at both very high and very low levels. Therefore, it must support the entire dynamic range and, due to the assumed component limitations, have narrowband implementation.

As a result, a GSM base station according to the proposal will contain three types of receivers, as shown in Fig. 2: traditional, narrowband receiver for the signaling channel, or a broadcast channel (BCCH) frequency, high gain wideband receiver for distant mobile stations (MSs) and a low gain wideband receiver for nearby terminals. Each of these receivers will cover a different frequency band. Thus, the detected signal level will be one of the criteria in radio channel allocation for a pending or active communication.

Other Signaling Channels

It is a common assumption in GSM network planning, that for every two carriers one physical channel must be reserved for signaling purposes. It is further stipulated that all signaling channels be located on the BCCH frequency, because they must be available anywhere within the cell coverage area. Therefore, the number of traffic channels on the BCCH frequency equals

$$N_{BCCH} = 8 - INT\left(\frac{N+1}{2}\right), \qquad (3)$$

where n equals the total number of carriers.

Since the combined amount of carriers in the wideband receivers equals $(n - 1)$, the number of traffic channels in the wideband receivers is given by

$$N_{WB} = 8^*(n - 1) \qquad (4)$$

If the low gain receiver is not necessary, N_{WB} equals the number of traffic channels in the wideband high gain receiver.

Received Signal Level Distributions

Simulated Environments

Signal propagation, cellular network layout, and terminal distribution are all dependent on the geographical environment. Therefore, it is beneficial to observe different environments in order to obtain more comprehensive information on the signal levels. With this aim, two rather dissimilar environments were considered:

- Suburban omnidirectional macrocell
- Urban omnidirectional microcell

The layout of the network in these environments is shown in Fig. 3.

In the next section signal propagation models will be discussed. At this point it appears prudent to mention that signal attenuation depends on the height and location of the terminal and base station antennas, as well as the geometry of the surroundings. Characteristic values for the relevant parameters were chosen and will be given in an ensuing summary.

Signal Level Variation

Electromagnetic theory teaches that in free space the field strength decreases in proportion to the second power of distance. Due to various obstacles encountered in terrestrial cells, the signal strength variation is more complex and the attenuation due to distance larger. The variation of signal strength is often modeled with three factors:

- Attenuation due to distance
- Slow variation due to obstacles or slow fading
- Fast variation due to multiple reflections or fast fading

In GSM systems, various techniques are utilized to combat the effect of multipath fading. These include data interleaving, channel coding, and frequency hopping. Since the combined effect of these features effectively averages out the fast fading, only the first two factors were considered in the simulations. Their nature is shown in Fig. 4.

Various propagation models have been devised to characterize distance-dependent attenuation. These include the macrocellular Okumura-Hata [15] and microcellular Walfish-Ikegami models [16], which were chosen for the study. Slow variation, or shadowing, is typically modeled as a zero-mean log-normally distributed random variable. The standard deviation of the distribution varies according to the physical environment. In the subsequent simulations, 6 dB was applied in the suburban cells and 8 dB in the microcellular environment.

Radio signal is also attenuated when it has to enter or exit buildings, a widely studied phenomenon. In the study, attenuation was modeled as another log-normally distributed random variable having 11 dB mean and 6 dB standard deviation, based on [17]. In the suburban case, indoor coverage was not stipulated.

A Multicellular Environment

In multicellular networks, cells are made to overlap slightly. One reason for this is the need to alleviate the effects of random signal level fluctuations, slow and fast fading. The overlap is built into the network plan by subtracting respective slow and fast fading margins from the ideal link budget. Adequate slow fading margin depends on the environment; a larger value is necessary in areas, such as dense urban, where shadow fading has large variations. In addition to noise, co-channel interference is also encountered in loaded networks. Hence, a typical interference degrading margin of 3 dB is sometimes included in the calculation.

Due to these parameters, the calculated cell range is reduced and overlap created. This is illustrated in Fig. 5. The magnitude of the margins is selectable and is reflected in the location probability, or the probability that adequate service quality is available at a particular location.

Another reason for the overlap is the need to hand over calls between cells. Successful handovers (HOs) call for adequate service from the target cell before connection to the original cell is lost. Furthermore, it is necessary to include some hysteresis, or HO margin, in the HO algorithm to prevent repeated HOs between neighboring cells. Without considerable cell overlap, implementation of the hysteresis is not possible. The effect of HO margin is different HO locations for departing and approaching terminals, which is demonstrated by Fig. 6.

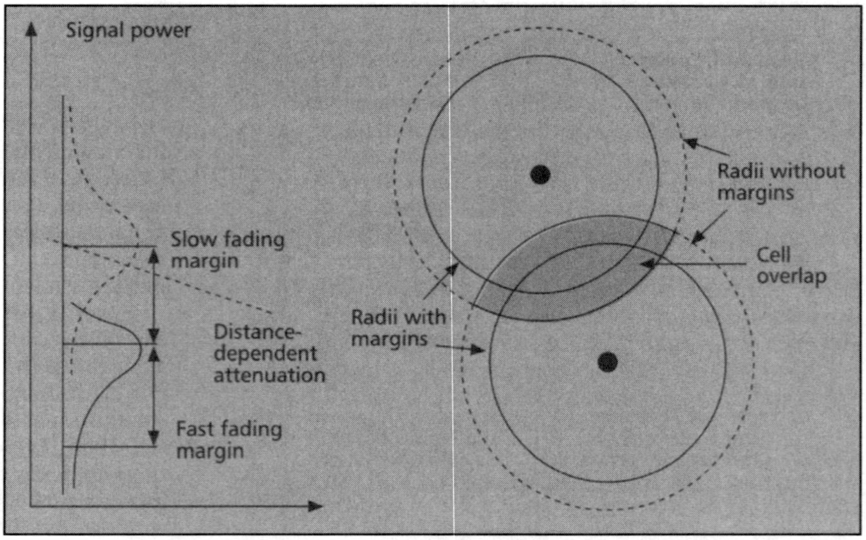

Figure 5. *Cell planning margins and resulting overlap.*

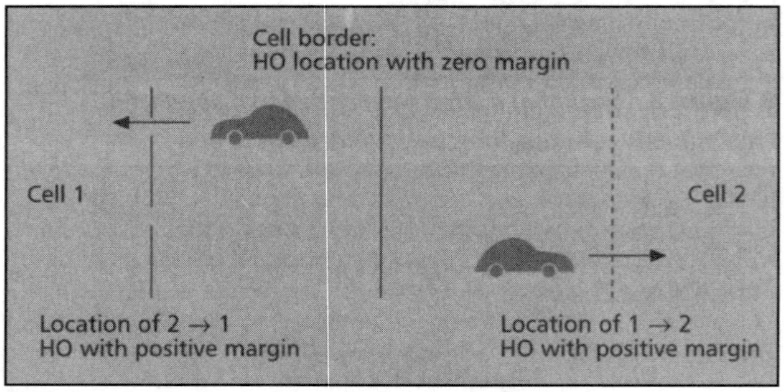

Figure 6. *Hysteresis due to the handover margin.*

In planning the simulated network, a slow fading margin of 4 dB was applied in the suburban environment and 5 dB in the urban network. In combination with the 3 dB interference degrading margin, these produce total margins of 7 and 8 dB. Since the specified base station sensitivity of –104 dBm [11] was used as the starting point, the margins lead to uplink cell edge target levels of –97 and –96 dBm. It should be noted that a –96 dBm indoor signal level corresponds to an outdoor level of –85 dBm, on average.

Terminals Considered in the Simulation

Initially, in the suburban case 390,625 terminals were placed uniformly, or in a regular square grid, over a network that embraced, in addition to the target cell, two tiers of neighboring cells. The path loss, consisting of the distance-dependent component and slow fading, was calculated from each terminal to each base station. Next, those terminals that were within the dominance area of the target cell were chosen. Thus, the amount of relevant mobiles became about 30,000.

In the urban environment the number of uniformly distributed terminals was 160,000. After consideration of the path losses to each base station, which additionally included the indoor loss, the number of considered terminals was reduced to 5400.

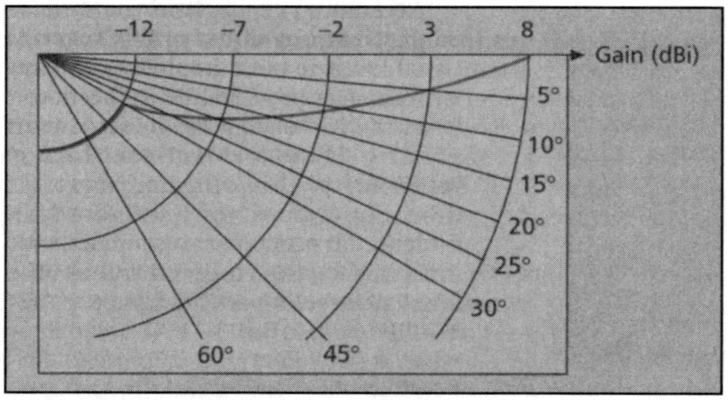

Figure 7. *A vertical radiation pattern of base station antennas.*

In defining the dominance area, the inter-cell handover margin was also considered. Inter-cell HO margin of 3 dB was chosen and taken into account in the following fashion:

- Departing terminals remain in the target cell until the strongest neighbor becomes 3 dB stronger than the target cell.
- Approaching terminals are handed over to the target cell as soon as it becomes 3 dB stronger than the strongest neighbor.

Fifty percent of the terminals were assumed approaching the target base station and 50 percent moving away from it. It should be noted that quality information was not considered in the intercell HO. The main reason for this was the desire to avoid modeling the adjacent and co-channel interference.

Base Station and Terminal Antenna Patterns

Gain and radiation pattern of base station antennas are chosen according to the network topology, coverage requirements, and base station location. In the simulations the antennas were omnidirectional in the horizontal plane. The vertical shape was designed to approximate that of commercial moderate-gain antennas. The radiation pattern is depicted in Fig. 7 and described by

$$G(\alpha)=\begin{cases}8\ \text{dBi} & \alpha=0^\circ\\ (e^{-0.0418\alpha+3.33}-20)\ \text{dBi} & \alpha\in(0^\circ,30^\circ)\\ -12\ \text{dBi} & \alpha\geq 30^\circ,\end{cases} \tag{5}$$

where α = clockwise angle from the horizontal plane.

The radiation pattern has the effect of reducing the received signal power from terminals very close to the base station. Hence, it somewhat reduces the probability of very high input levels.

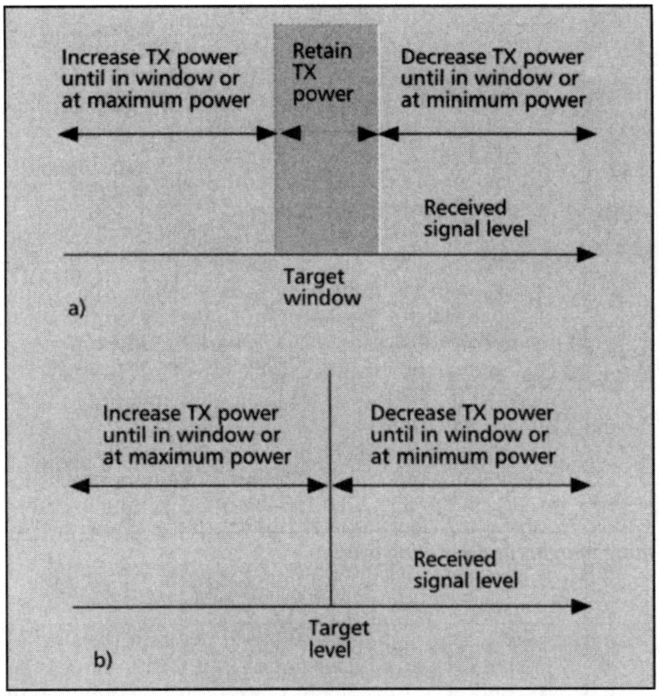

■ **Figure 8.** *Usage of a) a target window and b) a target level.*

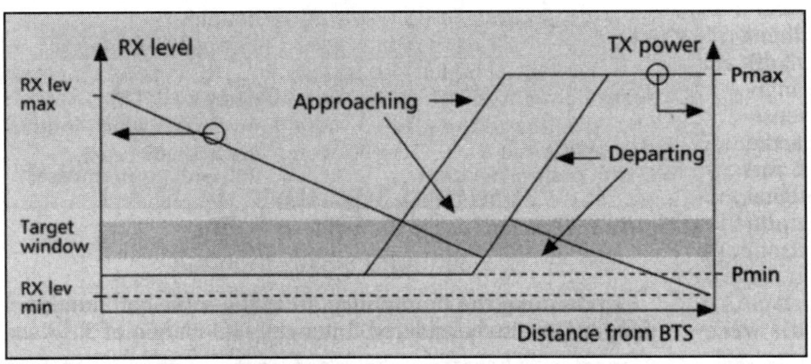

■ **Figure 9.** *The effect of a PC target window on transmit power and received signal level.*

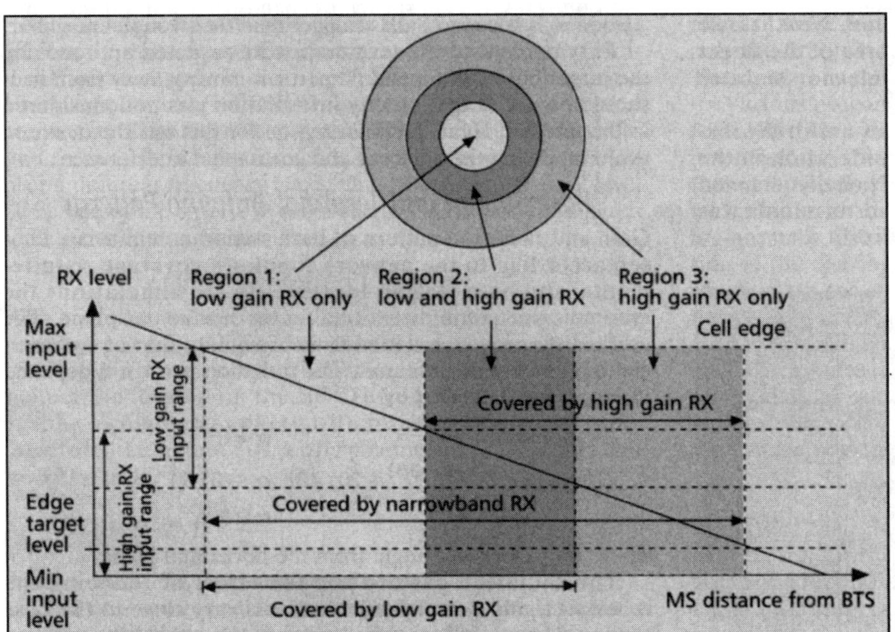

■ **Figure 10.** *Cell regions in the proposed BTS architecture.*

Dynamic Power Control

Dynamic power control, [18, pp. 342–46], is an important GSM functionality which has several aims. The most important are reduction of interference to neighbor cells and extension of MS battery life. Power control algorithms typically use the observed signal quality and level as adjustment criteria. However, in the simulations the only criterion was the received level, in order to retain the relative simplicity of the simulation model.

Several different level-based power control strategies can be envisioned, but the above-mentioned goals are best achieved with an algorithm that minimizes the transmitted power. In order to avoid loss of connection due to random fluctuations, it is typical to compromise and set the target received level somewhat above the receiver sensitivity threshold. Constant up and down adjustment of the transmission power level is normally avoided by using a target window instead of a single target value. The application of the power control target level and target window is shown in Fig. 8.

In the simulations the target window was assigned at –95 to –80 dBm and used in the following manner:

- The output power of departing terminals was kept at the minimum value until the received level equaled the lower edge of the window. Thereafter, the power was increased.
- The power of approaching terminals equaled the maximum value until the received level was at the upper edge of the window. Thereafter, the output power was decreased.

The effect of such a scheme is illustrated in Fig. 9.

A number of nonidealities are present in real-life power control. They include delay in the closed power control loop, which may be several seconds, receive level measurement error, transmit level setting error, and the power control step size. The combined effect of these phenomena is difficult to define precisely and is usually modeled as a zero-average, log-normally distributed random variable. In the simulations, PC error was assigned standard deviation of 6 dB in the suburban and 5.5 dB in the urban environment.

The power control range of a terminal is specified in [11]. Hence, notwithstanding the calculated power control error, the transmitted power cannot be above the maximum level or below the minimum level. This was taken into account in the simulations.

Simulation Results

The simulation parameters are summarized in Table 1.

Reference [19] presents the simulated distributions of the uplink received signal levels in both graphical and tabular format. It provides the proportion of levels within each 1 dB interval from –110 to –15 dBm as well the share of values below –110 dBm and above –15 dBm. For reasons which will shortly be evident, the proportion of mobiles within the following received signal level intervals were of particular interest:

- –110 to –45 dBm
- –45 to –40 dBm
- –40 to –15 dBm
- Above and below the extreme values

	Suburban macrocell	Urban microcell
Propagation model	Okumura-Hata	Walfish-Ikegami
Building height, separation, and width	N/A	20 m/20 m/20 m
Slow fading σ	6 dB	8 dB
BTS antenna height	30 m	10 m
Terminal height	1.5 m	
BTS antenna gain	–12 ... 8 dBi	
Indoor loss	N/A	Average 11 dB, σ 6 dB
MS TX power	33 dBm	
Target level from cell edge	–98 dBm	–96 dBm
Intercell HO margin	3 dB	
PC target window	–95 .... –80 dBm	
PC error σ	6 dB	5.5 dB
MS PC range	20 dB	
MS distribution	Uniform	Uniform, 2/3 indoors, 1/3 outdoors

Table 1. *Simulation parameters [19].*

The respective fractions are given for each of the considered environments in Table 2.

It is seen from the results that signal levels are somewhat higher in urban than in suburban environments. The reason is that due to higher attenuation, the cell radius is smaller, and therefore a larger fraction of the terminals is located very close to the base station. Furthermore, the existence of mobiles indoors spreads the received signal levels over a relatively wide range.

The actual distributions will further be somewhat affected by other phenomena, which have been omitted:

- Very close to a base station, the propagation loss may approach free-space loss with an exponent of two. This will increase the received level, but with the even distribution premise the fraction of such mobiles is small.
- There is a possibility that mobiles even further than the second tier of neighbors may select the target cell. These mobiles would be received at low levels, but their number is expected to be small.

Low/High Gain Receiver Dimensioning

Cell Regions

The partly overlapping input ranges of the three kinds of receivers will lead to a somewhat peculiar cell topology, as shown in Fig. 10. This is because each receiver type will support communications to mobiles in different geographical regions.

An additional feature of level-based intracell, intertransceiver handover must be designed. This is necessary in order to provide service through the correct receivers as the mobiles move across the regions. Prevention of ping-pong HOs near the region borders calls for HO margins, or hysteresis. Naturally, the HO thresholds must be located on the overlapping area of the low and high gain receivers.

Because of the topology, there will be a minimal requirement for both low and high gain capacity. The questions to be answered next are:

- What is the required number of low gain carriers in different base station configurations and environments?
- How would the reservation of low gain carriers affect the cell traffic capacity?

Traffic capacity in the low gain receiver can be minimized if the terminal power control can take full effect before a traffic channel is assigned. This would happen during the call setup signaling on the common control channel (CCCH) and stand-alone dedicated control channel (SDCCH) [20, 21], which are located on the BCCH frequency and therefore served by the narrowband receiver.

In calculating signaling channel capacity requirements, it is often assumed that a call stays seven seconds on the SDCCH. On the other hand, maximum output power adjustment of 20 dB has the duration:

$$T_{PC} = \left(delay + {}^{20}\!/_{step} * period\right)s, \tag{6}$$

where *delay* equals the interval between the start of a signal level measurement in a terminal and the reception of subsequent power control command in a base station, *step* is the maximum step size in power level reduction, and *period* is the duration between subsequent PC commands.

The parameter *period* has a value of 480 ms, as specified for GSM. *Delay* and *step* depend on power control parameterization, but typical values could be 3 s and 4 dB. With these values, Eq. 6 yields T_{PC} of 5.4 s, which is sufficiently fast and even allows for some margin.

Signal Ranges and Traffic Capacities of the Different Receivers

Although the cell edge target levels were chosen based on the specified base station sensitivity, it was decided to define the signal ranges so that input levels down to –110 dBm could be supported. Therefore, the range of the high gain receiver becomes –110 to –40 dBm and that of the low gain receiver –85 to –15 dBm.

Required low gain receiver capacity is again minimized if the HO region is placed at the upper end of the high gain receiver input range. Intertransceiver HO margin of 5 dB was selected, and hence the HO region equals –45 to –40 dBm. This arrangement is shown in Fig. 11.

Since, according to the simulations, the proportion of very high input levels is small, an assumption was now made that a dedicated low gain receiver is not needed. Instead, the nearby terminals are served by the narrowband receiver. Multicarrier receivers are potentially beneficial in base stations with a sufficiently large number of carriers. Therefore, the range of base station configurations was chosen as 5–16 carriers.

	Suburban macrocell	Urban microcell
< –110 dBm	5.09×10^{-3}	0
–110 to –45 dBm	9.95×10^{-1}	9.98×10^{-1}
–45 to 40 dBm	0	1.21×10^{-3}
–40 to –15 dBm	3.15×10^{-5}	7.21×10^{-4}
> –15 dBm	0	4.41×10^{-5}

Table 2. *Signal level fractions.*

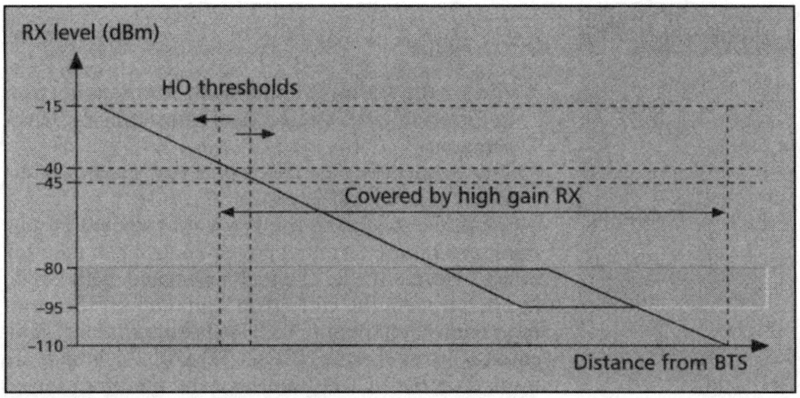

■ **Figure 11.** *Cell regions with the chosen parameters.*

The amount of traffic channels on the BCCH frequency was given by Eq. 3 and the number of traffic channels in the high gain receiver by Eq. 4. Using the Erlang B formula with 2 percent blocking probability it is possible to determine the traffic capacity of each receiver type. To simplify the analysis, it was further assumed that the BCCH carrier is reserved solely for terminals in region 1. The results of the traffic capacity calculation are given in Table 3, where:

- N_{BCCH} is the number of traffic channels in the narrowband receiver and T_{BCCH} is the respective traffic capacity.
- N_{WB} is the amount of traffic channels in the wideband high gain receiver, and T_{WB} is the resulting traffic capacity.

Applicability of the Selected Receiver Scheme

Table 2 summarized the share of input levels that fall within the various regions and outside the supported range. It is seen that about 0.5 percent of the offered traffic cannot be supported in the suburban case, due to inadequate signal strength. In the urban environment, some 0.005 percent of the input levels would exceed the maximum allowed level.

The 5 dB intertransceiver HO margin and the assumption that 50 percent of the terminals move in either radial direction lead to a situation where the traffic within the –45 to –40 dBm range is shared equally between the two receivers. Now it is straightforward to calculate from the information in Tables 2 and 3 the required and offered proportions of the total capacity in the narrowband receiver. The results are shown in Table 4.

It is seen that in the suburban cell the available traffic capacity for nearby mobiles is sufficient except in the 15- and 16-carrier base station configurations. In the urban area the offered capacity falls short in configurations of 13–16 carriers. The effect of this is higher blocking probability than the design value of 2 percent for those terminals located very close to the base station. It should be noted that based on the simulations, their proportion of the total terminal population is very small.

An interesting property of the presented architecture is the capability to extend or adjust the cell capacity according to the offered traffic, without addition or replacement of hardware. This can take place within the limits of maximum configurations given above.

The Effect of Dividing the Channels in Two Pools

In the previous section it was assumed that the traffic channels in the narrowband receiver would be reserved exclusively for nearby terminals. This action reduces cell capacity, because dividing the channels into smaller pools degrades channel utilization. This is easily seen by experimenting with the Erlang B formula. The magnitude of this phenomenon is shown in Table 5. Since the calculation only considers the number of traffic channels, the offered traffic has no influence and the values apply to any cellular environment.

The loss is largest in small base station configurations. However, in these cases there is substantial overcapacity in the narrowband receiver. Hence, a more intelligent scheme would also allocate channels in the narrowband receiver to more distant terminals whenever they are available. It is seen from Table 4 and Eq. 3 that one traffic channel is sufficient for the nearby mobiles in the suburban environment and two channels are required in the urban cell. For reference, Table 6 lists the capacity loss figures when only these are separated from the common channel pool.

At this point it should be noted that the necessary low gain capacity will be affected by a number of factors. These include the actual terminal distribution, path loss characteristics, cellular network plan, intertransceiver HO margin, and in particular by the wideband receiver dynamic range. Therefore, the conclusions only apply with the chosen parameters.

Implications on Radio Resource Management

As mentioned, there is a need for a novel intracell HO criterion, namely, received signal level. This measure is regularly used for deciding on intracell HOs, but for the purposes of the proposed architecture a new set of parameter values must be defined. These include separate HO thresholds for increasing and decreasing signal levels. The difference between these values, or the HO margin, must be selected appropriately. The dynamic range of the wideband receivers together with the cell edge target level will determine the position of the thresholds.

Since the coverage areas of the wideband receivers may overlap considerably beyond the HO region, it would be beneficial to consider their relative loads as well. The relative

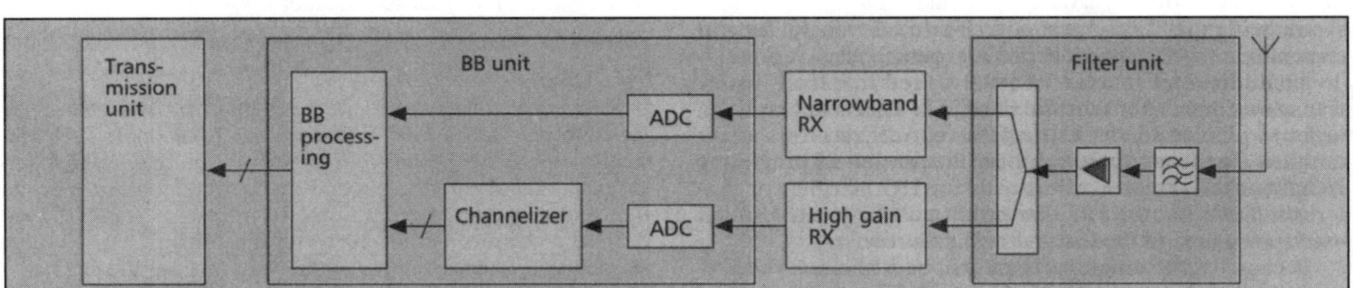

■ **Figure 12.** *The proposed architecture for GSM base stations.*

TRXs	5	6	7	8	9	10	11	12	13	14	15	16
*N*BCCH	5	5	4	4	3	3	2	2	1	1	0	0
*T*BCCH [Erl]	1.66	1.66	1.09	1.09	0.602	0.602	0.223	0.223	0.0204	0.0204	0	0
*N*WB	32	40	48	56	64	72	80	88	96	104	112	120
*T*WB (Erl)	23.7	31.0	38.4	45.9	53.4	61.0	68.7	76.4	84.1	91.9	99.6	107

■ **Table 3.** *Traffic capacities of the BCCH and high gain receivers.*

TRXs	5	6	7	8	9	10	11	12	13	14	15	16
Suburban macrocell Required traffic in NB RX 3.15×10^{-5}												
Traffic capacity in NB RX	6.53×10^{-2}	5.07×10^{-2}	2.77×10^{-2}	2.33×10^{-2}	1.11×10^{-2}	9.77×10^{-3}	3.24×10^{-3}	2.92×10^{-3}	2.43×10^{-4}	2.22×10^{-4}	0	0
Urban microcell Required traffic in NB RX 1.33×10^{-3}												
Traffic capacity in NB RX	6.53×10^{-2}	5.07×10^{-2}	2.77×10^{-2}	2.33×10^{-2}	1.11×10^{-2}	9.77×10^{-3}	3.24×10^{-3}	2.92×10^{-3}	2.43×10^{-4}	2.22×10^{-4}	0	0

■ **Table 4.** *Required and offered traffic capacity in the two receivers.*

TRXs	5	6	7	8	9	10	11	12	13	14	15	16
Traffic with two pools (Erl)	25.4	32.7	39.5	47.0	54.0	61.6	68.9	76.6	84.1	91.9	99.6	107
Traffic with one pool (Erl)	28.3	35.6	42.1	49.6	56.3	63.9	70.6	78.3	85.1	92.8	99.6	107
Loss of capacity (%)	10.2	8.3	6.3	5.4	4.0	3.5	2.4	2.2	1.1	1.0	0	0

■ **Table 5.** *Loss of traffic capacity when dividing traffic channels into two pools of 1/4-15 GSM carriers.*

TRXs	5	6	7	8	9	10	11	12	13	14	15	16
Suburban macrocell												
Loss of capacity (%)	3.2	2.5	2.2	1.9	1.7	1.5	1.3	1.2	1.1	1.0	0	0
Urban microcell												
Loss of capacity (%)	5.6	4.9	3.9	3.3	3.0	2.6	2.4	2.2	1.1	1.0	0	0

■ **Table 6.** *Loss of traffic capacity with a more intelligent channel allocation scheme.*

capacities of the receivers will be based on an assumed mobile distribution, uniform or otherwise. Occasionally, the traffic may be biased toward either the cell edge or the proximity of the base station. In such occurrences it would be advantageous to retain traffic in the less loaded receiver, although the signal level criteria may have been fulfilled.

These statements are also valid for the base station arrangement, where no wideband low gain receiver is present. As mentioned in the previous section, loss of traffic capacity may result from inappropriate channel allocation and HO schemes. Therefore, the importance of traffic load as an HO criterion and selection of adequate load-related parameters are emphasized.

It may be of interest to notice that an alternative solution would be to divide the channel pool in the time instead of the frequency domain. In practice, this could be achieved by applying different receiver gain during different time slots. Now the allocation of time slots for the mobile terminals should be done based on the received signal level. Naturally, a separate narrowband receiver would still be required for the BCCH frequency.

Finally, one should not forget that the proposed limited-range wideband transceivers are not compliant with the current GSM air interface specifications.

Conclusions

Wideband multicarrier receivers are an attractive technology for cellular base stations. However, due to technological constraints it may not be possible to achieve the required dynamic range. One method to circumvent this limitation is to divide the total dynamic range between two receivers. In GSM base stations a narrowband receiver with full dynamic range is also necessary in order to support random access capture.

The proposed solution creates regions within a cell that can only be supported by one of the receivers. Therefore, capacity dimensioning of the different receivers becomes important. In this study the target was to find out whether a separate low gain receiver is actually necessary in GSM base stations. The reason for questioning this is the indispensable narrowband receiver, which can be used to serve terminals in the proximity of the base station.

Toward these aims, the distribution of received signal levels in a base station was explored through simulation. The results indicate that in the suburban and urban environments, and with the chosen parameters, the proportion of high input levels is very small. Next, a GSM base station architecture was defined and the available traffic capacity in the different cell regions was analyzed. The results show that the narrowband receiver in fact provides sufficient capacity except in very large base station configurations of over 13 carriers. Hence, a separate wideband low gain receiver is not necessary, which simplifies the base station architecture; the proposed structure is shown in Fig. 12. Furthermore, the presented architecture allows flexible adjustment of traffic capacity without the addition or replacement of receiver hardware.

It was also noted that the novel architecture calls for new channel allocation and handover criteria. Unless the parameters are chosen carefully, some loss of traffic capacity may ensue.

References

[1] S. Phillips and D. Coons, "Extensive Use of DSPs Eases the Design of Wireless Systems," *Wireless Sys. Design*, Apr. 1997, pp. 41–44.

[2] J. Mitola, "Software Radio Technology Challenges and Opportunities," Presentation at EC/DG XIII-B Wksp. Software Radios, Brussels, Belgium, May 29, 1997.

[3] D. B. Chester, D. H. Damerow, and C. Olmstead, "Analog to digital converter requirements and implementations for narrowband channelization applications," *Proc. ICASSP '92*, Mar. 23-26, 1992, pp. 325–28.

[4] F. Goodenough and L. Goldberg, "Fast ADC Enables Digital Baseband RF Technology," *Elect. Design*, July 11, 1994, pp. 141–43.

[5] R. Baines, "The DSP Bottleneck," *IEEE Commun. Mag.*, May 1995, pp. 46–54.

[6] Analog Devices Inc., "Software Radios," *Microwave J.*, Feb. 1996, pp. 128–36.

[7] R. M. Lober, "Base-Station Technology Takes Software-Definable Approach," *Wireless Sys. Design*, Feb. 1998, pp. 19–30.

[8] T. Haynes, "DSP technology optimizes multi-channel digital receivers," *RF Receivers*, Apr. 1997, pp. 58–62.

[9] R. Carney, "Multichannel wideband digital receiver making use of multiple wideband tuners having individually selectable gains to extend overall system dynamic range," U.S. patent 5,590,156, Dec.31, 1996.

[10] ETSI TC-SMG, "Digital cellular telecommunications system (GSM Radio Access Phase 3): Radio transmission and reception (GSM 05.05)," v. 6.0.0, Jan. 1998.

[11] D. A. Hairfield *et al.*, "Design of a Wideband, High Dynamic Range Downconverter Used in Smart Antenna Applications," *Microwave J.*, Apr. 1998, pp. 100–8.

[12] J. A. Wepman, Analog-to-Digital converters and Their Applications in Radio Receivers, *IEEE Commun. Mag.*, May 1995, pp. 39–45.

[13] B. Burr, ADS852 preliminary data, June 1998.

[14] M. Hata, "Empirical Formula for Propagation Loss in Land Mobile Radio Services," *IEEE Trans. Vehic. Tech.*, vol. VT-29, Aug. 1980, pp. 317–20.

[15] COST 231 WG 2, "UHF Propagation, Urban Transmission Loss Models for Mobile Radio in the 900-1800-MHz Bands," Tech. rep. TD(90) 119 rev. 2, 1990.

[16] A. Davidson and C. Hill, "Measurement of Building Penetration Into Medium Buildings at 900 and 1500 MHz," *IEEE Trans. Vehic. Tech.*, vol. 46, no. 1, Feb. 1997.

[17] M. Mouly and M.-B. Pautet, *The GSM System for Mobile Communications*, 1992.

[18] R. Järvelä, "Distributions of Transmitted and Received Power Levels in Cellular Network Base Stations," M.Sc. thesis, Oulu Univ., 1998.

[19] ETSI TC-SMG, "Digital cellular telecommunication system (Phase 2+): Mobile radio interface layer 3 specification (GSM 04.08)," v. 5.8.0, Jan. 1998.

[20] ETSI TC-SMG, Digital cellular telecommunication system (GSM Radio Access Phase 3): Multiplexing and multiple access on the radio path (GSM 05.02)," v. 6.0.0, Jan. 1998.

Additional Readings

[1] J. R. Doner, "Designing a Broadband Software Radio," *Commun. Sys. Design*, Nov. 1996, pp. 23–30.

[2] P. Louzon, "Decipher High-Sample-Rate ADC Specs," *Elect. Design*, Mar. 20, 1995, pp. 91–100.

SOFTWARE RADIO ECONOMICS

Allan S. Margulies
The MITRE Corporation
Wireless Communications Department
Bedford, MA

Joseph Mitola,III
The MITRE Corporation
National Intelligence Department
McLean, VA

ABSTRACT

The non-recurring engineering (NRE) costs of software-defined radio (SDR) development programs were examined through interviews with knowledgeable government and industry sources. Starting with a technical staff with a strong background in programmable digital radios, it will take typically $25M and 22 months to produce a viable software radio product. The use of open architecture standards reduces the cost by about $15M and one year over proprietary designs of the same degree of flexibility as the open architecture SDR. There is a large variance in estimates, however. Some experts felt that the open architecture process would increase expense and not save any time, in some cases. This paper presents the raw data, assumptions, and viewpoints that seem to most strongly influence the near-term economics.

INTRODUCTION

One of the motivating factors behind developing software defined military radios with an open architecture and modular components is that they have the potential to provide enhanced performance, growth, and flexibility [1]. With a radio based on accepted industry standards and common commercial technology, performance and growth gains come from increased processing capabilities that can be installed as new technology generations develop, and flexibility comes from the ability to change the characteristics of waveforms through the use of software.

However, another major motivation for implementing software defined radios (SDR) is that they also have the potential to create substantial cost savings, and there has been considerable interest in those economic factors. Specifically it seems important to know whether or not a software-defined radio (SDR) with an open architecture would, in fact, be less expensive to acquire, own, and operate than would a radio design based on a proprietary design. This question addresses the effectiveness of open architectures for software-defined radios proposed by the Modular Multifunction Information Transfer Systems (MMITS) Forum [2]. In addition, the Joint Tactical Radio System (JTRS) Program Office is in the process of defining an architecture for this multi-billion dollar US Department of Defense (DoD) radio program. With DoD having over 200 radio programs currently on the books in the HF, VHF, and UHF bands and another 200 plus nomenclatured satellite communications terminals, even small incremental cost savings quickly develop into large aggregate savings.

Estimating the total life cycle cost for an SDR requires wide-ranging assumptions about the future, in turn, inevitably leading to questions about the validity of those assumptions. A more tractable approach to immediate economic issues is estimating cost savings in the non-recurring engineering (NRE) phases of a near term SDR program. Such an estimate requires the fewest assumptions and the fewest technical, marketing or other temporal projections. It also removes documentation costs, production, logistics, testing, maintenance, spares, and system integration from the equation. The NRE costs are commonly thought to represent less than 10% of the life-cycle cost of a radio program, but NRE generally represents the largest barrier to immediate application of a new technology. An understanding of costs and potential savings in NRE lays the foundation for future estimates of total program costs.

DoD policy for the acquisition of future tactical radios was outlined by the Programmable Modular Communications System (PMCS) Integrated Process Team (IPT), which met from February to September of 1997. Their deliberations indicated that some type of commercially available open architecture SDR would be in the best interests of the department. To quantify the potential benefits requires a first order estimate of the research, development and engineering costs for an SDR using an open architecture of the type now contemplated by the JTRS Program Office and the MMITS Forum. These costs then need to be compared to a development program based on conventional designs.

SURVEY METHODOLOGY

The authors constructed a simple Delphic survey. Ten experienced professionals from 7 vendors and a

Reprinted from the *IEEE MILCOM*, 1998.

government acquisition organization provided estimates for NRE in terms of time, staff, and/or investment requirements. Most of the survey participants are members of the MMITS Forum, an organization created for the purpose of promulgating and promoting standards for an open architecture SDR. They were promised anonymity so as not to compromise proprietary information.

The survey postulated the following three scenarios for a military mobile radio:

- Case 1. The radio is to be a completely new development with new waveforms, a new frequency band, and is being built for a new application, but some backward compatibility is required.
- Case 2. The radio is to be an existing radio requiring a few modifications in frequency band or bandwidth or a new waveform that is within the state of the art.
- Case 3. The radio is to be a limited production model built for a special application (multi-function or special form factor, or waveform, or limited capability, etc.)

The respondents were further asked to estimate the costs and time for NRE under each of the following two conditions:

- A. A JTRS or MMITS standard architecture exists; the required technology exists and the developer has experience using that technology.
- B. No architecture standard exists.

The assumption of developer experience eliminates learning curves from the NRE estimate. Condition B is intended to represent the historical status-quo in which each developer offers a different architecture based on the technology that is in-hand.

THE RESULTS

While the questionnaire asked for estimates of time and dollars for the two conditions for each of the 3 cases, the numbers of interest are the differences (deltas) between the two conditions, i.e., rather than try to correlate absolute values from different vendors, we looked at the percentage savings that the vendors expected as a result of using JTRS/MMITS technology instead of current conventional technology. The aggregate averages are presented in Table 1; the detailed results are contained in Table 2 for each of the three scenarios under the two postulated conditions.

Although, on average, there is an expectation that using an industry standard open-system technology for an SDR would result in roughly 40% savings in time and money for NRE in cases 1 and 2 which relate to the design of a new radio or the modification of an existing radio, the three most pessimistic respondents think that there may be no time savings at all and one of them feels that there may actually be a cost penalty associated with using the standard technology in case 2.

For the third scenario, the one in which the product is to be a special purpose design or form-factor, the potential for savings using the SDR standard technology is much more limited because requiring the use of a standard architecture or components can impose design constraints that might actually increase costs as the vendor struggles to reconcile the technology and the performance specifications.

DISCUSSION

Because the survey recipients are all participants in the MMITS Forum, they present two offsetting factors: on the one hand, they might be biased in favor of the SDR technology and they might over-estimate the benefit to be derived from using that technology. But, on the other hand, they are more knowledgeable about that technology and its potential for savings than would someone outside of the Forum.

However, based on these results and accounting for the possibility of positive bias, the uncertainty is not whether or not the use of a standard architecture and technology will save time and money on military radio acquisition, but rather how much time and money will they save. This implies that a reasonable investment in the development of a standard radio architecture and associated technology to facilitate an open architecture and plug and play, mix and match hardware and software suite for a military radio has large upside potential with a minimal downside risk.

Logistics and maintenance savings have the potential to generate the largest part of the cost savings resulting from using an open architecture and modular hardware and software elements for a military radio. But even without trying to extrapolate the NRE savings achieved into life cycle cost estimates to take advantage of those maintenance savings, it is apparent that an early investment to create the enabling technology will pay for itself within a relatively short period of time. With a $5 to 15 million per program saving in NRE, if only a small percentage of the scheduled programs use the technology, the technology investment costs will be quickly amortized.

ENGINEERING ANALYSIS

There are some other economic considerations, as shown in Figure 1, the major components of the software radio include the antenna, RF distribution, ADC/DAC, real-time software, on-line software and the development support environment. Some economic differences have to do with engineering assumptions. First, if one assumes that the antenna, RF, ADC, and DAC technology are sufficient for the application, then the development times and resources

Table 1
Base Numbers

CASE	Average Time with SDR Technology (months)	Average Cost with SDR Technology ($M)	Average Time without SDR Technology (months)	Average Cost without SDR Technology ($M)
1	22	24.8	34.2	40
2	10.4	5.8	17.3	10.1
3	14.7	9.5	20.5	12.3

Table 2
Survey Results

		Case 1A-1B		Case 2A-2B		Case 3A-3B	
Title	Years Experience	Delta time %	Delta $ %	Delta time %	Delta $ %	Delta time %	Delta $ %
Program Manager	20	0%	0%	0%	-33%	0%	-20%
Chief Scientist	20	25%	50%	8%	20%	0%	-33%
Chief Scientist	30	88%	82%	83%	91%	83%	90%
Program manager	20	41%	41%	53%	75%	0%	0%
Director, Program Development	35	22%	25%	11%	17%	25%	25%
Marketing Manager	9	58%	75%	67%	79%	67%	82%
Technical Director	37	0%	0%	0%	0%	0%	0%
Business Unit Manager	22	87%	91%	83%	71%	83%	90%
Government acquisition manager	20	43%	38%	38%	36%	11%	6%
Engineer/manager	16	0%	17%			0%	20%
AVERAGE CHANGE		36%	38%	40%	43%	28%	22%

are on the lower side. Most commercial developers realize that these technologies are not fully in hand and may require specialized designs such as dithering the ADC or voice channels to achieve dynamic range or speech quality. In addition, there are different assumptions about the ability of open architecture interfaces to cost-effectively support the real-time streams shown in the figure. For the technology to be worthwhile, it has to either reduce costs or enhance performance or both.

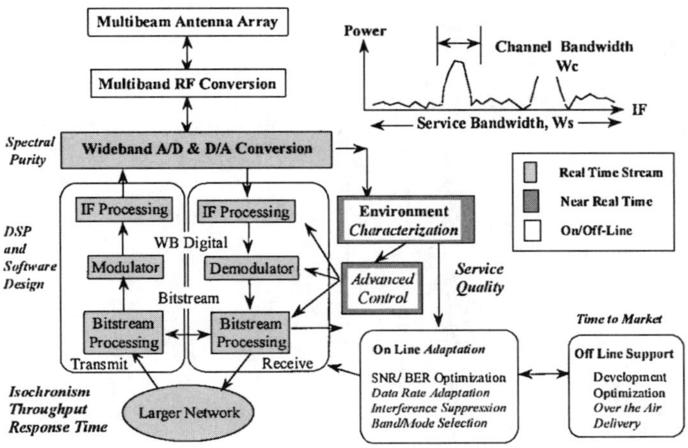

Figure 1. Functional Components Have Different Economic Assumptions

If the overhead of a layered software architecture forces the DSP to include more processing capacity than for a proprietary architecture, then the costs increase. Comparing a hand-optimized DSP architecture to an open architecture generally shows that the DSP costs may be 50% lower for the optimized but closed architecture. On the other hand, the initial acquisition costs of the software will increase by somewhere between 10 and 35% to pay for that optimization. The closed proprietary architecture sounds like a good idea based on these percentages before a closer examination of the actual costs. A nominal development program costing \$20M would include at least \$10M in labor for software development. That amount could change by from \$1 to \$3.5M, depending on whether the implementation uses open architecture software that is computationally inefficient, but quicker to implement or whether the software is optimized for a less capable DSP platform.

In production, the units should cost under \$30K. Only 10% of that cost is DSP, for a cost impact of about \$1500 per unit. In extremely cost-sensitive markets, this small amount can make all the difference. But considering the impact of reduced costs and increased capabilities of DSP chips, the production cost can be equalized within about 18 months, which is how long it will take to double the DSP capability of the unit if it is based on open architecture standards. If it is not based on open architecture standards, then the developer and service provider have to endure another round of \$1-\$3.5M of NRE to integrate the new technology into the optimized architecture.

The same economics affect the computer industry and it took some time for their impact to be felt. Initially, IBM safely dominated the mainframe market. Then DEC created the minicomputer market, costing IBM market share. The DEC minicomputer was based on a closed architecture. Then Sun created the workstation market, costing DEC market share. At the beginning, Sun was based on the VME open architecture hardware, Berkley UNIX and an aggressive third party software campaign. This had a lot to do with their early success. Then Intel/Microsoft (and a much wiser IBM) expanded the PC market costing all of the above market share and lost growth opportunities.

The wireless industry is following this same paradigm and is poised on the threshold of low cost plug-and-play wireless. The open architecture standards are nearly in place. Some risk-takers are already poised to challenge the incumbents. At least six start-up companies have been formed in the past 12 months alone, bent on becoming the PC of the wireless marketplace. This activity is all good for DoD, but it will require continuing assessments of the economics to determine when and how to expend significant resources on a new tactical radio program like JTRS. Industry is clearly moving in this direction. It may take 18 months for the de-facto architecture to emerge, but the economics are compelling. The standards will emerge and soon.

REFERENCES

[1] Programmable Modular Communication System (PMCS) Guidance Document, 31 July 1997 (Revision 2).

[2] MMITS Technical Report 1.1, Architecture and Elements of Software Defined Radio Systems as Related to Standards, January 1998.

Author Index

A

Abidi, A. A., 69
Antonesco, M., 225
Arnott, R., 258
Astier, L., 94
Athanas, P., 270
Auslander, E., 307
Aziz, P. M., 102

B

Baines, R., 154
Barron, K. S., 291
Bentzen, H. J., 424
Bonser, W., 326
Bose, V., 314
Boyle, R., 270
Brodersen, R., 163

C

Chaloupka, H., 258
Chen, T., 163
Chester, D. B., 148
Cook, P. G., 326
Cummings, J. P., 445
Cummings, M., 199, 421

E

Efstathiou, D., 193
Erben, H., 265
Evans, J. B., 375
Ewy, B., 375

F

Fettweiss, G., 389
Fridman, J., 193
Frost, V. S., 375
Fuchs, A., 434

G

Gass, W., 163
Gatherer, A., 307
Gershenfeld, N., 439
Gunn, J. E., 291
Guttag, J., 314

H

Haruyama, S., 199
Heath, S., 421
Hentschel, T., 389
Hutchings, B., 204

I

Ismert, M., 314

J

Järvelä, R., 461

K

Kennedy, J., 353
Kenington, P. B., 94
Koilpillai, R. D., 278
Kostic, Z., 179

L

Lackey, R. J., 229
Leppänen, P. P., 461
Likharev, K. K., 138
Liu, K. J. R., 360

M

Malinimohan, K., 375
Mandayam, N. B., 342
Maguire, G. Q. Jr., 413
Margulies, A., 470
Mateescu, M., 337
McMahan, M., 307
Meyr, H., 142, 235
Minden, G. J., 375
Mitola, J., 3, 12, 25, 38, 413, 470
Munro, A., 398
Murotake, D., 434

N

Nishitani, T., 163

O

Oates, J., 434
Oerder, M., 142

P

Pereira, J., 63
Petr, D., 375
Plumb, R., 375
Polydoros, A., 142
Ponnekanti, S., 258
Posti, H., 461
Prescott, G., 375
Prill, R. S., 225

R

Rabaey, J. M., 163
Rashid-Farrokhi, F., 360
Razavilar, J., 360
Reed, J. H., 270
Roberts, J., 375
Ruczczyk, W., 291

S

Sabatakakis, K., 265
Salthouse, C., 439
Sanchez, R., 375
Schuh, R. E., 337
Schuler, C., 337

Scott, N. L., 126
Seetharaman, S., 179
Semenov, V. K., 138
Seskar, I., 342
Sevenhans, J., 252
Shanmugan, K. S., 375
Shepherd, R., 408
Smith, J. R., 439
Sorensen, H. V., 102
Sparks, C., 375
Srikanteswara, S., 270
Stetzler, T., 307
Subramanian, R., 235
Sullivan, M. C., 353
Suzuki, Y., 246

T

Taraborrelli, S., 252
Taylor, C., 258
Tennenhouse, D., 424
Tsurumi, H., 246
Turletti, T., 424
Tuttlebee, W. H. W., 31

U

Upmal, D. W., 229

V

Van Der Spiegel, J., 102
Vaughan, R. G., 126
Verstraeten, B., 252
Villasenor, J., 204

W

Walden, R. H., 82
Welborn, M., 314
Wepman, J. A., 75
White, B. R., 126
Wikborg, E. B., 138

Z

Zangi, K. C., 278
Zvonar, Z., 193

Subject Index

1G
2G
3G, 32, 271, 277, 294, 309, 400

A

ACTS (Advanced communications technology and services), 25, 33, 63, 100, 261, 268
Ada, 55
ADC, 1, 4, 12, 17, 25, 30, 31, 44, 67, 82, 94, 102, 139, 248, 280, 298, 341, 346, 362, 387, 391, 464
ADC metrics, 75–80, 82–85, 263
ADC tradeoffs, 67, 75–80, 94–95, 104, 168, 261, 299
AGC, 58, 71, 150, 251, 464
Air interface, 5, 13. *See also* GSM
ALE (Adaptive link establishment), 55
Allocate resources. *See* resource allocation
AM (Amplitude modulation), 159, 305, 384, 461
Amateur bands, 382
AMPS (Analog mobile phone system), 4, 13, 41, 63, 253, 319, 369; (digital), 63, 94, 253, 278, 283; (supervisory audion tone) 354. *See also* IS-54, IS-136
Antenna bandwidth, 259
Antenna gain, 464
Antennas, 259, 355, 464
Antennas, programmable, 259
Anti-aliasing, 76–77
Aperture jitter, 5, 85, 89–91
API, 28, 36, 39, 417; modem, infosec, telephone 399; download: 421–423
Applications (Layer of protocol stack), 9
Architecture components, 1, 21, 268, 456
Architecture design rules, 1, 184
Architecture evolution, 12, 268
Architecture functions, 1, 51, 182, 186, 267, 329
Architecture tradeoffs, 31, 184, 201, 312
Architecture, layered virtual machine, 26, 271
Architecture, open. *See* Open architecture
ARIB, 179
ARM (Advanced RISC machines), 412
ARQ (Automatic repeat request), an error correction protocol, 18, 377
Array, 262
ASIC tradeoffs, 14, 67, 166, 199–200, 270, 293, 297, 307, 424
ASICs, 14, 25, 40, 42, 153, 181, 185, 241, 436
ASICS, digital filter, 17, 148–153, 279–282, 355, 385
ASICs, FEC, 18, 348, 426
ASICS, transceiver, 252
ATM, 21, 35, 59, 267, 375–377
Attributes (of an object), 403
AWACS, 448
AWGN (Additive white gaussian noise), 78, 96, 102, 134, 238, 342

B

Bandpass sampling, 41, 77, 126–137, 262, 320. *See also* Nyquist dynamic range
Baseband, 7, 16, 19, 21, 22, 46, 47, 69
Basestation, 258–262, 338, 366, 461–469. *See also* BSC and BSS
Beamforming, 379, 382–384
Behavior (of an object), 403
Benchmarking handsets, 27
Benchmarks, 5. *See also* Metrics, Processing capacity
BER (Bit error rate), 12, 15, 22, 237, 364, 382
BiCMOS, 252
B-ISDN (Broadband integrated services digital network), 124
BIT (Built in test), 458
Bit interleaving, 19, 427, 429
Bitstreams, 19, 22, 47, 409
Blocking probability, 369–370
BMW (Broadband RF, multichannel, wideband DSP)-SDR, 14, 15, 41, 246, 262, 288, 342, 344, 425, 435
BPF (Bandpass filter). *See* Digital filters
BPSK (Binary PSK), 18, 274, 376, 383
BSC (Base station controller), 186, 338
BSS (Base station subsystem), 186
BTS (Base transceiver station), 186, 195, 263, 424
Bus, 4. *See also* VME, PCI
Business models, 64, 173, 271

C

C40, (C44). *See also* TMS320
Canonical model, 12, 13, 17, 291
Capability profile, 421
Cascaded integrator comb (CIC) filter, 394
CASE (Computer-aided software engineering) tools, 10, 11, 60
Cathedral, 176
CCITT (International consultative committee on telephone and telegraphy), now the ITU-T
CDMA, 5, 17, 19, 41, 94, 154, 190, 196, 253, 259, 274, 337, 342, 360, 371, 412, 435, 439
CDPD (Cellular digital packet data), 179
Cellular, 13, 182, 195, 319, 463. *See* also MCR
CELP (Codebook excited linear prediction) vocoder, 21, 189

CEPT, *See* SDH
Channel coding, 186, 188
Channel control object, 403
Channel object, 59, 60, 428
Channel processing, 15, 304. *See also* digital filters, ASICs
Channelization, 29, 278, 279
Chip rate (DSSS), 299, 346, 443
CIR (Carrier to interference ratio), 353, 356–357, 360
CMOS, 69, 95
CNR (Carrier to noise ratio), 78–79
Cochannel interference, 360
Coded dynamic range, 84, 141. *See also* Dynamic range
Coding. *See* Source coding, Channel coding, Bit interleaving, Trellis coding
Cognitive radio, 413–418
Commanders tactical termina (CTT), 3
Commercialization, 271
Communications services (layer), 36
Comparitor ambiguity (of ADC), 85
Complexity, 19 (of modem software) 191, (of channel symbols), 186 (of equalizers), 186 (of decoding), (of data link), (QoS), 366 (of software) 19
Component view (UML). *See* UML
Components. *See* Architecture components
Computer-aided design (CAD), 3, 10, 236, 243
COMSEC (Communications security), 326
CONOPS, 447, 458. *See also* Scenario
Constant modulus, 259
Control objects, 403
CORBA, 16, 351, 405, 408. *See also* Real time
Cosite interference, 260, 343
COSSAP, 23, 181, 243
Costas loop, 152
COTS, 4, 152, 329, 448
CPU, 14
Cramer-Rao bound, 237
CT2 (Cordless telephone system 2), a European air interface, 94

D

DAC, 1, 25, 27, 42, 118, 259, 389, 425
DAMA (Satcom), 234, 446
DARPA (US Defense advanced research projects agency), 5, 25
DCS (Digital PCS), 235, 254, 337
DEC, formerly the Digital Equipment Corporation, now a division of Compaq, 426
Decimation, 393
DECT (Digital European cordless telephone), 37, 94, 179, 265, 399
Delay spread (Range of time delays in multipath propagation), 355
Delta-Sigma (ADC). *See* Sigma-Delta
Demand. *See* Processing demand
Demodulator, 11, 429, 442
Deployment view (UML). *See* UML
Design rules. *See* Architecture design rules
Despreader, 197
Digital down-conversion, 384–386, 390. *See also* Bandpass sampling
Digital filters, 17, 148, 157, 187, 365, 392
Digital processing tradeoffs. *see* DSP and Processing
Digital radio, 40. *See also* DSP and processing
Digital RF, 29
Direct conversion. *See* Homodyne receiver
Direct matrix inversion (DMI), 367
Direction finding. *See* Location
DISC (Dynamic instruction set computer), 212
Diversity (antenna), 36, 41, 45, 361
DMA, 55, 56, 317
Download, 31, 32, 33, 35, 39, 402, 421–423
DS0, 6
DSP, 5, 14, 20, 25, 32, 39, 94, 102, 124, 154–162, 165, 194, 231, 240, 258, 270, 291–306, 307, 314, 339
DSP core, 159, 294
DSP Motorola 56300, 159, 437
DSP pool, 39, 156
DSP tradeoffs, 5, 67, 154–162, 180–182, 346
DSSS (Direct sequence spread spectrum), 81, 292, 440. *See also* Spread spectrum
Dynamic range, 78, 84. *See also* SFDR

E

EC (European Community), 31, 66
Economics, 23, 24, 470
Efficiency (computational), 315, 425–429. *See also* MIPS, MFLOPS, Power (consumption)
EIRP (Effective radiated power relative to isotoropic), 372, 465
EMI (Electro-magnetic interference), 17, 45
Entity reference model, 40, 447
EPLRS, 292
Equalization, 188, 417
Erlang (Traffic intensity), 368, 466
Estimation, 142, 237
ETSI (European Telecommunications Standards Institute), 45, 54, 179, 252, 310
Exact decorrelator, 343

F

Facility utilization, 20
Facsimile, 314
Fading, 238
Fast transversal filter (FTF), 368
FDDI, 6, 21
FDM (Frequency division multiplexing), 278
FDMA (Frequency division multiple access), 260, 424
FEC (Forward error control), 7, 15, 18, 47, 55, 430
FFT, 4, 8, 16, 82, 135, 151–152, 155, 157, 168, 180, 279, 303, 306, 310, 321, 326, 436
FH (frequency hopping), 55, 292, 299, 323
Filters (e.g., FIR, IIR), 180, 202, 268, 280, 320–321, 365. *See also* Digital filters
Finite state automaton. *See* state machines
FIRST, 29, 34, 97, 259, 265, 269
Fixed point processors, 241, 312
FM (frequency modulation), 11, 459
FPGA, 25, 26, 28, 39, 41, 59, 67, 153, 172, 199–203, 246, 258, 276, 298, 339, 343, 345, 348, 394, 436
FPGA tradeoffs, 153, 223, 329
FPLMTS (Future public land mobile telephone service), early name for IMT-2000, 80, 179
FRAMES, 29, 34, 95, 259, 269
Framing, 186. *See also* Synchronization
Frequency performance, 357
FSK (Frequency shift keyed) modulation format, 319, 323, 380

G

General purpose (GP) processor, 312
GFLOPS, 44, 230
GIPS, 1000 MIPS, 34
GloMo, DARPA Global Mobile Project, 25, 29, 60
GMSK, 18, 19, 55, 196, 254, 427
GPRS, 405, 415
GPS (Global positioning satellite), 23, 252, 292, 375, 457
Graychip, 22, 152
Group IV (CCITT fax standard), 314
GSM, 5, 12, 17, 25, 33, 70, 94, 99, 149, 154, 186, 242, 279, 307, 337, 351, 395, 398, 401, 408, 419, 424–430, 436, 461–469
GUI. *See* User interface object

H

Handset (architecture), 27, 34, 246–251, 312
Harris Corp., 22, 143, 155
HAVE QUICK, a US slow-FH air interface used in aircraft, 26, 55, 234, 292, 328, 331, 445, 449
HDLC, 376
HDTV, 5
Heterogeneous multiprocessing, 20
HF (High frequency), 28, 353

HiperLAN, 400
HM (Hardware modularity), 225–226. *See also* Components
Hogenaur, 151, 155
HOL (High order language). *See* Java, Ada
Homodyne receiver, 41, 69, 71, 235, 240, 248, 254, 295
Hop sets. *See* FH
Host processors, 312
Hybrid fiber radio (HFR), 337

I

I/O, 315–317
IBMS, 34
ICNIA (Integrated communications, navigation, and identification architecture), 326
Identity (of an object), 403
IDL (Interface definition language), 415
IF filtering, 153, 293. *See also* ASICs
IF processing, 4, 5, 7, 16, 23, 34, 38, 45, 148–153, 239, 395
IF waveform, 41, 300
IFF (Identification, friend or foe), 228
IMT-2000 (International mobile telecommunications), 32, 34, 247, 434
Inductor, 256, 262
Inference, 417
INFOSEC (Information security), 26, 38, 45, 55, 233, 327, 427, 460
INFOSEC object, 59
Infrastructure. *See* Radio Infrastructure, DII, GII, NII
Infrastructure software, 408. *See also* Radio infrastructure
Intel (488, Pentium, MMX, etc.), 426, MMX
Interconnect, 160
Interface points, 454
Interference suppression, 224, 342–350, 360. *See also* EMI, Filters, Partial interference cancellation
Internetworking, 376 (API)
IP (Intellectual property), 203, 256
IS-136 (US TDMA system), 186, 292
IS-54 (Digital AMPS standard), 28, 242, 292
IS-95 (US CDMA air interface standard), 13, 41, 185, 197, 265, 292, 396
ISA (Industry standard architecture), e.g., of the IBM PC, 26, 44
ISO OSI protocol stack, 3, 9, 408–413
Isochronous streams, 57, 271–273
ISR (Interrupt service routine), 56
ITU (International telecommunications union), 16, 34, 45, 54, 63, 179, 399, 414, 434
IVHS (Intelligent vehicle highway systems), 353

J

Java, 33, 47, 351, 399, 405
Jini, 351, 405
JTIDS, 13, 19, 29, 44, 292
JTRS, 29, 419, 445, 447, 471

K

KLOC (Thousands of LOC). *See* LOC
KQML (Knowledge query manipulation language), 415
KRL (Knowledge representation language), 46, 413

L

LAN, 3, 21, 38, 323–324
Latency, 6, 322
Layered architecture. *See* Architecture, layered virtual machine
Least mean square (LMS), 368
Linearity, 347. *See also* Dynamic range
LMDS (Local multipoint distribution system), 337–338
LNA (Low noise amplifier), 70, 249, 252, 261 (dual mode)
LO (local oscillator), 29, 71, 151, 196, 301, 364, 383
LOC (lines of code), 3
Location (aware, finding, etc.), 312, 354, 378
Logical view (UML). *See* UML
LPI (Low probability of intercept), 326

M

MAC (Media access controller), 402
MAC (Multiply accumulate), 191, 198, 390
Management entity, 411
Matched filter, 343
Maximum likelihood receiver, 144
MCM (Multi chip module), 6, 231
MCR (Mobile cellular radio), 12, 15, 23
MEDIAN, 34
Memory (capacity, architecture), 22, 43, 150, 160, 165, 411
MEMS, 28, 29, 67
Mercury (Corp.), 23, 434
Metrics, processing, 14. *See also* MIPS, MOPS, MFLOPS, GFLOPS, Canonical model
MExE Protocol, 399, 423
MFLOPS, 14, 304–305
Middleware, 16, 332, 398–403. *See also* CORBA
MIMD (Multiple instructions, multiple data), 6, 16
MIPS, 14, 17, 155, 193, 258, 307, 346, 364, 427–432. *See also* MFLOPS
MMITS (Now the SDR Forum), 25, 39, 63, 199, 470
Mobile infrastructure, 29, 402
Mobile IP, 29
Models of computation, 48, 49
Modem, 5, 23, 39, 55, 300 (object model)
Modem software, 442, 454
Modes. *See* Multimode radio
Modulation, 11, 19, 56, 427. *See also* AM, FM, PSK, QAM
Modules. *See* Components
MOPS, 20, 21
Motorola (US telecommunications company), 308, 437
MPEG, 173
MPI (Message passing interface), 60
MSK, 380. *See also* GMSK
Multibeam antennas, 355, 360, 379. *See also* Smart antennas
Multimedia. *See* source coding
Multimode radio, 38, 229
Multithreaded. *See* Threads

N

Narrowband, 233
Network interfaces (See T- and E-Carrier), 47, 376
Networking, 31, 46. *See also* Internetworking
Noise floor matching, 79–81, 286
Noise power ratio (NPR), 79–81, 82–83
Nonlinear distortion, etc., 83, 112, 116, 250, 348, 362
NRT (near real time), 22
NTT (Nippon Telephone and Telegraph), 16
Nyquist dynamic range, 8, 79–81, 89. *See also* Dynamic range
Nyquist sampling, 17, 76, 81, 87, 105, 389

O

OA&M, 18, 20, 65
Object-oriented technology, 403
OC-(Optical carrier)-12, etc., 21
ODP (Open distributed processing, X.900), 57
Open architecture, 3, 21, 65. *See also* Standards, Bus
Optical fiber (RF transport over), 338–340
Oracle, 4
ORB (Object request broker), 404–405. *See also* CORBA
OSS (GSM operation subsystem), 27
OTA, Over The Air (Download) 35, 421–423
Oversampling, 5, 11, 30, 77, 102–125, 395, 436

P

Parallelism, 7, 276 (processing), 16, 58 (channels) 338, 358, 360, 384, 435
Partial interference cancellation, 348

Partitioning (of a radio into horizonal and vertical aspects), 184, 193, 231, 239, 428
Pave pillar, 225
PCI (A hardware bus standard), 425, 438
PCM (Pulse code modulation), 4, 103
PCS (Personal communications systems), 12, 23, 33, 37, 95, 414
PDA (personal digital assistant), 23, 351, 414. *See also* BMW-SDR
PDR (Programmable digital radio), 1, 225
Pentek, 345
Phase locked loop (PLL), 187, 302
PHS (Personal handyphone system), a Japanese standard, 94, 179, 265
Pipeline, 274
Platform, digital. *See* Radio platform
Platform, RF. *See* Radio platform
Plug-and-play, 39, 47
PMCS, 445, 453, 470
Point set topology. *See* Topology
Polyphase filters, 17
Power amplifier (PA), 248, 252, 263
Power (consumption), 30, 87, 94, 98, 170, 255, 291–306, 310
Processing capacity, 6, 19, 193. *See also* Metrics, MIPS
Processing demand, 5, 19–21
PROMURA, 34
Propagation modeling, 11
Protocols, 22, 29, 332
PSK (Phase shift keyed), 11, 301
PSTN, 13, 33, 406
Push down automaton. *See* State machines

Q

QAM, 5, 19, 56, 169, 181, 248, 301, 337, 382
QoS, 39, 402
QPSK (Quaternary PSK), 18, 196, 248, 273, 301, 337, 376, 385
Quadrature mixers, 253, 254, 384
Quadrature sampling, 77, 136. *See also* ADC
Queuing models, 11

R

Raceway. *See* Interconnect
Radar, 81
Radio applications (layer), 26
Radio infrastructure, 26
Radio noise, 238
Radio platform (reference model), 40
Radio resources, 48, 467
RAINBOW, 34
Rake receiver, 34, 238, 268, 299, 349, 434
RAM (Random Access Machine), 48, 57
Real time (RT CORBA etc.), 22, 405
Real time streams, 410
Receive (operation), 21
Reconfigurable [radio] networks, 64–65, 375–388, 422
Reconfigurable processors, 63, 67, 200, 204–221, 270, 273, 347, 387. *See also* FPGA
Recursive functions, 48, 49
Regulatory issues, 64
Rental of spectrum, 65
Resolution (of ADC), 83, 95, 295. *See also* ADC
Resource management, *see* Performance management
RF, 38
RF conversion, 16, 23, 67, 262, 265, 293, 339, 390–391, 456
RF LAN. *See* WLAN
RF MEMS. *See* MEMS
RFIC (RF integrated circuit), 252–256
RKRL (Radio KRL), 46, 54, 413
RL (Rome laboratory) US Air Force, sponsors of SPEAKeasy, 326
ROM (Rough order of magnitude price), 173
RPC (remote procedure call), 57
RPE-LTP (GSM vocoder), 18, 189, 307, 426, 431
RSFQ, 138
RSSI, 399

S

Sample rate (of ADC), 95, 151, 294, conversion: 389. *See also* Nyquist dynamic range
Satellite communications (Satcom), 23, 65, 233, 292, 450, 459
SAW (Surface acoustic wave), 255
SCA (Software communications architecture). *See* JTRS, middleware, CORBA
Scenario, 401
SDH (synchronous digital hierarchy), 7, 38
SDL (ITU Z.100), 54, 414
SDMA (Space division multiple access), 16
SDR forum, 25, 422
SDR joint control, 39
SDR personality, 38, 186
SDR (Software-defined radio), 1, 63, 138, 223, 337, 339, 342, 434, 470
SFDR, 78–80, 82, 140, 156, 196, 338, 436, 462. *See also* Dynamic range
SHARC, 21, 156, 197, 309, 344
SHF, 5, 19
SIC (Sequential interference cancellation), 349, 435
Sigma-Delta (ADC), 67, 75, 87, 102–125
Signaling, 19, 463
SIM (Subscriber identity module), 33, 35, 400
SIMD (Single instruction, multiple data processor), 8
Simplilcial complex, 53
SINCGARS (Single channel ground air radio system), 3, 27, 292, 445, 449
SINUS, 34
SLATS, 34
Smart antennas, 353–359, 360–374
SNR (Signal to noise ratio), 15, 22, 78–79, 96, 134, 361
Soft radio, 270–274
Software architecture, 322, 330–336, 408
Software-defined radio, 278
Software libraries , 186, 232, 243, 246, 322, 332–336
Software radio (Ideal), 3, 12, 138, 265, 278, 298
SONET (Synchronous optical network). *See* OC
SORT, 34
Source coding, 18, 26. *See also* Vocoder, Speech, VTC
Source segment software, 18
Space-time adaptive processing. *See* STAP
SPEAKeasy, 13, 25, 44, 223, 292, 326
SPEAKeasy I, 25, 49, 55, 223, 326 (Software)
SPEAKeasy II, 25, 223, 229–234, 326–336 (Software), (API), (Infosec API)
SpecFP, 424–429
SpecINT, 424–429
SpecMark, 5, 22, 156, 424
Spectrum management, 37, 64, 357, 434. *See also* Rental of spectrum
SpectrumWare (Spectra), 318, 432
Speech, 8, 18, 46, 174, 308. *See also* vocoder
Speech processing object, 8, 186
Spice, 255
SPOX (OS), 4, 22
Spread spectrum, 19. *See also* CDMA
SPW, Signal Processing Workbench, 23
Squelch, 38
SS7, 47, 351
Standards. *See* CORBA, Open architecture, SDR forum
Standards setting process, 174
STAP (Space-time adaptive processing), 361, 363, 434
State machines, 55, 58. *See also* SDL
STR-2000, 44
STREAMS, 410-412
Subscriber parameters, 347
Sun Microcomputer, 426
SUNBEAM, 34, 63
Superconducting (filters, ADCs, etc.), 261–263
Superheterodyne, 71, 183, 239, 295
SWR, (Software radio), 25, 27
Sybase (Data base), 4
Synchronization, 187
Synthesizer, 73, 227, 240. *See also* LO

T

T-Carrier, 13
T1 (1.544 Mbps), 5, 21, 233. *See also* T-Carrier
T1 Demultiplexer, 6
TCP/IP protocol, 22, 59, 406
TDD (Time domain duplexing), 258
TDM (Time division multiplexing), 17, 261, 278
TDMA (Time division multiple access), 17, 18, 250, 259, 376, 439
TDOA, 354
TETRA (Trunked radio), 94
Thermal noise, 96. *See also* Radio noise
Timing (of processes), 1, 18. *See also* performance management
Timing (of waveforms), 18, 58. *See also* Synchronization
TINA, 405–406
TMS 321 C30. *See* DSPs, C40
TMS320, 20, 44, 156, 344, 348, 358
TMS320 C6x, C80, 300, 309, 344
Top down design. *See* Object-oriented technology
Topology, 8–11, 40–43, 52–55
Transmultiplexer, 260, 261, 278
TRANSEC, 38, 299
Trellis coding, 18
Turing computability, 1, 7

U

UHF Band, 5, 19, 331, 446
UML (Unified modeling language), 59, 414. *See* Objects
UMTS (Universal mobile telephone service), European 3G, 32, 65, 80, 97, 100, 253, 258, 265, 271, 337, 340, 351, 396, 398, 406
UNIX, 22, 408
Utilization, *See* Facilty utilization
UTRA, 34, 412

V

VCELP (Vector codebook excited linear prediction),
VCO (Voltage controlled oscillator), 255, 256, 302
VELP (Vector excited linear prediction) vocoder, 189
VHDL, 16, 23, 209, 244
VHF, 5, 19
Virtual Radio, 44, 45, 314–325
Viterbi, 390, 429
VLF, 25, 31
VLIW, 165, 171, 185, 198, 310
VLSI, 163, 190, 235
VME (A hardware bus standard), 4, 21, 22, 231
Vocoder, 42, 189. *See also* CELP, VELP

W

WAP (Wireless applications protocol), 399
Waveform. *See* Air interface
W-CDMA, 26–28, 149, 266, 292, 310, 346
WLAN (Wireless LAN), 65, 337, 406

X

Xilinx, 349

Z

Zero-IF. *See* Homodyne receiver

About the Editors

Joseph Mitola III was introduced by Joao daSilva as keynote speaker of the First International Workshop on Software Radio as "the Godfather of the Software Radio." Dr. Mitola coined the term "software radio" in 1991, published the first technical paper on software radio, has been the most cited author in the field, and has contributed over two dozen technical papers on software radio architecture. He edited the landmark May 1995 special issue of the *IEEE Communications Magazine* and the April 1999 *Journal on Selected Areas in Communications on Software Radio*. Also, he was elected as the founding chair of the Global Software-Defined Radio Forum. His recent text *Software Radio Architecture* is the first interdisciplinary treatment of the subject.

Dr. Mitola is presently a consulting scientist with the MITRE Corporation, where he advises the U.S. DoD on telecommunications and information processing technology and policy. He also assisted the Executive Office of the President of the United States on Iran-Contra E-mail recovery technology. Previously, he was chief scientist of Electronic Systems with E-Systems. Dr. Mitola has held positions of technical responsibility with DoD, DARPA, Harris Corporation, Advanced Decision Systems, and ITT, having begun his career with the U.S. DoD in 1967. He holds a B.S. in electrical engineering from Northeastern University, Boston, MA, an M.S.E. from Johns Hopkins University, Baltimore, MD, and the Licentiate and Doctorate in Teleinformatics from KTH, the Royal Institute of Technology, Stockholm. Dr. Mitola's E-mail address is jmitola@compuserve.com.

Zoran Zvonar is systems development manager in the Communications Division, Analog Devices Inc., currently involved in the design of algorithms and architectures for wireless communications. He received Dipl.Ing. and M.S. degrees in electrical engineering from the University of Belgrade, Yugoslavia, and the Ph.D. degree from Northeastern University, Boston, MA.

Dr. Zvonar was a guest editor of the *International Journal of Wireless Information Networks*, *ACM/Baltzer Wireless Networks* and the *IEEE Transactions on Vehicular Technology*. He was coeditor of *GSM: Evolution Towards 3rd Generation Systems* (Kluwer Academic Publishers, 1998) and *Wireless Multimedia Networks Technologies* (Kluwer Academic Publishers, 1999). Currently, he is an associate editor of the *IEEE Communications Letters* and a feature editor of the series on Software and DSP in Radio in the *IEEE Communications Magazine*.